COMPUTATIONAL FLUID MECHANICS AND HEAT TRANSFER

Series in Computational and Physical Processes in Mechanics and Thermal Sciences

W. J. Minkowycz and E. M. Sparrow, *Editors*

Anderson, Tannehill, and Pletcher, Computational Fluid Mechanics and Heat Transfer
Aziz and Na, Perturbation Methods in Heat Transfer
Baker, Finite Element Computational Fluid Mechanics
Beck, Cole, Haji-Shiekh, and Litkouhi, Heat Conduction Using Green's Functions
Chung, Editor, Numerical Modeling in Combustion
Jaluria and Torrance, Computational Heat Transfer
Patankar, Numerical Heat Transfer and Fluid Flow
Pepper and Heinrich, The Finite Element Method: Basic Concepts and Applications
Shih, Numerical Heat Transfer
Tannehill, Anderson, and Pletcher, Computational Fluid Mechanics and Heat Transfer, Second Edition

PROCEEDINGS

Chung, Editor, Finite Elements in Fluids: Volume 8
Haji–Sheikh, Editor, Integral Methods in Science and Engineering–90
Shih, Editor, Numerical Properties and Methodologies in Heat Transfer: Proceedings of the Second National Symposium

IN PREPARATION

Heinrich and Pepper, The Finite Element Method: Advances Concepts and Applications

COMPUTATIONAL FLUID MECHANICS AND HEAT TRANSFER

Second Edition

John C. Tannehill
Professor of Aerospace Engineering and Engineering Mechanics
Iowa State University

Dale A. Anderson
Professor of Aerospace Engineering
University of Texas at Arlington

Richard H. Pletcher
Professor of Mechanical Engineering
Iowa State University

USA	Publishing Office:	Taylor & Francis 325 Chestnut Street, Suite 800 Philadelphia, PA 19106 Tel: (215) 625-8900 Fax: (215) 625-2940
	Distribution Center:	Taylor & Francis 47 Runway Road, Suite G Levittown, PA 19057 Tel: (215) 269-0400 Fax: (215) 269-0363
UK		Taylor & Francis Ltd. 11 New Fetter Lane London EC4P 4EE Tel: +44 (0) 171 583 9855 Fax: +44 (0) 171 842 2298

COMPUTATIONAL FLUID MECHANICS AND HEAT TRANSFER, Second Edition

2 3 4 5 6 7 8 9 0 B R B R 9 8 7

This book was set in Times Roman. The editors were Christine Williams and Carol Edwards. Cover design by Michelle Fleitz.

A CIP catalog record for this book is available from the British Library.

∞ The paper in this publication meets the requirements of the ANSI Standard Z39.48–1984 (Permanence of Paper)

Library of Congress Cataloging-in-Publication Data

Tannehill, John C.
Computational fluid mechanics and heat transfer / John C. Tannehill, Dale A. Anderson, Richard H. Pletcher. —2nd ed.
p. cm.—(Series in computational and physical processes in mechanics and thermal sciences)
Anderson's name appears first on the earlier ed.
Includes bibliographical references and index.
1. Fluid mechanics. 2. Heat—Transmission. I. Anderson, Dale A. (Dale Arden). II Pletcher, Richard H. III. Title. IV. Series.
QA901.A53 1997
532'.05'01515353—dc20

96-41097
CIP

ISBN 1-56032-046-X (case)

To our wives and children: Marcia, Michelle, and John Tannehill
Marleen, Greg, and Lisa Anderson
Carol, Douglas, Laura, and Cynthia Pletcher

CONTENTS

PREFACE

Almost fifteen years have passed since the first edition of this book was written. During the intervening years the literature in computational fluid dynamics (CFD) has expanded manyfold. Due in part to greatly enhanced computer power, the general understanding of the capabilities and limitations of algorithms has increased. A number of new ideas and methods have appeared. The authors have attempted to include new developments in this second edition while preserving those fundamental ideas covered in the first edition that remain important for mastery of the discipline. Ninety-five new homework problems have been added. The two part, ten chapter format of the book remains the same, although a shift in emphasis is evident in some of the chapters. The book is still intended to serve as an introductory text for advanced undergraduates and/or first-year graduate students. The major emphasis of the text is on finite-difference/finite-volume methods.

The first part, consisting of Chapters 1–4, presents basic concepts and introduces the reader to the fundamentals of finite-difference/finite-volume methods. The second part of the book, Chapters 5–10, is devoted to applications involving the equations of fluid mechanics and heat transfer. Chapter 1 serves as an introduction and gives a historical perspective of the discipline. This chapter has been brought up to date by reflecting the many changes that have occurred since the introduction of the first edition. Chapter 2 presents a brief review of those aspects of partial differential equation theory that have important implications for numerical solution schemes. This chapter has been revised for improved clarity and completeness. Coverage of the basics of discretization methods begins in Chapter 3. The second edition provides a more thorough introduction to the finite-volume method in this chapter. Chapter 4 deals with the application of numerical methods to selected model equations. Several additions have been made to this chapter. Treatment of methods for solving the wave equation now includes a discussion of Runge-Kutta schemes. The Keller box and modified box methods for solving parabolic equations are now included in Chapter 4. The method of approximate factorization is explained and demonstrated. The material on solution strategies for Laplace's equation has been revised and now contains an introduction to the multigrid method for both linear and nonlinear equations. Coloring schemes that can take advantage of vectorization are introduced. The material on discretization methods for the inviscid Burgers equation has been substantially revised in order to reflect the many developments, particularly with regard to upwind methods, that have occurred since the material for the first edition was drafted. Schemes due to Godunov, Roe, and Enquist and Osher are introduced. Higher-order upwind and total variation diminishing (TVD) schemes are also discussed in the revised Chapter 4.

The governing equations of fluid mechanics and heat transfer are presented in Chapter 5. The coverage has been expanded in several ways. The equations necessary to treat chemically reacting flows are discussed. Introductory information on direct and large-eddy simulation of turbulent flows is included. The filtered equations used in large-eddy simulation are presented as well as the Reynolds-averaged equations. The material on turbulence modeling has been augmented and now includes more details on one- and two-equation and Reynolds stress models as well as an introduction to the subgrid-scale modeling required for large-eddy simulation. A section has been added on the finite-volume formulation, a discretization procedure that proceeds from conservation equations in integral form.

Chapter 6 on methods for the inviscid flow equations is probably the most extensively revised chapter in the second edition. The revised chapter contains major new sections on flux splitting schemes, flux difference splitting schemes, the multidimensional case in generalized coordinates, and boundary conditions for the Euler equations. The chapter includes a discussion on implementing the integral form of conservation statements for arbitrarily shaped control volumes, particularly triangular cells, for two-dimensional applications.

Chapter 7 on methods for solving the boundary-layer equations includes new example applications of the inverse method, new material on the use of generalized coordinates, and a useful coordinate transformation for internal flows. In Chapter 8 methods are presented for solving simplified forms of the Navier-Stokes equations including the thin-layer Navier-Stokes (TLNS) equations, the parabolized Navier-Stokes (PNS) equations, the reduced Navier-Stokes (RNS) equations, the partially-parabolized Navier-Stokes (PPNS) equations, the viscous shock layer (VSL) equations, and the conical Navier-Stokes (CNS) equations. New material includes recent developments on pressure relaxation, upwind methods, coupled methods for solving the partially parabolized equations for subsonic flows, and applications.

Chapter 9 on methods for the "complete" Navier-Stokes equations has undergone substantial revision. This is appropriate because much of the research and development in CFD since the first edition appeared has been concentrated on solving these equations. Upwind methods that were first introduced in the context of model and Euler equations are described as they extend to the full Navier-Stokes equations. Methods to efficiently solve the compressible equations at very low Mach numbers through low Mach number preconditioning are described. New developments in methods based on derived variables, such as the dual potential method, are discussed. Modifications to the method of artificial compressibility required to achieve time accuracy are developed. The use of space-marching methods to solve the steady Navier-Stokes equations is described. Recent advances in pressure-correction (segregated) schemes for solving the Navier-Stokes equations such as the use of non-staggered grids and the pressure-implicit with splitting of operators (PISO) method are included in the revised chapter.

Grid generation, addressed in Chapter 10, is another area in which much activity has occurred since the appearance of the first edition. The coverage has been broadened to include introductory material on both structured and unstructured approaches. Coverage now includes algebraic and differential equation methods for constructing structured grids and the point insertion and advancing front methods for obtaining unstructured grids composed of triangles. Concepts employed in constructing hybrid grids composed of both quadrilateral cells (structured) and triangles, solution adaptive grids, and domain decomposition schemes are discussed.

We are grateful for the help received from many colleagues, users of the first edition and others, while this revision was being developed. We especially thank our colleagues Ganesh Rajagopalan, Alric Rothmayer, and Ijaz Parpia. We also continue to be indebted to our students, both past and present, for their contributions. We would like to acknowledge the skillful preparation of several new figures by Lynn Ekblad. Finally, we would like to thank our families for their patience and continued encouragement during the preparation of this second edition.

This text continues to be a collective work by the three of us. There is no junior or senior author. A coin flip determined the order of authors for the first edition, and a new coin flip has determined the order of authors for this edition.

John C. Tannehill
Dale A. Anderson
Richard H. Pletcher

PREFACE TO THE FIRST EDITION

This book is intended to serve as a text for introductory courses in computational fluid mechanics and heat transfer [or, synonymously, computational fluid dynamics (CFD)] for advanced undergraduates and/or first-year graduate students. The text has been developed from notes prepared for a two-course sequence taught at Iowa State University for more than a decade. No pretense is made that every facet of the subject is covered, but it is hoped that this book will serve as an introduction to this field for the novice. The major emphasis of the text is on finite-difference methods.

The material has been divided into two parts. The first part, consisting of Chapters 1–4, presents basic concepts and introduces the reader to the fundamentals of finite-difference methods. The second part of the book, consisting of Chapters 5–10, is devoted to applications involving the equations of fluid mechanics and heat transfer. Chapter 1 serves as an introduction, while a brief review of partial differential equations is given in Chapter 2. Finite-difference methods and the notions of stability, accuracy, and convergence are discussed in the third chapter.

Chapter 4 contains what is perhaps the most important information in the book. Numerous finite-difference methods are applied to linear and nonlinear model partial differential equations. This provides a basis for understanding the results produced when different numerical methods are applied to the same problem with a known analytic solution.

Building on an assumed elementary background in fluid mechanics and heat transfer, Chapter 5 reviews the basic equations of these subjects, emphasizing forms most suitable for numerical formulations of problems. A section on turbulence modeling is included in this chapter. Methods for solving inviscid flows using both conservative and nonconservative forms are presented in Chapter 6. Techniques for solving the boundary-layer equations for both laminar and turbulent flows are discussed in Chapter 7. Chapter 8 deals with equations of a class known as the "parabolized" Navier-Stokes equations which are useful for flows not adequately modeled by the boundary-layer equations, but not requiring the use of the full Navier-Stokes equations. Parabolized schemes for both subsonic and supersonic flows over external surfaces and in confined regions are included in this chapter. Chapter 9 is devoted to methods for the complete Navier-Stokes equations, including the Reynolds averaged form. A brief introduction to methods for grid generation is presented in Chapter 10 to complete the text.

At Iowa State University, this material is taught to classes consisting primarily of aerospace and mechanical engineers, although the classes often include students from other branches of engineering and earth sciences. It is our experience that Part I (Chapters 1–4) can be adequately covered in a one-semester, three-credit-hour course. Part II of the book contains more information than can be covered in great detail in most one-semester, three-credit-hour courses. This permits Part 2 to be used for courses with different objectives. Although we have found that the major thrust of each of Chapters 5 through 10 can be covered in one semester, it would also be possible to use only parts of this material for more specialized courses. Obvious modules would be Chapters 5, 6 and 10 for a course emphasizing inviscid flows or Chapters 5, 7–9, (and perhaps 10) for a course emphasizing viscous flows. Other combinations are clearly possible. If only one course can be offered in the subject, choices also exist. Part I of the text can be covered in detail in the single course or, alternatively, only selected material from Chapters 1–4 could be covered as well as some material on applications of particular interest from Part II. The material in the text is reasonably broad and should be appropriate for courses having a variety of objectives.

For background, students should have at least one basic course in fluid dynamics, one course in ordinary differential equations, and some familiarity with partial differential equations. Of course, some programming experience is also assumed.

The philosophy used throughout the CFD course sequence at Iowa State and embodied in this text is to encourage students to construct their own computer programs. For this reason, "canned" programs for specific problems do not appear in the text. Use of such programs does not enhance basic understanding necessary for algorithm development. At the end of each chapter, numerous problems are listed that necessitate numerical implementation of the text material. It is assumed that students have access to a high-speed digital computer.

We wish to acknowledge the contributions of all of our students, both past and present. We are deeply indebted to F. Blottner, S. Chakravarthy, G. Christoph, J. Daywitt, T. Holst, M. Hussaini, J. Ievalts, D. Jespersen, O. Kwon, M. Malik, J. Rakich, M. Salas, V. Shankar, R. Warming, and many others for helpful suggestions for improving the text. We would like to thank Pat Fox and her associates for skillfully preparing the illustrations. A special thanks to Shirley Riney for typing and editing the manuscript. Her efforts were a constant source of encouragement. To our wives and children, we owe a debt of gratitude for all of the hours stolen from them. Their forbearance is greatly appreciated.

Finally, a few words about the order in which the authors' names appear. This text is a collective work by the three of us. There is no junior or senior author. The final order was determined by a coin flip. Despite the emphasis of finite-difference methods in the text, we resorted to a "Monte Carlo" method for this determination.

Dale A. Anderson
John C. Tannehill
Richard H. Pletcher

PART ONE

FUNDAMENTALS

CHAPTER

ONE

INTRODUCTION

1.1 GENERAL REMARKS

The development of the high-speed digital computer during the twentieth century has had a great impact on the way principles from the sciences of fluid mechanics and heat transfer are applied to problems of design in modern engineering practice. Problems that would have taken years to work out with the computational methods and computers available 30 years ago can now be solved at very little cost in a few seconds of computer time. The ready availability of previously unimaginable computing power has stimulated many changes. These were first noticeable in industry and research laboratories, where the need to solve complex problems was the most urgent. More recently, changes brought about by the computer have become evident in nearly every facet of our daily lives. In particular, we find that computers are widely used in the educational process at all levels. Many a child has learned to recognize shapes and colors from mom and dad's computer screen before they could walk. To take advantage of the power of the computer, students must master certain fundamentals in each discipline that are unique to the simulation process. It is hoped that the present textbook will contribute to the organization and dissemination of some of this information in the fields of fluid mechanics and heat transfer.

Over the past half century, we have witnessed the rise to importance of a new methodology for attacking the complex problems in fluid mechanics and heat transfer. This new methodology has become known as computational fluid dynamics (CFD). In this computational (or numerical) approach, the equations (usually in partial differential form) that govern a process of interest are solved numerically. Some of the ideas are very old. The evolution of numerical methods, especially finite-difference methods for solving ordinary and partial differential equations, started approximately with the beginning of the twentieth century. The automatic digital computer was invented by Atanasoff in the late 1930s (see Gardner, 1982; Mollenhoff, 1988) and was used from nearly the beginning to solve problems in fluid dynamics. Still, these events alone did not revolutionize engineering practice. The explosion in computational activity did not begin until a third ingredient, general availability of high-speed digital computers, occurred in the 1960s.

Traditionally, both experimental and theoretical methods have been used to develop designs for equipment and vehicles involving fluid flow and heat transfer. With the advent of the digital computer, a third method, the numerical approach, has become available. Although experimentation continues to be important, especially when the flows involved are very complex, the trend is clearly toward greater reliance on computer-based predictions in design.

This trend can be largely explained by economics (Chapman, 1979). Over the years, computer speed has increased much more rapidly than computer costs. The net effect has been a phenomenal decrease in the cost of performing a given calculation. This is illustrated in Figure 1.1, where it is seen that the cost of performing a given calculation has been reduced by approximately a factor of 10 every 8 years. (Compare this with the trend in the cost of peanut butter in the past 8 years.) This trend in the cost of computations is based on the use of the best serial or vector computers available. It is true not every user will have easy access to the most recent computers, but increased access to very capable computers is another trend that started with the introduction of personal computers and workstations in the 1980s. The cost of performing a calculation on a desktop machine has probably dropped even more than a factor of 10 in an 8-year period, and the best of these "personal" machines are more capable than the best "mainframe" machines of a decade ago, achieving double-digit megaflops (millions of floating point operations per second). There seems to be

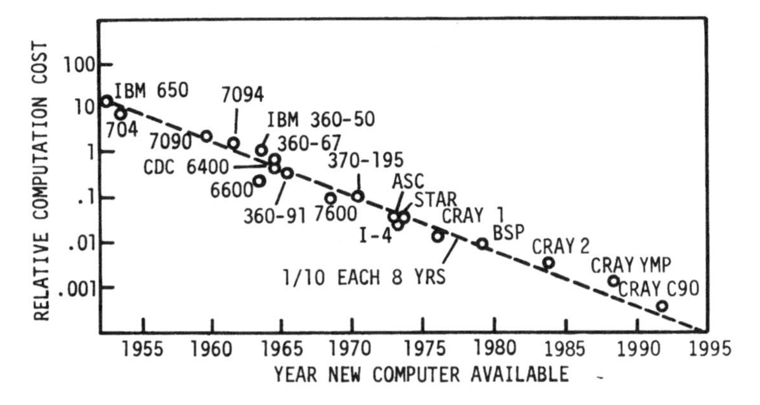

Figure 1.1 Trend of relative computation cost for a given flow and algorithm (based on Chapman, 1979; Kutler et al., 1987; Holst et al., 1992; Simon, 1995).

no real limit in sight to the computation speed that can be achieved when massively parallel computers are considered. Work is in progress toward achieving the goal of performance at the level of teraflops (10^{12} floating point operations per second) by the twenty-first century. This represents a 1000-fold increase in the computing speed that was achievable at the start of the 1990s. The increase in computing power per unit cost since the 1950s is almost incomprehensible. It is now possible to assign a homework problem in CFD, the solution of which would have represented a major breakthrough or could have formed the basis of a Ph.D. dissertation in the 1950s or 1960s. On the other hand, the costs of performing experiments have been steadily increasing over the same period of time.

The suggestion here is not that computational methods will soon completely replace experimental testing as a means to gather information for design purposes. Rather, it is believed that computer methods will be used even more extensively in the future. In most fluid flow and heat transfer design situations it will still be necessary to employ some experimental testing. However, computer studies can be used to reduce the range of conditions over which testing is required.

The need for experiments will probably remain for quite some time in applications involving turbulent flow, where it is presently not economically feasible to utilize computational models that are free of empiricism for most practical configurations. This situation is destined to change eventually, since it has become clear that the time-dependent Navier-Stokes equations can be solved numerically to provide accurate details of turbulent flow. Thus, as computer hardware and algorithms improve, the frontier will be pushed back continuously allowing flows of increasing practical interest to be computed by direct numerical simulation. The prospects are also bright for the increased use of large-eddy simulations, where modeling is required for only the smallest scales.

In applications involving multiphase flows, boiling, or condensation, especially in complex geometries, the experimental method remains the primary source of design information. Progress is being made in computational models for these flows, but the work remains in a relatively primitive state compared to the status of predictive methods for laminar single-phase flows over aerodynamic bodies.

1.2 COMPARISON OF EXPERIMENTAL, THEORETICAL, AND COMPUTATIONAL APPROACHES

As mentioned in the previous section, there are basically three approaches or methods that can be used to solve a problem in fluid mechanics and heat transfer. These methods are

1. Experimental
2. Theoretical

3. Computational (CFD)

The theoretical method is often referred to as an analytical approach, while the terms computational and numerical are used interchangeably. In order to illustrate how these three methods would be used to solve a fluid flow problem, let us consider the classical problem of determining the pressure on the front surface of a circular cylinder in a uniform flow of air at a Mach number (M_∞) of 4 and a Reynolds number (based on the diameter of the cylinder) of 5×10^6.

In the experimental approach, a circular cylinder model would first need to be designed and constructed. This model must have provisions for measuring the wall pressures, and it should be compatible with an existing wind tunnel facility. The wind tunnel facility must be capable of producing the required free stream conditions in the test section. The problem of matching flow conditions in a wind tunnel can often prove to be quite troublesome, particularly for tests involving scale models of large aircraft and space vehicles. Once the model has been completed and a wind tunnel selected, the actual testing can proceed. Since high-speed wind tunnels require large amounts of energy for their operation, the wind tunnel test time must be kept to a minimum. The efficient use of wind tunnel time has become increasingly important in recent years with the escalation of energy costs. After the measurements have been completed, wind tunnel correction factors can be applied to the raw data to produce the final wall pressure results. The experimental approach has the capability of producing the most realistic answers for many flow problems; however, the costs are becoming greater every day.

In the theoretical approach, simplifying assumptions are used in order to make the problem tractable. If possible, a closed-form solution is sought. For the present problem, a useful approximation is to assume a Newtonian flow (see Hayes and Probstein, 1966) of a perfect gas. With the Newtonian flow assumption, the shock layer (region between body and shock) is infinitesimally thin, and the bow shock lies adjacent to the surface of the body, as seen in Fig. 1.2(a). Thus the normal component of the velocity vector becomes zero after passing through the shock wave, since it immediately impinges on the body surface. The normal momentum equation across a shock wave (see Chapter 5) can be written as

$$p_1 + \rho_1 u_1^2 = p_2 + \rho_2 u_2^2 \tag{1.1}$$

where p is the pressure, ρ is the density, u is the normal component of velocity, and the subscripts 1 and 2 refer to the conditions immediately upstream and downstream of the shock wave, respectively. For the present problem [see Fig. 1.2(b)], Eq. (1.1) becomes

$$p_\infty + \rho_\infty V_\infty^2 \sin^2\sigma = p_{\text{wall}} + \rho_{\text{wall}} \cancelto{0}{u_{\text{wall}}^2} \tag{1.2}$$

or

$$p_{\text{wall}} = p_\infty \left(1 + \frac{\rho_\infty}{p_\infty} V_\infty^2 \sin^2\sigma \right) \tag{1.3}$$

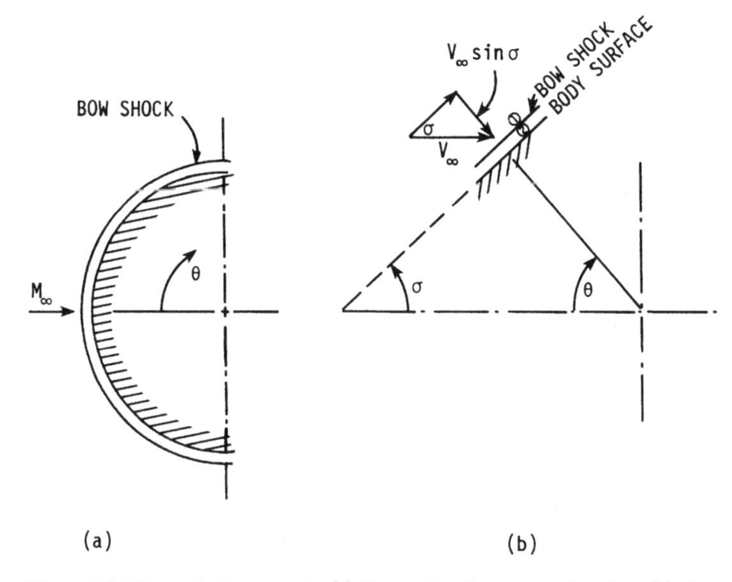

Figure 1.2 Theoretical approach. (a) Newtonian flow approximation. (b) Geometry at shock.

For a perfect gas, the speed of sound in the free stream is

$$a_\infty = \sqrt{\frac{\gamma p_\infty}{\rho_\infty}} \tag{1.4}$$

where γ is the ratio of specific heats. Using the definition of Mach number

$$M_\infty = \frac{V_\infty}{a_\infty} \tag{1.5}$$

and the trigonometric identity

$$\cos\theta = \sin\sigma \tag{1.6}$$

Eq. (1.3) can be rewritten as

$$p_{\text{wall}} = p_\infty(1 + \gamma M_\infty^2 \cos^2\theta) \tag{1.7}$$

At the stagnation point, $\theta = 0°$, so that the wall pressure becomes

$$p_{\text{stag}} = p_\infty(1 + \gamma M_\infty^2) \tag{1.8}$$

After inserting the stagnation pressure into Eq. (1.7), the final form of the equation is

$$p_{\text{wall}} = p_\infty + (p_{\text{stag}} - p_\infty)\cos^2\theta \tag{1.9}$$

The accuracy of this theoretical approach can be greatly improved if, in place of Eq. (1.8), the stagnation pressure is computed from Rayleigh's pitot formula

(Shapiro, 1953):

$$p_{\text{stag}} = p_\infty \left[\frac{(\gamma + 1) M_\infty^2}{2} \right]^{\gamma/(\gamma - 1)} \left[\frac{\gamma + 1}{2\gamma M_\infty^2 - (\gamma - 1)} \right]^{1/(\gamma - 1)} \tag{1.10}$$

which assumes an isentropic compression between the shock and body along the stagnation streamline. The use of Eq. (1.9) in conjunction with Eq. (1.10) is referred to as the modified Newtonian theory. The wall pressures predicted by this theory are compared in Fig. 1.3 to the results obtained using the experimental approach (Beckwith and Gallagher, 1961). Note that the agreement with the experimental results is quite good up to about $\pm 35°$. The big advantage of the theoretical approach is that "clean," general information can be obtained, in many cases, from a simple formula, as in the present example. This approach is quite useful in preliminary design work, since reasonable answers can be obtained in a minimum amount of time.

In the computational approach, a limited number of assumptions are made and a high-speed digital computer is used to solve the resulting governing fluid dynamic equations. For the present high Reynolds number problem, inviscid flow can be assumed, since we are only interested in determining wall pressures on the forward portion of the cylinder. Hence the Euler equations are the appropriate governing fluid dynamic equations. In order to solve these equations, the region between the bow shock and body must first be subdivided into a computational grid, as seen in Fig. 1.4. The partial derivatives appearing in the

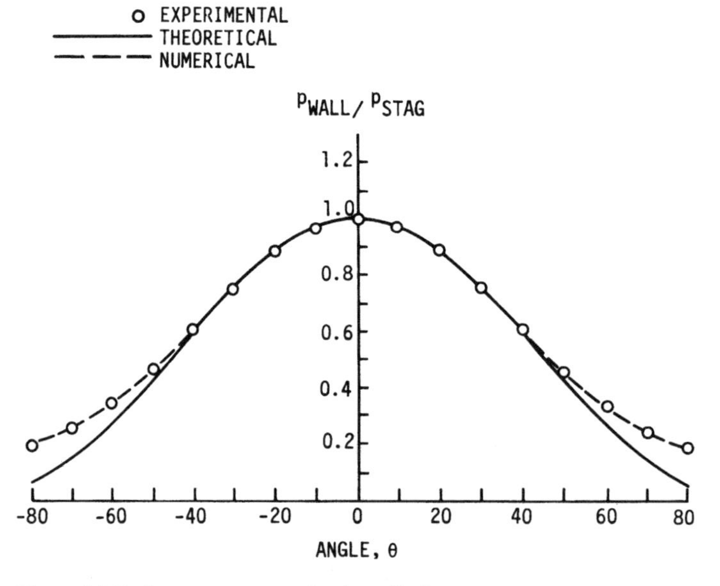

Figure 1.3 Surface pressure on circular cylinder.

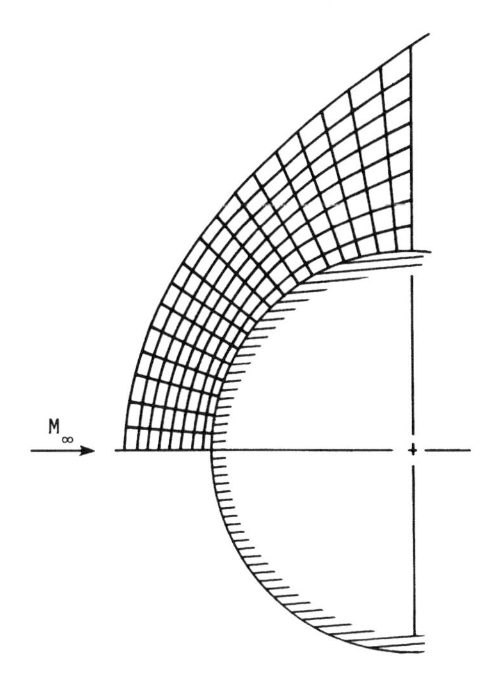

Figure 1.4 Computational grid.

unsteady Euler equations can be replaced by appropriate finite differences at each grid point. The resulting equations are then integrated forward in time until a steady-state solution is obtained asymptotically after a sufficient number of time steps. The details of this approach will be discussed in forthcoming chapters. The results of this technique (Daywitt and Anderson, 1974) are shown in Fig. 1.3. Note the excellent agreement with experiment.

In comparing the methods, we note that a computer simulation is free of some of the constraints imposed on the experimental method for obtaining information upon which to base a design. This represents a major advantage of the computational method, which should be increasingly important in the future. The idea of experimental testing is to evaluate the performance of a relatively inexpensive small-scale version of the prototype device. In performing such tests, it is not always possible to simulate the true operating conditions of the prototype. For example, it is very difficult to simulate the large Reynolds numbers of aircraft in flight, atmospheric reentry conditions, or the severe operating conditions of some turbomachines in existing test facilities. This suggests that the computational method, which has no such restrictions, has the potential of providing information not available by other means. On the other hand, computational methods also have limitations; among these are computer storage and speed. Other limitations arise owing to our inability to understand and mathematically model certain complex phenomena. None of these limitations of the computational method are insurmountable in principle, and current trends show reason for optimism about the role of the computational

Table 1.1 Comparison of approaches

Approach	Advantages	Disadvantages
Experimental	1. Capable of being most realistic	1. Equipment required 2. Scaling problems 3. Tunnel corrections 4. Measurement difficulties 5. Operating costs
Theoretical	1. Clean, general information, which is usually in formula form	1. Restricted to simple geometry and physics 2. Usually restricted to linear problems
Computational	1. No restriction to linearity 2. Complicated physics can be treated 3. Time evolution of flow can be obtained	1. Truncation errors 2. Boundary condition problems 3. Computer costs

method in the future. As seen in Fig. 1.1, the relative cost of computing a given flow field has decreased by almost 3 orders of magnitude during the past 20 years, and this trend is expected to continue in the near future. As a consequence, wind tunnels have begun to play a secondary role to the computer for many aerodynamic problems, much in the same manner as ballistic ranges perform secondary roles to computers in trajectory mechanics (Chapman, 1975). There are, however, many flow problems involving complex physical processes that still require experimental facilities for their solution.

Some of the advantages and disadvantages of the three approaches are summarized in Table 1.1. It should be mentioned that it is sometimes difficult to distinguish between the different methods. For example, when numerically computing turbulent flows, the eddy viscosity models that are frequently used are obtained from experiments. Likewise, many theoretical techniques that employ numerical calculations could be classified as computational approaches.

1.3 HISTORICAL PERSPECTIVE

As one might expect, the history of CFD is closely tied to the development of the digital computer. Most problems were solved using methods that were either analytical or empirical in nature until the end of World War II. Prior to this time, there were a few pioneers using numerical methods to solve problems. Of course, the calculations were performed by hand, and a single solution represented a monumental amount of work. Since that time, the digital computer has been developed, and the routine calculations required in obtaining a numerical solution are carried out with ease.

The actual beginning of CFD or the development of methods crucial to CFD is a matter of conjecture. Most people attribute the first definitive work of importance to Richardson (1910), who introduced point iterative schemes for numerically solving Laplace's equation and the biharmonic equation in an address to the Royal Society of London. He actually carried out calculations for the stress distribution in a masonry dam. In addition, he clearly defined the difference between problems that must be solved by a relaxation scheme and those that we refer to as marching problems.

Richardson developed a relaxation technique for solving Laplace's equation. His scheme used data available from the previous iteration to update each value of the unknown. In 1918, Liebmann presented an improved version of Richardson's method. Liebmann's method used values of the dependent variable both at the new and old iteration level in each sweep through the computational grid. This simple procedure of updating the dependent variable immediately reduced the convergence times for solving Laplace's equation. Both the Richardson method and Liebmann's scheme are usually used in elementary heat transfer courses to demonstrate how apparently simple changes in a technique greatly improve efficiency.

Sometimes the beginning of modern numerical analysis is attributed to a famous paper by Courant, Friedrichs, and Lewy (1928). The acronym CFL, frequently seen in the literature, stands for these three authors. In this paper, uniqueness and existence questions were addressed for the numerical solutions of partial differential equations. Testimony to the importance of this paper is evidenced in its re-publication in 1967 in the *IBM Journal of Research and Development*. This paper is the original source for the CFL stability requirement for the numerical solution of hyperbolic partial differential equations.

In 1940, Southwell introduced a relaxation scheme that was extensively used in solving both structural and fluid dynamic problems where an improved relaxation scheme was required. His method was tailored for hand calculations, in that point residuals were computed and these were scanned for the largest value. The point where the residual was largest was always relaxed as the next step in the technique. During the decades of the 1940s and 1950s. Southwell's methods were generally the first numerical techniques introduced to engineering students. Allen and Southwell (1955) applied Southwell's scheme to solve the incompressible, viscous flow over a cylinder. This solution was obtained by hand calculation and represented a substantial amount of work. Their calculation added to the existing viscous flow solutions that began to appear in the 1930s.

During World War II and immediately following, a large amount of research was performed on the use of numerical methods for solving problems in fluid dynamics. It was during this time that Professor John von Neumann developed his method for evaluating the stability of numerical methods for solving time-marching problems. It is interesting that Professor von Neumann did not publish a comprehensive description of his methods. However, O'Brien, Hyman, and Kaplan (1950) later presented a detailed description of the von Neumann method. This paper is significant because it presents a practical way of evaluating stability that can be understood and used reliably by scientists and engineers. The von Newman method is the most widely used technique in CFD for determining stability. Another of the important contributions appearing at about the same time was due to Peter Lax (1954). Lax developed a technique for computing fluid flows including shock waves that represent discontinuities in the flow variables. No special treatment was required for computing the shocks. This special feature developed by Lax was due to the use of the conservation-law form of the governing equations and is referred to as shock capturing.

At the same time, progress was being made on the development of methods for both elliptic and parabolic problems. Frankel (1950) presented the first version of the successive overrelaxation (SOR) scheme for solving Laplace's equation. This provided a significant improvement in the convergence rate. Peaceman and Rachford (1955) and Douglas and Rachford (1956) developed a new family of implicit methods for parabolic and elliptic equations in which sweep directions were alternated and the allowed step size was unrestricted. These methods are referred to as alternating direction implicit (ADI) schemes and were extended to the equations of fluid mechanics by Briley and McDonald (1973) and Beam and Warming (1976, 1978). This implementation provided fast efficient solvers for the solution of the Euler and Navier-Stokes equations.

Research in CFD continued at a rapid pace during the decade of the sixties. Early efforts at solving flows with shock waves used either the Lax approach or an artificial viscosity scheme introduced by von Neumann and Richtmyer (1950). Early work at Los Alamos included the development of schemes like the particle-in-cell (PIC) method, which used the dissipative nature of the finite-difference scheme to smear the shock over several mesh intervals (Evans and Harlow, 1957). In 1960, Lax and Wendroff introduced a method for computing flows with shocks that was second-order accurate and avoided the excessive smearing of the earlier approaches. The MacCormack (1969) version of this technique became one of the most widely used numerical schemes. Gary (1962) presented early work demonstrating techniques for fitting moving shocks, thus avoiding the smearing associated with the previous shock-capturing schemes. Moretti and Abbett (1966) and Moretti and Bleich (1968) applied shock-fitting procedures to multidimensional supersonic flow over various configurations. Even today, we see either shock-capturing or shock-fitting methods used to solve problems with shock waves.

Godunov (1959) proposed solving multidimensional compressible fluid dynamics problems by using a solution to a Riemann problem for flux calculations at cell faces. This approach was not vigorously pursued until van Leer (1974, 1979) showed how higher-order schemes could be constructed using the same idea. The intensive computational effort necessary with this approach led Roe (1980) to suggest using an approximate solution to the Riemann problem (flux-difference splitting) in order to improve the efficiency. This substantially reduced the work required to solve multidimensional problems and represents the current trend of practical schemes employed on convection-dominated flows. The concept of flux splitting was also introduced as a technique for treating convection-dominated flows. Steger and Warming (1979) introduced splitting where fluxes were determined using an upwind approach. Van Leer (1982) also proposed a new flux splitting technique to improve on the existing methods. These original ideas are used in many of the modern production codes, and improvements continue to be made on the basic concept.

As part of the development of modern numerical methods for computing flows with rapid variations such as those occurring through shock waves, the concept of limiters was introduced. Boris and Book (1973) first suggested this approach, and it has formed the basis for the nonlinear limiting subsequently used in most codes. Harten (1983) introduced the idea of total variation diminishing (TVD) schemes. This generalized the limiting concept and has led to substantial advances in the way the nonlinear limiting of fluxes is implemented. Others that also made substantial contributions to the development of robust methods for computing convection-dominated flows with shocks include Enquist and Osher (1980, 1981), Osher (1984), Osher and Chakravarthy (1983), Yee (1985a, 1985b), and Yee and Harten (1985). While this is not an all-inclusive list, the contributions of these and others have led to the addition of nonlinear dissipation with limiting as a major factor in state-of-the-art schemes in use today.

Other contributions were made in algorithm development dealing with the efficiency of the numerical techniques. Both multigrid and preconditioning techniques were introduced to improve the convergence rate of iterative calculations. The multigrid approach was first applied to elliptic equations by Fedorenko (1962, 1964) and was later extended to the equations of fluid mechanics by Brandt (1972, 1977). At the same time, strides in applying reduced forms of the Euler and Navier-Stokes equations were being made. Murman and Cole (1971) made a major contribution in solving the transonic small-disturbance equation by applying type-dependent differencing to the subsonic and supersonic portions of the flow field. The thin-layer Navier-Stokes equations have been extensively applied to many problems of interest, and the paper by Pulliam and Steger (1978) is representative of these applications. Also, the parabolized Navier-Stokes (PNS) equations were introduced by Rudman and Rubin (1968), and this approximate form of the Navier-Stokes equations has been used to solve many supersonic viscous flow fields. The correct treatment of the streamwise pressure gradient when solving the PNS equations was examined in detail by Vigneron et al. (1978a), and a new method of limiting the streamwise pressure gradient in subsonic regions was developed and is in prominent use today.

In addition to the changes in treating convection terms, the control-volume or finite-volume point of view as opposed to the finite-difference approach was applied to the construction of difference methods for the fluid dynamic equations. The finite-volume approach provides an easy way to apply numerical techniques to unstructured grids, and many codes presently in use are based on unstructured grids. With the development of methods that are robust for general problems, large-scale simulations of complete vehicles are now a common occurrence. Among the many researchers who have made significant contributions in this effort are Jameson and Baker (1983), Shang and Scherr (1985), Jameson et al. (1986), Flores et al. (1987), Obayashi et al. (1987), Yu et al. (1987), and Buning et al. (1988). At this time, the simulation of flow about a complete aircraft using the Euler equations is viewed as a reasonable tool for the analysis and design of these vehicles. Most simulations of this nature are still performed on serial vector computers. In the future, the full Navier-Stokes equations will be used, but the application of these equations to entire vehicles will only become an everyday occurrence when large parallel computers are available to the industry.

The progress in CFD over the past 25 years has been enormous. For this reason, it is impossible, with the short history given here, to give credit to all who have contributed. A number of review and history papers that provide a more precise state of the art may be cited and include those by Hall (1981), Krause (1985), Diewert and Green (1986), Jameson (1987), Kutler (1993), Rubin and Tannehill (1992), and MacCormack (1993). In addition, the Focus '92 issues of *Aerospace America* are dedicated to a review of the state of the art. The appearance of text materials for the study of CFD should also be mentioned in any brief history. The development of any field is closely paralleled by the appearance of books dealing with the subject. Early texts dealing with CFD include books by Roache (1972), Holt (1977), Chung (1978), Chow (1979), Patankar (1980), Baker (1983), Peyret and Taylor (1983), and Anderson et al. (1984). More recent books include those by Sod (1985), Thompson et al. (1985), Oran and Boris (1987), Hirsch (1988), Fletcher (1988), Hoffmann (1989), and Anderson (1995). The interested reader will also note that occasional writings appear in the popular literature that discuss the application of digital simulation to engineering problems. These applications include CFD but do not usually restrict the range of interest to this single discipline.

CHAPTER

TWO

PARTIAL DIFFERENTIAL EQUATIONS

2.1 INTRODUCTION

Many important physical processes in nature are governed by partial differential equations (PDEs). For this reason, it is important to understand the physical behavior of the model represented by the PDE. In addition, knowledge of the mathematical character, properties, and solution of the governing equations is required. In this chapter we will discuss the physical significance and the mathematical behavior of the most common types of PDEs encountered in fluid mechanics and heat transfer. Examples are included to illustrate important properties of the solutions of these equations. In the last sections we extend our discussion to systems of PDEs and present a number of model equations, many of which are used in Chapter 4 to demonstrate the application of various discretization methods.

2.2 PHYSICAL CLASSIFICATION

2.2.1 Equilibrium Problems

Equilibrium problems are problems in which a solution of a given PDE is desired in a closed domain subject to a prescribed set of boundary conditions (see Fig. 2.1). Equilibrium problems are boundary value problems. Examples of such problems include steady-state temperature distributions, incompressible inviscid flows, and equilibrium stress distributions in solids.

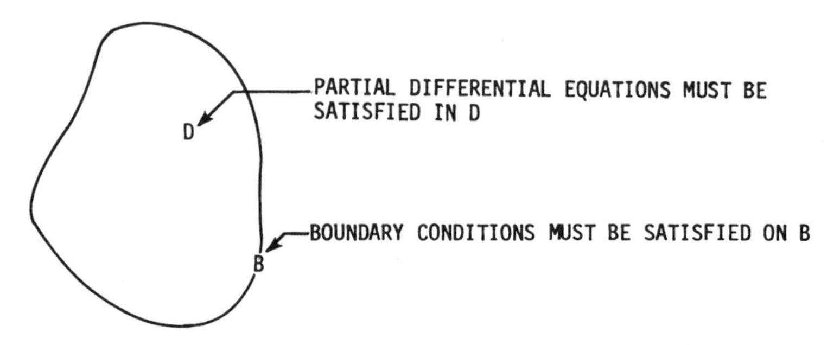

Figure 2.1 Domain for an equilibrium problem.

Sometimes equilibrium problems are referred to as jury problems. This is an apt name, since the solution of the PDE at every point in the domain depends upon the prescribed boundary condition at every point on B. In this sense the boundary conditions are certainly the jury for the solution in D. Mathematically, equilibrium problems are governed by elliptic PDEs.

Example 2.1 The steady-state temperature distribution in a conducting medium is governed by Laplace's equation. A typical problem requiring the steady-state temperature distribution in a two-dimensional (2-D) solid with the boundaries held at constant temperatures is defined by the equation

$$\nabla^2 T = \frac{\partial^2 T}{\partial x^2} + \frac{\partial^2 T}{\partial y^2} = 0 \qquad 0 \leqslant x \leqslant 1 \qquad 0 \leqslant y \leqslant 1 \tag{2.1}$$

with boundary conditions

$$T(0, y) = 0$$
$$T(1, y) = 0$$
$$T(x, 0) = T_0$$
$$T(x, 1) = 0$$

The 2-D configuration is shown in Fig. 2.2.

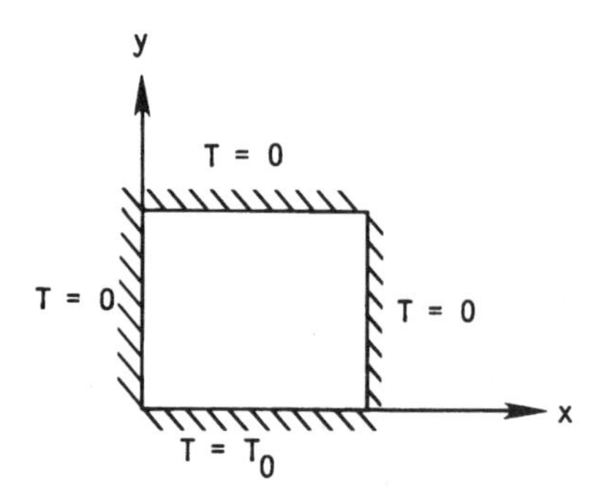

Figure 2.2 Unit square with fixed boundary temperatures.

Solution One of the standard techniques used to solve a linear PDE is separation of variables (Greenspan, 1961). This technique assumes that the unknown temperature can be written as the product of a function of x and a function of y, i.e.,

$$T(x, y) = X(x)Y(y)$$

If a solution of this form can be found that satisfies both the PDE and the boundary conditions, then it can be shown (Weinberger, 1965) that this is the one and only solution to the problem. After this form of the temperature is substituted into Laplace's equation, two ordinary differential equations (ODEs) are obtained. The resulting equations and homogeneous boundary conditions are

$$\begin{aligned} X'' + \alpha^2 X &= 0 \qquad Y'' - \alpha^2 Y = 0 \\ X(0) &= 0 \\ X(1) &= 0 \qquad Y(1) = 0 \end{aligned} \tag{2.2}$$

The prime denotes differentiation, and the factor α^2 arises from the separation process and must be determined as part of the solution to the problem. The solutions of the two differential equations given in Eq. (2.2) may be written

$$X(x) = A \sin(n\pi x) \qquad Y(y) = C \sinh[n\pi(y - 1)]$$

the boundary conditions enter the solution in the following way:

1. $$T(0, y) = 0 \rightarrow X(0) = 0$$

$$T(x, 1) = 0 \rightarrow Y(1) = 0$$

These two conditions determine the kinds of functions allowed in the expression for $T(x, y)$. The boundary condition $T(0, y) = 0$ is satisfied if the solution of the separated ODE satisfies $X(0) = 0$. Since the solution in general contains sine and cosine terms, this boundary condition eliminates the cosine terms. A similar behavior is observed by satisfying $T(x, 1) = 0$ through $Y(1) = 0$ for the separated equation.

2. $$T(1, y) = 0 \rightarrow X(1) = 0$$

This condition identifies the eigenvalues, i.e., the particular values of α that generate eigenfunctions satisfying this required boundary condition. Since the solution of the first separated equation, Eq. (2.2), was

$$X(x) = A \sin(\alpha x)$$

a nontrivial solution for $X(x)$ exists that satisfies $X(1) = 0$ only if $\alpha = n\pi$, where $n = 1, 2, \ldots$.

3. $$T(x, 0) = T_0$$

The prescribed temperature on the x axis determines the manner in which the eigenfunctions are combined to yield the correct solution to the problem.

The solution of the present problem is written

$$T(x, y) = \sum_{n=1}^{\infty} A_n \sin(n\pi x) \sinh[n\pi(y - 1)] \tag{2.3}$$

In this case, functions of the form $\sin(n\pi x)\sinh[n\pi(y - 1)]$ satisfy the PDE and three of the boundary conditions. In general, an infinite series composed of products of trigonometric sines and cosines and hyperbolic sines and cosines is required to satisfy the boundary conditions. For this problem, the fourth boundary condition along the lower boundary of the domain is given as

$$T(x, 0) = T_0$$

We use this to determine the coefficients A_n of Eq. (2.3). Thus we find (see Prob. 2.1)

$$A_n = \frac{2T_0}{n\pi} \frac{[(-1)^n - 1]}{\sinh(n\pi)}$$

The solution $T(x, y)$ provides the steady temperature distribution in the solid. It is clear that the solution at any point interior to the domain of interest depends upon the specified conditions at all points on the boundary. This idea is fundamental to all equilibrium problems.

Example 2.2 The irrotational flow of an incompressible inviscid fluid is governed by Laplace's equation. Determine the velocity distribution around the 2-D cylinder shown in Fig. 2.3 in an incompressible inviscid fluid flow. The flow is governed by

$$\nabla^2 \phi = 0$$

where ϕ is defined as the velocity potential, i.e., $\nabla\phi = \mathbf{V} =$ velocity vector. The boundary condition on the surface of the cylinder is

$$\mathbf{V} \cdot \nabla F = 0 \tag{2.4}$$

where $F(r, \theta) = 0$ is the equation of the surface of the cylinder. In addition, the *velocity must* approach the free stream value as distance from the body becomes large, i.e., as $(x, y) \to \infty$,

$$\nabla\phi = \mathbf{V}_\infty \tag{2.5}$$

Solution This problem is solved by combining two elementary solutions of Laplace's equation that satisfy the boundary conditions. This superposition of two elementary solutions is an acceptable way of obtaining a third solution only because Laplace's equation is linear. For a linear PDE, any linear combination of solutions is also a solution (Churchill, 1941). In this case, the flow around a cylinder can be simulated by adding the velocity potential for a uniform flow to

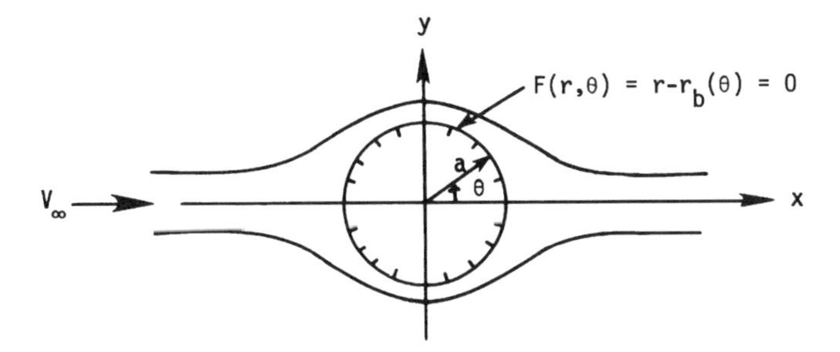

Figure 2.3 Two-dimensional flow around a cylinder.

that for a doublet (Karamcheti, 1966). The resulting solution becomes

$$\phi = V_\infty x + \frac{K \cos \theta}{\sqrt{x^2 + y^2}} = V_\infty x + \frac{Kx}{x^2 + y^2} \tag{2.6}$$

where the first term is the uniform oncoming flow, and the second term is a solution for a doublet of strength $2\pi K$.

2.2.2 Marching Problems

Marching or propagation problems are transient or transient-like problems where the solution of a PDE is required on an open domain subject to a set of initial conditions and a set of boundary conditions. Figure 2.4 illustrates the domain and marching direction for this case. Problems in this category are initial value or initial boundary value problems. The solution must be computed by marching outward from the initial data surface while satisfying the boundary conditions. Mathematically, these problems are governed by either hyperbolic or parabolic PDEs.

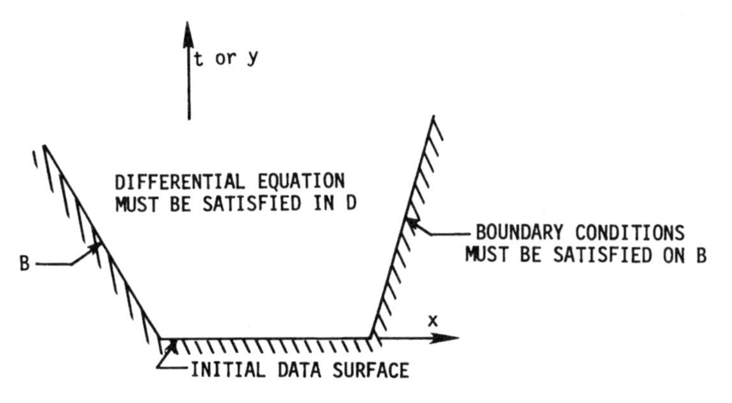

Figure 2.4 Domain for a marching problem.

Example 2.3 Determine the transient temperature distribution in a 1-D solid (Fig. 2.5) with a thermal diffusivity α if the initial temperature in the solid is 0° and if at all subsequent times, the temperature of the left side is held at 0° while the right side is held at T_0.

Solution The governing differential equation is the 1-D heat equation

$$\frac{\partial T}{\partial t} = \alpha \frac{\partial^2 T}{\partial x^2} \tag{2.7}$$

with boundary conditions

$$T(0, t) = 0 \qquad T(1, t) = T_0$$

and initial condition

$$T(x, 0) = 0$$

Again, for this linear equation, separation of variables will lead to a solution. Because of the nonhomogeneous boundary conditions in this problem, it is helpful to use the principle of superposition to determine the solution as the sum of the solution to the steady problem that results as the time becomes very large and a transient solution that dies out at large times. Thus we let $T(x, t) = u(x) + v(x, t)$. Substituting this decomposition into the governing PDE, we find that because u is independent of time,

$$\frac{d^2 u}{dx^2} = 0 \tag{2.8}$$

with boundary conditions

$$u(0) = 0 \qquad u(1) = T_0$$

The solution for the steady problem is thus $u(x) = T_0 x$. We find also that the transient solution must satisfy

$$\frac{\partial v}{\partial t} = \alpha \frac{\partial^2 v}{\partial x^2} \tag{2.9}$$

with associated boundary conditions

$$v(0, t) = v(1, t) = 0$$

and initial condition

$$v(x, 0) = -T_0 x$$

The initial condition for v is required in order that the sum of u and v satisfy the initial conditions of the problem. Separation of variables may be used to solve Eq. (2.9), and the solution is written in the form

$$v(x, t) = V(t)X(x)$$

If we denote the separation constant by $-\beta^2$, it is necessary to solve the ODEs

$$V' + \alpha\beta^2 V = 0 \qquad X'' + \beta^2 X = 0$$

$$X(0) = X(1) = 0$$

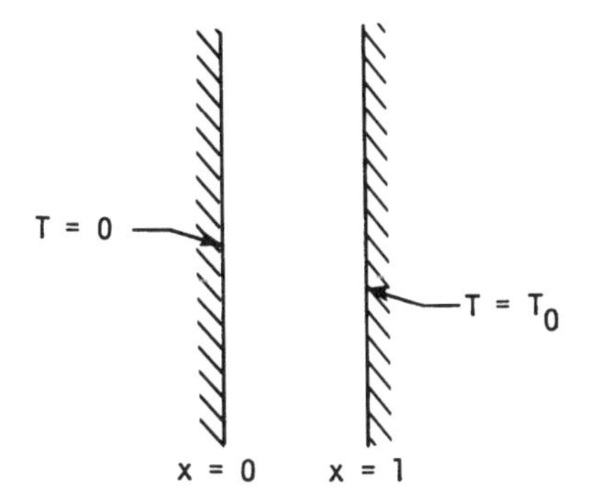

Figure 2.5 One-dimensional solid.

with the initial distribution on v as noted above. The general solution for V is readily obtained as

$$V(t) = e^{-\alpha \beta^2 t}$$

A solution for X that satisfies the boundary conditions is of the form

$$X(x) = \sin \beta x$$

where β must equal $n\pi$ ($n = 1, 2, \ldots$), so that the boundary conditions on X are met. The general solution that satisfies the PDE for v and the boundary conditions is then of the form

$$v(x,t) = e^{-\alpha n^2 \pi^2 t} \sin(n\pi x)$$

The orthogonality properties of the trigonometric functions (Weinberger, 1965) are used to meet the initial conditions as a Fourier sine series. This leads to the final solution for T, obtained by adding the solutions for u and v together:

$$T = T_0 x + \sum_{n=1}^{\infty} \frac{2T_0(-1)^n}{n\pi} e^{-n^2 \pi^2 \alpha t} \sin(n\pi x) \tag{2.10}$$

Example 2.4 Find the displacement $y(x,t)$ of a string of length l stretched between $x = 0$ and $x = l$ if it is displaced initially into position $y(x,0) = \sin \pi x/l$ and released from rest. Assume no external forces act on the string.

Solution In this case the motion of the string is governed by the wave equation

$$\frac{\partial^2 y}{\partial t^2} = a^2 \frac{\partial^2 y}{\partial x^2} \tag{2.11}$$

where a is a positive constant. The boundary conditions are

$$y(0,t) = y(l,t) = 0 \tag{2.12}$$

and initial conditions

$$y(x,0) = \sin \frac{\pi x}{l} \qquad \frac{\partial}{\partial t} y(x,t)|_{t=0} = 0 \tag{2.13}$$

The solution for this particular example is

$$y(x,t) = \sin\left(\pi \frac{x}{l}\right) \cos\left(a\pi \frac{t}{l}\right) \tag{2.14}$$

Solutions for problems of this type usually require an infinite series to correctly approximate the initial data. In this case, only one term of this series survives because the initial displacement requirement is exactly satisfied by one term.

The physical phenomena governed by the heat equation and the wave equation are different, but both are classified as marching problems. The behavior of the solutions to these equations and methods used to obtain these solutions are also quite different. This will become clear as the mathematical character of these equations is studied.

Typical examples of marching problems include unsteady inviscid flow, steady supersonic inviscid flow, transient heat conduction, and boundary-layer flow.

2.3 MATHEMATICAL CLASSIFICATION

The classification of PDEs is based on the mathematical concept of characteristics that are lines (in two dimensions) or surfaces (in three dimensions) along which certain properties remain constant or certain derivatives may be discontinuous. Such *characteristic* lines or surfaces are related to the directions in which "information" can be transmitted in physical problems governed by PDEs. Equations (single or system) that admit wave-like solutions are known as hyperbolic. If the equations admit solutions that correspond to damped waves, they are designated parabolic. If solutions are not wave-like, the equation or system is designated as elliptic. Although first-order equations or a system of first-order equations can be classified as indicated above, it is instructive at this point to develop classification concepts through consideration of the following general second-order PDE:

$$a\phi_{xx} + b\phi_{xy} + c\phi_{yy} + d\phi_x + e\phi_y + f\phi = g(x,y) \tag{2.15a}$$

where a, b, c, d, e, and f are functions of (x, y), i.e., we consider a linear equation. While this restriction is not essential, this form is convenient to use. Frequently, consideration is given to quasi-linear equations, which are defined as equations that are linear in the highest derivative. In terms of Eq. (2.15*a*), this means that a, b, and c could be functions of x, y, ϕ, ϕ_x, and ϕ_y. For our discussion, however, we assume that Eq. (2.15*a*) is linear and the coefficients depend only upon x and y.

We will indicate how equations having the general form of Eq. (2.15*a*) can be classified as hyperbolic, parabolic, or elliptic and how a standard or canonical form can be identified for each class by making use of the characteristic curves associated with the PDE. This will be discussed for equations with two independent variables, but the concepts can be extended to equations involving more independent variables, such as would be encountered in 3-D unsteady physical problems.

The classification of a second-order PDE depends only on the second-derivative terms of the equation, so we may rearrange Eq. (2.15*a*) as

$$a\phi_{xx} + b\phi_{xy} + c\phi_{yy} = -(d\phi_x + e\phi_y + f\phi - g) = H \tag{2.15b}$$

The characteristics, if they exist and are real curves within the solution domain, represent the locus of points along which the second derivatives may not be continuous. Along such curves, discontinuities in the solution, such as shock waves in supersonic flow, may appear. To identify such curves, we proceed as follows. For the general second-order PDE under consideration, the initial and boundary conditions are specified in terms of the function ϕ and first derivatives of ϕ. Assuming that ϕ and first derivatives of ϕ are continuous, we inquire if there may be any locations where this information would not uniquely determine the solution. In other words, may there be locations where the second derivatives are discontinuous?

Let τ be a parameter that varies along a curve C in the x-y plane. That is, on C, $x = x(\tau)$ and $y = y(\tau)$. The curve C may be on the boundary. For convenience, on C, we define

$$\begin{aligned} \phi_x &= p(\tau) \qquad & \phi_{xx} &= u(\tau) \\ \phi_y &= q(\tau) \qquad & \phi_{xy} &= v(\tau) \\ & & \phi_{yy} &= w(\tau) \end{aligned}$$

We suppose that ϕ, p, and q are given along C, as they might be given as boundary or initial conditions. With these definitions, Eq. (2.15*b*) becomes

$$au(\tau) + bv(\tau) + cw(\tau) = H \tag{2.15c}$$

Using the chain rule, we observe that

$$\frac{dp}{d\tau} = u\frac{dx}{d\tau} + v\frac{dy}{d\tau} \tag{2.15d}$$

$$\frac{dq}{d\tau} = v\frac{dx}{d\tau} + w\frac{dy}{d\tau} \tag{2.15e}$$

Equations (2.15*c*)–(2.15*e*) can be considered a system of three equations from which the second derivatives (u, v, and w) might be determined from the specified values of ϕ and the first derivatives of ϕ along C. These can be written in matrix form ($[A]\mathbf{x} = \mathbf{c}$) as

$$\begin{bmatrix} a & b & c \\ \dfrac{dx}{d\tau} & \dfrac{dy}{d\tau} & 0 \\ 0 & \dfrac{dx}{d\tau} & \dfrac{dy}{d\tau} \end{bmatrix} \begin{bmatrix} u \\ v \\ w \end{bmatrix} = \begin{bmatrix} H \\ \dfrac{dp}{d\tau} \\ \dfrac{dq}{d\tau} \end{bmatrix}$$

If the determinant of the coefficient matrix is zero, then there may be no unique solution for the second derivatives u, v, w along C for the given values of ϕ and its first derivatives. Thus we can write the condition for discontinuity (or nonuniqueness) in the highest order derivatives as

$$a\left(\frac{dy}{d\tau}\right)^2 - b\left(\frac{dx}{d\tau}\right)\left(\frac{dy}{d\tau}\right) + c\left(\frac{dx}{d\tau}\right)^2 = 0$$

or

$$a(dy)^2 - b\,dx\,dy + c(dx)^2 = 0 \tag{2.16}$$

Letting $h = dy/dx$, we can write Eq. (2.16) as

$$ah^2(dx)^2 - bh(dx)^2 + c(dx)^2 = 0$$

which, after division by $(dx)^2$, reduces to a quadratic equation in h:

$$ah^2 - bh + c = 0 \tag{2.17}$$

Solving for $h = dy/dx$ gives

$$h = \frac{dy}{dx} = \frac{b \pm \sqrt{b^2 - 4ac}}{2a} \tag{2.18}$$

The curves $y(x)$ that satisfy Eq. (2.18) are called the characteristics of the PDE. Along these curves, the second derivatives are not uniquely determined by specified values of ϕ and first derivatives of ϕ, and discontinuities in the highest order derivatives may exist. Note that when the coefficients a, b, and c are constants, the solution has a particularly simple form. In passing, we note that other useful relationships, known as the *compatibility relations*, can be developed from the system Eqs. (2.15*c*–2.15*e*). These are discussed in Chapter 6. See also Hirsch (1988).

We notice that the parameter $(b^2 - 4ac)$ plays a major role in the nature of the characteristic curves. If $(b^2 - 4ac)$ is positive, two distinct families of real characteristic curves exist. If $(b^2 - 4ac)$ is zero, only a single family of characteristic curves exist. If $(b^2 - 4ac)$ is negative, the right-hand side of Eq. (2.18) is complex, and no real characteristics exist. As in the classification of general second-degree equations in analytic geometry, the PDE is classified as (1) hyperbolic if $(b^2 - 4ac)$ is positive, (2) parabolic if $(b^2 - 4ac)$ is zero, and (3) elliptic if $(b^2 - 4ac)$ is negative. Note that if a, b, c are not constants, the classification may change from point to point in the problem domain.

Equations of each class can be reduced to a representative *canonical* or *characteristic coordinate* form by a coordinate transformation that makes use of the characteristic curves. We state these forms here and illustrate the transformations needed to obtain them in examples to follow.

Two characteristic coordinate forms exist for a hyperbolic PDE:

$$\phi_{\xi\xi} - \phi_{\eta\eta} = h_1(\phi_\xi, \phi_\eta, \phi, \xi, \eta) \tag{2.19}$$

$$\phi_{\xi\eta} = h_2(\phi_\xi, \phi_\eta, \phi, \xi, \eta) \tag{2.20}$$

The canonical form for a parabolic PDE can be written as either

$$\phi_{\xi\xi} = h_3(\phi_\xi, \phi_\eta, \phi, \xi, \eta) \tag{2.21}$$

or

$$\phi_{\eta\eta} = h_4(\phi_\xi, \phi_\eta, \phi, \xi, \eta) \tag{2.22}$$

For elliptic PDEs the canonical form is

$$\phi_{\xi\xi} + \phi_{\eta\eta} = h_5(\phi_\xi, \phi_\eta, \phi, \xi, \eta) \tag{2.23}$$

In the preceding equations, the coordinates ξ and η are functions of x and y. In a coordinate transformation of the form $(x, y) \rightarrow (\xi, \eta)$, a one-to-one relationship must exist between points specified by (x, y) and (ξ, η). We are assured of a nonsingular mapping, provided that the Jacobian of the transformation

$$J = \frac{\partial(\xi, \eta)}{\partial(x, y)} = \xi_x \eta_y - \xi_y \eta_x \tag{2.24}$$

is nonzero (Taylor, 1955). In order to apply this transformation to Eq. (2.15*a*), each derivative is replaced by repeated application of the chain rule. For example,

$$\begin{aligned} \frac{\partial \phi}{\partial x} &= \xi_x \frac{\partial \phi}{\partial \xi} + \eta_x \frac{\partial \phi}{\partial \eta} \\ \frac{\partial^2 \phi}{\partial x^2} &= \xi_x^2 \frac{\partial^2 \phi}{\partial \xi^2} + 2\xi_x \eta_x \frac{\partial^2 \phi}{\partial \xi\, \partial \eta} + \eta_x^2 \frac{\partial^2 \phi}{\partial \eta^2} + \xi_{xx} \frac{\partial \phi}{\partial \xi} + \eta_{xx} \frac{\partial \phi}{\partial \eta} \end{aligned} \tag{2.25}$$

Substitution into Eq. (2.15*a*) yields

$$A\phi_{\xi\xi} + B\phi_{\xi\eta} + C\phi_{\eta\eta} + \cdots = g(\xi, \eta)$$

where $A = a\xi_x^2 + b\xi_x \xi_y + c\xi_y^2$
$B = 2a\xi_x \eta_x + b\xi_x \eta_y + b\xi_y \eta_x + 2c\xi_y \eta_y$
$C = a\eta_x^2 + b\eta_x \eta_y + c\eta_y^2$

An important result of applying this transformation is immediately clear. The discriminant of the transformed equation becomes

$$B^2 - 4AC = (b^2 - 4ac)(\xi_x \eta_y - \xi_y \eta_x)^2 \tag{2.26}$$

where

$$\xi_x \eta_y - \xi_y \eta_x = J = \frac{\partial(\xi, \eta)}{\partial(x, y)}$$

Therefore, any real nonsingular transformation does not change the type of PDE.

2.3.1 Hyperbolic PDEs

From Eq. (2.18), we observe that two distinct families of characteristics exist for a hyperbolic equation. These can be found by first writing Eq. (2.18) as

$$\frac{dy}{dx} = \lambda_1 \qquad \frac{dy}{dx} = \lambda_2 \tag{2.27}$$

where the λ represent the right-hand side of Eq. (2.18) and a, b, and c are assumed constant. Upon solving the ODEs for the characteristic curves, we obtain

$$y - \lambda_1 x = k_1 \qquad y - \lambda_2 x = k_2 \tag{2.28}$$

A hyperbolic PDE in (x, y) can be written in canonical form,

$$\phi_{\xi\eta} = f(\xi, \eta, \phi, \phi_\xi, \phi_\eta) \tag{2.29}$$

by using the characteristic curves as the transformed coordinates $\xi(x, y)$ and $\eta(x, y)$. That is, we let

$$\xi = y - \lambda_1 x \qquad \eta = y - \lambda_2 x \tag{2.30}$$

In order to obtain the alternative canonical form for a hyperbolic equation,

$$\phi_{\bar{\xi}\bar{\xi}} - \phi_{\bar{\eta}\bar{\eta}} = f\left(\bar{\xi}, \bar{\eta}, \phi, \phi_{\bar{\xi}}, \phi_{\bar{\eta}}\right) \tag{2.31}$$

we can introduce linear combinations of ξ and η:

$$\bar{\xi} = \frac{\xi + \eta}{2} \qquad \bar{\eta} = \frac{\xi - \eta}{2}$$

An example utilizing the second-order wave equation is instructive.

Example 2.5 Solve the second-order wave equation

$$u_{tt} = c^2 u_{xx} \tag{2.32}$$

on the interval

$$-\infty < x < +\infty$$

with initial data

$$u(x,0) = f(x)$$
$$u_t(x,0) = g(x)$$

Solution The transformation to characteristic coordinates permits simple integration of the wave equation

$$u_{\xi\eta} = 0$$

where $\xi = x + ct$, $\eta = x - ct$.

We integrate to obtain the solution

$$u(x,t) = F_1(x + ct) + F_2(x - ct) \tag{2.33}$$

This is called the D'Alembert (Wylie, 1951) solution of the wave equation. The particular forms for F_1 and F_2 are determined from the initial data:

$$u(x,0) = f(x) = F_1(x) + F_2(x)$$
$$u_t(x,0) = g(x) = cF_1'(x) - cF_2'(x)$$

This results in a solution of the form

$$u(x,t) = \frac{f(x + ct) + f(x - ct)}{2} + \frac{1}{2c}\int_{x-ct}^{x+ct} g(\tau)\,d\tau \tag{2.34}$$

A distinctive property of hyperbolic PDEs can be deduced from the solution of Eq. (2.32) and the geometry of the physical domain of interest. Figure 2.6 shows the characteristics that pass through the point (x_0, t_0). The right running characteristic has a slope $+(1/c)$, while the left running one has slope $-(1/c)$. The solution $u(x,t)$ at (x_0, t_0) depends only upon the initial data contained in the interval

$$x_0 - ct_0 \leqslant x \leqslant x_0 + ct_0$$

The first term of the solution given by Eq. (2.34) represents propagation of the initial data along the characteristics, while the second term represents the effect of the data within the closed interval at $t = 0$.

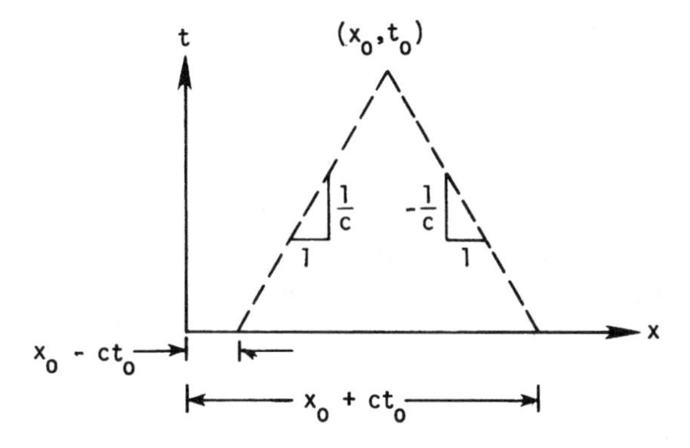

Figure 2.6 Characteristics for the wave equation.

A fundamental property of hyperbolic PDEs is the limited domain of dependence exhibited in Example 2.5. This domain of dependence is bounded by the characteristics that pass through the point (x_0, t_0). Clearly, the solution $u(x_0, t_0)$ depends only upon information in the interval bounded by these characteristics. This means that any disturbance that occurs outside of this interval can never influence the solution at (x_0, t_0). This behavior is common to all hyperbolic equations and is nicely demonstrated through the solution of the second-order wave equation. The basis for the term "initial value" or "marching problem" is clear. Initial conditions are specified, and the solution is marched outward in time or in a time-like direction.

The term "pure initial value problem" is frequently encountered in the study of hyperbolic PDEs. Example 2.5 is a pure initial value problem, i.e., there are no boundary conditions that must be applied at $x = \text{const}$. The solution at (x_0, t_0) depends only upon initial data.

In the classification of PDEs, many well-known names are associated with the specific problem types. The most well-known problem in the hyperbolic class is the Cauchy problem. This problem requires that one obtain a solution u to a hyperbolic PDE with initial data specified along a curve C. A very important theorem in mathematics assures us that a solution to the Cauchy problem exists. This is the Cauchy-Kowalewsky theorem. This theorem asserts that if the initial data are analytic in the neighborhood of (x_0, y_0) and the function u_{xx} (applied to our second-order wave equation of Example 2.5) is analytic there, a unique analytic solution for u exists in the neighborhood of (x_0, y_0).

Some discussion is warranted regarding the type of problem specification that is allowed for hyperbolic equations. For our second-order wave equation, initial conditions are required on the unknown function and its first derivatives along some curve C. It is important to observe that the curve C must not coincide with a characteristic of the differential equation. If an attempt is made to solve an initial value problem with characteristic initial data, a unique solution cannot be obtained (see Example 2.6). As is discussed further in Section 2.4, the problem is said to be "ill-posed."

Example 2.6 Solve the second-order wave equation in characteristic coordinates,

$$u_{\xi\eta} = 0$$

subject to initial data

$$u(0, \eta) = \phi(\eta) \qquad u_\xi(0, \eta) = \psi(\eta)$$

Solution The characteristics of the governing PDE are defined by $\xi = \text{const}$ and $\eta = \text{const}$. In this case the initial data are prescribed along a characteristic.

Suppose we attempt to write a Taylor-series expansion in ξ to obtain a solution for u in the neighborhood of the initial data surface $\xi = 0$. Our solution must be in the form

$$u(\xi, \eta) = u(0, \eta) + \xi u_\xi(0, \eta) + \frac{\xi^2}{2} u_{\xi\xi}(0, \eta) + \cdots$$

From the given initial data, $u(0, \eta)$ and $u_\xi(0, \eta)$ are known. It remains to determine $u_{\xi\xi}(0, \eta)$.

The governing differential equation requires

$$u_{\xi\eta}(0, \eta) = 0$$

However, we already have the condition that

$$u_{\xi\eta}(0, \eta) = \psi'(\eta) = 0$$

Therefore

$$\psi(\eta) = \text{const} = c_1$$

We may also write

$$\frac{\partial u_{\xi\eta}}{\partial \xi} = \frac{\partial u_{\xi\xi}}{\partial \eta} = 0$$

Integration of this equation yields

$$u_{\xi\xi} = f(\xi)$$

In view of the given initial data, we conclude that

$$u_{\xi\xi}(0, \eta) = \text{const} = c_2$$

and

$$u(\xi, \eta) = \phi(\eta) + \xi c_1 + \frac{\xi^2}{2} c_2$$

or

$$u(\xi, \eta) = \phi(\eta) + g(\xi)$$

We are unable to uniquely determine the function $g(\xi)$ when the initial data are given along the characteristic $\xi = 0$.

Proper specification of initial data or boundary conditions is very important in solving a PDE. Hadamard (1952) provided insight in noting that a well-posed problem is one in which the solution depends continuously upon the initial data. The concept of the well-posed problem is equally appropriate for elliptic and parabolic PDEs. An example for an elliptic problem is presented in Section 2.4.

2.3.2 Parabolic PDEs

A study of the solution of a simple hyperbolic PDE provided insight on the behavior of the solution of that type of equation. In a similar manner, we will now study the solution to parabolic equations. Referring to Eq. (2.15*a*), the parabolic case occurs when

$$b^2 - 4ac = 0$$

For this case the characteristic differential equation is given by

$$\frac{dy}{dx} = \frac{b}{2a} \tag{2.35}$$

The canonical form for the parabolic case is

$$\phi_{\xi\xi} = g(\phi_\xi, \phi_\eta, \phi, \xi, \eta) \tag{2.36}$$

If a and b are constant, this form may be obtained by identifying ξ and η as

$$\eta = y - \lambda_1 x \qquad \xi = y - \lambda_2 x$$

where λ_1 is given by the right-hand side of Eq. (2.35). In view of Eq. (2.35), we obtain only one characteristic. We must choose λ_2 to ensure linear independence of ξ and η. This requires that the Jacobian be nonzero:

$$\frac{\partial(\xi, \eta)}{\partial(x, y)} = f(\lambda_1, \lambda_2) \neq 0 \tag{2.37}$$

When λ_2 is selected, satisfying this requirement, and the transformation to (ξ, η) coordinates is completed, the canonical form given by Eq. (2.36) is obtained.

Parabolic PDEs are associated with diffusion processes. The solutions of parabolic equations clearly show this behavior. While the PDEs controlling diffusion are marching problems, i.e., we solve them starting at some initial data plane and march forward in time or in a time-like direction, they do not exhibit the limited zones of influence that hyperbolic equations have. In contrast, the solution of a parabolic equation at time t_1 depends upon the entire physical domain ($t \leqslant t_1$), including any side boundary conditions. To illustrate further, Example 2.3 required that we solve the heat equation for transient conduction in a 1-D solid. The initial temperature distribution was specified, as were the temperatures at the boundaries. Figure 2.7 illustrates the domain of dependence for this parabolic problem at t_1.

This shows that the solution at $t = t_1$ depends upon everything that occurred in the physical domain at all earlier times. The solution given by Eq. (2.10) also exhibits this behavior. Another example illustrating the behavior of a solution of a parabolic equation is of value.

Example 2.7 The unsteady motion due to the impulsive acceleration of an infinite flat plate in a viscous incompressible fluid is known as the Rayleigh problem and may be solved exactly. If the flow is 2-D, only the velocity component parallel to the plate will be nonzero. Let y be the coordinate normal to the plate and x be the coordinate along the plate. The equation that governs the velocity distribution is

$$\frac{\partial u}{\partial t} = v \frac{\partial^2 u}{\partial y^2} \tag{2.38}$$

where v is the kinematic viscosity. The time derivative term is the local acceleration of the fluid, while the right-hand side is the resisting force provided by the shear stress in the fluid ($\tau = v\rho\, \partial u / \partial y$). This equation is subject to the boundary conditions

$$u(0, y) = 0$$
$$u(t, 0) = U \qquad t > 0$$
$$u(t, \infty) = 0$$

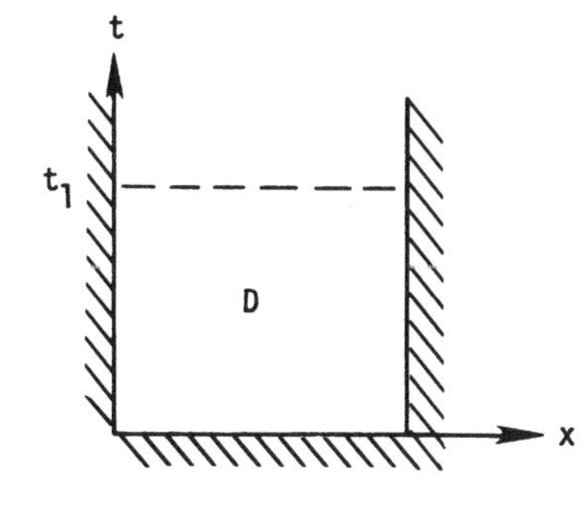

Figure 2.7 Domain of dependence for a simple parabolic problem.

Solution The solution of this problem provides the velocity distribution on a 2-D flat plate impulsively accelerated to a velocity U from rest. An interesting method frequently used in solving parabolic equations is to seek a similarity solution (Hansen, 1964). In finding a similarity solution, we introduce a change in variables, which results in reducing the number of independent variables in the original PDE (Churchill, 1974). In this case we attempt to reduce the PDE in (y, t) to an ODE in a new independent variable η. For this problem, let

$$f(\eta) = \frac{u}{U}$$

and

$$\eta = \frac{y}{2\sqrt{vt}}$$

The governing differential equation becomes

$$\frac{d^2f}{d\eta^2} + 2\eta\frac{df}{d\eta} = 0$$

with boundary conditions

$$f(0) = 1$$
$$f(\infty) = 0$$

This ODE may be solved directly to yield the solution

$$u = U\left(1 - \frac{2}{\sqrt{\pi}}\int_0^{\eta} e^{-\eta^2}\,d\eta\right) \tag{2.39}$$

Using the definition of the error function

$$\operatorname{erf}(\eta) = \frac{2}{\sqrt{\pi}}\int_0^{\eta} e^{-\eta^2}\,d\eta \tag{2.40}$$

the solution becomes

$$u = U[1 - \operatorname{erf}(\eta)]$$

This shows that the layer of fluid that is influenced by the moving plate increases in thickness with time. In fact, the layer of fluid has thickness proportional to $\sqrt{\nu t}$. This indicates that the growth of this layer is controlled by the kinematic viscosity ν and the velocity change in the layer is induced by diffusion of the plate velocity into the initially undisturbed fluid. We see that this is a diffusion process, as is 1-D transient heat conduction.

2.3.3 Elliptic PDEs

The third type of PDE is elliptic. As we previously noted, jury problems are governed by elliptic PDEs. If Eq. (2.15*a*) is elliptic, the discriminant is negative, i.e.,

$$b^2 - 4ac < 0 \tag{2.41}$$

and the characteristic differential equation has no real solution. For this case, the solutions to Eq. (2.18) take the form (assuming a, b, and c are constant)

$$\begin{aligned} y - c_1 x + i c_2 x &= k_1 \\ y - c_1 x - i c_2 x &= k_2 \end{aligned} \tag{2.42}$$

The transformation to the canonical form

$$\phi_{\xi\xi} + \phi_{\eta\eta} = h_5(\phi_\xi, \phi_\eta, \phi, \xi, \eta) \tag{2.43}$$

can be achieved by selecting ξ and η to be the real and imaginary parts of the complex conjugate functions in Eqs. (2.42). This gives

$$\xi = y - c_1 x \qquad \eta = c_2 x \tag{2.44}$$

The dependence of the solution upon the boundary conditions for elliptic PDEs has been previously discussed and demonstrated in Example 2.1. However, another example is presented here to reinforce this basic idea.

Example 2.8 Given Laplace's equation on the unit disk

$$\nabla^2 u = 0 \qquad 0 \leqslant r < 1 \qquad -\pi \leqslant \theta \leqslant \pi$$

subject to boundary conditions

$$\frac{\partial u}{\partial r}(1, \theta) = f(\theta) \qquad -\pi \leqslant \theta \leqslant \pi$$

what is the solution $u(r, \theta)$?

Solution This problem can be solved by assuming a solution of the form

$$u(r, \theta) = \frac{a_0}{2} + \sum_{n=1}^{\infty} r^n (a_n \cos n\theta + b_n \sin n\theta)$$

The correct expressions for a_n and b_n can be developed using standard techniques (Garabedian, 1964). For this example, the expressions for a_n and b_n depend upon the boundary conditions at all points on the unit disk. This dependence on the boundary conditions should be expected for all elliptic problems. The important point of this example is that a solution of this problem exists only if

$$\int f(\theta)\, dl = 0$$

over the boundary of the unit disk (Zachmanoglou and Thoe, 1976). This may be demonstrated by applying Green's theorem to the unit disk. In this problem the boundary conditions are not arbitrarily chosen but must satisfy the integral constraint shown above.

2.4 THE WELL-POSED PROBLEM

The previous section discussed the mathematical character of the different PDEs. The examples illustrated the dependence of the solution of a particular problem upon the initial data and boundary conditions. In our discussion of hyperbolic PDEs, it was noted that a unique solution to a hyperbolic PDE cannot be obtained if the initial data are given on a characteristic. Similar examples showing improper use of boundary conditions can be constructed for elliptic and parabolic equations.

The difficulty encountered in solving our hyperbolic equation subject to characteristic initial data had to do with the question of whether or not the problem was "well-posed." In order for a problem involving a PDE to be well-posed, the solution to the problem must exist, must be unique, and must depend continuously upon the initial or boundary data. Example 2.6 led to a uniqueness question. Hadamard (1952) has constructed a simple example that demonstrates the problem of continuous dependence on boundary data.

Example 2.9 A solution of Laplace's equation

$$u_{xx} + u_{yy} = 0 \qquad -\infty < x < \infty \qquad y \geqslant 0$$

is desired subject to the boundary conditions ($y = 0$)

$$u(x,0) = 0$$

$$u_y(x,0) = \frac{1}{n} \sin(nx) \qquad n > 0$$

Solution Using separation of variables, we obtain

$$u = \frac{1}{n^2} \sin(nx) \sinh(ny)$$

If our problem is well-posed, we expect the solution to depend continuously upon the boundary conditions. For the data given, we must have

$$u_y(x,0) = \frac{1}{n}\sin(nx)$$

We see that u_y becomes small for large values of n. The solution behaves in a different fashion for large n. As n becomes large, u approaches e^{ny}/n^2 and grows without bound even for small y. However, $u(x,0) = 0$, so that continuity with the initial data is lost. Thus we have an ill-posed problem. This is evident from our earlier discussions. Since Laplace's equation is elliptic, the solution depends upon conditions on the entire boundary of the closed domain. The problem given in this example requires the solution of an elliptic differential equation on an open domain. Boundary conditions were given only on the $y = 0$ line.

Problems requiring the solution of Laplace's equation subject to different types of boundary conditions are identified with specific names. The first of these is the *Dirichlet problem* (Fig. 2.8). In this problem, a solution of Laplace's equation is required on a closed domain subject to boundary conditions that require the solution to take on prescribed values on the boundary. The *Neumann problem* also requires the solution of Laplace's equation in D. However, the normal derivative of u is specified on B rather than the function u. If s is the arc length along B, then

$$\nabla^2 u = 0 \qquad \text{in D}$$

$$\frac{\partial u}{\partial n} = g(s) \qquad \text{on B}$$

The specification of the Dirichlet and Neumann problems leads one to speculate about the existence of a boundary value problem requiring specification of a combination of the function u and its normal derivative on the boundary. This is called the mixed or third boundary value problem (Zachmanoglou and Thoe, 1976) and is also referred to as *Robin's problem*. Mathematically, this problem may be written as

$$\nabla^2 u = 0$$

in D and

$$a_1(s)\frac{\partial u}{\partial n} + a_2(s)u = h(s)$$

on B. The assignment of the names Dirichlet, Neumann, and Robin to the three boundary value problems noted here is generally used to define types of boundary or initial data specified for any PDE. For example, if the comment "Dirichlet boundary data" is used, it is understood that the unknown, u, is prescribed on the boundary in question. This is accepted regardless of the type of differential equation.

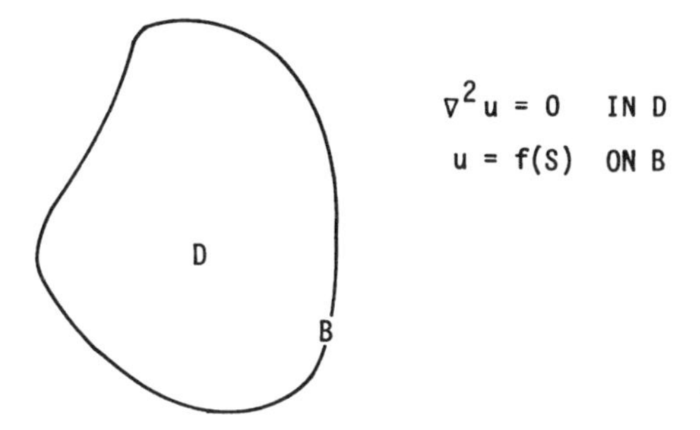

Figure 2.8 Dirichlet problem.

2.5 SYSTEMS OF EQUATIONS

In applying numerical methods to physical problems, systems of equations are frequently encountered. It is the exceptional case when a physical process is governed by a single equation. In those cases where the process is governed by a higher-order PDE, the PDE can usually be converted to a system of first-order equations. This can be most easily demonstrated by two simple examples.

The wave equation [Eq. (2.32)] can be written as a system of two first-order equations. Let

$$v = \frac{\partial u}{\partial t} \qquad w = c\frac{\partial u}{\partial x}$$

Then we may write

$$\begin{aligned} \frac{\partial v}{\partial t} &= c\frac{\partial w}{\partial x} \\ \frac{\partial w}{\partial t} &= c\frac{\partial v}{\partial x} \end{aligned} \tag{2.45}$$

If we introduce u as one of the variables in place of either w or v, then u can be seen to satisfy the second-order wave equation.

Many physical processes are governed by Laplace's equation [Eq. (2.1)]. As in the previous example, Laplace's equation can be replaced by a system of first-order equations. In this case, let u and v represent the unknown dependent variables. We require that

$$\begin{aligned} \frac{\partial u}{\partial x} &= +\frac{\partial v}{\partial y} \\ \frac{\partial u}{\partial y} &= -\frac{\partial v}{\partial x} \end{aligned} \tag{2.46}$$

These are the famous Cauchy-Riemann equations (Churchill, 1960). These equations are extensively used in conformally mapping one region onto another.*

* It should be noted that some differences exist in solving Laplace's equation and the Cauchy-Riemann equations. A solution of the Cauchy-Riemann equations is a solution of Laplace's equation, but the converse is not necessarily true.

The equations most frequently encountered in CFD may be written as first-order systems. We must be able to classify systems of first-order equations in order to correctly treat them. Consider the linear system of equations

$$\frac{\partial \mathbf{u}}{\partial t} + [A]\frac{\partial \mathbf{u}}{\partial x} + [B]\frac{\partial \mathbf{u}}{\partial y} + \mathbf{r} = 0 \tag{2.47}$$

We assume for simplicity that the coefficient matrices $[A]$ and $[B]$ are functions of t, x, y, and we restrict our attention to two space dimensions. The dependent variable $\mathbf{u}$ is a column vector of unknowns, and $\mathbf{r}$ depends upon $\mathbf{u}$, x, y.

According to Zachmanoglou and Thoe (1976), there are two cases that can be definitely identified for first-order systems. The system given in Eq. (2.47) is said to be hyperbolic at a point in (x, t) if the eigenvalues of $[A]$ are all real and distinct. Richtmyer and Morton (1967) define a system to be hyperbolic if the eigenvalues are all real and $[A]$ can be written as $[T][\lambda][T]^{-1}$, where $[\lambda]$ is a diagonal matrix of eigenvalues of $[A]$ and $[T]^{-1}$ is the matrix of left eigenvectors. The same can be said of the behavior of the system in (y, t) with respect to the eigenvalues of the B matrix.

This point can be illustrated by writing the system of equations given in Eq. (2.45) as

$$\frac{\partial \mathbf{u}}{\partial t} + [A]\frac{\partial \mathbf{u}}{\partial x} = 0 \tag{2.48a}$$

where

$$\mathbf{u} = \begin{bmatrix} v \\ w \end{bmatrix}$$

$$[A] = \begin{bmatrix} 0 & -c \\ -c & 0 \end{bmatrix}$$

The eigenvalues λ of the $[A]$ matrix are found from

$$\det \|[A] - \lambda[I]\| = 0$$

Thus

$$\begin{vmatrix} -\lambda & -c \\ -c & -\lambda \end{vmatrix} = 0$$

or

$$\lambda^2 - c^2 = 0$$

The roots of this characteristic equation are

$$\lambda_1 = +c$$

$$\lambda_2 = -c$$

These are the characteristic differential equations for the wave equation, i.e.,

$$\left(\frac{dx}{dt}\right)_1 = +c$$

$$\left(\frac{dx}{dt}\right)_2 = -c$$

The system of equations in this example is hyperbolic, and we see that the eigenvalues of the $[A]$ matrix represent the characteristic differential equations of the wave equation.

The second case that can be identified for the system given in Eq. (2.47) is elliptic. Equation (2.47) is said to be elliptic at a point in (x, t) if the eigenvalues of $[A]$ are all complex. An example illustrating this behavior is given by the Cauchy-Riemann equations.

Example 2.10 Classify the system given in Eq. (2.46), which may be written as

$$\frac{\partial \mathbf{w}}{\partial x} + [A]\frac{\partial \mathbf{w}}{\partial y} = 0$$

where

$$\mathbf{w} = \begin{bmatrix} u \\ v \end{bmatrix}$$

and

$$[A] = \begin{bmatrix} 0 & -1 \\ 1 & 0 \end{bmatrix}$$

Solution The eigenvalues of $[A]$ are

$$\lambda_1 = +i$$

$$\lambda_2 = -i$$

Since both eigenvalues of $[A]$ are complex, we identify the system as elliptic. Again, this is consistent with the behavior we are familiar with in Laplace's equation.

The first-order system represented by Eq. (2.47) can exhibit hyperbolic behavior in (x, t) space and elliptic behavior in (y, t) space, depending upon the eigenvalue structure of the A and B matrices. This is a result of evaluating the behavior of the PDE by examining the eigenvalues in (x, t) or (y, t) independently.

Note that a single first-order equation can be considered as a special case of the above development. That is, we can let $[A]$ and $[B]$ in Eq. (2.47) be real scalars a and b and the vector $\mathbf{u}$ be a scalar variable u. The conclusion is that such a single first-order equation would be classified as hyperbolic because there is only one root and it is real.

Since second-order PDEs can be represented as a system of first-order equations, one might wonder if such systems can also be identified as hyperbolic, parabolic, or elliptic by using a procedure that inquires about the continuity of the highest order derivatives. This seems reasonable, since discontinuities in second derivatives would show up as discontinuities in first derivatives in any first-order system that was developed from a second-order equation.

Consider a system of two first-order equations in two independent variables of the form

$$
\begin{aligned}
a_1 \frac{\partial u}{\partial x} + b_1 \frac{\partial v}{\partial x} + c_1 \frac{\partial u}{\partial y} + d_1 \frac{\partial v}{\partial y} &= f_1 \\
a_2 \frac{\partial u}{\partial x} + b_2 \frac{\partial v}{\partial x} + c_2 \frac{\partial u}{\partial y} + d_2 \frac{\partial v}{\partial y} &= f_2
\end{aligned}
\tag{2.48b}
$$

This system may be written as a matrix system of the form

$$
[A] \frac{\partial \mathbf{w}}{\partial x} + [C] \frac{\partial \mathbf{w}}{\partial y} = \mathbf{F} \tag{2.49a}
$$

where

$$
\mathbf{w} = \begin{bmatrix} u \\ v \end{bmatrix} \qquad \mathbf{F} = \begin{bmatrix} f_1 \\ f_2 \end{bmatrix}
$$

and

$$
[A] = \begin{bmatrix} a_1 & b_1 \\ a_2 & b_2 \end{bmatrix} \qquad [C] = \begin{bmatrix} c_1 & d_1 \\ c_2 & d_2 \end{bmatrix}
$$

As before, we consider curves C on which all but the highest order derivatives are specified (in this case, we consider u and v specified) and inquire about conditions that will indicate that the highest derivatives are not uniquely determined. Again, we let a parameter τ vary along curves C and use the chain rule to write

$$
\begin{aligned}
\frac{du}{d\tau} &= \frac{\partial u}{\partial x}\frac{dx}{d\tau} + \frac{\partial u}{\partial y}\frac{dy}{d\tau} \\
\frac{dv}{d\tau} &= \frac{\partial v}{\partial x}\frac{dx}{d\tau} + \frac{\partial v}{\partial y}\frac{dy}{d\tau}
\end{aligned}
\tag{2.49b}
$$

Writing the four equations (2.48b) and (2.49b) in matrix form with the prescribed data on the right-hand side gives

$$
\begin{bmatrix}
a_1 & b_1 & c_1 & d_1 \\
a_2 & b_2 & c_2 & d_2 \\
\frac{dx}{d\tau} & 0 & \frac{dy}{d\tau} & 0 \\
0 & \frac{dx}{d\tau} & 0 & \frac{dy}{d\tau}
\end{bmatrix}
\begin{bmatrix}
\frac{\partial u}{\partial x} \\ \frac{\partial v}{\partial x} \\ \frac{\partial u}{\partial y} \\ \frac{\partial v}{\partial y}
\end{bmatrix}
=
\begin{bmatrix}
f_1 \\ f_2 \\ \frac{du}{d\tau} \\ \frac{dv}{d\tau}
\end{bmatrix}
$$

A unique solution for the first derivatives of u and v with respect to x and y does not exist if the determinant of the coefficient matrix is zero. We can write this determinant in different ways. However, a Laplace development of the determinant on the elements of the last row followed by another development on the last rows of the third-order determinants allows the determinant to be written as

$$-\left(\frac{dy}{d\tau}\right)^2\begin{vmatrix} a_1 & b_1 \\ a_2 & b_2 \end{vmatrix} + \frac{dx}{d\tau}\frac{dy}{d\tau}\left(\begin{vmatrix} a_1 & d_1 \\ a_2 & d_2 \end{vmatrix} + \begin{vmatrix} c_1 & b_1 \\ c_2 & b_2 \end{vmatrix}\right) - \left(\frac{dx}{d\tau}\right)^2\begin{vmatrix} c_1 & d_1 \\ c_2 & d_2 \end{vmatrix}$$

Letting

$$|A| = \begin{vmatrix} a_1 & b_1 \\ a_2 & b_2 \end{vmatrix}$$

$$|B| = \begin{vmatrix} a_1 & d_1 \\ a_2 & d_2 \end{vmatrix} + \begin{vmatrix} c_1 & b_1 \\ c_2 & b_2 \end{vmatrix}$$

$$|C| = \begin{vmatrix} c_1 & d_1 \\ c_2 & d_2 \end{vmatrix}$$

and setting the determinant equal to zero gives the conditions under which first partial derivatives are not uniquely determined on C:

$$|A|\left(\frac{dy}{dx}\right)^2 - |B|\frac{dy}{dx} + |C| = 0$$

Notice that this expression has the same form as Eq. (2.17) except that a, b, and c have now become determinants. The classification of the first-order system is also similar to that of the second-order PDE. Letting

$$D = |B|^2 - 4|A|\,|C|$$

we find that the system is hyperbolic if $D > 0$, parabolic if $D = 0$, and elliptic if $D < 0$.

Several questions now appear regarding behavior of systems of equations with coefficient matrices where the roots of the characteristic equations contain both real and complex parts. In those cases, the system is mixed and may exhibit hyperbolic, parabolic, and elliptic behavior. The physical system under study usually provides information that is very useful in understanding the physical behavior represented by the governing PDE. Experience gained in solving mixed problems provides the best guidance in their correct treatment.

The classification of systems of second-order PDEs is very complex. It is difficult to determine the mathematical behavior of these systems except for simple cases. For example, the system of equations given by

$$\mathbf{u}_t = [A]\mathbf{u}_{xx}$$

is parabolic if all the eigenvalues of $[A]$ are real. The same uncertainties present in classifying mixed systems of first-order equations are also encountered in the classification of second-order systems.

2.6 OTHER DIFFERENTIAL EQUATIONS OF INTEREST

Our discussion in this chapter has centered on the second-order equations given by the wave equation, the heat equation, and Laplace's equation. In addition, systems of first-order equations were examined. A number of other very important equations should be mentioned, since they govern common physical phenomena or they are used as simple models for more complex problems. In many cases, exact analytical solutions for these equations exist.

1. The first-order, linear wave equation

$$\frac{\partial u}{\partial t} + c\frac{\partial u}{\partial x} = 0 \tag{2.50}$$

   governs propagation of a wave moving to the right at a constant speed c. This is called the advection equation in meteorology.
2. The inviscid Burgers equation

$$\frac{\partial u}{\partial t} + u\frac{\partial u}{\partial x} = 0 \tag{2.51}$$

   is also called the nonlinear first-order wave equation. This equation governs propagation of nonlinear waves for the simple 1-D case.
3. Burgers' equation

$$\frac{\partial u}{\partial t} + u\frac{\partial u}{\partial x} = \nu\frac{\partial^2 u}{\partial x^2} \tag{2.52}$$

   is the nonlinear wave equation [Eq. (2.51)] with diffusion added. This particular form is very similar to the equations governing fluid flow and can be used as a simple nonlinear model for numerical experiments.
4. The Tricomi equation

$$y\frac{\partial^2 u}{\partial x^2} + \frac{\partial^2 u}{\partial y^2} = 0 \tag{2.53}$$

   governs problems of the mixed type such as inviscid transonic flows. The properties of the Tricomi equation include a change from elliptic to hyperbolic character, depending upon the sign of y.
5. Poisson's equation

$$\frac{\partial^2 u}{\partial x^2} + \frac{\partial^2 u}{\partial y^2} = f(x, y) \tag{2.54}$$

   governs the temperature distribution in a solid with heat sources described by the function $f(x, y)$. Poisson's equation also determines the electric field in a region containing a charge density $f(x, y)$.

6. The advection-diffusion equation

$$\frac{\partial \xi}{\partial t} + u\frac{\partial \xi}{\partial x} = \alpha\frac{\partial^2 \xi}{\partial x^2} \tag{2.55}$$

represents the advection of a quantity ξ in a region with velocity u. The quantity α is a diffusion or viscosity coefficient.

7. The Korteweg-de Vries equation

$$\frac{\partial u}{\partial t} + u\frac{\partial u}{\partial x} + \frac{\partial^3 u}{\partial x^3} = 0 \tag{2.56}$$

governs the motion of nonlinear dispersive waves.

8. The Helmholtz equation

$$\frac{\partial^2 u}{\partial x^2} + \frac{\partial^2 u}{\partial y^2} + k^2 u = 0 \tag{2.57}$$

governs the motion of time-dependent harmonic waves, where k is a frequency parameter. Applications include the propagation of acoustic waves.

9. The biharmonic equation

$$\frac{\partial^4 u}{\partial x^4} + \frac{\partial^4 u}{\partial y^4} = 0 \tag{2.58}$$

determines the stream function for a very low Reynolds number viscous (Stokes) flow and is also a governing relation in the theory of elasticity.

10. The telegraph equation

$$\frac{\partial^2 u}{\partial t^2} + a\frac{\partial u}{\partial t} + bu = c^2\frac{\partial^2 u}{\partial x^2} \tag{2.59}$$

governs the transmission of electrical impulses in a long wire with distributed capacitance, inductance, and resistance. If $b = 0$, the equation is called the *damped wave equation*. Applications include the motion of a string with a damping force proportional to the velocity and heat conduction with a finite thermal propagation speed.

Many of the equations cited here will be used to demonstrate the application of discretization methods in subsequent chapters. While the list of equations is not exhaustive, examples of the various types of PDEs are included.

PROBLEMS

2.1 The solution of Laplace's equation for Example 2.1 is given in Eq. (2.3). Show that the expression for the Fourier coefficients A_n is correct as given in the example. Hint: Multiply Eq. (2.3) by $\sin(m\pi x)$ and integrate over the interval $0 \leqslant x \leqslant 1$ to obtain your answer after using the boundary condition $T(x, 0) = T_0$.

2.2 Show that the velocity field represented by the potential function in Eq. (2.6) satisfies the surface boundary condition given in Eq. (2.4).
2.3 Demonstrate that Eq. (2.14) is the solution of the wave equation as required in Example 2.4. Use the separation of variables technique.
2.4 Show that the type of PDE is unchanged when any nonsingular, real transformation is used.
2.5 Derive the canonical form for hyperbolic equations [Eq. (2.29)] by applying the transformations given in Eq. (2.30) to Eq. (2.15*a*).
2.6 Show that the canonical form for parabolic equations given in Eq. (2.36) is correct.
2.7 Show that a solution to Example 2.8 exists only if

$$\int f(\theta)\, dl = 0$$

on the unit circle.
2.8 Consider the equation

$$y^2 u_{xx} - x^2 u_{yy} = 0$$

(*a*) Discuss the mathematical character of this equation for *all* real values of x and y.
(*b*) Obtain the new coordinates ξ and η that will transform the given equation in the *first* quadrant to its canonical form.
2.9 (*a*) Classify the equation

$$2u_{xx} - 4u_{xy} + 2u_{yy} + 3u = 0$$

(*b*) Obtain the transformation variables required to transform the equation to its canonical form.
(*c*) Convert the equation into an equivalent system of first-order equations and write them as a matrix system.
(*d*) Apply the method for classification of a system of equations to the system determined in Prob. 2.9(*c*).
2.10 Classify the following system of equations:

$$\frac{\partial u}{\partial t} + 8\frac{\partial v}{\partial x} = 0$$

$$\frac{\partial u}{\partial t} + 2\frac{\partial v}{\partial x} = 0$$

2.11 The following system of equations is elliptic. Determine the possible range of values for a.

$$\frac{\partial u}{\partial x} - a\frac{\partial v}{\partial y} = 0$$

$$\frac{\partial v}{\partial y} + a\frac{\partial u}{\partial x} = 0$$

2.12 Determine the mathematical character of the equations given by

$$\beta^2\frac{\partial u}{\partial x} - \frac{\partial v}{\partial y} = 0$$

$$\frac{\partial v}{\partial x} - \frac{\partial u}{\partial y} = 0$$

2.13 Classify the following PDEs:

$$\frac{\partial^2 u}{\partial t^2} + \frac{\partial^2 u}{\partial x^2} + \frac{\partial u}{\partial x} = -e^{-kt}$$

$$\frac{\partial^2 u}{\partial x^2} - \frac{\partial^2 u}{\partial x\, \partial y} + \frac{\partial u}{\partial y} = 4$$

2.14 Classify the behavior of the following system of PDEs in (t, x) and (t, y) space:

$$\frac{\partial u}{\partial t} + \frac{\partial v}{\partial x} - \frac{\partial u}{\partial y} = 0$$

$$\frac{\partial v}{\partial t} - \frac{\partial u}{\partial x} + \frac{\partial v}{\partial y} = 0$$

2.15 (*a*) Write the Fourier cosine series for the function

$$f(x) = \sin(x) \qquad 0 < x < \pi$$

(*b*) Write the Fourier cosine series for the function

$$f(x) = \cos(x) \qquad 0 < x < \pi$$

2.16 Find the characteristics of each of the following PDEs:

(*a*)
$$\frac{\partial^2 u}{\partial x^2} + 3\frac{\partial^2 u}{\partial x\, \partial y} + 2\frac{\partial^2 u}{\partial y^2} = 0$$

(*b*)
$$\frac{\partial^2 u}{\partial x^2} - 2\frac{\partial^2 u}{\partial x\, \partial y} + \frac{\partial^2 u}{\partial y^2} = 0$$

2.17 Transform the PDEs given in Prob. 2.16 into canonical form.

2.18 Obtain the canonical form for the following elliptic PDEs:

(*a*)
$$\frac{\partial^2 u}{\partial x^2} + \frac{\partial^2 u}{\partial x\, \partial y} + \frac{\partial^2 u}{\partial y^2} = 0$$

(*b*)
$$\frac{\partial^2 u}{\partial x^2} - 2\frac{\partial^2 u}{\partial x\, \partial y} + 5\frac{\partial^2 u}{\partial y^2} + \frac{\partial u}{\partial y} = 0$$

2.19 Transform the following parabolic PDEs to canonical form:

(*a*)
$$\frac{\partial^2 u}{\partial x^2} - 6\frac{\partial^2 u}{\partial x\, \partial y} + 9\frac{\partial^2 u}{\partial y^2} + \frac{\partial u}{\partial x} - e^{xy} = 1$$

(*b*)
$$\frac{\partial^2 u}{\partial x^2} + 2\frac{\partial^2 u}{\partial x\, \partial y} + \frac{\partial^2 u}{\partial y^2} + 7\frac{\partial u}{\partial x} - 8\frac{\partial u}{\partial y} = 0$$

2.20 Find the solution of the wave equation

$$\frac{\partial^2 u}{\partial x^2} - \frac{\partial^2 u}{\partial y^2} = 0 \qquad y \geqslant 0$$

with initial data

$$u(x, 0) = 1$$
$$u_y(x, 0) = 0$$

2.21 Solve Laplace's equation,

$$\nabla^2 u = 0 \qquad 0 \leqslant x \leqslant \pi \qquad 0 \leqslant y \leqslant \pi$$

subject to boundary conditions

$$u(x, 0) = \sin x + 2\sin 2x$$
$$u(\pi, y) = 0$$
$$u(x, \pi) = 0$$
$$u(0, y) = 0$$

2.22 Repeat Prob. 2.21 with

$$u(x,0) = -\pi^2 x^2 + 2\pi x^3 - x^4$$

2.23 Determine the solution of the heat equation

$$\frac{\partial u}{\partial t} = \frac{\partial^2 u}{\partial x^2} \qquad 0 \leqslant x \leqslant 1$$

with boundary conditions

$$u(t,0) = 0$$
$$u(t,1) = 0$$

and an initial distribution

$$u(0,x) = \sin(2\pi x)$$

2.24 Repeat Prob. 2.23 if the initial distribution is given by

$$u(0,x) = 1 - \cos(4\pi x)$$

CHAPTER

THREE

BASICS OF DISCRETIZATION METHODS

3.1 INTRODUCTION

In this chapter, basic concepts and techniques needed in the formulation of finite-difference and finite-volume representations are developed. In the finite-difference approach, the continuous problem domain is "discretized," so that the dependent variables are considered to exist only at discrete points. Derivatives are approximated by differences, resulting in an algebraic representation of the partial differential equation (PDE). Thus a problem involving calculus has been transformed into an algebraic problem.

The nature of the resulting algebraic system depends on the character of the problem posed by the original PDE (or system of PDEs). Equilibrium problems usually result in a system of algebraic equations that must be solved simultaneously throughout the problem domain in conjunction with specified boundary values. Marching problems result in algebraic equations that usually can be solved one at a time (although it is often convenient to solve them several at a time). Several considerations determine whether the solution so obtained will be a good approximation to the exact solution of the original PDE. Among these considerations are truncation error, consistency, and stability, all of which will be discussed in the present chapter.

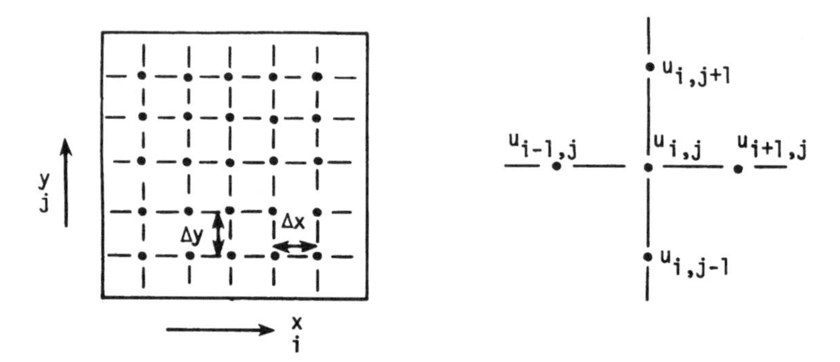

Figure 3.1 A typical finite-difference grid.

3.2 FINITE DIFFERENCES

One of the first steps to be taken in establishing a finite-difference procedure for solving a PDE is to replace the continuous problem domain by a finite difference mesh or grid. As an example, suppose that we wish to solve a PDE for which $u(x, y)$ is the dependent variable in the square domain $0 \leqslant x \leqslant 1$, $0 \leqslant y \leqslant 1$. We establish a grid on the domain by replacing $u(x, y)$ by $u(i\,\Delta x, j\,\Delta y)$. Points can be located according to values of i and j, so difference equations are usually written in terms of the general point (i, j) and its neighbors. This labeling is illustrated in Fig. 3.1. Thus, if we think of $u_{i,j}$ as $u(x_0, y_0)$, then

$$u_{i+1,j} = u(x_0 + \Delta x, y_0) \qquad u_{i-1,j} = u(x_0 - \Delta x, y_0)$$

$$u_{i,j+1} = u(x_0, y_0 + \Delta y) \qquad u_{i,j-1} = u(x_0, y_0 - \Delta y)$$

Often in the treatment of marching problems, the variation of the marching coordinate is indicated by a superscript, such as u_j^{n+1}, rather than a subscript. Many different finite-difference representations are possible for any given PDE and it is usually impossible to establish a "best" form on an absolute basis. First, the accuracy of a difference scheme may depend on the exact form of the equation and problem being solved, and second, our selection of a best scheme will be influenced by the aspect of the procedure that we are trying to optimize, i.e., accuracy, economy, or programming simplicity.

The idea of a finite-difference representation for a derivative can be introduced by recalling the definition of the derivative for the function $u(x, y)$ at $x = x_0$, $y = y_0$:

$$\frac{\partial u}{\partial x} = \lim_{\Delta x \to 0} \frac{u(x_0 + \Delta x, y_0) - u(x_0, y_0)}{\Delta x} \tag{3.1}$$

Here, if u is continuous, it is expected that $[u(x_0 + \Delta x, y_0) - u(x_0, y_0)]/\Delta x$ will be a "reasonable" approximation to $\partial u/\partial x$ for a "sufficiently" small but finite Δx. In fact, the mean-value theorem assures us that the difference representation is exact for some point within the Δx interval. The difference approximation can be put on a more formal basis through the use of either a Taylor-series expansion or Taylor's formula with a remainder. Developing a Taylor-series expansion for $u(x_0 + \Delta x, y_0)$ about (x_0, y_0) gives

$$u(x_0 + \Delta x, y_0) = u(x_0, y_0) + \left.\frac{\partial u}{\partial x}\right)_0 \Delta x + \left.\frac{\partial^2 u}{\partial x^2}\right)_0 \frac{(\Delta x)^2}{2!} + \cdots$$
$$+ \left.\frac{\partial^{n-1} u}{\partial x^{n-1}}\right)_0 \frac{(\Delta x)^{n-1}}{(n-1)!} + \left.\frac{\partial^n u}{\partial x^n}\right)_\xi \frac{(\Delta x)^n}{n!}$$
$$x_0 \leqslant \xi \leqslant (x_0 + \Delta x) \qquad (3.2)$$

where the last term can be identified as the remainder. Thus we can form the "forward" difference by rearranging Eq. (3.2):

$$\left.\frac{\partial u}{\partial x}\right)_{x_0, y_0} = \frac{u(x_0 + \Delta x, y_0) - u(x_0, y_0)}{\Delta x} - \left.\frac{\partial^2 u}{\partial x^2}\right)_0 \frac{\Delta x}{2!} - \cdots \qquad (3.3)$$

Switching now to the i, j notation for brevity, we consider

$$\left.\frac{\partial u}{\partial x}\right)_{i,j} = \frac{u_{i+1,j} - u_{i,j}}{\Delta x} + \text{T.E.} \qquad (3.4)$$

where $(u_{i+1,j} - u_{i,j})/\Delta x$ is obviously the finite-difference representation for $\partial u/\partial x)_{i,j}$. The truncation error (T.E.) is the difference between the partial derivative and its finite-difference representation. We can characterize the limiting behavior of the T.E. by using the order of (O) notation, whereby we write

$$\left.\frac{\partial u}{\partial x}\right)_{i,j} = \frac{u_{i+1,j} - u_{i,j}}{\Delta x} + O(\Delta x)$$

where $O(\Delta x)$ has a precise mathematical meaning. Here, when the T.E. is written as $O(\Delta x)$, we mean $|\text{T.E.}| \leqslant K|\Delta x|$ for $\Delta x \to 0$ (sufficiently small Δx), and K is a positive real constant. As a practical matter, the order of the T.E. in this case is found to be Δx raised to the largest power that is common to all terms in the T.E.

To give a more general definition of the O notation, when we say $f(x) = O[\phi(x)]$, we mean that there exists a positive constant K, independent of x, such that $|f(x)| \leqslant K|\phi(x)|$ for all x in S, where f and ϕ are real or complex functions defined in S. We often restrict S by $x \to \infty$ (sufficiently large x) or, as is most common in finite-difference applications, $x \to 0$ (sufficiently small x). More details on the O notation can be found in the work by Whittaker and Watson (1927).

Note that $O(\Delta x)$ tells us nothing about the exact size of the T.E., but rather how it behaves as Δx tends toward zero. If another difference expression had a T.E. $= O[(\Delta x)^2]$, we might expect or hope that the T.E. of the second representation would be smaller than the first for a convenient Δx, but we could only be *sure* that this would be true if we refined the mesh "sufficiently," and "sufficiently" is a quantity that is hard to estimate.

An infinite number of difference representations can be found for $\partial u/\partial x)_{i,j}$. For example, we could expand "backward":

$$u(x_0 - \Delta x, y_0) = u(x_0, y_0) - \left.\frac{\partial u}{\partial x}\right)_0 \Delta x + \left.\frac{\partial^2 u}{\partial x^2}\right)_0 \frac{(\Delta x)^2}{2} - \left.\frac{\partial^3 u}{\partial x^3}\right)_0 \frac{(\Delta x)^3}{6} + \cdots \tag{3.5}$$

and obtain the backward-difference representation

$$\left.\frac{\partial u}{\partial x}\right)_{i,j} = \frac{u_{i,j} - u_{i-1,j}}{\Delta x} + O(\Delta x) \tag{3.6}$$

We can subtract Eq. (3.5) from Eq. (3.2), rearrange, and obtain the "central" difference

$$\left.\frac{\partial u}{\partial x}\right)_{i,j} = \frac{u_{i+1,j} - u_{i-1,j}}{2\,\Delta x} + O[(\Delta x)^2] \tag{3.7}$$

We can also add Eq. (3.2) and Eq. (3.5) and rearrange to obtain an approximation to the second derivative:

$$\left.\frac{\partial^2 u}{\partial x^2}\right)_{i,j} = \frac{u_{i+1,j} - 2u_{i,j} + u_{i-1,j}}{(\Delta x)^2} + O[(\Delta x)^2] \tag{3.8}$$

It should be emphasized that these are only a few examples of the possible ways in which first and second derivatives can be approximated.

It is convenient to utilize difference operators to represent finite differences when particular forms are used repetitively. Here we define the first forward difference of $u_{i,j}$ with respect to x at the point i, j as

$$\Delta_x u_{i,j} = u_{i+1,j} - u_{i,j} \tag{3.9}$$

Thus we can express the forward finite-difference approximation for the first partial derivative as

$$\left.\frac{\partial u}{\partial x}\right)_{i,j} = \frac{u_{i+1,j} - u_{i,j}}{\Delta x} + O(\Delta x) = \frac{\Delta_x u_{i,j}}{\Delta x} + O(\Delta x) \tag{3.10}$$

Similarly, derivatives with respect to other variables such as y can be represented by

$$\frac{\Delta_y u_{i,j}}{\Delta y} = \frac{u_{i,j+1} - u_{i,j}}{\Delta y}$$

The first backward difference of $u_{i,j}$ with respect to x at i, j is denoted by

$$\nabla_x u_{i,j} = u_{i,j} - u_{i-1,j} \tag{3.11}$$

It follows that the first backward-difference approximation to the first derivative can be written as

$$\left.\frac{\partial u}{\partial x}\right)_{i,j} = \frac{u_{i,j} - u_{i-1,j}}{\Delta x} + O(\Delta x) = \frac{\nabla_x u_{i,j}}{\Delta x} + O(\Delta x) \tag{3.12}$$

The central-difference operators $\bar{\delta}$, δ and δ^2 will bc defined as

$$\bar{\delta}_x u_{i,j} = u_{i+1,j} - u_{i-1,j} \tag{3.13}$$

$$\delta_x u_{i,j} = u_{i+1/2,j} - u_{i-1/2,j} \tag{3.14}$$

$$\delta_x^2 u_{i,j} = \delta_x(\delta_x u_{i,j}) = u_{i+1,j} - 2u_{i,j} + u_{i-1,j} \tag{3.15}$$

and an averaging operator μ as

$$\mu_x u_{i,j} = \frac{u_{i+1/2,j} + u_{i-1/2,j}}{2} \tag{3.16}$$

Other convenient operators include the identity operator I and the shift operator E. The identity operator provides no operation, i.e., $Iu_{i,j} = u_{i,j}$. The shift operator advances the index associated with the subscripted variable by an amount indicated by the superscript. For example, $E_x^{-1} u_{i,j} = u_{i-1,j}$. When the superscript on E is $+1$, it is usually omitted. Difference representations can by indicated by using combinations of E and I, as for example,

$$\Delta_x u_{i,j} = (E_x - I)u_{i,j} = u_{i+1,j} - u_{i,j}$$

It is convenient to have specific operators for certain common central differences, although two of them can be easily expressed in terms of first-difference operators:

$$\bar{\delta}_x u_{i,j} = \Delta_x u_{i,j} + \nabla_x u_{i,j} \tag{3.17}$$

$$\delta_x^2 u_{i,j} = \Delta_x u_{i,j} - \nabla_x u_{i,j} = \Delta_x \nabla_x u_{i,j} \tag{3.18}$$

Using the newly defined operators, the central-difference representation for the first partial derivative can be written as

$$\left.\frac{\partial u}{\partial x}\right)_{i,j} = \frac{u_{i+1,j} - u_{i-1,j}}{2\,\Delta x} + O[(\Delta x)^2] = \frac{\bar{\delta}_x u_{i,j}}{2\,\Delta x} + O[(\Delta x)^2] \tag{3.19}$$

and the central-difference representation of the second derivative as

$$\left.\frac{\partial^2 u}{\partial x^2}\right)_{i,j} = \frac{u_{i+1,j} - 2u_{i,j} + u_{i-1,j}}{(\Delta x)^2} + O[(\Delta x)^2] = \frac{\delta_x^2 u_{i,j}}{(\Delta x)^2} + O[(\Delta x)^2] \tag{3.20}$$

Higher-order forward- and backward-difference operators are defined as

$$\Delta_x^n u_{i,j} = \Delta_x\left(\Delta_x^{n-1} u_{i,j}\right) \tag{3.21}$$

and

$$\nabla_x^n u_{i,j} = \nabla_x\left(\nabla_x^{n-1} u_{i,j}\right) \tag{3.22}$$

As an example, a forward second-derivative approximation is given by

$$\frac{\Delta_x^2 u_{i,j}}{(\Delta x)^2} = \frac{\Delta_x(u_{i+1,j} - u_{i,j})}{(\Delta x)^2} = \frac{u_{i+2,j} - u_{i+1,j} - u_{i+1,j} + u_{i,j}}{(\Delta x)^2}$$

$$= \frac{u_{i+2,j} - 2u_{i+1,j} + u_{i,j}}{(\Delta x)^2} = \left.\frac{\partial^2 u}{\partial x^2}\right)_{i,j} + O(\Delta x) \tag{3.23}$$

We can show that forward- and backward-difference approximations to derivatives of any order can be obtained from

$$\left.\frac{\partial^n u}{\partial x^n}\right)_{i,j} = \frac{\Delta_x^n u_{i,j}}{(\Delta x)^n} + O(\Delta x) \tag{3.24}$$

and

$$\left.\frac{\partial^n u}{\partial x^n}\right)_{i,j} = \frac{\nabla_x^n u_{i,j}}{(\Delta x)^n} + O(\Delta x) \tag{3.25}$$

Central-difference representations of derivatives of orders greater than the second can be expressed in terms of Δ and ∇ or δ. A more complete development on the use of difference operators can be found in many textbooks on numerical analysis such as that by Hildebrand (1956).

Most of the PDEs arising in fluid mechanics and heat transfer involve only first and second partial derivatives, and generally, we strive to represent these derivatives using values at only two or three grid points. Within these restrictions, the most frequently used first-derivative approximations on a grid for which $\Delta x = h = \text{const}$ are

$$\left.\frac{\partial u}{\partial x}\right)_{i,j} = \frac{u_{i+1,j} - u_{i,j}}{h} + O(h) \tag{3.26}$$

$$\left.\frac{\partial u}{\partial x}\right)_{i,j} = \frac{u_{i,j} - u_{i-1,j}}{h} + O(h) \tag{3.27}$$

$$\left.\frac{\partial u}{\partial x}\right)_{i,j} = \frac{u_{i+1,j} - u_{i-1,j}}{2h} + O(h^2) \tag{3.28}$$

$$\left.\frac{\partial u}{\partial x}\right)_{i,j} = \frac{-3u_{i,j} + 4u_{i+1,j} - u_{i+2,j}}{2h} + O(h^2) \tag{3.29}$$

$$\left.\frac{\partial u}{\partial x}\right)_{i,j} = \frac{3u_{i,j} - 4u_{i-1,j} + u_{i-2,j}}{2h} + O(h^2) \tag{3.30}$$

$$\left.\frac{\partial u}{\partial x}\right)_{i,j} = \frac{1}{2h}\left(\frac{\bar{\delta}_x u_{i,j}}{1 + \delta_x^2/6}\right) + O(h^4) \tag{3.31}$$

The most common three-point second-derivative approximations for a uniform grid, $\Delta x = h = \text{const}$, are

$$\left.\frac{\partial^2 u}{\partial x^2}\right)_{i,j} = \frac{u_{i,j} - 2u_{i+1,j} + u_{i+2,j}}{h^2} + O(h) \tag{3.32}$$

$$\left.\frac{\partial^2 u}{\partial x^2}\right)_{i,j} = \frac{u_{i,j} - 2u_{i-1,j} + u_{i-2,j}}{h^2} + O(h) \tag{3.33}$$

$$\left.\frac{\partial^2 u}{\partial x^2}\right)_{i,j} = \frac{u_{i+1,j} - 2u_{i,j} + u_{i-1,j}}{h^2} + O(h^2) \tag{3.34}$$

$$\left.\frac{\partial^2 u}{\partial x^2}\right)_{i,j} = \frac{\delta_x^2 u_{i,j}}{h^2(1 + \delta_x^2/12)} + O(h^4) \tag{3.35}$$

The compact, three-point schemes given by Eqs. (3.31) and (3.35) having fourth-order T.E.s deserve a further word of explanation (see also Orszag and Israeli, 1974). Letting $\partial u/\partial x)_{i,j} = v_{i,j}$, Eq. (3.31) is to be interpreted as

$$\left(1 + \frac{\delta_x^2}{6}\right) v_{i,j} = \frac{\bar{\delta}_x u_{i,j}}{2h}$$

or

$$\frac{1}{6}(v_{i+1,j} + 4v_{i,j} + v_{i-1,j}) = \frac{\bar{\delta}_x u_{i,j}}{2h} \tag{3.36}$$

which provides an *implicit* formula for the derivative of interest, $v_{i,j}$. The $v_{i,j}$ can be determined from the $u_{i,j}$ by solving a tridiagonal system of simultaneous algebraic equations, which can usually be accomplished quite efficiently. Tridiagonal systems commonly occur in connection with the use of implicit difference schemes for second-order PDEs arising from marching problems and are defined and discussed in some detail in Chapter 4. For now it is sufficient to think of a tridiagonal system as the arrangement of unknowns that would occur if each difference equation in a system only involved a single unknown variable evaluated at three adjacent grid locations. The interpretation of Eq. (3.35) proceeds in a similar manner, providing an *implicit* representation of $\partial^2 u/\partial x^2)_{i,j}$. Some difference approximations for derivatives that involve more than three grid points are given in Table 3.1. For completeness, a few common difference representations for mixed partial derivatives are presented in Table 3.2. These will prove useful for schemes discussed in subsequent chapters. The mixed-derivative approximations in Table 3.2 can be verified by using the

Table 3.1 Difference approximations using more than three points

Derivative	Finite-difference representation	Equation
$\left.\dfrac{\partial^3 u}{\partial x^3}\right)_{i,j} =$	$\dfrac{u_{i+2,j} - 2u_{i+1,j} + 2u_{i-1,j} - u_{i-2,j}}{2h^3} + O(h^2)$	(3.38)
$\left.\dfrac{\partial^4 u}{\partial x^4}\right)_{i,j} =$	$\dfrac{u_{i+2,j} - 4u_{i+1,j} + 6u_{i,j} - 4u_{i-1,j} + u_{i-2,j}}{h^4} + O(h^2)$	(3.39)
$\left.\dfrac{\partial^2 u}{\partial x^2}\right)_{i,j} =$	$\dfrac{-u_{i+3,j} + 4u_{i+2,j} - 5u_{i+1,j} + 2u_{i,j}}{h^2} + O(h^2)$	(3.40)
$\left.\dfrac{\partial^3 u}{\partial x^3}\right)_{i,j} =$	$\dfrac{-3u_{i+4,j} + 14u_{i+3,j} - 24u_{i+2,j} + 18u_{i+1,j} - 5u_{i,j}}{2h^3} + O(h^2)$	(3.41)
$\left.\dfrac{\partial^2 u}{\partial x^2}\right)_{i,j} =$	$\dfrac{2u_{i,j} - 5u_{i-1,j} + 4u_{i-2,j} - u_{i-3,j}}{h^2} + O(h^2)$	(3.42)
$\left.\dfrac{\partial^3 u}{\partial x^3}\right)_{i,j} =$	$\dfrac{5u_{i,j} - 18u_{i-1,j} + 24u_{i-2,j} - 14u_{i-3,j} + 3u_{i-4,j}}{2h^3} + O(h^2)$	(3.43)
$\left.\dfrac{\partial u}{\partial x}\right)_{i,j} =$	$\dfrac{-u_{i+2,j} + 8u_{i+1,j} - 8u_{i-1,j} + u_{i-2,j}}{12h} + O(h^4)$	(3.44)
$\left.\dfrac{\partial^2 u}{\partial x^2}\right)_{i,j} =$	$\dfrac{-u_{i+2,j} + 16u_{i+1,j} - 30u_{i,j} + 16u_{i-1,j} - u_{i-2,j}}{12h^2} + O(h^4)$	(3.45)

Taylor-series expansion for two variables:

$$
\begin{aligned}
u(x_0 + \Delta x, y_0 + \Delta y) &= u(x_0, y_0) + \left(\Delta x \frac{\partial}{\partial x} + \Delta y \frac{\partial}{\partial y}\right) u(x_0, y_0) \\
&+ \frac{1}{2!}\left(\Delta x \frac{\partial}{\partial x} + \Delta y \frac{\partial}{\partial y}\right)^2 u(x_0, y_0) \\
&+ \cdots + \frac{1}{n!}\left(\Delta x \frac{\partial}{\partial x} + \Delta y \frac{\partial}{\partial y}\right)^n u(x_0 + \theta\,\Delta x, y_0 + \theta\,\Delta y) \\
&\qquad 0 \leqslant \theta \leqslant 1 \quad (3.37)
\end{aligned}
$$

3.3 DIFFERENCE REPRESENTATION OF PARTIAL DIFFERENTIAL EQUATIONS

3.3.1 Truncation Error

As a starting point in our study of T.E., let us consider the heat equation

$$\frac{\partial u}{\partial t} = \alpha \frac{\partial^2 u}{\partial x^2} \tag{3.55}$$

Table 3.2 Difference approximations for mixed partial derivatives

Derivative	Finite-difference representation	Equation
$\left.\dfrac{\partial^2 u}{\partial x\,\partial y}\right)_{i,j} =$	$\dfrac{1}{\Delta x}\left(\dfrac{u_{i+1,j}-u_{i+1,j-1}}{\Delta y}-\dfrac{u_{i,j}-u_{i,j-1}}{\Delta y}\right)+O(\Delta x,\Delta y)$	(3.46)
$\left.\dfrac{\partial^2 u}{\partial x\,\partial y}\right)_{i,j} =$	$\dfrac{1}{\Delta x}\left(\dfrac{u_{i,j+1}-u_{l,j}}{\Delta y}-\dfrac{u_{i-1,j+1}-u_{i-1,j}}{\Delta y}\right)+O(\Delta x,\Delta y)$	(3.47)
$\left.\dfrac{\partial^2 u}{\partial x\,\partial y}\right)_{i,j} =$	$\dfrac{1}{\Delta x}\left(\dfrac{u_{i,j}-u_{i,j-1}}{\Delta y}-\dfrac{u_{i-1,j}-u_{i-1,j-1}}{\Delta y}\right)+O(\Delta x,\Delta y)$	(3.48)
$\left.\dfrac{\partial^2 u}{\partial x\,\partial y}\right)_{i,j} =$	$\dfrac{1}{\Delta x}\left(\dfrac{u_{i+1,j+1}-u_{i+1,j}}{\Delta y}-\dfrac{u_{i,j+1}-u_{i,j}}{\Delta y}\right)+O(\Delta x,\Delta y)$	(3.49)
$\left.\dfrac{\partial^2 u}{\partial x\,\partial y}\right)_{i,j} =$	$\dfrac{1}{\Delta x}\left(\dfrac{u_{i+1,j+1}-u_{i+1,j-1}}{2\,\Delta y}-\dfrac{u_{i,j+1}-u_{i,j-1}}{2\,\Delta y}\right)+O[\Delta x,(\Delta y)^2]$	(3.50)
$\left.\dfrac{\partial^2 u}{\partial x\,\partial y}\right)_{i,j} =$	$\dfrac{1}{\Delta x}\left(\dfrac{u_{i,j+1}-u_{i,j-1}}{2\,\Delta y}-\dfrac{u_{i-1,j+1}-u_{i-1,j-1}}{2\,\Delta y}\right)+O[\Delta x,(\Delta y)^2]$	(3.51)
$\left.\dfrac{\partial^2 u}{\partial x\,\partial y}\right)_{i,j} =$	$\dfrac{1}{2\,\Delta x}\left(\dfrac{u_{i+1,j+1}-u_{i+1,j-1}}{2\,\Delta y}-\dfrac{u_{i-1,j+1}-u_{i-1,j-1}}{2\,\Delta y}\right)+O[(\Delta x)^2,(\Delta y)^2]$	(3.52)
$\left.\dfrac{\partial^2 u}{\partial x\,\partial y}\right)_{i,j} =$	$\dfrac{1}{2\,\Delta x}\left(\dfrac{u_{i+1,j+1}-u_{i+1,j}}{\Delta y}-\dfrac{u_{i-1,j+1}-u_{i-1,j}}{\Delta y}\right)+O[(\Delta x)^2,\Delta y]$	(3.53)
$\left.\dfrac{\partial^2 u}{\partial x\,\partial y}\right)_{i,j} =$	$\dfrac{1}{2\,\Delta x}\left(\dfrac{u_{i+1,j}-u_{i+1,j-1}}{\Delta y}-\dfrac{u_{i-1,j}-u_{i-1,j-1}}{\Delta y}\right)+O[(\Delta x)^2,\Delta y]$	(3.54)

Using a forward-difference representation for the time derivative ($t = n\Delta t$) and a central-difference representation for the second derivative, we can approximate the heat equation by

$$\frac{u_j^{n+1}-u_j^n}{\Delta t} = \frac{\alpha}{(\Delta x)^2}\left(u_{j+1}^n - 2u_j^n + u_{j-1}^n\right) \tag{3.56a}$$

However, we noted in Section 3.2 that T.E.s were associated with the forward- and central-difference representations used in Eq. (3.56*a*). If we rearrange Eq. (3.55) to put zero on the right-hand side and include the T.E.s associated with the difference representation of the derivatives, we obtain

$$\underbrace{\frac{\partial u}{\partial t} - \alpha\frac{\partial^2 u}{\partial x^2}}_{\text{PDE}} = \underbrace{\frac{u_j^{n+1}-u_j^n}{\Delta t} - \frac{\alpha}{(\Delta x)^2}\left(u_{j+1}^n - 2u_j^n + u_{j-1}^n\right)}_{\text{FDE}}$$

$$+ \underbrace{\left[-\left.\frac{\partial^2 u}{\partial t^2}\right)_{n,j}\frac{\Delta t}{2} + \alpha\left.\frac{\partial^4 u}{\partial x^4}\right)_{n,j}\frac{(\Delta x)^2}{12} + \cdots\right]}_{\text{T.E.}} \tag{3.56b}$$

where PDE is the partial differential equation and FDE is the finite-difference equation. The T.E.s associated with all derivatives in any one PDE should be obtained by expanding about the same point (n, j in the above discussion).

The difference representation given by Eq. (3.56a) will be referred to as the *simple explicit scheme* for the heat equation. An *explicit* scheme is one for which only one unknown appears in the difference equation in a manner that permits evaluation in terms of known quantities. Since the parabolic heat equation governs a marching problem for which an initial distribution of u must be specified, u at the time level n can be considered as known. If the second-derivative term in the heat equation was approximated by u at the $n + 1$ time level, three unknowns would appear in the difference equation, and the procedure would be known as *implicit*, indicating that the algebraic formulation would require the simultaneous solution of several equations involving the unknowns. The differences between implicit and explicit schemes are discussed further in Chapter 4.

The quantity in brackets (note that only the leading terms have been written out utilizing Taylor-series expansions) in Eq. (3.56b) is identified as the *truncation error* for this finite-difference representation of the heat equation and is defined as the difference between the PDE and the difference approximation to it. That is, T.E. = PDE − FDE. The *order* of the T.E. in this case is $O(\Delta t) + O[(\Delta x)^2]$, which is frequently expressed in the form $O[\Delta t, (\Delta x)^2]$. Naturally, we solve only the finite-difference equations and hope that the T.E. is small. If we do not feel a little uneasy at this point, perhaps we should. How do we know that our difference representation is acceptable and that a marching solution technique will work in the sense of giving us an approximate solution to the PDE? In order to be acceptable, our difference representation for this marching problem needs to meet the conditions of *consistency* and *stability*.

3.3.2 Round-Off and Discretization Errors

Any computed solution, including sometimes an "exact" analytic solution to a PDE, may be affected by rounding to a finite number of digits in the arithmetic operations. These errors are called *round-off errors*, and we are especially aware of their existence in obtaining machine solutions to finite-difference equations because of the large number of dependent, repetitive operations that are usually involved. In some types of calculations, the magnitude of the round-off error is proportional to the number of grid points in the problem domain. In these cases, refining the grid may decrease the T.E. but increase the round-off error.

Discretization error is the error in the solution to the PDE caused by replacing the continuous problem by a discrete one and is defined as the difference between the exact solution of the PDE (round-off free) and the exact solution of the FDEs (round-off free). In terms of the definitions developed thus far, the difference between the exact solution of the PDE and the computer solution to the FDEs would be equal to the sum of the discretization error and the round-off error associated with the finite-difference calculation. We can also observe that the discretization error is the error in the solution that is caused by the T.E. in the difference representation of the PDE plus any errors introduced by the treatment of boundary conditions.

3.3.3 Consistency

Consistency deals with the extent to which the FDEs approximate the PDEs. The difference between the PDE and the finite-difference approximation has already been defined as the T.E. of the difference representation. A finite-difference representation of a PDE is said to be consistent if we can show that the difference between the PDE and its difference representation vanishes as the mesh is refined, i.e., $\lim_{\text{mesh} \to 0}(\text{PDE} - \text{FDE}) = \lim_{\text{mesh} \to 0}(\text{T.E.}) = 0$. This should always be the case if the order of the T.E. vanishes under grid refinement. An example of a questionable scheme would be one for which the T.E. was $O(\Delta t/\Delta x)$, where the scheme would not formally be consistent unless the mesh were refined in a manner such that $\Delta t/\Delta x \to 0$. The DuFort-Frankel (DuFort and Frankel, 1953) differencing of the heat equation,

$$\frac{u_j^{n+1} - u_j^{n-1}}{2\,\Delta t} = \frac{\alpha}{(\Delta x)^2}\left(u_{j+1}^n - u_j^{n+1} - u_j^{n-1} + u_{j-1}^n\right) \tag{3.57}$$

for which the leading terms in the T.E. are

$$+\left.\frac{\alpha}{12}\frac{\partial^4 u}{\partial x^4}\right)_{n,j}(\Delta x)^2 - \left.\alpha\frac{\partial^2 u}{\partial t^2}\right)_{n,j}\left(\frac{\Delta t}{\Delta x}\right)^2 - \left.\frac{1}{6}\frac{\partial^3 u}{\partial t^3}\right)_{n,j}(\Delta t)^2$$

serves as an example. All is well if

$$\lim_{\Delta t,\,\Delta x \to 0}\left(\frac{\Delta t}{\Delta x}\right) = 0$$

but if Δt and Δx were to approach zero at the same rate, such that $\Delta t/\Delta x = \beta$, then the DuFort-Frankel scheme is consistent with the hyperbolic equation

$$\frac{\partial u}{\partial t} + \alpha\beta^2\frac{\partial^2 u}{\partial t^2} = \alpha\frac{\partial^2 u}{\partial x^2}$$

3.3.4 Stability

Numerical stability is a concept applicable in the strict sense only to marching problems. A stable numerical scheme is one for which errors from any source (round-off, truncation, mistakes) are not permitted to grow in the sequence of numerical procedures as the calculation proceeds from one marching step to the next. Generally, concern over stability occupies much more of our time and energy than does concern over consistency. Consistency is relatively easy to check, and most schemes that are conceived will be consistent just owing to the methodology employed in their development. Stability is much more subtle, and usually a bit of hard work is required in order to establish analytically that a scheme is stable. More detail is presented in Section 3.6, and some very workable methods will be developed for establishing the stability limits for linear PDEs. It will be possible to extend these guidelines to nonlinear equations in an approximate sense.

Using these guidelines, the DuFort-Frankel scheme, Eq. (3.57), for the heat equation would be found to be unconditionally stable, whereas the simple explicit scheme would be stable only if $r = [\alpha \Delta t/(\Delta x)^2] \leqslant \frac{1}{2}$. This restriction would limit the size of the marching step permitted for any specific spatial mesh.

A scheme using a central time difference and having a more favorable T.E. of $O[(\Delta t)^2, (\Delta x)^2]$,

$$\frac{u_j^{n+1} - u_j^{n-1}}{2\,\Delta t} = \frac{\alpha}{(\Delta x)^2}(u_{j+1}^n - 2u_j^n + u_{j-1}^n) \tag{3.58}$$

is unconditionally unstable and therefore cannot be used for real calculations despite the fact that it looks to be more accurate, in terms of T.E., than the ones given previously that will work.

Sometimes instability can be identified with a physical implausibility. That is, conditions that would result in an unstable numerical procedure would also imply unacceptable modeling of physical processes. To illustrate this, we rearrange the simple explicit representation of the heat equation, Eq. (3.56*a*), so that the unknown appears on the left. Letting $r = \alpha \Delta t/(\Delta x)^2$, our difference equation becomes

$$u_j^{n+1} = r(u_{j+1}^n + u_{j-1}^n) + (1 - 2r)u_j^n \tag{3.59}$$

Suppose that at time t, $u_{j+1}^n = u_{j-1}^n = 100°\mathrm{C}$ and $u_j^n = 0°\mathrm{C}$. This arrangement is shown in Fig. 3.2. If $r > \frac{1}{2}$, we see that the temperature at point j at time level $n + 1$ will exceed the temperature at the two surrounding points at time level n. This seems unreasonable, since we expect heat to flow from the warmer region to a colder region but not vice versa. The maximum temperature that we would expect to find at point j at time level $n + 1$ is 100°C. If $r = 1$, for example, u_j^{n+1} would equal 200°C by Eq. (3.59).

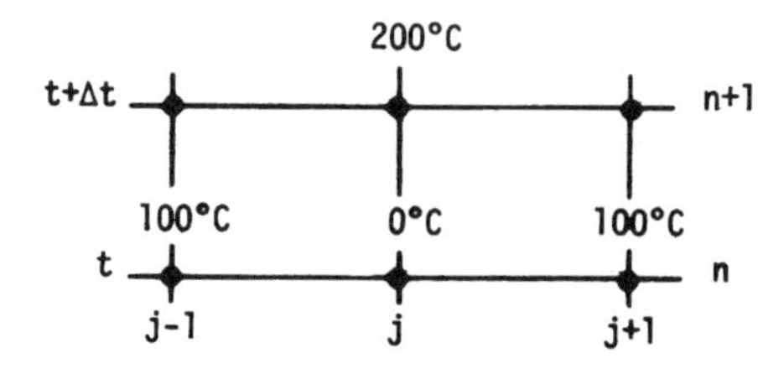

Figure 3.2 Physical implausibility resulting from $r = 1$.

3.3.5 Convergence for Marching Problems

Generally, we find that a consistent, stable scheme is convergent. Convergence here means that the solution to the finite-difference equation approaches the true solution to the PDE having the same initial and boundary conditions as the mesh is refined. A proof of this is available for initial value (marching) problems governed by linear PDEs. The theorem, due to Lax (see Richtmyer and Morton, 1967) is stated here without proof.

> **Lax's equivalence theorem:** Given a properly posed initial value problem and a finite-difference approximation to it that satisfies the consistency condition, stability is the necessary and sufficient condition for convergence.

We might add that most computational work proceeds as though this theorem applies also to nonlinear PDEs, although the theorem has never been proven for this more general category of equations.

3.3.6 A Comment on Equilibrium Problems

Throughout our discussion of stability and convergence, the focus was on marching problems (parabolic and hyperbolic PDEs). Despite this emphasis on initial value problems, most of the material presented in this chapter also applies to equilibrium problems. The exception is the concept of stability. We should observe, however, that the important concept of consistency applies to difference representations of PDEs of all classes.

The "convergence" of the solution of the difference equation to the exact solution of the PDE might be aptly termed truncation or discretization convergence. The solution to equilibrium problems (elliptic equations) leads us to a system of simultaneous algebraic equations that needs to be solved only once, rather than in a marching manner. Thus the concept of stability developed previously is not directly applicable as stated. To achieve "truncation convergence" for equilibrium problems, it would seem that it is only necessary to devise a solution scheme in which the error in solving the simultaneous algebraic equations can be controlled as the mesh size is refined without limit. Many common schemes are iterative (Gauss-Seidel iteration is one example) in nature, and for these we want to ensure that the iterative process converges. Here convergence means that the iterative process is repeated until the magnitude of the difference between the function at the $k+1$ and the k iteration levels is as small as we wish for each grid point, i.e., $|u_{i,j}^{k+1} - u_{i,j}^{k}| < \epsilon$. This is known as *iteration convergence*. It would appear that (no proof can be cited) truncation convergence will be assumed for a consistent representation to an equilibrium problem if it can be shown that the iterative method of solution converges even for arbitrarily small choices of mesh sizes.

It is possible to use direct (noniterative) methods to solve the algebraic equations associated with equilibrium problems. For these methods we would want to be sure that the errors inherent in the method, especially round-off errors, do not get out of control as the mesh is refined and the number of points tends toward infinity.

In closing this section, we should mention that there are aspects to the iterative solution of equilibrium problems that resemble the marching process in initial value problems and a sense in which stability concerns in the marching problems correspond to iterative convergence concerns in the solution to equilibrium problems.

3.3.7 Conservation Form and Conservative Property

Two different ideas will be discussed in this section. The first has to do with the PDEs themselves. The terms "conservation form," "conservation-law form," "conservative form," and "divergence form" are all equivalent, and PDEs possessing this form have the property that the coefficients of the derivative terms are either constant or, if variable, their derivatives appear nowhere in the equation. Normally, for the PDEs that represent a physical conservation statement, this means that the divergence of a physical quantity can be identified in the equation. If all spatial derivative terms of an equation can be identified as divergence terms, the equation is said to be in "strong conservation-law form." As an example, the conservative form of the equation for mass conservation (continuity equation) is

$$\frac{\partial \rho}{\partial t} + \frac{\partial \rho u}{\partial x} + \frac{\partial \rho v}{\partial y} + \frac{\partial \rho w}{\partial z} = 0 \tag{3.60}$$

which can be written in vector notation as

$$\frac{\partial \rho}{\partial t} + \nabla \cdot \rho \mathbf{V} = 0$$

A nonconservative or nondivergence form would be

$$\frac{\partial \rho}{\partial t} + u\frac{\partial \rho}{\partial x} + \rho\frac{\partial u}{\partial x} + v\frac{\partial \rho}{\partial y} + \rho\frac{\partial v}{\partial y} + w\frac{\partial \rho}{\partial z} + \rho\frac{\partial w}{\partial z} = 0 \tag{3.61}$$

As a second example, we consider the one-dimensional (1-D) heat conduction equation for a substance whose density ρ, specific heat c, and thermal conductivity k all vary with position. The conservative form of this equation is

$$\rho c\frac{\partial T}{\partial t} = \frac{\partial}{\partial x}\left(k\frac{\partial T}{\partial x}\right) \tag{3.62}$$

whereas a nonconservative form would be

$$\rho c\frac{\partial T}{\partial t} = k\frac{\partial^2 T}{\partial x^2} + \frac{\partial k}{\partial x}\frac{\partial T}{\partial x} \tag{3.63}$$

In Eq. (3.62) the right-hand side can be identified as the negative of the divergence of the heat flux vector specialized for 1-D conduction. A difference formulation based on a PDE in nondivergence form may lead to numerical difficulties in situations where the coefficients may be discontinuous, as in flows containing shock waves.

The second idea to be developed in this section deals with the *conservative property of a finite-difference representation.* The PDEs of interest in this book all have their basis in physical laws, such as the conservation of mass, momentum, and energy. Such a PDE represents a conservation statement at a point. We strive to construct finite-difference representations that provide a good approximation to the PDE in a small, local neighborhood involving a few grid points. The same conservation principles that gave rise to the PDEs also apply to arbitrarily large regions (control volumes). In fact, in deriving the PDEs, we usually start with the control-volume form of the conservation statement. If our finite-difference representation approximates the PDE closely in the neighborhood of each grid point, then we have reason to expect that the related conservation statement will be approximately enforced over a larger control volume containing a large number of grid points in the interior. Those finite-difference schemes that maintain the discretized version of the conservation statement exactly (except for round-off errors) for any mesh size over an arbitrary finite region containing any number of grid points is said to have the *conservative property*. For some problems this property is crucial.

The key word in the definition above is "exactly." All consistent schemes should approximately enforce the appropriate conservation statement over large regions, but schemes having the conservative property do so exactly (except for round-off errors) because of exact cancellation of terms. To illustrate this concept, we will consider a problem requiring the solution of the continuity equation for steady flow. The PDE can be written as

$$\nabla \cdot \rho \mathbf{V} = 0$$

We will assume that the PDE is approximated by a suitable finite-difference representation and solved throughout the flow. For an arbitrary control volume that could include the entire problem domain or any fraction of it, conservation of mass for steady flow requires that the net mass efflux be zero (mass flow rate in equals mass flow rate out). This is observed formally by applying the divergence theorem to the governing PDE,

$$\iiint_R \nabla \cdot \rho \mathbf{V} \, dR = \iint_S \rho \mathbf{V} \cdot \mathbf{n} \, dS = 0$$

To see if the finite-difference representation for the PDE has the conservative property, we must establish that the discretized version of the divergence theorem is satisfied. We normally check this for a control volume consisting of the entire problem domain. To do this, the integral on the left is evaluated by summing the difference representation of the PDE at all grid points. If the difference scheme has the conservative property, all terms will cancel except those that represent fluxes at the boundaries. This is sometimes referred to as the "telescoping property." It should be possible to rearrange the remaining terms to obtain identically a finite-difference representation of the integral on the right. For this example the result will be a verification that the mass flux into the control volume equals the mass flux out. If the difference scheme used for the PDE is not conservative, the numerical solution may permit the existence of small mass sources or sinks.

Schemes having the conservative property occur in a natural way when differencing starts with the divergence form of the PDE. For some equations and problems, the divergence form is not an appropriate starting point. For these situations, use of a control-volume method (Section 3.4.4) for obtaining the difference scheme is helpful. This difference representation will usually have the conservative property if care is taken to ensure that the expressions used to represent fluxes across the interface of two adjacent control volumes are the same in the difference form of the conservation statement for each of the two control volumes.

The conservative property issue has been actively discussed and debated over the short history of computational fluid mechanics and heat transfer. However, the conservative property is not the only important figure of merit for a difference representation. PDEs represent more than a conservation statement at a point. As shown by solution forms in Chapter 2, PDEs also contain information on characteristic directions and domains of dependence. Proper representation of this information is also important. Many useful finite-difference equations do not have the conservative property and, in a few instances, prove to be more accurate in some sense than those that do. The importance of maintaining the conservation statement with high accuracy over a finite region is highly problem dependent. All consistent formulations, whether or not they have the conservative property, can provide an adequate representation for most problems if the grid is refined sufficiently.

3.4 FURTHER EXAMPLES OF METHODS FOR OBTAINING FINITE-DIFFERENCE EQUATIONS

As we start with a given PDE and a finite-difference mesh, several procedures are available to us for developing finite-difference equations. Among these are

1. Taylor-series expansions
2. polynomial fitting
3. integral method (called the micro-integral method by some)
4. finite-volume (control-volume) approach

It is sometimes possible to obtain exactly the same finite-difference representation by using all four methods. In our introduction to the subject, we will lean most heavily on the use of Taylor-series expansions, utilizing polynomial fitting on occasion in treating boundary conditions.

3.4.1 Use of Taylor Series

We now demonstrate how one might proceed on a slightly more formal basis with Taylor-series expansions to develop difference expressions satisfying specified constraints. Suppose we want to develop a difference approximation for $\partial u/\partial x)_{i,j}$ having a T.E. of $O[(\Delta x)^2]$ using at most values $u_{i-2,j}$, $u_{i-1,j}$, and $u_{i,j}$.

With these constraints and objectives, it would appear logical to write Taylor-series expressions for $u_{i-2,j}$ and $u_{i-1,j}$ expanding about the point (i,j) and attempt to solve for $\partial u/\partial x)_{i,j}$ from the resulting equations in such a way as to obtain a T.E. of $O[(\Delta x)^2]$:

$$u_{i-2,j} = u_{i,j} + \left.\frac{\partial u}{\partial x}\right)_{i,j}(-2\,\Delta x) + \left.\frac{\partial^2 u}{\partial x^2}\right)_{i,j}\frac{(2\,\Delta x)^2}{2!} + \left.\frac{\partial^3 u}{\partial x^3}\right)_{i,j}\frac{(-2\,\Delta x)^3}{3!} + \cdots \tag{3.64}$$

$$u_{i-1,j} = u_{i,j} + \left.\frac{\partial u}{\partial x}\right)_{i,j}(-\Delta x) + \left.\frac{\partial^2 u}{\partial x^2}\right)_{i,j}\frac{(\Delta x)^2}{2!} + \left.\frac{\partial^3 u}{\partial x^3}\right)_{i,j}\frac{(-\Delta x)^3}{3!} + \cdots \tag{3.65}$$

It is often possible to determine the required form of the difference representation by inspection or simple substitution. To proceed by substitution, we will rearrange Eq. (3.64) to put $\partial u/\partial x)_{i,j}$ on the left-hand side, such that

$$\left.\frac{\partial u}{\partial x}\right)_{i,j} = \frac{u_{i,j}}{2\,\Delta x} - \frac{u_{i-2,j}}{2\,\Delta x} + \frac{\partial^2 u}{\partial x^2}\Delta x + O[(\Delta x)^2]$$

As is, the representation is $O(\Delta x)$ because of the term $(\partial^2 u/\partial x^2)\,\Delta x$. We can substitute for $\partial^2 u/\partial x^2$ in the above equation using Eq. (3.65) to obtain the desired result. A more formal procedure to obtain the desired expression is sometimes useful. To proceed more formally, we first multiply Eq. (3.64) by a and Eq. (3.65) by b and add the two equations. If $-2a - b = 1$, then the coefficient of $\partial u/\partial x)_{i,j}\,\Delta x$ will be 1 after the addition, and if $2a + b/2 = 0$, then the terms involving $\partial^2 u/\partial x^2)_{i,j}$, which would contribute a T.E. of $O(\Delta x)$ to the final result, will be eliminated. A solution to the equations

$$-2a - b = 1 \qquad 2a + \frac{b}{2} = 0$$

is given by $a = \frac{1}{2}$, $b = -2$. Thus, if we multiply Eq. (3.64) by $\frac{1}{2}$, Eq. (3.65) by -2, add the results, and solve for $\partial u/\partial x)_{i,j}$, we obtain

$$\left.\frac{\partial u}{\partial x}\right)_{i,J} = \frac{u_{i-2,j} - 4u_{i-1,j} + 3u_{i,j}}{2\,\Delta x} + O[(\Delta x)^2]$$

which can be recognized as Eq. (3.30). A careful check on the details of this example will reveal that it was really necessary to include terms involving $\partial^3 u/\partial x^3)_{i,j}$ in the Taylor-series expansions in order to determine whether or not these terms would cancel in the algebraic operations and reduce the T.E. even further to $O[(\Delta x)^3]$. Fortuitous cancellation of terms occurs frequently enough to warrant close attention to this point.

We should observe that it is sometimes necessary to carry out the inverse of the above process. That is, suppose we had obtained the approximation represented by Eq. (3.30) by some other means and we wanted to investigate the consistency and T.E. of such an expression. For this, the use of Taylor-series expansions would be invaluable, and the recommended procedure would be to substitute the Taylor-series expressions from Eq. (3.64) and Eq. (3.65) above for $u_{i-2,j}$ and $u_{i-1,j}$ into the difference representation to obtain an expression of the form $\partial u/\partial x)_{i,j}$ + T.E. on the right-hand side. At this point, the T.E. has been identified, and if $\lim_{\Delta x \to 0}(\text{T.E.}) = 0$, the difference representation is consistent.

As a slightly more complex example, we will develop a finite-difference approximation with T.E. of $O[(\Delta y)^2]$ for $\partial u/\partial y$ at point (i, j) using at most $u_{i,j}, u_{i,j+1}, u_{i,j-1}$ when the grid spacing is not uniform. We will adopt the notation that $\Delta y_+ = y_{i,j+1} - y_{i,j}$ and $\Delta y_- = y_{i,j} - y_{i,j-1}$, as indicated in Fig. 3.3.

We recall that for equal spacing, the central-difference representation for a first derivative was equivalent to the arithmetic average of a forward and backward representation. That is, for $\Delta y_+ = \Delta y_- = \Delta y$,

$$\left.\frac{\partial u}{\partial y}\right)_{i,j} = \frac{\bar{\delta}_y u_{i,j}}{2\,\Delta y} = \frac{\Delta_y u_{i,j} + \nabla_y u_{i,j}}{2\,\Delta y} + O\left[(\Delta y)^2\right]$$

We might wonder if, for unequal spacing, use of a geometrically weighted average will preserve the second-order accuracy:

$$\left.\frac{\partial u}{\partial y}\right)_{i,j} \stackrel{?}{=} \frac{\Delta_y u_{i,j}}{\Delta y_+}\left(\frac{\Delta y_-}{\Delta y_+ + \Delta y_-}\right) + \frac{\nabla_y u_{i,j}}{\Delta y_-}\left(\frac{\Delta y_+}{\Delta y_+ + \Delta y_-}\right) + O\left[(\Delta y)^2\right] \quad (3.66)$$

The truth of the above statement may be evident to some, but it can be verified from basics by use of Taylor-series expansions about point (i, j). Letting $\Delta y_+/\Delta y_- = \alpha$, and adopting the more compact subscript notation to denote differentiation, $u_y = \partial u/\partial y)_{i,j}$, $u_{yy} = \partial^2 u/\partial y^2)_{i,j}$, etc., we obtain

$$u_{i,j+1} = u_{i,j} + u_y \alpha \Delta y_- + u_{yy}\frac{(\alpha\,\Delta y_-)^2}{2!} + u_{yyy}\frac{(\alpha\,\Delta y_-)^3}{3!} + u_{yyyy}\frac{(\alpha\,\Delta y_-)^4}{4!} + \cdots \quad (3.67)$$

$$u_{i,j-1} = u_{i,j} + u_y(-\Delta y_-) + u_{yy}\frac{(-\Delta y_-)^2}{2!} + u_{yyy}\frac{(-\Delta y_-)^3}{3!} + u_{yyyy}\frac{(-\Delta y_-)^4}{4!} + \cdots \quad (3.68)$$

As before, we will multiply Eq. (3.67) by a and Eq. (3.68) by b, add the results, and solve for $\partial u/\partial y)_{i,j}$. Requiring that the coefficient of $\partial u/\partial y)_{i,j}\Delta y_-$

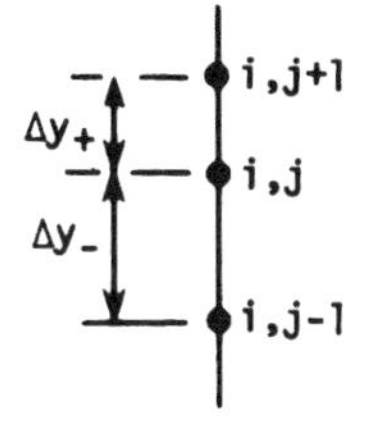

Figure 3.3 Notation for unequal y spacing.

be equal to 1 after the addition, gives $a\alpha - b = 1$. For the final result to have a T.E. of $O[(\Delta y)^2]$ or better, the coefficient of u_{yy} must be zero after the addition, which requires that $\alpha^2 a + b = 0$. A solution to these two algebraic equations can be obtained readily as $a = 1/\alpha(\alpha + 1)$, $b = -\alpha/(\alpha + 1)$. Thus

$$\left.\frac{\partial u}{\partial y}\right)_{i,j} = \frac{a \times \text{Eq. (3.67)} + b \times \text{Eq. (3.68)}}{\Delta y_-} + O\left[(\Delta y)^2\right]$$

The final result can be written as

$$\left.\frac{\partial u}{\partial y}\right)_{i,j} = \frac{u_{i,j+1} + (\alpha^2 - 1)u_{i,j} - \alpha^2 u_{i,j-1}}{\alpha(\alpha + 1)\Delta y_-} \tag{3.69}$$

which can be rearranged further into the form given by Eq. (3.66).

Our Taylor-series examples thus far have illustrated procedures for obtaining a finite-difference approximation to a single derivative. However, our main interest is in correctly approximating an entire PDE at an arbitrary point in the problem domain. For this reason, we must be careful to use the same expansion point in approximating all derivatives in the PDE by the Taylor-series method. If this is done, then the T.E. for the entire equation can be obtained by adding the T.E. for each derivative.

There is no requirement that the expansion point be (i, j), as indicated by the following examples, where the order of the T.E. and the most convenient expansion points are indicated. The geometric arrangement of points used in the difference equation is indicated by the sketch of the difference "molecule."

Fully implicit form for the heat equation, Eq. (3.55):

$$\frac{u_j^{n+1} - u_j^n}{\Delta t} = \frac{\alpha}{(\Delta x)^2}\left(u_{j+1}^{n+1} - 2u_j^{n+1} + u_{j-1}^{n+1}\right) \qquad \text{T.E.} = O[\Delta t, (\Delta x)^2] \tag{3.70}$$

The difference molecule for this scheme is shown in Fig. 3.4, and point $(n + 1, j)$ is indicated as the most convenient expansion point.

Crank-Nicholson form for the heat equation:

$$\frac{u_j^{n+1} - u_j^n}{\Delta t} = \frac{\alpha}{2(\Delta x)^2}\left[u_{j+1}^{n+1} + u_{j+1}^n - 2\left(u_j^{n+1} + u_j^n\right) + u_{j-1}^{n+1} + u_{j-1}^n\right] \tag{3.71a}$$

$$\text{T.E.} = O[(\Delta t)^2, (\Delta x)^2]$$

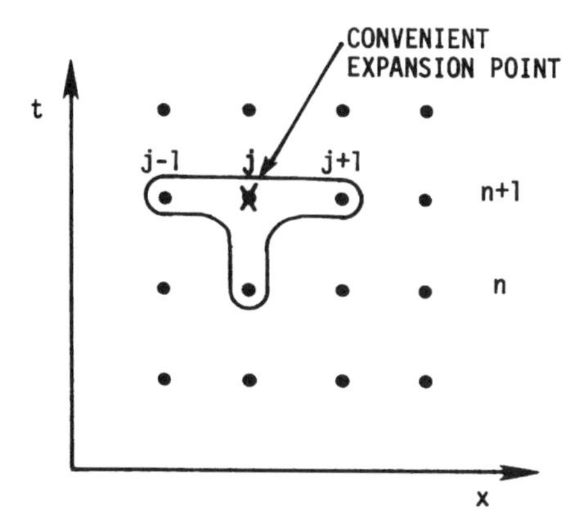

Figure 3.4 Difference molecule, fully implicit form for heat equation.

The difference molecule for the Crank-Nicolson scheme is shown in Fig. 3.5, and point $(n + \frac{1}{2}, j)$ is designated as the most convenient expansion point.

It is interesting to note that the *order* of the T.E. for difference representations of a complete PDE (not a single derivative term, however) is not dependent upon the choice of expansion point in the evaluation of this error by the Taylor-series method. We will demonstrate this point by considering the Crank-Nicolson scheme. The T.E. for the Crank-Nicolson scheme was most conveniently determined by expanding about the point $(n + \frac{1}{2}, j)$ to obtain the results stated above. Using this point resulted in the elimination of the maximum number of terms from the Taylor series by cancellation. Had we used point (n, j) or even $(n - 1, j)$ as the expansion point, the conclusion on the order of the T.E. would have been the same. To reach this conclusion, however, we often must examine the T.E. very carefully. To illustrate, evaluating the T.E. of the Crank-Nicolson scheme by using expansions for $u_{j-1}^n, u_{j+1}^n, u_{j-1}^{n+1}, u_{j+1}^{n+1}, u_j^{n+1}$ about point (n, j) in Eq. (3.71*a*) gives, after rearrangement,

$$u_t - \alpha u_{xx} = -u_{tt}\frac{\Delta t}{2} + \alpha u_{txx}\frac{\Delta t}{2} + O[(\Delta x)^2] + O[(\Delta t)^2] \qquad (3.71b)$$

At first glance, we are tempted to conclude that the T.E. for the Crank-Nicolson scheme becomes $O(\Delta t) + O[(\Delta x)^2]$, when evaluated by expanding about point (n, j), because of the appearance of the terms $-u_{tt}\,\Delta t/2$ and $\alpha u_{txx}\,\Delta t/2$. However, we can recognize these two terms as $-(\Delta t/2)(\partial/\partial t)(u_t - \alpha u_{xx})$, where the quantity in the second set of parentheses is the left-hand side of Eq. (3.71*b*). Thus we can differentiate Eq. (3.71*b*) with respect to t and multiply both sides by $-\Delta t/2$ to learn that $-(\Delta t/2)(\partial/\partial t)(u_t - \alpha u_{xx}) = O[(\Delta t)^2] + O[(\Delta x)^2]$. From this, we conclude that the T.E. for the Crank-Nicolson scheme is $O[(\Delta t)^2] + O[(\Delta x)^2]$ when evaluated about either point (n, j) or point $(n + \frac{1}{2}, j)$. Use of other points will give the same results for the order of the T.E. This example illustrates that the leading terms in the T.E. should be examined very carefully to see if they can be identified as a multiple of a derivative of the original PDE. If they can, they should be replaced by expressions of higher order.

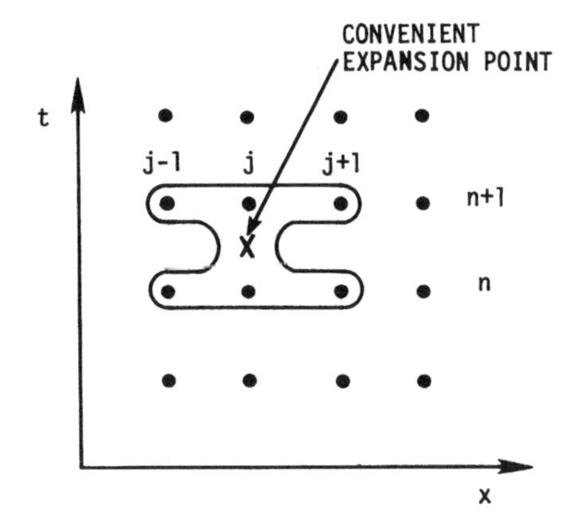

Figure 3.5 Difference molecule, Crank-Nicholson form for heat equation.

3.4.2 Use of Polynomial Fitting

Many applications of polynomial fitting are observed in computational fluid mechanics and heat transfer. The technique can be used to develop the entire finite-difference representation for a PDE. However, the technique is perhaps most commonly employed in the treatment of boundary conditions or in gleaning information from the solution in the neighborhood of the boundary. Consider some specific examples.

Example 3.1 In this example, the derivative approximations needed to represent a PDE will be obtained by assuming that the solution to the PDE can be approximated locally by a polynomial. The polynomial is then "fitted" to the points surrounding the general point (i, j), utilizing values of the function at the grid points. A sufficient number of points can be used to determine the coefficients in the polynomial exactly. The polynomial can then be differentiated to obtain the desired approximation to the derivatives. Consider Laplace's equation, which governs the 2-D temperature distribution in a solid under steady-state conditions:

$$\frac{\partial^2 T}{\partial x^2} + \frac{\partial^2 T}{\partial y^2} = 0 \tag{3.72}$$

Solution We suppose that both the x and y dependency of temperature can be expressed by a second-degree polynomial. For example, holding y fixed, we assume that temperatures at various x locations in the neighborhood of point (i,j) can be determined from

$$T(x, y_0) = a + bx + cx^2$$

For convenience, we let $x = 0$ at point (i, j), and $\Delta x = \text{const}$. Clearly,

$$\left.\frac{\partial T}{\partial x}\right)_{i,j} = b$$

$$\left.\frac{\partial^2 T}{\partial x^2}\right)_{i,j} = 2c$$

The coefficients a, b, and c can be evaluated in terms of temperatures at specific grid points and Δx. To do so, we must make some choices as to which neighboring grid points to use, and this choice determines the geometric arrangement of the difference molecule, that is, whether the resulting derivative approximations are central, forward, or backward differences. Here we will choose points $(i - 1, j)$, (i, j), and $(i + 1, j)$ and obtain

$$T(i, j) = a$$

$$T(i + 1, j) = a + b\,\Delta x + c(\Delta x)^2$$

$$T(i - 1, j) = a - b\,\Delta x + c(\Delta x)^2$$

from which we determine that

$$b = \left.\frac{\partial T}{\partial x}\right)_{i,j} = \frac{T_{i+1,j} - T_{i-1,j}}{2\,\Delta x}$$

$$c = \left.\frac{1}{2}\frac{\partial^2 T}{\partial x^2}\right)_{i,j} = \frac{T_{i+1,j} - 2T_{i,j} + T_{i-1,j}}{2(\Delta x)^2}$$

Thus

$$\left.\frac{\partial^2 T}{\partial x^2}\right)_{i,j} = \frac{T_{i+1,j} - 2T_{i,j} + T_{i-1,j}}{(\Delta x)^2} \tag{3.73}$$

This represents an exact result if indeed a second-degree polynomial expresses the correct variation of temperature with x. In the general case, we only suppose that the second-degree polynomial is a good approximation to the solution. The T.E. of the expression, Eq. (3.73), can be determined by substituting Taylor-series expansions about point (i, j) for $T_{i+1,j}$ and $T_{i-1,j}$ into Eq. (3.73). The T.E. is found to be $O[(\Delta x)^2]$ and will involve only fourth-order and higher derivatives, which are equal to zero when the temperature variation is given by a second-degree polynomial.

A finite-difference approximation for $\partial^2 T/\partial y^2$ can be found in a like manner. We notice that arbitrary decisions need to be made in the process of polynomial fitting, which will influence the form and T.E. of the result: in particular, these decisions influence which of the neighboring points will appear in the difference expression. We also observe that there is nothing unique about the procedure of polynomial fitting that guarantees that the difference approximation for the PDE is the best in any sense or that the numerical scheme is stable (when used for a marching problem).

Example 3.2 Suppose we have solved the finite-difference form of the energy equation for the temperature distribution near a solid boundary and we need to estimate the heat flux at the location. Our finite-difference solution gives us only the temperature at discrete grid points. From Fourier's law, the boundary heat flux is given by $q_w = -k\, \partial T/\partial y)_{y=0}$. Thus, we need to approximate $\partial T/\partial y)_{y=0}$ by a difference representation that uses the temperature obtained from the finite-difference solution to the energy equation.

Solution One way to proceed is to assume that the temperature distribution near the boundary is a polynomial and to "fit" such a polynomial, i.e., straight line, parabola, or third-degree polynomial, to the finite-difference solution that has been determined at discrete points. By requiring that the polynomial match the finite-difference solution for T at certain discrete points, the unknown coefficients in the polynomial can be determined.

For example, if we assume that the temperature distribution near the boundary is again a second-degree polynomial of the form $T = a + by + cy^2$, then referring to Fig. 3.6, we note that $\partial T/\partial y)_{y=0} = b$. Further, for equally spaced mesh points we can write

$$T_1 = a$$

$$T_2 = a + b\,\Delta y + c(\Delta y)^2$$

$$T_3 = a + b(2\,\Delta y) + c(2\,\Delta y)^2$$

from which we can determine that

$$a = T_1$$

$$b = \frac{-3T_1 + 4T_2 - T_3}{2\,\Delta y}$$

$$c = \frac{T_1 - 2T_2 + T_3}{2(\Delta y)^2}$$

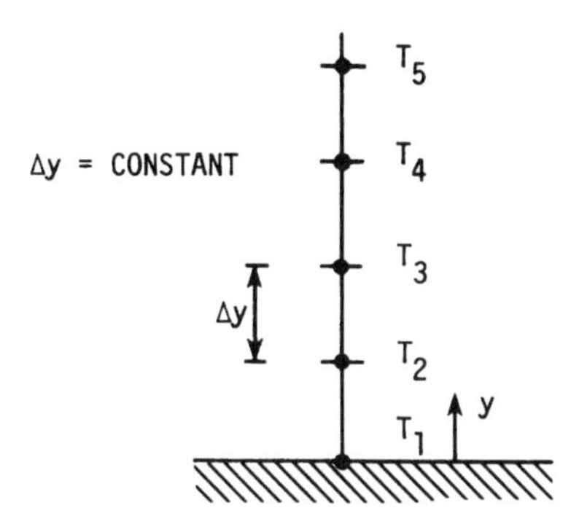

Figure 3.6 Finite-difference grid near wall.

Thus we can evaluate the wall heat flux by the approximation

$$q_w = -k\left.\frac{\partial T}{\partial y}\right)_{y=0} \simeq -kb = \frac{k}{2\,\Delta y}(3T_1 - 4T_2 + T_3)$$

It is natural to inquire about the T.E. of this approximation for $\partial T/\partial y)_{y=0}$. This may be established by expressing T_2 and T_3 in terms of Taylor-series expansions about the boundary point and substituting these evaluations into the difference expression for $\partial T/\partial y)_{y=0}$. Alternatively, we can identify the second-degree polynomial as a truncated Taylor-series expansion about $y = 0$.

Second-degree polynomial:

$$T = a + by + cy^2$$

Taylor series:

$$T = T(0) + \left.\frac{\partial T}{\partial y}\right)_0 y + \left.\frac{\partial^2 T}{\partial y^2}\right)_0 \frac{y^2}{2!} + \underbrace{\left.\frac{\partial^3 T}{\partial y^3}\right)_0 \frac{y^3}{3!} + \cdots}_{\text{T.E.}}$$

Thus the approximation $T \simeq a + by + cy^2$ is equivalent to utilizing the first three terms of a Taylor-series expansion with the resulting T.E. in the expression for T being $O[(\Delta y)^3]$. Solving the Taylor series for an expression for $\partial T/\partial y)_{y=0}$ involves division by Δy, which reduces the T.E. in the expression for $\partial T/\partial y)_{y=0}$ to $O[(\Delta y)^2]$.

Example 3.3 Suppose that the energy equation is being solved for the temperature distribution near the wall as in Example 3.2, but now the wall heat flux is specified as a boundary condition. We may then want to use polynomial fitting to obtain an expression for the boundary temperature that is called for in the difference equations for internal points. In other words, if $q_w = -k\,\partial T/\partial y)_{y=0}$ is given, how can we evaluate T at $y = 0$, i.e., (T_1) in terms of q_w/k and T_2, T_3, etc.?

Solution Here we might assume that $T = a + by + cy^2 + dy^3$ near the wall and that $\partial T/\partial y)_{y=0} = b = -q_w/k$ (given). Our objective is to evaluate T_1, which in this case equals a. Referring to Fig. 3.6, we can write

$$T_2 = a - \frac{q_w}{k}\,\Delta y + c(\Delta y)^2 + d(\Delta y)^3$$

$$T_3 = a - \frac{q_w}{k}(2\,\Delta y) + c(2\,\Delta y)^2 + d(2\,\Delta y)^3$$

$$T_4 = a - \frac{q_w}{k}(3\,\Delta y) + c(3\,\Delta y)^2 + d(3\,\Delta y)^3$$

These three equations can be solved for a, c, and d in terms of T_2, T_3, T_4, q_w/k, and Δy. The desired result, T_1 as a function of T_2, T_3, q_w/k, and Δy, follows directly from $T_1 = a$ and is given by

$$T_1 = \frac{1}{11}\left(18T_2 - 9T_3 + 2T_4 + \frac{6\,\Delta y\, q_w}{k}\right) + O\left[(\Delta y)^4\right] \tag{3.74}$$

The T.E. in Eq. (3.74) can be established by substituting Taylor-series expansions about (i, j) for the temperatures on the right-hand side or by identifying the polynomial as a truncated series by inspection. We will close this discussion on polynomial fitting by listing some expressions for wall values of a function and its first derivative in terms of values of the function. These expressions are useful, for example, in extracting a value of the function at the wall, if the wall value of the first derivative is specified. The results in Table 3.3 were obtained from polynomial fitting, assuming that $T(y)$ can be expressed as a polynomial of degree up to the fourth, and that $\Delta y = h = \text{const}$.

3.4.3 Integral Method

The integral method provides yet another means for developing difference approximations to PDEs. We consider again the heat equation as the specimen

Table 3.3 Some useful results from polynomial fitting

Polynomial degree	Wall value of function or derivative	Equation
1	$\left.\frac{\partial T}{\partial y}\right)_{i,j} = \frac{T_{i,j+1} - T_{i,j}}{h} + O(h)$	(3.75)
1	$T_{i,j} = T_{i,j+1} - h\left.\frac{\partial T}{\partial y}\right)_{i,j} + O(h^2)$	(3.76)
2	$\left.\frac{\partial T}{\partial y}\right)_{i,j} = \frac{1}{2h}(-3T_{i,j} + 4T_{i,j+1} - T_{i,j+2}) + O(h^2)$	(3.77)
2	$T_{i,j} = \frac{1}{3}\left[4T_{i,j+1} - T_{i,j+2} - 2h\left.\frac{\partial T}{\partial y}\right)_{i,j}\right] + O(h^3)$	(3.78)
3	$\left.\frac{\partial T}{\partial y}\right)_{i,j} = \frac{1}{6h}(-11T_{i,j} + 18T_{i,j+1} - 9T_{i,j+2} + 2T_{i,j+3}) + O(h^3)$	(3.79)
3	$T_{i,j} = \frac{1}{11}\left[18T_{i,j+1} - 9T_{i,j+2} + 2T_{i,j+3} - 6h\left.\frac{\partial T}{\partial y}\right)_{i,j}\right] + O(h^4)$	(3.80)
4	$\left.\frac{\partial T}{\partial y}\right)_{i,j} = \frac{1}{12h}(-25T_{i,j} + 48T_{i,j+1} - 36T_{i,j+2} + 16T_{i,j+3} - 3T_{i,j+4}) + O(h^4)$	(3.81)
4	$T_{i,j} = \frac{1}{25}\left[48T_{i,j+1} - 36T_{i,j+2} + 16T_{i,j+3} - 3T_{i,j+4} - 12h\left.\frac{\partial T}{\partial y}\right)_{i,j}\right] + O(h^5)$	(3.82)

equation:

$$\frac{\partial u}{\partial t} = \alpha \frac{\partial^2 u}{\partial x^2} \tag{3.83}$$

The strategy is to develop an algebraic relationship among the values of u at neighboring grid points by integrating the heat equation with respect to the independent variables t and x over the local neighborhood of point (n, j). The point (n, j) will also be identified as point (t_0, x_0). Grid points are spaced at intervals of Δx and Δt. We arbitrarily decide to integrate both sides of the equation over the interval t_0 to $t_0 + \Delta t$ and $x_0 - \Delta x/2$ to $x_0 + \Delta x/2$. Choosing $t_0 - \Delta t/2$ to $t_0 + \Delta t/2$ would lead to an inherently unstable difference equation. Unfortunately, at this point we have no way of knowing which choice for the integration interval would be the right or wrong one relative to stability of the solution method. This can only be determined by a trial calculation or application of the methods for stability analysis, presented in Section 3.6. The order of integration is chosen for each side in a manner to take advantage of exact differentials:

$$\int_{x_0 - \Delta x/2}^{x_0 + \Delta x/2} \left(\int_{t_0}^{t_0 + \Delta t} \frac{\partial u}{\partial t}\, dt \right) dx = \alpha \int_{t_0}^{t_0 + \Delta t} \left(\int_{x_0 - \Delta x/2}^{x_0 + \Delta x/2} \frac{\partial^2 u}{\partial x^2}\, dx \right) dt \tag{3.84}$$

The inner level of integration can be done exactly, giving

$$\int_{x_0 - \Delta x/2}^{x_0 + \Delta x/2} [u(t_0 + \Delta t, x) - u(t_0, x)]\, dx$$
$$= \alpha \int_{t_0}^{t_0 + \Delta t} \left[\frac{\partial u}{\partial x}\left(t, x_0 + \frac{\Delta x}{2}\right) - \frac{\partial u}{\partial x}\left(t, x_0 - \frac{\Delta x}{2}\right) \right] dt \tag{3.85}$$

For the next level of integration, we take advantage of the mean-value theorem for integrals, which assures us that for a continuous function $f(y)$,

$$\int_{y_1}^{y_1 + \Delta y} f(y)\, dy = f(\bar{y})\, \Delta y \tag{3.86}$$

where $\bar{y}$ is some value of y in the interval $y_1 \leqslant \bar{y} \leqslant y_1 + \Delta y$. Thus, any value of y on the interval will provide an approximation to the integral, and we can write

$$\int_{y_1}^{y_1 + \Delta y} f(y)\, dy \simeq f(\tilde{y})\, \Delta y \qquad y_1 \leqslant \tilde{y} \leqslant y_1 + \Delta y$$

As we invoke the mean-value theorem to further simplify Eq. (3.85), we arbitrarily select x_0 on the left-hand side and $t_0 + \Delta t$ on the right-hand side as the locations within the intervals of integration at which to evaluate the integrands:

$$[u(t_0 + \Delta t, x_0) - u(t_0, x_0)]\, \Delta x$$
$$= \alpha \left[\frac{\partial u}{\partial x}\left(t_0 + \Delta t, x_0 + \frac{\Delta x}{2}\right) - \frac{\partial u}{\partial x}\left(t_0 + \Delta t, x_0 - \frac{\Delta x}{2}\right) \right] \Delta t \tag{3.87}$$

To express the result in purely algebraic terms requires that the first derivatives, $\partial u/\partial x$, on the right-hand side be approximated by finite differences. We could achieve this by falling back on our experience to date and simply utilizing central differences. Alternatively, we can continue to pursue a purely integral approach and invoke the mean-value theorem for integrals, again observing that

$$\begin{aligned} u(t_0 + \Delta t, x_0 + \Delta x) &= u(t_0 + \Delta t, x_0) + \int_{x_0}^{x_0+\Delta x} \frac{\partial u}{\partial x}(t_0 + \Delta t, x)\, dx \\ &\simeq u(t_0 + \Delta t, x_0) + \frac{\partial u}{\partial x}\left(t_0 + \Delta t, x_0 + \frac{\Delta x}{2}\right)\Delta x \end{aligned} \tag{3.88}$$

from which we can write

$$\frac{\partial u}{\partial x}\left(t_0 + \Delta t, x_0 + \frac{\Delta x}{2}\right) \simeq \frac{u(t_0 + \Delta t, x_0 + \Delta x) - u(t_0 + \Delta t, x_0)}{\Delta x} \tag{3.89}$$

In evaluating the integral in Eq. (3.88) through the mean-value theorem, we have arbitrarily evaluated the integrand at the midpoint of the interval. Hence the final result is only an approximation. Treating the other first derivative in a similar manner permits the approximation to the heat equation to be written as

$$\begin{aligned} [u(t_0 + \Delta t, x_0) - u(t_0, x_0)]\,\Delta x = \frac{\alpha}{\Delta x}[u(t_0 + \Delta t, x_0 + \Delta x) - 2u(t_0 + \Delta t, x_0) \\ + u(t_0 + \Delta t, x_0 - \Delta x)]\,\Delta t \end{aligned} \tag{3.90}$$

Reverting back to the n, j notation, whereby n denotes time (t) and j denotes space (x), we can rearrange the above in the form

$$\frac{u_j^{n+1} - u_j^n}{\Delta t} = \frac{\alpha}{(\Delta x)^2}\left(u_{j+1}^{n+1} - 2u_j^{n+1} + u_{j-1}^{n+1}\right) \tag{3.91}$$

which can be recognized as the fully implicit representation of the heat equation, Eq. (3.70), given in Section 3.4.1. The choice of $t_0 + \Delta t$ as the location to use in utilizing the mean-value theorem for the second integration on the right-hand side is responsible for the implicit form. If t_0 had been chosen instead, an explicit formulation would have resulted. We note that a statement of the T.E. does not evolve naturally as part of this method for developing difference equations but must be determined as a separate step.

3.4.4 Finite-Volume (Control-Volume) Approach

In developing what has become known as the *finite-volume* method, the conservation principles are applied to a fixed region in space known as a *control volume*. Some authorities also refer to such a procedure as a control-volume method, so that the two terms, finite volume and control volume, are used somewhat interchangeably in the literature. In the finite-volume approach a point of view is taken that is distinctly different from that taken with any of the other methods considered thus far. In the Taylor-series and integral methods, we accepted the PDE as the correct and appropriate form of the conservation principle (physical law) governing our problem and merely turned to mathematical tools to develop algebraic approximations to derivatives. We never again considered the physical law represented by the PDE. The Taylor-series and integral methods then proceed in a rather formal, mechanical way, operating on the PDE, which represents the conservation statement (physical law) at a point.

In the finite-volume method the conservation statement is applied in a form applicable to a region in space (control volume). This integral form of the conservation statement is usually well known from first principles, or it can in most cases, be developed from the PDE form of the conservation law. In this approach, we are recognizing the discrete nature of the computational model at the outset. This feature is shared in common with finite-element methods. The finite-volume procedure can, in fact, be considered as a variant of the finite-element method (Hirsch, 1988), although it is, from another point of view, just a particular type of finite-difference scheme.

As an example, consider unsteady 2-D heat conduction in a rectangular-shaped solid. The problem domain is to be divided up into control volumes with associated grid points. We can establish the control volumes first and place grid points in the centers of the volumes (cell-centered method) or establish the grid first and then fix the boundaries of the control volumes (cell-vertex method) by, for example, placing the boundaries halfway between grid points. When the mesh spacing varies, the points will not be in the geometric center of the control volumes in the cell-vertex method. In the present example, equal spacing will be used, so that the two approaches will result in identical grid and control-volume arrangements.

We first consider the control volume labeled A in Fig. 3.7, which is representative of all internal (nonboundary) points. The appropriate form of the conservation statement for the control volume (namely, that the time rate of increase of energy stored in the volume is equal to the net rate at which energy is conducted into the volume) can be represented mathematically as

$$\iiint_R \rho c \frac{\partial T}{\partial t}\, dR + \oiint_S \mathbf{q} \cdot \mathbf{n}\, dS = 0$$

The first term in this equation, an integral over the control volume, represents the time rate of increase in the energy stored in the volume. The second term, an integral over the surface of the volume, represents the net rate at which energy is conducted out through the surface of the volume. This is the *integral* or control-volume form of the conservation law that we are applying in this case and is the usual starting point for the derivation of the conservation law in partial differential form. On the other hand, if the PDE form of the conservation law is available to us, we can usually work backward with the aid of the

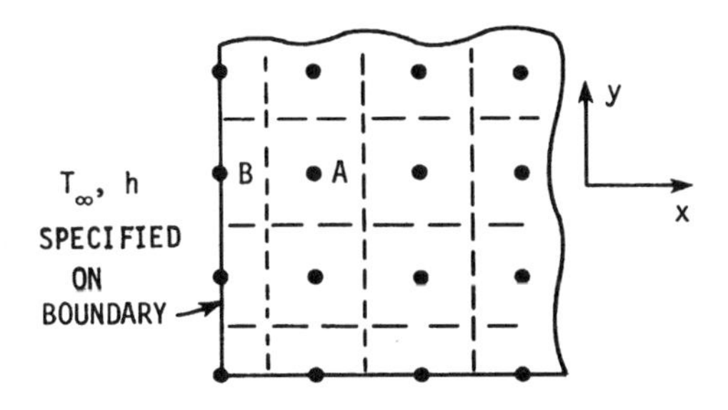

Figure 3.7 Finite-difference grid for control-volume method.

divergence theorem to obtain the appropriate integral form. For example, with constant properties, this problem is governed by the 2-D heat equation, an extension of Eq. (3.62), which can be written in the form

$$\rho c \frac{\partial T}{\partial t} = \frac{\partial}{\partial x}\left(k \frac{\partial T}{\partial x}\right) + \frac{\partial}{\partial y}\left(k \frac{\partial T}{\partial y}\right) = \boldsymbol{\nabla} \cdot (k\, \boldsymbol{\nabla} T) \tag{3.92a}$$

where k is the thermal conductivity, ρ is the density, c is the specific heat, and the heat flux vector $\mathbf{q}$ is given by $\mathbf{q} = -k\, \boldsymbol{\nabla} T$. We can integrate Eq. (3.92a) over the control volume to obtain

$$\iiint_R \left(\rho c \frac{\partial T}{\partial t} + \boldsymbol{\nabla} \cdot \mathbf{q}\right) dR = 0 \tag{3.92b}$$

Applying the divergence theorem gives

$$\iiint_R \rho c \frac{\partial T}{\partial t}\, dR + \oiint_S \mathbf{q} \cdot \mathbf{n}\, dS = 0$$

the integral form of the conservation law. The PDE form of the law is derived from the integral form by observing that Eq. (3.92a) must hold for all volumes regardless of size or shape. Therefore the integrand itself must be identically zero at every point. Of course, representing conservation of energy by Eqs. (3.92a–3.92b) assumes the existence of continuous derivatives that appear in the divergence term.

For a 2-D problem, the "volume" employs a unit depth. In two dimensions, we can represent $\mathbf{n}\, dS$ as $\mathbf{i}\, dy - \mathbf{j}\, dx$ for an integration path around the boundary in a *counterclockwise* direction. Thus the surface integral on the right, representing the net flow of heat out through the surface of the volume, can be evaluated as

$$\oint_S (q_x\, dy - q_y\, dx)$$

where q_x and q_y are components of the heat flux in the x and y directions, respectively. The conservation statement then becomes

$$\iiint_R \rho c \frac{\partial T}{\partial t}\, dR + \oiint_S (q_x\, dy - q_y\, dx) = 0 \tag{3.93}$$

It should be noted that Eq. (3.93) is valid for volumes of any shape. No assumption was necessary about the shape of the volume in order to obtain Eq. (3.93).

The term on the left containing the time derivative can be evaluated by assuming that the temperature at point (i, j) is the mean value for the volume and then using a forward time difference to obtain

$$\rho c \frac{\left(T_{i,j}^{n+1} - T_{i,j}^{n}\right)}{\Delta t} \Delta x \, \Delta y$$

The time level at which the term on the right, representing the net heat flow out of the volume, is evaluated determines whether the scheme will be explicit or implicit. Reasonable choices include time levels n, $n + 1$, or an average of the two. Fourier's law can be used to represent the heat flux components in terms of the temperature:

$$q_x = -k \frac{\partial T}{\partial x} \qquad q_y = -k \frac{\partial T}{\partial y}$$

The second integral in Eq. (3.93), representing the flow of heat out of the four boundaries of the control volume about point (i, j), can be represented by

$$-k \, \Delta y \frac{\partial T}{\partial x}\bigg)_{i+\frac{1}{2},j} \quad -k \, \Delta x \frac{\partial T}{\partial y}\bigg)_{i,j+\frac{1}{2}} \quad +k \, \Delta y \frac{\partial T}{\partial x}\bigg)_{i-\frac{1}{2},j} \quad +k \, \Delta x \frac{\partial T}{\partial y}\bigg)_{i,j-\frac{1}{2}}$$

The $\frac{1}{2}$ in the subscripts refers to evaluation at the boundaries of the control volume that are halfway between mesh points. The expression for the net flow of heat out of the volume is exact if the derivatives represent suitable average values for the boundaries concerned. Approximating the spatial derivatives by central differences at time level n and combining with the time-derivative representation yields

$$\rho c \frac{\left(T_{i,j}^{n+1} - T_{i,j}^{n}\right)}{\Delta t} \Delta x \, \Delta y + k \, \Delta y \frac{\left(T_{i,j}^{n} - T_{i+1,j}^{n}\right)}{\Delta x} + k \, \Delta x \frac{\left(T_{i,j}^{n} - T_{i,j+1}^{n}\right)}{\Delta y}$$
$$+ k \, \Delta y \frac{\left(T_{i,j}^{n} - T_{i-1,j}^{n}\right)}{\Delta x} + k \, \Delta x \frac{\left(T_{i,j}^{n} - T_{i,j-1}^{n}\right)}{\Delta y} = 0$$

Dividing by $\rho c \, \Delta x \, \Delta y$ and rearranging gives

$$\frac{T_{i,j}^{n+1} - T_{i,j}^{n}}{\Delta t} = \alpha \left(\frac{T_{i+1,j}^{n} - 2T_{i,j}^{n} + T_{i-1,j}^{n}}{(\Delta x)^2} + \frac{T_{i,j+1}^{n} - 2T_{i,j}^{n} + T_{i,j-1}^{n}}{(\Delta y)^2} \right) \tag{3.94}$$

where $\alpha = k/\rho c$. Equation (3.94) corresponds to the explicit finite-difference representation of the 2-D heat equation.

This equation was derived by approximating spatial derivatives at control volume boundaries by central differences; however, it is possible to develop appropriate representations for such derivatives by integral methods in a manner that is not restricted to Cartesian or even orthogonal grids (see Appendix D).

Now consider the control volume on the boundary, labeled B in Fig. 3.7. In this example we will assume that the boundary conditions are convective. For the continuous (nondiscrete) problem, this is formulated mathematically by $h(T_\infty - T_{i,j}) = -k \, \partial T/\partial x)_{i,j}$, where the point (i, j) is the point on the physical boundary associated with control volume B. If we were to proceed with the Taylor-series approach to this boundary condition, we would likely next seek a difference representation for $\partial T/\partial x)_{i,j}$. If a simple forward difference is used, the difference equation governing the boundary temperature would be

$$h(T_\infty - T^n_{i,j}) = \frac{k}{\Delta x}(T^n_{i,j} - T^n_{i+1,j}) \tag{3.95}$$

In the control-volume approach, however, we are forced to observe that there is some material associated with the boundary point so that conduction may occur along the boundary, and energy can be stored within the volume. The energy balance on the control volume will account for possible transfer across all four boundaries as well as storage. Applying Eq. (3.93) to volume B gives

$$\rho c \frac{\left(T^{n+1}_{i,j} - T^n_{i,j}\right)}{\Delta t}\frac{\Delta x\,\Delta y}{2} - k\,\Delta y \frac{\partial T}{\partial x}\bigg)_{i+\frac{1}{2},j} - k\frac{\Delta x}{2}\frac{\partial T}{\partial y}\bigg)_{i,j+\frac{1}{2}} + k\frac{\Delta x}{2}\frac{\partial T}{\partial y}\bigg)_{i,j-\frac{1}{2}}$$
$$+ h\,\Delta y(T^n_{i,j} - T_\infty) = 0$$

Using the same discretization strategy here as was used for volume A, we can write

$$\rho c \frac{\left(T^{n+1}_{i,j} - T^n_{i,j}\right)}{2\,\Delta t}\Delta x\,\Delta y + k\,\Delta y \frac{(T^n_{i,j} - T^n_{i+1,j})}{\Delta x} + \frac{k\,\Delta x}{2}\frac{(T^n_{i,j} - T^n_{i,j+1})}{\Delta y}$$
$$+ \frac{k\,\Delta x}{2}\frac{(T^n_{i,j} - T^n_{i,j-1})}{\Delta y} + h\,\Delta y(T^n_{i,j} - T_\infty) = 0$$

Dividing through by $\rho c\,\Delta x\,\Delta y$, we can write the result as

$$\frac{T^{n+1}_{i,j} - T^n_{i,j}}{2\,\Delta t} = \alpha\left[\frac{T^n_{i+1,j} - T^n_{i,j}}{(\Delta x)^2} + \frac{T^n_{i,j+1} - 2T^n_{i,j} + T^n_{i,j-1}}{2(\Delta y)^2}\right] + \frac{h(T_\infty - T^n_{i,j})}{\rho c\,\Delta x} \tag{3.96}$$

which is somewhat different from Eq. (3.95), which followed from the most obvious application of the Taylor-series method to approximate the mathematical statement of the boundary condition.

Looking back over the methodology of the finite-volume and Taylor-series methods, we can note that the Taylor-series method readily provided difference approximations to derivatives and the representation for the complete PDE was made up from the addition of several such representations. In contrast, the finite-volume method employs the conservation statement or physical law (usually invoked in integral form) corresponding to the entire PDE. The distinctive characteristic of the finite-volume approach is that a "balance" of some physical quantity is made on the region (control volume) in the neighborhood of a grid point. The discrete nature of the problem domain is always taken into account in the finite-volume approach, which ensures that the physical law is satisfied over a finite region rather than only at a point as the mesh is shrunk to zero. It would appear that the discretization developed by the finite-volume approach would almost certainly have the conservative property.

It is difficult to appreciate the subtle differences that may occur in the difference representations obtained for the same PDE by using the four different methods discussed in this section without working a large number of examples. In many cases, and especially for simple, linear equations, the resulting difference equations can be identical. That is, four different approaches can give the same result. There is no guarantee that difference equations developed by any of the methods will be numerically stable, so that the same difference scheme developed by all four methods could turn out to be worthless. The differences in the results obtained from using the different methods are more likely to become evident in coordinate systems other than rectangular.

3.5 INTRODUCTION TO THE USE OF IRREGULAR MESHES

Clearly, it is convenient to let the mesh increments such as Δx and Δy be constant throughout the computational domain. However, in many instances this is not possible because of domain boundaries that do not coincide with the regular mesh lines or because of the need to reduce the mesh spacing in certain regions in order to maintain the desired level of accuracy. These irregularities occur frequently enough in physical problems to command a significant amount of attention from workers in computational fluid mechanics and heat transfer. In fact, efficiently dealing with irregular geometries that cannot be defined in terms of coordinate lines from a known orthogonal coordinate system is one of the important practical problems challenging computational fluid dynamics at the present time. This problem is complex and has no optimum solution for all cases. Some of the ideas are introduced in this chapter, but the general issue of irregular meshes is addressed at various points throughout the remainder of the book, particularly in Chapters 5 and 10.

3.5.1 Irregular Mesh Due to Shape of a Boundary

Here we address those cases in which some portion of the boundary consists of a curve (in two dimensions) that does not coincide with a coordinate line (for an orthogonal coordinate system) that is satisfactory for the remainder of the boundaries. An example of this would arise in solving Laplace's equations in a rectangular region containing a circular interior "hole." This could also occur in solving for the inviscid flow in a channel containing a circular cylinder, or in a rectangular conduction medium containing a circular pipe. A square mesh, $\Delta x = \Delta y = \text{const}$, would be adequate except near the cylinder, where the spacing between some of the boundary points and the internal points is unequal, as illustrated in Fig. 3.8. If the boundary conditions are Dirichlet (u specified), the following three simple procedures may provide an adequate approximation.

1. Use an especially fine but regular mesh near the boundary and define the

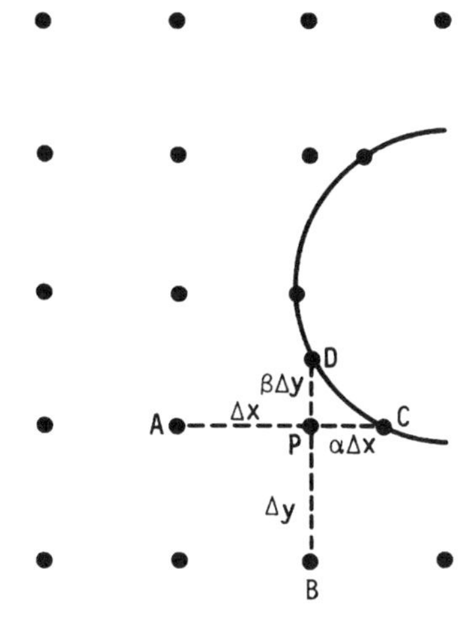

Figure 3.8 Irregular mesh caused by the shape of a boundary.

point closest to the actual boundary as the boundary point for computational purposes. This results in the boundary taking on a "zig-zag" appearance. Unless a coarser mesh is used away from the boundary, resulting in irregular mesh problems where the transition in spacing is made, this method could require a very large number of grid points to achieve reasonable accuracy.

2. Use linear (or bilinear) interpolation to assign values of u to any internal point that is less than a regular mesh increment from the boundary. The interpolation is between the specified boundary values of u and values of u determined at neighboring points by the finite-difference equations applicable to internal points in the regular mesh. This procedure may work but is not strongly recommended. Usually, we can do much better than this with very little additional effort, as indicated below.
3. Develop a finite-difference approximation to the governing PDE that is valid at internal points even when the mesh is irregular. Such a difference representation for Laplace's equation valid on a Cartesian grid with irregular spacing (Δx and Δy not constant) can be developed quite readily through the integral method by integrating about point x_0, y_0 and letting each integration interval extend halfway to a neighboring point. The mesh notation used is defined in Fig. 3.9. The starting point for the integral development of the difference expression is

$$\int_{y_0-\Delta y_-/2}^{y_0+\Delta y_+/2}\left(\int_{x_0-\Delta x_-/2}^{x_0+\Delta x_+/2}\frac{\partial^2 u}{\partial x^2}\,dx\right)dy+\int_{x_0-\Delta x_-/2}^{x_0+\Delta x_+/2}\left(\int_{y_0-\Delta y_-/2}^{y_0+\Delta y_+/2}\frac{\partial^2 u}{\partial y^2}\,dy\right)dx=0$$

Using the definition of an exact differential, this can be written as

$$\int_{y_0-\Delta y_-/2}^{y_0+\Delta y_+/2}\left[\frac{\partial u}{\partial x}\left(x_0+\frac{\Delta x_+}{2},y\right)-\frac{\partial u}{\partial x}\left(x_0-\frac{\Delta x_-}{2},y\right)\right]dy$$
$$+\int_{x_0-\Delta x_-/2}^{x_0+\Delta x_+/2}\left[\frac{\partial u}{\partial y}\left(x,y_0+\frac{\Delta y_+}{2}\right)-\frac{\partial u}{\partial y}\left(x,y_0-\frac{\Delta y_-}{2}\right)\right]dx=0$$

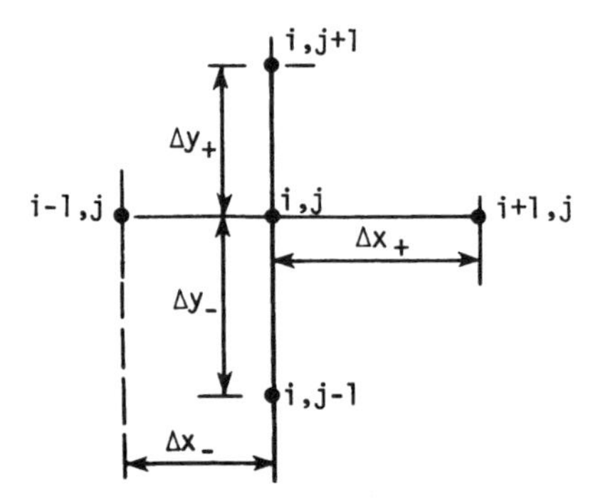

Figure 3.9 Notation for arbitrary irregular mesh.

Employing the mean-value theorem for integrals and using the central point of the interval to evaluate the integrands gives

$$\left[\frac{\partial u}{\partial x}\left(x_0 + \frac{\Delta x_+}{2}, y_0\right) - \frac{\partial u}{\partial x}\left(x_0 - \frac{\Delta x_-}{2}, y_0\right)\right]\frac{\Delta y_+ + \Delta y_-}{2}$$
$$+\left[\frac{\partial u}{\partial y}\left(x_0, y_0 + \frac{\Delta y_+}{2}\right) - \frac{\partial u}{\partial y}\left(x_0, y_0 - \frac{\Delta y_-}{2}\right)\right]\frac{\Delta x_+ + \Delta x_-}{2} = 0$$

Approximating these derivatives centrally, as was done in Section 3.4.3 gives, after rearrangement, the following approximation for Laplace's equation in subscript notation:

$$\frac{2}{\Delta x_+ + \Delta x_-}\left(\frac{u_{i+1,j} - u_{i,j}}{\Delta x_+} - \frac{u_{i,j} - u_{i-1,j}}{\Delta x_-}\right)$$
$$+\frac{2}{\Delta y_+ + \Delta y_-}\left(\frac{u_{i,j+1} - u_{i,j}}{\Delta y_+} - \frac{u_{i,j} - u_{i,j-1}}{\Delta y_-}\right) = 0 \qquad (3.97)$$

When the above is specialized to the points near the irregular boundary depicted in Fig. 3.8, the derivative approximations appear as

$$\left.\frac{\partial^2 u}{\partial x^2}\right)_P \cong \frac{2}{\Delta x(1+\alpha)}\left(\frac{u_C - u_P}{\alpha\,\Delta x} - \frac{u_P - u_A}{\Delta x}\right)$$

$$\left.\frac{\partial^2 u}{\partial y^2}\right)_P \cong \frac{2}{\Delta y(1+\beta)}\left(\frac{u_D - u_P}{\beta\,\Delta y} - \frac{u_P - u_B}{\Delta y}\right)$$

Equation (3.97) can also be developed by the control-volume method or by utilizing Taylor-series expansions. However, the unequal spacing makes the Taylor-series method noticeably more laborious, whereas the integral approach proceeds for unequal spacing with no increase in effort. Likewise, using the control-volume method would require little additional effort. However, Taylor-series expansions about (i, j) should be substituted into Eq. (3.97) to establish the consistency and T.E. of these approximations. This will be left as an exercise for the reader. As a note of warning, we recall that our second-derivative approximations on a regular mesh acquired second-order accuracy only through fortuitous cancellation of terms from the forward and backward Taylor-series expansions. This cancellation will not occur if the mesh increments are unequal.

When approximately the same number of grid points are being used, we might expect this third method of treating irregular points near boundaries to be the most accurate because the governing PDE is being approximated at each internal point (not the case for procedure 2), and the location of the boundary is not being altered as was done in procedure 1.

The above approximate procedures can be useful when solving a single equation for a problem in which Dirichlet boundary conditions are specified on an irregular boundary. However, when a system of equations is being solved or the boundary conditions involve derivatives (Neumann), the simple procedures given above are usually not adequate. Better ways of dealing with this problem usually add significantly to the complexity of the problem formulation. One common way of handling this type of problem is through the use of generalized body-fitted coordinates. This procedure is discussed in Chapters 5 and 10. The finite-volume method can also be extended to provide a satisfactory representation. An example of how the finite-volume approach can be applied to obtain satisfactory difference representations for control volumes associated with irregular boundaries is given below.

Finite-volume treatment of irregular boundary. As before, the governing PDE is Laplace's equation, and we are considering the effect of an irregular boundary on a computational domain that is otherwise discretized with an orthogonal coordinate system. In particular, we will consider the configuration depicted in Fig. 3.10, where control volumes will be rectangles except near the irregular boundary. In this case, application of the finite-volume methodology will result

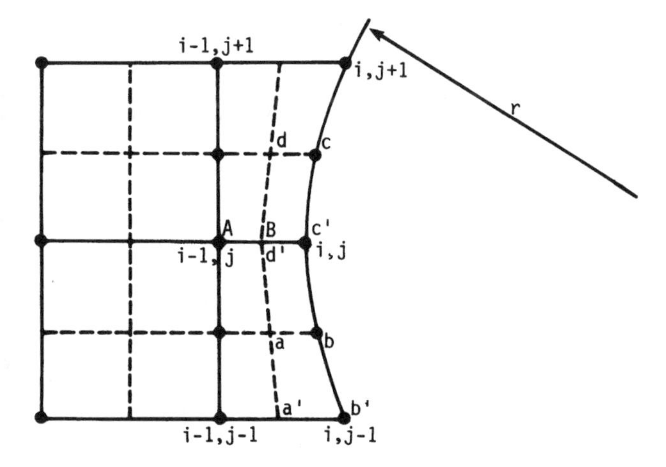

Figure 3.10 Finite-volume treatment of irregular boundary.

in Eq. (3.97) if the volume faces form a rectangle (i.e., if adjacent faces are orthogonal). The only exceptions to this will be for those volumes on the irregular boundary and their immediate internal neighbors. Some immediate internal neighbors to boundary cells, like volume A in Fig. 3.10, will have sufficient geometric symmetry so that Eq. (3.97) will be obtained from the finite-volume analysis.

If the boundary conditions at the irregular boundary are Dirichlet, no unknowns exist in the volumes that reside on the boundary, so a heat balance on those cells is not necessary. However, if the boundary condition is Neumann, corresponding to a specified value of boundary heat flux (q_w), the boundary temperature is unknown, and it is appropriate to apply the integral form of the conservation statement developed in Section 3.4.4:

$$\iiint_R \rho c \frac{\partial T}{\partial t} \, dR + \oiint_S (q_x \, dy - q_y \, dx) = 0$$

to the volume (labeled B in Fig. 3.10) on the boundary. The dashed lines in Fig. 3.10 denote the boundaries of the control volume. The corner points, located halfway between the specified nodal points, are labeled a, b, c, d. It is assumed that the coordinates of the nodal points are known. The coordinates of points a and d are given by

$$\begin{aligned} x_a &= (x_{i,j-1} + x_{i,j} + x_{i-1,j} + x_{i-1,j-1})/4 \\ y_a &= (y_{i,j-1} + y_{i,j} + y_{i-1,j} + y_{i-1,j-1})/4 \\ x_d &= (x_{i,j+1} + x_{i,j} + x_{i-1,j} + x_{i-1,j+1})/4 \\ y_d &= (y_{i,j+1} + y_{i,j} + y_{i-1,j} + y_{i-1,j+1})/4 \end{aligned}$$

Since we want points b and c to lie exactly on the boundary, we will fix $y_b = (y_{i,j} + y_{i,j-1})/2$ and $y_c = (y_{i,j+1} + y_{i,j})/2$ and establish the x coordinates so that the points are on the boundary. This is easily done, since the equation for the boundary curve is known in the form $(x - x_0)^2 + (y - y_0)^2 = r^2$, where x_0, y_0 are the coordinates of the center of the circle and r is the radius. We can separate the integral around the boundaries of B into line integrals over the four component line segments, a-b, b-c, c-d, d-a:

$$\begin{aligned} \oiint_S \mathbf{q} \cdot \mathbf{n} \, ds = \int_a^b (q_x \, dy - q_y \, dx) + q_w \, \Delta s_{bc} \\ + \int_c^d (q_x \, dy - q_y \, dx) + \int_d^a (q_x \, dy - q_y \, dx) \end{aligned} \tag{3.98}$$

where Δs_{bc} can be computed exactly in this case, making use of the known equation for a circle or approximated by a straight-line segment between the fixed points as is done for the other boundaries of the control volume. We now start at point a to evaluate the line integral [Eq. (3.98)] around the boundaries of control volume B in Fig. 3.10. Along the boundary from a to b the heat flux components can be evaluated by Fourier's law as

$$k\int_a^b\left(-\frac{\partial T}{\partial x}\,dy+\frac{\partial T}{\partial y}\,dx\right)\cong k\left(\frac{\partial T}{\partial y}\right)_{i-\frac{1}{4},j-\frac{1}{2}}\Delta x_{ab}-k\left(\frac{\partial T}{\partial x}\right)_{i-\frac{1}{4},j-\frac{1}{2}}\Delta y_{ab}$$

where $\Delta x_{ab}=x_b-x_a$ and $\Delta y_{ab}=y_b-y_a$. The values of $\partial T/\partial y)_{i-\frac{1}{4},j-\frac{1}{2}}$ and $\partial T/\partial x)_{i-\frac{1}{4},j-\frac{1}{2}}$ are approximately in the center of the region $a'b'c'd'$ denoted in Fig. 3.10. It is assumed that these derivatives can be approximated by averages over $a'b'c'd'$.

$$\left.\frac{\partial T}{\partial y}\right)_{i-\frac{1}{4},j-\frac{1}{2}}\cong\frac{1}{A'}\left(\iint_{A'}\frac{\partial T}{\partial y}\,dy\,dx\right)$$

$$\left.\frac{\partial T}{\partial x}\right)_{i-\frac{1}{4},j-\frac{1}{2}}\cong\frac{1}{A'}\left(\iint_{A'}\frac{\partial T}{\partial x}\,dy\,dx\right)$$

where A' denotes the area of the region $a'b'c'd'$. Using the Gauss divergence theorem again, the integrals over the area $a'b'c'd'$ can be evaluated by line integrals around the boundary of $a'b'c'd'$. This allows the heat flux across the a-b portion of the boundary of control volume B to be represented as

$$\int_a^b(q_x\,dy-q_y\,dx)$$

$$\cong-\frac{k}{A'}\left(\oint_{A'}T\,dx\,\Delta x_{ab}+\oint_{A'}T\,dy\,\Delta y_{ab}\right)$$

$$\cong\frac{-k}{A'}\Big[(T_{i-\frac{1}{4},j-1}\,\Delta x_{a'b'}+T_b\,\Delta x_{b'c'}+T_{i-\frac{1}{4},j}\,\Delta x_{c'd'}+T_a\,\Delta x_{d'a'})\,\Delta x_{ab}$$

$$+(T_{i-\frac{1}{4},j-1}\Delta y_{a'b'}+T_b\,\Delta y_{b'c'}+T_{i-\frac{1}{4},j}\,\Delta y_{c'd'}+T_a\,\Delta y_{d'a'})\,\Delta y_{ab}\Big] \quad (3.99)$$

Because $\Delta y=0$ along path a-b, half of the terms on the right-hand side of Eq. (3.99) vanish, so that the expression simplifies to

$$\frac{-k}{A'}\Big[T_{i-\frac{1}{4},j-1}(\Delta x_{a'b'}\,\Delta x_{ab})+T_b(\Delta x_{b'c'}\,\Delta x_{ab})$$

$$+T_{i-\frac{1}{4},j}(\Delta x_{c'd'}\,\Delta x_{ab})+T_a(\Delta x_{d'a'}\,\Delta x_{ab})\Big] \quad (3.100)$$

We note that further simplifications would occur if Δx were zero along paths b'-c' and d'-a' (i.e., the paths were parallel to the y axis). The temperatures required in Eq. (3.100) must be obtained by interpolation from values at nodal points. For this configuration, bilinear interpolation yields

$$T_a=0.25(T_{i,j}+T_{i,j-1}+T_{i-1,j-1}+T_{i-1,j})$$

$$T_b=0.5(T_{i,j}+T_{i,j-1})$$

$$T_{i-\frac{1}{4},j}=0.25T_{i-1,j}+0.75T_{i,j}$$

$$T_{i-\frac{1}{4},j-1}=0.25T_{i-1,j-1}+0.75T_{i,j-1}$$

The area A' can be approximated in several ways, one of which is by assuming that $a'b'c'd'$ forms a quadrilateral and computing its area as one-half the cross product of the diagonals of the quadrilateral region:

$$A' = 0.5(\Delta x_{d'b'} \Delta y_{a'c'} - \Delta y_{d'b'} \Delta x_{a'c'})$$

where $x_{a'} = 0.5(x_{i-1,j-1} + x_{i,j-1})$
$x_{b'} = x_{i,j-1}$
$x_{c'} = x_{i,j}$
$x_{d'} = 0.5(x_{i-1,j} + x_{i,j})$

In this formulation, care must be exercised in order to obtain a positive value for the area. This can be assured by employing the right-hand rule or by taking the absolute value of the cross product. The y coordinates of points a', b', c', d' are found by replacing x with y in the expressions above. The fluxes across control-volume boundaries *c-d* and *d-a* can be evaluated by extending the methodology illustrated above for boundary *a-b* appropriately.

Although the irregular shape of the boundary volumes clearly adds significant complexity to the solution procedure, the techniques needed to deal with this can be generalized and implemented reasonably systematically and efficiently. On the other hand, it is correct to conclude that when the boundaries of the domain of interest do not coincide with grid lines of an orthogonal coordinate system and the boundary conditions are not Dirichlet, a major escalation in the effort required to formulate the solution procedure seems to follow.

3.5.2 Irregular Mesh Not Caused by Shape of a Boundary

Here we assume that the boundaries of the problem domain conform to grid lines in an orthogonal coordinate system. The use of variable grid spacing may still be desirable in this situation because it is often necessary to employ very small grid spacings in regions where gradients of the dependent variables are especially large in order to obtain the desired accuracy or "resolution." However, in the interest of computational economy, we strive to use a coarser grid away from these critical regions. This requires that the mesh spacings vary. We can cite at least two ways to proceed:

1. We can employ a coordinate transformation so that unequal spacing in the original coordinate system becomes equal spacing in the new system but the PDE becomes altered somewhat in form. This procedure is described in detail in Chapter 5.
2. The difference equation can be formulated in such a way that it remains valid when the spacing is irregular (grid lines remain orthogonal, but the increments in each coordinate direction vary instead of remaining constant). Actually, this is the same as procedure 3 used above in connection with the irregular mesh caused by curved boundaries. Such a formulation for Laplace's equation is given as Eq. (3.97).

3.5.3 Concluding Remarks

The purpose of this section has been to introduce some of the problems and applicable solution procedures associated with irregular boundaries and unequal mesh spacing in general. Coverage of the topic has been by no means complete. More advanced considerations on this topic tend to quickly become quite specialized and detailed. Good pedagogy suggests that we move on and see more of the forest before we spend any more time studying this tree. Some ideas on this topic will be developed further in Chapters 5 and 10 and in connection with specific example problems in fluid mechanics and heat transfer.

3.6 STABILITY CONSIDERATIONS

A finite-difference approximation to a PDE may be consistent, but the solution will not necessarily converge to the solution of the PDE. The Lax Equivalence theorem (see Section 3.3.5) states that a stable numerical method must also be used. We will address the question of stability in this section.

The problem of stability in numerical analysis is similar to the problem of stability encountered in a modern control system. The transfer function in a control system plays the role of the difference operator. Consider a marching problem in which initial values at time level n are known and values of the unknown at time level $n + 1$ are required. The difference operator may be viewed as a "black box" that has a certain transfer function. A schematic representation would appear as shown in Fig. 3.11. The stability of such a system depends upon the operations performed by the black box on the input data. A control systems engineer would require that the transfer function have no poles in the right-half plane. Without this requirement, input signals would be falsely amplified, and the output would be useless; in fact, it would grow without bound. Similarly, the way in which the difference operator alters the input information to produce the solution at the next time level is the central concern of stability analysis.

As a starting point for stability analysis, consider the simple explicit approximation to the heat equation:

$$\frac{u_j^{n+1} - u_j^n}{\Delta t} = \frac{\alpha}{(\Delta x)^2}(u_{j+1}^n - 2u_j^n + u_{j-1}^n)$$

This may be solved for u_j^{n+1} to yield

$$u_j^{n+1} = u_j^n + \alpha\frac{\Delta t}{(\Delta x)^2}(u_{j+1}^n - 2u_j^n + u_{j-1}^n) \tag{3.101}$$

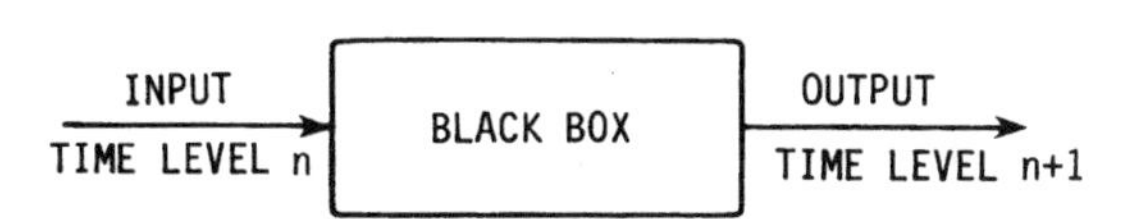

Figure 3.11 Schematic diagram of stability.

Let the exact solution of this equation be denoted by D. This is the solution that would be obtained using a computer with infinite accuracy. Similarly, denote by N the numerical solution of Eq. (3.101) computed using a real machine with finite accuracy. If the analytical solution of the PDE is A, then we may write

$$\text{Discretization error} = A - D$$
$$\text{Round-off error} = N - D$$

The question of stability of a numerical method examines the error growth while computations are being performed. O'Brien et al. (1950) pose the question of stability in the following manner:

1. Does the overall error due to round-off

$$\begin{bmatrix} \text{Grow} \\ \text{Not grow} \end{bmatrix} \Rightarrow \text{strong} \begin{bmatrix} \text{instability} \\ \text{stability} \end{bmatrix}$$

2. Does a single general round-off error

$$\begin{bmatrix} \text{Grow} \\ \text{Not grow} \end{bmatrix} \Rightarrow \text{weak} \begin{bmatrix} \text{instability} \\ \text{stability} \end{bmatrix}$$

The second question is the one most frequently answered because it can be treated much more easily from a practical point of view. The question of weak stability is usually answered by using a Fourier analysis. This method is also referred to as a von Neumann analysis. It is assumed that proof of weak stability using this method implies strong stability.

3.6.1 Fourier or von Neumann Analysis

Consider the finite-difference equation, Eq. (3.101). Let ϵ represent the error in the numerical solution due to round-off errors. The numerical solution actually computed may be written

$$N = D + \epsilon \tag{3.102}$$

This computed numerical solution must satisfy the difference equation. Substituting Eq. (3.102) into the difference equation, Eq. (3.101), yields

$$\frac{D_j^{n+1} + \epsilon_j^{n+1} - D_j^n - \epsilon_j^n}{\Delta t} = \alpha \left(\frac{D_{j+1}^n + \epsilon_{j+1}^n - 2D_j^n - 2\epsilon_j^n + D_{j-1}^n + \epsilon_{j-1}^n}{\Delta x^2} \right)$$

Since the exact solution D must satisfy the difference equation, the same is true of the error, i.e.,

$$\frac{\epsilon_j^{n+1} - \epsilon_j^n}{\Delta t} = \alpha \left(\frac{\epsilon_{j+1}^n - 2\epsilon_j^n + \epsilon_{j-1}^n}{\Delta x^2} \right) \tag{3.103}$$

In this case, the exact solution D and the error ϵ must both satisfy the same difference equation. This means that the numerical error and the exact numerical solution both possess the same growth property in time and either could be used

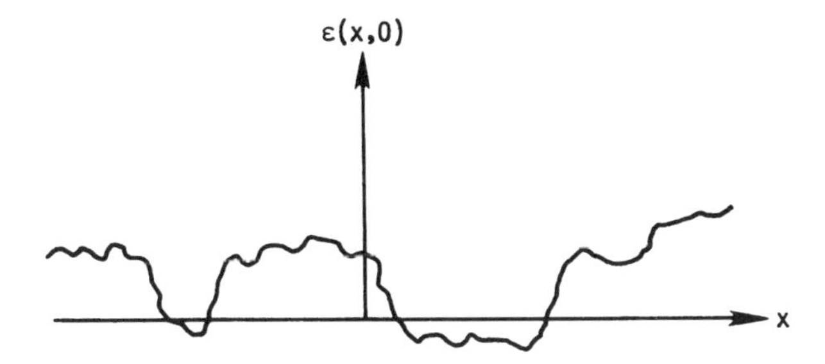

Figure 3.12 Initial error distribution.

to examine stability. Any perturbation of the input values at the nth time level will either be prevented from growing without bound for a stable system or will grow larger for an unstable system.

Consider a distribution of errors at any time in a mesh. We choose to view this distribution at time $t = 0$ for convenience. This error distribution is shown schematically in Fig. 3.12. We assume the error $\epsilon(x, t)$ can be written as a series of the form

$$\epsilon(x, t) = \sum_m b_m(t) e^{ik_m x} \tag{3.104}$$

where the period of the fundamental frequency ($m = 1$) is assumed to be $2L$. For the interval $2L$ units in length, the wave number may be written

$$k_m = \frac{2\pi m}{2L} \qquad m = 0, 1, 2, \ldots, M$$

where M is the number of increments Δx units long contained in length L. For instance, if an interval of length $2L$ is subdivided using five points, the value of M is 2, and the corresponding frequencies are

$$f_m = \frac{k_m}{2\pi} = \frac{m}{2L}$$

$$f_0 = 0 \qquad m = 0$$

$$f_1 = \frac{1}{2L} \qquad m = 1$$

$$f_2 = \frac{1}{L} \qquad m = 2$$

The frequency measures the number of wavelengths in each $2L$ units of length. The lowest frequency ($m = 0$, $f_0 = 0$) corresponds to a steady term in the assumed expansion. The highest frequency ($m = M$) has a wave number of $\pi/\Delta x$ and corresponds to the minimum number of points (3) required to approximately represent a sine or cosine wave between 0 and 2π.

Since the difference equation is linear, superposition may be used, and we may examine the behavior of a single term of the series given in Eq. (3.104). Consider the term

$$\epsilon_m(x,t) = b_m(t)e^{ik_m x}$$

We seek solutions of the form

$$z^n e^{ik_m x}$$

which reduces to $e^{ik_m x}$ when $t = 0$ ($n = 0$). Toward this end, let

$$z = e^{a\,\Delta t}$$

so that

$$\begin{aligned} z^n &= e^{an\,\Delta t} = e^{at} \\ \epsilon_m(x,t) &= e^{at}e^{ik_m x} \end{aligned} \tag{3.105}$$

where k_m is real but a may be complex.

If Eq. (3.105) is substituted into Eq. (3.103), we obtain

$$e^{a(t+\Delta t)}e^{ik_m x} - e^{at}e^{ik_m x} = r(e^{at}e^{ik_m(x+\Delta x)} - 2e^{at}e^{ik_m x} + e^{at}e^{ik_m(x-\Delta x)})$$

where $r = \alpha\,\Delta t/(\Delta x)^2$. If we divide by $e^{at}e^{ik_m x}$ and utilize the relation

$$\cos\beta = \frac{e^{i\beta} + e^{-i\beta}}{2}$$

the above expression becomes

$$e^{a\,\Delta t} = 1 + 2r(\cos\beta - 1)$$

where $\beta = k_m\,\Delta x$. Employing the trigonometric identity

$$\sin^2\frac{\beta}{2} = \frac{1 - \cos\beta}{2}$$

the final expression is

$$e^{a\,\Delta t} = 1 - 4r\sin^2\frac{\beta}{2} \tag{3.106}$$

Furthermore, since $\epsilon_j^{n+1} = e^{a\,\Delta t}\epsilon_j^n$ for each frequency present in the solution for the error, it is clear that if $|e^{a\,\Delta t}|$ is less than or equal to 1, a general component of the error will not grow from one time step to the next. This requires that

$$\left|1 - 4r\sin^2\frac{\beta}{2}\right| \leqslant 1 \tag{3.107}$$

The factor $1 - 4r\sin^2\beta/2$ (representing $\epsilon_j^{n+1}/\epsilon_j^n$) is called the *amplification factor* and will be denoted by G. Clearly, the influence of boundary conditions is not included in this analysis. In general, the Fourier stability analysis assumes that we have imposed periodic boundary conditions.

In evaluating the inequality Eq. (3.107), two possible cases must be considered:

1. Suppose $(1 - 4r \sin^2 \beta/2) \geqslant 0$; then $4r \sin^2 \beta/2 \geqslant 0$.
2. Suppose $(1 - 4r \sin^2 \beta/2) < 0$; then $4r \sin^2 \beta/2 - 1 \leqslant 1$.

The first condition is always satisfied if $r \geqslant 0$. The second inequality is satisfied only if $r \leqslant \frac{1}{2}$, which is the stability requirement for this method. This numerically places a constraint on the size of the time step relative to the size of the mesh spacing. The reason for the physically implausible temperatures calculated in the example at the end of Section 3.3.4 is now very clear. The step size Δt selected was too large by a factor of 2, and the solution began to diverge immediately. The stability of the calculation with $\alpha(\Delta t/\Delta x^2) = \frac{1}{2}$ can easily be verified. It should be noted that the amplification factor given by Eq. (3.106) could have been deduced by substituting a general form given by Eq. (3.104) into the difference equation. The proof is left as an exercise for the reader.

Example 3.4 The simple implicit scheme applied to the heat equation is given by

$$\frac{u_j^{n+1} - u_j^n}{\Delta t} = \frac{\alpha}{(\Delta x)^2}\left(u_{j+1}^{n+1} - 2u_j^{n+1} + u_{j-1}^{n+1}\right)$$

Determine the stability restrictions (if any) for this algorithm.

Solution After substituting Eq. (3.105) into this algorithm, we obtain

$$e^{a\,\Delta t}(1 + 2r - 2r\cos\beta) = 1$$

Using the trigonometric identity,

$$\sin^2\frac{\beta}{2} = \frac{1 - \cos\beta}{2}$$

the amplification factor becomes

$$G = \frac{1}{1 + 4r\sin^2\beta/2}$$

The condition for stability $|G| \leqslant 1$ is satisfied for all $r \geqslant 0$. Hence there is no upper limit on step size because of stability. However, there is a practical limit on step size because of T.E.

The application of the von Neumann or Fourier stability method is equally straightforward for hyperbolic equations. As an example, the first-order wave equation in one dimension is

$$\frac{\partial u}{\partial t} + c\frac{\partial u}{\partial x} = 0 \tag{3.108}$$

where c is the wave speed. This equation has one characteristic given by a solution of $x_t = c$. The solution of Eq. (3.108) is given by

$$u(x - ct) = \text{const}$$

This solution requires the initial data prescribed at $t = 0$ to be propagated along the characteristics.

Lax (1954) proposed the following first-order method for solving equations of this form:

$$u_j^{n+1} = \frac{u_{j+1}^n + u_{j-1}^n}{2} - c\frac{\Delta t}{\Delta x}\left(\frac{u_{j+1}^n - u_{j-1}^n}{2}\right) \tag{3.109}$$

The first term on the right-hand side represents an average value of the unknown at the previous time level, while the second term is the difference form of the spatial derivative. If a term of the form

$$u_j^n = e^{at}e^{ik_m x}$$

is substituted into the difference equation, the amplification factor becomes

$$e^{a\,\Delta t} = \cos\beta - i\nu\sin\beta$$

The stability requirement is $|G| \leqslant 1$ or

$$|\cos\beta - i\nu\sin\beta| \leqslant 1$$

where $\nu = c\,\Delta t/\Delta x$ is called the Courant number. Since the square of the absolute value of a complex number is the sum of the squares of the real and imaginary parts, the method is stable if

$$|\nu| \leqslant 1 \tag{3.110}$$

Again, a conditional stability requirement must be placed on the time step and the spatial mesh spacing. This is called the Courant-Friedrichs-Lewy (CFL) condition and was discussed at length relative to the concepts of convergence and stability in an historically important paper by Courant et al. (1928). Some authorities consider this paper to be the starting point for the development of modern numerical methods for PDEs.

The amplification factor or growth factor for a particular numerical method depends upon mesh size and wave number or frequency. The amplification factor for the Lax finite-difference method may be written

$$G = \cos\beta - i\nu\sin\beta = |G|e^{i\phi} = \sqrt{\cos^2\beta + \nu^2\sin^2\beta}\; e^{i\tan^{-1}(-\nu\tan\beta)} \tag{3.111}$$

where ϕ is the phase angle. Clearly, the magnitude of G changes with Courant number ν and frequency parameter β, which varies between 0 and π. A good understanding of the amplification factor can be obtained from a polar plot. Figure 3.13 is a plot of Eq. (3.111) for several different Courant numbers. Several interesting results can be deduced by a careful examination of this plot. The phase angle for the Lax method varies from 0 for the low frequencies to $-\pi$ for the high frequencies. This may be seen by computing the phase for both cases. For a Courant number of 1, all frequency components are propagated

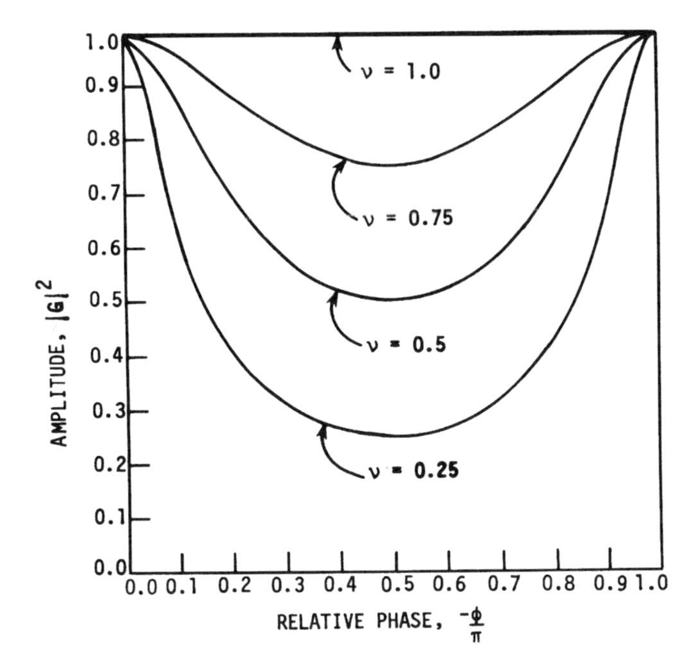

Figure 3.13 Amplitude-phase plot for the amplification factor of the Lax scheme.

without attenuation in the mesh. For Courant numbers less than 1, the low- and high-frequency components are only mildly altered, while the midrange frequency signal content is severely attenuated. The phase is also shown, and we can determine the phase error for any frequency from these curves.

A physical interpretation of the results provided by Eq. (3.110) for hyperbolic equations is important. Consider the second-order wave equation:

$$u_{tt} - c^2 u_{xx} = 0 \tag{3.112}$$

This equation has characteristics

$$x + ct = \text{const} = c_1$$
$$x - ct = \text{const} = c_2$$

A solution at a point (x, t) depends upon data contained between the characteristics that intersect that point, as sketched in Fig. 3.14. The analytic solution at (x, t) is influenced only by information contained between c_1 and c_2.

The numerical stability requirement for many explicit numerical methods for solving hyperbolic PDEs is the CFL condition, which, for the wave equation, is

$$\left| c \frac{\Delta t}{\Delta x} \right| \leqslant 1$$

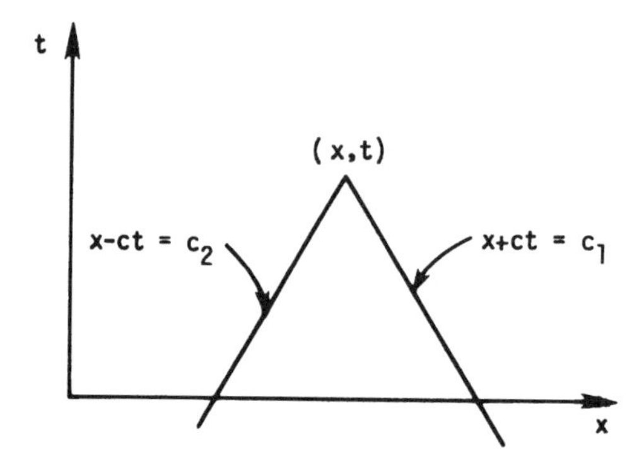

Figure 3.14 Characteristics of the second-order wave equation.

This is the same as given in Eq. (3.110) and may be written as

$$\left(\frac{\Delta t}{\Delta x}\right)^2 \leqslant \frac{1}{c^2}$$

The characteristic slopes are given by $dt/dx = \pm 1/c$. The CFL condition requires that the analytic domain of influence lie within the numerical domain of influence. The numerical domain may include more than, but not less than, the analytical zone. Another interpretation is that the slope of the lines connecting $(j \pm 1, n)$ and $(j, n+1)$ must be smaller in absolute value (flatter) than the characteristics. The CFL requirement makes sense from a physical point of view. One would also expect the numerical solution to be degraded if too much unnecessary information is included by allowing $c(\Delta t/\Delta x)$ to become greatly different from unity. This is, in fact, what occurs numerically. The best results for hyperbolic systems using the most common explicit methods are obtained with Courant numbers near unity. This is consistent with our observations about attenuation associated with the Lax method, as shown in Fig. 3.13.

Before we begin our study of stability for systems of equations, an example demonstrating the application of the von Neumann method to higher dimensional problems is in order.

Example 3.5 A solution of the 2-D heat equation

$$\frac{\partial u}{\partial t} = \alpha \frac{\partial^2 u}{\partial x^2} + \alpha \frac{\partial^2 u}{\partial y^2}$$

is desired using the simple explicit scheme. What is the stability requirement for the method?

Solution The finite-difference equation for this problem is

$$u_{j,k}^{n+1} = u_{j,k}^n + r_x(u_{j+1,k}^n - 2u_{j,k}^n + u_{j-1,k}^n) + r_y(u_{j,k+1}^n - 2u_{j,k}^n + u_{j,k-1}^n)$$

where $r_x = \alpha[\Delta t/(\Delta x)^2]$ and $r_y = \alpha[\Delta t/(\Delta y)^2]$. In this case, a Fourier component of the form

$$u_{j,k}^n = e^{at}e^{ik_x x}e^{ik_y y}$$

is assumed. If $\beta_1 = k_x \, \Delta x$ and $\beta_2 = k_y \, \Delta y$, we obtain

$$e^{a \, \Delta t} = 1 + 2r_x(\cos \beta_1 - 1) + 2r_y(\cos \beta_2 - 1)$$

If the identity $\sin^2(\beta/2) = (1 - \cos \beta)/2$ is used, the amplification factor is

$$G = 1 - 4r_x \sin^2 \frac{\beta_1}{2} - 4r_y \sin^2 \frac{\beta_2}{2}$$

Thus for stability, $|1 - 4r_x \sin^2(\beta_1/2) - 4r_y \sin^2(\beta_2/2)| \leqslant 1$, which is true only if $(4r_x \sin^2 \beta_1/2 + 4r_y \sin^2 \beta_2/2) \leqslant 2$. The stability requirement is then $(r_x + r_y) \leqslant \frac{1}{2}$ or $\alpha \, \Delta t[1/(\Delta x)^2 + 1/(\Delta y)^2] \leqslant \frac{1}{2}$. This is similar to the analysis of the same method for the 1-D case but shows that the effective time step in two dimensions is reduced. This example was easily completed, but in general, a stability analysis in more than a single space dimension and time is difficult. Frequently, the stability must be determined by computing the magnitude of the amplification factor for different values of r_x and r_y.

3.6.2 Stability Analysis for Systems of Equations

The previous discussion illustrates how the von Neumann analysis can be used to evaluate stability for a single equation. The basic idea used in this technique also provides a useful method of viewing stability for systems of equations. Systems of equations encountered in fluid mechanics and heat transfer can often be written in the form

$$\frac{\partial \mathbf{E}}{\partial t} + \frac{\partial \mathbf{F}}{\partial x} = 0 \tag{3.113}$$

where **E** and **F** are vectors and $\mathbf{F} = \mathbf{F}(\mathbf{E})$. In general, this system of equations is nonlinear. In order to perform a linear stability analysis, we rewrite the system as

$$\frac{\partial \mathbf{E}}{\partial t} + \left[\frac{\partial \mathbf{F}}{\partial \mathbf{E}}\right]\frac{\partial \mathbf{E}}{\partial x} = 0 \tag{3.114}$$

or

$$\frac{\partial \mathbf{E}}{\partial t} + [A]\frac{\partial \mathbf{E}}{\partial x} = 0$$

where $[A]$ is the Jacobian matrix $[\partial \mathbf{F}/\partial \mathbf{E}]$. We locally linearize the system by holding $[A]$ constant while the **E** vector is advanced through a single time step. A similar linearization is used for a single nonlinear equation, permitting the application of the von Neumann method of the previous section.

For the sake of discussion, let us apply the Lax method to this system. The result is

$$\mathbf{E}_j^{n+1} = \frac{1}{2}\left([I] + \frac{\Delta t}{\Delta x}[A]^n\right)\mathbf{E}_{j-1}^n + \frac{1}{2}\left([I] - \frac{\Delta t}{\Delta x}[A]^n\right)\mathbf{E}_{j+1}^n \quad (3.115)$$

where the notation is as previously defined and $[I]$ is the identity matrix. The stability of the difference equation can again be evaluated by applying the Fourier or von Neumann method. If a typical term of a Fourier series is substituted into Eq. (3.115), the following expression is obtained,

$$\mathbf{e}^{n+1}(k) = [G(\Delta t, k)]\mathbf{e}^n(k) \quad (3.116)$$

where

$$[G] = [I]\cos\beta - i\frac{\Delta t}{\Delta x}[A]\sin\beta \quad (3.117)$$

and $\mathbf{e}^n$ represents the Fourier coefficients of the typical term. The $[G]$ matrix is called the amplification matrix. This matrix is now dependent upon step size and frequency or wave number, i.e., $[G] = [G(\Delta t, k)]$. For a stable finite-difference calculation, the largest eigenvalue of $[G]$, $\sigma_{\max}$, must obey

$$|\sigma_{\max}| \leqslant 1 \quad (3.118)$$

This leads to the requirement that

$$\left|\lambda_{\max}\frac{\Delta t}{\Delta x}\right| \leqslant 1 \quad (3.119)$$

where $\lambda_{\max}$ is the largest eigenvalue of the $[A]$ matrix, i.e., the Jacobian matrix of the system. A simple example to demonstrate this is of value.

***Example** 3.6* Determine the stability requirement necessary for solving the system of first-order equations

$$\frac{\partial u}{\partial t} + c\frac{\partial v}{\partial x} = 0$$

$$\frac{\partial v}{\partial t} + c\frac{\partial u}{\partial x} = 0$$

using the Lax method.

Solution In this problem

$$\mathbf{E} = \begin{bmatrix} u \\ v \end{bmatrix}$$

and

$$\frac{\partial \mathbf{E}}{\partial t} + [A]\frac{\partial \mathbf{E}}{\partial x} = 0$$

where

$$[A] = \begin{bmatrix} 0 & c \\ c & 0 \end{bmatrix}$$

Thus, the maximum eigenvalue of [A] is c, and the stability requirement is the usual CFL condition

$$\left| c\frac{\Delta t}{\Delta x} \right| \leqslant 1$$

It should be noted that the stability analysis presented above does not include the effect of boundary conditions even though a matrix notation for the system is used. The influence of boundary conditions is easily included for systems of difference equations.

Equation (3.116) shows that the stability of a finite-difference operator is related to the amplification matrix. We may also write Eq. (3.116) as

$$\mathbf{e}^{n+1}(k) = [G(\Delta t, k)]^n [\mathbf{e}^1(k)] \tag{3.120}$$

The stability condition (Richtmyer and Morton, 1967) requires that for some positive τ, the matrices $[G(\Delta t, k)]^n$ be uniformly bounded for

$$0 < \Delta t < \tau$$
$$0 \leqslant n\,\Delta t \leqslant T$$

for all k, where T is the maximum time. This leads to the *von Neumann necessary condition* for stability, which is

$$|\sigma_i(\Delta t, k)| \leqslant 1 + O(\Delta t) \qquad 0 < \Delta t < \tau \tag{3.121}$$

for each eigenvalue and wave number, where σ_i represents the eigenvalues of $[G(\Delta t, k)]$. For a scalar equation, Eq. (3.121) reduces to

$$|G| \leqslant 1 + O(\Delta t)$$

The stability requirement used in previous examples required that the maximum eigenvalue have a modulus less than or equal to 1. Clearly, that requirement is more stringent than Eq. (3.121). The von Neumann necessary condition provides that local growth $c\,\Delta t$ can be acceptable and, in fact, must be possible in many physical problems. The classical example illustrating this point is the heat equation with a source term.

Example 3.7 Suppose we wish to solve the heat equation with a source term

$$\frac{\partial u}{\partial t} = \alpha \frac{\partial^2 u}{\partial x^2} + cu$$

using the simple explicit finite-difference method. Determine the stability requirement.

Solution If a Fourier stability analysis is performed, the amplification factor is

$$G = 1 - 4r \sin^2 \frac{\beta}{2} + c\,\Delta t$$

This shows that the solution of the difference equation may grow with time and still satisfy the von Neumann necessary condition. Physical insight must be used when the stability of a finite-difference method is investigated. One must recognize that for hyperbolic systems the strict condition less than or equal to 1 should be used. Hyperbolic equations are wave-like and do not possess solutions that increase exponentially with time.

We have investigated stability of various finite-difference methods by using the von Neumann method. If the influence of boundary conditions on stability is desired, we must use the *matrix method.* This is most easily demonstrated by applying the Lax method to solve the 1-D linear wave equation:

$$\frac{\partial u}{\partial t} + c\frac{\partial u}{\partial x} = 0$$

Assume that an array of m points is used to solve this problem and that the boundary conditions are periodic, i.e.,

$$u_{m+1}^n = u_1^n \tag{3.122}$$

If the Lax method is applied to this problem, a system of algebraic equations is generated that has the form

$$\mathbf{u}^{n+1} = [X]\mathbf{u}^n \tag{3.123}$$

where

$$\mathbf{u}^n = [u_1, u_2, \ldots, u_m]^T \tag{3.124}$$

and

$$[X] = \begin{bmatrix} 0 & \frac{1-\nu}{2} & 0 & \cdot & \cdot & \cdot & \frac{1+\nu}{2} \\ \frac{1+\nu}{2} & 0 & \frac{1-\nu}{2} & \cdot & \cdot & \cdot & 0 \\ 0 & \frac{1+\nu}{2} & \ddots & \ddots & & & \cdot \\ \cdot & 0 & \ddots & \ddots & \ddots & & \cdot \\ \cdot & \cdot & & \ddots & \ddots & \ddots & \cdot \\ \cdot & \cdot & & & \ddots & \ddots & \frac{1-\nu}{2} \\ \frac{1-\nu}{2} & 0 & \cdot & \cdot & \cdot & \frac{1+\nu}{2} & 0 \end{bmatrix} \tag{3.125}$$

The stability of the finite-difference calculation in Eq. (3.123) is governed by the eigenvalue structure of $[X]$. Since $[X]$ was formed assuming periodic boundary conditions, only the three diagonals noted in Eq. (3.125) and the two corner elements contribute to the calculation. This matrix is called an aperiodic matrix.

For matrices of the form

$$\begin{bmatrix} a_1 & a_2 & 0 & \cdot & \cdot & \cdot & a_0 \\ a_0 & a_1 & a_2 & 0 & \cdot & \cdot & 0 \\ 0 & a_0 & \ddots & \ddots & \ddots & & \cdot \\ \cdot & \cdot & & \ddots & \ddots & \ddots & \cdot \\ \cdot & \cdot & & & \ddots & \ddots & \cdot \\ \cdot & \cdot & & & & \ddots & a_2 \\ a_2 & 0 & \cdot & \cdot & \cdot & a_0 & a_1 \end{bmatrix} \tag{3.126}$$

the eigenvalues are given by

$$\lambda_j = a_1 + (a_0 + a_2)\cos\frac{2\pi}{m}(j-1) + i(a_0 - a_2)\sin\frac{2\pi}{m}(j-1)$$

In this case a_0, a_1, and a_2 have the values

$$a_0 = \frac{1+\nu}{2} \qquad a_1 = 0 \qquad a_2 = \frac{1-\nu}{2}$$

and the eigenvalues are

$$\lambda_j = \cos\frac{2\pi}{m}(j-1) + i\nu\sin\frac{2\pi}{m}(j-1) \tag{3.127}$$

The numerical method is thus stable if $|\nu| \leqslant 1$, i.e., if the CFL condition is satisfied. This shows that an analysis based upon the matrix operator associated with the Lax method yields the same stability requirement as previously derived for the simple wave equation. For periodic boundary conditions, the Fourier and matrix method yield virtually identical results. Another example is needed in order to demonstrate the effect of boundary conditions and the discreteness of the mesh.

Example 3.8 As in the previous example, assume that the Lax method is used to solve the first-order linear wave equation. If a four-point mesh is used, special treatment is needed to enforce the boundary conditions at the first and fourth points. For simplicity we set u at the first point equal to a constant value for all time, so the equation for the first point reads

$$u_1^{n+1} = u_1^n$$

Since we are computing a solution to the wave equation, the value of u_4 cannot be arbitrarily chosen. It must be consistent with the way the solution is propagated. We elect to set

$$u_4^{n+1} = u_3^n$$

which determines the boundary value from the interior solution.

Solution For the present boundary condition treatment the $[X]$ matrix becomes

$$[X] = \begin{bmatrix} 1 & 0 & 0 & 0 \\ \dfrac{1+\nu}{2} & 0 & \dfrac{1-\nu}{2} & 0 \\ 0 & \dfrac{1+\nu}{2} & 0 & \dfrac{1-\nu}{2} \\ 0 & 0 & 1 & 0 \end{bmatrix}$$

The eigenvalues are easily computed and are

$$\lambda_1 = 1$$

$$\lambda_2 = 0$$

$$\lambda_{3,4} = \pm\tfrac{1}{2}\sqrt{(1-\nu)(3+\nu)}$$

Using the requirement that $|\lambda| \leqslant 1$ for stability, the restriction on ν is not the usual CFL condition but is

$$(-\sqrt{8} - 1) \leqslant \nu \leqslant (\sqrt{8} - 1)$$

The CFL condition is altered by the boundary conditions in this example, as is normally the case.

It is clear that the boundary conditions on the mesh are included in the matrix method. This means that the influence of boundary conditions on stability is automatically included if the matrix analysis is used. Unfortunately, a closed-form solution for the eigenvalues is usually not available for arbitrary end boundary conditions.

The treatment of stability presented in this section has included the Fourier (von Neumann) method and the matrix method of analysis. These two techniques are probably the most widely used to determine the stability of numerical schemes. Other methods of analyzing stability have been devised and are frequently very convenient to use. The works of Hirt (1968) and Warming and Hyett (1974) are typical of these techniques. A more comprehensive mathematical analysis of stability including many theorems and proofs is contained in the book by Richtmyer and Morton (1967).

PROBLEMS

3.1 Verify that

$$\left.\frac{\partial^3 u}{\partial x^3}\right)_{i,j} = \frac{\Delta_x^3 u_{i,j}}{(\Delta x)^3} + O(\Delta x)$$

3.2 Consider the function $f(x) = e^x$. Using a mesh increment $\Delta x = 0.1$, determine $f'(x)$ at $x = 2$ with the forward-difference formula, Eq. (3.26), the central-difference formula, Eq. (3.28), and the second-order three-point formula, Eq. (3.29). Compare the results with the exact value. Repeat the comparisons for $\Delta x = 0.2$. Have the order estimates for truncation errors been a reliable guide? Discuss this point.

3.3 Verify whether or not the following difference representation for the continuity equation for a 2-D steady incompressible flow has the conservation property:

$$\frac{(u_{i+1,j} + u_{i+1,j-1} - u_{i,j} - u_{i,j-1})}{2\,\Delta x} + \frac{(v_{i+1,j} - v_{i+1,j-1})}{\Delta y} = 0$$

where u and v are the x and y components of velocity, respectively.
3.4 Repeat Prob. 3.3, for the following difference representation for the continuity equation:

$$\frac{(u_{i+1,j} - u_{i-1,j})}{2\,\Delta x} + \frac{(v_{i,j+1} - v_{i,j-1})}{2\,\Delta y} = 0$$

3.5 Consider the nonlinear equation

$$u\frac{\partial u}{\partial x} = \mu\frac{\partial^2 u}{\partial y^2}$$

where μ is a constant.

(*a*) Is this equation in conservative form? If not, can you suggest a conservative form for the equation?

(*b*) Develop a finite-difference formulation for this equation using the integral approach.

3.6 Verify the approximation to $\partial^2 u/\partial x\,\partial y$ given by Eq. (3.50) in Table 3.2.
3.7 Verify the approximation to $\partial^2 u/\partial x^2$ given by Eq. (3.40) in Table 3.1.
3.8 Verify Eq. (3.79) in Table 3.3.
3.9 Verify Eq. (3.80) in Table 3.3.
3.10 Verify the following finite-difference approximation for use in two dimensions at the point (i, j). Assume $\Delta x = \Delta y = h$.

$$\frac{\partial^2 u}{\partial x^2} + \frac{\partial^2 u}{\partial y^2} = \frac{u_{i+1,j-1} + u_{i+1,j+1} + u_{i-1,j-1} + u_{i-1,j+1} - 4u_{i,j}}{2h^2} + O(h^2)$$

3.11 Develop a finite-difference approximation with T.E. of $O(\Delta y)$ for $\partial^2 u/\partial y^2$ at point (i, j) using $u_{i,j}, u_{i,j+1}, u_{i,j-1}$ when the grid spacing is *not* uniform. Use the Taylor-series method. Can you devise a three-point scheme with second-order accuracy with unequal spacing? Before you draw your final conclusions, consider the use of compact implicit representations.
3.12 Establish the T.E. of the following finite-difference approximation to $\partial u/\partial y$ at the point (i, j) for a uniform mesh:

$$\frac{\partial u}{\partial y} \simeq \frac{-3u_{i,j} + 4u_{i,j+1} - u_{i,j+2}}{2\,\Delta y}$$

What is the order of the T.E.?
3.13 Investigate the T.E. of the following finite-difference approximation for a uniform mesh:

$$\left.\frac{\partial u}{\partial x}\right)_{i,j} \simeq \frac{1}{2h}\frac{\bar{\delta}_x u_{i,j}}{1 + \delta_x^2/6}$$

3.14 Utilize Taylor-series expansions about the point $(n + \frac{1}{2}, j)$ to determine the T.E. of the Crank-Nicolson representation of the heat equation, Eq. (3.71*a*). Compare these results with the T.E. obtained from Taylor-series expansions about point n, j.
3.15 Develop a finite-difference approximation with T.E. of $O(\Delta y)^2$ for $\partial T/\partial y$ at point (i, j) using $T_{i,j}$, $T_{i,j+1}$, and $T_{i,j+2}$ when the grid spacing is *not* uniform.
3.16 Determine the T.E. of the following finite-difference approximation for $\partial u/\partial x$ at point (i, j) when the grid spacing is *not* uniform:

$$\left.\frac{\partial u}{\partial x}\right)_{i,j} \simeq \frac{u_{i+1,j} - (\Delta x_+/\Delta x_-)^2 u_{i-1,j} - [1 - (\Delta x_+/\Delta x_-)^2]u_{i,j}}{\Delta x_-(\Delta x_+/\Delta x_-)^2 + \Delta x_+}$$

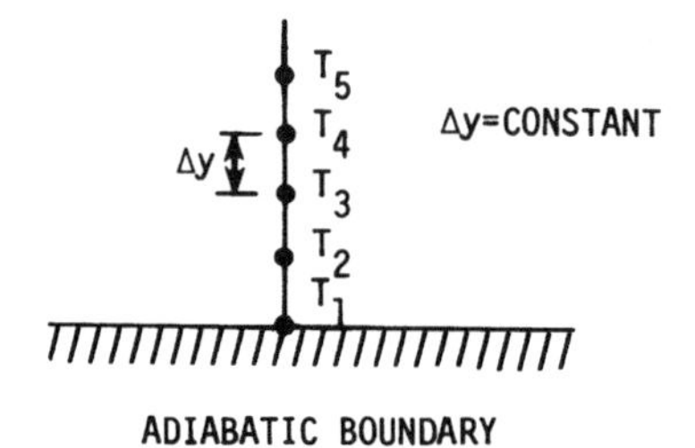

Figure P3.1

3.17 Suppose that a finite-difference solution has been obtained for the temperature T, near but not at an adiabatic boundary (i.e., $\partial T/\partial y = 0$ at the boundary) (Fig. P3.1). In most instances, it would be necessary or desirable to evaluate the temperature at the boundary point itself. For this case of an adiabatic boundary, develop expressions for the temperature at the boundary T_1, in terms of temperatures at neighboring points T_2, T_3, etc., by assuming that the temperature distribution in the neighborhood of the boundary is

(a) a straight line

(b) a second-degree polynomial

(c) a cubic polynomial (you only need to indicate how you would derive this one).

Indicate the order of the T.E. in each of the above approximations used to evaluate T_1.

3.18 Consider a steady-state conduction problem governed by Laplace's equation with convective boundary conditions (see Fig. P3.2). The formal statement of the boundary condition is $-k\,\partial T/\partial y)_{\text{bdy}} = h(T_w - T_\infty)$, which can be readily cast into finite-difference form as $-k[(T_0 - T_\infty)/\Delta y] + O(\Delta y) = h(T_0 - T_\infty)$. Use the control-volume approach to develop an expression for the boundary condition at point 0. Evaluate the T.E. in this expression assuming that Laplace's equation applies at the boundary point.

3.19 Consider a heat conduction problem governed by $\partial T/\partial t = \alpha(\partial^2 T/\partial x^2)$. Develop a finite-difference representation for this equation by the control-volume approach. Do not assume that the grid is uniform.

3.20 For 2-D steady-state conduction in a solid, apply the control-volume method to derive an appropriate difference expression for the boundary temperature in control volume B in Fig. 3.7 for *adiabatic wall* boundary conditions.

3.21 Solve the 1-D heat equation using forward-time centered-space differences with $\alpha(\Delta t/\Delta x^2) = \frac{1}{2}$. Let the grid consist of five points, including three interior and two boundary points. Assume a constant unity wall temperature and a zero initial temperature on the interior. Complete this calculation for 10 integration steps. Compare your results with those obtained in the example of Section 3.3.4.

3.22 Refer to Fig. 3.10. Following the methodology illustrated in the text material associated with Fig. 3.10, develop an appropriate finite-volume expression for the heat flux across the boundary from c to d.

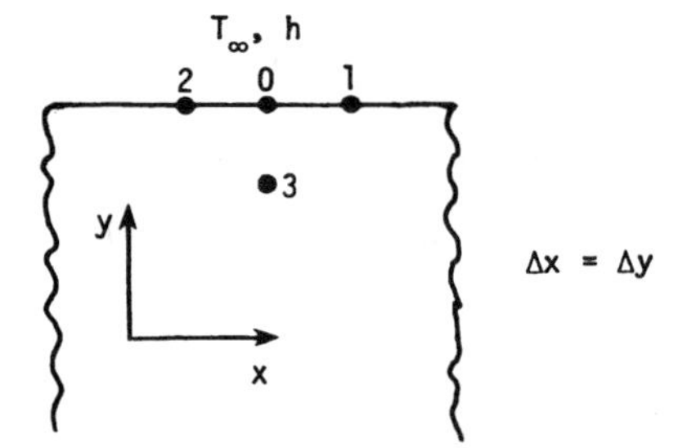

Figure P3.2

3.23 Refer to Fig. 3.10 and the associated text material. In an example in Section 3.5, an expression was developed for the heat flux across boundary *a-b* for control volume B (Eq. 3.100). Following a similar methodology, develop an appropriate finite-volume expression for the heat flow into volume B across the boundary from d to a. If the difference scheme is to be conservative, this heat flow should be equal in magnitude but opposite in direction to the inflow computed for volume A across boundary *d-a*. Check to see if this is true.

3.24 Show that the amplification factor derived for the finite-difference solution of the heat equation, Eq. (3.101), could be obtained by direct substitution of a solution of the form

$$u_j^n = \sum_{-\infty}^{+\infty} C_m g_m^n e^{ik_x x}$$

In this form C_m represent the Fourier coefficients of the initial error distribution and g_m is the amplification factor. Identify g_m with Eq. (3.106). Discuss the convergence of the solution and relate your conclusions to the Lax equivalence theorem.

3.25 Use a von Neumann stability analysis to show for the wave equation that a simple explicit Euler predictor using central differencing in space is unstable. The difference equation is

$$u_j^{n+1} = u_j^n - c\frac{\Delta t}{\Delta x}\left(\frac{u_{j+1}^n - u_{j-1}^n}{2}\right)$$

Now show that the same difference method is stable when written as the implicit formula

$$u_j^{n+1} = u_j^n - c\frac{\Delta t}{\Delta x}\left(\frac{u_{j+1}^{n+1} - u_{j-1}^{n+1}}{2}\right)$$

3.26 The DuFort-Frankel method for solving the heat equation requires solution of the difference equation

$$\frac{u_j^{n+1} - u_j^{n-1}}{2\,\Delta t} = \frac{\alpha}{(\Delta x)^2}(u_{j+1}^n - u_j^{n+1} - u_j^{n-1} + u_{j-1}^n)$$

Develop the stability requirements necessary for the solution of this equation.

3.27 Prove that the CFL condition is the stability requirement when the Lax-Wendroff method is applied to solve the simple 1-D wave equation. The difference equation is of the form

$$u_j^{n+1} = u_j^n - \frac{c\,\Delta t}{2\,\Delta x}(u_{j+1}^n - u_{j-1}^n) + \frac{c^2(\Delta t)^2}{2(\Delta x)^2}(u_{j+1}^n - 2u_j^n + u_{j-1}^n)$$

3.28 An implicit scheme for solving the heat equation is given by

$$u_j^{n+1} = u_j^n + \frac{\alpha\,\Delta t}{(\Delta x)^2}\left[\frac{1}{3}(u_{j+1}^{n+1} - 2u_j^{n+1} + u_{j-1}^{n+1}) + \frac{2}{3}(u_{j+1}^n - 2u_j^n + u_{j-1}^n)\right]$$

Apply the Fourier stability analysis to this scheme and determine the stability restrictions, if any.

3.29 An implicit scheme for solving the first-order wave equation is given by

$$u_j^{n+1} = u_j^n - \frac{c\,\Delta t}{\Delta x}(u_{j+1}^{n+1} - u_j^{n+1})$$

Apply the Fourier stability analysis to this scheme and determine the stability restrictions, if any.

3.30 The leap frog method for solving the 1-D wave equation is given by

$$\frac{u_j^{n+1} - u_j^{n-1}}{2\,\Delta t} + c\frac{u_{j+1}^n - u_{j-1}^n}{2\,\Delta x} = 0$$

Apply the Fourier stability analysis to this method, and determine the stability restrictions, if any.

3.31 Determine the stability requirement necessary to solve the 1-D heat equation with a source term

$$\frac{\partial u}{\partial t} = \alpha \frac{\partial^2 u}{\partial x^2} + ku$$

Use the central-space, forward-time difference method. Does the von Neumann necessary condition, Eq. (3.121), make physical sense for this type of computational problem?

3.32 Use the matrix method to determine the stability of the Lax method used to solve the first-order wave equation on a mesh with two interior points and two boundary points. Assume the boundaries are held at constant values $u_{\text{left}} = 1$, $u_{\text{right}} = 0$.

3.33 Use the matrix method and evaluate the stability of the numerical method used in Prob. 3.21 for the heat equation using a five-point mesh. How many frequencies must one be concerned with in this case?

3.34 In attempting to solve a simple PDE, a system of finite-difference equations of the form $u_j^{n+1} = [A]u_j^n$ has evolved, where

$$[A] = \begin{bmatrix} 1+\nu & \nu & 0 \\ 0 & 1+\nu & \nu \\ -\nu & 0 & 1+\nu \end{bmatrix}$$

Investigate the stability of this scheme.

3.35 The application of a finite-difference scheme to the heat equation on a three-point grid results in the following system of equations:

$$u_j^{n+1} = [A]u_j^n$$

where

$$[A] = \begin{bmatrix} 1 & 0 & 0 \\ r & 1-2r & r \\ 0 & 0 & 1 \end{bmatrix}$$

and $r = \alpha\,\Delta t/(\Delta x)^2$. Determine the stability of this scheme.

3.36 The upstream scheme

$$\frac{u_j^{n+1} - u_j^n}{\Delta t} + c\frac{u_j^n - u_{j-1}^n}{\Delta x} = 0$$

is used to solve the wave equation on a four-point grid for the boundary conditions

$$u_1 = 1 \qquad u_4^{n+1} = u_3^n$$

and the initial conditions ($n = 1$)

$$u_1^1 = 1 \qquad u_2^1 = u_3^1 = u_4^1 = 0$$

Use the matrix method to determine the stability restrictions for this method.

CHAPTER

FOUR

APPLICATION OF NUMERICAL METHODS TO SELECTED MODEL EQUATIONS

In this chapter we examine in detail various numerical schemes that can be used to solve simple model partial differential equations (PDEs). These model equations include the first-order wave equation, the heat equation, Laplace's equation, and Burgers' equation. These equations are called model equations because they can be used to "model" the behavior of more complicated PDEs. For example, the heat equation can serve as a model equation for other parabolic PDEs such as the boundary-layer equations. All of the present model equations have exact solutions for certain boundary and initial conditions. We can use this knowledge to quickly evaluate and compare numerical methods that we might wish to apply to more complicated PDEs. The various methods discussed in this chapter were selected because they illustrate the basic properties of numerical algorithms. Each of the methods exhibits certain distinctive features that are characteristic of a class of methods. Some of these features may not be desirable, but the method is included anyway for pedagogical reasons. Other very useful methods have been omitted because they are similar to those that are included. Space does not permit a discussion of all possible methods that could be used.

4.1 WAVE EQUATION

The one-dimensional (1-D) wave equation is a second-order hyperbolic PDE given by

$$\frac{\partial^2 u}{\partial t^2} = c^2 \frac{\partial^2 u}{\partial x^2} \tag{4.1}$$

This equation governs the propagation of sound waves traveling at a wave speed c in a uniform medium. A first-order equation that has properties similar to those of Eq. (4.1) is given by

$$\frac{\partial u}{\partial t} + c \frac{\partial u}{\partial x} = 0 \qquad c > 0 \tag{4.2}$$

Note that Eq. (4.1) can be obtained from Eq. (4.2). We will use Eq. (4.2) as our model equation and refer to it as the first-order 1-D wave equation, or simply the "wave equation." This linear hyperbolic equation describes a wave propagating in the x direction with a velocity c, and it can be used to "model" in a rudimentary fashion the nonlinear equations governing inviscid flow. Although we will refer to Eq. (4.2) as the wave equation, the reader is cautioned to be aware of the fact that Eq. (4.1) is the classical wave equation. More appropriately, Eq. (4.2) is often called the 1-D linear convection equation.

The exact solution of the wave equation [Eq. (4.2)] for the pure initial value problem with initial data

$$u(x, 0) = F(x) \qquad -\infty < x < \infty \tag{4.3}$$

is given by

$$u(x, t) = F(x - ct) \tag{4.4}$$

Let us now examine some schemes that could be used to solve the wave equation.

4.1.1 Euler Explicit Methods

The following simple explicit one-step methods,

$$\frac{u_j^{n+1} - u_j^n}{\Delta t} + c \frac{u_{j+1}^n - u_j^n}{\Delta x} = 0 \qquad c > 0 \tag{4.5}$$

$$\frac{u_j^{n+1} - u_j^n}{\Delta t} + c \frac{u_{j+1}^n - u_{j-1}^n}{2\,\Delta x} = 0 \tag{4.6}$$

have truncation errors (T.E.s) of $O[\Delta t, \Delta x]$ and $O[\Delta t, (\Delta x)^2]$, respectively. We refer to these schemes as being first-order accurate, since the lowest-order term in the T.E. is first order, i.e., Δt and Δx for Eq. (4.5) and Δt for Eq. (4.6). These schemes are explicit, since only one unknown u_j^{n+1} appears in each equation. Unfortunately, when the von Neumann stability analysis is applied to these schemes, we find that they are unconditionally unstable. These simple schemes, therefore, prove to be worthless in solving the wave equation. Let us now proceed to look at methods that have more utility.

4.1.2 Upstream (First-Order Upwind or Windward) Differencing Method

The simple Euler method, Eq. (4.5), can be made stable by replacing the forward space difference by a backward space difference, provided that the wave speed c is positive. If the wave speed is negative, a forward difference must be used to assure stability. This point is discussed further at the end of the present section. For a positive wave speed, the following algorithm results:

$$\frac{u_j^{n+1} - u_j^n}{\Delta t} + c\frac{u_j^n - u_{j-1}^n}{\Delta x} = 0 \qquad c > 0 \tag{4.7}$$

This is a first-order accurate method with T.E. of $O[\Delta t, \Delta x]$. The von Neumann stability analysis shows that this method is stable, provided that

$$0 \leqslant \nu \leqslant 1 \tag{4.8}$$

where $\nu = c\,\Delta t/\Delta x$.

Let us substitute Taylor-series expansions into Eq. (4.7) for u_j^{n+1} and u_{j-1}^n. The following equation results:

$$\frac{1}{\Delta t}\left\{\left[u_j^n + \Delta t\, u_t + \frac{(\Delta t)^2}{2}u_{tt} + \frac{(\Delta t)^3}{6}u_{ttt} + \cdots\right] - u_j^n\right\}$$
$$+ \frac{c}{\Delta x}\left\{u_j^n - \left[u_j^n - \Delta x\, u_x + \frac{(\Delta x)^2}{2}u_{xx} - \frac{(\Delta x)^3}{6}u_{xxx} + \cdots\right]\right\} = 0 \tag{4.9}$$

Equation (4.9) simplifies to

$$u_t + cu_x = -\frac{\Delta t}{2}u_{tt} + \frac{c\,\Delta x}{2}u_{xx} - \frac{(\Delta t)^2}{6}u_{ttt} - c\frac{(\Delta x)^2}{6}u_{xxx} + \cdots \tag{4.10}$$

Note that the left-hand side of this equation corresponds to the wave equation and the right-hand side is the T.E., which is generally not zero. The significance of terms in the T.E. can be more easily interpreted if the time-derivative terms are replaced by spatial derivatives. In order to replace u_{tt} by a spatial-derivative term, we take the partial derivative of Eq. (4.10) with respect to time, to obtain

$$u_{tt} + cu_{xt} = -\frac{\Delta t}{2}u_{ttt} + \frac{c\,\Delta x}{2}u_{xxt} - \frac{(\Delta t)^2}{6}u_{tttt} - \frac{c(\Delta x)^2}{6}u_{xxxt} + \cdots \tag{4.11}$$

and take the partial derivative of Eq. (4.10) with respect to x and multiply by $-c$:

$$-cu_{tx} - c^2 u_{xx} = \frac{c\,\Delta t}{2} u_{ttx} - \frac{c^2\,\Delta x}{2} u_{xxx} + \frac{c(\Delta t)^2}{6} u_{tttx} + \frac{c^2(\Delta x)^2}{6} u_{xxxx} + \cdots \tag{4.12}$$

Adding Eqs. (4.11) and (4.12) gives

$$u_{tt} = c^2 u_{xx} + \Delta t\left(\frac{-u_{ttt}}{2} + \frac{c}{2} u_{ttx} + O[\Delta t]\right) + \Delta x\left(\frac{c}{2} u_{xxt} - \frac{c^2}{2} u_{xxx} + O[\Delta x]\right) \tag{4.13}$$

In a similar manner, we can obtain the following expressions for u_{ttt}, u_{ttx}, and u_{xxt}:

$$\begin{aligned} u_{ttt} &= -c^3 u_{xxx} + O[\Delta t, \Delta x] \\ u_{ttx} &= c^2 u_{xxx} + O[\Delta t, \Delta x] \\ u_{xxt} &= -c u_{xxx} + O[\Delta t, \Delta x] \end{aligned} \tag{4.14}$$

Combining Eqs. (4.10), (4.13), and (4.14) leaves

$$\begin{aligned} u_t + cu_x = {} & \frac{c\,\Delta x}{2}(1 - \nu) u_{xx} - \frac{c(\Delta x)^2}{6}(2\nu^2 - 3\nu + 1) u_{xxx} \\ & + O[(\Delta x)^3, (\Delta x)^2 \Delta t, \Delta x(\Delta t)^2, (\Delta t)^3] \end{aligned} \tag{4.15}$$

An equation, such as Eq. (4.15), is called a *modified equation* (Warming and Hyett, 1974). It can be thought of as the PDE that is actually solved (if sufficient boundary conditions were available) when a finite-difference method is applied to a PDE. It is important to emphasize that the equation obtained after substitution of the Taylor-series expansions, i.e., Eq. (4.10), must be used to eliminate the higher-order time derivatives rather than the original PDE, Eq. (4.2). This is due to the fact that a solution of the original PDE does not in general satisfy the difference equation, and since the modified equation represents the difference equation, it is obvious that the original PDE should not be used to eliminate the time derivatives.

The process of eliminating time derivatives can be greatly simplified if a table is constructed (Table 4.1). The coefficients of each term in Eq. (4.10) are placed in the first row of the table. Note that all terms have been moved to the left-hand side of the equation. The u_{tt} term is then eliminated by multiplying Eq. (4.10) by the operator

$$-\frac{\Delta t}{2}\frac{\partial}{\partial t}$$

Table 4.1 Procedure for determining modified equation

	u_t	u_x	u_{tt}	u_{tx}	u_{xx}	u_{ttt}	u_{ttx}	u_{txx}	u_{xxx}	u_{tttt}	u_{tttx}	u_{ttxx}	u_{txxx}	u_{xxxx}
Coefficients of Eq. (4.10)	1	c	$\frac{\Delta t}{2}$	0	$-\frac{c\,\Delta x}{2}$	$\frac{\Delta t^2}{6}$	0	0	$\frac{c\,\Delta x^2}{6}$	$\frac{\Delta t^3}{24}$	0	0	0	$-\frac{c\,\Delta x^3}{24}$
$-\frac{\Delta t}{2}\frac{\partial}{\partial t}$ Eq. (4.10)			$-\frac{\Delta t}{2}$	$-\frac{c\,\Delta t}{2}$	0	$-\frac{\Delta t^2}{4}$	0	$\frac{c\,\Delta t\,\Delta x}{4}$	0	$-\frac{\Delta t^3}{12}$	0	0	$-\frac{c\,\Delta t\,\Delta x^2}{12}$	0
$\frac{c}{2}\Delta t\frac{\partial}{\partial x}$ Eq. (4.10)				$\frac{c\,\Delta t}{2}$	$\frac{c^2}{2}\Delta t$	0	$\frac{c\,\Delta t^2}{4}$	0	$-\frac{c^2\,\Delta t\,\Delta x}{4}$	0	$\frac{c\,\Delta t^3}{12}$	0	0	$\frac{c^2\,\Delta t\,\Delta x^2}{12}$
$\frac{1}{12}\Delta t^2\frac{\partial^2}{\partial t^2}$ Eq. (4.10)						$\frac{1}{12}\Delta t^2$	$\frac{c\,\Delta t^2}{12}$	0	0	$\frac{1}{24}\Delta t^3$	0	$-\frac{c}{24}\Delta x\,\Delta t^2$	0	0
$-\frac{1}{3}c\,\Delta t^2\frac{\partial^2}{\partial t\,\partial x}$ Eq. (4.10)							$-\frac{1}{3}c\,\Delta t^2$	$-\frac{1}{3}c^2\,\Delta t^2$	0	0	$-\frac{1}{6}c\,\Delta t^3$	0	$+\frac{c^2}{6}\Delta x\,\Delta t^2$	0
$\left(\frac{1}{3}c^2\,\Delta t^2 - \frac{c\,\Delta t\,\Delta x}{4}\right)\frac{\partial^2}{\partial x^2}$ Eq. (4.10)								$\frac{1}{3}c^2\,\Delta t^2$ $-\frac{c\,\Delta t\,\Delta x}{4}$	$\frac{1}{3}c^3\,\Delta t^2$ $-\frac{c^2\,\Delta t\,\Delta x}{4}$	0	0	$\frac{1}{6}c^2\,\Delta t^3$ $-\frac{c\,\Delta t^2\,\Delta x}{8}$	0	$-\frac{1}{6}c^3\,\Delta t^2\,\Delta x$ $+\frac{c^2}{8}\Delta t\,\Delta x^2$
$\frac{1}{12}c\,\Delta t^3\frac{\partial^3}{\partial t^2\,\partial x}$ Eq. (4.10)											$\frac{1}{12}c\,\Delta t^3$	$\frac{c^2}{12}\Delta t^3$	0	0
$\left(\frac{1}{6}c\,\Delta x\,\Delta t^2 - \frac{1}{4}c^2\,\Delta t^3\right)\frac{\partial^3}{\partial t\,\partial x^2}$ Eq. (4.10)												$\frac{1}{6}c\,\Delta x\,\Delta t^2$ $-\frac{1}{4}c^2\,\Delta t^3$	$\frac{1}{6}c^2\Delta x\,\Delta t^2$ $-\frac{1}{4}c^3\,\Delta t^3$	0
$\left(\frac{c}{12}\Delta t\,\Delta x^2 - \frac{1}{3}c^2\,\Delta x\,\Delta t^2 + \frac{1}{4}c^3\,\Delta t^3\right)\frac{\partial^3}{\partial x^3}$ Eq. (4.10)													$\frac{c}{12}\Delta t\,\Delta x^2$ $-\frac{1}{3}c^2\,\Delta x\,\Delta t^2$ $+\frac{1}{4}c^3\,\Delta t^3$	$\frac{c^2}{12}\Delta t\,\Delta x^2$ $-\frac{1}{3}c^3\,\Delta x\,\Delta t^2$ $+\frac{1}{4}c^4\,\Delta t^3$
...														
Sum of coefficients	1	c	0	0	$\frac{c\,\Delta x}{2}(\nu - 1)$	0	0	0	$\frac{c\,\Delta x^2}{6}(2\nu^2 - 3\nu + 1)$	0	0	0	0	$\frac{c\,\Delta x^3}{24}(6\nu^3 - 12\nu^2 + 7\nu - 1)$

and adding the result to the first row, i.e., Eq. (4.10). This introduces the term $-(c\,\Delta t/2)u_{tx}$, which is eliminated by multiplying Eq. (4.10) by the operator

$$\frac{c\,\Delta t}{2}\frac{\partial}{\partial x}$$

and adding the result to the first two rows of the table. This procedure is continued until the desired time derivatives are eliminated. Each coefficient in the modified equation is then obtained by simply adding the coefficients in the corresponding column of the table. The algebra required to derive the modified equation can be programmed on a digital computer using an algebraic manipulation code.

The right-hand side of the modified equation [Eq. (4.15)] is the T.E., since it represents the difference between the original PDE and the finite-difference approximation to it. Consequently, the lowest order term on the right-hand side of the modified equation gives the order of the method. In the present case, the method is first-order accurate, since the lowest order term is $O[\Delta t, \Delta x]$. If $\nu = 1$, the right-hand side of the modified equation becomes zero, and the wave equation is solved exactly. In this case, the upstream differencing scheme reduces to

$$u_j^{n+1} = u_{j-1}^n$$

which is equivalent to solving the wave equation exactly using the method of characteristics. Finite-difference algorithms that exhibit this behavior are said to satisfy the *shift condition* (Kutler and Lomax, 1971).

The lowest order term of the T.E. in the present case contains the partial derivative u_{xx}, which makes this term similar to the viscous term in 1-D fluid flow equations. For example, the viscous term in the 1-D Navier-Stokes equation (see Chapter 5) may be written as

$$\frac{\partial}{\partial x}(\tau_{xx}) = \frac{4}{3}\mu u_{xx} \tag{4.16}$$

if a constant coefficient of viscosity is assumed. Thus, when $\nu \neq 1$, the upstream differencing scheme introduces an *artificial viscosity* into the solution. This is often called implicit artificial viscosity, as opposed to explicit artificial viscosity, which is purposely added to a difference scheme. Artificial viscosity tends to reduce all gradients in the solution whether physically correct or numerically induced. This effect, which is the direct result of even derivative terms in the T.E., is called *dissipation.*

Another quasi-physical effect of numerical schemes is called *dispersion.* This is the direct result of the odd derivative terms that appear in the T.E. As a result of dispersion, phase relations between various waves are distorted. The combined effect of dissipation and dispersion is sometimes referred to as *diffusion.* Diffusion tends to spread out sharp dividing lines that may appear in the computational region. Figure 4.1 illustrates the effects of dissipation and dispersion on the computation of a discontinuity. In general, if the lowest order

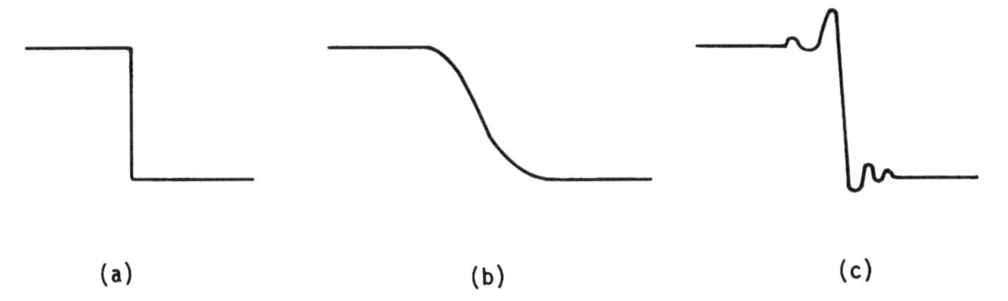

Figure 4.1 Effects of dissipation and dispersion. (a) Exact solution. (b) Numerical solution distorted primarily by dissipation errors (typical of first-order methods). (c) Numerical solution distorted primarily by dispersion errors (typical of second-order methods).

term in the T.E. contains an even derivative, the resulting solution will predominantly exhibit dissipative errors. On the other hand, if the leading term is an odd derivative, the resulting solution will predominantly exhibit dispersive errors.

In Chapter 3 we discussed a technique for finding the relative errors in both amplitude (dissipation) and phase (dispersion) from the amplification factor. At this point it seems natural to ask if the amplification factor is related to the modified equation. The answer is definitely yes! Warming and Hyett (1974) have developed a "heuristic" stability theory based on the even derivative terms in the modified equation and have determined the phase shift error by examining the odd derivative terms. However, the analysis of Warming and Hyett has been shown by Chang (1987) to be restricted to schemes involving only two time levels $(n, n+1)$. Before showing the correspondence between the modified equation and the amplification factor, let us first examine the amplification factor of the present upstream differencing scheme:

$$G = (1 - \nu + \nu \cos \beta) - i(\nu \sin \beta) \tag{4.17}$$

The modulus of this amplification factor,

$$|G| = \left[(1 - \nu + \nu \cos \beta)^2 + (-\nu \sin \beta)^2\right]^{1/2}$$

is plotted in Fig. 4.2 for several values of ν. It is clear from this plot that ν must be less than or equal to 1 if the von Neumann stability condition $|G| \leqslant 1$ is to be met.

The amplification factor, Eq. (4.17), can also be expressed in the exponential form for a complex number:

$$G = |G|e^{i\phi}$$

where ϕ is the phase angle given by

$$\phi = \tan^{-1}\left[\frac{\operatorname{Im}(G)}{\operatorname{Re}(G)}\right] = \tan^{-1}\left(\frac{-\nu \sin \beta}{1 - \nu + \nu \cos \beta}\right)$$

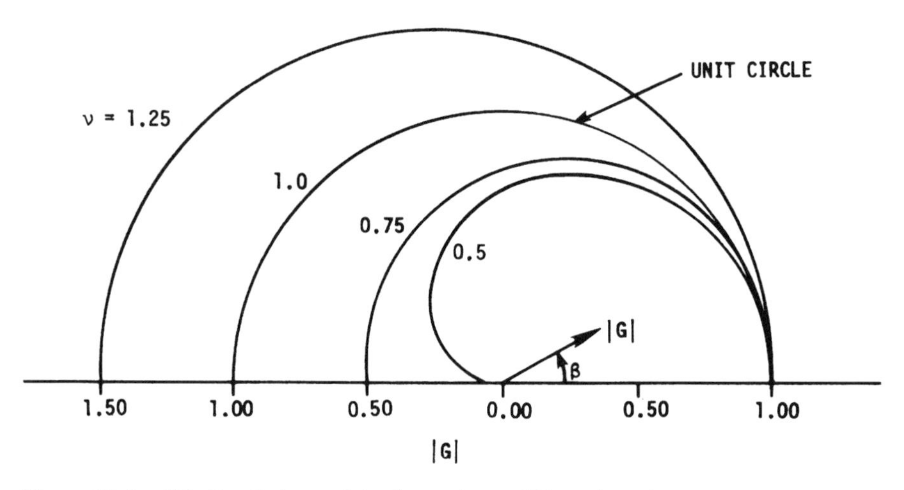

Figure 4.2 Amplification factor modulus for upstream differencing scheme.

The phase angle for the exact solution of the wave equation (ϕ_e) is determined in a similar manner once the amplification factor of the exact solution is known. In order to find the exact amplification factor we substitute the elemental solution

$$u = e^{\alpha t} e^{ik_m x}$$

into the wave equation and find that $\alpha = -ik_m c$, which gives

$$u = e^{ik_m(x-ct)}$$

The exact amplification factor is then

$$G_e = \frac{u(t+\Delta t)}{u(t)} = \frac{e^{ik_m[x-c(t+\Delta t)]}}{e^{ik_m(x-ct)}}$$

which reduces to

$$G_e = e^{-ik_m c\,\Delta t} = e^{i\phi_e}$$

where

$$\phi_e = -k_m c\,\Delta t = -\beta\nu$$

and

$$|G_e| = 1$$

Thus the total dissipation (amplitude) error that accrues from applying the upstream differencing method to the wave equation for N steps is given by

$$(1 - |G|^N)A_0$$

where A_0 is the initial amplitude of the wave. Likewise, the total dispersion (phase) error can be expressed as $N(\phi_e - \phi)$. The relative phase shift error after one time step is given by

$$\frac{\phi}{\phi_e} = \frac{\tan^{-1}[(-\nu \sin \beta)/(1 - \nu + \nu \cos \beta)]}{-\beta\nu} \tag{4.18}$$

and is plotted in Fig. 4.3 for several values of ν. For small wave numbers (i.e., small β) the relative phase error reduces to

$$\frac{\phi}{\phi_e} \cong 1 - \frac{1}{6}(2\nu^2 - 3\nu + 1)\beta^2 \tag{4.19}$$

If the relative phase error exceeds 1 for a given value of β, the corresponding Fourier component of the numerical solution has a wave speed greater than the exact solution, and this is a *leading phase error*. If the relative phase error is less than 1, the wave speed of the numerical solution is less than the exact wave speed, and this is a *lagging phase error*. The upstream differencing scheme has a leading phase error for $0.5 < \nu < 1$ and a lagging phase error for $\nu < 0.5$.

Example 4.1 Suppose the upstream differencing scheme is used to solve the wave equation ($c = 0.75$) with the initial condition

$$u(x,0) = \sin(6\pi x) \qquad 0 \leqslant x \leqslant 1$$

and periodic boundary conditions. Determine the amplitude and phase errors after 10 steps if $\Delta t = 0.02$ and $\Delta x = 0.02$.

Solution In this problem a unique value of β can be determined because the exact solution of the wave equation (for the present initial condition) is

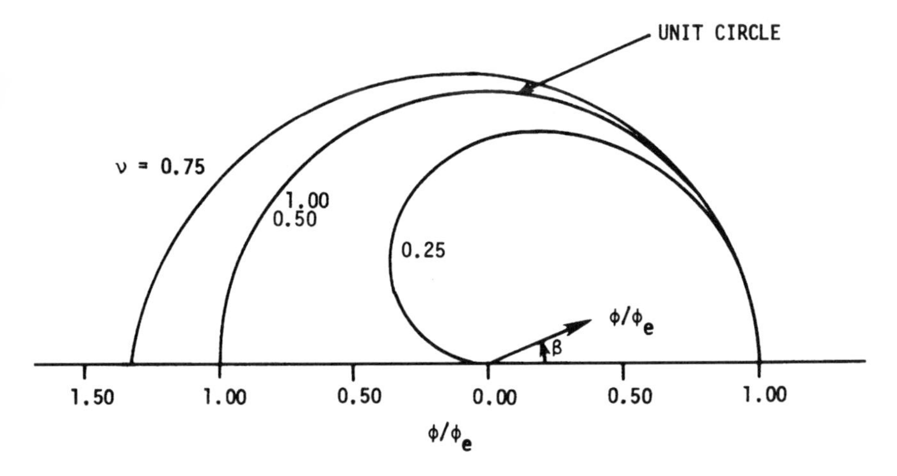

Figure 4.3 Relative phase error of upstream differencing scheme.

represented by a single term of a Fourier series. Since the amplification factor is also determined using a single term of a Fourier series that satisfies the wave equation, the frequency of the exact solution is identical to the frequency associated with the amplification factor, i.e., $f_m = k_m/2\pi$. Thus the wave number for the present problem is given by

$$k_m = \frac{2m\pi}{2L} = \frac{6\pi}{1} = 6\pi$$

and β can be calculated as

$$\beta = k_m \, \Delta x = (6\pi)(0.02) = 0.12\pi$$

Using the Courant number,

$$\nu = \frac{c\,\Delta t}{\Delta x} = \frac{(0.75)(0.02)}{(0.02)} = 0.75$$

the modulus of the amplification factor becomes

$$|G| = \left[(1 - \nu + \nu \cos \beta)^2 + (-\nu \sin \beta)^2\right]^{1/2} = 0.986745$$

and the resulting amplitude error after 10 steps is

$$(1 - |G|^N) A_0 = (1 - |G|^{10})(1) = 1 - 0.8751 = 0.1249$$

The phase angle (ϕ) after one step,

$$\phi = \tan^{-1}\left[\frac{-\nu \sin \beta}{1 - \nu + \nu \cos \beta}\right] = -0.28359$$

can be compared with the exact phase angle (ϕ_e) after one step,

$$\phi_e = -\beta\nu = -0.28274$$

to give the phase error after 10 steps:

$$10(\phi_e - \phi) = 0.0084465$$

Let us now compare the exact and numerical solutions after 10 steps where the time is

$$t = 10\,\Delta t = 0.2$$

The exact solution is given by

$$u(x, 0.2) = \sin\left[6\pi(x - 0.15)\right]$$

and the numerical solution that results after applying the upstream differencing scheme for 10 steps is

$$u(x, 0.2) = (0.8751) \sin\left[6\pi(x - 0.15) - 0.0084465\right]$$

In order to show the correspondence between the amplification factor and the modified equation, we write the modified equation [Eq. (4.15)] in the following form:

$$u_t + cu_x = \sum_{n=1}^{\infty} \left(C_{2n} \frac{\partial^{2n} u}{\partial x^{2n}} + C_{2n+1} \frac{\partial^{2n+1} u}{\partial x^{2n+1}} \right) \tag{4.20}$$

where C_{2n} and C_{2n+1} represent the coefficients of the even and odd spatial-derivative terms, respectively. Warming and Hyett (1974) have shown that a necessary condition for stability is

$$(-1)^{l-1} C_{2l} > 0 \tag{4.21}$$

where C_{2l} represents the coefficient of the lowest order even derivative term. This is analogous to the requirement that the coefficient of viscosity in viscous flow equations be greater than zero. In Eq. (4.15) the coefficient of the lowest order even derivative term is

$$C_2 = \frac{c\,\Delta x}{2}(1 - \nu) \tag{4.22}$$

and therefore the stability condition becomes

$$\frac{c\,\Delta x}{2}(1 - \nu) > 0 \tag{4.23}$$

or $\nu < 1$, which was obtained earlier from the amplification factor. It should be remembered that the "heuristic" stability analysis, i.e., Eq. (4.21), can only provide a necessary condition for stability. Thus, for some finite-difference algorithms, only partial information about the complete stability bound is obtained, and for others (such as algorithms for the heat equation) a more complete theory must be employed.

Warming and Hyett have also shown that the relative phase error for difference schemes applied to the wave equation is given by

$$\frac{\phi}{\phi_e} = 1 - \frac{1}{c} \sum_{n=1}^{\infty} (-1)^n (k_m)^{2n} C_{2n+1} \tag{4.24}$$

where $k_m = \beta/\Delta x$ is the wave number. For small wave numbers, we need only retain the lowest order term. For the upstream differencing scheme, we find that

$$\frac{\phi}{\phi_e} \cong 1 - \frac{1}{c}(-1)\left(\frac{\beta}{\Delta x}\right)^2 C_3 = 1 - \frac{1}{6}(2\nu^2 - 3\nu + 1)\beta^2 \tag{4.25a}$$

which is identical to Eq. (4.19). Thus we have demonstrated that the amplification factor and the modified equation are directly related.

The upstream method given by Eq. (4.7) may be written in a more general form to account for either positive or negative wave speeds. The method is normally written separately for these two cases as

$$u_j^{n+1} = u_j^n - c\frac{\Delta t}{\Delta x}(u_j^n - u_{j-1}^n) \qquad c > 0$$

$$u_j^{n+1} = u_j^n - c\frac{\Delta t}{\Delta x}(u_{j+1}^n - u_j^n) \qquad c < 0$$

However, if we make use of the following definitions,

$$c^+ = \tfrac{1}{2}(c + |c|)$$

$$c^- = \tfrac{1}{2}(c - |c|)$$

the upstream scheme may be written as the single expression

$$u_j^{n+1} = u_j^n - \frac{\Delta t}{\Delta x}\left[c^+(u_j^n - u_{j-1}^n) + c^-(u_{j+1}^n - u_j^n)\right]$$

Upon substituting for the values of c^+ and c^-, the final form becomes

$$u_j^{n+1} = u_j^n - c\frac{\Delta t}{2\,\Delta x}(u_{j+1}^n - u_{j-1}^n) + \frac{|c|\,\Delta t}{2\,\Delta x}(u_{j+1}^n - 2u_j^n + u_{j-1}^n) \quad (4.25b)$$

It is interesting to note that this form of the upstream scheme gives the impression that it is a centered method. We recognize the first difference term as a central-difference approximation and interpret the last term as an artificial viscosity term. The function of this last term is to add the appropriate dissipation to produce the upstream scheme when c is either positive or negative.

4.1.3 Lax Method

The Euler method, Eq. (4.6), can be made stable by replacing u_j^n with the averaged term $(u_{j+1}^n + u_{j-1}^n)/2$. The resulting algorithm is the well-known Lax method (Lax, 1954), which was presented earlier:

$$\frac{u_j^{n+1} - (u_{j+1}^n + u_{j-1}^n)/2}{\Delta t} + c\frac{u_{j+1}^n - u_{j-1}^n}{2\,\Delta x} = 0 \qquad (4.26)$$

This explicit one-step scheme is first-order accurate with T.E. of $O[\Delta t, (\Delta x)^2/\Delta t]$ and is stable if $|\nu| \leqslant 1$. The modified equation is given by

$$u_t + cu_x = \frac{c\,\Delta x}{2}\left(\frac{1}{\nu} - \nu\right)u_{xx} + \frac{c(\Delta x)^2}{3}(1 - \nu^2)u_{xxx} + \cdots \qquad (4.27)$$

Note that this method is not uniformly consistent, since $(\Delta x)^2/\Delta t$ may not approach zero in the limit as Δt and Δx go to zero. However, if ν is held constant as Δt and Δx approach zero, the method is consistent. The Lax method is known for its large dissipation error when $\nu \neq 1$. This large dissipation is readily apparent when we compare the coefficient of the u_{xx} term in Eq. (4.27) with the same coefficient in the modified equation of the upstream differencing scheme for various values of ν. The large dissipation can also be observed in the amplification factor

$$G = \cos\beta - i\nu\sin\beta \tag{4.28}$$

which is described in Section 3.6.1. The modulus of the amplification factor is plotted in Fig. 4.4(a). The relative phase error is given by

$$\frac{\phi}{\phi_e} = \frac{\tan^{-1}(-\nu\tan\beta)}{-\beta\nu}$$

which produces a leading phase error, as seen in Fig. 4.4(b).

4.1.4 Euler Implicit Method

The algorithms discussed previously for the wave equation have all been explicit. The following implicit scheme,

$$\frac{u_j^{n+1} - u_j^n}{\Delta t} + \frac{c}{2\,\Delta x}\left(u_{j+1}^{n+1} - u_{j-1}^{n+1}\right) = 0 \tag{4.29}$$

is first-order accurate with T.E. of $O[\Delta t, (\Delta x)^2]$ and, according to a Fourier stability analysis, is unconditionally stable for all time steps. However, a system of algebraic equations must be solved at each new time level. To illustrate this, let us rewrite Eq. (4.29) so that the unknowns at time level $(n+1)$ appear on the left-hand side of the equation and the known quantity u_j^n appears on the right-hand side. This gives

$$\frac{\nu}{2}u_{j+1}^{n+1} + (1)u_j^{n+1} - \frac{\nu}{2}u_{j-1}^{n+1} = u_j^n \tag{4.30}$$

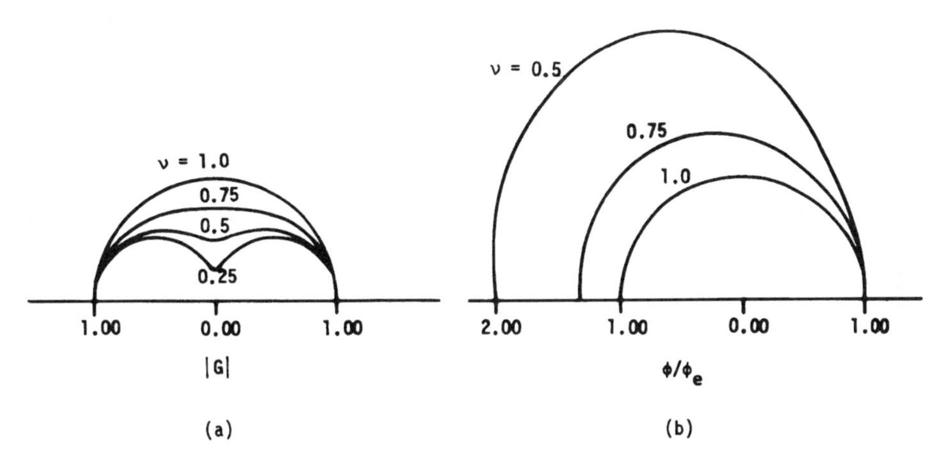

Figure 4.4 Lax method. (a) Amplification factor modulus. (b) Relative phase error.

or

$$a_j u_{j+1}^{n+1} + d_j u_j^{n+1} + b_j u_{j-1}^{n+1} = C_j \tag{4.31}$$

where $a_j = \nu/2$, $d_j = 1$, $b_j = -\nu/2$, and $C_j = u_j^n$. Consider the computational mesh shown in Fig. 4.5, which contains $M + 2$ grid points in the x direction and known initial conditions at $n = 0$. Along the left boundary, u_0^{n+1} has a fixed value of u_0. Along the right boundary, u_{M+1}^{n+1} can be computed as part of the solution using characteristic theory. For example, if $\nu = 1$, then $u_{M+1}^{n+1} = u_M^n$. Applying Eq. (4.31) to the grid shown in Fig. 4.5, we find that the following system of M linear algebraic equations must be solved at each $(n + 1)$ time level:

$$\underset{[A]}{\begin{bmatrix} d_1 & a_1 & 0 & \cdot & \cdot & \cdot & \cdot & \cdot & 0 \\ b_2 & d_2 & a_2 & & & & & & \cdot \\ 0 & b_3 & d_3 & a_3 & & & & & \cdot \\ \cdot & & \ddots & \ddots & \ddots & & & & \cdot \\ \cdot & & & & & & & & \cdot \\ \cdot & & & & & & & & 0 \\ \cdot & & & & & & b_{M-1} & d_{M-1} & a_{M-1} \\ 0 & \cdot & \cdot & \cdot & \cdot & & 0 & b_M & d_M \end{bmatrix}} \underset{[u]}{\begin{bmatrix} u_1^{n+1} \\ u_2^{n+1} \\ \cdot \\ \cdot \\ \cdot \\ \cdot \\ u_{M-1}^{n+1} \\ u_M^{n+1} \end{bmatrix}} = \underset{[C]}{\begin{bmatrix} C_1 \\ C_2 \\ \cdot \\ \cdot \\ \cdot \\ \cdot \\ C_{M-1} \\ C_M \end{bmatrix}} \tag{4.32}$$

In Eq. (4.32), C_1 and C_M are given by

$$\begin{aligned} C_1 &= u_1^n - bu_0^{n+1} \\ C_M &= u_M^n - au_{M+1}^{n+1} \end{aligned} \tag{4.33}$$

where u_0^{n+1} and u_{M+1}^{n+1} are the known boundary conditions.

Matrix $[A]$ in Eq. (4.32) is a tridiagonal matrix. A technique for rapidly solving a tridiagonal system of linear algebraic equations is due to Thomas (1949) and is called the Thomas algorithm. In this algorithm, the system of equations is first put into upper triangular form by replacing the diagonal

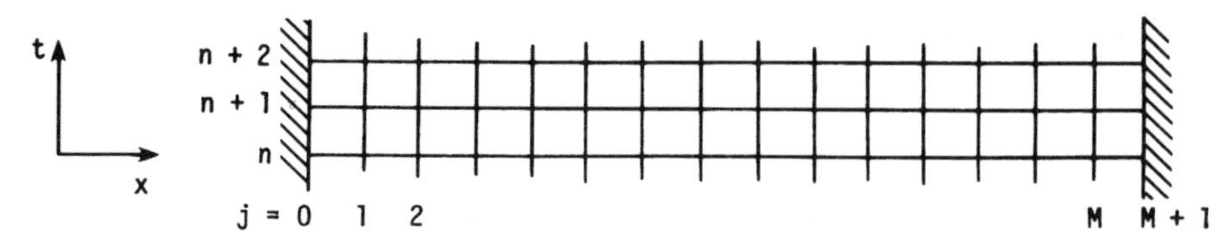

Figure 4.5 Computational mesh.

elements d_i with

$$d_i - \frac{b_i}{d_{i-1}} a_{i-1} \qquad i = 2, 3, \ldots, M$$

and the C_i with

$$C_i - \frac{b_i}{d_{i-1}} C_{i-1} \qquad i = 2, 3, \ldots, M$$

The unknowns are then computed using back substitution starting with

$$u_M^{n+1} = \frac{C_M}{d_M}$$

and continuing with

$$u_j^{n+1} = \frac{C_j - a_j u_{j+1}^{n+1}}{d_j} \qquad j = M-1, M-2, \ldots, 1$$

Further details of the Thomas algorithm are given in Section 4.3.3.

In general, implicit schemes require more computation time per time step but, of course, permit a larger time step, since they are usually unconditionally stable. However, the solution may become meaningless if too large a time step is taken. This is due to the fact that a large time step produces large T.E.s. The modified equation for the Euler implicit scheme is

$$u_t + cu_x = \left(\tfrac{1}{2}c^2 \Delta t\right) u_{xx} - \left[\tfrac{1}{6}c(\Delta x)^2 + \tfrac{1}{3}c^3(\Delta t)^2\right] u_{xxx} + \cdots \tag{4.34}$$

which does not satisfy the shift condition. The amplification factor

$$G = \frac{1 - i\nu \sin \beta}{1 + \nu^2 \sin^2 \beta} \tag{4.35}$$

and the relative phase error

$$\frac{\phi}{\phi_e} = \frac{\tan^{-1}(-\nu \sin \beta)}{-\beta\nu} \tag{4.36}$$

are plotted in Fig. 4.6. The Euler implicit scheme is very dissipative for intermediate wave numbers and has a large lagging phase error for high wave numbers.

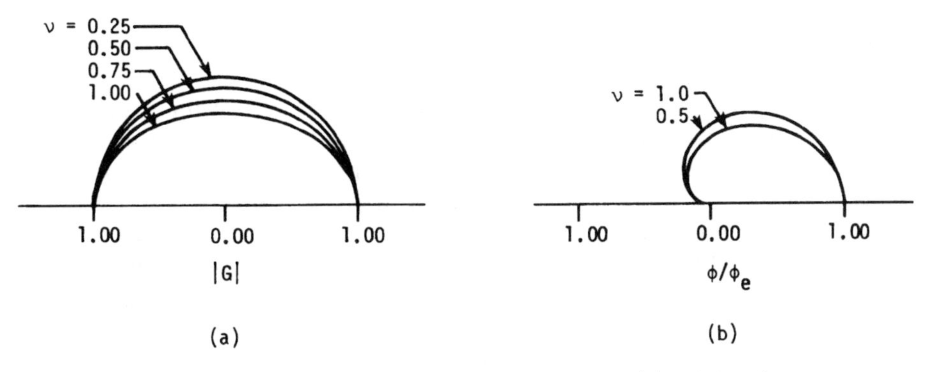

Figure 4.6 Euler implicit method. (a) Amplification factor modulus. (b) Relative phase error.

4.1.5 Leap Frog Method

The numerical schemes presented so far in this chapter for solving the linear wave equation are all first-order accurate. In most cases, first-order schemes are not used to solve PDEs because of their inherent inaccuracy. The leap frog method is the simplest second-order accurate method. When applied to the first-order wave equation, this explicit one-step three-time-level scheme becomes

$$\frac{u_j^{n+1} - u_j^{n-1}}{2\,\Delta t} + c\frac{u_{j+1}^n - u_{j-1}^n}{2\,\Delta x} = 0 \tag{4.37}$$

The leap frog method is referred to as a three-time-level scheme, since u must be known at time levels n and $n-1$ in order to find u at time level $n+1$. This method has a T.E. of $O[(\Delta t)^2, (\Delta x)^2]$ and is stable whenever $|\nu| \leqslant 1$. The modified equation is given by

$$u_t + cu_x = \frac{c(\Delta x)^2}{6}(\nu^2 - 1)u_{xxx} - \frac{c(\Delta x)^4}{120}(9\nu^4 - 10\nu^2 + 1)u_{xxxxx} + \cdots \tag{4.38}$$

The leading term in the T.E. contains the odd derivative u_{xxx}, and hence the solution will predominantly exhibit dispersive errors. This is typical of second-order accurate methods. In this case, however, there are no even derivative terms in the modified equation, so that the solution will not contain any dissipation error. As a consequence, the leap frog algorithm is neutrally stable, and errors caused by improper boundary conditions or computer round-off will not be damped (assuming periodic boundary conditions and $|\nu| \leqslant 1$). The amplification factor

$$G = \pm(1 - \nu^2 \sin^2 \beta)^{1/2} - i\nu \sin \beta \tag{4.39}$$

and the relative phase error

$$\frac{\phi}{\phi_e} = \frac{\tan^{-1}\left[-\nu \sin \beta / \pm(1 - \nu^2 \sin^2 \beta)^{1/2}\right]}{-\beta\nu} \tag{4.40}$$

are plotted in Fig. 4.7.

The leap frog method, while being second-order accurate with no dissipation error, does have its disadvantages. First, initial conditions must be specified at two-time levels. This difficulty can be circumvented by using a two-time-level scheme for the first time step. A second disadvantage is due to the "leap frog" nature of the differencing (i.e., u_j^{n+1} does not depend on u_j^n), so that two independent solutions develop as the calculation proceeds. And finally, the leap frog method may require additional computer storage because it is a three-time-level scheme. The required computer storage is reduced considerably if a simple overwriting procedure is employed, whereby u_j^{n-1} is overwritten by u_j^{n+1}.

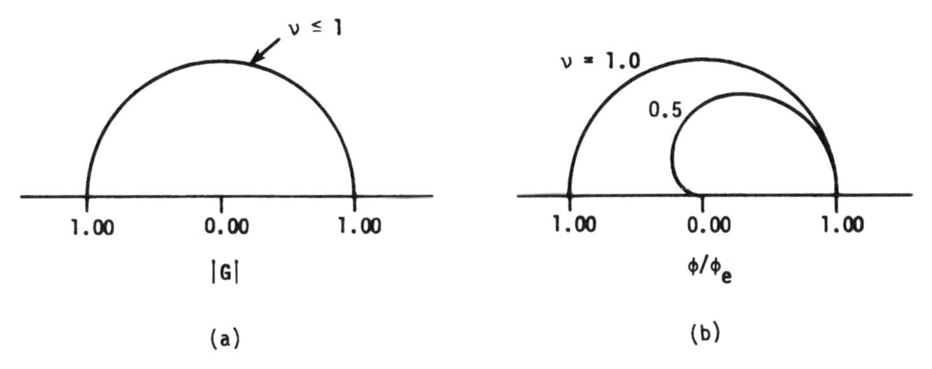

Figure 4.7 Leap frog method. (a) Amplification factor modulus. (b) Relative phase error.

4.1.6 Lax-Wendroff Method

The Lax-Wendroff finite-difference scheme (Lax and Wendroff, 1960) can be derived from a Taylor-series expansion in the following manner:

$$u_j^{n+1} = u_j^n + \Delta t u_t + \tfrac{1}{2}(\Delta t)^2 u_{tt} + O[(\Delta t)^3] \tag{4.41}$$

Using the wave equations

$$\begin{aligned} u_t &= -cu_x \\ u_{tt} &= c^2 u_{xx} \end{aligned} \tag{4.42}$$

Equation (4.41) may be written as

$$u_j^{n+1} = u_j^n - c\,\Delta t u_x + \tfrac{1}{2}c^2(\Delta t)^2 u_{xx} + O[(\Delta t)^3] \tag{4.43}$$

And finally, if u_x and u_{xx} are replaced by second-order accurate central-difference expressions, the well-known Lax-Wendroff scheme is obtained:

$$u_j^{n+1} = u_j^n - \frac{c\,\Delta t}{2\,\Delta x}(u_{j+1}^n - u_{j-1}^n) + \frac{c^2(\Delta t)^2}{2(\Delta x)^2}(u_{j+1}^n - 2u_j^n + u_{j-1}^n) \tag{4.44}$$

This explicit one-step scheme is second-order accurate with a T.E. of $O[(\Delta x)^2, (\Delta t)^2]$ and is stable whenever $|\nu| \leqslant 1$. The modified equation for this method is

$$u_t + cu_x = -c\frac{(\Delta x)^2}{6}(1 - \nu^2)u_{xxx} - \frac{c(\Delta x)^3}{8}\nu(1 - \nu^2)u_{xxxx} + \cdots \tag{4.45}$$

The amplification factor

$$G = 1 - \nu^2(1 - \cos\beta) - i\nu\sin\beta \tag{4.46}$$

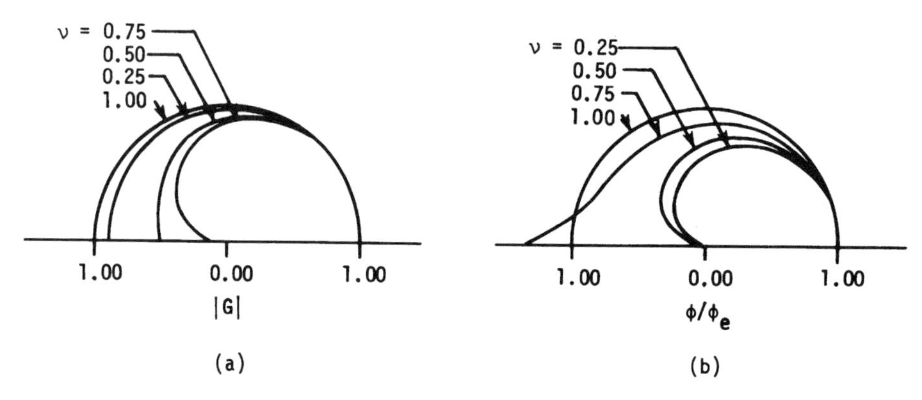

Figure 4.8 Lax-Wendroff method. (a) Amplification factor modulus. (b) Relative phase error.

and the relative phase error

$$\frac{\phi}{\phi_e} = \frac{\tan^{-1}\{-\nu \sin \beta/[1 - \nu^2(1 - \cos \beta)]\}}{-\beta\nu} \tag{4.47}$$

are plotted in Fig. 4.8. The Lax-Wendroff scheme has a predominantly lagging phase error except for large wave numbers with $\sqrt{0.5} < \nu < 1$.

4.1.7 Two-Step Lax-Wendroff Method

For nonlinear equations such as the inviscid flow equations, a two-step variation of the original Lax-Wendroff method can be used. When applied to the wave equation, this explicit two-step three-time-level method becomes

Step 1:
$$\frac{u_{j+1/2}^{n+1/2} - (u_{j+1}^n + u_j^n)/2}{\Delta t/2} + c\frac{u_{j+1}^n - u_j^n}{\Delta x} = 0 \tag{4.48}$$

Step 2:
$$\frac{u_j^{n+1} - u_j^n}{\Delta t} + c\frac{u_{j+1/2}^{n+1/2} - u_{j-1/2}^{n+1/2}}{\Delta x} = 0 \tag{4.49}$$

This scheme is second-order accurate with a T.E. of $O[(\Delta x)^2, (\Delta t)^2]$ and is stable whenever $|\nu| \leqslant 1$. Step 1 is the Lax method applied at the midpoint $j + \frac{1}{2}$ for a half time step, and step 2 is the leap frog method for the remaining half time step. When applied to the linear wave equation, the two-step Lax-Wendroff scheme is equivalent to the original Lax-Wendroff scheme. This can be readily shown by substituting Eq. (4.48) into Eq. (4.49). Since the two schemes are equivalent, it follows that the modified equation and the amplification factor are the same for the two methods.

4.1.8 MacCormack Method

The MacCormack method (MacCormack, 1969) is a widely used scheme for solving fluid flow equations. It is a variation of the two-step Lax-Wendroff scheme that removes the necessity of computing unknowns at the grid points $j+\frac{1}{2}$ and $j-\frac{1}{2}$. Because of this feature, the MacCormack method is particularly useful when solving nonlinear PDEs, as is shown in Section 4.4.3. When applied to the linear wave equation, this explicit, predictor-corrector method becomes

Predictor: $$u_j^{\overline{n+1}} = u_j^n - c\frac{\Delta t}{\Delta x}(u_{j+1}^n - u_j^n) \tag{4.50}$$

Corrector: $$u_j^{n+1} = \frac{1}{2}\left[u_j^n + u_j^{\overline{n+1}} - c\frac{\Delta t}{\Delta x}\left(u_j^{\overline{n+1}} - u_{j-1}^{\overline{n+1}}\right)\right] \tag{4.51}$$

The term $u_j^{\overline{n+1}}$ is a temporary "predicted" value of u at the time level $n+1$. The corrector equation provides the final value of u at the time level $n+1$. Note that in the predictor equation a forward difference is used for $\partial u/\partial x$, while in the corrector equation a backward difference is used. This differencing can be reversed, and in some problems it is advantageous to do so. This is particularly true for problems involving moving discontinuities. For the present linear wave equation, the MacCormack scheme is equivalent to the original Lax-Wendroff scheme. Hence the truncation error, stability limit, modified equation, and amplification factor are identical with those of the Lax-Wendroff scheme.

4.1.9 Second-Order Upwind Method

The second-order upwind method (Warming and Beam, 1975) is a variation of the MacCormack method, which uses backward (upwind) differences in both the predictor and corrector steps for $c > 0$:

Predictor: $$u_j^{\overline{n+1}} = u_j^n - \frac{c\,\Delta t}{\Delta x}(u_j^n - u_{j-1}^n) \tag{4.52}$$

Corrector:

$$u_j^{n+1} = \frac{1}{2}\left[u_j^n + u_j^{\overline{n+1}} - \frac{c\,\Delta t}{\Delta x}\left(u_j^{\overline{n+1}} - u_{j-1}^{\overline{n+1}}\right) - \frac{c\,\Delta t}{\Delta x}(u_j^n - 2u_{j-1}^n + u_{j-2}^n)\right] \tag{4.53}$$

The addition of the second backward difference in Eq. (4.53) makes this scheme second-order accurate with T.E. of $O[(\Delta t)^2, (\Delta t)(\Delta x), (\Delta x)^2]$. If Eq. (4.52) is substituted into Eq. (4.53), the following one-step algorithm is obtained:

$$u_j^{n+1} = u_j^n - \nu(u_j^n - u_{j-1}^n) + \tfrac{1}{2}\nu(\nu-1)(u_j^n - 2u_{j-1}^n + u_{j-2}^n) \tag{4.54}$$

The modified equation for this scheme is

$$u_t + cu_x = \frac{c(\Delta x)^2}{6}(1-\nu)(2-\nu)u_{xxx} - \frac{(\Delta x)^4}{8\,\Delta t}\nu(1-\nu)^2(2-\nu)u_{xxxx} + \cdots \tag{4.55}$$

The second-order upwind method satisfies the shift condition for both $\nu = 1$ and $\nu = 2$. The amplification factor is

$$G = 1 - 2\nu\left(\nu + 2(1-\nu)\sin^2\frac{\beta}{2}\right)\sin^2\frac{\beta}{2} - i\nu\sin\beta\left(1 + 2(1-\nu)\sin^2\frac{\beta}{2}\right) \tag{4.56}$$

and the resulting stability condition becomes $0 \leqslant \nu \leqslant 2$. The modulus of the amplification factor and the relative phase error are plotted in Fig. 4.9. The second-order upwind method has a predominantly leading phase error for $0 < \nu < 1$ and a predominantly lagging phase error for $1 < \nu < 2$. We observe that the second-order upwind method and the Lax-Wendroff method have opposite phase errors for $0 < \nu < 1$. This suggests that a considerable reduction in dispersive error would occur if a linear combination of the two methods were used. Fromm's method of zero-average phase error (Fromm, 1968) is based on this observation.

4.1.10 Time-Centered Implicit Method (Trapezoidal Differencing Method)

A second-order accurate implicit scheme can be obtained if the two Taylor-series expansions

$$u_j^{n+1} = u_j^n + \Delta t(u_t)_j^n + \frac{(\Delta t)^2}{2}(u_{tt})_j^n + \frac{(\Delta t)^3}{6}(u_{ttt})_j^n + \cdots$$

$$u_j^n = u_j^{n+1} - \Delta t(u_t)_j^{n+1} + \frac{(\Delta t)^2}{2}(u_{tt})_j^{n+1} - \frac{(\Delta t)^3}{6}(u_{ttt})_j^{n+1} + \cdots \tag{4.57}$$

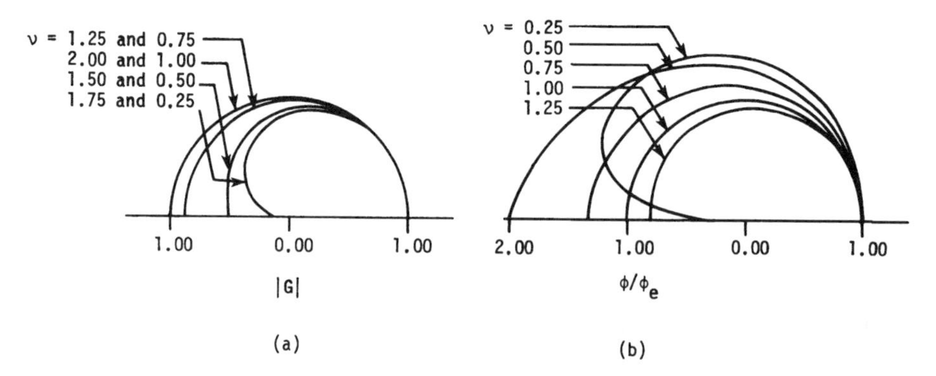

Figure 4.9 Second-order upwind method. (a) Amplification factor modulus. (b) Relative phase error.

are subtracted and $(u_{tt})_j^{n+1}$ is replaced with

$$(u_{tt})_j^{n+1} = (u_{tt})_j^n + \Delta t (u_{ttt})_j^n + \cdots$$

The resulting expression becomes

$$u_j^{n+1} = u_j^n + \frac{\Delta t}{2}\left[(u_t)^n + (u_t)^{n+1}\right]_j + O[(\Delta t)^3] \tag{4.58}$$

The time differencing in this equation is known as trapezoidal differencing or Crank-Nicolson differencing. Upon substituting the linear wave equation $u_t = -cu_x$, we obtain

$$u_j^{n+1} = u_j^n - \frac{c\,\Delta t}{2}\left[(u_x)^n + (u_x)^{n+1}\right]_j + O[(\Delta t)^3] \tag{4.59}$$

And finally, if the u_x terms are replaced by second-order central differences, the time-centered implicit method results:

$$u_j^{n+1} = u_j^n - \frac{\nu}{4}\left(u_{j+1}^{n+1} + u_{j+1}^n - u_{j-1}^{n+1} - u_{j-1}^n\right) \tag{4.60}$$

This method has second-order accuracy with T.E. of $O[(\Delta x)^2, (\Delta t)^2]$ and is unconditionally stable for all time steps. However, a tridiagonal matrix must be solved at each new time level. The modified equation for this scheme is

$$u_t + cu_x = -\left[\frac{c^3(\Delta t)^2}{12} + \frac{c(\Delta x)^2}{6}\right]u_{xxx} - \left[\frac{c(\Delta x)^4}{120} + \frac{c^3(\Delta t)^2(\Delta x)^2}{24} + \frac{c^4(\Delta t)^4}{80}\right]u_{xxxxx} + \cdots \tag{4.61}$$

Note that the modified equation contains no even derivative terms, so that the scheme has no implicit artificial viscosity. When this scheme is applied to the nonlinear fluid dynamic equations, it often becomes necessary to add some explicit artificial viscosity to prevent the solution from "blowing up." The addition of explicit artificial viscosity (i.e., "smoothing" term) to this scheme will be discussed in Section 4.4.7. The modulus of the amplification factor,

$$G = \frac{1 - (i\nu/2)\sin\beta}{1 + (i\nu/2)\sin\beta} \tag{4.62}$$

and the relative phase error are plotted in Fig. 4.10.

The time-centered implicit method can be made fourth-order accurate in space if the difference approximation given by Eq. (3.31) is used for u_x:

$$(u_x)_j = \frac{1}{2\,\Delta x}\frac{\bar{\delta}_x}{1 + \delta_x^2/6}u_j + O[(\Delta x)^4] \tag{4.63}$$

The modified equation and phase error diagram for the resulting scheme can be found in the work by Beam and Warming (1976).

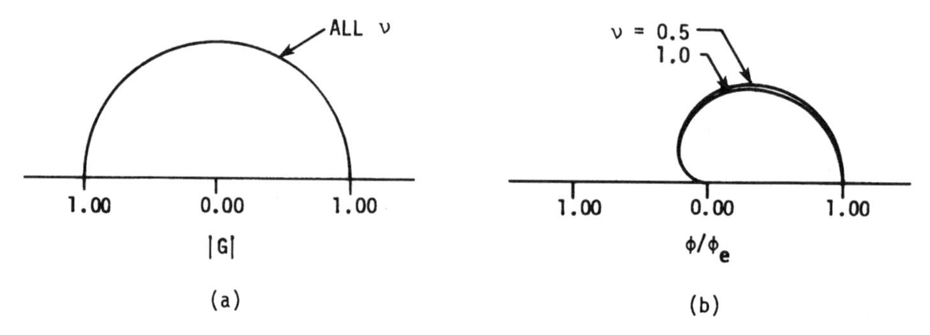

Figure 4.10 Time-centered implicit method. (a) Amplification factor modulus. (b) Relative phase error.

4.1.11 Rusanov (Burstein-Mirin) Method

The methods presented thus far for solving the wave equation have either been first-order or second-order accurate. Only a small number of third-order methods have appeared in the literature. Rusanov (1970) and Burstein and Mirin (1970) simultaneously developed the following explicit three-step method:

Step 1:
$$u_{j+1/2}^{(1)} = \tfrac{1}{2}(u_{j+1}^n + u_j^n) - \tfrac{1}{3}\nu(u_{j+1}^n - u_j^n)$$

Step 2:
$$u_j^{(2)} = u_j^n - \tfrac{2}{3}\nu\left(u_{j+1/2}^{(1)} - u_{j-1/2}^{(1)}\right)$$

Step 3:
$$\begin{aligned} u_j^{n+1} = u_j^n &- \frac{1}{24}\nu(-2u_{j+2}^n + 7u_{j+1}^n - 7u_{j-1}^n + 2u_{j-2}^n) \\ &- \frac{3}{8}\nu\left(u_{j+1}^{(2)} - u_{j-1}^{(2)}\right) \\ &- \frac{\omega}{24}(u_{j+2}^n - 4u_{j+1}^n + 6u_j^n - 4u_{j-1}^n + u_{j-2}^n) \end{aligned} \tag{4.64}$$

Step 3 contains the fourth-order difference term

$$\delta_x^4 u_j^n = u_{j+2}^n - 4u_{j+1}^n + 6u_j^n - 4u_{j-1}^n + u_{j-2}^n$$

which is multiplied by a free parameter ω. This term has been added to make the scheme stable. The need for this term is apparent when we examine the stability requirements for the scheme:

$$|\nu| \leqslant 1$$
$$4\nu^2 - \nu^4 \leqslant \omega \leqslant 3 \tag{4.65}$$

If the fourth-order difference term were not present (i.e., $\omega = 0$), we could not satisfy Eq. (4.65) for $0 < \nu \leqslant 1$. The modified equation for this method is

$$\begin{aligned} u_t + cu_x = &-\frac{c(\Delta x)^3}{24}\left(\frac{\omega}{\nu} - 4\nu + \nu^3\right)u_{xxxx} \\ &+ \frac{c(\Delta x)^4}{120}(-5\omega + 4 + 15\nu^2 - 4\nu^4)u_{xxxxx} + \cdots \end{aligned} \tag{4.66}$$

In order to reduce the dissipation of this scheme, we can make the coefficient of the fourth derivative equal to zero by letting

$$\omega = 4\nu^2 - \nu^4 \tag{4.67}$$

In a like manner, we can reduce the dispersive error by setting the coefficient of the fifth derivative to zero, which gives

$$\omega = \frac{(4\nu^2 + 1)(4 - \nu^2)}{5} \tag{4.68}$$

The amplification factor for this method is

$$G = 1 - \frac{\nu^2}{2}\sin^2\beta - \frac{2\omega}{3}\sin^4\frac{\beta}{2} - i\nu\sin\beta\left[1 + \frac{2}{3}(1 - \nu^2)\sin^2\frac{\beta}{2}\right] \tag{4.69}$$

The modulus of the amplification factor and the relative phase error are plotted in Fig. 4.11. This figure shows that the Rusanov method has a leading or a lagging phase error, depending on the value of the free parameter ω.

4.1.12 Warming-Kutler-Lomax Method

Warming et al. (1973) developed a third-order method that uses MacCormack's method for the first two steps and has the same third step as the Rusanov method. This so-called WKL method is given by

Step 1: $$u_j^{(1)} = u_j^n - \tfrac{2}{3}\nu(u_{j+1}^n - u_j^n)$$

Step 2: $$u_j^{(2)} = \tfrac{1}{2}\left[u_j^n + u_j^{(1)} - \tfrac{2}{3}\nu\left(u_j^{(1)} - u_{j-1}^{(1)}\right)\right]$$

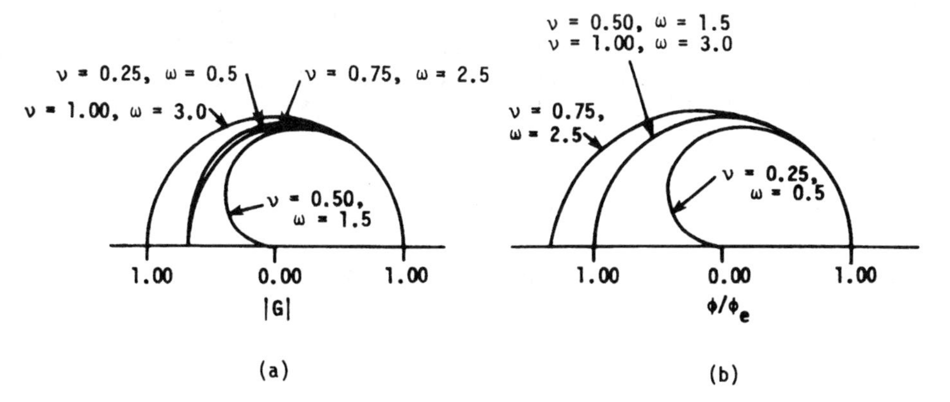

Figure 4.11 Rusanov method. (a) Amplification factor modulus. (b) Relative phase error.

Step 3:
$$u_j^{n+1} = u_j^n - \frac{1}{24}\nu(-2u_{j+2}^n + 7u_{j+1}^n - 7u_{j-1}^n + 2u_{j-2}^n) \tag{4.70a}$$
$$- \frac{3}{8}\nu\left(u_{j+1}^{(2)} - u_{j-1}^{(2)}\right)$$
$$- \frac{\omega}{24}(u_{j+2}^n - 4u_{j+1}^n + 6u_j^n - 4u_{j-1}^n + u_{j-2}^n)$$

This method has the same stability bounds as the Rusanov method. In addition, the modified equation is identical to Eq. (4.66) for the present linear wave equation. The WKL method has the same advantage over the Rusanov method that the MacCormack method has over the two-step Lax-Wendroff method.

4.1.13 Runge-Kutta Methods

Runge-Kutta methods are frequently employed to solve ordinary differential equations (ODEs). They can also be applied to solve PDEs (Lomax et al., 1970; and Jameson et al., 1981, 1983). In fact, several of the methods described previously in this section can be derived using Runge-Kutta methodology. The first step in this process is to convert the PDE into a "pseudo-ODE." This is accomplished by separating out a partial derivative with respect to a single independent variable in the marching direction and placing the remaining partial derivatives into a term that is a function of the dependent variable. For example, the linear wave equation can be written as

$$u_t = R(u) \tag{4.70b}$$

where $R(u) = -cu_x$. This pseudo-ODE is a time-continuous equation, and any integration scheme applicable to ODEs, including Runge-Kutta methods, may be used. Once the time differencing is completed, the partial derivatives contained in $R(u)$ can be differenced using appropriate spatial differences. To illustrate this approach, let us apply the second-order Runge-Kutta method, also referred to as the improved Euler's method (Carnahan et al., 1969), to Eq. (4.70b), which gives

Step 1:
$$u^{(1)} = u^n + \Delta t R^n$$

Step 2:
$$u^{n+1} = u^n + \frac{\Delta t}{2}(R^n + R^{(1)})$$

where

$$R^n \equiv R(u^n) = -cu_x^n$$

The term $R^{(1)}$ in step 2 can be evaluated by making use of step 1 in the following manner:

$$\begin{aligned} R^{(1)} &= -cu_x^{(1)} \\ &= -c(u_x^n + \Delta t R_x^n) \\ &= -cu_x^n + c^2 \Delta t u_{xx}^n \end{aligned}$$

Substituting this expression for $R^{(1)}$ into step 2 yields

$$u^{n+1} = u^n + \frac{\Delta t}{2}(-2cu_x^n + c^2 \Delta t u_{xx}^n)$$

If second-order accurate central differences are then used to approximate the spatial derivatives, the resulting scheme becomes

$$u_j^{n+1} = u_j^n - \frac{c\,\Delta t}{2\,\Delta x}(u_{j+1}^n - u_{j-1}^n) + \frac{c^2(\Delta t)^2}{2(\Delta x)^2}(u_{j+1}^n - 2u_j^n + u_{j-1}^n)$$

which is the second-order accurate Lax-Wendroff scheme, Eq. (4.44).

Procedures and equations for obtaining nth-order Runge-Kutta methods can be found in the works by Carnahan et al. (1969), Luther (1966), and Yu et al. (1992). A fourth-order Runge-Kutta method, attributed to Kutta, is given by

Step 1: $$u^{(1)} = u^n + \frac{\Delta t}{2} R^n$$

Step 2: $$u^{(2)} = u^n + \frac{\Delta t}{2} R^{(1)}$$

Step 3: $$u^{(3)} = u^n + \Delta t R^{(2)}$$

Step 4: $$u^{n+1} = u^n + \frac{\Delta t}{6}(R^n + 2R^{(1)} + 2R^{(2)} + R^{(3)})$$

where $R^{(\,)} = -cu_x^{(\,)}$ for the linear wave equation. If second-order accurate spatial differences are inserted into this algorithm, the resulting scheme will have a T.E. of $O[(\Delta t)^4, (\Delta x)^2]$. In order to obtain higher-order spatial accuracy, it is convenient to employ compact difference schemes (Yu et al., 1992) with the Runge-Kutta time stepping.

4.1.14 Additional Comments

The improved accuracy of higher-order methods is at the expense of added computer time and additional complexity. These factors must be considered carefully when choosing a scheme to solve a PDE. In general, second-order accurate methods provide enough accuracy for most practical problems.

For the 1-D linear wave equation, the second-order accurate explicit schemes such as the Lax-Wendroff scheme give excellent results with a minimum of computational effort. An implicit scheme may not be the optimum choice in this case because the solution is unsteady and intermediate results are typically desired at relatively small time intervals.

4.2 HEAT EQUATION

The 1-D heat equation (diffusion equation),

$$\frac{\partial u}{\partial t} = \alpha \frac{\partial^2 u}{\partial x^2} \tag{4.71}$$

is a parabolic PDE. In its present form, it is the governing equation for heat conduction or diffusion in a 1-D isotropic medium. It can be used to "model" in a rudimentary fashion the parabolic boundary-layer equations. The exact solution of the heat equation for the initial condition

$$u(x,0) = f(x)$$

and boundary conditions

$$u(0,t) = u(1,t) = 0$$

is

$$u(x,t) = \sum_{n=1}^{\infty} A_n e^{-\alpha k^2 t} \sin(kx) \tag{4.72}$$

where

$$A_n = 2\int_0^1 f(x) \sin(kx)\, dx$$

and $k = n\pi$. Let us now examine some of the more important finite-difference algorithms that can be used to solve the heat equation.

4.2.1 Simple Explicit Method

The following explicit one-step method,

$$\frac{u_j^{n+1} - u_j^n}{\Delta t} = \alpha \frac{u_{j+1}^n - 2u_j^n + u_{j-1}^n}{(\Delta x)^2} \tag{4.73}$$

is first-order accurate with T.E. of $O[\Delta t, (\Delta t)^2]$. At steady-state the accuracy is $O[(\Delta x)^2]$. As we have shown earlier, this scheme is stable whenever

$$0 \leqslant r \leqslant \tfrac{1}{2} \tag{4.74}$$

where

$$r = \frac{\alpha\, \Delta t}{(\Delta x)^2} \tag{4.75}$$

The modified equation is given by

$$u_t - \alpha u_{xx} = \left[-\frac{1}{2}\alpha^2 \Delta t + \frac{\alpha(\Delta x)^2}{12} \right] u_{xxxx} + \left[\frac{1}{3}\alpha^3(\Delta t)^2 - \frac{1}{12}\alpha^2 \Delta t(\Delta x)^2 + \frac{1}{360}\alpha(\Delta x)^4 \right] u_{xxxxxx} + \cdots \tag{4.76}$$

We note that if $r = \frac{1}{6}$, the T.E. becomes of $O[(\Delta t)^2, (\Delta x)^4]$. It is also interesting to note that no odd derivative terms appear in the T.E. As a consequence, this scheme, as well as almost all other schemes for the heat equation, has no dispersive error. This fact can also be ascertained by examining the amplification factor for this scheme:

$$G = 1 + 2r(\cos \beta - 1) \tag{4.77}$$

which has no imaginary part and hence no phase shift. The amplification factor is plotted in Fig. 4.12 for two values of r and is compared with the exact amplification factor of the solution. The exact amplification (decay) factor is obtained by substituting the elemental solution

$$u = e^{-\alpha k_m^2 t} e^{i k_m x}$$

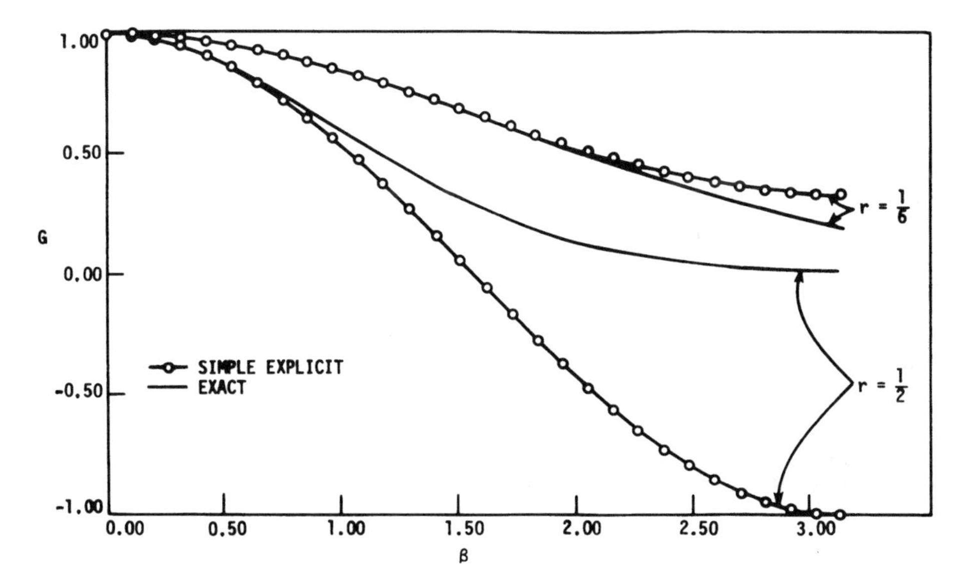

Figure 4.12 Amplification factor for simple explicit method.

into

$$G_e = \frac{u(t + \Delta t)}{u(t)}$$

which gives

$$G_e = e^{-\alpha k_m^2 \Delta t} \tag{4.78}$$

or

$$G_e = e^{-r\beta^2} \tag{4.79}$$

where $\beta = k_m \Delta x$. Hence the amplitude of the exact solution decreases by the factor $e^{-r\beta^2}$ during one time step, assuming no boundary condition influence.

In Fig. 4.12, we observe that the simple explicit method is highly dissipative for large values of β when $r = \frac{1}{2}$. As expected, the amplification factor is in closer agreement with the exact decay factor when $r = \frac{1}{6}$.

The present simple explicit scheme marches the solution outward from the initial data line in much the same manner as the explicit schemes of the previous section. This is illustrated in Fig. 4.13. In this figure we see that the unknown u can be calculated at point P without any knowledge of the boundary conditions along AB and CD. We know, however, that point P should depend on the boundary conditions along AB and CD, since the parabolic heat equation has the characteristic $t =$ const. From this we conclude that the present explicit scheme (with a finite Δt) does not properly model the physical behavior of the parabolic PDE. It would appear that an implicit method would be the more appropriate method for solving a parabolic PDE, since an implicit method normally assimilates information from all grid points located on or below the characteristic $t =$ const. On the other hand, explicit schemes seem to provide a

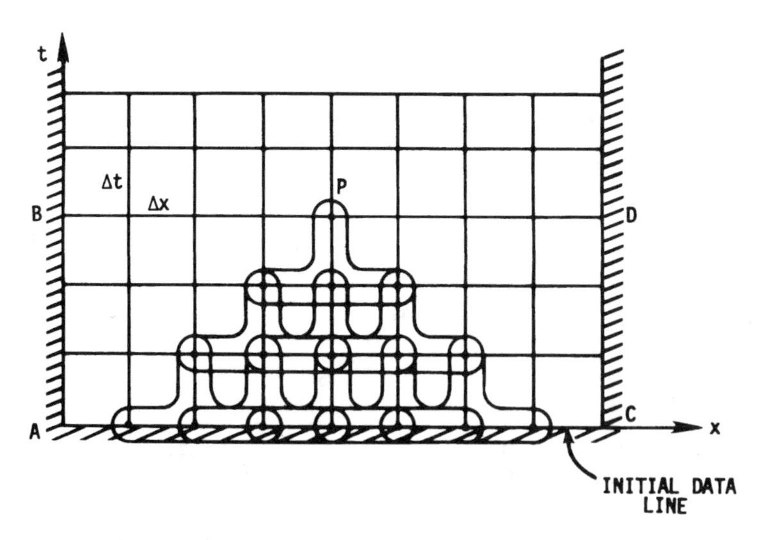

Figure 4.13 Zone of influence of simple explicit scheme.

more natural finite-difference approximation for hyperbolic PDEs that possess limited zones of influence.

Example 4.2 Suppose the simple explicit method is used to solve the heat equation ($\alpha = 0.05$) with the initial condition

$$u(x,0) = \sin(2\pi x) \qquad 0 \leqslant x \leqslant 1$$

and periodic boundary conditions. Determine the amplitude error after 10 steps if $\Delta t = 0.1$ and $\Delta x = 0.1$.

Solution A unique value of β can be determined in this problem for the same reason that was given in Example 4.1. Thus the value of β becomes

$$\beta = k_m \, \Delta x = (2\pi)(0.1) = 0.2\pi$$

After computing r,

$$r = \frac{\alpha \, \Delta t}{(\Delta x)^2} = \frac{(0.05)(0.1)}{(0.1)^2} = 0.5$$

the amplification factor for the simple explicit method is given by

$$G = 1 + 2r(\cos \beta - 1) = 0.809017$$

while the exact amplification factor becomes

$$G_e = e^{-r\beta^2} = 0.820869$$

As a result, the amplitude error is

$$A_0 |G_e^{10} - G^{10}| = (1)(0.1389 - 0.1201) = 0.0188$$

Using Eq. (4.72), the exact solution after 10 steps ($t = 1.0$) is given by

$$u(x,1) = e^{-\alpha 4\pi^2} \sin(2\pi x) = 0.1389 \sin(2\pi x)$$

which can be compared to the numerical solution:

$$u(x,1) = 0.1201 \sin(2\pi x)$$

4.2.2 Richardson's Method

Richardson (1910) proposed the following explicit one-step three-time-level scheme for solving the heat equation:

$$\frac{u_j^{n+1} - u_j^{n-1}}{2 \, \Delta t} = \alpha \frac{u_{j+1}^n - 2u_j^n + u_{j-1}^n}{(\Delta x)^2} \tag{4.80}$$

This scheme is second-order accurate with T.E. of $O[(\Delta t)^2, (\Delta x)^2]$. Unfortunately, this method proves to be unconditionally unstable and cannot be used to solve the heat equation. It is presented here for historic purposes only.

4.2.3 Simple Implicit (Laasonen) Method

A simple implicit scheme for the heat equation was proposed by Laasonen (1949). The algorithm for this scheme is

$$\frac{u_j^{n+1} - u_j^n}{\Delta t} = \alpha \frac{u_{j+1}^{n+1} - 2u_j^{n+1} + u_{j-1}^{n+1}}{(\Delta x)^2} \tag{4.81}$$

If we make use of the central-difference operator

$$\delta_x^2 u_j^n = u_{j+1}^n - 2u_j^n + u_{j-1}^n$$

we can rewrite Eq. (4.81) in the simpler form:

$$\frac{u_j^{n+1} - u_j^n}{\Delta t} = \alpha \frac{\delta_x^2 u_j^{n+1}}{(\Delta x)^2} \tag{4.82}$$

This scheme has first-order accuracy with a T.E. of $O[\Delta t, (\Delta x)^2]$ and is unconditionally stable. Upon examining Eq. (4.82), it is apparent that a tridiagonal system of linear algebraic equations must be solved at each time level $n + 1$.

The modified equation for this scheme is

$$u_t - \alpha u_{xx} = \left[\frac{1}{2}\alpha^2 \Delta t + \frac{\alpha(\Delta x)^2}{12}\right] u_{xxxx} + \left[\frac{1}{3}\alpha^3(\Delta t)^2 + \frac{1}{12}\alpha^2 \Delta t(\Delta x)^2 + \frac{1}{360}\alpha(\Delta x)^4\right] u_{xxxxxx} + \cdots \tag{4.83}$$

It is interesting to observe that in this modified equation, the terms in the coefficient of u_{xxxx} are of the same sign, whereas they are of opposite sign in the modified equation for the simple explicit scheme, Eq. (4.76). This observation can explain why the simple explicit scheme is generally more accurate than the simple implicit scheme when used within the appropriate stability limits. The amplification factor for the simple implicit scheme,

$$G = [1 + 2r(1 - \cos \beta)]^{-1} \tag{4.84}$$

is plotted in Fig. 4.14 for $r = \frac{1}{2}$ and is compared with the exact decay factor.

4.2.4 Crank-Nicolson Method

Crank and Nicolson (1947) used the following implicit algorithm to solve the heat equation:

$$\frac{u_j^{n+1} - u_j^n}{\Delta t} = \alpha \frac{\delta_x^2 u_j^n + \delta_x^2 u_j^{n+1}}{2(\Delta x)^2} \tag{4.85}$$

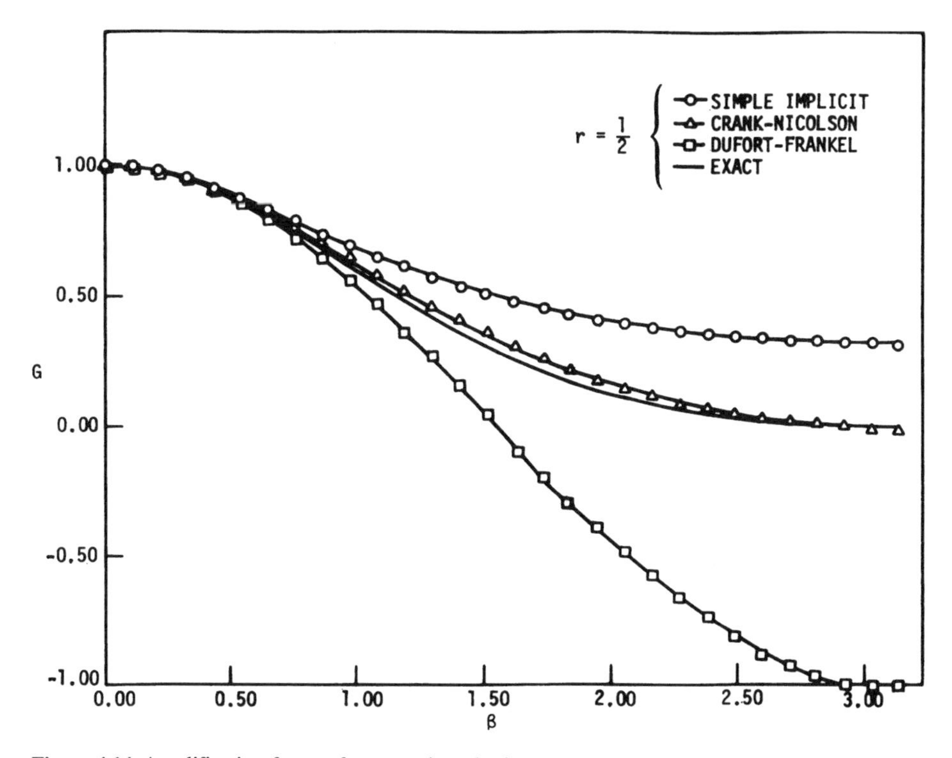

Figure 4.14 Amplification factors for several methods.

This unconditionally stable algorithm has become very well known and is referred to as the Crank-Nicolson scheme. This scheme makes use of trapezoidal differencing to achieve second-order accuracy with a T.E. of $O[(\Delta t)^2, (\Delta x)^2]$. Once again, a tridiagonal system of linear algebraic equations must be solved at each time level $n + 1$. The modified equation for the Crank-Nicolson method is

$$u_t - \alpha u_{xx} = \frac{\alpha(\Delta x)^2}{12} u_{xxxx} + \left[\frac{1}{12}\alpha^3(\Delta t)^2 + \frac{1}{360}\alpha(\Delta x)^4\right] u_{xxxxxx} + \cdots \tag{4.86}$$

The amplification factor

$$G = \frac{1 - r(1 - \cos\beta)}{1 + r(1 - \cos\beta)} \tag{4.87}$$

is plotted in Fig. 4.14 for $r = \frac{1}{2}$.

4.2.5 Combined Method A

The simple explicit, the simple implicit, and the Crank-Nicolson methods are special cases of a general algorithm given by

$$\frac{u_j^{n+1} - u_j^n}{\Delta t} = \alpha \frac{\theta\, \delta_x^2 u_j^{n+1} + (1 - \theta)\, \delta_x^2 u_j^n}{(\Delta x)^2} \tag{4.88}$$

where θ is a constant ($0 \leqslant \theta \leqslant 1$). The simple explicit method corresponds to $\theta = 0$, the simple implicit method corresponds to $\theta = 1$, and the Crank-Nicolson method corresponds to $\theta = \frac{1}{2}$. This combined method has first-order accuracy with T.E. of $O[\Delta t, (\Delta x)^2]$ except for special cases such as

1. $\theta = \frac{1}{2}$(Crank-Nicolson method) $\quad$ T.E. $= O[(\Delta t)^2, (\Delta x)^2]$

2. $$\theta = \frac{1}{2} - \frac{(\Delta x)^2}{12 \alpha \Delta t} \qquad \text{T.E.} = O[(\Delta t)^2, (\Delta x)^4]$$

3. $$\theta = \frac{1}{2} - \frac{(\Delta x)^2}{12 \alpha \Delta t} \quad \text{and} \quad \frac{(\Delta x)^2}{\alpha \Delta t} = \sqrt{20} \qquad \text{T.E.} = O[(\Delta t)^2, (\Delta x)^6]$$

The T.E.s of these special cases can be obtained by examining the modified equation

$$u_t - \alpha u_{xx} = \left[\left(\theta - \frac{1}{2}\right)\alpha^2 \Delta t + \frac{\alpha (\Delta x)^2}{12}\right] u_{xxxx} + \left[\left(\theta^2 - \theta + \frac{1}{3}\right)\alpha^3 (\Delta t)^2 + \frac{1}{6}\left(\theta - \frac{1}{2}\right)\alpha^2 \Delta t (\Delta x)^2 + \frac{1}{360}\alpha (\Delta x)^4\right] u_{xxxxxx} + \cdots \tag{4.89}$$

The present combined method is unconditionally stable if $\frac{1}{2} \leqslant \theta \leqslant 1$. However, when $0 \leqslant \theta < \frac{1}{2}$, the method is stable only if

$$0 \leqslant r \leqslant \frac{1}{2 - 4\theta} \tag{4.90}$$

4.2.6 Combined Method B

Richtmyer and Morton (1967) present the following general algorithm for a three-time-level implicit scheme:

$$(1 + \theta)\frac{u_j^{n+1} - u_j^n}{\Delta t} - \theta \frac{u_j^n - u_j^{n-1}}{\Delta t} = \alpha \frac{\delta_x^2 u_j^{n+1}}{(\Delta x)^2} \tag{4.91}$$

This general algorithm has first-order accuracy with T.E. of $O[\Delta t, (\Delta x)^2]$ except for special cases:

1. $\theta = \frac{1}{2}$ $\quad$ T.E. $= O[(\Delta t)^2, (\Delta x)^2]$

2. $$\theta = \frac{1}{2} + \frac{(\Delta x)^2}{12 \alpha \Delta t} \qquad \text{T.E.} = O[(\Delta t)^2, (\Delta x)^4]$$

which can be verified by examining the modified equation

$$u_t - \alpha u_{xx} = \left[-(\theta - \tfrac{1}{2})\alpha^2 \Delta t + \tfrac{1}{12}\alpha(\Delta x)^2\right] u_{xxxx} + \cdots$$

4.2.7 DuFort-Frankel Method

The unstable Richardson method [Eq. (4.80)] can be made stable by replacing u_j^n with the time-averaged expression $(u_j^{n+1} + u_j^{n-1})/2$. The resulting explicit three-time-level scheme,

$$\frac{u_j^{n+1} - u_j^{n-1}}{2\,\Delta t} = \alpha \frac{u_{j+1}^n - u_j^{n+1} - u_j^{n-1} + u_{j-1}^n}{(\Delta x)^2} \tag{4.92}$$

was first proposed by DuFort and Frankel (1953). Note that Eq. (4.92) can be rewritten as

$$u_j^{n+1}(1 + 2r) = u_j^{n-1} + 2r\left(u_{j+1}^n - u_j^{n-1} + u_{j-1}^n\right) \tag{4.93}$$

so that only one unknown, u_j^{n+1}, appears in the scheme, and therefore it is explicit. The T.E. for the DuFort-Frankel method is $O[(\Delta t)^2, (\Delta x)^2, (\Delta t/\Delta x)^2]$. Consequently, if this method is to be consistent, then $(\Delta t/\Delta x)^2$ must approach zero as Δt and Δx approach zero. As pointed out in Chapter 3, if $\Delta t/\Delta x$ approaches a constant value γ, instead of zero, the DuFort-Frankel scheme is consistent with the hyperbolic equation

$$\frac{\partial u}{\partial t} + \alpha\gamma^2 \frac{\partial^2 u}{\partial t^2} = \alpha \frac{\partial^2 u}{\partial x^2}$$

If we let r remain constant as Δt and Δx approach zero, the term $(\Delta t/\Delta x)^2$ becomes formally a first-order term of $O(\Delta t)$. The modified equation is given by

$$u_t - \alpha u_{xx} = \left[\frac{1}{12}\alpha(\Delta x)^2 - \alpha^3 \frac{(\Delta t)^2}{(\Delta x)^2}\right] u_{xxxx}$$
$$+ \left[\frac{1}{360}\alpha(\Delta x)^4 - \frac{1}{3}\alpha^3(\Delta t)^2 + 2\alpha^5 \frac{(\Delta t)^4}{(\Delta x)^4}\right] u_{xxxxxx} + \cdots$$

The amplification factor

$$G = \frac{2r\cos\beta \pm \sqrt{1 - 4r^2\sin^2\beta}}{1 + 2r}$$

is plotted in Fig. 4.14 for $r = \frac{1}{2}$. The explicit DuFort-Frankel scheme has the unusual property of being unconditionally stable for $r \geqslant 0$. In passing, we note that the DuFort-Frankel method can be extended to two and three dimensions without any unexpected penalties. The scheme remains unconditionally stable.

4.2.8 Keller Box and Modified Box Methods

The Keller box method (Keller, 1970) for parabolic PDEs is an implicit scheme with second-order accuracy in both space and time. This formulation allows for the spatial and temporal steps to vary without causing deterioration in the formal second-order accuracy. The scheme differs from others considered thus far, in that second and higher derivatives are replaced by first derivatives through the introduction of additional variables as discussed in Section 2.5. Thus a system of first-order equations results. For the 1-D heat equation,

$$\frac{\partial u}{\partial t} = \alpha \frac{\partial^2 u}{\partial x^2}$$

we can define

$$v = \frac{\partial u}{\partial x}$$

so that the second-order heat equation can be written as a system of two first-order equations:

$$\frac{\partial u}{\partial x} = v$$

$$\frac{\partial u}{\partial t} = \alpha \frac{\partial v}{\partial x}$$

Now we endeavor to approximate these equations using only central differences, making use of the four points at the corners of a "box" about $(n + \frac{1}{2}, j - \frac{1}{2})$ (see Fig. 4.15). The resulting difference equations are

$$\frac{u_j^{n+1} - u_{j-1}^{n+1}}{\Delta x_j} = v_{j-\frac{1}{2}}^{n+1} \tag{4.94a}$$

$$\frac{u_{j-\frac{1}{2}}^{n+1} - u_{j-\frac{1}{2}}^{n}}{\Delta t_{n+1}} = \frac{\alpha\left(v_j^{n+\frac{1}{2}} - v_{j-1}^{n+\frac{1}{2}}\right)}{\Delta x_j} \tag{4.94b}$$

where the difference molecules are shown in Figs. 4.16 and 4.17. The mesh functions that contain a subscript or superscript $\frac{1}{2}$ are defined as averages, as for example,

$$u_{j-\frac{1}{2}}^{n+1} = \frac{u_j^{n+1} + u_{j-1}^{n+1}}{2}$$

$$v_j^{n+\frac{1}{2}} = \frac{v_j^n + v_j^{n+1}}{2}$$

After substituting the averaged expressions into Eqs. (4.94*a*) and (4.94*b*), the new difference equations become

$$\frac{u_j^{n+1} - u_{j-1}^{n+1}}{\Delta x_j} = \frac{v_j^{n+1} + v_{j-1}^{n+1}}{2} \tag{4.95a}$$

$$\frac{u_j^{n+1} + u_{j-1}^{n+1}}{\Delta t_{n+1}} = \alpha \frac{v_j^{n+1} - v_{j-1}^{n+1}}{\Delta x_j} + \frac{u_j^n + u_{j-1}^n}{\Delta t_{n+1}} + \alpha \frac{v_j^n - v_{j-1}^n}{\Delta x_j} \tag{4.95b}$$

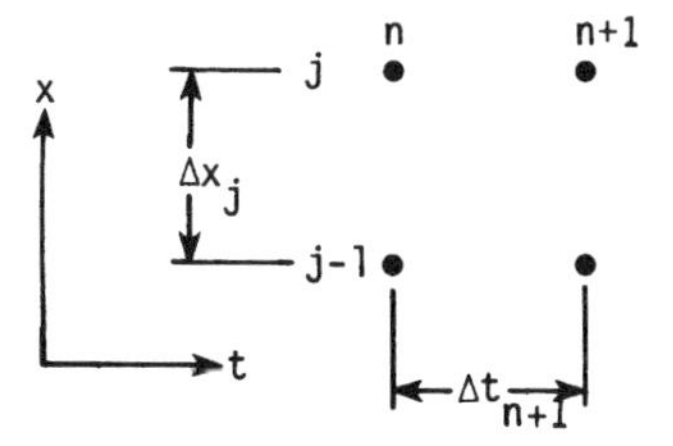

Figure 4.15 Grid for box scheme.

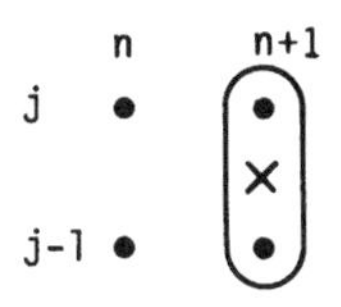

Figure 4.16 Difference molecule for evaluation of $v_{j-1/2}^{n+1}$.

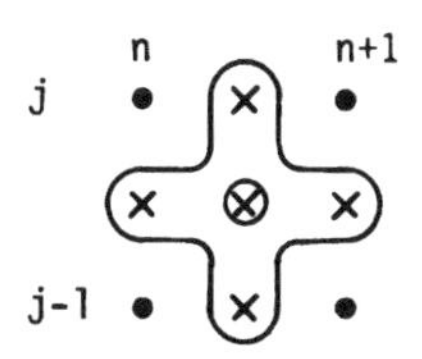

Figure 4.17 Difference molecule for Eq. (4.94*b*).

The unknowns (u^{n+1}, v^{n+1}) in the above equations are located at grid points j and $j - 1$. However, when boundary conditions are included, the unknowns may also occur at grid point $j + 1$. Thus the algebraic system resulting from the Keller box scheme for the general point can be represented in matrix form as

$$[B]\mathbf{F}_{j-1}^{n+1} + [D]\mathbf{F}_j^{n+1} + [A]\mathbf{F}_{j+1}^{n+1} = \mathbf{C}$$

where

$$\mathbf{F} = [u, v]^T$$

and $[B]$, $[D]$, and $[A]$ are 2×2 matrices and $\mathbf{C}$ is a two-component vector. When the entire system of equations for a given problem is assembled and boundary conditions are taken into account, the algebraic problem can be expressed in the general form $[M]\mathbf{x} = \mathbf{c}$, where the "elements" of the coefficient matrix $[M]$ are now 2×2 matrices, and each "component" of the column vectors becomes the two components of $\mathbf{F}_j^{n+1}$ and $\mathbf{C}_j$ associated with point j. This system of equations is a block tridiagonal system and can be solved with the general block tridiagonal algorithm given in Appendix B or with a special-purpose algorithm specialized to take advantage of zeros that may be present in the coefficient matrices. The block algorithm actually proceeds with the same operations as for the scalar tridiagonal algorithm with matrix and matrix-vector multiplications replacing scalar operations. When division by a matrix would be indicated by this analogy, premultiplication by the inverse of the matrix is carried out.

The work required to solve the algebraic system resulting from the box-difference stencil can be reduced by combining the difference representations at two adjacent grid points to eliminate one variable. This system can then be solved with the simple scalar Thomas algorithm. This revision of the box method, which simplifies the final algebraic formulation, will be referred to as the *modified box method.*

Modified box method. The strategy in the development of the modified box method is to express the v's in terms of u's. The term v_{j-1}^{n+1} can be eliminated from Eq. (4.95*b*) by a simple substitution using Eq. (4.95*a*). Similarly, v_{j-1}^{n} can be eliminated through substitution by evaluating Eq. (4.95*a*) at time level n. This gives

$$\frac{u_j^{n+1} + u_{j-1}^{n+1}}{\Delta t_{n+1}} = 2\alpha\frac{v_j^{n+1}}{\Delta x_j} - 2\alpha\frac{u_j^{n+1} - u_{j-1}^{n+1}}{(\Delta x_j)^2} + \frac{u_j^n + u_{j-1}^n}{\Delta t_{n+1}} + 2\alpha\frac{v_j^n}{\Delta x_j} - 2\alpha\frac{u_j^n - u_{j-1}^n}{(\Delta x_j)^2} \tag{4.96a}$$

To eliminate v_j^{n+1} and v_j^n, Eqs. (4.95*a*) and (4.95*b*) can first be rewritten with the j index advanced by 1 and combined. The result is

$$\frac{u_{j+1}^{n+1} + u_j^{n+1}}{\Delta t_{n+1}} = \frac{-2\alpha v_j^{n+1}}{\Delta x_{j+1}} + \frac{2\alpha\left(u_{j+1}^{n+1} - u_j^{n+1}\right)}{(\Delta x_{j+1})^2} + \frac{u_{j+1}^n + u_j^n}{\Delta t_{n+1}} + \frac{-2\alpha v_j^n}{\Delta x_{j+1}} + 2\alpha\frac{u_{j+1}^n - u_j^n}{(\Delta x_{j+1})^2} \tag{4.96b}$$

The terms v_j^{n+1} and v_j^n can then be eliminated by multiplying Eq. (4.96*a*) by Δx_j and Eq. (4.96*b*) by Δx_{j+1} and adding the two products. The result can be written in the tridiagonal format

$$B_j u_{j-1}^{n+1} + D_j u_j^{n+1} + A_j u_{j+1}^{n+1} = C_j$$

where

$$B_j = \frac{\Delta x_j}{\Delta t_{n+1}} - \frac{2\alpha}{\Delta x_j} \qquad A_j = \frac{\Delta x_{j+1}}{\Delta t_{n+1}} - \frac{2\alpha}{\Delta x_{j+1}}$$

$$D_j = \frac{\Delta x_j}{\Delta t_{n+1}} + \frac{\Delta x_{j+1}}{\Delta t_{n+1}} + \frac{2\alpha}{\Delta x_j} + \frac{2\alpha}{\Delta x_{j+1}}$$

$$C_j = 2\alpha\frac{u_{j-1}^n - u_j^n}{\Delta x_j} + 2\alpha\frac{u_{j+1}^n - u_j^n}{\Delta x_{j+1}} + (u_j^n + u_{j-1}^n)\frac{\Delta x_j}{\Delta t_{n+1}} + (u_{j+1}^n + u_j^n)\frac{\Delta x_{j+1}}{\Delta t_{n+1}}$$

The above equations can be simplified somewhat if the spacing in the x direction is uniform. Even then, a few more algebraic operations per time step are required than for the Crank-Nicolson scheme, which is also second-order accurate for a uniformly spaced mesh. A conceptual advantage of schemes based on the box-difference molecule is that formal second-order accuracy is achieved even when the mesh is nonuniform. The Crank-Nicolson scheme can be extended to cases of nonuniform grid spacing by representing the second derivative term as indicated for Laplace's equation in Eq. (3.97). Formally, the T.E. for that representation is reduced to first order for arbitrary grid spacing. Blottner (1974) has shown that if the variable grid spacing used is one that could be established through a coordinate stretching transformation, then the Crank-Nicolson scheme is also second-order accurate for that variable grid arrangement.

4.2.9 Methods for the Two-Dimensional Heat Equation

The 2-D heat equation is given by

$$\frac{\partial u}{\partial t} = \alpha \left(\frac{\partial^2 u}{\partial x^2} + \frac{\partial^2 u}{\partial y^2} \right) \tag{4.97}$$

Since this PDE is different from the 1-D equation, caution must be exercised when attempting to apply the previous finite-difference methods to this equation. The following two examples illustrate some of the difficulties. If we apply the simple explicit method to the 2-D heat equation, the following algorithm results:

$$\frac{u_{i,j}^{n+1} - u_{i,j}^{n}}{\Delta t} = \alpha \left[\frac{u_{i+1,j}^{n} - 2u_{i,j}^{n} + u_{i-1,j}^{n}}{(\Delta x)^2} + \frac{u_{i,j+1}^{n} - 2u_{i,j}^{n} + u_{i,j-1}^{n}}{(\Delta y)^2} \right] \tag{4.98}$$

where $x = i\,\Delta x$ and $y = j\,\Delta y$. As shown in Chapter 3, the stability condition is

$$\alpha\,\Delta t \left[\frac{1}{(\Delta x)^2} + \frac{1}{(\Delta y)^2} \right] \leqslant \frac{1}{2}$$

If $(\Delta x)^2 = (\Delta y)^2$, the stability condition reduces to $r \leqslant \frac{1}{4}$, which is twice as restrictive as the 1-D constraint $r \leqslant \frac{1}{2}$ and makes this method even more impractical.

When we apply the Crank-Nicolson scheme to the 2-D heat equation, we obtain

$$\frac{u_{i,j}^{n+1} - u_{i,j}^{n}}{\Delta t} = \frac{\alpha}{2} \left(\hat{\delta}_x^2 + \hat{\delta}_y^2 \right) \left(u_{i,j}^{n+1} + u_{i,j}^{n} \right) \tag{4.99}$$

where the 2-D central-difference operators $\hat{\delta}_x^2$ and $\hat{\delta}_y^2$ are defined by

$$\begin{aligned} \hat{\delta}_x^2 u_{i,j}^n &= \frac{u_{i+1,j}^{n} - 2u_{i,j}^{n} + u_{i-1,j}^{n}}{(\Delta x)^2} = \frac{\delta_x^2 u_{i,j}^n}{(\Delta x)^2} \\ \hat{\delta}_y^2 u_{i,j}^n &= \frac{u_{i,j+1}^{n} - 2u_{i,j}^{n} + u_{i,j-1}^{n}}{(\Delta y)^2} = \frac{\delta_y^2 u_{i,j}^n}{(\Delta y)^2} \end{aligned} \tag{4.100}$$

As with the 1-D case, the Crank-Nicolson scheme is unconditionally stable when applied to the 2-D heat equation with periodic boundary conditions. Unfortunately, the resulting system of linear algebraic equations is no longer tridiagonal because of the five unknowns $u_{i,j}^{n+1}$, $u_{i+1,j}^{n+1}$, $u_{i-1,j}^{n+1}$, $u_{i,j+1}^{n+1}$, and $u_{i,j-1}^{n+1}$. The same is true for all the implicit schemes we have studied previously. In order to examine this further, let us rewrite Eq. (4.99) as

$$au_{i,j-1}^{n+1} + bu_{i-1,j}^{n+1} + cu_{i,j}^{n+1} + bu_{i+1,j}^{n+1} + au_{i,j+1}^{n+1} = d_{i,j}^{n} \tag{4.101}$$

where

$$a = -\frac{\alpha\,\Delta t}{2(\Delta y)^2} = -\frac{1}{2}r_y$$

$$b = -\frac{\alpha\,\Delta t}{2(\Delta x)^2} = -\frac{1}{2}r_x$$

$$c = 1 + r_x + r_y$$

$$d_{i,j}^{n} = u_{i,j}^{n} + \frac{\alpha\,\Delta t}{2}\left(\hat{\delta}_x^2 + \hat{\delta}_y^2\right)u_{i,j}^{n}$$

If we apply Eq. (4.101) to the 2-D (6 × 6) computational mesh shown in Fig. 4.18, the following system of 16 linear algebraic equations must be solved at each $(n + 1)$ time level:

$$\begin{bmatrix}
c & b & 0 & 0 & a & 0 & & & & & & & & & & 0 \\
b & c & b & & & a & & & & & & & & & & \\
0 & b & c & b & & & a & & & & & & & & & \\
0 & & b & c & 0 & & & a & & & & & & & & \\
a & & & 0 & c & b & & & a & & & & & & & \\
0 & a & & & b & c & b & & & a & & & & & & \\
 & & a & & & b & c & b & & & a & & & & & \\
 & & & a & & & b & c & 0 & & & a & & & & \\
 & & & & a & & & 0 & c & b & & & a & & & \\
 & & & & & a & & & b & c & b & & & a & & \\
 & & & & & & a & & & b & c & b & & & a & 0 \\
 & & & & & & & a & & & b & c & 0 & & & a \\
 & & & & & & & & a & & & 0 & c & b & & 0 \\
 & & & & & & & & & a & & & b & c & b & 0 \\
 & & & & & & & & & & a & & & b & c & b \\
0 & & & & & & & & & & 0 & a & 0 & 0 & b & c
\end{bmatrix}
\begin{bmatrix}
u_{2,2}^{n+1} \\ u_{3,2}^{n+1} \\ u_{4,2}^{n+1} \\ u_{5,2}^{n+1} \\ u_{2,3}^{n+1} \\ u_{3,3}^{n+1} \\ u_{4,3}^{n+1} \\ u_{5,3}^{n+1} \\ u_{2,4}^{n+1} \\ u_{3,4}^{n+1} \\ u_{4,4}^{n+1} \\ u_{5,4}^{n+1} \\ u_{2,5}^{n+1} \\ u_{3,5}^{n+1} \\ u_{4,5}^{n+1} \\ u_{5,5}^{n+1}
\end{bmatrix}
=
\begin{bmatrix}
d_{2,2}''' \\ d_{3,2}' \\ d_{4,2}' \\ d_{5,2}''' \\ d_{2,3}'' \\ d_{3,3} \\ d_{4,3} \\ d_{5,3}'' \\ d_{2,4}'' \\ d_{3,4} \\ d_{4,4} \\ d_{5,4}'' \\ d_{2,5}''' \\ d_{3,5}' \\ d_{4,5}' \\ d_{5,5}'''
\end{bmatrix} \tag{4.102}$$

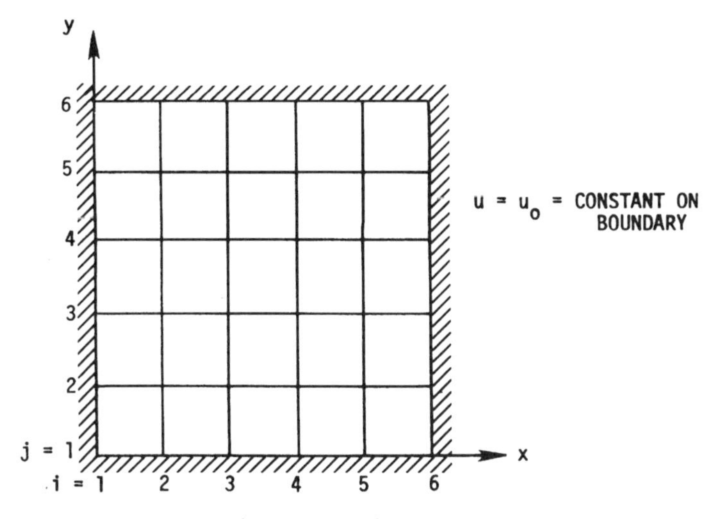

Figure 4.18 Two-dimensional computational mesh.

where $d' = d - au_0$
$d'' = d - bu_0$
$d''' = d - (a + b)u_0$

A system of equations like Eq. (4.102) requires substantially more computer time to solve than does a tridiagonal system. In fact, equations of this type are often solved by iterative methods. These methods are discussed in Section 4.3.

4.2.10 ADI Methods

The difficulties described above, which occur when attempting to solve the 2-D heat equation by conventional algorithms, led to the development of alternating-direction implicit (ADI) methods by Peaceman and Rachford (1955) and Douglas (1955). The usual ADI method is a two-step scheme given by

$$\text{Step 1:} \qquad \frac{u_{i,j}^{n+1/2} - u_{i,j}^{n}}{\Delta t/2} = \alpha\left(\hat{\delta}_x^2 u_{i,j}^{n+1/2} + \hat{\delta}_y^2 u_{i,j}^{n}\right)$$

$$\text{Step 2:} \qquad \frac{u_{i,j}^{n+1} - u_{i,j}^{n+1/2}}{\Delta t/2} = \alpha\left(\hat{\delta}_x^2 u_{i,j}^{n+1/2} + \hat{\delta}_y^2 u_{i,j}^{n+1}\right) \tag{4.103}$$

As a result of the "splitting" that is employed in this algorithm, only tridiagonal systems of linear algebraic equations must be solved. During step 1, a tridiagonal matrix is solved for each j row of grid points, and during step 2, a tridiagonal matrix is solved for each i row of grid points. This procedure is illustrated in Fig. 4.19. The ADI method is second-order accurate with a T.E. of

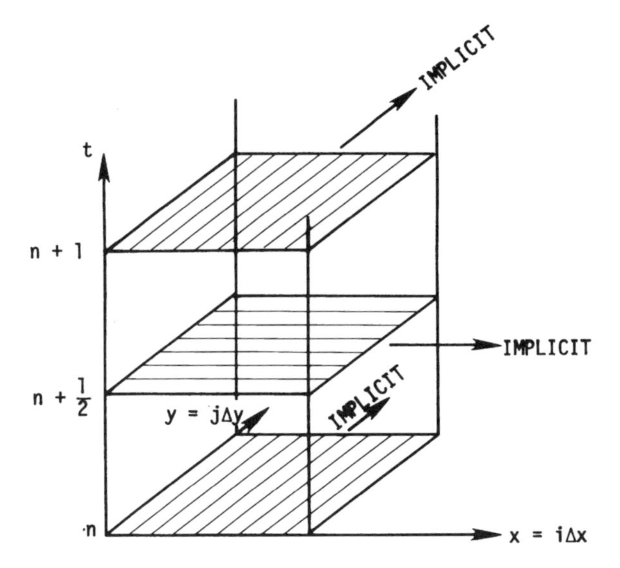

Figure 4.19 ADI calculation procedure.

$O[(\Delta t)^2, (\Delta x)^2, (\Delta y)^2]$. Upon examining the amplification factor

$$G = \frac{[1 - r_x(1 - \cos \beta_x)][1 - r_y(1 - \cos \beta_y)]}{[1 + r_x(1 - \cos \beta_x)][1 + r_y(1 - \cos \beta_y)]}$$

where

$$r_x = \alpha\, \Delta t / (\Delta x)^2 \qquad \beta_x = k_m\, \Delta x$$

$$r_y = \alpha\, \Delta t / (\Delta y)^2 \qquad \beta_y = k_m\, \Delta y$$

we find this method to be unconditionally stable. The obvious extension of this method to three dimensions (making use of the time levels $n, n + \frac{1}{3}, n + \frac{2}{3}, n + 1$) leads to a conditionally stable method with T.E. of $O[(\Delta t, (\Delta x)^2, (\Delta y)^2, (\Delta z)^2]$.

In order to circumvent these problems, Douglas and Gunn (1964) developed a general method for deriving ADI schemes that are unconditionally stable and retain second-order accuracy. Their method of derivation is commonly called *approximate factorization*. When an implicit procedure such as the Crank-Nicolson scheme is cast into residual or *delta* form, the motivation for factoring becomes evident. The delta form is obtained by defining $\Delta u_{i,j}$ as

$$\Delta u_{i,j} \equiv u_{i,j}^{n+1} - u_{i,j}^{n}$$

Substituting this into the 2-D Crank-Nicolson scheme [Eq. (4.99)] gives

$$\Delta u_{i,j} = \frac{\alpha\,\Delta t}{2}\left[\frac{\delta_x^2\,\Delta u_{i,j}}{(\Delta x)^2} + \frac{\delta_y^2\,\Delta u_{i,j}}{(\Delta y)^2} + \frac{2\,\delta_x^2 u_{i,j}^n}{(\Delta x)^2} + \frac{2\,\delta_y^2 u_{i,j}^n}{(\Delta y)^2}\right]$$

After rearranging this equation to put all of the unknowns on the left-hand side of the equation and inserting r_x and r_y, we obtain

$$\left(1 - \frac{r_x}{2}\delta_x^2 - \frac{r_y}{2}\delta_y^2\right)\Delta u_{i,j} = \left(r_x\delta_x^2 + r_y\delta_y^2\right)u_{i,j}^n$$

If the quantity in parentheses on the left can be arranged as the product of two operators, one involving x-direction differences and the other involving y-direction differences, then the algorithm can proceed in two steps. One such *factorization* is

$$\left(1 - \frac{r_x}{2}\delta_x^2\right)\left(1 - \frac{r_y}{2}\delta_y^2\right)\Delta u_{i,j} = \left(r_x\delta_x^2 + r_y\delta_y^2\right)u_{i,j}^n$$

In order to achieve this factorization, the quantity $r_x r_y\,\delta_x^2\delta_y^2\,\Delta u_{i,j}/4$ must be added to the left-hand side. The T.E. is thus augmented by the same amount. The factored equation can now be solved in the two steps:

Step 1: $$\left(1 - \frac{r_x}{2}\delta_x^2\right)\Delta u_{i,j}^* = \left(r_x\delta_x^2 + r_y\delta_y^2\right)u_{i,j}^n$$

Step 2: $$\left(1 - \frac{r_y}{2}\delta_y^2\right)\Delta u_{i,j} = \Delta u_{i,j}^*$$

where the superscript asterisk denotes an intermediate value. The unknown $u_{i,j}^{n+1}$ is then obtained from

$$u_{i,j}^{n+1} = u_{i,j}^n + \Delta u_{i,j}$$

Using the preceding factorization approach and starting with the three-dimensional (3-D) Crank-Nicolson scheme, Douglas and Gunn also developed an algorithm to solve the 3-D heat equation:

Step 1: $$\left(1 - \frac{r_x}{2}\delta_x^2\right)\Delta u^* = \left(r_x\delta_x^2 + r_y\delta_y^2 + r_z\delta_z^2\right)u^n$$

Step 2: $$\left(1 - \frac{r_y}{2}\delta_y^2\right)\Delta u^{**} = \Delta u^* \tag{4.104}$$

Step 3: $$\left(1 - \frac{r_z}{2}\delta_z^2\right)\Delta u = \Delta u^{**}$$

where the superscript asterisks and double asterisks denote intermediate values and the subscripts i, j, k have been dropped from each term.

4.2.11 Splitting or Fractional-Step Methods

The ADI methods are closely related and in some cases identical to the method of fractional steps or methods of splitting, which were developed by Soviet mathematicians at about the same time as the ADI methods were developed in the United States. The basic idea of these methods is to split a finite-difference algorithm into a sequence of 1-D operations. For example, the simple implicit scheme applied to the 2-D heat equation could be split in the following manner:

$$\text{Step 1:} \qquad \frac{u_{i,j}^{n+1/2} - u_{i,j}^{n}}{\Delta t} = \alpha\, \hat{\delta}_x^2 u_{i,j}^{n+1/2}$$

$$\text{Step 2:} \qquad \frac{u_{i,j}^{n+1} - u_{i,j}^{n+1/2}}{\Delta t} = \alpha\, \hat{\delta}_y^2 u_{i,j}^{n+1} \tag{4.105}$$

to give a first-order accurate method with a T.E. of $O[\Delta t, (\Delta x)^2, (\Delta y)^2]$. For further details on the method of fractional steps, the reader is urged to consult the book by Yanenko (1971).

4.2.12 ADE Methods

Another way of solving the 2-D heat equation is by means of an alternating-direction explicit (ADE) method. Unlike the ADI methods, the ADE methods do not require tridiagonal matrices to be "inverted." Since the ADE methods can also be used to solve the 1-D heat equation, we will apply the ADE algorithms to this equation, for simplicity.

The first ADE method was proposed by Saul'yev (1957). His two-step scheme is given by

$$\text{Step 1:} \qquad \frac{u_j^{n+1} - u_j^{n}}{\Delta t} = \alpha \frac{u_{j-1}^{n+1} - u_j^{n+1} - u_j^{n} + u_{j+1}^{n}}{(\Delta x)^2}$$

$$\text{Step 2:} \qquad \frac{u_j^{n+2} - u_j^{n+1}}{\Delta t} = \alpha \frac{u_{j-1}^{n+1} - u_j^{n+1} - u_j^{n+2} + u_{j+1}^{n+2}}{(\Delta x)^2} \tag{4.106}$$

In the application of this method, step 1 marches the solution from the left boundary to the right boundary. By marching in this direction, u_{j-1}^{n+1} is always known, and consequently, u_j^{n+1} can be determined "explicitly." In a like manner, step 2 marches the solution from the right boundary to the left boundary, again resulting in an "explicit" formulation, since u_{j+1}^{n+2} is always known. We assume that u is known on the boundaries. Although this scheme involves three time levels, only one storage array is required for u because of the unique way in which the calculation procedure sweeps through the mesh. This scheme is unconditionally stable, and the T.E. is $O[(\Delta t)^2, (\Delta x)^2, (\Delta t/\Delta x)^2]$. The scheme is formally first-order accurate (if r is constant) owing to the presence of the inconsistent term $(\Delta t/\Delta x)^2$ in the T.E.

Another ADE method was proposed by Barakat and Clark (1966). In this method the calculation procedure is simultaneously "marched" in both directions, and the resulting solutions (p_j^{n+1} and q_j^{n+1}) are averaged to obtain the final value of u_j^{n+1}:

$$\frac{p_j^{n+1} - p_j^n}{\Delta t} = \alpha \frac{p_{j-1}^{n+1} - p_j^{n+1} - p_j^n + p_{j+1}^n}{(\Delta x)^2}$$

$$\frac{q_j^{n+1} - q_j^n}{\Delta t} = \alpha \frac{q_{j-1}^n - q_j^n - q_j^{n+1} + q_{j+1}^{n+1}}{(\Delta x)^2} \tag{4.107}$$

$$u_j^{n+1} = \tfrac{1}{2}\left(p_j^{n+1} + q_j^{n+1}\right)$$

This method is unconditionally stable, and the T.E. is approximately $O[(\Delta t)^2, (\Delta x)^2]$ because the simultaneous marching tends to cancel the $(\Delta t/\Delta x)^2$ terms. It has been observed that this method is about 18/16 times faster than the ADI method for the 2-D heat equation.

Larkin (1964) proposed a slightly different algorithm, which replaces the p and q with u whenever possible. His algorithm is

$$\frac{p_j^{n+1} - u_j^n}{\Delta t} = \alpha \frac{p_{j-1}^{n+1} - p_j^{n+1} - u_j^n + u_{j+1}^n}{(\Delta x)^2}$$

$$\frac{q_j^{n+1} - u_j^n}{\Delta t} = \alpha \frac{u_{j-1}^n - u_j^n - q_j^{n+1} + q_{j+1}^{n+1}}{(\Delta x)^2} \tag{4.108}$$

$$u_j^{n+1} = \tfrac{1}{2}\left(p_j^{n+1} + q_j^{n+1}\right)$$

Numerical tests indicate that this method is usually less accurate than the Barakat and Clark scheme.

4.2.13 Hopscotch Method

As our final algorithm for solving the 2-D heat equation, let us examine the hopscotch method. This method is an explicit procedure that is unconditionally stable. The calculation procedure, illustrated in Fig. 4.20, involves two sweeps through the mesh. For the first sweep, $u_{i,j}^{n+1}$ is computed at each grid point (for which $i + j + n$ is even) by the simple explicit scheme

$$\frac{u_{i,j}^{n+1} - u_{i,j}^n}{\Delta t} = \alpha\left(\hat{\delta}_x^2 u_{i,j}^n + \hat{\delta}_y^2 u_{i,j}^n\right) \tag{4.109}$$

For the second sweep, $u_{i,j}^{n+1}$ is computed at each grid point (for which $i + j + n$ is odd) by the simple implicit scheme

$$\frac{u_{i,j}^{n+1} - u_{i,j}^n}{\Delta t} = \alpha\left(\hat{\delta}_x^2 u_{i,j}^{n+1} + \hat{\delta}_y^2 u_{i,j}^{n+1}\right) \tag{4.110}$$

The second sweep appears to be implicit, but no simultaneous algebraic equations must be solved because $u_{i+1,j}^{n+1}$, $u_{i-1,j}^{n+1}$, $u_{i,j+1}^{n+1}$, and $u_{i,j-1}^{n+1}$ are known from the first sweep; hence the algorithm is explicit. The T.E. for the hopscotch method is of $O[\Delta t, (\Delta x)^2, (\Delta y)^2]$.

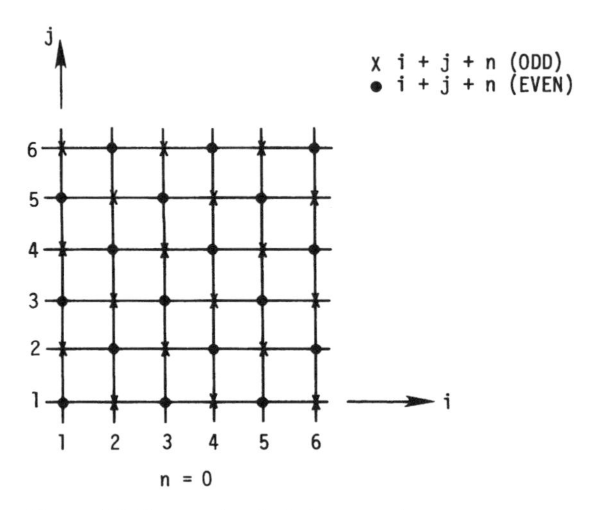

Figure 4.20 Hopscotch calculation procedure.

4.2.14 Additional Comments

The selection of a best method for solving the heat equation is made difficult by the large variety of acceptable methods. In general, implicit methods are considered more suitable than explicit methods. For the 1-D heat equation, the Crank-Nicolson method is highly recommended because of its second-order temporal and spatial accuracy. For the 2-D and 3-D heat equations, both the ADI schemes of Douglas and Gunn and the modified Keller box method give excellent results.

4.3 LAPLACE'S EQUATION

Laplace's equation is the model form for elliptic PDEs. For 2-D problems in Cartesian coordinates, Laplace's equation is

$$\frac{\partial^2 u}{\partial x^2} + \frac{\partial^2 u}{\partial y^2} = 0 \tag{4.111}$$

Some of the important practical problems governed by a single elliptic equation include the steady-state temperature distribution in a solid and the incompressible irrotational ("potential") flow of a fluid.

The incompressible Navier-Stokes equations are an example of a more complicated system of equations that has an elliptic character. The steady incompressible Navier-Stokes equations are elliptic but in a coupled and complicated fashion, since the pressure derivatives as well as velocity derivatives are sources of elliptic behavior. The elliptic equation arising in many physical problems is a Poisson equation of the form

$$\frac{\partial^2 u}{\partial x^2} + \frac{\partial^2 u}{\partial y^2} = f(x, y) \tag{4.112}$$

Thus elliptic PDEs will be found to frequently govern important problems in heat transfer and fluid mechanics. For this reason, we will give serious attention to ways of solving a model elliptic equation.

4.3.1 Finite-Difference Representations for Laplace's Equation

The differences between "methods" for Laplace's equation and elliptic equations in general are not so much differences in the finite-difference representations, (although these will vary) but more often, differences in the techniques used for solving the resulting system of linear algebraic equations.

Five-point formula. By far the most common difference scheme for the 2-D Laplace equation is the five-point formula first used by Runge in 1908:

$$\frac{u_{i+1,j} - 2u_{i,j} + u_{i-1,j}}{(\Delta x)^2} + \frac{u_{i,j+1} - 2u_{i,j} + u_{i,j-1}}{(\Delta y)^2} = 0 \tag{4.113}$$

which has a T.E. of $O[(\Delta x)^2, (\Delta y)^2]$. The modified equation is

$$u_{xx} + u_{yy} = -\tfrac{1}{12}\left[u_{xxxx}(\Delta x)^2 + u_{yyyy}(\Delta y)^2\right] + \cdots$$

Nine-point formula. The nine-point formula appears to be a logical choice when greater accuracy is desired for Laplace's equation in the Cartesian coordinate system. Letting $\Delta x = h$ and $\Delta y = k$, the formula becomes

$$u_{i+1,j+1} + u_{i-1,j+1} + u_{i+1,j-1} + u_{i-1,j-1} - 2\frac{h^2 - 5k^2}{h^2 + k^2}(u_{i+1,j} + u_{i-1,j})$$
$$+ 2\frac{5h^2 - k^2}{h^2 + k^2}(u_{i,j+1} + u_{i,j-1}) - 20u_{i,j} = 0 \tag{4.114}$$

The T.E. for this scheme is $O(h^2, k^2)$ but becomes $O(h^6)$ on a square mesh. Details of the T.E. and modified equation for this scheme are left as an exercise. Although the nine-point formula appears to be very attractive for Laplace's equation because of the favorable T.E., this error may be only $O(h^2, k^2)$ when applied to a more general elliptic equation (including the Poisson equation) containing other terms. More details on the nine-point scheme can be found in the work by Lapidus and Pinder (1982).

Other finite-difference schemes for Laplace's equation can be found in the literature (see, for example, Thom and Apelt, 1961), but none seems to offer significant advantages over the five- and nine-point schemes given here. To obtain smaller formal T.E. in these schemes, more grid points must be used in the difference molecules. High accuracy is difficult to maintain near boundaries with such schemes.

Residual form of the difference equations. In some solution schemes it is advantageous to solve the difference equations in delta or residual form. We will illustrate the residual form by way of an example based on the five-point stencil given in Eq. (4.113). We let L be a difference operator giving the five-point difference representation. Thus, $Lu_{i,j} = 0$ is equivalent to Eq. (4.113). A *delta* (change in the variable) is defined by $u_{i,j} = \tilde{u}_{i,j} + \Delta u_{i,j}$, where $\tilde{u}_{i,j}$ represents a provisional solution such as might occur at some point in an iterative process before convergence, and $u_{i,j}$ represents the exact numerical solution to the difference equation. (Readers should note that all deltas that arise in computational fluid dynamics are not defined in the same way. The deltas may have slightly different meanings, depending upon the algorithm or application. Deltas denote a change in something, but be alert to exactly how the delta is defined.) We can substitute $\tilde{u}_{i,j} + \Delta u_{i,j}$ for $u_{i,j}$ in the difference equation $Lu_{i,j} = 0$ and obtain

$$Lu_{i,j} = L\tilde{u}_{i,j} + L\,\Delta u_{i,j} = 0 \tag{4.115}$$

The *residual* is defined as the number that results when the difference equation, written in a form giving zero on the right-hand side, is evaluated for an intermediate or provisional solution. For Laplace's equation the residual can be evaluated as $R_{i,j} = L\tilde{u}_{i,j}$. If the provisional solution satisfies the difference equation exactly, the residual vanishes. With this definition, Eq. (4.115) can be written as

$$L\,\Delta u_{i,j} = -R_{i,j} \tag{4.116}$$

Equation (4.116) is an alternate and equivalent form of the difference equation for Laplace's equation. Starting with any provisional solution or, in fact, a simple guess, allows the residual to be computed at each grid point. From there, Eq. (4.116) can be solved for the deltas that are then added to the provisional solution. In some iterative schemes, the residuals are updated at each iteration, and new deltas are computed. To update the solution for u, the delta values are added to the provisional solution used to compute the residual. An example below in connection with the multigrid method will help clarify these ideas.

4.3.2 Simple Example for Laplace's Equation

Consider how we might determine a function satisfying

$$\frac{\partial^2 u}{\partial x^2} + \frac{\partial^2 u}{\partial y^2} = 0$$

on the square domain

$$0 \leqslant x \leqslant 1 \qquad 0 \leqslant y \leqslant 1$$

subject to Dirichlet boundary conditions. Series solutions can be obtained for this problem (most readily by separation of variables) satisfying certain distributions of u at the boundaries. These are available in most textbooks that cover conduction heat transfer (Chapman, 1974) and can be used as test cases to verify the finite-difference formulation. In this example, we will use the five-point scheme, Eq. (4.113), and let $\Delta x = \Delta y = 0.1$, resulting in a uniform 11×11 grid over the square problem domain (see Fig. 4.21). With $\Delta x = \Delta y$, the difference equation can be written as

$$u_{i+1,j} + u_{i-1,j} + u_{i,j+1} + u_{i,j-1} - 4u_{i,j} = 0 \tag{4.117}$$

for each point where u is unknown. In this example problem with Dirichlet boundary conditions, we have 81 grid points where u is unknown. For each one of those points, we can write the difference equation so that our problem is one of solving the system of 81 simultaneous linear algebraic equations for the 81 unknown $u_{i,j}$. Mathematically, our problem can be expressed as

$$\begin{aligned}
a_{11}u_1 + a_{12}u_2 + \cdots\cdots\cdots a_{1n}u_n &= c_1 \\
a_{21}u_1 + a_{22}u_2 + \cdots\cdots\cdots a_{2n}u_n &= c_2 \\
\vdots \qquad\qquad\qquad\qquad \vdots \quad &\quad \vdots \\
a_{n1}u_1 + \cdots\cdots\cdots\cdots\cdots\cdots\cdots a_{nn}u_n &= c_n
\end{aligned} \tag{4.118}$$

or more compactly as $[A]\mathbf{u} = \mathbf{C}$, where $[A]$ is the matrix of known coefficients, $\mathbf{u}$ is the column vector of unknowns, and $\mathbf{C}$ is a column vector of known quantities. It is worth noting that the matrix of coefficients will be very sparse, since about 76 of the 81 a's in each row will be zero. To make our example algebraically as simple as possible, we have let $\Delta x = \Delta y$. If $\Delta x \neq \Delta y$, the

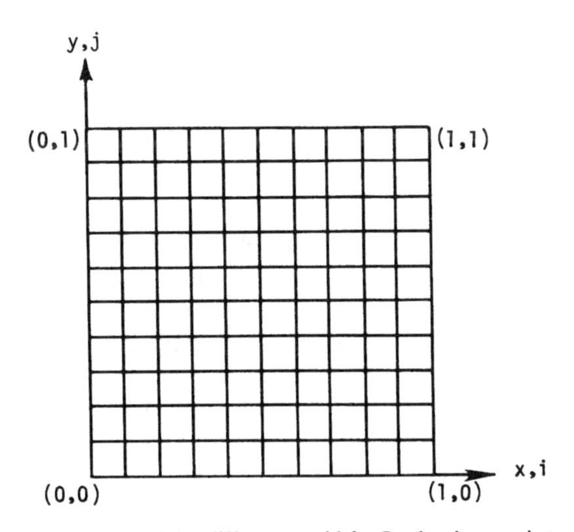

Figure 4.21 Finite-difference grid for Laplace's equation.

coefficients will be a little more involved, but the algebraic equations will still be linear and can be represented by the general $[A]\mathbf{u} = \mathbf{C}$ system given above.

Methods for solving systems of linear algebraic equations can be readily classified as either direct or iterative. Direct methods are those that give the solution (exactly if round-off error does not exist) in a finite and predeterminable number of operations using an algorithm that is often quite complicated. Iterative methods consist of a repeated application of an algorithm that is usually quite simple. They yield the exact answer only as a limit of a sequence, but if the iterative procedure converges, we can come within ϵ of the answer in a finite but usually not predeterminable number of operations. Some examples of both types of methods will be given.

4.3.3 Direct Methods for Solving Systems of Linear Algebraic Equations

Cramer's rule. Cramer's rule is one of the most elementary methods. All students have certainly heard of it, and most are familiar with the workings of the procedure. Unfortunately, the algorithm is immensely time consuming, the number of operations being approximately proportional to $(n + 1)!$, where n is the number of unknowns. A number of horror stories have been told about the large computation time required to solve systems of equations by Cramer's rule. The number of operations (multiplications and divisions) required to solve a system of algebraic equations by Gauss elimination (described below) is approximately $n^3/3$. The example problem discussed above for an 11×11 grid involved 81 unknowns. The operation count for solving this problem by Cramer's rule is 82!, which is a very large number indeed (4.75×10^{122}), dwarfing even the national debt. By comparison, solving the problem by Gauss elimination would require only about 177,147 operations. If we applied Cramer's rule to the example problem with 81 unknowns using a machine capable of performing 100 million floating point operations per second (100 megaflops), the calculation would require about 3.20×10^{101} years. This is not worth waiting for! Using Gauss elimination would only require a fraction of a second on the same machine. Cramer's rule should never be used for more than about three unknowns, since it rapidly becomes very inefficient as the number of unknowns increases.

Gaussian elimination. Gaussian elimination is a very useful and efficient tool for solving many systems of algebraic equations, particularly for the special case of a tridiagonal system of equations. However, the method is not as fast as some others to be considered for more general systems of algebraic equations that arise in solving PDEs. Approximately $n^3/3$ multiplications and divisions are required in solving n equations. Also, round-off errors, which can accumulate through the many algebraic operations, sometimes cause deterioration of accuracy when n is large. Actually, the accuracy of the method depends on the specific system of equations, and the matter is too complex to resolve by a simple general statement. Rearranging the equations to the extent possible in order to put the coefficients that are largest in magnitude on the main diagonal (known as "pivoting") will tend to improve accuracy. However, since we will want to use an elimination scheme for tridiagonal systems of equations that arise in implicit difference schemes for marching problems, it would be well to gain some notion of how the basic Gaussian elimination procedure works.

Consider the equations

$$\begin{aligned} a_{11}u_1 + a_{12}u_2 + \cdots\cdots\cdots &= c_1 \\ a_{21}u_1 + a_{22}u_2 + \cdots\cdots\cdots &= c_2 \\ \vdots \qquad\qquad\qquad & \quad \vdots \\ a_{n1}u_n + \cdots\cdots\cdots\cdots\cdots &= c_n \end{aligned} \tag{4.119}$$

The objective is to transform the system into an upper triangular array by eliminating some of the unknowns from some of the equations by algebraic operations. To illustrate, we choose the first equation (row) as the "pivot" equation and use it to eliminate the u_1 term from each equation below it. This is done by multiplying the first equation by a_{21}/a_{11}† and subtracting it from the second equation to eliminate u_1 from the second equation. Multiplying the pivot equation by a_{31}/a_{11} and subtracting it from the third equation eliminates the first term from the third equation. This procedure can be continued to eliminate the u_1 from equations 2 through n. The system now appears in Fig. 4.22.

Next, the second equation (as altered by the above procedure) is used as the pivot equation to eliminate u_2 from all equations below it, leaving the system in the form shown in Fig. 4.23. The third equation in the altered system is then used as the next pivot equation and the process continues until only an upper triangular form remains:

$$\begin{aligned} a_{11}u_1 + a_{12}u_2 + \cdots\cdots\cdots\cdots &= c_1 \\ a'_{22}u_2 + a'_{23}u_3 + \cdots\cdots\cdots &= c'_2 \\ a'_{33}u_3 + \cdots\cdots\cdots &= c'_3 \\ &\ \vdots \\ a'_{nn}u_n &= c'_n \end{aligned} \tag{4.120}$$

† We must always interchange rows if necessary to avoid division by zero.

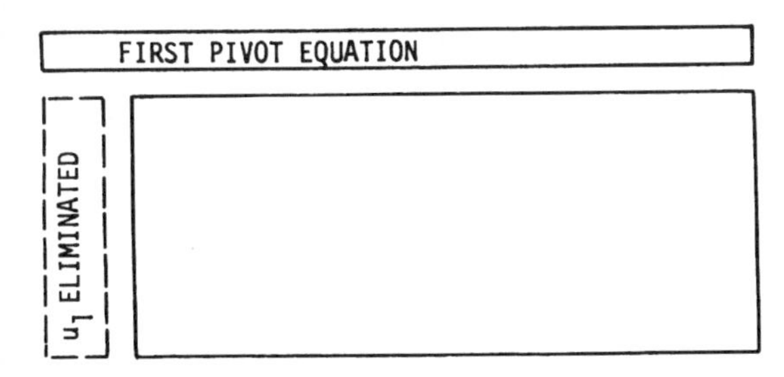

Figure 4.22 Gaussian elimination, u_1 eliminated below main diagonal.

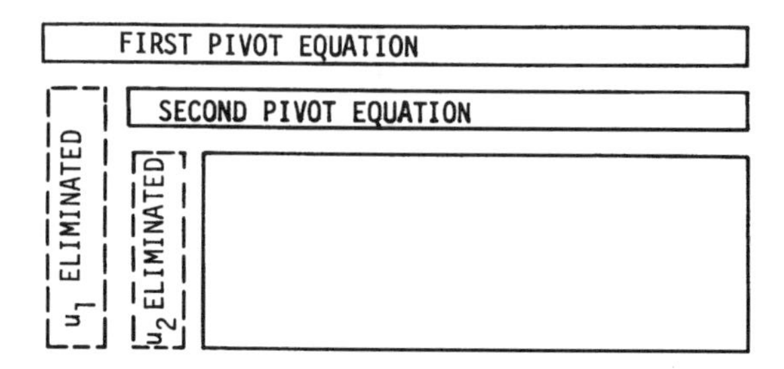

Figure 4.23 Gaussian elimination, u_1 and u_2 eliminated below main diagonal.

At this point, only one unknown appears in the last equation, two in the next to last equation, etc., so a solution can be obtained by back substitution.

Consider the following system of three equations as a specific numerical example:

$$\begin{aligned} U_1 + 4U_2 + U_3 &= 7 \\ U_1 + 6U_2 - U_3 &= 13 \\ 2U_1 - U_2 + 2U_3 &= 5 \end{aligned}$$

Using the top equation as a pivot, we can eliminate U_1 from the lower two equations:

$$\begin{aligned} U_1 + 4U_2 + U_3 &= 7 \\ 2U_2 - 2U_3 &= 6 \\ -9U_2 + 0 &= -9 \end{aligned}$$

Now using the second equation as a pivot, we obtain the upper triangular form:

$$\begin{aligned} U_1 + 4U_2 + U_3 &= 7 \\ 2U_2 - 2U_3 &= 6 \\ -9U_3 &= 18 \end{aligned}$$

Back substitution yields $U_3 = -2$, $U_2 = 1$, $U_1 = 5$.

Block-iterative methods for Laplace's equation (Section 4.3.4) lead to systems of simultaneous algebraic equations which have a tridiagonal matrix of coefficients. This was also observed in Sections 4.1 and 4.2 for implicit formulations of PDEs for marching problems. To illustrate how Gaussian elimination can be efficiently modified to take advantage of the tridiagonal form of the coefficient matrix, we will consider the simple implicit scheme for the heat equation as an example:

$$\frac{\partial u}{\partial t} = \alpha \frac{\partial^2 u}{\partial x^2}$$

$$\frac{u_j^{n+1} - u_j^n}{\Delta t} = \frac{\alpha}{(\Delta x)^2}\left(u_{j+1}^{n+1} + u_{j-1}^{n+1} - 2u_j^{n+1}\right)$$

In terms of the format used before for algebraic equations, this can be rewritten as

$$b_j u_{j-1}^{n+1} + d_j u_j^{n+1} + a_j u_{j+1}^{n+1} = c_j$$

where

$$a_j = b_j = -\frac{\alpha\,\Delta t}{(\Delta x)^2} \qquad c_j = u_j^n \qquad d_j = 1 + \frac{2\alpha\,\Delta t}{(\Delta x)^2}$$

For Dirichlet boundary conditions, u_{j-1}^{n+1} is known at one boundary and u_{j+1}^{n+1} at the other. All known u are collected into the c_j term, so our system looks like

$$\begin{bmatrix} d_1 & a_1 & 0 & \cdot & \cdot & \cdot & \cdot & 0 \\ b_2 & d_2 & a_2 & 0 & \cdot & \cdot & \cdot & \cdot \\ 0 & b_3 & d_3 & a_3 & 0 & \cdot & \cdot & \cdot \\ 0 & 0 & b_4 & d_4 & a_4 & 0 & \cdot & \cdot \\ \cdot & \cdot & \cdot & \ddots & \ddots & \ddots & \cdot & \cdot \\ \cdot & \cdot & \cdot & & \ddots & \ddots & \ddots & 0 \\ \cdot & \cdot & \cdot & \cdot & \cdot & \ddots & \ddots & a_{NJ-1} \\ 0 & \cdot & \cdot & \cdot & \cdot & 0 & b_{NJ} & d_{NJ} \end{bmatrix} \begin{bmatrix} u_1^{n+1} \\ u_2^{n+1} \\ \cdot \\ \cdot \\ \cdot \\ \cdot \\ \cdot \\ u_{NJ}^{n+1} \end{bmatrix} = \begin{bmatrix} c_1 \\ c_2 \\ \cdot \\ \cdot \\ \cdot \\ \cdot \\ \cdot \\ c_{NJ} \end{bmatrix}$$

Even when other boundary conditions apply, the system can be cast into the above form, although the first and last equations in the array may result from auxiliary relationships related to the boundary conditions and not to the original difference equation, which applies to nonboundary points.

For this tridiagonal system it is easy to modify the Gaussian elimination procedure to take advantage of the zeros in the matrix of coefficients. This modified procedure, suggested by Thomas (1949), is discussed briefly in Section 4.1.4.

Thomas algorithm. Referring to the tridiagonal matrix of coefficients above, the system is put into an upper triangular form by computing the new d_j by

$$d_j = d_j - \frac{b_j}{d_{j-1}} a_{j-1} \qquad j = 2, 3, \ldots, NJ$$

and the new c_j by

$$c_j = c_j - \frac{b_j}{d_{j-1}} c_{j-1} \qquad j = 2, 3, \ldots, NJ$$

then computing the unknowns from back substitution according to $u_{NJ} = c_{NJ}/d_{NJ}$, and then

$$u_k = \frac{c_k - a_k u_{k+1}}{d_k} \qquad k = NJ - 1, NJ - 2, \ldots, 1$$

In the above equations, the equals sign means "is replaced by," as in the FORTRAN programming language. A FORTRAN program for this procedure is given in Appendix A.

Some flexibility exists in the way in which boundary conditions are handled when the Thomas algorithm is used to solve for the unknowns. It is best that the reader develop an appreciation for these details through experience; however, a comment or two will be offered here by way of illustration. The main purpose of the elimination scheme is to determine the unknowns; therefore, for Dirichlet boundary conditions, the u's at the boundary need not be included in the list of unknowns. That is, u_1 in the elimination algorithm could correspond to the u at the first nonboundary point, and u_{NJ} to the u at the last nonboundary point. However, no harm is done, and programming may be made easier, by specializing a_1, d_1, b_{NJ}, and d_{NJ} to provide a redundant statement of the boundary conditions. That is, if we let $d_1 = 1$, $a_1 = 0$, $d_{NJ} = 1$, $b_{NJ} = 0$, $c_1 = u_1^{n+1}$ (given), and $c_{NJ} = d_{NJ}^{n+1}$ (given), the first and last algebraic equations become just a statement of the boundary conditions. As an example of how other boundary conditions fall easily into the tridiagonal format, consider convective (mixed) boundary conditions for the heat equation:

$$h(u_\infty - u_{\text{bdy}}) = -k \left.\frac{\partial u}{\partial x}\right)_{\text{bdy}}$$

A control-volume analysis at the boundary where $j = 1$ leads to a difference equation that can be written as

$$d_1 u_1^{n+1} + a_1 u_2^{n+1} = c_1$$

where

$$d_1 = 1 + \frac{2\alpha\,\Delta t}{(\Delta x)^2}\left(1 + \frac{h\,\Delta x}{k}\right)$$

$$a_1 = \frac{-2\alpha\,\Delta t}{(\Delta x)^2} \qquad c_1 = \frac{2\alpha(\Delta t)h(\Delta x)}{(\Delta x)^2 k}u_\infty + u_1^n$$

which obviously fits the tridiagonal form for the first row.

Advanced-direct methods. Direct methods for solving systems of algebraic equations that are faster than Gaussian elimination certainly exist. Unfortunately, none of these methods is completely general. That is, they are applicable only to the algebraic equations arising from a special class of difference equations and associated boundary conditions. Many of these methods are "field size limited" (limited in applicability to relatively small systems of algebraic equations) owing to the accumulation of round-off errors. As a class, the algorithms for fast direct procedures tend to be rather complicated and not easily adapted to irregular problem domains or complex boundary conditions. Somewhat more computer storage is usually required than for an iterative method suitable for a given problem. In addition, the best iterative methods developed in recent years (mainly multigrid methods) are actually faster than the direct methods. It seems that the simplest of the direct methods suffer from the field size limitations and are relatively restricted in their range of application, and those that are not field size limited have algorithms with very involved details that are beyond the scope of this text. Consequently, only a few of these methods will be mentioned here, and none will be discussed in detail.

One of the simplest of the advanced, direct methods is the error vector propagation (EVP) method developed for the Poisson equation by Roache (1972). This method is field size limited; however, the concepts are straightforward. Two fast direct methods for the Poisson equation that are not limited, owing to accumulation of round-off errors, are the "odd-even reduction" method of Buneman (1969) and the fast Fourier transform method of Hockney (1965, 1970). Swartztrauber (1977) discusses optimal combinations of the two schemes. More recent developments, including implementation details for supercomputers, are discussed by Hockney and Jesshope (1981). Clearly, the fast direct methods should be considered for problems where the geometry and boundary conditions will permit their use and when computer execution time is an overriding consideration. These methods can be 10–30 times faster than the simpler iterative methods but are not expected to be faster than the multigrid iterative methods, which will be introduced below.

Another type of advanced direct method is known as a *sparse matrix method*, the most well known of which is the Yale sparse matrix package (Eisenstadt et al., 1977). Unlike the "fast" solvers discussed above, the sparse matrix methods can be quite general. The methods are essentially "smart" elimination methods that take advantage of the sparseness of the coefficient matrix. The methods generally require extensive computer memory and are not more efficient than the better iterative methods. Examples of the use of such methods to solve the Navier-Stokes equations can be found in the works of Bender and Khosla (1988) and Venkatakrishnan and Barth (1989).

4.3.4 Iterative Methods for Solving Systems of Linear Algebraic Equations

Iterative methods are also known as "relaxation" methods, a term derived from the residual relaxation method introduced by Southwell many years ago. This class of methods can be further broken down into point- (or explicit) iterative methods and block- (or implicit) iterative methods. In brief, for point-iterative methods, the same simple algorithm is applied to each point where the unknown function is to be determined in successive iterative sweeps, whereas in block-iterative methods, subgroups of points are singled out for solution by elimination (direct) schemes in an overall iterative procedure.

Gauss-Seidel iteration. Although many different iterative methods have been suggested over the years, Gauss-Seidel iteration (often called Liebmann iteration when applied to the algebraic equation resulting from the differencing of an elliptic PDE) is one of the most efficient and useful point-iterative procedures for large systems of equations. The method is extremely simple but only converges under certain conditions related to "diagonal dominance" of the matrix of coefficients. Fortunately, the differencing of many steady-state conservation statements provides this diagonal dominance. The method makes explicit use of the sparseness of the matrix of coefficients.

The simplicity of the procedure will be demonstrated by an example prior to a concise statement regarding the sufficient condition for convergence. When the method can be used, the procedure for a general system of algebraic equations would be to (1) make initial guesses for all unknowns (a guessed value for one unknown will not be needed, as seen in example below); (2) solve each equation for the unknown whose coefficient is largest in magnitude, using guessed values initially and the most recently computed values thereafter for the other unknowns in each equation; (3) repeat iteratively the solution of the equations in this manner until changes in the unknowns become "small," remembering to use the most recently computed value for each unknown when it appears on the right-hand side of an equation. As an example, consider the system

$$\begin{aligned} 4x_1 - x_2 + x_3 &= 4 \\ x_1 + 6x_2 + x_3 &= 9 \\ -x_1 + 2x_2 + 5x_3 &= 2 \end{aligned}$$

We would first rewrite the equations as

$$\begin{aligned} x_1 &= \tfrac{1}{4}(4 + x_2 - x_3) \\ x_2 &= \tfrac{1}{6}(9 - x_1 - x_3) \\ x_3 &= \tfrac{1}{5}(2 + x_1 - 2x_2) \end{aligned}$$

then make initial guesses for x_2 and x_3 (a guess for x_1 is not needed) and compute x_1, x_2, x_3 iteratively as indicated above.

Referring to the five-point stencil for Laplace's equation, we observe that the unknown having the coefficient largest in magnitude is $u_{i,j}$. Letting $\beta = \Delta x/\Delta y$, the grid aspect ratio, the general equation for the Gauss-Seidel procedure for Laplace's equation can be written as

$$u_{i,j}^{k+1} = \frac{u_{i+1,j}^{k} + u_{i-1,j}^{k+1} + \beta^2\left(u_{i,j+1}^{k} + u_{i,j-1}^{k+1}\right)}{2(1+\beta^2)} \tag{4.121}$$

where k denotes the iterative level, i denotes the column, and j the row. In Eq. (4.121), the sweep direction is assumed to be from low values of i and j to large values. Thus, at least two unknowns in each equation would already have been calculated at the $k+1$ level. In terms of a general system of equations,

$[A]\mathbf{x} = \mathbf{b}$, the Gauss-Seidel scheme can be written as

$$x_i^{k+1} = \frac{1}{a_{ii}}\left(b_i - \sum_{j=1}^{i-1} a_{ij}x_j^{k+1} - \sum_{j=i+1}^{n} a_{ij}x_i^k\right) \tag{4.122}$$

where it is understood that the system of equations has been reordered, if necessary, so that the coefficients largest in magnitude are on the main diagonal.

In passing, we shall mention that an iterative process can also be performed without continuously updating values on the right-hand side. If the unknowns on the right-hand side are updated only after each iterative sweep through the entire field, the process is known as *Jacobi iteration*. For model problems, Jacobi iteration requires approximately twice as many iterations for convergence as Gauss-Seidel iteration.

Sufficient condition for convergence of the Gauss-Seidel procedure. In order to provide a compact notation, as above, we will order the equations, if possible, so that the coefficient largest in magnitude in each row is on the main diagonal. Then if the system is irreducible (cannot be arranged so that some of the unknowns can be determined by solving less than n equations) and if

$$|a_{ii}| \geqslant \sum_{\substack{j=1 \\ j\neq i}}^{n} |a_{ij}| \tag{4.123}$$

for all i and if

$$|a_{ii}| > \sum_{\substack{j=1 \\ j\neq i}}^{n} |a_{ij}| \tag{4.124}$$

for at least one i, then the Gauss-Seidel iteration will converge. This is a *sufficient condition*, which means that convergence may sometimes be observed when the above condition is not met. A necessary condition can be stated, but it is impractical to evaluate. Stated in words, the sufficient condition can be interpreted as requiring for each equation that the magnitude of the coefficient on the diagonal be greater than or equal to the sum of the magnitudes of the other coefficients in the equation, with the "greater than" holding for at least one (usually corresponding to a point near a boundary for a physical problem) equation.

Perhaps we should now relate the above iterative convergence criteria to the system of equations that results from differencing Laplace's equation according to Eq. (4.113). First we observe that the coefficient largest in magnitude belongs to $u_{i,j}$. Since we apply Eq. (4.113) to each point where $u_{i,j}$ is unknown, we could clearly arrange all the equations in the system so that the coefficient largest in magnitude appeared on the diagonal. With the exercise of proper care in establishing difference representations, this type of diagonal dominance can normally be achieved for all elliptic equations. In terms of a linear difference equation for u, we would expect the Gauss-Seidel iterative procedure to converge if the finite-difference equation applicable to each point i, j, where $u_{i,j}$ is unknown, is such that the magnitudes of the coefficient of $u_{i,j}$ is greater than or equal to the sum of the magnitudes of the coefficients of the other unknowns in the equation. The "greater than" must hold for at least one equation.

We will not offer a proof for this sufficient condition for the convergence of the Gauss-Seidel iteration, but hopefully, a simple example will suggest why it is true. If we look back to our simple three-equation example for Gauss-Seidel iteration and consider that at any point our intermediate values of x are the exact solution plus some ϵ, i.e., $x_1 = (x_1)_{\text{exact}} + \epsilon_1$, then our condition of diagonal dominance is forcing the ϵ to become smaller and smaller as the iteration is repeated cyclically. For one run through the iteration, we could observe

$$|\epsilon_1^2| \leqslant \tfrac{1}{4}|\epsilon_2^1| + \tfrac{1}{4}|\epsilon_3^1|$$

$$|\epsilon_2^2| \leqslant \tfrac{1}{6}|\epsilon_1^2| + \tfrac{1}{6}|\epsilon_3^1|$$

$$|\epsilon_3^2| \leqslant \tfrac{1}{5}|\epsilon_1^2| + \tfrac{2}{5}|\epsilon_2^2|$$

If ϵ_2^1 and ϵ_3^1 were initially each 10, $|\epsilon_1^2|$ would be $\leqslant 5$ and $|\epsilon_1^3| \leqslant 1.446$. Here, superscripts denote iterative level.

Finally, we note for a general system of equations, the multiplications per iteration could be as great as n^2 but could be much less if the matrix was sparse.

Successive overrelaxation. Successive overrelaxation (SOR) is a technique that can be used in an attempt to accelerate any iterative procedure, but we will propose it here primarily as a refinement to the Gauss-Seidel method. As to the origins of the method, one story (probably inaccurate) being passed around is that the method was suggested by a duck hunter, who finally learned that if he pointed his gun ahead of the duck, he would score more hits than if he pointed the gun right at the duck. The duck is a moving target, and if we anticipate its motion, we are more likely to hit it with the shot pattern. The duck hunter told his story to his neighbor, who was a numerical analyst, and SOR was born—or so the story goes.

As we apply Gauss-Seidel iteration to a system of simultaneous algebraic equations, we expect to make several recalculations or iterations before convergence to an acceptable level is achieved. Suppose that during this process we observe the change in the value of the unknown at a point between two successive iterations, note the direction of change, and anticipate that the same trend will continue on to the next iteration. Why not go ahead and make a correction to the variable in the anticipated direction *before* the next iteration, thereby, hopefully, accelerating the convergence? An arbitrary correction to the intermediate values of the unknowns from *any* iterative procedure (Gauss-Seidel iteration is of most interest to us at this point, so we will use it as the representative iterative scheme), according to the form

$$u_{i,j}^{k+1'} = u_{i,j}^{k'} + \omega\left(u_{i,j}^{k+1} - u_{i,j}^{k'}\right) \tag{4.125}$$

is known as overrelaxation or successive overrelaxation. Here, k denotes iteration level, $u_{i,j}^{k+1}$ is the most recent value of $u_{i,j}$ calculated from the Gauss-Seidel procedure, $u_{i,j}^{k'}$ is the value from the previous iteration as adjusted by previous application of this formula if the overrelaxation is being applied successively (at each iteration), and $u_{i,j}^{k+1'}$ is the newly adjusted or "better guess" for $u_{i,j}$ at the $k+1$ iteration level. That is, we expect $u_{i,j}^{k+1'}$ to be closer to the final solution than the unaltered value $u_{i,j}^{k+1}$ from the Gauss-Seidel calculation. The formula is applied immediately at each point after $u_{i,j}^{k+1}$ has been obtained, and $u_{i,j}^{k+1'}$ replaces $u_{i,j}^{k+1}$ in all subsequent calculations in the cycle. Here, ω is the relaxation parameter, and when $1 < \omega < 2$, *overrelaxation* is being employed. Overrelaxation can be likened to linear extrapolation based on values $u_{i,j}^{k'}$ and $u_{i,j}^{k+1}$. In some problems, *underrelaxation* $0 < \omega < 1$ is employed. Underrelaxation appears to be most appropriate when the convergence at a point is taking on an oscillatory pattern and tending to "overshoot" the apparent final solution. For underrelaxation the adjusted value, $u_{i,j}^{k+1'}$ is between $u_{i,j}^{k'}$ and $u_{i,j}^{k+1}$. Overrelaxation is usually appropriate for numerical solutions to Laplace's equation with Dirichlet boundary conditions. Underrelaxation is sometimes called for in elliptic problems, it seems, when the equations are nonlinear. Occasionally, for nonlinear problems, underrelaxation is even observed to be necessary for convergence.

We note that the relaxation parameter should be restricted to the range $0 < \omega < 2$. For convergence, we require that the magnitude of the changes in u from one iteration to the next become smaller. Use of $\omega \geqslant 2$ forces these changes to remain the same or to increase, in contradiction to convergent behavior.

Two important remaining questions are, how can we properly determine a good or even the best value for ω, and by how much does this procedure accelerate the convergence? No completely general answers to these questions are available, but some guidelines can be drawn.

For Laplace's equation on a rectangular domain with Dirichlet boundary conditions, theories pioneered by Young (1954) and Frankel (1950) lead to an expression for the optimum ω (hereafter denoted by ω_{opt}). First, defining σ as

$$\sigma = \frac{1}{1+\beta^2}\left(\cos\frac{\pi}{p} + \beta^2\cos\frac{\pi}{q}\right) \tag{4.126}$$

the optimum ω is given by

$$\omega_{\text{opt}} = \frac{2}{1 + (1-\sigma^2)^{1/2}} \tag{4.127}$$

where β is the grid aspect ratio as defined previously, p is the number of Δx increments, and q is the number of Δy increments along the sides of the rectangular region. The formula can also be used for problems in rectangular regions with certain combinations of Dirichlet and Neumann boundary conditions that permit an equivalent Dirichlet problem to be recognized by identifying the Neumann boundaries as lines of symmetry in a Dirichlet problem. In general, however, for more complex elliptic problems it is not possible to determine ω_{opt} in advance. In these cases, some numerical experimentation should be helpful in identifying useful values for ω. Numerical examples and theory generally indicate that it is better to guess on the high side of ω_{opt} than on the low side. Hageman and Young (1981) discuss considerations in the search for ω_{opt} in some detail.

Is the ω search worthwhile? The answer is emphatically yes. In some problems it is possible to reduce the computation time by a factor of 30. This is significant! Occasionally, SOR may be found not to be of much help in accelerating convergence, but it should always be considered and evaluated. The potential for savings in computation time is too great to ignore.

Since overrelaxation can be viewed as applying a correction to the values obtained from the Gauss-Seidel procedure based on extrapolation from previous iterates, it is natural to wonder if other, perhaps more accurate (in terms of T.E. of the extrapolation formula) extrapolation schemes can be used to accelerate the convergence of iterative procedures. In fact, other schemes such as Aitken and Richardson extrapolation have been used in this application. The details of these extrapolation schemes are covered in standard texts on numerical analysis, but as is perhaps expected, any advantage in accelerating the convergence of the iterative process by using more complex extrapolation schemes has to be weighed against any added computation costs due to requirements of additional storage or algebraic operations. SOR has simplicity in its favor, and it can be programmed so that no additional arrays need to be stored.

In the SOR scheme the calculations normally proceed in a systematic way with sweeps from the lower left-hand corner of the domain to the upper right-hand side (in two dimensions from low values of i and j to high values of i and j). This scheme has a bias in terms of sweep direction that may permit the largest errors to accumulate at the high values of i and j. A modification to SOR known as symmetric successive overrelaxation (SSOR) attempts to improve upon this condition. In SSOR, one alternates the sweep direction. A pass from low values of i and j to high values of i and j is followed by a sweep from high i and j to low i and j.

Coloring schemes. Supercomputers in common use today employ *vector processing* to obtain greater execution speeds. The *vectorization* occurs in FORTRAN DO loops (only in the innermost loop if the loops are nested) and can be thought of as a simultaneous execution of the statements in the DO loop for all values of the DO parameter. If the statements in the DO loop are recursive in nature, i.e., the right-hand side of the statement contains results previously computed in the loop, then the compiler rejects that loop for vectorization because an apparent error would occur if the statements were executed simultaneously. Vectorization naturally speeds up the algorithm and is a desirable feature. An example of a vectorizable FORTRAN DO loop is

```
   do 10 j = 1, nj
   a(j) = b(j) * c + d
10 continue
```

and an example of a recursive loop that cannot be performed as given in a vector manner is

```
   do 10 j = 1, nj
   a(j) = a(j - 1) * c + d
10 continue
```

The Gauss-Seidel algorithm is recursive in appearance because of the preference to use the most recent (updated) values of the unknown function on the right-hand side. Thus the algorithm is not vectorizable. Simple Jacobi iteration is vectorizable but converges more slowly than Gauss-Seidel iteration. A variation of the Gauss-Seidel procedure known as the *red-black* or *checkerboard* scheme has approximately the same convergence properties as the Gauss-Seidel procedure but is vectorizable. Imagine that the nodes are colored like a checkerboard, every other point red, alternate points black, as indicated in Fig. 4.24. The red-black scheme updates the variables in two sweeps, much as was done for the hopscotch scheme. This can be thought of as performing a Jacobi iteration on every other point. Sweep 1 updates red points (points for which $i + j$ is even in two dimensions). At this point, black points are surrounded by nodes for which the unknown has been updated (see Fig. 4.24). Sweep 2 updates the black points (points for which $i + j$ is odd). The two sweeps constitute one iteration. The DO loops proceed in strides of 2 (every other point), and a compiler directive may be needed to confirm the vectorization because the appearance of the right-hand sides may give the impression that the

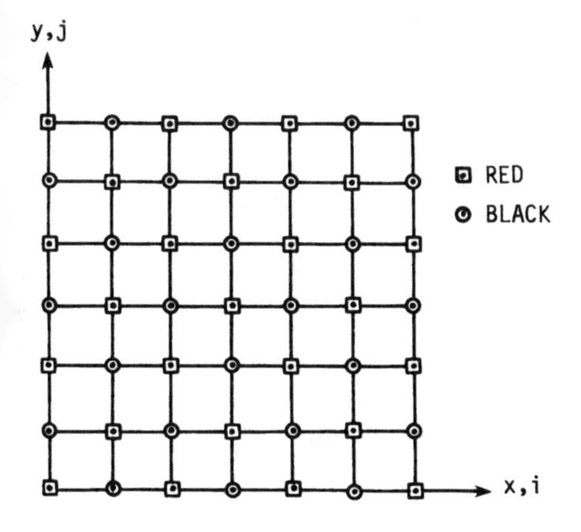

Figure 4.24 Red-black (checkerboard) ordering.

scheme is recursive. The favorable convergence properties arise because some of the updates use information that has been obtained within the same iteration (but not the same DO loop).

For Laplace's equation in two dimensions in the Cartesian coordinate system, only two subdivisions of points (red and black) are necessary to recover the convergence properties of the Gauss-Seidel scheme on a vector machine. The question might arise as to whether more than two colors (or two subdivisions of points) are ever needed. This question is most likely to arise in the use of unstructured grids, for which the cell shape and the ordering of nodes in the solution algorithm may vary greatly. Unstructured grids will be discussed in Chapters 6 and 10. If we restrict our concern to the nearest neighbors and wish to color the domain so that adjacent nodes (or cells) are of a different color, a famous theorem from graph theory states that four colors are sufficient to do this in two dimensions. This fact has been generally accepted for more than 100 years but was only proven by Appel and Haken in 1976. Four colors mean that four DO loops would be used to "visit" every node or cell. Clearly, less than four colors is sometimes adequate, as is the case for a 2-D grid formed in the Cartesian coordinate system.

Block-iterative methods. The Gauss-Seidel iteration method with SOR stands as the best all-around method for the finite-difference solution of elliptic equations discussed in detail thus far in this chapter. The number of iterations can usually be reduced even further by use of block-iterative concepts, but the number of algebraic operations required per iterative cycle generally increases, and whether the reduction in number of required iterative cycles compensates for the extra computation time per cycle is a matter that must be studied for each problem. However, several cases can be cited where the use of block-iterative methods has resulted in a net saving of computation time, so that these procedures warrant serious attention. Ames (1977) and Lapidus and Pinder (1982) present useful discussions that compare the rates of convergence for several point- and block-iterative methods.

In block- (or group) iterative methods, subgroups of the unknowns are singled out, and their values modified simultaneously by obtaining a solution to the simultaneous algebraic equations by elimination methods. Thus the block-iterative methods have an implicit nature and are sometimes known as implicit-iterative methods. In the most common block-iterative methods the unknowns in the subgroups (to be modified simultaneously) are set up so that the matrix of coefficients will be tridiagonal in form, permitting the Thomas algorithm to be used. The simplest block procedure is SOR by lines.

SOR by lines (SLOR). Although SLOR is workable with almost any iterative algorithm, it makes the most sense within the framework of the Gauss-Seidel method with SOR. We can choose either rows or columns for grouping with equal ease. To illustrate the procedure, consider again the solution to Laplace's equation on a square domain with Dirichlet boundary conditions using the five-point scheme. If we agree to start at the bottom of the square and sweep up by rows, we could write, for the general point

$$u_{i,j}^{k+1} = \frac{u_{i+1,j}^{k+1} + u_{i-1,j}^{k+1} + \beta^2\left(u_{i,j+1}^{k} + u_{i,j-1}^{k+1}\right)}{2(1+\beta^2)} \tag{4.128}$$

If we study this equation carefully, we observe that only three unknowns are present, since $u_{i,j-1}^{k+1}$ would be known from either the lower boundary conditions, if we were applying the equation to the first row of unknowns, or from the solution already obtained at the $k+1$ level from the row below. We have chosen to evaluate $u_{i,j+1}$ at the k iteration level rather than the $k+1$ level in order to obtain just three unknowns in the equation, so that the efficient Thomas algorithm can be used. This configuration can be seen in Fig. 4.25.

The procedure is then to solve the system of $I-2$ simultaneous algebraic equations for the $I-2$ unknowns representing the values of $u_{i,j}$ at the $k+1$ iteration level. SOR can now be applied in the same manner as indicated previously before moving on to the next row. Some flexibility exists in the way SOR is applied. After the Thomas algorithm is used to solve Eq. (4.128) for each row, the newly calculated values can be simply overrelaxed, as indicated by Eq. (4.125) before the calculation is advanced to the next row.

Alternatively, the overrelaxation parameter ω can be introduced prior to solution of the simultaneous algebraic equations. This is accomplished by substituting the right-hand side of Eq. (4.128) into the right-hand side of Eq. (4.125) to replace $u_{i,j}^{k+1}$. The resulting equation

$$u_{i,j}^{k+1} = (1-\omega)u_{i,j}^{k} + \frac{\omega}{2(1+\beta^2)}\left[u_{i+1,j}^{k+1} + u_{i-1,j}^{k+1} + \beta^2\left(u_{i,j+1}^{k} + u_{i,j-1}^{k+1}\right)\right]$$

is then solved for each row by the Thomas algorithm. The overrelaxation has been accomplished as part of the row solution and not as a separate step. Since

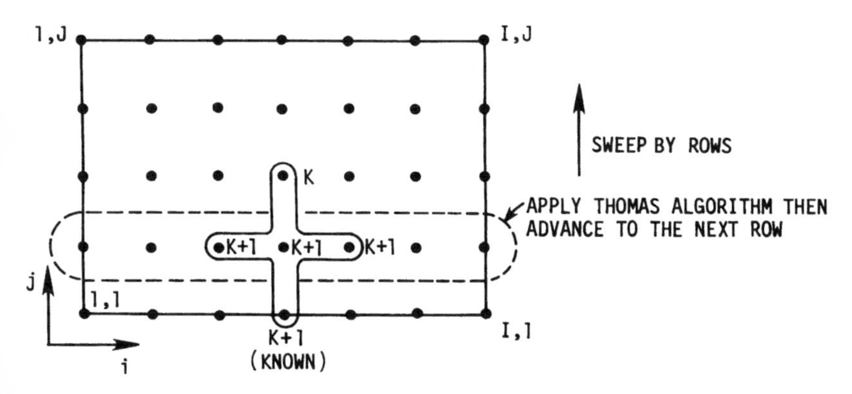

Figure 4.25 SOR by lines.

it is highly desirable to maintain diagonal dominance in the application of the Thomas algorithm, care should be taken when this latter procedure is used to ensure that $\omega \leqslant 1 + \beta^2$.

In the SLOR procedure, one iterative cycle is completed when the tridiagonal inversion has been applied to all the rows. The process is then repeated until convergence has been achieved. In applying the method to a standard example problem with Dirichlet boundary conditions, Ames (1977) indicates that only $1/\sqrt{2}$ as many iterations would be required as for a Gauss-Seidel iteration with SOR to reduce the initial errors by the same amount. On the other hand, use of the Thomas algorithm is expected to increase the computation time per iteration cycle somewhat.

The improved convergence rates observed for block-iterative methods compared with point-iterative methods might be thought of as being due to the greater influence exerted by the boundary values in each iterative pass. For example, in SOR by rows, the unknowns in each row are determined simultaneously, so it is possible for the boundary values to influence all the unknowns in the row in one iteration. This is not the case for point-iterative methods such as the Gauss-Seidel procedure, where in the first pass, at least one of the boundary points (details depend on sequence used in sweeping) only influences adjacent points.

ADI methods. The SLOR method proceeds by taking all lines in the same direction in a repetitive manner. The convergence rate can often be improved by following the sequence by rows, say, by a second sequence in the column direction. Thus a complete iteration cycle would consist of a sweep over all rows followed by a sweep over the columns. Several closely related ADI forms are observed in practice. Perhaps the simplest procedure is to first employ Eq. (4.128) to sweep by rows. We will designate the values so determined by a $k + \frac{1}{2}$ superscript. This is followed by a sweep by columns using

$$u_{i,j}^{k+1} = \frac{u_{i+1,j}^{k+1/2} + u_{i-1,j}^{k+1} + \beta^2\left(u_{i,j+1}^{k+1} + u_{i,j}^{k+1}\right)}{2(1+\beta^2)}$$

This completes one iteration, and overrelaxation can be achieved by applying Eq. (4.125) to all grid points as a second step before another iterative sweep is carried out. Alternatively, we can include the overrelaxation as part of the row and column sweeps using first for rows,

$$u_{i,j}^{k+1/2} = (1-\omega)u_{i,j}^{k} + \frac{\omega}{2(1+\beta^2)}\left[u_{i+1,j}^{k+1/2} + u_{i-1,j}^{k+1/2} + \beta^2\left(u_{i,j+1}^{k} + u_{i,j-1}^{k+1/2}\right)\right] \tag{4.129a}$$

and then for columns

$$u_{i,j}^{k+1} = (1-\omega)u_{i,j}^{k+1/2} + \frac{\omega}{2(1+\beta^2)}\left[u_{i+1,j}^{k+1/2} + u_{i-1,j}^{k+1} + \beta^2\left(u_{i,j+1}^{k+1} + u_{i,j-1}^{k+1}\right)\right] \tag{4.129b}$$

To preserve diagonal dominance in the Thomas algorithm requires $\omega \leqslant 1 + \beta^2$ in the sweep by rows and $\omega \leqslant (1 + \beta^2)/\beta^2$ in the sweep by columns.

Schemes patterned after the ADI procedures for the 2-D heat equation [Eq. (4.97)] are also very commonly used for obtaining solutions to Laplace's equation. If the boundary conditions for an unsteady problem governed by Eq. (4.97) are independent of time, the solution will asymptotically approach a steady-state distribution that satisfies Laplace's equation. Since we are only interested in the "steady-state" solution, the size of the time step can be selected with a view toward speeding convergence of the iterative process. Letting $\alpha\,\Delta t/2 = \rho_k$ in Eq. (4.103), we can write the Peaceman-Rachford ADI scheme for solving Laplace's equation as the two-step procedure:

Step 1: $$u_{i,j}^{k+1/2} = u_{i,j}^{k} + \rho_k\left(\hat{\delta}_x^2 u_{i,j}^{k+1/2} + \hat{\delta}_y^2 u_{i,j}^{k}\right) \tag{4.130a}$$

Step 2: $$u_{i,j}^{k+1} = u_{i,j}^{k+1/2} + \rho_k\left(\hat{\delta}_x^2 u_{i,j}^{k+1/2} + \hat{\delta}_y^2 u_{i,j}^{k+1}\right) \tag{4.130b}$$

where $\hat{\delta}_x^2$ and $\hat{\delta}_y^2$ are defined by Eq. (4.100).

Step 1 proceeds using the Thomas algorithm by rows, and step 2 completes the iteration cycle by applying the Thomas algorithm by columns. The ρ_k are known as iteration parameters, and Mitchell and Griffiths (1980) show that the Peaceman-Rachford iterative procedure for solving Laplace's equation in a square is convergent for any fixed value of ρ_k. On the other hand, for maximum computational efficiency, the iteration parameters should be varied with k, but the same ρ_k should be used in both steps of the iterative cycle. The key to using the ADI method most efficiently for elliptic problems lies in the proper choice of ρ_k. Peaceman and Rachford (1955) suggested one procedure, and another in common usage was suggested by Wachspress (1966). Although the evidence is not conclusive, some studies have suggested that the Wachspress parameters are superior to those suggested by Peaceman and Rachford. The reader is encouraged to study the literature regarding the selection of ρ_k prior to using the Peaceman-Rachford ADI method.

It is difficult to compare the computation times required by point- and block-iterative methods with SOR because of the difficulty in establishing the optimum value of the overrelaxation factor. Conclusions are also very much dependent upon the specific problem considered, the boundary conditions, and the number of grid points involved. The block-iterative methods as a class require fewer iterations than point-iterative methods, but as was mentioned earlier, more computational effort is required by each iteration. Experience suggests that SLOR will require very close to the same computation time as the Gauss-Seidel procedure with SOR for convergence to the same level for most problems. Use of an ADI procedure with SOR (fixed parameter) often provides a savings in computer time of 20-40% over that required by the Gauss-Seidel procedure with SOR. A greater savings can normally be observed if the iteration parameters are suitably varied in the ADI procedure.

Strongly implicit methods. In recent years, another type of block-iterative procedure has been gaining favor as an efficient method for solving the algebraic equations arising from the numerical solution of elliptic PDEs. To illustrate this approach, let us consider the system of algebraic equations arising from the use of the five-point difference scheme for Laplace's equations as

$$[A]\mathbf{u} = \mathbf{C}$$

where $[A]$ is the relatively sparse matrix of known coefficients, $\mathbf{u}$ is the column vector of unknowns, and $\mathbf{C}$ is a column vector of known quantities. It is well known that if the matrix $[A]$ could be factored into the product of upper and lower triangular matrices, the solution for $\mathbf{u}$ could proceed in two sweeps, involving only forward and backward substitution. To do this exactly, however, requires approximately the same effort as solving the system by Gaussian elimination. On the other hand, a number of investigators have explored the merits of obtaining an approximate (or incomplete) $[L][U]$ factorization, which requires less effort than the complete factorization, and then solving for $\mathbf{u}$ iteratively. The strongly implicit procedure (SIP) proposed by Stone (1968) is one example of this factorization strategy. The objective is to replace the sparse matrix $[A]$ by a modified form $[A + P]$ such that the modified matrix can be decomposed into upper and lower triangular sparse matrices denote by $[U]$ and $[L]$, respectively. If the $[L]$ and $[U]$ matrices are not sparse, then very little will be gained in computational efficiency over the use of Gaussian elimination. Thus the key to any computational advantage of the SIP procedure lies in the manner in which $[P]$ is selected. It is essential that the elements of $[P]$ be small in magnitude and permit the set of equations to remain more strongly implicit than for the ADI procedure. An iterative procedure is defined by writing $[A]\mathbf{u} = \mathbf{C}$ as

$$[A + P]\mathbf{u}^{n+1} = \mathbf{C} + [P]\mathbf{u}^n$$

Decomposing $[B] = [A + P]$ into the upper and lower triangular matrices $[U]$ and $[L]$ permits our system to be written as

$$[L][U]\mathbf{u}^{n+1} = \mathbf{C} + [P]\mathbf{u}^n$$

Defining an intermediate vector as $\mathbf{V}^{n+1} = [U]\mathbf{u}^{n+1}$, we form the following two-step algorithm:

Step 1: $$[L]\mathbf{V}^{n+1} = \mathbf{C} + [P]\mathbf{u}^n \tag{4.131a}$$

Step 2: $$[U]\mathbf{u}^{n+1} = \mathbf{V}^{n+1} \tag{4.131b}$$

which is repeated iteratively. Step 1 consists simply of a forward substitution. This is followed by the backward substitution indicated by step 2.

Stone (1968) selected $[P]$ so that $[L]$ and $[U]$ have only three nonzero diagonals, the principal diagonal of $[U]$ being the unity diagonal. Furthermore, the elements of $[L]$ and $[U]$ were determined such that the coefficients in the $[B]$ matrix in the locations of the nonzero entries of matrix $[A]$ were identical with those in $[A]$. Two additional nonzero diagonals appear in $[B]$. The elements of $[L]$, $[U]$, and $[P]$ can be determined from the defining equations established by forming the $[L][U]$ product. The details of this are given by Stone (1968). The procedure is implicit in both the x and y directions. Studies have indicated that, for a solution to Laplace's equations, the method requires only on the order of 50-60% of the computation time required for ADI schemes.

Schneider and Zedan (1981) proposed an alternative procedure for establishing the $[L][U]$ matrices, which is reported to reduce the computational cost for a converged solution to Laplace's equation by a factor of 2–4 over the procedures proposed by Stone (1968). They refer to their alternative procedure as the modified strongly implicit (MSI) procedure. The basic two-step iterative sequence remains the same as given in Eqs. (4.131a) and (4.131b). The improvement apparently results from extending the approach of Stone to a nine-point formulation. The MSI procedure then easily treats five-point difference representations as a special case, and the great reduction in computational cost mentioned above (a factor of 2–4) applies to use of the five-point representation. This new scheme appears to hold great promise as a very efficient and general procedure. Further details on the MSI procedure are given in Appendix C.

Despite the recursive steps in the SIP procedure, it is possible to vectorize the algorithm by structuring DO loops to move along diagonals. The scheme has also been successfully extended to solve coupled systems of equations, i.e., a "block" version of SIP has been developed (see, for example, Zedan and Schneider, 1985; Chen and Pletcher, 1991).

Other iterative methods. The quest continues for more economical and memory efficient iterative methods. Generally, methods that are economical in terms of iteration count or computer time require a relatively large amount of memory. The cost of computer memory has been decreasing rapidly, but memory can still be a limiting factor for 3-D problems involving the Navier-Stokes equations. Another iterative method that appears promising for use with large systems of equations is the generalized minimum residual (GMRES) algorithm introduced by Saad and Schultz (1986). The GMRES algorithm is closely related to the conjugate gradient (see, for example, Golub and van Loan, 1989) procedure but is applicable to problems in which the coefficient matrix may be nonsymmetric. Examples of the application of the GMRES algorithm to flow problems can be found in the works by Wigton et al. (1985), Venkatakrishnan and Mavriplis (1991), and Hixon and Sankar (1992).

4.3.5 Multigrid Method

The Gauss-Seidel method with and without SOR and the block-iterative methods just discussed provide excellent smoothing of the local error. However, because the difference stencil for Laplace's equation is relatively compact, on fine grids a very large number of iterations is often required for the influence of boundary conditions to propagate throughout the grid. Convergence often becomes painfully slow. This violates the "golden rule" of computational physics: "The amount of computational work should be proportional to the amount of real physical changes in the simulated system."

It is the removal of the low-frequency component of the error that usually slows convergence of iterative schemes on a fixed grid. However, a low-frequency component on a fine grid becomes a high-frequency component on a coarse grid. Therefore it makes good sense to use coarse grids to remove the low-frequency errors and propagate boundary information throughout the domain in combination with fine grids to improve accuracy. The strategy known as *multigrid* can do this (Brandt, 1977).

The multigrid method is one of the most efficient general iterative methods known today. The key word here is "general." More efficient schemes can be found for certain problems or certain choices of grids, but it is difficult to find a method more efficient than multigrid for the general case. The multigrid technique can be applied using any of the iterative schemes discussed in this chapter as the "smoother," although the Gauss-Seidel procedure will be used to illustrate the main points of this technique in the introductory material presented here. The objective of the multigrid technique is to accelerate the convergence of an iterative scheme.

To take full advantage of multigrid, several mesh levels are typically used. Normally, the mesh size is increased by a factor of 2 with each coarsening. For many problems the coarsening may continue until the grid consists of one internal point. It would be instructive, however, to illustrate the method first using a two-level scheme applied to Laplace's equation.

The standard Gauss-Seidel scheme will be used based on the five-point stencil. For convenience, let the operator L be defined such that $Lu_{i,j}$ becomes the standard difference representation for the left-hand side of Laplace's equation. That is,

$$Lu_{i,j} = \frac{u_{i+1,j} - 2u_{i,j} + u_{i-1,j}}{(\Delta x)^2} + \frac{u_{i,j+1} - 2u_{i,j} + u_{i,j-1}}{(\Delta y)^2} \tag{4.132}$$

The residual, $R_{i,j}$, has been defined as the number that results when the difference equation, written in a form giving zero on the right-hand side, is evaluated for an intermediate solution. Thus for the present application, $R_{i,j} = Lu_{i,j}$, where it is understood that at convergence, $R_{i,j} = 0$. Let the final converged solution of the difference equations be $u_{i,j}$ and define the corrections, $\Delta u_{i,j}$ by $u_{i,j} = \Delta u_{i,j} + u_{i,j}^k$, where the superscript k denotes iteration level. Thus the correction is the value that must be added to an intermediate solution in order to obtain the final converged solution. Since the difference equation to be solved is $Lu_{i,j} = 0$, we can write

$$L\,\Delta u_{i,j} + Lu_{i,j}^k = 0$$

but

$$Lu_{i,j}^k = R_{i,j}$$

so that

$$L\,\Delta u_{i,j} + R_{i,j} = 0 \tag{4.133}$$

This is known as the residual or delta form of the equation, as discussed earlier. This equation can be solved iteratively for the $\Delta u_{i,j}$ until convergence. If $u_{i,j}$ and $R_{i,j}$ are updated after each iteration, the delta variables vanish upon convergence. Alternatively, if $R_{i,j}$ is held fixed, the iterations will converge, yielding generally finite values for the $\Delta u_{i,j}$, which can be added to the $u_{i,j}^k$ in $R_{i,j}$ to obtain the final value for the solution.

The key idea in multigrid is to improve the fine-grid solution. We do not seek a solution to the original problem on the coarse grid. The coarse grid or grids are only used to obtain corrections to the fine-grid solution. For the present linear PDE (Laplace's equation), we can "transfer the problem" to a coarser grid by interpolating the fine-grid residual to the coarser grid and then solving Eq. (4.133) for the corrections. This form of the multigrid scheme that is applicable to linear PDEs is known as the *coarse-grid correction* scheme or the *correction storage* (CS) scheme. The residuals, of course, would be treated as known, so we would normally rearrange Eq. (4.133) for numerical solution as

$$L\,\Delta u_{i,j} = -R_{i,j} \tag{4.134}$$

and solve for the $\Delta u_{i,j}$ by the Gauss-Seidel procedure. For a two-level scheme, we would proceed through the following steps:

1. Do n iterations on the fine grid, solving Laplace's equation, $Lu_{i,j} = 0$, for $u_{i,j}$ using a "smoother" like the Gauss-Seidel scheme. The value of n would be 3 or 4 in most cases. Do not overrelax—use "pure" Gauss-Seidel. If the solution has not converged after n iterations, compute and store the residual at each fine-grid point.
2. Interpolate the residual onto the coarse grid by using a *restriction operator*. The most common way to do this on a uniform grid is by "injection," which means using the values of the residual at the fine-grid points that coincide with the coarse-grid points. That is, every second fine-grid point will be a coarse-grid point. We have $R_{i,j}$ at these points, so we use it. The chore in practice is in defining or setting up the coarse grid. Solve Eq. (4.134), the residual form of Laplace's equation, for the corrections using zero as the initial guess. [A common mistake here is to neglect to change the grid size in implementing Eq. (4.134); the corrections are being computed on the coarse grid, so the grid increments in the difference equation need to be adjusted accordingly.] The residuals are not updated during the iterations because we want the computed deltas to represent corrections to the fine-grid solution. It is common to iterate the corrections to some predetermined convergence level on the coarsest grid.
3. The corrections are interpolated onto the fine grid using a *prolongation operator*. The simplest procedure is to use bilinear interpolation. This can be

carried out as follows:

(a) Sweep through the coarse-grid rows adding values needed on the fine grid by simply averaging values of the correction existing to the right and left, i.e., simply average the neighboring values of the correction in the row.
(b) Sweep through the coarse-grid columns, adding values needed on the fine grid by averaging values above and below, much as was done for the rows in step 3(a).
(c) Examining Fig. 4.26, we note that to fill out the fine grid, we still need values at the locations marked x. We can obtain these values by sweeping either by rows or columns and filling in with averages of the neighbors above and below or to the right and the left. The result for either method will be identical because of the previous averaging.

The corrections at the fine-grid points are now added to the intermediate solution obtained from step 1. One cycle has now been completed. We would be finished except for errors associated with the interpolation (down to the coarse grid and back up to the fine grid). We now go back to step 1 and iterate n times to obtain an improved solution. If convergence is not indicated, the new residuals are interpolated to the coarse grid, and another cycle is implemented. This continues until convergence is observed on the fine grid.

Most features of the multigrid strategy can be demonstrated by use of the two-level scheme. However, significant improvements in computational economy can be expected by use of additional levels. The extension to additional levels is straightforward in principle but requires careful interpretation of the multigrid concept in order to correctly implement the procedure. Here we will discuss only the simple V cycle, in which the calculation proceeds from the finest grid down to the coarsest and then back up to the finest. Many variations in cycles are possible, and reference to more complete works on multigrid, such as Briggs (1987), Hackbusch and Trottenberg (1982), and Brandt (1977), is recommended in order to obtain more complete information on the multigrid concept. As before, the scheme will be applied to solve Laplace's equation on a square

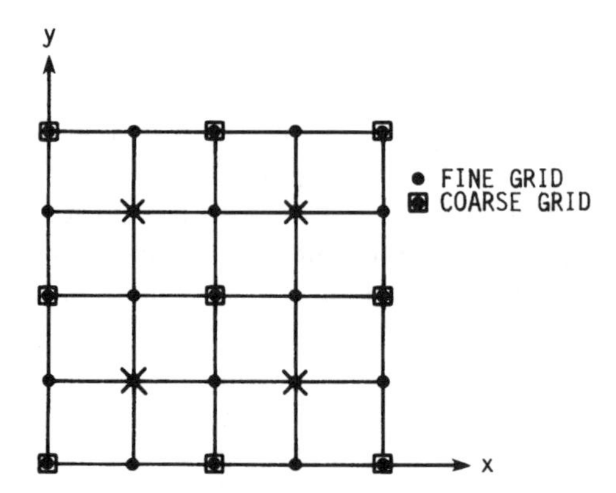

Figure 4.26 Prolongation details.

domain. The main steps are as follows:

1. The general multilevel scheme begins in the same manner as the two-level scheme discussed previously. The difference scheme, $Lu_{i,j} = 0$, is iterated n times (again n is a small number like 3 or 4, or alternatively, a "stalling factor" can be used to determine when to transfer to the coarser grids) on the finest grid.
2. The residual, $Lu_{i,j}^k = R_{i,j}^1$ is computed and stored at each point. This residual is then restricted by injection to the next coarsest grid. The restricted residual is denoted as $I_1^2 R_{i,j}^1$, where I is the transfer operator, the subscript indicates the level of origin, and the superscript the level of the destination. The superscript on the R indicates the grid upon which the residual was computed. The grids will be numbered from the finest (level 1) to coarsest.
3. The equation $L\Delta(u_2)_{i,j} = -I_1^2 R_{i,j}^1$ is iterated ("relaxed") n times on grid level 2 using zero as the initial guesses while keeping the residual fixed at each grid point. The solution after n iterations, $\Delta(u_2)_{i,j}^k$, represents a correction to the fine-grid solution. This solution, as well as the residual used to obtain it, are stored for future use in the prolongation phase. In order to transfer the problem to a coarser grid, an updated residual needs to be computed on grid level 2. It is about at this point that beginners typically lose their concentration and make mistakes. The updated residual at level 2 is $R_{i,j}^2 = I_1^2 R_{i,j}^1 + L\Delta(u_2)_{i,j}^k$, where $\Delta(u_2)_{i,j}^k$ is the solution obtained on grid level 2 after n iterations. The newly updated residual is then restricted to the next coarsest grid (level 3) as $I_2^3 R_{i,j}^2$.
4. The equation $L\Delta(u_3)_{i,j} = -I_2^3 R_{i,j}^2$ is iterated n times on grid level 3 using zero as the initial guess. The solution after n iterations can be thought of as a correction to the correction obtained on grid level 2, which of course, represents a further correction to the fine-grid solution. This solution and the residual used to obtain it are stored for use in the prolongation phase. The transfer to coarser grids, relaxation sweeps, and creation of new corrections continues following the residual update and restriction steps described above until the coarsest grid is reached. The coarsest grid may consist of one grid point in the interior. The solution is usually iterated to convergence on the coarsest grid. With one grid point, this solution can be obtained analytically, although the iterative scheme will normally reflect convergence in two passes, depending of course, upon how the convergence criterion is applied.
5. The corrections obtained on the coarsest grid are prolongated (interpolated) onto the next finer grid following the steps outlined in the description of the two-level scheme. Let us assume that the coarsest grid is grid level 4, in order to provide specific notation. These prolongated corrections are then $I_4^3 \Delta(u_4)_{i,j}^k$. These are added to the corrections obtained earlier at level 3 in the restriction phase, $\Delta(u_3)_{i,j}^k$. The sum of the two corrections is used as the initial guess at each point, as we continue to solve the problem we started to solve on grid level 3 on the way down in the restriction phase. That is, the problems to be solved as we move up the sequence of finer grids are the

continuation of the problems started on the way down. In other words, on grid level 3 in the prolongation phase, we continue to solve the problem

$$L\Delta(u_3)_{i,j} = -I_2^3 R_{i,j}^2$$

but the computation is started with the corrections identified above as the initial guesses. As in the restriction phase, n sweeps are made at level 3. The solution represents improved corrections.

6. The corrections from level 3 are prolongated onto the next finer grid at level 2. These corrections are added to the values of $\Delta(u_2)_{i,j}^k$ obtained at level 2 in the restriction phase, and the sums are used as the initial guesses for continuing the computation of the same problem solved at level 2 on the way down from finer to coarser grids, $L\Delta(u_2)_{i,j} = -I_1^2 R_{i,j}^1$. Again, n sweeps are made, and the solution after n sweeps represents improved corrections. Note that no new residuals are computed in the prolongation phase of moving up from coarser to finer grids. The solution is being improved as we move up toward finer grids because additional sweeps are being made that start with improved guesses.
7. The corrections from level 2 are prolongated onto grid level 1, the finest grid, and added to the last solution obtained on the fine grid, $u_{i,j}$. The corrected solution is then iterated through n sweeps unless convergence is detected before n sweeps are completed. If convergence has not occurred, new residuals are computed after n sweeps, and the cycle down to the coarsest grid and back up is repeated.

Example using multigrid. Here we will apply the multigrid technique to obtain the solution to Laplace's equation with Dirichlet boundary conditions. The computational effort for the multigrid scheme will be compared with that required for the simple Gauss-Seidel scheme and for the Gauss-Seidel scheme with optimum overrelaxation. A square domain will be utilized. The Gauss-Seidel scheme will also be used as the smoother for the multigrid scheme. Results will be presented for both the two-level and the multilevel V-cycle procedure. For simplicity, each side of the square will be set at a fixed value of u, as indicated in Fig. (4.27). Although use of discontinuous boundary conditions is physically somewhat unrealistic, the points of discontinuity do not enter into the finite-difference calculation. The primary purpose of this example is to compare the computational effort required by the four procedures, so detailed results of the solution values themselves will not be given. As a rough check, the reader could confirm that the solution at the center is the arithmetic average of the temperatures of the four boundaries. An analytic series solution can actually be obtained for this problem by superposition and could be used to check the numerical solution.

Use of Dirichlet boundary conditions will make it easy to determine an optimum overrelaxation factor to use with the Gauss-Seidel scheme for purposes of comparison. The overrelaxation factors are obtained from the formula given

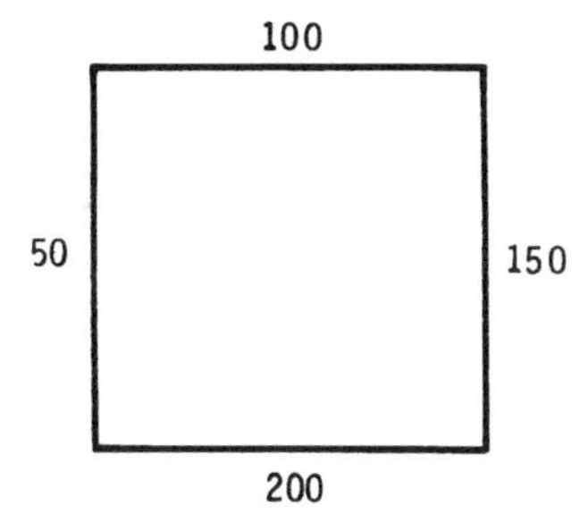

Figure 4.27 Boundary conditions.

previously in this chapter. When the coarsest grid in a multilevel scheme is to be simply one interior grid point (i.e., a 3×3 grid), it is convenient to let 2^n be the number of mesh increments into which each side of the square is divided. Hence grids of 9×9, 17×17, 33×33, 65×65, and 129×129 will be used. The overrelaxation factors computed for these grids are 1.45, 1.67, 1.82, 1.91, and 1.95. These will be used with the Gauss-Seidel scheme without multigrid.

Overrelaxation was not used with multigrid, since it did not seem to improve the convergence rate. The computational effort will be reported in terms of equivalent fine-grid sweeps that are usually referred to as work units. That is, in the multigrid calculations the total number of times the Gauss-Seidel smoother was applied was determined, and the total was then divided by the number of calculation (internal) grid points in the finest grid. As is customary in such comparisons, other operations in the multigrid algorithm such as computation of the residual, the restriction and prolongation operations, and the addition of the corrections were not counted, as such effort is generally considered to be a fairly negligible percentage of the total effort. The convergence parameter used in the calculations was the maximum change in the computed variable (u or Δu) between two successive sweeps divided by the maximum value of the dependent variable on the boundary. In the present example, that maximum boundary value was 200.

Using the same reference in the denominator of the convergence parameter for both u and Δu appeared to be important in order to obtain results that had the property that the number of iterations to convergence was independent of whether the variable itself or a correction was being computed. Convergence was declared when this parameter was less than 10^{-5}.

The number of iterations (in terms of equivalent fine-grid sweeps or work units) required for convergence for the four schemes is given in Table 4.2. For the multigrid results shown, three sweeps were made on the fine and all intermediate grids before a transfer was made and convergence was achieved on the coarsest grid for each cycle. The first column, labeled GS, gives results obtained with the conventional Gauss-Seidel scheme with no overrelaxation. The second column labeled GS ω_{opt} gives results obtained with the Gauss-Seidel scheme using the optimum overrelaxation factor. The column labeled MG2 gives results obtained with the two-level multigrid, and the column labeled

Table 4.2 Number of equivalent fine-grid iterations required for convergence

Grid size	GS	GS ω_{opt}	MG2	MGMAX
9×9	62	19	19	17
17×17	215	40	40	19
33×33	715	75	95	20
65×65	2282	137	258	20
129×129	6826	282	732	21

MGMAX provides multigrid results obtained using the maximum number of levels, i.e., taking the calculation down to one internal grid point. This results in use of seven, six, five, four, and three levels for the 129×129, 65×65, 33×33, 17×17, and 9×9 grids, respectively.

A number of interesting points can be made from the results shown in Table 4.2. The number of iterative sweeps required by the standard Gauss-Seidel scheme can be seen to be almost proportional to the number of grid points used. Of course, the computational effort per sweep is also proportional to the number of points used. The use of the optimum overrelaxation factor reduces the computational effort substantially, especially as the number of grid points increases. For the finest grid, use of overrelaxation reduces the computational effort by a factor of about 24. This is significant. The two-level multigrid is seen to provide a significant reduction in computational effort, but it does not perform quite as well as the Gauss-Seidel scheme with optimum overrelaxation. However, it is general, whereas the optimum overrelaxation factor can only be computed in advance for special cases.

The performance of the multigrid with the maximum number of levels (hereafter referred to as the n-level scheme) is truly amazing. The number of sweeps is seen to be nearly independent of the number of grid points used. Only 21 sweeps were required for the finest grid compared to 6826 for the conventional Gauss-Seidel scheme. This is a reduction in effort by a factor of 325! It requires only 1/13th as much effort as the Gauss-Seidel scheme with the optimum overrelaxation. This reduction in effort or "speed-up factor" is shown graphically in Fig. 4.28. Both multigrid schemes required four cycles for convergence for this problem, which utilized only 13 or 14 sweeps through the finest grid.

For the 129×129 grid, it was observed that using four or five levels gave nearly the same performance as using seven levels. Specifically, it was observed that 732, 82, 25, 21, 21, and 21 work units (equivalent fine-grid sweeps) were required when using 2, 3, 4, 5, 6, and 7 levels, respectively. This trend is illustrated in Fig. 4.29. The especially large improvement observed when moving from two to three levels is noteworthy.

The results were fairly insensitive to the number of sweeps for the fine and intermediate grids. The following trend was observed for the n-level scheme on the 129×129 grid. For use of 2, 3, 4, and 5 sweeps, the number or work units

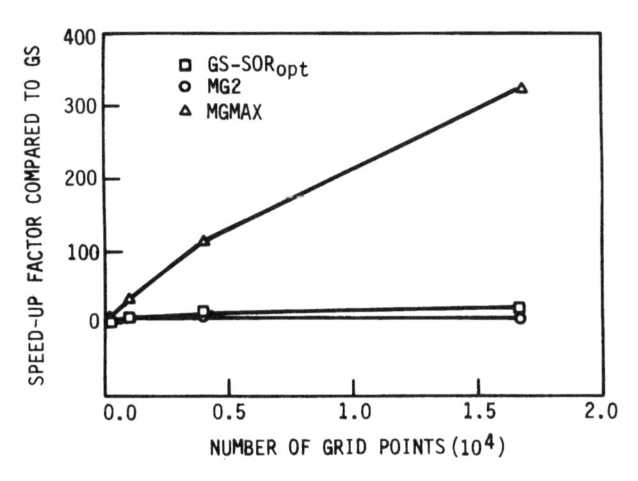

Figure 4.28 Comparison of effort for Dirichlet problem. GS-SOR$_{opt}$, Gauss-Seidel with optimum overrelaxation; MG2, two-level multigrid; MGMAX, multigrid using maximum number of levels.

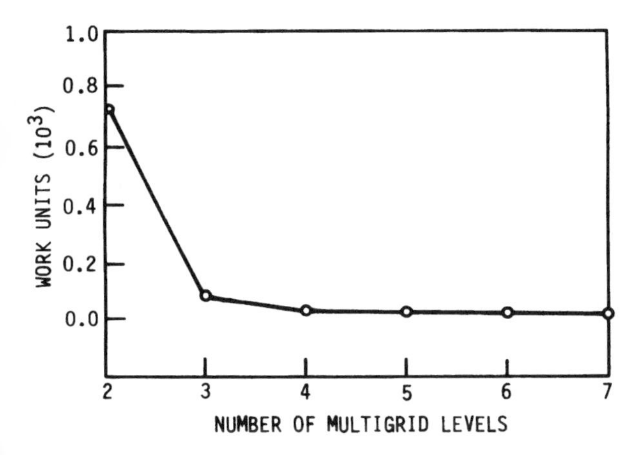

Figure 4.29 Effect of multigrid levels on Dirichlet problem.

required was 21, 21, 21, and 25, respectively. It was assumed that performance would continue to deteriorate as the number of sweeps was increased above 5.

For the two-level scheme, performance was actually improved by not converging on the coarse grid. For the 129 × 129 grid the number of equivalent fine-grid sweeps required for convergence was reduced to 495, 471, 474, and 479 if the maximum number of coarse-grid sweeps was limited to 75, 100, 125, and 150, respectively. This represents a reduction of about one-third in the computational effort for the two-level scheme. No effort was made to determine a general criterion for determining the optimum number of coarse-grid sweeps.

Multigrid for nonlinear equations. For a linear equation it is possible to "transfer the problem" from one grid to another by merely transferring the residual. This is the correction storage scheme. If the equation is nonlinear, this transfer by residual alone is generally not possible, and we must transfer (restrict and prolongate) the solution as well as the residual. Such a version of multigrid is known as the *full approximation storage* (FAS) method. It is not really much more complicated than the coarse-grid correction scheme, but it may seem like it initially because the problem formulation itself for a nonlinear equation is more complicated. Consider a 1-D problem governed by the following nonlinear equation:

$$\frac{\partial u^2}{\partial x} + \frac{\partial^2 u}{\partial x^2} = 0 \tag{4.135}$$

Choosing a central-difference discretization, we wish to find a way to satisfy the difference equation

$$\frac{u_{i+1}^2 - u_{i-1}^2}{2\,\Delta x} + \frac{u_{i+1} - 2u_i + u_{i-1}}{(\Delta x)^2} = 0 \tag{4.136}$$

Let us introduce a nonlinear difference operator N such that the difference equation above can be represented as $Nu_i = 0$. the residual R_i will be the N operator operating on any intermediate solution Nu_i^k. This definition of the residual is consistent with the terminology used for the linear problem. It will be convenient to solve the problem using a delta form for the equation on all grids. In this nonlinear case the delta will be an iteration delta, $\Delta u_i = u_i^{k+1} - u_i^k$. Suppose we wish Eq. (4.136) to be satisfied at the $k + 1$ level and substitute $u_i^k + \Delta u_i$ for u_i in Eq. (4.136). The result can be written as

$$\frac{2u_{i+1}^k\,\Delta u_{i+1} - 2u_{i-1}^k\,\Delta u_{i-1}}{2\,\Delta x} + \frac{\Delta u_{i+1} - 2\,\Delta u_i + \Delta u_{i-1}}{(\Delta x)^2}$$
$$+ \frac{(\Delta u_{i+1})^2 - (\Delta u_{i-1})^2}{2\,\Delta x} = -\left(\frac{(u_{i+1}^k)^2 - (u_{i-1}^k)^2}{2\,\Delta x} + \frac{u_{i+1}^k - 2u_i^k + u_{i-1}^k}{(\Delta x)^2}\right) \tag{4.137}$$

Observe that the last term on the left-hand side involves the change in u to the second power. Dropping this term, which is small near convergence, linearizes the difference equation and is equivalent to a Newton linearization (see Section 7.3.3), which can be developed through the use of a Taylor-series expansion, neglecting terms involving derivatives of higher order than the first. Other ways of linearizing the algebraic equation can obviously be found. The final working equation for the fine grid can be written

$$\frac{2u_{i+1}^k \Delta u_{i+1} - 2u_{i-1}^k \Delta u_{i-1}}{2\,\Delta x} + \frac{\Delta u_{i+1} - 2\,\Delta u_i + \Delta u_{i-1}}{(\Delta x)^2}$$
$$= -\left(\frac{(u_{i+1}^k)^2 - (u_{i-1}^k)^2}{2\,\Delta x} + \frac{u_{i+1}^k - 2u_i^k + u_{i-1}^k}{(\Delta x)^2} \right) \tag{4.138}$$

where the right-hand side can be recognized as the negative of the residual evaluated at the kth iteration level. Notice that as the iterative process converges, the delta terms on the left-hand side vanish, and the nonlinear difference equation is satisfied exactly as the residual goes to zero.

Only a two-level FAS multigrid scheme will be discussed here. The sweeping strategy is the same as before: n sweeps on the fine grid and then a transfer to the coarse grid, where normally, the solution would be iterated to convergence and then the changes transferred to the fine grid. The equation being solved on the fine grid is Eq. (4.138). Note that because the residual or delta form of the governing equation is being solved on the fine grid, the residual in Eq. (4.138) is updated at every iteration. After n sweeps, the most recent residual and the current solution u_i are restricted to the coarse grid.

As before, on the coarse grid we wish to compute changes to the solution that will annihilate the residual on the fine grid. However, because the equation is nonlinear, we continue to solve for the solution itself on the coarse grid even though it will be the changes or corrections to the fine-grid solution that will be prolongated to the fine grid. Thus Eq. (4.138) would be appropriate to use on the coarse grid, provided the right-hand side (residual) would vanish if and only if the residual on the fine grid vanished. This would happen if we modified the residual by adding the difference between the restricted fine-grid residual and the residual computed on the coarse grid using the restricted fine-grid solution. This modification to the residual can be thought of as compensation for the difference in T.E.s associated with the solution on meshes of different sizes. Thus, letting M be the operator that gives the left-hand side of Eq. (4.138), we can write the difference equation solved on the coarse grid as

$$M\,\Delta u_i = -R_i^2 - I_1^2 R_i^1 + R_i^2(I_1^2 u_i^1) \tag{4.139}$$

where the first term on the right-hand side is the residual computed from the coarse-grid solution [evaluated as indicated by the right-hand side of Eq. (4.138)] at each coarse-grid iteration. The second term on the right is the restricted fine-grid residual. This is the same term that was used in the correction scheme in the linear example. The third term is the residual computed on the coarse grid using the restricted fine-grid solution. The second and third terms are source terms that remain fixed during the iterative process on the coarse grid.

The restricted fine-grid solution is taken as the starting value for u on the coarse grid. Thus we notice that the first and third terms cancel for the first coarse-grid sweep. This leaves the restricted residual from the fine grid as the source term driving the changes. Notice also that no changes would be computed if the residual on the fine grid were zero. This formulation for the right-hand side properly takes into account discrepancies that arise owing to the restriction. If no "errors" were introduced in the restriction, the second and third terms would cancel. The formulation also ensures that it is the residual on the fine grid that drives the multigrid process. At the conclusion of the coarse-grid iterations (usually signaled by convergence on the coarse grid), the *changes* in u computed over the duration of the coarse-grid iterations are prolongated onto the fine grid. An additional n iterations are performed on the fine grid, and if the solution has not converged, the cycle is repeated.

4.4 BURGERS' EQUATION (INVISCID)

We have discussed finite-difference methods and have applied them to simple linear problems. This has provided an understanding of the various techniques and acquainted us with the peculiarities of each approach. Unfortunately, the usual fluid mechanics problem is highly nonlinear. The governing PDEs form a nonlinear system that must be solved for the unknown pressures, densities, temperatures, and velocities.

A single equation that could serve as a nonlinear analog of the fluid mechanics equations would be very useful. This single equation must have terms that closely duplicate the physical properties of the fluid equations, i.e., the model equation should have a convective term, a diffusive or dissipative term, and a time-dependent term. Burgers (1948) introduced a simple nonlinear equation that meets these requirements:

$$\underbrace{\frac{\partial u}{\partial t}}_{\substack{\text{Unsteady}\\ \text{term}}} + \underbrace{u\frac{\partial u}{\partial x}}_{\substack{\text{Convective}\\ \text{term}}} = \underbrace{\mu\frac{\partial^2 u}{\partial x^2}}_{\substack{\text{Viscous}\\ \text{term}}} \tag{4.140}$$

Equation (4.140) is parabolic when the viscous term is included. If the viscous term is neglected, the remaining equation is composed of the unsteady term and a nonlinear convection term. The resulting hyperbolic equation

$$\frac{\partial u}{\partial t} + u\frac{\partial u}{\partial x} = 0 \tag{4.141}$$

may be viewed as a simple analog of the Euler equations for the flow of an inviscid fluid. Equation (4.141) is a nonlinear convection equation and possesses properties that need to be examined in some detail. Methods for solving the inviscid Burgers equation will be presented in this section. Typical results for a number of commonly used finite-difference/finite-volume methods are included, and the effects of the nonlinear terms are discussed. A discussion of the viscous Burgers equation follows in Section 4.5.

Equation (4.141) may be viewed as a nonlinear wave equation, where each point on the wave front can propagate with a different speed. In contrast, the speed of propagation of all signals or waves was constant for the linear, 1-D convection equation, Eq. (4.2). A consequence of the changing wave speed is the coalescence of characteristics and the formation of discontinuous solutions similar to shock waves in fluid mechanics. This means the class of solutions that include discontinuities can be studied with this simple 1-D model.

Nonlinear hyperbolic PDEs exhibit two types of solutions according to Lax (1954). For simplicity, we consider a simple scalar equation,

$$\frac{\partial u}{\partial t} + \frac{\partial F}{\partial x} = 0 \tag{4.142}$$

For the general case, both the unknown u and the variable $F(u)$ are vectors. We may write Eq. (4.142) as

$$\frac{\partial u}{\partial t} + A\frac{\partial u}{\partial x} = 0 \tag{4.143}$$

where $A = A(u)$ is the Jacobian matrix $\partial F_i/\partial u_j$ for the general case and is dF/du for our simple equation. Our equation or system of equations is hyperbolic, which means that the eigenvalues of the matrix A are all real. A *genuine solution* of Eq. (4.143) is one in which u is continuous but bounded discontinuities in the derivatives of u may occur (Lipschitz continuous). A *weak solution* of Eq. (4.143) is a solution that is genuine except along a surface in (x, t) space, across which the function u may be discontinuous. A constraint is placed upon the jump in u across the discontinuity in the domain of interest. If w is a test vector that is continuous and has continuous first derivatives but vanishes outside some bounded set, then u is termed a weak solution of Eq. (4.142) if

$$\iint_D (w_t u + w_x F)\, dx\, dt + \int w(x,0)\phi(x)\, dx = 0 \tag{4.144}$$

where $\phi(x) = u(x, 0)$. A genuine solution is a weak solution, and a weak solution that is continuous is a genuine solution. A complete discussion of the weak solution concept may be found in the excellent texts by Whitham (1974) and Jeffrey and Taniuti (1964). The mathematical theory of weak solutions for hyperbolic equations is a relatively recent development. Clearly, the existence of shock waves in inviscid supersonic flow is an example of a weak solution. It is interesting to recognize that the shock solutions in inviscid supersonic flow were known 50–100 years before the theory of weak solutions for hyperbolic systems was developed.

Let us return to the study of the inviscid Burgers equation and develop the requirements for a weak solution, i.e., the requirements necessary for the existence of a solution with a discontinuity such as that shown in Fig. 4.30.

Let $w(x, t)$ be an arbitrary test function that is continuous and has continuous first derivatives. Let $w(x, t)$ vanish on the boundary B of the domain

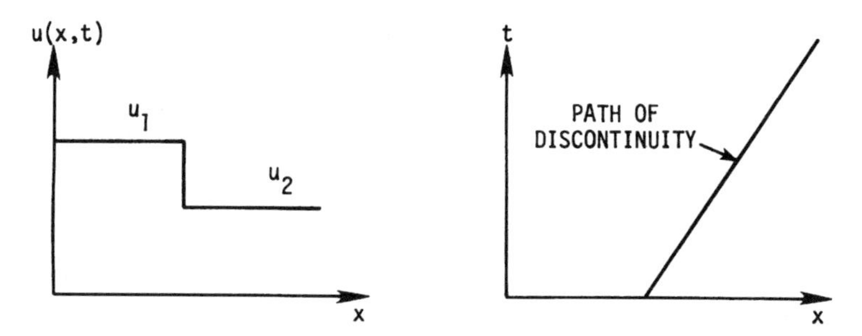

Figure 4.30 Typical traveling discontinuity problem for Burgers' equation.

D and everywhere outside D (complement of D). D is an arbitrary rectangular domain in the (x, t) plane. We may write

$$\iint_{D}\left(\frac{\partial u}{\partial t}+\frac{\partial F}{\partial x}\right)w(x,t)\,dx\,dt=0 \tag{4.145}$$

or

$$\iint_{D}(uw_t+Fw_x)\,dx\,dt=0 \tag{4.146}$$

Equations (4.145) and (4.146) are equivalent when both u and F are continuous and have continuous first derivatives. The second integral of Eq. (4.144) does not appear, since the function w vanishes on the boundary. Functions $u(x, t)$, which satisfy Eq. (4.146) for all test functions w, are called weak solutions of the inviscid Burgers equation. We do not require that u be differentiable in order to satisfy Eq. (4.146).

Suppose our domain D is now a rectangular region in the (x, t) plane, which is separated by a curve $\tau(x, t) = 0$, across which u is discontinuous. We assume that u is continuous and has continuous derivatives to the left of $\tau(D_1)$ and to the right of $\tau(D_2)$. Let the test function vanish on the boundary of D and outside of D. With these restrictions, Eq. (4.146) can be integrated by parts to yield

$$\iint_{D_1}\left(\frac{\partial u}{\partial t}+\frac{\partial F}{\partial x}\right)w\,dx\,dt+\iint_{D_2}\left(\frac{\partial u}{\partial t}+\frac{\partial F}{\partial x}\right)w\,dx\,dt$$
$$+\int_{\tau}([u]\cos\alpha_1+[F]\cos\alpha_2)\,ds=0 \tag{4.147}$$

The last integrand is evaluated along the curve $\tau(x, t) = 0$ separating the two regions D_1 and D_2. This integral occurs through the limits of the integration by parts on the discontinuity surface $\tau(x, t) = 0$. The square brackets denote the jump in the quantity across the discontinuity, and $\cos\alpha_1, \cos\alpha_2$ are the cosines of the angles between the normal to $\tau(x, t) = 0$ and the t and x directions, respectively. The problem is illustrated in Fig. 4.31.

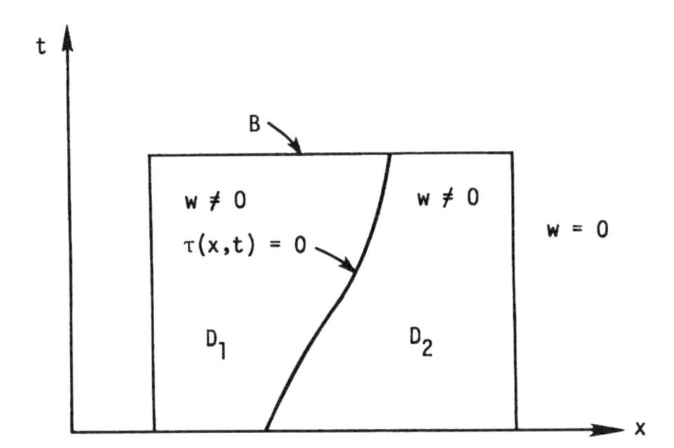

Figure 4.31 Schematic representation of an arbitrary domain with a discontinuity.

The integrals in Eq. (4.147) over D_1 and D_2 are zero by Eq. (4.145). We conclude that since the last integral vanishes for all test functions w with the required properties, we must have

$$[u]\cos\alpha_1 + [F]\cos\alpha_2 = 0 \tag{4.148}$$

This is the condition that u be a weak solution for Burgers' equation. Let us apply this condition to a moving discontinuity. Suppose initial data are prescribed for $u(x,0)$ as shown in Fig. 4.30, where u_1 and u_2 denote the values to the left and to the right of the discontinuity. In one dimension, we may write the equation of the surface $\tau(x,t) = 0$ as $t - t_1(x) = 0$. The direction cosines as required in Eq. (4.148) become

$$\cos\alpha_1 = \frac{1}{\left[1 + t_1'^2\right]^{1/2}} \qquad \cos\alpha_2 = -\frac{t_1'}{\left[1 + t_1'^2\right]^{1/2}}$$

where the prime denotes differentiation with respect to x. Thus

$$\frac{[u]}{[1 + t'^2]^{1/2}} - \frac{[F]t'}{[1 + t'^2]^{1/2}} = 0$$

or

$$u_2 - u_1 = \frac{u_2^2 - u_1^2}{2}\frac{dt}{dx}$$

Therefore

$$\frac{dx}{dt} = \frac{u_1 + u_2}{2} \tag{4.149}$$

which shows that the discontinuity travels at the average value of the u function across the wave front. Since we now see that a discontinuity in u simply propagates at constant speed $(u_1 + u_2)/2$ with uniform states on each side, a

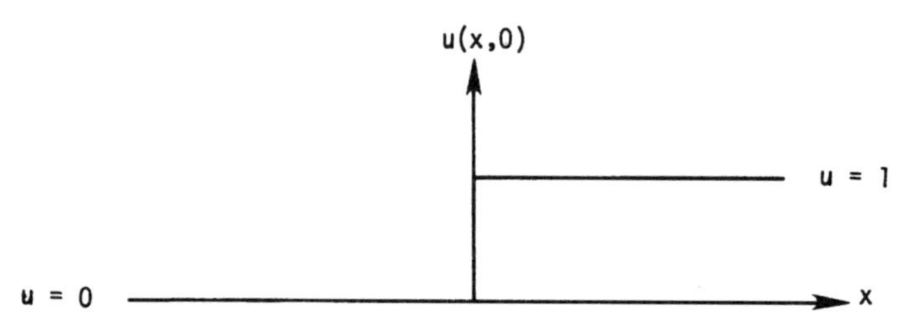

Figure 4.32 Initial data for rarefaction wave.

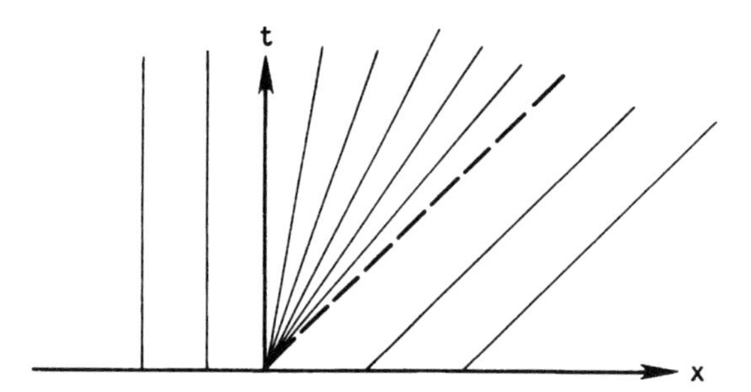

Figure 4.33 Characteristics for centered expansion.

numerical solution of a similar problem for a discontinuity can be compared with the exact solution. These comparisons are presented for a number of finite-difference/finite-volume methods in this section.

Rarefactions are as prevalent in high-speed flows as shock waves, and the exact solution of Burgers' equation for a rarefaction is known. Consider initial data $u(x,0)$ as shown in Fig. 4.32. The characteristic for Burgers' equation is given by

$$\frac{dt}{dx} = \frac{1}{u} \tag{4.150}$$

Figure 4.33 shows the characteristic diagram plotted in the (x,t) plane. In the left half-plane, the characteristics are simply vertical lines, while they are lines at an angle of $\pi/4$ radians to the right of the characteristic that bounds the expansion. This particular problem is similar to a centered expansion wave in compressible flow. Here the expansion is bounded by the $x = 0$ axis and the characteristic originating at the origin denoted by the dashed line. The solution for this problem may be written

$$\begin{aligned} u &= 0 & x &\leqslant 0 \\ u &= \frac{x}{t} & 0 &< x < t \\ u &= 1 & x &\geqslant t \end{aligned}$$

The initial distribution of u forms a centered expansion, where the width of the expansion grows linearly with time.

We have examined two problems, shocks and rarefactions, which are frequently encountered in high-speed flows by using the simple analog provided by Burgers' equation. Clearly, these types of solutions can occur in systems of nonlinear equations of the hyperbolic type. Armed with simple analytic solutions for these two important cases, let us examine the application of some numerical algorithms to the nonlinear, inviscid Burgers equation.

4.4.1 Lax Method

First-order methods for solving hyperbolic equations are infrequently used. The Lax (1954) method is presented as a typical first-order method to demonstrate the application to a nonlinear equation and the dissipative character of the result.

The conservation form of the basic PDE,

$$\frac{\partial u}{\partial t} + \frac{\partial F}{\partial x} = 0$$

is used for all examples that follow. For the Lax method, we expand in a Taylor series about the point (x, t), retaining only the first two terms

$$u(x, t + \Delta t) = u(x, t) + \Delta t\left(\frac{\partial u}{\partial t}\right)_{x,t} + \cdots$$

and substitute for the time derivative

$$u(x, t + \Delta t) = u(x, t) - \Delta t\left(\frac{\partial F}{\partial x}\right)_{x,t} + \cdots$$

Using centered differences and averaging the first term yields the Lax method (see Section 4.1.3):

$$u_j^{n+1} = \frac{u_{j+1}^n + u_{j-1}^n}{2} - \frac{\Delta t}{\Delta x}\frac{F_{j+1}^n - F_{j-1}^n}{2} \tag{4.151}$$

In Burgers' equation, $F = u^2/2$. The amplification factor in this case is

$$G = \cos\beta - i\frac{\Delta t}{\Delta x}A\sin\beta \tag{4.152}$$

where A is the Jacobian dF/du, which is just the single element u for Burgers' equation. The stability requirement for this method is

$$\left|\frac{\Delta t}{\Delta x}u_{\max}\right| \leqslant 1 \tag{4.153}$$

because $u_{\max}$ is the maximum eigenvalue of the A matrix with the single element u.

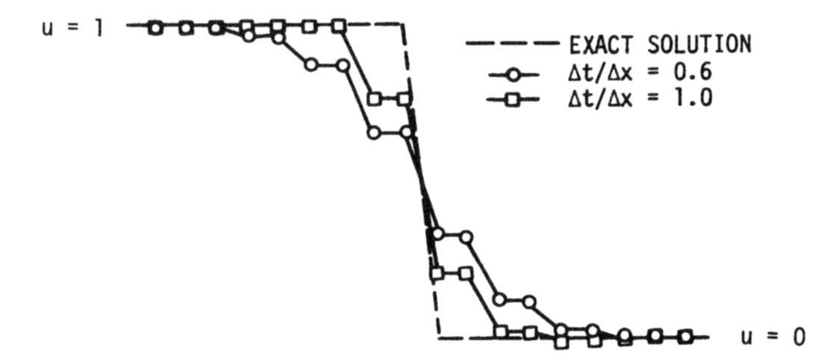

Figure 4.34 Numerical solution of Burgers' equation using Lax method.

The Lax method applied to a 1-0 right-moving discontinuity produces the solutions shown in Fig. 4.34. The location of the moving discontinuity is correctly predicted, but the dissipative nature of the method is evident in the smearing of the discontinuity over several mesh intervals. As previously noted, this smearing becomes worse as the Courant number decreases. It is of interest to note that the application of the Lax method to Burgers' equation with a discontinuity produces the double-point solutions as shown. A further comment on these results is in order. Notice that the computed solutions are monotone, i.e., the solution does not oscillate. Godunov (1959) has shown that monotone behavior of a solution cannot be assured for finite-difference methods with more than first-order accuracy. This monotone property is very desirable when discontinuities are computed as part of the solution. Unfortunately, the desirability of monotone behavior must be reconciled with the highly dissipative character of the results. The relative importance of these properties must be carefully evaluated for each case.

The finite-volume equivalent of the Lax method can be readily developed by noting that first-order integration (in time) over a control volume (see Fig. 4.35) provides the expression

$$u_j^{n+1} = u_j^n - \frac{\Delta t}{\Delta x}\left[f_{j+\frac{1}{2}}^n - f_{j-\frac{1}{2}}^n\right] \tag{4.154}$$

In this expression the control volume may be considered to be centered at the point $(j, n + \frac{1}{2})$. The flux terms, $f_{j\pm\frac{1}{2}}^n$ are referred to as the numerical fluxes, since they represent the flux at the surface of the control volume in the finite-volume formulation. Functionally, the numerical flux is written

$$f_{j+\frac{1}{2}} = f(u_j, u_{j+1})$$

The numerical flux must be consistent with the analytical flux F, so that

$$f(u_j, u_{j+1}) = F(u_j) \tag{4.155}$$

when

$$u_j = u_{j+1} = u$$

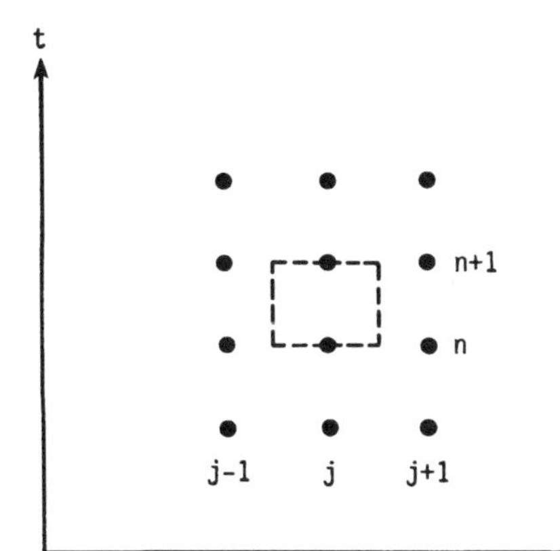

Figure 4.35 Control volume for Lax method.

The problem of finding the flux function is very important because it represents the control-volume boundary flux. This is needed in constructing methods for solving the conservative form of the equations of fluid dynamics.

For the Lax method the numerical flux becomes

$$f_{j+\frac{1}{2}} = \frac{1}{2}\left[F_j + F_{j+1} - \frac{\Delta x}{\Delta t}(u_{j+1} - u_j)\right] \tag{4.156}$$

The last term in this expression can be viewed as a dissipation term. The order of the numerical scheme can be altered by using a different form for this term. It will be of value to compare Eq. (4.156) with the numerical flux terms of other methods in this chapter.

Another observation can be made regarding the Lax method. Notice that the computer solutions in Fig. 4.34 are monotone, i.e., the solution does not oscillate. Godunov (1959) studied numerical methods applied to the simple 1-D wave equation (Eq. 4.2) having the linear form

$$u_j^{n+1} = \sum_k a_k u_{j+k}^n \tag{4.157}$$

If the right-hand side is expanded in a Taylor series, second-order accuracy for such a method is established if

$$\sum_k a_k = 1 \tag{4.158}$$

$$\sum_k k a_k = -\nu \tag{4.159}$$

$$\sum_k k^2 a_k = \nu^2 \tag{4.160}$$

This may be verified by examining the second-order schemes previously considered in Section 4.1.

If solutions produced by Eq. (4.157) do not oscillate, what are the conditions that guarantee monotone behavior? A necessary and sufficient condition is that all of the coefficients a_k be positive. Consider the change in the solution between points j and $j+1$:

$$u_{j+1}^{n+1} - u_j^{n+1} = \sum_k a_k (u_{j+k+1}^n - u_{j+k}^n) \tag{4.161}$$

If the solution at level n is monotone, then every difference on the right-hand side is of the same sign. Thus, if the a_k are all positive, then the differences on the right- and left-hand sides carry the same sign. This is also a necessary condition because for at least one value of a_k of opposite sign, monotone initial data may be constructed that produce an oscillation at the next level.

Godunov's (1959) theorem states that second-order schemes are not monotone. This may be proved by letting $a_k = (e_k)^2$ and substituting into Eqs. (4.158)–(4.160) to obtain

$$\sum_k (e_k)^2 \Big(\sum k e_k^2 \Big)^2 = \sum_k k^2 e_k^2$$

This equation violates the Cauchy inequality (Taylor, 1955) and shows that no monotone second-order schemes exist. This result provided a major difficulty that needed to be overcome in the development of methods for solving hyperbolic PDEs. In regions where discontinuities develop, measures must be taken in order to avoid oscillations.

4.4.2 Lax-Wendroff Method

The Lax-Wendroff method (Lax and Wendroff, 1960) was one of the first second-order finite-difference methods for hyperbolic PDEs. The development of the Lax-Wendroff scheme for nonlinear equations again follows from a Taylor series:

$$u(x, t+\Delta t) = u(x,t) + \Delta t \left(\frac{\partial u}{\partial t} \right)_{x,t} + \frac{(\Delta t)^2}{2} \left(\frac{\partial^2 u}{\partial t^2} \right)_{x,t} + \cdots$$

The first time derivative can be directly replaced using the differential equation, but we need to examine the second-derivative term in more detail. We consider the original equation in the form

$$\frac{\partial u}{\partial t} = -\frac{\partial F}{\partial x}$$

Taking the time derivative of this expression yields

$$\frac{\partial^2 u}{\partial t^2} = -\frac{\partial^2 F}{\partial t\, \partial x} = -\frac{\partial^2 F}{\partial x\, \partial t} = -\frac{\partial}{\partial x}\left(\frac{\partial F}{\partial t} \right)$$

where the order of differentiation on F has been interchanged. Now $F = F(u)$, which permits us to write

$$\frac{\partial u}{\partial t} = -\frac{\partial F}{\partial x} = -\frac{\partial F}{\partial u}\frac{\partial u}{\partial x} = -A\frac{\partial u}{\partial x}$$

and

$$\frac{\partial F}{\partial t} = \frac{\partial F}{\partial u}\frac{\partial u}{\partial t} = A\frac{\partial u}{\partial t}$$

Hence we may replace $\partial F/\partial t$ with

$$\frac{\partial F}{\partial t} = -A\frac{\partial F}{\partial x}$$

so that

$$\frac{\partial^2 u}{\partial t^2} = \frac{\partial}{\partial x}\left(A\frac{\partial F}{\partial x}\right)$$

The Jacobian A contains a single element for Burgers' equation. It is clear that A is a matrix when u and F are vectors in treating a system of equations. Making the appropriate substitution in the Taylor-series expansion for u, we obtain

$$u(x, t + \Delta t) = u(x, t) - \Delta t\frac{\partial F}{\partial x} + \frac{(\Delta t)^2}{2}\frac{\partial}{\partial x}\left(A\frac{\partial F}{\partial x}\right) + \cdots$$

After using central differencing, the Lax-Wendroff method is obtained:

$$u_j^{n+1} = u_j^n - \frac{\Delta t}{\Delta x}\frac{F_{j+1}^n - F_{j-1}^n}{2} + \frac{1}{2}\left(\frac{\Delta t}{\Delta x}\right)^2 \times\left[A_{j+1/2}^n(F_{j+1}^n - F_j^n) - A_{j-1/2}^n(F_j^n - F_{j-1}^n)\right] \tag{4.162}$$

The Jacobian matrix is evaluated at the half interval, i.e.,

$$A_{j+1/2} = A\left(\frac{u_j + u_{j+1}}{2}\right)$$

In Burgers' equation, $F = u^2/2$ and $A = u$. In this case $A_{j+1/2} = (u_j + u_{j+1})/2$ and $A_{j-1/2} = (u_j + u_{j-1})/2$. The amplification factor for this method is

$$G = 1 - 2\left(\frac{\Delta t}{\Delta x}A\right)^2(1 - \cos\beta) - 2i\frac{\Delta t}{\Delta x}A\sin\beta \tag{4.163}$$

and the stability requirement reduces to $|(\Delta t/\Delta x)u_{max}| \leqslant 1$.

The results obtained when the Lax-Wendroff method is applied to our example problem are shown in Fig. 4.36. The right-moving discontinuity is correctly positioned and is sharply defined. The dispersive nature of this method is evidenced through the presence of oscillations near the discontinuity. Even though the method uses central differences, some asymmetry will occur, since the wave is moving. The solution shows more oscillations when a Courant

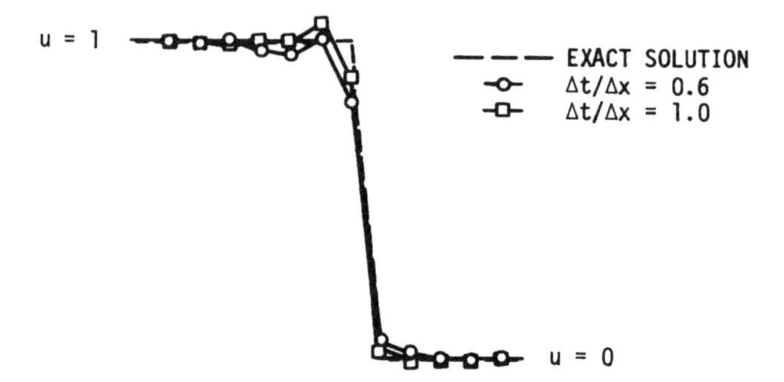

Figure 4.36 Application of the Lax-Wendroff method to the inviscid Burgers equation.

number of 0.6 is used than for a Courant number of 1.0. In general, as the Courant number is reduced, the quality of the solution will be degraded (see Section 4.1.6).

The numerical flux for the Lax-Wendroff scheme consistent with Eq. (4.156) may be written as

$$f_{j+\frac{1}{2}} = \frac{1}{2}(F_j + F_{j+1}) - \frac{1}{2}\frac{\Delta t}{\Delta x}(\lambda_{j+\frac{1}{2}})^2(u_{j+1} - u_j) \tag{4.164}$$

where $\lambda_{j+\frac{1}{2}}$ is defined as the eigenvalue of the Jacobian $A_{j+\frac{1}{2}}$, which is simply $u_{j+\frac{1}{2}}$ for Burgers' equation.

A comparison of Eq. (4.164) with Eq. (4.156) can yield valuable insight into the order and the behavior of the numerical methods. As noted in the previous section, the Lax scheme is monotone, while Fig. 4.36 shows that the second-order Lax-Wendroff scheme is not. If the numerical flux terms are compared, we see that the difference in these terms is

$$f^{LW}_{j+\frac{1}{2}} - f^{L}_{j+\frac{1}{2}} = \frac{\Delta x}{\Delta t}\left(\frac{u_{j+1} - u_j}{2}\right)\left[1 - \left(\lambda_{j+\frac{1}{2}}\frac{\Delta t}{\Delta x}\right)^2\right] \tag{4.165}$$

This difference is a correction that may be added to the Lax method to provide second-order accuracy and, in fact, modify the solution to the form of the Lax-Wendroff scheme.

However, if oscillations at discontinuities occur as seen in Fig. 4.36, the correction term should be suppressed in the region where discontinuities appear. This will have the effect of reducing the order of the method and have the potential of providing monotone or near-monotone profiles through discontinuities. The control of the correction term is usually accomplished by the use of *limiters*. With this idea in mind, the flux may be written

$$f_{j+\frac{1}{2}} = \frac{1}{2}\left[F_j + F_{j+1} - \frac{\Delta x}{\Delta t}(u_{j+1} - u_j)\right] + \phi\frac{\Delta x}{\Delta t}\left(\frac{u_{j+1} - u_j}{2}\right)\left[1 - \left(\lambda_{j+\frac{1}{2}}\frac{\Delta t}{\Delta x}\right)^2\right] \tag{4.166}$$

where ϕ is a function that can be adjusted to limit the addition of the second-order terms. The hybrid method presented by Harten and Zwas (1972) used this idea. The concept of using limiters to improve the effectiveness of solution techniques is discussed in detail in Section 4.4.12.

4.4.3 MacCormack Method

MacCormack's (1969) method is a predictor-corrector version of the Lax-Wendroff scheme, as has been discussed in Section 4.1.8. This method is much easier to apply than the Lax-Wendroff scheme because the Jacobian does not appear. When applied to the inviscid Burgers equation, the MacCormack method becomes

$$
\begin{aligned}
u_j^{\overline{n+1}} &= u_j^n - \frac{\Delta t}{\Delta x}(F_{j+1}^n - F_j^n) \\
u_j^{n+1} &= \frac{1}{2}\left[u_j^n + u_j^{\overline{n+1}} - \frac{\Delta t}{\Delta x}\left(F_j^{\overline{n+1}} - F_{j-1}^{\overline{n+1}}\right)\right]
\end{aligned}
\tag{4.167}
$$

The amplification factor and stability requirement are the same as presented for the Lax-Wendroff method. The results of applying this method are shown in Fig. 4.37. Again the right-moving wave is well defined. We note that the solutions obtained for the same problem at the same Courant number are different from those obtained using the Lax-Wendroff scheme. This is due both to the switched differencing in the predictor and the corrector and the nonlinear nature of the governing PDE. One should expect results that show some differences, even though both methods are equivalent for linear problems.

In general, the MacCormack method provides good resolution at discontinuities. It should be noted in passing that reversing the differencing in the predictor and corrector steps leads to quite different results. The best resolution of discontinuities occurs when the difference in the predictor is in the

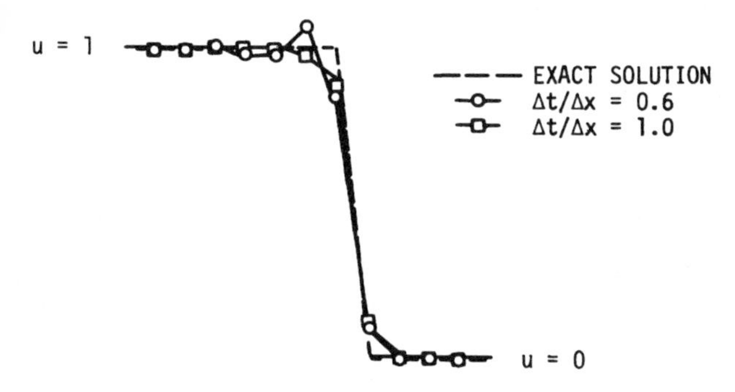

Figure 4.37 Solution of Burgers' equation using MacCormack's method.

direction of propagation of the discontinuity. This will be apparent when problems at the end of the chapter are completed.

4.4.4 Rusanov (Burstein-Mirin) Method

The third-order Rusanov or Burstein-Mirin method was discussed in Section 4.1.11. This method uses central differencing and, when applied to Eq. (4.142), becomes

$$u_{j+1/2}^{(1)} = \frac{1}{2}(u_{j+1}^n + u_j^n) - \frac{1}{3}\frac{\Delta t}{\Delta x}(F_{j+1}^n - F_j^n)$$

$$u_j^{(2)} = u_j^n - \frac{2}{3}\frac{\Delta t}{\Delta x}\left(F_{j+1/2}^{(1)} - F_{j-1/2}^{(1)}\right)$$

$$u_j^{n+1} = u_j^n - \frac{1}{24}\frac{\Delta t}{\Delta x}(-2F_{j+2}^n + 7F_{j+1}^n - 7F_{j-1}^n + 2F_{j-2}^n) - \frac{3}{8}\frac{\Delta t}{\Delta x}\left(F_{j+1}^{(2)} - F_{j-1}^{(2)}\right)$$
$$- \frac{\omega}{24}(u_{j+2}^n - 4u_{j+1}^n + 6u_j^n - 4u_{j-1}^n + u_{j-2}^n) \tag{4.168}$$

The last term in the third step represents a fourth-derivative term,

$$(\Delta x)^4 \frac{\partial^4 u}{\partial x^4}$$

and is added for stability. The third-order accuracy of the method is unaffected, since this added term is $O[(\Delta x)^4]$. A stability analysis of this method shows that the amplification factor is

$$G = 1 - \left(\frac{\Delta t}{\Delta x}u\right)^2 \frac{\sin^2 \beta}{2} - \frac{\omega}{6}(1 - \cos \beta) + i\frac{\Delta t}{\Delta x}u \sin \beta$$
$$\times \left\{1 + \frac{1}{3}(1 - \cos \beta)\left[1 - \left(\frac{\Delta t}{\Delta x}u\right)^2\right]\right\} \tag{4.169}$$

It follows that stability is assured for Burgers' equation if

$$|\nu| \leqslant 1 \qquad \text{or} \qquad \left|\frac{\Delta t}{\Delta x}u_{\max}\right| \leqslant 1$$

and

$$4\nu^2 - \nu^4 \leqslant \omega \leqslant 3 \tag{4.170}$$

Application of this method to Burgers' equation for a right-moving shock produces the results shown in Fig. 4.38. The magnitude and position of the discontinuity are correctly produced, but the results show an overshoot on both sides of the shock front. A schematic showing the numerical solution as it is computed from the base points is shown in Fig. 4.39.

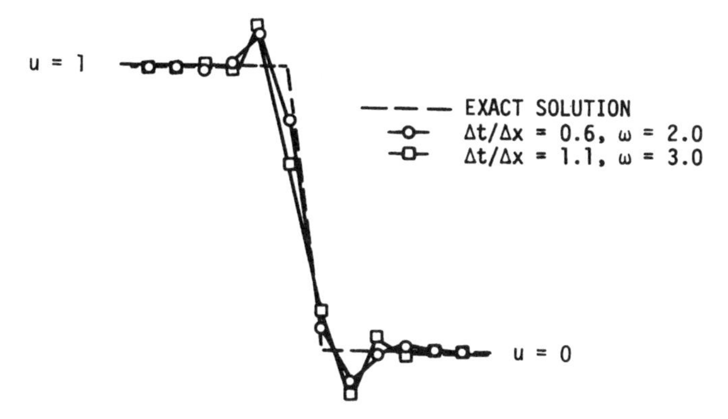

Figure 4.38 Rusanov method applied to Burgers' equation.

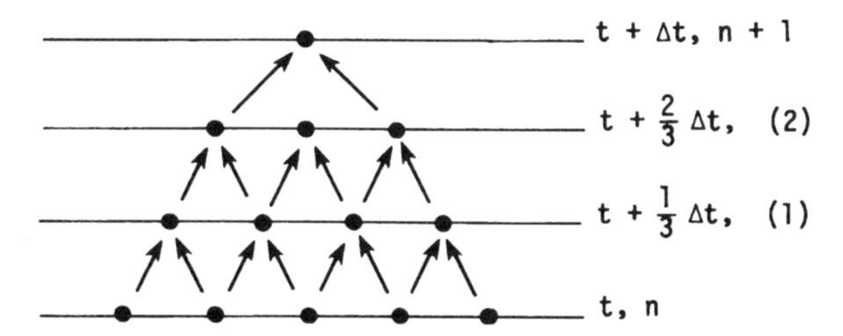

Figure 4.39 Point pyramid for the Rusanov method.

4.4.5 Warming-Kutler-Lomax Method

Warming et al. (1973) developed a third-order scheme using noncentered differences. This technique uses the MacCormack method for the first two levels evaluated at $\frac{2}{3}\Delta t$. The advantage of this method over the Rusanov technique is that only values at integral mesh points are required in the calculation.

The Warming-Kutler-Lomax (WKL) method applied to Eq. (4.142) becomes

$$u_j^{(1)} = u_j^n - \frac{2}{3}\frac{\Delta t}{\Delta x}(F_{j+1}^n - F_j^n)$$

$$u_j^{(2)} = \frac{1}{2}\left[u_j^n + u_j^{(1)} - \frac{2}{3}\frac{\Delta t}{\Delta x}\left(F_j^{(1)} - F_{j-1}^{(1)}\right)\right]$$

$$u_j^{n+1} = u_j^n - \frac{1}{24}\frac{\Delta t}{\Delta x}(-2F_{j+2}^n + 7F_{j+1}^n - 7F_{j-1}^n + 2F_{j-2}^n) - \frac{3}{8}\frac{\Delta t}{\Delta x}\left(F_{j+1}^{(2)} - F_{j-1}^{(2)}\right)$$
$$- \frac{\omega}{24}(u_{j+2}^n - 4u_{j+1}^n + 6u_j^n - 4u_{j-1}^n + u_{j-2}^n) \tag{4.171}$$

The third level for the WKL scheme is exactly the same as that used in the

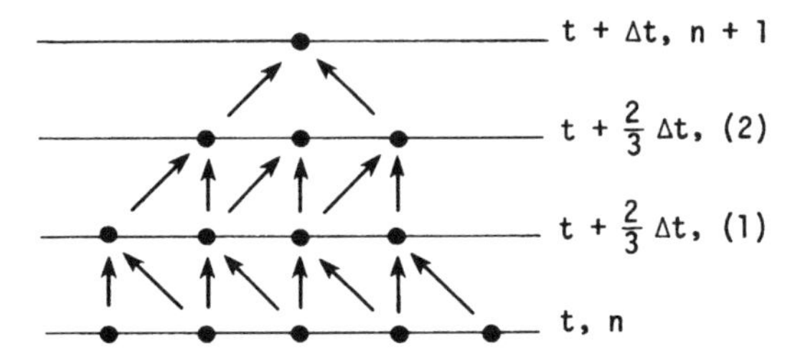

Figure 4.40 Point schematic for WKL method.

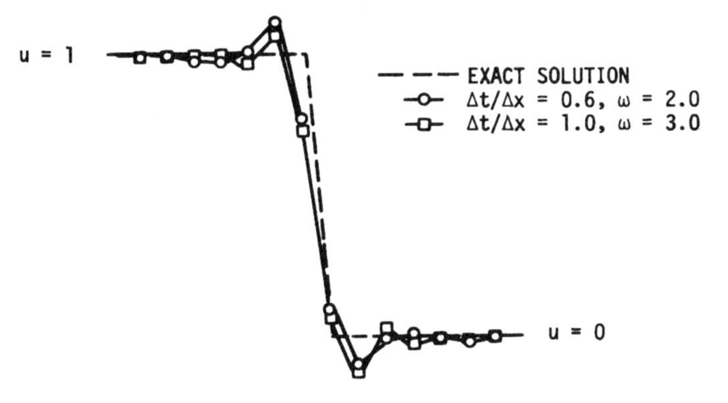

Figure 4.41 Burgers' equation solution using the WKL method.

Rusanov technique. We should note that different third-order schemes can be generated by altering the first two steps. Burstein and Mirin have shown that any second-order method may be used to generate $u_j^{(2)}$. The linear stability bound for the WKL and the Rusanov methods is the same as given in Eq. (4.170). The schematic illustrating the grid points used in the WKL method is presented in Fig. 4.40. Notice that the preferential treatment of the first two levels is readily apparent in this diagram. The differencing in the first two levels can be reversed or even cycled from time step to time step.

The results of using the WKL method to solve Burgers' equation for a right-moving discontinuity are shown in Fig. 4.41. The solution is nearly the same as that obtained in the previous section. Based upon the calculated results, either of the third-order methods may be used with approximately equal accuracy.

4.4.6 Tuned Third-Order Methods

The parameter ω, which appears in the third level of the methods of the previous two sections, can be chosen arbitrarily, as long as the stability bound is not violated. Once ω is selected at the beginning of a calculation, it retains the same value throughout the mesh. However, if the numerical damping term is written in conservation-law form for the third level, i.e.,

$$\frac{\partial}{\partial x}\left(\omega \frac{\partial^3 u}{\partial x^3}\right)$$

then ω may be altered from point to point in the calculation, and correct flux conservation in the mesh is assured. Using this approach, the ω term in the last level of either the Rusanov or WKL method can be written as

$$\begin{aligned} -\frac{\omega_{j+1/2}^n}{24}&(u_{j+2}^n - 3u_{j+1}^n + 3u_j^n - u_{j-1}^n) \\ &+ \frac{\omega_{j-1/2}^n}{24}(u_{j+1}^n - 3u_j^n + 3u_{j-1}^n - u_{j-2}^n) \end{aligned} \tag{4.172}$$

The $\omega_{j\pm1/2}^n$ values are now varied according to the effective Courant number in the mesh. Warming et al. (1973) suggest that these parameters be calculated at each point in the mesh to minimize either the dispersive error or the dissipative error.

A discussion of the modified equation for third-order methods is presented in Section 4.1.11. If the minimum dispersive error is desired, then according to Eq. (4.68), we should choose

$$\omega_{j\pm1/2}^n = \frac{\left(4\nu_{j\pm1/2}^2 + 1\right)\left(4 - \nu_{j\pm1/2}^2\right)}{5} \tag{4.173}$$

It remains to arrive at a rational method to determine the effective Courant numbers, $\nu_{j\pm1/2}$. Warming et al. (1973) suggest that the effective Courant numbers, used to determine the $\omega_{j\pm1/2}$ parameters, be the average value at the mesh points used in the difference formula. Since the term containing $\omega_{j+1/2}$ involves points $j+2$, $j+1$, j, and $j-1$, we can write

$$\nu_{j+1/2} = \frac{1}{4}(\lambda_{j+2} + \lambda_{j+1} + \lambda_j + \lambda_{j-1})\frac{\Delta t}{\Delta x} \tag{4.174}$$

and similarly,

$$\nu_{j-1/2} = \frac{1}{4}(\lambda_{j+1} + \lambda_j + \lambda_{j-1} + \lambda_{j-2})\frac{\Delta t}{\Delta x}$$

where λ is the local eigenvalue. For Burgers' equation, λ is just the unknown u. Results obtained using this variable ω or tuned approach are shown in Fig. 4.42. This shows that both third-order methods provide satisfactory solutions for the minimum dispersion case. A slightly larger overshoot occurs at the left of the discontinuity, but a nearly exact solution is obtained on the right. The minimum dissipative method of computing $\omega_{j\pm1/2}$ is not recommended. The ω parameter was added to provide stability, and when the dissipation is minimized, stability problems can occur. Even for stable solutions, large oscillations may be present. It should be noted that the parameters $\omega_{j\pm1/2}$ may be computed using any

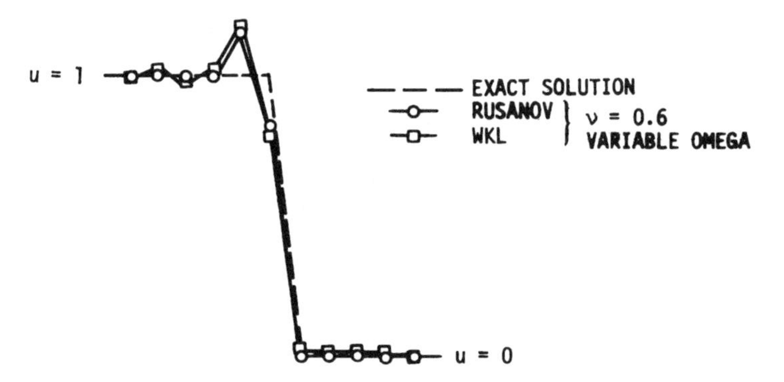

Figure 4.42 Tuned or variable ω method applied to Burgers' equation.

technique that does not violate the stability bound. Clearly, a different computed solution will be obtained for each way of computing these parameters.

4.4.7 Implicit Methods

The time-centered implicit method (trapezoidal method) is presented in Section 4.1.10. This scheme is based upon Eq. (4.58). If we substitute into Eq. (4.58) for the time derivatives using our model equation, we obtain

$$u_j^{n+1} = u_j^n - \frac{\Delta t}{2}\left[\left(\frac{\partial F}{\partial x}\right)^n + \left(\frac{\partial F}{\partial x}\right)^{n+1}\right] \tag{4.175}$$

It is immediately apparent that we now have a nonlinear problem, and some sort of linearization or iteration technique must be used. Beam and Warming (1976) have suggested that we write

$$F^{n+1} \approx F^n + \left(\frac{\partial F}{\partial u}\right)^n (u^{n+1} - u^n) = F^n + A^n(u^{n+1} - u^n)$$

Thus

$$u_j^{n+1} = u_j^n - \frac{\Delta t}{2}\left\{2\left(\frac{\partial F}{\partial x}\right)^n + \frac{\partial}{\partial x}\left[A\left(u_j^{n+1} - u_j^n\right)\right]\right\}$$

If the x derivatives are replaced by second-order central differences, then

$$-\frac{\Delta t A_{j-1}^n}{4\,\Delta x}u_{j-1}^{n+1} + u_j^{n+1} + \frac{\Delta t A_{j+1}^n}{4\,\Delta x}u_{j+1}^{n+1}$$
$$= -\frac{\Delta t}{\Delta x}\frac{F_{j+1}^n - F_{j-1}^n}{2} - \frac{\Delta t A_{j-1}^n u_{j-1}^n}{4\,\Delta x} + u_j^n + \frac{\Delta t A_{j+1}^n}{4\,\Delta x}u_{j+1}^n \tag{4.176}$$

The Jacobian A has the single element u for Burgers' equation, and a further simplification of the right side is possible. We see that the linearization applied

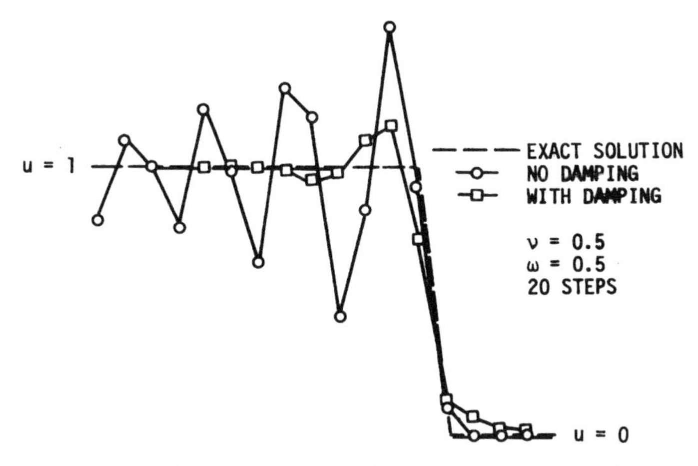

Figure 4.43 Solution of Burgers' equation using Beam-Warming (trapezoidal) method.

by Beam and Warming leads to a linear system of algebraic equations at the next time level. This is a tridiagonal system and may be solved using the Thomas algorithm.

As pointed out in Section 4.1.10, this method is stable for any time step. It should be noted that the roots of the characteristic equation always lie on the unit circle. This is consistent with the fact that the modified equation contains no even derivative terms. Consequently, artificial smoothing is added to the scheme. The usual fourth difference,

$$-\frac{\omega}{8}(u_{j+2}^n - 4u_{j+1}^n + 6u_j^n - 4u_{j-1}^n + u_{j-2}^n) \tag{4.177}$$

may be added to Eq. (4.176), and the formal accuracy of the method is unaltered. According to Beam and Warming, the implicit formula Eq. (4.176) with explicit damping added is stable if

$$0 < \omega \leqslant 1 \tag{4.178}$$

Figure 4.43 shows the results of applying the time-centered implicit formula to a right-moving discontinuity. The solution with no damping is clearly unacceptable. When explicit damping given by Eq. (4.177) is added, better results are obtained.

In addition to the trapezoidal formula just presented, Beam and Warming (1976) developed a three-point-backward implicit and an Euler implicit method as part of a family of techniques. The Beam and Warming version of the Euler implicit scheme follows from the backward Euler formula:

$$u^{n+1} = u^n + \Delta t\left(\frac{\partial u}{\partial t}\right)^{n+1}$$

which for our nonlinear equation becomes

$$u^{n+1} = u^n - \Delta t \left(\frac{\partial F}{\partial x} \right)^{n+1}$$

If the same linearization is applied, we obtain

$$-\frac{\Delta t A_{j-1}^n}{2\,\Delta x} u_{j-1}^{n+1} + u_j^{n+1} + \frac{\Delta t A_{j+1}^n}{2\,\Delta x} u_{j+1}^{n+1}$$

$$= -\frac{\Delta t}{\Delta x} \frac{F_{j+1}^n - F_{j-1}^n}{2} - \frac{\Delta t A_{j-1}^n}{2\,\Delta x} u_{j-1}^n + u_j^n + \frac{\Delta t A_{j+1}^n u_{j+1}^n}{2\,\Delta x} \quad (4.179)$$

This is again a tridiagonal system and is easily solved. We note that this scheme is unconditionally stable, but damping must be added such as that given in Eq. (4.177), to ensure a usable numerical result.

A simpler form of the implicit algorithms presented in this section can be obtained if they are written in "delta" form. This form uses the increments in the conserved variables and fluxes. In multidimensional problems it has the advantage of providing a steady-state solution that is independent of the time step in problems that possess a steady-state solution. Let us develop the time-centered implicit method using the delta form. Let $\Delta u_j = u_j^{n+1} - u_j^n$. The trapezoidal formula Eq. (4.175) may be written

$$\Delta u_j = -\frac{\Delta t}{2}\left[\left(\frac{\partial F}{\partial x} \right)^n + \left(\frac{\partial F}{\partial x} \right)^{n+1} \right]$$

Again a local linearization is used to obtain

$$F_j^{n+1} = F_j^n + A_j^n \, \Delta u_j$$

The final form of the difference equation becomes

$$-\frac{\Delta t A_{j-1}^n}{4\,\Delta x} \Delta u_{j-1} + \Delta u_j + \frac{\Delta t A_{j+1}^n}{4\,\Delta x} \Delta u_{j+1} = -\frac{\Delta t}{2\,\Delta x}(F_{j+1}^n - F_{j-1}^n) \quad (4.180)$$

This is much simpler than Eq. (4.176). The tridiagonal form is still retained, but the right side does not require the multiplications of the original algorithm. This can be important for systems of equations where the operation count is large. The solution of Eq. (4.180) provides the incremental changes in the unknowns between two time levels. As noted previously, the stability of the delta form is unrestricted, but the usual higher-order damping terms must be added. Results obtained using the delta form for our simple right-moving shock are shown in Fig. 4.44. The solutions with and without damping are essentially identical to those obtained using the expanded form, as should be expected. The delta form of the time-centered implicit scheme is recommended over the expanded version. In problems with time asymptotic solutions, the Δu terms approach zero, and in all cases, matrix multiplications are reduced.

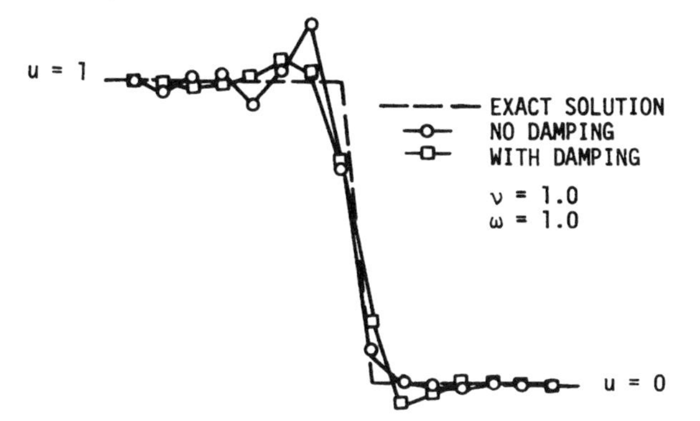

Figure 4.44 Solution for right-moving discontinuity time-centered implicit method, delta form.

Solutions of the inviscid Burgers equation computed with an implicit scheme are generally inferior to those calculated with explicit techniques, and more computational effort per integration step is required. In addition, transient results are usually desired, and the larger step sizes permitted by implicit schemes are not of major significance. When discontinuities are present, results produced with explicit methods are superior to those produced with implicit techniques using central differences. For these reasons, explicit numerical methods are recommended for solving the inviscid Burgers equation.

4.4.8 Godunov Scheme

The numerical methods applied to Burgers' equation in this chapter have used Taylor series to establish appropriate expressions for the values of the dependent variables at the next time level. Differences in the spatial directions were also based upon the requirement of having a certain accuracy using a series approximation. Taylor-series expansions work very well when conditions for convergence of the series are met. In fact, the series will converge everywhere, provided the function that is approximated is sufficiently smooth. In the case of a finite-difference method, we assume that a series expansion is an appropriate means of obtaining a difference approximation and the functions are continuous and have continuous derivatives at least through the order of the difference approximation. This is certainly not true when shock waves or other discontinuities are present. Godunov (1959) recognized this basic problem and proposed to avoid the requirement of differentiability by using a finite-volume approximation in solving the conservation equations and evaluating the flux terms at the cell interfaces by the solution of a Riemann problem. In this section we will describe the Godunov scheme specifically applied to Burgers' equation and see how this method leads to a numerical technique that treats the problem of discontinuous solutions in a very specific way.

Consider the inviscid Burgers equation, Eq. (4.142), and a finite-volume approximation with a control volume as shown in Fig. 4.35. For an explicit method the control volume extends from t to $t + \Delta t$ and from $x - \Delta x/2$ to $x + \Delta x/2$. If a control volume centered at $(j, n + \frac{1}{2})$ is selected, the resulting numerical approximation for the dependent variable may be written

$$\bar{u}_j^{n+1} = \bar{u}_j^n - \frac{\Delta t}{\Delta x}\left[f(u_{j+\frac{1}{2}}) - f(u_{j-\frac{1}{2}})\right] \tag{4.181}$$

In this equation the value of u is averaged over the volume element, i.e.,

$$\bar{u}_j = \frac{1}{\Delta x}\int_{x-\Delta x/2}^{x+\Delta x/2} u(x,t)\,dx$$

and the flux term is the time-averaged value of the flux at the control-volume interface:

$$f = \frac{1}{\Delta t}\int_t^{t+\Delta t} f\,dt$$

The Godunov method solves a local Riemann problem at each cell interface in order to obtain a value of the flux necessary to advance the solution. The Riemann problem specifically for Burgers' equation is

$$\frac{\partial u}{\partial t} + \frac{\partial}{\partial x}\left(\frac{u^2}{2}\right) = 0 \tag{4.182}$$

with initial conditions

$$u(x,0) = \begin{cases} u_j & x \leqslant 0 \\ u_{j+1} & x > 0 \end{cases}$$

The geometry of the problem is shown in Fig. 4.45. The averaged values of the dependent variable give the appearance of a slab-like variation in distribution,

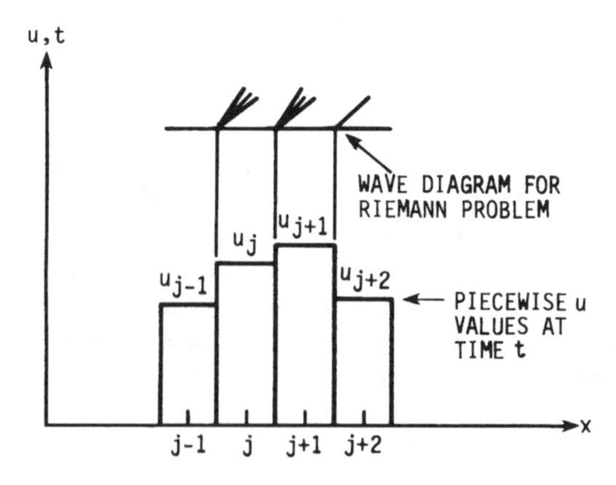

Figure 4.45 Wave diagram for Godunov's method.

leading to a discontinuity at each cell interface. At each cell interface, either a shock or an expansion is initiated and propagates in time. As noted in earlier discussions of solutions for the Riemann problem for Burgers' equation, the solution is self-similar where the similarity variable is x/t. In this sense, the value of x is assumed to be zero at the interface. There are several cases to consider in solving this problem.

With the notation

$$c_{j+\frac{1}{2}} = \left(\frac{dx}{dt}\right)_{j+\frac{1}{2}} = \frac{u_j + u_{j+1}}{2}$$

the solution of the Riemann problem for this equation may be written for the following cases:

Case 1: Shock waves

$$u_j > u_{j+1}$$

$$u = \begin{cases} u_j & x/t < c_{j+\frac{1}{2}} \\ u_{j+1} & x/t > c_{j+\frac{1}{2}} \end{cases}$$

$$f_{j+\frac{1}{2}} = \begin{cases} \frac{1}{2}u_j^2 & c_{j+\frac{1}{2}} > 0 \\ \frac{1}{2}u_{j+1}^2 & c_{j+\frac{1}{2}} < 0 \end{cases} \tag{4.183}$$

Case 2: Expansion waves

$$u_j < u_{j+1}$$

$$u = \begin{cases} u_j & x/t < u_j \\ x/t & u_j < x/t < u_{j+1} \\ u_{j+1} & x/t > u_{j+1} \end{cases}$$

$$f_{j+\frac{1}{2}} = \begin{cases} 0 & u_j < 0 < u_{j+1} & \\ \frac{1}{2}u_j^2 & c_{j+\frac{1}{2}} > 0 & u_{j+1} > u_j > 0 \\ \frac{1}{2}u_{j+1}^2 & c_{j+\frac{1}{2}} < 0 & u_j < u_{j+1} < 0 \end{cases} \tag{4.184}$$

An assumption implicit in the above discussion is that the waves from adjacent cells do not interact. This is required in order to write a simple solution for the state variables at the cell boundaries and is assured only if the wave can travel, at most, half of one cell in distance. As a consequence, the stability restriction placed on the Godunov scheme is that

$$\left| u_{\max} \frac{\Delta t}{\Delta x} \right| \leqslant \frac{1}{2}$$

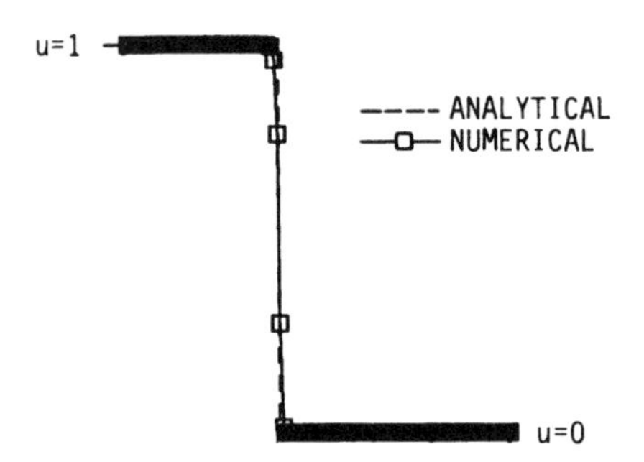

Figure 4.46 Godunov method applied to shock problem.

Using Eq. (4.181), we can now integrate to obtain a solution of Burgers' equation where the flux is evaluated using the solution of the Riemann problem. A typical result is shown in Fig. 4.46. The results of this calculation show that the solution is superior for shock propagation. The results for the case of an expansion wave are nearly as good. It should be remembered that these results are only for a 1-D calculation and that higher-dimensional applications lead to a much more rigorous test of the Godunov idea. In multiple dimensions the flux at the control-volume boundaries is still determined by solving the 1-D Riemann problem and the influence of this 1-D view is shown in Chapter 6.

4.4.9 Roe Scheme

The solution of Burgers' equation using the Godunov method is easily accomplished, and the results are excellent. However, when this method is applied to the solution of the equations that govern fluid flow, it is necessary to employ a computationally inefficient iterative technique. One idea that has been used with good success in computing the solution to nonlinear systems is to solve an approximate Riemann problem rather than having to deal with the exact nonlinear iterative scheme. One of the most popular approximate Riemann solvers was proposed by Roe (1980, 1981). Roe suggested solving the linear problem

$$\frac{\partial u}{\partial t} + \bar{A}\frac{\partial u}{\partial x} = 0 \tag{4.185}$$

where $\bar{A}$ is a constant matrix that is dependent on local conditions. In the case of Burgers' equation, the $\bar{A}$ matrix is, of course, a single scalar element. The $\bar{A}$ matrix is constructed to satisfy what Roe termed property U, which includes the following conditions:

1. For any u_j, u_{j+1},

$$F_{j+1} - F_j = \bar{A}(u_{j+1} - u_j) \tag{4.186}$$

2. When $u = u_j = u_{j+1}$, then

$$\bar{A}(u_j, u_{j+1}) = \bar{A}(u, u) = \frac{\partial F}{\partial u} = u \tag{4.187}$$

The first of these conditions ensures that the correct jump is recovered when a discontinuity is encountered. The second condition provides that in smooth regions, the Jacobian, in this case the nonlinear wave speed, reduces to the correct value.

In applying Roe's scheme to Burgers' equation, $\bar{u}$ is a constant and denotes the averaged value of $\bar{A}$. We then consider the linear problem

$$\frac{\partial u}{\partial t} + \bar{u}\frac{\partial u}{\partial x} = 0 \tag{4.188}$$

where the appropriate value of $\bar{u}$ for cells j and $j+1$ is determined from the first of the requirements noted above. This gives

$$\bar{u} = \bar{u}_{j+\frac{1}{2}} = \frac{F_{j+1} - F_j}{u_{j+1} - u_j} \tag{4.189}$$

which for Burgers' equation reduces to

$$\bar{u}_{j+\frac{1}{2}} = \begin{cases} \dfrac{u_j + u_{j+1}}{2} & u_j \neq u_{j+1} \\ u_j & u_j = u_{j+1} \end{cases} \tag{4.190}$$

With this information, the numerical flux can be developed. Since the original Riemann problem has been reduced to a linearized form, the Rankine-Hugoniot relations directly provide the relationship between the jump in the flux and the jump in the dependent variable u across a wave, i.e.,

$$F_{j+1} - F_j = \bar{u}_{j+\frac{1}{2}}(u_{j+1} - u_j) \tag{4.191}$$

As a consequence, this approximate Riemann solver only recognizes discontinuities and cannot distinguish between expansions that are discontinuous and shock waves. An alteration will be added below to correct for this fact.

Consider the approximate Riemann problem for Burgers' equation as depicted in Fig. 4.47. For this case, we see that a single wave emanates from the cell interface. This single wave travels either in the positive or negative direction, depending upon $\bar{u}_{j+\frac{1}{2}} = dx/dt$. Utilizing the definition of the jump across this wave, we may write either

$$f_{j+\frac{1}{2}} - F_j = \bar{u}^-_{j+\frac{1}{2}}(u_{j+1} - u_j)$$

or

$$F_{j+1} - f_{j+\frac{1}{2}} = \bar{u}^+_{j+\frac{1}{2}}(u_{j+1} - u_j)$$

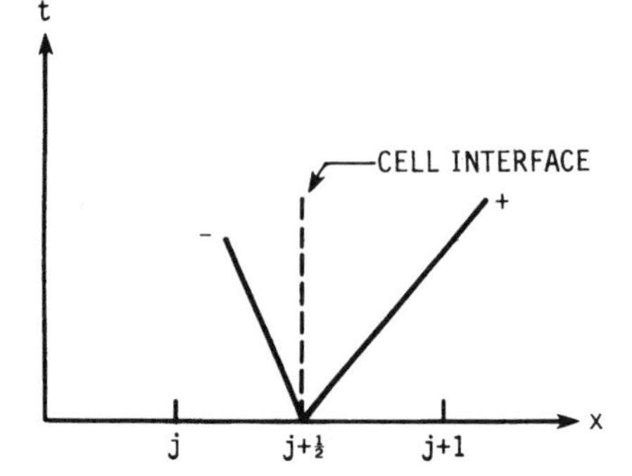

Figure 4.47 Wave diagram for Roe scheme applied to Burgers' equation.

The numerical flux may then be written in the symmetric form

$$f_{j+\frac{1}{2}} = \frac{F_j + F_{j+1}}{2} + \frac{1}{2}\left(\bar{u}^-_{j+\frac{1}{2}} - \bar{u}^+_{j+\frac{1}{2}}\right)(u_{j+1} - u_j)$$

Consider the individual contributions that occur when the wave travels either from the right or left. If each case is evaluated separately, the numerical flux can be represented in a single equation of the form

$$f_{j+\frac{1}{2}} = \frac{F_j + F_{j+1}}{2} - \frac{1}{2}|\bar{u}_{j+\frac{1}{2}}|(u_{j+1} - u_j) \tag{4.192}$$

Figure 4.48 illustrates the calculation of a propagating discontinuity using Roe's scheme, and the results show excellent agreement with the known analytical solution.

Roe's scheme has the stability bound common to explicit methods, i.e., the Courant number must be less than 1. The Roe formulation propagates the difference in dependent variables between two points as if a discontinuity existed between these points. A consequence of this is that expansion shocks that are nonphysical may appear. In Burgers' equation this is a problem when $\bar{u}_{j+\frac{1}{2}}$ vanishes. This is referred to as a sonic transition. The classical example is a solution computed using Roe's scheme with initial data

$$u = \begin{cases} -1 & 0 \leqslant x \leqslant x_0 \\ +1 & x_0 < x \leqslant L \end{cases}$$

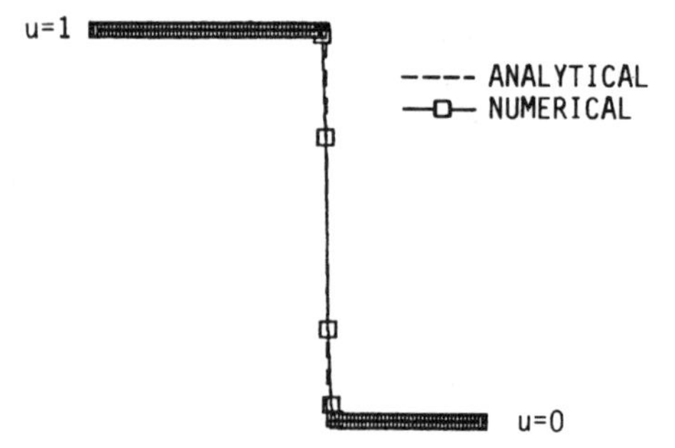

Figure 4.48 Discontinuity propagated using Roe's scheme.

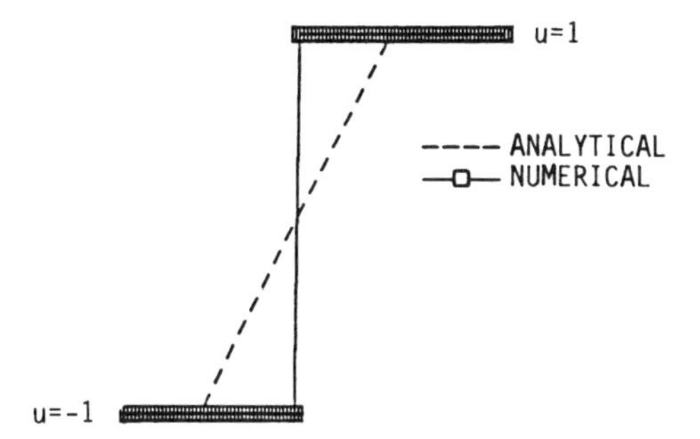

Figure 4.49 Expansion shock using Roe's scheme.

The analytic solution is a centered expansion about $x = x_0$. The solution computed using the Roe scheme is shown in Fig. 4.49. The initial data are faithfully reproduced, showing an incorrect stationary expansion shock.

This nonphysical behavior is due to the fact that the scheme cannot distinguish between an expansion shock and a compression shock. Each is a valid solution for this formulation. The existence of the expansion shock is said to violate the entropy condition, allowing incorrect physical behavior. Oleinik (1957) and, later, Lax (1973) developed conditions that must be satisfied by discontinuous solutions of hyperbolic equations. The simplest statement of this condition applied to Burgers' equation may be written

$$u_{\mathrm{R}} < \left(\frac{dx}{dt}\right)_{x_0} < u_{\mathrm{L}} \tag{4.193}$$

where u_{R} and u_{L} are the values to the right and the left of the discontinuity. The initial data for the expansion violate this condition, eliminating expansion shocks. In the example just cited, no information is present regarding the sonic transition ($u = 0$), and Roe's scheme interprets the two-point expansion in the initial data as a discontinuity. Unlike the Godunov scheme, where admissible solutions incorporate the correct physical behavior, a modification of the numerics is necessary. A number of techniques to accomplish this have been proposed. Harten and Hyman (1983) proposed the following modification of $\bar{u}_{j+\frac{1}{2}}$. Let

$$\epsilon = \max\left(0, \frac{u_{j+1} - u_j}{2}\right)$$

then

$$\bar{u}_{j+\frac{1}{2}} = \begin{cases} \bar{u}_{j+\frac{1}{2}} & \bar{u}_{j+\frac{1}{2}} \geqslant \epsilon \\ \epsilon & \bar{u}_{j+\frac{1}{2}} < \epsilon \end{cases} \tag{4.194}$$

The compression case ($\epsilon = 0$) uses the unaltered scheme to propagate discontinuities, while the case of an expansion fan [$\epsilon = (u_{j+1} - u_j)/2$] requires the modification. Figure 4.50 shows the results computed for an expansion $(-1, 1)$ using Roe's scheme with the entropy fix included. The agreement between the numerical and analytical solutions is good and is approximately

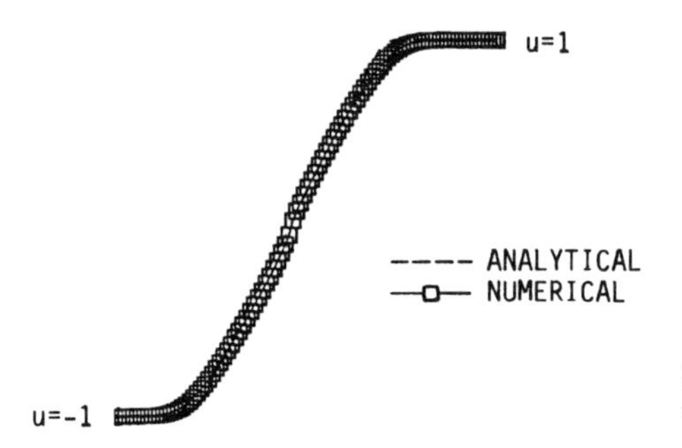

Figure 4.50 Expansion using Roe's scheme with entropy correction.

what one would expect for a first-order method. The modification of Roe's scheme may be viewed as a way of adding an appropriate amount of dissipation to the basic method when the dissipation at the sonic transition goes to zero. The alteration given here is equivalent to the introduction of a small expansion in the approximate Riemann solution when an expansion appears through a sonic point.

4.4.10 Enquist-Osher Scheme

In the previous section, Roe's scheme was seen to replace both shock and expansion waves by discontinuities. In order to correctly reproduce expansions and not produce expansion shocks, a modification of the numerical flux was introduced to correct the physical behavior. The scheme introduced by Enquist and Osher (1980, 1981) treats the change in the dependent variables across waves as a continuous transition in state space and produces a method that is monotone and conservative and that correctly treats both shock and expansion waves. In this scheme, the shock discontinuities in the exact Riemann solution are replaced by smooth compression waves. The discontinuities are resolved with at most two interior points. The stability bound of this approximate Riemann solver is the usual Courant-Friedrichs-Lewy (CFL) condition for explicit methods.

The flux at any location is written as the sum of contributions from crossing waves with slopes in (x, t) space that are positive and negative. Thus

$$F = F^{+} + F^{-} \tag{4.195}$$

where the individual components are obtained by an integration in phase space defined by

$$F^{+}(u) = \int_0^u \mu(\tau) F'(\tau)\, d\tau \tag{4.196a}$$

$$F^{-}(u) = \int_0^u [1 - \mu(\tau)] F'(\tau)\, d\tau \tag{4.196b}$$

and the switch is defined as

$$\mu(\tau) = \begin{cases} 1 & A = \partial F/\partial u \geqslant 0 \\ 0 & A < 0 \end{cases}$$

In this evaluation of the flux, the sonic transition is specifically identified through the switch in the integration. With this notation, we may write the cell interface flux by starting at either side, as we did in Roe's scheme:

$$f_{j+\frac{1}{2}} = F_j + \int_{u_j}^{u_{j+1}} [1 - \mu(\tau)] F'(\tau)\, d\tau \tag{4.197a}$$

$$f_{j+\frac{1}{2}} = F_{j+1} - \int_{u_j}^{u_{j+1}} \mu(\tau) F'(\tau)\, d\tau \tag{4.197b}$$

The numerical flux is usually written in the symmetric form

$$f_{j+\frac{1}{2}} = \frac{F_j + F_{j+1}}{2} - \frac{1}{2}\int_{u_j}^{u_{j+1}} |u|\, du \tag{4.198}$$

where the flux derivative representing the Jacobian and the switch have been replaced by $|u|$ in the integral. With this definition of the interface flux, the correct form for Burgers' equation may now be written

$$f_{j+\frac{1}{2}} = \begin{cases} u_{j+1}^2/2 & u_j, u_{j+1} < 0 \\ u_j^2/2, & u_j, u_{j+1} > 0 \end{cases} \tag{4.199}$$

If u_j and u_{j+1} are of opposite sign, then

$$f_{j+\frac{1}{2}} = \begin{cases} 0 & u_j < 0 < u_{j+1} \\ \left(u_j^2 + u_{j+1}^2\right)/2 & u_j > 0 > u_{j+1} \end{cases} \tag{4.200}$$

While the Godunov scheme treats shocks as discontinuities, the Enquist-Osher scheme replaces shocks by what van Leer (1984) describes as overturned centered compression waves. Discontinuities are excluded owing to the smooth transitions in the phase space integrals. The treatment of sonic transitions leads to a small deviation in the slope of expansions when such points are present.

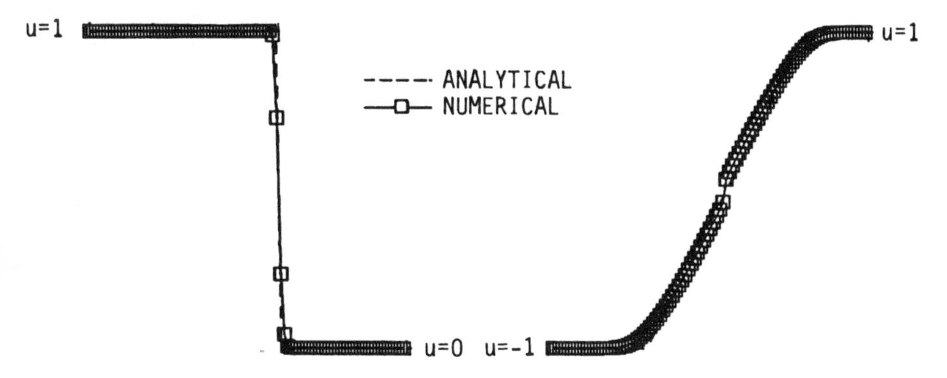

Figure 4.51 Results using the Enquist-Osher scheme. (a) Shock wave. (b) Expansion wave.

Such a case is shown in Fig. 4.51, where the results produced by applying the Enquist-Osher scheme to both a propagating discontinuity and expansion are depicted. According to Chakravarthy and Osher (1985), this small deviation at the sonic location in the expansion is bounded and does not lead to erroneous results. The results shown in Fig. 4.51 were computed using the explicit scheme given in Eq. (4.154). For either the Roe or Enquist-Osher scheme, the stability bound is the usual requirement of Courant number less than 1. Since only an approximate Riemann problem is considered, the smaller limit imposed by the Godunov method is avoided.

4.4.11 Higher-Order Upwind Schemes

In applying finite-difference methods for the solution of PDEs, a higher-order approximation is obtained by introducing more points in the stencil. For upwind schemes a first derivative can be approximated to first order using two points, while three points are necessary for a second-order expression. In the Riemann or approximate Riemann solvers, a higher-order approximation must be interpreted in terms of flux values at control-volume boundaries.

In the original Godunov approach, state variables were assumed to be constant in control volumes. For the first-order schemes, this assumption was sufficient. Van Leer (1979) extended this idea by assuming that the state variables, as projected on each cell, can have a variation. The cell-averaged values from the Riemann solution are used to reconstruct the assumed variation in each control volume. This idea essentially leads to a higher-order extrapolation of the flux or state variables at the cell boundaries. For the variable extrapolation approach, van Leer coined the term "monotone upstream-centered schemes for conservation laws." This is referred to as the MUSCL approach, or sometimes MUSCL differencing. It should be noted that the flux can be used in the formulas for extrapolation to the cell boundaries. In this case, the flux terms do not need to be recalculated, since the extrapolation directly provides the needed information.

For the first-order Godunov scheme, the average value of the dependent variable is set equal to the constant value of that variable in the cell. For higher-order schemes, the values assigned to each cell are the averages over each cell. Depending on the accuracy desired, the dependent variables may be curve fit with a polynomial of arbitrary order in each cell. Consider the piecewise linear representation shown in Fig. 4.52. For cell j, the expression for the variation of u is a function of position in the cell, and the average value of u is assigned to the cell. In order to arrive at an extrapolation for the control-volume boundary value, a Taylor-series representation is employed:

$$u_{j+\frac{1}{2}} = u_j + \frac{\Delta x}{2}(u_x)_j + \frac{(\Delta x)^2}{8}(u_{xx})_j + \cdots \tag{4.201}$$

where the derivatives are evaluated by differences using cell averages.

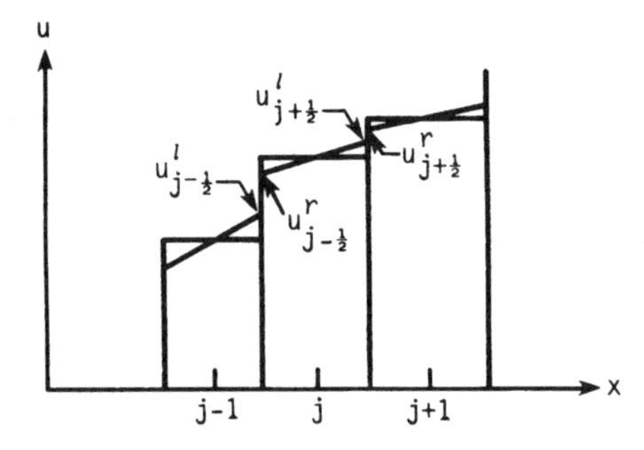

Figure 4.52 Assumed linear variation in each cell.

For cell j the expressions for the independent variables will have a right and left side. The usual expressions for the right and left extrapolations are

$$u^l_{j+\frac{1}{2}} = u_j + \frac{1-\kappa}{4}\,\delta u_{j-\frac{1}{2}} + \frac{1+\kappa}{4}\,\delta u_{j+\frac{1}{2}} \tag{4.202}$$

$$u^r_{j+\frac{1}{2}} = u_{j+1} - \frac{1+\kappa}{4}\,\delta u_{j+\frac{1}{2}} - \frac{1-\kappa}{4}\,\delta u_{j+\frac{3}{2}} \tag{4.203}$$

These expressions are Taylor-series representations for the state variables, with difference approximations for the derivatives. When fluxes are used instead of the independent variables, the fluxes are introduced in the extrapolation to the cell boundaries. This has the advantage of providing the flux directly, and the flux does not require reformulation from the independent variable extrapolation. Care must be exercised using this approach because it may not provide a solution that is as good as that using the independent variable extrapolation and subsequent reformulation of the fluxes at the cell boundaries (Anderson et al., 1987). In the extrapolation the value of κ determines the type of method (van Leer, 1979; Yee, 1989). For example,

$$\kappa = \begin{cases} -1 & \text{upwind scheme} \\ 0 & \text{Fromm's (1968) method} \\ 1 & \text{central difference} \end{cases} \tag{4.204}$$

It is of value to consider the case of $\kappa = -1$ and the relationship of this upwind extrapolation to an upwind finite-difference approximation. The second-order upwind scheme studied in Section 4.1.9 can be recovered using the one-sided extrapolation with $\kappa = -1$ (see Prob. 4.60). The values for the extrapolated variables on both sides of the interface become

$$u^l_{j+\frac{1}{2}} = u_j + \frac{1}{2}\,\delta u_{j-\frac{1}{2}}$$

$$u^r_{j+\frac{1}{2}} = u_{j+1} - \frac{1}{2}\,\delta u_{j+\frac{3}{2}}$$

while for $\kappa = 1$, a central approximation becomes

$$u^l_{j+\frac{1}{2}} = u_j + \frac{1}{2}\,\delta u_{j+\frac{1}{2}}$$

$$u^r_{j-\frac{1}{2}} = u_j - \frac{1}{2}\,\delta u_{j-\frac{1}{2}}$$

Both the upwind and central differences are linear approximations to the cell boundary values of u. The difference between them is the different approximation to the slope. The interface values for the $\kappa = 1$ case reduce to the same average between adjacent cells, indicating the central-difference scheme.

Second-order flux values are formed from the dependent variables by replacement. Consider Roe's scheme, where the first-order flux was given by Eq. (4.192). Functionally, we may write

$$f^{(1)}_{j+\frac{1}{2},Roe} = f(u_j, u_{j+1}) \tag{4.205}$$

The second-order flux is obtained by replacing u_j, u_{j+1} in the flux equation with right and left values:

$$f^{(2)}_{j+\frac{1}{2},Roe} = f\left(u^l_{j+\frac{1}{2}}, u^r_{j+\frac{1}{2}}\right) \tag{4.206}$$

The second-order Roe flux may then be written

$$f^{(2)}_{j+\frac{1}{2},Roe} = \frac{1}{2}\left[F\left(u^l_{j+\frac{1}{2}}\right) + F\left(u^r_{j+\frac{1}{2}}\right) - \left|\overline{u_{j+\frac{1}{2}}}\right|\left(u^r_{j+\frac{1}{2}} - u^l_{j+\frac{1}{2}}\right)\right] \tag{4.207}$$

where

$$\overline{u_{j+\frac{1}{2}}} = \frac{F\left(u^r_{j+\frac{1}{2}}\right) - F\left(u^l_{j+\frac{1}{2}}\right)}{u^r_{j+\frac{1}{2}} - u^l_{j+\frac{1}{2}}} \tag{4.208}$$

and is defined in the same manner as the equivalent term in the first-order flux.

An explicit predictor-corrector method similar to the two-step Lax-Wendroff method, producing a second-order solution in space and time, may now be constructed. The predictor step is

$$u^{n+\frac{1}{2}}_j = u^n_j - \frac{\Delta t}{2\,\Delta x}\left(f^n_{j+\frac{1}{2}} - f^n_{j-\frac{1}{2}}\right) \tag{4.209}$$

where f represents a first-order flux. Variable extrapolations are then employed to obtain the left and right interface values of the dependent variables, yielding

$$u^{l,n+\frac{1}{2}}_{j+\frac{1}{2}} = u^{n+\frac{1}{2}}_j + \frac{1-\kappa}{4}\,\delta u^{n+\frac{1}{2}}_{j-\frac{1}{2}} + \frac{1+\kappa}{4}\,\delta u^{n+\frac{1}{2}}_{j+\frac{1}{2}} \tag{4.210}$$

$$u^{r,n+\frac{1}{2}}_{j+\frac{1}{2}} = u^{n+\frac{1}{2}}_j - \frac{1+\kappa}{4}\,\delta u^{n+\frac{1}{2}}_{j+\frac{1}{2}} - \frac{1-\kappa}{4}\,\delta u^{n+\frac{1}{2}}_{j+\frac{3}{2}} \tag{4.211}$$

The corrector step is written

$$u_j^{n+1} = u_j^n - \frac{\Delta t}{\Delta x}\left(f_{j+\frac{1}{2}}^{n+\frac{1}{2}} - f_{j-\frac{1}{2}}^{n+\frac{1}{2}}\right) \tag{4.212}$$

where

$$f_{j+\frac{1}{2}}^{n+\frac{1}{2}} = f\left(u_{j+\frac{1}{2}}^{l,n+\frac{1}{2}}, u_{j+\frac{1}{2}}^{r,n+\frac{1}{2}}\right) \tag{4.213}$$

The stability of this scheme depends on the extrapolation used to construct the flux terms. Upwind schemes of higher order have a less restrictive stability constraint than central schemes, as we have seen in previous sections. A Courant number less than 1 is necessary for the central scheme, while the less restrictive value of 2 may be used for the upwind case. Figure 4.53 shows the results of applying this scheme to Burgers' equation for a shock propagating to the right. As is characteristic of second-order schemes, oscillations are present in the calculation. It is desirable to eliminate these wiggles and maintain the monotone character of the initial profile. In Section 4.4.1 the concept of a monotone solution was introduced when the Godunov theorem was presented. It is apparent that the nonmonotone behavior of the present solutions must be modified in some way if a better result is to be obtained. It should also be apparent that the behavior of the present solution obtained with the second-order upwind scheme does little to suggest an improvement over a central-difference scheme. In both cases the solution oscillates, and some way of controlling this behavior is desired. In the case of the upwind scheme, one can argue that the physics of the problem is more closely represented, even though the solution does not show any marked improvement over the central schemes. Improvements in the results can be achieved with the introduction of limiters that avoid the overshoots and undershoots shown in typical solutions.

4.4.12 TVD Schemes

The results of the previous section showed that even though upwind schemes appear to account for physics in a more appropriate way, as compared to central-difference schemes, higher-order methods share the same deficiencies

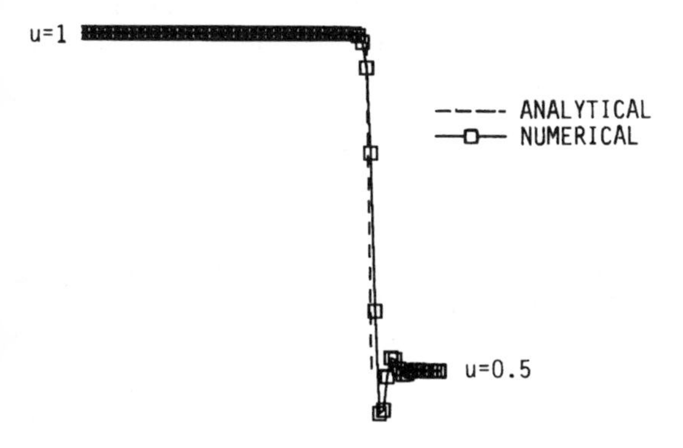

Figure 4.53 Second-order Roe scheme applied to Burgers' equation without limiting.

when discontinuities are encountered. The problem can be demonstrated computationally by starting with a monotone profile for a compression wave. As the compression front steepens, higher-order methods produce numerical solutions with new extrema. Consequently, oscillations are introduced that are undesirable. This numerical experiment (see Prob. 4.64) is a computational verification of Godunov's theorem given in Section 4.4.1.

Lax (1973) showed that for scalar conservation laws of the form given by Burgers' equation, Eq. (4.142), the total variation of physically possible solutions does not increase in time. The total variation (TV) is given by

$$\mathrm{TV} = \int \left| \frac{\partial u}{\partial x} \right| dx \tag{4.214}$$

and the total variation for the discrete case is

$$\mathrm{TV}(u) = \sum_j |u_{j+1} - u_j| \tag{4.215}$$

A numerical method is said to be total variation diminishing, or TVD, if

$$\mathrm{TV}(u^{n+1}) \leqslant \mathrm{TV}(u^n) \tag{4.216}$$

Harten (1983) proved that

1. a monotone scheme is TVD, and
2. a TVD scheme is monotonicity preserving.

Thus, if higher-order TVD schemes can be constructed, these schemes will be monotonicity preserving.

The central idea in constructing a TVD scheme is to attempt to develop a higher-order method that will avoid oscillations and exhibit properties similar to those of a monotone scheme. For such schemes the solution is first order near discontinuities and higher order in smooth regions. The transition to higher order is accomplished by the use of slope limiters on the dependent variables or flux limiters. Boris and Book (1973) developed the first nonlinear limiting by adding a limited difference between the first- and second-order fluxes to prevent oscillations associated with second-order schemes. The work of van Leer (1974) was based upon limiting the extrapolation of the dependent variables to the cell boundaries in order to prevent overshoots in the solution. Sweby (1984) and Roe (1985) developed limiters based on the TVD models of the Lax-Wendroff scheme, while Harten (1978) applied the concept of a modified flux to achieve the same effect. It is important to understand that methods that are either central differenced or upwind can be made TVD by suitable modification with limiters. When such higher-order schemes are constructed, a central-difference scheme may appear to be upwind in many regions, and upwind schemes may sometimes appear to be similar to a central-difference scheme.

The problem of constructing a solution of Burgers' equation that does not include unwanted oscillations when MUSCL differencing is employed can now be considered. In this case, unwanted oscillations can be avoided if the slopes of the variables used in the extrapolations are limited in such a way that the end point values do not create a new maximum or minimum. For the case of $u_{j-1} < u_{j+1}$, it is desired that

$$u^{\mathrm{r}}_{j-\frac{1}{2}} \geqslant u_{j-1} \tag{4.217a}$$

$$u^{\mathrm{l}}_{j+\frac{1}{2}} \leqslant u_{j+1} \tag{4.217b}$$

For this specification of the extrapolated values of the dependent variables, an easy control is devised by the introduction of slope limiters in the following way:

$$u^{\mathrm{l}}_{j+\frac{1}{2}} = u_j + \frac{1-\kappa}{4}\overline{\delta^{+}u_{j-\frac{1}{2}}} + \frac{1+\kappa}{4}\overline{\delta^{-}u_{j+\frac{1}{2}}} \tag{4.218a}$$

$$u^{\mathrm{r}}_{j+\frac{1}{2}} = u_{j+1} - \frac{1+\kappa}{4}\overline{\delta^{+}u_{j+\frac{1}{2}}} - \frac{1-\kappa}{4}\overline{\delta^{-}u_{j+\frac{3}{2}}} \tag{4.218b}$$

In these equations, the quantities $\overline{\delta^{\pm}u_j}$ are limited slopes. A number of different limiters may be used. One choice is the minmod limiter defined by

$$\overline{\delta^{-}u_{j+\frac{1}{2}}} = \operatorname{minmod}(\delta u_{j+\frac{1}{2}}, \omega\, \delta u_{j-\frac{1}{2}}) \tag{4.219a}$$

$$\overline{\delta^{+}u_{j+\frac{1}{2}}} = \operatorname{minmod}(\delta u_{j+\frac{1}{2}}, \omega\, \delta u_{j+\frac{3}{2}}) \tag{4.219b}$$

The minmod limiter is used extensively in TVD numerical methods. This is a function that selects the smallest number from a set when all have the same sign but is zero if they have different signs. For example,

$$\operatorname{minmod}(x, y) = \begin{cases} x & \text{if} \quad |x| < |y| \quad \text{and} \quad xy > 0 \\ y & \text{if} \quad |x| > |y| \quad \text{and} \quad xy > 0 \\ 0 & \text{if} \quad \qquad\qquad\qquad\qquad xy < 0 \end{cases} \tag{4.220}$$

or written in another form,

$$\operatorname{minmod}(x, \omega y) = \operatorname{sgn}(x)\max\{0, \min[|x|, \omega y \operatorname{sgn}(x)]\} \tag{4.221}$$

with the limits on ω given as

$$1 \leqslant \omega \leqslant \frac{3-\kappa}{1-\kappa} \tag{4.222}$$

and the values of κ not equal to 1. The usual notation of sgn represents the sign of the argument indicated. The values of the limited slopes in these expressions are selected to satisfy the end point conditions given by Eqs. (4.217). The higher-order fluxes are computed as indicated in the previous section, and the slopes are limited to prevent the occurrence of nonphysical oscillations. The same calculation presented in Fig. 4.53 without the use of limiters is repeated in Fig. 4.54 including limiting. The improvement in the solution is apparent, and the MUSCL approach does provide a good way of calculating a higher-order solution that is dramatically better when using limiters.

In general, higher-order schemes that are either upwind or central differenced can be modified to have the TVD property, thus avoiding the

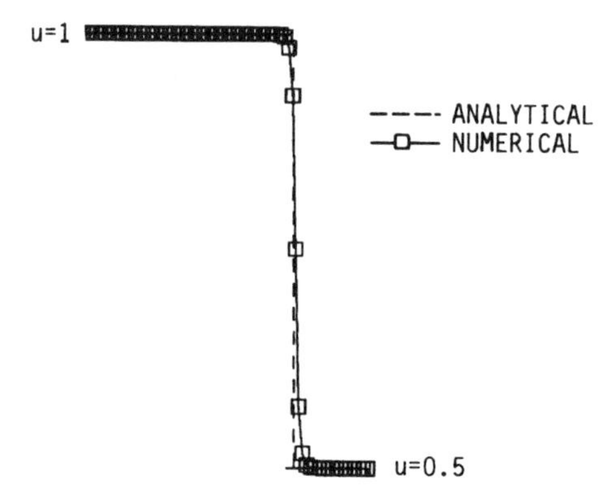

Figure 4.54 Second-order Roe scheme applied to Burgers' equation using the minmod limiter.

deficiencies of the classical shock-capturing numerical methods. Insight into the process of designing limiters is gained by utilizing the general class of numerical schemes of Yee (1987) and following the analysis of Harten (1984). Consider the one-parameter family of difference schemes written in conservative form:

$$u_j^{n+1} + \frac{\Delta t}{\Delta x}\theta\left(f_{j+\frac{1}{2}}^{n+1} - f_{j-\frac{1}{2}}^{n+1}\right) = u_j^n - \frac{\Delta t}{\Delta x}(1-\theta)\left(f_{j+\frac{1}{2}}^n - f_{j-\frac{1}{2}}^n\right) \quad (4.223)$$

where θ varies between 0 and 1 and

$$f_{j+\frac{1}{2}}^n = f(u_{j-1}^n, u_j^n, u_{j+1}^n, u_{j+2}^n) \quad (4.224)$$

This scheme represents several variations. If $\theta = 0$, this is an explicit scheme, while it is implicit if $\theta \neq 0$. If $\theta = \frac{1}{2}$, the differencing is trapezoidal and the technique is second order in time. This one-parameter family of methods can be more easily recognized if the average flux is defined as

$$\bar{f}_{j+\frac{1}{2}} = (1-\theta)f_{j+\frac{1}{2}}^n + \theta f_{j+\frac{1}{2}}^{n+1}$$

With this notation, Eq. (4.223) may be written in the familiar form

$$u_j^{n+1} = u_j^n - \frac{\Delta t}{\Delta x}\left(\bar{f}_{j+\frac{1}{2}} - \bar{f}_{j-\frac{1}{2}}\right)$$

The average flux is consistent with the conservation statement, in that

$$\bar{f}(u,u,u,u) = f(u)$$

The general scheme may be written in terms of implicit and explicit operators as

$$L(u^{n+1}) = R(u^n) \quad (4.225)$$

where the L and R represent the operators defined by

$$L(u_j) = u_j + \theta \frac{\Delta t}{\Delta x}(f_{j+\frac{1}{2}} - f_{j-\frac{1}{2}}) \tag{4.226}$$

$$R(u_j) = u_j - \frac{\Delta t}{\Delta x}(1-\theta)(f_{j+\frac{1}{2}} - f_{j-\frac{1}{2}}) \tag{4.227}$$

In order to ensure that the numerical schemes represented by Eq. (4.223) are TVD, it is required that

$$\mathrm{TV}(u^{n+1}) \leqslant \mathrm{TV}(u^n)$$

Harten (1984) showed that the sufficient conditions for this are

$$\mathrm{TV}[R(u^n)] \leqslant \mathrm{TV}(u^n) \tag{4.228a}$$

$$\mathrm{TV}[L(u^{n+1})] \geqslant \mathrm{TV}(u^{n+1}) \tag{4.228b}$$

Rewriting Eq. (4.223) in the form

$$u_j^{n+1} - \frac{\Delta t}{\Delta x}\theta\left(\overline{C}_{j+\frac{1}{2}}^{-}\,\delta u_{j+\frac{1}{2}} - \overline{C}_{j-\frac{1}{2}}^{+}\,\delta u_{j-\frac{1}{2}}\right)^{n+1}$$

$$= u_j^n + \frac{\Delta t}{\Delta x}(1-\theta)\left(\overline{C}_{j+\frac{1}{2}}^{-}\,\delta u_{j+\frac{1}{2}} - \overline{C}_{j-\frac{1}{2}}^{+}\,\delta u_{j-\frac{1}{2}}\right)^{n} \tag{4.229}$$

Harten proved that the sufficient conditions are

$$C_{j+\frac{1}{2}}^{\pm} = \frac{\Delta t}{\Delta x}(1-\theta)\overline{C}_{j+\frac{1}{2}}^{\pm} \geqslant 0 \tag{4.230}$$

and

$$C_{j+\frac{1}{2}}^{+} + C_{j+\frac{1}{2}}^{-} = \frac{\Delta t}{\Delta x}(1-\theta)\left(\overline{C}_{j+\frac{1}{2}}^{+} + \overline{C}_{j+\frac{1}{2}}^{-}\right) \leqslant 1 \tag{4.231}$$

with

$$-\infty < C \leqslant -\frac{\Delta t}{\Delta x}\theta\overline{C}_{j+\frac{1}{2}}^{\pm} \leqslant 0 \tag{4.232}$$

where C is some positive constant. For values of $\theta = 0$ and 1, the method is first order and the TVD conditions are satisfied. In fact, the Lax scheme and all first-order upwind schemes can be shown to have the TVD property (see Prob. 4.65). If the value of θ is taken to be 0.5, the method is a trapezoidal scheme, and the TVD requirements are not satisfied. Consequently, the scheme must be modified with the limiter concept to avoid oscillations. The construction of appropriate limiters is possible using the requirements given in the preceding equations for guidance.

Jameson and Lax (1984) generalized the idea of TVD schemes for multipoint methods. In the case of higher-order schemes, the terms $C^{\pm}$ are written as

$$C_{j+\frac{1}{2}}^{-} = C^{-}(u_{j+2}, u_{j+1}, u_j, u_{j-1}) \tag{4.233a}$$

$$C_{j-\frac{1}{2}}^{+} = C^{+}(u_{j+1}, u_j, u_{j-1}, u_{j-2}) \tag{4.233b}$$

The general explicit scheme

$$u_j^{n+1} = u_j^n + \frac{\Delta t}{\Delta x} \sum_{k=1}^{J} \left(C^{-(k)}_{j+k-\frac{1}{2}}\, \delta u_{j+k-\frac{1}{2}} - C^{+(k)}_{j-k+\frac{1}{2}}\, \delta u_{j-k+\frac{1}{2}} \right)^n \tag{4.234}$$

satisfies the TVD conditions when

$$C^{+(1)}_{j+\frac{1}{2}} \geqslant C^{+(2)}_{j+\frac{1}{2}} \geqslant \cdots \geqslant C^{+(k)}_{j+\frac{1}{2}} \geqslant 0 \tag{4.235a}$$

$$C^{-(1)}_{j+\frac{1}{2}} \geqslant C^{-(2)}_{j+\frac{1}{2}} \geqslant \cdots \geqslant C^{-(k)}_{j+\frac{1}{2}} \geqslant 0 \tag{4.235b}$$

$$\frac{\Delta t}{\Delta x}\left(C^{+(1)}_{j+\frac{1}{2}} + C^{-(1)}_{j+\frac{1}{2}} \right) \leqslant 1 \tag{4.235c}$$

which is the same condition derived by Harten for the three-point schemes. The general implicit scheme

$$u_j^{n+1} - \frac{\Delta t}{\Delta x} \sum_{k=1}^{J} B^{-(k)}_{j+k-\frac{1}{2}}\, \delta u_{j+k-\frac{1}{2}} - B^{+(k)}_{j-k+\frac{1}{2}}\, \delta u_{j-k+\frac{1}{2}} \Big)^{n+1}$$

$$= u_j^n + \frac{\Delta t}{\Delta x} \sum_{k=1}^{J} \left(C^{-(k)}_{j+k-\frac{1}{2}}\, \delta u_{j+k-\frac{1}{2}} - C^{+(k)}_{j-k+\frac{1}{2}}\, \delta u_{j-k+\frac{1}{2}} \right)^n \tag{4.236}$$

satisfies the TVD conditions if and only if the coefficients of the implicit and explicit operators obey the expressions

$$\begin{aligned} B^{+(1)}_{j+\frac{1}{2}} \geqslant B^{+(2)}_{j+\frac{1}{2}} \geqslant \cdots \geqslant B^{+(k)}_{j+\frac{1}{2}} \geqslant 0 \\ B^{-(1)}_{j+\frac{1}{2}} \geqslant B^{-(2)}_{j+\frac{1}{2}} \geqslant \cdots \geqslant B^{-(k)}_{j+\frac{1}{2}} \geqslant 0 \end{aligned} \tag{4.237}$$

and

$$\begin{aligned} C^{+(1)}_{j+\frac{1}{2}} \geqslant C^{+(2)}_{j+\frac{1}{2}} \geqslant \cdots \geqslant C^{+(k)}_{j+\frac{1}{2}} \geqslant 0 \\ C^{-(1)}_{j+\frac{1}{2}} \geqslant C^{-(2)}_{j+\frac{1}{2}} \geqslant \cdots \geqslant C^{-(k)}_{j+\frac{1}{2}} \geqslant 0 \end{aligned} \tag{4.238}$$

This condition is also consistent with the earlier development for the general three-point scheme.

Armed with the information gleaned from the TVD conditions, we can now test the second-order upwind scheme (Section 4.1.9) to determine if it is a TVD method. The second-order upwind method applied to the first-order wave equation yields

$$u_j^{n+1} = u_j^n - \nu(u_j^n - u_{j-1}^n) + \frac{1}{2}\nu(\nu - 1)(u_j^n - 2u_{j-1}^n + u_{j-2}^n) \tag{4.239}$$

This may be written as

$$u_j^{n+1} = u_j^n - \frac{\nu}{2}(3 - \nu)\, \delta u^n_{j-\frac{1}{2}} - \frac{\nu}{2}(\nu - 1)\, \delta u^n_{j-\frac{3}{2}} \tag{4.240}$$

showing that

$$C^{+(1)}_{j-\frac{1}{2}} = \frac{c}{2}(3 - \nu) \qquad C^{+(2)}_{j-\frac{3}{2}} = \frac{c}{2}(\nu - 1) \tag{4.241}$$

In this case the coefficient $C^{+(2)}_{j-3/2}$ has the wrong sign when the CFL condition is satisfied, and oscillations will occur at discontinuities.

As another example, let us test the Lax-Wendroff scheme, Eq. (4.44), to see if it satisfies the TVD condition. This scheme is written

$$u_j^{n+1} = u_j^n - \frac{\nu}{2}(u_{j+1}^n - u_{j-1}^n) + \frac{\nu^2}{2}(u_{j+1}^n - 2u_j^n + u_{j-1}^n) \qquad (4.242)$$

or equivalently,

$$u_j^{n+1} = u_j^n - \frac{\nu}{2}(1-\nu)\,\delta u_{j+\frac{1}{2}}^n - \frac{\nu}{2}(1+\nu)\,\delta u_{j-\frac{1}{2}}^n \qquad (4.243)$$

For this example,

$$C^+_{j-\frac{1}{2}} = \frac{c}{2}(1+\nu) \qquad C^-_{j+\frac{1}{2}} = -\frac{c}{2}(1-\nu) \qquad (4.244)$$

which indicates that the TVD conditions are not satisfied in the range of Courant numbers where this scheme is stable.

The idea of limiting to control oscillations is demonstrated effectively by considering the second-order upwind scheme in the form of Eq. (4.239). This may be rewritten as

$$u_j^{n+1} = u_j^n - \nu\,\delta u_{j-\frac{1}{2}}^n + \frac{\nu}{2}(\nu-1)\left(\delta u_{j-\frac{1}{2}}^n - \delta u_{j-\frac{3}{2}}^n\right) \qquad (4.245)$$

The terms following the positive sign represent contributions from the second-order corrections to the first-order difference. The idea is to restrict the corrections in regions of rapid change to avoid undesirable behavior by limiting the magnitude of the difference in u or, more generally, the flux or variable gradients. In this sense, the scheme may be written

$$u_j^{n+1} = u_j^n - \nu\,\delta u_{j-\frac{1}{2}}^n + \frac{\nu}{2}(\nu-1)\left(\bar{\delta} u_{j-\frac{1}{2}}^n - \bar{\delta} u_{j-\frac{3}{2}}^n\right) \qquad (4.246)$$

In this expression the quantity $\bar{\delta}$ is defined as

$$\bar{\delta} u_j = \psi\,\delta u_j$$

where ψ is a limiter function. Equation (4.246) may be written in a form showing the dependence of the limited portion on successive changes in the dependent variable in the following way:

$$u_j^{n+1} = u_j^n - \nu\,\delta u_{j-\frac{1}{2}}^n + \frac{\nu}{2}(\nu-1)\left(\psi^+_{j-\frac{1}{2}}\,\delta u_{j-\frac{1}{2}}^n - \psi^+_{j-\frac{3}{2}}\,\delta u_{j-\frac{3}{2}}^n\right) \quad (4.247)$$

In order to arrive at the conditions to ensure the TVD property is contained in the difference formulation, this expression is written as

$$u_j^{n+1} = u_j^n - \nu\,\delta u_{j-\frac{1}{2}}^n + \frac{\nu}{2}(\nu-1)\left(\psi^+_{j-\frac{1}{2}} - \frac{\psi^+_{j-\frac{3}{2}}}{r^+_{j-\frac{3}{2}}}\right)\delta u_{j-\frac{1}{2}}^n \qquad (4.248)$$

where the ratio $r^+_{j+1/2}$ is defined as

$$r^+_{j+\frac{1}{2}} = \frac{u_{j+2} - u_{j+1}}{u_{j+1} - u_j} \tag{4.249}$$

Limiters are customarily written in terms of the successive variations in the dependent variables or the flux. In addition to the r^+ definition, one may also define a quantity r^- of the form

$$r^-_{j+\frac{1}{2}} = \frac{u_j - u_{j-1}}{u_{j+1} - u_j} \tag{4.250}$$

Limiters are then written in terms of these ratios. While they may be written in general as functions of any number of successive variations, for simplicity, they are usually written showing a dependence on only the single ratio at the local point in question:

$$\psi^+_{j+\frac{1}{2}} = \psi\left(r^+_{j+\frac{1}{2}}\right) \tag{4.251}$$

If a wave propagating in the negative direction is present, limiting on the opposite family can be achieved using

$$\psi^-_{j+\frac{1}{2}} = \psi\left(r^-_{j+\frac{1}{2}}\right) \tag{4.252}$$

The TVD conditions that must be satisfied by the upwind method are

$$0 \leqslant \nu\left[1 + \frac{1}{2}(1-\nu)\left(\psi^+_{j-\frac{1}{2}} - \frac{\psi^+_{j-\frac{3}{2}}}{r^+_{j-\frac{3}{2}}}\right)\right] \leqslant 1 \tag{4.253}$$

where

$$C^-_{j+\frac{1}{2}} = 0$$

The TVD condition shows that the limiter must satisfy an equation of the form

$$\frac{\psi(r_1)}{r_1} - \psi(r_2) \leqslant \frac{2}{\nu} \tag{4.254}$$

The stability bounds for this formulation are normally selected to satisfy the standard CFL condition. This provides a bound on the allowable values of ν, and the TVD condition can be simplified to

$$\frac{\psi(r_1)}{r_1} - \psi(r_2) \leqslant 2 \tag{4.255}$$

Many different formulations for ψ can be written to satisfy this inequality. Several constraints are imposed on the construction of the different forms. These include the fact that ψ must be a positive function, leading to the condition

$$\psi(r) \geqslant 0 \quad \text{for} \quad r \geqslant 0 \tag{4.256}$$

and when r is negative, ψ is set equal to zero, i.e.,

$$\psi(r) = 0 \quad \text{for} \quad r = 0 \tag{4.257}$$

This results in the following general constraints:

$$0 \leqslant \psi(r) \leqslant 2r \tag{4.258}$$

If an additional constraint on the magnitude of the limiter is imposed, i.e.,

$$\psi(r) \leqslant 2 \tag{4.259}$$

the TVD requirement may be written as

$$0 \leqslant \psi(r) \leqslant \min(2r, 2) \tag{4.260}$$

The second-order upwind scheme will then satisfy the TVD condition at any point in the shaded area of Fig. 4.55. If the limiters are set equal to 1, the original unlimited upwind, or Beam-Warming (1976) explicit upwind scheme, is recovered. Roe (1985) has developed weaker conditions for the limiters of the form

$$\frac{\psi(r)}{r} \leqslant \frac{2}{1 - \nu} \tag{4.261}$$

and

$$\psi(r) \leqslant \frac{2}{\nu} \tag{4.262}$$

These conditions are the TVD conditions for the basic Lax-Wendroff method (see Prob. 4.65). We have used a linear equation to develop the TVD conditions. The corrections necessary to avoid the oscillations associated with the method are alterations to the second-order terms that appear as nonlinear terms in the difference formulas. It should be clear that the corrections that make a method TVD are always associated with nonlinear limiting even for linear convection problems.

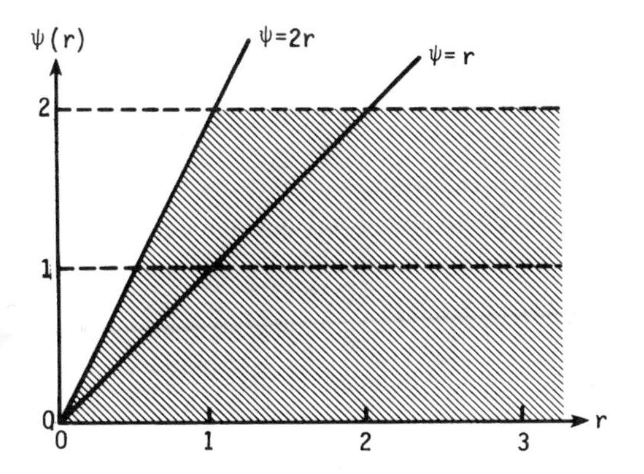

Figure 4.55 Limiter range for second-order TVD schemes.

Other limiters that are useful in computations have been developed and published in the open literature. One of the early examples is the van Leer (1979) limiter, which is of the form

$$\psi(r) = \frac{r + |r|}{1 + r^2} \tag{4.263}$$

The van Albada (1982) limiter is of the form

$$\psi(r) = \frac{r + r^2}{1 + r^2} \tag{4.264}$$

The minmod limiter is also used extensively in TVD numerical methods. As previously noted, this is a function that selects the smallest number from a set when all have the same sign but is zero if they have different signs. This limiter may be written

$$\psi(r) = \text{minmod}(1, r) \tag{4.265}$$

Another limiter that has received use, particularly when contact surfaces are of importance, is the "Superbee" limiter of Roe (1985), which is written

$$\psi(r) = \max[0, \min(2r, 1), \min(r, 2)] \tag{4.266}$$

All of these limiters satisfy a symmetry condition written in the form

$$\frac{\psi(r)}{r} = \psi\left(\frac{1}{r}\right) \tag{4.267}$$

which ensures that the limits for forward and backward gradients are treated the same way. If this condition is not satisfied, the treatment afforded gradients by the limiters is not symmetric. Other TVD schemes may be constructed and are available for both upwind and central-difference methods. The Roe-Sweby scheme (Roe, 1984; Sweby, 1984) begins with a first-order upwind scheme with a numerical flux from Eq. (4.192) written in the form

$$f_{j+\frac{1}{2}} = \frac{1}{2}\left[F_{j+1} + F_j - \text{sgn}(\bar{u}_{j+\frac{1}{2}})(F_{j+1} - F_j)\right] \tag{4.268}$$

A second-order correction is added to this flux, which is a limiter multiplied by the difference in the first-order upwind flux and the flux for the Lax-Wendroff scheme. Thus

$$f^{\text{RS}}_{j+\frac{1}{2}} = f_{j+\frac{1}{2}} + \psi(r)\left(f^{\text{LW}}_{j+\frac{1}{2}} - f_{j+\frac{1}{2}}\right) \tag{4.269}$$

This results in an expression for the flux

$$f^{\text{RS}}_{j+\frac{1}{2}} = f_{j+\frac{1}{2}} + \frac{\psi(r)}{2}\left[\text{sgn}(\bar{u}_{j+\frac{1}{2}}) - \frac{\Delta t}{\Delta x}\bar{u}_{j+\frac{1}{2}}\right](F_{j+1} - F_j) \tag{4.270}$$

where the notation for r is

$$r = \frac{u_{j+1+\sigma} - u_{j+\sigma}}{\delta u_{j+\frac{1}{2}}} \tag{4.271}$$

and σ may take on integer values defining the r values as previously used for $r^{\pm}$. The limiters that may be applied with success in this expression are the same as previously given in Eqs. (4.261)–(4.266). The Roe-Sweby scheme produces results for the 1-D case that are similar to the Burgers equation solution shown previously for the MUSCL scheme (see Prob. 4.66).

Another technique for creating a TVD scheme is Harten's modified flux method (Harten, 1984). In this approach, the T.E. of a first-order upwind scheme is developed, and the idea is to subtract this away like an antidiffusive flux term. This is accomplished by modifying the flux used in the original approximation. The modifications to the flux are due to the T.E., and as a result must necessarily be limited to obtain a TVD scheme. This technique is discussed in more detail in Chapter 6.

4.5 BURGERS' EQUATION (VISCOUS)

The complete nonlinear Burgers equation

$$\frac{\partial u}{\partial t} + u\frac{\partial u}{\partial x} = \mu\frac{\partial^2 u}{\partial x^2} \tag{4.272}$$

is a parabolic PDE, which can serve as a model equation for the boundary-layer equations, the "parabolized" Navier-Stokes equations, and the complete Navier-Stokes equations. In order to better model the steady boundary-layer and "parabolized" Navier-Stokes equations, the independent variables t and x can be replaced by x and y to give

$$\frac{\partial u}{\partial x} + u\frac{\partial u}{\partial y} = \mu\frac{\partial^2 u}{\partial y^2} \tag{4.273}$$

where x is the marching direction.

As with previous model equations, Burgers' equation has exact analytical solutions for certain boundary and initial conditions. These exact solutions are useful when comparing different numerical algorithms. The exact steady-state solution [i.e., $\lim_{t\to\infty} u(x,t)$] of Eq. (4.272) for the boundary conditions

$$u(0,t) = u_0 \tag{4.274}$$

$$u(L,t) = 0 \tag{4.275}$$

is given by

$$u = u_0\bar{u}\left\{\frac{1 - \exp[\bar{u}\,\mathrm{Re}_L(x/L - 1)]}{1 + \exp[\bar{u}\,\mathrm{Re}_L(x/L - 1)]}\right\} \tag{4.276}$$

where

$$\mathrm{Re}_L = \frac{u_0 L}{\mu} \tag{4.277}$$

and $\bar{u}$ is a solution of the equation

$$\frac{\bar{u} - 1}{\bar{u} + 1} = \exp(-\bar{u}\,\mathrm{Re}_L) \tag{4.278}$$

For simplicity, the linearized Burgers equation

$$\frac{\partial u}{\partial t} + c\frac{\partial u}{\partial x} = \mu\frac{\partial^2 u}{\partial x^2} \tag{4.279}$$

is often used in place of Eq. (4.272). Note that if $\mu = 0$, the wave equation is obtained. If $c = 0$, the heat equation is obtained. The exact steady-state solution of Eq. (4.279) for the boundary conditions given by Eqs. (4.274) and (4.275) is

$$u = u_0\left\{\frac{1 - \exp\left[R_L(x/L - 1)\right]}{1 - \exp\left(-R_L\right)}\right\} \tag{4.280}$$

where

$$R_L = \frac{cL}{\mu}$$

The exact unsteady solution of Eq. (4.279) for the initial condition

$$u(x,0) = \sin(kx)$$

and periodic boundary conditions is

$$u(x,t) = \exp(-k^2\mu t)\sin k(x - ct) \tag{4.281}$$

This latter exact solution is useful in evaluating the temporal accuracy of a method.

Equations (4.272) and (4.279) can be combined into a generalized equation (Rakich, 1978):

$$u_t + (c + bu)u_x = \mu u_{xx} \tag{4.282}$$

where c and b are free parameters. If $b = 0$, the linearized Burgers equation is obtained and if $c = 0$ and $b = 1$, the nonlinear Burgers equation is obtained. If $c = \frac{1}{2}$ and $b = -1$, the generalized Burgers equation has the stationary solution

$$u = -\frac{c}{b}\left[1 + \tanh\frac{c(x - x_0)}{2\mu}\right] \tag{4.283}$$

which is shown in Fig. 4.56 for $\mu = \frac{1}{4}$. Hence, if the initial distribution of u is given by Eq. (4.283), the exact solution does not vary with time but remains fixed at the initial distribution. Additional exact solutions of Burgers' equation can be found in the paper by Benton and Platzman (1972), which describes 35 different exact solutions.

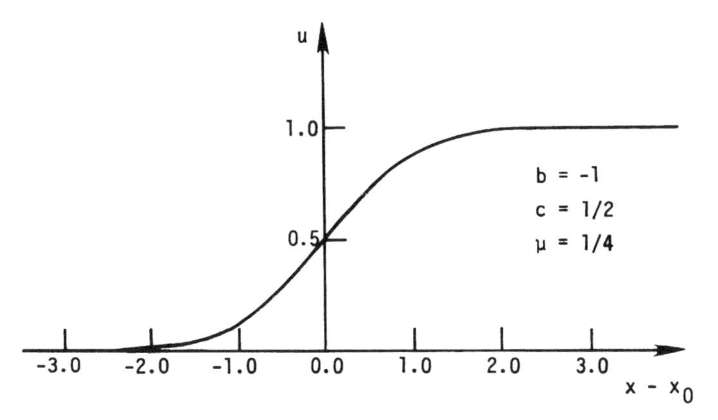

Figure 4.56 Exact solution of Eq. (4.283).

Equation (4.282) can be put into conservation form:

$$u_t + \bar{F}_x = 0 \tag{4.284}$$

where $\bar{F}$ is defined by

$$\bar{F} = cu + \frac{bu^2}{2} - \mu u_x \tag{4.285}$$

Alternatively, Eq. (4.282) can be rewritten as

$$u_t + F_x = \mu u_{xx} \tag{4.286}$$

where F is defined by

$$F = cu + \frac{bu^2}{2} \tag{4.287}$$

For the linearized case ($b = 0$), F reduces to

$$F = cu$$

If we let $A = \partial F/\partial u$, then Eq. (4.286) becomes

$$u_t + Au_x = \mu u_{xx} \tag{4.288}$$

where $A = u$ for the nonlinear Burgers equation ($c = 0$, $b = 1$) and $A = c$ for the linear Burgers equation ($b = 0$). We will use either Eq. (4.286) or Eq. (4.288) to represent Burgers' equation in the following discussion of applicable finite-difference/finite-volume schemes.

The various schemes described previously for the inviscid Burgers equation can also be applied to the complete Burgers equation. This is accomplished by simply adding a second-order (or higher-order) central-difference expression for the viscous term u_{xx}. We will describe these methods as well as other methods for solving the complete Burgers equation in the following sections.

4.5.1 FTCS Method

Roache (1972) has given the name FTCS method to the scheme obtained by applying forward-time and centered-space differences to the linearized Burgers equation [i.e., Eq. (4.288) with $A = c$]. The resulting algorithm is

$$\frac{u_j^{n+1} - u_j^n}{\Delta t} + c\frac{u_{j+1}^n - u_{j-1}^n}{2\,\Delta x} = \mu\frac{u_{j+1}^n - 2u_j^n + u_{j-1}^n}{(\Delta x)^2} \tag{4.289}$$

This is a first-order, explicit, one-step scheme with a T.E. of $O[\Delta t, (\Delta x)^2]$. The modified equation can be written as

$$\begin{aligned} u_t + cu_x &= \left(\mu - \frac{c^2\,\Delta t}{2}\right)u_{xx} + \frac{c(\Delta x)^2}{3}\left(3r - \nu^2 - \frac{1}{2}\right)u_{xxx} \\ &\quad + \frac{c(\Delta x)^3}{12}\left(\frac{r}{\nu} - \frac{3r^2}{\nu} - 2\nu + 10\nu r - 3\nu^3\right)u_{xxxx} + \cdots \end{aligned} \tag{4.290}$$

where r is defined as $\mu\,\Delta t/(\Delta x)^2$ for the viscous Burgers equation and $\nu = c\,\Delta t/\Delta x$. Note that if $r = \frac{1}{2}$ and $\nu = 1$, the coefficients of the first two terms on the right-hand side of the modified equation become zero. Unfortunately, this eliminates the viscous term (μu_{xx}) that appears in the PDE we wish to solve. Thus the FTCS method with $r = \frac{1}{2}$ and $\nu = 1$, which incidentally reduces to $u_j^{n+1} = u_{j-1}^u$, is an unacceptable difference representation for Burgers' equation.

The "heuristic" stability analysis, described in Section 4.1.2, requires that the coefficient on u_{xx} be greater than zero. Hence

$$\frac{c^2\,\Delta t}{2} \leqslant \mu$$

or

$$\frac{c^2(\Delta t)^2}{(\Delta x)^2} \leqslant 2\mu\frac{\Delta t}{(\Delta x)^2}$$

which can be rewritten as

$$\nu^2 \leqslant 2r \tag{4.291}$$

A very useful parameter that arises naturally when solving Burgers' equation is the *mesh Reynolds number*, which is defined by

$$\mathrm{Re}_{\Delta x} = \frac{c\,\Delta x}{\mu} \tag{4.292}$$

This nondimensional parameter gives the ratio of convection to diffusion and plays an important role in determining the character of the solution for Burgers' equation. The mesh (cell) Reynolds number (also called Peclet number) can be expressed in terms of ν and r in the following manner:

$$\mathrm{Re}_{\Delta x} = \frac{c\,\Delta x}{\mu} = \frac{c\,\Delta t}{\Delta x}\frac{(\Delta x)^2}{\mu\,\Delta t} = \frac{\nu}{r}$$

Thus the stability condition given by Eq. (4.291) becomes

$$\mathrm{Re}_{\Delta x} \leqslant \frac{2}{\nu} \tag{4.293}$$

As pointed out earlier, the "heuristic" stability analysis does not always give the complete stability restrictions for a given numerical scheme, and this happens in the present case. In order to obtain all of the stability conditions it is necessary to use the Fourier stability analysis. For the FTCS method, the amplification factor is

$$G = 1 + 2r(\cos\beta - 1) - i\nu(\sin\beta) \tag{4.294}$$

which is plotted in Fig. 4.57(a) for a given ν and r. The equation for G describes an ellipse that is centered on the positive real axis at $(1 - 2r)$ and has semimajor and semiminor axes given by $2r$ and ν, respectively. In addition, the ellipse is tangent to the unit circle at the point where the positive real axis intersects the unit circle. For stability, it is necessary that $|G| \leqslant 1$, which requires that the ellipse be entirely within the unit circle. This leads to the following necessary stability restrictions, which are based on the lengths of the semimajor and semiminor axes:

$$\nu \leqslant 1 \qquad 2r \leqslant 1 \tag{4.295}$$

It is possible, however, for these restrictions to be satisfied and the solution to still be unstable, as can be seen in Fig. 4.57(b). Of course, the complete stability

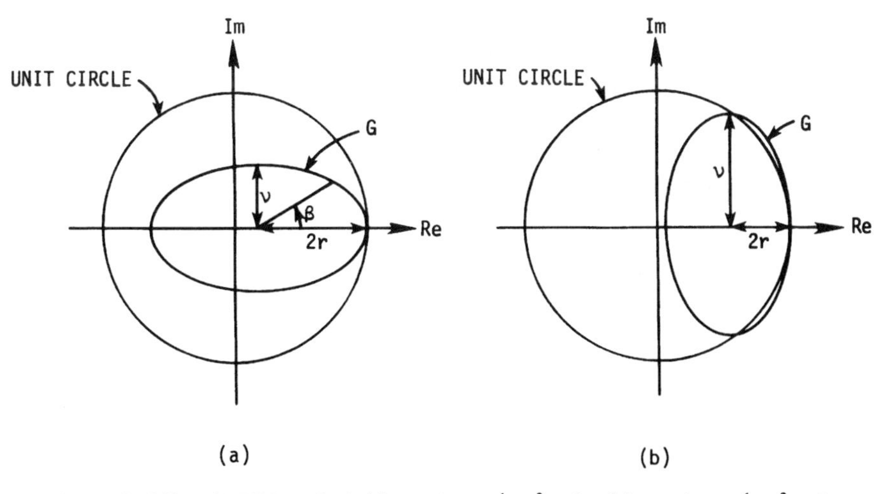

Figure 4.57 Stability of FTCS method: (a) $\nu < 1$, $r < \frac{1}{2}$, $\nu^2 < 2r$; (b) $\nu < 1$, $r < \frac{1}{2}$, $\nu^2 > 2r$.

limitations can be obtained by examining the modulus of the amplification factor in the usual manner. This analysis yields

$$\nu^2 \leqslant 2r \qquad r \leqslant \tfrac{1}{2} \tag{4.296}$$

Note that the first restriction was obtained previously by the heuristic stability analysis and that the two inequalities can be combined to yield a third inequality,

$$\nu \leqslant 1$$

which was obtained by graphical considerations. In terms of the mesh Reynolds number, the stability restrictions become

$$2\nu \leqslant \text{Re}_{\Delta x} \leqslant \frac{2}{\nu} \tag{4.297}$$

It should be mentioned that the right-hand inequality is incorrectly given as $\text{Re}_{\Delta x} \leqslant 2$ in some references.

An important characteristic of finite-difference schemes that are used to solve Burgers' equation is whether they produce oscillations (wiggles) in the solution. Obviously, we do not want these oscillations to occur in our solutions of fluid flow problems. The FTCS method will produce oscillations in the solution of Burgers' equation for mesh Reynolds numbers in the range

$$2 \leqslant \text{Re}_{\Delta x} \leqslant \frac{2}{\nu}$$

For mesh Reynolds numbers slightly above $2/\nu$, the oscillations will eventually cause the solution to "blow up," as expected from our previous stability analysis. In order to explain the origin of the wiggles, let us rewrite Eq. (4.289) in the following form:

$$u_j^{n+1} = \left(r - \frac{\nu}{2}\right)u_{j+1}^n + (1 - 2r)u_j^n + \left(r + \frac{\nu}{2}\right)u_{j-1}^n \tag{4.298}$$

which is equivalent to

$$u_j^{n+1} = \frac{r}{2}(2 - \text{Re}_{\Delta x})u_{j+1}^n + (1 - 2r)u_j^n + \frac{r}{2}(2 + \text{Re}_{\Delta x})u_{j-1}^n \tag{4.299}$$

Furthermore, assume we are trying to find the steady-state solution of Burgers' equation for the initial condition,

$$u(x,0) = 0 \qquad 0 \leqslant x < 1$$

and boundary conditions,

$$u(0,t) = 0$$
$$u(1,t) = 1$$

using an 11-point mesh. For the first time step the values of u at time level $n + 1$ are all zero except at $j = 10$, where

$$u_{10}^{n+1} = \frac{r}{2}(1 - \text{Re}_{\Delta x})(1) + (1 - 2r)(0) + \frac{r}{2}(2 + \text{Re}_{\Delta x})(0) = \frac{r}{2}(2 - \text{Re}_{\Delta x})$$

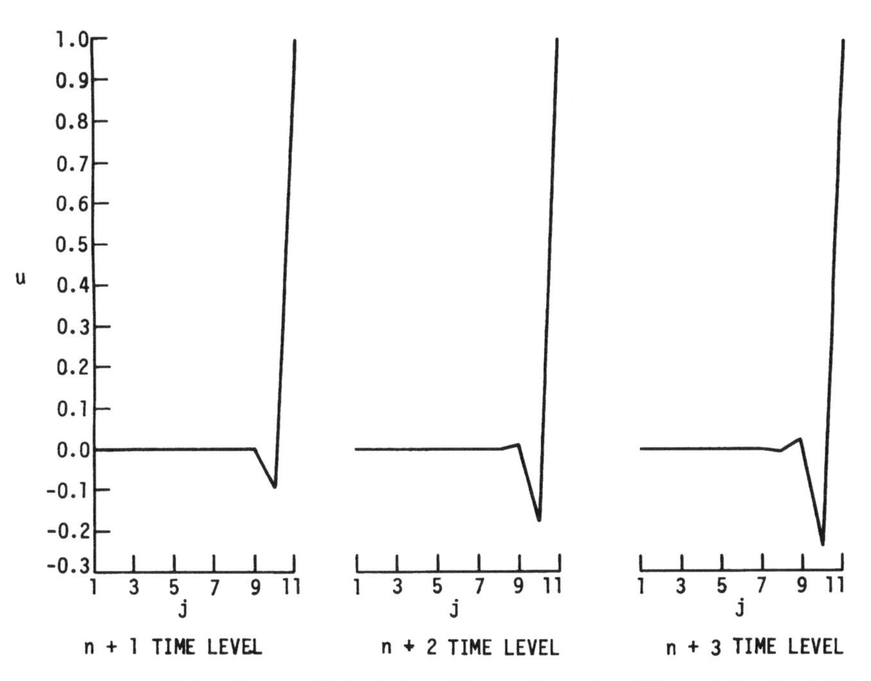

Figure 4.58 Oscillations in numerical solution of Burgers' equation: (a) $n + 1$ time level; (b) $n + 2$ time level; (c) $n + 3$ time level.

and at the boundary ($j = 11$), where u_{11} is fixed at 1. If $\mathrm{Re}_{\Delta x} > 2$, the value of u_{10}^{n+1} will be negative, which will initiate an oscillation as shown in Fig. 4.58a. This figure is drawn for the conditions

$$\nu = 0.4$$
$$r = 0.1$$
$$\mathrm{Re}_{\Delta x} = 4 < \frac{2}{\nu}$$

which make $u_{10}^{n+1} = -0.1$. During the next time step, the oscillation propagates one grid point further from the right-hand boundary. The values of u at $j = 9$ and $j = 10$ become

$$u_9^{n+2} = +0.01$$
$$u_{10}^{n+2} = -0.18$$

and the resulting solution is shown in Fig. 4.58b. The wiggles will eventually propagate to the other boundary but will remain bounded throughout the iteration to steady state. The oscillations that occur in this case are similar to the oscillations that appear when a second-order (or higher) scheme is used to solve the inviscid Burgers equation for a propagating discontinuity.

Additional insight into the origin of the wiggles can be obtained by examining the coefficients in Eq. (4.299) from a physical standpoint. We observe that when $\text{Re}_{\Delta x} > 2$, the coefficient in front of u_{j+1}^n becomes negative. Hence, the larger the value for u_{j+1}^n, the smaller the value for u_j^{n+1}. This represents a nonphysical behavior for a viscous problem, since we would expect a greater "pull" on u_j^{n+1} (i.e., increased value) because of viscosity as u_{j+1}^n is increased. As a consequence of this nonphysical behavior, oscillations are produced in the solution.

The oscillations of the FTCS method can be eliminated if the second-order central difference used for the convective term cu_x is replaced by a first-order upwind difference. The resulting algorithm for $c > 0$ becomes

$$\frac{u_j^{n+1} - u_j^n}{\Delta t} + c\frac{u_j^n - u_{j-1}^n}{\Delta x} = \mu\frac{u_{j+1}^n - 2u_j^n + u_{j-1}^n}{(\Delta x)^2} \tag{4.300}$$

This first-order scheme eliminates the oscillations by adding additional dissipation to the solution. Unfortunately, the amount of dissipation causes the resulting solution to be sufficiently inaccurate to exclude Eq. (4.300) as a viable difference scheme for Burgers' equation. The large amount of dissipation is evident when we examine the modified equation for the scheme

$$u_t + cu_x = \left[\mu\left(1 + \frac{\text{Re}_{\Delta x}}{2}\right) - \frac{c^2\,\Delta t}{2}\right]u_{xx} + \cdots \tag{4.301}$$

and compare it to the modified equation of the FTCS method. Equation (4.301) has the additional term $\mu\,\text{Re}_{\Delta x}/2$ appearing in the coefficient of u_{xx}. Hence, if $\text{Re}_{\Delta x} > 2$, this additional term produces more dissipation (diffusion) than is present in the original problem governed by Burgers' equation. In order to reduce dispersive errors without adding a large amount of artificial viscosity, Leonard (1979a, 1979b) has suggested using a third-order upstream (upwind) difference for the convective term. The resulting algorithm for $c > 0$ is

$$\frac{u_j^{n+1} - u_j^n}{\Delta t} + c\left(\frac{u_{j+1}^n - u_{j-1}^n}{2\,\Delta x} - \frac{u_{j+1}^n - 3u_j^n + 3u_{j-1}^n - u_{j-2}^n}{6\,\Delta x}\right) = \mu\frac{u_{j+1}^n - 2u_j^n + u_{j-1}^n}{(\Delta x)^2} \tag{4.302}$$

and for $c < 0$ the algorithm becomes

$$\frac{u_j^{n+1} - u_j^n}{\Delta t} + c\left(\frac{u_{j+1}^n - u_{j-1}^n}{2\,\Delta x} - \frac{u_{j+2}^n - 3u_{j+1}^n + 3u_j^n - u_{j-1}^n}{6\,\Delta x}\right) = \mu\frac{u_{j+1}^n - 2u_j^n + u_{j-1}^n}{(\Delta x)^2} \tag{4.303}$$

4.5.2 Leap Frog/DuFort-Frankel Method

We have noted earlier that the linearized Burgers equation is a combination of the first-order wave equation and the heat equation. This suggests that we might be able to combine some of the algorithms given previously for the wave equation and the heat equation. The leap frog/DuFort-Frankel method is onc such example. When applied to Eq. (4.288), this method becomes

$$\frac{u_j^{n+1} - u_j^{n-1}}{2\,\Delta t} + A_j^n \frac{u_{j+1}^n - u_{j-1}^n}{2\,\Delta x} = \mu \frac{u_{j+1}^n - u_j^{n+1} - u_j^{n-1} + u_{j-1}^n}{(\Delta x)^2} \tag{4.304}$$

This explicit, one-step scheme is first-order accurate with a T.E. of $O[(\Delta t/\Delta x)^2, (\Delta t)^2, (\Delta x)^2]$. The modified equation for the linear case ($A = c$) can be written as

$$u_t + cu_x = \mu(1 - \nu^2)u_{xx} + \left[\frac{2\mu^2 c(\Delta t)^2}{(\Delta x)^2} - \frac{1}{6}c(\Delta x)^2 + \frac{1}{6}c^3(\Delta t)^2 - \frac{2\mu^2 c^3(\Delta t)^4}{(\Delta x)^4}\right]u_{xxx} + \cdots \tag{4.305}$$

Also for the linear case, a Fourier stability analysis can be performed, which gives the stability condition

$$\nu \leqslant 1$$

Note that this stability condition is independent of the viscosity coefficient μ because of the DuFort-Frankel type of differencing used for the viscous term. However, because consistency requires that $(\Delta t/\Delta x)^2$ approach zero as Δt and Δx approach zero, a much smaller time step than allowed by $\nu \leqslant 1$ is implied. For this reason, the leap frog/DuFort-Frankel scheme seems better suited for the calculation of steady solutions (where time accuracy is unimportant) than for unsteady solutions according to Peyret and Viviand (1975). In the nonlinear case, this scheme is unstable if $\mu = 0$.

4.5.3 Brailovskaya Method

The following two-step explicit method for Eq. (4.286) was proposed by Brailovskaya (1965):

$$\begin{aligned}
&\text{Predictor:} \quad \overline{u_j^{n+1}} = u_j^n - \frac{\Delta t}{2\,\Delta x}(F_{j+1}^n - F_{j-1}^n) + r(u_{j+1}^n - 2u_j^n + u_{j-1}^n) \\
&\text{Corrector:} \quad u_j^{n+1} = u_j^n - \frac{\Delta t}{2\,\Delta x}\left(\overline{F_{j+1}^{n+1}} - \overline{F_{j-1}^{n+1}}\right) + r(u_{j+1}^n - 2u_j^n + u_{j-1}^n)
\end{aligned} \tag{4.306}$$

This scheme is formally first-order accurate with a T.E. of $O[\Delta t, (\Delta x)^2]$. If only a steady-state solution is desired, the first-order temporal accuracy is not important. For the linear Burgers equation, the von Neumann necessary condition for stability is

$$|G|^2 = 1 - \{\nu^2 \sin^2 \beta(1 - \nu^2 \sin^2 \beta) + 4r(1 - \cos \beta) \times [1 - r(1 - \cos \beta)(1 + \nu^2 \sin^2 \beta)]\} \leqslant 1 \quad (4.307)$$

If we ignore viscous effects (i.e., set $r = 0$), the stability condition becomes

$$\nu \leqslant 1$$

On the other hand, if we ignore the convection term, i.e., set $\nu = 0$, the stability condition becomes

$$r \leqslant \tfrac{1}{2}$$

Based on these observations, Carter (1971) has suggested the following stability criterion for the Brailovskaya scheme:

$$\Delta t \leqslant \min\left[\frac{(\Delta x)^2}{2\mu}, \frac{(\Delta x)}{|A|}\right] \quad (4.308)$$

An attractive feature of this scheme is that the viscous term remains the same in both predictor and corrector steps and needs to be computed only once.

4.5.4 Allen-Cheng Method

Allen and Cheng (1970) modified the Brailovskaya scheme to eliminate the stability restriction on r. Their scheme is given by

Predictor:
$$\overline{u_j^{n+1}} = u_j^n - \frac{\Delta t}{2\,\Delta x}(F_{j+1}^n - F_{j-1}^n) + r\left(u_{j+1}^n - 2\overline{u_j^{n+1}} + u_{j-1}^n\right)$$

Corrector:
$$u_j^{n+1} = u_j^n - \frac{\Delta t}{2\,\Delta x}\left(\overline{F_{j+1}^{n+1}} - \overline{F_{j-1}^{n+1}}\right) + r\left(\overline{u_{j+1}^{n+1}} - 2u_j^{n+1} + \overline{u_{j-1}^{n+1}}\right) \quad (4.309)$$

The unconventional differencing of the viscous term eliminates the stability restriction on r, so that the stability condition becomes

$$\nu \leqslant 1$$

for the linear Burgers equation. As a result, when μ is large, this method permits a much larger time step to be taken than does the Brailovskaya scheme. The Allen-Cheng method is formally first-order accurate with a T.E. of $O[\Delta t, (\Delta x)^2]$.

4.5.5 Lax-Wendroff Method

We have previously applied the two-step Lax-Wendroff method to the wave equation. When applied to the complete Burgers equation, several different variations of the method are possible, including the following:

Step 1:
$$u_j^{n+1/2} = \frac{1}{2}(u_{j+1/2}^n - u_{j-1/2}^n) - \frac{\Delta t}{\Delta x}(F_{j+1/2}^n - F_{j-1/2}^n) + r\left[(u_{j-3/2}^n - 2u_{j-1/2}^n + u_{j+1/2}^n) + (u_{j+3/2}^n - 2u_{j+1/2}^n + u_{j-1/2}^n)\right] \tag{4.310}$$

Step 2:
$$u_j^{n+1} = u_j^n - \frac{\Delta t}{\Delta x}\left(F_{j+1/2}^{n+1/2} - F_{j-1/2}^{n+1/2}\right) + r(u_{j+1}^n - 2u_j^n + u_{j-1}^n)$$

This version is based on the Lax-Wendroff scheme used by Thommen (1966) to solve the Navier-Stokes equations. An alternate version has been proposed by Palumbo and Rubin (1972), which computes provisional values at time level $n+1$ instead of $n+\frac{1}{2}$. The present version is formally first-order accurate with a T.E. of $O[\Delta t, (\Delta x)^2]$. The exact linear stability condition is

$$\frac{\Delta t}{(\Delta x)^2}(A^2 \Delta t + 2\mu) \leqslant 1 \tag{4.311}$$

4.5.6 MacCormack Method

The original MacCormack method (1969) applied to the complete Burgers equation (4.286) is

Predictor:
$$\overline{u_j^{n+1}} = u_j^n - \frac{\Delta t}{\Delta x}(F_{j+1}^n - F_j^n) + r(u_{j+1}^n - 2u_j^n + u_{j-1}^n)$$

Corrector:
$$u_j^{n+1} = \frac{1}{2}\left[u_j^n + \overline{u_j^{n+1}} - \frac{\Delta t}{\Delta x}\left(\overline{F_j^{n+1}} - \overline{F_{j-1}^{n+1}}\right) + r\left(\overline{u_{j+1}^{n+1}} - 2\overline{u_j^{n+1}} + \overline{u_{j-1}^{n+1}}\right)\right] \tag{4.312}$$

which is second-order accurate in both time and space. In this version of the MacCormack scheme, a forward difference is employed in the predictor step for $\partial F/\partial x$ and a backward difference is used in the corrector step. The alternate version of the MacCormack scheme employs a backward difference in the predictor step and a forward difference in the corrector step. Both variants of the MacCormack scheme are second-order accurate. It is not possible to obtain a simple stability criterion for the MacCormack scheme applied to the Burgers equation. However, either the condition given by Eq. (4.308) or the empirical formula (Tannehill et al., 1975)

$$\Delta t \leqslant \frac{(\Delta x)^2}{|A|\,\Delta x + 2\mu} \tag{4.313}$$

can be used with an appropriate safety factor. The latter formula reduces to the usual viscous condition $r \leqslant \frac{1}{2}$ when $|A|$ is set equal to zero, and reduces to the usual inviscid condition $|A|\,\Delta t/\Delta x \leqslant 1$ when μ is set equal to zero. The MacCormack method has been widely used to solve not only the Euler equations but also the Navier-Stokes equations for laminar flow. For multidimensional problems, a time-split version of the MacCormack scheme has been developed and will be described in Section 4.5.8. For high-Reynolds-number problems, MacCormack has devised several newer methods, which will be discussed in Chapter 9.

An interesting variation of the original MacCormack scheme is obtained when overrelaxation is applied to both predicted and corrected values (Désidéri and Tannehill, 1977a) in the following manner:

$$\text{Predictor:} \quad v_j^{\overline{n+1}} = u_j^n - \frac{\Delta t}{\Delta x}(F_{j+1}^n - F_j^n) + r(u_{j+1}^n - 2u_j^n + u_{j-1}^n)$$

$$u_j^{\overline{n+1}} = u_j^{\overline{n}} + \overline{\omega}\left(v_j^{\overline{n+1}} - u_j^{\overline{n}}\right) \tag{4.314}$$

$$\text{Corrector:} \quad v_j^{n+1} = u_j^{\overline{n+1}} - \frac{\Delta t}{\Delta x}\left(F_j^{\overline{n+1}} - F_{j-1}^{\overline{n+1}}\right) + r\left(u_{j+1}^{\overline{n+1}} - 2u_j^{\overline{n+1}} + u_{j-1}^{\overline{n+1}}\right)$$

$$u_j^{n+1} = u_j^n + \omega\left(v_j^{n+1} - u_j^n\right) \tag{4.315}$$

In these equations, the v's are intermediate quantities, the u's denote final predictions, $\overline{\omega}$ and ω are overrelaxation parameters, and $u_j^{\overline{n}}$ represents the predicted value for u_j from the previous step. The original MacCormack scheme is obtained by setting $\overline{\omega} = 1$ and $\omega = \frac{1}{2}$. In general, the overrelaxed MacCormack method is first-order accurate with a T.E. of $O[\Delta t, (\Delta x)^2]$. However, it can be shown (Désidéri and Tannehill, 1977b) that if

$$\omega\overline{\omega} = |\overline{\omega} - \omega| \tag{4.316}$$

the method is second-order accurate in time when applied to the linearized Burgers equation. The overrelaxation scheme accelerates the convergence over that of the original MacCormack scheme by an approximate factor Ω, given by

$$\Omega = \frac{2\overline{\omega}\omega}{1 - (\overline{\omega} - 1)(\omega - 1)} \tag{4.317}$$

A Fourier stability analysis applied to the linearized Burgers equation does not yield a necessary and sufficient stability condition in the form of an algebraic relation between the parameters ν, r, $\overline{\omega}$, and ω. However, a necessary condition of stability is

$$|(\overline{\omega} - 1)(\omega - 1)| \leqslant 1 \tag{4.318}$$

In general, the stability limitation must be computed numerically, and it is usually more restrictive than the conditions $\overline{\omega} \leqslant 2$ and $\omega \leqslant 2$.

4.5.7 Briley-McDonald Method

The Briley-McDonald (1974) method is an implicit scheme which is often based on the following time differencing (Euler implicit) of Eq. (4.286):

$$\frac{u_j^{n+1} - u_j^n}{\Delta t} + \left(\frac{\partial F}{\partial x}\right)_j^{n+1} = \mu\left(\frac{\partial^2 u}{\partial x^2}\right)_j^{n+1} \tag{4.319}$$

The term $(\partial F/\partial x)_j^{n+1}$ is expanded as

$$\left(\frac{\partial F}{\partial x}\right)_j^{n+1} = \left(\frac{\partial F}{\partial x}\right)_j^n + \Delta t\left[\frac{\partial}{\partial t}\left(\frac{\partial F}{\partial x}\right)\right]_j^n + O[(\Delta t)^2] \tag{4.320}$$

thereby introducing $\partial/\partial t(\partial F/\partial x)$, which can be replaced by

$$\frac{\partial}{\partial t}\left(\frac{\partial F}{\partial x}\right) = \frac{\partial}{\partial x}\left(\frac{\partial F}{\partial t}\right) = \frac{\partial}{\partial x}\left(\frac{\partial F}{\partial u}\frac{\partial u}{\partial t}\right) = \frac{\partial}{\partial x}\left(A\frac{\partial u}{\partial t}\right) \tag{4.321}$$

Finally, if we combine Eqs. (4.319), (4.320), and (4.321) and employ forward-time differences and centered-spatial differences, the Briley-McDonald method is obtained:

$$\frac{u_j^{n+1} - u_j^n}{\Delta t} + \frac{F_{j+1}^n - F_{j-1}^n}{2\,\Delta x} + \frac{A_{j+1}^n\left(u_{j+1}^{n+1} - u_{j+1}^n\right) - A_{j-1}^n\left(u_{j-1}^{n+1} - u_{j-1}^n\right)}{2\,\Delta x}$$
$$= \mu\,\hat{\delta}_x^2 u_j^{n+1} \tag{4.322}$$

This scheme is formally first-order accurate with a T.E. of $O[\Delta t, (\Delta x)^2]$. However, at steady-state the accuracy is $O[(\Delta x)^2]$. The temporal accuracy can be increased by using trapezoidal differencing or by using additional time levels in the same manner as discussed earlier for the Beam-Warming scheme. For example, if we apply trapezoidal time differencing to Eq. (4.286), the following equation is obtained:

$$\frac{u_j^{n+1} - u_j^n}{\Delta t} + \frac{1}{2}\left[\left(\frac{\partial F}{\partial x}\right)_j^n + \left(\frac{\partial F}{\partial x}\right)_j^{n+1}\right] = \frac{1}{2}\mu\left[\left(\frac{\partial^2 u}{\partial x^2}\right)_j^n + \left(\frac{\partial^2 u}{\partial x^2}\right)_j^{n+1}\right] \tag{4.323}$$

Proceeding as before, we find the resulting second-order accurate scheme to be

$$\frac{u_j^{n+1} - u_j^n}{\Delta t} + \frac{F_{j+1}^n - F_{j-1}^n}{2\,\Delta x} + \frac{A_{j+1}^n\left(u_{j+1}^{n+1} - u_{j+1}^n\right) - A_{j-1}^n\left(u_{j-1}^{n+1} - u_{j-1}^n\right)}{4\,\Delta x}$$
$$= \frac{\mu}{2(\Delta x)^2}\left[(\delta_x^2 u)_j^n + (\delta_x^2 u)_j^{n+1}\right] \tag{4.324}$$

Both of these schemes, Eq. (4.322) and Eq. (4.324), are unconditionally stable and produce tridiagonal systems of linear algebraic equations that can be solved using the Thomas algorithm.

The Briley-McDonald method is directly related to the method developed by Beam and Warming (1978) to solve the Navier-Stokes equations. In fact, when the two methods are applied to Burgers' equation, they can be reduced to the same form. In order to do this, the delta terms in the Beam-Warming method must be replaced by their equivalent expressions [i.e., Δu_j^n is replaced by $(u_j^{n+1} - u_j^n)$]. The Beam-Warming method for the Navier-Stokes equations is discussed in Chapter 9.

4.5.8 Time-Split MacCormack Method

In order to illustrate methods that are designed specifically for multidimensional problems, we introduce the 2-D Burgers equation

$$\frac{\partial u}{\partial t} + \frac{\partial F}{\partial x} + \frac{\partial G}{\partial y} = \mu\left(\frac{\partial^2 u}{\partial x^2} + \frac{\partial^2 u}{\partial y^2}\right) \tag{4.325}$$

If we let $A = \partial F/\partial u$ and $B = \partial G/\partial u$, Eq. (4.325) can be rewritten as

$$u_t = Au_x + Bu_y = \mu(u_{xx} + u_{yy}) \tag{4.326}$$

The exact steady-state solution (derived by Rai, 1982) of the 2-D linearized Burgers equation,

$$u_t + cu_x + du_y = \mu(u_{xx} + u_{yy}) \tag{4.327}$$

for the boundary conditions $(0 \leqslant t \leqslant \infty)$,

$$\begin{aligned} u(x,0,t) &= \frac{1 - \exp[(x-1)c/\mu]}{1 - \exp(-c/\mu)} \qquad & u(x,1,t) &= 0 \\ u(0,y,t) &= \frac{1 - \exp[(y-1)d/\mu]}{1 - \exp(-d/\mu)} \qquad & u(1,y,t) &= 0 \end{aligned} \tag{4.328}$$

and the initial condition $(0 < x \leqslant 1,\ 0 < y \leqslant 1)$,

$$u(x,y,0) = 0$$

is given by

$$u(x,y) = \left\{\frac{1 - \exp[(x-1)c/\mu]}{1 - \exp(-c/\mu)}\right\}\left\{\frac{1 - \exp[(y-1)d/\mu]}{1 - \exp(-d/\mu)}\right\} \tag{4.329}$$

Note that the extension of this form of solution to the 3-D linearized Burgers equation is straightforward. All of the methods we have discussed for the 1-D Burgers equation can be readily extended to the 2-D Burgers equation. However, because of the more restrictive stability conditions of the explicit methods and the desire to maintain tridiagonal matrices in the implicit schemes, it is usually necessary to modify the previous algorithms for multidimensional problems. As an example, let us first consider the explicit time-split MacCormack method.

The time-split MacCormack method (MacCormack, 1971; MacCormack and Baldwin, 1975) "splits" the original MacCormack scheme into a sequence of 1-D operations, thereby achieving a less restrictive stability condition. In other words, the splitting makes it possible to advance the solution in each direction with the maximum allowable time step. This is particularly advantageous if the allowable time steps $(\Delta t_x, \Delta t_y)$ are much different because of differences in the mesh spacings $(\Delta x, \Delta y)$. In order to explain this method, we will make use of the 1-D difference operators $L_x(\Delta t_x)$ and $L_y(\Delta t_y)$. The $L_x(\Delta t_x)$ operator applied to $u_{i,j}^n$,

$$u_{i,j}^* = L_x(\Delta t_x) u_{i,j}^n \tag{4.330}$$

is by definition equivalent to the two-step formula:

$$\begin{aligned} \overline{u_{i,j}^*} &= u_{i,j}^n - \frac{\Delta t_x}{\Delta x}(F_{i+1,j}^n - F_{i,j}^n) + \mu\,\Delta t_x \hat{\delta}_x^2 u_{i,j}^n \\ u_{i,j}^* &= \frac{1}{2}\left[u_{i,j}^n + \overline{u_{i,j}^*} - \frac{\Delta t_x}{\Delta x}\left(\overline{F_{i,j}^*} - \overline{F_{i-1,j}^*}\right) + \mu\,\Delta t_x \hat{\delta}_x^2 \overline{u_{i,j}^*}\right] \end{aligned} \tag{4.331}$$

These expressions make use of a dummy time index, which is denoted by the asterisk. The $L_y(\Delta t_y)$ operator is defined in a similar manner, that is,

$$u_{i,j}^* = L_y(\Delta t_y) u_{i,j}^n \tag{4.332}$$

is equivalent to

$$\begin{aligned} \overline{u_{i,j}^*} &= u_{i,j}^n - \frac{\Delta t_y}{\Delta y}(G_{i,j+1}^n - G_{i,j}^n) + \mu\,\Delta t_y \hat{\delta}_j^2 u_{i,j}^n \\ u_{i,j}^* &= \frac{1}{2}\left[u_{i,j}^n + \overline{u_{i,j}^*} - \frac{\Delta t_y}{\Delta y}\left(\overline{G_{i,j}^*} - \overline{G_{i,j-1}^*}\right) + \mu\,\Delta t_y \hat{\delta}_y^2 \overline{u_{i,j}^*}\right] \end{aligned} \tag{4.333}$$

A second-order accurate scheme can be constructed by applying the L_x and L_y operators to $u_{i,j}^n$ in the following manner:

$$u_{i,j}^{n+1} = L_y\left(\frac{\Delta t}{2}\right) L_x(\Delta t) L_y\left(\frac{\Delta t}{2}\right) u_{i,j}^n \tag{4.334}$$

This scheme has a T.E. of $O[(\Delta t)^2, (\Delta x)^2, (\Delta y)^2]$. In general, a scheme formed by a sequence of these operators is (1) stable, if the time step of each operator does not exceed the allowable step size for that operator; (2) consistent, if the sums of the time steps for each of the operators are equal; and (3) second-order accurate, if the sequence is symmetric. Other sequences that satisfy these criteria are given by

$$\begin{aligned} u_{i,j}^{n+1} &= L_y\left(\frac{\Delta t}{2}\right) L_x\left(\frac{\Delta t}{2}\right) L_x\left(\frac{\Delta t}{2}\right) L_y\left(\frac{\Delta t}{2}\right) u_{i,j}^n \\ u_{i,j}^{n+1} &= \left[L_y\left(\frac{\Delta t}{2m}\right)\right]^m L_x(\Delta t)\left[L_y\left(\frac{\Delta t}{2m}\right)\right]^m u_{i,j}^n \qquad m = \text{integer} \end{aligned} \tag{4.335}$$

The last sequence is quite useful for the case where $\Delta y \ll \Delta x$.

4.5.9 ADI Methods

Polezhaev (1967) used an adaptation of the Peaceman-Rachford ADI scheme to solve the compressible Navier-Stokes equations. When applied to the 2-D Burgers equation, Eq. (4.326), this scheme becomes

$$\left[1 + \frac{\Delta t}{2}\left(A_{i,j}^n \frac{\bar{\delta}_x}{2\,\Delta x} - \mu \hat{\delta}_x^2\right)\right] u_{i,j}^* = \left[1 - \frac{\Delta t}{2}\left(B_{i,j}^n \frac{\bar{\delta}_y}{2\,\Delta y} - \mu \hat{\delta}_y^2\right)\right] u_{i,j}^n$$

$$\left[1 + \frac{\Delta t}{2}\left(B_{i,j}^* \frac{\bar{\delta}_y}{2\,\Delta y} - \mu \hat{\delta}_y^2\right)\right] u_{i,j}^{n+1} = \left[1 - \frac{\Delta t}{2}\left(A_{i,j}^n \frac{\bar{\delta}_x}{2\,\Delta x} - \mu \hat{\delta}_x^2\right)\right] u_{i,j}^* \tag{4.336}$$

This method is first-order accurate with a T.E. of $O[\Delta t, (\Delta x)^2, (\Delta y)^2]$ and is unconditionally stable for the linear case. Obviously, a tridiagonal system of algebraic equations must be solved during each step.

When the Briley-McDonald scheme, Eq. (4.322), is applied directly to the 2-D Burgers equation, a tridiagonal system of algebraic equations is no longer obtained. This difficulty can be avoided by applying the two-step ADI procedure of Douglas and Gunn (1964):

$$\left[1 + \Delta t\left(\frac{\bar{\delta}_x}{2\,\Delta x} A_{i,j}^n - \mu \hat{\delta}_x^2\right)\right] u_{i,j}^* = \left[1 - \Delta t\left(\frac{\bar{\delta}_y}{2\,\Delta y} B_{i,j}^n - \mu \hat{\delta}_y^2\right)\right] u_{i,j}^n + (\Delta t) S_{i,j}^n \tag{4.337}$$

$$\left[1 + \Delta t\left(\frac{\bar{\delta}_y}{2\,\Delta y} B_{i,j}^n - \mu \hat{\delta}_x^2\right)\right] u_{i,j}^{n+1} = u_{i,j}^n - \Delta t\left(\frac{\bar{\delta}_x}{2\,\Delta x} A_{i,j}^n - \mu \hat{\delta}_x^2\right) u_{i,j}^* + \Delta t\, S_{i,j}^n \tag{4.338}$$

where

$$S_{i,j}^n = -\frac{\bar{\delta}_x}{2\,\Delta x} F_{i,j}^n - \frac{\bar{\delta}_y}{2\,\Delta y} G_{i,j}^n + \frac{\bar{\delta}_x}{2\,\Delta x}(A_{i,j}^n u_{i,j}^n) + \frac{\bar{\delta}_y}{2\,\Delta y}(B_{i,j}^n u_{i,j}^n)$$

4.5.10 Predictor-Corrector, Multiple-Iteration Method

Rubin and Lin (1972) devised a predictor-corrector, multiple-iteration method to solve the 3-D "parabolized" Navier-Stokes equations. Their scheme eliminates cross coupling of grid points in the normal (y) and lateral (z) directions and uses an iterative procedure to recover acceptable accuracy. In order to illustrate this method, let us use the following 3-D linear Burgers equation,

$$u_x + c u_y + d u_z = \mu(u_{yy} + u_{zz}) \tag{4.339}$$

as a model for the "parabolized" Navier-Stokes equations. The predictor-corrector, multiple-iteration method applied to this model equation is

$$\begin{aligned} u_{i+1,j,k}^{m+1} = u_{i,j,k} &- \frac{c\,\Delta x}{2\,\Delta y}\left(u_{i+1,j+1,k}^{m+1} - u_{i+1,j-1,k}^{m+1}\right) \\ &- \frac{d\,\Delta x}{2\,\Delta z}\left(u_{i+1,j,k+1}^{m} - u_{i+1,j,k-1}^{m}\right) \\ &+ \frac{\mu\,\Delta x}{(\Delta y)^2}\left(u_{i+1,j+1,k}^{m+1} - 2u_{i+1,j,k}^{m+1} + u_{i+1,j-1,k}^{m+1}\right) \\ &+ \frac{\mu\,\Delta x}{(\Delta z)^2}\left(u_{i+1,j,k+1}^{m} - 2u_{i+1,j,k}^{m+1} + u_{i+1,j,k-1}^{m}\right) \end{aligned} \tag{4.340}$$

where the superscript m indicates the iteration level and $x = i\,\Delta x$, $y = j\,\Delta y$, and $z = k\,\Delta z$. For the first iteration, m is set equal to zero and the corresponding terms are approximated by either linear replacement,

$$u_{i+1,j,k}^{0} = u_{i,j,k}$$

or by Taylor-series expansions such as

$$u_{i+1,j,k}^{0} = 2u_{i,j,k} - u_{i-1,j,k} + O[(\Delta x)^2]$$

As a result, Eq. (4.340) has three unknowns

$$m = 0 \quad \begin{cases} u_{i+1,j+1,k}^{1} \\ u_{i+1,j,k}^{1} \\ u_{i+1,j-1,k}^{1} \end{cases} \tag{4.341}$$

which produces a tridiagonal system of algebraic equations. The computation in the $i+1$ plane proceeds outward from the known boundary conditions at $k = 1$ to the last k column of grid points. This completes the first iteration. For the next iteration ($m = 1$) the three unknowns in Eq. (4.340) are

$$m = 1 \quad \begin{cases} u_{i+1,j+1,k}^{2} \\ u_{i+1,j,k}^{2} \\ u_{i+1,j-1,k}^{2} \end{cases} \tag{4.342}$$

This iteration procedure is continued until the solution is converged in the $i+1$ plane. Usually, only two iterations ($m = 0$, $m = 1$) are required to recover acceptable accuracy. The computation then advances to the $i+2$ plane.

4.5.11 Roe Method

The Roe (1981) scheme was previously applied to the inviscid Burgers equation in Section 4.4.9. When applied to the complete Burgers' equation (Eq. 4.286), the algorithm becomes

$$u_j^{n+1} = u_j^n - \frac{\Delta t}{2\,\Delta x}\left[(F_{j+1}^n - F_{j-1}^n) - |\bar{u}_{j+\frac{1}{2}}^n|(u_{j+1}^n - u_j^n) + |\bar{u}_{j-\frac{1}{2}}^n|(u_j^n - u_{j-1}^n)\right]$$
$$+ r(u_{j+1}^n - 2u_j^n + u_{j-1}^n) \tag{4.343}$$

where $F = u^2/2$ and

$$\bar{u}_{j+\frac{1}{2}}^n = \frac{u_j^n + u_{j+1}^n}{2}$$

This explicit one-step method is first-order accurate with a T.E. of $O[\Delta t, (\Delta x)^2]$. Higher-order versions of Roe's scheme can be obtained using the techniques described in Sections 4.4.11 and 4.4.12.

4.6 CONCLUDING REMARKS

In this chapter an attempt has been made to introduce basic numerical methods for solving simple model PDEs. It has not been the intent to include all numerical techniques that have been proposed for these equations. Some very useful methods have not been included. However, those that have been presented should provide a reasonable background for the more complex applications that follow in Chapters 6–9.

Based on the information presented on the various techniques, it is clear that many different numerical methods can be used to solve the same problem. The differences in the quality of the solutions produced using the applicable methods are frequently small, and the selection of an optimum technique becomes difficult. However, the selection process can be aided by the experience gained in programming the various methods to solve the model equations presented in this chapter.

PROBLEMS

4.1 Derive Eq. (4.19).

4.2 Derive the modified equation for the Lax method applied to the wave equation. Retain terms up to and including u_{xxxx}.

4.3 Repeat Prob. 4.2 for the Euler implicit scheme.

4.4 Derive the modified equation for the leap frog method. Retain terms up to and including u_{xxxxx}.

4.5 Repeat Prob. 4.4 for the Lax-Wendroff method.

4.6 Determine the errors in amplitude and phase for $\beta = 90°$ if the Lax method is applied to the wave equation for 10 time steps with $\nu = 0.5$.

4.7 Repeat Prob. 4.6 for the MacCormack scheme.

4.8 Suppose the Lax scheme is used to solve the wave equation ($c = 0.75$) for the initial condition

$$u(x,0) = 2\sin(2\pi x - 0.4\pi) \qquad 0 \leqslant x \leqslant 2$$

and periodic boundary conditions with $\Delta x = 0.02$ and $\Delta t = 0.02$.

(*a*) Use the amplification factor to find the amplitude and phase errors after 10 steps.

(*b*) Use a truncated version of the modified equation to determine (approximately) the amplitude and phase errors after 10 steps.
Hint: The exact solution for the PDE

$$u_t + cu_x = \mu\mu_{xx} + du_{xxx}$$

with initial condition

$$u(x,0) = A_0 \sin(kx)$$

and periodic boundary conditions is

$$u(x,t) = A_0 \exp(-k^2\mu t) \sin\{k[x - (c + k^2 d)t]\}$$

4.9 Suppose the Lax-Wendroff scheme is used to solve the wave equation ($c = 0.75$) for the initial condition

$$u(x,0) = 2\sin(\pi x) \qquad 0 \leqslant x \leqslant 2$$

and periodic boundary conditions with $\Delta x = 0.1$ and $\Delta t = 0.1$.

(*a*) Use the amplification factor to find the amplitude and phase errors after 10 steps.

(*b*) Use a truncated version of the modified equation to determine (approximately) the amplitude and phase errors after 10 steps.
Hint: The exact solution for the PDE

$$u_t + cu_x = du_{xxx} + \mu u_{xxxx}$$

with initial condition

$$u(x,0) = A_0 \sin(kx)$$

and periodic boundary conditions is given by

$$u(x,t) = A_0 \exp(k^4\mu t) \sin\{k[x - (c + k^2 d)t]\}$$

4.10 Repeat Prob. 4.9 for the leap-frog method.
4.11 Suppose the Rusanov scheme is used to solve the wave equation ($c = 0.5$) for the initial condition

$$u(x,0) = 2\sin(\pi x - 0.3\pi) \qquad 0 \leqslant x \leqslant 2$$

and periodic boundary conditions with $\Delta x = 0.1$, $\Delta t = 0.1$, and $\omega = 1.0$.

(*a*) Use the amplification factor to find the amplitude and phase errors after 10 steps.

(*b*) Use a truncated version of the modified equation to determine (approximately) the amplitude and phase errors after 10 steps, if the exact solution for the PDE

$$u_t + cu_x = \mu u_{xxxx} + du_{xxxxx}$$

with initial condition

$$u(x,0) = A_0 \sin(kx)$$

and periodic boundary conditions is given by

$$u(x,t) = A_0 \exp(k^4\mu t) \sin\{k[x - (c - k^4 d)t]\}$$

4.12 Derive the amplification factor for the leap frog method applied to the wave equation and determine the stability restriction for this scheme.
4.13 Repeat Prob. 4.12 for the second-order upwind method.
4.14 Show that the Rusanov method applied to the wave equation is equivalent to the following one-step scheme:

$$u_j^{n+1} = u_j^n - \nu(\mu_x \delta_x)\left(1 - \frac{\delta_x^2}{6}\right)u_j^n + \nu^2\delta_x^2\left(\frac{1}{2} + \frac{\delta_x^2}{8}\right)u_j^n - \frac{\nu^3}{6}(\mu_x \delta_x^3)u_j^n - \frac{\omega}{24}\delta_x^4 u_j^n$$

4.15 Evaluate the stability of the Rusanov method applied to the wave equation using the Fourier stability analysis. Hint: See Prob. 4.14.

4.16 The following second-order accurate explicit scheme for the wave equation was proposed by Crowley (1967):

$$u_j^{n+1} = u_j^n - \nu(\mu_x \delta_x)u_j^n + \frac{\nu^2}{2}(\mu_x^2 \delta_x^2)u_j^n - \frac{1}{8}\nu^3(\mu_x \delta_x^3)u_j^n$$

(a) Derive the modified equation for this scheme. Retain terms up to and including u_{xxxxx}.
(b) Evaluate the necessary condition for stability.
(c) Determine the errors in amplitude and phase for $\beta = 90°$ if this scheme is applied to the wave equation for 10 time steps with $\nu = 1$.

4.17 Solve the wave equation $u_t + u_x = 0$ on a digital computer using
(a) Lax scheme
(b) Lax-Wendroff scheme
for the initial condition

$$u(x,0) = \sin 2n\pi\left(\frac{x}{40}\right) \qquad 0 \leqslant x \leqslant 40$$

and periodic boundary conditions. Choose a 41 grid point mesh with $\Delta x = 1$ and compute to $t = 18$. Solve this problem for $n = 1, 3$ and $\nu = 1.0, 0.6, 0.3$ and compare graphically with the exact solution. Determine β for $n = 1$ and $n = 3$, and calculate the errors in amplitude and phase for each scheme with $\nu = 0.6$. Compare these errors with the errors appearing on the graphs.

4.18 Repeat Prob. 4.17 using the following schemes:
(a) Windward differencing scheme
(b) MacCormack scheme

4.19 Repeat Prob. 4.17 using the following schemes:
(a) MacCormack scheme
(b) Rusanov scheme ($\omega = 3$)

4.20 Solve the wave equation $u_t + u_x = 0$ on a digital computer using
(a) Windward differencing scheme
(b) MacCormack scheme
for the initial conditions

$$u(x,0) = 1 \qquad x \leqslant 10$$
$$u(x,0) = 0 \qquad x > 10$$

and Dirichlet boundary conditions. Choose a 41 grid point mesh with $\Delta x = 1$ and compute to $t = 18$. Solve this problem for $\nu = 1.0, 0.6$, and 0.3 and compare graphically with the exact solution.

4.21 Apply the windward differencing scheme to the two-dimensional wave equation

$$u_t + c(u_x + u_y) = 0$$

and determine the stability of the resulting scheme.

4.22 Derive the modified equation for the simple implicit method applied to the 1-D heat equation. Retain terms up to and including u_{xxxxxx}.

4.23 Evaluate the stability of the combined method B applied to the 1-D heat equation.

4.24 Determine the amplification factor of the ADE method of Saul'yev and examine the stability.

4.25 For the grid points $(i + j + n)$ even, show that the hopscotch method reduces to

$$u_{i,j}^{n+2} = 2u_{i,j}^{n+1} - u_{i,j}^n$$

4.26 Use the simple explicit method to solve the 1-D heat equation on the computational grid (Fig. P4.1) with boundary conditions

$$u_1^n = 2 = u_3^n$$

and initial conditions

$$u_1^1 = 2 = u_3^1 \qquad u_2^1 = 1$$

Show that if $r = \frac{1}{4}$, the steady-state value of u along $j = 2$ becomes

$$u_2^\infty = \lim_{n \to \infty} \sum_{k=1}^{n} \frac{1}{2^{k-1}}$$

Note that this infinite series is a geometric series that has a known sum.

4.27 Apply the ADI scheme to the 2-D heat equation and find u^{n+1} at the internal grid points in the mesh shown in Fig. P4.2 for $r_x = r_y = 2$. The initial conditions are

$$u^n = 1 - \frac{x}{3\,\Delta x} \qquad \text{along } y = 0$$

$$u^n = 1 - \frac{y}{2\,\Delta y} \qquad \text{along } x = 0$$

$$u^n = 0 \qquad \text{everywhere else}$$

and the boundary conditions remain fixed at their initial values.

4.28 Solve the heat equation $u_t = 0.2u_{xx}$ on a digital computer using

(*a*) Simple explicit method

(*b*) Barakat and Clark ADE method

for the initial condition

$$u(x,0) = 100 \sin \frac{\pi x}{L} \qquad L = 1$$

and boundary conditions

$$u(0,t) = u(L,t) = 0$$

Compute to $t = 0.5$ using the parameters in Table P4.1 (if possible) and compare graphically with the exact solution.

4.29 Repeat Prob. 4.28 using the Crank-Nicolson scheme.

4.30 Repeat Prob. 4.28 using the DuFort-Frankel scheme.

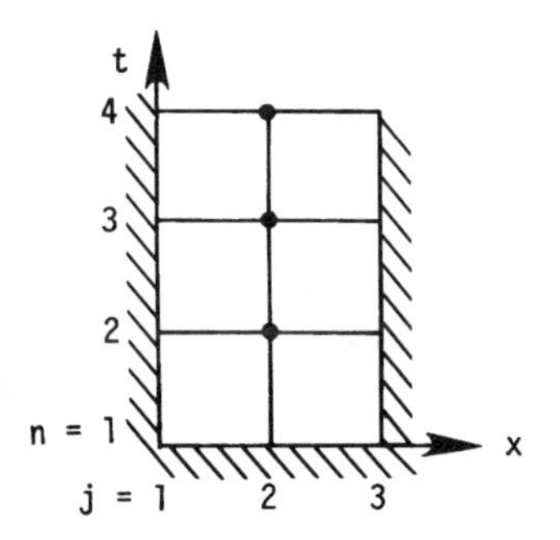

Figure P4.1

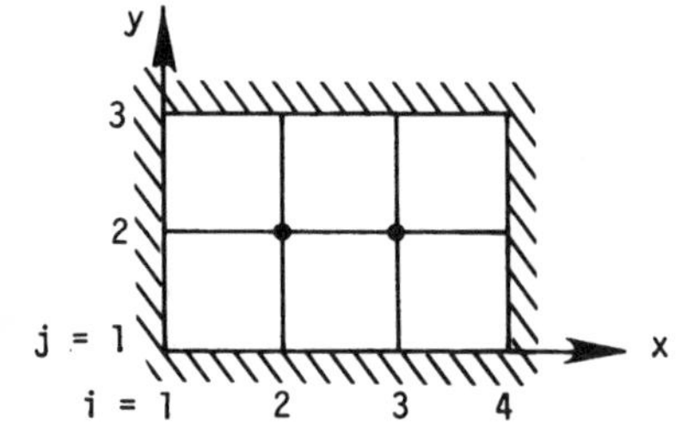

Figure P4.2

Table P4.1

Case	Number of grid points	r
1	11	0.25
2	11	0.50
3	16	0.50
4	11	1.00
5	11	2.00

4.31 The heat equation

$$\frac{\partial T}{\partial t} = \alpha \frac{\partial^2 T}{\partial x^2}$$

governs the time-dependent temperature distribution in a homogeneous constant property solid under conditions where the temperature varies only in one space dimension. Physically, this may be nearly realized in a long thin rod or very large (infinite) wall of finite thickness.

Consider a large wall of thickness L whose initial temperature is given by $T(t, x) = c \sin \pi x/L$. If the faces of the wall continue to be held at 0°, then a solution for the temperature at $t > 0$, $0 \leqslant x \leqslant L$ is

$$T(t, x) = c \exp\left(\frac{-\alpha \pi^2 t}{L^2}\right) \sin \frac{\pi x}{L}$$

For this problem let $c = 100°\text{C}$, $L = 1$ m, $\alpha = 0.02\ \text{m}^2/\text{h}$. We will consider two explicit methods of solution, A. Simple explicit method, Eq. (4.73). Stability requires that $\alpha\,\Delta t/(\Delta x)^2 \leqslant \frac{1}{2}$ for this method, B. Alternating direction explicit (ADE) method, Eq. (4.107). This particular version of the ADE method was suggested by Barakat and Clark (1966). In this algorithm, the equation for p_j^{n+1} can be solved explicitly starting from the boundary at $x = 0$, whereas the equation for q_j^{n+1} should be solved starting at the boundary at $x = L$. There is no stability constraint on the size of the time step for this method. Develop computer programs to solve the problem described above by methods A and B. Also, you will want to provide a capability for evaluating the exact solution for purposes of comparison. Make at least the following comparisons:

1. For $\Delta x = 0.1$, $\Delta t = 0.1$ [resulting in $\alpha\,\Delta t/(\Delta x)^2 = 0.2$], compare the results from methods A and B and the exact solution for $t = 10$ h. A graphical comparison is suggested.
2. Repeat the above comparison after refining the space grid, i.e., let $\Delta x = 0.066667$ (15 increments). Is the reduction in error as suggested by $O[(\Delta x)^2]$?
3. For $\Delta x = 0.1$ choose Δt such that $\alpha\,\Delta t/(\Delta x)^2 = 0.5$ and compare the predictions of methods A and B and the exact solution for $t \approx 10$ h.
4. Demonstrate that method A does become unstable as $\alpha\,\Delta t/(\Delta x)^2$ exceeds 0.5. One suggestion is to plot the centerline temperature vs. time for $\alpha\,\Delta t/(\Delta x)^2 \simeq 0.6$ for 10–20 hours of problem time.
5. For $\Delta x = 0.1$, choose Δt such that $\alpha\,\Delta t/(\Delta x)^2 = 1.0$ and compare the results of method B and the exact solution for $t \approx 10$ h.
6. Increment $\alpha\,\Delta t/(\Delta x)^2$ to 2, then 3, etc., and repeat comparison 5 above until the agreement with the exact solution becomes noticeably poor.

4.32 Work Prob. 4.31 letting method B be the Crank-Nicolson scheme.

4.33 Work Prob. 4.31 letting method B be the simple implicit scheme.

4.34 Derive a way to solve the problem described in Prob. 4.31 utilizing the fourth-order accurate representation of the second derivative given by Eq. (3.35).

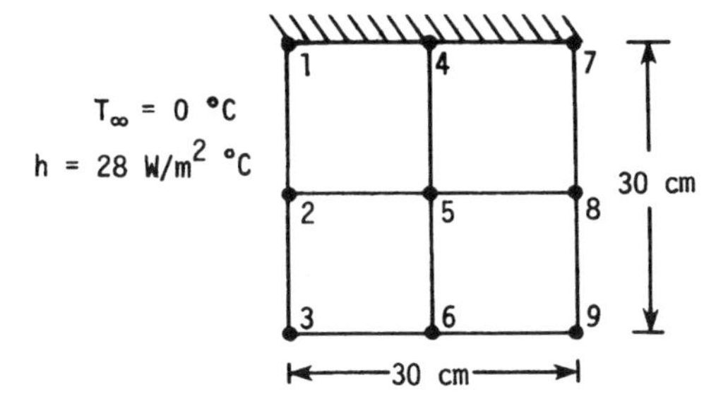

Figure P4.3

4.35 Use the difference scheme of Eq. (3.35) for second derivatives

$$\frac{\partial^2 u}{\partial x^2} \approx \frac{\delta_x^2 u_{i,j}}{h^2(1 + \delta_x^2/12)}$$

to develop a finite-difference representation for Laplace's equation where $\Delta x = \Delta y$. Write out the scheme explicitly in terms of u on the finite-difference mesh. What is the T.E. of this representation?

4.36 Evaluate the T.E. of the difference scheme of Eq. (4.114) for Laplace's equation (a) when $\Delta x = \Delta y$, (b) when $\Delta x \neq \Delta y$.

4.37 What is the T.E. for the difference equation employing the nine-point scheme of Eq. (4.114) with $\Delta x = \Delta y$ for the Poisson equation $u_{xx} + u_{yy} = x + y$?

4.38 In the cross section illustrated in Fig. P4.3, the surface 1-4-7 is insulated (adiabatic). The convective heat transfer coefficient at surface 1-2-3 is 28 W/m^2°C. The thermal conductivity of the solid material is 3.5 W/m°C. The temperature at nodes 3, 6, 7, 8, 9 is held constant at 100°C. Using Gauss-Seidel iteration, compute the temperature at nodes 1, 2, 4, and 5.

4.39 A cylindrical pin fin (Fig. P4.4) is attached to a 200°C wall while its surface is exposed to a gas at 30°C. The convection heat transfer coefficient is 300 W/m^2°C. The fin is made of stainless steel with a thermal conductivity of 18 W/m°C. Use five subdivisions and find the steady-state nodal temperatures by Gauss-Seidel iteration. Compute the total rate at which heat is transferred from the fin. You may neglect the heat loss from the outer end of the fin (i.e., assume end is adiabatic).

4.40 Determine the inviscid (ideal) flow in a 2-D channel containing a cylinder. Use a stream function formulation in which

$$u = \frac{\partial \psi}{\partial y} \qquad v = -\frac{\partial \psi}{\partial x}$$

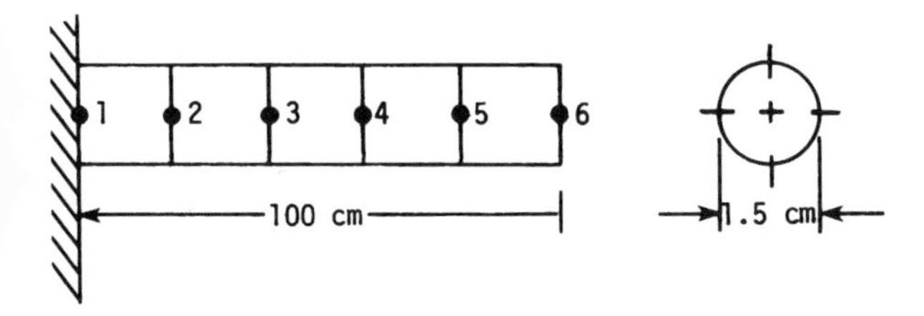

Figure P4.4

Without viscous effects, the rotation of fluid particles cannot be changed, leading to

$$\frac{\partial^2 \psi}{\partial x^2} + \frac{\partial^2 \psi}{\partial y^2} = 0$$

The flow domain is sketched in Fig. P4.5. Use a grid of $\Delta x = \Delta y = 0.5$ cm wherever possible. Unequal grid spacing will be needed near the cylinder. At the inlet (along A-B) the flow is uniform at 1 m/s. Along B-C the stream function remains constant. Along C-D, $\partial \psi / \partial x = 0$. The stream function is zero along D-E-A.

(*a*) Determine the values of ψ throughout the flow.

(*b*) Determine the velocity distribution along C-D.

(*c*) Sketch as best you can the streamline pattern for the flow.

(*d*) From the computed results, estimate the pressure coefficient at the top of the cylinder (point D) and compare this with the pressure coefficient from the "exact" analytical solution for inviscid flow around a cylinder in a stream of infinite extent.

4.41 Solve the steady-state, 2-D heat conduction equation in the unit square, $0 < x < 1$, $0 < y < 1$, by finite-differences using mesh increments $\Delta x = \Delta y = 0.1$ and 0.05. Compare the center temperatures with the exact solution. Use boundary conditions

$$T = 0 \qquad \text{at } x = 0,\ x = 1$$

$$\frac{\partial T}{\partial y} = 0 \qquad \text{at } y = 0$$

$$T = \sin(\pi x) \qquad \text{at } y = 1$$

4.42 For the conditions of Prob. 4.41,

(*a*) Use the Gauss-Seidel iterative procedure with SOR. Establish a convergence criteria. For each mesh size, use $\omega = 1$ and at least three other values of ω between 1 and 2 in an attempt to determine an appropriate ω_{opt}. Compare this with the value predicted by the Young-Frankel theory. For each calculation, use the same initial guess. Make a plot of the number of iterations required for convergence to the same specified tolerance vs. ω. Also compare the computer solution at the center with the exact solution.

(*b*) Devise and explain an SOR point iterative algorithm based on the red-black (checkerboard) strategy. Use this algorithm for the same series of computations and comparisons as indicated for Prob. 4.42(*a*). Compare the number of iterations required with the results in Prob. 4.42(*a*).

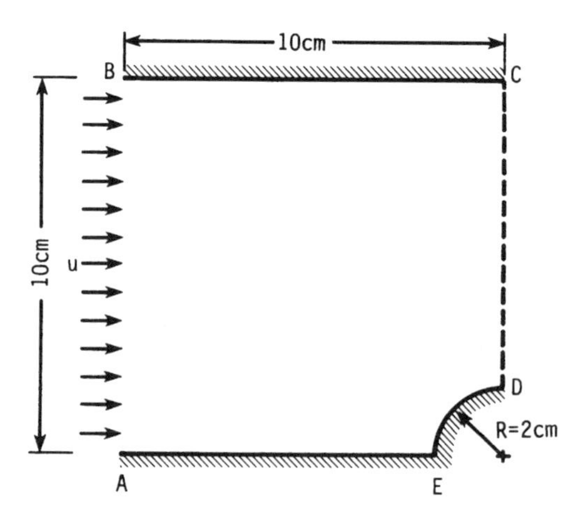

Figure P4.5

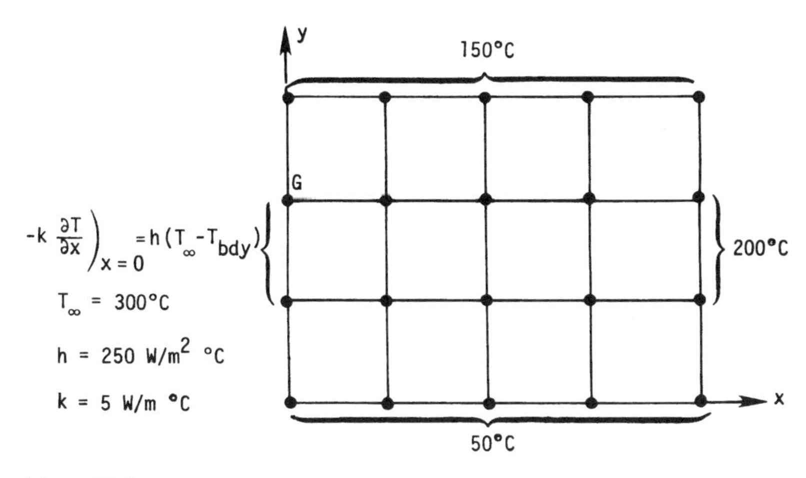

Figure P4.6

(*c*) Using the same initial guess and convergence criterion as in Prob. 4.42(*a*), solve the problem for $\Delta x = \Delta y = 0.1$ using the line iterative procedure with SOR (SLOR). Solve the problem for $\omega = 1$ and at least three other values of ω. Does this scheme appear to have the same ω_{opt} for this problem? Compare the number of iterations required with the results in Prob. 4.42(*a*) and 4.42(*b*).

4.43 Consider steady-state conduction governed by Laplace's equation in the 2-D domain shown in Fig. P4.6. The boundary conditions are shown in the figure. The mesh is square, i.e., $\Delta x = \Delta y =$ 0.02 m.

(*a*) Develop an approximate difference equation for the boundary temperature at point G using the control-volume approach.

(*b*) After obtaining a suitable finite-difference representation for Laplace's equation, use Gauss-Seidel iteration to obtain the steady-state temperature distribution.

4.44 Solve Prob. 4.43 using the line iterative method.

4.45 It is required to estimate the temperature distribution in the two-dimensional wall of a combustion chamber at steady state. The geometry has been simplified for this preliminary analysis and is given in Fig. P4.7. Write a computer program using Gauss-Seidel iteration with SOR to solve this problem. Give careful attention to the equations at the boundaries. Use grid spacing of 2 cm ($\Delta x = \Delta y$), resulting in a 6×11 mesh, and use a thermal conductivity of 20 W/m^2°C.

(*a*) Compute the steady-state temperature distribution.

(*b*) Compute the rate of heat transfer to the top, and check to see how closely it matches the heat removed by the coolant.

(*c*) For the same convergence criteria, repeat the calculation for at least three values of the relaxation parameter ω. If sufficient computer time is available, make a more detailed search for ω_{opt}.

4.46 Solve Prob. 4.45 using the line iterative method.

4.47 Solve Prob. 4.45 using the ADI method.

4.48 Write a computer program to solve Laplace's equation with Dirichlet boundary conditions on a unit square using the Gauss-Seidel procedure with

(*a*) SOR (using both $\omega = 1$ and $\omega = \omega_{opt}$)

(*b*) a two-level multigrid scheme

Compare the relative efficiencies of the three procedures for side temperatures (going counterclockwise around the square) of 20, 40, 80, and 100 using three different grids, 9×9, 17×17, and 33×33. Start all calculations with initial guesses of zero. Indicate the solution

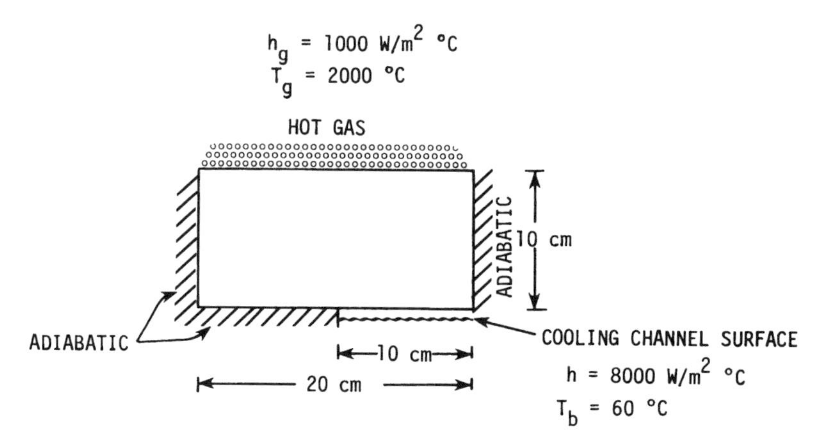

Figure P4.7

obtained at the center. Determine the multigrid effort in terms of "work units" of equivalent fine-grid iterations. Four coarse-grid sweeps approximately equal one fine-grid sweep. Three or four fine-grid sweeps per cycle should work well. Converge the coarse-grid calculation. A number of convergence criteria are workable. Monitoring the magnitude of the maximum change from one iterative sweep to the next, normalized with the maximum or average boundary value, is easy to implement:

$$\frac{|f^{k+1} - f^k|}{|f_{\text{ref}}|}$$

where f denotes either u or Δu and f_{ref} is the maximum or average value of u on the boundary. A convergence level of 10^{-5} on that basis is suggested.

4.49 Use the Lax method to solve the inviscid Burgers equation using a mesh with 51 points in the x direction. Solve this equation for a right propagating discontinuity with initial data $u = 1$ on the first 11 mesh points and $u = 0$ at all other points. Repeat your calculations for Courant numbers of 1.0, 0.6, and 0.3 and compare your numerical solutions with the analytical solution at the same time.

4.50 Repeat Prob. 4.49 using MacCormack's method. Use both a forward-backward and a backward-forward predictor-corrector sequence.

4.51 Repeat Prob. 4.49 using the WKL method.

4.52 Repeat Prob. 4.49 using the Beam-Warming method.

4.53 Solve the inviscid Burgers equation for an expansion with initial data $u = 0$ for the first 21 mesh points and $u = 1$ elsewhere. Use MacCormack's method with both forward-backward and backward-forward predictor-corrector sequences. Compare your results at two different Courant numbers with the analytic solution.

4.54 Repeat Prob. 4.53 using the Beam-Warming method (trapezoidal) and the Euler implicit scheme.

4.55 Solve the inviscid Burgers equation for a standing discontinuity. Initialize using $u = 1$ at the left end point and $u = -1$ at the right end point and zero everywhere else. Apply MacCormack's method to this problem.

4.56 Repeat Prob. 4.55 using the Beam-Warming scheme.

4.57 Determine the solution of the inviscid Burgers equation for the double-shock profile given by $u_l = 1.0$ and $u_r = 0.5$. Compute this solution using MacCormack's scheme and the Godunov method. What is the analytic solution for this set of initial conditions? Compare your numerical calculation with the analytical solution.

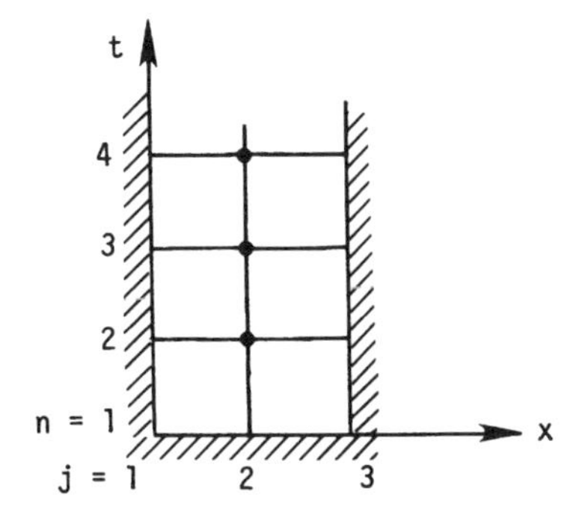

Figure P4.8

4.58 Repeat Prob. 4.49 using the schemes listed below. Remember that the stability bound for Godunov's method is $\upsilon \leqslant 0.5$.

(*a*) Godunov scheme

(*b*) Roe scheme (first order)

(*c*) Enquist-Osher scheme

4.59 Repeat Prob. 4.53 using the methods listed below:

(*a*) Godunov scheme

(*b*) Roe scheme (first-order) with and without the entropy fix

(*c*) Enquist-Osher scheme

4.60 Verify that the extrapolation formulas given in Section 4.4.11 give central or upwind differences for $\kappa = \pm 1$.

4.61 Solve Prob. 4.49 using the second-order Roe scheme with and without the use of limiters. Use the minmod limiter and the van Leer limiter in your calculations. Use an initial profile of $u_l = 1.0$ and $u_r = 0.5$ for this problem.

4.62 A numerical method is said to be monotone if it does not produce oscillations in the numerical solution. Schemes for solving the inviscid Burgers equation may be written in the form

$$u^{n+1} = H(u^n_{j-k}, \ldots, u^n_{j+k})$$

The condition for monotonicity requires H to be a monotone increasing function of its arguments, i.e., $\partial H / \partial u_j \geqslant 0$. Show that the Lax method is monotone and that the first-order upwind scheme is monotone. What conditions must be satisfied to meet this condition?

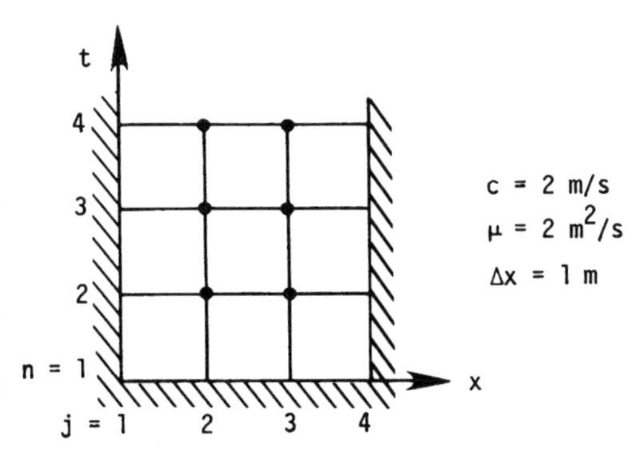

Figure P4.9

4.63 In the MUSCL approach the dependent variables are extrapolated directly to the cell boundaries, and the flux terms are recalculated. In the non-MUSCL approach, the fluxes are directly extrapolated to the interface boundaries. Repeat Prob. 4.49 using Roe's scheme with both the MUSCL and non-MUSCL approaches. Can you draw any conclusions from your results?
4.64 Solve the inviscid Burgers equation using the Lax-Wendroff method with an initial profile that is linear between the left and right boundary values of 1 and -1. This profile will become steeper as the solution progresses until a shock wave ultimately results. Perform the same experiment with the Lax method, and compare your results. Does this provide any insight into the Godunov theorem?
4.65 In Section 4.4.12 the sufficient conditions for a scheme to be TVD were given. Show that the Lax scheme satisfies these conditions, while the Lax-Wendroff scheme does not.
4.66 Solve Prob. 4.49 using the Roe-Sweby scheme. Compare your results with those obtained earlier with Roe's method using the MUSCL approach (Prob. 4.63), and comment on any differences.
4.67 Show graphically the exact steady-state solution of Eq. (4.272) for the boundary conditions

$$u(0,t) = 1$$
$$u(1,t) = 0$$

and $\mu = 0.1$.
4.68 Verify that Eq. (4.283) is an exact stationary solution of Eq. (4.282).
4.69 Derive stability conditions for the FTCS method applied to the 1-D linearized Burgers equation.
4.70 Derive the stability conditions for the upwind difference scheme given by Eq. (4.300).
4.71 Suppose the FTCS method is used to solve the linearized Burgers equation ($c = 0.5$, $\mu = 0.01$) for the initial condition

$$u(x,0) = \sin(2\pi x) \qquad 0 \leqslant x \leqslant 2$$

and periodic boundary conditions with $\Delta x = 0.02$ and $\Delta t = 0.02$. Find the amplitude error and the phase error after 20 steps.
Hint: The exact solution for the linearized Burgers equation for the above initial and boundary conditions is

$$u(x,t) = \exp(-4\pi^2 \mu t) \sin[2\pi(x - ct)]$$

4.72 Use the FTCS method to solve the linearized Burgers equation for the initial condition

$$u(x,0) = 0 \qquad 0 \leqslant x \leqslant 1$$

and the boundary conditions

$$u(0,t) = 100$$
$$u(1,t) = 0$$

on a 21 grid point mesh. Find the steady-state solution for the conditions
(*a*) $r = 0.50$, $\nu = 0.25$
(*b*) $r = 0.50$, $\nu = 1.00$
(*c*) $r = 0.10$, $\nu = 0.40$
(*d*) $r = 0.05$, $\nu = 0.50$
and compare the numerical solutions with the exact solution.
4.73 Repeat Prob. 4.72 using the scheme proposed by Leonard.
4.74 Repeat Prob. 4.72 using the leap frog/DuFort-Frankel method.
4.75 Repeat Prob. 4.72 using the Allen-Cheng method.
4.76 Use the Fourier stability analysis to determine the stability limitations of the scheme proposed by Leonard, Eq. (4.302).
4.77 Determine the modified equation for the Allen-Cheng method. Retain terms up to and including u_{xxx}.

4.78 Apply the Brailovskaya scheme to the linearized Burgers equation on the computational grid shown in Fig. P4.8 and show that the steady-state value for u at $j = 2$ is

$$u_2^\infty = \lim_{n \to \infty} \sum_{i=1}^{n} \frac{1}{3^{n-i}} = \frac{3}{2}$$

Boundary conditions are $u_1^n = \frac{3}{2} = u_3^n$, and the initial condition is $u_2^1 = 1$. Do not use a digital computer to solve this problem.

4.79 Apply the Beam-Warming scheme with Euler implicit time differencing to the linearized Burgers equation on the computational grid shown in Fig. P4.9, and determine the steady-state values for u at $j = 2$ and $j = 3$. The boundary conditions are $u_1^n = 1$, $u_4^n = 4$, and the initial conditions are $u_2^1 = 0 = u_3^1$. Do not use a digital computer to solve this problem.

4.80 Apply the two-step Lax-Wendroff method to the PDE

$$u_t + F_x + uu_{xxx} = 0$$

where $F = F(u)$. Develop the final finite-difference equations.

4.81 Solve the linearized Burgers equation using

(*a*) FTCS method

(*b*) Upwind method, Eq. (4.300)

(*c*) Leonard method, Eq. (4.302)

for the initial condition

$$u(x, 0) = 0 \qquad 0 \leqslant x \leqslant 1$$

and the boundary conditions

$$u(0, t) = 100$$
$$u(1, t) = 0$$

on a 21 grid point mesh. Find the steady-state solution for $r = 0.10$ and $\nu = 0.40$, and compare the numerical solutions with the exact solution.

4.82 Repeat Prob. 4.81 using the following methods

(*a*) Leap frog/DuFort-Frankel method

(*b*) Allen-Cheng method

(*c*) MacCormack method, Eq. (4.312)

4.83 Repeat Prob. 4.81 using the Briley-McDonald method with Euler implicit time differencing.

4.84 Solve the generalized Burgers equation

$$u_t + \left(\frac{1}{2} - u\right)u_x = 0.001u_{xx}$$

using

(*a*) MacCormack's scheme

(*b*) Roe's scheme

for the initial condition

$$u = \frac{1}{2}\{1 + \tanh[250(x - 20)]\} \qquad 0 \leqslant x \leqslant 40$$

and exact Dirichlet boundary conditions. Choose a 41 grid point mesh with $\Delta x = 1$, and compute to $t = 18$. Solve this problem for $\Delta t = 1.0$ and 0.5, and compare graphically with the exact stationary solution.

PART TWO

APPLICATION OF NUMERICAL METHODS TO THE EQUATIONS OF FLUID MECHANICS AND HEAT TRANSFER

CHAPTER

FIVE

GOVERNING EQUATIONS OF FLUID MECHANICS AND HEAT TRANSFER

In this chapter, the governing equations of fluid mechanics and heat transfer (i.e., fluid dynamics) are described. Since the reader is assumed to have some background in this field, a complete derivation of the governing equations is not included. The equations are presented in order of decreasing complexity. For the most part, only the classical forms of the equations are given. Other forms of the governing equations, which have been simplified primarily for computational purposes, are presented in later chapters. Also included in this chapter is an introduction to turbulence modeling.

5.1 FUNDAMENTAL EQUATIONS

The fundamental equations of fluid dynamics are based on the following universal laws of conservation:

Conservation of Mass
Conservation of Momentum
Conservation of Energy

The equation that results from applying the Conservation of Mass law to a fluid flow is called the continuity equation. The Conservation of Momentum law is nothing more than Newton's Second Law. When this law is applied to a fluid flow, it yields a vector equation known as the momentum equation. The

Conservation of Energy law is identical to the First Law of Thermodynamics, and the resulting fluid dynamic equation is named the energy equation. In addition to the equations developed from these universal laws, it is necessary to establish relationships between fluid properties in order to close the system of equations. An example of such a relationship is the equation of state, which relates the thermodynamic variables pressure p, density ρ, and temperature T.

Historically, there have been two different approaches taken to derive the equations of fluid dynamics: the phenomenological approach and the kinetic theory approach. In the phenomenological approach, certain relations between stress and rate of strain and heat flux and temperature gradient are postulated, and the fluid dynamic equations are then developed from the conservation laws. The required constants of proportionality between stress and rate of strain and heat flux and temperature gradient (which are called transport coefficients) must be determined experimentally in this approach. In the kinetic theory approach (also called the mathematical theory of nonuniform gases), the fluid dynamic equations are obtained with the transport coefficients defined in terms of certain integral relations, which involve the dynamics of colliding particles. The drawback to this approach is that the interparticle forces must be specified in order to evaluate the collision integrals. Thus a mathematical uncertainty takes the place of the experimental uncertainty of the phenomenological approach. These two approaches will yield the same fluid dynamic equations if equivalent assumptions are made during their derivations.

The derivation of the fundamental equations of fluid dynamics will not be presented here. The derivation of the equations using the phenomenological approach is thoroughly treated by Schlichting (1968), and the kinetic theory approach is described in detail by Hirschfelder et al. (1954). The fundamental equations given initially in this chapter were derived for a uniform, homogeneous fluid without mass diffusion or finite-rate chemical reactions. In order to include these later effects it is necessary to consider extra relations, called the species continuity equations, and to add terms to the energy equation to account for diffusion. These additional equations and terms are given in Section 5.1.5. Further information on reacting flows can be found in the works by Dorrance (1962) and Anderson (1989).

5.1.1 Continuity Equation

The Conservation of Mass law applied to a fluid passing through an infinitesimal, fixed control volume (see Fig. 5.1) yields the following equation of continuity:

$$\frac{\partial \rho}{\partial t} + \nabla \cdot (\rho \mathbf{V}) = 0 \tag{5.1}$$

where ρ is the fluid density and $\mathbf{V}$ is the fluid velocity. The first term in this equation represents the rate of increase of the density in the control volume, and the second term represents the rate of mass flux passing out of the control surface (which surrounds the control volume) per unit volume. It is convenient

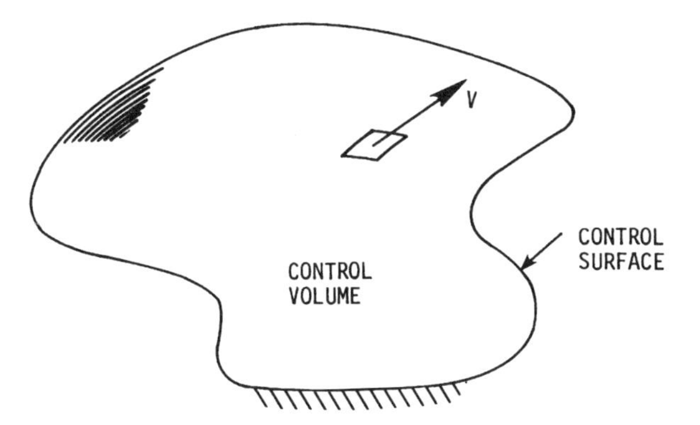

Figure 5.1 Control volume for Eulerian approach.

to use the substantial derivative

$$\frac{D(\ )}{Dt} \equiv \frac{\partial(\ )}{\partial t} + \mathbf{V} \cdot \nabla(\ ) \tag{5.2}$$

to change Eq. (5.1) into the form

$$\frac{D\rho}{Dt} + \rho(\nabla \cdot \mathbf{V}) = 0 \tag{5.3}$$

Equation (5.1) was derived using the *Eulerian approach*. In this approach, a fixed control volume is utilized, and the changes to the fluid are recorded as the fluid passes through the control volume. In the alternative *Lagrangian approach*, the changes to the properties of a fluid element are recorded by an observer moving with the fluid element. The Eulerian viewpoint is commonly used in fluid mechanics.

For a Cartesian coordinate system, where u, v, w represent the x, y, z components of the velocity vector, Eq. (5.1) becomes

$$\frac{\partial \rho}{\partial t} + \frac{\partial}{\partial x}(\rho u) + \frac{\partial}{\partial y}(\rho v) + \frac{\partial}{\partial z}(\rho w) = 0 \tag{5.4}$$

Note that this equation is in conservation-law (divergence) form.

A flow in which the density of each fluid element remains constant is called *incompressible*. Mathematically, this implies that

$$\frac{D\rho}{Dt} = 0 \tag{5.5}$$

which reduces Eq. (5.3) to

$$\nabla \cdot \mathbf{V} = 0 \tag{5.6}$$

or

$$\frac{\partial u}{\partial x} + \frac{\partial v}{\partial y} + \frac{\partial w}{\partial z} = 0 \tag{5.7}$$

for the Cartesian coordinate system. For steady air flows with speed $V < 100$ m/s or $M < 0.3$ the assumption of incompressibility is a good approximation.

5.1.2 Momentum Equation

Newton's Second Law applied to a fluid passing through an infinitesimal, fixed control volume yields the following momentum equation:

$$\frac{\partial}{\partial t}(\rho \mathbf{V}) + \boldsymbol{\nabla} \cdot \rho \mathbf{V}\mathbf{V} = \rho \mathbf{f} + \boldsymbol{\nabla} \cdot \boldsymbol{\Pi}_{ij} \tag{5.8}$$

The first term in this equation represents the rate of increase of momentum per unit volume in the control volume. The second term represents the rate of momentum lost by convection (per unit volume) through the control surface. Note that $\rho \mathbf{V}\mathbf{V}$ is a tensor, so that $\boldsymbol{\nabla} \cdot \rho \mathbf{V}\mathbf{V}$ is not a simple divergence. This term can be expanded, however, as

$$\boldsymbol{\nabla} \cdot \rho \mathbf{V}\mathbf{V} = \rho \mathbf{V} \cdot \boldsymbol{\nabla} \mathbf{V} + \mathbf{V}(\boldsymbol{\nabla} \cdot \rho \mathbf{V}) \tag{5.9}$$

When this expression for $\boldsymbol{\nabla} \cdot \rho \mathbf{V}\mathbf{V}$ is substituted into Eq. (5.8), and the resulting equation is simplified using the continuity equation, the momentum equation reduces to

$$\rho \frac{D\mathbf{V}}{Dt} = \rho \mathbf{f} + \boldsymbol{\nabla} \cdot \boldsymbol{\Pi}_{ij} \tag{5.10}$$

The first term on the right-hand side of Eq. (5.10) is the body force per unit volume. Body forces act at a distance and apply to the entire mass of the fluid. The most common body force is the gravitational force. In this case, the force per unit mass ($\mathbf{f}$) equals the acceleration of gravity vector $\mathbf{g}$:

$$\rho \mathbf{f} = \rho \mathbf{g} \tag{5.11}$$

The second term on the right-hand side of Eq. (5.10) represents the surface forces per unit volume. These forces are applied by the external stresses on the fluid element. The stresses consist of normal stresses and shearing stresses and are represented by the components of the stress tensor $\boldsymbol{\Pi}_{ij}$.

The momentum equation given above is quite general and is applicable to both continuum and noncontinuum flows. It is only when approximate expressions are inserted for the shear-stress tensor that Eq. (5.8) loses its generality. For all gases that can be treated as a continuum, and most liquids, it has been observed that the stress at a point is linearly dependent on the rates of strain (deformation) of the fluid. A fluid that behaves in this manner is called a *Newtonian fluid*. With this assumption, it is possible to derive (Schlichting, 1968) a general deformation law that relates the stress tensor to the pressure and velocity components. In compact tensor notation, this relation becomes

$$\mathbf{\Pi}_{ij} = -p\delta_{ij} + \mu\left(\frac{\partial u_i}{\partial x_j} + \frac{\partial u_j}{\partial x_i}\right) + \delta_{ij}\mu'\frac{\partial u_k}{\partial x_k} \qquad i,j,k = 1,2,3 \quad (5.12)$$

where δ_{ij} is the Kronecker delta function ($\delta_{ij} = 1$ if $i = j$ and $\delta_{ij} = 0$ if $i \neq j$); u_1, u_2, u_3 represent the three components of the velocity vector $\mathbf{V}$; x_1, x_2, x_3 represent the three components of the position vector; μ is the coefficient of viscosity (dynamic viscosity), and μ' is the second coefficient of viscosity. The two coefficients of viscosity are related to the coefficient of bulk viscosity κ by the expression

$$\kappa = \tfrac{2}{3}\mu + \mu' \qquad (5.13)$$

In general, it is believed that κ is negligible except in the study of the structure of shock waves and in the absorption and attenuation of acoustic waves. For this reason, we will ignore bulk viscosity for the remainder of the text. With $\kappa = 0$, the second coefficient of viscosity becomes

$$\mu' = -\tfrac{2}{3}\mu \qquad (5.14)$$

and the stress tensor may be written as

$$\mathbf{\Pi}_{ij} = -p\delta_{ij} + \mu\left[\left(\frac{\partial u_i}{\partial x_j} + \frac{\partial u_j}{\partial x_i}\right) - \frac{2}{3}\delta_{ij}\frac{\partial u_k}{\partial x_k}\right] \qquad i,j,k = 1,2,3 \quad (5.15)$$

The stress tensor is frequently separated in the following manner:

$$\mathbf{\Pi}_{ij} = -p\delta_{ij} + \boldsymbol{\tau}_{ij} \qquad (5.16)$$

where $\boldsymbol{\tau}_{ij}$ represents the viscous stress tensor given by

$$\boldsymbol{\tau}_{ij} = \mu\left[\left(\frac{\partial u_i}{\partial x_j} + \frac{\partial u_j}{\partial x_i}\right) - \frac{2}{3}\delta_{ij}\frac{\partial u_k}{\partial x_k}\right] \qquad i,j,k = 1,2,3 \qquad (5.17)$$

Upon substituting Eq. (5.15) into Eq. (5.10), the famous *Navier-Stokes equation* is obtained:

$$\rho\frac{D\mathbf{V}}{Dt} = \rho\mathbf{f} - \nabla p + \frac{\partial}{\partial x_j}\left[\mu\left(\frac{\partial u_i}{\partial x_j} + \frac{\partial u_j}{\partial x_i}\right) - \frac{2}{3}\delta_{ij}\mu\frac{\partial u_k}{\partial x_k}\right] \qquad (5.18)$$

For a Cartesian coordinate system, Eq. (5.18) can be separated into the following three scalar Navier-Stokes equations:

$$\rho\frac{Du}{Dt} = \rho f_x - \frac{\partial p}{\partial x} + \frac{\partial}{\partial x}\left[\frac{2}{3}\mu\left(2\frac{\partial u}{\partial x} - \frac{\partial v}{\partial y} - \frac{\partial w}{\partial z}\right)\right] + \frac{\partial}{\partial y}\left[\mu\left(\frac{\partial u}{\partial y} + \frac{\partial v}{\partial x}\right)\right]$$
$$+ \frac{\partial}{\partial z}\left[\mu\left(\frac{\partial w}{\partial x} + \frac{\partial u}{\partial z}\right)\right]$$
$$\rho\frac{Dv}{Dt} = \rho f_y - \frac{\partial p}{\partial y} + \frac{\partial}{\partial x}\left[\mu\left(\frac{\partial v}{\partial x} + \frac{\partial u}{\partial y}\right)\right] + \frac{\partial}{\partial y}\left[\frac{2}{3}\mu\left(2\frac{\partial v}{\partial y} - \frac{\partial u}{\partial x} - \frac{\partial w}{\partial z}\right)\right]$$
$$+ \frac{\partial}{\partial z}\left[\mu\left(\frac{\partial v}{\partial z} + \frac{\partial w}{\partial y}\right)\right]$$
$$\rho\frac{Dw}{Dt} = \rho f_z - \frac{\partial p}{\partial z} + \frac{\partial}{\partial x}\left[\mu\left(\frac{\partial w}{\partial x} + \frac{\partial u}{\partial z}\right)\right] + \frac{\partial}{\partial y}\left[\mu\left(\frac{\partial v}{\partial z} + \frac{\partial w}{\partial y}\right)\right]$$
$$+ \frac{\partial}{\partial z}\left[\frac{2}{3}\mu\left(2\frac{\partial w}{\partial z} - \frac{\partial u}{\partial x} - \frac{\partial v}{\partial y}\right)\right] \tag{5.19}$$

Utilizing Eq. (5.8), these equations can be rewritten in conservation-law form as

$$\frac{\partial \rho u}{\partial t} + \frac{\partial}{\partial x}(\rho u^2 + p - \tau_{xx}) + \frac{\partial}{\partial y}(\rho uv - \tau_{xy})$$
$$+ \frac{\partial}{\partial z}(\rho uw - \tau_{xz}) = \rho f_x$$
$$\frac{\partial \rho v}{\partial t} + \frac{\partial}{\partial x}(\rho uv - \tau_{xy}) + \frac{\partial}{\partial y}(\rho v^2 + p - \tau_{yy})$$
$$+ \frac{\partial}{\partial z}(\rho vw - \tau_{yz}) = \rho f_y \tag{5.20}$$
$$\frac{\partial \rho w}{\partial t} + \frac{\partial}{\partial x}(\rho uw - \tau_{xz}) + \frac{\partial}{\partial y}(\rho vw - \tau_{yz})$$
$$+ \frac{\partial}{\partial z}(\rho w^2 + p - \tau_{zz}) = \rho f_z$$

where the components of the viscous stress tensor τ_{ij} are given by

$$\tau_{xx} = \frac{2}{3}\mu\left(2\frac{\partial u}{\partial x} - \frac{\partial v}{\partial y} - \frac{\partial w}{\partial z}\right)$$
$$\tau_{yy} = \frac{2}{3}\mu\left(2\frac{\partial v}{\partial y} - \frac{\partial u}{\partial x} - \frac{\partial w}{\partial z}\right)$$

$$\tau_{zz} = \frac{2}{3}\mu\left(2\frac{\partial w}{\partial z} - \frac{\partial u}{\partial x} - \frac{\partial v}{\partial y}\right)$$

$$\tau_{xy} = \mu\left(\frac{\partial u}{\partial y} + \frac{\partial v}{\partial x}\right) = \tau_{yx}$$

$$\tau_{xz} = \mu\left(\frac{\partial w}{\partial x} + \frac{\partial u}{\partial z}\right) = \tau_{zx}$$

$$\tau_{yz} = \mu\left(\frac{\partial v}{\partial z} + \frac{\partial w}{\partial y}\right) = \tau_{zy}$$

The Navier-Stokes equations form the basis upon which the entire science of viscous flow theory has been developed. Strictly speaking, the term Navier-Stokes equations refers to the components of the viscous momentum equation [Eq. (5.18)]. However, it is common practice to include the continuity equation and the energy equation in the set of equations referred to as the Navier-Stokes equations.

If the flow is incompressible and the coefficient of viscosity (μ) is assumed constant, Eq. (5.18) will reduce to the much simpler form

$$\rho\frac{D\mathbf{V}}{Dt} = \rho\mathbf{f} - \nabla p + \mu\nabla^2\mathbf{V} \tag{5.21}$$

It should be remembered that Eq. (5.21) is derived by assuming a constant viscosity, which may be a poor approximation for the nonisothermal flow of a liquid whose viscosity is highly temperature dependent. On the other hand, the viscosity of gases is only moderately temperature dependent, and Eq. (5.21) is a good approximation for the incompressible flow of a gas.

5.1.3 Energy Equation

The First Law of Thermodynamics applied to a fluid passing through an infinitesimal, fixed control volume yields the following energy equation:

$$\frac{\partial E_t}{\partial t} + \nabla \cdot E_t\mathbf{V} = \frac{\partial Q}{\partial t} - \nabla \cdot \mathbf{q} + \rho\mathbf{f} \cdot \mathbf{V} + \nabla \cdot (\mathbf{\Pi}_{ij} \cdot \mathbf{V}) \tag{5.22}$$

where E_t is the total energy per unit volume given by

$$E_t = \rho\left(e + \frac{V^2}{2} + \text{potential energy} + \cdots\right) \tag{5.23}$$

and e is the internal energy per unit mass. The first term on the left-hand side of Eq. (5.22) represents the rate of increase of E_t in the control volume, while the second term represents the rate of total energy lost by convection (per unit volume) through the control surface. The first term on the right-hand side of Eq. (5.22) is the rate of heat produced per unit volume by external agencies, while the second term ($\nabla \cdot \mathbf{q}$) is the rate of heat lost by conduction (per unit volume) through the control surface. Fourier's law for heat transfer by conduction will be assumed, so that the heat transfer $\mathbf{q}$ can be expressed as

$$\mathbf{q} = -k\,\nabla T \tag{5.24}$$

where k is the coefficient of thermal conductivity and T is the temperature. The third term on the right-hand side of Eq. (5.22) represents the work done on the control volume (per unit volume) by the body forces, while the fourth term represents the work done on the control volume (per unit volume) by the surface forces. It should be obvious that Eq. (5.22) is simply the First Law of Thermodynamics applied to the control volume. That is, the increase of energy in the system is equal to heat added to the system plus the work done on the system.

For a Cartesian coordinate system, Eq. (5.22) becomes

$$\begin{aligned}\frac{\partial E_t}{\partial t} - \frac{\partial Q}{\partial t} - \rho(f_x u + f_y v + f_z w) + \frac{\partial}{\partial x}(E_t u + pu - u\tau_{xx} - v\tau_{xy} - w\tau_{xz} + q_x)\\ + \frac{\partial}{\partial y}(E_t v + pv - u\tau_{xy} - v\tau_{yy} - w\tau_{yz} + q_y)\\ + \frac{\partial}{\partial z}(E_t w + pw - u\tau_{xz} - v\tau_{yz} - w\tau_{zz} + q_z) = 0\end{aligned} \tag{5.25}$$

which is in conservation-law form. Using the continuity equation, the left-hand side of Eq. (5.22) can be replaced by the following expression:

$$\rho\frac{D(E_t/\rho)}{Dt} = \frac{\partial E_t}{\partial t} + \nabla \cdot E_t\mathbf{V} \tag{5.26}$$

which is equivalent to

$$\rho\frac{D(E_t/\rho)}{Dt} = \rho\frac{De}{Dt} + \rho\frac{D(V^2/2)}{Dt} \tag{5.27}$$

if only internal energy and kinetic energy are considered significant in Eq. (5.23). Forming the scalar dot product of Eq. (5.10) with the velocity vector $\mathbf{V}$

allows one to obtain

$$\rho \frac{D\mathbf{V}}{Dt} \cdot \mathbf{V} = \rho \mathbf{f} \cdot \mathbf{V} - \nabla p \cdot \mathbf{V} + (\nabla \cdot \boldsymbol{\tau}_{ij}) \cdot \mathbf{V} \tag{5.28}$$

Now if Eqs. (5.26), (5.27), and (5.28) are combined and substituted into Eq. (5.22), a useful variation of thc original energy equation is obtained:

$$\rho \frac{De}{Dt} + p(\nabla \cdot \mathbf{V}) = \frac{\partial Q}{\partial t} - \nabla \cdot \mathbf{q} + \nabla \cdot (\boldsymbol{\tau}_{ij} \cdot \mathbf{V}) - (\nabla \cdot \boldsymbol{\tau}_{ij}) \cdot \mathbf{V} \tag{5.29}$$

The last two terms in this equation can be combined into a single term, since

$$\tau_{ij} \frac{\partial u_i}{\partial x_j} = \nabla \cdot (\boldsymbol{\tau}_{ij} \cdot \mathbf{V}) - (\nabla \cdot \boldsymbol{\tau}_{ij}) \cdot \mathbf{V} \tag{5.30}$$

This term is customarily called the *dissipation function* Φ and represents the rate at which mechanical energy is expended in the process of deformation of the fluid due to viscosity. After inserting the dissipation function, Eq. (5.29) becomes

$$\rho \frac{De}{Dt} + p(\nabla \cdot \mathbf{V}) = \frac{\partial Q}{\partial t} - \nabla \cdot \mathbf{q} + \Phi \tag{5.31}$$

Using the definition of enthalpy,

$$h = e + \frac{p}{\rho} \tag{5.32}$$

and the continuity equation, Eq. (5.31) can be rewritten as

$$\rho \frac{Dh}{Dt} = \frac{Dp}{Dt} + \frac{\partial Q}{\partial t} - \nabla \cdot \mathbf{q} + \Phi \tag{5.33}$$

For a Cartesian coordinate system, the dissipation function, which is always positive if $\mu' = -(2/3)\mu$, becomes

$$\Phi = \mu \left[2\left(\frac{\partial u}{\partial x}\right)^2 + 2\left(\frac{\partial v}{\partial y}\right)^2 + 2\left(\frac{\partial w}{\partial z}\right)^2 + \left(\frac{\partial v}{\partial x} + \frac{\partial u}{\partial y}\right)^2 + \left(\frac{\partial w}{\partial y} + \frac{\partial v}{\partial z}\right)^2 + \left(\frac{\partial u}{\partial z} + \frac{\partial w}{\partial x}\right)^2 - \frac{2}{3}\left(\frac{\partial u}{\partial x} + \frac{\partial v}{\partial y} + \frac{\partial w}{\partial z}\right)^2 \right] \tag{5.34}$$

If the flow is incompressible, and if the coefficient of thermal conductivity is assumed constant, Eq. (5.31) reduces to

$$\rho \frac{De}{Dt} = \frac{\partial Q}{\partial t} + k \nabla^2 T + \Phi \tag{5.35}$$

5.1.4 Equation of State

In order to close the system of fluid dynamic equations it is necessary to establish relations between the thermodynamic variables (p, ρ, T, e, h) as well as to relate the transport properties (μ, k) to the thermodynamic variables. For example, consider a compressible flow without external heat addition or body forces and use Eq. (5.4) for the continuity equation, Eqs. (5.19) for the three momentum equations, and Eq. (5.25) for the energy equation. These five scalar equations contain seven unknowns ρ, p, e, T, u, v, w, provided that the transport coefficients μ, k can be related to the thermodynamic properties in the list of unknowns. It is obvious that two additional equations are required to close the system. These two additional equations can be obtained by determining relations that exist between the thermodynamic variables. Relations of this type are known as equations of state. According to the *state principle* of thermodynamics, the local thermodynamic state is fixed by any two independent thermodynamic variables, provided that the chemical composition of the fluid is not changing owing to diffusion or finite-rate chemical reactions. Thus for the present example, if we choose e and ρ as the two independent variables, then equations of state of the form

$$p = p(e, \rho) \qquad T = T(e, \rho) \tag{5.36}$$

are required.

For most problems in gas dynamics, it is possible to assume a *perfect gas*. A perfect gas is defined as a gas whose intermolecular forces are negligible. A perfect gas obeys the perfect gas equation of state,

$$p = \rho RT \tag{5.37}$$

where R is the gas constant. The intermolecular forces become important under conditions of high pressure and relatively low temperature. For these conditions, the gas no longer obeys the perfect gas equation of state, and an alternative equation of state must be used. An example is the Van der Waals equation of state,

$$(p + a\rho^2)\left(\frac{1}{\rho} - b\right) = RT$$

where a and b are constants for each type of gas.

For problems involving a perfect gas at relatively low temperatures, it is possible to also assume a *calorically perfect gas*. A calorically perfect gas is defined as a perfect gas with constant specific heats. In a calorically perfect gas, the specific heat at constant volume c_v, the specific heat at constant pressure c_p, and the ratio of specific heats γ all remain constant, and the following relations exist:

$$e = c_v T \qquad h = c_p T \qquad \gamma = \frac{c_p}{c_v} \qquad c_v = \frac{R}{\gamma - 1} \qquad c_p = \frac{\gamma R}{\gamma - 1}$$

For air at standard conditions, $R = 287\ \text{m}^2/(\text{s}^2\ \text{K})$ and $\gamma = 1.4$. If we assume that the fluid in our example is a calorically perfect gas, then Eqs. (5.36) become

$$p = (\gamma - 1)\rho e \qquad T = \frac{(\gamma - 1)e}{R} \tag{5.38}$$

For fluids that cannot be considered calorically perfect, the required state relations can be found in the form of tables, charts, or curve fits.

The coefficients of viscosity and thermal conductivity can be related to the thermodynamic variables using kinetic theory. For example, Sutherland's formulas for viscosity and thermal conductivity are given by

$$\mu = C_1 \frac{T^{\frac{3}{2}}}{T + C_2} \qquad k = C_3 \frac{T^{\frac{3}{2}}}{T + C_4}$$

where C_1–C_4 are constants for a given gas. For air at moderate temperatures, $C_1 = 1.458 \times 10^{-6}$ kg/(m s $\sqrt{\text{K}}$), $C_2 = 110.4$ K, $C_3 = 2.495 \times 10^{-3}$ (kg m)/(s^{3}K$^{\frac{3}{2}}$), and $C_4 = 194$ K. The Prandtl number

$$\text{Pr} = \frac{c_p \mu}{k}$$

is often used to determine the coefficient of thermal conductivity k once μ is known. This is possible because the ratio (c_p/Pr), which appears in the expression

$$k = \frac{c_p}{\text{Pr}} \mu$$

is approximately constant for most gases. For air at standard conditions, Pr = 0.72.

5.1.5 Chemically Reacting Flows

The assumption of a calorically perfect gas is valid if the intermolecular forces are negligible and the temperature is relatively low. The equations governing a calorically perfect gas are given in the previous section. As the temperature of the gas increases to higher values, the gas can no longer be considered calorically perfect. At first, the vibrational energy of the molecules becomes excited and the specific heats c_p and c_v are no longer constant but are functions of temperature. For air, this occurs at temperatures above 800 K, where the air first becomes *thermally perfect*. By definition, a thermally perfect gas is a perfect gas whose specific heats are functions only of temperature. As the temperature of the gas is increased further, chemical reactions begin to take place, and the gas is no longer thermally perfect. For air at sea level pressure, the dissociation of molecular oxygen ($O_2 \rightarrow 2O$) starts at about 2000 K, and the molecular oxygen is totally dissociated at about 4000 K. The dissociation of molecular nitrogen ($N_2 \rightarrow 2N$) then begins, and total dissociation occurs at about 9000 K. Above 9000 K, ionization of the air takes place ($N \rightarrow N^+ + e^-$ and $O \rightarrow O^+ + e^-$), and the gas becomes a partially ionized plasma (Anderson, 1989).

For most chemically reacting gases, it is possible to assume that the intermolecular forces are negligible, and hence each individual species obeys the perfect gas equation of state. In addition, each individual species can be assumed to be thermally perfect. In this case, the gas is a chemically reacting mixture of thermally perfect gases, and this assumption will be employed for the remainder of this section. The equation of state for a mixture of perfect gases can be written as

$$p = \rho \frac{\mathscr{R}}{\mathscr{M}} T \tag{5.39}$$

where $\mathscr{R}$ is the universal gas constant [8314.34 J/(kg mol K)] and $\mathscr{M}$ is the molecular weight of the mixture of gases. The molecular weight of the mixture can be calculated using

$$\mathscr{M} = \left(\sum_{i=1}^{n} \frac{c_i}{\mathscr{M}_i} \right)^{-1}$$

where c_i is the mass fraction of species i and $\mathscr{M}_i$ is the molecular weight of each species.

The species mass fractions in a reacting mixture of gases are determined by solving the species continuity equations, which are given by

$$\frac{\partial \rho_i}{\partial t} + \nabla \cdot [\rho_i(\mathbf{V} + \mathbf{U}_i)] = \dot{\omega}_i \qquad i = 1, 2, \ldots, n \tag{5.40}$$

where ρ_i is the species density, $\mathbf{U}_i$ is the species diffusion velocity, and $\dot{\omega}_i$ is the rate of production of species i due to chemical reactions. The species mass fraction is related to the species density by

$$c_i = \rho_i / \rho$$

If ρ_i is replaced with ρc_i and the global continuity equation, Eq. (5.1), is employed, the species continuity equation can be rewritten as

$$\rho \left(\frac{\partial c_i}{\partial t} + \mathbf{V} \cdot \nabla c_i \right) + \nabla \cdot (\rho_i \mathbf{U}_i) = \dot{\omega}_i \qquad i = 1, 2, \ldots, n \tag{5.41}$$

The mass flux of species $i(\rho_i \mathbf{U}_i)$ due to diffusion can be approximated for most applications using Fick's law:

$$\rho_i \mathbf{U}_i = -\rho \mathscr{D}_{\text{im}} \nabla c_i$$

where $\mathscr{D}_{\text{im}}$ is the multicomponent diffusion coefficient for each species. The multicomponent diffusion coefficient is often replaced with a binary diffusion coefficient $\mathscr{D}$ for mixtures of gases like air. The binary diffusion coefficient is assumed to be the same for all species in the mixture.

The rate of production of each species $\dot{\omega}_i$ is evaluated by using an appropriate chemistry model to simulate the reacting mixture. A chemistry model consists of m reactions, n species, and n_t reactants and can be symbolically represented as

$$\sum_{i=1}^{n_t} v'_{l,i} A_i \rightleftarrows \sum_{i=1}^{n_t} v''_{l,i} A_i \qquad l = 1, 2, \ldots, m$$

where $v'_{l,i}$ and $v''_{l,i}$ are the stoichiometric coefficients and A_i is the chemical symbol of the ith reactant. For example, a widely used chemistry model for air is due to Blottner et al. (1971) and consists of molecular oxygen (O_2), atomic oxygen (O), molecular nitrogen (N_2), nitric oxide (NO), nitric oxide ion (NO^+), atomic nitrogen (N), and electrons (e^-), which are reacting according to the chemical reactions

$$\begin{aligned}
O_2 + M_1 &\rightleftarrows 2O + M_1 \\
N_2 + M_2 &\rightleftarrows 2N + M_2 \\
N_2 + N &\rightleftarrows 2N + N \\
NO + M_3 &\rightleftarrows N + O + M_3 \\
NO + O &\rightleftarrows O_2 + N \\
N_2 + O &\rightleftarrows NO + N \\
N + O &\rightleftarrows NO^+ + e^-
\end{aligned}$$

where M_1, M_2, and M_3 are catalytic third bodies. This chemistry model involves 7 reactions ($m = 7$), 6 species ($n = 6$) excluding electrons, and 10 reactants ($n_t = 10$), which include species, electrons, and catalytic third bodies.

Once the chemistry model is specified, the rate of production of species i can be computed using the Law of Mass Action (Vincenti and Kruger, 1965):

$$\dot{\omega}_i = \mathscr{M}_i \sum_{l=1}^{m} (\upsilon''_{l,i} - \upsilon'_{l,i}) \left[K_{f_l} \prod_{j=1}^{n_t} (\rho\gamma_j)^{\upsilon'_{l,j}} - K_{b_l} \prod_{j=1}^{n_t} (\rho\gamma_j)^{\upsilon''_{l,j}} \right] \tag{5.42}$$

where K_{f_l} and K_{b_l} are the forward and backward reaction rates for the lth reaction and γ_j is the mole-mass ratios of the reactants, defined by

$$\gamma_j = \begin{cases} c_j/\mathscr{M}_j & j = 1, 2, \ldots, n \\ \displaystyle\sum_{i=1}^{n} Z_{j-n,i}\gamma_i & j = n+1, n+2, \ldots, n_t \end{cases}$$

where $Z_{j-n,i}$ are the third-body efficiencies. The forward and backward reaction rates are functions of temperature and can be expressed in the modified Arrhenius form as

$$K_{f_l} = \exp\left(\ln K_1 + \frac{K_2}{T} + K_3 \ln T \right)$$

$$K_{b_l} = \exp\left(\ln K_4 + \frac{K_5}{T} + K_6 \ln T \right)$$

For the present air chemistry model, the constants for each reaction ($K_1, K_2, K_3, K_4, K_5, K_6$) and the third-body efficiencies ($Z_{j-n,i}$) are given by Blottner et al. (1971).

A chemically reacting mixture of gases can be classified as being *frozen*, in *equilibrium*, or in *nonequilibrium*, depending on the reaction rates. If the reaction rates are essentially zero, the mixture is said to be frozen, and the rate of production of species i ($\dot{\omega}_i$) is zero. If the reaction rates approach infinity, the mixture is said to be in *chemical equilibrium.* If the reaction rates are finite, the mixture is in *chemical nonequilibrium,* and Eq. (5.42) can be used to find $\dot{\omega}_i$. At high velocities and rarefield conditions, the mixture of gases may also be in *thermal nonequilibrium.* In this case, the translational, rotational, vibrational, and electronic modes of the thermal energy are not in equilibrium. As a consequence, the modeling of the chemistry will require a multitemperature approach as opposed to the usual single-temperature formulation. For the present discussion, the mixture will be assumed to be in thermal equilibrium. See Anderson (1989), Park (1990), and Vincenti and Kruger (1965) for information on flows in thermal nonequilibrium.

The thermodynamic properties for a reacting mixture in chemical nonequilibrium are functions of both temperature and the mass fractions. For example, the enthalpy and internal energy of the mixture can be expressed as

$$h = h(T, c_1, c_2, \ldots, c_n)$$

$$e = e(T, c_1, c_2, \ldots, c_n)$$

If the mixture is in chemical equilibrium, the thermodynamic properties are a unique function of any two thermodynamic variables, such as temperature and pressure. In this case, h and e can be expressed as

$$h = h(T, p)$$

$$e = e(T, p)$$

Computer programs are available (i.e., Gordon and McBride, 1971) that can be used to compute the composition and thermodynamic properties of equilibrium mixtures of gases. Also available are computer programs that obtain properties by interpolating values from tables of equilibrium data or by using simplified curve fits of the data. Included in the latter approach are the curve fits of Srinivasan et al. (1987) for the thermodynamic properties of equilibrium air. These curve fits include correlations for $p(e, \rho)$, $a(e, \rho)$, $T(e, \rho)$, $s(e, \rho)$, $h(p, \rho)$, $T(p, \rho)$, $\rho(p, s)$, $e(p, s)$, and $a(p, s)$. The curve fits are based on the data from the NASA RGAS (Real GAS) program (Bailey, 1967) and are valid for temperatures up to 25,000 K and densities from 10^{-7} to 10^3 amagats (ρ/ρ_0). In addition, Srinivasan and Tannehill (1987) have developed simplified curve fits for the transport properties of equilibrium air. These curve fits include correlations for $\mu(e, \rho)$ $k(e, \rho)$, $\mu(T, \rho)$, and $\Pr(T, \rho)$. The curve fits are based on the data of Peng and Pindroh (1962) and are valid for temperatures up to 15,000 K and densities from 10^{-5} to 10^3 amagats.

The thermodynamic properties for a reacting mixture in chemical nonequilibrium can be determined once the mass fractions of each species are known. If each species is assumed to be thermally perfect, the species enthalpy and specific heat at constant pressure are given by

$$h_i = C_{1,i} T + h_i^0$$

$$c_{p_i} = C_{2,i}$$

where h_i^0 is the enthalpy of formation for species i. The coefficients $C_{1,i}$ and $C_{2,i}$ are functions of temperature and can be interpolated from the tabulated data of Blottner et al. (1971) or McBride et al. (1963). The mixture enthalpy and frozen specific heat at constant pressure are then given by

$$h = \sum_{i=1}^{n} c_i h_i$$

$$c_{p_f} = \sum_{i=1}^{n} c_i c_{p_i}$$

The transport properties for a chemically reacting mixture can be determined in a similar manner. The viscosity for each species is given by Svehla (1962) in the form of curve fits. Using the viscosity for each species, the species thermal conductivity can be evaluated using Eucken's semi-empirical formula (Prabhu et al., 1987a, 1987b). The mixture viscosity and thermal conductivity can then be determined using Wilke's mixing rule (Wilke, 1950). Further details on these methods can be found in the works by Prabhu et al. (1987a, 1987b) and Buelow et al. (1991).

For chemically reacting flows, it is also necessary to modify the energy equation to include the effect of mass diffusion. This effect is accounted for by adding the following component to the heat flux vector **q**:

$$\rho \sum_{i=1}^{n} c_i h_i \mathbf{U}_i$$

where $\mathbf{U}_i$ is the diffusion velocity for each species.

5.1.6 Vector Form of Equations

Before applying a numerical algorithm to the governing fluid dynamic equations, it is often convenient to combine the equations into a compact vector form. For example, the compressible Navier-Stokes equations in Cartesian coordinates without body forces, mass diffusion, finite-rate chemical reactions, or external heat addition can be written as

$$\frac{\partial \mathbf{U}}{\partial t} + \frac{\partial \mathbf{E}}{\partial x} + \frac{\partial \mathbf{F}}{\partial y} + \frac{\partial \mathbf{G}}{\partial z} = 0 \tag{5.43}$$

where **U**, **E**, **F**, and **G** are vectors given by

$$\mathbf{U} = \begin{bmatrix} \rho \\ \rho u \\ \rho v \\ \rho w \\ E_t \end{bmatrix}$$

$$\mathbf{E} = \begin{bmatrix} \rho u \\ \rho u^2 + p - \tau_{xx} \\ \rho u v - \tau_{xy} \\ \rho u w - \tau_{xz} \\ (E_t + p)u - u\tau_{xx} - v\tau_{xy} - w\tau_{xz} + q_x \end{bmatrix}$$

$$\mathbf{F} = \begin{bmatrix} \rho v \\ \rho u v - \tau_{xy} \\ \rho v^2 + p - \tau_{yy} \\ \rho v w - \tau_{yz} \\ (E_t + p)v - u\tau_{xy} - v\tau_{yy} - w\tau_{yz} + q_y \end{bmatrix} \tag{5.44}$$

$$\mathbf{G} = \begin{bmatrix} \rho w \\ \rho u w - \tau_{xz} \\ \rho v w - \tau_{yz} \\ \rho w^2 + p - \tau_{zz} \\ (E_t + p)w - u\tau_{xz} - v\tau_{yz} - w\tau_{zz} + q_z \end{bmatrix}$$

The first row of the vector Eq. (5.43) corresponds to the continuity equation as given by Eq. (5.4). Likewise, the second, third, and fourth rows are the momentum equations, Eqs. (5.20), while the fifth row is the energy equation, Eq. (5.25). With the Navier-Stokes equations written in this form, it is often easier to code the desired numerical algorithm. Other fluid dynamic equations that are written in conservation-law form can be placed in a similar vector form.

5.1.7 Nondimensional Form of Equations

The governing fluid dynamic equations are often put into nondimensional form. The advantage in doing this is that the characteristic parameters such as Mach number, Reynolds number, and Prandtl number can be varied independently. Also, by nondimensionalizing the equations, the flow variables are "normalized," so that their values fall between certain prescribed limits such as 0 and 1. Many different nondimensionalizing procedures are possible. An example of one such procedure is

$$x^* = \frac{x}{L} \qquad y^* = \frac{y}{L} \qquad z^* = \frac{z}{L} \qquad t^* = \frac{t}{L/V_\infty}$$

$$u^* = \frac{u}{V_\infty} \qquad v^* = \frac{v}{V_\infty} \qquad w^* = \frac{w}{V_\infty} \qquad \mu^* = \frac{\mu}{\mu_\infty}$$

$$\rho^* = \frac{\rho}{\rho_\infty} \qquad p^* = \frac{p}{\rho_\infty V_\infty^2} \qquad T^* = \frac{T}{T_\infty} \qquad e^* = \frac{e}{V_\infty^2}$$

where the nondimensional variables are denoted by an asterisk, free stream conditions are denoted by ∞, and L is the reference length used in the Reynolds number:

$$\mathrm{Re}_L = \frac{\rho_\infty V_\infty L}{\mu_\infty}$$

If this nondimensionalizing procedure is applied to the compressible Navier-Stokes equations given previously by Eqs. (5.43) and (5.44), the following nondimensional equations are obtained:

$$\frac{\partial \mathbf{U}^*}{\partial t^*} + \frac{\partial \mathbf{E}^*}{\partial x^*} + \frac{\partial \mathbf{F}^*}{\partial y^*} + \frac{\partial \mathbf{G}^*}{\partial z^*} = 0 \tag{5.45}$$

where $\mathbf{U}^*$, $\mathbf{E}^*$, $\mathbf{F}^*$, and $\mathbf{G}^*$ are the vectors

$$\mathbf{U}^* = \begin{bmatrix} \rho^* \\ \rho^* u^* \\ \rho^* v^* \\ \rho^* w^* \\ E_t^* \end{bmatrix}$$

$$\mathbf{E}^* = \begin{bmatrix} \rho^* u^* \\ \rho^* u^{*2} + p^* - \tau_{xx}^* \\ \rho^* u^* v^* - \tau_{xy}^* \\ \rho^* u^* w^* - \tau_{xz}^* \\ (E_t^* + p^*)u^* - u^*\tau_{xx}^* - v^*\tau_{xy}^* - w^*\tau_{xz}^* + q_x^* \end{bmatrix}$$

$$\mathbf{F}^* = \begin{bmatrix} \rho^* v^* \\ \rho^* u^* v^* - \tau_{xy}^* \\ \rho^* v^{*2} + p^* - \tau_{yy}^* \\ \rho^* v^* w^* - \tau_{yz}^* \\ (E_t^* + p^*)v^* - u^*\tau_{xy}^* - v^*\tau_{yy}^* - w^*\tau_{yz}^* + q_y^* \end{bmatrix} \tag{5.46}$$

$$\mathbf{G}^* = \begin{bmatrix} \rho^* w^* \\ \rho^* u^* w^* - \tau_{xz}^* \\ \rho^* v^* w^* - \tau_{yz}^* \\ \rho^* w^{*2} + p^* - \tau_{zz}^* \\ (E_t^* + p^*)w^* - u^*\tau_{xz}^* - v^*\tau_{yz}^* - w^*\tau_{zz}^* + q_z^* \end{bmatrix}$$

and

$$E_t^* = \rho^* \left(e^* + \frac{u^{*2} + v^{*2} + w^{*2}}{2} \right)$$

The components of the shear-stress tensor and the heat flux vector in nondimensional form are given by

$$\tau_{xx}^* = \frac{2\mu^*}{3\,\mathrm{Re}_L}\left(2\frac{\partial u^*}{\partial x^*} - \frac{\partial v^*}{\partial y^*} - \frac{\partial w^*}{\partial z^*}\right)$$

$$\tau_{yy}^* = \frac{2\mu^*}{3\,\mathrm{Re}_L}\left(2\frac{\partial v^*}{\partial y^*} - \frac{\partial u^*}{\partial x^*} - \frac{\partial w^*}{\partial z^*}\right)$$

$$
\begin{aligned}
\tau_{zz}^* &= \frac{2\mu^*}{3\,\mathrm{Re}_L}\left(2\frac{\partial w^*}{\partial z^*} - \frac{\partial u^*}{\partial x^*} - \frac{\partial v^*}{\partial y^*}\right) \\
\tau_{xy}^* &= \frac{\mu^*}{\mathrm{Re}_L}\left(\frac{\partial u^*}{\partial y^*} + \frac{\partial v^*}{\partial x^*}\right) \\
\tau_{xz}^* &= \frac{\mu^*}{\mathrm{Re}_L}\left(\frac{\partial u^*}{\partial z^*} + \frac{\partial w^*}{\partial x^*}\right) \\
\tau_{yz}^* &= \frac{\mu^*}{\mathrm{Re}_L}\left(\frac{\partial v^*}{\partial z^*} + \frac{\partial w^*}{\partial y^*}\right) \\
q_x^* &= -\frac{\mu^*}{(\gamma - 1)M_\infty^2\,\mathrm{Re}_L\,\mathrm{Pr}}\frac{\partial T^*}{\partial x^*} \\
q_y^* &= -\frac{\mu^*}{(\gamma - 1)M_\infty^2\,\mathrm{Re}_L\,\mathrm{Pr}}\frac{\partial T^*}{\partial y^*} \\
q_z^* &= -\frac{\mu^*}{(\gamma - 1)M_\infty^2\,\mathrm{Re}_L\,\mathrm{Pr}}\frac{\partial T^*}{\partial z^*}
\end{aligned} \tag{5.47}
$$

where M_∞ is the free stream Mach number,

$$M_\infty = \frac{V_\infty}{\sqrt{\gamma R T_\infty}}$$

and the perfect gas equations of state [Eqs. (5.38)] become

$$p^* = (\gamma - 1)\rho^* e^*$$

$$T^* = \frac{\gamma M_\infty^2 p^*}{\rho^*}$$

Note that the nondimensional forms of the equations given by Eqs. (5.45) and (5.46) are identical (except for the asterisks) to the dimensional forms given by Eqs. (5.43) and (5.44). For convenience, the asterisks can be dropped from the nondimensional equations, and this is usually done.

5.1.8 Orthogonal Curvilinear Coordinates

The basic equations of fluid dynamics are valid for any coordinate system. We have previously expressed these equations in terms of a Cartesian coordinate system. For many applications it is more convenient to use a different orthogonal coordinate system. Let us define x_1, x_2, x_3 to be a set of generalized orthogonal curvilinear coordinates whose origin is at point P and let $\mathbf{i}_1, \mathbf{i}_2, \mathbf{i}_3$ be the corresponding unit vectors (see Fig. 5.2). The rectangular Cartesian coordinates are related to the generalized curvilinear coordinates by

$$
\begin{aligned}
x &= x(x_1, x_2, x_3) \\
y &= y(x_1, x_2, x_3) \\
z &= z(x_1, x_2, x_3)
\end{aligned} \tag{5.48}
$$

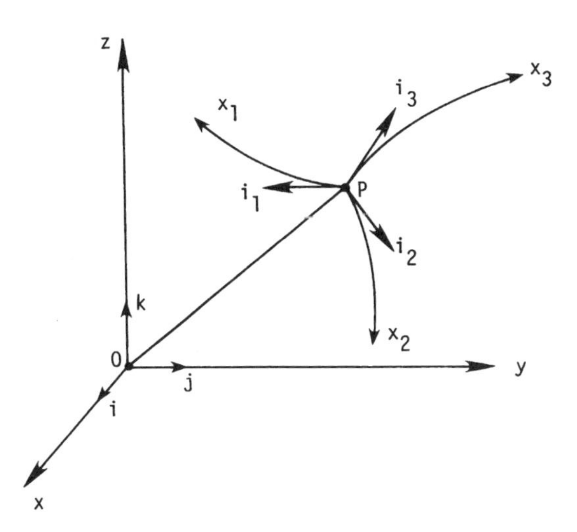

Figure 5.2 Orthogonal curvilinear coordinate system.

so that if the Jacobian

$$\frac{\partial(x, y, z)}{\partial(x_1, x_2, x_3)}$$

is nonzero, then

$$\begin{aligned} x_1 &= x_1(x, y, z) \\ x_2 &= x_2(x, y, z) \\ x_3 &= x_3(x, y, z) \end{aligned} \tag{5.49}$$

The elemental arc length ds in Cartesian coordinates is obtained from

$$(ds)^2 = (dx)^2 + (dy)^2 + (dz)^2 \tag{5.50}$$

If Eq. (5.48) is differentiated and substituted into Eq. (5.50), the following result is obtained:

$$(ds)^2 = (h_1\, dx_1)^2 + (h_2\, dx_2)^2 + (h_3\, dx_3)^2 \tag{5.51}$$

where

$$\begin{aligned} (h_1)^2 &= \left(\frac{\partial x}{\partial x_1}\right)^2 + \left(\frac{\partial y}{\partial x_1}\right)^2 + \left(\frac{\partial z}{\partial x_1}\right)^2 \\ (h_2)^2 &= \left(\frac{\partial x}{\partial x_2}\right)^2 + \left(\frac{\partial y}{\partial x_2}\right)^2 + \left(\frac{\partial z}{\partial x_2}\right)^2 \\ (h_3)^2 &= \left(\frac{\partial x}{\partial x_3}\right)^2 + \left(\frac{\partial y}{\partial x_3}\right)^2 + \left(\frac{\partial z}{\partial x_3}\right)^2 \end{aligned}$$

If ϕ is an arbitrary scalar and $\mathbf{A}$ is an arbitrary vector, the expressions for the gradient, divergence, curl, and Laplacian operator in the generalized curvilinear coordinates become

$$\nabla\phi = \frac{1}{h_1}\frac{\partial\phi}{\partial x_1}\mathbf{i}_1 + \frac{1}{h_2}\frac{\partial\phi}{\partial x_2}\mathbf{i}_2 + \frac{1}{h_3}\frac{\partial\phi}{\partial x_3}\mathbf{i}_3 \tag{5.52}$$

$$\nabla\cdot\mathbf{A} = \frac{1}{h_1h_2h_3}\left[\frac{\partial}{\partial x_1}(h_2h_3A_1) + \frac{\partial}{\partial x_2}(h_3h_1A_2) + \frac{\partial}{\partial x_3}(h_1h_2A_3)\right] \tag{5.53}$$

$$\nabla\times\mathbf{A} = \frac{1}{h_1h_2h_3}\left\{h_1\left[\frac{\partial(h_3A_3)}{\partial x_2} - \frac{\partial(h_2A_2)}{\partial x_3}\right]\mathbf{i}_1 + h_2\left[\frac{\partial(h_1A_1)}{\partial x_3} - \frac{\partial(h_3A_3)}{\partial x_1}\right]\mathbf{i}_2 + h_3\left[\frac{\partial(h_2A_2)}{\partial x_1} - \frac{\partial(h_1A_1)}{\partial x_2}\right]\mathbf{i}_3\right\} \tag{5.54}$$

$$\nabla^2\phi = \frac{1}{h_1h_2h_3}\left[\frac{\partial}{\partial x_1}\left(\frac{h_2h_3}{h_1}\frac{\partial\phi}{\partial x_1}\right) + \frac{\partial}{\partial x_2}\left(\frac{h_3h_1}{h_2}\frac{\partial\phi}{\partial x_2}\right) + \frac{\partial}{\partial x_3}\left(\frac{h_1h_2}{h_3}\frac{\partial\phi}{\partial x_3}\right)\right] \tag{5.55}$$

The expression $\mathbf{V}\cdot\nabla\mathbf{V}$, which is contained in the momentum equation term $D\mathbf{V}/Dt$, can be evaluated as

$$\mathbf{V}\cdot\nabla\mathbf{V} = \left(\frac{u_1}{h_1}\frac{\partial}{\partial x_1} + \frac{u_2}{h_2}\frac{\partial}{\partial x_2} + \frac{u_3}{h_3}\frac{\partial}{\partial x_3}\right)(u_1\mathbf{i}_1 + u_2\mathbf{i}_2 + u_3\mathbf{i}_3)$$

where u_1, u_2, u_3 are the velocity components in the x_1, x_2, x_3 coordinate directions. After taking into account the fact that the unit vectors are functions of the coordinates, the final expanded form becomes

$$\begin{aligned}\mathbf{V}\cdot\nabla\mathbf{V} = &\left(\frac{u_1}{h_1}\frac{\partial u_1}{\partial x_1} + \frac{u_2}{h_2}\frac{\partial u_1}{\partial x_2} + \frac{u_3}{h_3}\frac{\partial u_1}{\partial x_3} + \frac{u_1u_2}{h_1h_2}\frac{\partial h_1}{\partial x_2}\right.\\ &\left.+\frac{u_1u_3}{h_1h_3}\frac{\partial h_1}{\partial x_3} - \frac{u_2^2}{h_1h_2}\frac{\partial h_2}{\partial x_1} - \frac{u_3^2}{h_1h_3}\frac{\partial h_3}{\partial x_1}\right)\mathbf{i}_1\\ &+\left(\frac{u_1}{h_1}\frac{\partial u_2}{\partial x_1} + \frac{u_2}{h_2}\frac{\partial u_2}{\partial x_2} + \frac{u_3}{h_3}\frac{\partial u_2}{\partial x_3} - \frac{u_1^2}{h_1h_2}\frac{\partial h_1}{\partial x_2}\right.\\ &\left.+\frac{u_1u_2}{h_1h_2}\frac{\partial h_2}{\partial x_1} + \frac{u_2u_3}{h_2h_3}\frac{\partial h_2}{\partial x_3} - \frac{u_3^2}{h_2h_3}\frac{\partial h_3}{\partial x_2}\right)\mathbf{i}_2\\ &+\left(\frac{u_1}{h_1}\frac{\partial u_3}{\partial x_1} + \frac{u_2}{h_2}\frac{\partial u_3}{\partial x_2} + \frac{u_3}{h_3}\frac{\partial u_3}{\partial x_3} - \frac{u_1^2}{h_1h_3}\frac{\partial h_1}{\partial x_3}\right.\\ &\left.-\frac{u_2^2}{h_2h_3}\frac{\partial h_2}{\partial x_3} + \frac{u_1u_3}{h_1h_3}\frac{\partial h_3}{\partial x_1} + \frac{u_2u_3}{h_2h_3}\frac{\partial h_3}{\partial x_2}\right)\mathbf{i}_3\end{aligned}$$

The components of the stress tensor given by Eq. (5.15) can be expressed in terms of the generalized curvilinear coordinates as

$$\begin{aligned}
\Pi_{x_1x_1} &= -p + \tfrac{2}{3}\mu(2e_{x_1x_1} - e_{x_2x_2} - e_{x_3x_3}) \\
\Pi_{x_2x_2} &= -p + \tfrac{2}{3}\mu(2e_{x_2x_2} - e_{x_1x_1} - e_{x_3x_3}) \\
\Pi_{x_3x_3} &= -p + \tfrac{2}{3}\mu(2e_{x_3x_3} - e_{x_1x_1} - e_{x_2x_2}) \\
\Pi_{x_2x_3} &= \Pi_{x_3x_2} = \mu e_{x_2x_3} \\
\Pi_{x_1x_3} &= \Pi_{x_3x_1} = \mu e_{x_1x_3} \\
\Pi_{x_1x_2} &= \Pi_{x_2x_1} = \mu e_{x_1x_2}
\end{aligned} \tag{5.56}$$

where the expressions for the strains are

$$\begin{aligned}
e_{x_1x_1} &= \frac{1}{h_1}\frac{\partial u_1}{\partial x_1} + \frac{u_2}{h_1h_2}\frac{\partial h_1}{\partial x_2} + \frac{u_3}{h_1h_3}\frac{\partial h_1}{\partial x_3} \\
e_{x_2x_2} &= \frac{1}{h_2}\frac{\partial u_2}{\partial x_2} + \frac{u_3}{h_2h_3}\frac{\partial h_2}{\partial x_3} + \frac{u_1}{h_1h_2}\frac{\partial h_2}{\partial x_1} \\
e_{x_3x_3} &= \frac{1}{h_3}\frac{\partial u_3}{\partial x_3} + \frac{u_1}{h_1h_3}\frac{\partial h_3}{\partial x_1} + \frac{u_2}{h_2h_3}\frac{\partial h_3}{\partial x_2} \\
e_{x_2x_3} &= \frac{h_3}{h_2}\frac{\partial}{\partial x_2}\left(\frac{u_3}{h_3}\right) + \frac{h_2}{h_3}\frac{\partial}{\partial x_3}\left(\frac{u_2}{h_2}\right) \\
e_{x_1x_3} &= \frac{h_1}{h_3}\frac{\partial}{\partial x_3}\left(\frac{u_1}{h_1}\right) + \frac{h_3}{h_1}\frac{\partial}{\partial x_1}\left(\frac{u_3}{h_3}\right) \\
e_{x_1x_2} &= \frac{h_2}{h_1}\frac{\partial}{\partial x_1}\left(\frac{u_2}{h_2}\right) + \frac{h_1}{h_2}\frac{\partial}{\partial x_2}\left(\frac{u_1}{h_1}\right)
\end{aligned} \tag{5.57}$$

The components of $\mathbf{\nabla} \cdot \mathbf{\Pi}_{ij}$ are

$$x_1: \frac{1}{h_1h_2h_3}\left[\frac{\partial}{\partial x_1}(h_2h_3\Pi_{x_1x_1}) + \frac{\partial}{\partial x_2}(h_1h_3\Pi_{x_1x_2}) + \frac{\partial}{\partial x_3}(h_1h_2\Pi_{x_1x_3})\right]$$
$$+ \Pi_{x_1x_2}\frac{1}{h_1h_2}\frac{\partial h_1}{\partial x_2} + \Pi_{x_1x_3}\frac{1}{h_1h_3}\frac{\partial h_1}{\partial x_3} - \Pi_{x_2x_2}\frac{1}{h_1h_2}\frac{\partial h_2}{\partial x_1} - \Pi_{x_3x_3}\frac{1}{h_1h_3}\frac{\partial h_3}{\partial x_1}$$

$$x_2: \frac{1}{h_1h_2h_3}\left[\frac{\partial}{\partial x_1}(h_2h_3\Pi_{x_1x_2}) + \frac{\partial}{\partial x_2}(h_1h_3\Pi_{x_2x_2}) + \frac{\partial}{\partial x_3}(h_1h_2\Pi_{x_2x_3})\right]$$
$$+ \Pi_{x_2x_3}\frac{1}{h_2h_3}\frac{\partial h_2}{\partial x_3} + \Pi_{x_1x_2}\frac{1}{h_1h_2}\frac{\partial h_2}{\partial x_1} - \Pi_{x_3x_3}\frac{1}{h_2h_3}\frac{\partial h_3}{\partial x_2} - \Pi_{x_1x_1}\frac{1}{h_1h_2}\frac{\partial h_1}{\partial x_2}$$

$$x_3: \frac{1}{h_1 h_2 h_3}\left[\frac{\partial}{\partial x_1}(h_2 h_3 \Pi_{x_1 x_3}) + \frac{\partial}{\partial x_2}(h_1 h_3 \Pi_{x_2 x_3}) + \frac{\partial}{\partial x_3}(h_1 h_2 \Pi_{x_3 x_3})\right]$$

$$+ \Pi_{x_1 x_3}\frac{1}{h_1 h_3}\frac{\partial h_3}{\partial x_1} + \Pi_{x_2 x_3}\frac{1}{h_2 h_3}\frac{\partial h_3}{\partial x_2} - \Pi_{x_1 x_1}\frac{1}{h_1 h_3}\frac{\partial h_1}{\partial x_3} - \Pi_{x_2 x_2}\frac{1}{h_2 h_3}\frac{\partial h_2}{\partial x_3} \tag{5.58}$$

In generalized curvilinear coordinates, the dissipation function becomes

$$\Phi = \mu\Big[2\big(e_{x_1x_1}^2 + e_{x_2x_2}^2 + e_{x_3x_3}^2\big) + e_{x_2x_3}^2 + e_{x_1x_3}^2 + e_{x_1x_2}^2 - \tfrac{2}{3}(e_{x_1x_1} + e_{x_2x_2} + e_{x_3x_3})^2\Big] \tag{5.59}$$

The above formulas can now be used to derive the fluid dynamic equations in any orthogonal curvilinear coordinate system. Examples include

Cartesian coordinates

$$x_1 = x \qquad h_1 = 1 \qquad u_1 = u$$
$$x_2 = y \qquad h_2 = 1 \qquad u_2 = v$$
$$x_3 = z \qquad h_3 = 1 \qquad u_3 = w$$

Cylindrical coordinates

$$x_1 = r \qquad h_1 = 1 \qquad u_1 = u_r$$
$$x_2 = \theta \qquad h_2 = r \qquad u_2 = u_\theta$$
$$x_3 = z \qquad h_3 = 1 \qquad u_3 = u_z$$

Spherical coordinates

$$x_1 = r \qquad h_1 = 1 \qquad u_1 = yu_r$$
$$x_2 = \theta \qquad h_2 = r \qquad u_2 = u_\theta$$
$$x_3 = \phi \qquad h_3 = r \sin\theta \qquad u_3 = u_\phi$$

Two-dimensional (2-D) or axisymmetric body intrinsic coordinates

$$x_1 = \xi \qquad h_1 = 1 + K(\xi)\eta \qquad u_1 = u$$
$$x_2 = \eta \qquad h_2 = 1 \qquad u_2 = v$$
$$x_3 = \phi \qquad h_3 = [r(\xi) + \eta\cos\alpha(\xi)]^m \qquad u_3 = mw$$

where $K(\xi)$ is the local body curvature, $r(\xi)$ is the cylindrical radius, and

$$m = \begin{cases} 0 & \text{for 2-D flow} \\ 1 & \text{for axisymmetric flow} \end{cases}$$

These coordinate systems are illustrated in Fig. 5.3.

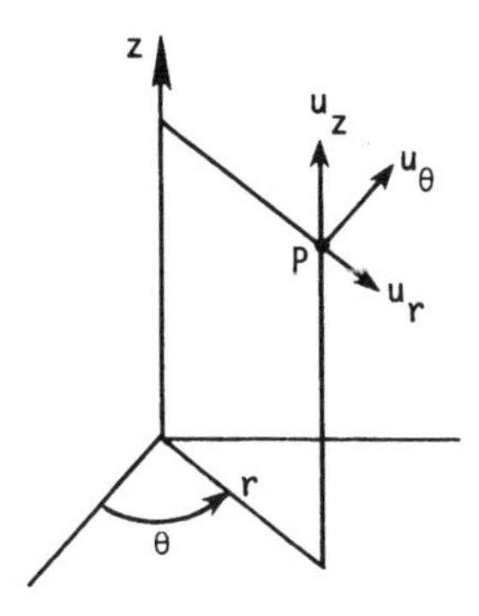

(a) CYLINDRICAL COORDINATES (r, θ, z)

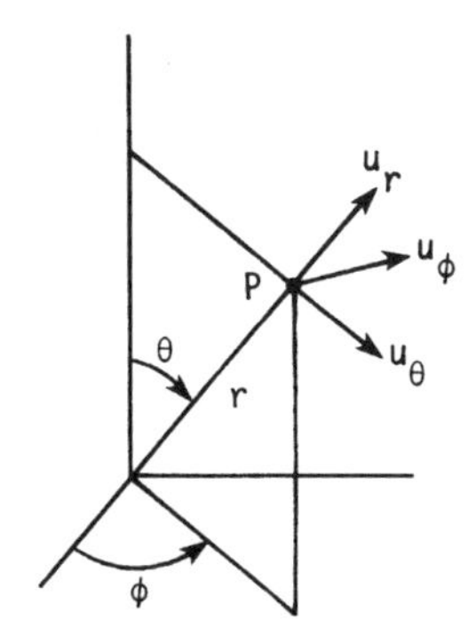

(b) SPHERICAL COORDINATES (r, θ, φ)

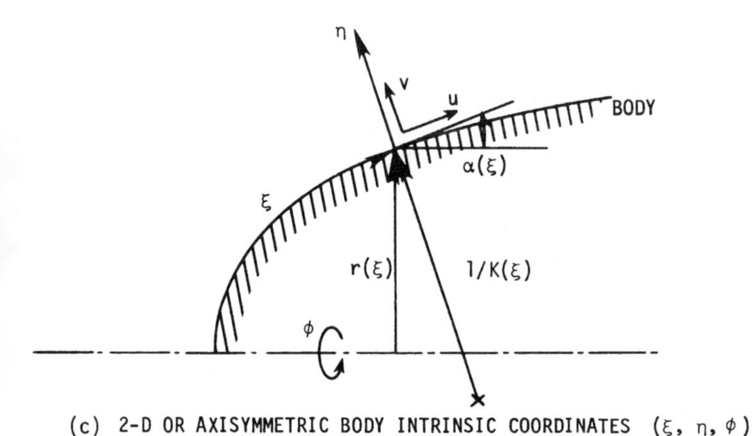

(c) 2-D OR AXISYMMETRIC BODY INTRINSIC COORDINATES (ξ, η, φ)

Figure 5.3 Curvilinear coordinate systems. (a) Cylindrical coordinates (r, θ, z); (b) spherical coordinates (r, θ, ϕ); (c) 2-D or axisymmetric body intrinsic coordinates (ξ, η, ϕ).

5.2 AVERAGED EQUATIONS FOR TURBULENT FLOWS

5.2.1 Background

For more than 60 years it has been recognized that our understanding of turbulent flows is incomplete. A quotation attributed to Sir Horace Lamb in 1932 might still be appropriate: "I am an old man now, and when I die and go to Heaven there are two matters on which I hope for enlightenment. One is quantum electrodynamics and the other is the turbulent motion of fluids. And about the former I am rather optimistic."

According to Hinze (1975), "Turbulent fluid motion is an irregular condition of flow in which the various quantities show a random variation with time and space coordinates so that statistically distinct average values can be discerned."

We are all familiar with some of the differences between laminar and turbulent flows. Usually, higher values of friction drag and pressure drop are associated with turbulent flows. The diffusion rate of a scalar quantity is usually greater in a turbulent flow than in a laminar flow (increased "mixing"), and turbulent flows are usually noisier. A turbulent boundary layer can normally negotiate a more extensive region of unfavorable pressure gradient prior to separation than can a laminar boundary layer. Users of dimpled golf balls are well aware of this.

The unsteady Navier-Stokes equations are generally considered to govern turbulent flows in the continuum regime. If this is the case, then we might wonder why turbulent flows cannot be solved numerically as easily as laminar flows. Perhaps the wind tunnels can be dismantled once and for all. This is indeed a possibility, but it is not likely to happen very soon. To resolve a turbulent flow by *direct numerical simulation* (DNS) requires that all relevant length scales be resolved from the smallest eddies to scales on the order of the physical dimensions of the problem domain. The computation needs to be 3-D even if the time-mean aspects of the flow are 2-D, and the time steps must be small enough that the small-scale motion can be resolved in a time-accurate manner even if the flow is steady in a time-mean sense. Such requirements place great demands on computer resources, to the extent that only relatively simple flows at low Reynolds numbers can be computed directly with present-day machines.

The computations of Kim et al. (1987) provide an example of required resources. They computed a nominally fully developed incompressible channel flow at a Reynolds number based on channel height of about 6000 using grids of 2 and 4 million points. For the finer grid, 250 hours of Cray XMP time were required. For channel flow the number of grid points needed can be estimated from the expression (Wilcox, 1993)

$$N_{\mathrm{DNS}} = (0.088\,\mathrm{Re}_h)^{9/4}$$

where Re_h is the Reynolds number based on the mean channel velocity and channel height. Turbulent wall shear flows that have been successfully simulated directly include planar and square channel flows, flow over a rearward facing step, and flow over a flat plate. Much useful information has been gained from such simulations, since many of the statistical quantities of interest cannot be measured experimentally, but can be evaluated from the simulations. Recently, simulations have been extended to include compressible and transitional flows.

Another promising approach is known as *large-eddy simulation* (LES), in which the large-scale structure of the turbulent flow is computed directly and only the effects of the smallest (subgrid-scale) and more nearly isotropic eddies are modeled. This is accomplished by "filtering" the Navier-Stokes equations to obtain a set of equations that govern the "resolved" flow. This filtering, to be defined below, is a type of space averaging of the flow variables over regions approximately the size of the computational control volume (cell). The computational effort required for LES is less than that of DNS by approximately a factor of 10 using present-day methods. Clearly, with present-day computers, it is not possible to simulate directly or on a large-eddy basis most of the turbulent flows arising in engineering applications. However, the frontier is advancing relentlessly as computer hardware and algorithms improve. With each advance in computer capability, it becomes possible to apply DNS and LES to more and more flows of increasing complexity. Sometime during the twenty-first century it is highly likely that DNS and LES will replace the more approximate modeling methods currently used as the primary design procedure for engineering applications.

The main thrust of present-day research in computational fluid mechanics and heat transfer in turbulent flows is through the time-averaged Navier-Stokes equations. These equations are also referred to as the Reynolds equations of motion or the *Reynolds averaged Navier-Stokes* (RANS) equations. Time averaging the equations of motion gives rise to new terms, which can be interpreted as "apparent" stress gradients and heat flux quantities associated with the turbulent motion. These new quantities must be related to the mean flow variables through turbulence models. This process introduces further assumptions and approximations. Thus this attack on the turbulent flow problem through solving the Reynolds equations of motion does not follow entirely from first principles, since additional assumptions must be made to "close" the system of equations.

The Reynolds equations are derived by decomposing the dependent variables in the conservation equations into time-mean (obtained over an appropriate time interval) and fluctuating components and then time averaging the entire equation. Two types of averaging are presently used, the classical Reynolds averaging and the mass-weighted averaging suggested by Favre (1965). For flows in which density fluctuations can be neglected, the two formulations become identical.

5.2.2 Reynolds Averaged Navier-Stokes Equations

In the conventional averaging procedure, following Reynolds, we define a time-averaged quantity $\bar{f}$ as

$$\bar{f} \equiv \frac{1}{\Delta t} \int_{t_0}^{t_0 + \Delta t} f\, dt \tag{5.60}$$

We require that Δt be large compared to the period of the random fluctuations associated with the turbulence, but small with respect to the time constant for any slow variations in the flow field associated with ordinary unsteady flows. The Δt is sometimes indicated to approach infinity as a limit, but this should be interpreted as being relative to the characteristic fluctuation period of the turbulence. For practical measurements, Δt must be finite.

In the conventional Reynolds decomposition, the randomly changing flow variables are replaced by time averages plus fluctuations (see Fig. 5.4) about the average. For a Cartesian coordinate system, we may write

$$\begin{array}{llll} u = \bar{u} + u' & v = \bar{v} + v' & w = \bar{w} + w' & \rho = \bar{\rho} + \rho' \\ p = \bar{p} + p' & h = \bar{h} + h' & T = \bar{T} + T' & H = \bar{H} + H' \end{array} \tag{5.61}$$

where total enthalpy H is defined by $H = h + u_i u_i/2$. Fluctuations in other fluid properties such as viscosity, thermal conductivity, and specific heat are usually small and will be neglected here.

By definition, the time average of a fluctuating quantity is zero:

$$\overline{f'} = \frac{1}{\Delta t} \int_{t_0}^{t_0 + \Delta t} f' \, dt \equiv 0 \tag{5.62}$$

It should be clear from these definitions that for symbolic flow variables f and g, the following relations hold:

$$\overline{\bar{f} g'} = 0 \qquad \overline{\bar{f} g} = \bar{f} \bar{g} \qquad \overline{f + g} = \bar{f} + \bar{g} \tag{5.63}$$

It should also be clear that, whereas $\overline{f'} \equiv 0$, the time average of the product of two fluctuating quantities is, in general, not equal to zero, i.e., $\overline{f'f'} \neq 0$. In fact, the root mean square of the velocity fluctuations is known as the turbulence intensity.

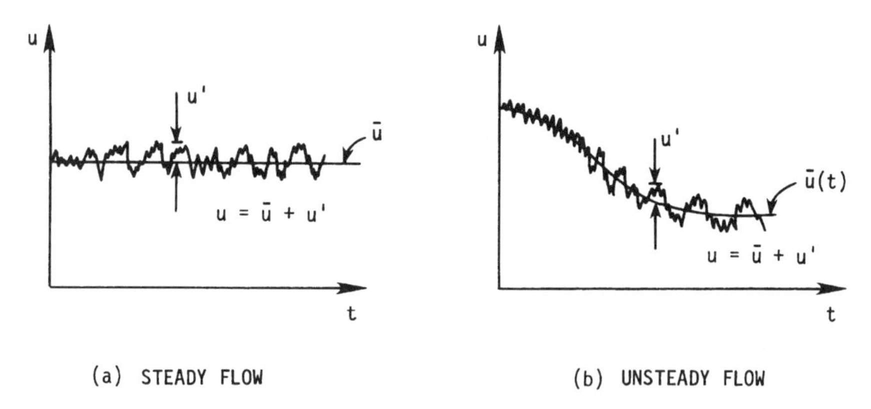

Figure 5.4 Relation between u, $\bar{u}$, and u'. (a) Steady flow. (b) Unsteady flow.

For treatment of compressible flows and mixtures of gases in particular, mass-weighted averaging is convenient. In this approach we define mass-averaged variables according to $\tilde{f} = \overline{\rho f}/\bar{\rho}$. This gives

$$\tilde{u} = \frac{\overline{\rho u}}{\bar{\rho}} \qquad \tilde{v} = \frac{\overline{\rho v}}{\bar{\rho}} \qquad \tilde{w} = \frac{\overline{\rho w}}{\bar{\rho}} \qquad \tilde{h} = \frac{\overline{\rho h}}{\bar{\rho}} \qquad \tilde{T} = \frac{\overline{\rho T}}{\bar{\rho}} \qquad \tilde{H} = \frac{\overline{\rho H}}{\bar{\rho}} \tag{5.64}$$

We note that only the velocity components and thermal variables are mass averaged. Fluid properties such as density and pressure are treated as before.

To substitute into the conservation equations, we define new fluctuating quantities by

$$u = \tilde{u} + u'' \qquad v = \tilde{v} + v'' \qquad w = \tilde{w} + w'' \qquad h = \tilde{h} + h'' \qquad T = \tilde{T} + T''$$
$$H = \tilde{H} + H'' \tag{5.65}$$

It is very important to note that the time averages of the doubly primed fluctuating quantities ($\overline{u''}, \overline{v''}$, etc.) *are not* equal to zero, in general, unless $\rho' = 0$. In fact, it can be shown that $\overline{u''} = -\overline{\rho' u'}/\bar{\rho}$, $\overline{v''} = -\overline{\rho' v'}/\bar{\rho}$, etc. Instead, the time average of the doubly primed fluctuation multiplied by the density is equal to zero:

$$\overline{\rho f''} \equiv 0 \tag{5.66}$$

The above identity can be established by expanding $\overline{\rho f} = \overline{\rho(\tilde{f} + f'')}$ and using the definition of $\tilde{f}$.

5.2.3 Reynolds Form of the Continuity Equation

Starting with the continuity equation in the Cartesian coordinate system as given by Eq. (5.4), we first decompose the variables into the conventional time-averaged variables plus fluctuating components as given by Eqs. (5.61).

The entire equation is then time averaged, yielding in summation notation

$$\frac{\partial \bar{\rho}}{\partial t} + \overbrace{\frac{\partial \overline{\rho'}}{\partial t}}^{0} + \frac{\partial}{\partial x_j}(\overline{\bar{\rho}\bar{u}_j}) + \frac{\partial}{\partial x_j}\overbrace{(\overline{\rho'\bar{u}_j})}^{0} + \frac{\partial}{\partial x_j}\overbrace{(\overline{\bar{\rho}u'_j})}^{0} + \frac{\partial}{\partial x_j}(\overline{\rho' u'_j}) = 0 \tag{5.67}$$

Three of the terms are identically zero as indicated because of the identity given by Eq. (5.62). Finally, the Reynolds form of the continuity equation in conventionally averaged variables can be written

$$\frac{\partial \bar{\rho}}{\partial t} + \frac{\partial}{\partial x_j}(\bar{\rho}\bar{u}_j + \overline{\rho' u'_j}) = 0 \tag{5.68}$$

Substituting the mass-weighted averaged variables plus the doubly primed fluctuations given by Eqs. (5.65) into Eq. (5.4) and time averaging the entire equation gives

$$\frac{\partial \bar{\rho}}{\partial t} + \overset{0}{\cancel{\frac{\partial \overline{\rho'}}{\partial t}}} + \frac{\partial}{\partial x_j}\overline{(\bar{\rho}\tilde{u}_j)} + \frac{\partial}{\partial x_j}\overset{0}{\cancel{\overline{(\rho'\tilde{u}_j)}}} + \frac{\partial}{\partial x_j}\overline{(\bar{\rho}u_j'')} + \frac{\partial}{\partial x_j}\overline{(\rho' u_j'')} = 0 \tag{5.69}$$

Two of the terms in Eq. (5.69) are obviously identically zero as indicated. In addition, the last two terms can be combined, i.e.,

$$\frac{\partial}{\partial x_j}\overline{(\bar{\rho}u_j'')} + \frac{\partial}{\partial x_j}\overline{(\rho' u_j'')} = \frac{\partial}{\partial x_j}\overline{\rho u_j''}$$

which is equal to zero by Eq. (5.66). This permits the continuity equation in mass-weighted variables to be written as

$$\frac{\partial \bar{\rho}}{\partial t} + \frac{\partial}{\partial x_j}(\bar{\rho}\tilde{u}_j) = 0 \tag{5.70}$$

We note that Eq. (5.70) is more compact in form than Eq. (5.68). For incompressible flows, $\rho' = 0$, and the differences between the conventional and mass-weighted variables vanish, so that the continuity equation can be written as

$$\frac{\partial \bar{u}_j}{\partial x_j} = 0 \tag{5.71}$$

5.2.4 Reynolds Form of the Momentum Equations

The development of the Reynolds form of the momentum equations proceeds most easily when we start with the Navier-Stokes momentum equations in divergence or conservation-law form as in Eq. (5.20). Working first with the conventionally averaged variables, we replace the dependent variables in Eq. (5.20) with the time averages plus fluctuations according to Eq. (5.61). As an example, the resulting x component of Eq. (5.20), after neglecting body forces, becomes

$$\begin{aligned}
&\frac{\partial}{\partial t}[(\bar{\rho} + \rho')(\bar{u} + u')] + \frac{\partial}{\partial x}[(\bar{\rho} + \rho')(\bar{u} + u')(\bar{u} + u') + (\bar{p} + p') - \tau_{xx}] \\
&\quad + \frac{\partial}{\partial y}\left[(\bar{\rho} + \rho')(\bar{u} + u')(\bar{v} + v') - \tau_{yx}\right] \\
&\quad + \frac{\partial}{\partial z}[(\bar{\rho} + \rho')(\bar{u} + u')(\bar{w} + w') - \tau_{zx}] = 0
\end{aligned}$$

Next, the entire equation is time averaged. Terms that are linear in fluctuating quantities become zero when time averaged, as they did in the continuity equation. Several terms disappear in this manner, while others can be grouped together and found to be zero through use of the continuity equation. The resulting Reynolds x-momentum equation can be written as

$$\frac{\partial}{\partial t}(\bar{\rho}\bar{u} + \overline{\rho' u'}) + \frac{\partial}{\partial x}(\bar{\rho}\bar{u}\bar{u} + \bar{u}\overline{\rho' u'})$$
$$+ \frac{\partial}{\partial y}(\bar{\rho}\bar{u}\bar{v} + \bar{u}\overline{\rho' v'}) + \frac{\partial}{\partial z}(\bar{\rho}\bar{u}\bar{w} + \bar{u}\overline{\rho' w'})$$
$$= -\frac{\partial \bar{p}}{\partial x} + \frac{\partial}{\partial x}\left[\mu\left(2\frac{\partial \bar{u}}{\partial x} - \frac{2}{3}\frac{\partial \bar{u}_k}{\partial x_k}\right) - \bar{u}\overline{\rho' u'} - \bar{\rho}\overline{u' u'} - \overline{\rho' u' u'}\right]$$
$$+ \frac{\partial}{\partial y}\left[\mu\left(\frac{\partial \bar{u}}{\partial y} + \frac{\partial \bar{v}}{\partial x}\right) - \bar{v}\overline{\rho' u'} - \bar{\rho}\overline{u' v'} - \overline{\rho' u' v'}\right]$$
$$+ \frac{\partial}{\partial z}\left[\mu\left(\frac{\partial \bar{u}}{\partial z} + \frac{\partial \bar{w}}{\partial x}\right) - \bar{w}\overline{\rho' u'} - \bar{\rho}\overline{u' w'} - \overline{\rho' u' w'}\right] \tag{5.72}$$

The complete Reynolds momentum equations (all three components) can be written

$$\frac{\partial}{\partial t}(\bar{\rho}\bar{u}_i + \overline{\rho' u_i'}) + \frac{\partial}{\partial x_j}(\bar{\rho}\bar{u}_i\bar{u}_j + \bar{u}_i\overline{\rho' u_j'})$$
$$= -\frac{\partial \bar{p}}{\partial x_i} + \frac{\partial}{\partial x_j}\left(\bar{\tau}_{ij} - \bar{u}_j\overline{\rho' u_i'} - \bar{\rho}\overline{u_i' u_j'} - \overline{\rho' u_i' u_j'}\right) \tag{5.73}$$

where

$$\bar{\tau}_{ij} = \mu\left[\left(\frac{\partial \bar{u}_i}{\partial x_j} + \frac{\partial \bar{u}_j}{\partial x_i}\right) - \frac{2}{3}\delta_{ij}\frac{\partial \bar{u}_k}{\partial x_k}\right] \tag{5.74}$$

To develop the Reynolds momentum equation in mass-weighted variables, we again start with Eq. (5.20) but use the decomposition indicated by Eq. (5.65) to represent the instantaneous variables. As an example, the resulting x component of Eq. (5.20) becomes

$$\frac{\partial}{\partial t}[(\bar{\rho} + \rho')(\tilde{u} + u'')] + \frac{\partial}{\partial x}[(\bar{\rho} + \rho')(\tilde{u} + u'')(\tilde{u} + u'') + (\bar{p} + p') - \tau_{xx}]$$
$$+ \frac{\partial}{\partial y}[(\bar{\rho} + \rho')(\tilde{u} + u'')(\tilde{v} + v'') - \tau_{yx}]$$
$$+ \frac{\partial}{\partial z}[(\bar{\rho} + \rho')(\tilde{u} + u'')(\tilde{w} + w'') - \tau_{zx}] = 0 \tag{5.75}$$

Next, the entire equation is time averaged, and the identity given by Eq. (5.66) is used to eliminate terms. The complete Reynolds momentum equation in mass-weighted variables becomes

$$\frac{\partial}{\partial t}(\bar{\rho}\tilde{u}_i) + \frac{\partial}{\partial x_j}(\bar{\rho}\tilde{u}_i\tilde{u}_j) = -\frac{\partial \bar{p}}{\partial x_i} + \frac{\partial}{\partial x_j}(\bar{\tau}_{ij} - \overline{\rho u_i'' u_j''}) \tag{5.76}$$

where, neglecting viscosity fluctuations, $\bar{\tau}_{ij}$ becomes

$$\bar{\tau}_{ij} = \mu\left[\left(\frac{\partial \tilde{u}_i}{\partial x_j} + \frac{\partial \tilde{u}_j}{\partial x_i}\right) - \frac{2}{3}\delta_{ij}\frac{\partial \tilde{u}_k}{\partial x_k}\right] + \mu\left[\left(\frac{\partial \overline{u_i''}}{\partial x_j} + \frac{\partial \overline{u_j''}}{\partial x_i}\right) - \frac{2}{3}\delta_{ij}\frac{\partial \overline{u_k''}}{\partial x_k}\right] \tag{5.77}$$

The momentum equation, Eq. (5.76), in mass-weighted variables is simpler in form than the corresponding equation using conventional variables. We note, however, that even when viscosity fluctuations are neglected, $\bar{\tau}_{ij}$ is more complex in Eq. (5.77) than the $\bar{\tau}_{ij}$ that appeared in the conventionally averaged equation [Eq. (5.74)]. In practice, the viscous terms involving the doubly primed fluctuations are expected to be small and are likely candidates for being neglected on the basis of order of magnitude arguments.

For incompressible flows the momentum equation can be written in the simpler form

$$\frac{\partial}{\partial t}(\rho\bar{u}_i) + \frac{\partial}{\partial x_j}(\rho\bar{u}_i\bar{u}_j) = -\frac{\partial \bar{p}}{\partial x_i} + \frac{\partial}{\partial x_j}(\bar{\tau}_{ij} - \rho\overline{u_i' u_j'}) \tag{5.78}$$

where $\bar{\tau}_{ij}$ takes on the reduced form

$$\bar{\tau}_{ij} = \mu\left(\frac{\partial \bar{u}_i}{\partial x_j} + \frac{\partial \bar{u}_j}{\partial x_i}\right) \tag{5.79}$$

As we noted in connection with the continuity equation, there is no difference between the mass-weighted and conventional variables for incompressible flow.

5.2.5 Reynolds Form of the Energy Equation

The thermal variables H, h, and T are all related, and the energy equation takes on different forms, depending upon which one is chosen to be the transported thermal variable. To develop one common form, we start with the energy equation as given by Eq. (5.22). The generation term, $\partial Q/\partial t$, will be neglected. Assuming that the total energy is composed only of internal energy and kinetic energy, and replacing E_t by $\rho H - p$, we can write Eq. (5.22) in summation notation as

$$\frac{\partial}{\partial t}(\rho H) + \frac{\partial}{\partial x_j}(\rho u_j H + q_j - u_i\tau_{ij}) = \frac{\partial p}{\partial t} \tag{5.80}$$

To obtain the Reynolds energy equation in conventionally averaged variables, we replace the dependent variables in Eq. (5.80) with the decomposition indicated by Eq. (5.61). After time averaging, the equation becomes

$$\frac{\partial}{\partial t}(\bar{\rho}\bar{H} + \overline{\rho' H'}) + \frac{\partial}{\partial x_j}\left(\bar{\rho}\bar{u}_j\bar{H} + \bar{\rho}\overline{u'_j H'} + \overline{\rho' u'_j}\bar{H} + \overline{\rho' u'_j H'} + \bar{u}_j\overline{\rho' H'} - k\frac{\partial \bar{T}}{\partial x_j}\right)$$
$$= \frac{\partial \bar{p}}{\partial t} + \frac{\partial}{\partial x_j}\left[\bar{u}_i\left(\frac{2}{3}\mu\delta_{ij}\frac{\partial \bar{u}_k}{\partial x_k}\right) + \mu\bar{u}_i\left(\frac{\partial \bar{u}_j}{\partial x_i} + \frac{\partial \bar{u}_i}{\partial x_j}\right)\right.$$
$$\left. - \frac{2}{3}\mu\delta_{ij}\overline{u'_i\frac{\partial u'_k}{\partial x_k}} + \mu\left(\overline{u'_i\frac{\partial u'_j}{\partial x_i}} + \overline{u'_i\frac{\partial u'_i}{\partial x_j}}\right)\right] \tag{5.81}$$

It is frequently desirable to utilize static temperature as a dependent variable in the energy equation. We will let $h = c_p T$ and write Eq. (5.33) in conservative form to provide a convenient starting point for the development of the Reynolds averaged form:

$$\frac{\partial}{\partial t}(\rho c_p T) + \frac{\partial}{\partial x_j}\left(\rho c_p u_j T - k\frac{\partial T}{\partial x_j}\right) = \frac{\partial p}{\partial t} + u_j\frac{\partial p}{\partial x_j} + \Phi \tag{5.82}$$

The dissipation function Φ [see Eq. (5.34)] can be written in terms of the velocity components using summation convention as

$$\Phi = \tau_{ij}\frac{\partial u_i}{\partial x_j} = \mu\left[-\frac{2}{3}\left(\frac{\partial u_k}{\partial x_k}\right)^2 + \frac{1}{2}\left(\frac{\partial u_j}{\partial x_i} + \frac{\partial u_i}{\partial x_j}\right)^2\right] \tag{5.83}$$

The variables in Eq. (5.83) are then replaced with the decomposition indicated by Eq. (5.61), and the resulting equation is time averaged. After eliminating terms known to be zero, the Reynolds energy equation in terms of temperature becomes

$$\frac{\partial}{\partial t}\left(c_p\bar{\rho}\bar{T} + c_p\overline{\rho' T'}\right) + \frac{\partial}{\partial x_j}\left(\bar{\rho}c_p\bar{T}\bar{u}_j + c_p\bar{T}\overline{\rho' u'_j}\right)$$
$$= \frac{\partial \bar{p}}{\partial t} + \bar{u}_j\frac{\partial \bar{p}}{\partial x_j} + \overline{u'_j\frac{\partial p'}{\partial x_j}}$$
$$+ \frac{\partial}{\partial x_j}\left(k\frac{\partial \bar{T}}{\partial x_j} - \bar{\rho}c_p\overline{T' u'_j} - c_p\overline{\rho' T' u'_j} - \bar{u}_j c_p\overline{\rho' T'}\right) + \bar{\Phi} \tag{5.84}$$

where

$$\bar{\Phi} = \overline{\tau_{ij}\frac{\partial u_i}{\partial x_j}} = \bar{\tau}_{ij}\frac{\partial \bar{u}_i}{\partial x_j} + \overline{\tau'_{ij}\frac{\partial u'_i}{\partial x_j}} \tag{5.85}$$

The $\bar{\tau}_{ij}$ in Eq. (5.85) should be evaluated as indicated by Eq. (5.74).

To develop the Reynolds form of the energy equation in mass-weighted variables, we replace the dependent variables in Eq. (5.80) with the decomposition indicated by Eq. (5.65) and time average the entire equation. The result can be written

$$\frac{\partial}{\partial t}(\bar{\rho}\tilde{H}) + \frac{\partial}{\partial x_j}\left(\bar{\rho}\tilde{u}_j\tilde{H} + \overline{\rho u_j'' H''} - k\frac{\partial \bar{T}}{\partial x_j}\right) = \frac{\partial \bar{p}}{\partial t} + \frac{\partial}{\partial x_j}\left(\tilde{u}_i\bar{\tau}_{ij} + \overline{u_i''\tau_{ij}}\right) \tag{5.86}$$

where $\bar{\tau}_{ij}$ can be evaluated as given by Eq. (5.77) in terms of mass-weighted variables.

In terms of static temperature, the Reynolds energy equation in mass-weighted variables becomes

$$\frac{\partial}{\partial t}(\bar{\rho}c_p\tilde{T}) + \frac{\partial}{\partial x_j}(\bar{\rho}c_p\tilde{T}\tilde{u}_j) = \frac{\partial \bar{p}}{\partial t} + \tilde{u}_j\frac{\partial \bar{p}}{\partial x_j} + \overline{u_j''\frac{\partial p}{\partial x_j}}$$
$$+ \frac{\partial}{\partial x_j}\left(k\frac{\partial \tilde{T}}{\partial x_j} + k\frac{\partial \overline{T''}}{\partial x_j} - c_p\overline{\rho T'' u_j''}\right) + \overline{\Phi} \tag{5.87}$$

where

$$\overline{\Phi} = \overline{\tau_{ij}\frac{\partial u_i}{\partial x_j}} = \bar{\tau}_{ij}\frac{\partial \tilde{u}_i}{\partial x_j} + \overline{\tau_{ij}\frac{\partial u_i''}{\partial x_j}} \tag{5.88}$$

For incompressible flows the energy equation can be written in terms of total enthalpy as

$$\frac{\partial \rho\overline{H}}{\partial t} + \frac{\partial}{\partial x_j}\left(\rho u_j\overline{H} + \rho\overline{u_j' H'} - k\frac{\partial \overline{T}}{\partial x_j}\right)$$
$$= \frac{\partial \bar{p}}{\partial t} + \frac{\partial}{\partial x_j}\left[\mu\bar{u}_i\left(\frac{\partial \bar{u}_j}{\partial x_i} + \frac{\partial \bar{u}_i}{\partial x_j}\right) + \mu\left(\overline{u_i'\frac{\partial u_j'}{\partial x_i}} + \overline{u_i'\frac{\partial u_i'}{\partial x_j}}\right)\right] \tag{5.89}$$

and in terms of static temperature as

$$\frac{\partial}{\partial t}(\rho c_p\overline{T}) + \frac{\partial}{\partial x_j}(\rho c_p\overline{T}\bar{u}_j) = \frac{\partial \bar{p}}{\partial t} + \bar{u}_j\frac{\partial \bar{p}}{\partial x_j} + \overline{u_j'\frac{\partial p'}{\partial x_j}}$$
$$+ \frac{\partial}{\partial x_j}\left(k\frac{\partial \overline{T}}{\partial x_j} - \rho c_p\overline{T' u_j'}\right) + \overline{\Phi} \tag{5.90}$$

where $\overline{\Phi}$ is reduced slightly in complexity owing to the vanishing of the volumetric dilatation term in $\bar{\tau}_{ij}$ for incompressible flow.

5.2.6 Comments on the Reynolds Equations

At first glance, the Reynolds equations are likely to appear quite complex, and we are tempted to question whether or not we have made any progress toward solving practical problems in turbulent flow. Certainly, a major problem in fluid mechanics is that more equations can be written than can be solved. Fortunately, for many important flows, the Reynolds equations can be simplified. Before we turn to the task of simplifying the equations, let us examine the Reynolds equations further.

We will consider an incompressible turbulent flow first and interpret the Reynolds momentum equation in the form of Eq. (5.78). The equation governs the time-mean motion of the fluid, and we recognize some familiar momentum flux and laminar-like stress terms plus some new terms involving fluctuations that must represent apparent turbulent stresses. These apparent turbulent stresses originated in the momentum flux terms of the Navier-Stokes equations. To put this another way, the equations of mean motion relate the particle acceleration to stress gradients, and since we know how acceleration for the time-mean motion is expressed, anything new in these equations must be apparent stress gradients due to the turbulent motion. To illustrate, we will utilize the continuity equation to arrange Eq. (5.78) in a form in which the particle (substantial) derivative appears on the left-hand side,

$$\underbrace{\rho \frac{D\bar{u}_i}{Dt}}_{\substack{\text{Particle}\\ \text{acceleration}\\ \text{of mean motion}}} = \underbrace{-\frac{\partial \bar{p}}{\partial x_i}}_{\substack{\text{Mean pressure}\\ \text{gradient}}} + \underbrace{\frac{\partial (\bar{\tau}_{ij})_{\text{lam}}}{\partial x_j}}_{\substack{\text{Laminar-like}\\ \text{stress gradients}\\ \text{for the mean motion}}} + \underbrace{\frac{\partial (\bar{\tau}_{ij})_{\text{turb}}}{\partial x_j}}_{\substack{\text{Apparent stress}\\ \text{gradients due to}\\ \text{transport of}\\ \text{momentum by}\\ \text{turbulent fluctuations}}} \tag{5.91}$$

where $(\bar{\tau}_{ij})_{\text{lam}}$ is the same as Eq. (5.79) and has the same form in terms of the time-mean velocities as the stress tensor for a laminar incompressible flow. The apparent turbulent stresses can be written as

$$(\bar{\tau}_{ij})_{\text{turb}} = -\rho\overline{u_i' u_j'} \tag{5.92}$$

These apparent stresses are commonly called the Reynolds stresses.

For compressible turbulent flow, labeling the terms according to the acceleration of the mean motion and apparent stresses becomes more of a challenge. Using conventional averaging procedures, the presence of terms like $\overline{\rho' u_i'}$ can result in the flux of momentum across mean flow streamlines, frustrating our attempts to categorize terms. The use of mass-weighted averaging eliminates the $\overline{\rho' u_i'}$ terms and provides a compact expression for the particle acceleration but complicates the separation of stresses into purely laminar-like and apparent turbulent categories. When conventionally averaged variables are used, the fluctuating components of $\bar{\tau}_{ij}$ vanish when the equations are time averaged. They do not vanish, however, when mass-weighted averaging is used. To illustrate, we will arrange Eq. (5.76) (using the continuity equation) in a form that utilizes the substantial derivative and label the terms as follows:

$$\underbrace{\bar{\rho}\frac{D\tilde{u}_i}{Dt}}_{\substack{\text{Particle}\\\text{acceleration}\\\text{of mean motion}}} = \underbrace{-\frac{\partial \bar{p}}{\partial x_i}}_{\substack{\text{Mean}\\\text{pressure}\\\text{gradient}}} + \underbrace{\frac{\partial(\bar{\tau}_{ij})_{\text{lam}}}{\partial x_j}}_{\substack{\text{Laminar-like}\\\text{stress gradients}\\\text{for the mean}\\\text{motion}}} + \underbrace{\frac{\partial(\bar{\tau}_{ij})_{\text{turb}}}{\partial x_j}}_{\substack{\text{Apparent stress gradients}\\\text{due to transport of}\\\text{momentum by turbulent}\\\text{fluctuations and deformations}\\\text{attributed to fluctuations}}} \tag{5.93a}$$

The form of Eq. (5.93*a*) is identical to that of Eq. (5.91) except that $\tilde{u}_i$ replaces the $\bar{u}_i$ used in Eq. (5.91). If we insist that $(\bar{\tau}_{ij})_{\text{lam}}$ have the same *form* as for a laminar flow, then the second half of the $\bar{\tau}_{ij}$ of Eq. (5.77) should be attributed to turbulent transport, resulting in

$$(\bar{\tau}_{ij})_{\text{lam}} = \mu\left[\left(\frac{\partial \tilde{u}_i}{\partial x_j} + \frac{\partial \tilde{u}_j}{\partial x_i}\right) - \frac{2}{3}\delta_{ij}\frac{\partial \tilde{u}_k}{\partial x_k}\right] \tag{5.93b}$$

and

$$(\bar{\tau}_{ij})_{\text{turb}} = -\overline{\rho u_i'' u_j''} + \mu\left[\left(\frac{\partial \overline{u_i''}}{\partial x_j} + \frac{\partial \overline{u_j''}}{\partial x_i}\right) - \frac{2}{3}\delta_{ij}\frac{\partial \overline{u_k''}}{\partial x_k}\right] \tag{5.93c}$$

As before, viscosity fluctuations have been neglected in obtaining Eq. (5.93*a*). The second term in the expression for $(\bar{\tau}_{ij})_{\text{turb}}$ involving the molecular viscosity is expected to be much smaller than the $-\overline{\rho u_i'' u_j''}$ component.

We can perform a similar analysis on the Reynolds form of the energy equation and identify certain terms involving temperature or enthalpy fluctuations as apparent heat flux quantities. For example, in Eq. (5.84) the molecular "laminar-like" heat flux term is

$$-(\nabla \cdot \mathbf{q})_{\text{lam}} = \frac{\partial}{\partial x_j}\left(k\frac{\partial \bar{T}}{\partial x_j}\right) \tag{5.94a}$$

and the apparent turbulent (Reynolds) heat flux component is

$$-(\nabla \cdot \mathbf{q})_{\text{turb}} = \frac{\partial}{\partial x_j}\left(-\bar{\rho}c_p\overline{T'u_j'} - c_p\overline{\rho'T'u_j'} - \bar{u}_j c_p\overline{\rho'T'}\right) \tag{5.94b}$$

Further examples illustrating the form of the Reynolds stress and heat flux terms will be given in subsequent sections that consider reduced forms of the Reynolds equations.

The Reynolds equations cannot be solved in the form given because the new apparent turbulent stresses and heat flux quantities must be viewed as new unknowns. To proceed further, we need to find additional equations involving the new unknowns or make assumptions regarding the relation between the new apparent turbulent quantities and the time-mean flow variables. This is known as the closure problem, which is most commonly handled through *turbulence modeling*, which is discussed in Section 5.4.

5.2.7 Filtered Navier-Stokes Equations for Large-Eddy Simulation

At the present time, DNS and LES require such enormous computer resources that it is not feasible to use them for design calculations. However, during the professional careers of the present generation of students, it is very likely that the use of LES as a design tool will become as commonplace as the more refined closure schemes used today for the Reynolds averaged equations. In light of these anticipated advances, it seems appropriate to describe the LES approach briefly in this section. Because the goal is to only introduce the LES approach, the incompressible equations will be considered for the sake of brevity.

The methodology for LES parallels that employed for computing turbulent flows through closure of the Reynolds averaged equations in many respects. First, a set of equations is derived from the Navier-Stokes equations by performing a type of averaging. For LES a spatial average is employed instead of the temporal average used in deriving the Reynolds equations. The averaged equations contain stress terms that must be evaluated through modeling to achieve closure. The equations are then solved numerically. The basic difference between the RANS and LES approaches arises in the choice of quantities to be resolved.

The equations solved in LES are formally developed by "filtering" the Navier-Stokes equations to remove the small spatial scales. The resulting equations describe the evolution of the large eddies and contain the subgrid-scale stress tensor that represents the effects of the unresolved small scales.

Following Leonard (1974), flow variables are decomposed into large (filtered, resolved) and subgrid (residual) scales, as follows:

$$u_i = \bar{u}_i + u'_i \tag{5.94c}$$

Note that the bar and prime have a different meaning here than when previously used in connection with the Reynolds equations. The filtered variable is defined in the general case by the convolution integral

$$\bar{u}_i(x_1, x_2, x_3) = \iiint_D \left[\prod_{j=1}^{3} G_j(x_j, x'_j) \right] u_i(x'_1, x'_2, x'_3)\, dx'_1\, dx'_2\, dx'_3 \tag{5.94d}$$

over the entire flow domain, where x_i and x'_i are position vectors and G is the general filter function. To return the correct value when u is constant, G is normalized by requiring that

$$\iiint_D \left[\prod_{j=1}^{3} G_j(x_j, x'_j) \right] dx'_1\, dx'_2\, dx'_3 = 1$$

Although several filter functions, G, have been employed (Aldama, 1990), the volume-averaged "box" (also known as "top-hat") filter is most frequently used with finite-difference and finite-volume methods. The box filter is given by

$$G_j(x_j - x_j') = \begin{cases} 1/\Delta_j & |x_j - x_j'| \leqslant \Delta_j/2 \\ 0 & \text{otherwise} \end{cases} \tag{5.94e}$$

This gives

$$\bar{u}_i(\mathbf{x}, t) = \frac{1}{\Delta^3} \int_{x_1 - \Delta x_1/2}^{x_1 + \Delta x_1/2} \int_{x_2 - \Delta x_2/2}^{x_2 + \Delta x_2/2} \times \int_{x_3 - \Delta x_3/2}^{x_3 + \Delta x_3/2} u(x_1 - x_1', x_2 - x_2', x_3 - x_3')\, dx_1'\, dx_2'\, dx_3' \tag{5.94f}$$

where $\Delta = (\Delta_1\, \Delta_2\, \Delta_3)^{1/3}$ and $\Delta_1, \Delta_2, \Delta_3$ are increments in x_1, x_2, x_3, respectively. Filtering the Navier-Stokes continuity and momentum equations gives

$$\frac{\partial \bar{u}_j}{\partial x_j} = 0 \tag{5.94g}$$

$$\frac{\partial \bar{u}_i}{\partial t} + \frac{\partial \overline{u_i u_j}}{\partial x_j} = -\frac{\partial \bar{p}}{\partial x_i} + \nu \frac{\partial^2 \bar{u}_i}{\partial x_k\, \partial x_k} \tag{5.94h}$$

However, we cannot solve the system for both $\bar{u}_i$ and $\overline{u_i u_j}$, so we represent the convective flux in terms of decomposed variables, as follows:

$$\overline{u_i u_j} = \overline{\bar{u}_i \bar{u}_j} + \tau_{ij}$$

resulting in

$$\frac{\partial \bar{u}_i}{\partial t} + \frac{\partial \overline{\bar{u}_i \bar{u}_j}}{\partial x_j} = -\frac{\partial \bar{p}}{\partial x_i} + \nu \frac{\partial^2 \bar{u}_i}{\partial x_k\, \partial x_k} - \frac{\partial \tau_{ij}}{\partial x_j} \tag{5.94i}$$

where τ_{ij} is the subgrid-scale stress tensor,

$$\tau_{ij} = \left(\overline{\bar{u}_i \bar{u}_j} - \bar{u}_i \bar{u}_j\right) + \left(\overline{u_i' \bar{u}_j} + \overline{\bar{u}_i u_j'}\right) + \overline{(u_i' u_j')} \tag{5.94j}$$

The first term in parentheses on the right-hand side is known as the Leonard stress, the second term, the cross-term stress, and the third term, the Reynolds stress. Note that if time averaging were being employed instead of filtering, the first two terms would be zero, leaving only the Reynolds stress. Thus an important difference between time averaging and filtering is that for filtering, $\bar{\bar{u}}_i \neq \bar{u}_i$. That is, a second averaging yields a different result from the first averaging.

Although the Leonard term can be computed from the resolved flow, it is not easily done with most finite-difference and finite-volume schemes. Furthermore, the Leonard stresses can be shown to be (Shaanan et al., 1975) of the same order as the truncation error for second-order schemes. As a result, most present-day finite-difference and finite-volume approaches consider that the subgrid-scale model accounts for the important effects of all three terms in Eq. (5.94*i*). Another point in defense of this approach is that the sum of the cross-term stress and the Reynolds stress is not Galilean invariant (Speziale, 1985), whereas the sum of all three terms is.

Solving the filtered Navier-Stokes equations gives the time-dependent solution for the resolved variables. In applications that operate nominally at steady state, we are usually interested in the time-mean motion of the flow and the drag, lift, etc., in a steady sense. Thus the solution to the filtered equations must be averaged in time (and/or averaged spatially in directions in which the flow is assumed to be homogeneous) to obtain the time-averaged values of the variables. If it is desired to compute turbulence statistics to compare with experimental measurements, the fluctuations in the time-averaged sense are computed from the difference between the resolved flow and the time-averaged results, as for example, $\bar{u}_i'' = \bar{u}_i - \langle \bar{u}_i \rangle$, where the double prime designates a time-basis fluctuation and the angle brackets indicate a time- or ensemble-averaged quantity. Of course, the solution to the filtered equations resolves the motion of the large eddies and is not quite the same as the true instantaneous solution to the full Navier-Stokes equations. Thus we cannot expect perfect agreement when turbulence statistics computed from LES are compared with DNS results or experimental data.

Modeling for the subgrid-scale stress is addressed in Section 5.4.8. Information on the extension of LES to compressible flows can be found in the works of Erlebacher et al. (1992) and Moin et al. (1991).

5.3 BOUNDARY-LAYER EQUATIONS

5.3.1 Background

The concept of a boundary layer originated with Ludwig Prandtl in 1904 (Prandtl, 1926). Prandtl reasoned from experimental evidence that for sufficiently large Reynolds numbers a thin region existed near a solid boundary where viscous effects were at least as important as inertia effects no matter how small the viscosity of the fluid might be. Prandtl deduced that a much reduced form of the governing equations could be used by systematically employing two constraints. These were that the viscous layer must be thin relative to the characteristic streamwise dimension of the object immersed in the flow, $\delta/L \ll 1$, and that the largest viscous term must be of the same approximate magnitude as any inertia (particle acceleration) term. Prandtl used what we now call an order of magnitude analysis to reduce the governing equations. Essentially, his conclusions were that second derivatives of the velocity components in the streamwise direction were negligible compared to corresponding derivatives transverse to the main flow direction and that the entire momentum equation for the transverse direction could be neglected.

In the years since 1904, we have found that a similar reduction can often be made in the governing equations for other flows for which a primary flow direction can be identified. These flows include jets, wakes, mixing layers, and the developing flow in pipes and other internal passages. Thus the terminology "boundary-layer flow" or "boundary-layer approximation" has taken on a more general meaning, which refers to circumstances that permit the neglect of the transverse momentum equation and the streamwise second-derivative term in the remaining momentum equation (or equations in the case of 3-D flow). It is increasingly common to refer to these reduced equations as the "thin-shear-layer" equations. This terminology seems especially appropriate in light of the applicability of the equations to free-shear flows such as jets and wakes as well as flows along a solid boundary. We will use both designations, boundary layer and thin-shear layer, interchangeably in this book.

5.3.2 Boundary-Layer Approximation for Steady Incompressible Flow

It is useful to review the methodology used to obtain the boundary-layer approximations to the Navier-Stokes and Reynolds equations for steady 2-D incompressible constant-property flow along an isothermal surface at temperature T_w. First, we define the nondimensional variables (much as was done in Section 5.1.7):

$$u^* = \frac{u}{u_\infty} \qquad v^* = \frac{v}{u_\infty} \qquad x^* = \frac{x}{L} \qquad y^* = \frac{y}{L} \qquad p^* = \frac{p}{\rho u_\infty^2} \qquad \theta = \frac{T - T_\infty}{T_w - T_\infty} \tag{5.95}$$

and introduce them into the Navier-Stokes equations by substitution. After rearrangement, the results can be written as

continuity:

$$\frac{\partial u^*}{\partial x^*} + \frac{\partial v^*}{\partial y^*} = 0 \tag{5.96}$$

x momentum:

$$u^* \frac{\partial u^*}{\partial x^*} + v^* \frac{\partial u^*}{\partial y^*} = -\frac{\partial p^*}{\partial x^*} + \frac{1}{\mathrm{Re}_L}\left(\frac{\partial^2 u^*}{\partial x^{*2}} + \frac{\partial^2 u^*}{\partial y^{*2}}\right) \tag{5.97}$$

y momentum:

$$u^* \frac{\partial v^*}{\partial x^*} + v^* \frac{\partial v^*}{\partial y^*} = -\frac{\partial p^*}{\partial y^*} + \frac{1}{\mathrm{Re}_L}\left(\frac{\partial^2 v^*}{\partial x^{*2}} + \frac{\partial^2 v^*}{\partial y^{*2}}\right) \tag{5.98}$$

energy:

$$u^* \frac{\partial \theta}{\partial x^*} + v^* \frac{\partial \theta}{\partial y^*} = \frac{1}{\mathrm{Re}_L \Pr}\left(\frac{\partial^2 \theta}{\partial x^{*2}} + \frac{\partial^2 \theta}{\partial y^{*2}}\right) + \mathrm{Ec}\left(u^* \frac{\partial p^*}{\partial x^*} + v^* \frac{\partial p^*}{\partial y^*}\right)$$
$$+ \frac{\mathrm{Ec}}{\mathrm{Re}_L}\left[2\left(\frac{\partial u^*}{\partial x^*}\right)^2 + 2\left(\frac{\partial v^*}{\partial y^*}\right)^2 + \left(\frac{\partial v^*}{\partial x^*} + \frac{\partial u^*}{\partial y^*}\right)^2\right] \tag{5.99}$$

In the above,

$$\text{Re}_L = \text{Reynolds number} = \frac{\rho u_\infty L}{\mu}$$

$$\text{Pr} = \text{Prandtl number} = \frac{c_p \mu}{k}$$

$$\text{Ec} = \text{Eckert number} = 2\frac{T_0 - T_\infty}{T_w - T_\infty}$$

and u_∞, T_∞ are the free stream velocity and temperature, respectively, and T_0 is the stagnation temperature. The product Re Pr is also known as the Peclet number Pe.

Following Prandtl, we assume that the thicknesses of the viscous and thermal boundary layers are small relative to a characteristic length in the primary flow direction. That is, $\delta/L \ll 1$ and $\delta_t/L \ll 1$ (see Fig. 5.5). For convenience, we let $\epsilon = \delta/L$ and $\epsilon_t = \delta_t/L$. Since ϵ and ϵ_t are both assumed to be small, we will take them to be of the same order of magnitude. We are assured that ϵ and ϵ_t are small over L if $\partial\delta/\partial x$ and $\partial\,\delta_t/\partial x$ are everywhere small. At a distance L from the origin of the boundary layer, we now estimate typical or expected sizes of terms in the equation.

As a general rule, we estimate sizes of derivatives by using the "mean value" provided by replacing the derivative by a finite difference over the expected range of the variables in the boundary-layer flow. For example, we estimate the size of $\partial u^*/\partial x^*$ by noting that for flow over a flat plate in a uniform stream u^* ranges between 1 and 0 as x^* ranges between 0 and 1; thus we say that we expect $\partial u^*/\partial x^*$ to be of the order of magnitude of 1. That is,

$$\left|\frac{\partial u^*}{\partial x^*}\right| \approx \left|\frac{0 - 1}{1 - 0}\right| = 1$$

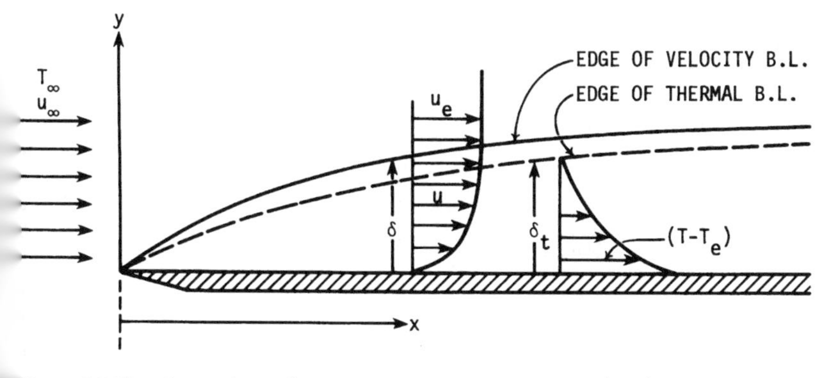

Figure 5.5 Notation and coordinate system for a boundary layer (B.L.) on a flat plate.

A factor of 2 or so does not matter in our estimates, but a factor of 10–100 does and represents an order of magnitude. It should be noted that the velocity at the outer edge of the boundary layer may deviate somewhat from u_∞ (as would be the case for flows with a pressure gradient) without changing the order of magnitude of $\partial u^*/\partial x^*$. Having established $(\partial u^*/\partial x^*) \cong 1$, we now consider the $\partial v^*/\partial y^*$ term in the continuity equation. We require that this term be of the same order of magnitude as $\partial u^*/\partial x^*$, so that mass can be conserved. Since y^* ranges between 0 and ϵ in the boundary layer, we expect from the continuity equation that v^* will also range between 0 and ϵ. Thus, $v^* \cong \epsilon$. If $\partial\delta/\partial x$ should locally become large owing to some perturbation, then the continuity equation suggests that v^* could also become large locally. The nondimensional thermal variable θ clearly ranges between 0 and 1 for incompressible constant-property flow.

We are now in a position to establish the order of magnitudes for the terms in the Navier-Stokes equations. The estimates are labeled underneath the terms in Eqs. (5.100)–(5.103).

continuity:

$$\frac{\partial u^*}{\partial x^*} + \frac{\partial v^*}{\partial y^*} = 0 \tag{5.100}$$

$$1 \qquad 1$$

x momentum:

$$u^* \frac{\partial u^*}{\partial x^*} + v^* \frac{\partial u^*}{\partial y^*} = -\frac{\partial p^*}{\partial x^*} + \frac{1}{\mathrm{Re}_L}\left(\frac{\partial^2 u^*}{\partial x^{*2}} + \frac{\partial^2 u^*}{\partial y^{*2}}\right) \tag{5.101}$$

$$1 \quad 1 \qquad \epsilon \quad \frac{1}{\epsilon} \qquad 1 \qquad \epsilon^2 \qquad 1 \qquad \frac{1}{\epsilon^2}$$

y momentum:

$$u^* \frac{\partial v^*}{\partial x^*} + v^* \frac{\partial v^*}{\partial y^*} = -\frac{\partial p^*}{\partial y^*} + \frac{1}{\mathrm{Re}_L}\left(\frac{\partial^2 v^*}{\partial x^{*2}} + \frac{\partial^2 v^*}{\partial y^{*2}}\right) \tag{5.102}$$

$$1 \quad \epsilon \qquad \epsilon \quad 1 \qquad \epsilon \qquad \epsilon^2 \qquad \epsilon \qquad \frac{1}{\epsilon}$$

energy:

$$u^* \frac{\partial \theta}{\partial x^*} + v^* \frac{\partial \theta}{\partial y^*} = \frac{1}{\mathrm{Re}_L \,\mathrm{Pr}}\left(\frac{\partial^2 \theta}{\partial x^{*2}} + \frac{\partial^2 \theta}{\partial y^{*2}}\right) + \mathrm{Ec}\left(u^* \frac{\partial p^*}{\partial x^*} + v^* \frac{\partial p^*}{\partial y^*}\right)$$

$$1 \quad 1 \qquad \epsilon \quad \frac{1}{\epsilon} \qquad \epsilon^2 \qquad 1 \qquad \frac{1}{\epsilon^2} \qquad 1 \quad 1 \quad 1 \qquad \epsilon \quad \epsilon$$

$$+ \frac{\mathrm{Ec}}{\mathrm{Re}_L}\left[2\left(\frac{\partial u^*}{\partial x^*}\right)^2 + 2\left(\frac{\partial v^*}{\partial y^*}\right)^2 + \left(\frac{\partial v^*}{\partial x^*} + \frac{\partial u^*}{\partial y^*}\right)^2\right]$$

$$\epsilon^2 \qquad 1 \qquad 1 \qquad \epsilon^2 \quad 1 \quad \frac{1}{\epsilon^2}$$

$$\tag{5.103}$$

Some comments are in order. In Eq. (5.101), the order of magnitude of the pressure gradient was established by the observation that the Navier-Stokes equations reduce to the Euler equations (see Section 5.5) at the outer edge of the viscous region. The pressure gradient must be capable of balancing the inertia terms. Hence the pressure gradient and the inertia terms must be of the same order of magnitude. We are also requiring that the largest viscous term be of the same order of magnitude as the inertia terms. For this to be true, Re_L must be of the order of magnitude of $1/\epsilon^2$, as can be seen from Eq. (5.101).

The order of magnitude of all terms in Eq. (5.102) can be established in a straight-forward manner except for the pressure gradient. Since the pressure gradient must be balanced by other terms in the equation, its order of magnitude cannot be greater than any of the others in Eq. (5.102). Accordingly, its maximum order of magnitude must be ϵ, as recorded in Eq. (5.102).

In the energy equation, we have assumed somewhat arbitrarily that the Eckert number Ec was of the order of magnitude of 1. This should be considered a typical value. Ec can become an order of magnitude larger or smaller in certain applications. The order of magnitude of the Peclet number, Re Pr, was set at $1/\epsilon^2$. Since we have already assumed $\mathrm{Re}_L \cong (1/\epsilon^2)$ in dealing with the momentum equations, this suggests that $\mathrm{Pr} \cong 1$. This is consistent with our original hypothesis that ϵ and ϵ_t were both small, i.e., $\delta \cong \delta_t$. In other words, we are assuming that the Pe is of the same order of magnitude as Re. We expect that the present results will be applicable to flows in which Pr does not vary from 1 by more than an order of magnitude. The exact limitation of the analysis must be determined by comparisons with experimental data. Three orders are specified for the last term in parentheses in Eq. (5.103) to account for the cross-product term that results from squaring the quantity in parentheses.

Carrying out the multiplication needed to establish the order of each term in Eqs. (5.101)–(5.103), we observe that all terms in the x-momentum equation are of the order of 1 in magnitude except for the streamwise second-derivative (diffusion) term, which is of the order of ϵ^2. No term in the y-momentum equation is larger than ϵ in estimated magnitude. Several terms in the energy equation are of the order 1 in magnitude, although several in the compression work and viscous dissipation terms are smaller. Keeping terms whose order of magnitude estimates are equal to 1 gives the boundary-layer equations. These are recorded below in terms of dimensional variables.

continuity:

$$\frac{\partial u}{\partial x} + \frac{\partial v}{\partial y} = 0 \tag{5.104}$$

momentum:

$$u\frac{\partial u}{\partial x} + v\frac{\partial u}{\partial y} = -\frac{1}{\rho}\frac{dp}{dx} + \nu\frac{\partial^2 u}{\partial y^2} \tag{5.105}$$

energy:

$$u\frac{\partial T}{\partial x} + v\frac{\partial T}{\partial y} = \alpha\frac{\partial^2 T}{\partial y^2} + \frac{\beta T u}{\rho c_p}\frac{dp}{dx} + \frac{\mu}{\rho c_p}\left(\frac{\partial u}{\partial y}\right)^2 \tag{5.106}$$

where ν is the kinematic viscosity μ/ρ and α is the thermal diffusivity $k/\rho c_p$.

The form of the energy equation has been generalized to include nonideal gas behavior through the introduction of β, the volumetric expansion coefficient:

$$\beta = -\frac{1}{\rho}\frac{\partial \rho}{\partial T}\bigg)_p$$

For an ideal gas, $\beta = 1/T$, where T is absolute temperature. It should be pointed out that the last two terms in Eq. (5.106) were retained from the order of magnitude analysis on the basis that Ec $\sim$ 1. Should Ec become of the order of ϵ or smaller for a particular flow, neglecting these terms should be permissible.

To complete the mathematical formulation, initial and boundary conditions must be specified. The steady boundary-layer momentum and energy equations are parabolic with the streamwise direction being the marching direction. Initial distributions of u and T must be provided. The usual boundary conditions are

$$\begin{gathered} u(x,0) = v(x,0) = 0 \\ T(x,0) = T_w(x) \qquad \text{or} \qquad -\frac{\partial T}{\partial y}\bigg)_{y=0} = \frac{q(x)}{k} \\ \lim_{y\to\infty} u(x,y) = u_e(x) \qquad \lim_{y\to\infty} T(x,y) = T_e(x) \end{gathered} \tag{5.107}$$

where the subscript e refers to conditions at the edge of the boundary layer. The pressure gradient term in Eqs. (5.105) and (5.106) is to be evaluated from the given boundary information. With $u_e(x)$ specified, dp/dx can be evaluated from an application of the equations that govern the inviscid outer flow (Euler's equations), giving $dp/dx = -\rho u_e\, du_e/dx$.

It is not difficult to extend the boundary-layer equations to variable-property and/or compressible flows. The constant-property restriction was made only as a convenience as we set about the task of illustrating principles that can frequently be used to determine a reduced but approximate set of governing equations for a flow of interest. The compressible form of the boundary-layer equations, which will also account for property variations, is presented in Section 5.3.3.

Before moving on from our order of magnitude deliberations for laminar flows, we should raise the question as to which terms neglected in the boundary-layer approximation should first become important as δ/L becomes larger and larger. Terms of the order of ϵ will next become important and then eventually, terms of the order of ϵ^2. We note that the second-derivative term neglected in the streamwise momentum equation is of the order of ϵ^2, whereas most terms in the y-momentum equation are of the order of ϵ. This means that contributions through the transverse momentum equation are expected to become important before additional terms need to be considered in the x-momentum equation. The set of equations that results from retaining terms of both orders of 1 and ϵ in the order of magnitude analysis, while neglecting terms of the order of ϵ^2 and higher, has proven to be useful in computational fluid dynamics. Such steady flow equations, which neglect all streamwise second-derivative terms, are known as the "parabolized" Navier-Stokes equations for supersonic applications and are known as the "partially parabolized" Navier-Stokes equations for subsonic applications. These are but two examples from a category of equations frequently called parabolized Navier-Stokes equations. These equations are intermediate in complexity between the Navier-Stokes and boundary-layer equations and are discussed in Chapter 8.

We next consider extending the boundary-layer approximation to an incompressible constant-property 2-D turbulent flow. Under our incompressible assumption, $\rho' = 0$ and the Reynolds equations simplify considerably. We will nondimensionalize the incompressible Reynolds equations very much in the same manner as for the Navier-Stokes equations, letting

$$u^* = \frac{\bar{u}}{u_\infty} \qquad v^* = \frac{\bar{v}}{u_\infty} \qquad x^* = \frac{x}{L} \qquad y^* = \frac{y}{L} \qquad p^* = \frac{\bar{p}}{\rho u_\infty^2} \qquad (u')^* = \frac{u'}{u_\infty}$$

$$(v')^* = \frac{v'}{u_\infty} \qquad \theta = \frac{\bar{T} - T_\infty}{T_w - T_\infty} \qquad H^* = \frac{\bar{H} - H_\infty}{H_w - H_\infty} \qquad \theta' = \frac{T'}{T_w - T_\infty} \tag{5.108}$$

Asterisks appended to parentheses indicate that all quantities within the parentheses are dimensionless; that is, instead of $\overline{u'^* v'^*}$, we will use the more convenient notation $(\overline{u'v'})^*$.

As before, we assume that $\delta/L \ll 1$, $\delta_t/L \ll 1$, and let $\epsilon = \delta/L \approx \delta_t/L$. We rely on experimental evidence to guide us in establishing the magnitude estimates for the Reynolds stress and heat flux terms. Experiments indicate that the Reynolds stresses can be at least as large as the laminar counterparts. This requires that $(\overline{u'v'})^* \sim \epsilon$. Measurements suggest that $(\overline{u'^2})^*, (\overline{u'v'})^*, (\overline{v'^2})^*$, while differing in magnitudes and distribution somewhat, are nevertheless of the same order of magnitude in the boundary layer. That is, we cannot stipulate that the magnitudes of any of these terms are different by a factor of 10 or more. A similar observation can be made for the energy equation, leading to the conclusion that $(\overline{\theta' v'})^*$ and $(\overline{\theta' u'})^*$ are of the order of magnitude of ϵ. Triple correlations such as $(\overline{u'u'u'})^*$ are clearly expected to be smaller than double correlations, and they will be taken to be of the order ϵ^2 (Schubauer and Tchen, 1959). It will be expedient to invoke the boundary-layer approximation to the form of the energy equation that employs the total enthalpy as the transported thermal variable, Eq. (5.89). We will, however, substitute for H' according to

$$H' = c_p T' + u_i' \bar{u}_i + \frac{u_i' u_i'}{2} - \frac{1}{2}\overline{u_i' u_i'}$$

The above expression for H' may look suspicious because a time-averaged expression appears on the right-hand side. However, this is necessary so that $\overline{H'} = 0$. The correct expression for H' can be derived by expanding H in terms of decomposed temperature and velocity variables and then subtracting $\overline{H}$. That is, let H' be defined as $H' = H - \overline{H}$, where

$$H = c_p T + \frac{u_i u_i}{2} = c_p \overline{T} + c_p T' + \frac{(\bar{u}_i + u_i')(\bar{u}_i + u_i')}{2}$$

The incompressible nondimensional Reynolds equations are given below along with the order of magnitude estimates for the individual terms:

continuity:

$$\frac{\partial u^*}{\partial x^*} + \frac{\partial v^*}{\partial y^*} = 0 \tag{5.109}$$

$$1 \qquad 1$$

x momentum:

$$u^* \frac{\partial u^*}{\partial x^*} + v^* \frac{\partial u^*}{\partial y^*} = -\frac{\partial p^*}{\partial x^*} + \frac{1}{\mathrm{Re}_L}\left(\frac{\partial^2 u^*}{\partial x^{*2}} + \frac{\partial^2 u^*}{\partial y^{*2}}\right)$$

$$1 \quad 1 \qquad \epsilon \quad \frac{1}{\epsilon} \qquad\qquad 1 \qquad \epsilon^2 \qquad 1 \qquad\qquad \frac{1}{\epsilon^2}$$

$$-\frac{\partial}{\partial y^*}(\overline{u'v'})^* - \frac{\partial}{\partial x^*}\left(\overline{u'^2}\right)^* \tag{5.110}$$

$$\frac{\epsilon}{\epsilon} \qquad\qquad \epsilon$$

y momentum:

$$u^* \frac{\partial v^*}{\partial x^*} + v^* \frac{\partial v^*}{\partial y^*} = -\frac{\partial p^*}{\partial y^*} + \frac{1}{\mathrm{Re}_L}\left(\frac{\partial^2 v^*}{\partial x^{*2}} + \frac{\partial^2 v^*}{\partial y^{*2}}\right) - \frac{\partial(\overline{v'u'})^*}{\partial x^*} - \frac{\partial\left(\overline{v'^2}\right)^*}{\partial y^*}$$

$$1 \quad \epsilon \qquad \epsilon \quad 1 \qquad\quad 1 \qquad \epsilon^2 \qquad \epsilon \qquad \frac{1}{\epsilon} \qquad\quad \epsilon \qquad\qquad 1 \tag{5.111}$$

energy:

$$u^* \frac{\partial H^*}{\partial x^*} + v^* \frac{\partial H^*}{\partial y^*} = \frac{T_w - T_\infty}{T_w - T_0}\left[-\frac{\partial}{\partial x^*}(\overline{u'\theta'})^* - \frac{\partial}{\partial y^*}(\overline{v'\theta'})^*\right]$$

$$1 \quad 1 \qquad \epsilon \quad \frac{1}{\epsilon} \qquad\qquad 1 \qquad\qquad \epsilon \qquad\qquad \frac{\epsilon}{\epsilon}$$

$$+\frac{1}{\mathrm{Re}_L \,\mathrm{Pr}}\left(\frac{\partial^2\theta}{\partial x^{*2}} + \frac{\partial^2\theta}{\partial y^{*2}}\right) + \mathrm{Ec}\left[-\frac{\partial}{\partial x^*}(u\overline{u'u'})^* - \frac{\partial}{\partial x^*}(v\overline{v'u'})^*\right.$$

$$\epsilon^2 \qquad 1 \qquad \frac{1}{\epsilon^2} \qquad\quad 1 \qquad\qquad \epsilon \qquad\qquad\qquad \epsilon^2$$

$$-\frac{\partial}{\partial y^*}(\overline{uu'v'})^* - \frac{\partial}{\partial y^*}(\overline{vv'v'})^* - \frac{1}{2}\frac{\partial}{\partial x^*}(\overline{u'u'u'})^* - \frac{1}{2}\frac{\partial}{\partial x^*}(\overline{v'v'u'})^*$$

$$\frac{\epsilon}{\epsilon} \qquad\qquad \frac{\epsilon^2}{\epsilon} \qquad\qquad \epsilon^2 \qquad\qquad \epsilon^2$$

$$-\frac{1}{2}\frac{\partial}{\partial y^*}(\overline{u'u'v'})^* - \frac{1}{2}\frac{\partial}{\partial y^*}(\overline{v'v'v'})^*\Bigg] + \frac{\text{Ec}}{\text{Re}_L}\left[2\frac{\partial}{\partial x^*}\left(u^*\frac{\partial u^*}{\partial x^*}\right)\right.$$

$$\frac{\epsilon^2}{\epsilon} \qquad\qquad \frac{\epsilon^2}{\epsilon} \qquad\qquad \epsilon^2 \qquad\qquad 1$$

$$+\frac{\partial}{\partial x^*}\left(v^*\frac{\partial u^*}{\partial y^*}\right) + \frac{\partial}{\partial x^*}\left(v^*\frac{\partial v^*}{\partial x^*}\right) + \frac{\partial}{\partial y^*}\left(u^*\frac{\partial v^*}{\partial y^*}\right) + \frac{\partial}{\partial y^*}\left(u^*\frac{\partial u^*}{\partial y^*}\right)$$

$$\frac{\epsilon}{\epsilon} \qquad\qquad \epsilon^2 \qquad\qquad \frac{\epsilon}{\epsilon^2} \qquad\qquad \frac{1}{\epsilon^2}$$

$$+2\frac{\partial}{\partial y^*}\left(v^*\frac{\partial v^*}{\partial y^*}\right) + 2\frac{\partial}{\partial x^*}\left(\overline{u'\frac{\partial u'}{\partial x}}\right)^* + \frac{\partial}{\partial x^*}\left(\overline{v'\frac{\partial u'}{\partial y}}\right)^* + \frac{\partial}{\partial x}\left(\overline{v'\frac{\partial v'}{\partial x}}\right)^*$$

$$\frac{\epsilon^2}{\epsilon^2} \qquad\qquad \epsilon \qquad\qquad \frac{\epsilon}{\epsilon} \qquad\qquad \epsilon$$

$$\left.+\frac{\partial}{\partial y^*}\left(\overline{u'\frac{\partial v'}{\partial x}}\right)^* + \frac{\partial}{\partial y^*}\left(\overline{u'\frac{\partial u'}{\partial y}}\right)^* + 2\frac{\partial}{\partial y^*}\left(\overline{v'\frac{\partial v'}{\partial y}}\right)^*\right] \tag{5.112}$$

$$\frac{\epsilon}{\epsilon} \qquad\qquad \frac{\epsilon}{\epsilon^2} \qquad\qquad \frac{\epsilon}{\epsilon^2}$$

Again we assume that Pr and Ec are near 1 in order of magnitude. The 2-D boundary-layer equations are obtained by retaining only terms of the order of 1. They can be written in dimensional variables as follows:

continuity:

$$\frac{\partial \bar{u}}{\partial x} + \frac{\partial \bar{v}}{\partial y} = 0$$

momentum:

$$\rho\bar{u}\frac{\partial \bar{u}}{\partial x} + \rho\bar{v}\frac{\partial \bar{u}}{\partial y} = -\frac{d\bar{p}}{dx} + \mu\frac{\partial^2 \bar{u}}{\partial y^2} - \rho\frac{\partial}{\partial y}(\overline{u'v'}) \tag{5.113}$$

energy:

$$\rho\bar{u}\frac{\partial \overline{H}}{\partial x} + \rho\bar{v}\frac{\partial \overline{H}}{\partial y} = k\frac{\partial^2 \overline{T}}{\partial y^2} - \rho c_p\frac{\partial}{\partial y}(\overline{v'T'}) - \rho\frac{\partial}{\partial y}(\overline{\bar{u}v'u'}) + \mu\frac{\partial}{\partial y}\left(\bar{u}\frac{\partial \bar{u}}{\partial y}\right) \tag{5.114}$$

It should be noted that terms of the order of 1 do remain in the y-momentum equation for turbulent flow, namely,

$$\frac{1}{\rho}\frac{\partial \bar{p}}{\partial y} = -\frac{\partial}{\partial y}\left(\overline{v'^2}\right)$$

These terms have not been listed above with the boundary-layer equations because they contribute no information about the mean velocities. The pressure variation across the boundary layer is of the order of ϵ (negligible in comparison with the streamwise variation). The boundary-layer energy equation can be easily written in terms of static temperature by substituting

$$c_p \bar{T} + \frac{\bar{u}^2}{2}$$

for $\bar{H}$ in Eq. (5.114). In doing this, we are neglecting $\bar{v}^2$ compared to $\bar{u}^2$ in the kinetic energy of the mean motion. Examination of the way in which $\bar{H}$ appears in Eq. (5.114) reveals that this is permissible in the boundary-layer approximation. Utilizing Eq. (5.113) to eliminate the kinetic energy terms permits the boundary-layer form of the energy equation to be written as

$$\rho \bar{u} c_p \frac{\partial \bar{T}}{\partial x} + \rho \bar{v} c_p \frac{\partial \bar{T}}{\partial y} = k \frac{\partial^2 \bar{T}}{\partial y^2} - \rho c_p \frac{\partial}{\partial y}\left(\overline{v'T'}\right) + \bar{u}\frac{d\bar{p}}{dx} + \left(\mu \frac{\partial \bar{u}}{\partial y} - \rho \overline{v'u'}\right)\frac{\partial \bar{u}}{\partial y} \tag{5.115}$$

The last two terms on the right-hand side of Eq. (5.115) can be neglected in some applications. However, it is not correct to categorically neglect these terms for incompressible flows. The last term on the right-hand side, for example, represents the viscous dissipation of energy, which obviously is important in incompressible lubrication applications, where the major heat transfer concern is to remove the heat generated by the viscous dissipation. In some instances, it is also possible to neglect one or both of the last two terms on the right-hand side of Eq. (5.114). Both Eq. (5.114) and (5.115) can be easily treated with finite-difference/finite-volume methods in their entirety, so the temptation to impose further reductions should generally be resisted unless it is absolutely clear that the terms neglected will indeed be negligible. The boundary conditions remain unchanged for turbulent flow.

In closing this section on the development of the thin-shear-layer approximation, it is worthwhile noting that for turbulent flow, the largest term neglected in the streamwise momentum equation, the Reynolds normal stress term, was estimated to be an order of magnitude larger, ϵ, than the largest term neglected in the laminar flow analysis. We note also that only one Reynolds stress term and one Reynolds heat flux term remain in the governing equations after the boundary-layer approximation is invoked.

For any steady internal flow application of the thin-shear-layer equations, it is possible to develop a global channel mass flow constraint. This permits the pressure gradient to be computed rather than requiring that it be given, as in the case for external flow. This point is developed further in Chapter 7.

5.3.3 Boundary-Layer Equations for Compressible Flow

The order of magnitude reduction of the Reynolds equations to boundary-layer form is a lengthier process for compressible flow. Only the results will be presented here. Details of the arguments for elimination of terms are given by Schubauer and Tchen (1959), van Driest (1951), and Cebeci and Smith (1974). As was the case for incompressible flow, guidance must be obtained from experimental observations in assessing the magnitudes of turbulence quantities. An estimate must be made for $\rho'/\bar{\rho}$ for compressible flows.

Measurements in gases for Mach numbers (M) less than about 5 indicate that temperature fluctuations are nearly isobaric for adiabatic flows. This suggests that $T'/\bar{T} \approx -\rho'/\bar{\rho}$. However, there is evidence that appreciable pressure fluctuations exist (8–10% of the mean wall static pressure) at $M = 5$ and it is speculated that $p'/\bar{p}$ increases with increasing M. In the absence of specific experimental evidence to the contrary, it is common to base the order of magnitude estimates of fluctuating terms on the assumption that the pressure fluctuations are small. This appears to be a safe assumption for $M \leqslant 5$, and good predictions based on this assumption have been noted for M as high as 7.5. We will adopt the isobaric assumption here. It is primarily the correlation terms involving the density fluctuations that may increase in magnitude with increasing Mach number above $M \approx 5$.

We find that the difference between $\tilde{u}$ and $\bar{u}$ vanishes under the boundary-layer approximation. This follows because $\overline{\rho' u'}$ is expected to be small compared to $\bar{\rho}\bar{u}$ and can be neglected in the momentum equation. We also find $\bar{T} = \tilde{T}$ and $\bar{H} = \tilde{H}$ to be consistent with the boundary-layer approximation. On the other hand, $\overline{\rho' v'}$ and $\bar{\rho}\bar{v}$ are both of about the same order of magnitude in a thin shear layer. Thus $\bar{v} \neq \tilde{v}$. Below, the unsteady boundary-layer equations for a compressible fluid are written in a form applicable to both 2-D and axisymmetric turbulent flow. For convenience, we will drop the use of bars over time-mean quantities and make use of $\tilde{v} = (\bar{\rho}\bar{v} + \overline{\rho' v'})/\bar{\rho}$. The equations are also valid for laminar flow when the terms involving fluctuating quantities are set equal to zero. The coordinate system is indicated in Fig. 5.6. The equations are as follows.

continuity:

$$\frac{\partial \rho}{\partial t} + \frac{\partial}{\partial x}(r^m \rho u) + \frac{\partial}{\partial y}(r^m \rho \tilde{v}) = 0 \tag{5.116}$$

momentum:

$$\rho\frac{\partial u}{\partial t} + \rho u\frac{\partial u}{\partial x} + \rho\tilde{v}\frac{\partial u}{\partial y} = -\frac{dp}{dx} + \frac{1}{r^m}\frac{\partial}{\partial y}\left[r^m\left(\mu\frac{\partial u}{\partial y} - \overline{\rho u' v'}\right)\right] \tag{5.117}$$

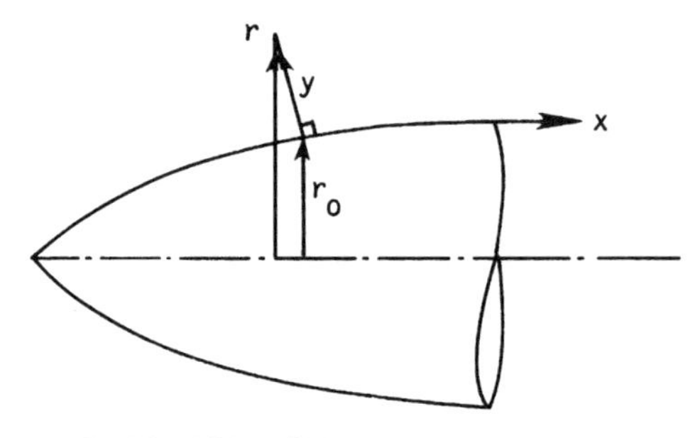

(a) EXTERNAL BOUNDARY LAYER

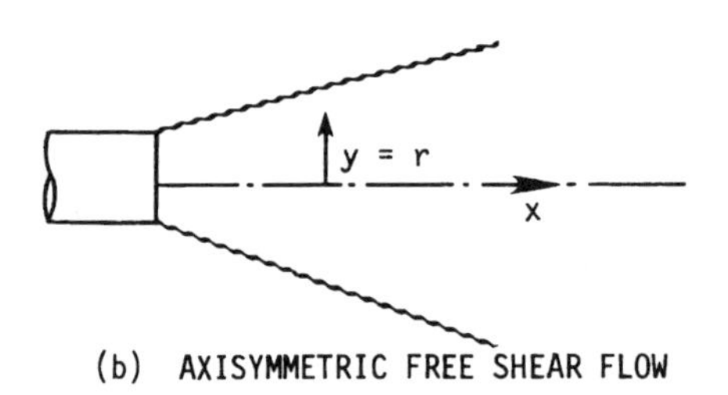

(b) AXISYMMETRIC FREE SHEAR FLOW

(c) CONFINED AXISYMMETRIC FLOW

Figure 5.6 Coordinate system for axisymmetric thin-shear-layer equations. (a) External boundary layer; (b) axisymmetric free-shear flow; (c) confined axisymmetric flow.

energy:

$$\rho\frac{\partial H}{\partial t} + \rho u\frac{\partial H}{\partial x} + \rho\tilde{v}\frac{\partial H}{\partial y}$$
$$= \frac{1}{r^m}\frac{\partial}{\partial y}\left(r^m\left\{\frac{\mu}{\mathrm{Pr}}\frac{\partial H}{\partial y} - \rho c_p\overline{v'T'} + u\left[\left(1 - \frac{1}{\mathrm{Pr}}\right)\mu\frac{\partial u}{\partial y} - \rho\overline{v'u'}\right]\right\}\right) + \frac{\partial p}{\partial t} \tag{5.118}$$

state:

$$\rho = \rho(p, T) \tag{5.119}$$

In the preceding equations, m is a flow index equal to unity for axisymmetric flow ($r^m = r$) and equal to zero for 2-D flow ($r^m = 1$). Other forms of the energy equation will be noted in subsequent sections.

We note that the boundary-layer equations for compressible flow are not significantly more complex than for incompressible flow. Only one Reynolds stress and one heat flux term appear, regardless of whether the flow is compressible or incompressible. As for purely laminar flows, the main difference is in the property variations of μ, k, and ρ for the compressible case, which nearly always requires that a solution be obtained for some form of the energy equation. When properties can be assumed constant (as for many incompressible flows), the momentum equation is independent of the energy equation, and as a result, the energy equation need not be solved for many problems of interest.

The boundary-layer approximation remains valid for a flow in which the turning of the main stream results in a 3-D flow as long as velocity derivatives with respect to only one coordinate direction are large. That is, the 3-D boundary layer is a flow that remains "thin" with respect to only one coordinate direction. The 3-D unsteady boundary-layer equations in Cartesian coordinates, applicable to a compressible turbulent flow, are given below. The y direction is normal to the wall.

continuity:

$$\frac{\partial \rho}{\partial t} + \frac{\partial \rho u}{\partial x} + \frac{\partial \rho \tilde{v}}{\partial y} + \frac{\partial \rho w}{\partial z} = 0 \tag{5.120}$$

x-momentum:

$$\frac{\partial u}{\partial t} + \rho u \frac{\partial u}{\partial x} + \rho \tilde{v} \frac{\partial u}{\partial y} + \rho w \frac{\partial u}{\partial z} = -\frac{\partial p}{\partial x} + \frac{\partial}{\partial y}\left(\mu \frac{\partial u}{\partial y} - \rho \overline{u'v'}\right) \tag{5.121}$$

z-momentum:

$$\frac{\partial w}{\partial t} + \rho u \frac{\partial w}{\partial x} + \rho \tilde{v} \frac{\partial w}{\partial y} + \rho w \frac{\partial w}{\partial z} = -\frac{\partial p}{\partial z} + \frac{\partial}{\partial y}\left(\mu \frac{\partial w}{\partial y} - \rho \overline{w'v'}\right) \tag{5.122}$$

energy:

$$\begin{aligned}
&\frac{\partial H}{\partial t} + \rho u \frac{\partial H}{\partial x} + \rho \tilde{v} \frac{\partial H}{\partial y} + \rho w \frac{\partial H}{\partial z} \\
&\quad = \frac{\partial}{\partial y}\left[\frac{\mu}{\Pr}\frac{\partial H}{\partial y} - \rho c_p \overline{v'T'} + \mu\left(1 - \frac{1}{\Pr}\right)\left(u \frac{\partial u}{\partial y} + w \frac{\partial w}{\partial y}\right)\right. \\
&\qquad \left. - \rho u \overline{v'u'} - \rho w \overline{v'w'}\right]
\end{aligned} \tag{5.123}$$

For a 3-D flow, the boundary-layer approximation permits H to be written as

$$H = c_p T + \frac{u^2}{2} + \frac{w^2}{2}$$

The 3-D boundary-layer equations are used primarily for external flows. This usually permits the pressure gradient terms to be evaluated from the solution to the inviscid flow (Euler) equations. Three-dimensional internal flows are normally computed from slightly different equations, discussed in Chapter 8.

It is common to employ body intrinsic curvilinear coordinates to compute the 3-D boundary layers occurring on wings and other shapes of practical interest. Often, this curvilinear coordinate system is nonorthogonal. An example of this can be found in the work by Cebeci et al. (1977). The orthogonal system is somewhat more common (see, for example, Blottner and Ellis, 1973). One coordinate, x_2, is almost always taken to be orthogonal to the body surface. This convention will be followed here. Below we record the 3-D boundary-layer equations in the orthogonal curvilinear coordinate system described in Section 5.1.8. Typically, x_1 will be directed roughly in the primary flow direction and x_3 will be in the crossflow direction. The metric coefficients (h_1, h_2, h_3) are as defined in Section 5.1.8; however, h_2 will be taken as unity as a result of the boundary-layer approximation. In addition, we will make use of the geodesic curvatures of the surface coordinate lines,

$$K_1 = \frac{1}{h_1 h_3} \frac{\partial h_1}{\partial x_3} \qquad K_3 = \frac{1}{h_1 h_3} \frac{\partial h_3}{\partial x_1} \tag{5.124}$$

With this notation, the boundary-layer form of the conservation equations for a compressible turbulent flow can be written as follows:

continuity:

$$\frac{\partial}{\partial x_1}(\rho h_3 u_1) + \frac{\partial}{\partial x_2}(h_1 h_3 \rho \tilde{u}_2) + \frac{\partial}{\partial x_3}(\rho h_1 u_3) = 0 \tag{5.125}$$

x_1-momentum:

$$\frac{\rho u_1}{h_1} \frac{\partial u_1}{\partial x_1} + \rho \tilde{u}_2 \frac{\partial u_1}{\partial x_2} + \frac{\rho u_3}{h_3} \frac{\partial u_1}{\partial x_3} + \rho u_1 u_3 K_1 - \rho u_3^2 K_3$$
$$= -\frac{1}{h_1} \frac{\partial p}{\partial x_1} + \frac{\partial}{\partial x_2}\left(\mu \frac{\partial u_1}{\partial x_2} - \rho \overline{u_1' u_2'} \right) \tag{5.126}$$

x_3-momentum:

$$\frac{\rho u_1}{h_1} \frac{\partial u_3}{\partial x_1} + \rho \tilde{u}_2 \frac{\partial u_3}{\partial x_2} + \frac{\rho u_3 \partial u_3}{h_3 \partial x_3} + \rho u_1 u_3 K_3 - \rho u_1^2 K_1$$
$$= -\frac{1}{h_3} \frac{\partial p}{\partial x_3} + \frac{\partial}{\partial x_2}\left(\mu \frac{\partial u_3}{\partial x_2} - \rho \overline{u_3' u_2'} \right) \tag{5.127}$$

energy:

$$\frac{\rho u_1}{h_1}\frac{\partial H}{\partial x_1} + \rho\tilde{u}_2\frac{\partial H}{\partial x_2} + \frac{\rho u_3}{h_3}\frac{\partial H}{\partial x_3}$$

$$= \frac{\partial}{\partial x_2}\left[\frac{\mu}{\text{Pr}}\frac{\partial H}{\partial x_2} - \rho c_p\overline{u_2'T'} + \mu\left(1 - \frac{1}{\text{Pr}}\right)\left(u_1\frac{\partial u_1}{\partial x_2} + u_3\frac{\partial u_3}{\partial x_2}\right)\right.$$

$$\left. - \rho u_1\overline{u_2'u_1'} - \rho u_3\overline{u_2'u_3'}\right] \quad (5.128)$$

As always, an equation of state, $\rho = \rho(p, T)$, is needed to close the system of equations for a compressible flow. The above equations remain valid for a laminar flow when the fluctuating quantities are set equal to zero.

5.4 INTRODUCTION TO TURBULENCE MODELING

5.4.1 Background

The need for turbulence modeling was pointed out in Section 5.2. In order to predict turbulent flows by numerical solutions to the Reynolds equations, it becomes necessary to make closing assumptions about the apparent turbulent stress and heat flux quantities. All presently known turbulence models have limitations: the ultimate turbulence model has yet to be developed. Some argue philosophically that we have a system of equations for turbulent flows that is both accurate and general in the Navier-Stokes set, and therefore, to hope to develop an alternative system having the same accuracy and generality (but being simpler to solve) through turbulence modeling is being overly optimistic. If this premise is accepted, then our expectations in turbulence modeling are reduced from seeking the ultimate to seeking models that have reasonable accuracy over a limited range of flow conditions.

It is important to remember that turbulence models must be verified by comparing predictions with experimental measurements. Care must be taken in interpreting predictions of models outside the range of conditions over which they have been verified by comparisons with experimental data.

The purpose of this section is to introduce the methodology commonly used in turbulence modeling. The intent is not to present all models in sufficient detail that they can be used without consulting the original references, but rather to outline the rationale for the evolution of modeling strategy. Simpler models will be described in sufficient detail to enable the reader to formulate a "baseline" model applicable to simple thin shear layers.

5.4.2 Modeling Terminology

Boussinesq (1877) suggested, more than 100 year ago, that the apparent turbulent shearing stresses might be related to the rate of mean strain through an apparent scalar turbulent or "eddy" viscosity. For the general Reynolds stress tensor, the Boussinesq assumption gives

$$-\overline{\rho u_i' u_j'} = 2\mu_T S_{ij} - \frac{2}{3}\delta_{ij}\left(\mu_T \frac{\partial u_k}{\partial x_k} + \rho\bar{k}\right) \tag{5.129a}$$

where μ_T is the turbulent viscosity, $\bar{k}$ is the kinetic energy of turbulence, $\bar{k} = \overline{u_i' u_i'}/2$, and the rate of the mean strain tensor S_{ij} is given by

$$S_{ij} = \frac{1}{2}\left(\frac{\partial u_i}{\partial x_j} + \frac{\partial u_j}{\partial x_i}\right) \tag{5.129b}$$

Following the convention introduced in Section 5.3.2, we are omitting bars over the time-mean variables.

By analogy with kinetic theory, by which the molecular viscosity for gases can be evaluated with reasonable accuracy, we might expect that the turbulent viscosity can be modeled as

$$\mu_T = \rho v_T l \tag{5.130}$$

where v_T and l are characteristic velocity and length scales of the turbulence, respectively. The problem, of course, is to find suitable means for evaluating v_T and l.

Turbulence models to close the Reynolds equations can be divided into two categories, according to whether or not the Boussinesq assumption is used. Models using the Boussinesq assumption will be referred to as Category I, or turbulent viscosity models. These are also known as first-order models. Most models currently employed in engineering calculations are of this type. Experimental evidence indicates that the turbulent viscosity hypothesis is a valid one in many flow circumstances. There are exceptions, however, and there is no physical requirement that it hold. Models that affect closure to the Reynolds equations without this assumption will be referred to as Category II models and include those known as Reynolds stress or stress-equation models. The stress-equation models are also referred to as second-order or second-moment closures.

The other common classification of models is according to the number of supplementary partial differential equations that must be solved in order to supply the modeling parameters. This number ranges from zero for the simplest algebraic models to 12 for the most complex of the Reynolds stress models (Donaldson and Rosenbaum, 1968).

Category III models will be defined as those that are not based entirely on the Reynolds equations. Large-eddy simulations fall into this category, since it is a filtered set of conservation equations that is solved instead of the Reynolds equations.

As we turn to examples of specific turbulence models, it will be helpful to keep in mind an example set of conservation equations for which turbulence modeling is needed. The thin-shear-layer equations, Eqs. (5.116)–(5.119), will serve this purpose reasonably well. In the incompressible 2-D or axisymmetric thin-shear-layer equations, the modeling task reduces to finding expressions for $-\rho\overline{v'u'}$ and $\rho c_p\overline{v'T'}$.

5.4.3 Simple Algebraic or Zero-Equation Models

Algebraic turbulence models invariably utilize the Boussinesq assumption. One of the most successful of this type of model was suggested by Prandtl in the 1920s:

$$\mu_T = \rho l^2 \left| \frac{\partial u}{\partial y} \right| \tag{5.131a}$$

where l, a "mixing length," can be thought of as a transverse distance over which particles maintain their original momentum, somewhat on the order of a mean free path for the collision or mixing of globules of fluid. The product $l|\partial u/\partial y|$ can be interpreted as the characteristic velocity of turbulence, v_T. In Eq. (5.131*a*), u is the component of velocity in the primary flow direction, and y is the coordinate transverse to the primary flow direction.

For 3-D thin shear layers, Prandtl's formula is usually interpreted as

$$\mu_T = \rho l^2 \left[\left(\frac{\partial u}{\partial y} \right)^2 + \left(\frac{\partial w}{\partial y} \right)^2 \right]^{1/2} \tag{5.131b}$$

This formula treats the turbulent viscosity as a scalar and gives qualitatively correct trends, especially near the wall. There is increasing experimental evidence, however, that in the outer layer, the turbulent viscosity should be treated as a tensor (i.e., dependent upon the direction of strain) in order to provide the best agreement with measurements. For flows in corners or in other geometries where a single "transverse" direction is not clearly defined, Prandtl's formula must be modified further (see, for example, Patankar et al., 1979).

The evaluation of l in the mixing-length model varies with the type of flow being considered, wall boundary layer, jet, wake, etc. For flow along a solid surface (internal or external flow), good results are observed by evaluating l according to

$$l_i = \kappa y(1 - e^{-y^+/A^+}) \tag{5.132}$$

in the inner region closest to the solid boundaries and switching to

$$l_0 = C_1 \delta \tag{5.133}$$

when l_i predicted by Eq. (5.132) first exceeds l_0. The constant C_1 in Eq. (5.133) is usually assigned a value close to 0.089, and δ is the velocity boundary-layer thickness.

In Eq. (5.132), κ is the von Kármán constant, usually taken as 0.41, and A^+ is the damping constant, most commonly evaluated as 26. The quantity in parentheses is the van Driest damping function (van Driest, 1956) and is the most common expression used to bridge the gap between the fully turbulent region where $l = \kappa y$ and the viscous sublayer where $l \to 0$. The parameter y^+ is defined as

$$y^+ = \frac{y(|\tau_w|/\rho_w)^{1/2}}{\nu_w}$$

Numerous variations on the exponential function have been utilized in order to account for effects of property variations, pressure gradients, blowing, and surface roughness. A discussion of modifications to account for several of these effects can be found in the work by Cebeci and Smith (1974). It appears reasonably clear from comparisons in the literature, however, that the inner layer model as stated [Eq. (5.132)] requires no modification to accurately predict the variable-property flow of gases with moderate pressure gradients on smooth surfaces.

The expression for l_i, Eq. (5.132), is responsible for producing the inner, "law-of-the-wall" region of the turbulent flow, and l_0 [Eq. (5.133)] produces the outer "wake-like" region. These two zones are indicated in Fig. 5.7, which depicts a typical velocity distribution for an incompressible turbulent boundary

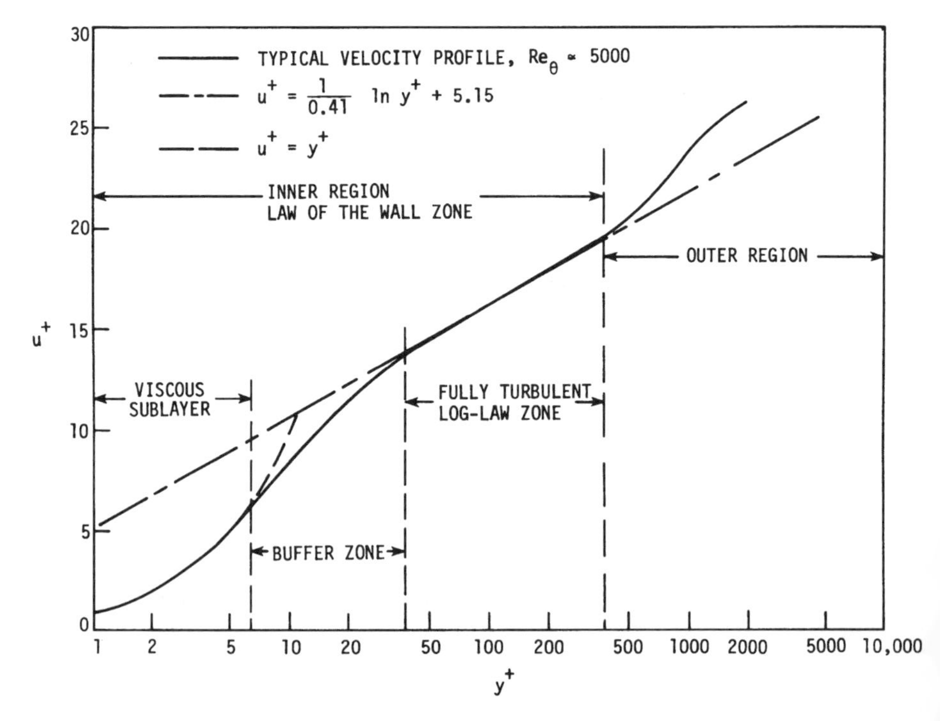

Figure 5.7 Zones in the turbulent boundary layer for a typical incompressible flow over a smooth flat plate.

layer on a smooth impermeable plate using law-of-the-wall coordinates. In Fig. 5.7, Re_θ is the Reynolds number based on momentum thickness, $\rho_e u_e \theta/\mu_e$, where for 2-D flow, the momentum thickness is defined as

$$\theta = \int_0^\infty \frac{\rho u}{\rho_e u_e}\left(1 - \frac{u}{u_e}\right) dy$$

The nondimensional velocity u^+ is defined as $u^+ = u/(|\tau_w|/\rho_w)^{1/2}$. The inner and outer regions are indicated on the figure. Under normal conditions, the inner law-of-the-wall zone only includes about 20% of the boundary layer. The log-linear zone is the characteristic "signature" of a turbulent wall boundary layer, although the law-of-the-wall plot changes somewhat in general appearance as Re and M are varied.

It is worth noting that for low Re_θ, i.e., relatively near the origin of the turbulent boundary layer, both inner and outer regions are tending toward zero, and problems might be expected with the two-region turbulence model employing Eqs. (5.132) and (5.133). The difficulty occurs because the smaller δ occurring near the origin of the turbulent boundary layer are causing the switch to the outer model to occur before the wall damping effect has permitted the fully turbulent law-of-the-wall zone to develop. This causes the numerical scheme using such a model to underpredict the wall shear stress. The discrepancy is nearly negligible for incompressible flow, but the effect is more serious for compressible flows, persisting at higher and higher Re as M increases owing to the relative thickening of the viscous sublayer from thermal effects (Pletcher, 1976). Naturally, details of the effect are influenced by the level of wall cooling present in the compressible flow.

Predictions can be brought into good agreement with measurements at low Re by simply delaying the switch from the inner model. Eq. (5.132), to the outer, Eq. (5.133), until $y^+ \geqslant 50$. If, at $y^+ = 50$ in the flow, $l/\delta \leqslant 0.089$, then no adjustment is necessary. On the other hand, if Eq. (5.132) predicts $l/\delta > 0.089$, then the mixing length becomes constant in the outer region at the value computed at $y^+ = 50$ by Eq. (5.132). This simple adjustment ensures the existence of the log-linear region in the flow, which is in agreement with the preponderance of measurements.

Other modeling procedures have been used successfully for the inner and outer regions. Some workers advocate the use of wall functions based on a Couette flow assumption (Patankar and Spalding, 1970) in the near-wall region. This approach probably has not been quite as well refined as the van Driest function to account for variable properties, transpiration, and other near-wall effects.

An alternative treatment to Eq. (5.133) is often used to evaluate the turbulent viscosity in the outer region (Cebeci and Smith, 1974). This follows the Clauser formulation,

$$\mu_{T(\text{outer})} = \alpha \rho u_e |\delta_k^*| \tag{5.134}$$

where α accounts for low-Re effects. Cebeci and Smith (1974) recommend

$$\alpha = 0.0168 \frac{(1.55)}{1 + \pi} \tag{5.135}$$

where $\pi = 0.55[1 - \exp(-0.243 z^{1/2} - 0.298 z)]$ and $z = (\mathrm{Re}_\theta/425) - 1$. For $\mathrm{Re}_\theta > 5000$, $\alpha \cong 0.0168$. The parameter δ_k^* is the kinematic displacement thickness, defined as

$$\delta_k^* = \int_0^\infty \left(1 - \frac{u}{u_e}\right) dy \tag{5.136}$$

Closure for the Reynolds heat flux term, $\rho c_p \overline{v'T'}$, is usually handled in algebraic models by a form of the Reynolds analogy, which is based on the similarity between the transport of heat and momentum. The Reynolds analogy is applied to the apparent turbulent conductivity in the assumed Boussinesq form:

$$\rho c_p \overline{v'T'} = -k_T \frac{\partial T}{\partial y}$$

In turbulent flow this additional transport of heat is caused by the turbulent motion. Experiments confirm that the ratio of the diffusivities for the turbulent transport of heat and momentum, called the turbulent Prandtl number, $\mathrm{Pr}_T = \mu_T c_p / k_T$, is a well-behaved function across the flow. Most algebraic turbulence models do well by letting the Pr_T be a constant near 1; most commonly, $\mathrm{Pr}_T = 0.9$. Experiments indicate that for wall shear flows, Pr_T varies somewhat from about 0.6–0.7 at the outer edge of the boundary layer to about 1.5 near the wall, although the evidence is not conclusive. Several semi-empirical distributions for Pr_T have been proposed; a sampling is found in the works by Cebeci and Smith (1974), Kays (1972), and Reynolds (1975). Using Pr_T, the apparent turbulent heat flux is related to the turbulent viscosity and mean flow variables as

$$-\rho c_p \overline{v'T'} = \frac{c_p \mu_T}{\mathrm{Pr}_T} \frac{\partial T}{\partial y} \tag{5.137a}$$

and closure has been completed.

For other than thin shear flows, it may be necessary to model other Reynolds heat flux terms. To do so, the turbulent conductivity, $k_T = c_p \mu_T / \mathrm{Pr}_T$, is normally considered as a scalar, and the Boussinesq-type approximation is extended to other components of the temperature gradient. As an example, we would evaluate $-\rho c_p \overline{u'T'}$ as

$$-\rho c_p \overline{u'T'} = \frac{c_p \mu_T}{\mathrm{Pr}_T} \frac{\partial T}{\partial x}$$

To summarize, a recommended baseline algebraic model for wall boundary layers consists of evaluating the turbulent viscosity by Prandtl's mixing-length formula, Eq. (5.131a), where l is given by Eq. (5.132) for the inner region, and is given by Eq. (5.133) for the outer region. Alternatively, the Clauser formulation, Eq. (5.134), could be used in the outer region. The apparent turbulent heat flux can be evaluated through Eq. (5.137*a*) using $\Pr_T = 0.9$. This simplest form of modeling has employed four empirical, adjustable constants as given in Table 5.1.

Note that outer-region models used with the boundary-layer equations typically require the determination of the boundary-layer thickness or displacement thickness. Such measures are satisfactory for the boundary-layer equations because the streamwise velocity from the boundary-layer solution smoothly reaches the specified outer-edge velocity at the outer edge of the viscous region. As a consequence, the edge of the viscous region is reasonably well defined when solving the boundary-layer equations. This is not the case when solving more complete forms of the conservation equations, as for example, the full Reynolds averaged Navier-Stokes (RANS) equations, because the solution domain extends well outside the viscous region and the solution to the remaining Euler terms often results in a streamwise velocity that varies with distance from the solid boundary. This makes it difficult to identify a viscous layer thickness. A widely used algebraic model introduced by Baldwin and Lomax (1978) avoids the use of boundary-layer parameters such as displacement thickness or boundary-layer thickness. Basically, the Baldwin-Lomax model is a two-region algebraic model similar to the Cebeci-Smith model but formulated with the requirements of RANS solution schemes in mind.

The inner region is resolved as for many other algebraic models but with the use of the full measure of vorticity:

Inner:

$$\mu_{T_i} = \rho l^2 |\omega| \tag{5.137b}$$

$$|\omega| = \sqrt{\left(\frac{\partial v}{\partial x} - \frac{\partial u}{\partial y}\right)^2 + \left(\frac{\partial w}{\partial y} - \frac{\partial v}{\partial z}\right)^2 + \left(\frac{\partial u}{\partial z} - \frac{\partial w}{\partial x}\right)^2} \tag{5.137c}$$

where $l = \kappa y(1 - e^{-(y^+/A^+)})$.

Table 5.1 Empirical constants employed in algebraic turbulence models for wall boundary layers

Symbol	Description
κ	von Kármán constant, used for inner layer ≈ 0.41
A^+	van Driest constant for damping function ≈ 26, but frequently modified to account for complicating effects
C_1 or α	constant for outer-region model $C_1 \approx 0.089$, $\alpha \approx 0.0168$, but usually includes $f(\mathrm{Re}_\theta)$ in α
$\Pr_T$	turbulent Prandtl number, $= 0.9$ *is most common*

Outer:

$$\mu_{T_0} = 0.0168\rho V_{T_0} L_0 \tag{5.137d}$$

where

$$V_{T_0} = \min(F_{\max}, 0.25 q_{\text{Diff}}^2 / F_{\max}) \tag{5.137e}$$

$$q_{\text{Diff}}^2 = (u^2 + v^2 + w^2)_{\max} - (u^2 + v^2 + w^2)_{\min} \tag{5.137f}$$

$$F_{\max} = \max[y|\omega|(1 - e^{-(y^+/A^+)})] \tag{5.137g}$$

$$L_0 = 1.6 y_{\max} I^k \tag{5.137h}$$

$$I^k = \left[1 + 5.5\left(\frac{0.3y}{y_{\max}}\right)^6\right]^{-1} \tag{5.137i}$$

and $y_{\max}$ = value of y for which $F_{\max}$ occurs.

Algebraic models have accumulated an impressive record of good performance for relatively simple viscous flows but need to be modified in order to accurately predict flows with "complicating" features. It should be noted that compressible flows do not represent a "complication" in general. The turbulence structure of the flow appears to remain essentially unchanged for Mach numbers up through at least 5. Naturally, the variation of density and other properties must be accounted for in the form of the conservation equations used with the turbulence model. Table 5.2 lists several flow conditions requiring alterations or extensions to the simplest form of algebraic models cited above. Some key references are also tabulated where such model modifications are discussed.

The above discussion of algebraic models for wall boundary layers is by no means complete. Over the years, dozens of slightly different algebraic models have been suggested. Eleven algebraic models were compared in a study by

Table 5.2 Effects requiring alterations or additions to simplest form of algebraic turbulence models

Effect	References
Low Reynolds number	Cebeci and Smith (1974), Pletcher (1976), Bushnell et al. (1975), Herring and Mellor (1968), McDonald (1970)
Roughness	Cebeci and Smith (1974), Bushnell et al. (1976), McDonald and Fish (1973), Healzer et al. (1974), Adams and Hodge (1977)
Transpiration	Cebeci and Smith (1974), Bushnell et al. (1976), Pletcher (1974), Baker and Launder (1974), Kays and Moffat (1975)
Strong pressure gradients	Cebeci and Smith (1974), Bushnell et al. (1976), Adams and Hodge (1977), Pletcher (1974), Baker and Launder (1974), Kays and Moffat (1975), Jones and Launder (1972), Kreskovsky et al. (1974), Horstman (1977)
Merging shear layers	Bradshaw et al. (1973), Stephenson (1976), Emery and Gessner (1976), Cebeci and Chang (1978), Malik and Pletcher (1978)

McEligot et al. (1970) for turbulent pipe flow with heat transfer. None were found superior to the van Driest–damped mixing-length model presented here.

Somewhat less information is available in the literature on algebraic turbulence models for free-shear flows. This category of flows has historically been more difficult to model than wall boundary layers, especially if model generality is included as a measure of merit. Some discussion of the status of the simple models for round jets can be found in the works of Madni and Pletcher (1975b, 1977a). The initial mixing region of round jets can be predicted fairly well using Prandtl's mixing-length formulation, Eq. (5.131*a*), with

$$l = 0.0762\,\delta_m \tag{5.138}$$

where δ_m is the width of the mixing zone. This model does not perform well after the shear layers have merged, and a switch at that point to models of the form (Hwang and Pletcher, 1978)

$$\nu_T = \mu_T/\rho = \gamma F y_{1/2}(u_{\max} - u_{\min}) \tag{5.139}$$

or (Madni and Pletcher, 1975b)

$$\nu_T = \frac{2F}{a}\int_y^{\infty} |u_e - u|\, y\, dy \tag{5.140}$$

has provided good agreement with measurements for round co-flowing jets. Equation (5.139) is a modification of the model suggested for jets by Prandtl (1926). In the above equations, a is the jet discharge radius and γ is an intermittency function,

$$\gamma = 1 \qquad 0 \leqslant \frac{y}{y_{1/2}} \leqslant 0.8$$

$$\gamma = (0.5)^z \qquad \frac{y}{y_{1/2}} > 0.8 \qquad \text{where} \qquad z = \left(\frac{y}{y_{1/2}} - 0.8\right)^{2.5} \tag{5.141}$$

F is a function of the ratio (R) of the stream velocity to the jet discharge velocity given by $F = 0.015(1 + 2.13R^2)$. The distance y is measured from the jet centerline, and $y_{1/2}$ is the "velocity half width," the distance from the centerline to the point at which the velocity has decreased to the average of the centerline and external stream velocities.

Philosophically, the strongest motivation for turning to more complex models is the observation that the algebraic model evaluates the turbulent viscosity only in terms of *local* flow parameters, yet we feel that a turbulence model ought to provide a mechanism by which effects upstream can influence the turbulence structure (and viscosity) downstream. Further, with the simplest models, ad hoc additions and corrections are frequently required to handle specific effects, and constants need to be changed to handle different classes of shear flows. To many investigators, it is appealing to develop a model general enough that specific modifications to the constants are not required to treat different classes of flows.

If we accept the general form for the turbulent viscosity, $\mu_T = \rho v_T l$, then a logical way to extend the generality of turbulent viscosity models is to permit v_T and perhaps l to be more complex (and thus more general) functions of the flow capable of being influenced by upstream (historic) effects. This rationale serves to motivate several of the more complex turbulence models.

5.4.4 One-Half-Equation Models

A *one-half-equation model* will be defined as one in which the value of one model parameter (v_T, l, or μ_T itself) is permitted to vary with the primary flow direction in a manner determined by the solution to an *ordinary* differential equation (ODE). The ODE usually results from either neglecting or assuming the variation of the model parameter with one coordinate direction. Extended mixing-length models and relaxation models fall into this category. A *one-equation model* is one in which an additional *partial* differential equation (PDE) is solved for a model parameter. The main features of several one-half-equation models are tabulated in Table 5.3.

The first three models in Table 5.3 differ in detail, although all three utilize an integral form of a transport equation for turbulence kinetic energy as a basis for letting flow history influence the turbulent viscosity. Models of this type have been refined to allow prediction of transition, roughness effects, transpiration, pressure gradients, and qualitative features of relaminarization. Most of the test cases reported for the models have involved external rather than channel flows.

Although models 5D, 5E, and 5F appear to be purely empirical relaxation or lag models, Birch (1976) shows that models of this type are actually equivalent to 1-D versions of transport PDEs for the quantities concerned except that these transport equations are not generally derivable from the Navier-Stokes equations. This is no serious drawback, since transport equations cannot be

Table 5.3 Some one-half-equation models

Model	Transport equation used as basis for ODE	Model parameter determined by ODE solution	References
5A	Turbulence kinetic energy	l_∞	McDonald and Camerata (1968), Kreskovsky et al. (1974), McDonald and Kreskovsky (1974)
5B	Turbulence kinetic energy	l_∞	Chan (1972)
5C	Turbulence kinetic energy	l_∞	Adams and Hodge (1977)
5D	Empirical ODE for $\mu_{T(\text{outer})}$	$\mu_{T(\text{outer})}$	Shang and Hankey (1975)
5E	Empirical ODE for $\mu_{T(\text{outer})}$	$\mu_{T(\text{outer})}$	Reyhner (1968)
5F	Empirical ODE for l_∞	l_∞	Malik and Pletcher (1978). Pletcher (1978)
5G	Empirical ODE for $\tau_{\max}$	$\tau_{\max}$	Johnson and King (1985)

solved without considerable empirical simplification and modeling of terms. In the end, these transport equations tend to have similar forms characterized by generation, dissipation, diffusion, and convection terms, regardless of the origin of the equation.

To illustrate, for wall boundary layers, model 5F utilizes the expression for mixing length given by Eq. (5.132) for the inner region. In the outer part of the flow, the mixing length is calculated according to

$$l_0 = 0.12L \tag{5.142}$$

where L is determined by the solution to an ODE. For a shear layer of constant width, the natural value of L is presumed to be the thickness δ of the shear layer. When δ is changing with the streamwise direction, x, L will lag δ in a manner controlled by the relaxation time for the large-eddy structure, which is assumed to be equal to $\delta/\bar{u}_\tau$, where $\bar{u}_\tau$ is a characteristic turbulence velocity. If it is further assumed that the fluid in the outer part of the shear layer travels at velocity u_e, then the streamwise distance traversed by the flow during the relaxation time is $L^* = C_2 u_e \delta/\bar{u}_\tau$. A rate equation can be developed by assuming that L will tend toward δ, according to

$$\frac{dL}{dx} = \frac{\delta - L}{L^*} \tag{5.143}$$

This model has been extended to free-shear flows (Minaie and Pletcher, 1982) by interpreting δ as the distance between the location of the maximum shear stress and the outer edge of the shear flow and replacing u_e by the average streamwise velocity over the shear layer. The optimum evaluation for $\bar{u}_\tau$ appears unsettled. The expression $\bar{u}_\tau = (L/\delta)(|\tau_w|/\rho_w)^{1/2}$ has been employed successfully for flows along solid surfaces, whereas $\bar{u}_\tau = (\tau_{\max}/\rho_w)^{1/2}$ has proven satisfactory for free-shear flows predicted to date. It might be speculated that this latter evaluation would work reasonably well for wall boundary layers also. The final form of the transport ODE for L used for separating wall boundary layers (Pletcher, 1978) and merging shear layers in annular passages (Malik and Pletcher, 1978) can be written as

$$u_e \frac{dL}{dx} = 1.25 \left| \frac{\tau_w}{\rho_w} \right|^{1/2} \left[\frac{L}{\delta} - \left(\frac{L}{\delta} \right)^2 \right] \tag{5.144}$$

Although the simple one-half-equation model described above provides remedies for shortcomings in zero-equation models for several flows, it performs badly in predicting shock-separated flows. The one-half-equation model proposed by Johnson and King (1985), and modified by Johnson and Coakley (1990), 5G in Table 5.3, was developed to excel especially for the nonequilibrium conditions present in transonic separated flows. Basically, it provides an extension of algebraic models to nonequilibrium flows and effectively reduces to the Cebeci-Smith model under equilibrium conditions.

5.4.5 One-Equation Models

One obvious shortcoming of algebraic viscosity models that normally evaluate v_T in the expression $\mu_T = \rho v_T l$ by $v_T = l|\partial u/\partial y|$ is that $\mu_T = k_T = 0$ whenever $\partial u/\partial y = 0$. This would suggest that μ_T and k_T would be zero at the centerline of a pipe, in regions near the mixing of a wall jet with a main stream and in flow through an annulus or between parallel plates where one wall is heated and the other cooled. Measurements (and common sense) indicate that μ_T and k_T are *not* zero under all conditions whenever $\partial u/\partial y = 0$. The mixing-length models can be "fixed up" to overcome this deficiency, but this conceptual shortcoming provides motivation for considering other interpretations for μ_T and k_T. In fairness to the algebraic models, we should mention that this defect is not always crucial because Reynolds stresses and heat fluxes are frequently small when $\partial u/\partial y = 0$. For some examples illustrating this point, see Malik and Pletcher (1981).

It was the suggestion of Prandtl and Kolmogorov in the 1940s to let v_T in $\mu_T = \rho v_T l$ be proportional to the square root of the kinetic energy of turbulence, $\bar{k} = \frac{1}{2}\overline{u_i' u_i'}$. Thus the turbulent viscosity can be evaluated as

$$\mu_T = C_k\, \rho l (\bar{k})^{1/2} \tag{5.145}$$

and μ_T no longer becomes equal to zero when $\partial u/\partial y = 0$. The kinetic energy of turbulence is a measurable quantity and is easily interpreted physically. We naturally inquire how we might predict $\bar{k}$.

A transport PDE can be developed (Prob. 5.22) for $\bar{k}$ from the Navier-Stokes equations. For incompressible flows, the equation takes the form

$$\rho\frac{\partial \bar{k}}{\partial t} + \rho u_j \frac{\partial \bar{k}}{\partial x_j} = \frac{\partial}{\partial x_j}\left(\mu\frac{\partial \bar{k}}{\partial x_j} - \frac{1}{2}\overline{\rho u_i' u_i' u_j'} - \overline{p' u_j'}\right) - \overline{\rho u_i' u_j'}\frac{\partial u_i}{\partial x_j} - \mu\overline{\frac{\partial u_i'}{\partial x_k}\frac{\partial u_i'}{\partial x_k}} \tag{5.146}$$

The term $-\frac{1}{2}\overline{\rho u_i' u_i' u_j'} - \overline{p' u_j'}$ is typically modeled as a gradient diffusion process,

$$-\frac{1}{2}\overline{\rho u_i' u_i' u_j'} - \overline{p' u_j'} = \frac{\mu_T}{\Pr_k}\frac{\partial \bar{k}}{\partial x_j}$$

where $\Pr_k$ is a turbulent Prandtl number for turbulent kinetic energy and, as such, is purely a closure constant. Using the Boussinesq eddy viscosity assumption, the second term on the right-hand side readily becomes

$$-\overline{\rho u_i' u_j'}\frac{\partial u_i}{\partial x_j} = \left(2\mu_T S_{ij} - \frac{2}{3}\rho\bar{k}\delta_{ij}\right)\frac{\partial u_i}{\partial x_j}$$

The last term on the right-hand side of Eq. (5.146) is the dissipation rate of turbulent kinetic energy per unit volume, $\rho\varepsilon$. Based on dimensional arguments, the dissipation rate of turbulent kinetic energy is given by $\varepsilon = C_D \bar{k}^{3/2}/l$. Thus the modeled form of the turbulent kinetic energy equation becomes

$$\underbrace{\rho \frac{D\bar{k}}{Dt}}_{\substack{\text{Particle rate}\\ \text{of increase}\\ \text{of } \bar{k}}} = \underbrace{\frac{\partial}{\partial x_j}\left[(\mu + \mu_T/\text{Pr}_k)\frac{\partial \bar{k}}{\partial x_j}\right]}_{\substack{\text{Diffusion rate}\\ \text{for } \bar{k}}}$$

$$+ \underbrace{\left(2\mu_T S_{ij} - \frac{2}{3}\rho \bar{k}\delta_{ij}\right)\frac{\partial u_i}{\partial x_j}}_{\substack{\text{Generation}\\ \text{rate for } \bar{k}}} - \underbrace{C_D \rho \bar{k}^{3/2}/l}_{\substack{\text{Dissipation}\\ \text{rate for } \bar{k}}} \tag{5.147}$$

The physical interpretation of the various terms is indicated for Eq. (5.147). This modeled transport equation is then added to the system of PDEs to be solved for the problem at hand. Note that a length parameter, l, needs to be specified algebraically. In the above, $\text{Pr}_k \approx 1.0$ and $C_D \simeq 0.164$ if l is taken as the ordinary mixing length.

The above modeling for the $\bar{k}$ transport equation is only valid in the fully turbulent regime, i.e., away from any wall damping effects. For typical wall flows, this means for y^+ greater than about 30. Inner boundary conditions for the $\bar{k}$ equations are often supplied through the use of wall functions (Launder and Spalding, 1974).

Wall functions are based on acceptance of the law of the wall as the link between the near-wall velocity and the wall shear stress. The logarithmic portion of the law of the wall follows exactly from Prandtl's mixing-length hypothesis and is confirmed by experiments under a fairly wide range of conditions. In the law of the wall region, experiments also indicate that convection and diffusion of $\bar{k}$ are negligible. Thus generation and dissipation of $\bar{k}$ are in balance, and it can be shown (Prob. 5.23) that the turbulence kinetic energy model for the turbulent viscosity reduces to Prandtl's mixing-length formulation, Eq. (5.131*a*), under these conditions if we take $C_k = (C_D)^{1/3}$ in Eq. (5.145). At the location where the diffusion and convection are first neglected, we can establish (Prob. 5.26) an inner boundary condition for $\bar{k}$ in a 2-D flow as

$$\bar{k}(x, y_c) = \frac{\tau(y_c)}{\rho C_D^{2/3}} \tag{5.148a}$$

where y_c is a point within the region where the logarithmic law of the wall is expected to be valid. For $y < y_c$ the Prandtl-type algebraic inner-region model [Eqs. (5.131*a*) and (5.132)] can be used.

Other one-equation models have been suggested that deviate somewhat from the Prandtl-Kolmogorov pattern. One of the most successful of these is by Bradshaw et al. (1967). The turbulence energy equation is used in the Bradshaw model, but the modeling is different in both the momentum equation, where the turbulent shearing stress is assumed proportional to $\bar{k}$, and in the turbulence energy equation. The details will not be given here, but an interesting feature of the Bradshaw method is that, as a consequence of the form of modeling used for the turbulent transport terms, the system of equations becomes hyperbolic and can be solved by a procedure similar to the method of characteristics. The Bradshaw method has enjoyed good success in the prediction of wall boundary layers. Even so, the predictions have not been notably superior to those of the algebraic models, one-half-equation models, or other one-equation models.

Not all one-equation models have been based on the turbulent kinetic energy equation. Nee and Kovasznay (1968) and, more recently, Baldwin and Barth (1990) and Spalart and Allmaras (1992) have devised model equations for the transport of the turbulent viscosity or a parameter proportional to the turbulent viscosity. To illustrate the approach, the model of Spalart and Allmaras will be outlined below.

The Spalart and Allmaras turbulent kinematic viscosity is given by

$$\nu_T = \tilde{\nu} f_{\nu 1} \tag{5.148b}$$

The parameter $\tilde{\nu}$ is obtained from the solution of the transport equation,

$$\frac{\partial \tilde{\nu}}{\partial t} + u_j \frac{\partial \tilde{\nu}}{\partial x_j} = \frac{1}{\sigma} \frac{\partial}{\partial x_k}\left[(\nu + \tilde{\nu})\frac{\partial \tilde{\nu}}{\partial x_k}\right] + c_{b1}(1 - f_{\nu 2})\tilde{S}\tilde{\nu} - c_{w1} f_w \left(\frac{\tilde{\nu}}{d}\right)^2 + \frac{c_{b2}}{\sigma}\frac{\partial \tilde{\nu}}{\partial x_k}\frac{\partial \tilde{\nu}}{\partial x_k} \tag{5.148c}$$

where the closure coefficients and functions are given by

$$c_{b1} = 0.1355 \qquad c_{b2} = 0.622 \qquad c_{\nu 1} = 7.1 \qquad \sigma = 2/3$$

$$c_{w1} = \frac{c_{b1}}{\kappa^2} + \frac{(1 + c_{b2})}{\sigma} \qquad c_{w2} = 0.3 \qquad c_{w3} = 2 \qquad \kappa = 0.41$$

$$f_{\nu 1} = \frac{\chi^3}{\chi^3 + c_{\nu 1}^3} \qquad f_{\nu 2} = 1 - \frac{\chi}{1 + \chi f_{\nu 1}} \qquad f_w = g\left(\frac{1 + c_{w3}^6}{g^6 + c_{w3}^6}\right) \qquad \chi = \frac{\tilde{\nu}}{\nu}$$

$$g = r + c_{w2}(r^6 - r) \qquad r = \frac{\tilde{\nu}}{\tilde{S}\kappa^2 d^2} \qquad \tilde{S} = S + \frac{\tilde{\nu}}{\kappa^2 d^2} f_{\nu 2}$$

$$S = \sqrt{2\Omega_{ij}\Omega_{ij}} \qquad \Omega_{ij} = \frac{1}{2}\left(\frac{\partial u_i}{\partial x_j} - \frac{\partial u_j}{\partial x_i}\right)$$

and d is the distance from the closest surface. Corrections for transition can be found in the work by Spalart and Allmaras (1992). A comparison of the performance of the Baldwin-Barth and Spalart-Allmaras models has been reported by Mani et al. (1995).

The one-equation model has been extended to compressible flow (Rubesin, 1976) and appears to provide a definite improvement over algebraic models. The recent one-equation models of (Baldwin-Barth and Spalart-Allmaras) have provided better agreement with experimental data for some separated flows than has generally been possible with algebraic models. On the whole, however, the performance of most one-equation models (for both incompressible and compressible flows) has been disappointing, in that relatively few cases have been observed in which these models offer an improvement over the predictions of the algebraic models.

5.4.6 One-and-One-Half- and Two-Equation Models

One conceptual advance made by moving from a purely algebraic mixing-length model to a one-equation model was that the latter permitted one model parameter to vary throughout the flow, being governed by a PDE of its own. In most one-equation models, a length parameter still appears that is generally evaluated by an algebraic expression dependent upon only *local* flow parameters. Researchers in turbulent flow have long felt that the length scale in turbulence models should also depend upon the upstream "history" of the flow and not just on local flow conditions. An obvious way to provide more complex dependence of l on the flow is to derive a transport equation for the variation of l. If the equation for l added to the system is an ODE, such as given by Eq. (5.144) for model 5F, the resulting model might well be termed a *one-and-one-half-equation model*. Such a model has been employed to predict separating external turbulent boundary layers (Pletcher, 1978), flow in annular passages with heat transfer (Malik and Pletcher, 1981), and plane and round jets (Minaie and Pletcher, 1982).

Frequently, the equation from which the length scale is obtained is a PDE, and the model is then referred to as a *two-equation turbulence model*. Although a transport PDE can be developed for a length scale, the terms in this equation are not easily modeled, and some workers have experienced better success by solving a transport equation for a length-scale-related parameter rather than the length scale itself.

Several combinations of variables have been used as the transported quantity, including the dissipation rate ε, $\bar{k}l$, ω, τ, and $\bar{k}\tau$, where $\omega = \varepsilon/\bar{k}$ is known as the specific dissipation rate and τ is a dissipation time, $\tau = 1/\omega$. Thus the length scale needed in the expression for turbulent viscosity, $\mu_T = C_D^{1/3}\rho l\bar{k}^{1/2}$, can be obtained from the solution for any of the combinations listed above. For example,

$$l = C_D\bar{k}^{3/2}/\varepsilon \qquad l = C_D\bar{k}^{1/2}/\omega \qquad l = C_D\bar{k}^{1/2}\tau$$

The most commonly used variable for a second transport equation is the dissipation rate ε. The versions of the $\bar{k}$-ε model used today can largely be traced back to the early work of Harlow and Nakayama (1968) and Jones and Launder (1972). The description here follows the work of Jones and Launder (1972) and Launder and Spalding (1974).

In terms of $\bar{k}$ and ε, the turbulent viscosity can be evaluated as $\mu_T = C_\mu \rho \bar{k}^2/\varepsilon$, where in terms of C_D introduced earlier, $C_\mu = C_D^{4/3}$. Although an exact equation can be derived for the transport of ε (see, for example, Wilcox, 1993), it provides relatively little guidance for modeling except to support the idea that the modeled equation should allow for the production, dissipation, diffusion, and convective transport of ε. The turbulent kinetic energy equation used in the $\bar{k}$-ε equation is given by Eq. (5.147) except that ε is maintained as an unknown. The transport equations used in the "standard" $\bar{k}$-ε model are as follows:

turbulent kinetic energy:

$$\rho \frac{D\bar{k}}{Dt} = \frac{\partial}{\partial x_j}\left[(\mu + \mu_T/\mathrm{Pr}_k)\frac{\partial \bar{k}}{\partial x_j}\right] + \left(2\mu_T S_{ij} - \frac{2}{3}\rho \bar{k}\delta_{ij}\right)\frac{\partial u_i}{\partial x_j} - \rho\varepsilon \quad (5.149)$$

dissipation rate:

$$\rho \frac{D\varepsilon}{Dt} = \frac{\partial}{\partial x_j}\left[(\mu + \mu_T/\mathrm{Pr}_\varepsilon)\frac{\partial \varepsilon}{\partial x_j}\right] + C_{\varepsilon 1}\frac{\varepsilon}{\bar{k}}\left(2\mu_T S_{ij} - \frac{2}{3}\rho \bar{k}\delta_{ij}\right)\frac{\partial u_i}{\partial x_j} - C_{\varepsilon 2}\rho\frac{\varepsilon^2}{\bar{k}} \quad (5.150a)$$

The terms on the right-hand side of Eq. (5.150*a*) from left to right can be interpreted as the diffusion, generation, and dissipation rates of ϵ. Typical values of the model constants are tabulated in Table 5.4.

The most common $\bar{k}$-ε closure for the Reynolds heat flux terms utilizes the same turbulent Prandtl number formulation as used with algebraic models, Eq. (5.137*a*). A number of modifications or adjustments to the $\bar{k}$-ε model have been suggested to account for effects not accounted for in the standard model such as buoyancy and streamline curvature.

Numerous other two-equation models have been proposed and evaluated. Several of these are described and discussed by Wilcox (1993). Of these other two-equation models, the $\bar{k}$-ω model in the form prescribed by Wilcox (1988) has probably been developed and tested the most extensively.

The standard $\bar{k}$-ε model given above is not appropriate for use in the viscous sublayer because the damping effect associated with solid boundaries has not been included in the model. Closure can be achieved with the use of the standard model by assuming that the law of the wall holds in the inner region and either using wall functions of a form described by Launder and Spalding (1974) or using a traditional damped mixing-length algebraic model and matching

Table 5.4 Model constants for $\bar{k}$-ϵ two-equation model

C_μ	$C_{\varepsilon 1}$	$C_{\varepsilon 2}$	Pr_k	Pr_ϵ	Pr_T
0.09	1.44	1.92	1.0	1.3	0.9

with the two-equation model by neglecting the convection and diffusion of $\bar{k}$ and ε. This provides a boundary condition for $\bar{k}$ at a point y_c within the law of the wall region given by Eq. (5.148a). Applying the strategy to ε gives

$$\varepsilon = \frac{C_D \bar{k}^{3/2}}{l} = \frac{C_D \left[\bar{k}(y_c)\right]^{3/2}}{\kappa y} \tag{5.150b}$$

A frequently used alternative is to employ a model based on transport equations for $\bar{k}$ and ε that have been modified by the addition of damping terms to extend applicability to the near-wall region. The models proposed by Jones and Launder (1972), Launder and Sharma (1974), Lam and Bremhorst (1981), and Chien (1982) are among the most commonly used of these *low Reynolds number* $\bar{k}$-ε *models*. The Re that is "low" in these models is the Re of turbulence, $\mathrm{Re}_T = \bar{k}^2/\varepsilon\nu$, where $\bar{k}^{3/2}/\varepsilon$ is used as a length scale for dissipation of kinetic energy. As an example, the low Reynolds number $\bar{k}$-ε model of Chien (1982) is given below with the modified or added terms identified by labels underneath.

turbulent kinetic energy:

$$\rho \frac{D\bar{k}}{Dt} = \frac{\partial}{\partial x_j}\left[(\mu + \mu_T/\mathrm{Pr}_k)\frac{\partial \bar{k}}{\partial x_j}\right] + \left(2\mu_T S_{ij} - \frac{2}{3}\rho\bar{k}\,\delta_{ij}\right)\frac{\partial u_i}{\partial x_j} - \rho\varepsilon - \underbrace{2\mu\frac{\bar{k}}{y^2}}_{\substack{\text{added}\\ \text{term}}} \tag{5.150c}$$

dissipation rate:

$$\rho \frac{D\varepsilon}{Dt} = \frac{\partial}{\partial x_j}\left[(\mu + \mu_T/\mathrm{Pr}_\varepsilon)\frac{\partial \varepsilon}{\partial x_j}\right] + C_{\varepsilon 1}\frac{\varepsilon}{\bar{k}}\left(2\mu_T S_{ij} - \frac{2}{3}\rho\bar{k}\,\delta_{ij}\right)\frac{\partial u_i}{\partial x_j} - \underbrace{C_{\varepsilon 2} f_1 \rho \frac{\varepsilon^2}{\bar{k}}}_{\substack{\text{modified}\\ \text{term}}} - \underbrace{f_2}_{\substack{\text{added}\\ \text{term}}} \tag{5.150d}$$

where the turbulent viscosity is evaluated as

$$\mu_T = C_\mu f_\mu \rho \bar{k}^2/\varepsilon \tag{5.150e}$$

and the additional functions are given by

$$f_\mu = 1 - e^{-0.0115y^+} \qquad f_1 = 1 - 0.22e^{-(\mathrm{Re}_T/6)^2} \qquad f_2 = -2\mu\frac{\varepsilon}{y^2}e^{-y^+/2}$$

The constants C_μ, Pr_k, Pr_ε have the same values as in the standard $\bar{k}$-ε model, but $C_{\varepsilon 1}$ and $C_{\varepsilon 2}$ are assigned slightly different values of 1.35 and 1.80, respectively.

Only a few of the large number of two-equation models proposed in the literature have been discussed in this section. A number of those omitted, particularly those based on renormalization group theory (Yakhot and Orszag, 1986), show much promise.

The literature on the application of two-equation turbulence models to compressible flows is relatively sparse (see Viegas et al., 1985; Viegas and Horstman, 1979; Coakley, 1983a; Wilcox, 1993). Details of modeling considerations for the compressible RANS equations will not be given here, but the compressible form of the $\bar{k}$-ε model equations (Coakley, 1983a) will be listed below in Favre variables for completeness.

turbulent kinetic energy:

$$\frac{\partial(\bar{\rho}\tilde{k})}{\partial t} + \frac{\partial}{\partial x_j}\left(\bar{\rho}\tilde{u}_j\tilde{k}\right) = \frac{\partial}{\partial x_j}\left[(\mu + \mu_T/\text{Pr}_k)\frac{\partial \tilde{k}}{\partial x_j}\right] + \left[2\mu_T\left(S_{ij} - \frac{1}{3}\delta_{ij}\frac{\partial \tilde{u}_k}{\partial x_k}\right) - \frac{2}{3}\bar{\rho}\tilde{k}\,\delta_{ij}\right]\frac{\partial \tilde{u}_i}{\partial x_j} - \bar{\rho}\tilde{\varepsilon} \tag{5.150f}$$

dissipation rate:

$$\frac{\partial}{\partial t}(\bar{\rho}\tilde{\varepsilon}) + \frac{\partial}{\partial x_j}\left(\bar{\rho}\tilde{u}_j\tilde{\epsilon}\right) = \frac{\partial}{\partial x_j}\left[(\mu + \mu_T/\text{Pr}_\varepsilon)\frac{\partial \tilde{\varepsilon}}{\partial x_j}\right] + C_{\varepsilon 1}\frac{\tilde{\varepsilon}}{\tilde{k}}\left[2\mu_T\left(S_{ij} - \frac{1}{3}\delta_{ij}\frac{\partial \tilde{u}_k}{\partial x_k}\right) - \frac{2}{3}\bar{\rho}\tilde{k}\,\delta_{ij}\right]\frac{\partial \tilde{u}_i}{\partial x_j} - C_{\varepsilon 2}\,\bar{\rho}\frac{\tilde{\varepsilon}^2}{\tilde{k}} \tag{5.150g}$$

Despite the enthusiasm that is noted from time to time over two-equation models, it is perhaps appropriate to point out again the two major restrictions on this type of model. First, two-equation models of the type discussed herein are merely turbulent *viscosity* models, which assume that the Boussinesq approximation [Eq. (5.129*a*)] holds. In algebraic models, μ_T is a local function, whereas in two-equation models, μ_T is a more general and complex function governed by two additional PDEs. If the Boussinesq approximation fails, then even two-equation models fail. Obviously, in many flows the Boussinesq approximation models reality closely enough for engineering purposes.

The second shortcoming of two-equation models is the need to make assumptions in evaluating the various terms in the model transport equations, especially in evaluating the third-order turbulent correlations. This same shortcoming, however, plagues all higher-order closure attempts. These model equations contain no magic; they only reflect the best understanding and intuition of the originators. We can be optimistic, however, that the models can be improved by improved modeling of these terms.

5.4.7 Reynolds Stress Models

By Reynolds stress models (sometimes called stress-equation models), we are referring to those Category II (second-order) closure models that *do not* assume that the turbulent shearing stress is proportional to the rate of mean strain. That is, for a 2-D incompressible flow,

$$-\rho\overline{u'v'} \neq \mu_T\left(\frac{\partial u}{\partial y} + \frac{\partial v}{\partial x}\right)$$

Such models are more general than those based on the Boussinesq assumption and can be expected to provide better predictions for flows with sudden changes in the mean strain rate or with effects such as streamline curvature or gradients in the Reynolds normal stresses. For example, without specific ad hoc adjustments, two-equation viscosity models are not able to predict the existence of the secondary flow patterns observed experimentally in turbulent flow through channels having a noncircular cross section.

Exact transport equations can be derived (Prob. 5.21) for the Reynolds stresses. This is accomplished for 2-D incompressible flow by multiplying the Navier-Stokes form of the ith momentum equation by the fluctuating velocity component u'_j and adding to it the product of u'_i and the jth momentum equation and then time averaging the result. That is, if we write the Navier-Stokes momentum equations in a form equal to zero, we form

$$\overline{u'_i N_j + u'_j N_i} = 0$$

where the N_i and N_j denote the ith and jth components of the Navier-Stokes momentum equation, respectively. Before the time averaging is carried out, the variables in the Navier-Stokes equations are replaced by the usual decomposition, $u_i = \bar{u}_i + u'_i$, etc. Intermediate steps are outlined by Wilcox (1993). The result can be written in the form

$$\rho\frac{D(\overline{-u'_i u'_j})}{Dt} = -\rho\overline{u'_i u'_k}\frac{\partial u_j}{\partial x_k} - \rho\overline{u'_j u'_k}\frac{\partial u_i}{\partial x_k} + \varepsilon_{ij} - \Pi_{ij} + \frac{\partial}{\partial x_k}\left(\mu\frac{\partial(\overline{-u'_i u'_j})}{\partial x_k} + C_{ijk}\right) \tag{5.150h}$$

where

$$\Pi_{ij} = \overline{p'\left(\frac{\partial u'_i}{\partial x_j} + \frac{\partial u'_j}{\partial x_i}\right)} \qquad \varepsilon_{ij} = 2\mu\overline{\frac{\partial u'_i}{\partial x_k}\frac{\partial u'_j}{\partial x_k}}$$

$$C_{ijk} = \rho\overline{u'_i u'_j u'_k} + \overline{p'u'_i}\,\delta_{jk} + \overline{p'u'_j}\,\delta_{ik}$$

Thus, in principle, transport equations can be solved for the six components of the Reynolds stress tensor. However, the equations contain 22 new unknowns that must be modeled first. Thus, we see again that in turbulence modeling, transport equations can be written for nearly anything of interest but none of them can be solved exactly. Models that employ transport PDEs for the Reynolds stresses are often referred to as second-order or second-moment closures.

Modeling for stress transport equations has followed the pioneering work of Rotta (1951). A commonly used example of the stress equation approach is the Reynolds stress model proposed by Launder et al. (1975), although numerous variations have since been suggested. Because of their computational complexity, the Reynolds stress models have not been widely used for engineering applications. Because they are not restricted by the Boussinesq approximation and because the closure contains the greatest number of model PDEs and constants of all the models considered, it would seem that the Reynolds stress models would have the best chance of emerging as "ultimate" turbulence models. Such ultimate turbulence models may eventually appear, but after more than 20 years of serious numerical research with these models, the results have been somewhat disappointing, considering the computational effort needed to implement the models.

Another Category II approach that has shown considerable promise recently is known as the *algebraic Reynolds stress model* (ASM). The idea here is to allow a nonlinear constitutive relationship between the Reynolds stresses and the rate of mean strain while avoiding the need to solve full PDEs for each of the six stresses. To do this, of course, requires modeling assumptions, just as modeling assumptions are needed to close the transport equations that arise in the development of a full Reynolds stress model. Many of the ASM models employ $\bar{k}$ and ε as parameters, and the working forms of the models often have the appearance of a two-equation model but with the constant in the expression for the turbulent viscosity being replaced by a function. Thus the computational effort required for such models is only slightly greater than that for a traditional two-equation model.

Two approaches have been followed in developing ASM models. Several researchers have simply proposed nonlinear relationships between the Reynolds stresses and the rate of mean strain, usually in the form of a series expansion having the Boussinesq approximation as the leading term (see for example, Lumley, 1970; Speziale, 1987, 1991). Such constitutive relationships are usually required to satisfy several physical and mathematical principles, such as Galilean invariance and "realizability." Realizability (Schumann, 1977; Lumley, 1978) requires that the turbulent normal stresses be positive and that Schwarz's inequality hold for fluctuating quantities, as, for example, $|\overline{ab}/\sqrt{\overline{a^2}\,\overline{b^2}}| \leqslant 1$, where a and b are fluctuating quantities. Without the realizability constraint, models (even the standard $\bar{k}$-ϵ model) may lead to nonphysical results, such as negative values for the turbulent kinetic energy. As an example, the ASM model of Shih et al. (1994) for incompressible flow will be summarized below.

The Reynolds stress in the Shih et al. (1994) ASM model is evaluated as

$$-\overline{u_i u_j} = 2C_\mu \frac{\bar{k}^2}{\varepsilon} S_{ij}^* - \frac{2}{3}\bar{k}\,\delta_{ij} - 2C_2 \frac{\bar{k}^3}{\varepsilon^2}(-S_{ik}^*\Omega_{kj} + \Omega_{ik}S_{kj}^*) \quad (5.150i)$$

where

$$S_{ij}^* = S_{ij} - \frac{1}{3}S_{kk}\,\delta_{ij} \qquad S_{ij} = \frac{1}{2}\left(\frac{\partial u_i}{\partial x_j} + \frac{\partial u_j}{\partial x_i}\right) \qquad \Omega_{ij} = \frac{1}{2}\left(\frac{\partial u_i}{\partial x_j} - \frac{\partial u_j}{\partial x_i}\right)$$

The constants in Eq. (5.150i) are given by

$$C_\mu = \frac{1}{6.5 + A_s^* U^* \bar{k}/\varepsilon} \qquad C_2 = \frac{\sqrt{1 - 9C_\mu^2 (S^* \bar{k}/\varepsilon)^2}}{1.0 + 6S^*\Omega^*\bar{k}^2/\varepsilon^2}$$

where

$$S^* = \sqrt{S_{ij}^* S_{ij}^*} \qquad \Omega^* = \sqrt{\Omega_{ij}\Omega_{ij}} \qquad U^* = \sqrt{S_{ij}^* S_{ij}^* + \Omega_{ij}\Omega_{ij}}$$

and

$$A_s^* = \sqrt{6}\cos\phi \qquad \phi = \cos^{-1}(\sqrt{6}\,W^*) \qquad W^* = \frac{S_{ij}^* S_{jk}^* S_{ki}^*}{(S^*)^3}$$

The values of $\bar{k}$ and ε are determined according to the standard $\bar{k}$-ε model given by Eqs. (5.149) and (5.150a).

The second approach, and the one followed by Rodi (1976), is to deduce a nonlinear algebraic equation for the Reynolds stresses by simplifying the full Reynolds stress PDE, Eq. (5.150h). Rodi argued that the convection minus the diffusion of Reynolds stress was proportional to the convection minus the diffusion of turbulent kinetic energy, with the proportionality factor being the ratio of the stress to the turbulent kinetic energy. That is,

$$\begin{aligned}
\rho \frac{D(-\overline{u_i' u_j'})}{Dt} &- \frac{\partial}{\partial x_k}\left(\mu \frac{\partial(-\overline{u_i' u_j'})}{\partial x_k} + C_{ijk}\right) \\
&= \frac{-\rho\overline{u_i' u_j'}}{\rho\bar{k}}\left[\rho\frac{D\bar{k}}{Dt} - \frac{\partial}{\partial x_j}\left(\mu\frac{\partial \bar{k}}{\partial x_j} - \frac{1}{2}\rho\overline{u_i' u_i' u_j'} - \overline{p' u_j'}\right)\right] \\
&= \frac{-\rho\overline{u_i' u_j'}}{\rho\bar{k}}\left(-\rho\overline{u_i' u_j'}\frac{\partial u_i}{\partial x_j} - \mu\overline{\frac{\partial u_i'}{\partial x_k}\frac{\partial u_i'}{\partial x_k}}\right)
\end{aligned} \quad (5.150j)$$

Simply put, this relates the difference between production and dissipation of Reynolds stresses to the difference between production and dissipation of turbulent kinetic energy. Thus, by defining

$$P = -\rho\overline{u_i' u_j'}\frac{\partial u_i}{\partial x_j} \qquad P_{ij} = -\rho\overline{u_i' u_k'}\frac{\partial u_j}{\partial x_k} - \rho\overline{u_j' u_k'}\frac{\partial u_i}{\partial x_k}$$

and recalling the definitions of ε, ε_{ij}, and Π_{ij}, we can directly write the following nonlinear algebraic relation:

$$\frac{-\rho\overline{u_i'u_j'}}{\rho\bar{k}}(P-\rho\varepsilon)=P_{ij}+\varepsilon_{ij}-\Pi_{ij}$$

With closure approximations for ε_{ij} and Π_{ij}, Rodi (1976) developed a nonlinear constitutive equation relating Reynolds stresses and rates of mean strain that contained $\bar{k}$ and ε as parameters. For thin shear layers the constitutive relation reduces to

$$-\overline{u'v'}=\frac{2}{3}\,\frac{(1-C_2)}{C_1}\,\frac{C_1-1+C_2(P/\varepsilon)}{[C_1-1+(P/\varepsilon)]^2}\,\frac{\bar{k}^2}{\varepsilon}\,\frac{\partial u}{\partial y} \tag{5.150k}$$

where $C_1 \cong 1.5$ and $C_2 \cong 0.6$. The above can be viewed as a variation of the $\bar{k}$-ε eddy viscosity model in which the "constant" (C_μ) is replaced by a function of P/ε. Notice that when $P=\varepsilon$ (production and dissipation of turbulent kinetic energy are in balance), the function takes on the standard value of C_μ in the $\bar{k}$-ε model, 0.09.

Because the Boussinesq assumption is not invoked directly in the ASM models, they are considered to belong in Category II. However, they are not generally considered to be second-order or second-moment models because they are not based on solutions to modeled forms of the full-transport PDEs for Reynolds stresses.

The algebraic Reynolds stress models have provided a means of accounting for a number of effects, including streamline curvature, rotation, and buoyancy that cannot be predicted without ad hoc adjustments to standard two-equation models. They can predict the Reynolds-stress-driven secondary flow patterns in noncircular ducts, for example. However, they have generally not performed as well as full Reynolds stress models for flows with sudden changes in mean strain rate.

5.4.8 Subgrid-Scale Models for Large-Eddy Simulation

In LES the effect of the subgrid-scale (SGS) stresses must be modeled. Because the small-scale motion tends to be fairly isotropic and universal, there is hope that a relatively simple model will suffice. The earliest and simplest model was proposed by Smagorinsky (1963). It takes the form of a mixing-length or gradient diffusion model with the length, $l_s = C_s\Delta$, being proportional to the filter width. Thus the SGS stress tensor is represented by

$$\tau_{ij}=2\mu_T S_{ij} \tag{5.150l}$$

where S_{ij} is the rate of strain tensor,

$$S_{ij}=\frac{1}{2}\left(\frac{\partial\bar{u}_i}{\partial x_j}+\frac{\partial\bar{u}_j}{\partial x_i}\right) \tag{5.150m}$$

as before, and

$$\mu_T = \rho(C_s\Delta)^2\sqrt{2S_{ij}S_{ij}} \tag{5.150n}$$

The Smagorinsky constant C_s, unfortunately, is not universal. Values ranging from 0.1 to 0.24 have been reported. For wall shear flows it is common to multiply the mixing length by a Van Driest–type exponential damping function (Moin and Kim, 1982; Piomelli, 1988) to force the length scale to approach zero at the wall.

A number of more complex models have been proposed for LES. One of the most promising model variations is the dynamic SGS model proposed by Germano et al. (1991). In its most basic form, the dynamic model follows the Smagorinsky mixing-length format but provides a basis for the value of the modeling constant C_s to be computed as part of the solution by using filters of two sizes. This approach has been extended to the transport of a scalar and compressible flow (Moin et al., 1991). When applied to flows with heat transfer, the dynamic model permits the turbulent Prandtl number to be computed as part of the solution rather than being specified a priori, and the van Driest damping function is not required to provide the correct limiting behavior near solid boundaries. Very good results have been reported with the use of the model to date (see, for example, Yang and Ferziger, 1993; Akselvoll and Moin, 1993; Wang and Pletcher 1995a, 1995b).

5.5 EULER EQUATIONS

Prandtl discovered in 1904 (see Section 5.3.1) that for sufficiently large Re the important viscous effects are confined to a thin boundary layer near the surface of a solid boundary. As a consequence of this discovery, the inviscid (non-viscous, nonconducting) portion of the flow field can be solved independently of the boundary layer. Of course, this is only true if the boundary layer is very thin compared to the characteristic length of the flow field, so that the interaction between the boundary layer and the inviscid portion of the flow field is negligible. For flows in which the interaction is not negligible, it is still possible to use separate sets of equations for the two regions, but the equations must be solved in an iterative fashion. This iterative procedure can be computationally inefficient, and as a result, it is sometimes desirable to use a single set of equations that remain valid throughout the flow field. Equations of this latter type are discussed in Chapter 8.

In the present section, a reduced set of equations will be discussed that are valid only in the inviscid portion of the flow field. These equations are obtained by dropping both the viscous terms and the heat-transfer terms from the complete Navier-Stokes equations. The resulting equations can be numerically solved (see Chapter 6) using less computer time than is required for the complete Navier-Stokes equations. We will refer to these simplified equations as the *Euler equations*, although strictly speaking, Euler's name should be attached only to the inviscid momentum equation. In addition to the assumption of inviscid flow, it will also be assumed that there is no external heat transfer, so that the heat-generation term $\partial Q/\partial t$ in the energy equation can be dropped.

5.5.1 Continuity Equation

The continuity equation does not contain viscous terms or heat transfer terms, so that the various forms of the continuity equation given in Section 5.1.1 cannot be simplified for an inviscid flow. However, if the steady form of the continuity equation reduces to two terms for a given coordinate system, it becomes possible to discard the continuity equation by introducing the so-called *stream function* ψ. This holds true whether the flow is viscous or nonviscous. For example, the continuity equation for a 2-D steady compressible flow in Cartesian coordinates is

$$\frac{\partial}{\partial x}(\rho u) + \frac{\partial}{\partial y}(\rho v) = 0 \tag{5.151}$$

If the stream function ψ is defined such that

$$\begin{aligned} \rho u &= \frac{\partial \psi}{\partial y} \\ \rho v &= -\frac{\partial \psi}{\partial x} \end{aligned} \tag{5.152}$$

it can be seen by substitution that Eq. (5.151) is satisfied. Hence the continuity equation does not need to be solved, and the number of dependent variables is reduced by 1. The disadvantage is that the velocity derivatives in the remaining equations are replaced using Eqs. (5.152), so that these remaining equations will now contain derivatives that are one order higher. The physical significance of the stream function is obvious when we examine

$$\begin{aligned} d\psi &= \frac{\partial \psi}{\partial x}\,dx + \frac{\partial \psi}{\partial y}\,dy = -\rho v\,dx + \rho u\,dy \\ &= \rho \mathbf{V} \cdot d\mathbf{A} = d\dot{m} \end{aligned} \tag{5.153}$$

We see that lines of constant $\psi (d\psi = 0)$ are lines across which there is no mass flow ($d\dot{m} = 0$). A *streamline* is defined as a line in the flow field whose tangent at any point is in the same direction as the flow at that point. Hence lines of constant ψ are streamlines, and the difference between the values of ψ for any two streamlines represents the mass flow rate per unit width between those streamlines.

For an incompressible 2-D flow the continuity equation in Cartesian coordinates is

$$\frac{\partial u}{\partial x} + \frac{\partial v}{\partial y} = 0 \tag{5.154}$$

and the stream function is defined by

$$u = \frac{\partial \psi}{\partial y}$$
$$v = -\frac{\partial \psi}{\partial x} \tag{5.155}$$

For a steady, axially symmetric compressible flow in cylindrical coordinates (see Section 5.1.8), the continuity equation is given by

$$\frac{1}{r}\frac{\partial}{\partial r}(r\rho u_r) + \frac{\partial}{\partial z}(\rho u_z) = 0 \tag{5.156}$$

and the stream function is defined by

$$\rho u_r = \frac{1}{r}\frac{\partial \psi}{\partial z}$$
$$\rho u_z = -\frac{1}{r}\frac{\partial \psi}{\partial r} \tag{5.157}$$

For the case of 3-D flows, it is possible to use stream functions to replace the continuity equation. However, the complexity of this approach usually makes it less attractive than using the continuity equation in its original form.

5.5.2 Inviscid Momentum Equations

When the viscous terms are dropped from the Navier-Stokes equations [Eq. (5.18)], the following equation results:

$$\rho\frac{D\mathbf{V}}{Dt} = \rho\mathbf{f} - \boldsymbol{\nabla} p \tag{5.158}$$

This equation was first derived by Euler in 1755 and has been named Euler's equation. If we neglect body forces and assume steady flow, Euler's equation reduces to

$$\mathbf{V}\cdot\boldsymbol{\nabla}\mathbf{V} = -\frac{1}{\rho}\boldsymbol{\nabla} p \tag{5.159}$$

Integrating this equation along a line in the flow field gives

$$\int(\mathbf{V}\cdot\boldsymbol{\nabla}\mathbf{V})\cdot d\mathbf{r} = -\int\frac{1}{\rho}\boldsymbol{\nabla} p\cdot d\mathbf{r} \tag{5.160}$$

where $d\mathbf{r}$ is the differential length of the line. For a Cartesian coordinate system, $d\mathbf{r}$ is defined by

$$d\mathbf{r} = dx\,\mathbf{i} + dy\,\mathbf{j} + dz\,\mathbf{k} \tag{5.161}$$

Let us assume that the line is a streamline. Hence $\mathbf{V}$ has the same direction as $d\mathbf{r}$, and we can simplify the integrand on the left side of Eq. (5.160) in the following manner:

$$(\mathbf{V} \cdot \nabla \mathbf{V}) \cdot d\mathbf{r} = V \frac{\partial \mathbf{V}}{\partial r} \cdot d\mathbf{r} = V \frac{\partial V}{\partial r} dr = V dV = d\left(\frac{V^2}{2}\right)$$

Likewise, the integrand on the right-hand side becomes

$$\frac{1}{\rho} \nabla p \cdot d\mathbf{r} = \frac{dp}{\rho}$$

and Eq. (5.160) reduces to

$$\frac{V^2}{2} + \int \frac{dp}{\rho} = \text{const} \tag{5.162}$$

The integral in this equation can be evaluated if the flow is assumed *barotropic*. A barotropic fluid is one in which ρ is a function only of p (or a constant), i.e., $\rho = \rho(p)$. Examples of barotropic flows are as follows.

1. steady incompressible flow:

$$\rho = \text{const} \tag{5.163}$$

2. isentropic (constant entropy) flow (see Section 5.5.4):

$$\rho = (\text{const}) p^{1/\gamma} \tag{5.164}$$

Thus for an incompressible flow, the integrated Euler's equation [Eq. (5.162)] becomes

$$p + \tfrac{1}{2}\rho V^2 = \text{const} \tag{5.165}$$

which is called *Bermoulli's equation*. For an isentropic, compressible flow, Eq. (5.162) can be expressed as

$$\frac{V^2}{2} + \frac{\gamma}{\gamma - 1} \frac{p}{\rho} = \text{const} \tag{5.166}$$

which is sometimes referred to as the *compressible Bernoulli equation*. It should be remembered that Eqs. (5.165) and (5.166) are valid only along a given streamline, since the constants appearing in these equations can vary between streamlines.

We will now show that Eqs. (5.165) and (5.166) can be made valid everywhere in the flow field if the flow is assumed *irrotational*. An irrotational flow is one in which the fluid particles do not rotate about their axes. From the study of kinematics (see, for example, Owczarek, 1964), the vorticity $\boldsymbol{\zeta}$, which is defined by

$$\boldsymbol{\zeta} = \nabla \times \mathbf{V} \tag{5.167}$$

is equivalent to twice the angular velocity of a fluid particle. Thus for an irrotational flow,

$$\boldsymbol{\zeta} = \nabla \times \mathbf{V} = 0 \tag{5.168}$$

and as a result, we can express $\mathbf{V}$ as the gradient of a single-valued point function ϕ, since

$$\nabla \times \mathbf{V} = \nabla \times (\nabla\phi) = 0 \tag{5.169}$$

The scalar ϕ is called the *velocity potential.* Also, from kinematics, the acceleration of a fluid particle, $D\mathbf{V}/Dt$, is given by

$$\frac{D\mathbf{V}}{Dt} = \frac{\partial \mathbf{V}}{\partial t} + \nabla\left(\frac{V^2}{2}\right) - \mathbf{V} \times \boldsymbol{\zeta} \tag{5.170}$$

which is called *Lagrange's acceleration formula*. For an irrotational flow, this equation reduces to

$$\frac{D\mathbf{V}}{Dt} = \frac{\partial \mathbf{V}}{\partial t} + \nabla\left(\frac{V^2}{2}\right)$$

which can be substituted into Euler's equation to give

$$\frac{\partial \mathbf{V}}{\partial t} + \nabla\left(\frac{V^2}{2}\right) = \mathbf{f} - \frac{1}{\rho}\nabla p \tag{5.171}$$

If we again neglect body forces and assume steady flow, Eq. (5.171) can be rewritten as

$$\nabla\left(\frac{V^2}{2} + \int \frac{dp}{\rho}\right) = 0 \tag{5.172}$$

since

$$\nabla \int \frac{dp}{\rho} \cdot d\mathbf{r} = \frac{\nabla p}{\rho} \cdot d\mathbf{r}$$

Integrating Eq. (5.172) along any arbitrary line in the flow field yields

$$\frac{V^2}{2} + \int \frac{dp}{\rho} = \text{const} \tag{5.173}$$

The constant in this equation now has the same value everywhere in the flow field, since Eq. (5.173) was integrated along any arbitrary line. The incompressible Bernoulli equation [Eq. (5.165)] and the compressible Bernoulli equation [Eq. (5.166)] follow directly from Eq. (5.173) in the same manner as before. The only difference is that the resulting equations are now valid everywhere in the inviscid flow field because of our additional assumption of irrotationality.

For the special case of an inviscid incompressible irrotational flow, the continuity equation

$$\nabla \cdot \mathbf{V} = 0 \tag{5.174}$$

can be combined with

$$\mathbf{V} = \nabla\phi \tag{5.175}$$

to give Laplace's equation

$$\nabla^2\phi = 0 \tag{5.176}$$

5.5.3 Inviscid Energy Equations

The inviscid form of the energy equation given by Eq. (5.22) becomes

$$\frac{\partial E_t}{\partial t} + \nabla \cdot E_t\mathbf{V} = \rho\mathbf{f} \cdot \mathbf{V} - \nabla \cdot (p\mathbf{V}) \tag{5.177}$$

which is equivalent to

$$\frac{\partial}{\partial t}(\rho H) + \nabla \cdot (\rho H \mathbf{V}) = \rho\mathbf{f} \cdot \mathbf{V} + \frac{\partial p}{\partial t} \tag{5.178}$$

Additional forms of the inviscid energy equation can be obtained from Eq. (5.29),

$$\rho\frac{De}{Dt} + p(\nabla \cdot \mathbf{V}) = 0 \tag{5.179}$$

and from Eq. (5.33),

$$\rho\frac{Dh}{Dt} = \frac{Dp}{Dt} \tag{5.180}$$

If we use the continuity equation and ignore the body force term, Eq. (5.178) can be written as

$$\frac{DH}{Dt} = \frac{1}{\rho}\frac{\partial p}{\partial t} \tag{5.181}$$

which for a steady flow becomes

$$\mathbf{V} \cdot \nabla H = 0 \tag{5.182}$$

This equation can be integrated along a streamline to give

$$H = h + \frac{V^2}{2} = \text{const} \tag{5.183}$$

The constant will remain the same throughout the inviscid flow field for the special case of an isoenergetic (homenergic) flow.

For an incompressible flow, Eq. (5.179) reduces to

$$\frac{De}{Dt} = 0 \tag{5.184}$$

which, for a steady flow, implies that the internal energy is constant along a streamline.

5.5.4 Additional Equations

The conservation equations for an inviscid flow have been presented in this section. It is possible to derive additional relations that prove to be quite useful in particular applications. In some cases, these auxiliary equations can be used to replace one or more of the conservation equations. Several of the auxiliary equations are based on the First and Second Laws of Thermodynamics, which provide the relation

$$T\,ds = de + p\,d\left(\frac{1}{\rho}\right) \tag{5.185}$$

where s is the entropy. Using the definition of enthalpy,

$$h = e + \frac{p}{\rho}$$

it is possible to rewrite Eq. (5.185) as

$$T\,ds = dh - \frac{dp}{\rho} \tag{5.186}$$

This latter equation can also be written as

$$T\,\nabla s = \nabla h - \frac{\nabla p}{\rho}$$

since at any given instant, a fluid particle can change its state to that of a neighboring particle. Upon combining this equation with Eqs. (5.170) and (5.158) and ignoring body forces, we obtain

$$\frac{\partial \mathbf{V}}{\partial t} - \mathbf{V} \times \boldsymbol{\zeta} = T\,\nabla s - \nabla h - \nabla\left(\frac{V^2}{2}\right)$$

or

$$\frac{\partial \mathbf{V}}{\partial t} - \mathbf{V} \times \boldsymbol{\zeta} = T\,\nabla s - \nabla H \tag{5.187}$$

which is called *Crocco's equation.* This equation provides a relation between vorticity and entropy. For a steady flow it becomes

$$\mathbf{V} \times \boldsymbol{\zeta} = \nabla H - T\,\nabla s \tag{5.188}$$

We have shown earlier that for a steady inviscid adiabatic flow,

$$\mathbf{V} \cdot \nabla H = 0$$

which if combined with Eq. (5.188), gives

$$\mathbf{V} \cdot \nabla s = 0$$

since $\mathbf{V} \times \boldsymbol{\zeta}$ is normal to $\mathbf{V}$. Thus we have proved that entropy remains constant along a streamline for a steady, nonviscous, nonconducting, adiabatic flow. This is called an *isentropic* flow. If we also assume that the flow is irrotational and isoenergetic, then Crocco's equation tells us that the entropy remains constant everywhere (i.e., homentropic flow).

The thermodynamic relation given by Eq. (5.185) involves only changes in properties, since it does not contain path-dependent functions. For the isentropic flow of a perfect gas it can be written as

$$T\,ds = 0 = c_p\,dT - RT\frac{dp}{p}$$

or

$$\frac{dp}{p} = \frac{\gamma}{\gamma - 1}\frac{dT}{T}$$

The latter equation can be integrated to yield

$$\frac{p}{T^{\gamma/(\gamma-1)}} = \text{const}$$

which becomes

$$\frac{p}{\rho^{\gamma}} = \text{const} \tag{5.189}$$

after substituting the perfect gas equation of state. The latter isentropic relation was used earlier to derive the compressible Bernoulli equation [Eq. (5.166)]. It is interesting to note that the integrated energy equation, given by Eq. (5.183), can be made identical to Eq. (5.166) if the flow is assumed to be isentropic.

The speed of sound is given by

$$a = \sqrt{\left(\frac{\partial p}{\partial \rho}\right)_s} \tag{5.190}$$

where the subscript s indicates a constant entropy process. At a point in the flow of a perfect gas, Eqs. (5.189) and (5.190) can be combined to give

$$a = \sqrt{\frac{dp}{d\rho}} = \sqrt{\frac{\gamma p}{\rho}} = \sqrt{\gamma RT} \tag{5.191}$$

5.5.5 Vector Form of Euler Equations

The compressible Euler equations in Cartesian coordinates without body forces or external heat addition can be written in vector form as

$$\frac{\partial \mathbf{U}}{\partial t} + \frac{\partial \mathbf{E}}{\partial x} + \frac{\partial \mathbf{F}}{\partial y} + \frac{\partial \mathbf{G}}{\partial z} = 0 \tag{5.192}$$

where **U**, **E**, **F**, and **G** are vectors given by

$$\mathbf{U} = \begin{bmatrix} \rho \\ \rho u \\ \rho v \\ \rho w \\ E_t \end{bmatrix} \qquad \mathbf{E} = \begin{bmatrix} \rho u \\ \rho u^2 + p \\ \rho u v \\ \rho u w \\ (E_t + p)u \end{bmatrix}$$

$$\mathbf{F} = \begin{bmatrix} \rho v \\ \rho u v \\ \rho v^2 + p \\ \rho v w \\ (E_t + p)v \end{bmatrix} \qquad \mathbf{G} = \begin{bmatrix} \rho w \\ \rho u w \\ \rho v w \\ \rho w^2 + p \\ (E_t + p)w \end{bmatrix}$$

For a steady isoenergetic flow of a perfect gas, it becomes possible to remove the energy equation from the vector set and use, instead, the algebraic form of the equation given by Eq. (5.166). This reduces the overall computation time, since one less PDE needs to be solved.

5.5.6 Simplified Forms of Euler Equations

The Euler equations can be simplified by making additional assumptions. If the flow is steady, irrotational, and isentropic, the Euler equations can be combined into a single equation called the *velocity potential equation*. The velocity potential equation is derived in the following manner. In a Cartesian coordinate system, the continuity equation may be written as

$$\frac{\partial}{\partial x}(\rho\phi_x) + \frac{\partial}{\partial y}(\rho\phi_y) + \frac{\partial}{\partial z}(\rho\phi_z) = 0 \tag{5.193}$$

where the velocity components have been replaced by

$$u = \frac{\partial \phi}{\partial x} \qquad v = \frac{\partial \phi}{\partial y} \qquad w = \frac{\partial \phi}{\partial z} \tag{5.194}$$

The momentum (and energy) equations reduce to Eq. (5.162) with the assumptions of steady, irrotational, and isentropic flow. In differential form this equation becomes

$$dp = -\rho d\left(\frac{V^2}{2}\right) = -\rho d\left(\frac{\phi_x^2 + \phi_y^2 + \phi_z^2}{2}\right) \tag{5.195}$$

Combining Eqs. (5.190) and (5.195) yields the equation

$$d\rho = -\frac{\rho}{a^2} d\left(\frac{\phi_x^2 + \phi_y^2 + \phi_z^2}{2}\right) \tag{5.196}$$

which may be used to find the derivatives of ρ in each direction. After substituting these expressions for ρ_x, ρ_y, and ρ_z into Eq. (5.193) and simplifying, the velocity potential equation is obtained:

$$\left(1 - \frac{\phi_x^2}{a^2}\right)\phi_{xx} + \left(1 - \frac{\phi_y^2}{a^2}\right)\phi_{yy} + \left(1 - \frac{\phi_z^2}{a^2}\right)\phi_{zz} - \frac{2\phi_x\phi_y}{a^2}\phi_{xy} - \frac{2\phi_x\phi_z}{a^2}\phi_{xz} - \frac{2\phi_y\phi_z}{a^2}\phi_{yz} = 0 \tag{5.197}$$

Note that for an incompressible flow ($a \to \infty$), the velocity potential equation reduces to Laplace's equation.

The Euler equations can be further simplified if we consider the flow over a slender body where the free stream is only slightly disturbed (perturbed). An example is the flow over a thin airfoil. An analysis of this type is an example of small-perturbation theory. In order to demonstrate how the velocity potential equation can be simplified for flows of this type, we assume that a slender body is placed in a 2-D flow. The body causes a disturbance of the uniform flow, and the velocity components are written as

$$\begin{aligned} u &= U_\infty + u' \\ v &= v' \end{aligned} \tag{5.198}$$

where the prime denotes perturbation velocity. If we let ϕ' be the perturbation velocity potential, then

$$\begin{aligned} u &= \frac{\partial \phi}{\partial x} = U_\infty + \frac{\partial \phi'}{\partial x} \\ v &= \frac{\partial \phi}{\partial y} = \frac{\partial \phi'}{\partial y} \end{aligned} \tag{5.199}$$

Substituting these expressions along with Eq. (5.191) into Eq. (5.166) gives

$$a^2 = a_\infty^2 - \frac{\gamma - 1}{2}\left[2u'U_\infty + (u')^2 + (v')^2\right] \tag{5.200}$$

which can then be combined with the velocity potential equation to yield

$$\begin{aligned} (1 - M_\infty^2)\frac{\partial u'}{\partial x} + \frac{\partial v'}{\partial y} &= M_\infty^2\left[(\gamma + 1)\frac{u'}{U_\infty} + \left(\frac{\gamma + 1}{2}\right)\frac{(u')^2}{U_\infty^2} + \left(\frac{\gamma - 1}{2}\right)\frac{(v')^2}{U_\infty^2}\right]\frac{\partial u'}{\partial x} \\ &+ M_\infty^2\left[(\gamma - 1)\frac{u'}{U_\infty} + \left(\frac{\gamma + 1}{2}\right)\frac{(v')^2}{U_\infty^2} + \left(\frac{\gamma - 1}{2}\right)\frac{(u')^2}{U_\infty^2}\right]\frac{\partial v'}{\partial y} \\ &+ M_\infty^2\frac{v'}{U_\infty}\left(1 + \frac{u'}{U_\infty}\right)\left(\frac{\partial u'}{\partial y} + \frac{\partial v'}{\partial x}\right) \end{aligned} \tag{5.201}$$

Since the flow is only slightly disturbed from the free stream, we assume

$$\frac{u'}{U_\infty}, \frac{v'}{U_\infty} \ll 1$$

As a result, Eq. (5.200) simplifies to

$$a^2 = a_\infty^2 - (\gamma - 1)u'U_\infty \tag{5.202}$$

and Eq. (5.201) becomes

$$\left[\frac{1 - M_\infty^2}{M_\infty^2} - (\gamma + 1)\frac{u'}{U_\infty}\right] M_\infty^2 \phi'_{xx} + \phi'_{yy} = 0 \tag{5.203}$$

The latter equation is called the *transonic small-disturbance equation.* This nonlinear equation is either elliptic or hyperbolic, depending on whether the flow is subsonic or supersonic.

For flows in the subsonic or supersonic regimes, the magnitude of the term $M_\infty^2(\gamma + 1)(u'/U_\infty)\phi'_{xx}$ is small in comparison with $(1 - M_\infty^2)\phi'_{xx}$, and Eq. (5.203) reduces to the linear *Prandtl-Glauert equation*:

$$(1 - M_\infty^2)\phi'_{xx} + \phi'_{yy} = 0 \tag{5.204}$$

Once the perturbation velocity potential is known, the pressure coefficient can be determined from

$$C_p = \frac{p - p_\infty}{\frac{1}{2}\rho U_\infty^2} = \frac{2}{\gamma M_\infty^2}\left(\frac{p}{p_\infty} - 1\right) = -\frac{2u'}{U_\infty} \tag{5.205}$$

which is derived using Eqs. (5.166), (5.189), (5.198), and the binomial expansion theorem.

5.5.7 Shock Equations

A shock wave is a very thin region in a supersonic flow, across which there is a large variation in the flow properties. Because these variations occur in such a short distance, viscosity and heat conductivity play dominant roles in the structure of the shock wave. However, unless one is interested in studying the structure of the shock wave, it is usually possible to consider the shock wave to be infinitesimally thin (i.e., a mathematical discontinuity) and use the Euler equations to determine the changes in flow properties across the shock wave. For example, let us consider the case of a stationary straight shock wave oriented perpendicular to the flow direction (i.e., a normal shock). The flow is in the positive x direction, and the conditions upstream of the shock wave are designated with a subscript 1, while the conditions downstream are designated with a subscript 2. Since a shock wave is a weak solution to the hyperbolic Euler equations, we can apply the theory of weak solutions, described in Section 4.4, to Eq. (5.192). For the present discontinuity, this gives

$$[\mathbf{E}] = 0$$

or

$$\mathbf{E}_1 = \mathbf{E}_2$$

Thus

$$\begin{aligned} \rho_1 u_1 &= \rho_2 u_2 \\ p_1 + \rho_1 u_1^2 &= p_2 + \rho_2 u_2^2 \\ \rho_1 u_1 v_1 &= \rho_2 u_2 v_2 \\ (E_{t_1} + p_1) u_1 &= (E_{t_2} + p_2) u_2 \end{aligned}$$

Upon simplifying the above shock relations, we find that

$$\begin{aligned} \rho_1 u_1 &= \rho_2 u_2 \\ p_1 + \rho_1 u_1^2 &= p_2 + \rho_2 u_2^2 \\ v_1 &= v_2 \\ h_1 + \frac{u_1^2}{2} &= h_2 + \frac{u_2^2}{2} \end{aligned} \tag{5.206}$$

Solving these equations for the pressure ratio across the shock, we obtain

$$\frac{p_2}{p_1} = \frac{(\gamma + 1)\rho_2 - (\gamma - 1)\rho_1}{(\gamma + 1)\rho_1 - (\gamma - 1)\rho_2} \tag{5.207}$$

Equation (5.207) relates thermodynamic properties across the shock wave and is called the *Rankine-Hugoniot equation.* The label "Rankine-Hugoniot relations" is frequently applied to all equations that relate changes across shock waves.

For shock waves inclined to the free stream (i.e., oblique shocks) the shock relations becomes

$$\begin{aligned} \rho_1 V_{n_1} &= \rho_2 V_{n_2} \\ p_1 + \rho_1 V_{n_1}^2 &= p_2 + \rho_2 V_{n_2}^2 \\ V_{t_1} &= V_{t_2} \\ h_1 + \frac{V_1^2}{2} &= h_2 + \frac{V_2^2}{2} \end{aligned} \tag{5.208}$$

where V_n and V_t are the normal and tangential components of the velocity vector, respectively. These equations also apply to moving shock waves if the velocity components are measured with respect to the moving shock wave. In this case, the normal component of the flow velocity ahead of the shock (measured with respect to the shock) can be related to the pressure behind the shock by manipulating the above equations to form

$$V_{n_1}^2 = \frac{\gamma + 1}{2} \frac{p_1}{\rho_1} \left(\frac{p_2}{p_1} + \frac{\gamma - 1}{\gamma + 1} \right) \tag{5.209}$$

This latter equation is useful when attempting to numerically treat moving shock waves as discontinuities, as seen in Chapter 6. A comprehensive listing of shock relations is available in NACA Report 1135 (Ames Research Staff, 1953).

5.6 TRANSFORMATION OF GOVERNING EQUATIONS

The classical governing equations of fluid dynamics have been presented in this chapter. These equations have been written in either vector or tensor form. In Section 5.1.8, it was shown how these equations can be expressed in terms of any generalized orthogonal curvilinear coordinate system. For many applications, however, a nonorthogonal coordinate system is desirable. In this section, we will show how the governing equations can be transformed from a Cartesian coordinate system to any general nonorthogonal (or orthogonal) coordinate system. In the process, we will demonstrate how simple transformations can be used to cluster grid points in regions of large gradients such as boundary layers and how to transform a nonrectangular computational region in the physical plane into a rectangular uniformly-spaced grid in the computational plane. These latter transformations are simple examples from a very important topic of computational fluid dynamics called grid generation. A complete discussion of grid generation is presented in Chapter 10.

5.6.1 Simple Transformations

In this section, simple independent variable transformations are used to illustrate how the governing fluid dynamic equations are transformed. As a first example, we will consider the problem of clustering grid points near a wall. Refinement of the mesh near a wall is mandatory, in most cases, if the details of the boundary layer are to be properly resolved. Figure 5.8(a) shows a mesh above a flat plate in which grid points are clustered near the plate in the normal direction (y), while the spacing in the x direction is uniform. Because the spacing is not uniform in the y direction, it is convenient to apply a transformation to the y coordinate, so that the governing equations can be solved on a uniformly spaced grid in the computational plan $(\bar{x}, \bar{y})$ as seen in Fig. 5.8(b). A suitable

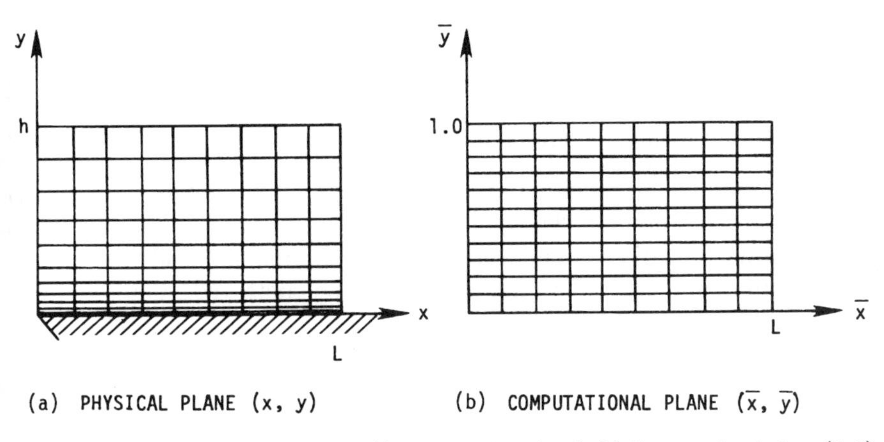

Figure 5.8 Grid clustering near a wall. (a) Physical plane (x, y). (b) Computational plane $(\bar{x}, \bar{y})$.

transformation for a 2-D boundary-layer type of problem is given by the following.

Transformation 1:

$$\bar{x} = x$$
$$\bar{y} = 1 - \frac{\ln\{[\beta + 1 - (y/h)]/[\beta - 1 + (y/h)]\}}{\ln[(\beta + 1)/(\beta - 1)]} \qquad 1 < \beta < \infty \tag{5.210}$$

This stretching transformation clusters more points near $y = 0$ as the stretching parameter β approaches 1.

In order to apply this transformation to the governing fluid dynamic equations, the following partial derivatives are formed:

$$\frac{\partial}{\partial x} = \frac{\partial \bar{x}}{\partial x}\frac{\partial}{\partial \bar{x}} + \frac{\partial \bar{y}}{\partial x}\frac{\partial}{\partial \bar{y}}$$
$$\frac{\partial}{\partial y} = \frac{\partial \bar{x}}{\partial y}\frac{\partial}{\partial \bar{x}} + \frac{\partial \bar{y}}{\partial y}\frac{\partial}{\partial \bar{y}} \tag{5.211}$$

where

$$\frac{\partial \bar{x}}{\partial x} = 1 \qquad \frac{\partial \bar{y}}{\partial x} = 0$$
$$\frac{\partial \bar{x}}{\partial y} = 0 \qquad \frac{\partial \bar{y}}{\partial y} = \frac{2\beta}{h\left\{\beta^2 - [1 - (y/h)]^2\right\}\ln[(\beta + 1)/(\beta - 1)]}$$

As a result, the partial derivatives simplify to

$$\frac{\partial}{\partial x} = \frac{\partial}{\partial \bar{x}}$$
$$\frac{\partial}{\partial y} = \left(\frac{\partial \bar{y}}{\partial y}\right)\frac{\partial}{\partial \bar{y}} \tag{5.212}$$

If we now apply this transformation to the steady 2-D incompressible continuity equation written in Cartesian coordinates,

$$\frac{\partial u}{\partial x} + \frac{\partial v}{\partial y} = 0 \tag{5.213}$$

the following transformed equation is obtained:

$$\frac{\partial u}{\partial \bar{x}} + \left(\frac{\partial \bar{y}}{\partial y}\right)\frac{\partial v}{\partial \bar{y}} = 0 \tag{5.214}$$

This transformed equation can now be differenced on the uniformly spaced grid in the computational plane. The grid spacing can be computed from

$$\Delta \bar{x} = \frac{L}{NI - 1}$$
$$\Delta \bar{y} = \frac{1}{NJ - 1} \tag{5.215}$$

where NI and NJ are the number of grid points in the x and y directions, respectively. We note that the expression for the metric $\partial \bar{y}/\partial y$ contains y, so that we must be able to express y as a function of $\bar{y}$. This is referred to as the inverse of the transformation. For the present transformation, given by Eqs. (5.210), the inverse can be readily found as

$$x = \bar{x}$$
$$y = h\frac{(\beta + 1) - (\beta - 1)\left\{[(\beta + 1)/(\beta - 1)]^{1-\bar{y}}\right\}}{[(\beta + 1)/(\beta - 1)]^{1-\bar{y}} + 1} \tag{5.216}$$

The stretching transformation discussed here is from the family of general stretching transformations proposed by Roberts (1971). Another transformation from this family refines the mesh near walls of a duct, as seen in Fig. 5.9. This transformation is given by the following.

Transformation 2:

$$\bar{x} = x$$
$$\bar{y} = \alpha + (1 - \alpha) \tag{5.217}$$
$$\times \frac{\ln(\{\beta + [y(2\alpha + 1)/h] - 2\alpha\}/\{\beta - [y(2\alpha + 1)/h] + 2\alpha\})}{\ln[(\beta + 1)/(\beta - 1)]}$$

For this transformation, if $\alpha = 0$, the mesh will be refined near $y = h$ only, whereas if $\alpha = \frac{1}{2}$, the mesh will be refined equally near $y = 0$ and $y = h$. Roberts has shown that the stretching parameter β is related (approximately) to the nondimensional boundary-layer thickness (δ/h) by

$$\beta = \left(1 - \frac{\delta}{h}\right)^{-1/2} \qquad 0 < \frac{\delta}{h} < 1 \tag{5.218}$$

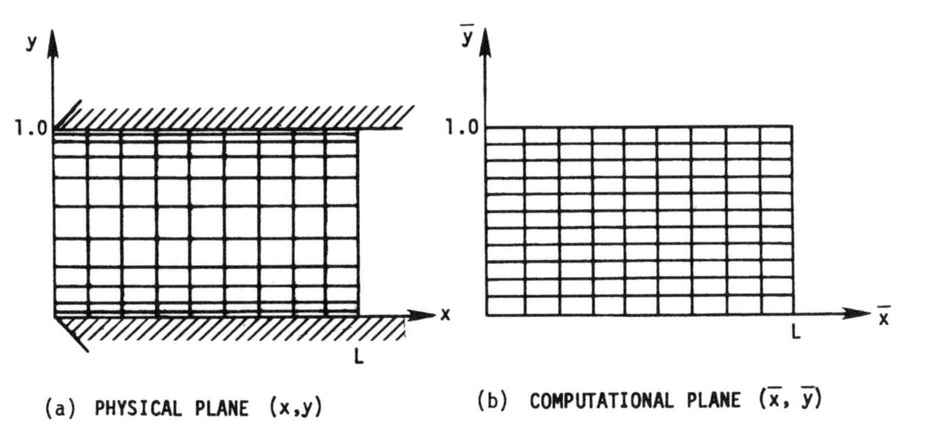

Figure 5.9 Grid clustering in a duct. (a) Physical plane (x, y). (b) Computational plane $(\bar{x}, \bar{y})$.

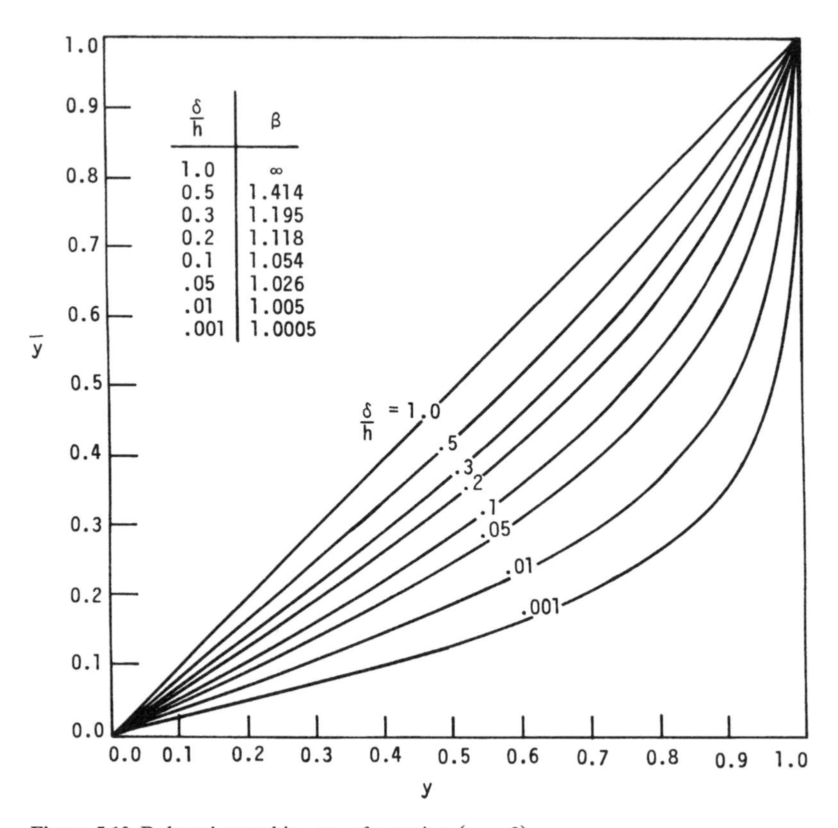

Figure 5.10 Roberts' stretching transformation ($\alpha = 0$).

where h is the height of the mesh. The amount of stretching for various values of δ/h is illustrated in Fig. 5.10 for the case where $\alpha = 0$. For the transformation given by Eqs. (5.217), the metric $\partial \bar{y}/\partial y$ is

$$\frac{\partial \bar{y}}{\partial y} = \frac{2\beta(1-\alpha)(2\alpha+1)}{h\left\{\beta^2 - [y(2\alpha+1)/h - 2\alpha]^2\right\} \ln[(\beta+1)/(\beta-1)]} \tag{5.219}$$

and the inverse transformation becomes

$$x = \bar{x}$$

$$y = h\frac{(\beta+2\alpha)[(\beta+1)/(\beta-1)]^{(\bar{y}-\alpha)/(1-\alpha)} - \beta + 2\alpha}{(2\alpha+1)\left\{1 + [(\beta+1)/(\beta-1)]^{(\bar{y}-\alpha)/(1-\alpha)}\right\}} \tag{5.220}$$

A useful transformation for refining the mesh about some interior point y_c (see Fig. 5.11) is given by the following:

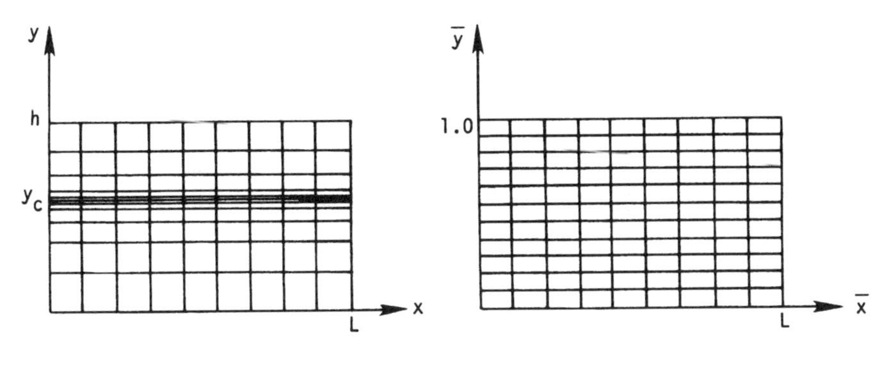

Figure 5.11 Grid clustering near an interior point. (a) Physical plane (x, y). (b) Computational plane $(\bar{x}, \bar{y})$.

Transformation 3:

$$\begin{aligned} \bar{x} &= x \\ \bar{y} &= B + \frac{1}{\tau} \sinh^{-1}\left[\left(\frac{y}{y_c} - 1\right) \sinh(\tau B)\right] \end{aligned} \tag{5.221}$$

where

$$B = \frac{1}{2\tau} \ln\left[\frac{1 + (e^{\tau} - 1)(y_c/h)}{1 + (e^{-\tau} - 1)(y_c/h)}\right] \qquad 0 < \tau < \infty$$

In this transformation, τ is the stretching parameter, which varies from zero (no stretching) to large values that produce the most refinement near $y = y_c$. The metric $\partial \bar{y}/\partial y$ and y become

$$\frac{\partial \bar{y}}{\partial y} = \frac{\sinh(\tau B)}{\tau y_c \sqrt{1 + [(y/y_c) - 1]^2 \sinh^2(\tau B)}} \tag{5.222}$$

$$y = y_c\left\{1 + \frac{\sinh[\tau(\bar{y} - B)]}{\sinh(\tau B)}\right\} \tag{5.223}$$

For our final transformation, we will examine a simple transformation that can be used to transform a nonrectangular region in the physical plane into a rectangular region in the computational plane, as seen in Fig. 5.12. The required transformation is as follows:

Transformation 4:

$$\begin{aligned} \bar{x} &= x \\ \bar{y} &= \frac{y}{h(x)} \end{aligned} \tag{5.224}$$

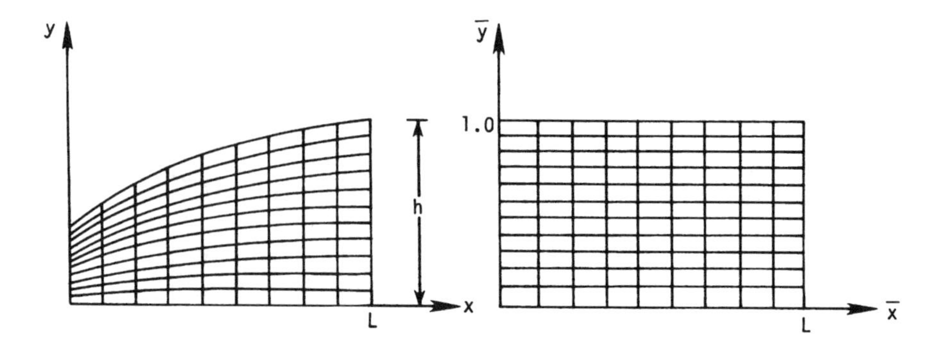

Figure 5.12 Rectangularization of computational grid. (a) Physical plane (x, y); (b) Computational plane $(\bar{x}, \bar{y})$.

The known distance between the lower boundary and the upper boundary (measured along a x = constant line) is designated by $h(x)$. The required partial derivatives are

$$\begin{aligned} \frac{\partial}{\partial x} &= \frac{\partial}{\partial \bar{x}} - \bar{y}\frac{h'(x)}{h(x)}\frac{\partial}{\partial \bar{y}} \\ \frac{\partial}{\partial y} &= \frac{1}{h(x)}\frac{\partial}{\partial \bar{y}} \end{aligned} \tag{5.225}$$

where $h'(x) = dh(x)/dx$. Hence the steady 2-D incompressible continuity equation in Cartesian coordinates is transformed to

$$\frac{\partial u}{\partial \bar{x}} - \bar{y}\frac{h'(\bar{x})}{h(\bar{x})}\frac{\partial u}{\partial \bar{y}} + \frac{1}{h(\bar{x})}\frac{\partial v}{\partial \bar{y}} = 0 \tag{5.226}$$

5.6.2 Generalized Transformation

In the preceding section, we examined simple independent variable transformations that make it possible to solve the governing equations on a uniformly spaced computational grid. Let us now consider a completely general transformation of the form

$$\begin{aligned} \xi &= \xi(x, y, z) \\ \eta &= \eta(x, y, z) \\ \zeta &= \zeta(x, y, z) \end{aligned} \tag{5.227}$$

which can be used to transform the governing equations from the physical domain (x, y, z) to the computational domain (ξ, η, ζ). Using the chain rule of partial differentiation, the partial derivatives become

$$\begin{aligned}\frac{\partial}{\partial x} &= \xi_x \frac{\partial}{\partial \xi} + \eta_x \frac{\partial}{\partial \eta} + \zeta_x \frac{\partial}{\partial \zeta} \\ \frac{\partial}{\partial y} &= \xi_y \frac{\partial}{\partial \xi} + \eta_y \frac{\partial}{\partial \eta} + \zeta_y \frac{\partial}{\partial \zeta} \\ \frac{\partial}{\partial z} &= \xi_z \frac{\partial}{\partial \xi} + \eta_z \frac{\partial}{\partial \eta} + \zeta_z \frac{\partial}{\partial \zeta}\end{aligned} \tag{5.228}$$

The metrics ($\xi_x, \eta_x, \zeta_x, \xi_y, \eta_y, \zeta_y, \xi_z, \eta_z, \zeta_z$) appearing in these equations can be determined in the following manner. We first write the differential expressions

$$\begin{aligned}d\xi &= \xi_x\,dx + \xi_y\,dy + \xi_z\,dz \\ d\eta &= \eta_x\,dx + \eta_y\,dy + \eta_z\,dz \\ d\zeta &= \zeta_x\,dx + \zeta_y\,dy + \zeta_z\,dz\end{aligned} \tag{5.229}$$

which in matrix form become

$$\begin{bmatrix} d\xi \\ d\eta \\ d\zeta \end{bmatrix} = \begin{bmatrix} \xi_x & \xi_y & \xi_z \\ \eta_x & \eta_y & \eta_z \\ \zeta_x & \zeta_y & \zeta_z \end{bmatrix} \begin{bmatrix} dx \\ dy \\ dz \end{bmatrix} \tag{5.230}$$

In a like manner, we can write

$$\begin{bmatrix} dx \\ dy \\ dz \end{bmatrix} = \begin{bmatrix} x_\xi & x_\eta & x_\zeta \\ y_\xi & y_\eta & y_\zeta \\ z_\xi & z_\eta & z_\zeta \end{bmatrix} \begin{bmatrix} d\xi \\ d\eta \\ d\zeta \end{bmatrix} \tag{5.231}$$

Therefore

$$\begin{aligned}\begin{bmatrix} \xi_x & \xi_y & \xi_z \\ \eta_x & \eta_y & \eta_z \\ \zeta_x & \zeta_y & \zeta_z \end{bmatrix} &= \begin{bmatrix} x_\xi & x_\eta & x_\zeta \\ y_\xi & y_\eta & y_\zeta \\ z_\xi & z_\eta & z_\zeta \end{bmatrix}^{-1} \\ &= J \begin{bmatrix} y_\eta z_\zeta - y_\zeta z_\eta & -(x_\eta z_\zeta - x_\zeta z_\eta) & x_\eta y_\zeta - x_\zeta y_\eta \\ -(y_\xi z_\zeta - y_\zeta z_\xi) & x_\xi z_\zeta - x_\zeta z_\xi & -(x_\xi y_\zeta - x_\zeta y_\xi) \\ y_\xi z_\eta - y_\eta z_\xi & -(x_\xi z_\eta - x_\eta z_\xi) & x_\xi y_\eta - x_\eta y_\xi \end{bmatrix}\end{aligned} \tag{5.232}$$

Thus the metrics are

$$\begin{aligned}\xi_x &= J(y_\eta z_\zeta - y_\zeta z_\eta) \\ \xi_y &= -J(x_\eta z_\zeta - x_\zeta z_\eta) \\ \xi_z &= J(x_\eta y_\zeta - x_\zeta y_\eta)\end{aligned}$$

$$\begin{aligned}
\eta_x &= -J(y_\xi z_\zeta - y_\zeta z_\xi) \\
\eta_y &= J(x_\xi z_\zeta - x_\zeta z_\xi) \\
\eta_z &= -J(x_\xi y_\zeta - x_\zeta y_\xi) \\
\zeta_x &= J(y_\xi z_\eta - y_\eta z_\xi) \\
\zeta_y &= -J(x_\xi z_\eta - x_\eta z_\xi) \\
\zeta_z &= J(x_\xi y_\eta - x_\eta y_\xi)
\end{aligned} \tag{5.233}$$

where J is the Jacobian of the transformation,

$$J = \frac{\partial(\xi, \eta, \zeta)}{\partial(x, y, z)} = \begin{vmatrix} \xi_x & \xi_y & \xi_z \\ \eta_x & \eta_y & \eta_z \\ \zeta_x & \zeta_y & \zeta_z \end{vmatrix} \tag{5.234}$$

which can be evaluated in the following manner:

$$J = 1/J^{-1} = 1 \Bigg/ \frac{\partial(x, y, z)}{\partial(\xi, \eta, \zeta)} = 1 \Bigg/ \begin{vmatrix} x_\xi & x_\eta & x_\zeta \\ y_\xi & y_\eta & y_\zeta \\ z_\xi & z_\eta & z_\zeta \end{vmatrix}$$

$$= 1/\left[x_\xi(y_\eta z_\zeta - y_\zeta z_\eta) - x_\eta(y_\xi z_\zeta - y_\zeta z_\xi) + x_\zeta(y_\xi z_\eta - y_\eta z_\xi)\right] \tag{5.235}$$

The metrics can be readily determined if analytical expressions are available for the inverse of the transformation:

$$\begin{aligned}
x &= x(\xi, \eta, \zeta) \\
y &= y(\xi, \eta, \zeta) \\
z &= z(\xi, \eta, \zeta)
\end{aligned} \tag{5.236}$$

For cases where the transformation is the direct result of a grid generation scheme, the metrics can be computed numerically using central differences in the computational plane. A brief discussion on the proper way to compute metrics is presented in Chapter 10.

If we apply the generalized transformation to the compressible Navier-Stokes equations written in vector form [Eqs. (5.43)], the following transformed equation is obtained:

$$\mathbf{U}_t + \xi_x \mathbf{E}_\xi + \eta_x \mathbf{E}_\eta + \zeta_x \mathbf{E}_\zeta + \xi_y \mathbf{F}_\xi + \eta_y \mathbf{F}_\eta + \zeta_y \mathbf{F}_\zeta + \xi_z \mathbf{G}_\xi + \eta_z \mathbf{G}_\eta + \zeta_z \mathbf{G}_\zeta = 0 \tag{5.237}$$

Viviand (1974) and Vinokur (1974) have shown that the gas dynamic equations can be put back into strong conservation-law form after a transformation has been applied. In order to do this, the transformed equation is first divided by the Jacobian and is then rearranged into conservation-law form by adding and subtracting like terms. When this procedure is applied to Eq. (5.237), the following equation results:

$$\left(\frac{\mathbf{U}}{J}\right)_t + \left(\frac{\mathbf{E}\xi_x + \mathbf{F}\xi_y + \mathbf{G}\xi_z}{J}\right)_\xi + \left(\frac{\mathbf{E}\eta_x + \mathbf{F}\eta_y + \mathbf{G}\eta_z}{J}\right)_\eta$$
$$+ \left(\frac{\mathbf{E}\zeta_x + \mathbf{F}\zeta_y + \mathbf{G}\zeta_z}{J}\right)_\zeta - \mathbf{E}\left[\left(\frac{\xi_x}{J}\right)_\xi + \left(\frac{\eta_x}{J}\right)_\eta + \left(\frac{\zeta_x}{J}\right)_\zeta\right]$$
$$- \mathbf{F}\left[\left(\frac{\xi_y}{J}\right)_\xi + \left(\frac{\eta_y}{J}\right)_\eta + \left(\frac{\zeta_y}{J}\right)_\zeta\right] - \mathbf{G}\left[\left(\frac{\xi_z}{J}\right)_\xi + \left(\frac{\eta_z}{J}\right)_\eta + \left(\frac{\zeta_z}{J}\right)_\zeta\right] = 0 \tag{5.238}$$

The last three terms in brackets are all equal to zero and can be dropped. This can be verified by substituting the metrics given by Eqs. (5.233) into these terms. If we now define the quantities

$$\begin{aligned}
\mathbf{U}_1 &= \frac{\mathbf{U}}{J} \\
\mathbf{E}_1 &= \frac{1}{J}(\mathbf{E}\xi_x + \mathbf{F}\xi_y + \mathbf{G}\xi_z) \\
\mathbf{F}_1 &= \frac{1}{J}(\mathbf{E}\eta_x + \mathbf{F}\eta_y + \mathbf{G}\eta_z) \\
\mathbf{G}_1 &= \frac{1}{J}(\mathbf{E}\zeta_x + \mathbf{F}\zeta_y + \mathbf{G}\zeta_z)
\end{aligned} \tag{5.239}$$

and substitute them into Eq. (5.238), the final equation is in strong conservation-law form:

$$\frac{\partial \mathbf{U}_1}{\partial t} + \frac{\partial \mathbf{E}_1}{\partial \xi} + \frac{\partial \mathbf{F}_1}{\partial \eta} + \frac{\partial \mathbf{G}_1}{\partial \zeta} = 0 \tag{5.240}$$

It should be kept in mind that the vectors $\mathbf{E}_1$, $\mathbf{F}_1$, and $\mathbf{G}_1$ contain partial derivatives in the viscous and heat-transfer terms. These partial derivative terms are to be transformed using Eqs. (5.228). For example, the shearing stress term, τ_{xy}, would be transformed to

$$\tau_{xy} = \mu\left(\xi_y\frac{\partial u}{\partial \xi} + \eta_y\frac{\partial u}{\partial \eta} + \zeta_y\frac{\partial u}{\partial \zeta} + \xi_x\frac{\partial v}{\partial \xi} + \eta_x\frac{\partial v}{\partial \eta} + \zeta_x\frac{\partial v}{\partial \zeta}\right) \tag{5.241}$$

The strong conservation-law form of the governing equations is a convenient form for applying finite-difference schemes. However, when using this form of the equations, caution must be exercised if the grid is changing. In this case, a constraint on the way the metrics are differenced, called the *geometric conservation law* (Thomas and Lombard, 1978), must be satisfied in order to prevent additional errors from being introduced into the solution.

5.7 FINITE-VOLUME FORMULATION

The governing equations of fluid dynamics have been mathematically expressed in differential form in this chapter. When a numerical scheme is applied to these differential equations, the computational domain is subdivided into grid points, and the finite-difference equations are solved at each point. An alternative approach is to solve the integral form of the governing equations. In this approach, the physical domain is subdivided into small volumes (or areas for a 2-D case), and the dependent variables are evaluated either at the centers of the volumes (cells) or at the corners of the volumes.

The integral approach includes both the finite-volume and finite-element methods, but only the finite-volume method will be discussed here. The finite-volume method has an obvious advantage over a finite-difference method if the physical domain is highly irregular and complicated, since arbitrary volumes can be utilized to subdivide the physical domain. Also since the integral equations are solved directly in the physical domain, no coordinate transformation is required. Another advantage of the finite-volume method is that mass, momentum, and energy are automatically conserved, since the integral forms of the governing equations are solved.

5.7.1 Two-Dimensional Finite-Volume Method

In order to explain the finite-volume method, consider the following 2-D model equation:

$$\frac{\partial U}{\partial t} + \frac{\partial E}{\partial x} + \frac{\partial F}{\partial y} = 0 \tag{5.242}$$

Integrating this equation over the finite volume abcd (with unit depth) shown in Fig. 5.13 gives

$$\iiint_{\text{abcd}} \left(\frac{\partial U}{\partial t} + \frac{\partial E}{\partial x} + \frac{\partial F}{\partial y} \right) d\mathscr{V} = 0 \tag{5.243}$$

where the differential volume $d\mathscr{V}$ is $dx\,dy$ (1). After applying Green's theorem, this equation becomes

$$\frac{\partial}{\partial t} \iiint_{\text{abcd}} U(1)\,dx\,dy + \oiint_{\text{abcd}} \mathbf{H} \cdot \mathbf{n}\,dS = 0 \tag{5.244}$$

where $\mathbf{n}$ is the unit normal to the surface S of the finite volume and $\mathbf{H}$ can be expressed in Cartesian coordinates as

$$\mathbf{H} = E\mathbf{i} + F\mathbf{j}$$

For the present 2-D geometry,

$$\mathbf{H} \cdot \mathbf{n}\,dS = (E\,dy - F\,dx)(1) \tag{5.245}$$

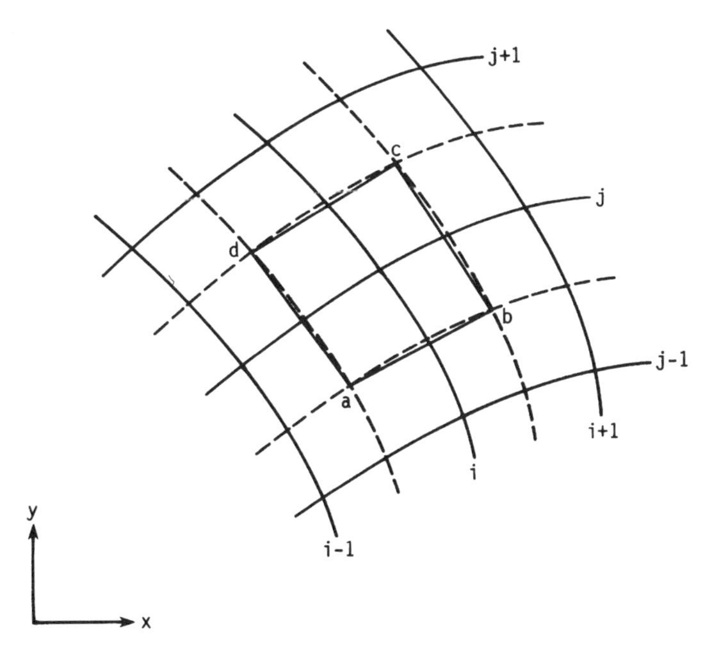

Figure 5.13 Two-dimensional finite volume.

which can be substituted into Eq. (5.244) to yield

$$\frac{\partial}{\partial t}\iint_{\text{abcd}} U\,dx\,dy + \oint_{\text{abcd}} (E\,dy - F\,dx) = 0 \tag{5.246}$$

This expression can then be approximated as

$$\left(\frac{U_{i,j}^{n+1} - U_{i,j}^{n}}{\Delta t}\right) S_{\text{abcd}} + \left(E_{i,j-\frac{1}{2}}\,\Delta y_{\text{ab}} + E_{i+\frac{1}{2},j}\,\Delta y_{\text{bc}} + E_{i,j+\frac{1}{2}}\,\Delta y_{\text{cd}} + E_{i-\frac{1}{2},j}\,\Delta y_{\text{da}}\right)$$
$$-\left(F_{i,j-\frac{1}{2}}\,\Delta x_{\text{ab}} + F_{i+\frac{1}{2},j}\,\Delta x_{\text{bc}} + F_{i,j+\frac{1}{2}}\,\Delta x_{\text{cd}} + F_{i-\frac{1}{2},j}\,\Delta x_{\text{da}}\right) = 0 \tag{5.247}$$

where S_{abcd} is the area (which is assumed constant) of the quadrilateral abcd and $U_{i,j}$ is the average value of U in the quadrilateral or cell. This formulation is referred to as a *cell-centered* finite-volume scheme. An alternate approach would be to evaluate the dependent variables at the vertices of the cell, and this is called a *nodal-point* finite-volume scheme.

The increments in x and y are given by

$$\Delta x_{\text{ab}} = x_{\text{b}} - x_{\text{a}} \qquad \Delta x_{\text{bc}} = x_{\text{c}} - x_{\text{b}} \qquad \Delta x_{\text{cd}} = x_{\text{d}} - x_{\text{c}} \qquad \Delta x_{\text{da}} = x_{\text{a}} - x_{\text{d}}$$
$$\Delta y_{\text{ab}} = y_{\text{b}} - y_{\text{a}} \qquad \Delta y_{\text{bc}} = y_{\text{c}} - y_{\text{b}} \qquad \Delta y_{\text{cd}} = y_{\text{d}} - y_{\text{c}} \qquad \Delta y_{\text{da}} = y_{\text{a}} - y_{\text{d}} \tag{5.248}$$

The fluxes E and F can be evaluated at time level n or $n+1$ to provide either an explicit or implicit scheme. In addition, the spatial values of the fluxes can be determined in a variety of ways, which will lead to the various algorithms discussed in Chapter 4. As an example, let us evaluate the fluxes using average values given by

$$\begin{aligned} E_{i,j-\frac{1}{2}} &= 0.5(E_{i,j-1} + E_{i,j}) & F_{i,j-\frac{1}{2}} &= 0.5(F_{i,j-1} + F_{i,j}) \\ E_{i+\frac{1}{2},j} &= 0.5(E_{i+1,j} + E_{i,j}) & F_{i+\frac{1}{2},j} &= 0.5(F_{i+1,j} + F_{i,j}) \\ E_{i,j+\frac{1}{2}} &= 0.5(E_{i,j+1} + E_{i,j}) & F_{i,j+\frac{1}{2}} &= 0.5(F_{i,j+1} + F_{i,j}) \\ E_{i-\frac{1}{2},j} &= 0.5(E_{i-1,j} + E_{i,j}) & F_{i-\frac{1}{2},j} &= 0.5(F_{i-1,j} + F_{i,j}) \end{aligned} \tag{5.249}$$

Substituting these expressions into Eq. (5.247) yields

$$\begin{aligned} &\left(\frac{U_{i,j}^{n+1} - U_{i,j}^{n}}{\Delta t}\right) S_{\text{abcd}} + 0.5(E_{i,j-1} + E_{i,j})\,\Delta y_{\text{ab}} - 0.5(F_{i,j-1} + F_{i,j})\,\Delta x_{\text{ab}} \\ &\quad + 0.5(E_{i+1,j} + E_{i,j})\,\Delta y_{\text{bc}} - 0.5(F_{i+1,j} + F_{i,j})\,\Delta x_{\text{bc}} \\ &\quad + 0.5(E_{i,j+1} + E_{i,j})\,\Delta y_{\text{cd}} - 0.5(F_{i,j+1} + F_{i,j})\,\Delta x_{\text{cd}} \\ &\quad + 0.5(E_{i-1,j} + E_{i,j})\,\Delta y_{\text{da}} - 0.5(F_{i-1,j} + F_{i,j})\,\Delta x_{\text{da}} = 0 \end{aligned} \tag{5.250}$$

If the quadrilateral abcd is rectangular in shape and if the sides coincide with lines of constant x and y, Eq. (5.250) reduces to

$$\frac{U_{i,j}^{n+1} - U_{i,j}^{n}}{\Delta t} + \frac{E_{i+1,j} - E_{i-1,j}}{2\,\Delta x} + \frac{F_{i,j+1} - F_{i,j-1}}{2\,\Delta y} = 0 \tag{5.251}$$

which we recognize as the FTCS scheme applied to our model equation. Other schemes, such as upwind algorithms, can be obtained by using appropriate expressions for the fluxes at the cell faces.

The finite-volume method described thus far in this section has been applied to a model PDE containing only first derivatives. In order to show how the finite-volume method can be applied to equations containing second derivatives, let us consider the 2-D heat equation,

$$\frac{\partial T}{\partial t} = \alpha\left(\frac{\partial^2 T}{\partial x^2} + \frac{\partial^2 T}{\partial y^2}\right) \tag{5.252}$$

where α is assumed constant. Integrating this equation over the finite volume abcd (with unit depth) shown in Fig. 5.14 gives

$$\iiint_{\text{abcd}} \frac{\partial T}{\partial t}(1)\,dx\,dy = \alpha \iiint_{\text{abcd}} \left(\frac{\partial^2 T}{\partial x^2} + \frac{\partial^2 T}{\partial y^2}\right)(1)\,dx\,dy \tag{5.253}$$

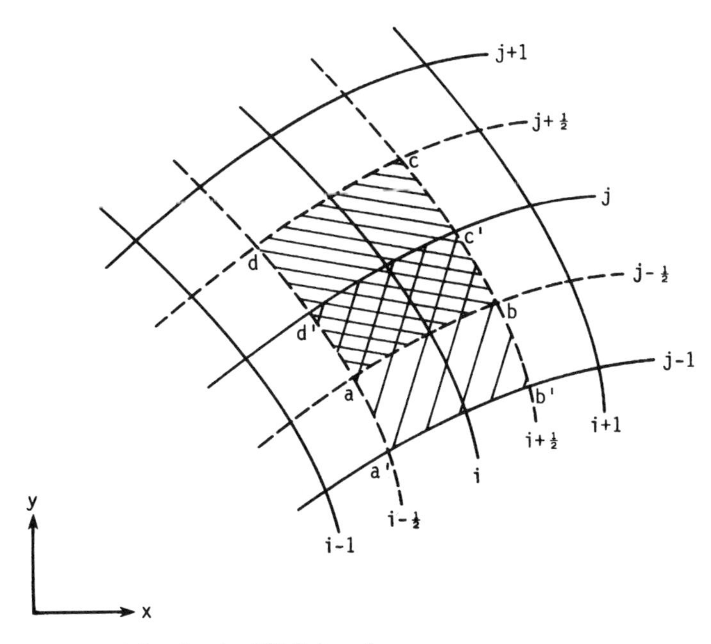

Figure 5.14 Overlapping 2-D finite volumes.

After applying Green's theorem, this equation becomes

$$\frac{\partial}{\partial t}\iiint_{\text{abcd}} T(1)\, dx\, dy = \alpha \oiint_{\text{abcd}} \mathbf{H} \cdot \mathbf{n}\, dS \tag{5.254}$$

where $\mathbf{H}$ can be expressed in Cartesian coordinates as

$$\mathbf{H} = \frac{\partial T}{\partial x}\mathbf{i} + \frac{\partial T}{\partial y}\mathbf{j}$$

For the present 2-D geometry,

$$\mathbf{H} \cdot \mathbf{n}\, dS = \left(\frac{\partial T}{\partial x}\, dy - \frac{\partial T}{\partial y}\, dx\right)(1)$$

which can be substituted into Eq. (5.254) to yield

$$\frac{\partial}{\partial t}\iint_{\text{abcd}} T\, dx\, dy = \alpha \oint_{\text{abcd}} \left(\frac{\partial T}{\partial x}\, dy - \frac{\partial T}{\partial y}\, dx\right) \tag{5.255}$$

Equation (5.255) can then be approximated, as before, to obtain

$$\left(\frac{T_{i,j}^{n+1} - T_{i,j}^{n}}{\Delta t}\right) S_{\text{abcd}}$$
$$= \alpha\left[\left(\frac{\partial T}{\partial x}\right)_{i,j-\frac{1}{2}} \Delta y_{\text{ab}} + \left(\frac{\partial T}{\partial x}\right)_{i+\frac{1}{2},j} \Delta y_{\text{bc}} + \left(\frac{\partial T}{\partial x}\right)_{i,j+\frac{1}{2}} \Delta y_{\text{cd}}\right.$$
$$+ \left(\frac{\partial T}{\partial x}\right)_{i-\frac{1}{2},j} \Delta y_{\text{da}} - \left(\frac{\partial T}{\partial y}\right)_{i,j-\frac{1}{2}} \Delta x_{\text{ab}} - \left(\frac{\partial T}{\partial y}\right)_{i+\frac{1}{2},j} \Delta x_{\text{bc}}$$
$$\left. - \left(\frac{\partial T}{\partial y}\right)_{i,j+\frac{1}{2}} \Delta x_{\text{cd}} - \left(\frac{\partial T}{\partial y}\right)_{i-\frac{1}{2},j} \Delta x_{\text{da}}\right] \tag{5.256}$$

where the increments in x and y are given in Eq. (5.248). Different techniques (see Peyret and Taylor, 1983) can be used to evaluate the derivatives in Eq. (5.256). A common approach is to evaluate the derivatives as a mean value over the appropriate area. For example, the derivatives $(\partial T/\partial x)_{i,j-\frac{1}{2}}$ and $(\partial T/\partial y)_{i,j-\frac{1}{2}}$ can be evaluated as their mean values over the finite volume a′b′c′d′ in Fig. 5.14. Thus

$$\left(\frac{\partial T}{\partial x}\right)_{i,j-\frac{1}{2}} = \frac{1}{S_{\text{a}'\text{b}'\text{c}'\text{d}'}} \iint \left(\frac{\partial T}{\partial x}\right) dx\,dy = \frac{1}{S_{\text{a}'\text{b}'\text{c}'\text{d}'}} \oint T\,dy \tag{5.257}$$

where the line integral can be approximated by

$$\oint T\,dy \cong T_{i,j-1}\,\Delta y_{\text{a}'\text{b}'} + T_b\,\Delta y_{\text{b}'\text{c}'} + T_{i,j}\,\Delta y_{\text{c}'\text{d}'} + T_a\,\Delta y_{\text{d}'\text{a}'} \tag{5.258}$$

The temperatures T_a and T_b are evaluated as the average of the four surrounding temperatures:

$$\begin{aligned} T_a &= \tfrac{1}{4}(T_{i,j} + T_{i-1,j} + T_{i-1,j-1} + T_{i,j-1}) \\ T_b &= \tfrac{1}{4}(T_{i,j} + T_{i+1,j} + T_{i+1,j-1} + T_{i,j-1}) \end{aligned} \tag{5.259}$$

In a like manner,

$$\left(\frac{\partial T}{\partial y}\right)_{i,j-\frac{1}{2}} = \frac{1}{S_{\text{a}'\text{b}'\text{c}'\text{d}'}} \iint \left(\frac{\partial T}{\partial y}\right) dx\,dy = -\frac{1}{S_{\text{a}'\text{b}'\text{c}'\text{d}'}} \oint T\,dx \tag{5.260}$$

and the line integral can be approximated as

$$\oint T\,dx = T_{i,j-1}\,\Delta x_{\text{a}'\text{b}'} + T_b\,\Delta x_{\text{b}'\text{c}'} + T_{i,j}\,\Delta x_{\text{c}'\text{d}'} + T_a\,\Delta x_{\text{d}'\text{a}'} \tag{5.261}$$

The other derivatives appearing in Eq. (5.256) can be determined in a similar manner.

5.7.2 Three-Dimensional Finite-Volume Method

The finite-volume formulation can readily be extended to three dimensions, although it does become more complicated. Consider the 3-D Navier-Stokes (or Euler) equations [Eq. (5.43)]:

$$\frac{\partial \mathbf{U}}{\partial t} + \frac{\partial \mathbf{E}}{\partial x} + \frac{\partial \mathbf{F}}{\partial y} + \frac{\partial \mathbf{G}}{\partial z} = 0 \tag{5.262}$$

These equations can be expressed in integral form as

$$\frac{\partial}{\partial t} \iiint_{\mathscr{V}} \mathbf{U}\, d\mathscr{V} + \oiint_S (\overline{\mathbf{H}} \cdot \mathbf{n})\, dS = 0 \tag{5.263}$$

where the finite volume $\mathscr{V}$ is bounded by the surface S and the tensor $\overline{\mathbf{H}}$ is given in Cartesian coordinates as

$$\overline{\mathbf{H}} = \mathbf{E}\mathbf{i} + \mathbf{F}\mathbf{j} + \mathbf{G}\mathbf{k} \tag{5.264}$$

If we utilize the cell-face surface-area vector $d\mathbf{S}$ (defined as $dS\,\mathbf{n}$) and assume that the volume $\mathscr{V}$ is constant, then Eq. (5.263) can be written in discrete form for a finite volume l as

$$\mathscr{V}_l \frac{\partial}{\partial t}(\mathbf{U}_l) + \sum_{\text{sides}} \overline{\mathbf{H}} \cdot \mathbf{S} = 0 \tag{5.265}$$

where $\mathbf{U}_l$ is the value of $\mathbf{U}$ associated with the finite volume l and the summation is applied to all exterior sides of the finite volume.

In three dimensions the computational region is usually subdivided using six-sided hexahedrons such as the one shown in Fig. 5.15. The edges of the hexahedron (cell) are taken to be straight-line segments, so that a cell face can

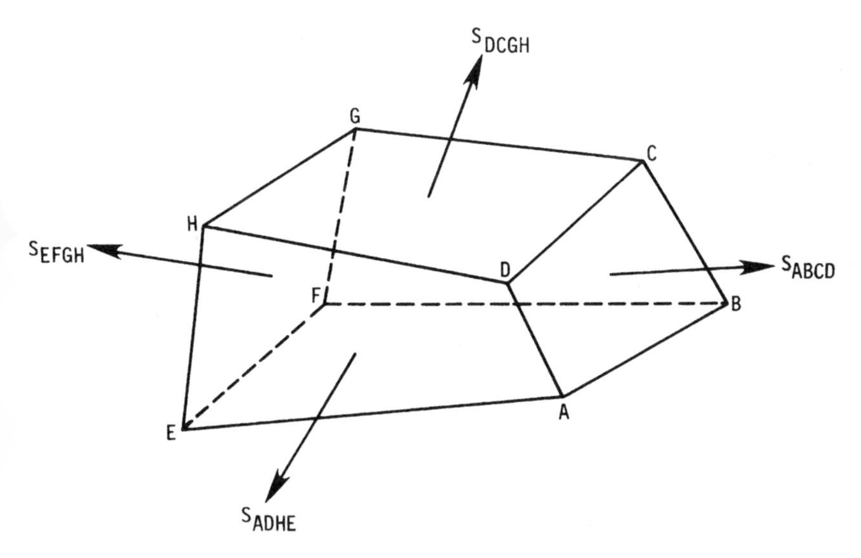

Figure 5.15 Three-dimensional hexahedral finite volume.

be considered to consist of two planar triangles. For example, the face ABCD in Fig. 5.15 can be subdivided into triangles ABC and ADC or, alternately, triangles ABD and BCD. The cell-face surface-area vector **S** is then obtained by summing the triangular area vectors. This vector is not dependent on which diagonal is used to separate the face into two triangles. Expressions for determining **S** in an efficient manner are given by Vinokur (1986). For face ABCD the surface-area vector can be determined from

$$\mathbf{S}_{\mathrm{ABCD}} = \tfrac{1}{2}(\mathbf{r}_{\mathrm{AC}} \times \mathbf{r}_{\mathrm{BD}}) \tag{5.266}$$

where $\mathbf{r}_{\mathrm{AC}}$ and $\mathbf{r}_{\mathrm{BD}}$ are given by

$$\mathbf{r}_{\mathrm{AC}} = \mathbf{r}_{\mathrm{C}} - \mathbf{r}_{\mathrm{A}}$$
$$\mathbf{r}_{\mathrm{BD}} = \mathbf{r}_{\mathrm{D}} - \mathbf{r}_{\mathrm{B}}$$

and $\mathbf{r}_{\mathrm{A}}, \mathbf{r}_{\mathrm{B}}, \mathbf{r}_{\mathrm{C}}, \mathbf{r}_{\mathrm{D}}$ are the position vectors of the points A, B, C, D, respectively. Similar expressions can be written for the other faces. For example,

$$\begin{aligned} \mathbf{S}_{\mathrm{ADHE}} &= \tfrac{1}{2}(\mathbf{r}_{\mathrm{ED}} \times \mathbf{r}_{\mathrm{AH}}) \\ \mathbf{S}_{\mathrm{DCGH}} &= \tfrac{1}{2}(\mathbf{r}_{\mathrm{DG}} \times \mathbf{r}_{\mathrm{CH}}) \\ \mathbf{S}_{\mathrm{BFGC}} &= \tfrac{1}{2}(\mathbf{r}_{\mathrm{CF}} \times \mathbf{r}_{\mathrm{BG}}) \end{aligned} \tag{5.267}$$

The volume of a hexahedral cell can be determined in several different ways. The usual approach is to subdivide the hexahedron into tetrahedrons or pyramids. Vinokur (1986) has devised a relatively simple expression for the volume of a hexahedron, which can be expressed in terms of the notation of Fig. 5.15 as

$$\mathscr{V} = \tfrac{1}{3}(\mathbf{S}_{\mathrm{ABCD}} + \mathbf{S}_{\mathrm{DCGH}} + \mathbf{S}_{\mathrm{BFGC}}) \cdot (\mathbf{r}_{\mathrm{C}} - \mathbf{r}_{\mathrm{E}}) \tag{5.268}$$

This formula is derived by breaking the hexahedron into three pyramids that share the main diagonal as a common edge.

PROBLEMS

5.1 Verify Eq. (5.9).
5.2 Show that for an incompressible constant-property flow, Eq. (5.18) reduces to Eq. (5.21).
5.3 Verify Eq. (5.30).
5.4 Using the nondimensionalization procedure described in Section 5.1.7, derive Eqs. (5.47).
5.5 Write the energy equation [Eq. (5.33)] in terms of axisymmetric body intrinsic coordinates.
5.6 Write the incompressible Navier-Stokes equation [Eq. (5.21)] in a spherical coordinate system.
5.7 Show that $\overline{\rho' u''} = \overline{\rho' u'}$.
5.8 Show that $\tilde{u} - \bar{u} = \overline{\rho' u'}/\bar{\rho}$.
5.9 Verify that $\overline{u''} = -\overline{\rho' u''}/\bar{\rho}$.
5.10 Starting with Eq. (5.80), show the steps in the development of Eq. (5.81).
5.11 Develop Eq. (5.84) by substitution (i.e., using $c_p\overline{T} = \overline{H} - \bar{u}_i\bar{u}_i/2 - \overline{u_i' u_i'}/2$) starting with Eq. (5.81).
5.12 Show the steps in the derivation of Eq. (5.76) starting with the Navier-Stokes equations.
5.13 Using the decomposition indicated in Section 5.2.7 for large-eddy simulation, verify the expression for τ_{ij} given in Eq. (5.94*j*).

5.14 Apply an order of magnitude analysis to the incompressible 2-D Navier-Stokes equations for the case of a planar 2-D laminar jet. Indicate which terms in the Navier-Stokes equations can be neglected in this flow.
5.15 Verify that $H' = c_p T' + u_i' \bar{u}_i + u_i' u_i'/2 - \overline{u_i' u_i'}/2$.
5.16 Explain why the boundary-layer equations may be applicable to the developing flow in a tube.
5.17 Determine the proper boundary conditions to apply to the thin-shear-layer equations for the 2-D shear layer formed by the merging of two infinite streams at uniform velocities U_a and U_b.
5.18 In the boundary-layer equations for a compressible turbulent flow, explain why $\overline{\rho u} + \overline{\rho' u'}$ has been replaced by $\overline{\rho u}$ but $\overline{\rho v} + \overline{\rho' v'}$ has been left intact.
5.19 In a flow governed by the incompressible boundary-layer equations, it is often said that the Reynolds number is of the order of $1/\varepsilon^2$. What is the basis for this statement?
5.20 The boundary-layer equations, Eqs. (5.104)-(5.106), were developed for Prandtl numbers of the order of magnitude of 1. For a laminar flow over a heated flat plate, indicate what alterations should be made in these equations to properly treat flows in which the Prandtl number becomes of the order of magnitude of (a) ϵ, (b) ϵ^2, (c) $1/\epsilon$, (d) $1/\epsilon^2$.
5.21 Using the Navier-Stokes equations, develop an exact Reynolds stress transport equation applicable to an incompressible turbulent boundary layer, i.e., obtain an expression for $\rho D\overline{u_i' u_j'}/Dt$. Show the steps in your development.
5.22 Using the expression for the transport of Reynolds stresses from Prob. 5.21, let $i = j$ to obtain an expression for the transport of turbulence kinetic energy.
5.23 Using the modeled form of the turbulence kinetic energy equation, Eq. (5.147), show that when convection and diffusion of turbulence kinetic energy are negligible, the kinetic energy turbulence model reduces to the Prandtl mixing-length formula.
5.24 Assuming that convection and diffusion of turbulence kinetic energy are negligible within the log-law region of a turbulent wall boundary layer, find an expression for the turbulence kinetic energy at the outer edge of the log-law region in terms of the wall shear stress. Compare this estimate with experimental measurements of $\bar{k}$ such as those of Klebanoff (see Hinze, 1975).
5.25 Assuming the validity of the Prandtl mixing-length formula for a turbulent wall boundary layer, obtain an expression for the ratio of the apparent turbulent viscosity to the molecular viscosity in the log-law region.
5.26 Verify the inner boundary condition for $\bar{k}$ stated in Eq. (5.148a).
5.27 Verify that when $P = \varepsilon$ in Eq. (5.150k), the representation for $\overline{u'v'}$ becomes equivalent to that used in the standard $\bar{k}$-ε model.
5.28 In a 2-D body intrinsic coordinate system, define the stream function for a steady compressible flow.
5.29 Obtain Eq. (5.220).
5.30 Verify Eqs. (5.222) and (5.223).
5.31 Transform Laplace's equation

$$\frac{\partial^2 u}{\partial x^2} + \frac{\partial^2 u}{\partial y^2} = 0$$

into the (ξ, η) computational space using the transformation

$$\xi = x$$
$$\eta = y/h(x)$$

Note that x and y (as well as the partial derivatives with respect to these variables) should not appear in the final transformed equation.
5.32 Transform the steady 2-D incompressible continuity equation

$$u_x + v_y = 0$$

to (ξ, η) computational space using the transformation

$$\xi = x \qquad \eta = \frac{y}{x^2}$$

and display the results in strong conservation-law form using the technique of Viviand.

5.33 The 2-D physical space (x, y) is transformed to the computational space (ξ, η) by the following transformation:

$$\xi = x$$

$$\eta = \frac{y}{(x+1) - x^2}$$

(*a*) Find the Jacobian of this transformation.

(*b*) Using this transformation, transform the 2-D steady incompressible continuity equation in Cartesian coordinates. The transformed equation should contain ξ, η as the only independent variables.

5.34 Transform the 2-D incompressible Navier-Stokes equation [Eq. (5.21)] using the transformation defined by Eqs. (5.217).

5.35 Show that the transformation defined by

$$x = r \cos \theta$$

$$y = r \sin \theta$$

$$z = z$$

will transform the 3-D compressible continuity equation expressed in cylindrical coordinates into the compressible continuity equation in Cartesian coordinates.

5.36 Apply in a successive manner the transformations given by Eqs. (5.224) and Eqs. (5.210) to the inviscid energy equation [Eq. (5.179)] written for a 2-D steady flow.

5.37 Transform the 2-D continuity equation

$$\frac{\partial \rho}{\partial t} + \frac{\partial \rho u}{\partial x} + \frac{\partial \rho v}{\partial y} = 0$$

to the (τ, ξ, η) computation domain using the transformation

$$\tau = t$$

$$\xi = \xi(t, x, y)$$

$$\eta = \eta(t, x, y)$$

Use the technique of Viviand to write the transformed equation in conservation-law form. Show all intermediate steps.

5.38 Transform the steady form of Euler's equations [Eqs. (5.192)] to the (ξ, η, ζ) computational domain using the transformation

$$\xi = x$$

$$\eta = \eta(x, y, z)$$

$$\zeta = \zeta(x, y, z)$$

Using the technique of Viviand, write the transformed equations in conservation-law form.

5.39 Consider the generalized transformation

$$\tau = t$$

$$\xi = \xi(t, x, y, z)$$

$$\eta = \eta(t, x, y, z)$$

$$\zeta = \zeta(t, x, y, z)$$

(*a*) Determine suitable expressions for the Jacobian of the transformation as well as the metrics.

(*b*) Apply this transformation to the compressible Navier-Stokes equations written in vector form [Eqs. (5.43)].

CHAPTER

SIX

NUMERICAL METHODS FOR INVISCID FLOW EQUATIONS

6.1 INTRODUCTION

The Navier-Stokes equations govern the flows commonly encountered in both internal and external applications. Computing a solution of the Navier-Stokes equations is often difficult or at least impractical and, in many of these applications, unnecessary. Results obtained from a solution of the Euler equations are particularly useful in preliminary design work, where information on pressure alone is desired. In problems where heat transfer and skin friction are required, a solution of the boundary-layer equations usually provides an adequate approximation. However, the outer-edge conditions, including the pressure, must be specified from the inviscid solution as the first step in such an analysis. The Euler equations are also of interest because many of the major elements of fluid dynamics are incorporated in them. For example, fluid flows frequently have internal discontinuities such as shock waves or contact surfaces. Solutions relating the end states across a shock are given by the Rankine-Hugoniot relations; these relations are contained in solutions of the Euler equations.

The Euler equations govern the motion of an inviscid nonheat-conducting gas and have a different character in different flow regimes. If the time-dependent terms are retained, the resulting unsteady equations are hyperbolic for all Mach numbers, and solutions can be obtained using time-marching procedures. The situation is very different when a steady flow is assumed. In this case, the Euler equations are elliptic when the flow is subsonic, and hyperbolic when the flow is supersonic. This change in character of the governing equations is the reason that the development of methods for solving steady transonic flows has required many years. Many simplified versions of the Euler equations are used for inviscid fluid flows. When studying incompressible flows, it is often to our advantage to assume irrotationality. Under these conditions, a solution of Laplace's equation for the velocity potential or stream function provides the flow field information. Associated with the Euler equations is the companion set of small-perturbation equations. In subsonic and supersonic flow, we observe that the Prandtl-Glauert equation provides the first-order theory for the potential function. In transonic flow the equation obtained for small perturbations is still a nonlinear equation. The classification of the various forms of the inviscid equations of motion is given in Table 6.1.

Many different methods are used to obtain solutions to the Euler equations or any of their reduced forms. The main goal of this chapter is to present the most commonly used methods for solving inviscid flow problems. While many techniques may be used to solve the partial differential equations (PDEs) governing such flows, our attention will be restricted to finite-difference and finite-volume methods.

The methods presented in this chapter are selected to illustrate the basic ideas as well as give information on useful solution schemes. In a textbook, only fundamental methods that appear to have some measure of permanence should be included. Although some question always exists regarding the long-term survivability of "current techniques," it is hoped that those selected for discussion will stand the test of time.

6.2 METHOD OF CHARACTERISTICS

Closed-form solutions of nonlinear hyperbolic PDEs do not exist for general cases. In order to obtain solutions to such equations we are required to use numerical methods. The method of characteristics is the oldest and most nearly exact method that can be used to solve hyperbolic PDEs. Even though this technique has been replaced by newer, more easily implemented finite-difference/finite-volume methods, a background in characteristic theory and its application is essential.

In our discussion in Chapter 2, we observed that certain directions or surfaces that bound the zones of influence are associated with hyperbolic equations. Signals are propagated along these particular surfaces influencing the solution at other points within the zone of influence. The method of

Table 6.1 Classification of the Euler equations

	Subsonic, $M < 1$	Sonic, $M = 1$	Supersonic, $M > 1$
Steady	Elliptic	Parabolic	Hyperbolic
Unsteady	Hyperbolic	Hyperbolic	Hyperbolic

characteristics is a technique that utilizes the known physical behavior of the solution at each point in the flow. A clear understanding of the essential elements of the method of characteristics can be obtained by studying a second-order linear PDE.

6.2.1 Linear Systems of Equations

Consider the steady supersonic flow of an inviscid, nonheat-conducting perfect gas. Suppose the free stream flow is only slightly disturbed by a thin body, so the fluid motion satisfies the small perturbation assumptions given by (see Section 5.5.6)

$$\frac{u}{U_\infty} \ll 1 \qquad \frac{v}{U_\infty} \ll 1$$

where u and v are perturbation velocity components. If transonic and hypersonic flows are not considered, the governing PDEs reduce to the Prandtl-Glauert equation for supersonic flow. If the x axis is aligned with the free stream, this equation may be written

$$(1 - M_\infty^2)\phi_{xx} + \phi_{yy} = 0 \tag{6.1}$$

The free stream Mach number is denoted by M_∞, and the perturbation potential is denoted by ϕ. Initial data are specified along a smooth curve, C, which we choose to be $x = \text{const}$ in this case. Boundary conditions are prescribed at $y = 0$.

$$\begin{aligned} \frac{\partial \phi}{\partial y}(x,0) &= U_\infty\left(\frac{dy}{dx}\right)_{\text{wall}} \\ \phi(0,y) &= 0 \end{aligned} \tag{6.2}$$

In order to present the formulation for a system of equations, it is advantageous to consider the similar formulation introduced in Chapter 2. Using the perturbation velocity components

$$u = \frac{\partial \phi}{\partial x} \qquad v = \frac{\partial \phi}{\partial y}$$

and denoting $M_\infty^2 - 1$ by β^2, Eq. (6.1) may be written as the system

$$\begin{aligned} \beta^2\frac{\partial u}{\partial x} - \frac{\partial v}{\partial y} &= 0 \\ \frac{\partial v}{\partial x} - \frac{\partial u}{\partial y} &= 0 \end{aligned} \tag{6.3}$$

with associated initial data and boundary conditions

$$\begin{aligned} &\left.\begin{aligned} u(0,y) &= 0 \\ v(0,y) &= 0 \end{aligned}\right\} \qquad y > 0 \\ &v(x,0) = v_{\text{wall}} \qquad y = 0 \end{aligned} \tag{6.4}$$

In order to use the method of characteristics, the system given by Eq. (6.3) is written along the characteristics. The differential equations of the characteristics are developed as the first step in this procedure.

Suppose the initial data for this problem are prescribed along an arbitrary smooth curve, C, and we consider methods for constructing a solution of Eq. (6.3) in the neighborhood of this curve. If the solution is sufficiently smooth, the first method that might be considered is to write a Taylor series about a point on C. Assume that our interest is in a small neighborhood, and only terms through the first derivatives need to be retained. The solution for either u or v may then be written in the form

$$u(x + \Delta x, y + \Delta y) = u(x, y) + \Delta x \frac{\partial u}{\partial x}(x, y) + \Delta y \frac{\partial u}{\partial y}(x, y) + \cdots \tag{6.5}$$

In this expression, the coordinates (x, y) are on the initial data curve where u and v are known. However, we need to compute the first derivatives in the Taylor series. If s represents arc length along the curve C, we may write

$$\begin{aligned} \frac{du}{ds} &= \frac{\partial u}{\partial x}\frac{dx}{ds} + \frac{\partial u}{\partial y}\frac{dy}{ds} \\ \frac{dv}{ds} &= \frac{\partial v}{\partial x}\frac{dx}{ds} + \frac{\partial v}{\partial y}\frac{dy}{ds} \end{aligned} \tag{6.6}$$

The system of four equations in the unknown derivatives given by Eqs. (6.3) and (6.6) may be solved by any standard method, such as Cramer's rule. It is clear that the determinant of the coefficients of the system must not vanish. (If the determinant of the coefficients vanishes, the direction of curve C is along the characteristics of the system and, consistent with our discussion in Chapter 2, the derivatives may not be uniquely determined.) The differential equations of the characteristics are obtained by setting the determinant of this system equal to zero:

$$\begin{vmatrix} \beta^2 & 0 & 0 & -1 \\ 0 & -1 & 1 & 0 \\ \dfrac{dx}{ds} & \dfrac{dy}{ds} & 0 & 0 \\ 0 & 0 & \dfrac{dx}{ds} & \dfrac{dy}{ds} \end{vmatrix} = 0 \tag{6.7}$$

Expanding this determinant and solving the characteristic equation yields the expressions

$$\frac{dy}{dx} = \pm \frac{1}{\beta} \tag{6.8}$$

which are differential equations of the characteristics as illustrated in Fig. 6.1. Since β is constant, the characteristics can be obtained by integration and are

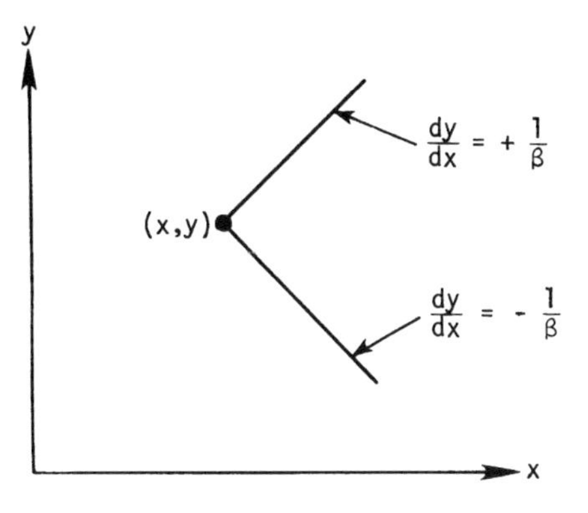

Figure 6.1 Characteristics of the Prandtl-Glauert equation.

given by

$$\begin{aligned} \xi &= x - \beta y \\ \eta &= x + \beta y \end{aligned} \tag{6.9}$$

The original differential equations written along the characteristics are called the compatibility equations. These compatibility equations may be derived by continuing to solve the original system of equations for the first derivatives. Along the characteristic directions, the determinant of the coefficients vanishes. If we solve for any of the first derivatives, for instance, $\partial u/\partial x$, and require that they are at least bounded, the determinant forming the numerator must also vanish. This may be written

$$\begin{vmatrix} 0 & 0 & 0 & -1 \\ 0 & -1 & 1 & 0 \\ \dfrac{du}{ds} & \dfrac{dy}{ds} & 0 & 0 \\ \dfrac{dv}{ds} & 0 & \dfrac{dx}{ds} & \dfrac{dy}{ds} \end{vmatrix} = 0 \tag{6.10}$$

If this determinant is expanded, the compatibility equations are given by

$$\frac{du}{ds} = \left(\frac{dy}{dx}\right)\frac{dv}{ds}$$

or

$$\frac{d}{ds}(\beta u + v) = 0 \tag{6.11}$$

along a right running characteristic, where

$$\frac{dy}{dx} = -\frac{1}{\beta}$$

and

$$\frac{d}{ds}(\beta u - v) = 0 \tag{6.12}$$

along the left running characteristic

$$\frac{dy}{dx} = \frac{1}{\beta}$$

A more general procedure for deriving the characteristics is given by Whitham (1974). We will repeat the details of the procedure here and omit the derivation of the technique. In order to find the characteristics of the system [Eq. (6.3)], we write these equations in the vector form:

$$\frac{\partial \mathbf{w}}{\partial x} + [A]\frac{\partial \mathbf{w}}{\partial y} = 0 \tag{6.13}$$

where

$$\mathbf{w} = \begin{bmatrix} u \\ v \end{bmatrix}$$

and

$$[A] = \begin{bmatrix} 0 & -\dfrac{1}{\beta^2} \\ -1 & 0 \end{bmatrix} \tag{6.14}$$

The eigenvalues of this system are the eigenvalues of $[A]$. These are obtained by extracting the roots of the characteristic equation of $[A]$. Thus we write

$$\|[A] - \lambda[I]\| = 0$$

or

$$\begin{vmatrix} -\lambda & -\dfrac{1}{\beta^2} \\ -1 & -\lambda \end{vmatrix} = 0$$

This produces the quadratic equation

$$\lambda^2 - \frac{1}{\beta^2} = 0$$

The roots of this equation are

$$\lambda_1 = \frac{1}{\beta}$$

$$\lambda_2 = -\frac{1}{\beta}$$

This pair of roots form the differential equations of the characteristics we have already derived in Eq. (6.8). Since our original Prandtl-Glauert equation for supersonic flow is just a wave equation in ϕ, we could have written the characteristic differential equations using the results from our discussion of the second-order PDE [Eq. (2.15*a*)]. The next step is to determine the compatibility equations. Following Whitham, these equations are obtained by premultiplying the system given by Eq. (6.13) by the left eigenvector of $[A]$. This effectively provides a method of writing the equations along the characteristics.

Let $\mathbf{L}^1$ represent the left eigenvector of $[A]$ corresponding to λ_1 and $\mathbf{L}^2$ represent the left eigenvector corresponding to λ_2. We derive the eigenvectors of $[A]$, by writing

$$[L^i]^T[A - \lambda_i I] = 0 \tag{6.15}$$

If we let

$$\mathbf{L}^1 = \begin{bmatrix} l_1 \\ l_2 \end{bmatrix}$$

then

$$[l_1^1, l_2^1]\begin{bmatrix} -\dfrac{1}{\beta} & -\dfrac{1}{\beta^2} \\ -1 & -\dfrac{1}{\beta} \end{bmatrix} = 0$$

This provides the equations

$$\frac{l_1^1}{\beta} + l_2^1 = 0 \qquad \frac{l_1^1}{\beta^2} + \frac{l_2^1}{\beta} = 0$$

which are equivalent as expected. Since we are only able to obtain the normalized components of $\mathbf{L}^1$, assume $l_1^1 = -\beta$. Then the solution for l_2^1 is

$$l_2^1 = 1$$

and

$$\mathbf{L}^1 = \begin{bmatrix} -\beta \\ 1 \end{bmatrix}$$

In a similar manner, the solution for $\mathbf{L}^2$ is

$$\mathbf{L}^2 = \begin{bmatrix} \beta \\ 1 \end{bmatrix}$$

The compatibility equations are now obtained by writing our system [Eq. (6.13)] along the characteristics. To do this, we multiply Eq. (6.13) by the transpose of the left eigenvector:

$$[\mathbf{L}^i]^T\left[\mathbf{w}_x + [A]\mathbf{w}_y\right] = 0 \tag{6.16}$$

The term $[\mathbf{L}^i]^T[A]$ may be replaced by $[\mathbf{L}^i]^T\lambda_i[I]$ by substituting from Eq. (6.15). Thus, we may write Eq. (6.16) as

$$[\mathbf{L}^i]^T[\mathbf{w}_x + \lambda_i \mathbf{w}_y] = 0$$

The compatibility equation along λ_1 is obtained from

$$[-\beta, 1]\begin{bmatrix} u_x + \dfrac{1}{\beta}u_y \\ v_x + \dfrac{1}{\beta}v_y \end{bmatrix} = 0$$

Thus

$$\frac{\partial}{\partial x}(\beta u - v) + \frac{1}{\beta}\frac{\partial}{\partial y}(\beta u - v) = 0 \tag{6.17a}$$

In a similar manner, the compatibility equation along the right running characteristic in partial derivative form is

$$\frac{\partial}{\partial x}(\beta u + v) - \frac{1}{\beta}\frac{\partial}{\partial y}(\beta u + v) = 0 \tag{6.17b}$$

Equation (6.17*a*) is valid along the positive or left running characteristic. It expresses the fact that the quantity $(\beta u - v)$ is constant along λ_1. This can be demonstrated by letting s represent distance along the characteristic and writing

$$\frac{d}{ds}(\beta u - v) = \frac{\partial}{\partial x}(\beta u - v)\frac{dx}{ds} + \frac{\partial}{\partial y}(\beta u - v)\frac{dy}{ds}$$

However, if $(\beta u - v)$ is constant along the characteristic, we may write

$$\frac{d}{ds}(\beta u - v) = 0$$

or

$$\frac{\partial}{\partial x}(\beta u - v) + \left(\frac{dy}{dx}\right)\frac{\partial}{\partial y}(\beta u - v) = 0$$

which is the same as Eq. (6.17*a*). Therefore we conclude that $(\beta u - v)$ is constant along λ_1, and $(\beta u + v)$ is constant along λ_2. The quantities $(\beta u - v)$ and $(\beta u + v)$ are called *Riemann invariants* (Garabedian, 1964). Since these two quantities are constant along opposite pairs of characteristics, it is easy to determine u and v at a given point. If at a point (x, y) we know $(\beta u + v)$ and $(\beta u - v)$, we can immediately compute both u and v. An example illustrating this is in order.

Example 6.1 A uniform inviscid supersonic flow ($M_\infty = \sqrt{2}$) encounters a one-period sine wave wrinkle in the metal skin of a wind tunnel. The geometry of this configuration is shown in Fig. 6.2. The maximum amplitude of the sine wave is ϵ/L, where $\epsilon/L \ll 1$. Determine the solution for the perturbation velocities u and v using the method of characteristics.

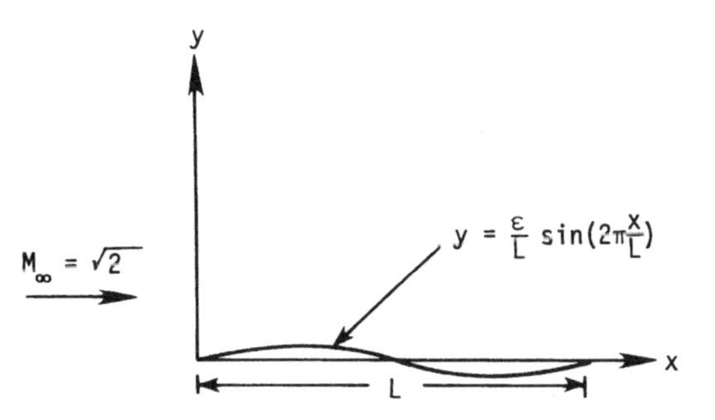

Figure 6.2 Wavy wall geometry.

Solution Since the flow is assumed to satisfy the small-perturbation assumption, the Prandtl-Glauert equation can be used. We choose to solve the system of equations [Eq. (6.3)] for the perturbation components u and v. In this case, $\beta^2 = 1$, and we solve the system of PDEs,

$$\frac{\partial u}{\partial x} - \frac{\partial v}{\partial y} = 0$$

$$\frac{\partial v}{\partial x} - \frac{\partial u}{\partial y} = 0$$

with initial data specified along $x = 0$, $y > 0$

$$u = 0$$
$$v = 0$$

subject to the surface boundary condition (see Section 6.7),

$$v = 2\pi U_\infty \frac{\epsilon}{L^2} \cos\left(2\pi \frac{x}{L}\right) \qquad 0 \leqslant x \leqslant L$$

Since the problem is two-dimensional and obeys the small-perturbation assumptions, we may apply the boundary conditions in the $y = 0$ plane. This makes the problem much easier.

We begin our characteristic solution by sketching the characteristics that originate at the initial data surface $x = 0$. Along the left running characteristics, we know that

$$\frac{dy}{dx} = 1 \qquad u - v = P = \text{const}$$

while along the other characteristic,

$$\frac{dy}{dx} = -1 \qquad u + v = Q = \text{const}$$

Therefore we determine u and v at any point as

$$u = \frac{P + Q}{2} \qquad v = \frac{Q - P}{2}$$

Since the right running characteristics that strike the surface originate in the free stream, the Q variable is initially zero. It is also true that $P = 0$ for those characteristics that originate in the free stream (see Fig. 6.3).

Consider the characteristic that strikes the wavy wall. An up or left running characteristic is introduced at that point in such a way that the surface boundary condition is satisfied. Thus at any station x_1, we have

$$Q = u + v = 0$$

$$v = \frac{2\pi\epsilon}{L^2} U_\infty \cos\left(2\pi\frac{x_1}{L}\right)$$

Therefore

$$u = -\frac{2\pi\epsilon}{L^2} U_\infty \cos\left(2\pi\frac{x_1}{L}\right)$$

and

$$P = u - v = -\frac{4\pi\epsilon}{L^2} U_\infty \cos\left(2\pi\frac{x_1}{L}\right)$$

The solution for u and v is constructed by marching outward from the initial data surface in the x direction. A grid with indices and the corresponding characteristics is shown in Fig. 6.4. The solution can now be obtained at the intersections of the characteristics. At point (1, 3),

$$P = 0$$
$$Q = 0$$
$$u = 0$$
$$v = 0$$

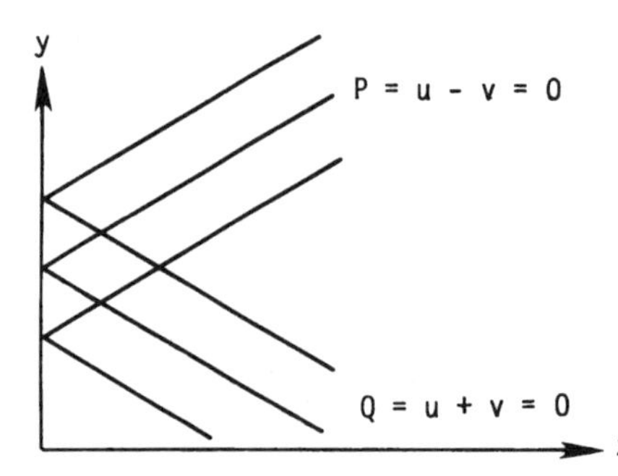

Figure 6.3 Initial data line.

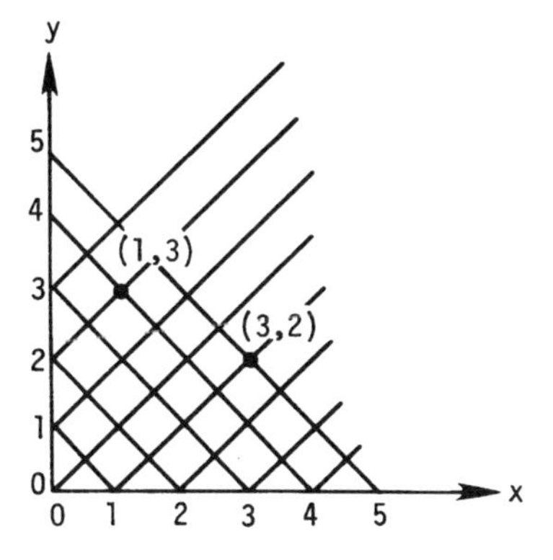

Figure 6.4 Characteristic net.

At (3, 2),

$$P = -\frac{4\pi\epsilon}{L^2} U_\infty \cos\left(2\pi \frac{x_1}{L}\right)$$

$$Q = 0$$

$$u = -\frac{2\pi\epsilon}{L^2} U_\infty \cos\left(2\pi \frac{x_1}{L}\right)$$

$$v = \frac{2\pi\epsilon}{L^2} \cos\left(2\pi \frac{x_1}{L}\right)$$

The solution is known everywhere in the domain of interest. The results of this example may be verified by solving the Prandtl-Glauert equation directly for the velocity potential and then computing the solution for u and v.

6.2.2 Nonlinear Systems of Equations

The development presented thus far is for a system of two linear equations and was chosen for its simplicity. In more complex nonlinear problems, the results are not as easily obtained. In the general case, the characteristic slopes are not constant but must vary as the fluid properties change. The governing PDEs may be nonhomogeneous. Clearly, the compatibility equations cannot be directly integrated in closed form along the characteristics in that case. For the general nonlinear problem, both the compatibility equations and the characteristic equations must be integrated numerically to obtain a complete flow field solution. Not only are the flow variables unknown, but the location in the field along the characteristics must be computed.

In order to illustrate the difference in applying the method of characteristics to a linear and a nonlinear problem, we consider the two-dimensional supersonic flow of a perfect gas over a flat surface. For simplicity, we choose a rectangular coordinate system and write the Euler equations (see Section 5.5) governing this inviscid flow as the matrix system:

$$\frac{\partial \mathbf{w}}{\partial x} + [A]\frac{\partial \mathbf{w}}{\partial y} = 0 \tag{6.18}$$

where

$$\mathbf{w} = \begin{bmatrix} u \\ v \\ p \\ \rho \end{bmatrix}$$

and

$$[A] = \frac{1}{u^2 - a^2}\begin{bmatrix} uv & -a^2 & -\dfrac{v}{\rho} & 0 \\ 0 & \dfrac{v}{u}(u^2 - a^2) & \dfrac{u^2 - a^2}{\rho u} & 0 \\ -\rho v a^2 & \rho u a^2 & uv & 0 \\ -\rho v & \rho u & \dfrac{v}{u} & \dfrac{v}{u}(u^2 - a^2) \end{bmatrix}$$

The initial data, **I**, are prescribed and may be written as

$$\mathbf{w}(0, y) = \mathbf{I}(y) \qquad 0 \leqslant y \leqslant h$$

and the boundary conditions are

$$\begin{aligned} v(x, 0) &= 0 \\ u(x, h) &= u_\infty \\ v(x, h) &= u_\infty \\ p(x, h) &= p_\infty \\ \rho(x, h) &= \rho_\infty \end{aligned}$$

The eigenvalues of $[A]$ determine the characteristic directions and must be found as the first step. These eigenvalues are

$$\lambda_1 = \frac{v}{u} \qquad \lambda_2 = \frac{v}{u}$$

$$\lambda_3 = \frac{uv + a\sqrt{u^2 + v^2 - a^2}}{u^2 - a^2} \qquad \lambda_4 = \frac{uv - a\sqrt{u^2 + v^2 - a^2}}{u^2 - a^2} \tag{6.19a}$$

The matrix of left eigenvectors associated with these values of λ may be written

$$[T]^{-1} = \begin{bmatrix} \dfrac{\rho u}{a^2} & \dfrac{\rho v}{a^2} & 0 & 1 \\ \rho u & \rho v & 1 & 0 \\ -\dfrac{1}{\sqrt{u^2+v^2-a^2}} & +\dfrac{u}{v}\dfrac{1}{\sqrt{u^2+v^2-a^2}} & \dfrac{1}{\rho v a} & 0 \\ \dfrac{1}{\sqrt{u^2+v^2-a^2}} & -\dfrac{u}{v}\dfrac{1}{\sqrt{u^2+v^2-a^2}} & \dfrac{1}{\rho v a} & 0 \end{bmatrix} \tag{6.19b}$$

We obtain the compatibility relations by premultiplying the original system by $[T]^{-1}$. These relations along the wave fronts are given by

$$-v\frac{du}{ds_3} + u\frac{dv}{ds_3} + \frac{\beta}{\rho}\frac{dp}{ds_3} = 0 \tag{6.20}$$

along

$$\frac{dy}{dx} = \lambda_3$$

and

$$v\frac{du}{ds_4} - u\frac{dv}{ds_4} + \frac{\beta}{\rho}\frac{dp}{ds_4} = 0 \tag{6.21}$$

along

$$\frac{dy}{dx} = \lambda_4$$

In these expressions,

$$\beta = \sqrt{M^2 - 1} \qquad M^2 = \frac{u^2+v^2}{a^2}$$

Equation (6.20) is an ordinary differential equation, which holds along the characteristic with slope λ_3. Arc length along this characteristic is denoted by s_3. A similar result is expressed in Eq. (6.21). In contrast to our linear example using the Prandtl-Glauert equation, the analytic solution for the characteristics is not known for the general nonlinear problem. It is clear that we must numerically integrate to determine the shape of the characteristics in a step-by-step manner. Consider the characteristic defined by λ_3:

$$\frac{dy}{dx} = \frac{uv + a\sqrt{u^2+v^2-a^2}}{u^2-a^2}$$

Starting at an initial data surface, this expression can be integrated to obtain the coordinates of the next point on the curve. At the same time, the differential equation defining the other wave front characteristic can be integrated. For a simple first-order integration, this provides us with two equations for the wave front characteristics. From these expressions, we determine the coordinates of their intersection (point A in Fig. 6.5). Once the point A is known, the compatibility relations, Eqs. (6.20) and (6.21), are integrated along the characteristics to this point. This provides a system of equations for the unknowns at point A. Of course, auxiliary relationships are required to complete the problem, and these are provided by integrating the streamline compatibility equations or by using other valid equations relating the unknowns at A.

By using this procedure, a first-order estimate of both the location of point A and the associated flow variables can be obtained. These first-order estimates are usually used as a first step in a predictor-corrector scheme in calculating the solution to a system of hyperbolic PDEs using the method of characteristics. In the corrector step, a new intersection point B can be computed, which now includes the nonlinear nature of the characteristic curves. In a similar manner, the dependent variables at B are computed.

The calculation of the solution at point B presents an interesting problem. Because the problem is nonlinear, the final intersection point B does not necessarily appear at the same value of x for all solution points. Consequently, the solution is usually interpolated onto an x = const surface before the next integration step is started. This requires additional logic and adds considerably to the difficulty in structuring an accurate code.

The problem of integrating the compatibility equations and satisfying the boundary conditions at both permeable and impermeable boundaries is discussed in Section 6.7. It should be clear that the wall boundary condition is iterative in the sense that we attempt to satisfy a particular boundary condition at a point on a surface with an initially unknown x coordinate.

The two-dimensional (2-D) flow problem used in this section actually can be treated using characteristics in a much simpler setting (see Prob. 6.3). The main reason for this discussion is to present the ideas behind the numerical integration of the equations of motion using characteristic methods and to introduce some of the inherent difficulties in the general method. More complete descriptions are given by numerous authors, including Owczarek (1964), Shapiro (1953), and Courant and Friedrichs (1948).

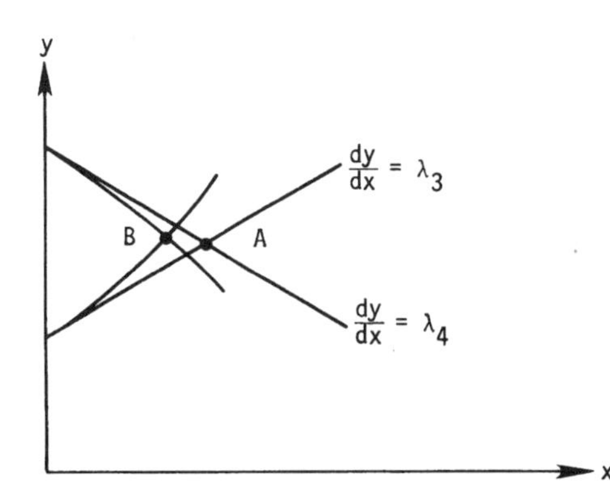

Figure 6.5 Characteristic solution point.

6.3 CLASSICAL SHOCK-CAPTURING METHODS

Shock-capturing schemes are the most widely used techniques for computing inviscid flows with shocks. In this approach the Euler equations are cast in conservation-law form, and any shock waves or other discontinuities are computed as part of the solution. The shock waves predicted by these methods are usually smeared over several mesh intervals, but the simplicity of the approach may outweight the slight compromise in results compared to the more elaborate shock-fitting schemes. Classical shock-capturing methods have the disadvantage that very strong shocks will cause the methods to fail. This failure is usually evidenced by oscillations. Computations in hypersonic flow with very strong shocks typically lead to the appearance of negative pressures and subsequent divergence of the solution during the time-dependent computation process. In addition to this problem, higher-order schemes tend to produce oscillations in the solution. However, these methods are useful and will be modified in later sections to avoid these difficulties. The alternative approach is to fit each shock wave as a discontinuity and solve for the discontinuity as part of the solution. This shock-fitting approach is very elegant and produces shocks that are truly discontinuous. Unfortunately, the procedure for general shock fitting in three dimensions with multiple shocks is extremely complex, and as a result, the use of shock fitting is usually limited to fitting shocks at boundaries.

In supersonic flow, when one boundary of the physical domain is a shock wave, shock fitting is frequently employed, and the shock shape is computed as part of the solution. Since boundary shocks can be fit with either the standard schemes discussed in this section or Section 6.7, the real advantage accrues when a complicated internal shock structure is captured and the special treatment of each shock wave is eliminated. This is a standard approach, where the outer boundary is fit when it is a shock wave and the internal shocks are captured. In this section, we will examine several simple shock-capturing schemes and apply them to example problems to gain experience in understanding the behavior of these numerical methods and interpreting the results that are produced when they are used.

Lax (1954) has shown that shock wave speed and strength are correctly predicted when the conservative form of the Euler equations is used. This means that the physically correct weak solution corresponding to the Rankine-Hugoniot equations for shocks is obtained if the conservation-law form is used and the equations are discretized in a conservative manner. In our study of Burgers' equation in Chapter 4, we saw that incorrect results were produced when the nonconservative form was used. While the nonconservative form of the Euler equations will have a weak solution, the solution depends upon the form of the equations used. In order that the solution satisfy the Rankine-Hugoniot equations, the conservative form must be used when we apply shock-capturing techniques.

As an example of conservation form, consider the supersonic flow of a perfect gas over a 2-D surface. If we assume the x axis forms the body surface and is also the marching direction, the equations are given by the steady 2-D version of Eq. (5.192) and may be written

$$\frac{\partial \rho u}{\partial x} + \frac{\partial \rho v}{\partial y} = 0 \tag{6.22}$$

$$\frac{\partial(p + \rho u^2)}{\partial x} + \frac{\partial(\rho u v)}{\partial y} = 0 \tag{6.23}$$

$$\frac{\partial(\rho u v)}{\partial x} + \frac{\partial(p + \rho v^2)}{\partial y} = 0 \tag{6.24}$$

For a steady isoenergetic flow, the total enthalpy is constant. In this case, the differential energy equation can be integrated to give

$$H = \frac{\gamma}{\gamma - 1}\frac{p}{\rho} + \frac{u^2 + v^2}{2} = \text{const} \tag{6.25}$$

The system formed by Eqs. (6.22)–(6.24) in conjunction with the constant total enthalpy equation is hyperbolic for supersonic flow, and a solution can be obtained by marching or integrating the equations in the x direction starting from an initial data surface. The geometry for such a marching problem is shown in Fig. 6.6. Initial data are prescribed along the line $x = 0$, and the solution is advanced in the x direction subject to wall boundary conditions and an appropriate condition at y_{max}.

Equations (6.22)–(6.24) are of the form

$$\frac{\partial \mathbf{E}}{\partial x} + \frac{\partial \mathbf{F}}{\partial y} = 0 \tag{6.26}$$

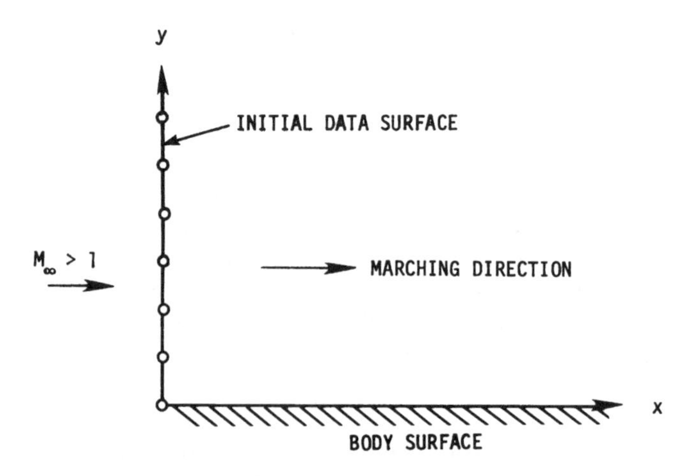

Figure 6.6 Coordinate system for marching problem.

where

$$\mathbf{E} = \begin{bmatrix} \rho u \\ p + \rho u^2 \\ \rho u v \end{bmatrix} \qquad \mathbf{F} = \begin{bmatrix} \rho v \\ \rho u v \\ p + \rho v^2 \end{bmatrix}$$

Equation (6.26) may be integrated with any of the methods presented in Chapter 4 for hyperbolic PDEs. If the forward predictor–backward corrector version of MacCormack's method is applied, Eq. (6.26) may be written

$$\overline{\mathbf{E}_j^{n+1}} = \mathbf{E}_j^n - \frac{\Delta x}{\Delta y}(\mathbf{F}_{j+1}^n - \mathbf{F}_j^n)$$

$$\mathbf{E}_j^{n+1} = \frac{1}{2}\left[\mathbf{E}_j^n + \overline{\mathbf{E}_j^{n+1}} - \frac{\Delta x}{\Delta y}\left(\overline{\mathbf{F}_j^{n+1}} - \overline{\mathbf{F}_{j-1}^{n+1}}\right)\right] \tag{6.27}$$

At the end of the predictor and corrector steps, **E** must be decoded to obtain the primitive variables. In this way, the new flux vector can be formed for the next integration step. After advancing the solution, the y component of velocity is immediately known as

$$v = \frac{E_3}{E_1}$$

where the subscripts denote elements of **E**. A quadratic equation must be solved for the x component of the velocity. If we combine E_2 with the energy equation to eliminate p, we have

$$\rho = \frac{E_2}{u^2 + [(\gamma - 1)/2\gamma](2H - u^2 - v^2)}$$

We now eliminate ρ in favor of u by using

$$\rho = \frac{E_1}{u} \tag{6.28}$$

This yields a quadratic equation for u, which has roots

$$u = \frac{\gamma}{\gamma + 1}\frac{E_2}{E_1} \pm \sqrt{\left(\frac{\gamma}{\gamma + 1}\frac{E_2}{E_1}\right)^2 - \frac{\gamma - 1}{\gamma + 1}(2H - v^2)} \tag{6.29}$$

The correct sign on the radical is typically positive. The density can now be computed from E_1, and the pressure from E_2, as

$$p = E_2 - \rho u^2 \tag{6.30}$$

Having completed this process, **F** can be recalculated, and the next step in the integration can be implemented.

Example 6.2 Compute the flow field produced by a 2-D wedge moving at a Mach number of 2.0 if the wedge half angle is 15°. Assume inviscid flow of a perfect gas.

Solution The problem requires that we determine the shock wave location and strength as well as internal flow detail. The wedge and associated flow are shown in Fig. 6.7. In a 2-D wedge flow with an attached shock wave, the flow is conical. This means that flow properties along rays from the vertex of the wedge are constant (Anderson, 1982). This results in a simplification of the problem.

For this problem, the governing PDEs are given by Eqs. (6.22)–(6.24) and the energy equation [Eq. (6.25)]. The boundary conditions are the surface tangency requirement at the wedge surface and free stream conditions outside the shock wave. We recognize that we can select the x axis along the wedge surface and march the equations in this direction so long as the shock layer Mach number is greater than 1. Unfortunately, the shock layer expands as we move downstream, and this eventually causes our outer boundary point (at $y = y_{\max}$) to interfere with the shock wave.

The problem can easily be solved utilizing the fact that the shock wave is straight and that the thickness of the shock layer grows linearly with x. We introduce the independent variable transformation given by

$$\xi = x \qquad \eta = \frac{y}{x} \tag{6.31}$$

This provides the grid shown in Fig. 6.8. We can solve the wedge-flow problem with no difficulty now because the constant η lines grow linearly with x. Since the governing equations are hyperbolic in the ξ direction, initial data must be prescribed along some noncharacteristic surface. The line $\xi = 1$ is an easy choice. The PDEs are integrated in the ξ direction using arbitrarily assigned initial data. Since the solution to 2-D wedge flow is conical, the conical solution will be obtained for large ξ (asymptotically).

If the governing PDEs are transformed from (x, y) into (ξ, η) coordinates, they become

$$\frac{\partial \overline{\mathbf{E}}}{\partial \xi} + \frac{\partial \overline{\mathbf{F}}}{\partial \eta} = 0 \tag{6.32}$$

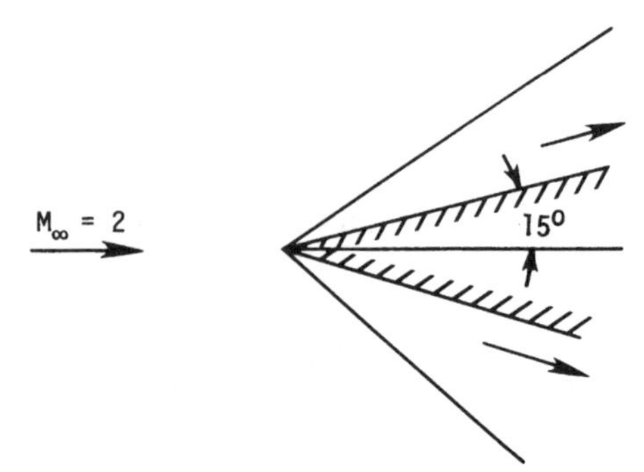

Figure 6.7 Wedge flow with attached shock.

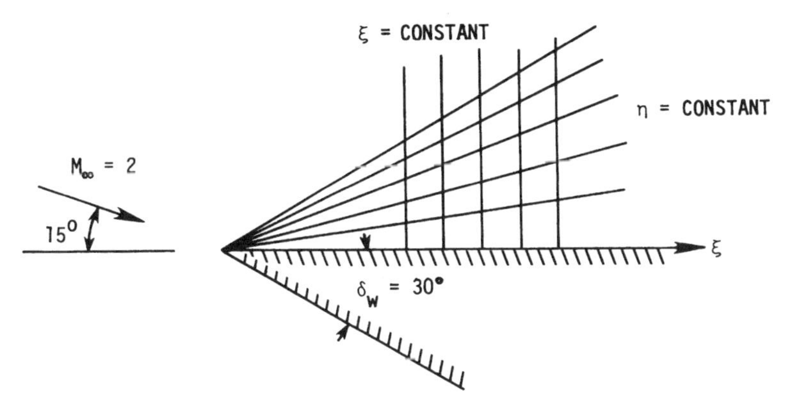

Figure 6.8 Wedge with transformed shock layer.

where

$$\bar{\mathbf{E}} = \xi \mathbf{E}$$
$$\bar{\mathbf{F}} = \mathbf{F} - \eta \mathbf{E}$$

An additional problem can be avoided by utilizing the conical flow property in this problem. The stability of the integration scheme used in solving Eq. (6.32) depends upon the eigenvalue structure of the $[A]$ matrix of the expanded system written in (ξ, η) coordinates.

$$\frac{\partial \mathbf{w}}{\partial \xi} + [A]\frac{\partial \mathbf{w}}{\partial \eta} + \mathbf{H} = 0 \tag{6.33}$$

In this expression, $\mathbf{w}$ is the vector of primitive variables and $\mathbf{H}$ is a source term that occurs in this expanded form. If the eigenvalues of $[A]$ are evaluated, they are found to depend explicitly on the ξ coordinate. That is, ξ appears in the expressions for the eigenvalues. As the solution is marched downstream in ξ, the allowable step size must change as ξ increases if an explicit method such as MacCormack's is used. If the step size did not change as ξ increased, a stability problem would occur. This problem can be avoided if we elect to integrate the equations from $\xi = 1$ to $\xi = 1 + \Delta\xi$ in an iterative manner until a converged solution is obtained.

The application of boundary conditions requires careful consideration. We must include enough points in the η direction so that the shock wave can form naturally and not be interfered with by the fixed free stream conditions, which are maintained at $\eta = \eta_{\max}$. For example, if our shock wave angle (measured from the wedge surface) is 20°, and we elect to use 10 points in the shock layer, then

$$\eta_{\text{shock}} = \tan(20°) = 0.3640$$

$$\Delta\eta = \frac{0.3640}{10 - 1} = 0.0404$$

Suppose we add an additional 5 points using this computed $\Delta\eta$; then

$$\eta_{\max} = 0.0404(15 - 1) = 0.5662$$

and the last mesh point is at an angle of 29.52°. This should provide sufficient freedom for the shock wave to form without interference from the fixed boundary condition at $\eta_{\max}$.

When predictor-corrector methods are used, one or both integration steps may require modification when applied at a solid boundary. For example, a MacCormack forward predictor can be directly applied at the wall but the backward corrector requires modification. One way to assure satisfaction of surface tangency is to also use a forward corrector and overwrite the decoded value of v at the wall with the boundary condition $v = 0$. While the use of forward differences in both the predictor and corrector is generally unstable, the wall boundary condition alters the stability in such a way as to provide a stable solution.

Typical shock-capturing pressure results for wedge flow are presented in Fig. 6.9. These results show an excellent solution, at a Courant number (ν) of one with a sharp shock wave, and very few oscillations. However, the same

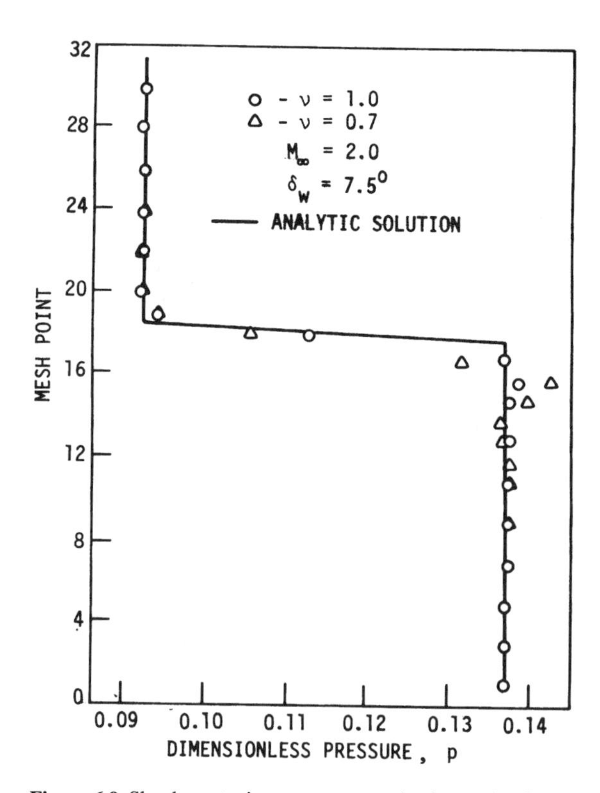

Figure 6.9 Shock-capturing pressure results for wedge flow.

calculation at a Courant number of 0.7 demonstrates the dispersive behavior of second-order methods previously discussed in Chapter 4.

Before we leave the wedge-flow problem, it is worthwhile to note that a solution could also have been obtained using a time-dependent formulation. If the governing PDEs are written in polar coordinates including the time terms, they are of the form

$$\frac{\partial \mathbf{E}}{\partial t} + \frac{\partial \mathbf{F}}{\partial \theta} + \frac{\partial \mathbf{G}}{\partial R} + \mathbf{H} = 0 \tag{6.34}$$

where the origin is at the vertex of the wedge and the vectors are the appropriate polar forms. If we assume a priori that the flow is conical, a solution can be computed in an $R =$ const plane if the radial derivatives are discarded. This requires a solution of the system

$$\frac{\partial \mathbf{E}}{\partial t} + \frac{\partial \mathbf{F}}{\partial \theta} + \mathbf{H} = 0 \tag{6.35}$$

This system is hyperbolic in time and can be integrated to attain a steady wedge-flow solution. In some ways, the time-dependent set is easier to use. For example, the decoding procedure is much simpler.

As in Example 6.2, the equations of motion are usually transformed into a computational domain. One of the more frequently used transformations is that of Viviand (1974) and Vinokur (1974). This transformation (see Section 5.6.2) assures us that a system of equations in a strong conservation-law form can be written in the same form after changing the independent variables. There may be disadvantages to Viviand's transformed equation form because the Jacobian of the mapping always appears in the denominator of the conservative variable terms. In order to avoid the introduction of errors through the geometry, special care must be taken in forming the metrics.

The difficulty encountered in using a simple rectangular mesh in Example 6.2 could have been eliminated if the shock wave was treated as a discontinuity. In fact, most shock-capturing codes fit boundary shock waves as discontinuities and capture interior shock waves as they develop. While the same philosophy of shock fitting holds for the steady flow marching problem as for time-dependent flows, a slightly different scheme is sometimes used to predict the interior or post-shock pressure when the conservative form of the original equations is used. Consider a system of PDEs of the form given in Eq. (6.26). Suppose we make use of a normalizing transformation,

$$(x, y) \rightarrow (\xi, \eta)$$

$$\xi = x \qquad \eta = \frac{y}{y_s(x)} \tag{6.36}$$

where $y - y_s(x) = 0$ is the equation for the position of the shock wave. As shown in Fig. 6.10, the physical domain is now transformed into a computational domain with the shock wave at $\eta = 1.0$. The conservation form for the governing

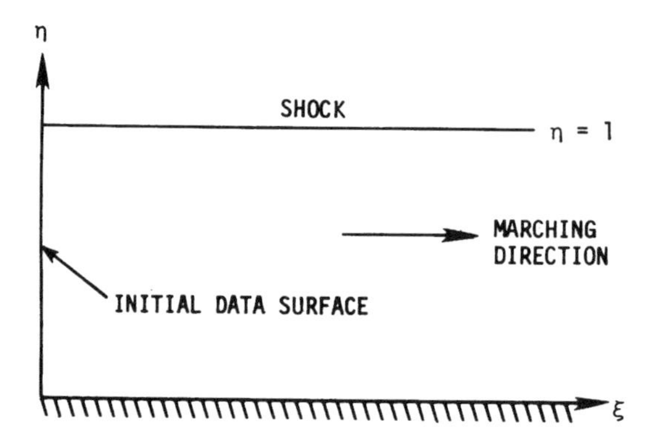

Figure 6.10 Normalizing transformation.

equations using such a transformation may be similar to Viviand's or any other form that conserves the appropriate flux terms. We again assume that the solution for the interior of the shock layer is advanced. At the shock wave, one-sided integration must be used to obtain an estimate for one of the variables. We assume initially that we know everything along an initial data surface, including the shock slope. We advance the solution on the interior, including the shock point. In addition, the shock slope equation (dy_s/dx) is integrated, providing an updated estimate of the new shock position. We now calculate the shock slope at the new location, and the dependent variables other than pressure can then be obtained.

If the pressure on the downstream side of the shock is known, we clearly can determine the density and both velocity components from the Rankine-Hugoniot equations. Our requirement is to develop the expression for shock slope. We write the surface equation of the shock wave as

$$y - y_s(x) = 0 \tag{6.37}$$

The shock normal is then written

$$\mathbf{n}_s = \frac{1}{\left[1 + (dy_s/dx)^2\right]^{1/2}}\left(-\mathbf{i}\frac{dy_s}{dx} + \mathbf{j}\right) \tag{6.38}$$

The normal component of velocity on the free stream side of the shock wave is given by

$$u_{\infty n} = \mathbf{n}_s \cdot \mathbf{V}_\infty = \frac{1}{\left[1 + (dy_s/dx)^2\right]^{1/2}}\left(-u_\infty \frac{dy_s}{dx} + v_\infty\right) \tag{6.39}$$

If this equation is solved for the shock slope, we obtain

$$(u_{\infty n}^2 - u_\infty^2)\frac{dy_s}{dx} = -u_\infty v_\infty \pm \sqrt{u_\infty^2 v_\infty^2 - (u_{\infty n}^2 - u_\infty^2)(u_{\infty n}^2 - v_\infty^2)} \tag{6.40}$$

The term $u_{\infty n}^2$ required in Eq. (6.40) is known from the pressure ratio across the shock as given in Eq. (5.209) and is

$$u_{\infty n}^2 = \frac{\gamma - 1}{2} \frac{p_\infty}{\rho_\infty} \left(1 + \frac{\gamma + 1}{\gamma - 1} \frac{p_2}{p_\infty}\right) \tag{6.41}$$

After the shock slope is computed, all quantities are known at the new location. The same procedure is repeated for both the predictor and corrector steps. We have again performed the shock fitting assuming the post-shock pressure (or other quantity) was known. This follows the approach suggested by Thomas et al. (1972).

Since we are examining methods for either time-dependent or steady supersonic inviscid flows, the governing equations are hyperbolic. Hyperbolic systems are often solved using explicit methods. However, the step size for most explicit schemes is limited by the CFL condition. This can lead to unreasonably long computation times for some problems. To overcome the step size limitation, implicit methods can be used. Examples of implicit algorithms that have been developed for the Euler equations include those of Lindemuth and Killeen (1973), Briley and McDonald (1973), and Beam and Warming (1976). The advantage of implicit methods lies in the unrestricted stability limit. Although more computational effort is required per time step compared to an explicit method, the overall time required to obtain a solution may be less. We will review the development of the basic scheme presented by Beam and Warming (1976) for the conservation form of the governing equations.

The basic system under consideration is of the form given in Eq. (5.192) and is repeated here for convenience:

$$\frac{\partial \mathbf{U}}{\partial t} + \frac{\partial \mathbf{E}}{\partial x} + \frac{\partial \mathbf{F}}{\partial y} = 0 \tag{6.42}$$

where $\mathbf{U}$ is the vector of conservative variables and $\mathbf{E}$ and $\mathbf{F}$ are vector functions of $\mathbf{U}$. If the trapezoidal rule given by Eq. (4.58) is used as the basic integration scheme, the value of $\mathbf{U}$ at the advanced time level is given by

$$\mathbf{U}^{n+1} = \mathbf{U}^n + \frac{\Delta t}{2}\left[\left(\frac{\partial \mathbf{U}}{\partial t}\right)^n + \left(\frac{\partial \mathbf{U}}{\partial t}\right)^{n+1}\right]$$

or

$$\mathbf{U}^{n+1} = \mathbf{U}^n - \frac{\Delta t}{2}\left[\left(\frac{\partial \mathbf{E}}{\partial x} + \frac{\partial \mathbf{F}}{\partial y}\right)^n + \left(\frac{\partial \mathbf{E}}{\partial x} + \frac{\partial \mathbf{F}}{\partial y}\right)^{n+1}\right] \tag{6.43}$$

This expression provides a second-order integration algorithm for the unknown vector $\mathbf{U}^{n+1}$ at the next time level. It is implicit because the derivatives of $\mathbf{U}$ as well as $\mathbf{U}$ appear at the advanced level, thus coupling the unknowns at neighboring grid points. A local Taylor-series expansion of the derivatives of $\mathbf{E}$

and $\mathbf{F}$ is used to obtain a linear equation that can be solved for $\mathbf{U}^{n+1}$. Let

$$\begin{aligned} \mathbf{E}^{n+1} &= \mathbf{E}^n + [A](\mathbf{U}^{n+1} - \mathbf{U}^n) \\ \mathbf{F}^{n+1} &= \mathbf{F}^n + [B](\mathbf{U}^{n+1} - \mathbf{U}^n) \end{aligned} \tag{6.44}$$

where $[A]$ and $[B]$ are defined:

$$[A] = \frac{\partial \mathbf{E}}{\partial \mathbf{U}} \qquad [B] = \frac{\partial \mathbf{F}}{\partial \mathbf{U}}$$

When the linearization given by Eq. (6.44) is substituted into Eq. (6.43), a linear system for $\mathbf{U}^{n+1}$ results and may be written as

$$\begin{aligned} &\left\{[I] + \frac{\Delta t}{2}\left(\frac{\partial}{\partial x}[A]^n + \frac{\partial}{\partial y}[B]^n\right)\right\}\mathbf{U}^{n+1} \\ &\quad = \left\{[I] + \frac{\Delta t}{2}\left(\frac{\partial}{\partial x}[A]^n + \frac{\partial}{\partial y}[B]^n\right)\right\}\mathbf{U}^n - \Delta t\left(\frac{\partial \mathbf{E}}{\partial x} + \frac{\partial \mathbf{F}}{\partial y}\right)^n \end{aligned} \tag{6.45}$$

This is a linear system for the unknown $\mathbf{U}^{n+1}$. Direct solution of Eq. (6.45) is usually avoided owing to the large operation count in treating multidimensional systems. The path usually chosen is to reduce the multidimensional problem into a sequence of one-dimensional inversions. This is done using the method of fractional steps (Yanenko, 1971) or the method of approximate factorization (Peaceman and Rachford, 1955; Douglas, 1955).

Equation (6.45) may be approximately factored into the equation

$$\begin{aligned} &\left([I] + \frac{\Delta t}{2}\frac{\partial}{\partial x}[A]^n\right)\left([I] + \frac{\Delta t}{2}\frac{\partial}{\partial y}[B]^n\right)\mathbf{U}^{n+1} \\ &\quad = \left([I] + \frac{\Delta t}{2}\frac{\partial}{\partial x}[A]^n\right)\left([I] + \frac{\Delta t}{2}\frac{\partial}{\partial y}[B]^n\right)\mathbf{U}^n - \Delta t\left(\frac{\partial \mathbf{E}}{\partial x} + \frac{\partial \mathbf{F}}{\partial y}\right)^n \end{aligned} \tag{6.46}$$

This expression differs from the original Eq. (6.45) by a term that is of $O[(\Delta t)^2]$, and the formal accuracy of our implicit algorithm is maintained as second order. This factored scheme may be written as the alternating direction sequence:

$$\begin{aligned} \left([I] + \frac{\Delta t}{2}\frac{\partial}{\partial x}[A]^n\right)\mathbf{U}' &= \text{RHS of Eq. (6.46)} \\ \left([I] + \frac{\Delta t}{2}\frac{\partial}{\partial y}[B]^n\right)\mathbf{U}^{n+1} &= \mathbf{U}' \end{aligned} \tag{6.47}$$

A simpler algorithm results if the delta form introduced in Chapter 4 is used. Since the operators on both sides of Eq. (6.46) are the same, define

$$\Delta\mathbf{U}^n = \mathbf{U}^{n+1} - \mathbf{U}^n$$

so that

$$\left([I] + \frac{\Delta t}{2}\frac{\partial}{\partial x}[A]^n\right)\left([I] + \frac{\Delta t}{2}\frac{\partial}{\partial y}[B]^n\right)\Delta \mathbf{U}^n = -\Delta t\left(\frac{\partial \mathbf{E}}{\partial x} + \frac{\partial \mathbf{F}}{\partial y}\right)^n \tag{6.48}$$

Again this may be replaced by the alternating direction sequence:

$$\begin{aligned}\left([I] + \frac{\Delta t}{2}\frac{\partial}{\partial x}[A]^n\right)\Delta \mathbf{U}' &= -\Delta t\left(\frac{\partial \mathbf{E}}{\partial x} + \frac{\partial \mathbf{F}}{\partial y}\right)^n \\ \left([I] + \frac{\Delta t}{2}\frac{\partial}{\partial y}[B]^n\right)\Delta \mathbf{U}^n &= \Delta \mathbf{U}'\end{aligned} \tag{6.49}$$

The solution of this system is not trivial. The x and y sweeps each require the solution of a block tridiagonal system of equations assuming the spatial derivatives are approximated by central differences. Each block is $m \times m$ if there are m elements in the unknown $\mathbf{U}$ vector (see Appendix B).

The implicit algorithm developed here used the trapezoidal rule. Generalized time differencing presented by Warming and Beam (1977) can be used to generate a number of implicit algorithms with varying accuracy. This point is discussed in Section 8.3.3. Additional consideration is presented on the required addition of artificial damping in conjunction with nondissipative schemes.

6.4 FLUX SPLITTING SCHEMES

In the previous section, classical shock-capturing methods using central differences were discussed. In this section the concept of *flux-vector splitting* is introduced. The underlying idea behind flux-vector splitting is to split the flux contributions into positive and negative components, where splitting is based on the eigenvalue structure of the system or some other appropriately assumed behavior. In presenting these methods, the view is taken that the fundamental problem that must be solved is to determine the correct flux at the boundaries of the control-volume faces. Interpretation of the numerical methods in terms of the control-volume surface fluxes for the various methods may also be considered in the sense of finite-difference schemes. Wherever this dual interpretation is appropriate, a comparison will be made.

To set the stage for the study of solutions of the Euler equation, consider a control volume as shown in Fig. 6.11. As previously discussed, the conservative form of the governing equations is integrated over the control volume. The 2-D Euler equations are given in Section 5.5.5 in the conservative form:

$$\frac{\partial \mathbf{U}}{\partial t} + \frac{\partial \mathbf{E}}{\partial x} + \frac{\partial \mathbf{F}}{\partial y} = 0 \tag{6.50}$$

where the conservative variables are defined in the usual way. Integrating this

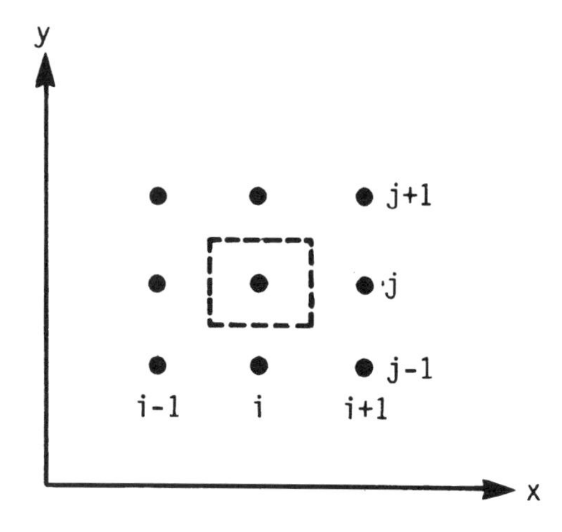

Figure 6.11 Control volume for Euler equations.

equation over the control volume yields the form

$$\int_{\delta v} \frac{\partial \mathbf{U}}{\partial t} dv + \int_{\delta v} \left(\frac{\partial \mathbf{E}}{\partial x} + \frac{\partial \mathbf{F}}{\partial y} \right) dv = 0 \tag{6.51}$$

Applying Green's Theorem (Taylor, 1955) to the second term converts this to a surface integral of the form

$$\int_{\delta v} \frac{\partial \mathbf{U}}{\partial t} dv + \oint_s (\mathbf{E}\, dy - \mathbf{F}\, dx) = 0 \tag{6.52}$$

where the subscript on the integral around the boundary is denoted by the small s. In discrete form the integration results in

$$\frac{\partial \mathbf{U}}{\partial t} \delta v + \sum_{\text{cell faces}} (\mathbf{E}\, \Delta y - \mathbf{F}\, \Delta x) = 0 \tag{6.53}$$

where the δv represents the volume of the cell and the Δx and Δy are the arc lengths of the cell sides for the 2-D case. The evaluation of the sum of the fluxes on the boundary requires that the flux values, i.e., the values of **E** and **F**, be known on the surface of the control volume. The evaluation of the flux terms on the control volume surfaces is the fundamental problem in the development of methods for solving the Euler equations.

6.4.1 Steger-Warming Splitting

Steger and Warming (1979) developed an implicit algorithm using a splitting of **E** and **F** in the governing equations. This is similar to the Beam scheme studied earlier in the work of Sanders and Prendergast (1974). In splitting the flux terms, the flux is assumed to be composed of a positive and a negative component. For illustration, consider a 1-D problem where the Euler system under investigation has the form

$$\frac{\partial \mathbf{U}}{\partial t} + \frac{\partial \mathbf{E}}{\partial x} = 0 \tag{6.54}$$

This system can also be written in the form

$$\frac{\partial \mathbf{U}}{\partial t} + [A]\frac{\partial \mathbf{U}}{\partial x} = 0 \tag{6.55}$$

where $[A]$ is the Jacobian $\partial \mathbf{E}/\partial \mathbf{U}$. This system is hyperbolic if a similarity transformation exists so that

$$[T]^{-1}[A][T] = [\lambda] \tag{6.56}$$

where $[\lambda]$ is a diagonal matrix of real eigenvalues of $[A]$ and $[T]^{-1}$ is the matrix whose rows are the left eigenvectors of $[A]$ taken in order.

According to Steger and Warming, if the equation of state is of the form

$$p = \rho f(e) \tag{6.57}$$

where e is the internal energy, then the flux vector $\mathbf{E}(\mathbf{U})$ is a homogeneous function of degree one in $\mathbf{U}$, which means that

$$\mathbf{E}(\alpha \mathbf{U}) = \alpha \mathbf{E}(\mathbf{U}) \tag{6.58}$$

for any α. This permits the flux vectors $\mathbf{E}$ and $\mathbf{F}$ of the Euler equations to be written in the form

$$\mathbf{E} = [A]\mathbf{U} \tag{6.59}$$

We can use this property and the fact that the system is hyperbolic to achieve the desired split flux form.

Combining Eqs. (6.56) and (6.59), $\mathbf{E}$ may be written

$$\mathbf{E} = [A]\mathbf{U} = [T][\lambda][T]^{-1}\mathbf{U} \tag{6.60}$$

The matrix of eigenvalues is divided into two matrices, one with only positive elements and the other with negative elements. We write the $[A]$ matrix as

$$[A] = [A^+] + [A^-] = [T][\lambda^+][T]^{-1} + [T][\lambda^-][T]^{-1} \tag{6.61}$$

and define

$$\mathbf{E} = \mathbf{E}^+ + \mathbf{E}^- \tag{6.62}$$

so that

$$\mathbf{E}^+ = [A^+]\mathbf{U} \qquad \mathbf{E}^- = [A^-]\mathbf{U} \tag{6.63}$$

The original conservation-law form written using the split-flux notation becomes

$$\frac{\partial \mathbf{U}}{\partial t} + \frac{\partial \mathbf{E}^+}{\partial x} + \frac{\partial \mathbf{E}^-}{\partial x} = 0 \tag{6.64}$$

where the plus and minus signs indicate that the flux components are associated with wave propagation in the positive and negative directions, respectively. The key point is that the flux vector $\mathbf{E}$ can be split into a positive part and a negative part, each associated with the signal propagation directions. The eigenvalues of

$\partial \mathbf{E}^{\pm}/\partial \mathbf{U}$ are not the same as $\lambda^{\pm}$, but the correct sign is preserved. For the 1-D case, the eigenvalues of $[A]$ are the familiar streamline and signal propagation terms written as

$$\lambda_1 = u$$
$$\lambda_2 = u + a$$
$$\lambda_3 = u - a$$

For the supersonic case, with u positive, $\lambda^+ = \lambda$ and $\lambda^- = 0$. For the subsonic case, both λ^+ and λ^- are nonzero. For subsonic flow,

$$[\lambda^+] = \begin{bmatrix} u & & \\ & u + a & \\ & & 0 \end{bmatrix} \tag{6.65a}$$

and

$$[\lambda^-] = \begin{bmatrix} 0 & & \\ & 0 & \\ & & u - a \end{bmatrix} \tag{6.65b}$$

The associated split-flux terms are

$$\mathbf{E}^- = \frac{1}{2}\frac{\rho}{\gamma}(u - a)\begin{bmatrix} 1 \\ u - a \\ \dfrac{1}{2}(u - a)^2 + \dfrac{1}{2}a^2\left(\dfrac{3 - \gamma}{\gamma - 1}\right) \end{bmatrix} \tag{6.66a}$$

and

$$\mathbf{E}^+ = \mathbf{E} - \mathbf{E}^- = \frac{1}{2}\frac{\rho}{\gamma}\begin{bmatrix} (2\gamma - 1)u + a \\ 2(\gamma - 1)u^2 + (u + a)^2 \\ (\gamma - 1)u^3 + \dfrac{1}{2}(u + a)^3 + \dfrac{1}{2}a^2\dfrac{3 - \gamma}{\gamma - 1}(u + a) \end{bmatrix} \tag{6.66b}$$

A first-order upwind scheme is easily constructed with this split-flux idea. A simple integration of the equations for a 1-D problem may be written

$$\mathbf{U}_i^{n+1} = \mathbf{U}_i^n - \frac{\Delta t}{\Delta x}(\mathbf{E}_{i+\frac{1}{2}} - \mathbf{E}_{i-\frac{1}{2}}) \tag{6.67}$$

In this setting, the cell-face values of the flux are composed of both + and − components according to the splitting, i.e.,

$$\mathbf{E}_{i+\frac{1}{2}} = (\mathbf{E}^+ + \mathbf{E}^-)_{i+\frac{1}{2}} \tag{6.68}$$

For a first-order calculation the flux components may be evaluated with an extrapolation consistent with the expressions given in Section 4.4.11 for the MUSCL scheme, where the primitive variables were extrapolated to the cell faces. In the Steger-Warming splitting, the fluxes are extrapolated to the cell faces. However, the MUSCL approach with primitive variables may also be used in this splitting. In the simplest case, the values of $\mathbf{E}^+_{i+\frac{1}{2}}$ are set equal to $\mathbf{E}^+_i$, and the values of $\mathbf{E}^-_{i+\frac{1}{2}}$ are set equal to $\mathbf{E}^-_{i+1}$. This produces a numerical algorithm of the form

$$\mathbf{U}_i^{n+1} = \mathbf{U}_i^n - \frac{\Delta t}{\Delta x}[(\mathbf{E}^+ + \mathbf{E}^-)_{i+\frac{1}{2}} - (\mathbf{E}^+ + \mathbf{E}^-)_{i-\frac{1}{2}}]$$

or

$$\mathbf{U}_i^{n+1} = \mathbf{U}_i^n - \frac{\Delta t}{\Delta x}(\nabla \mathbf{E}_i^+ + \Delta \mathbf{E}_i^-) \tag{6.69}$$

This is the finite-difference form of Eq. (6.64) when the $\mathbf{E}^+$ derivative is backward differenced and the $\mathbf{E}^-$ term is forward differenced. Based on earlier discussions, the equivalence of the finite-difference and the finite-volume formulations is clear.

A second-order algorithm may be developed by using the trapezoidal scheme given in Section 4.1.10. The integration in time is written

$$\mathbf{U}_i^{n+1} = \mathbf{U}_i^n + \frac{\Delta t}{2}\left[\left(\frac{\partial \mathbf{U}}{\partial t}\right)_i^n + \left(\frac{\partial \mathbf{U}}{\partial t}\right)_i^{n+1}\right] \tag{6.70}$$

where the term $\partial \mathbf{U}^{n+1}/\partial t$ is interpreted as a predicted value for an explicit scheme and is included as part of the computed solution for an implicit technique. In the application of this scheme the derivative at n is written in terms of the fluxes on the surface of the control volume just as in the first-order method. However, the flux terms evaluated at the control-volume surfaces must be second order in space if the result is to be used in the first term of the trapezoidal integration scheme. This can easily be accomplished by noting that the second-order flux is obtained by using the upwind extrapolation formula given in Section 4.4.11. A second-order spatial calculation requires that a linear extrapolation be used for the primitive variables. Here $\mathbf{u}$ represents the vector of primitive variables, and the extrapolation for the positive terms is written

$$\mathbf{u}^+_{i-\frac{1}{2}} = \mathbf{u}_{i-1} + \psi^+(r^+)\tfrac{1}{2}(\mathbf{u}_{i-1} - \mathbf{u}_{i-2}) \tag{6.71}$$

while for the negative terms,

$$\mathbf{u}^-_{i+\frac{1}{2}} = \mathbf{u}_{i+1} + \psi^-(r^-)\tfrac{1}{2}(\mathbf{u}_{i+2} - \mathbf{u}_{i+1}) \tag{6.72}$$

From these extrapolations, the split fluxes may be reformed for the terms in the integration. The $\psi^\pm$ terms are the same limiters presented in Section 4.4.12, and any of the limiters discussed may be used. Split flux schemes will also produce oscillations when higher-order algorithms are constructed, so limiting is necessary. The final algorithm for the second-order upwind scheme may be written in the following two-step sequence:

$$\overline{\mathbf{U}_i^{n+1}} = \mathbf{U}_i^n - \frac{\Delta t}{\Delta x}(\nabla \mathbf{E}_i^+ + \Delta \mathbf{E}_i^-) \tag{6.73}$$

$$\mathbf{U}_i^{n+1} = \frac{1}{2}\left[\mathbf{U}_i^n + \overline{\mathbf{U}_i^{n+1}} - \frac{\Delta t}{\Delta x}\left(\nabla^2 \mathbf{E}_i^{+n} + \nabla \overline{\mathbf{E}_i^{+n+1}}\right) + \frac{\Delta t}{\Delta x}\left(\Delta^2 \mathbf{E}_i^{-n} - \Delta \overline{\mathbf{E}_i^{-n+1}}\right)\right] \tag{6.74}$$

The 2-D version of this scheme follows with the addition of the appropriate terms.

An implicit algorithm using the trapezoidal rule is easily derived using the split-flux idea, and a first-order spatial scheme takes the form

$$\left([I] + \frac{\Delta t}{2\,\Delta x}(\nabla[A_i^+] + \Delta[A_i^-])\right)\Delta\,\mathbf{U}_i^n = -\frac{\Delta t}{\Delta x}(\nabla \mathbf{E}^+ + \Delta \mathbf{E}^-) \tag{6.75}$$

where

$$\Delta \mathbf{U}_i^n = \mathbf{U}_i^{n+1} - \mathbf{U}_i^n$$

This is written in the delta form introduced in Section 4.4.7. This algorithm is first-order accurate in space even though it is second-order accurate in time. The spatial accuracy can be improved by simply increasing the order of the spatial operators. Frequently, interest is in the steady-state solution. If this is the case, the right-hand side can be modified to obtain second-order accuracy in space for the steady-state result without altering the block tridiagonal structure of the left-hand side. It is interesting to note that an approximate factorization of the left-hand side of Eq. (6.75) is possible, resulting in the product of two operators

$$\left([I] + \frac{1}{2}\frac{\Delta t}{\Delta x}\nabla[A_i^+]\right)\left([I] + \frac{1}{2}\frac{\Delta t}{\Delta x}\Delta[A_i^-]\right)\Delta\,\mathbf{U}_i^n = \text{RHS of Eq. (6.75)} \tag{6.76}$$

This permits the algorithm to be implemented in the sequence

$$\left([I] + \frac{1}{2}\frac{\Delta t}{\Delta x}\nabla[A_i^+]\right)\Delta\,\mathbf{U}_i' = \text{RHS of Eq. (6.75)} \tag{6.77a}$$

$$\left([I] + \frac{1}{2}\frac{\Delta t}{\Delta x}\Delta[A_i^-]\right)\Delta\,\mathbf{U}_i^n = \Delta\mathbf{U}_i' \tag{6.77b}$$

If Eqs. (6.77*a*) and (6.77*b*) are used, each 1-D sweep requires the solution of two block bidiagonal systems. The original system [Eq. (6.75)] requires the solution of a single block tridiagonal system for each time step. It is important to note that savings expected in using Eqs. (6.77*a*) and (6.77*b*) may not be realized for all problems. Usually, the major advantage of using the split form with bidiagonal systems occurs in multidimensional cases.

The use of split-flux techniques for shock-capturing applications produces better results than central-difference methods, but some problems remain even for this formulation. Using the Steger-Warming splitting, the shock waves are well represented, but some oscillations are produced when a sonic condition is encountered. The problem is that the components of the split flux are not continuously differentiable at sonic and stagnation points. Figure 6.12 shows the split mass flux behavior as the sonic region is traversed. Steger and Warming (1981) attempted to eliminate this problem by modifying the eigenvalues when they change signs to be of the form

$$\lambda^{\pm} = \frac{\lambda \pm \sqrt{\lambda^2 + \epsilon^2}}{2} \tag{6.78}$$

where ϵ is viewed as a blending function to ensure a smooth transition when the λ's change sign. This modification was only moderately successful, and more appropriate schemes employing flux-vector splitting evolved later.

6.4.2 Van Leer Flux Splitting

The problems encountered at sonic transitions and at stagnation points with the Steger-Warming splitting were addressed by van Leer (1982). He proposed a different splitting, defined so that the flux terms were continuously differentiable through sonic and stagnation zones. The conditions van Leer imposed to accomplish a flux splitting were

1. $\mathbf{E} = \mathbf{E}^{+} + \mathbf{E}^{-}$,
2. $\partial \mathbf{E}^{+} / \partial \mathbf{U}$ must have all eigenvalues $\geqslant 0$, and
3. $\partial \mathbf{E}^{-} / \partial \mathbf{U}$ must have all eigenvalues $\leqslant 0$,

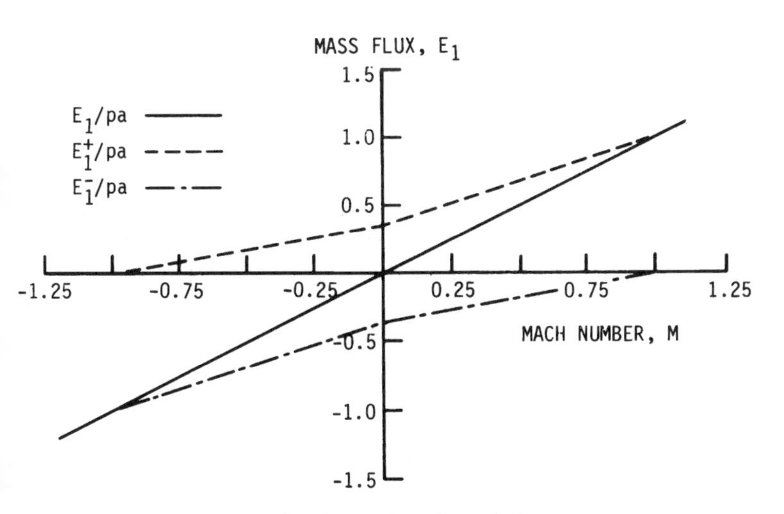

Figure 6.12 Split mass flux using Steger-Warming splitting.

with the following restrictions:

1. $\mathbf{E}^{\pm}$ must be continuous with

$$\mathbf{E}^{+} = \mathbf{E}(\mathbf{U}) \qquad M \geqslant 1$$
$$\mathbf{E}^{-} = \mathbf{E}(\mathbf{U}) \qquad M \leqslant -1$$

2. The components of $\mathbf{E}^{+}, \mathbf{E}^{-}$ must exhibit the same symmetry as $\mathbf{E}$ in terms of Mach number. For each component,

$$\mathbf{E}^{+} = \pm \mathbf{E}^{-}(-M)$$

if

$$\mathbf{E}(M) = \pm \mathbf{E}(-M)$$

3. The Jacobians $\partial \mathbf{E}^{\pm}/\partial \mathbf{U}$ must be continuous.
4. The Jacobians $\partial \mathbf{E}^{\pm}/\partial \mathbf{U}$ must have one eigenvalue vanish for $|M| < 1$.
5. $\mathbf{E}^{\pm}$ must be a polynomial in Mach number of lowest possible order.

Conditions 1 and 2 ensure that this splitting procedure produces the standard upwind differences in supersonic flow. Restrictions 1–3 provide symmetry and eliminate the problem at sonic and stagnation points associated with the Steger-Warming scheme. The fourth restriction ensures that a stationary shock with two interior zones can be constructed, while the last restriction is included to provide uniqueness of the splitting.

For a 1-D problem in (x, t) the van Leer fluxes may be written

$$\mathbf{E}^{+} = \frac{1}{4}\rho a(M+1)^2 \begin{bmatrix} 1 \\ \dfrac{2a}{\gamma}\left(1 + \dfrac{\gamma - 1}{2}M\right) \\ \dfrac{2a^2}{\gamma^2 - 1}\left(1 + \dfrac{\gamma - 1}{2}M\right)^2 \end{bmatrix} \tag{6.79a}$$

$$\mathbf{E}^{-} = \mathbf{E} - \mathbf{E}^{+} \tag{6.79b}$$

The flux terms as presented in Eqs. (6.79) satisfy the constraints of the van Leer splitting and are continuously differentiable at sonic and stagnation points. Splitting for the multidimensional case is accomplished in the same manner as in one dimension with the addition of the necessary flux components.

Both flux splitting schemes have proved to be dissipative (van Leer, 1990, 1992), and some questions have been raised about the various splitting ideas. Van Leer et al. (1987) pointed out that the splitting he proposed does not identify contact surfaces and, in some cases, leads to large dissipation. This is especially apparent in viscous regions, where large errors may occur (see Anderson et al., 1985). Hanel et al. (1987) and Hanel and Schwane (1989) observed that the original flux splitting of van Leer does not preserve total enthalpy in solutions of the steady Euler equations. They proposed a modification of the energy flux component by writing

$$\mathbf{E}^{\pm}_{\text{energy}} = \pm\tfrac{1}{4}\rho a(M \pm 1)^2 H(u^{\pm}) \tag{6.80}$$

where the total enthalpy H is now included in the split-flux terms.

An additional modification due to Hanel and Schwane (1989) employs upwinding of the transverse momentum flux. While these changes have been shown to improve the dissipation at contact surfaces, and prevent artificial thickening of boundary layers, some problems still appear in computing wall temperatures for viscous flows. Van Leer (1990) applied the Hanel and Schwane (1989) upwind idea to the energy flux in an attempt to resolve this issue.

6.4.3 Other Flux Splitting Schemes

Other ideas for splitting the Euler equations have been suggested. A recent example is provided by the work of Liou and Steffen (1991). They have presented a new scheme, where the pressure and convection terms are treated separately. The flux-vector splitting technique of Liou and Steffan has been dubbed by them the *advection upstream splitting method* (AUSM). The inviscid flux terms are viewed as a combination of scalar quantities convected by an appropriately defined cell interface velocity, and the pressure terms are treated as being governed only by the acoustic wave speeds. The inviscid 1-D flux term is written as

$$\mathbf{E} = \begin{bmatrix} \rho u \\ \rho u^2 \\ u(E_t + p) \end{bmatrix} + \begin{bmatrix} 0 \\ p \\ 0 \end{bmatrix} \tag{6.81}$$

or

$$\mathbf{E} = \mathbf{E}_c + \mathbf{E}_p \tag{6.82}$$

where the subscripts denote convection terms and pressure terms. These terms are discretized differently. The interface fluxes for supersonic flow are selected by taking either the left (L) and right (R) state, depending on the sign of the Mach number. The subsonic case needs a more careful evaluation. For this case, Liou and Steffan have suggested that

$$(\mathbf{E}_c)_{i+1/2} = u_{\frac{1}{2}} \begin{bmatrix} \rho \\ \rho u \\ E_t + p \end{bmatrix}_{L/R} \tag{6.83}$$

where

$$\frac{L}{R} = \begin{cases} L & u_{\frac{1}{2}} > 0 \\ R & \text{otherwise} \end{cases} \tag{6.84}$$

The quantity $u_{\frac{1}{2}}$ is referred to as the advective velocity, and numerous choices exist for defining this value to be used at the cell interface. Liou and Steffan have suggested that

$$u_{\frac{1}{2}} = a_{L/R} M_{\frac{1}{2}} \tag{6.85}$$

where

$$M_{\frac{1}{2}} = M_L^+ + M_R^- \tag{6.86}$$

and the split Mach numbers for the left and right states use the van Leer definitions

$$M^{\pm} = \pm \tfrac{1}{4}(M \pm 1)^2 \tag{6.87}$$

The interface convective flux terms become

$$(\mathbf{E}_c)_{i+1/2} = M_{\frac{1}{2}} \begin{bmatrix} \rho a \\ \rho u a \\ a(E_t + p) \end{bmatrix}_{L/R} \tag{6.88}$$

The pressure is written as the sum of left and right contributions:

$$p_{\frac{1}{2}} = p_L^+ + p_R^- \tag{6.89}$$

A number of representations can be used for the terms composing $p_{\frac{1}{2}}$. The simplest presented was given as

$$p^{\pm} = \tfrac{1}{2} p(1 \pm M) \tag{6.90}$$

In fact, the splittings used for both the pressure and convection terms can be accomplished in many ways. Unfortunately, the approach that proves to work best is not always clear, and numerous numerical experiments are necessary to explore the effectiveness of a numerical scheme. The fact that there is no unique way to accomplish the flux splitting provides ample opportunity to develop new ideas for solving the Euler equations. The flux splittings presented in this section each have some issues that are unresolved. The Steger-Warming scheme has been modified to eliminate the problems at the stagnation and sonic conditions, with limited success. The van Leer scheme seem to be too dissipative except for use in the solution of the Euler equations, and the AUSM scheme appears to be sensitive to the pressure evaluation. Van Leer (1992) has suggested that split-flux schemes only be applied to the Euler equations because of the dissipative nature of the methods currently available. Work on construction of new schemes that are improvements on existing methods is expected to continue. The work of Zha and Bilgen (1993) is an example.

In the construction of the second-order schemes, the MUSCL approach has been used to extrapolate the primitive variables to the cell interface rather than the fluxes. Anderson et al. (1985) has performed a series of numerical experiments with results that indicate that extrapolation of the primitive variables and reconstruction of the flux give better flow solutions. Based on these results, it is advisable to use primitive-variable extrapolation and flux reconstruction at the cell boundaries wherever practical.

6.4.4 Application for Arbitrarily Shaped Cells

In the previous sections, no information was given regarding the shapes of the control volumes. However, it is usually implicitly assumed that the mesh is structured and the cells in a physical domain are quadrilateral. A significant amount of interest has recently surfaced in applying solvers to cells with arbitrary shapes and in using unstructured grids.

The integration of the conservation law presented in Eq. (6.53) is general and applies to any cell. The conserved variable resulting from the application of the divergence theorem to a control volume is a cell-averaged value, as previously noted in Chapter 4. Calculating the flux values at the cell boundaries may require careful consideration when the cells are of arbitrary shape.

Recently, the use of triangular cells has gained popularity. In this case, the flow variables may be computed for a cell-centered scheme or a vertex (node)-centered scheme. When using finite-volume schemes, it is natural to consider a cell-centered approach. Consider the mesh shown in Fig. 6.13. Each cell is triangular, and the cell-centered conservative variables depend upon careful estimates of the boundary fluxes for accuracy.

If a first-order method is used to solve the Euler equations, the primitive variables at the cell face may be quickly obtained by simple extrapolation of the cell values to the faces using the appropriate upwind directions. Of course, the estimation of the cell-face values and techniques to determine these from the cell-averaged values must be applied not only for split-flux methods but for any numerical method applied on this type of grid. The determination of the required values is more complicated when a higher-order approximation is desired.

Consider a cell-centered scheme and assume that we seek a linear extrapolation of the primitive variables to the control-volume faces in order to obtain second-order accuracy. The centroidal values are known, but gradient information is needed to complete the extrapolation. Let u represent any scalar primitive variable, and consider the cell face value given by

$$u_{\mathrm{cf}} = u_i + \nabla u_i \cdot (\mathbf{r}_{\mathbf{cf}} - \mathbf{r}_{\mathbf{i}}) \tag{6.91}$$

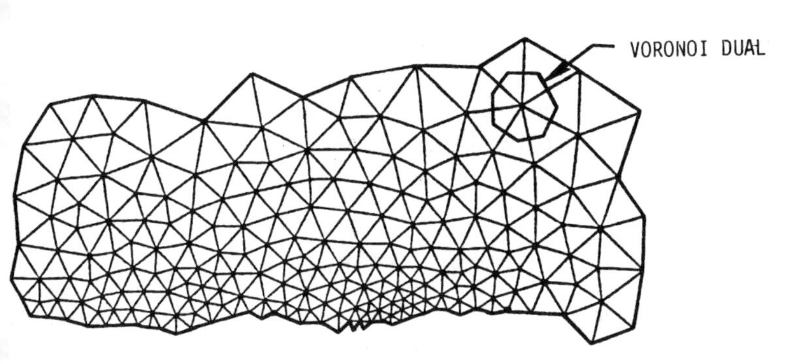

Figure 6.13 Unstructured mesh with triangular cells.

This is a first-order Taylor series expression representing the cell-face values in terms of the cell averages. The term $\mathbf{r}_{cf} - \mathbf{r}_i$ represents the distance from the cell centroid to the midpoint of the cell face, and ∇u_i is the gradient of u_i. The product of these two terms represents the directional derivative toward the cell face midpoint times the distance to the midpoint. In order to evaluate this expression, a way of computing the gradient is needed.

A common technique to establish the gradient (Barth, 1991) is to compute the gradients at the vertices of the cell and use these values to establish a value to use in the extrapolation. For triangular cells a simple average of the vertex gradients is one way of establishing the required value. The vertex values of the gradients are most easily established by using a mesh dual. A Delaunay mesh (Delaunay, 1934) is shown in Fig. 6.13, and the Voronoi dual (Voronoi, 1908) is an appropriate choice for this case. A description of the mesh construction and terminology is given in Chapter 10. It is sufficient for our purposes to use the mesh as given in Fig. 6.13. At any node, an evaluation of the gradient may be made by applying the identity

$$\int_{\delta v} \nabla u \, \delta v = \oint_s u\mathbf{n} \, dl \tag{6.92}$$

to the cell formed by the dual, where $\mathbf{n}$ represents the unit normal to the cell surface. When applied in a discrete sense, this takes the form

$$\nabla u = \frac{1}{\delta v} \sum_s u\mathbf{n} \, \Delta\mathbf{s} \tag{6.93}$$

With this expression, any component of the gradient may be evaluated at the nodal locations in the mesh. Once these values are known, the gradients at the cell centroids may be computed.

This procedure may be used for cell-centered schemes for arbitrary mesh configurations. In calculating higher-order solutions, the gradient terms are limited in the same way as previously indicated. Figure 6.14 shows an NACA 0012 airfoil with the associated unstructured mesh. The corresponding transonic solution for the pressure field is shown in Fig. 6.15. In these calculations a van Leer split-flux scheme was used, and the solver was second order. The pressure data are compared with the calculations of Anderson et al. (1985). In solving the Euler equations on an unstructured mesh, significantly more computational effort and a larger number of cells are probably required to produce the same solution quality when compared to a solution computed on a structured mesh. This is a problem of some concern, but increasing availability of low-cost memory and improved processor speed suggest that storage and speed may not continue to be major problems.

6.5 FLUX-DIFFERENCE SPLITTING SCHEMES

The main challenge in constructing methods for solving the Euler equations is to find ways of estimating the flux terms at the control-volume faces. Several flux-splitting schemes were reviewed in the previous section and were interpreted

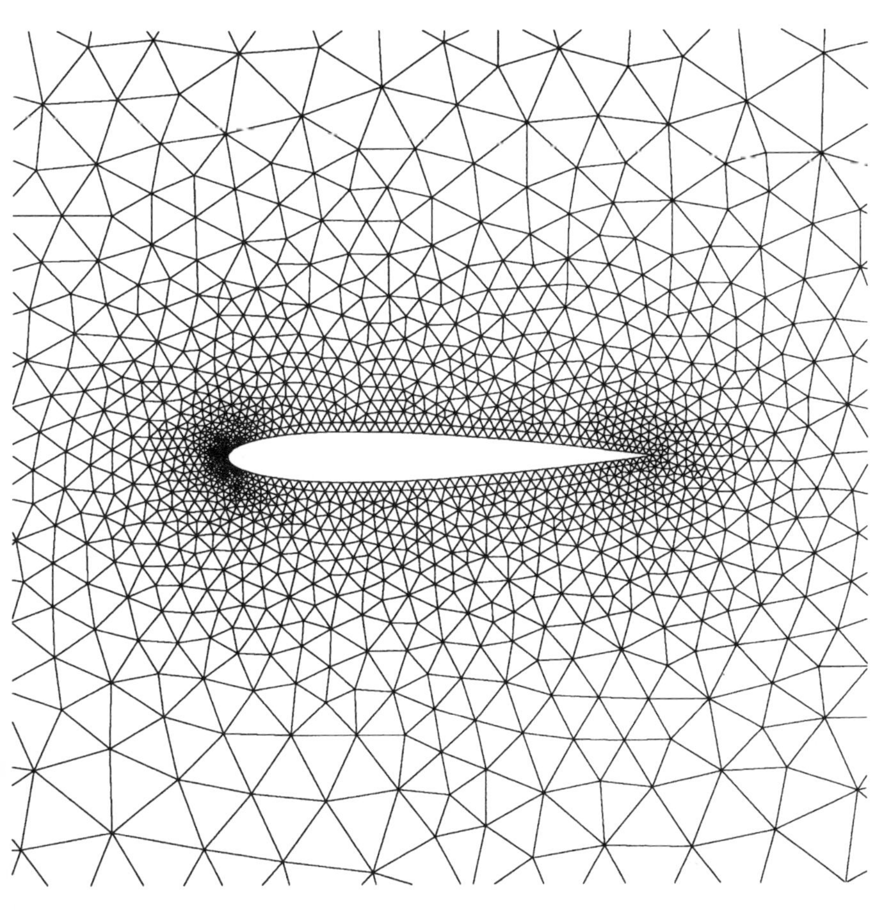

Figure 6.14 Unstructured grid for NACA 0012 airfoil.

as schemes that transport particles according to the characteristic information (van Leer, 1990). In contrast, the changes in the flux quantities at the cell interface using *flux-difference splitting* have been interpreted as being caused by a series of waves. The wave interpretation is derived from the characteristic field of the Euler equations. The problem of computing the cell-face fluxes for a control volume is viewed as a series of 1-D Riemann problems along the direction normal to the control-volume faces. One way of determining these fluxes is to solve the Riemann problem using Godunov's method as outlined in Section 4.4.8 for the 1-D Burgers equation. Of course, the solution in the present case would be for a generalized problem with arbitrary initial states. The original Godunov method has been substantially improved by employing a variety of techniques to accelerate the solution of the nonlinear wave problem (Gottlieb and Growth, 1988). Because some of the details of the exact solution, obtained at considerable cost, are lost in the cell-averaged representation of the

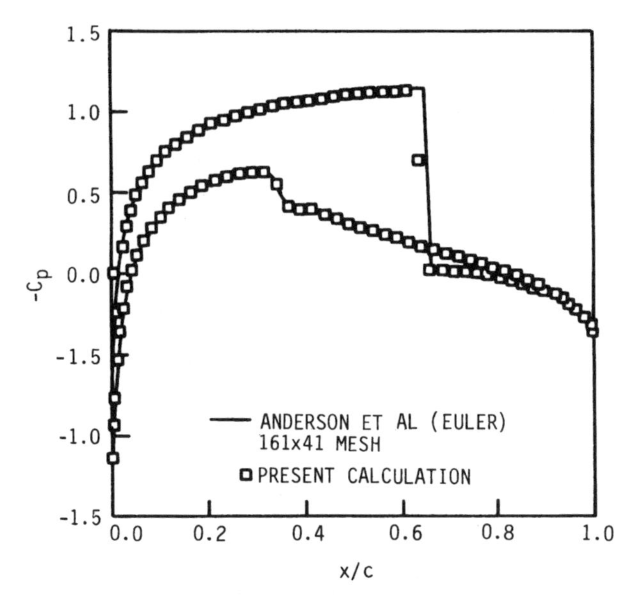

Figure 6.15 Pressure contours for NACA 0012 airfoil in transonic flow. $M = 0.80$, $\alpha = 1.25°$.

data, the solution of the full Riemann problem is usually replaced by methods referred to as approximate Riemann solvers. The Roe method (Roe, 1980) and the Osher scheme (Osher, 1984) are the most well known of these schemes. Owing to its simplicity, the Roe scheme and its many variations have evolved as the method of choice among flux-difference splitting schemes. In the next section, Roe's scheme will be discussed as applied to the Euler equations. This technique is another way of calculating the flux values at the control-volume boundaries in the finite-volume approach.

6.5.1 Roe Scheme

In view of the fact that the Riemann problem requires a solution of a nonlinear system, a significant gain in efficiency can be realized if a solution to a linear problem approximating the original Riemann problem can be obtained. This is the basis for Roe's scheme. Consider the original Riemann problem in the form

$$\frac{\partial \mathbf{U}}{\partial t} + \frac{\partial \mathbf{E}}{\partial x} = 0 \tag{6.94a}$$

$$\mathbf{U}(x,0) = \begin{cases} \mathbf{U}_L & x < 0 \\ \mathbf{U}_R & x > 0 \end{cases} \tag{6.94b}$$

The notation for the left and right states has been used previously in Chapter 4. Roe's linear approximation to the Riemann problem is written

$$\frac{\partial \mathbf{U}}{\partial t} + [\hat{A}]\frac{\partial \mathbf{U}}{\partial x} = 0 \tag{6.95}$$

where the initial conditions are the same as those in the nonlinear problem and $[\hat{A}]$ is Roe's averaged matrix and is assumed to be a constant in this formulation. Recall that the original Jacobian was defined by

$$[A] = \frac{\partial \mathbf{E}}{\partial \mathbf{U}} \tag{6.96}$$

The Jacobian matrix is replaced by $[\hat{A}]$ in this system. The components of the $[\hat{A}]$ matrix are evaluated using averaged values of $\mathbf{U}$ at the interface separating the two states. This is indicated by writing

$$[\hat{A}] = \left[\hat{A}(\mathbf{U}_L, \mathbf{U}_R)\right] \tag{6.97}$$

The Roe-averaged matrix $[\hat{A}]$ is chosen to satisfy certain conditions, so that a solution of the linear problem becomes an approximate solution of the nonlinear Riemann problem. These conditions include the following.

1. A linear mapping relates the vector space $\mathbf{U}$ to the vector space $\mathbf{E}$.
2. As $\mathbf{U}_L$ approaches $\mathbf{U}_R$, i.e., as an undisturbed state is reached,

   $$\left[\hat{A}(\mathbf{U}_L, \mathbf{U}_R)\right] \Rightarrow [A]$$

   when

   $$\mathbf{U}_L \to \mathbf{U}_R \to \mathbf{U}$$

   where $[A]$ is the Jacobian of the original system.
3. For any two values $\mathbf{U}_L, \mathbf{U}_R$, the jump condition across the interface must be correct, i.e.,

   $$\mathbf{E}_R - \mathbf{E}_L = [\hat{A}](\mathbf{U}_R - \mathbf{U}_L)$$

4. The eigenvalues of $[\hat{A}]$ are real and linearly independent.

Consider the system of equations given by Eq. (6.95). This is a hyperbolic system that may be diagonalized by writing the constant matrix $[\hat{A}]$ as

$$[\hat{A}] = [\hat{T}][\hat{\Lambda}][\hat{T}]^{-1} \tag{6.98}$$

The original equations can then be cast in the form

$$\frac{\partial \mathbf{U}}{\partial t} + [\hat{T}][\hat{\Lambda}][\hat{T}]^{-1}\frac{\partial \mathbf{U}}{\partial x} = 0 \tag{6.99}$$

Premultiplying by $[\hat{T}]^{-1}$ and defining the vector $\mathbf{W}$ as

$$\mathbf{W} = [\hat{T}]^{-1}\mathbf{U} \tag{6.100}$$

leads to the linear problem

$$\frac{\partial \mathbf{W}}{\partial t} + [\hat{\Lambda}]\frac{\partial \mathbf{W}}{\partial x} = 0 \tag{6.101}$$

where the matrix of eigenvalues $[\hat{\Lambda}]$ is a diagonal matrix. This produces an uncoupled hyperbolic system. The numerical method may be applied to each of the uncoupled equations of this system, and the result transformed back to the original variables. For a single linear equation, the value of W is constant along the characteristic defined by $dx/dt = \lambda_k$. As each of the waves associated with the eigenvalues of the system is crossed, the values of the dependent variables experience a jump. Consequently, the values of W_k are constant between each pair of waves in the domain. Mathematically, this can be stated as

$$W_k = \text{const}$$

when

$$\lambda_{k-1} \leqslant \frac{x}{t} \leqslant \lambda_k$$

Consequently, the value of $\mathbf{W}$ at any point may be written

$$W_k = W_1 + \sum_{j=2}^{k} (W_j - W_{j-1}) \tag{6.102}$$

Again, since $[\hat{A}]$ is a constant matrix, we may write

$$U_k = U_1 + \sum_{j=2}^{k} (U_j - U_{j-1}) \tag{6.103}$$

and the final result is that the flux changes may be written

$$E_k = E_1 + \sum_{j=2}^{k} \delta E_j \tag{6.104}$$

where the flux increments are associated with the crossing of each wave in the system.

If the entire wave system is traversed and the left and right states are identified with appropriate subscripts, then

$$\mathbf{E}_R = \mathbf{E}_L + [\hat{A}](\mathbf{U}_R - \mathbf{U}_L) \tag{6.105}$$

As shown in the previous section, the $[\hat{A}]$ matrix may be split, corresponding to changes that occur across negative and positive waves. Consequently, we may split the calculation of the fluxes into contributions across negative and positive waves to determine appropriate formulas for the cell-face fluxes in the linear Riemann problem. Referring to Fig. 6.16, one notes that the interface flux can be computed by starting at either the left or the right state.

Starting at the left state, we can write

$$\begin{aligned} \tilde{E}_{i+\frac{1}{2}} &= \mathbf{E}_L + [\hat{A}^-](\mathbf{U}_R - \mathbf{U}_L) \\ \mathbf{E}_R &= \tilde{E}_{i-\frac{1}{2}} + [\hat{A}^+](\mathbf{U}_R - \mathbf{U}_L) \end{aligned} \tag{6.106a}$$

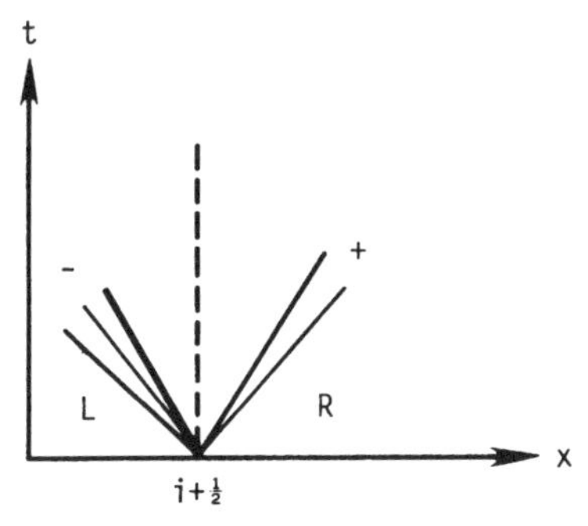

Figure 6.16 Decomposed flux for the linear Riemann problem.

or, as is usually written, the two expressions for the interface flux become

$$\tilde{E}_{i+\frac{1}{2}} = \mathbf{E}_L + [\hat{A}^-](\mathbf{U}_R - \mathbf{U}_L)$$
$$\tilde{E}_{i+\frac{1}{2}} = \mathbf{E}_R - [\hat{A}^+](\mathbf{U}_R - \mathbf{U}_L) \tag{6.106b}$$

A symmetric result is used in applications of computational fluid dynamics, and this may be obtained by averaging the cell-face flux formulas to obtain the following appropriate expression:

$$\tilde{E}_{i+\frac{1}{2}} + \tfrac{1}{2}\left\{(\mathbf{E}_R + \mathbf{E}_L) - [|\hat{A}|](\mathbf{U}_R - \mathbf{U}_L)\right\} \tag{6.107}$$

In this equation, $[|\hat{A}|] = [\hat{T}][|\hat{\Lambda}|][\hat{T}]^{-1}$ and $[|\hat{\Lambda}|]$ is the diagonal matrix whose entries are the absolute values of the eigenvalues. The numerical flux expression incorporates upwind influence through the addition of contributions across positive and negative waves. Condition 3 and the subsequent expressions for the interface flux given by Eqs. (6.106) show that the change across any wave depends upon the change in state variables across all waves. This point can be noted by recalling that a diagonalization of the system leads to uncoupled equations providing the changes across each wave in a modified set of variables derived by multiplication by $[\hat{T}]^{-1}$. When the flux values are recovered by multiplying by $[\hat{T}]$, the change in flux across each wave is seen to depend upon the change in **U** across the entire system of waves.

The Roe-averaged matrix may be constructed by noting that **U** and **E** are quadratic functions of the variable **z**, defined as

$$\mathbf{z} = \sqrt{\rho}\begin{bmatrix} 1 \\ u \\ H \end{bmatrix} \tag{6.108}$$

The conservative variables may be written in terms of the **z** variable as

$$\mathbf{U} = \begin{bmatrix} z_1^2 \\ z_1 z_2 \\ \dfrac{z_1 z_3}{\gamma} + \dfrac{1}{2}\dfrac{\gamma - 1}{\gamma} z_2^2 \end{bmatrix} \tag{6.109}$$

where the vector of conservative variables **U** is defined by Eq. (5.44) for the 1-D case where $v = w = 0$. The flux term may also be written as

$$\mathbf{E} = \begin{bmatrix} z_1 z_2 \\ \dfrac{\gamma - 1}{\gamma} z_1 z_3 + \dfrac{1}{2} \dfrac{\gamma + 1}{\gamma} z_2^2 \\ z_2 z_3 \end{bmatrix} \tag{6.110}$$

We define the arithmetic average of any quantity with an overbar symbol in the following manner

$$\bar{x}_{i+\frac{1}{2}} = \tfrac{1}{2}(x_i + x_{i+1})$$

and note the exact expansion formula:

$$\Delta(xy)_{i+\frac{1}{2}} = \bar{x}_{i+\frac{1}{2}} \Delta y_{i+\frac{1}{2}} + \bar{y}_{i+\frac{1}{2}} \Delta x_{i+\frac{1}{2}} \tag{6.111}$$

Applying this expansion formula results in conservative variable and flux formulas of the form

$$\mathbf{U}_{i+1} - \mathbf{U}_i = [B](\mathbf{z}_{i+1} - \mathbf{z}_i) \tag{6.112}$$

and

$$\mathbf{E}_{i+1} - \mathbf{E}_i = [C](\mathbf{z}_{i+1} - \mathbf{z}_i) \tag{6.113}$$

where

$$[B] = \begin{bmatrix} 2\bar{z}_1 & 0 & 0 \\ \bar{z}_2 & \bar{z}_1 & 0 \\ \dfrac{\bar{z}_3}{\gamma} & \dfrac{\gamma - 1}{\gamma} \bar{z}_2 & \dfrac{\bar{z}_1}{\gamma} \end{bmatrix} \tag{6.114}$$

and

$$[C] = \begin{bmatrix} \bar{z}_2 & \bar{z}_1 & 0 \\ \dfrac{\gamma - 1}{\gamma} \bar{z}_3 & \dfrac{\gamma + 1}{\gamma} \bar{z}_2 & \dfrac{\gamma - 1}{\gamma} \bar{z}_1 \\ 0 & \bar{z}_3 & \bar{z}_2 \end{bmatrix} \tag{6.115}$$

with the result that

$$\mathbf{E}_{i+1} - \mathbf{E}_i = [C][B]^{-1}(\mathbf{U}_{i+1} - \mathbf{U}_i) \tag{6.116}$$

The matrix $[C][B]^{-1}$ is identical to the Jacobian matrix $[A]$ if the original variables are replaced by an average weighted by the square root of the density. If

$$R_{i+\frac{1}{2}} = \sqrt{\frac{\rho_{i+1}}{\rho_i}} \tag{6.117}$$

then

$$\hat{\rho}_{i+\frac{1}{2}} = R_{i+\frac{1}{2}}\rho_i \tag{6.118}$$

$$\hat{u}_{i+\frac{1}{2}} = \frac{R_{i+\frac{1}{2}}u_{i+1} + u_i}{1 + R_{i+\frac{1}{2}}} \tag{6.119}$$

$$\hat{H}_{i+\frac{1}{2}} = \frac{R_{i+\frac{1}{2}}H_{i+1} + H_i}{1 + R_{i+\frac{1}{2}}} \tag{6.120}$$

where the quantity $\hat{H}$ is the averaged total enthalpy and H is defined by

$$H = \frac{E_t + p}{\rho} \tag{6.121}$$

The development of the averaged matrix has used the so-called parameter vector approach. This "Roe-averaged state" may be directly obtained by solving Eq. (6.105) for the state variables. This follows because the correct averaged matrix is the only one that will provide the correct relationship satisfying these equations. For further details, the reader should consult Roe and Pike (1985).

The numerical flux for the first-order Roe scheme is then written in the form

$$\tilde{\mathbf{E}}_{i+\frac{1}{2}} = \tfrac{1}{2}\left\{\mathbf{E}_i + \mathbf{E}_{i+1} - \left[\hat{T}_{i+\frac{1}{2}}\right]\left[|\hat{\Lambda}_{i+\frac{1}{2}}|\right]\left[\hat{T}_{i+\frac{1}{2}}\right]^{-1}(\mathbf{U}_{i+1} - \mathbf{U}_i)\right\} \tag{6.122}$$

This may be used to calculate a first-order solution using the standard explicit or implicit techniques for advancing the solution in time. In this formula, the problem of expansion shocks must be considered. By way of review, recall that the formulation of the Roe scheme admits an expansion shock as a perfectly appropriate solution of the approximate problem. As a consequence, stationary expansion shocks are not dissipated by this method. An appropriate entropy fix, but one that does not distinguish between shocks and expansions, is easily implemented by replacing the components of $[|\hat{\Lambda}|]$ by $\beta(\hat{\lambda}^l_{i+\frac{1}{2}})$, where

$$\left[|\hat{\Lambda}_{i+\frac{1}{2}}|\right] = \begin{bmatrix} |\hat{\lambda}^1_{i+\frac{1}{2}}| & 0 & 0 \\ 0 & |\hat{\lambda}^2_{i+\frac{1}{2}}| & 0 \\ 0 & 0 & |\hat{\lambda}^3_{i+\frac{1}{2}}| \end{bmatrix} \tag{6.123}$$

and

$$\beta(\hat{\lambda}) = \begin{cases} |\hat{\lambda}| & |\hat{\lambda}| \geqslant \epsilon \\ (\hat{\lambda}^2 + \epsilon^2)/2\epsilon & |\hat{\lambda}| < \epsilon \end{cases} \tag{6.124}$$

In this set of expressions, the Roe average is implied by the circumflex symbol with subscript $i + \frac{1}{2}$.

While the explicit methods described in previous sections may be used with the Roe scheme, more details on the implementation of implicit schemes using flux-difference splitting (FDS) are in order. Consider a simple Euler implicit scheme resulting in the expression

$$\mathbf{U}_i^{n+1} = \mathbf{U}_i^n - \frac{\Delta t}{\Delta x}\left(\tilde{\mathbf{E}}_{i+\frac{1}{2}}^{n+1} - \tilde{\mathbf{E}}_{i-\frac{1}{2}}^{n+1}\right) \tag{6.125}$$

Define the residual $\mathbf{R}$ as

$$\mathbf{R} = \frac{1}{\Delta x}\left(\tilde{\mathbf{E}}_{i+\frac{1}{2}} - \tilde{\mathbf{E}}_{i-\frac{1}{2}}\right) \tag{6.126}$$

In terms of the residual, the scheme may be written

$$\left(\frac{[I]}{\Delta t} + \frac{\partial \mathbf{R}}{\partial \mathbf{U}}\right)(\mathbf{U}_i^{n+1} - \mathbf{U}_i^n) = -\mathbf{R}_i^n \tag{6.127}$$

where the linearization of the equations is performed as in the previous section with flux-vector splitting (FVS) implicit schemes and $[I]$ represents the identity matrix. Implicit schemes result in coupled systems that must be solved simultaneously, and the structure of the coefficient matrices of the system is important. The elements of $\partial \mathbf{R}/\partial \mathbf{U}$ result in filling the ith row of the coefficient matrix with a bandwidth corresponding to the functional dependence of $\mathbf{R}_j$ on $\mathbf{U}_i$, i.e.,

$$\frac{\partial \mathbf{R}_j}{\partial \mathbf{U}_{i-1}}, \frac{\partial \mathbf{R}_j}{\partial \mathbf{U}_i}, \frac{\partial \mathbf{R}_j}{\partial \mathbf{U}_{i+1}}, \text{ etc.} \tag{6.128}$$

Barth (1987) has considered exact linearizations as indicated here and an additional linearization, leading to what he has called the "frozen" matrix scheme, given by

$$\left(\frac{[I]}{\Delta t} + [M]^n\right)(\mathbf{U}_i^{n+1} - \mathbf{U}_i^n) = -\mathbf{R}_i^n = -[M]^n \mathbf{U}_i^n \tag{6.129}$$

The idea of the frozen matrix scheme is that various alterations of the $[M]$ matrix may be used. If one is only interested in a steady-state result, any approach that causes the residual to vanish (as rapidly as possible) is appropriate.

For illustration, consider an FDS scheme with a first-order Roe flux:

$$\tilde{\mathbf{E}}_{i+\frac{1}{2}} = \tfrac{1}{2}\left\{\mathbf{E}_i + \mathbf{E}_{i+1} - [T][|\Lambda|][T]^{-1}\delta \mathbf{U}_{i+\frac{1}{2}}\right\} \tag{6.130}$$

As a simplification in the notation, let the dissipation term be given as

$$\Delta|\tilde{\mathbf{E}}_{i+\frac{1}{2}}| = [T][|\Lambda|][T]^{-1}\delta \mathbf{U}_{i+\frac{1}{2}} \tag{6.131}$$

With this notation, the residual vector for the Roe FDS may be written

$$\mathbf{R}_i = \frac{1}{2\,\Delta x}\left[\mathbf{E}_{i+1} - \mathbf{E}_{i-1} - \left(\Delta|\tilde{\mathbf{E}}_{i+\frac{1}{2}}| - \Delta|\tilde{\mathbf{E}}_{i-\frac{1}{2}}|\right)\right] \tag{6.132}$$

and the exact local linearization of the Jacobian produces elements of the form

$$\frac{\partial \mathbf{R}_i}{\partial \mathbf{U}_{i+1}} = \frac{1}{2\,\Delta x}\left([A_{i+1}] - \frac{\partial \Delta|\tilde{\mathbf{E}}_{i+\frac{1}{2}}|}{\partial \mathbf{U}_{i+1}}\right) \tag{6.133a}$$

$$\frac{\partial \mathbf{R}_i}{\partial \mathbf{U}_i} = \frac{1}{2\,\Delta x}\left(-\frac{\partial \Delta|\tilde{\mathbf{E}}_{i+\frac{1}{2}}|}{\partial \mathbf{U}_i} + \frac{\partial \Delta|\tilde{\mathbf{E}}_{i-\frac{1}{2}}|}{\partial \mathbf{U}_i}\right) \tag{6.133b}$$

$$\frac{\partial \mathbf{R}_i}{\partial \mathbf{U}_{i-1}} = \frac{1}{2\,\Delta x}\left(-[A_{i-1}] + \frac{\partial \Delta|\tilde{\mathbf{E}}_{i-\frac{1}{2}}|}{\partial \mathbf{U}_{i-1}}\right) \tag{6.133c}$$

FVS and FDS produce different evaluations for terms like $\partial \Delta|\mathbf{E}|/\partial \mathbf{U}$. In order to understand this difference, the FVS idea can be employed in a setting where the numerical flux is represented in a form similar to the FDS approach. The linearizations then require treatment of similar terms for both techniques. For the FDS scheme,

$$\frac{\partial \Delta|\tilde{\mathbf{E}}_{i+\frac{1}{2}}|}{\partial \mathbf{U}_{i+1}} = |A_{i+\frac{1}{2}}| + \frac{\partial |A_{i+\frac{1}{2}}|}{\partial \mathbf{U}_{i+1}}(\mathbf{U}_{i+1} - \mathbf{U}_i) \tag{6.134}$$

while the equivalent calculation for the FVS schemes is

$$\frac{\partial \Delta|\tilde{\mathbf{E}}_{i+\frac{1}{2}}|}{\partial \mathbf{U}_{i+1}} = |A_{i+1}| + \frac{\partial |A_{i+1}|}{\partial \mathbf{U}_{i+1}}\mathbf{U}_{i+1} \tag{6.135}$$

In both of these approaches, the second terms involve differentiation of matrix elements and then matrix multiplication. Note the second term of the FDS linearization is multiplied by the difference in $\mathbf{U}$. This would suggest that for reasonable changes in $\mathbf{U}$, the second term of the FDS linearization is smaller than that produced using FVS. Consequently, linearization errors produced will also be smaller. The effect of the use of these approximate linearizations on convergence rate has been examined by Barth (1987) and by Jesperson and Pulliam (1983). Their results show better convergence properties in time asymptotic calculations for a variety of steady-state problems using FDS schemes. Other linearizations have been considered by Harten (1984), Chakravarthy and Osher (1985), and numerous others. However, the linearizations presented here for either exact or approximate cases, where the second terms are neglected, are conservative and will provide satisfactory results.

6.5.2 Second-Order Schemes

Several different techniques are used to extend FDS methods to higher order. The MUSCL idea has been explored and used with success in Chapter 4 and in the section on flux-vector splitting. Yee (1989) has constructed a framework where both MUSCL and non-MUSCL approaches to higher-order schemes can be described through similar numerical flux functions. The numerical flux for the non-MUSCL approach may be written

$$\tilde{\mathbf{E}}_{i+\frac{1}{2}} = \tfrac{1}{2}\left\{\mathbf{E}_i + \mathbf{E}_{i+1} + \left[\hat{T}_{i+\frac{1}{2}}\right]\hat{\mathbf{\Phi}}_{i+\frac{1}{2}}\right\} \tag{6.136}$$

and the corresponding form for the MUSCL approach is

$$\tilde{\mathbf{E}}_{i+\frac{1}{2}} = \tfrac{1}{2}\left\{\mathbf{E}\left(\mathbf{U}^r_{i+\frac{1}{2}}\right) + \mathbf{E}\left(\mathbf{U}^l_{i+\frac{1}{2}}\right) + \left[\hat{T}_{i+\frac{1}{2}}\right]\hat{\mathbf{\Phi}}_{i+\frac{1}{2}}\right\} \tag{6.137}$$

The $[\hat{T}]$ matrix is evaluated at some average such as the Roe average, and the elements of $\hat{\mathbf{\Phi}}$ are the same as the scalar case, but each element is evaluated with the same symmetric average. In the MUSCL approach, the average state between i and $i+1$ is replaced by the right and left states, as indicated in the MUSCL extrapolation. Limiting is accomplished by reducing (limiting) the slopes in the extrapolation to cell faces. The dissipation term is written in terms of the limited variables and, in general, includes an entropy fix. The construction of the numerical flux in this form permits one to limit the wave strengths in the non-MUSCL approach (Roe, 1984) rather than restrict the slopes using the MUSCL idea. It is argued that this is a better interpretation of the physics. When this idea is implemented, the limiters are slightly different, but the general form of the flux function can be written the same way.

The form of $\hat{\mathbf{\Phi}}$ may be written

$$\hat{\mathbf{\Phi}}_{i+\frac{1}{2}} = \left[|\hat{\Lambda}_{i+\frac{1}{2}}|\right]\left[\hat{T}_{i+\frac{1}{2}}\right]^{-1}(\mathbf{U}_{i+1} - \mathbf{U}_i) \tag{6.138}$$

where the wave strengths are given by

$$\boldsymbol{\alpha}_{i+\frac{1}{2}} = \left[\hat{T}_{i+\frac{1}{2}}\right]^{-1}(\mathbf{U}_{i+1} - \mathbf{U}_i) \tag{6.139}$$

With this notation, the form of the last term in the numerical flux, using the idea of local characteristics, may be written

$$\left[\hat{T}_{i+\frac{1}{2}}\right]\hat{\mathbf{\Phi}}_{i+\frac{1}{2}} = \left[\hat{T}_{i+\frac{1}{2}}\right]\left[|\hat{\Lambda}_{i+\frac{1}{2}}|\right]\boldsymbol{\alpha}_{i+\frac{1}{2}} \tag{6.140}$$

The form of $\hat{\mathbf{\Phi}}$ now is in terms of the wave strengths, and the general description of this term for the various methods will appear in terms of the wave strengths and the appropriate limiters.

A second-order Roe scheme using the MUSCL idea was used to solve Burgers' equation in Section 4.4.11. In that case, the limiting was applied to the variables extrapolated to the cell faces. For the non-MUSCL approach the limiting is applied through the dissipation term. A non-MUSCL second-order Roe-Sweby scheme may be developed where the elements of $\hat{\mathbf{\Phi}}$ take the form

$$\left(\hat{\mathbf{\Phi}}^l_{i+\frac{1}{2}}\right)^R = -\left\{|\lambda^l_{i+\frac{1}{2}}| - \frac{\psi(r^l)}{2}\left[|\lambda^l_{i+\frac{1}{2}}| - \frac{\Delta t}{\Delta x}\left(\lambda^l_{i+\frac{1}{2}}\right)^2\right]\right\}\alpha^l_{i+\frac{1}{2}} \tag{6.141}$$

In this expression the definition of r^l is

$$r^l = \frac{w^l_{i+1+\zeta} - w^l_{i+\zeta}}{w^l_{i+1} - w^l_i} \tag{6.142}$$

where w^l are components of the characteristic variable vector $\mathbf{W}$ and ζ is defined as the $\operatorname{sgn}(\lambda)$. The limiter $\psi(r^l)$ may be any of the limiters previously discussed in Section 4.4.12. The limiter is expressed in terms of the characteristic variables with this formulation.

Another method arrived at by the characteristic variable extension was originally developed by Harten (1984) and later modified by Yee (1986). The elements of $\hat{\mathbf{\Phi}}$ for this second-order upwind scheme are written

$$\left(\phi^l_{i+\frac{1}{2}}\right)^{\text{HY}} = \sigma\left(\lambda^l_{i+\frac{1}{2}}\right)(g^l_{i+1} - g^l_i) - \psi\left(\lambda^l_{i+\frac{1}{2}} + \gamma^l_{i+\frac{1}{2}}\right)\alpha^l_{i+\frac{1}{2}} \tag{6.143}$$

The definition of σ is

$$\sigma(s) = \frac{1}{2}\left[\beta(s) - \frac{\Delta t}{\Delta x}s^2\right] \tag{6.144}$$

and

$$\gamma^l_{i+\frac{1}{2}} = \sigma\left(\lambda^l_{i+\frac{1}{2}}\right)\begin{cases}\left(g^l_{i+\frac{1}{2}} - g^l_i\right)/\alpha^l_{i+\frac{1}{2}} & \alpha^l_{i+\frac{1}{2}} \neq 0 \\ 0 & \alpha^l_{i+\frac{1}{2}} = 0\end{cases} \tag{6.145}$$

The limiter function in this case is denoted by g^l_i and may be written as

$$g^l_i = \operatorname{minmod}\left(\alpha^l_{i-\frac{1}{2}}, \alpha^l_{i+\frac{1}{2}}\right) \tag{6.146}$$

or

$$g^l_i = \left(\alpha^l_{i+\frac{1}{2}}\alpha^l_{i-\frac{1}{2}} + |\alpha^l_{i+\frac{1}{2}}\alpha^l_{i-\frac{1}{2}}|\right)/\left(\alpha^l_{i+\frac{1}{2}} + \alpha^l_{i-\frac{1}{2}}\right) \tag{6.147}$$

For other limiters that may be used, the reader is referred to the review by Yee (1989).

In Chapter 4 a general formulation was given to provide time integration of the equations for flux formulas similar to those given above. A similar general formulation for the integration can be written for systems of equations and appears in the following form:

$$\mathbf{U}^{n+1}_i + \theta\frac{\Delta t}{\Delta x}\left(\mathbf{E}^{n+1}_{i+\frac{1}{2}} - \mathbf{E}^{n+1}_{i-\frac{1}{2}}\right) = \mathbf{U}^n_i - (1-\theta)\frac{\Delta t}{\Delta x}\left(\mathbf{E}^n_{i+\frac{1}{2}} - \mathbf{E}^n_{i-\frac{1}{2}}\right) \tag{6.148}$$

In this expression, θ has the same meaning as used previously in Chapter 4. For $\theta = 0$ the scheme is explicit; the $\theta = 1$ case represents backward Euler differencing, and when $\theta = \frac{1}{2}$, this reduces to the trapezoidal scheme. The backward Euler method is first order in time, while the trapezoidal scheme is second order in time. If the explicit scheme is used, stability limitations will generally be encountered with first-order time differencing and second-order spatially accurate fluxes. With the use of limiters, the stability problem may be suppressed. However, this is not recommended. For time asymptotic calculations of steady-state flows, the backward Euler scheme with second-order spatial fluxes or the trapezoidal scheme are recommended. Depending upon the time accuracy desired, these implicit schemes may also be used for time-accurate calculations. If a time-accurate solution is desired, the flux Jacobians must also be correctly treated.

6.6 MULTIDIMENSIONAL CASE IN A GENERAL COORDINATE SYSTEM

In previous sections the basic concepts used to compute solutions of 1-D time-dependent flow were discussed. In practical applications, calculations are almost always multidimensional and are usually performed using a boundary conforming grid. This grid may be structured or unstructured, and the numerical method should be applicable to either case in a finite-volume formulation.

In Chapter 5 the general form of the Euler equations in conservative form was written as

$$\frac{\partial \mathbf{U}}{\partial t} + \frac{\partial \mathbf{E}}{\partial x} + \frac{\partial \mathbf{F}}{\partial y} = 0 \tag{6.149}$$

For simplicity, only the 2-D case will be studied, since the extension to three dimensions follows the same path. For a transformation to a general coordinate system, the conservative equation becomes

$$\frac{\partial \mathbf{U}_1}{\partial t} + \frac{\partial \mathbf{E}_1}{\partial \xi} + \frac{\partial \mathbf{F}_1}{\partial \eta} = 0 \tag{6.150}$$

where the subscript 1 is used to denote the altered conservative variables defined by

$$\mathbf{U}_1 = \mathbf{U}/J \tag{6.151a}$$

$$\mathbf{E}_1 = [\xi_x \mathbf{E} + \xi_y \mathbf{F}]/J \tag{6.151b}$$

$$\mathbf{F}_1 = [\eta_x \mathbf{E} + \eta_y \mathbf{F}]/J \tag{6.151c}$$

and the Jacobian J is

$$J = \xi_x \eta_y - \xi_y \eta_x$$

The flux terms and the corresponding limiters need to be more generally interpreted for application to complex geometries.

In order to proceed with the extension, let the Jacobian matrices of the fluxes be identified in the same manner as before, with the subscript 1 indicating the transformed coordinate system:

$$[A_1] = \xi_x [A] + \xi_y [B] \tag{6.152a}$$

$$[B_1] = \eta_x [A] + \eta_y [B] \tag{6.152b}$$

where

$$[A] = \partial \mathbf{E}/\partial \mathbf{U} \tag{6.153a}$$

$$[B] = \partial \mathbf{F}/\partial \mathbf{U} \tag{6.153b}$$

The eigenvalues of the $[A_1]$ and $[B_1]$ matrices will be denoted by λ_{a1} and λ_{b1}. The new matrices $[A_1]$ and $[B_1]$ may be written

$$[A_1] = [T_1][\Lambda_{a1}][T_1]^{-1} \tag{6.154a}$$

$$[B_1] = [S_1][\Lambda_{b1}][S_1]^{-1} \tag{6.154b}$$

The application of the ideas of the previous sections to multidimensional problems is based upon computing the cell-face flux terms as a series of 1-D calculations. The Riemann problem is solved as if the flow was normal to the cell face in each case. Considering this, severely skewed grids should be avoided wherever possible, as is true in any numerical calculation. The application of the idea of a 1-D Riemann problem at the cell interface in multidimensional problems has been carried out with good success. Much work continues to be done in creating a truly multidimensional Riemann solver, but success has been limited at best (Parpia and Michelak, 1993). A good review of this approach and the difficulties encountered is given in the paper by van Leer (1992). Another technique that has been explored to improve the applicability of the Riemann solver in the multidimensional case is to use the idea of a rotated difference stencil. The idea is to orient the local coordinates so the Riemann problem is solved along coordinate lines. This is similar to the work earlier reported by Davis (1984) using more classical methods. While some authors report good success with this approach (Kontinos and McRae (1991, 1994), it is not clear that the improvement in accuracy justifies the additional complexity. For a review of other work using the rotated difference idea, the papers of Deconick et al. (1992), Leck and Tannehill (1993), and Levy et al. (1993) are suggested.

At the beginning of Section 6.4, the equations for inviscid flow were integrated around a general control volume in physical space. The cell-face areas and volumes appeared explicitly. The metrics of the transformations contained in the conservation form of the general equations appear when the transformed equations are used. In either case, the Riemann problem is solved normal to the cell faces with the relationships between the geometry of the cells as given in Section 8.3.3.

The numerical flux function that must be treated in the generalized case may be written in a general form:

$$\hat{\mathbf{E}}_{i+\frac{1}{2},j} = \frac{1}{2}\Bigg[\left(\frac{\xi_x}{J}\right)_{i+\frac{1}{2},j}(\mathbf{E}_{i,j} + \mathbf{E}_{i+1,j}) + \left(\frac{\xi_y}{J}\right)_{i+\frac{1}{2},j}(\mathbf{F}_{i,j} + \mathbf{F}_{i+1,j}) + [T_{i+\frac{1}{2},j}][\mathbf{\Phi}_{i+\frac{1}{2},j}]/J_{i+\frac{1}{2},j}\Bigg] \tag{6.155}$$

In this formulation the matrix $[T_{i+\frac{1}{2},j}]$ is the eigenvector matrix associated with $[A_1]$. The computational coordinates are denoted by ξ and η, and the metric coefficients correspond to the projections of the cell-face areas as shown in Section 8.3.3. If $\mathbf{U}_{i+\frac{1}{2},j}$ denotes the symmetric average,

$$\mathbf{U}_{i+\frac{1}{2},j} = \tfrac{1}{2}(\mathbf{U}_{i,j} + \mathbf{U}_{i+1,j}) \tag{6.156}$$

or a Roe average, then the quantity $\lambda^l_{a,i+\frac{1}{2},j}$ represents the eigenvalues of $[A_1]$ evaluated at $\mathbf{U}_{i+\frac{1}{2},j}$, and a similar notation may be used for the eigenvalues of $[B_1]$. We may also define the wave strengths for the multidimensional case as

$$\alpha_{a,i+\frac{1}{2},j} = [T]^{-1}(\mathbf{U}_{i+1,j} - \mathbf{U}_{i,j}) \tag{6.157}$$

for the ξ direction, and a corresponding form for the η direction. The notation for the other terms in Eq. (6.155) is

$$\left(\frac{\xi_x}{J}\right)_{i+\frac{1}{2},j} = \frac{1}{2}\left[\left(\frac{\xi_x}{J}\right)_{i,j} + \left(\frac{\xi_x}{J}\right)_{i+1,j}\right] \tag{6.158}$$

The averaged values of the metrics and Jacobians are used in order to preserve the free stream. This is similar to the geometric conservation requirements that appear in finite-difference formulations (Hindman, 1981).

The $\mathbf{\Phi}$ term that contains the limiters for symmetric or upwind total variation diminishing methods may be written in the same form as previously given. The structure of the limiter must be interpreted in terms of the altered eigenvalues associated with the generalized directions.

The MUSCL approach leads to a slightly different formulation of the flux terms, but the general form is the same as previously defined:

$$\hat{\mathbf{E}}_{i+\frac{1}{2},j} = \frac{1}{2}\left\{\left(\frac{\xi_x}{J}\right)_{i+\frac{1}{2},j}\left[\mathbf{E}\left(\mathbf{U}^R_{i+\frac{1}{2},j}\right) + \mathbf{E}\left(\mathbf{U}^L_{i+\frac{1}{2},j}\right)\right]\right.$$
$$\left. + \left(\frac{\xi_y}{J}\right)_{i+\frac{1}{2},j}\left[\mathbf{F}\left(\mathbf{U}^R_{i+\frac{1}{2},j}\right) + \mathbf{F}\left(\mathbf{U}^L_{i+\frac{1}{2},j}\right)\right] + [T_{i+\frac{1}{2},j}]\mathbf{\Phi}_{i+\frac{1}{2},j}/J_{i+\frac{1}{2},j}\right\} \tag{6.159}$$

In the MUSCL formulation, some liberty exists in selecting the quantity that is limited. There are a number of choices that can be made that will provide acceptable results. These include limiting the primitive variables, the characteristic variables, or any other quantity that is used in the extrapolation. The appropriate form of the limiter depends upon the choice of the extrapolation variable. The primitive variables are normally the best choice.

An explicit scheme is easily constructed with the same predictor-corrector algorithm as used in Chapter 4. In this case, one may predict at time $t + (\Delta t/2)$ and use the results to form the fluxes needed to advance the solution to the next time level. Perhaps the most well-known explicit scheme that has been employed to solve hyperbolic problems in fluid mechanics is the MacCormack (1969) method. An approximate finite-volume form of this method may be constructed by using the standard MacCormack method for the first two steps with an added third step to provide the selective dissipation to control the oscillations present in conventional shock-capturing methods. This may be written in the form

$$\overline{\mathbf{U}}_{1,i,j}^{n+1} = \mathbf{U}_{1,i,j}^{n} - \frac{\Delta t}{\Delta \xi}(\mathbf{E}_{1,i+1,j}^{n} - \mathbf{E}_{1,i,j}^{n}) - \frac{\Delta t}{\Delta \eta}(\mathbf{F}_{1,i,j+1}^{n} - \mathbf{F}_{1,i,j}^{n})$$

$$\overline{\overline{\mathbf{U}}}_{1,i,j}^{n+1} = \frac{1}{2}\left[\overline{\mathbf{U}}_{1,i,j}^{n+1} + \mathbf{U}_{1,i,j}^{n} - \frac{\Delta t}{\Delta \xi}\left(\overline{\mathbf{E}}_{1,i,j}^{n+1} - \overline{\mathbf{E}}_{1,i-1,j}^{n+1}\right) - \frac{\Delta t}{\Delta \eta}\left(\overline{\mathbf{F}}_{1,i,j}^{n+1} - \overline{\mathbf{F}}_{1,i,j-1}^{n+1}\right)\right]$$

$$\mathbf{U}_{1,i,j}^{n+1} = \overline{\overline{\mathbf{U}}}_{1,i,j}^{n+1} + \frac{1}{2}\left(\frac{\overline{T}_{i+\frac{1}{2},j}\overline{\mathbf{\Phi}}_{i+\frac{1}{2},j}}{J_{i+\frac{1}{2},j}} - \frac{\overline{T}_{i-\frac{1}{2},j}\overline{\mathbf{\Phi}}_{i-\frac{1}{2},j}}{J_{i-\frac{1}{2},j}}\right)$$
$$+ \frac{1}{2}\left(\frac{\overline{T}_{i,j+\frac{1}{2}}\overline{\mathbf{\Phi}}_{i,j+\frac{1}{2}}}{J_{i,j+\frac{1}{2}}} - \frac{\overline{T}_{i,j-\frac{1}{2}}\overline{\mathbf{\Phi}}_{i,j-\frac{1}{2}}}{J_{i,j-\frac{1}{2}}}\right) \tag{6.160}$$

The third step is a postprocessing step appended to the classical MacCormack method in order to eliminate the deficiencies associated with calculations through strong shocks that are present in the classical methods. In this step the overbar quantities may be evaluated either at time level n or at the end of the second step.

Other techniques for generating explicit methods may be used. Many of the standard integration schemes for ordinary differential equations have been used to advance the solution in time. The most popular of these schemes is the Runge-Kutta (R-K) method described in detail for use on time-continuous systems in the work of Lomax et al. (1970) and applied extensively by Jameson et al. (1981) and Jameson (1987). The formulation of the R-K procedure is described in Section 4.1.13, where application is made to the linear wave equation.

Implicit schemes may be constructed using a number of techniques. Most common among these is the simple Euler implicit method or the trapezoidal scheme described in Chapter 4 as well as the Euler implicit scheme considered in the last section. As in the 1-D case, implicit methods create the necessity of calculating the flux Jacobians. The discussion presented in the previous section applies not only to the 1-D problem, but equally well to any number of dimensions. The linear system produced must be solved as efficiently as possible, and this becomes the major effort in computing a solution with an implicit method.

In Section 4.3, several techniques were presented for solving scalar systems. In solving systems of equations that result from the conservation laws, the same methods apply, and to some extent, the same philosophy may be followed. For example, one of the goals that is used in solving the 1-D equations is to try to create tridiagonal systems of equations. The Thomas algorithm is then applied to obtain a direct solution. The same methodology may be applied in the multi-equation case, where the goal is to have block tridiagonal systems. This naturally suggests the use of approximate factorization, as discussed earlier in this chapter. In 3-D problems the errors introduced by this method have historically proved difficult to control. However, through suitable preconditioning, the use of approximate factorization has been shown to be effective for any number of dimensions (Choi and Merkle, 1993). Many applications of the approximate factorization approach may be found in the literature (Beam and Warming, 1976); Steger, 1977; Pulliam, 1985). Other techniques used to solve these equations include LU factorizations as described in Section 4.3.4. The technique employed to solve the linear system of equations for an implicit scheme will also depend upon the application. If only the steady-state solution is of interest, the implicit operator may be modified in any way that accelerates convergence of the solution to the steady-state value. On the other hand, if the transient is of importance, the implicit as well as the explicit operator must be retained or approximated as closely as possible (in non-iterative schemes) to provide time-accurate results.

Approximations to the implicit side of the equations have been the most popular modification to the basic implicit algorithms. These modifications have the largest influence on reducing resources needed to compute solutions to the equations of fluid flow. The approximate factorization scheme employed by Beam and Warming (1976) was modified by a diagonalization procedure by Pulliam and Chaussee (1981), reducing the operation count for the solution of the system. Another idea is to use a lower-order implicit operator and a higher-order explicit operator when computing time-asymptotic solutions. Such procedures have been advocated in the work by Steger and Warming (1982) and Rai (1987). Other approaches to the problem of solving systems of equations, with the banded structure common to the problems of computational fluid dynamics, include the LU factorization schemes noted in Section 4.3.4 and the modifications to LU factorization with Gauss-Seidel iteration recently presented by Yoon and Kwak (1993). In recent applications of implicit methods, the concept of using local iteration schemes to compute solutions of arbitrary accuracy have been developed and used with success. In general, the problem of computing a solution could be viewed as having two separate issues. The first is that a banded matrix structure will probably result from the implicit scheme and some ingenuity will be required to solve the system in an efficient manner. The other issue is that the modification of the implicit operator in many cases can be accomplished by viewing changes as alterations of the dissipation term in the numerical flux terms. These two issues are obviously coupled but should be recognized as significant in accelerating computations for implicit schemes.

6.7 BOUNDARY CONDITIONS FOR THE EULER EQUATIONS

In computing solutions to PDEs, the application of boundary conditions is a key ingredient. The technique used in implementing the boundary conditions can have a major effect on the stability and convergence of the numerical solution. The way that boundary conditions are applied in solving analytical problems is usually well defined. This is not necessarily the case when applying boundary conditions for the discrete problem. In this case, reasonable approximations have been found to work, although the exact form may be different from that expected in comparison to the analytical case.

In the solution of flow field problems, appropriate conditions must be applied on the domain boundaries. Since only finite domains may be considered because of limitations in computer memory and speed, we are forced to specify conditions on boundaries at a finite distance from the airfoil, body, aircraft, or other geometry that is of interest in the simulation. These conditions depend upon the type of boundary and the type of flow regime. For hyperbolic problems the number of boundary conditions needed can be determined by a careful evaluation of the direction that information is carried by the characteristics. The number of boundary conditions that must be supplied is equal to the number of characteristics that are directed from the exterior of the region toward the boundary. This will become apparent below, when specific conditions and boundaries are considered. For elliptic and parabolic problems, the application of boundary conditions usually results in the specification of a set of dependent variables on the boundary or the specification of the normal derivatives. These problems require a different approach because of the different physics involved.

As an example, consider a supersonic flow over a body and the different types of domain boundaries that are present. At the inflow boundary the flow is entirely supersonic. If one examines the direction of signal propagation for these conditions, the characteristics carry information from the exterior of the domain toward the interior in all cases. For this steady case the solution for the flow over the representative body may be obtained using a marching procedure. Reference to the characteristics (see Fig. 6.17) shows that the signals are carried into the domain from the upstream region by both the streamline characteristics as well as the characteristics involving the acoustic speeds. This indicates that all information at the inflow boundary for a supersonic flow must be specified using the free stream conditions. There are no characteristics that carry information from the interior of the domain to the boundary. For a time-dependent flow, it should also be clear that a supersonic inflow will always carry information toward the boundary from the exterior [see Fig. 6.18(a)]. In this case the inflow conditions are always prescribed, as is the case for the steady marching problem. For subsonic flow the characteristics carry information toward the domain boundary both from the interior and the exterior [Fig. 6.18(b)]. The boundary conditions are used to replace the information carried to the boundary by the characteristics from the exterior.

In order to understand the development of boundary condition procedures in more than one dimension, consider the 2-D Euler equation written in (x, y) coordinates:

$$\frac{\partial \mathbf{U}}{\partial t} + \frac{\partial \mathbf{E}}{\partial x} + \frac{\partial \mathbf{F}}{\partial y} = 0$$

This equation can be expanded and written in the form

$$\frac{\partial \mathbf{U}}{\partial t} + [A]\frac{\partial \mathbf{U}}{\partial x} + [B]\frac{\partial \mathbf{U}}{\partial y} = 0 \tag{6.161}$$

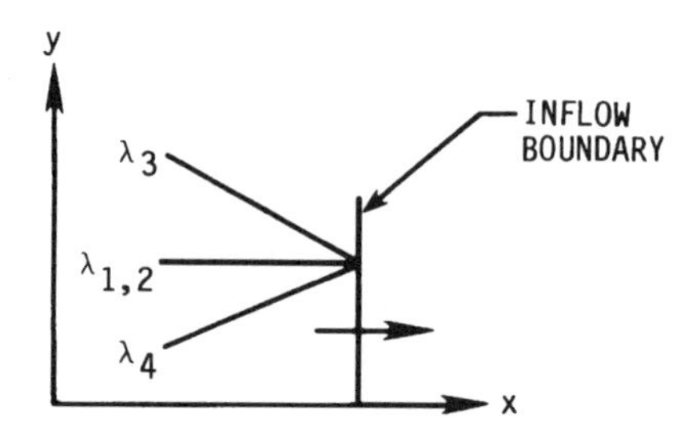

Figure 6.17 Inflow boundary conditions for steady supersonic flow.

It should be understood that this equation could have been written in generalized (ξ, η) coordinates and the ideas for applying boundary conditions would be developed in the same way. The differences would be that the contravariant velocities as well as the computational coordinates appear in the flux terms.

The Jacobian matrices $[A], [B]$ and the dependent variables are those associated with the equations written in the (x, y) system, while the Jacobians written in generalized coordinates would be in terms of the contravariant velocities. At an inflow boundary, assume that the positive x direction points from the free stream toward the interior of the domain. The y direction is assumed to lie along the boundary, with the boundary defined as a constant x surface. In the case where general (ξ, η) coordinates are used, the inflow boundary would be defined as a constant ξ surface with the η coordinate changing as one moves along the surface. Returning to the formulation in the (x, y) system, we write the governing differential equation in the form

$$\frac{\partial \mathbf{U}}{\partial t} + [T][\Lambda_a][T]^{-1} \frac{\partial \mathbf{U}}{\partial x} + [S][\Lambda_b][S]^{-1} \frac{\partial \mathbf{U}}{\partial y} = 0 \tag{6.162}$$

In this form the dependent variables may be either the primitive variables or the conservative variables denoted by **U**. Usually, primitive variables are used to develop the compatibility relations. The compatibility equations written in the (t, x) direction are desired for the development of boundary conditions on a constant x surface. If the governing system written in primitive variables is premultiplied by the $[T]^{-1}$ matrix, the compatibility equations are obtained.

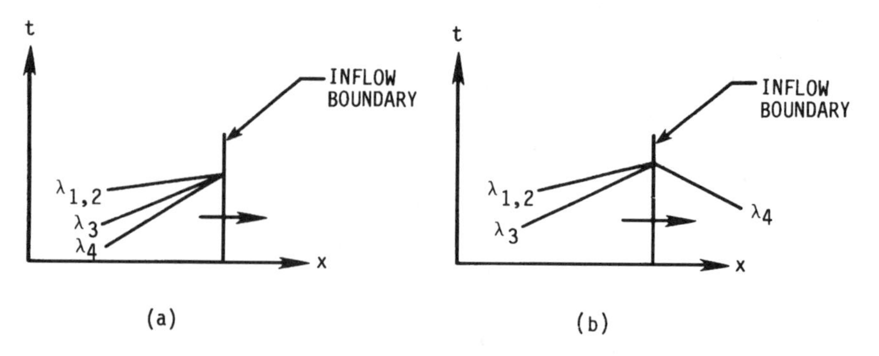

Figure 6.18 Inflow boundaries for time-dependent flow. (a) Supersonic flow. (b) Subsonic flow.

These equations may be integrated along with the boundary conditions to provide a solution at either inflow or outflow boundaries. The spatial terms must be differenced according to the direction of propagation of signals. If the elements of the $[T]^{-1}$ matrix are assumed to be constant and a new set of variables $\mathbf{W}$ is defined as $[T]^{-1}\mathbf{U}$, the compatibility equations may be written as

$$\frac{\partial \mathbf{W}}{\partial t} + [\Lambda_a]\frac{\partial \mathbf{W}}{\partial x} + [T]^{-1}[S][\Lambda_b][S]^{-1}[T]\frac{\partial \mathbf{W}}{\partial y} = 0 \tag{6.163}$$

The terms in the vector $\mathbf{W}$ are called characteristic variables, Riemann variables, or sometimes Riemann invariants in the literature. The use of characteristic variables seems to be the most appropriate choice of notation in the multidimensional case. The matrix $[\Lambda]$ denotes the matrix of eigenvalues and, for the case under consideration, consists of four elements corresponding to the repeated streamline characteristics and the wave fronts defining the subsonic or supersonic flow:

$$\begin{aligned} \lambda_{1,2} &= u, u \\ \lambda_{3,4} &= u \pm a \end{aligned} \tag{6.164}$$

For subsonic inflow the streamline characteristics are both positive. The remaining characteristics representing the acoustic wave fronts are composed of one characteristic that is positive and one that is negative. The one with the negative slope carries information from the interior to the boundary of the domain and therefore may be viewed as a valid equation to be retained in the evaluation of the inflow conditions. The other characteristics all have positive slopes, which indicates that information is carried from the free stream to the inflow boundary. Therefore these three equations must be discarded, and the boundary conditions must provide the missing three pieces of data required to close the problem at the inflow boundary.

The compatibility equations for the multidimensional case cannot be written in a perfectly diagonalized form as in the 1-D case. As a consequence, the transverse terms in this case, the terms involving derivatives in the y direction, must be retained when the integration of the compatibility equations is performed to establish the conditions at a boundary. In establishing the boundary conditions for an inflow, the assumption is often made that the transverse terms do not contribute measurably and may be neglected. The resulting equation is the 1-D compatibility equation, and it can be explicitly integrated along the characteristics. This integration yields the Riemann invariants in the 1-D case. These invariants at the inflow boundary are set by the free stream and are used as the inflow boundary conditions. The form of these conditions is given as

$$\begin{aligned} w_1 &= \left[\frac{p}{\rho^\gamma}\right] \\ w_2 &= v \\ w_3 &= u + \left[\frac{2a}{\gamma - 1}\right] \end{aligned} \tag{6.165}$$

These values of w at the inflow plane are evaluated using free stream conditions. The second value of w_2 has been set equal to the tangential velocity and is somewhat arbitrary, since this is a 1-D approximation. For supersonic flows the characteristics all carry the same sign, and free stream conditions are set at the inflow boundary, since no upstream influence is present.

At outflow boundaries the characteristics all carry the same sign for the supersonic case, and the solution must be determined entirely from conditions based on the interior. The compatibility conditions may be integrated using a suitable upwind approximation to appropriately account for the signals. If the outflow is subsonic, the analysis shows that one characteristic has a negative slope, indicating that one condition must be specified and that only three of the compatibility conditions are valid. This condition is not arbitrary but must be selected to ensure that the numerical problem is well posed (Yee, 1981). At an outflow boundary it is permissible to specify the pressure, the density, or the velocity normal to the cell. The choice of outflow boundary conditions usually depends on the way the flow solver is influenced by the boundary conditions. Hirsch (1990) suggests that the density should be among the imposed conditions for a subsonic inflow boundary. The choice of boundary conditions that work best will change owing to the manner of implementation, and the exact form of these conditions will be different for different codes.

Characteristic-based boundary conditions have been extensively used in computational fluid dynamics codes. While these methods give excellent results, they sometimes constitute a case of overkill, in the sense that a simpler boundary condition implementation would have worked equally as well. For example, simple extrapolation methods work well in many cases and are very easy to apply. The idea of extrapolation can be applied to either inflow or outflow boundaries, and any combination of primitive variables, characteristic variables, or conservative variables may be used. The easiest extrapolation to use is to set the adjacent cell values equal. This is equivalent to the enforcement of a zero slope condition on the variable. Higher-order extrapolation procedures may be employed as well.

In applying boundary conditions at inflow or outflow boundaries, the use of FVS and FDS schemes may simplify the accounting for signal propagation directions and the correct application of exterior information. Parpia (1994) has shown that these inflow boundaries may be treated without any special conditions if the exterior cells simply are assigned the correct free stream conditions and the standard split-scheme operator is then used as if the boundary cells were interior cells. Since the characteristic signal propagation directions are correctly accounted for in the scheme, the correct information is applied at the boundaries through the numerical method. This simplifies the boundary condition procedure by a significant amount.

Impermeable surfaces such as solid boundaries present another situation where a valid set of boundary conditions is needed. At a solid boundary, the correct boundary condition for the Euler equations is that the material derivative of the surface must vanish. Let the equation of the surface be given by

$$F(x, y) = y - f(x) = 0 \tag{6.166}$$

where the surface is assumed to be independent of time. The inviscid tangency condition for a steady flow simply states that the velocity component normal to the surface is zero and may be written in the form

$$v = u\frac{\partial f}{\partial x} = u\left(\frac{\partial y}{\partial x}\right)_{\text{surface}} \tag{6.167}$$

When the Euler equations are solved using a cell-centered finite-volume method, the surface tangency condition makes all fluxes on the surface vanish, and the only remaining term that must be evaluated is the surface pressure. An approximate value for the surface pressure may be established by simple extrapolation from the interior. A more common approach is to solve the normal momentum equation to find a suitable value at the surface. If the normal momentum equation is written at the body, the pressure gradient is balanced by the centrifugal force term and may be written in the form

$$\frac{\partial p}{\partial n} = \frac{\rho v_t^2}{R} \tag{6.168}$$

where v_t represents the tangential velocity and R the radius of curvature of the surface. With this information, an approximation to the pressure at the surface may be written. This is accomplished by using the pressure computed at the first cell center and employing a Taylor series to find the pressure at the cell face corresponding to the body surface. This approach is easily applied in an (x, y) system, where the body surface is represented by a constant y surface because the normal direction is the same as the y direction. However, it is a simple matter to write the series expansion to obtain the wall pressure in a general coordinate system. If the body surface is not a cell boundary, the boundary conditions required include the variables necessary to find the full array of primitive variables. For this case the equations of motion need to be solved at the body surface, and a method that permits integration at this location must be found. Sometimes the simplest boundary condition procedure for a solid surface is to use the idea of *reflection*. If the body is located on the cell boundary, the ghost-cell values are established by assigning the reflected values at its cell center (see Fig. 6.19). The tangential velocity in the ghost cell is set equal to the tangential velocity in the first cell, while the velocity in the ghost cell normal to the body is taken to be the negative of the first cell value. In addition, the density and the pressure values are equated to the first cell values. With the

Figure 6.19 Body-surface representation. (a) Cell face; (b) grid point.

ghost-cell conditions known, the correct flux values at the body surface needed to continue the calculations can be obtained in a number of ways. One way is to use characteristic ideas with information carried from the ghost cell and the first cell to the shared body-surface segment. In this case the body-surface segment would be a constant η curve in generalized coordinates, and the cell-centered values of the characteristic variables consistent with the propagation directions are then used to determine the surface conditions. The body-surface boundary condition (flow tangency) is also implemented in the surface-flux evaluation. Of course, the same procedure used in the interior flow field may also be used once the appropriate values of the dependent variables in the ghost cells are available.

Other ways of dealing with the body-surface conditions have been shown to be very useful. The compatibility equations can be written and discretized to obtain a system of equations for the body-surface conditions when the surface tangency requirement is imposed. In this case, the characteristic that carries information from the interior of the body is replaced by the body-surface condition. The discretized compatibility conditions include the equations along the streamline (body surface in this case) as well as the acoustic front equation. Kentzer (1970) proposed a scheme of applying the surface tangency condition in conjunction with the compatibility equations. In his approach, the surface tangency condition is used in differential form. If a time-dependent flow is considered, the time derivative of the tangency condition is employed. Others have implemented various forms of these conditions to compute surface solutions based on the characteristic information available at the body (Chakravarthy et al., 1980; Chakravarthy, 1983; Rai and Chaussee, 1994).

When the compatibility equations are used to determine the body boundary conditions, the equations are integrated along the characteristics. In this case, assuming that the (x, y, t) system previously employed can be a usable model, the body-surface boundary condition replaces the compatibility equation in the (y, t) direction, corresponding to the positive eigenvalue. This assumes that the body surface is at $y = 0$ with the interior of the body defined for $y < 0$. The same approach can be used in steady marching problems.

For codes where the numerical method is a cell-centered scheme the body surface is often a cell face. In this case the only remaining term in the inviscid flux is the body pressure. Consequently, a simple wave corrector may be used to evaluate the body-pressure term. This may be viewed as integration across the waves, as opposed to the ideas presented previously, where the equations were integrated along the characteristics. In this sense, the Roe scheme may be directly employed to obtain a simple wave corrector for hyperbolic problems. For the time-dependent case the characteristic variable form for the equations of motion show that the velocity normal to the cell face is proportional to the change in cell surface pressure. This permits the normal velocity to be corrected to satisfy the surface tangency condition by correcting the pressure with a weak expansion or compression wave. Abbett (1973) proposed that for problems where a steady supersonic flow was being computed, a test to evaluate the velocity misalignment with the body surface be made after the flow field solution had been advanced. The flow misalignment is then corrected by turning the flow back parallel to the body by introducing a simple wave. When a cell-centered scheme with a cell center located on the body is used, the velocity is computed on the body surface. Otherwise, the body-surface velocity is not computed. Consequently, the misalignment of the velocity vector must be approximated by using the cell-center value from the first cell in the flow field. The velocity vector can be divided into a component normal to the surface and a component tangent to the surface.

The velocity components in the respective directions are (u, v, w). Let the unit vector normal to the surface be

$$\mathbf{n} = \frac{\nabla F}{|\nabla F|}$$

where the body-surface equation is given by

$$F(x_1, x_2, x_3) = x_1 - f(x_2, x_3) = 0 \tag{6.169}$$

Therefore the body surface normal is

$$\mathbf{n} = \frac{\mathbf{i}_1/h_1 - [(\mathbf{i}_2/h_2)(\partial f/\partial x_2)] - [(\mathbf{i}_3/h_3)(\partial f/\partial x_3)]}{\left\{1/h_1^2 + [(1/h_2)(\partial f/\partial x_2)]^2 + [(1/h_3)(\partial f/\partial x_3)]^2\right\}^{1/2}} \tag{6.170}$$

The velocity vector can be divided into a component normal to the surface and a component tangent to the surface. If the normal velocity is computed as

$$\mathbf{u}_{\text{nor}} = (\mathbf{V} \cdot \mathbf{n})\mathbf{n} \tag{6.171}$$

the small misalignment angle $\Delta\theta$, representing the orientation of the velocity vector with respect to a surface tangent, becomes

$$|\sin(\Delta\theta)| = \frac{|\mathbf{u}_{\text{nor}}|}{|\mathbf{V}|} \tag{6.172}$$

We may write this as

$$\sin(\Delta\theta) = \frac{\mathbf{V} \cdot \mathbf{n}}{|\mathbf{V}|} \tag{6.173}$$

The geometry of this problem is shown in Fig. 6.20. The misalignment angle $\Delta\theta$ is clearly shown. The velocity vector $\mathbf{V}$ represents the velocity computed at the

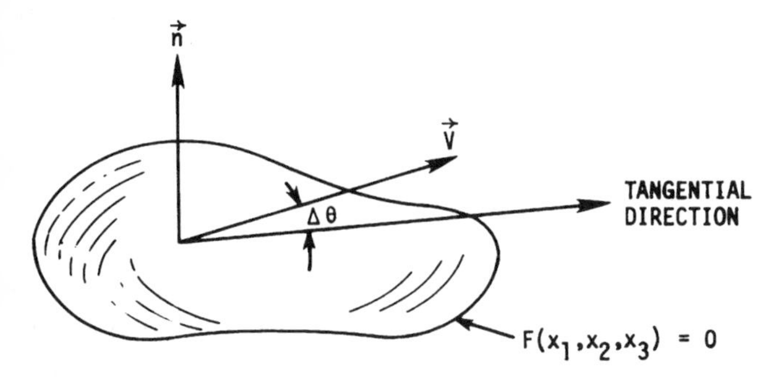

Figure 6.20 Velocity vector orientation on body surface.

body surface using the integration scheme. If MacCormack's method was used to solve the equations of motion, the velocity vector **V**, shown in Fig. 6.20, is the value at the end of the corrector step. Again, remember that a forward corrector must be used at the body.

In order to turn the velocity vector through an angle $\Delta\theta$ so that it is parallel to the body, a weak wave is introduced into the flow. If $\Delta\theta$ is positive, an expansion is required. As the flow turns through an angle $\Delta\theta$, the body-surface pressure must also change. For weak waves the pressure is related to the flow turning angle by the expression [see NACA Report 1135 (Ames Research Staff, 1953)].

$$\frac{p_2}{p_1} = 1 - \frac{\gamma M^2}{\sqrt{M^2 - 1}} \Delta\theta + \gamma M^2 \left[\frac{(\gamma + 1)M^4 - 4(M^2 - 1)}{4(M^2 - 1)^2} \right] (\Delta\theta)^2 + \cdots \tag{6.174}$$

In this expression, M and p_1 are the Mach number and pressure before turning, while p_2 represents the pressure after the turn takes place. With the pressure known from Eq. (6.174), the density change can now be computed. This is one point where Abbett's scheme requires additional information. It is assumed that the value of the surface entropy is known. At least along the streamline that wets the body, the value of p/ρ^γ is known. The new surface pressure (p_2) is used in conjunction with the surface entropy to calculate a new density ρ_2.

The magnitude of the velocity in the tangential direction is computed by use of the steady energy equation. If H is the total enthalpy, the velocity along the body surface is calculated as

$$|\mathbf{V}_2| = \sqrt{2\left(H - \frac{\gamma}{\gamma - 1}\frac{p_2}{\rho_2}\right)} \tag{6.175}$$

The velocity components must now be determined. The direction of the new velocity vector along the surface is obtained by subtracting the normal velocity from the original velocity computed using the integration routine. This produces the result

$$\mathbf{V}_T = \mathbf{V} - (\mathbf{V} \cdot \mathbf{n})\mathbf{n} \tag{6.176}$$

and represents the tangential component of the original velocity. It is assumed that the new surface velocity vector is in the same direction. The new velocity $\mathbf{V}_2$ is given by

$$\mathbf{V}_2 = |\mathbf{V}_2| \frac{\mathbf{V}_T}{|\mathbf{V}_T|} \tag{6.177}$$

This boundary condition routine is relatively easy to apply and provides excellent results (see Kutler et al., 1973). One of the major problems is that of determining the proper direction for the final velocity vector. Abbett's method assumes this final velocity vector lies in the tangent plane of the body in the direction of the intersection of the tangent plane and the plane formed by the unit normal and the original velocity vector. No out-of-plane correction is used.

In supersonic flow, it is common procedure to attempt to fit the outer boundary of the physical domain when a shock wave is coincident with the outer boundary. This procedure may be followed in either time-dependent or steady flow. The process of shock fitting saves memory and produces a precise description of the shock front that one obtains from the computational procedure. The process of fitting the shock wave is a matter of satisfying the Rankine-Hugoniot equations while simultaneously requiring that the solution on the downstream side of the shock be compatible with the rest of the flow field.

In the time-dependent problem the solution for the flow variables downstream of a shock is determined by the free stream conditions, the shock velocity, and the shock orientation. If we know the free stream conditions, the initial shock slope, and the shock velocity, the shock pressure can be considered as the primary unknown in the shock-fitting procedure. The normal procedure is to combine the Rankine-Hugoniot equations with one compatibility equation to provide the expression for shock acceleration and post-shock conditions. For example, once the downstream pressure has been determined from the integration on the interior, the other downstream flow variables may be computed using the Rankine-Hugoniot equations:

$$\mathbf{n}_s = \frac{\mathbf{i}\eta_x + \mathbf{j}\eta_y}{\sqrt{\eta_x^2 + \eta_y^2}}$$

$$u_{\infty n} = |\mathbf{V}_\infty \cdot \mathbf{n}_s|$$

$$M_s = \left\{ \frac{1}{2\gamma} \left[\frac{p_2}{p_\infty}(\gamma + 1) + (\gamma - 1) \right] \right\}^{1/2}$$

$$V_s = a_\infty M_s - u_{\infty n} \tag{6.178}$$

$$u_{2n} - u_{\infty n} = \frac{2a_\infty(1 - M_\infty^2)}{(\gamma + 1)M_s}$$

$$\rho_2 = \rho_\infty \left(\frac{p_2}{p_\infty} + \frac{\gamma - 1}{\gamma + 1} \right) \left[\frac{1}{1 + (\gamma - 1)(p_2/p_\infty)/(\gamma + 1)} \right]$$

$$\mathbf{V}_2 = \mathbf{V}_\infty + (u_{2n} - u_{\infty n})\mathbf{n}_s[\text{sign}\,(\mathbf{V}_\infty \cdot \mathbf{n}_s)]$$

The subscript ∞ refers to free stream conditions, subscript 2 denotes conditions immediately downstream of the shock wave, and subscript s indicates the shock surface, and n indicates the normal to this surface. Equations (6.178) can easily be derived from the relative velocity expression for the shock motion and the Rankine-Hugoniot equations. Figure 6.21 illustrates the notation and the orientation of the shock in physical space. Consistent with the discussion of boundary condition procedures, only one characteristic carries information from

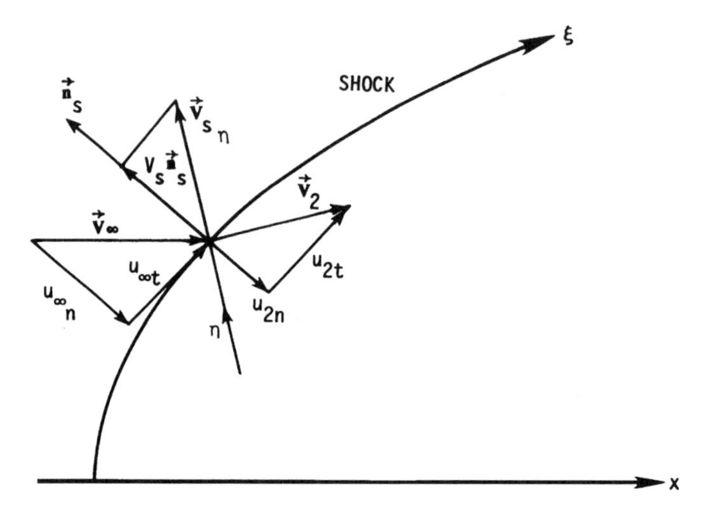

Figure 6.21 Shock geometry.

the interior to the shock wave. If this characteristic is λ_4, the corresponding compatibility equation (where τ is the transformed time) is

$$\sum_{i=1}^{4} s_{4i}(w_{i_\tau} + \lambda_4 w_{i_\eta} + g_i) = 0 \tag{6.179}$$

Since we have the shock wave as one boundary of our domain, we may write

$$\frac{\partial w_i}{\partial \tau} = \frac{\partial w_i}{\partial p}\frac{\partial p}{\partial \tau} \tag{6.180}$$

That is, we explicitly include the dependence of the w_i variables on the shock pressure. The derivative $\partial w_i/\partial p$ can be explicitly evaluated from Eq. (6.178). If we substitute Eq. (6.180) into Eq. (6.179), an expression for the time rate of change of pressure is obtained:

$$\frac{\partial p}{\partial \tau}\sum_{i=1}^{4} s_{4i}\frac{\partial w_i}{\partial p} = -\sum_{i=1}^{4} s_{4i}(\lambda_4 w_{i_\eta} + g_i) \tag{6.181}$$

The w_{i_η} derivatives in this expression are evaluated using backward differences, which is consistent with the fact that information is being carried along a positive characteristic. The expression given in Eq. (6.181) permits p_τ to be computed, and then time derivatives of the other variables can be obtained using Eq. (6.180). These expressions are then integrated to provide the updated dependent variables. The shock position is updated by integrating the known shock speed. Moretti (1974, 1975) prefers to use the shock speed as the dependent variable. This can be easily accomplished within the above analysis.

The dependence of the w_i variables given in Eq. (6.180) is replaced by

$$\frac{\partial w_i}{\partial \tau} = \frac{\partial w_i}{\partial V_s}\frac{\partial V_s}{\partial \tau} \tag{6.182}$$

where we again compute $\partial w_i/\partial V_s$ from the Rankine-Hugoniot equations. Substituting this expression into our compatibility equation yields an equation that may be solved for the shock acceleration

$$\frac{\partial V_s}{\partial \tau}\sum_{i=1}^{4} s_{4i}\frac{\partial w_i}{\partial V_s} = -\sum_{i=1}^{4} s_{4i}(\lambda_4 w_{i\eta} + g_i) \tag{6.183}$$

Once the shock acceleration is known, the velocity and position are obtained by integration in time. The new dependent variables are computed from the Rankine-Hugoniot equations using the new shock velocity.

Boundary shock fitting is also used when the solution of the Euler equations is obtained using a marching procedure. This was discussed in Section 6.3, and a general procedure was given to compute the shock wave shape as part of the solution.

The boundary condition procedures presented in this section have been successfully applied to solve the Euler equations. While a limited number of ideas have been presented here, the literature abounds with different boundary condition application procedures. However, most of the ideas follow the guidelines given here. As a final comment, it is frequently necessary to specify more on a boundary than is required in the analytic formulation of the problem when using a numerical method. This should not be surprising, since the modification of the continuous problem also has a major effect on the application of the boundary conditions. At least, one should expect the boundary conditions to enter the computational procedure in ways not anticipated.

6.8 METHODS FOR SOLVING THE POTENTIAL EQUATION

While solutions of the Euler equations are computed on a routine basis, finding the flow field for a complete configuration is still a time-consuming procedure. More rapid computational procedures that preserve accuracy are desirable in many applications where repetitive calculations are needed. A case in point is in preliminary design. Many geometric configurations and modifications are studied, and fast reliable methods are needed in order to evaluate each change in the proposed design. Potential methods are well suited for these situations.

As is well known in fluid mechanics, a hierarchy of equations exists based upon the order of the approximation attempted or the assumptions made in the derivation of the governing equation. As we consider reductions from the Euler equations, the next logical step is to consider the solution provided by a full potential formulation.

The *full potential equation* in conservative or nonconservative form is frequently used for solving transonic flow problems. In developing the full potential equation, the existence of the velocity potential requires that the flow be irrotational. Furthermore, Crocco's equation [Eq. (5.187)] requires that no entropy production occur. Thus no entropy changes are permitted across shocks in supersonic flows when a full potential formulation is used. At first glance, this appears to be a poor assumption. However, experience has shown that the full potential and Euler solutions do not differ significantly if the component of the Mach number normal to the shock is close to 1. The entropy production across a weak shock is dependent on the normal Mach number M_n, and is approximately (Liepmann and Roshko, 1957)

$$\frac{\Delta s}{R} \sim \frac{2\gamma}{\gamma+1}(M_n^2 - 1)^3 \tag{6.184}$$

This shows that the assumption of no entropy change across a shock is reasonable so long as the normal component of the Mach number is sufficiently close to 1. It is important to note that the restriction is on the normal component of the local Mach number and not the free stream Mach number.

If the irrotational flow assumption is valid, we expect solutions of the potential equation to yield results nearly as accurate as solutions to the Euler equations even in supersonic and transonic flows with shocks. Difficulties in solving the Euler equations are not completely circumvented by the potential formulation, since we retain the nonlinear fluid behavior even with this simplification. We will discuss the application of the potential equation to typical flow problems later in this section.

The full potential approximation to the Euler equations can be developed in either a nonconservative or conservative form. The nonconservative form of the steady potential equation may be written for two dimensions as [Eq. (5.197)]

$$\left(1 - \frac{u^2}{a^2}\right)\phi_{xx} - \frac{2uv}{a^2}\phi_{xy} + \left(1 - \frac{v^2}{a^2}\right)\phi_{yy} = 0 \tag{6.185}$$

where

$$u = \frac{\partial \phi}{\partial x} \qquad v = \frac{\partial \phi}{\partial y} \tag{6.186}$$

and a is the speed of sound, which may be obtained from the energy equation

$$\frac{a^2}{\gamma - 1} + \frac{u^2 + v^2}{2} = H = \text{const} \tag{6.187}$$

The nonconservative form of the potential equation is sometimes referred to as the quasi-linear form of the full potential equation. In our discussion of solutions of the Euler equations, the use of the nonconservative form did not produce acceptable results at the shocks. A similar condition holds here, and the solutions for flows with shocks obtained with the nonconservative form appear to have mass sources at the shock. Most recent techniques use a conservative formulation. The conservative form of the full potential equation is simply the continuity equation:

$$\frac{\partial \rho u}{\partial x} + \frac{\partial \rho v}{\partial y} = 0 \tag{6.188}$$

This equation is written in nondimensional form, where the asterisks denoting nondimensional variables have been omitted. The velocity components are related to the potential as noted above, and the density is calculated from the energy equation in the form

$$\rho = \left[1 - \frac{\gamma - 1}{2} M_\infty^2 (u^2 + v^2 - 1)\right]^{1/(\gamma-1)} \tag{6.189}$$

In this formulation, the velocity components and the density are nondimensionalized by the free stream values. We wish to solve Eqs. (6.188) and (6.189) subject to the surface tangency condition, written as

$$\frac{\partial \phi}{\partial n} = 0 \tag{6.190}$$

and appropriate boundary conditions at infinity.

When the full potential equation is solved, care must be exercised in correctly treating the spatial derivative terms, as was the case for the spatial derivatives for the Euler equations. Since the potential equation eliminates entropy changes, both expansion shocks and compression shocks are valid solutions, and the expansion shocks must be eliminated. In cases where this possibility exists, the addition of dissipation through upwinding is the most obvious choice. Artificial viscosity may be added either by explicit means or through the more widely used method of upwinding.

The evolution of methods for solving the potential equation is worth reviewing. Murman and Cole (1971), in a landmark paper treating transonic flow, pointed out that derivatives at each mesh point in the domain of interest must be correctly treated using type-dependent differencing. They were particularly interested in solving the transonic small-disturbance equation, but the same idea is applicable to the full potential equation. To illustrate type-dependent differencing used by Murman and Cole, consider the nonconservative equation [Eq. (6.185)]. This equation is hyperbolic at points where

$$\frac{u^2 + v^2}{a^2} - 1 > 0$$

and elliptic at points where

$$\frac{u^2 + v^2}{a^2} - 1 < 0$$

Consider the case when the flow is aligned with the x direction. If the flow is subsonic, the equation is elliptic, and central differences are used for the derivatives. If the flow is supersonic, the equation is hyperbolic at the point of interest, and the streamwise second derivative is retarded in the upstream direction. The expressions for the finite-difference representation of the second derivatives at point (i, j) become

$$\begin{aligned}
\phi_{xx} &= \frac{\phi_{i,j} - 2\phi_{i-1,j} + \phi_{i-2,j}}{(\Delta x)^2} \\
\phi_{xy} &= \frac{\phi_{i,j+1} - \phi_{i,j-1} - \phi_{i-1,j+1} + \phi_{i-1,j-1}}{2\,\Delta x\,\Delta y} \\
\phi_{yy} &= \frac{\phi_{i,j+1} - 2\phi_{i,j} + \phi_{i,j-1}}{(\Delta y)^2}
\end{aligned} \tag{6.191}$$

The grid points used for both supersonic and subsonic points are shown in Fig. 6.22.

The structure of the point clusters shown in Fig. 6.22 illustrates the correct type dependence for either supersonic or subsonic points. The location of the points used in the finite-difference representation of the steady potential equation shows that it is desirable to use an implicit scheme to compute solutions. If we consider only supersonic flow so that no elliptic points exist in the field, a solution can be obtained using an explicit formulation. This is not advisable if the flow is only slightly supersonic at some field points because the CFL stability criterion prohibits reasonable step sizes. In that case an explicit solution even for purely supersonic flow becomes impractical.

If we examine the truncation error in the finite-difference representation of ϕ_{xx} at hyperbolic points, we find the leading terms to be of the form

$$\Delta x(u^2 - a^2)\phi_{xxx} \tag{6.192}$$

This provides a positive artificial viscosity at all points where $u^2 > a^2$. If the differencing given in Eq. (6.191) is used at an elliptic point, the artificial viscosity becomes negative and a stability problem results. Jameson (1974) pointed out that difficulty arises in those cases where the flow is supersonic and the x component (u) of the velocity is less than the speed of sound. The problem can be understood by considering a case where the flow is not aligned with the x direction, as shown in Fig. 6.23. The proper domain of dependence for all points

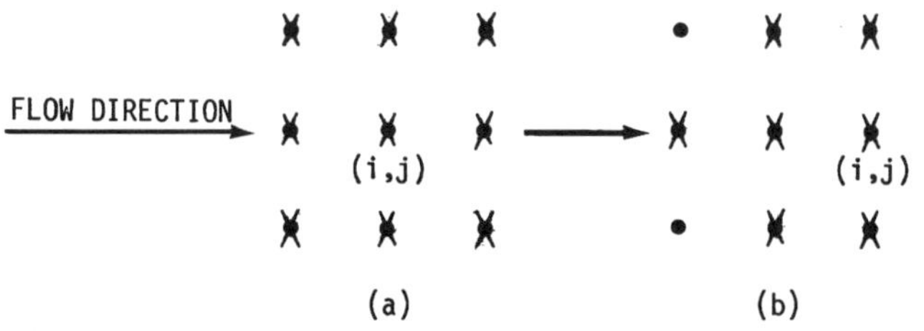

Figure 6.22 Type-dependent differencing. (a) Elliptic point; (b) hyperbolic point.

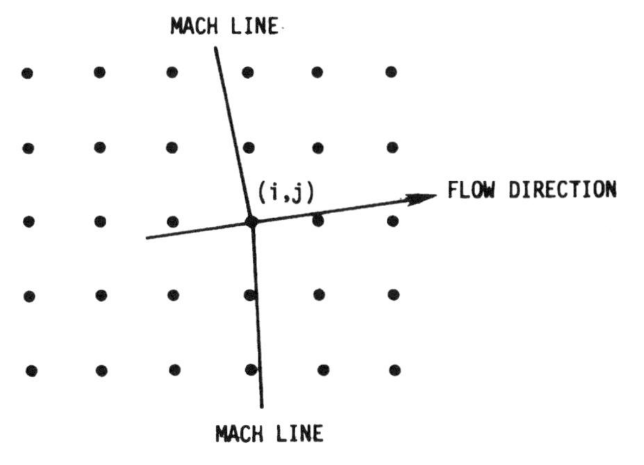

Figure 6.23 Flow with nonaligned mesh system.

is not included. One of the y coordinates of a point in the finite-difference molecule lies behind one of the characteristics passing through the point $(i\,\Delta x, j\Delta y)$. In order to remedy this problem, Jameson introduced his well-known rotated difference scheme. The idea is to write the potential equation in natural coordinates as

$$(a^2 - V^2)\phi_{ss} + a^2\phi_{nn} = 0 \tag{6.193}$$

where s and n are distances along and normal to the streamlines. By applying the chain rule for partial derivatives, the second derivatives may be written in terms of x and y as

$$\begin{aligned}\phi_{ss} &= \frac{1}{V^2}\left(u^2\phi_{xx} + 2uv\phi_{xy} + v^2\phi_{yy}\right)\\ \phi_{nn} &= \frac{1}{V^2}\left(v^2\phi_{xx} - 2uv\phi_{xy} + u^2\phi_{yy}\right)\end{aligned} \tag{6.194}$$

Both x and y derivative contributions to ϕ_{ss} are lagged or retarded, while central differences are used for the ϕ_{nn} term. When the flow is aligned with the grid, the rotated scheme reduces to that given in Eq. (6.191) and produces an artificial viscosity with leading term of the form

$$\left(1 - \frac{a^2}{V^2}\right)(\Delta s\, u^2\phi_{sss} + \cdots) \tag{6.195}$$

This provides us with a positive artificial viscosity for all points where the flow is supersonic, and we expect shock waves to form only as compressions. While the concept of artificial viscosity is used as a means of explaining the behavior of the solutions of the full potential equation, it should be understood that the same conclusions regarding proper treatment of the various terms can be reached by a careful analysis of the finite-difference equations.

Hafez et al. (1979) applied the idea of artificial compressibility in transonic flows in order to provide artificial viscosity in supersonic regions. This concept was originally introduced by Harten (1978) in attempting to devise better methods of shock capturing in supersonic flows. Holst and Ballhaus (1979) and Holst (1979) used an upwind density bias to provide the necessary artificial viscosity. The method presented below incorporates these ideas and is very useful for solving the full potential equation.

To understand the role of density biasing in providing an artificial viscosity, it is instructive to consider the 1-D form of the potential equation:

$$\frac{\partial}{\partial x}\left(\rho\frac{\partial\phi}{\partial x}\right) = 0 \tag{6.196}$$

This expression may be approximated to second order by writing

$$\nabla(\rho_{i+1/2}\,\Delta\phi_i) = 0 \tag{6.197}$$

where the notation is as previously defined. For elliptic points, Eq. (6.197) is satisfactory. For hyperbolic points, an artificial viscosity must be added such as that used by Jameson (1975):

$$-\Delta x(\mu\phi_{xx})_x \tag{6.198}$$

where

$$\mu = \min\begin{cases} 0 \\ \rho\left(1 - \dfrac{\phi_x^2}{a^2}\right)\end{cases} \tag{6.199}$$

As previously noted, this explicit addition of artificial viscosity is equivalent to the type-dependent differencing introduced by Murman and Cole (1971). Jameson (1975) has shown that Eq. (6.198) is equivalent to a term with the form

$$-\Delta x(\nu\rho_x\phi_x)_x \tag{6.200}$$

where

$$\nu = \max\begin{cases} 0 \\ 1 - \dfrac{a^2}{\phi_x^2}\end{cases} \tag{6.201}$$

This form is obtained by differentiation of the 1-D form of the energy equation. If this artificial viscosity form is incorporated into the potential equation, the finite-difference approximation to Eq. (6.196) becomes

$$\frac{\partial}{\partial x}\left(\rho\frac{\partial\phi}{\partial x}\right) \approx \nabla[\rho_{i+1/2}\,\Delta\phi_i] - \nabla\left[\nu_i(\rho_{i+1/2} - \rho_{i-1/2})\,\Delta\phi_i\right] = 0 \tag{6.202}$$

This expression, due to Holst and Ballhaus (1979), is second-order accurate and centrally differenced in subsonic regions. In supersonic regions this is a first-order upwind scheme due to the addition of the artificial viscosity. The differencing becomes more strongly biased in the upwind direction as the Mach number increases. In subsonic regions the density biasing is switched off.

The difference expression given by Eq. (6.202) can also be written

$$\frac{\partial}{\partial x}\left(\rho\frac{\partial\phi}{\partial x}\right) \approx \nabla(\tilde{\rho}_{i+1/2}\,\Delta\phi_i) = 0 \tag{6.203}$$

if the new density is identified by

$$\tilde{\rho}_{i+\frac{1}{2}} = (1 - \nu_i)\rho_{i+\frac{1}{2}} + \nu_i\,\rho_{i-\frac{1}{2}} \tag{6.204}$$

where the values at the cell midpoint are obtained from the energy equation [Eq. (6.189)]. In this expression for $\rho_{i+\frac{1}{2}}$, only u appears and is evaluated as $(\phi_{i+1} - \phi_i)/\Delta x$. Equations (6.203) and (6.204) show that the effect of adding artificial viscosity is equivalent to using a retarded density. In Jameson's (1975) method, the artificial viscosity is explicitly added, while in the scheme outlined here, the artificial viscosity is included in the treatment of the density. If the artificial viscosity ν is chosen as given in Eq. (6.199), the two techniques give identical results. If $\nu = 0$, the scheme is valid only in elliptic regions and is unstable for supersonic flow. However, if ν is set equal to a positive constant, the scheme can be used for both subsonic and supersonic flows. It should be noted that the resulting method is first order and highly dissipative when ν is set equal to a constant.

Other techniques for including the upwind or density-biasing effect have been developed that are more accurate. Shankar et al. (1985) used a streamwise flux-biasing approach that has proven to be effective. The value of $\tilde{\rho}_{i+\frac{1}{2}}$ is written in terms of mass flux values in the streamwise direction for the 1-D case as

$$\tilde{\rho} = \frac{1}{u}\left[\rho u \pm \Delta x\frac{\partial}{\partial x}(\rho u)^{-}\right] \tag{6.205}$$

where the negative sign is used with a backward difference and the positive with a forward difference. The term $(\rho u)^{-}$ is defined as

$$\begin{aligned} (\rho u)^{-} &= \rho u - \rho^* u^* \qquad & u > u^* \\ (\rho u)^{-} &= 0 \qquad & u \leqslant u^* \end{aligned} \tag{6.206}$$

and the starred quantities represent the sonic values of the density and the velocity. For steady flow these values are constant and generally depend only on the free stream Mach number. If the flow is unsteady, the values are computed at all points due to the unsteady behavior of the flow. In order to use flux biasing, four different cases must be evaluated, and these will be detailed below.

In solving the potential equation, methods for computing solutions of the steady equations using relaxation are popular. When relaxation methods are used, the behavior of the equations switches whenever sonic lines are encountered. If the time-dependent form of the governing equations is used for the general case, the solution procedure is valid for either steady or unsteady flow. However, the special treatment of the density still needs to be carried during the solution process.

The unsteady potential equation may be written

$$\left[\frac{\rho}{J}\right]_\tau + \left[\rho\frac{U}{J}\right]_\xi + \left[\rho\frac{V}{J}\right]_\eta = 0 \tag{6.207}$$

where U, V are the contravariant velocities given by

$$\begin{aligned} U &= \xi_t + a_{11}\phi_\xi + a_{12}\phi_\eta \\ V &= \eta_t + a_{21}\phi_\xi + a_{22}\phi_\eta \end{aligned} \tag{6.208}$$

with the usual definitions for the metric coefficients,

$$\begin{aligned} a_{11} &= \xi_x^2 + \xi_y^2 \qquad & a_{12} &= \xi_x\eta_y + \xi_y\eta_x \\ a_{21} &= \xi_x\eta_y + \xi_y\eta_x \qquad & a_{22} &= \eta_x^2 + \eta_y^2 \end{aligned} \tag{6.209}$$

and J is the Jacobian. In the unsteady formulation the density is given by

$$\rho = \left\{1 - \frac{\gamma - 1}{2}M_\infty^2\left[2\phi_\tau + (U + \xi_\tau)\phi_\xi + (v + \eta_\tau)\phi_\eta - 1\right]\right\}^{1/(\gamma-1)} \tag{6.210}$$

We seek a solution of Eq. (6.207), and any scheme that provides the desired accuracy may be used. Shankar et al. (1985) used a Newton method to solve this equation, and we present the basic idea of their approach.

The conservative form of the continuity equation may be written as a function of the velocity potential in the form

$$F[\phi] = 0 \tag{6.211}$$

In this expression the value of ϕ is the unknown at each mesh point at the $n + 1$ time level. The standard Newton iteration scheme for computing this value of ϕ is

$$F[\phi_*] + \left(\frac{\partial F}{\partial \phi}\right)_{\phi_*}(\phi - \phi_*) = 0 \tag{6.212}$$

The asterisk denotes the iteration value for ϕ. That is, the iteration procedes by starting with an assumed value of ϕ at the $n + 1$ level. Initially, this value of ϕ is assigned to ϕ_*. After the first Newton iteration, the new values of ϕ that result are then assigned to the ϕ_* array. In this manner, the iteration continues until ϕ approaches ϕ_* within the desired accuracy.

The solution procedure begins with a specific treatment of the time- and spatial-derivative terms. The details of this treatment are outlined below.

Treatment of the time derivatives. The time derivative may be formed in a number of ways. In order to provide flexibility in selecting the temporal accuracy, the term

$$\frac{\partial}{\partial \tau}\left[\frac{\rho}{J}\right]$$

may be written in the following form:

$$\frac{(a_1 - \theta b_1)\left[(\rho/J)^{n+1} - (\rho/J)^n\right] - \theta b_1\left[(\rho/J)^n - (\rho/J)^{n-1}\right]}{D_1} \tag{6.213}$$

where the denominator is defined as

$$D_1 = a_1\,\Delta\tau_1 - \theta b_1(\Delta\tau_1 + \Delta\tau_2) \tag{6.214}$$

and

$$\begin{aligned} a_1 &= (\Delta\tau_1 + \Delta\tau_2)^2 \\ b_1 &= \Delta\tau_1^2 \\ \Delta\tau_1 &= \tau^{n+1} - \tau^n \\ \Delta\tau_2 &= \tau^n - \tau^{n-1} \end{aligned} \tag{6.215}$$

The time accuracy of the method is controlled by θ, where a value of zero corresponds to first-order accuracy and one provides a second-order accurate scheme.

The unknown shown in the time discretization is the density. However, we write the density in terms of ϕ following the original Newton iteration procedure. Thus we write the density in the following form:

$$\rho(\phi_* + \Delta\phi) = \rho(\phi_*) + \Delta\rho \tag{6.216}$$

where

$$\Delta\rho = \frac{\partial\rho}{\partial\phi}\,\Delta\phi \tag{6.217}$$

and

$$\Delta\phi = \phi - \phi_* \tag{6.218}$$

The density derivative is evaluated by writing

$$\Delta\rho = \frac{\partial\rho}{\partial\phi_t}\,\Delta\phi_t + \frac{\partial\rho}{\partial\phi_\xi}\,\Delta\phi_\xi + \frac{\partial\rho}{\partial\phi_\eta}\,\Delta\phi_\eta \tag{6.219}$$

With the time derivative approximated by $1/\Delta\tau$, we obtain the density change by differentiation, resulting in the expression for $\Delta\rho$:

$$\Delta\rho = \left[-\frac{\rho}{a^2}\left(\frac{1}{\Delta\tau_1} + U\frac{\partial}{\partial\xi} + V\frac{\partial}{\partial\eta}\right)\right]_{\phi_*}\Delta\phi \tag{6.220}$$

Spatial derivatives. The density appearing in the spatial derivatives is treated in a manner similar to that outlined above. Consistent with the Newton iteration procedure for ϕ, the spatial derivatives are written as

$$\frac{\partial}{\partial \xi}\left(\rho \frac{U}{J}\right) = \frac{\partial}{\partial \xi}\left(f + \frac{\partial f}{\partial \phi}\Delta\phi\right) \tag{6.221}$$

where

$$f = \left(\rho \frac{U}{J}\right) \tag{6.222}$$

with

$$\frac{\partial f}{\partial \phi} = \frac{1}{J}\left(\rho \frac{\partial U}{\partial \phi} + U \frac{\partial \rho}{\partial \phi}\right) \tag{6.223}$$

The upwinding that must be used when the quantity $(a_{11} - U^2/a^2)$ is negative will create a pentadiagonal matrix. In the interest of computational efficiency, the term $U(\partial\rho/\partial\phi)$ is sometimes neglected because a tridiagonal form is recovered. Since the change in ϕ goes to zero when the solution converges, this should not produce errors. The final form assumed for the spatial derivative becomes

$$\frac{\partial}{\partial \xi}\left(\frac{\tilde{\rho}U}{J}\right) = \left(\tilde{\rho}\frac{U}{J}\right)_{i+\frac{1}{2},j} - \left(\tilde{\rho}\frac{U}{J}\right)_{i-\frac{1}{2},j} \tag{6.224}$$

where the expanded terms may be written

$$\left(\tilde{\rho}\frac{U}{J}\right)_{i+\frac{1}{2},j} = \left\{\frac{\tilde{\rho}}{J}(\xi_t + a_{11}[\phi_* + \Delta\phi]_\xi + a_{12}[\phi_* + \Delta\phi]_\eta)\right\}_{i+\frac{1}{2},j} \tag{6.225}$$

The density $\tilde{\rho}$ is given by $\tilde{\rho}(\phi_*)$, where ϕ_* is the initial guess in the Newton iteration.

In flows with shocks or where $M > 1$, artificial viscosity is added to the scheme by density biasing. This may be accomplished in a number of different ways. The density may be biased strictly in the coordinate direction, so that

$$\tilde{\rho}_{i+\frac{1}{2}} = \rho_{i+\frac{1}{2}} \pm \nu\Delta\xi\left(\frac{\partial\rho}{\partial\xi}\right)_{i+\frac{1}{2}} \tag{6.226}$$

where the coefficient ν takes the usual form,

$$\nu = \max\left(0, 1 - \frac{1}{M^2}\right)_{i+\frac{1}{2}} \tag{6.227}$$

For values of U that are positive, the negative sign and backward differencing are used, while the positive sign and forward differencing are used when U is negative.

Directional flux biasing can also be employed successfully and consists of writing the density in terms of the weighted mass flux. This is written in the form

$$\tilde{\rho} = \frac{1}{q}\left(\rho q \pm \Delta\xi\frac{\partial}{\partial\xi}(\rho q)^{-}\right) \tag{6.228}$$

The streamwise biasing approach weights the density by using the streamwise mass flux and is written

$$\tilde{\rho} = \frac{1}{q}\left(\rho q \pm \Delta s\frac{\partial}{\partial s}(\rho q)^{-}\right) \tag{6.229}$$

where s is the local streamwise coordinate and q represents the speed. This may be written in a form consistent with the previous notation as

$$\tilde{\rho} = \frac{1}{q}\left[\rho q \pm \left(\frac{U}{Q}\Delta\xi\frac{\partial}{\partial\xi} + \frac{V}{Q}\Delta\eta\frac{\partial}{\partial\eta}\right)(\rho q)^{-}\right] \tag{6.230}$$

where

$$Q = (U^2 + V^2)^{\frac{1}{2}} \tag{6.231}$$

with $(\rho q)^{-}$ defined as

$$\begin{aligned} (\rho u)^{-} &= \rho u - \rho^* u^* \qquad & u > u^* \\ (\rho u)^{-} &= 0 & u \leqslant u^* \end{aligned} \tag{6.232}$$

and the starred values represent sonic conditions. These sonic conditions are given by

$$(q^*)^2 = \frac{2}{M_\infty^2(\gamma+1)}\left[1 + \frac{\gamma-1}{2}M_\infty^2(1 - 2\phi_\tau - 2\xi_\tau\phi_\xi - 2\eta_\tau\phi_\eta)\right] \tag{6.223}$$

$$\rho^* = (q^* M_\infty)^{2/(\gamma-1)} \tag{6.234}$$

There are four cases that must be considered in the biasing of the density.

1. Subsonic flow. In the case of subsonic flow the velocity is less than the speed of sound at both points under consideration. For $q < q^*$ at $(i + \frac{1}{2}, j)$ and $(i - \frac{1}{2}, j)$, the density term becomes

$$\tilde{\rho}_{i+\frac{1}{2},j} = \frac{1}{q_{i+\frac{1}{2},j}}\left\{(\rho q)_{i+\frac{1}{2},j} - \left[(\rho q)^{-}_{i+\frac{1}{2},j} - (\rho q)^{-}_{i-\frac{1}{2},j}\right]\right\} \tag{6.235}$$

or

$$\tilde{\rho}_{i+\frac{1}{2},j} = \rho_{i+\frac{1}{2},j} \tag{6.236}$$

2. Supersonic flow. For the case of supersonic flow, the velocity is fully supersonic at both mesh half intervals, $(i + \frac{1}{2}, j), (i - \frac{1}{2}, j)$:

$$q > q^*$$

For $U > 0$, the density becomes

$$\tilde{\rho}_{i+\frac{1}{2},j} = \frac{1}{q_{i+\frac{1}{2},j}}\left\{(\rho q)_{i+\frac{1}{2},j} - \left[(\rho q - \rho^* q^*)_{i+\frac{1}{2},j} - (\rho q - \rho^* q^*)_{i-\frac{1}{2},j}\right]\right\} \tag{6.237}$$

When the flow is steady and supersonic, the value of the density is simplified considerably to

$$\tilde{\rho}_{i+\frac{1}{2},j} = \rho_{i-\frac{1}{2},j}\left(\frac{q_{i-\frac{1}{2},j}}{q_{i+\frac{1}{2},j}}\right) \tag{6.238}$$

3. Transition through a sonic point. Figure 6.24 shows a schematic of transition through a sonic point region along with the shock transition point. For the sonic transition, $q > q^*$ at $(i + \frac{1}{2}, j)$ and $q < q^*$ at $(i - \frac{1}{2}, j)$ for $U > 0$. The density is written

$$\tilde{\rho}_{i+\frac{1}{2},j} = \frac{1}{q_{i+\frac{1}{2},j}}\left\{(\rho q)^{-}_{i+\frac{1}{2},j} - \left[(\rho q - \rho^* q^*)_{i+\frac{1}{2},j} - (\rho q)^{-}_{i-\frac{1}{2},j}\right]\right\} \tag{6.239}$$

This may be more simply written as

$$\tilde{\rho}_{i+\frac{1}{2},j} = \frac{\rho^* q^*}{q_{i+\frac{1}{2},j}} \tag{6.240}$$

4. Transition through a shock. In the case of transition through a shock, $q > q^*$ at $(i - \frac{1}{2}, j)$ and $q < q^*$ at $(i + \frac{1}{2}, j)$. For $U > 0$,

$$\tilde{\rho}_{i+\frac{1}{2},j} = \frac{1}{q_{i+\frac{1}{2},j}}\left\{(\rho q)_{i+\frac{1}{2},j} - \left[(\rho q)^{-}_{i+\frac{1}{2},j} - (\rho q - \rho^* q^*)_{i-\frac{1}{2},j}\right]\right\} \tag{6.241}$$

or in a simpler form,

$$\tilde{\rho}_{i+\frac{1}{2},j} = \rho_{i+\frac{1}{2},j} + \frac{1}{q_{i+\frac{1}{2},j}}(\rho q - \rho^* q^*)_{i-\frac{1}{2},j} \tag{6.242}$$

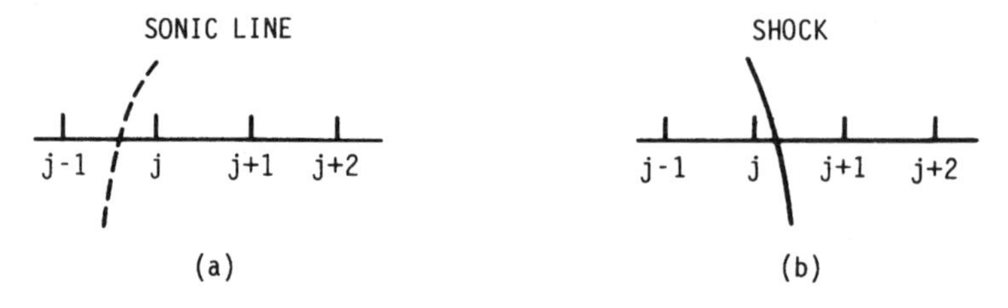

Figure 6.24 Transition. (a) Sonic point. (b) Shock point.

For steady flows it may be shown that flux biasing and the density biasing procedure employed by Holst [1980] give identical results at a purely supersonic point.

The remaining task that must be accomplished to make this full potential formulation applicable to both steady and unsteady problems is to correctly treat the circulation at the boundary of the computational region. In particular, the velocity potential jumps in value across the wake, and an appropriate technique to establish the correct jump is needed. The circulation is defined as

$$\Gamma = \oint \mathbf{V} \cdot d\mathbf{s} \tag{6.243}$$

When circulation is generated in the lifting case, the jump in the velocity potential across the airfoil wake is equal to the circulation, i.e.,

$$\phi_u - \phi_l = \Gamma \tag{6.244}$$

where the u and l subscripts indicate the values taken at the upper and lower sides of the airfoil wake. Kelvin's theorem (Karamcheti, 1966) states that the circulation around a fluid curve remains constant for all time if the curve moves with the fluid. This permits the conservation of circulation to be expressed as

$$\frac{D\Gamma}{Dt} = 0 \tag{6.245}$$

Integrating along the wake cut provides the correct Γ variation relating the values of ϕ across the wake. The circulation is simply convected with the fluid particles, and the most convenient form to use may be written

$$\frac{\partial \Gamma}{\partial t} + U\frac{\partial \Gamma}{\partial \xi} + V\frac{\partial \Gamma}{\partial \eta} = 0 \tag{6.246}$$

This equation may be substantially simplified with the proper choice of coordinates. When this expression is integrated to find the variation in Γ along the wake cut, the recommended practice is to assume that the upper and lower values of ϕ are correct at alternating time steps. This means that the upper and lower values are alternately determined by the solutions of the field equations or the integration of Kelvin's theorem. For the steady flow case, the value of Γ along the wake is set equal to a constant and is simply the jump in ϕ at the trailing edge point.

The unsteady formulation also requires a specification on the normal derivatives of ϕ along the wake cut when the governing equations for ϕ are written. This is usually accomplished by an extrapolation that also includes the calculated value of the circulation. The jump in the second derivative of ϕ across the wake may be written

$$[\phi_{\eta\eta}]_{ul} = -\frac{\rho a_{11}\Gamma_\xi/J}{\rho a_{22}/J} \tag{6.247}$$

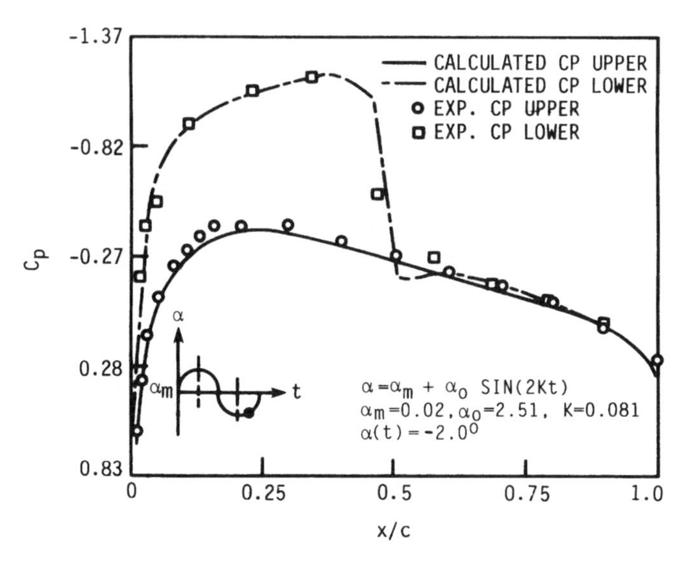

Figure 6.25 Comparison of the unsteady pressure coefficients (Cp) for a NACA 0012 airfoil (Shankar et al., 1985. Copyright © 1985 AIAA. Reprinted with permission).

The pressure is continuous across the wake, and the density is also assumed to be continuous from the lower to the upper side.

In addition to the wake treatment, the body-surface boundary conditions and conditions in the far field require specification. On the body surface the flow must satisfy the surface tangency condition for the steady case, or the more general statement for the unsteady case requires that the normal velocity of the moving surface be equal to the normal component of the fluid velocity. The far field is treated in the same way as the Euler equation boundary conditions, in

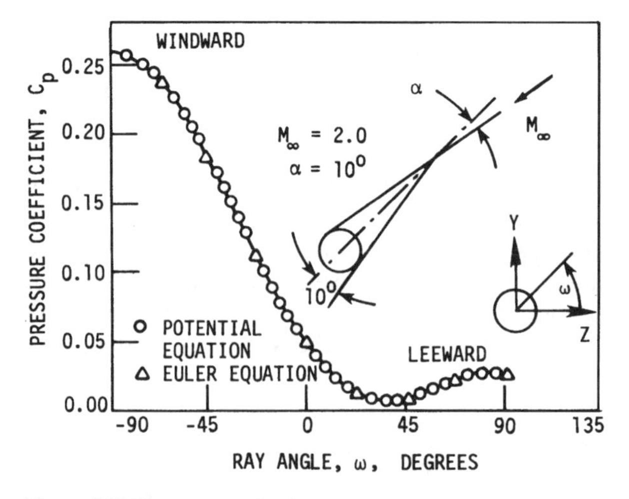

Figure 6.26 Flow over a circular cone.

the sense that the Reimann invariants may be used as outlined in the previous section.

The original unsteady equation (continuity) may be written in the form

$$H(\Delta\phi) - R(\phi^*, \phi^n, \phi^{n-1}, \dots) = 0 \tag{6.248}$$

In this equation the residual is denoted by R and the operator H may represent the particular scheme selected to compute a solution of the system. Approximate factorization, relaxation, or any other suitable method may be chosen. Figure 6.25 shows the unsteady pressure distribution on an oscillating NACA 0012 airfoil compared to experimental data from AGARD R-702. This is a transonic case and is representative of the unsteady results obtained with full potential formulations. Typical results from steady full potential calculations of supersonic flows from Shankar and Chakravarthy (1981) are shown in Figs. 6.26 and 6.27. In both cases, the agreement with solutions using the Euler equations is excellent. In applications where the assumptions inherent in a potential formulation are valid, the calculated solutions for the flow field will provide a quick, accurate result. In areas such as preliminary design, application of full potential codes

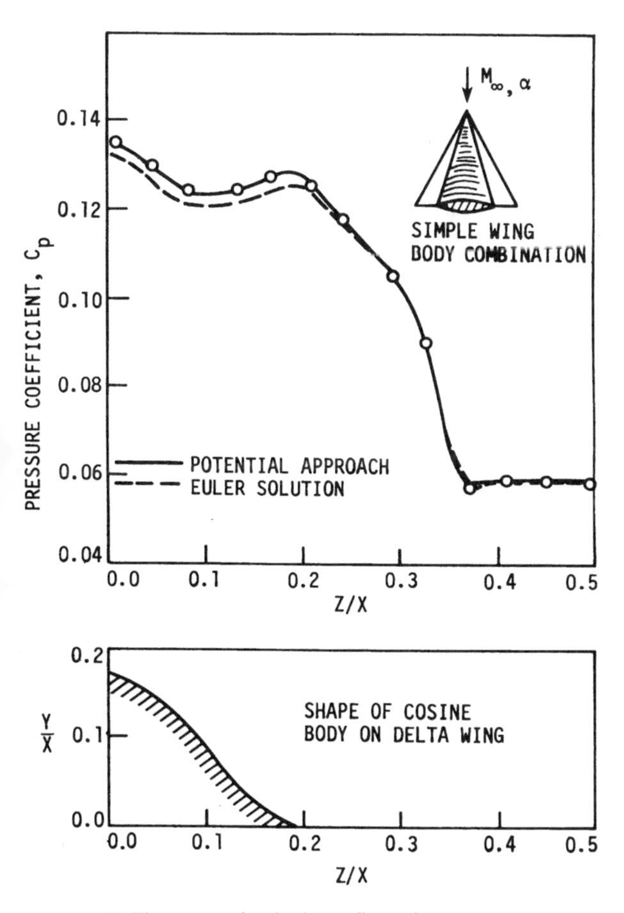

Figure 6.27 Flow over wing-body configuration.

can substantially reduce the computational effort expended in comparison with other approaches.

6.9 TRANSONIC SMALL-DISTURBANCE EQUATIONS

The use of the full potential equation for inviscid transonic flows was discussed in the previous section. Results obtained for airfoils and some 3-D body shapes compare very well with available experimental data. Methods for solving the full potential equation are very efficient and are being used extensively. However, we still find numerous applications where the sophistication provided by the full potential formulation is not required and the accuracy of the solution of the transonic small-disturbance equation is sufficient. In addition, a significant advantage accrues in the application of boundary conditions. Boundary conditions for 2-D problems are applied on the slit in two dimensions or on the plane for 3-D problems. The governing equations are greatly simplified, since complex body-aligned mappings are unnecessary for the application of boundary conditions. This can result in significant reductions in computer time and storage requirements, particularly in 3-D problems.

The transonic small-disturbance equations may be derived by a systematic expansion procedure. The details of this procedure are given by Cole and Messiter (1957) and Hayes (1966) and provide a means of systematically developing higher-order approximations to the Euler equations. In Chapter 5 the transonic small-disturbance equation [Eq. (5.203)] is derived using a perturbation procedure. This may be written in the nondimensional form

$$[K - (\gamma + 1)\phi_x]\phi_{xx} + \phi_{\tilde{y}\tilde{y}} = 0 \tag{6.249}$$

where K is the transonic similarity parameter given by

$$K = \frac{1 - M_\infty^2}{\delta^{2/3}} \tag{6.250}$$

with δ representing the maximum thickness ratio and f the shape function of an airfoil defined by the expression

$$y = \delta f(x) \tag{6.251}$$

The velocity potential used in Eq. (6.249) is the perturbation velocity potential defined in such a way that the x derivative of ϕ is the perturbation velocity in the x direction nondimensionalized with respect to the free stream velocity, and similarly in the y direction. The scaled coordinate $\tilde{y}$ is defined by

$$\tilde{y} = \delta^{1/3} y \tag{6.252}$$

Equation (6.249) is formally equivalent to Eq. (5.203), and both are forms of the Guderley–von Kármán transonic small-disturbance equations. The similarity form given in Eq. (6.249) is the equation originally treated by Murman and Cole (1971) in calculating the inviscid flow over a nonlifting airfoil. The pressure coefficient is the same as Eq. (5.205) and may be written

$$C_p = -2\phi_x$$

For flows that are not considered transonic, we obtain the Prandtl-Glauert equation for subsonic or supersonic flow. This expression has been used in numerous examples in previous chapters and takes the form

$$(1 - M_\infty^2)\phi_{xx} + \phi_{yy} = 0 \tag{6.253}$$

The main point to remember is that the transonic small-disturbance equation is nonlinear and switches from elliptic to hyperbolic in the same manner as the full potential and Euler equations.

In their original paper, Murman and Cole treated the inviscid transonic flow over a nonlifting airfoil and solved the transonic small-disturbance equation as given in Eq. (6.249). In addition to the governing PDE, the necessary body-surface boundary conditions for zero angle of attack are given by

$$\phi_{\tilde{y}}(x, 0) = f'(x) \tag{6.254}$$

applied in the plane $\tilde{y} = 0$ consistent with the theory. A boundary condition must also be applied at the outer boundary of the computational mesh. For this case, in the far field,

$$\phi \simeq \frac{1}{2\pi\sqrt{K}} \frac{Mx}{x^2 + K\tilde{y}^2} \tag{6.255}$$

where

$$M = 2\int_{-1}^{1} f(\xi)\, d\xi + \frac{\gamma + 1}{2} \iint_{-\infty}^{+\infty} d\xi\, d\eta \tag{6.256}$$

and the airfoil is confined to the interval

$$-1 \leqslant x \leqslant 1$$

In the lifting case, the circulation must be imposed and determined by satisfying the Kutta condition on the airfoil. The far-field boundary condition in this case takes the form of a vortex with the value of circulation determined by the Kutta condition. For development of the far-field boundary condition, the papers by Ludford (1951) and Klunker (1971) are recommended.

Murman and Cole solved Eq. (6.249) for a nonlifting transonic airfoil using line relaxation methods. Type-dependent differencing given in Eq. (6.191) was used for hyperbolic regions, and central differencing was used in the elliptic regions. This switched differencing used by Murman and Cole provides a method that is equivalent to a first-order Roe scheme. The airfoil now appears on the x axis as a line or slit, and boundary conditions are applied there. In this case the airfoil lies between two mesh points at the half-mesh interval, as shown

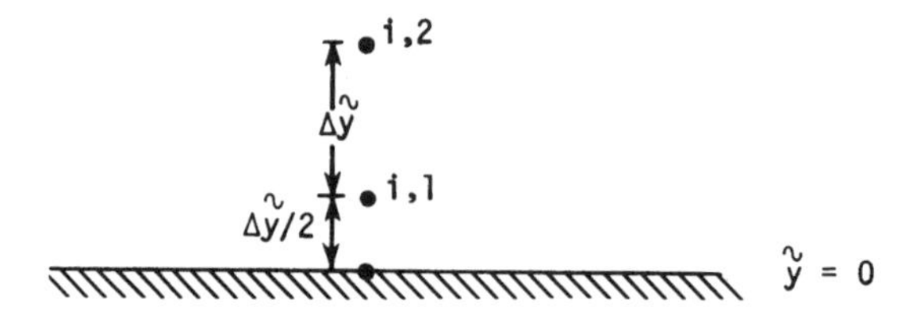

Figure 6.28 Surface boundary point distribution.

in Fig. 6.28. The boundary condition at $\tilde{y} = 0$ enters as a body slope or derivative of ϕ given in Eq. (6.254). At the $(i, 1)$ point, the $\phi_{\tilde{y}\tilde{y}}$ derivative is differenced as

$$\phi_{\tilde{y}\tilde{y}} = \frac{1}{\Delta \tilde{y}}\left(\phi_{\tilde{y}_{i,3/2}} - \phi_{\tilde{y}_{i,1/2}}\right) = \frac{1}{\Delta \tilde{y}}\left[\frac{\phi_{i,2} - \phi_{i,1}}{\Delta \tilde{y}} - \phi_{\tilde{y}}(x, 0)\right] \qquad (6.257)$$

The surface boundary condition explicitly enters the calculation through the $\phi_{\tilde{y}}$ term.

Figure 6.29 shows the pressure distribution for a circular arc airfoil obtained by solving the transonic small-disturbance equation. As can be seen, the experimental data of Knechtel (1959) and the computed results compare favorably for both the subcritical and supercritical cases. It is interesting that the

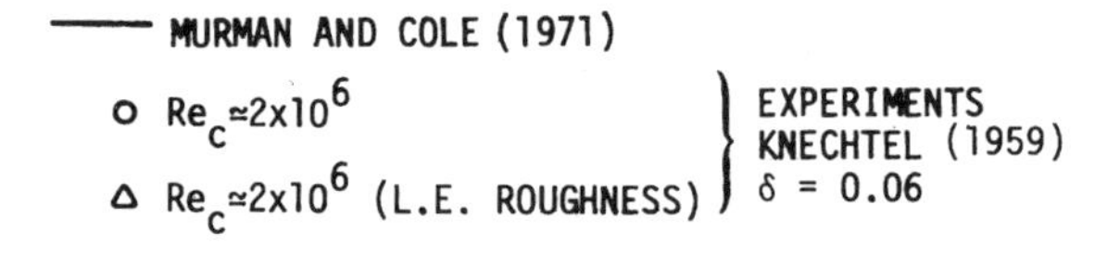

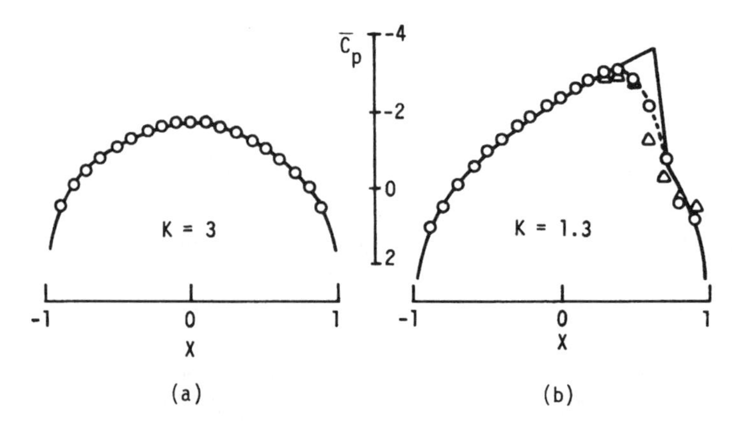

Figure 6.29 Pressure distribution for circular are airfoil. (a) Subcritical case. (b) Supercritical case.

shock location and strength for this example agree well with the experimental measurements. The nonconservative equations of small-disturbance theory for an inviscid flow underestimate shock strength and produce the same effect as a shock–boundary layer interaction on shock strength and location. Thus the nonconservative form has been popular even though the conservative form is mathematically appealing. Numerous applications of the technique, presented by Murman and Cole for solving the transonic small-disturbance equation, have been made since it was originally introduced, and many refinements of the basic method have been developed. However, the main point to remember is that a significant simplification over either the Euler equations or the full potential formulation is realized when this approach is used.

The simplified form of the transonic small-disturbance equation was used in developing solutions for 3-D wings by Bailey and Ballhaus (1972). Their work led to the development of a widely used 3-D code for transonic wing analysis. This code has been used extensively in designing improved wings for flight in the transonic speed regime. For those interested in 3-D transonic flow over wings, the paper by Bailey and Ballhaus is recommended reading.

Most current work in transonic flow is concentrated on developing Euler or Navier-Stokes equation solvers. One area where considerable effort is being expended using the transonic full potential or small-disturbance equations is in the development of design codes. In the inverse design problem the body pressure is prescribed, and the body shape is unknown. For this type of problem a simplified approach offers advantages.

6.10 METHODS FOR SOLVING LAPLACE'S EQUATION

The numerical techniques presented in the previous sections of this chapter were applied to the nonlinear equations governing inviscid fluid flow. Linear PDEs are often used to model both internal and external flows. Examples include Laplace's equation for incompressible inviscid irrotational flow and the Prandtl-Glauert equation, which is valid in compressible flow if the small-perturbation assumptions are satisfied. The methods for solving both of these equations are similar. Finite-difference/finite-volume methods for solving Laplace's equation are presented in Chapter 4 and will not be reviewed here. Instead, the basic idea underlying the use of panel methods will be discussed. These methods have received extensive use in industry.

The advantage of using panel methods is that a solution for the body-pressure distribution can be obtained without solving for the flow field throughout the domain. In this case the problem is reduced to the solution of a system of algebraic equations for source, doublet, or vortex strengths on the boundaries. Using the resulting solution, the body-surface pressures can be computed. Panel schemes require the solution of a large system of algebraic equations. For most practical configurations, the storage and speed capability of modern computers have been sufficient. However, judicious selection of the number of surface panels and correct placement are essential in obtaining a good solution for the body-surface pressure.

In studying panel methods, we will consider the flow of an incompressible inviscid irrotational fluid that is governed by a solution of Laplace's equation written in terms of the velocity potential. We require

$$\nabla^2\phi = 0 \tag{6.258}$$

in the domain of interest and specify either ϕ or $\partial\phi/\partial n$ on the boundary of the domain. For simplicity, we restrict our attention to the 2-D case, although the fully 3-D problem is conceptually the same. The geometry of the problem under consideration is shown in Fig. 6.30. The basic idea underlying all panel methods is to replace the required solution of Laplace's equation in the domain with a surface integral. This method is developed by the application of Green's second identity to the domain of interest. If u and v are two functions with continuous derivatives through second order (class C^{II}), then Green's second identity may be written

$$\iint_A (u\,\nabla^2 v - v\,\nabla^2 u)\,dA = \oint_s (v\,\nabla u - u\,\nabla v)\cdot\mathbf{n}\,ds$$

where $\mathbf{n}$ is the unit normal to the boundary and s is arc length along the boundary. Suppose we choose u to be the potential ϕ and v to be of the form

$$v = \ln(r)$$

where

$$r = \sqrt{(x-\xi)^2 + (y-\eta)^2}$$

We take (ξ, η) as the coordinates of the point P where ϕ is to be determined and (x, y) as the coordinates of the point Q on the boundary where a source is located. In evaluating the integrals in applying Green's identity, we must

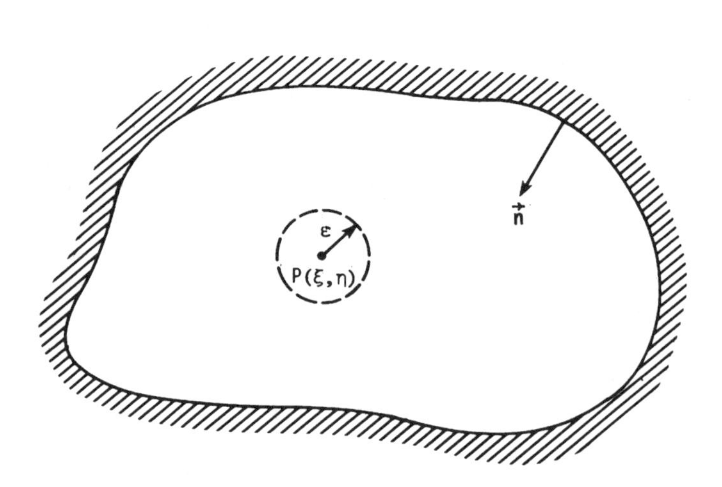

Figure 6.30 Physical domain for Laplace's equation.

exercise caution as (ξ, η) approach (x, y), that is, when $r \to 0$. In order to avoid this difficulty we think of enclosing the point $P(\xi, \eta)$ with a small circle of radius ϵ and apply Green's identity to the region enclosed by the original boundary (B) and that of the small circle enclosing P. Thus

$$0 = \oint_B (v\,\nabla u - u\,\nabla v) \cdot \mathbf{n}\, ds - \oint_\epsilon (v\,\nabla u - u\,\nabla v) \cdot \mathbf{n}\, ds$$

Consider the second integral with u, v replaced as noted above:

$$\oint_\epsilon [\ln(r)\,\nabla\phi - \phi\,\nabla \ln(r)] \cdot \mathbf{n}\, ds$$

On the boundary of the small circle, $r = \epsilon$, and we may write this integral as

$$\ln(\epsilon)\left(\oint \nabla\phi \cdot \mathbf{n}\, ds\right) - \oint \frac{\phi}{r}\, ds$$

By our original hypothesis, ϕ is a solution of Laplace's equation, and therefore the first term must vanish (see Prob. 2.7). The second term may be written

$$\frac{1}{\epsilon}\oint_\epsilon \phi\, ds$$

which, by the mean-value property of harmonic functions, becomes

$$\frac{1}{\epsilon}\oint \phi\, ds = 2\pi\phi(\xi, \eta)$$

We substitute this result into our original expression to obtain

$$\phi(\xi, \eta) = \frac{1}{2\pi}\oint\left[\ln(r)\frac{\partial\phi}{\partial n} - \phi\frac{\partial \ln(r)}{\partial n}\right] ds \tag{6.259}$$

Thus we have reduced the problem of computing a solution to Laplace's equation in the domain to solving an integral equation over the boundary. The first term represents a Neumann problem, where $\partial\phi/\partial n$ is given on the boundary, while the second is an example of the classical Dirichlet boundary value problem, where ϕ is specified. These integrals correspond to contributions to ϕ from sources and doublets. We could write

$$\phi = \frac{1}{2\pi}\oint\left[\mu\frac{\partial \ln(r)}{\partial n} + \sigma \ln(r)\right] ds \tag{6.260}$$

where we interpret σ as a source distribution and μ as a doublet distribution with axis normal to the bounding surface.

A surface source distribution with density σ per unit length produces a potential at an external point given by

$$\phi = \frac{1}{2\pi}\oint \sigma \ln(r)\, ds \tag{6.261}$$

where the integration is taken over the surface. If we have n surfaces or panels, the total potential at a point P is the sum of the contributions from each panel:

$$\phi_i = \sum_{j=1}^{n} \frac{1}{2\pi} \int_j \phi_j \ln(r_{ij})\, ds_j \tag{6.262}$$

A similar expression can be developed for a doublet distribution.

When a uniform stream is superimposed on the domain that includes the source panels, we include the velocity potential of the free stream and write

$$\phi_i = U_\infty x_i + \sum_{j=1}^{n} \frac{1}{2\pi} \int_j \sigma_j \ln(r_{ij})\, ds_j \tag{6.263}$$

The simplest panel representation to treat numerically is obtained when the source strength of each panel is assumed to be constant. Some advanced methods assume other distributions, and the representation of the velocity potential becomes correspondingly more complex. For a constant source strength per panel,

$$\phi_i = U_\infty x_i + \sum_{j=1}^{n} \frac{\sigma_j}{2\pi} \int_j \ln(r_{ij})\, ds_j \tag{6.264}$$

The geometry appropriate to the above potential distribution is shown in Fig. 6.31. The problem in using the source panel representation for a given body is to determine the source strengths σ_j. This is accomplished by selecting a control point on each panel and requiring that no flow cross the panel. The control point is selected at the midpoint of each panel. We now specify the point P to be at the control point of the ith panel. The boundary condition that no flow passes through the panel at this point is

$$\frac{\partial}{\partial n_i} \phi(x_i, y_i) = 0 \tag{6.265}$$

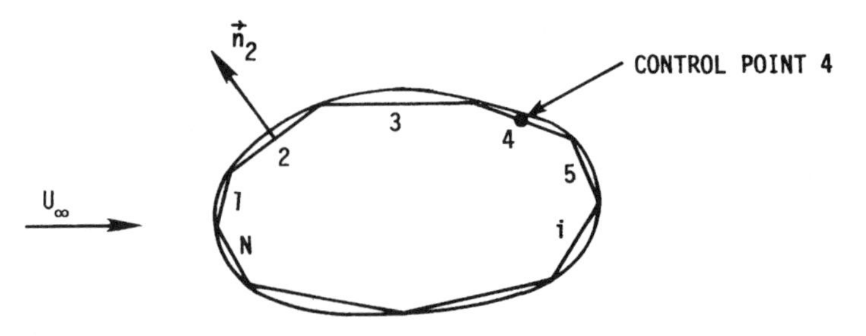

Figure 6.31 Panel representation for general shape.

Since ϕ is the velocity potential, this requires that the normal velocity at the control point of the ith panel vanish. Therefore

$$\sum_{i=1}^{n} \frac{\sigma_j}{2\pi} \int_j \frac{\partial}{\partial n_i} \ln (r_{ij})\, ds_j = -\mathbf{U}_\infty \cdot \mathbf{n}_i \tag{6.266}$$

The dot product is used because the velocity component normal to the surface is required. The velocity induced at the ith control point due to the ith panel is $\sigma_i/2$ and is usually taken out of the above summation. With this convention, we may write

$$\frac{\sigma_i}{2\pi} + \sum_{i \neq j}^{n} \frac{\sigma_j}{2\pi} \int \frac{\partial}{\partial n_i} \ln (r_{ij})\, ds_j = -\mathbf{U}_\infty \cdot \mathbf{n}_i \tag{6.267}$$

Application of this equation to each panel provides n algebraic equations for the n source strengths. Once the σ_j are computed, the pressure coefficients can be determined. When Eq. (6.267) is used to generate the required panel source strengths, the integrand function is most easily developed by using the vector dot product and may be written

$$\frac{\partial \ln (r_{ij})}{\partial n_i} = \mathbf{\nabla}_i \ln (r_{ij}) \cdot \mathbf{n}_i \tag{6.268}$$

An example demonstrating the procedure for generating the required algebraic equations is in order.

Example 6.3 Suppose we wish to solve for the pressure distribution on a cylinder of unit radius in an incompressible flow using the method of source panels. The cylinder is to be represented by eight panels, and the configuration is shown in Fig. 6.32.

Solution In order to determine the surface pressures, we must calculate the panel strengths required for all eight panels on the cylinder. This is done by solving the system of algebraic equations generated by the application of Eq.

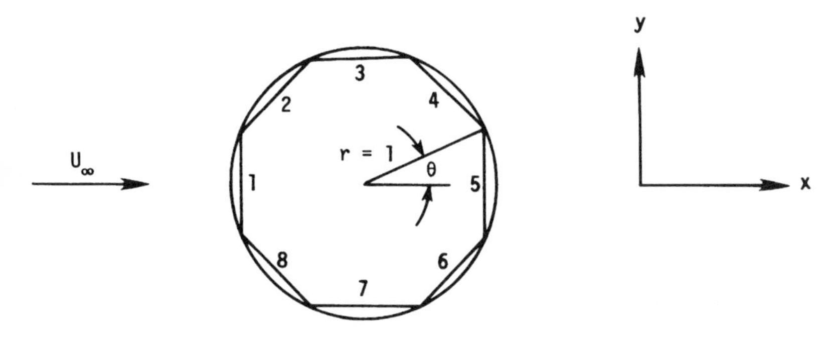

Figure 6.32 Panel representation of cylinder.

(6.267) to each panel. In applying Eq. (6.267) to any panel, the most difficult part is the evaluation of the integral term. In general, it is convenient to view the integral as an influence coefficient and write the system of governing equations in the form

$$[C]\frac{\sigma}{2\pi U_\infty} = -\frac{\mathbf{U}_\infty \cdot \mathbf{n}_i}{U_\infty} \tag{6.269}$$

Using the following notation, we may write the components of $[C]$ as

$$c_{ij} = \begin{cases} \int_j \boldsymbol{\nabla}_i \ln(r_{ij}) \cdot \mathbf{n}_i \, ds_j & i \neq j \\ \pi & i = j \end{cases} \tag{6.270}$$

To demonstrate the application of Eq. (6.270), we elect to compute c_{53}, which represents the normal velocity at the control point of panel 5 due to a constant source strength of magnitude $1/U_\infty$ on panel 3. For this case we write the radius as

$$r_{53} = \left[(x_5 - x_3)^2 + (y_5 - y_3)^2\right]^{1/2}$$

and

$$\boldsymbol{\nabla}_5 \ln(r_{53}) = \frac{(x_5 - x_3)\mathbf{i} + (y_5 - y_3)\mathbf{j}}{(x_5 - x_3)^2 + (y_5 - y_3)^2} \tag{6.271}$$

The unit normal on panel 5 is the unit vector in the positive x direction and

$$\boldsymbol{\nabla}_5 \ln(r_{53}) \cdot \mathbf{n}_5 = \frac{x_5 - x_3}{(x_5 - x_3)^2 + (y_5 - y_3)^2}$$

For this integral, $x_5 = 0.9239$, $y_5 = 0$, and $y_3 = 0.9239$, while x_3 is a variable on panel 3. This reduces the integral required to the form

$$c_{53} = \int_{-0.3827}^{+0.3827} \frac{0.9239 - x}{x^2 - 1.848x + 1.707} \, dx = 0.4018$$

In this expression the arc length along panel 3 is equal to $x - 0.3827$; therefore $ds_3 = dx$, and we use the x coordinates of the panel end points as the integration limits. Notice that the integration proceeds clockwise around the cylinder, which is the positive sense for the domain where a solution of Laplace's equation is required. The $[C]$ matrix is symmetric, i.e., $c_{ij} = c_{ji}$, and the solution for the σ_i must be such that

$$\sum_{i=1}^{n} \sigma_i = 0$$

This requirement is an obvious result of the requirement that we have a closed body.

A comparison of the eight-panel solution with the analytically derived pressure coefficient is shown in Fig. 6.33. Clearly, the panel scheme provides a very accurate numerical solution for this case.

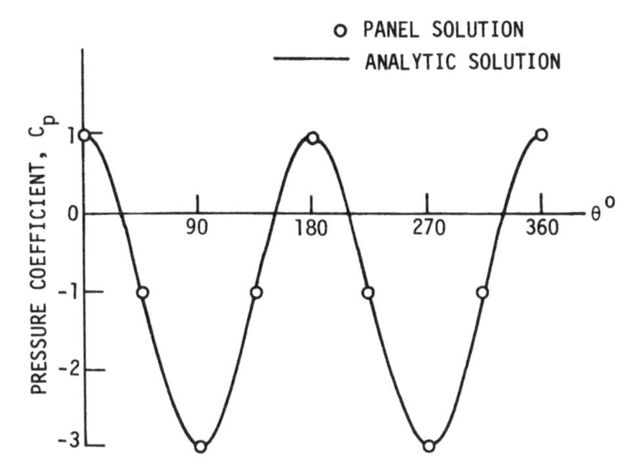

Figure 6.33 Pressure coefficient for a circular cylinder.

We have used the method of source panels in our example to demonstrate the mechanics of applying the technique. We could use doublets or dipoles to construct bodies as well as vortex panels. Clearly, we must include circulation if we are concerned with lifting airfoils. This may be done in a number of ways, but one technique is to use a vortex panel distribution along the mean camber line to provide circulation and satisfy the Kutta condition at a control point just aft of the trailing edge.

Panel methods represent a powerful approach for solving certain classes of flow problems. They have received extensive use and have resulted in a number of standard codes that are used industrywide. For more details on the development of these schemes, the paper by Hess and Smith (1967) provides basic details, while the papers by Rubbert and Saaris (1972) and Johnson and Rubbert (1975) present more advanced ideas.

PROBLEMS

6.1 In Example 6.1, we used a characteristic method to solve for supersonic flow over a wavy wall. Verify the velocity field obtained by solving the Prandtl-Glauert equation [Eq. (6.1)] directly.

6.2 Derive the differential equations of the characteristics of the nonlinear system of equations governing 2-D supersonic flow written in rectangular coordinates [Eq. (6.18)].

6.3 The differential equations of the characteristics obtained in Prob. 6.2 are written in rectangular coordinates. Transform these results using the streamline angle θ, and show that the characteristics are inclined at the local Mach angle, i.e.,

$$\tan(\theta \pm \mu) = \frac{dy}{dx}$$

6.4 Develop the compatibility equations for the nonlinear equations of Prob. 6.2

6.5 Use the results of Probs. 6.2 and 6.4 and solve Example 6.1 using the method of characteristics for the nonlinear equations.

6.6 Complete the derivation in Prob. 6.4 by computing $[T]$. The governing equations are then ready to be numerically integrated using an appropriate upwind scheme.
6.7 The 1-D unsteady Euler equations are given by

$$\mathbf{U}_t + [A]\mathbf{U}_x = 0$$

where

$$\mathbf{U} = [\rho, u, p]^T$$

and

$$[A] = \begin{bmatrix} u & \rho & 0 \\ 0 & u & 1/\rho \\ 0 & \rho a^2 & u \end{bmatrix}$$

Find the following:
- (a) eigenvalues
- (b) characteristics
- (c) left eigenvectors
- (d) $[T]^{-1}$
- (e) compatibility equations

6.8 Develop a code to solve for the supersonic flow over the 2-D wedge of Example 6.2. Use a shock-capturing approach and MacCormack's method in solving the steady flow equations. Use 26 grid points with the outer boundary located at an angle of 40° with respect to the wedge surface. Nondimensionalize the governing equations using the procedure given in Section 5.1.7. Let $V_\infty = 1$, $\rho_\infty = 1$, and $p_\infty = 1/(\gamma M_\infty^2)$. Compute this case by integrating from $\xi \cong 0$ to a relatively large value of ξ in order to asymptotically obtain the converged solution. Use the maximum allowable $\Delta\xi$, which can be determined by performing numerical experiments. Compare your numerical solution with the exact solution.
6.9 Solve the wedge-flow problem of Prob. 6.8 by using the unsteady (time-dependent) approach and conical flow. Use the maximum allowable Δt, which can be determined by performing numerical experiments. Compare your numerical solution with the exact solution.
6.10 Develop a code to solve the 2-D supersonic wedge problem of Prob. 6.8, but fit the shock wave as a discontinuity. Use either the conservative or nonconservative form.
6.11 Suppose that a solid boundary lies on a ray (θ = const) in a 2-D flow problem. Use reflection to establish a suitable means of determining the flow variables of the sublayer points. Use rectangular velocity components.
6.12 Develop the appropriate boundary condition procedure using Kentzer's method (see Section 6.7) for the supersonic wedge-flow problem.
6.13 The flux-vector splitting method of Steger and Warming "splits" the system of equations

$$\mathbf{U}_t + \mathbf{E}_x = 0$$

into the following form:

$$\mathbf{U}_t + \mathbf{E}_x^+ + \mathbf{E}_x^- = 0$$

If this method is applied to the system of equations

$$\mathbf{U} = \begin{bmatrix} u \\ v \end{bmatrix} \qquad \mathbf{E} = \begin{bmatrix} cv \\ cu \end{bmatrix}$$

where c is a constant, find the following quantities:
- (a) $[A]$
- (b) $[\lambda^+], [\lambda^-]$
- (c) $[T]^{-1}, [T]$
- (d) $[A^+], [A^-]$
- (e) $\mathbf{E}^+, \mathbf{E}^-$

6.14 Repeat Prob. 6.13 for the 1-D unsteady Euler equations given in Prob. 6.7 Assume that $(0 \ll u \ll a)$.
6.15 The split-coefficient matrix method (Chakravarthy, 1979) "splits" the system of equations

$$\mathbf{U}_t + [A]\mathbf{U}_x = 0$$

into the following nonconservative form:

$$\mathbf{U}_t + [A^+]\mathbf{U}_x + [A^-]\mathbf{U}_x = 0$$

where

$$[A^+] = [T][\lambda^+][T]^{-1}$$

$$[A^-] = [T][\lambda^-][T]^{-1}$$

If this method is applied to the system of equations

$$\mathbf{U} = \begin{bmatrix} u \\ v \end{bmatrix} \qquad [A] = \begin{bmatrix} 0 & c \\ c & 0 \end{bmatrix}$$

where c is a constant, find the following quantities:

(*a*) $[\lambda^+], [\lambda^-]$
(*b*) $[T], [T]^{-1}$
(*c*) $[A^+], [A^-]$

6.16 Repeat Prob. 6.15 for the 1-D unsteady Euler equations given in Prob. 6.7. Assume that $0 < u < a$.

6.17 In the CSCM (Conservative Supra-Characteristics Method) flux-difference splitting scheme (Lombard et al., 1983), the system of equations

$$\mathbf{U}_t + \mathbf{E}_x = 0$$

is "split" into the following form:

$$\mathbf{U}_t + [Q^+]\mathbf{E}_x + [Q^-]\mathbf{E}_x = 0$$

where

$$[Q^+] = [T][\lambda^+][\lambda]^{-1}[T]^{-1}$$

$$[Q^-] = [T][\lambda^-][\lambda]^{-1}[T]^{-1}$$

If this method is applied to the system of equations given in Prob. 6.13, find the following quantities:

(*a*) $[\lambda], [\lambda^+], [\lambda^-], [\lambda]^{-1}$
(*b*) $[T], [T]^{-1}$
(*c*) $[Q^+], [Q^-]$

6.18 Repeat Prob. 6.17 for the 1-D unsteady Euler equations given in Prob. 6.7. Assume that $0 < u < a$.

6.19 Apply Roe's scheme to the system of equations given in Prob. 6.13 and find

$$[|A|] = [T][|\Lambda|][T]^{-1}$$

6.20 Apply Roe's scheme to the 1-D unsteady Euler equation given in Prob. 6.7 and find

$$[|A|] = [T][|\Lambda|][T]^{-1}$$

if $0 < u < a$.

6.21 Compute the split-flux mass using Steger-Warming flux-vector splitting and show that the van Leer flux-vector splitting eliminates the discontinuities in derivatives as indicated in Fig. 6.12.

6.22 Verify Eqs. (6.114) and (6.115).

6.23 Develop a code to solve Prob. 6.9 using the Steger-Warming flux-vector splitting scheme.

6.24 Develop a code to solve Prob. 6.9 using van Leer flux-vector splitting.

6.25 Write a computer code to solve the 1-D shock tube problem using van Leer flux-vector splitting. Assume that the low-pressure end of the infinitely long shock tube is at standard atmospheric conditions and the high-pressure end has a pressure of 10 atm and standard atmospheric temperature.

6.26 Develop a code to solve Prob. 6.25 using the advection upstream splitting (AUSM) method.

6.27 Develop a code to solve Prob. 6.25 using a first- and second-order Roe scheme.

6.28 Show that Eq. (6.197) is a second-order representation of the 1-D potential equation.

6.29 Show that the retarded density formulation of Eq. (6.203) is equivalent to Eq. (6.202).

6.30 Derive Eq. (6.210).

6.31 Show that Eq. (6.260) is a valid representation for the potential in an incompressible fluid flow.

6.32 Compute c_{43} of Example 6.3.

CHAPTER
SEVEN

NUMERICAL METHODS FOR BOUNDARY-LAYER TYPE EQUATIONS

7.1 INTRODUCTION

It was pointed out in Chapter 5 that the equations that result from the boundary-layer (or thin-shear-layer) approximation provide a useful mathematical model for several important flows occurring in engineering applications. Among these are many jet and wake flows, two-dimensional or axisymmetric flows in channels and tubes, as well as the classical wall boundary layer. Certain three-dimensional flows can also be economically treated through the boundary-layer approximation. In addition, methods have been developed to extend the boundary-layer approximation to flows containing small regions of recirculation. Often, a small region exists near the streamwise starting plane of these flows in which the thin-shear-layer approximation is a poor one, but for moderate to large Reynolds numbers, this region is very (and usually negligibly) small.

In this chapter, methods and numerical considerations related to the numerical solution of these equations will be presented. The emphasis will be on the application of methods and principles covered in Chapters 3 and 4 rather than on the exposition of a single general-numerical procedure. Several finite-difference/finite-volume methods for these equations are described in detail elsewhere. Except as an aid in illustrating key principles, those details will not be repeated here.

The history of numerical methods for boundary-layer equations goes back to the 1930s and 1940s. Finite-difference methods in a form very similar to those now in use began emerging in the 1950s (Friedrich and Forstall, 1953; Rouleau and Osterle, 1955). We can think of numerical schemes for the boundary-layer equations as being well developed and tested as compared to methods for some other classes of flows. Despite this, new developments in the numerical treatment of these equations continue to appear regularly.

7.2 BRIEF COMPARISON OF PREDICTION METHODS

Before proceeding with a discussion of numerical methods for boundary-layer flows, it is well to remember that over the years, useful solutions have been obtained by other methods and for some simple flows, engineering results are available as simple formulas. These results are presented in standard textbooks on fluid mechanics, aerodynamics, and heat transfer. The books by Schlichting (1979) and White (1991) are especially valuable references for viscous flows.

Except for a few isolated papers based on similarity methods, the calculation methods for boundary-layer type problems that appear in the current literature can generally be categorized as (1) integral methods, (2) finite-difference/finite-volume methods, or (3) finite-element methods.

Integral methods can be applied to a wide range of both laminar and turbulent flows and, in fact, any problem that can be solved by a finite-difference/finite-volume method can also be solved by an integral method. Prior to the 1960s, integral methods were the primary "advanced" calculation method for solving complex problems in fluid mechanics and heat transfer. Loosely speaking, the method transforms the partial differential equations (PDEs) into one or more ordinary differential equations (ODEs) by integrating out the dependence of one independent variable (usually the normal coordinate) in advance by making assumptions about the general form of the velocity and temperature profiles (often functions of "N" parameters). Many of these procedures can be grouped as weighted residual methods. It can be shown that the solution by the method of weighted residuals approaches the exact solution of the PDE as N becomes very large. Modern versions of integral methods for complex problems make use of digital computers. In practice, it appears that implementing integral methods is not as straightforward (requiring more "intuition" about the problem) as for finite-difference/finite-volume methods. The integral methods are not as flexible or general in that more changes are generally required as boundary or other problem conditions are changed. In the 1970s the preference of the scientific community shifted in favor of using finite-difference/finite-volume methods over integral methods for computing the more complex boundary-layer flows. However, integral methods have at least a few strong advocates and can be used to solve important current problems.

Finite-element methodology has been applied to boundary-layer equations. Comments on this approach for boundary layers can be found in the work by Chung (1978). The objective of all three of these methods is to transform the problem posed through PDEs to one having an algebraic representation. The methods differ in the procedures used to implement this discretization.

7.3 FINITE-DIFFERENCE METHODS FOR TWO-DIMENSIONAL OR AXISYMMETRIC STEADY EXTERNAL FLOWS

7.3.1 Generalized Form of the Equations

The preferred form for the boundary-layer equations will vary from problem to problem. In the case of laminar flows, coordinate transformations are especially useful for maintaining a nearly constant number of grid points across the flow. The energy equation is usually written differently for compressible flow than it is for incompressible flow. In practice, it is frequently necessary to extend or alter a difference scheme established for one PDE to accommodate one that is similar but different in some detail. Optimizing the representation often requires a trial-and-error procedure.

The boundary-layer equations were given in Chapter 5 [Eqs. (5.116)–(5.119)] in physical coordinates. Here, we will utilize the Boussinesq approximation to evaluate the Reynolds shear-stress and heat flux quantities in terms of a turbulent viscosity μ_T and the turbulent Prandtl number $\Pr_T$. Specifically, we will let

$$-\rho\overline{u'v'} = \mu_T \frac{\partial u}{\partial y}$$

and

$$-\rho c_p \overline{v'T'} = \frac{c_p \mu_T}{\Pr_T} \frac{\partial T}{\partial y}$$

To solve the energy equation numerically using H as the primary thermal variable, it will be helpful to eliminate T in the expression for the Reynolds heat flux by using the definition of total enthalpy, $H = c_p T + u^2/2 + v^2/2$. The $v^2/2$ term can be neglected in keeping with the boundary-layer approximation. These substitutions permit the boundary-layer equations for a steady compressible 2-D or axisymmetric flow to be written as follows:

x momentum:

$$\rho u \frac{\partial u}{\partial x} + \rho \tilde{v} \frac{\partial u}{\partial y} = \rho_e u_e \frac{du_e}{dx} + \frac{1}{r^m} \frac{\partial}{\partial y}\left[r^m(\mu + \mu_T)\frac{\partial u}{\partial y}\right] \tag{7.1}$$

energy:

$$\rho u \frac{\partial H}{\partial x} + \rho \tilde{v} \frac{\partial H}{\partial y} = \frac{1}{r^m} \frac{\partial}{\partial y}\left(r^m\left\{\left(\frac{\mu}{\Pr} + \frac{\mu_T}{\Pr_T}\right)\frac{\partial H}{\partial y} + \left[\mu\left(1 - \frac{1}{\Pr}\right) + \mu_T\left(1 - \frac{1}{\Pr_T}\right)\right] u \frac{\partial u}{\partial y}\right\}\right) \tag{7.2}$$

continuity:

$$\frac{\partial}{\partial x}(r^m \rho u) + \frac{\partial}{\partial y}(r^m \rho \tilde{v}) = 0 \tag{7.3}$$

state:

$$\rho = \rho(T, p) \tag{7.4}$$

Property relationships are also needed to evaluate μ, k, c_p as a function (usually) of temperature.

As indicated in Chapter 5, m is a flow index equal to unity for axisymmetric flow and equal to zero for 2-D flow, and $\tilde{v} = (\bar{\rho}\bar{v} + \overline{\rho' v'})/\bar{\rho}$. When $m = 0$, $r^m = 1$ and the equations are in appropriate form for 2-D flows.

The primary dependent variable in the momentum equation is u, and it is useful to think of Eq. (7.1) as a "transport" equation for u, in which terms representing convection, diffusion, and "sources" of u can be recognized. Likewise, the energy equation can be viewed as a transport equation for H with similar categories of terms. This interpretation can also be extended to include the unsteady form of the boundary-layer momentum and energy equations.

Within the transport equation context, both Eqs. (7.1) and (7.2) can usually [an exception may occur with the use of some turbulence models, Bradshaw et al. (1967)] be cast into the general form

$$\underbrace{\rho u \frac{\partial \phi}{\partial x} + \rho \tilde{v} \frac{\partial \phi}{\partial y}}_{\text{Convection of } \phi} = \underbrace{\frac{1}{r^m}\frac{\partial}{\partial y}\left(r^m \lambda \frac{\partial \phi}{\partial y}\right)}_{\text{Diffusion of } \phi} + \underbrace{S}_{\substack{\text{Source}\\ \text{terms}}} \tag{7.5}$$

In Eq. (7.5), ϕ is a generalized variable, which would be u for the boundary-layer momentum equation and H for the boundary-layer energy equation; λ is a generalized diffusion coefficient and S represents the source terms. Source terms are those terms in the PDE that do not involve a derivative of ϕ. The term $\rho_e u_e \, du_e/dx$ in Eq. (7.1) and the term involving $u \, \partial u/\partial y$ in Eq. (7.2) are examples of source terms. Most of the transport equations for turbulence model parameters given in Chapter 5 also fit the form of Eq. (7.5).

The momentum and energy equations that can be cast into the general form of Eq. (7.5) are parabolic with x as the marching coordinate. By making appropriate assumptions regarding the evaluation of coefficients, it is possible to decouple the finite-difference representation of the equations, permitting the momentum, continuity, and energy equations to be marched one step in the x direction independently to provide new values of u_j, H_j, and $\tilde{v}_j$. This strategy is illustrated below:

Equation	*Marched to obtain*
x momentum	u_j^{n+1}
energy	H_j^{n+1}
equation of state + continuity	$\tilde{v}_j^{n+1}$

After each marching step, the coefficients in the equations are reevaluated (updated), so that the solutions of the three equations are in fact interdependent —the decoupling is in the algebraic system for one marching step at a time. In some solution schemes, the coupling is maintained so that at each marching step, a larger system of algebraic equations must be solved simultaneously for new values of $u_j, H_j, \tilde{v}_j$. Uncoupling the algebraic system is conceptually the simplest procedure and can usually be made to work satisfactorily for most flow problems.

7.3.2 Example of a Simple Explicit Procedure

Although the simplest explicit method is no longer widely used for boundary layers owing to the restrictive stability constraint associated with it, it will be used here for pedagogical purposes to demonstrate the general solution algorithm for boundary-layer flows (Wu, 1961). Consider a 2-D laminar incompressible flow without heat transfer. The governing equations in partial differential form are given as Eqs. (5.104) and (5.105).

The difference equations can be written as follows:

x momentum:

$$u_j^n \frac{\left(u_j^{n+1} - u_j^n\right)}{\Delta x} + \boxed{v_j^n \frac{\left(u_{j+1}^n - u_{j-1}^n\right)}{2\,\Delta y}}$$

$$= u_e^n \frac{\left(u_e^{n+1} - u_e^n\right)}{\Delta x} + \frac{\nu}{(\Delta y)^2}\left(u_{j+1}^n - 2u_j^n + u_{j-1}^n\right) + O(\Delta x) + O\left[(\Delta y)^2\right] \tag{7.6}$$

continuity:

$$\frac{v_j^{n+1} - v_{j-1}^{n+1}}{\Delta y} + \frac{u_j^{n+1} + u_{j-1}^{n+1} - u_j^n - u_{j-1}^n}{2\,\Delta x} = 0 + O(\Delta x) + O\left[(\Delta y)^2\right] \tag{7.7}$$

For flow over a flat plate (see Fig. 7.1), the computation is usually started by assuming that $u_j^n = u_\infty$ at the leading edge and $v_j^n = 0$. The value v_j^n is required in the explicit algorithm in order to advance the solution to the $n+1$ level. However, in the formal mathematical formulation of the PDE problem, it is not necessary to specify an initial distribution of v_j^n. A compatible initial distribution can be obtained for v_j^n (Ting, 1965) by first using the continuity equation to eliminate $\partial u/\partial x$ from the boundary-layer momentum equation. For a laminar, incompressible flow, this gives

$$-u\frac{\partial v}{\partial y} + v\frac{\partial u}{\partial y} = u_e\frac{du_e}{dx} + \nu\frac{\partial^2 u}{\partial y^2}$$

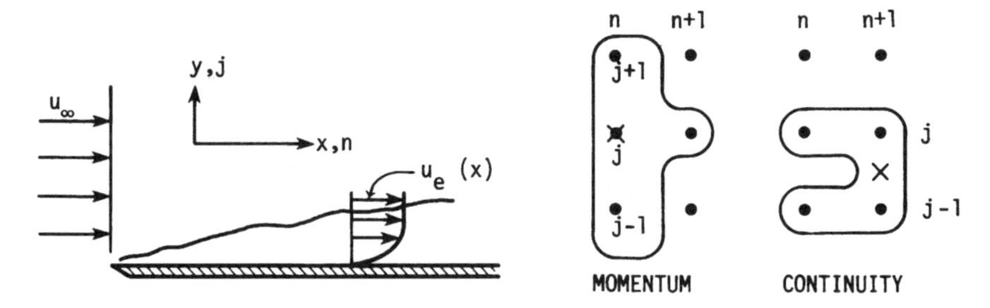

Figure 7.1 Simple explicit procedure.

We can observe that

$$-u\frac{\partial v}{\partial y} + v\frac{\partial u}{\partial y} = -u^2\frac{\partial}{\partial y}\left(\frac{v}{u}\right)$$

Thus

$$\frac{\partial}{\partial y}\left(\frac{v}{u}\right) = -\frac{1}{u^2}\left(u_e\frac{du_e}{dx} + \nu\frac{\partial^2 u}{\partial y^2}\right)$$

and using $v = 0$ at $y = 0$, we find

$$v(y) = -u\int_0^y \frac{1}{u^2}\left(u_e\frac{du_e}{dx} + \nu\frac{\partial^2 u}{\partial y^2}\right)dy \tag{7.8}$$

For the flat plate problem at hand, we would assume that $u_j^n = u_\infty$ at $x = 0$ (the leading edge) except at the wall, where $u_1^n = 0$. We can use a numerical evaluation of the integral in Eq. (7.8) to obtain an estimate of a compatible initial distribution of v_j^n to use in the explicit difference procedure. Employing the usual central-difference representation for $\partial^2 u/\partial y^2$ gives $v_j^n = 2\nu/\Delta y$ at all points except the point on the wall where $v_1^n = 0$ and the point adjacent to the wall where $v_2^n = \nu/\Delta y$. In practice, letting $v_j^n = 0$ initially throughout is also found to work satisfactorily.

Having initial values for u_j^n, the momentum equation, Eq. (7.6), can be solved for u_j^{n+1} explicitly, usually by starting from the wall and working outward until $u_j^{n+1}/u_e^{n+1} = 1 - \epsilon \approx 0.9995$; that is, owing to the asymptotic boundary condition, we *find* the location of the outer boundary as the solution proceeds. The values of v_j^{n+1} can now be computed from Eq. (7.7), starting with the point next to the lower boundary and computing outward. The difference formulation of the continuity equation and the solution procedure described is equivalent to integrating the continuity equation by the trapezoidal rule for v_j^{n+1}.

The *stability constraints* for this method are

$$\frac{2\nu\,\Delta x}{u_j^n(\Delta y)^2} \leqslant 1 \text{ and } \frac{(v_j^n)^2\,\Delta x}{u_j^n \nu} \leqslant 2$$

The second term in the momentum equation, Eq. (7.6), has been enclosed by a dashed box for two reasons. First, we should be aware that the presence of this term is mainly responsible for any difference between the stability constraints of Eq. (7.6) and the heat equation, and second, we will suggest an alternative treatment for this term below.

Alternative formulation for explicit method. In order to control the stability of the explicit method by checking only a single inequality, the boxed term in Eq. (7.6), can be expressed as

$$v_j^n \frac{u_j^n - u_{j-1}^n}{\Delta y}$$

when

$$v_j^n > 0$$

and

$$v_j^n \frac{u_{j+1}^n - u_j^n}{\Delta y}$$

when $v_j^n < 0$, whereby the stability constraint becomes

$$\Delta x \leqslant \frac{1}{2\nu\big/\left[u_j^n(\Delta y)^2\right] + |v_j^n|/(u_j^n\,\Delta y)}$$

The truncation error (T.E.) deteriorates to only $O(\Delta x) + O(\Delta y)$ when this treatment of $v\,\partial u/\partial y$ is used.

Note that the stability constraints for both methods depend upon the local values of u and v. This is typical for equations with variable coefficients. The von Neumann stability analysis has proven to be a reliable guide to stability for boundary-layer equations if the coefficients u and v that appear in the equations are treated as being locally constant. Treatment of μ_T for turbulent flow in the stability analysis requires further consideration. For some models, μ_T will contain derivatives whose difference representation could contribute to numerical instabilities. In the stability analysis, one can treat μ_T as simply a specified variable property and then by trial and error develop a stable difference representation for μ_T or one can express μ_T in terms of the dependent flow variables and attempt to determine the appropriate stability constraints by the usual methods.

7.3.3 Crank-Nicolson and Fully Implicit Methods

The characteristics of most implicit methods can be visualized by considering the following representation of the compressible laminar boundary-layer equations in physical coordinates on a mesh for which $\Delta y = \text{const}$.

momentum:

$$\frac{\left[\theta\left(\rho_j^{n+1}u_j^{n+1}\right)+(1-\theta)\left(\rho_j^n u_j^n\right)\right]\left(u_j^{n+1}-u_j^n\right)}{\Delta x}$$
$$+\frac{\theta\left(\rho_j^{n+1}v_j^{n+1}\right)\left(u_{j+1}^{n+1}-u_{j-1}^{n+1}\right)+(1-\theta)\left(\rho_j^n v_j^n\right)\left(u_{j+1}^n-u_{j-1}^n\right)}{2\,\Delta y}$$
$$=\frac{\left[\theta\left(\rho_e^{n+1}u_e^{n+1}\right)+(1-\theta)\left(\rho_e^n u_e^n\right)\right]\left(u_e^{n+1}-u_e^n\right)}{\Delta x}$$
$$+\frac{1}{(\Delta y)^2}\left\{\theta\left[\mu_{j+1/2}^{n+1}\left(u_{j+1}^{n+1}-u_j^{n+1}\right)-\mu_{j-1/2}^{n+1}\left(u_j^{n+1}-u_{j-1}^{n+1}\right)\right]\right.$$
$$\left.+(1-\theta)\left[\mu_{j+1/2}^n\left(u_{j+1}^n-u_j^n\right)-\mu_{j-1/2}^n\left(u_j^n-u_{j-1}^n\right)\right]\right\} \tag{7.9}$$

In the above, θ is a weighting factor. If

$\theta = 0$ Method is explicit, most convenient expansion point is (n, j), truncation error is $O(\Delta x) + O[(\Delta y)^2]$; the von Neumann stability constraint, given previously, presents a severe limitation on the marching step size.

$\theta = \frac{1}{2}$ Crank-Nicolson implicit; the most convenient expansion point is $(n + \frac{1}{2}, j)$; the T.E. is $O[(\Delta x)^2] + O[(\Delta y)^2]$ if coefficients (and properties) are evaluated at $(n + \frac{1}{2}, j)$. No stability constraint arises from the von Neumann analysis, but difficulties can arise if diagonal dominance is not maintained for the tridiagonal algorithm (Hirsh and Rudy, 1974).

$\theta = 1$ Fully implicit, expansion point $(n + 1, j)$, T.E. is $O(\Delta x) + O[(\Delta y)^2]$ [if properties and coefficients are evaluated at $(n + 1, j)$]. No stability constraint by the von Neumann method, but same comment as for $\theta = \frac{1}{2}$ applies for diagonal dominance.

We note that the above scheme becomes implicit if $\theta > 0$, and inherently stable if $\theta \geqslant \frac{1}{2}$. Values of θ between $\frac{1}{2}$ and 1 have been used successfully. The same form of the continuity equation can be used for both the fully implicit and the explicit methods.

continuity:

$$\frac{\rho_j^{n+1}v_j^{n+1}-\rho_{j-1}^{n+1}v_{j-1}^{n+1}}{\Delta y}+\frac{\rho_j^{n+1}u_j^{n+1}-\rho_j^n u_j^n+\rho_{j-1}^{n+1}u_{j-1}^{n+1}-\rho_{j-1}^n u_{j-1}^n}{2\,\Delta x}=0 \tag{7.10}$$

When $\theta = \frac{1}{2}$, we can consider ρ and v in the first term to be at the $n + \frac{1}{2}$ level and rewrite Eq. (7.10) accordingly. This results in a T.E. of $O[(\Delta x)^2] + O[(\Delta y)^2]$ for the continuity equation. Differencing of the energy equation follows the same general pattern as used for the momentum equation. Choosing T as the primary thermal variable as we might for low speed flow, we can write the energy equation as

energy:

$$\rho u c_p \frac{\partial T}{\partial x} + \rho v c_p \frac{\partial T}{\partial y} = \frac{\partial}{\partial y}\left(k \frac{\partial T}{\partial y}\right) + \beta T u \frac{dp}{dx} + \mu \left(\frac{\partial u}{\partial y}\right)^2 \tag{7.11}$$

which, utilizing the θ notation, can be written in difference form as

$$\begin{aligned}
&\left[\theta\left(\rho_j^{n+1} u_j^{n+1} c_{p_j}^{n+1}\right) + (1-\theta)\left(\rho_j^n u_j^n c_{p_j}^n\right)\right] \frac{T_j^{n+1} - T_j^n}{\Delta x} \\
&\quad + \frac{\theta\left(\rho_j^{n+1} v_j^{n+1} c_{p_j}^{n+1}\right)\left(T_{j+1}^{n+1} - T_{j-1}^{n+1}\right) + (1-\theta)\left(\rho_j^n v_j^n c_{p_j}^n\right)\left(T_{j+1}^n - T_{j-1}^n\right)}{2\,\Delta y} \\
&= \frac{1}{(\Delta y)^2}\left\{\theta\left[k_{j+1/2}^{n+1}\left(T_{j+1}^{n+1} - T_j^{n+1}\right) - k_{j-1/2}^{n+1}\left(T_j^{n+1} - T_{j-1}^{n+1}\right)\right]\right. \\
&\qquad \left. + (1-\theta)\left[k_{j+1/2}^n\left(T_{j+1}^n - T_j^n\right) - k_{j-1/2}^n\left(T_j^n - T_{j-1}^n\right)\right]\right\} \\
&\quad + \frac{\left[\theta\left(\beta_j^{n+1} T_j^{n+1} u_j^{n+1}\right) + (1-\theta)\left(\beta_j^n T_j^n u_j^n\right)\right]\left(p_j^{n+1} - p_j^n\right)}{\Delta x} \\
&\quad + \theta \mu_j^{n+1}\left(\frac{u_{j+1}^{n+1} - u_{j-1}^{n+1}}{2\,\Delta y}\right)^2 + (1-\theta)\mu_j^n\left(\frac{u_{j+1}^n - u_{j-1}^n}{2\,\Delta y}\right)^2
\end{aligned} \tag{7.12}$$

The T.E. for the energy equation is identical to that stated for the momentum equation for $\theta = 0, \frac{1}{2}, 1$.

The fully implicit ($\theta = 1$) scheme can be elevated to formal second-order accuracy by representing streamwise derivatives by three-level ($n-1, n, n+1$) second-order accurate differences, such as can be found in Chapter 3. Davis (1963) and Harris (1971) have demonstrated the feasibility of such a procedure.

For any implicit method ($\theta \neq 0$) the finite-difference forms of the momentum and energy equations [Eqs. (7.9) and (7.12)] are algebraically nonlinear in the unknowns owing to the appearance of quantities unknown at the $n+1$ level in the coefficients. Linearizing procedures that can and have been utilized are described in the following sections.

Lagging the coefficients. The simplest and most common strategy is to linearize the difference equations by evaluating all coefficients at the n level. This is known as "lagging" the coefficients. The procedure provides a consistent representation, since for a general function $\phi(x, y)$, $\phi(x_0 + \Delta x, y_0) = \phi(x_0, y_0) + O(\Delta x)$. This procedure causes the difference scheme to be no better than first-order accurate in the marching coordinate. Using the generalized form [Eq. (7.5)] for a transport equation, the linearized difference representation obtained by lagging the coefficients can be written

$$
\begin{aligned}
\rho_j^n u_j^n \frac{\phi_j^{n+1} - \phi_j^n}{\Delta x} &+ \frac{\rho_j^n v_j^n}{2\,\Delta y}\left[\theta\left(\phi_{j+1}^{n+1} - \phi_{j-1}^{n+1}\right) + (1-\theta)(\phi_{j+1}^n - \phi_{j-1}^n)\right] \\
&= \frac{1}{(\Delta y)^2}\Big\{\lambda_{j+1/2}^n\left[\theta\left(\phi_{j+1}^{n+1} - \phi_j^{n+1}\right) + (1-\theta)(\phi_{j+1}^n - \phi_j^n)\right] \\
&\qquad - \lambda_{j-1/2}^n\left[\theta\left(\phi_j^{n+1} - \phi_{j-1}^{n+1}\right) + (1-\theta)(\phi_j^n - \phi_{j-1}^n)\right]\Big\} \\
&\quad + \theta S_j^{n+1} + (1-\theta)S_j^n
\end{aligned}
\tag{7.13}
$$

The three conservation equations in difference form can now be solved in an uncoupled manner. The momentum equation can be solved for u_j^{n+1}, the energy equation for T_j^{n+1}, and an equation of state used to obtain ρ_j^{n+1}. Finally, the continuity equation can be solved for v_j^{n+1}. The matrix of unknowns in each equation (for momentum and energy) is tridiagonal, and the Thomas algorithm can be employed.

Simple iterative update of coefficients. The coefficients can be ultimately evaluated at the $n+1$ level as required in Eqs. (7.9), (7.10), and (7.12) by use of a simple iterative updating procedure. To do this, the coefficients are first evaluated at the n level (lagged) and the system solved for new values of u, T, v at the $n+1$ level. The coefficients can then be updated by utilizing the solution just obtained at the $n+1$ level and the calculation repeated to obtain "better" predictions at $n+1$.

This procedure can be repeated iteratively until changes are small. Usually only two or three iterations are used, although Blottner (1975a) points out that up to 19 iterations were required in a sample calculation with the Crank-Nicolson procedure before the solution obtained behaved like a second-order accurate scheme under grid refinement (see Section 3.2). Although the programming changes involved in advancing from the lagged procedure to the simple iterative update are minimal, the use of Newton linearization, to be described next, is more efficient and is recommended for that reason.

Use of Newton linearization to iteratively update coefficients. Newton linearization is another linearization procedure that can be used to iteratively update coefficients and, in fact, to provide a useful representation for most nonlinear expressions arising in computational fluid dynamics (CFD). The Newton procedure is actually more efficient (converges in fewer iterations) than the simple iterative update procedure described above. To be general, suppose we wish to linearize a function of several dependent variables in a conservation equation such as u, v, and p. These variables may, in turn, depend upon independent variables such as position and time. With an iteration sequence in mind, we may think of u, v, and p as being functions of a time-like parameter (pseudo-time) that is incremented as the iterative sequence proceeds. We let

Δu, Δv, and Δp equal the change in u, v, and p, respectively, between two iterative solutions to the difference equations. Thus $u_j^{n+1} = \hat{u}_j^{n+1} + \Delta u_j$, $v_j^{n+1} = \hat{v}_j^{n+1} + \Delta v_j$, $p_j^{n+1} = \hat{p}_j^{n+1} + \Delta p_j$, where the circumflex denotes an evaluation of the variable from a previous iteration level. For the first iteration with the steady boundary-layer equations, the variables with circumflexes will be assigned values from the previous marching station. We may expand the nonlinear function $F^{n+1}(u, v, p)$ in a Taylor series in the iteration parameter τ about the present state $\hat{F}^{n+1}$:

$$F^{n+1} = \hat{F}^{n+1} + \frac{\partial \hat{F}}{\partial \tau} \Delta \tau + \cdots \tag{7.14}$$

Linearization is enabled by truncating the series after the first-derivative term. Using the chain rule, we can represent the derivative on the right-hand side in terms of u, v, p:

$$\frac{\partial \hat{F}}{\partial \tau} = \frac{\partial \hat{F}}{\partial u}\frac{\partial u}{\partial \tau} + \frac{\partial \hat{F}}{\partial v}\frac{\partial v}{\partial \tau} + \frac{\partial \hat{F}}{\partial p}\frac{\partial p}{\partial \tau}$$

Substituting this result into Eq. (7.14) gives

$$F^{n+1} \cong \hat{F}^{n+1} + \frac{\partial \hat{F}}{\partial u}\Delta u + \frac{\partial \hat{F}}{\partial v}\Delta v + \frac{\partial \hat{F}}{\partial p}\Delta p \tag{7.15}$$

Note that the iteration parameter τ does not appear in this final working form of Newton linearization.

As an example, let the function to be linearized be uv/p. Applying the results indicated by Eq. (7.15), we obtain

$$F^{n+1} = \left(\frac{\hat{u}\hat{v}}{\hat{p}}\right)^{n+1} + \left(\frac{\hat{v}}{\hat{p}}\right)^{n+1} \Delta u + \left(\frac{\hat{u}}{\hat{p}}\right)^{n+1} \Delta v - \left(\frac{\hat{u}\hat{v}}{\hat{p}^2}\right)^{n+1} \Delta p$$

When solving the boundary-layer equations with an implicit scheme, we need to linearize $(u_j^{n+1})^2$. Here $F^{n+1} = F^{n+1}(u) = (u_j^{n+1})^2$ and $\partial \hat{F}/\partial u = 2\hat{u}^{n+1}$. Our representation after linearization is

$$\left(u_j^{n+1}\right)^2 \cong \left(\hat{u}_j^{n+1}\right)^2 + 2\Delta u_j \hat{u}_j^{n+1} \tag{7.16}$$

in which Δu_j is the only unknown. Alternatively, we can substitute for Δu_j according to $\Delta u_j = u_j^{n+1} - \hat{u}_j^{n+1}$ and rewrite Eq. (7.16) as

$$\left(u_j^{n+1}\right)^2 \cong 2u_j^{n+1}\hat{u}_j^{n+1} - \left(\hat{u}_j^{n+1}\right)^2 \tag{7.17}$$

This latter procedure will be employed in this chapter. However, both procedures, the one in which the delta quantities are treated as the unknowns [as in Eq. (7.16)] and the other in which the deltas are eliminated by substitution [as in Eq. (7.17)] are widely used in CFD. Note that upon iterative convergence, both representations are exact.

For a more specific example of the use of this procedure, consider a fully implicit ($\theta = 1$) application in which the conservation equations are to be solved in an uncoupled manner for an incompressible flow. The most obvious nonlinearity appears in the representation for the $\rho u\, \partial u/\partial x$ term. Applying Newton linearization to the fully implicit finite-difference representation of this term gives

$$\frac{\rho\left[2\hat{u}_j^{n+1}u_j^{n+1} - \left(\hat{u}_j^{n+1}\right)^2 - u_j^n u_j^{n+1}\right]}{\Delta x} \tag{7.18}$$

in which u_j^{n+1} is the only unknown. For the first iteration, $\hat{u}_j^{n+1}$ is evaluated as u_j^n. A slightly different final result is obtained if we apply the linearization procedure to this term in the mathematically equivalent form, $\rho\, \partial(u^2/2)/\partial x$.

If the conservation equations are to be solved in an uncoupled manner, i.e., one unknown is to be determined independently from each conservation equation, the other nonlinear terms

$$\rho v \frac{\partial u}{\partial y} \qquad \frac{\partial}{\partial y}\left(\mu \frac{\partial u}{\partial y}\right)$$

are usually evaluated by the simple iterative updating procedure described above.

Evaluating $\rho u\, \partial u/\partial x$ by the Newton linearization as indicated in Eq. (7.18) and using simple updating on other nonlinear terms results in a tridiagonal coefficient matrix, which permits use of the Thomas algorithm with no special modifications. The calculation is repeated two or more times at each streamwise location, updating variables as indicated.

Newton linearization with coupling. Several investigators have observed that convergence of the iterations to update coefficients at each streamwise step in the boundary-layer momentum equation can be accelerated greatly by solving the momentum and continuity equations in a coupled manner. Second-order accuracy for the Crank-Nicolson procedure has been observed using only one iteration at each streamwise station when the equations are solved in a coupled manner (Blottner, 1975a). According to Blottner (1975a), coupling was first suggested by R. T. Davis and used by Werle and co-workers (Werle and Bertke, 1972; Werle and Dwoyer, 1972). An example of the coupled procedure for a fully implicit formulation for incompressible, constant property flow follows.

The $u\, \partial u/\partial x$ term is treated as in Eq. (7.18). The $v\, \partial u/\partial y$ term is linearized by using $v_j^{n+1} = \hat{v}_j^{n+1} + \Delta v_j$ and $u_j^{n+1} = \hat{u}_j^{n+1} + \Delta u_j$. For the first iteration, $\hat{v}_j^{n+1}$ and $\hat{u}_j^{n+1}$ are most conveniently evaluated as v_j^n and u_j^n, respectively. Here we are considering F in Eq. (7.15) to be $F(u, v)$. After replacing the delta quantities by differences in variables at two iteration levels, the term $v\,\partial u/\partial y$ becomes

$$\left(v\frac{\partial u}{\partial y}\right)^{n+1} \approx \hat{v}^{n+1}\left(\frac{\partial u}{\partial y}\right)^{n+1} + v^{n+1}\left(\frac{\partial \hat{u}}{\partial y}\right)^{n+1} - \hat{v}^{n+1}\left(\frac{\partial \hat{u}}{\partial y}\right)^{n+1} \tag{7.19}$$

The continuity and momentum equations can then be written in difference form as

$$\frac{u_j^{n+1} - u_j^n + u_{j-1}^{n+1} - u_{j-1}^n}{2\,\Delta x} + \frac{v_j^{n+1} - v_{j-1}^{n+1}}{\Delta y} = 0 \tag{7.20}$$

$$\begin{aligned}
&\frac{2\hat{u}_j^{n+1}u_j^{n+1} - \left(\hat{u}_j^{n+1}\right)^2 - u_j^n u_j^{n+1}}{\Delta x} + \frac{\hat{v}_j^{n+1}\left(u_{j+1}^{n+1} - \hat{u}_{j+1}^{n+1} - u_{j-1}^{n+1} + \hat{u}_{j-1}^{n+1}\right)}{2\,\Delta y} \\
&\quad + \frac{v_j^{n+1}\left(\hat{u}_{j+1}^{n+1} - \hat{u}_{j-1}^{n+1}\right)}{2\,\Delta y} \\
&= \frac{\nu}{(\Delta y)^2}\left(u_{j+1}^{n+1} - 2u_j^{n+1} + u_{j-1}^{n+1}\right) + \frac{\left(u_e^{n+1}\right)^2 - u_e^{n+1}u_e^n}{\Delta x}
\end{aligned} \tag{7.21}$$

To clarify the algebraic formulation of the problem, the momentum equation can be written as

$$B_j u_{j-1}^{n+1} + D_j u_j^{n+1} + A_j u_{j+1}^{n+1} + a_j v_j^{n+1} + b_j v_{j-1}^{n+1} = C_j \tag{7.22}$$

where

$$B_j = -\frac{\hat{v}_j^{n+1}}{2\,\Delta y} - \frac{\nu}{(\Delta y)^2} \qquad D_j = \frac{2\hat{u}_j^{n+1} - u_j^n}{\Delta x} + \frac{2\nu}{(\Delta y)^2}$$

$$A_j = \frac{\hat{v}_j^{n+1}}{2\,\Delta y} - \frac{\nu}{(\Delta y)^2} \qquad a_j = \frac{\hat{u}_{j+1}^{n+1} - \hat{u}_{j-1}^{n+1}}{2\,\Delta y} \qquad b_j = 0$$

$$C_j = \frac{\left(\hat{u}_j^{n+1}\right)^2}{\Delta x} + \hat{v}_j^{n+1}\frac{\hat{u}_{j+1}^{n+1} - \hat{u}_{j-1}^{n+1}}{2\,\Delta y} + \frac{\left(u_e^{n+1}\right)^2 - u_e^{n+1}u_e^n}{\Delta x}$$

In this example, b_j could be dropped, since it is equal to zero. We will continue to develop the solution algorithm including b_j because the result will be useful to us for solving other difference equations in this chapter.

For any j value, *four* unknowns (five if $b_j \neq 0$) appear on the left-hand side of Eq. (7.22), u_{j-1}^{n+1}, u_j^{n+1}, u_{j+1}^{n+1}, and v_j^{n+1}. It is obvious that the matrix of coefficients is no longer tridiagonal. However, the continuity equation can be written as

$$v_j^{n+1} = v_{j-1}^{n+1} - e_j\left(u_{j-1}^{n+1} + u_j^{n+1}\right) + d_j \tag{7.23}$$

where

$$e_j = \frac{\Delta y}{2\,\Delta x} \qquad d_j = \frac{(u_{j-1}^n + u_j^n)\,\Delta y}{2\,\Delta x}$$

and Eqs. (7.22) and (7.23) together form a coupled system, which can be written in "block-tridiagonal" form (see Appendix B) with 2×2 blocks as

$$\begin{bmatrix} B_j & b_j \\ e_j & -1 \end{bmatrix} \begin{bmatrix} u_{j-1}^{n+1} \\ v_{j-1}^{n+1} \end{bmatrix} + \begin{bmatrix} D_j & a_j \\ e_j & 1 \end{bmatrix} \begin{bmatrix} u_j^{n+1} \\ v_j^{n+1} \end{bmatrix} + \begin{bmatrix} A_j & 0 \\ 0 & 0 \end{bmatrix} \begin{bmatrix} u_{j+1}^{n+1} \\ v_{j+1}^{n+1} \end{bmatrix} = \begin{bmatrix} C_j \\ d_j \end{bmatrix}$$

A solution algorithm has been developed (see also Werle et al., 1973, or Blottner, 1975a) for solving this coupled system of equations. In this procedure (often called the modified tridiagonal algorithm), the blocks above the main diagonal are first eliminated. This permits the velocities, u_j^{n+1}, to be calculated from the recursion formula $u_j^{n+1} = E_j u_{j-1}^{n+1} + F_j + G_j v_{j-1}^{n+1}$ after E_j, F_j, G_j, and v_{j-1}^{n+1} are computed as indicated below. At the upper boundary, corresponding to $j = J$, conditions are specified as

$$E_J = 0$$

$$F_J = u_J^{n+1} \quad \text{(specified boundary value)}$$

$$G_J = 0$$

Then for $j = J-1, J-2, \ldots, 2$ we compute

$$\overline{D}_j = D_j + A_j E_{j+1} - e_j(A_j G_{j+1} + a_j)$$

$$E_j = -\left(\frac{B_j - e_j(A_j G_{j+1} + a_j)}{\overline{D}_j}\right)$$

$$F_j = \frac{C_j - A_j F_{j+1} - d_j(A_j G_{j+1} + a_j)}{\overline{D}_j}$$

$$G_j = -\left(\frac{A_j G_{j+1} + a_j + b_j}{\overline{D}_j}\right)$$

Then the lower boundary conditions are utilized to compute $v_1^{n+1} = 0$, $u_1^{n+1} = 0$, after which the velocities can be computed for $j = 2, \ldots, J$ by utilizing $u_j^{n+1} = E_j u_{j-1}^{n+1} + F_j + G_j v_{j-1}^{n+1}$ and $v_j^{n+1} = v_{j-1}^{n+1} - e_j(u_{j-1}^{n+1} + u_j^{n+1}) + d_j$. The above procedure reduces to the Thomas algorithm (but with elements above the main diagonal being eliminated) for a scalar tridiagonal system whenever a_j, b_j, e_j, and d_j are all set to zero. This system of equations can also be solved by the general algorithm for a block tridiagonal system given in Appendix B. However, the algorithm given above is more efficient because it is specialized to systems exactly of the form given by Eqs. (7.22) and (7.23).

The procedure can be extended readily to compressible variable property flows (see Blottner, 1975a). The energy equation is nearly always solved in an uncoupled manner in this case.

Extrapolating the coefficients. Values of the coefficients can be obtained at the $n + 1$ level by extrapolation based on values already obtained from previous n levels. Formally, the T.E. of this procedure can be made as small as we wish. For example, we can write

$$u_j^{n+1} = u_j^n + \left.\frac{\partial u}{\partial x}\right)_j^n \Delta x_+ + O[(\Delta x)^2]$$

Approximating $(\partial u/\partial x)_j^n$ by only a first-order accurate representation such as

$$\left.\frac{\partial u}{\partial x}\right)_j^n = \frac{u_j^n - u_j^{n-1}}{\Delta x_-} + O(\Delta x)$$

gives the following representation for u_j^{n+1}, which formally has a T.E. of $O[(\Delta x)^2]$:

$$u_j^{n+1} = u_j^n + \frac{u_j^n - u_j^{n-1}}{\Delta x_-} \Delta x_+ + O[(\Delta x)^2]$$

A similar procedure can be used for other coefficients needed at the $n + 1$ level. This approach has been used satisfactorily for boundary-layer flows by Harris (1971).

A recommendation. For many calculations, the linearization introduced by simply lagging the coefficients u and v and the fluid properties (in cases with temperature variations) will cause no serious deterioration of accuracy. Errors associated with linearization of coefficients are simply truncation errors, which can be controlled by adjustment of the marching step size. Many investigators have used this procedure satisfactorily. For any problem in which this linearization causes special difficulties, extrapolation of coefficients or Newton linearization with coupling is recommended. The former procedure requires no iterations to update the coefficients and for that reason should be more economical in terms of computation time. Clearly, it is desirable to use a method that is consistent, so that the numerical errors can be reduced to any level required. For turbulent flow calculations in particular, the uncertainties in the experimental data that are used to guide and verify the calculations and the uncertainties introduced by turbulence modeling add up to several (at least three to five) percent, making extreme accuracy in the numerical procedures unrewarding. In this situation the merits of using a higher-order method (highly accurate in terms of order of truncation error) should be determined on the basis of computer time that can be saved through the use of the coarser grids permitted by the more accurate schemes.

A warming on stability. Implicit schemes are touted as being unconditionally stable (in the von Neumann sense) if $\theta \geqslant \frac{1}{2}$. The Crank-Nicolson scheme just barely satisfies the formal stability requirement in term of θ, and this requirement was based on a heuristic extension of von Neumann's analysis for linear equations to nonlinear ones.

For turbulent flows in particular, the Crank-Nicolson procedure has occasionally been found to become unstable. For this reason, the fully implicit scheme has become more widely used. Formal second-order accuracy can be achieved by use of a three-point representation of the streamwise derivative and extrapolation of the coefficients. As an example, for uniform grid spacing, the convective terms

$$u\frac{\partial u}{\partial x} + v\frac{\partial u}{\partial y}$$

can be represented by

$$u\frac{\partial u}{\partial x} + v\frac{\partial u}{\partial y} = \frac{\left(2u_j^n - u_j^{n-1}\right)\left(3u_j^{n+1} - 4u_j^n + u_j^{n-1}\right)}{2\,\Delta x} + \frac{\left(2v_j^n - v_j^{n-1}\right)\left(u_{j+1}^{n+1} - u_{j-1}^{n+1}\right)}{2\,\Delta y} + O\left[(\Delta x)^2\right] + O\left[(\Delta y)^2\right] \tag{7.24}$$

With a slight increase in algebraic complexity, these representations can be generalized to also provide second-order accurate representations when the mesh increments Δx and Δy are not constant (Harris, 1971).

There is still one very real constraint on the use of the implicit schemes given for boundary-layer flows. Though not detected by the von Neumann stability analysis, a behavior very much characteristic of numerical instability can occur if the choice of grid spacing permits the convective transport (of momentum or energy) to dominate the diffusive transport. Two sources of this difficulty can be identified. First, errors can grow out of hand in the tridiagonal elimination scheme if diagonal dominance is not maintained, that is, in terms of the notation being used for the Thomas algorithm, if $|D_j|$ is not greater than $|B_j| + |A_j|$. A second and equally important cause of these unacceptable solutions can be related to a physical implausibility that arises when the choice of grid size permits the algebraic model to be an inaccurate representation for a viscous flow. The same difficulty for the viscous Burgers equation was discussed in Chapter 4. It can be shown that satisfying the conditions required to keep the algebraic representation a physically valid one provides a sufficient condition for diagonal dominance in the elimination scheme.

To illustrate the basis for these difficulties, consider the fully implicit procedure applied to the boundary-layer momentum equation for constant property flow with the coefficients lagged. The finite-difference equation can be written as

$$B_j u_{j-1}^{n+1} + D_j u_j^{n+1} + A_j u_{j+1}^{n+1} = C_j \tag{7.25}$$

with

$$B_j = -\frac{v_j^n}{2\,\Delta y} - \frac{\nu}{(\Delta y)^2}$$

$$D_j = \frac{u_j^n}{\Delta x} + \frac{2\nu}{(\Delta y)^2}$$

$$A_j = \frac{v_j^n}{2\,\Delta y} - \frac{\nu}{(\Delta y)^2}$$

$$C_j = \frac{(u_j^n)^2}{\Delta x} + u_e^n \frac{(u_e^{n+1} - u_e^n)}{\Delta x}$$

By reflecting on the implications of Eq. (7.25) in terms of the predicted behavior of u_j^{n+1} relative to changes in u_{j-1}^{n+1} and u_{j+1}^{n+1}, we would expect both A_j and B_j to be negative to properly imply the expected behavior of a viscous fluid. The expected behavior would be such that a decrease in the velocity of the fluid below or above the point $n + 1, j$ would contribute toward a decrease in the velocity at point $n + 1, j$ through the effects of viscosity. We should be able to see that such would not be the case if either A_j or B_j would become positive. To keep A_j and B_j negative in value requires

$$\frac{|v_j^n|}{2\,\Delta y} - \frac{\nu}{(\Delta y)^2} < 0$$

or

$$\frac{|v_j^n|\,\Delta y}{\nu} \leqslant 2 \tag{7.26}$$

Equation (7.26) confirms our suspicion that the "correct" representation is one that permits viscous-like behavior, in that the inequality can be satisfied for a sufficiently fine mesh, which of course, is achieved at convergence. The term $|v_j^n|\,\Delta y/\nu$ can be identified as a mesh Reynolds number. Mesh Peclet number, a more general terminology, is also frequently used for this term.

Maintaining the inequality of Eq. (7.26) provides a sufficient (but not the necessary) condition for diagonal dominance of the algebraic system. It appears that keeping the coefficients A_j and B_j negative to provide correct simulation of viscous behavior should be the major concern.

For some flows the constraint of Eq. (7.26) tends to require the use of an excessively large number of grid points. This has motivated several investigators to consider ways of altering the difference scheme to eliminate the mesh Reynolds number constraint. Most of the studies on this problem have focused on the more complex Navier-Stokes equations where the motivation for computational economy is stronger. The simplest remedy to the problem of the mesh Reynolds number constraint is to replace the central-difference representation for $v\,\partial u/\partial y$ by an upstream (one-way) difference:

$$v\frac{\partial u}{\partial y} \cong \frac{v_j^n\left(u_j^{n+1} - u_{j-1}^{n+1}\right)}{\Delta y}$$

when

$$v_j^n > 0$$

and

$$\frac{v_j^n\left(u_{j+1}^{n+1} - u_j^{n+1}\right)}{\Delta y}$$

when

$$v_j^n < 0$$

The T.E. associated with the upstream (also called "upwind") scheme creates an "artificial viscosity," which tends to enhance viscous-like behavior, causing a deterioration in accuracy in some cases.

It is clearly possible to devise upstream weighted schemes having a more favorable T.E. (using two or more upstream grid points), but these can lead to coefficient matrices that are not tridiagonal in form—a distinct disadvantage. Most of the example calculations illustrating the detrimental effects of upstream differencing have been for the Navier-Stokes equations. Less specific information appears to be available for the boundary-layer equations. The tentative conclusion is that the use of upstream differencing for $v\,\partial u/\partial y$ (when mandated by the mesh Reynolds number) is a sufficient solution to the constraint of Eq. (7.26). Use of central differencing for this term is, of course, recommended whenever feasible.

Rather than switching abruptly from the central to the upwind scheme as the mesh Reynolds number exceeds 2, the use of a combination (hybrid) of central and upwind schemes is recommended. This concept was originally suggested by Allen and Southwell (1955). Others, apparently not aware of this early work, have proposed similar or identical forms (Spalding, 1972; Raithby and Torrance, 1974). To illustrate this principle, we let $R_{\Delta y} = |v_j^n|\,\Delta y/\nu$ and R_c equal the desired critical mesh Reynolds number for initiating the hybrid scheme, $R_c \leqslant 2$. Then for $R_{\Delta y} \geqslant R_c$ we represent $v\,\partial u/\partial y$ by

$$v\frac{\partial u}{\partial y} \cong \underbrace{\left(\frac{R_c}{R_{\Delta y}}\right) v_j^n \frac{\left(u_{j+1}^{n+1} - u_{j-1}^{n+1}\right)}{2\,\Delta y}}_{\text{Central-difference component}}$$
$$+ \underbrace{\left(1 - \frac{R_c}{R_{\Delta y}}\right)\left(\frac{\left(v_j^n + |v_j^n|\right)}{2}\frac{\left(u_j^{n+1} - u_{j-1}^{n+1}\right)}{\Delta y} + \frac{\left(v_j^n - |v_j^n|\right)}{2}\frac{\left(u_{j+1}^{n+1} - u_j^{n+1}\right)}{\Delta y}\right)}_{\text{Upwind component}} \tag{7.27}$$

We observe that as $R_{\Delta y}$ increases, the weighting shifts toward the upwind representation. As $R_{\Delta y} \to \infty$, the representation is entirely upwind. The hybrid scheme maintains negative values for A_j and B_j in Eq. (7.25) while permitting the maximum utilization of the central-difference representation.

The reader is referred to the work of Raithby (1976), Leonard (1979a, 1979b), and Chow and Tien (1978) for an introduction to the literature on the mesh Reynolds number problem.

It is interesting that nothing has been noted in the technical literature about the mesh Reynolds number constraint for boundary-layer equations when the equations are solved in a coupled manner, as with the Davis coupled scheme discussed in this section or the modified box method discussed in Section 7.3.5. When coupling is used, the v in $v\,\partial u/\partial y$ is treated algebraically as an unknown and not merely as a coefficient for the unknown u's. It is possible that the coupling eliminates the "wiggles" and nonphysical behavior observed when central differencing is used for large mesh Reynolds numbers.

Closing comment on Crank-Nicolson and fully implicit methods. The difference schemes presented in this section have been purposely applied to equations in physical coordinates and have been written assuming Δx and Δy were both constant. This has been done primarily to keep the equations as simple as possible as the fundamental characteristics of the schemes were being discussed. As familiarity is gained with the basic concepts involved with differencing the boundary-layer equations, ways of extending schemes to a nonuniform grid will be pointed out.

7.3.4 DuFort-Frankel Method

Another finite-difference procedure that has worked well for both laminar and turbulent boundary layers is an extension of the method proposed by DuFort and Frankel (1953) for the heat equation. The difference representation will be written in a form that will accommodate variable grid spacing. We let $\Delta x_+ = x^{n+1} - x^n$, $\Delta x_- = x^n - x^{n-1}$, $\Delta y_+ = y_{j+1} - y_j$, $\Delta y_- = y_j - y_{j-1}$. The implicit methods of the previous section can be extended in applicability to a nonuniform grid by following a similar procedure.

In presenting the DuFort-Frankel procedure for the momentum and energy equations, the generalized transport PDE, Eq. (7.5), will be employed with the dependent variable ϕ denoting velocity components, a turbulence model parameter, or a thermal variable such as temperature or enthalpy. In the DuFort-Frankel differencing, stability is promoted by eliminating the appearance of ϕ_j^n in the diffusion term through the use of an average of ϕ at the $n+1$ and $n-1$ levels. With unequal spacing, however, Dancey and Pletcher (1974) observed that accuracy was improved by use of a linearly interpolated value of ϕ between $n-1$ and $n+1$ levels instead of a simple average. Here we define the linearly interpolated value as $\overline{\phi}_j^n$ according to $\overline{\phi}_j^n = (\Delta x_+ \phi_j^{n-1} + \Delta x_- \phi_j^{n+1})/$

$(\Delta x_+ + \Delta x_-)$. As before, for turbulent flows it is understood that u and v are time-mean quantities. For a compressible flow, $v = \tilde{v}$. For generality, let $\bar{\lambda} = \lambda_T + \lambda$, where λ_T is a turbulent diffusion coefficient. The DuFort-Frankel representation of the generalized transport equation becomes

$$\frac{\rho_j^n u_j^n\left(\phi_j^{n+1} - \phi_j^{n-1}\right)}{\Delta x_+ + \Delta x_-} + \frac{\rho_j^n v_j^n\left(\phi_{j+1}^n - \phi_{j-1}^n\right)}{\Delta y_+ + \Delta y_-}$$

$$= \frac{2}{\Delta y_+ + \Delta y_-}\left[\frac{\bar{\lambda}_{j+1/2}^n\left(\phi_{j+1}^n - \bar{\phi}_j^n\right)}{\Delta y_+} - \frac{\bar{\lambda}_{j-1/2}^n\left(\bar{\phi}_j^n - \phi_{j-1}^n\right)}{\Delta y_-}\right] + S_j^n \quad (7.28)$$

In the above, S_j^n denotes the source terms. Examples of source terms that frequently occur include the pressure gradient dp/dx in the x-momentum equation, where

$$S_j^n = \frac{p_j^{n+1} - p_j^{n-1}}{\Delta x_+ + \Delta x_-}$$

a viscous dissipation term $\bar{\mu}(\partial u/\partial y)^2$ in the energy equation when T is used as the thermal variable,

$$S_j^n = \bar{\mu}_j^n\left(\frac{u_{j+1}^n - u_{j-1}^n}{\Delta y_+ + \Delta y_-}\right)^2$$

and a dissipation term $C_D\,\rho(\bar{k})^{3/2}/l$ in the modeled form of the turbulence kinetic energy equation

$$S_j^n = C_D\,\rho_j^n\left(\frac{\Delta y_+(\bar{k})_{j-1}^n + \Delta y_-(\bar{k})_{j+1}^n}{\Delta y_+ + \Delta y_-}\right)^{1/2}\left(\frac{\Delta x_+(\bar{k})_j^{n-1} + \Delta x_-(\bar{k})_j^{n+1}}{\Delta x_+ + \Delta x_-}\right)\Big/ l_j^n$$

Note that this latter representation avoids using the dependent variable $\bar{k}$ at (n, j). This is required by stability (see Malik and Pletcher, 1978), as might be expected in light of the special treatment required for the diffusion term noted above.

We recall (Chapter 4) that the DuFort-Frankel representation is explicit. Although ϕ_j^{n+1} appears in both the left and right sides (within $\bar{\phi}_j^n$) of the equation, the equation can be rearranged to isolate ϕ_j^{n+1}, so that we can write ϕ_j^{n+1} = (all known quantities at the n and $n-1$ levels). The formal T.E. for the equation with $\Delta x_+ = \Delta x_-$ and $\Delta y_+ = \Delta y_-$ is $O[(\Delta x)^2] + O[(\Delta y)^2] + O[(\Delta x/\Delta y)^2]$. However, the leading term in the T.E. represented by $O[(\Delta x/\Delta y)^2]$ is actually $(\Delta x/\Delta y)^2(\partial^2\phi/\partial x^2)$, and $\partial^2\phi/\partial x^2$ is presumed to be very small for boundary-layer flows. One can show that a deterioration in the formal T.E. is generally expected as the grid spacing becomes unequal, although a paper by Blottner (1974) points out exceptions. This deterioration would be observed in all methods presented thus far in this chapter. In practice, the increase in actual error due to the use of unequal spacing may be negligible. In nearly all cases, remedies can be found that will restore the original formal T.E. at the expense of algebraic operations. For example, Hong (1974) demonstrated that the streamwise derivative $\partial\phi/\partial x$ in the DuFort-Frankel method can be written as

$$\frac{(\Delta x_-)^2\phi_j^{n+1} - (\Delta x_+)^2\phi_j^{n-1} + \left[(\Delta x_+)^2 - (\Delta x_-)^2\right]\phi_j^n}{\Delta x_-\,(\Delta x_+)^2 + \Delta x_+\,(\Delta x_-)^2}$$

with second-order accuracy even when $\Delta x_+ \neq \Delta x_-$.

A consistent treatment of the continuity equation is given by

$$\frac{\rho_j^{n+1}v_j^{n+1} - \rho_{j-1}^{n+1}v_{j-1}^{n+1}}{\Delta y_-} + \frac{\rho_j^{n+1}u_j^{n+1} - \rho_j^{n-1}u_j^{n-1} + \rho_{j-1}^{n+1}u_{j-1}^{n+1} - \rho_{j-1}^{n-1}u_{j-1}^{n-1}}{2(\Delta x_+ + \Delta x_-)} = 0 \tag{7.29}$$

with T.E. of $O(\Delta x) + O[(\Delta y)^2]$.

A stability analysis for $\Delta y = \text{const}$ (Madni and Pletcher, 1975a, 1975b) suggests that

$$\Delta x_+ \leqslant \frac{\rho_j^n u_j^n\,\Delta y}{\left|\rho_j^n v_j^n + \left(\bar{\lambda}_{j-1}^n - \bar{\lambda}_{j+1}^n\right)/2\,\Delta y\right|} \tag{7.30}$$

It would appear that this constraint could also be used to provide a rough guide under variable Δy conditions. In practice, this condition has not proven to be especially restrictive on the marching step size, probably because v/u is generally very small and the other term in the denominator involves differences in the diffusion coefficient rather than the coefficient itself.

It is interesting to note that Eq. (7.30) follows essentially from the Courant-Friedrichs-Lewy (CFL) condition rather than the diffusion stability limit for the boundary-layer momentum equation. This becomes evident when the diffusion term $\partial/\partial y(\bar{\lambda}\,\partial\phi/\partial y)$ is expanded to two terms and the boundary-layer equation is rearranged as

$$\frac{\partial\phi}{\partial x} + \frac{1}{\rho u}\left(\rho v - \frac{\partial\bar{\lambda}}{\partial y}\right)\frac{\partial\phi}{\partial y} = \frac{\bar{\lambda}}{\rho u}\frac{\partial^2\phi}{\partial y^2} + \frac{S}{\rho u}$$

Now simply applying the CFL condition gives Eq. (7.30).

The boundary-layer calculation begins by utilizing an initial distribution for the ϕ variables. Since the DuFort-Frankel procedure requires information at *two* streamwise levels in order to advance the calculation, some other method must be used to obtain a solution for at least one streamwise station before the DuFort-Frankel scheme can be employed. A simple explicit scheme is most frequently used to provide these starting values. A typical calculation would require the solution to the momentum, continuity, and energy equations. The equations can be solved sequentially starting with the momentum equation in an uncoupled manner. The usual procedure is to solve first for the unknown streamwise velocities from the momentum equation starting with the point nearest the wall and working outward to the outer edge of the boundary layer. The outer edge of the boundary layer is located when the velocity from the solution is within a prescribed tolerance of the velocity specified as the outer boundary condition. The energy equation can be solved in a like manner for the thermal variable. The density at the new station can be evaluated from an equation of state. Finally, the continuity equation is used to obtain the normal component of velocity at the $n + 1$ level starting from the point adjacent to the wall and working outward.

The explicit nature of the DuFort-Frankel procedure is probably its most attractive feature. Those inexperienced in numerical methods are likely to feel more comfortable programming an explicit procedure than they are in applying an implicit scheme. A second significant feature of the scheme is that no additional linearizations, iterations, or assumptions are needed to evaluate coefficients in the equation, since these all appear at the n-level, where they are known values. Further details on the application of the DuFort-Frankel type schemes to wall boundary layers can be found in the work by Pletcher (1969, 1970, 1971).

7.3.5 Box Method

Keller and Cebeci (1972) applied the box-difference scheme (introduced in Section 4.2.8) to the boundary-layer momentum and continuity equations after they had first been transformed to a single third-order PDE using the Mangler and Levy-Lees transformations (see Cebeci and Smith, 1974). The third-order PDE is written as a system of three first-order PDEs using newly defined variables in a manner that parallels the procedure commonly employed in the numerical solution of third-order ODEs. The box-differencing scheme with Newton linearization is then applied to the three first-order PDEs, giving rise to a block tridiagonal system having 3×3 blocks, which is solved by a block elimination scheme. The corresponding treatment for the energy equation gives rise to a block tridiagonal system with 2×2 blocks.

The details of the Keller-Cebeci box method for the boundary-layer equations will not be given here but can be found in the work by Cebeci and Smith (1974). Instead, we will indicate how a modified box scheme can be developed that only requires the use of the same modified tridiagonal elimination scheme presented in Section 7.3.3 for the Davis coupled scheme. From reports in the literature (Blottner, 1975a; Wornom, 1977), the modified box scheme appears to require only on the order of one-half as much computer time as the standard box scheme for the boundary-layer equations.

The momentum and energy equations for a compressible flow can be written in the generalized form given by Eq. (7.5). For rectangular coordinates, this becomes

$$\rho u \frac{\partial \phi}{\partial x} + \rho \tilde{v} \frac{\partial \phi}{\partial y} = \frac{\partial}{\partial y}\left(\bar{\lambda} \frac{\partial \phi}{\partial y}\right) + S \qquad (7.31)$$

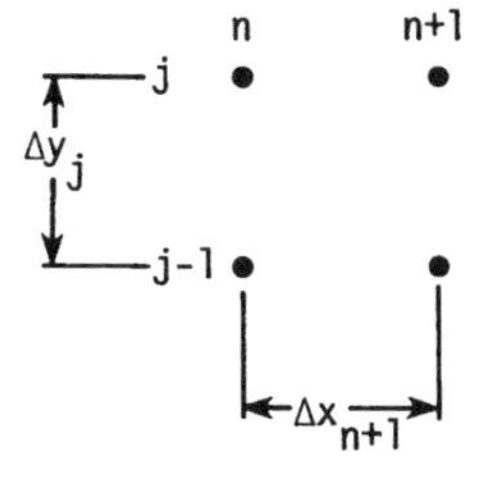

Figure 7.2 Grid arrangement for the modified box scheme.

where

$$\bar{\lambda} = \lambda_T + \lambda$$

The continuity equation can be written as

$$\frac{\partial \rho u}{\partial x} + \frac{\partial \rho \tilde{v}}{\partial y} = 0 \tag{7.32}$$

The grid nomenclature is given in Fig. 7.2. If we let

$$\bar{\lambda}\frac{\partial \phi}{\partial y} = q$$

$$D = S + \frac{\partial q}{\partial y} - \rho\tilde{v}\frac{\partial \phi}{\partial y}$$

Equation (7.31) can then be written as

$$\rho u \frac{\partial \phi}{\partial x} = D \tag{7.33}$$

Centering on the box grid gives

$$\begin{aligned} &\frac{(\rho u)_{j-1/2}^{n+1} + (\rho u)_{j-1/2}^{n}}{2} \frac{\phi_{j-1/2}^{n+1} - \phi_{j-1/2}^{n}}{\Delta x_{n+1}} \\ &\quad = S_{j-1/2}^{n+1/2} - \frac{(\rho\tilde{v})_{j-1/2}^{n+1/2}\left(\phi_j^{n+1/2} - \phi_{j-1}^{n+1/2}\right)}{\Delta y_j} + \frac{q_j^{n+1/2} - q_{j-1}^{n+1/2}}{\Delta y_j} \end{aligned} \tag{7.34}$$

Utilizing the definition of $q_{j-1/2}^{n+1/2}$,

$$\bar{\lambda}_{j-1/2}^{n+1/2} \frac{\phi_j^{n+1/2} - \phi_{j-1}^{n+1/2}}{\Delta y_j} = \frac{q_j^{n+1/2} + q_{j-1}^{n+1/2}}{2} \tag{7.35}$$

we can eliminate $q_{j-1}^{n+1/2}$ from Eq. (7.34). This gives

$$\frac{(\rho u)_{j-1/2}^{n+1} + (\rho u)_{j-1/2}^{n}}{2} \frac{\phi_{j-1/2}^{n+1} - \phi_{j-1/2}^{n}}{\Delta x_{n+1}}$$
$$= S_{j-1/2}^{n+1/2} - (\rho \tilde{v})_{j-1/2}^{n+1/2} \frac{\phi_j^{n+1/2} - \phi_{j-1}^{n+1/2}}{\Delta y_j}$$
$$+ \frac{2q_j^{n+1/2}}{\Delta y_j} - 2\bar{\lambda}_{j-1/2}^{n+1/2} \frac{\phi_j^{n+1/2} - \phi_{j-1}^{n+1/2}}{(\Delta y_j)^2} \tag{7.36}$$

In a similar manner, the momentum equation can be put into difference form centering about the point $(n + \frac{1}{2}, j + \frac{1}{2})$, and $q_{j+1}^{n+1/2}$ can be eliminated from that equation using the definition of $q_{j+1/2}^{n+1/2}$. The result is

$$\frac{(\rho u)_{j+1/2}^{n+1} + (\rho u)_{j+1/2}^{n}}{2} \frac{\phi_{j+1/2}^{n+1} - \phi_{j+1/2}^{n}}{\Delta x_{n+1}}$$
$$= S_{j+1/2}^{n+1/2} - (\rho \tilde{v})_{j+1/2}^{n+1/2} \frac{\phi_{j+1}^{n+1/2} - \phi_j^{n+1/2}}{\Delta y_{j+1}}$$
$$+ 2\bar{\lambda}_{j+1/2}^{n+1/2} \frac{\phi_{j+1}^{n+1/2} - \phi_j^{n+1/2}}{(\Delta y_{j+1})^2} - \frac{2q_j^{n+1/2}}{\Delta y_{j+1}} \tag{7.37}$$

Equations (7.36) and (7.37) can be combined to eliminate $q_j^{n+1/2}$. This is accomplished by multiplying Eq. (7.36) by Δy_j, and Eq. (7.37) by Δy_{j+1} and adding the two products. After replacing quantities specified for evaluation at grid midpoints by averages from adjacent grid points, the results can be written as

$$\frac{\Delta y_j}{2\,\Delta x_{n+1}} \left[(\rho u)_j^{n+1} + (\rho u)_{j-1}^{n+1} + (\rho u)_j^{n} + (\rho u)_{j-1}^{n} \right]$$
$$\times \left(\phi_j^{n+1} + \phi_{j-1}^{n+1} - \phi_j^{n} - \phi_{j-1}^{n} \right)$$
$$+ \frac{\Delta y_{j+1}}{2\,\Delta x_{n+1}} \left[(\rho u)_{j+1}^{n+1} + (\rho u)_j^{n+1} + (\rho u)_{j+1}^{n} + (\rho u)_j^{n} \right]$$
$$\times \left(\phi_{j+1}^{n+1} + \phi_j^{n+1} - \phi_{j+1}^{n} - \phi_j^{n} \right)$$
$$+ \tfrac{1}{2} \left[(\rho \tilde{v})_j^{n+1} + (\rho \tilde{v})_{j-1}^{n+1} + (\rho \tilde{v})_j^{n} + (\rho \tilde{v})_{j-1}^{n} \right]$$
$$\times \left(\phi_j^{n+1} + \phi_j^{n} - \phi_{j-1}^{n+1} - \phi_{j-1}^{n} \right)$$
$$+ \tfrac{1}{2} \left[(\rho \tilde{v})_{j+1}^{n+1} + (\rho \tilde{v})_j^{n+1} + (\rho \tilde{v})_{j+1}^{n} + (\rho \tilde{v})_j^{n} \right]$$

$$\times\left(\phi_{j+1}^{n+1}+\phi_{j+1}^{n}-\phi_{j}^{n+1}-\phi_{j}^{n}\right)$$
$$=\Delta y_j\left(S_j^{n+1}+S_{j-1}^{n+1}+S_j^n+S_{j-1}^n\right)+\Delta y_{j+1}\left(S_{j+1}^{n+1}+S_j^{n+1}+S_{j+1}^n+S_{j-1}^n\right)$$
$$+\frac{\left(\bar{\lambda}_{j+1}^{n+1}+\bar{\lambda}_j^{n+1}+\bar{\lambda}_{j+1}^n+\bar{\lambda}_j^n\right)\left(\phi_{j+1}^{n+1}+\phi_{j+1}^n-\phi_j^{n+1}-\phi_j^n\right)}{\Delta y_{j+1}}$$
$$-\frac{\left(\bar{\lambda}_j^{n+1}+\bar{\lambda}_{j-1}^{n+1}+\bar{\lambda}_j^n+\bar{\lambda}_{j-1}^n\right)\left(\phi_j^{n+1}+\phi_j^n-\phi_{j-1}^{n+1}-\phi_{j-1}^n\right)}{\Delta y_j} \tag{7.38}$$

Equation (7.38) can be expressed in the tridiagonal format for the unknown ϕ's, but as for all implicit methods, some scheme must be devised for treating the algebraic nonlinearities arising through the coefficients. Conceptually, any of the procedures presented in Section 7.3.3 can be employed. The most suitable representation of the continuity equation may depend upon the procedure used to accomplish the linearization of the momentum equation. To date, Newton linearization with coupling (Blottner, 1975a) has been the most commonly used procedure. For this, the continuity equation can be written as

$$\frac{(\rho u)_j^{n+1}+(\rho u)_{j-1}^{n+1}-(\rho u)_j^n-(\rho u)_{j-1}^n}{2\,\Delta x_{n+1}}$$
$$+\frac{(\rho\tilde{v})_j^{n+1}+(\rho\tilde{v})_j^n-(\rho\tilde{v})_{j-1}^{n+1}-(\rho\tilde{v})_{j-1}^n}{2\,\Delta y_j}=0 \tag{7.39}$$

The momentum equation involves $(\rho\tilde{v})_{j+1}^{n+1}$, $(\rho\tilde{v})_j^{n+1}$, and $(\rho\tilde{v})_{j-1}^{n+1}$. To employ the modified tridiagonal elimination scheme, the continuity equation (Eq. 7.39) can be written between the j and $j+1$ levels and $(\rho\tilde{v})_{j+1}^{n+1}$ eliminated from the momentum equation by substitution. After employing Newton linearization in a manner that parallels the procedures illustrated in Section 7.3.3 for the fully implicit Davis coupled method, the coupled momentum and continuity equations can be solved with the modified tridiagonal elimination scheme. The energy equation is usually solved in an uncoupled manner, and properties (including the turbulent viscosity) are updated iteratively as desired or required by accuracy constraints.

With the box and modified box schemes, the wall shear stress and heat flux are usually determined by evaluating q at the wall ($j=1$). For the modified box scheme this is done after the solutions for ϕ, $\tilde{v}$, and ρ have been determined. An expression for $q_1^{n+1/2}$ can be obtained by writing Eqs. (7.34) and (7.35) for $j=2$ and eliminating $q_2^{n+1/2}$ by a simple substitution.

7.3.6 Other Methods

Exploratory studies of limited scope have indicated that the Barakat and Clark ADE method can be used for solving the boundary-layer equations (R. G. Hindman, 1975, private communication: S. S. Hwang, 1975, private communication). These results indicate that ADE methods are roughly equivalent in accuracy and computation time to the more conventional implicit methods for boundary-layer problems. Higher-order schemes (up through fourth order) have also been applied to the boundary-layer equations. A critical study of some of these schemes was reported by Wornom (1977). It is worthwhile to note that the accuracy of lower-order methods can also be improved through the use of Richardson extrapolation (Ralston, 1965; Cebeci and Smith, 1974).

It is believed that the most commonly used difference schemes for 2-D or axisymmetric boundary layers have been described in this section. No attempt has been made to cover all known methods in detail.

7.3.7 Coordinate Transformations for Boundary Layers

The general subject of coordinate transformations has been treated in Chapter 5. In the present chapter the focus has been on the difference schemes themselves, and to illustrate these in the simplest possible manner, the equations have been presented in rectangular Cartesian, "physical coordinates."

It is well to point out that there may be advantages to numerically solving the equations in alternative forms. Two approaches are observed. One proceeds by introducing new dependent and independent variables analytically to transform the mathematical representation of the conservation principles before the equations are discretized. We shall refer to this strategy approach as the *analytical transformation approach.* A second strategy employs an independent variable transformation very much along lines introduced in Chapter 5 and will be referred to as the *generalized coordinate approach.*

The main objective of the transformations is generally to obtain a coordinate frame for computation in which the boundary-layer thickness remains as constant as possible and to remove the singularity in the equations at the leading edge or stagnation point. Unfortunately, for complex turbulent flows, the optimum transformation leading to a constant boundary-layer thickness in the transformed plane has not been identified, although the transformation suggested by Carter et al. (1980) shows promise.

Analytical transformation approach. The most commonly used analytical transformation makes use of the transverse similarity variable η employed in the Blasius similarity solution to the laminar boundary layer. We will give an example of such a transformation applied to the constant property laminar boundary-layer equations. We start with

continuity:

$$\frac{\partial u}{\partial x} + \frac{\partial v}{\partial y} = 0 \tag{7.40}$$

momentum:

$$u\frac{\partial u}{\partial x} + v\frac{\partial u}{\partial y} = u_e\frac{du_e}{dx} + \nu\frac{\partial^2 u}{\partial y^2} \tag{7.41}$$

The crucial element of the transformation is the introduction of

$$\eta = \frac{y}{x}\left(\frac{u_e x}{\nu}\right)^{1/2}$$

From this point on, several variations are possible, but a common procedure is to let $x = x$ (no stretching of x) and $F = u/u_e$. Using the chain rule, we note that

$$\left.\frac{\partial}{\partial x}\right)_y = \left.\frac{\partial}{\partial x}\right)_\eta + \left.\frac{\partial \eta}{\partial x}\right)_y \left.\frac{\partial}{\partial \eta}\right)_x = \left.\frac{\partial}{\partial x}\right)_\eta + \left(\frac{\eta}{2u_e}\frac{du_e}{dx} - \frac{\eta}{2x}\right)\left.\frac{\partial}{\partial \eta}\right)_x$$

and

$$\left.\frac{\partial}{\partial y}\right)_x = \left.\frac{\partial x}{\partial y}\right)_x \left.\frac{\partial}{\partial x}\right)_\eta + \left.\frac{\partial \eta}{\partial y}\right)_x \left.\frac{\partial}{\partial \eta}\right)_x = \left(\frac{u_e}{x\nu}\right)^{1/2} \left.\frac{\partial}{\partial \eta}\right)_x$$

Replacing the x and y derivatives in Eqs. (7.40) and (7.41) as indicated, and utilizing F, results in the transformed momentum and continuity equations:

momentum:

$$xF\frac{\partial F}{\partial x} + V\frac{\partial F}{\partial \eta} = \beta(1 - F^2) + \frac{\partial^2 F}{\partial \eta^2} \tag{7.42}$$

continuity:

$$x\frac{\partial F}{\partial x} + \frac{\partial V}{\partial \eta} + F\frac{\beta + 1}{2} = 0 \tag{7.43}$$

where

$$V = \frac{\beta - 1}{2}F\eta + \left(\frac{x}{\nu u_e}\right)^{1/2} v$$

$$\beta = \frac{x}{u_e}\frac{du_e}{dx}$$

When $x = 0$, the streamwise derivatives vanish from the transformed equations, and a system of two ODEs remains. It is common to solve these equations with a slightly modified version of the marching technique employed for the rest of the flow domain, i.e., for $x > 0$, although special numerical procedures applicable to ODEs could be used.

There is no singular behavior at $x = 0$ in the new coordinate system, since the troublesome streamwise derivatives have been eliminated. In fact, for laminar flow over a plate, the solution for $x = 0$ is the well-known Blasius similarity solution. Naturally, for a zero pressure gradient flow, the marching solution for $x > 0$ should reproduce essentially the same solution downstream, and the boundary-layer thickness should remain constant. When pressure gradients or wall boundary conditions force a nonsimilar laminar flow solution, the boundary-layer thickness will change somewhat along the flow. For nearly similar laminar flows, we would expect the solution of the transformed equations to provide greater and more uniform accuracy near the leading edge than the solution in physical coordinates because of the tendency of the former procedure to divide the boundary layer into a more nearly constant number of points in the transverse direction. For turbulent flows, we observe the boundary-layer thickness growing along the surface, generally quite significantly, even with the use of the above transformed variables.

Analytical transformations have also proven useful for solving the compressible boundary-layer equations. The Levy-Lees and Mangler transformations extend the similarity variable approach to compressible 2-D and axisymmetric boundary layers and have been successfully utilized in finite-difference methods by Blottner (1975b) and Christoph and Pletcher (1983).

For external laminar boundary-layer calculations, the use of transformed coordinates of the similarity type is recommended. For turbulent flows the advantages of transformations suggested to date are less certain.

Generalized coordinate approach. Generalized nonorthogonal coordinates can be introduced into the boundary-layer equations if care is taken to be consistent with the boundary-layer approximation. In particular, the x-axis should align with the main flow direction; however, the axis need not be straight. A second condition is that the grid lines that intersect the x axis should be straight (negligible curvature) and should be orthogonal to the x axis. A grid system that meets these conditions will be referred to as the generalized boundary-layer grid and can be established by the intersection of lines of constant ξ and η, where ξ is $\xi(x)$ and η is $\eta(x, y)$. Two examples of grids that meet the above conditions are illustrated in Fig. 7.3. The grid shown in Fig. 7.3(a) would be convenient for external flow applications, while the grid of Fig. 7.3(b) would be particularly well suited for solving a flow in a straight, symmetric, 2-D channel of varying cross-sectional area. Notice that this grid allows the spacing between lines of constant η to vary in the x direction, permitting grid points to be packed more closely in regions where the boundary layer is thin. Letting $\eta = (y/x)\sqrt{\text{Re}_x}$ gives the well-known Blasius similarity variable as a special case of the generalized boundary-layer grid. The generalized boundary-layer grid can be implemented by introducing ξ and η into the governing equations as new independent variables using the chain rule to replace derivatives with respect to x and y:

$$\frac{\partial}{\partial x} = \xi_x \frac{\partial}{\partial \xi} + \eta_x \frac{\partial}{\partial \eta}$$

$$\frac{\partial}{\partial y} = \eta_y \frac{\partial}{\partial \eta}$$

The corresponding Jacobian is $J = \xi_x \eta_y$.

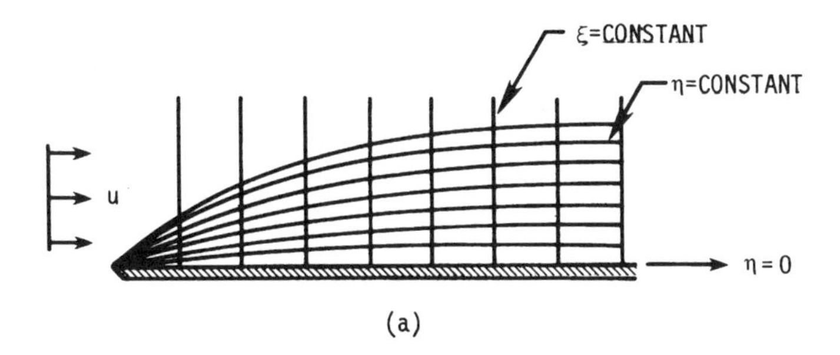

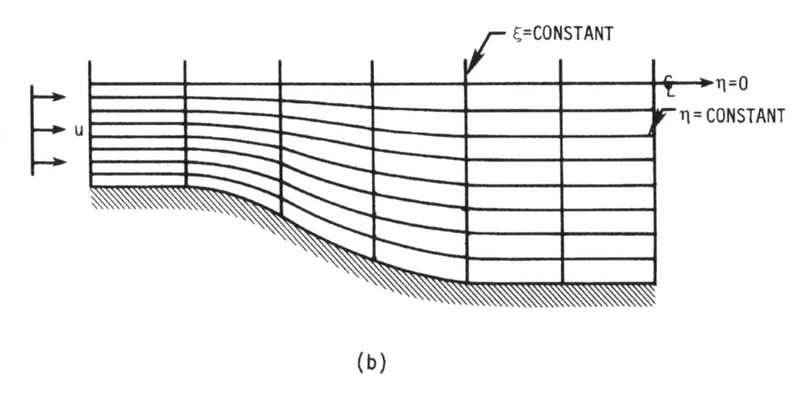

Figure 7.3 Examples of generalized boundary layer grids. (a) External flow. (b) Internal flow.

The metric terms in the equations are most conveniently represented as

$$\begin{aligned} \xi_x &= Jy_\eta = \frac{1}{x_\xi} \\ \eta_x &= -Jy_\xi \qquad \eta_y = Jx_\xi = \frac{1}{y_\eta} \end{aligned} \tag{7.44}$$

After applying this independent variable transformation to the steady incompressible boundary-layer momentum and continuity equations in nondimensional form (as in Section 5.3.2), the following are obtained:

$$\begin{aligned} &\frac{\partial}{\partial \xi}(x_\xi y_\eta u) + \frac{\partial}{\partial \eta}(x_\xi y_\eta V) = 0 \\ &\frac{u}{x_\xi}\frac{\partial u}{\partial \xi} + V\frac{\partial u}{\partial \eta} = -\frac{1}{x_\xi}\frac{\partial p}{\partial \xi} + \frac{1}{y_\eta \operatorname{Re}}\frac{\partial}{\partial \eta}\left(\frac{1}{y_\eta}\frac{\partial u}{\partial \eta}\right) \end{aligned} \tag{7.45}$$

The contravariant velocity component $V = \eta_x u + \eta_y v$ has been introduced. The u velocity component is orthogonal to lines of constant ξ and thus also represents a contravariant velocity. Care must be taken in representing the metric quantities when discretizing the equations. When the fully implicit formulation of the finite-difference equations is used, the best results have been obtained (Ramin and Pletcher, 1993) by representing the diffusion term as

$$\frac{1}{y_\eta}\frac{\partial}{\partial \eta}\left(\frac{1}{y_\eta}\frac{\partial u}{\partial \eta}\right) \cong \frac{1}{\left[(y_\eta)_{i,j} + (y_\eta)_{i+1,j}\right]}\left[\frac{1}{(y_\eta)_{i+1,j+1/2}}(u_{i+1,j+1} - u_{i+1,j}) - \frac{1}{(y_\eta)_{i+1,j-1/2}}(u_{i+1,j} - u_{i+1,j-1})\right] \tag{7.46}$$

where $(y_\eta)_{i,j}$ is to be interpreted as $(y_{i,j+1} - y_{i,j-1})/2\,\Delta\eta$, and $(y_\eta)_{i+1,j+1/2}$ as $(y_{i+1,j+1} - y_{i+1,j})/\Delta\eta$. A qualifying test for the correct representation of the metric quantities is that the scheme should reproduce the Blasius laminar boundary-layer solution exactly when the grid is specified according to $\eta = (y/x)\sqrt{\text{Re}_x}$.

An important advantage of using the analytical transformed coordinates discussed previously was that leading edge and stagnation point singularities could be removed. This permitted very accurate solutions to be obtained near these singular points, in contrast to what was possible when simple physical (generally Cartesian) coordinates were used. It is also possible to obtain very accurate starting solutions and to resolve regions near leading edges and stagnation points using the generalized boundary-layer grid. Since it is well known that similarity solutions are valid in such regions, it is only necessary to connect the first two marching stations by the use of a grid that is constructed from the similarity variables, such as $\xi = x$ and $\eta = (y/x)\sqrt{\text{Re}_x}$ for an incompressible 2-D flow. It is best to omit the points at $x = 0$, since the boundary layer is vanishingly thin at that location and only exists in the sense of a limit. It is sufficient to resolve the boundary layer at the first Δx beyond $x = 0$, since that is the first location where the boundary-layer thickness is finite. A characteristic of the similarity solution is that it is independent of the marching coordinate. Thus the finite-difference approximation to it can be obtained by iteratively marching from station 2 to 3. With each sweep, the solution most recently obtained at station 3 is used at station 2. The iterative sweeps from stations 2 to 3 continue until the two solutions agree. This indicates that a solution locally independent of the marching coordinate has been found.

7.3.8 Special Considerations for Turbulent Flows

The accurate solution of the boundary-layer equations for turbulent flow using models that evaluate the turbulent viscosity at all points within the flow requires that grid points be located within the viscous sublayer, $y^+ \leqslant 4.0$ for incompressible flow, and perhaps $y^+ \leqslant 1.0$ or 2.0 for flows in which a solution to the energy equation is also being obtained. The use of equal grid spacing for the transverse coordinate would require several thousand grid points across the boundary layer for a typical calculation at moderate Reynolds numbers. This at least provides motivation for considering ways to reduce the number of grid points required to span the boundary layer. The techniques that have been used successfully fall into three categories: use of wall functions, unequal grid spacing, and coordinate transformations.

Use of wall functions. For many turbulent wall boundary layers the inner portion of the flow appears to have a "universal" character captured by the logarithmic "law-of-the-wall" discussed previously (see Fig. 5.7). Basically, this inner region is a zone in which convective transport is relatively unimportant. The law-of-the-wall can be roughly thought of as a solution to the boundary-layer momentum equation using Prandtl's mixing-length turbulence model when convective and pressure gradient terms are unimportant. Corresponding nearly universal behavior has been observed for the temperature distribution for many turbulent flows, and wall functions can be used to provide an inner boundary condition for solutions to the energy equation. Thus, with the wall function approach, the boundary-layer equations are solved using a turbulence model in the outer region on a relatively coarse grid, and the near-wall region is "patched in" through the use of a form of the law-of-the-wall, which in fact, represents an approximate solution for the near-wall region. In this approach, the law-of-the-wall is usually assumed to be valid in the range $30 < y^+ < 200$, and the first computational point away from the wall is located in this interval. Boundary conditions are developed for the dependent variables in the transport equations being solved (u, T, $\bar{k}$, ϵ, etc.) at this point from the wall functions. Many variations in this procedure are possible, and details depend upon the turbulence model and difference scheme being used. The procedure has been well developed for use with the $\bar{k}$-ϵ turbulence model, and recommended wall functions for u, T, $\bar{k}$, and ϵ can be found in the work by Launder and Spalding (1974).

Like turbulence models themselves, wall functions need modifications to accurately treat effects such as wall blowing and suction and surface roughness. Their use does, however, circumvent the need for many closely spaced points near the wall.

Use of unequal grid spacing. Almost without exception, turbulent boundary-layer calculations that have applied the difference scheme right down to the wall have utilized either a variable grid scheme or what is often equivalent, a coordinate transformation. Arbitrary spacing will work. Pletcher (1969) used Δy corresponding to Δy^+ [defined as $\Delta y(\tau_w/\rho)^{1/2}/\nu_w] \cong 1.0$ for several mesh increments nearest the wall and then approximately doubled every few points until Δy^+ reached 100 in the outer part of the flow.

Another commonly used (Cebeci and Smith, 1974) and very workable scheme maintains a constant ratio between two adjacent increments:

$$\frac{\Delta y_+}{\Delta y_-} = \frac{\Delta y_{j+1}}{\Delta y_j} = K \tag{7.47}$$

In this constant ratio scheme, each grid spacing is increased by a fixed percentage from the wall outward. This results in a geometric progression in the size of the spacing. K is usually a number between 1.0 and 2.0 for turbulent flows. For the constant ratio scheme it follows that

$$\Delta y_j = K^{j-1}\,\Delta y_1 \qquad y_j = \Delta y_1 \frac{K^{j-1} - 1}{K - 1} \tag{7.48}$$

The accuracy (and occasionally the stability) of some schemes appears sensitive to the value of K being used. Most methods appear to give satisfactory results for $K \leqslant 1.15$. For a typical calculation using $\Delta y_1^+ \cong 1.5$, $K = 1.04$, and $y_e^+ \cong 3000$, Eqs. (7.47) and (7.48) can be used to determine that about 113 grid points in the transverse direction would be required.

As a difference scheme is being generalized to accommodate variable grid spacing, the truncation error should be reevaluated, since a deterioration in the formal T.E. is common in these circumstances. For example, the treatment previously recommended for the transverse shear stress derivative is

$$\frac{\partial}{\partial y}\left(\mu \frac{\partial u}{\partial y}\right)_j^n = \frac{2}{\Delta y_+ + \Delta y_-}\left(\mu_{j+1/2}^n \frac{u_{j+1}^n - u_j^n}{\Delta y_+} - \mu_{j-1/2}^n \frac{u_j^n - u_{j-1}^n}{\Delta y_-}\right)$$
$$+ O(\Delta y_+ - \Delta y_-) + O\left[(\Delta y_+ + \Delta y_-)^2\right]$$

which at first appears to be first order accurate unless there is a way to show that $O(\Delta y_+ - \Delta y_-) = O[(\Delta y)^2]$ for a particular scheme. Blottner (1974) has shown that the treatment of derivatives indicated above in his Crank-Nicolson scheme using the constant ratio arrangement for mesh spacing is locally second order accurate. To prove this, Blottner interpreted the constant ratio scheme in terms of a coordinate transformation (see below) and verified his findings by calculations that indicated that his scheme behaved as though the T.E.s were second order as the mesh was refined.

Use of coordinate transformations. The general topic of coordinate transformations was treated in Chapter 5. Here we are considering the use of a coordinate transformation for the purpose of providing unequal grid spacing in the physical plane. Transformation 1 of Section 5.6 provides a good example of this concept (see also Fig. 5.8). Such a transformation permits the use of standard equal-increment differencing of the governing equations in terms of the transformed coordinates. Thus the clustering of points near the wall can be achieved without deterioration in the *order* of the T.E. On the other hand, the equations generally become more complex in terms of the transformed variables, and new variable coefficients always appear. The actual *magnitude* of the T.E. will be influenced by the new coefficients.

Transformations 1 and 2 in Section 5.6 of Chapter 5 are representative of those that can be readily used with the boundary-layer equations.

7.3.9 Example Applications

For laminar flows in which the boundary-layer approximation is valid, finite-difference predictions can easily be made to agree with results of more exact theories to several significant figures. Even with only modest attention to mesh size, agreement to within ± 1–2% of some "exact" standard is relatively common. Figure 7.4 compares the velocity profile computed by a DuFort-Frankel type difference scheme (Pletcher, 1971) with the analytical results of van Driest (1952) for laminar flow at a Mach number of 4 and $T_w/T_e = 4$. The temperature profiles are compared for the same flow conditions in Fig. 7.5. The agreement is excellent and typical of what can be expected for laminar boundary-layer flows.

The prediction of turbulent boundary-layer flows is another matter. The issue of turbulence modeling adds complexity and uncertainty to the prediction. Turbulence models can be adjusted to give good predictions for a limited class of flows, but when applied to other flows containing conditions not accounted for by the model, poor agreement is often noted. Because of the usual level of uncertainty in both the experimental measurements and turbulence models, agreement to within ± 3–4% is generally considered good for turbulent flows.

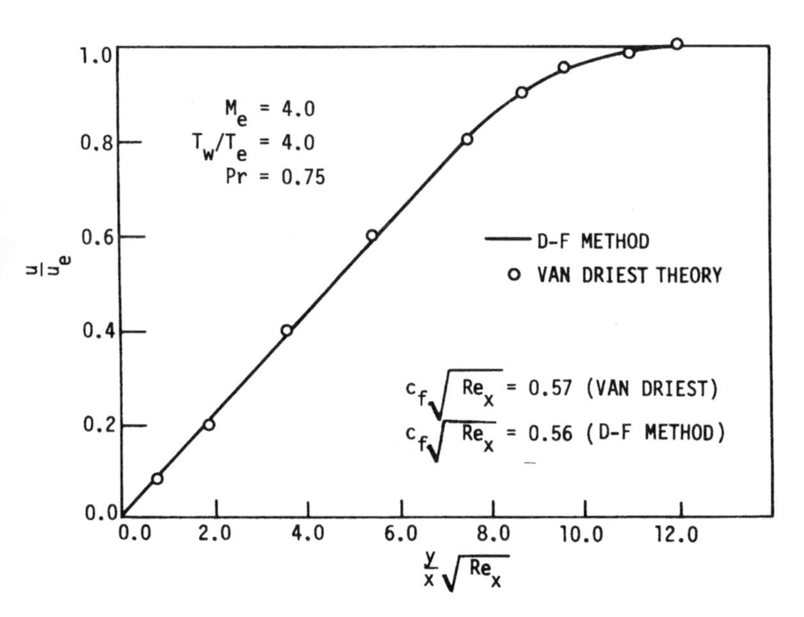

Figure 7.4 Velocity profile comparisons for a laminar compressible boundary layer. Solid line represents predictions from the DuFort-Frankel finite-difference scheme (Pletcher, 1971).

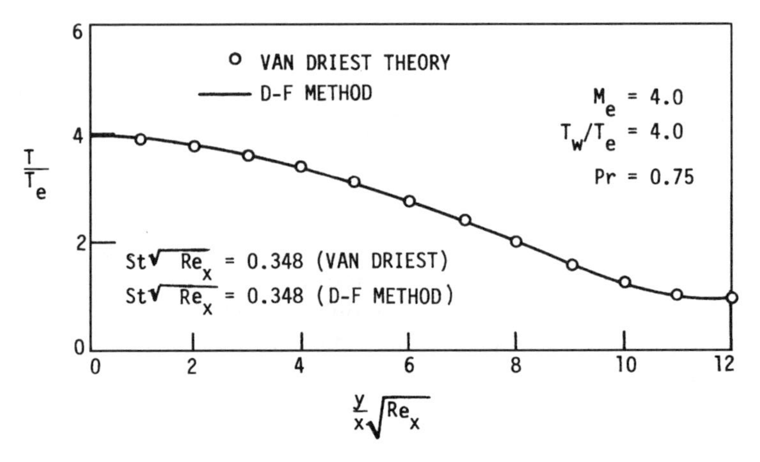

Figure 7.5 Temperature profile comparisons for a laminar compressible boundary layer. Solid line represents predictions from the DuFort-Frankel finite-difference scheme (Pletcher, 1971).

Even a simple algebraic turbulence model can give good predictions over a wide range of Mach numbers for turbulent boundary-layer flows in zero or mild pressure gradients. Figure 7.6 compares the prediction of a DuFort-Frankel finite-difference method with the measurements of Coles (1953) for a turbulent boundary layer on an adiabatic plate at a free stream Mach number of 4.554. The agreement is excellent.

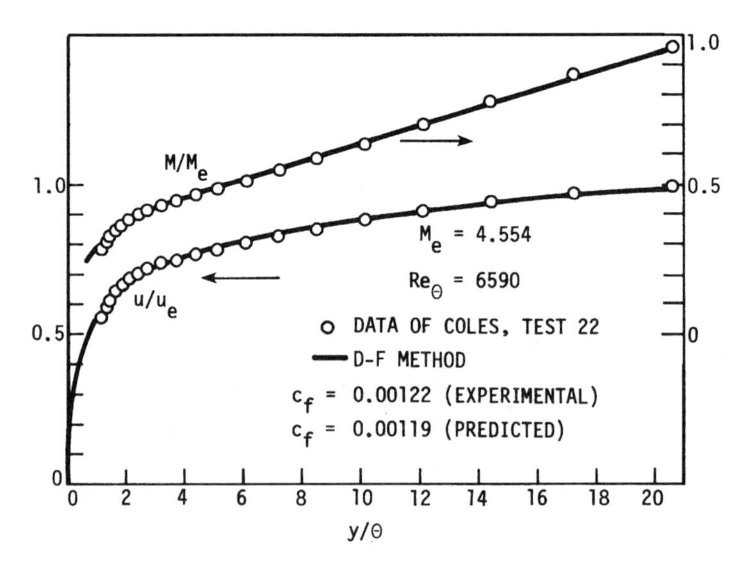

Figure 7.6 Comparison for a compressible flat plate flow measured by Coles (1953). Solid line represents predictions from the DuFort-Frankel finite-difference scheme (Pletcher, 1970).

Finite-difference methods easily accommodate step changes in boundary conditions, permitting solutions to be obtained for conditions under which simple correlations are especially unreliable. Figure 7.7 compares predictions of an algebraic mixing-length turbulence model used with a DuFort-Frankel type finite-difference procedure with the measurements of Moretti and Kays (1965) for low-speed flow over a cooled flat plate with a step change in wall temperature and a favorable pressure gradient. The Stanton number (St) in Fig. 7.7 is defined as $k(\partial T/\partial y)_w/[\rho_e u_e(H_{aw} - H_w)]$, where H_{aw} is the total enthalpy of the wall under adiabatic conditions.

Examples of cases where predictions of the simplest algebraic turbulence models fail to agree with experimental data abound in the technical literature. Several effects, which are not well predicted by the simplest models, were cited in Chapter 5. One of these was flow at low Re, especially at supersonic Mach numbers. This low Re effect is demonstrated in Fig. 7.8, where it can be seen that the point at which the simplest algebraic model (Model A) begins to fail

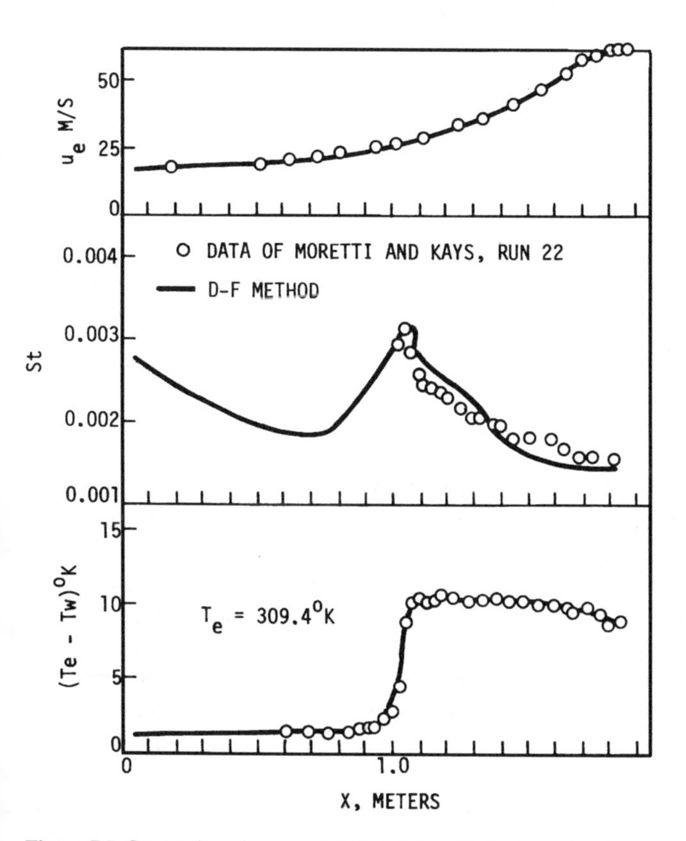

Figure 7.7 Comparison for a cooled flat plate with flow acceleration measured by Moretti and Kays (1965). Solid lines represent predictions from the DuFort-Frankel finite-difference scheme (Pletcher, 1970).

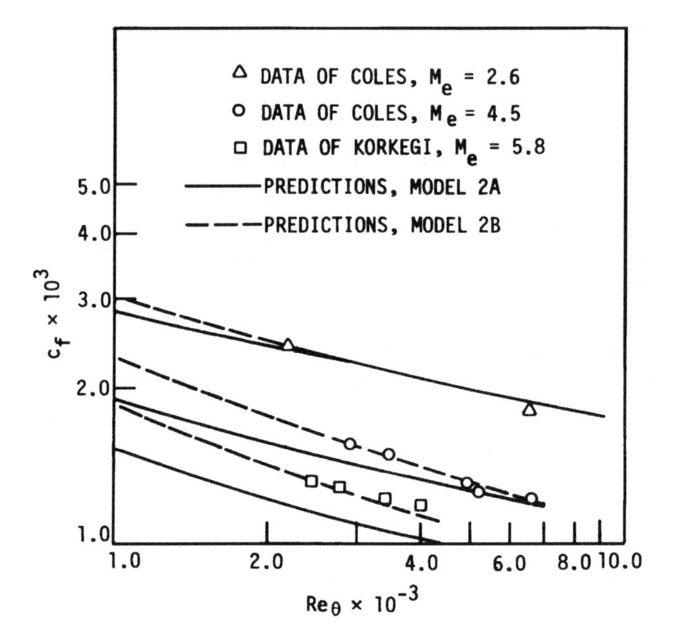

Figure 7.8 Comparison of predicted skin-friction coefficients with the measurements of Coles (1953) and Korkegi (1956) for the compressible turbulent boundary layer on a flat plate at low Reynolds number.

shifts to higher and higher Re as the Mach number of the flow increases. Predictions of a model containing the simple modification discussed in Section 5.4.3 for low Re are included in Fig. 7.8 as Model B.

7.3.10 Closure

This section has discussed several topics that are important in the finite-difference solution of the boundary-layer equations for two-dimensional and axisymmetric flows. Several difference schemes have been described. In computational work, as in many endeavors, "hands on" participation or "practice" is important. Accordingly, several example problems should be solved using the schemes discussed in order to develop an appreciation of the concepts and issues involved. Just as an engineer could hardly be considered an experimentalist without running an experiment, one likewise should not be considered a computational fluid dynamicist until some computations have been completed.

Which finite-difference scheme is best for the boundary-layer equations? The question is a logical one to raise at this point, but we need to establish measures by which "best" can be identified. All consistent difference schemes should provide numerical results as accurate as needed with sufficient grid refinement. With ultimate accuracy no longer an issue, the remaining concerns are computational costs and, to a lesser extent, ease of programming. In the present discussion it is assumed that the user insists on understanding all algebraic operations. We will include the time and effort needed to *understand* a given algorithm as part of the programming effort category. Programming effort then will measure, not the number of statements in a computer program, but the implied algebraic complexity of the steps in the algorithm and the difficulty of following the various steps for the beginner.

A review of the technical literature suggests that the schemes listed in Table 7.1 have been satisfactorily employed in the solution of the 2-D or axisymmetric boundary-layer equations for both laminar and turbulent flow and are recommended on the basis of their well-established performance.

The computation time for a typical calculation for all of the above schemes is expected to be modest (only a few seconds) on present-day computers. More details can be found in the literature cited above within the discussion of those methods. Only a few studies have been reported in which the computer times for several schemes have been compared for the boundary-layer equations. The work of Blottner (1975a) suggests that the Crank-Nicolson scheme with coupling requires about the same time as the modified box scheme for comparable accuracy. Again, for comparable accuracy, Blottner (1975a) found that the box scheme requires 2 to 3 times more computer time than the modified box scheme.

For the beginner wishing to establish a general purpose boundary-layer computer program, a reasonable way to start would be with the fully implicit scheme. The scheme is only first-order accurate in the marching direction, but second-order accuracy does not appear to be crucial for most boundary-layer calculations. This may be due partly to the fact that the $O(\Delta x)$ term in the T.E. usually includes the second streamwise derivative, which is relatively small when the boundary-layer approximation is valid. If second-order accuracy becomes desirable in the streamwise coordinate, it can be achieved with only minor changes through the use of a three-point, second-order representation of the streamwise derivative or by switching to the Crank-Nicolson representation. In increasing order of programming complexity, the logical choices for linearizing the coefficients are lagging, extrapolation, and Newton linearization with

Table 7.1 Recommended finite-difference schemes for the boundary-layer equations listed in estimated order of increasing programming effort

1	DuFort-Frankel
2	Fully implicit
3	Crank-Nicolson implicit
4	Fully implicit with continuity equation coupling
5	Crank-Nicolson implicit with continuity equation coupling
6	Modified box scheme
7	Box scheme

coupling. If lagging is adopted as standard, it would be advisable to program one of the latter two more accurate (for the same mesh increment) procedures as an option to provide periodic checks.

7.4 INVERSE METHODS, SEPARATED FLOWS, AND VISCOUS-INVISCID INTERACTION

7.4.1 Introduction

Thus far we have only considered the conventional or "direct" boundary-layer solution methods for the standard equations and boundary conditions given in Section 5.3. An "inverse" calculation method for the boundary-layer equations is a scheme whereby a solution is obtained that satisfies boundary conditions that differ from the standard ones. The usual procedure in an inverse method is to replace the outer boundary condition,

$$\lim_{y \to \infty} u(x, y) = u_e(x)$$

by the specification of a displacement thickness or wall shear stress that must be satisfied by the solution. The pressure gradient [or $u_e(x)$] is determined as part of the solution. It should be noted clearly that it is the *boundary conditions* that differ between the conventional direct methods and the inverse methods. It is perhaps more correct to think of the problem specification as being direct or inverse rather than the method. However, we will yield to convention and refer to the solution method as being direct or inverse.

The inverse methods are not merely an alternative way to solve the boundary-layer equations. The successful development of inverse calculation methods has permitted an expansion of the range of usefulness for the boundary-layer approximation.

Clearly, some design applications can be envisioned where it is desirable to calculate the boundary-layer pressure distribution that will accompany a specified distribution of displacement thickness or wall shear stress. This has provided some of the motivation for the development of inverse methods for the boundary-layer equations. Perhaps the most interesting applications of inverse methods have been in connection with separated flow. The computation of separated flows has long been thought to require the solution of the full Navier-Stokes equations. Thus, any suggestion that these flows, which are very important in applications, can be adequately treated with a much simpler mathematical model has been received with great interest. For this reason, the present discussion of inverse methods will emphasize applications to flows containing separated regions. The ability to remove the separation point singularity (Goldstein, 1948) is one of the most unique characteristics of the inverse methods.

7.4.2 Comments on Computing Separated Flows Using the Boundary-Layer Equations

Originally, it was thought that the usefulness of the boundary-layer approximation ended as the flow separation point was approached. This was because of the well-known singularity (Goldstein, 1948) of the standard boundary-layer formulation at separation and because the entire boundary-layer approximation is subject to question as the layer thickens and the normal component of velocity becomes somewhat larger (relative to u) than in the usual high Reynolds number flow. It is now known that the inverse formulation is regular at separation (Klineberg and Steger, 1974) and the evidence suggests (Williams, 1977; Kwon and Pletcher, 1979) that the boundary-layer equations provide a useful approximation for flows containing small, confined (bubble) separated regions. In support of the validity of the boundary-layer approximation, it is noted that the formation of a separation bubble normally does not cause the thickness of the viscous region to increase by an order of magnitude; that is, the boundary-layer measure of thinness, $\delta/L \ll 1$, is still met. The "triple-deck theory" of Lighthill (1953) and Stewartson (1974) (see Section 7.4.4) also provides analytical support for the validity of the boundary-layer approximation for large Reynolds number flows containing small separated regions. On the other hand, large local values of $d\delta/dx$ may occur and are expected to induce rather large values of v/u. At best, it should be conceded that the boundary-layer model is a weaker approximation for flow containing recirculation, even though it may provide estimates of flow parameters accurate enough for many purposes. The full range of applicability of the boundary layer equations for separated flows is still under study.

Flow separation presents two obstacles to a straightforward space-marching solution procedure using conventional boundary conditions with the boundary-layer equations; these are (1) the singularity at separation and (2) the flow reversal, which prohibits marching the solution in the direction of the external flow (see Fig. 7.9) unless the convection terms in the equations are altered. When the pressure gradient is fixed near separation by the conventional boundary conditions, the normal component of the velocity and $d\tau_w/dx$ tend

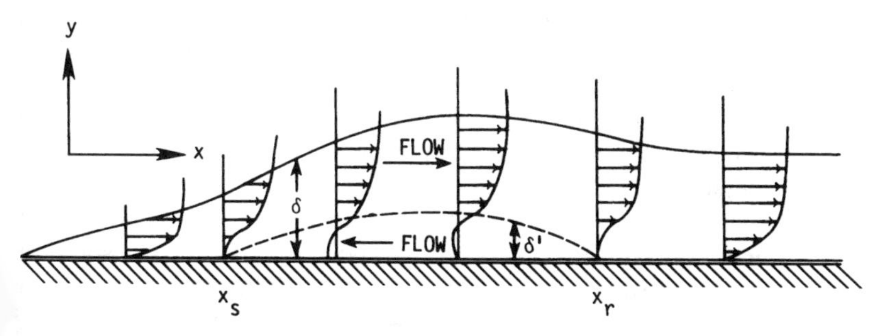

Figure 7.9 Flow containing a separation bubble.

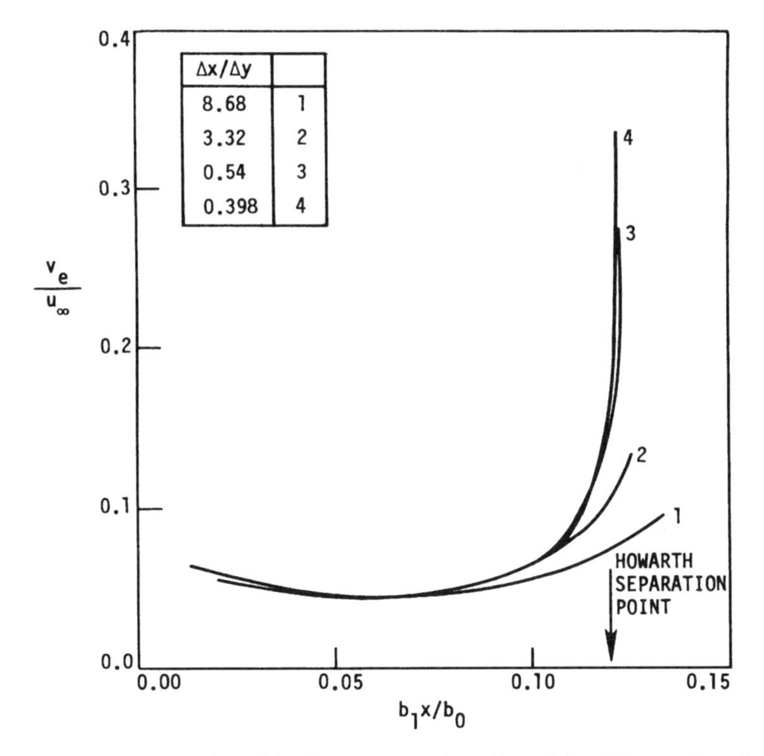

Figure 7.10 Effect of x-grid refinement on v_e for a direct finite-difference boundary-layer calculation (Pletcher and Dancey, 1976) near the separation point for a linearly retarded stream: $u_e = b_0 - b_1 x$, $b_0 = 30.48$ m/s, $b_1 = 300\ \text{s}^{-1}$, $\nu = 1.49 \times 10^{-4}\ \text{m}^2/\text{s}$.

toward infinity at the point of separation. A detailed discussion of this singularity can be found in the works of Goldstein (1948) and Brown and Stewartson (1969). This phenomenon appears in finite-difference solutions, where $u_e(x)$ is prescribed, as the tendency for v to increase without limit as the streamwise step size is reduced. This is illustrated in Fig. 7.10 for the Howarth linearly retarded flow (Howarth, 1938). Naturally, a finite v will be obtained for a finite step size, but the solution will not be unique. This singular behavior, which is mathematical rather than physical, can be overcome by the use of an auxiliary pressure interaction relationship with direct methods (Reyhner and Flügge-Lotz, 1968; Napolitano et al., 1978) or by the use of inverse procedures. In this section we will concentrate on the inverse procedures that require no auxiliary relationships to eliminate the singular behavior.

The difficulty with the convective terms can be viewed as follows. We recall that the steady boundary-layer equations are parabolic. For $u > 0$ the solution can be marched in the positive x direction. Physically, information is carried downstream from the initial plane by the flow. In regions of reversed flow, however, the "downstream" direction is in the negative x direction (Fig. 7.9).

Mathematically, we observe that when $u < 0$, the boundary-layer momentum equation remains parabolic, but the correct marching direction is in the negative x direction.

It would seem, then, that a solution procedure might be devised to overcome the problem associated with the "correct" marching direction by making initial guesses or approximations for the velocities in the reversed flow portion of a flow with a separation bubble, storing these velocities, and correcting them by successive iterative calculation sweeps over the entire flow field. To do this requires using a difference representation that honors the appropriate marching direction, forward or backward, depending on the direction of flow. To follow this iterative procedure means abandoning the once-through simplicity of the usual boundary-layer approach. Computer storage must also be provided for velocities in and near the region of reversed flow. Such multiple-pass procedures have been employed by Klineberg and Steger (1974), Carter and Wornom (1975), and Cebeci (1976). Some crucial aspects of differencing for multiple-pass procedures will become apparent from the material presented in Chapter 8.

Reyhner and Flügge-Lotz (1968) suggested a simpler alternative to the multiple-pass procedure. Noting that the reversed flow velocities are generally quite small for confined regions of recirculation, they suggested that the convective term $u\,\partial u/\partial x$ in the boundary-layer momentum equation be represented in the reversed flow regions by $C|u|\,\partial u/\partial x$, where C is zero or a small positive constant. This representation has become known as the FLARE approximation and permits the boundary-layer solution to proceed through separated regions by a simple forward-marching procedure. It should be clear that the FLARE procedure introduces an additional approximation (or assumption) into the boundary-layer formation, namely, that the $u\,\partial u/\partial x$ term is small relative to other terms in the momentum equation in the region of reversed flow. On the other hand, the FLARE approximation appears to give smooth and plausible solutions for many flows with separation bubbles. Example solutions are presented in Section 7.4.3. Experimental and computational evidence accumulated to date indicates that for naturally occurring separation bubbles, the u component of velocity in reversed flow regions is indeed fairly small in magnitude, usually less than about 10% of the maximum velocity found in the viscous region.

It should be noted that, although ways of satisfactorily treating the $u\,\partial u/\partial x$ convective term have been presented, it still does not appear possible to obtain a unique convergent solution of the steady boundary-layer equations alone by a direct marching procedure. Direct calculation procedures reported to date have always employed an interaction relation whereby the pressure gradient specified becomes dependent upon the displacement thickness (or related parameter) of the viscous regions, usually in a time-dependent manner (Napolitano et al., 1978). This is not necessarily a disadvantage. Viscous-inviscid interaction usually needs to be considered ultimately in obtaining the solution for the complete flow field containing a separated region, if the boundary-layer equations are used for the viscous regions. Viscous-inviscid interaction is treated further in

Section 7.4.4. On the other hand, we should note that a unique convergent solution can be obtained for the steady boundary-layer equations alone using inverse methods.

7.4.3 Inverse Finite-Difference Methods

Two procedures will be illustrated. The first is conceptually the simplest and is especially useful for illustrating the concept of the inverse method. It appears to work very well when the flow is attached (no reversed flow region), but gives rise to small controlled oscillations in the skin friction when reversed flow is present. This oscillatory behavior is overcome by the second method, which solves the boundary-layer equations in a coupled manner. The FLARE approximation will be employed in both of these methods. For simplicity, the methods will be illustrated for incompressible flows.

Inverse Method A. The boundary-layer equations are written as follows.

continuity:

$$\frac{\partial u}{\partial x} + \frac{\partial v}{\partial y} = 0 \tag{7.49}$$

momentum:

$$C|u|\frac{\partial u}{\partial x} + v\frac{\partial u}{\partial y} = u_e\frac{du_e}{dx} + \frac{1}{\rho}\frac{\partial \tau}{\partial y} \tag{7.50}$$

In the above, $C = 1.0$ when $u > 0$, and C is a small ($\leqslant 0.2$) positive constant when $u \leqslant 0$ and

$$\tau = \mu\frac{\partial u}{\partial y} - \rho\overline{u'v'} = (\mu + \mu_T)\frac{\partial u}{\partial y} \tag{7.51}$$

The above equations are in a form applicable to either laminar or turbulent flow. For laminar flow the primed velocities and μ_T are zero, and for turbulent flow the unprimed velocities are time-mean quantities.

The boundary conditions for the inverse procedure are

$$u(x,0) = v(x,0) = 0 \tag{7.52}$$

and

$$\int_0^\infty \left(1 - \frac{u}{u_e}\right) dy = \delta^*(x) \tag{7.53}$$

where δ^* is a prescribed function. Alternatively, $\tau_w(x)$ can be specified as a boundary condition. Clearly Eqs. (7.49) and (7.50) can be solved by a direct method utilizing the conventional boundary condition

$$\lim_{y\to\infty} u(x,y) = u_e(x) \tag{7.54}$$

in place of Eq. (7.53) for attached portions of the flow. It is possible to start a boundary-layer calculation in the direct mode and switch to the inverse procedure when desired.

The boundary-layer equations are cast into a fully implicit difference form, and the coefficients are lagged. Such a difference representation is discussed in Section 7.3 and will not be repeated here. The inverse treatment of boundary conditions is implemented by varying u_e in successive iterations at each streamwise calculation station until the solution satisfies the specified value of $\delta^*(x)$. In each of these iterations the numerical formulation and implementation of boundary conditions are the same as for a direct method. The displacement thickness is evaluated from the computed velocity distribution by numerical integration (use of either Simpson's rule or the trapezoidal rule is suggested). The appropriate value of u_e needed to satisfy the boundary condition on δ^* (δ^*_{BC}) is determined by considering $\delta^* - \delta^*_{BC}$ to be a function of u_e at each streamwise station, $\delta^* - \delta^*_{BC} = F(u_e)$, and seeking the value of u_e required to establish $F = 0$ by a variable secant (Fröberg, 1969) procedure. In the above, δ^* is the δ^* actually obtained from the solution for a specified value of u_e. Two initial guesses are required for this procedure, which usually converges in three or four iterations (Pletcher, 1978).

The variable secant procedure can be thought of as a generalization of Newton's method (also known as the Newton-Raphson method) for finding the root of $F(x) = 0$. In Newton's method we expand $F(x)$ in a Taylor series about a reference point x_n:

$$F(x_n + \Delta x) = F(x_n) + F'(x_n)\,\Delta x + \cdots$$

We truncate the series after the first derivative term and compute the value of Δx required to establish $F(x_n + \Delta x) = 0$. For Newton's method this gives

$$x_{n+1} - x_n = \Delta x = -\frac{F(x_n)}{F'(x_n)} \tag{7.55}$$

Thus, starting with an initial guess, x_n, an improved approximation, x_{n+1}, can be computed from Eq. (7.55). The process is repeated iteratively until $|(x_{n+1} - x_n)| < \epsilon$.

Newton's method is a simple and effective procedure. Its use does, however, require that $F'(x)$ be evaluated analytically. When this is not possible, the variable secant generalization of the Newton procedure represents a reasonable alternative.

In the variable secant procedure, the derivative is replaced by a secant line approximation through two points:

$$F'(x_n) \cong \frac{F(x_n) - F(x_{n-1})}{x_n - x_{n-1}}$$

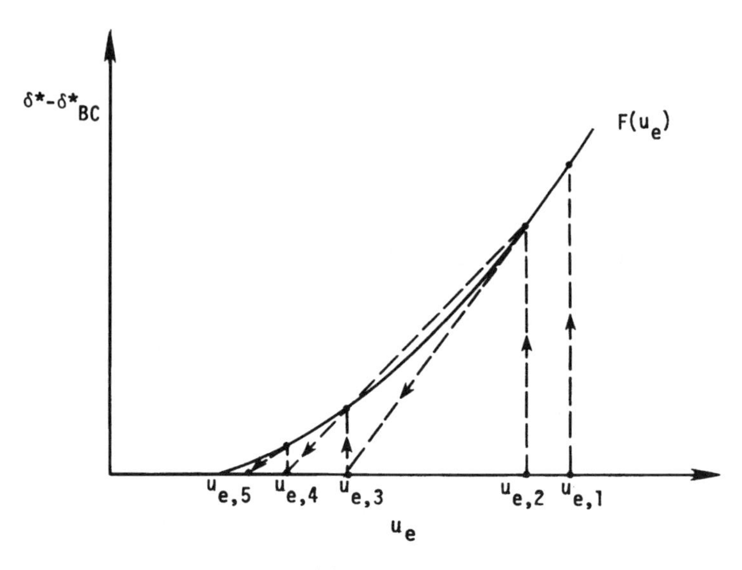

Figure 7.11 Determination of $u_e(x)$ through the use of the variable secant procedure.

After two initial guesses for x, the third approximation to the root is obtained from

$$x_{n+1} = x_n - \frac{F(x_n)}{F(x_n) - F(x_{n-1})}(x_n - x_{n-1}) \tag{7.56}$$

In the application of the variable secant method to the inverse boundary-layer calculation, x_n becomes $(u_e)_n$ and $F = \delta^* - \delta^*_{BC}$. The iterative process is illustrated in Fig. 7.11.

When the iterative search for the $u_e(x)$ that provides the specified $\delta^*(x)$ is completed, the solution may be advanced to another streamwise station in the usual manner for parabolic equations. The simplicity of inverse method A is obvious. Apart from small changes to implement the FLARE representation, the difference equations are solved the same as for the standard direct method for boundary layers. The method performs reasonably well (Pletcher, 1978; Kwon and Pletcher, 1979) but does predict small oscillations in the wall shear stress when separation is present. These oscillations can be eliminated by suitable coupling of the momentum and continuity equations and are not present in solutions obtained by the method described below.

Inverse Method B. Here we will describe the method developed by Kwon and Pletcher (1981). More recently, Truman et al. (1993) have described a similar strategy in a more general context. The overall strategy of this method is to couple all unknowns and the boundary conditions in one simultaneous system of algebraic equations to be solved at each streamwise station. To accomplish this, it is convenient to introduce the stream function ψ. Accordingly,

$$u = \frac{\partial \psi}{\partial y}$$

$$v = -\frac{\partial \psi}{\partial x}$$

The conservation equations for mass and momentum are written as

$$u = \frac{\partial \psi}{\partial y} \tag{7.57}$$

$$Cu\frac{\partial u}{\partial x} - \frac{\partial \psi}{\partial x}\frac{\partial u}{\partial y} = u_e\frac{du_e}{dx} + \frac{1}{\rho}\frac{\partial \tau}{\partial y} \tag{7.58}$$

where

$$\tau = \bar{\mu}\frac{\partial u}{\partial y}$$

$$\bar{\mu} = \mu + \mu_T$$

The boundary conditions are

$$u(x,0) = \psi(x,0) = 0 \tag{7.59}$$

$$\psi_e = u_e[y_e - \delta^*(x)] \tag{7.60}$$

where $\delta^*(x)$ is a prescribed function. The boundary condition for ψ_e follows from the definition of δ^*:

$$\delta^* = \int_0^\infty \left(1 - \frac{u}{u_e}\right) dy$$

The upper limit of this integral can be replaced by y at the outer edge of the boundary layer, y_e, since the integrand is equal to zero for $y > y_e$. Multiplying by u_e gives

$$u_e \delta^* = u_e y_e - \int_0^{y_e} u\, dy$$

Expressing u in terms of the stream function permits the integral to be evaluated as ψ_e. Rearranging gives Eq. (7.60). When the difference equations that follow are solved in a direct mode, the outer boundary condition becomes the conventional one given by Eq. (7.54) instead of that specified by Eq. (7.60).

Equations (7.57) and (7.58) are first represented in finite-difference form as

$$\frac{u_j^{n+1} + u_{j-1}^{n+1}}{2} = \frac{\psi_j^{n+1} - \psi_{j-1}^{n+1}}{\Delta y_-} \tag{7.61}$$

$$Cu_j^{n+1}\frac{u_j^{n+1} - u_j^n}{\Delta x} - \frac{\psi_j^{n+1} - \psi_j^n}{\Delta x}\frac{u_{j+1}^{n+1} - u_{j-1}^{n+1}}{\Delta y_+ + \Delta y_-}$$

$$= \chi^{n+1} + \frac{2}{\rho(\Delta y_+ + \Delta y_-)}$$

$$\times\left(\bar{\mu}_{j+1/2}\frac{u_{j+1}^{n+1} - u_j^{n+1}}{\Delta y_+} - \bar{\mu}_{j-1/2}\frac{u_j^{n+1} - u_{j-1}^{n+1}}{\Delta y_-}\right) \tag{7.62}$$

In the above,

$$C = 1 \quad \text{when} \quad u_j^{n+1} > 0$$

$$C = 0 \quad \text{when} \quad u_j^{n+1} < 0$$

$$\chi = -\frac{1}{\rho}\frac{dp}{dx}$$

Newton linearization is next applied to the above nonlinear convective terms following the procedures presented in Section 7.3.3. We let $u_j^{n+1} = \hat{u}_j^{n+1} + \Delta u_j$ and $\psi_j^{n+1} = \hat{\psi}_j^{n+1} + \Delta\psi_j$, where the circumflexes indicate provisional values of the variables in an iterative process. The quantities Δu_j and $\Delta\psi_j$ are the changes in the variables between two iterative sweeps, i.e., $\Delta\phi_j = \phi_j^{n+1} - \hat{\phi}_j^{n+1}$ for a general variable ϕ. The resulting difference equations can be written in the form

$$\psi_{j-1}^{n+1} - \psi_j^{n+1} + b_j\left(u_{j-1}^{n+1} + u_j^{n+1}\right) = 0 \tag{7.63}$$

$$B_j u_{j-1}^{n+1} + D_j u_j^{n+1} + A_j u_{j+1}^{n+1} + E_j \psi_j^{n+1} = H_j \chi^{n+1} + C_j \tag{7.64}$$

where

$$A_j = -\frac{\hat{\psi}_j^{n+1} - \psi_j^n}{\Delta x(\Delta y_+ + \Delta y_-)} - \frac{2\bar{\mu}_{j+1/2}}{\rho\Delta y_+(\Delta y_+ + \Delta y_-)}$$

$$B_j = \frac{\hat{\psi}_j^{n+1} - \psi_j^n}{\Delta x(\Delta y_+ + \Delta y_-)} - \frac{2\bar{\mu}_{j-1/2}}{\rho\Delta y_-(\Delta y_+ + \Delta y_-)}$$

$$C_j = \frac{C\left(\hat{u}_j^{n+1}\right)^2}{\Delta x} - \frac{\hat{\psi}_j^{n+1}\left(\hat{u}_{j+1}^{n+1} - \hat{u}_{j-1}^{n+1}\right)}{\Delta x(\Delta y_+ + \Delta y_-)}$$

$$D_j = \frac{C\left(2\hat{u}_j^{n+1} - u_j^n\right)}{\Delta x} + \frac{2}{\rho(\Delta y_+ + \Delta y_-)}\left(\frac{\bar{\mu}_{j+1/2}}{\Delta y_+} + \frac{\bar{\mu}_{j-1/2}}{\Delta y_-}\right)$$

$$E_j = -\frac{\hat{u}_{j+1}^{n+1} - \hat{u}_{j-1}^{n+1}}{\Delta x(\Delta y_+ + \Delta y_-)}$$

$$H_j = 1$$

$$b_j = \frac{\Delta y_-}{2}$$

The above algebraic formulation is similar to that presented in Section 7.3 in connection with the Davis coupled scheme and solved by the modified Thomas algorithm. Equations (7.63) and (7.64) form a block-tridiagonal system with 2×2 blocks and require the simultaneous solution of $2(NJ) - 2$ equations for $2(NJ) - 2$ unknowns at each streamwise marching step. The parameter NJ is the number of grid points across the flow, including boundary points. One difference between this formulation and the algebraic equations arising from the Davis coupled scheme is the appearance of the new term, $H_j \chi^{n+1}$, on the right-hand side of Eq. (7.64). The pressure gradient parameter χ^{n+1} is one of the unknowns in the inverse formulation. The outer boundary conditions are also different. These facts preclude the use of the modified tridiagonal algorithm presented in Section 7.3. However, the blocks below the main diagonal can be eliminated, and a recursion formula can be developed (Kwon and Pletcher, 1981) for the back substitution. Before the back substitution is carried out, however, the parameter χ^{n+1} must be determined by a special procedure to be indicated subsequently.

The unknowns can be computed from

$$u_j^{n+1} = A'_j u_{j+1}^{n+1} + H'_j \chi^{n+1} + C'_j \tag{7.65}$$

$$\psi_j^{n+1} = B'_j u_{j+1}^{n+1} + D'_j \chi^{n+1} + E'_j \tag{7.66}$$

providing the coefficients A'_j, H'_j, C'_j, B'_j, D'_j, E'_j and the quantities u_{j+1}^{n+1} and χ^{n+1} are known a priori. The coefficients are given by

$$A'_j = -\frac{A_j}{R_1}$$

$$B'_j = A'_j R_2$$

$$C'_j = \frac{C_j - B_j C'_{j-1} - E_j(b_j C'_{j-1} + E'_{j-1})}{R_1}$$

$$D'_j = b_j H'_{j-1} + D'_{j-1} + H'_j R_2$$

$$E'_j = b_j C'_{j-1} + E'_{j-1} + C'_j R_2$$

$$H'_j = \frac{H_j - B_j H'_{j-1} - E_j(b_j H'_{j-1} + D'_{j-1})}{R_1}$$

$$R_1 = D_j + (B_j + E_j b_j)A'_{j-1} + E_j(B'_{j-1} + b_j)$$

$$R_2 = b_j(1 + A'_{j-1}) + B'_{j-1}$$

Since the inner ($j = 1$) boundary conditions on u_j^{n+1} and ψ_j^{n+1} are zero, the coefficients A'_1, B'_1, C'_1, D'_1, E'_1, H'_1, are also zero and the coefficients above can be computed starting from $j = 2$ and continuing to the outer boundary ($j = NJ$).

The pressure gradient parameter χ^{n+1} is evaluated by simultaneously solving the equations obtained from Eqs. (7.65) and (7.66) by replacing j with $NJ - 1$ and the boundary conditions in the following manner. At $j = NJ - 1$, Eqs. (7.65) and (7.66) become

$$u_{NJ-1}^{n+1} = A'_{NJ-1} u_{NJ}^{n+1} + H'_{NJ-1} \chi^{n+1} + C'_{NJ-1} \tag{7.67}$$

$$\psi_{NJ-1}^{n+1} = B'_{NJ-1} u_{NJ}^{n+1} + D'_{NJ-1} \chi^{n+1} + E'_{NJ-1} \tag{7.68}$$

The boundary conditions are written as

$$\psi_{NJ}^{n+1} = u_{NJ}^{n+1}\left(y_{NJ} - \delta^{*n+1}\right) \tag{7.69}$$

and

$$\chi^{n+1} = \frac{1}{\Delta x}\left[(2\hat{u}_{NJ}^{n+1} - u_{NJ}^{n})u_{NJ}^{n+1} - (\hat{u}_{NJ}^{n+1})^2\right] \tag{7.70}$$

Equation (7.61) is written as

$$\psi_{NJ}^{n+1} = \psi_{NJ-1}^{n+1} + \frac{\Delta y_-}{2}(u_{NJ}^{n+1} + u_{NJ-1}^{n+1}) \tag{7.71}$$

Solving Eqs. (7.67)–(7.71) for χ^{n+1} gives

$$\chi^{n+1} = \frac{(F_3/F_1)(2\hat{u}_{NJ}^{n+1} - u_{NJ}^{n}) - (u_{NJ}^{n+1})^2}{\Delta x - (F_2/F_1)(2\hat{u}_{NJ}^{n+1} - u_{NJ}^{n})} \tag{7.72}$$

where

$$F_1 = y_{NJ} - \delta^{*n+1} - B'_{NJ-1} - \frac{\Delta y_-}{2}(1 + A'_{NJ-1})$$

$$F_2 = D'_{NJ-1} + \frac{\Delta y_-}{2} H'_{NJ-1}$$

$$F_3 = E'_{NJ-1} + \frac{\Delta y_-}{2} C'_{NJ-1}$$

Once the pressure gradient parameter χ^{n+1} is determined, the edge velocity u_{NJ}^{n+1} can be calculated using Eqs. (7.67)–(7.71) as

$$u_{NJ}^{n+1} = \frac{F_2}{F_1}\chi^{n+1} + \frac{F_3}{F_1} \tag{7.73}$$

Then, ψ_{NJ}^{n+1} can be computed directly from Eq. (7.69). Now the back substitution process can be initiated using Eqs. (7.65) and (7.66) to compute u_j^{n+1} and ψ_j^{n+1} from the outer edge to the wall. The Newton linearization requires that the system of equations be solved iteratively, with $\hat{u}_j^{n+1}$ and $\hat{\psi}_j^{n+1}$ being updated between iterations. The iterative process is continued at each streamwise location until the maximum change, in u's and ψ's between two successive iterations is less than some predetermined tolerance. The calculation is initiated at each streamwise station by setting $\hat{u}_j^{n+1} = u_j^n$ and $\hat{\psi}_j^{n+1} = \psi_j^n$. In previous applications of this method, only two or three iterations were generally required for the maximum fractional change in the variables, i.e., $\Delta\phi/\phi$, to be reduced to 5×10^{-4}.

Details of other somewhat different coupled inverse boundary-layer finite-difference procedures employing the FLARE approximation can be found in the works of Cebeci (1976) and Carter (1978).

7.4.4 Viscous-Inviscid Interaction

In design it is common to obtain the pressure distribution about aerodynamic bodies from an inviscid flow solution. The inviscid flow solution then provides the edge velocity distribution needed as a boundary condition for solving the boundary-layer equations to obtain the viscous drag on the body. In many cases, the presence of the viscous boundary layer only slightly modifies the flow pattern over the body. It is possible to obtain an improved inviscid flow solution by augmenting the physical thickness of the body by the boundary-layer displacement thickness. The definition of δ^* is such that the new inviscid flow solution properly accounts for the displacement of the inviscid flow caused by the viscous flow near the body. The improved inviscid edge velocity distribution can then be used to obtain yet another viscous flow solution. In principle, this viscous-inviscid interaction procedure can be continued iteratively until changes are small. In practice, however, severe underrelaxation of the changes from one iterative cycle to another is often required for convergence.

Fortunately, for most flows involving an attached boundary layer, the changes that arise from accounting for the viscous-inviscid interaction are negligibly small, and it suffices for engineering design purposes to compute the inviscid and viscous flows independently (i.e., without considering viscous-inviscid interaction). Flows that separate or contain separation bubbles are a notable exception.

The displacement effect of the separated regions locally alters the pressure distribution in a significant manner. A rapid thickening of the boundary layer under the influence of an adverse pressure gradient even without separation can also alter the pressure distribution locally, to the extent that a reasonable flow solution cannot be obtained without accounting for the displacement effect of the viscous flow. Often under such conditions, a boundary-layer calculation obtained using the edge velocity distribution from an inviscid flow solution that neglects the displacement effect will predict separation when the real flow does not separate at all.

It is often possible to confine the region where viscous-inviscid interaction effects are important to the local neighborhood of the "bulge" in the displacement surface. Such a local interaction region is depicted in Fig. 7.12.

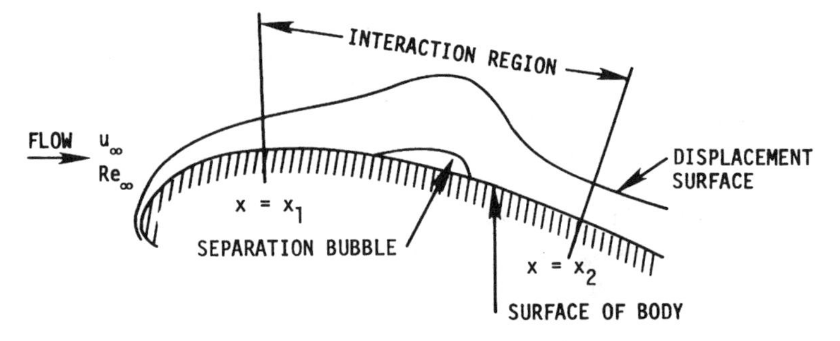

Figure 7.12 Local interaction region on a two-dimensional body.

The inverse boundary procedure described in Section 7.4.3 is particularly well suited for flows in which separation may occur.

The essential elements of a viscous-inviscid interaction calculation procedure are the following:

1. A method for obtaining an improved inviscid flow solution that provides a pressure distribution or edge velocity distribution that accounts for the viscous flow displacement effect. In principle, any inviscid flow "solver" could be used, but it is also frequently possible to employ a greatly simplified inviscid flow calculation scheme based on a small-disturbance approximation.
2. A technique for obtaining a solution to the boundary-layer equations suitable for the problem at hand. For a flow that may separate, an inverse boundary-layer procedure would be appropriate.
3. A procedure for relating the inviscid and viscous flow solutions in a manner that will drive the changes from one iterative cycle to the next toward zero.

Over the years, numerous viscous-inviscid interaction schemes have been proposed. It will not be possible to discuss all of these here. Instead, we will summarize a suitable approach for predicting the flow in the neighborhood of a separation bubble on an airfoil in incompressible flow. This configuration is illustrated in Fig. 7.12.

For this case, a good estimate of the effect of the displacement correction for the inviscid flow solution can be obtained by the use of a small-disturbance approximation. We let $u_{e,o}$ denote the tangential component of velocity of the inviscid flow over the solid body (neglecting all effects of the viscous flow) and u_c be the velocity on the displacement surface induced only by the sources and sinks distributed on the surface of the body due to the displacement effect of the viscous flow in the interaction region. Then, the x component of velocity of a fluid particle on the displacement surface can be written as

$$u_e = u_{e,o} + u_c \tag{7.74}$$

Following Lighthill (1958), the intensity of the line source or sinks displacing a streamline at a displacement surface of the viscous flow can be evaluated as

$$q = \frac{d(u_e \delta^*)}{dx} \tag{7.75}$$

For small values of δ^*, u_c can be evaluated from the Hilbert integral

$$u_c(x) = \frac{1}{\pi} \int_{-\infty}^{\infty} \frac{d(u_e \delta^*)}{dx'} \frac{dx'}{x - x'} \tag{7.76}$$

In the numerical computation of u_c, it is usually assumed that strong interaction is limited to the region $x_1 \leqslant x \leqslant x_2$ shown in Fig. 7.12. The intensity of the source or sink caused by the viscous displacement is assumed to approach zero as x approaches $\pm\infty$. Consequently, $d(u_e \delta^*/dx)$ is normally only computed in the region $x_1 < x < x_2$ using the boundary-layer solution. An arbitrary extrapolation of the form (Kwon and Pletcher, 1979)

$$q'(x) = \frac{b}{x^2} \tag{7.77}$$

is often used for the regions $x < x_1$ and $x > x_2$ in order to evaluate the integral in Eq. (7.76). The constant b is chosen to match the q obtained from the boundary-layer solution at x_1 and x_2. Equation (7.76) can now be written as

$$u_c(x) = \frac{1}{\pi} \left[\int_{-\infty}^{x_1} \frac{q'(x')}{x - x'} dx' + \int_{x_1}^{x_2} \frac{q(x')}{x - x'} dx' + \int_{x_2}^{\infty} \frac{q'(x')}{x - x'} dx' \right] \tag{7.78}$$

The first and third integrals can be evaluated analytically. The second integral is evaluated numerically, normally using the trapezoidal rule. The singularity at $x = x'$ can be isolated using the procedure found in the work by Jobe (1974). Some authors have found it possible to evaluate the integral numerically with no special attention given to the singularity as long as $(x - x')$ remained finite (Briley and McDonald, 1975).

The inviscid surface velocity on the solid body (neglecting the boundary layer), $u_{e,o}$, can be obtained by the methods cited in Chapter 6 [as, for example, the Hess and Smith (1967) method], or from experimental data. The Hess and Smith procedure could be used iteratively for all of the inviscid flow calculations. However, the relatively simple small-disturbance procedure requires significantly less computer time and has been found to provide sufficient accuracy for incompressible viscous-inviscid interaction calculations of a type that permits the use of the boundary-layer equations for the viscous flow.

The inverse boundary-layer procedures discussed in Section 7.4.3 are quite suitable for computing the viscous portion of the flow, which may include separated regions. The iterative updating of the solutions can be effectively carried out by the method successfully demonstrated by Carter (1978) and Kwon and Pletcher (1979).

The interaction strategy known as a semi-inverse (inviscid solution proceeds directly, the viscous, inversely) method proceeds in the following way. First, u_{eo} is obtained for the body of interest, and the viscous flow is computed up to the beginning of the interaction region by a conventional direct method. These two solutions do not change. Next, an initial $\delta^*(x)$ distribution is chosen over the region $x_1 < x < x_2$ (see Fig. 7.12). The initial guess is purely arbitrary but should match the $\delta^*(x)$ of the boundary layer computed by the direct method at $x = x_1$. The boundary-layer solution is next obtained by an inverse procedure using this $\delta^*(x)$ as a boundary condition. An edge velocity distribution $u_{e,\mathrm{BL}}(x)$ is obtained as an output.

Now the small-disturbance inviscid flow procedure, Eq. (7.78), is used to compute the correction to the inviscid flow velocity. This establishes a new distribution for the edge (surface) velocity $u_{e,\mathrm{inv}}(x)$. The $u_e(x)$ from the two calculations, boundary layer and inviscid, will not agree until convergence has been achieved. The difference between $u_e(x)$ calculated both ways can be used as a potential to calculate an improved distribution for $\delta^*(x)$. To do this formally, one would seek to determine the way in which a change in u_e would influence δ^*. A suitable scheme has been developed for subsonic flows by noting that a response to small excursions in local u_e tends to preserve the volume flow rate per unit width in the boundary layer, i.e., $u_e \delta^* \cong$ const. This implies that a local decrease in $u_e(x)$ (associated with a more adverse pressure gradient) causes an increase in $\delta^*(x)$ and a local increase in $u_e(x)$ (associated with a more favorable pressure gradient) causes a decrease in $\delta^*(x)$. This concept is put into practice by computing the appropriate new distribution of δ^* (Carter, 1978) to use for a new pass through the boundary-layer calculation by

$$\delta^*_{k+1} = \delta^*_k \left(\frac{u_{e,\mathrm{BL}_k}}{u_{e,\mathrm{inv}_k}} \right) \tag{7.79}$$

where k denotes iteration level. It is important to note that Eq. (7.79) only serves as a basis for correcting δ^* between iterative passes so that no formal justification for its use is required so long as the iterative process converges. At convergence $u_{e,\mathrm{BL}} = u_{e,\mathrm{inv}}$; thus, Eq. (7.79) represents an identity, thereby having no effect on the final solution. In this sense, the use of Eq. (7.79) is somewhat like the use of an arbitrary overrelaxation factor in the numerical solution of an elliptic equation by successive overrelaxation (SOR). Carter (1978) has given a somewhat more formal justification of Eq. (7.79) based on the von Kármán momentum integral.

The viscous-inviscid interaction calculation is completed by making successive passes first through the inverse boundary-layer scheme, then through the inviscid flow procedure with δ^* being computed by Eq. (7.79) prior to each boundary-layer calculation. When $|u_{e,\mathrm{BL}} - u_{e,\mathrm{inv}}|$ is less than a prescribed tolerance, convergence is considered to have been achieved. In some applications of this matching procedure, overrelaxation of δ^* in Eq. (7.79) has been observed to speed convergence. An illuminating discussion of several matching procedures can be found in the paper by Wigton and Holt (1981).

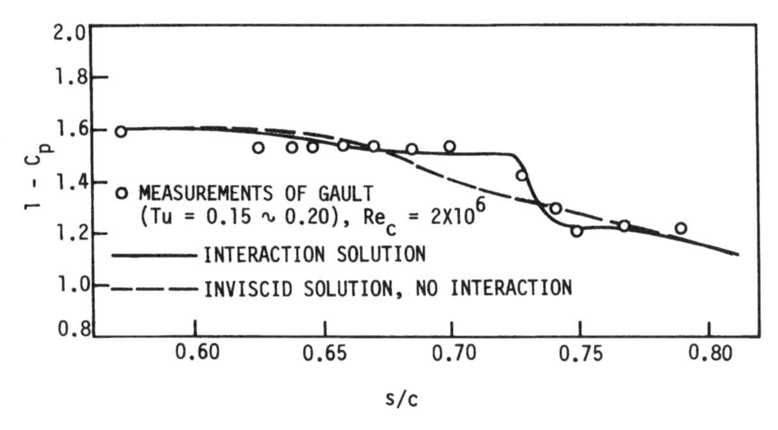

Figure 7.13 Comparison of the predicted pressure distribution with experimental data (Gault, 1955) for a NACA 66_3-0.18 airfoil at zero angle of incidence.

Some example predictions (Kwon and Pletcher, 1979) are shown in Figs. 7.13 and 7.14 for the flow in the neighborhood of a transitional separation bubble on a NACA 66_3-018 airfoil. The parameter Tu in the figures is the free stream turbulence level, and Re_c is the Reynolds number based on the airfoil chord. Figure 7.13 compares the predicted pressure coefficient with measurements. The dashed line in Fig. 7.13 indicates the pressure coefficient predicted by inviscid flow theory neglecting the presence of the boundary layer. In the neighborhood of the separation bubble centered at $s/c \simeq 0.7$ (s is the distance along the airfoil surface measured from the leading edge, and c is the chord), this predicted pressure coefficient is seen to be considerably in error compared to the measurements. The solid line indicates the prediction of a viscous-inviscid interaction procedure, which is seen to follow the trend of the measurements fairly closely. Seventeen passes through the viscous-inviscid procedure were required for convergence in this case. Velocity profiles are compared in Fig. 7.14. Reversed flow is evident from the profiles in the vicinity of $s/c \simeq 0.7$. The predicted results are quite sensitive to the model used for laminar-turbulent transition.

The same general strategy outlined above for viscous-inviscid interaction calculations has also been found to work well for compressible flows, including transonic and supersonic applications (Carter, 1981; Werle and Verdon, 1979). However, there has been some evidence that the semi-inverse coupling procedure described above becomes unstable for large separated flow regions in a supersonic stream. This has led to the development of quasi-simultaneous (Le Balleur, 1984; Houwink and Veldman, 1984; Bartels and Rothmayer, 1994) and simultaneous (Lee and Pletcher, 1988) methods for achieving the coupling between viscous and inviscid solutions. When the flow becomes compressible, the boundary-layer form of the energy equation is solved in the viscous flow region, usually with the use of the FLARE approximation. The solution

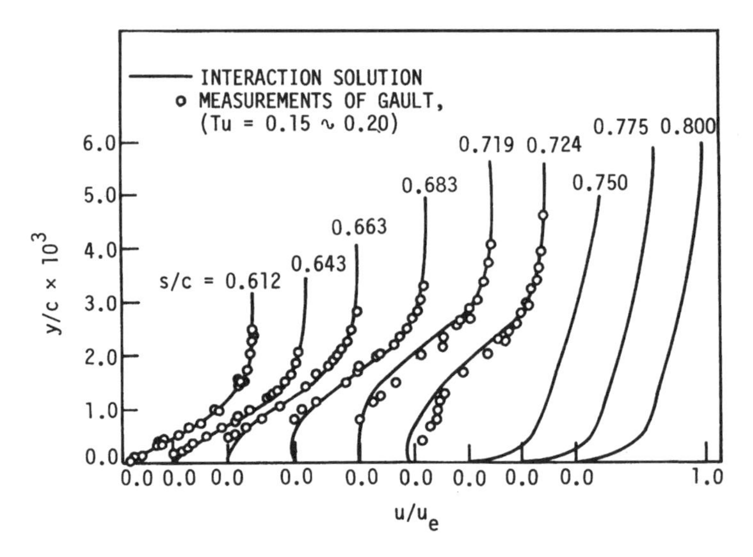

Figure 7.14 Comparison of the predicted mean velocity profiles with experimental data (Gault, 1955) for a NACA 66_3-018 airfoil at zero angle of attack.

procedure used for the inviscid flow normally varies as the flow regime changes. A relaxation solution of the full potential equation for inviscid flow was used by Carter (1981) in his transonic viscous-inviscid interaction calculations. For fully supersonic streams the concept of a small-disturbance approximation (linearized theory) is again useful, and the component of the pressure gradient attributed to viscous displacement can be related to the second derivative of the boundary-layer displacement thickness in a very simple manner. The exact form of the appropriate pressure gradient relation varies somewhat with the application considered. The reader is referred to the works of Werle and Vatsa (1974) and Burggraf et al. (1979) for specific examples. Despite the fact that the pressure gradient depends upon local quantities in the case of a supersonic external stream, a downstream condition must be imposed (usually it is on δ^*) in order to obtain a unique solution. Various time-dependent interaction schemes have also been successfully applied to both subsonic and supersonic flows (Briley and McDonald, 1975; Werle and Vatsa, 1974).

Mention is often made of "triple-deck theory" or "triple-deck structure" in connection with viscous-inviscid interactions. It is natural to wonder if this theory introduces something that ought to be taken into account by those applying finite-difference methods to viscous-inviscid interaction problems. The theory itself is based on a multistructured asymptotic expansion valid as $\text{Re} \rightarrow \infty$ for laminar flow in the neighborhood of a perturbation to a boundary-layer flow such as would occur owing to small separated regions or near the trailing edge of a flat plate. We will primarily concentrate on the application of triple-deck theory to the small-separation problem.

Several individuals have contributed to the theory. Some of the early concepts were introduced by Lighthill (1958). Stewartson and co-workers have made several contributions. An excellent review of developments in the theory up through 1974 is given by Stewartson (1974).

The theory is applicable if the streamwise length of the disturbance is relatively short. Thus, the theory would be applicable to small separation bubbles, but not to catastrophic separation. The length of the perturbation region where the triple-deck analysis would be applicable is of the order of $\mathrm{Re}^{-3/8}$, where Re is the Reynolds number based on the origin of the boundary layer. The "decks" in the theory are flow regions measured normal to the wall. The thickness of the lower deck is of order $\mathrm{Re}^{-5/8}$. The flow in this thin lower region has very little inertia, so that it responds quite readily to disturbances transmitted by the pressure gradient. The thickness of the middle (main) deck is of the order $\mathrm{Re}^{-1/2}$. The flow in this region is essentially a streamwise continuation of the upstream boundary-layer flow and is predominantly rotational and inviscid. All flow quantities in this region are only perturbed slightly from those in a conventional noninteracting boundary layer. The disturbances being transmitted by the lower deck displace the main deck boundary layer outward. The upper deck is of order $\mathrm{Re}^{-3/8}$ in thickness. The upper deck flow is the perturbed part of the inviscid irrotational flow.

Triple-deck theory provides the equations and boundary conditions needed to match the solutions in each of the three regions. The results are only valid for laminar flows where $\mathrm{Re} \to \infty$ so in a sense are of limited practical value. These equations are frequently solved numerically using viscous-inviscid interaction procedures (Jobe and Burggraf, 1974).

To the computational fluid dynamicist, the most important ideas and conclusions that come from the development of the triple-deck theory to date are as follows.

1. The equations that result from the triple-deck theory applied to flows containing small perturbations (such as small closed separated regions and trailing edge flow) contain no terms that are not present in the boundary-layer viscous-inviscid interaction model. This tends to confirm that the boundary-layer viscous-inviscid interaction model is correct in the limit as $\mathrm{Re} \to \infty$. Normal pressure gradients are neglected in the triple-deck theory when applied to the class of flows being considered here.
2. Triple-deck theory identifies length scales that can prove useful in finite-difference computations for laminar flows. The theory predicts that the lower deck is of order $\mathrm{Re}^{-5/8}$ in thickness. Although this conclusion is only strictly valid in the limit as $\mathrm{Re} \to \infty$, it would appear prudent to use a mesh near the wall sufficiently fine to resolve this lower deck region, where pressure variations can have a fairly drastic effect on the flow. The importance of honoring this scaling is confirmed by the finite-difference study made by Burggraf et al. (1979).

3. The theory provides clear evidence that the supersonic separation problem is boundary value in nature, requiring a downstream boundary condition in order to select a unique solution from the branching solutions that might otherwise be obtained. This requirement is not immediately obvious in the supersonic case because the boundary-layer equations themselves are parabolic and, according to linearized theory, the pressure depends only on the *local* slope of the displacement body. The downstream boundary condition is usually invoked as a prescribed value of the displacement thickness.

The paper by Burggraf et al. (1979) is helpful in clarifying the differences between the application of boundary-layer viscous-inviscid interaction schemes and the numerical solution of the triple-deck equations. At very large Re ($\geqslant 10^9$) the boundary-layer viscous-inviscid interaction calculation agreed very well with the triple-deck results for separating supersonic flow past a compression ramp. As Re decreased, the predictions of the boundary-layer viscous-inviscid interaction procedure and the triple-deck results differed very noticeably.

7.5 METHODS FOR INTERNAL FLOWS

7.5.1 Introduction

The thin-shear-layer equations provide a reasonably accurate mathematical model for two-dimensional and axisymmetric internal flows. These include the developing flow in straight tubes and in the annulus formed between two concentric straight tubes. In addition, the flow in the central portion of a large aspect ratio straight rectangular channel ("parallel plate duct") is often found to be reasonably two-dimensional. These flow configurations are illustrated in Fig. 7.15. The flow cross-sectional area does not change with axial distance in these standard geometries. The boundary-layer model also provides a good approximation for some internal flows in channels having abrupt expansions in cross-sectional area, which cause regions of flow reversal. These new areas of possible applicability of the thin-shear-layer equations are discussed further in Section 7.5.3.

The finite-difference/finite-volume approach is particularly useful in analyzing the flow from the inlet to the region of fully developed flow. The flow is said to be hydrodynamically fully developed when the velocity distribution is no longer changing with the axial distance along the flow passage. The hydrodynamic fully developed idealization is generally only realized for flows in which fluid property variations in the main flow direction are negligible. The thermal development of the flow is also of interest and can be predicted for the class of flow mentioned above by solving the thin-shear-layer form of the energy equation simultaneously with the momentum and continuity equations. Under constant property assumptions and with either constant wall temperature or uniform wall heat flux thermal boundary conditions, it is possible for the nondimensional temperature distribution to become independent of the axial

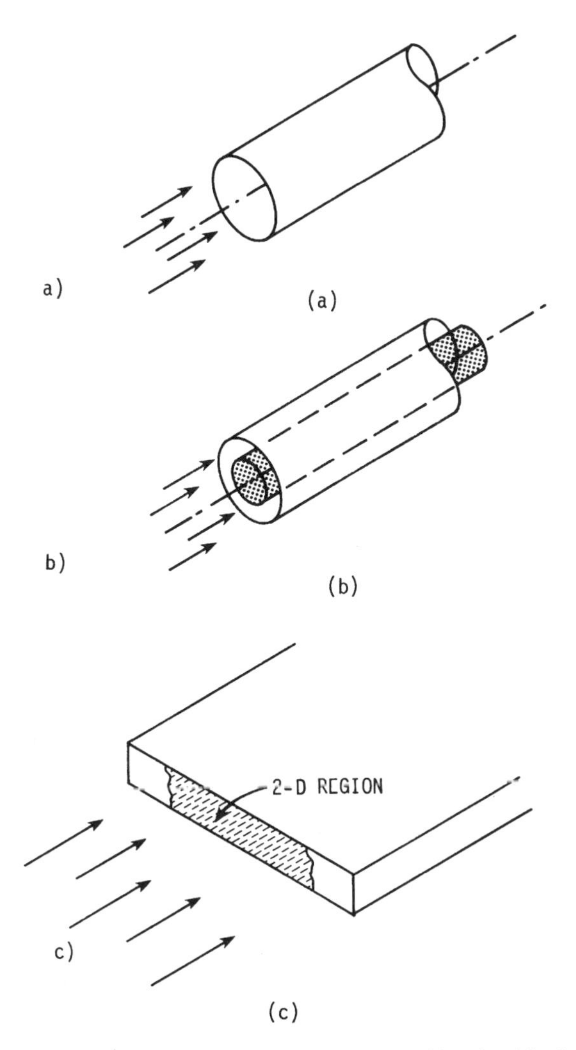

Figure 7.15 Internal flow configurations in which the thin-shear-layer equations are applicable. (a) Circular duct. (b) Annular passage. (c) Large aspect ratio rectangular channel.

direction. Shah and London (1978) provide an excellent discussion of the thermal aspects of internal flows.

The finite-difference/finite-volume approach is of less interest for treating constant-property fully developed flow, since the governing PDEs reduce to ODEs under these conditions. Laminar fully developed flow in a tube is the well-known Hagen-Poiseuille flow (White, 1991). If a relatively simple algebraic turbulence model is used, even turbulent fully developed flow can be treated by numerical methods appropriate for ODEs. With heat transfer present, it becomes more likely that the property variations will prevent the flow from reaching a fully developed state.

The characteristic Re for internal flows makes use of the channel hydraulic diameter, D_H, as the characteristic length. The hydraulic diameter is evaluated as $4A/P$, where A is the flow cross-sectional area and P is the wetted perimeter. For circular ducts, D_H reduces to the duct diameter.

We expect to find a small region near the channel inlet where the boundary-layer approximation is poor. This corresponds to the low Re leading edge region in external flows. For channel Re greater than about 75, this region is negligibly small. Comparisons of various numerical models for very low Re channel inlet flows can be found in the works of McDonald et al. (1972) and Chilukuri and Pletcher (1980).

7.5.2 Coordinate Transformation for Internal Flows

For steady laminar two-dimensional and axisymmetric constant property flows in straight channels (including pipes of circular cross section), Re can be removed from the governing equations by a simple nondimensionalization. As a result, it becomes apparent that only one solution of the boundary-layer equations is required for each geometry. The scaling factors for this nondimensionalization are given by

$$u^* = \frac{u}{\bar{u}_i} \qquad v^* = \frac{v\,\mathrm{Re}}{\bar{u}_i} \qquad x^* = \frac{x}{D\,\mathrm{Re}} \qquad y^* = \frac{y}{D} \qquad p^* = \frac{p}{\rho \bar{u}_i^2} \tag{7.80}$$

where D is the channel diameter and Re is the Reynolds number based on the inlet average velocity $\bar{u}_i$ and diameter D. The x axis is at the center of the straight channel. Specializing Eqs. (7.1) and (7.3) for constant property laminar flow and introducing the variables defined above yields

$$\frac{\partial u^*}{\partial x^*} + \frac{1}{r^m}\frac{\partial r^m v^*}{\partial y^*} = 0 \tag{7.81}$$

$$u^*\frac{\partial u^*}{\partial x^*} + v^*\frac{\partial u^*}{\partial y^*} = -\frac{dp^*}{dx^*} + \frac{1}{r^m}\frac{\partial}{\partial y^*}\left(r^m\frac{\partial u^*}{\partial y^*}\right) \tag{7.82}$$

where r is the distance from the center of the channel and m is a flow index equal to unity for axisymmetric flow and equal to zero for 2-D flow, as before. It should be easy to see that if one has a solution to Eqs. (7.81) and (7.82) for a developing flow, the results can be stretched (scaled) to be applicable for a flow at any specific Re by using Eqs. (7.80). Note again that this simple state of affairs is for constant-property laminar flow and, to date, an appropriate general scaling for variable-property and turbulent flows has not been identified.

7.5.3 Computational Strategies for Internal Flows

It is very important to observe that for steady channel flow, the flux of mass across any plane perpendicular to the channel axis is constant in the absence of wall blowing or suction. Since an initial velocity and temperature distribution must be given as part of the problem specification for the parabolic equations, the mass flow rate can also be considered as specified. If wall blowing or suction occurs, the normal component of velocity at the walls would be required as part of the boundary conditions for the boundary-layer equations; hence the changing mass flow rate through the channel can be computed from the problem specifications. For simplicity, in the discussion to follow, we will assume that no flow passes through the channel walls. However, computation procedures can easily be modified to account for these effects. This additional information about the global or overall mass flow in the channel permits a constraint to be placed on the solution from which the pressure gradient can be determined. In a sense, this mass flow constraint serves the same purpose as the simple relation between $u_e(x)$ and dp/dx, which can be obtained from the steady Euler momentum equation for external flows. In the usual treatment for external flows, the flow outside the boundary layer is assumed to be inviscid, and at the outer edge we specialize the Euler equation to $dp/dx = -\rho u_e \, du_e/dx$. Thus for external flows, we usually think of the pressure gradient being specified, meaning that dp/dx is either given or easily calculated from $u_e(x)$. When we apply the boundary-layer equations alone to calculate steady internal flows, no information is available from an inviscid flow solution, and "outer" boundary conditions on u are established from geometric considerations. In general, viscous effects may be important throughout the flow, so that the Euler momentum equations cannot be used in any manner to obtain the pressure gradient. Instead, the global mass flow constraint is used. Thus, in steady internal flows, *the pressure gradient is determined from the solution* (with the help of the global mass flow constraint) rather than being "specified" as for external flows. This is the primary difference between the numerical treatment of internal and external flows.

The thin-shear-layer equations can be written in a form applicable to 2-D internal flows as follows:

momentum:

$$\rho u \frac{\partial u}{\partial x} + \rho \tilde{v} \frac{\partial u}{\partial y} = -\frac{dp}{dx} + \frac{1}{r^m} \frac{\partial}{\partial y}(r^m \tau) \tag{7.83}$$

energy:

$$\rho u c_p \frac{\partial T}{\partial x} + \rho \tilde{v} c_p \frac{\partial T}{\partial y} = \frac{1}{r^m} \frac{\partial}{\partial y}(-r^m q_y) + \beta T u \frac{dp}{dx} + \tau \frac{\partial u}{\partial y} \tag{7.84}$$

mass:

$$\frac{\partial}{\partial x}(\rho u r^m) + \frac{\partial}{\partial y}(\rho \tilde{v} r^m) = 0 \tag{7.85}$$

global mass:

$$\dot{m} = \int_A \rho u \, dA = \text{const} \tag{7.86}$$

In the above, A is the cross-sectional area perpendicular to the channel axis. In addition, an equation of state is normally used to relate density to temperature and pressure. When $m = 0$, the above equations are applicable to 2-D flows, and when $m = 1$, they apply to axisymmetric flows. For turbulence models utilizing the Boussinesq assumption, we find

$$\tau = \mu \frac{\partial u}{\partial y} - \rho \overline{u'v'} = (\mu + \mu_T) \frac{\partial u}{\partial y} \tag{7.87}$$

$$q_y = -k \frac{\partial T}{\partial y} + \rho c_p \overline{v'T'} = \left(-k + \frac{c_p \mu_T}{\mathrm{Pr}_T} \right) \frac{\partial T}{\partial y} \tag{7.88}$$

The governing equations reduce to a form applicable to laminar flows whenever the fluctuating Reynolds terms above are equal to zero.

The wall boundary conditions remain the same as for external flows. For flows in straight tubes and parallel plate channels, a symmetry line or plane exists, and outer boundary conditions of the form

$$\left. \frac{\partial u}{\partial y} \right)_{r=0} = \left. \frac{\partial T}{\partial y} \right)_{r=0} = 0 \tag{7.89}$$

are used. For tube flow, the shear-stress and heat flux terms in Eqs. (7.83) and (7.84) are singular at $r = 0$. A correct representation can be found from an application of L'Hospital's rule, from which we find

$$\lim_{r \to 0} \frac{1}{r} \frac{\partial}{\partial y} \left(\mu r \frac{\partial \phi}{\partial y} \right) = 2 \frac{\partial}{\partial y} \left(\mu \frac{\partial \phi}{\partial y} \right)$$

Except for the treatment of the pressure gradient, the differencing of the governing equations proceeds in the same manner as for external boundary-layer flow. The pressure gradient is treated as an unknown in the internal flow case, its value to be determined with the aid of the global mass flow constraint, as indicated previously. This can be done in several ways.

When explicit difference schemes are used, the pressure gradient can be determined as follows. The finite-difference form of the momentum equation can be written in the form

$$u_j^{n+1} = Q_j^n + \frac{dp}{dx} R_j^n \tag{7.90}$$

where Q_j^n and R_j^n contain quantities that are all known. Equation (7.90) is then multiplied by the density $\hat{\rho}_j^{n+1}$, and the resulting equation integrated numerically over the channel cross section by Simpson's or the trapezoidal rule. This gives

$$\int_A \hat{\rho}_j^{n+1} u_j^{n+1} \, dA = \dot{m} = \int_A \hat{\rho}_j^{n+1} Q_j^n \, dA + \frac{dp}{dx} \int_A \hat{\rho}_j^{n+1} R_j^n \, dA \tag{7.91}$$

The density $\hat{\rho}_j^{n+1}$ is not known a priori at the $n + 1$ level at the time the pressure gradient is being determined. The circumflex indicates the provisional nature of this one variable. Very good results have been obtained by simply letting $\hat{\rho}_j^{n+1} = \rho_j^n$. In fact, this has been the most common procedure. An alternative is to evaluate $\hat{\rho}_j^{n+1}$ from ρ_j^n and ρ_j^{n-1} by a second-order accurate extrapolation. Since $\dot{m}$ is specified by the problem initial conditions and the integrals in Eq. (7.91) contain all known quantities, dp/dx can be determined as

$$\frac{dp}{dx} = \frac{\dot{m} - \int_A \hat{\rho}_j^{n+1} Q_j^n \, dA}{\int_A \hat{\rho}_j^{n+1} R_j^n \, dA} \tag{7.92}$$

Once dp/dx has been evaluated, the finite-difference form of the momentum, continuity, and energy equations can be solved just as for external flows. The most widely used explicit scheme for internal flow appears to be of the DuFort-Frankel type. The DuFort-Frankel scheme was given for the thin-shear-layer equations in Section 7.3.4. A typical comparison between the predictions of the DuFort-Frankel scheme and experimental measurements of Barbin and Jones (1963) is shown in Fig. 7.16 for the turbulent flow of air in a tube. In the figure, u_b denotes the bulk velocity in the tube and r_w is the radius of the tube. Even very near the inlet ($x/D = 1.5$), the predictions are seen to be in good agreement with the measurements. A simple algebraic turbulence model was used in the predictions.

The internal flow problem is conceptually very similar to the inverse boundary-layer problem discussed in Section 7.4 for external flows. This is most evident when implicit difference schemes are used. For internal flows the "correct" pressure gradient must be determined that will give velocities that

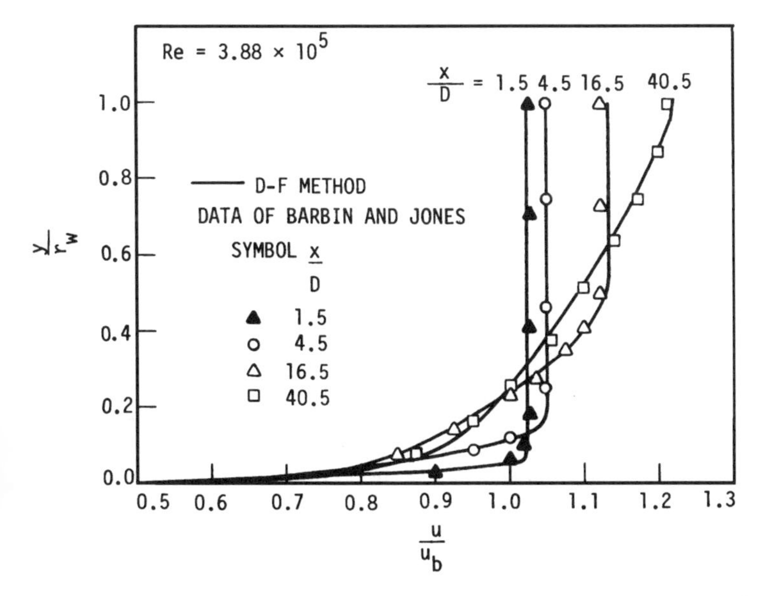

Figure 7.16 Comparison of predicted and measured turbulent velocity profiles in the entrance region of a pipe (Nelson and Pletcher, 1974). D-F denotes DuFort-Frankel method.

satisfy the global mass flow constraint. This corresponds to adjusting the pressure gradient (or edge velocity) until the velocity distribution satisfies the specified displacement thickness in inverse methods for external flows. Several different procedures have been used with implicit methods to determine the pressure gradient. A number of these are briefly discussed below.

Variable secant iteration. The pressure gradient can be varied iteratively at each streamwise location until the global mass flow constraint is met (Briley, 1974) by employing the variable secant procedure discussed in Section 7.4.3 in connection with imposing the δ^* boundary condition for the inverse boundary-layer method. For fixed coefficients, velocities vary linearly with the pressure gradient, so that convergence is usually obtained with three iterations.

Lagging the pressure adjustment. Patankar and Spalding (1970) pointed out that iterating at each streamwise station is uneconomical and have suggested that a value for the pressure gradient be guessed to advance the solution and then let the knowledge of any resulting error in mass flow rate guide our choice of pressure gradient for the *next* step. That is, in analogy with the way an automobile is steered, adjustments are made to correct the course without going back to retrace the path. This was the common-sense approach included in the early versions of the Patankar-Spalding finite-difference method for confined flows. Although the common-sense aspect of this logic cannot be denied, the algorithm appears a bit too approximate by present-day standards and is not recommended. In concert with the trend toward lower computer costs observed over the past decade, there has been a cost-equalizing trend toward the use of algorithms that are potentially more accurate.

Newton's method. Raithby and Schneider (1979) proposed a scheme suitable for incompressible flows that requires one-third less effort than the minimum (three iterations) variable secant calculation. The scheme assumes that the coefficients in the difference equations will remain constant, i.e., no form of updating is employed as the pressure gradient is adjusted until the global mass flow constraint is satisfied. The idea is that once an initial guess for dp/dx is made and a provisional solution obtained for the difference equations, a correction can be obtained by employing a form of Newton's method. With "frozen" coefficients, the velocities will vary linearly with the pressure gradient, and it follows that one Newton-type correction should provide the correct pressure gradient. To illustrate, we will let $S = dp/dx$. We make an initial guess for $dp/dx = (dp/dx)^*$ and calculate provisional velocities $(u_j^{n+1})^*$ and a provisional mass flow rate $\dot{m}^*$. Due to the linearity of the momentum equation with frozen coefficients, we observe from an application of Newton's method (see Section 7.4.2) that the correct velocity at each point would be

$$u_j^{n+1} = \left(u_j^{n+1}\right)^* + \frac{\partial u_j^{n+1}}{\partial S} \Delta S \tag{7.93}$$

where ΔS is the change in the pressure gradient required to satisfy the global mass flow constraint. We define $u_{p,j}^{n+1} = \partial u_j^{n+1}/\partial S$. The difference equations are actually differentiated with respect to the pressure gradient (S) to obtain difference equations for $u_{p,j}^{n+1}$ that are tridiagonal in form. The coefficients for the unknowns in these equations will be the same as for the original implicit difference equations. The Thomas algorithm is used to solve the system of algebraic equations for $u_{p,j}^{n+1}$. The boundary conditions on $u_{p,j}^{n+1}$ must be consistent with the velocity boundary conditions. On boundaries where the velocity is specified, $u_{p,j}^{n+1} = 0$, whereas on boundaries where the velocity gradient is specified, $\partial u_{p,j}^{n+1}/\partial n = 0$ (n normal to boundary). The solution for $u_{p,j}^{n+1}$ is then used to compute ΔS by noting that $u_{p,j}^{n+1}\,\Delta S$ is the correction in velocity at each point required to satisfy the global mass flow constraint. Thus we can write

$$\dot{m} - \dot{m}^* = \Delta S \int_A \rho u_{p,j}^{n+1}\, dA \tag{7.94}$$

where the integral is evaluated by numerical means. The $\dot{m}$ in Eq. (7.94) is the known value specified by the initial conditions. The required value of ΔS is determined from Eq. (7.94). The correct values of velocity u_j^{n+1} can then be determined from Eq. (7.93). The continuity equation is then used to determine v_j^{n+1}. The computational effort of this procedure is roughly equivalent to two iterations of the method employing the variable secant procedure.

Treating the pressure gradient as a dependent variable. In all of the procedures discussed above, the pressure gradient is treated as a known quantity whenever the simultaneous algebraic equations for the new velocities are solved. The standard Thomas algorithm can be used for the three methods above. Here we consider schemes in which the pressure gradient is treated as an unknown in the algebraic formulation. The coefficient matrix is no longer tridiagonal. Early methods of this type (Hornbeck, 1963) tended to employ conventional Gaussian elimination. More recent procedures (Blottner, 1977; Cebeci and Chang, 1978; Kwon and Pletcher, 1981) have used more efficient block elimination procedures. The method of Kwon and Pletcher (1981) is a modification of inverse method B presented in Section 7.4.3. The procedure employs the FLARE approximation to permit the calculation of separated regions in internal flows. The changes that must be made in inverse method B in order to treat internal incompressible flows in a 2-D channel will now be described. The flow is assumed to be symmetric about the channel centerline located at $y = H/2$, where y is measured from the channel wall. The channel height is H. Equations (7.57) and (7.58) apply. The outer boundary conditions become

$$\left.\frac{\partial u}{\partial y}\right)_{y=H/2} = 0 \qquad \psi\left(x, \frac{H}{2}\right) = \frac{\dot{m}}{2\rho} \tag{7.95}$$

where $\dot{m}$ is the mass flow rate per unit width for a 2-D channel. The difference equations, Eqs. (7.61)-(7.64), are applicable, and χ^{n+1} represents the unknown pressure gradient

$$-\frac{1}{\rho}\frac{dp}{dx}$$

as before. The procedures for internal and external flow differ in the way in which χ^{n+1} and u_{NJ}^{n} are determined from the outer boundary conditions, which are different in the two cases. We express $\partial u/\partial y)_{H/2}$ in terms of a one-sided second-order accurate difference representation,

$$\left(\frac{\partial u}{\partial y}\right)_{NJ}^{n+1} \cong \frac{u_{NJ}^{n+1}}{2}\left(\frac{4}{\Delta y_-} - \frac{1}{\Delta y_{--}}\right) - \frac{2u_{NJ-1}^{n+1}}{\Delta y_-} + \frac{u_{NJ-2}^{n+1}}{2\,\Delta y_{--}} \tag{7.96}$$

where

$$\Delta y_- = y_{NJ} - y_{NJ-1}$$
$$\Delta y_{--} = y_{NJ-1} - y_{NJ-2}$$

The outer boundary conditions, Eq. (7.95) can now be written as

$$u_{NJ}^{n+1} = c_1 u_{NJ-1}^{n+1} - c_2 u_{NJ-2}^{n+1} \tag{7.97}$$

$$\psi_{NJ}^{n+1} = \frac{\dot{m}}{2\rho} \tag{7.98}$$

where

$$c_1 = \frac{4}{4-K}$$
$$c_2 = \frac{K}{4-K}$$
$$K = \frac{\Delta y_-}{\Delta y_{--}}$$

Equations (7.97) and (7.98) are to be solved with Eqs. (7.67), (7.68), and (7.71). However, one additional relationship is needed, since five unknowns appear $(u_{NJ}^{n+1}, u_{NJ-1}^{n+1}, u_{NJ-2}^{n+1}, \psi_{NJ-1}^{n+1}, \chi^{n+1})$ and only four independent relationships among them have been identified thus far [Eqs. (7.97), (7.67), (7.68), and (7.71)]. The additional equation can be obtained by specializing Eq. (7.65) for u_{NJ-2}^{n+1} as

$$u_{NJ-2}^{n+1} = A'_{NJ-2}u_{NJ-1}^{n+1} + H'_{NJ-2}\,\chi^{n+1} + C'_{NJ-2} \tag{7.99}$$

This system of equations can be solved for χ^{n+1} by defining

$$\alpha_1 = 1 - A'_{NJ-1}(c_1 - c_2 A'_{NJ-2})$$
$$\alpha_2 = (c_1 - c_2 A'_{NJ-2})H'_{NJ-1} - c_2 H'_{NJ-2}$$
$$\alpha_3 = (c_1 - c_2 A'_{NJ-2})C'_{NJ-1} - c_2 C'_{NJ-2}$$
$$\alpha_4 = 1 + \frac{2}{\Delta y_-}B'_{NJ-1} + A'_{NJ-1}$$

$$\alpha_5 = -\left(H'_{NJ-1} + \frac{2}{\Delta y_-} D'_{NJ-1}\right)$$

$$\alpha_6 = \frac{\dot{m}}{\rho \Delta y_-} - \frac{2}{\Delta y_-} E'_{NJ-1} - C'_{NJ-1}$$

Then

$$\chi^{n+1} = \frac{\alpha_1 \alpha_6 - \alpha_3 \alpha_4}{\alpha_2 \alpha_4 - \alpha_1 \alpha_5} \tag{7.100}$$

The axial component of velocity at the line of symmetry can be found from

$$u_{NJ}^{n+1} = \frac{\alpha_2}{\alpha_1} \chi^{n+1} + \frac{\alpha_3}{\alpha_1} \tag{7.101}$$

At this point the back substitution process can be initiated using Eqs. (7.65) and (7.66) to compute u_j^{n+1} and ψ_j^{n+1} from the outer boundary to the wall. The remaining portions of the algorithm are as discussed in Section 7.4.3. The only differences between inverse method B and the related procedure for internal flows are due to the minor differences in the boundary conditions for the two cases. This requires that slightly different algebraic procedures be used to evaluate χ^{n+1} and u_{NJ}^{n+1} prior to the back substitution step in the block tridiagonal solution procedure.

An interesting application of this method has been made to laminar channel flows having a sudden symmetric expansion that creates a region of recirculation downstream of the expansion. The general pattern of such a flow is illustrated in Fig. 7.17. The predictions were obtained by the boundary-layer method described above utilizing a fully developed velocity profile at the step. Re_h is the Reynolds

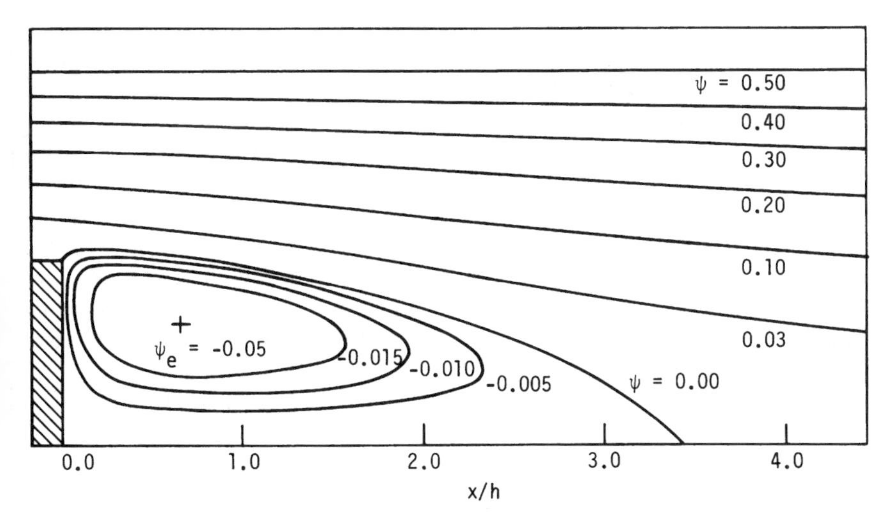

Figure 7.17 Streamline contours predicted from boundary-layer equations (Kwon et al., 1984) for a laminar flow in a channel with a symmetric sudden expansion, $\mathrm{Re}_h = 50$, $H_1/H_2 = 0.5$.

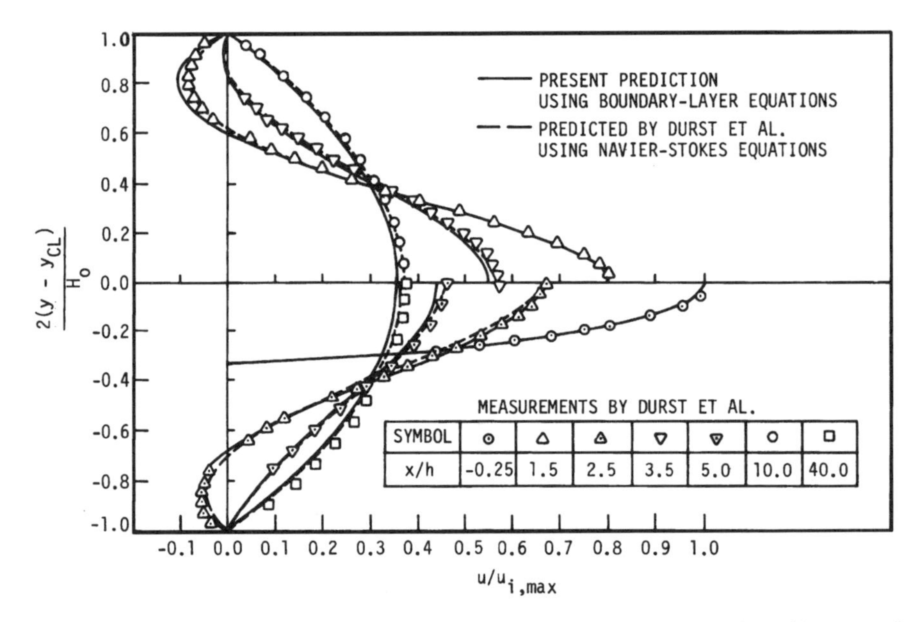

Figure 7.18 Velocity profiles for a laminar flow in a channel with a symmetric sudden expansion. Re_h (based on $u_{\max}$) = 56, $H_1/H_2 = \frac{1}{3}$ (Kwon et al., 1984).

number based on step height, and H_1/H_2 is the ratio of the channel height before and after the expansion. Such flows have customarily been predicted by solving the full Navier-Stokes equations.

Figure 7.18 compares velocity profiles predicted by the same method with experimental data and Navier-Stokes solutions for a symmetric sudden expansion flow. In the figure, $u_{i,\max}$ denotes the maximum velocity just upstream of the expansion (step), H_0 is the channel height downstream of the expansion, and y_{CL} is the distance from the wall to the channel centerline. The symbols h, H_1, H_2 are as defined previously. The method based on boundary-layer equations requires an order of magnitude less computer time than required for the solution of the Navier-Stokes equations.

In Fig. 7.19 the reattachment length and distance to the vortex center predicted by the boundary-layer solutions (Lewis and Pletcher, 1986) are compared with the Navier-Stokes solutions and experimental data obtained by Macagno and Hung (1967) for a 1:2 pipe expansion. The Reynolds number (Re) and l/d in the figure are based on the diameter upstream of the expansion. The agreement is quite good for Re greater than about 20. Note that the boundary-layer results yield straight lines in Fig. 7.19 due to the scaling laws of Eqs. (7.80), which are applicable even in the case of fully developed flow undergoing a sudden expansion in cross-sectional flow area. Although the solutions to the boundary-layer equations are in close agreement with experimental data and solutions to the full Navier-Stokes equations for velocity profiles and the distance to the reattachment point, some limitations exist. For example, details

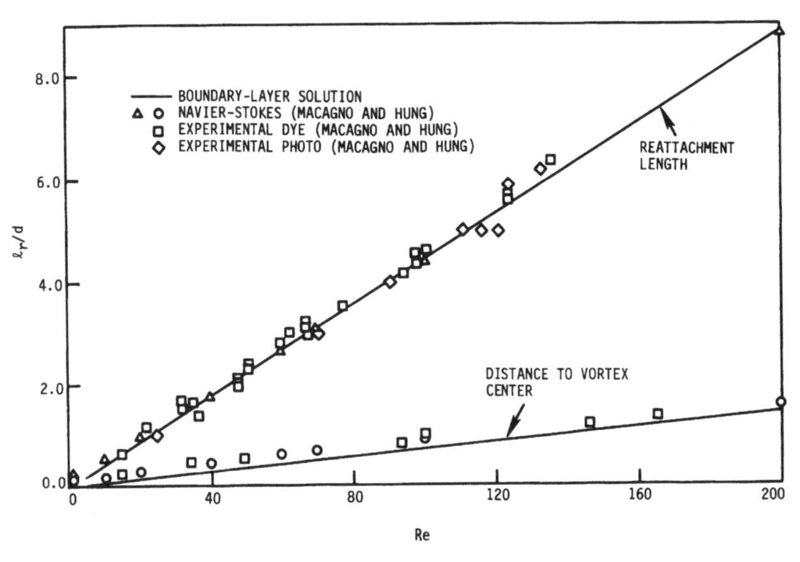

Figure 7.19 Comparison of the distances to flow reattachment and vortex center for experiments and Navier-Stokes solutions (Macagno and Hung, 1967) and solution of boundary-layer equations with FLARE (Lewis and Pletcher, 1986) for a 1:2 pipe expansion.

of the eddy structure are not well predicted at low Re, as can be seen in Fig. 7.20, where the magnitude of the minimum nondimensional stream function is plotted against Re for the 1:2 pipe and planar expansions. The minimum stream function measures the volume rate of flow in the recirculating eddy. Note that the boundary-layer values are independent of Re (indicating that the dimensional

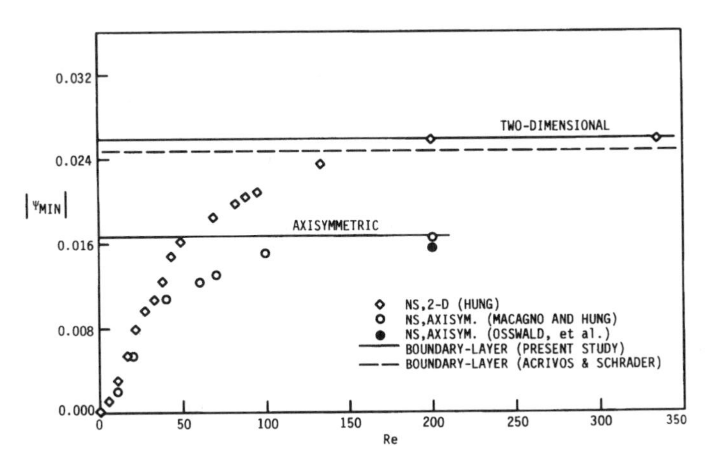

Figure 7.20 Comparison of the stream function at the center of the trapped vortex predicted by boundary-layer equations (Lewis and Pletcher, 1986; Acrivos and Schrader, 1982) and the Navier-Stokes equations (Macagno and Hung, 1967; Osswald et al., 1984) for a 1:2 expansion.

volume flow reduces linearly with Re) but are only correct for Re greater than about 200. At lower Re the volume rate of flow in the eddy actually reduces more rapidly with decreasing Re as indicated by the solutions to the full Navier-Stokes equations.

7.5.4 Additional Remarks

Details of several difference schemes suitable for the thin-shear-layer equations were presented in Section 7.3 as they applied to ordinary external boundary-layer flows. As methods for confined flows have been considered in the present section, numerical details that remain the same as for external flows have not been repeated. However, an attempt has been made to clearly point out and emphasize those details that change and are unique to internal flows.

Coverage in this section has been limited to flows in straight channels. Blottner (1977) demonstrated that the thin-shear-layer approximation (also known as the "slender channel" approximation) can be extended to curved 2-D channels with varying channel height. The equations solved are the boundary-layer equations with longitudinal curvature (Van Dyke, 1969). The normal pressure gradient induced by the channel curvature is accounted for by

$$\frac{\partial p}{\partial n} = \frac{\kappa \rho u^2}{1 + \kappa n} = 0$$

where n is the coordinate normal to the channel centerline and κ is the curvature of the centerline.

Viscous-inviscid interaction schemes can be applied to internal flows in which an inviscid core region can be identified. The interaction effect is expected to be negligibly small except very near the inlet for low Reynolds number flows and under conditions in which the channel cross-sectional area changes abruptly. Viscous-inviscid interaction permits information to be transmitted upstream and can give improved predictions for flows in which the pressure field at a point is expected to be influenced by conditions farther downstream. The incompressible inviscid flow in channels is conveniently determined through a numerical solution of Laplace's equation for the stream function. Inverse method B discussed in Section 7.4.3 can be used for the viscous portion of the flow. Such a combination has been employed interactively to predict the flow over a rearward facing step in a channel (Kwon and Pletcher, 1986a, 1986b).

7.6 APPLICATION TO FREE-SHEAR FLOWS

The thin-shear-layer equations provide a fairly accurate mathematical model for a number of free-shear flows. These include the plane or axisymmetric jet discharging to a quiescent or co-flowing ambient, the planar mixing layer, and simple wake flows. The majority of free-shear layers encountered in engineering applications are turbulent. To date, turbulence models for free-shear flows have not exhibited nearly the degree of generality as those used for wall boundary layers. It is still a major challenge to find models that can provide accurate predictions for the development of both the planar and axisymmetric jet without requiring adjustments in the model parameters.

A complete treatise on the subject of the numerical prediction of free-shear flows might devote 60% of its content to turbulence modeling, 25% to coverage of the physics of various categories of free-shear flows, and 15% to numerical procedures. The numerical procedures, which are our main concern here, are the least troublesome aspect of the problem of obtaining accurate predictions for turbulent free-shear flows.

The round jet has been studied extensively both experimentally and analytically and provides a representative example of a free-shear flow. The thin-shear-layer equations will provide a good mathematical model for the round jet following a straight trajectory if pressure in the interior of the jet can be assumed to be equal to that of the surrounding medium. This requires that the surface tension of the jet be negligible and that the jet be fully expanded, i.e., the pressure at the discharge plane equals the pressure in the surrounding medium. A subsonic jet discharging from a tube can always be considered as fully expanded. For the jet cross section to remain round and the trajectory to remain straight, it is necessary that no forces act on the jet in the lateral direction. This requires that the medium into which the jet is injected be at rest or flowing in the same direction as the discharging jet (co-flowing) and that body forces (such as buoyancy) be negligible. Under these conditions, the form of the thin-shear-layer equations given by Eqs. (5.116)-(5.119) are applicable. These equations are specialized further below for the steady incompressible flow of a round jet in the absence of a pressure gradient:

continuity:

$$\frac{\partial(yu)}{\partial x} + \frac{\partial(yv)}{\partial y} = 0 \tag{7.102}$$

momentum:

$$u\frac{\partial u}{\partial x} + v\frac{\partial u}{\partial y} = \frac{1}{y}\frac{\partial}{\partial y}\left[y\left(\nu\frac{\partial u}{\partial y} - \overline{u'v'}\right)\right] \tag{7.103}$$

Numerically, the primary difference between the wall boundary layer and the round jet is in the specification of the boundary conditions. Figure 7.21 illustrates the round jet flow configuration. Due to the symmetry that exists about the jet centerline, the appropriate boundary conditions at $y = 0$ are $(\partial u/\partial y)_{y=0} = 0$ and $v(x, 0) = 0$. The outer boundary condition is identical to that for a wall boundary layer,

$$\lim_{y\to\infty} u(x, y) = u_e$$

Initial conditions are also needed for the finite-difference/finite-volume

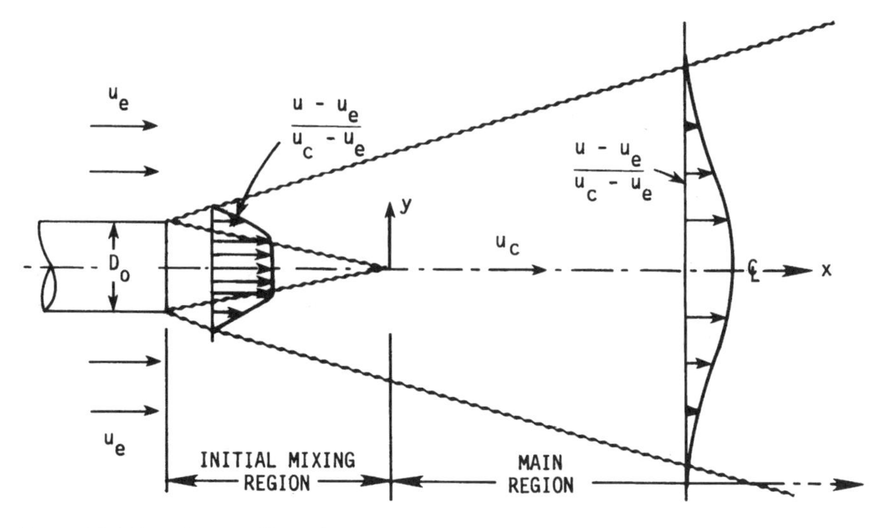

Figure 7.21 Round jet configuration.

calculation. For turbulent jets especially, the initial streamwise velocity distribution is usually taken as a uniform stream at the discharge velocity, u_0. Naturally, this cannot be completely correct, in that the velocities in a small region must exhibit the retarding effects of the tube walls. On the other hand, the boundary-layer equations are not expected to provide an extremely accurate solution very near the discharge plane, i.e., for x/D_0 less than about one, where D_0 is the diameter of the jet at discharge. Using a uniform velocity distribution at discharge for the turbulent jet appears to provide fairly accurate results for $x/D_0 > 1$, which includes the region of most interest in engineering applications. Some finite-difference schemes applied to the jet problem in the Cartesian coordinate system will also require an initial distribution for v. As was mentioned in Section 7.3 for wall boundary layers, this is a requirement of the numerical procedure and not a requirement in the mathematical specification of the problem. When an initial distribution for v is required, using $v(0, y) = 0$ is recommended. Taking several very small streamwise steps near the starting plane helps confine the effects of the starting singularity (which is due to the very large values of $\partial u/\partial x$ associated with the vanishingly small initial mixing zone) to a small region. This starting singularity is similar to that observed at the leading edge of a flat plate for the wall boundary layer when the equations are solved in the Cartesian coordinate system.

For a turbulent jet discharging to a quiescent ambient, the initial mixing region, indicated in Fig. 7.21, extends to an x/D of about 5. For jets discharging to a co-flowing stream, the initial mixing region is even longer. This initial mixing region is characterized by the fact that the fluid at the centerline is moving at the jet discharge velocity. Beyond the initial mixing region, the velocities throughout the jet are influenced by the ambient stream velocity, u_e. The growth properties of the jet differ in the two regions, initial and main, and when algebraic turbulence models are used, it is expected that somewhat different models (or values for the constants in the models) will be required in the two regions.

Most of the finite-difference schemes discussed in Section 7.3 have been observed to work well for jets. Several methods are described in the *Proceedings of the Langley Working Conference on Free Turbulent Shear Flows* (NASA, 1972). This reference should provide a good starting point for obtaining a background on the special problems associated with achieving accurate predictions for several types of turbulent free-shear flows. Many numerical details can also be found in the works of Hornbeck (1973), Madni and Pletcher (1975a, 1975b, 1977a), and Hwang and Pletcher (1978). This latter work gives the difference equations used in evaluating the fully implicit, Crank-Nicolson implicit, DuFort-Frankel, Larkin alternating direction explicit (ADE), Saul'yev ADE, and Barakat and Clark ADE methods for the round jet. A useful evaluation of available experimental data on uniform-density turbulent free-shear layers has been provided by Rodi (1975).

The boundary-layer form of the energy equation is also applicable to free-shear flows. For a heated jet discharging vertically into a quiescent ambient, with or without thermal stratification, the trajectory of the jet is straight, and no special difficulties arise with the boundary-layer model. For the heated jet discharging at other angles or discharging at any angle with a cross flow, the jet is expected to follow a curved trajectory. Such flows have been treated by solving the fully 3-D Navier-Stokes equations by Patankar et al. (1977) and others and by more approximate parabolic finite-difference models that assume the flow remains axisymmetric (Madni and Pletcher, 1977b; Hwang and Pletcher, 1978). In these latter axisymmetric models, the momentum equation in the transverse direction is treated in a lumped manner, which yields an ODE for the angle between the tangent to the trajectory of the jet centerline and the horizontal direction. This approach requires only slightly more computational effort than solving the axisymmetric boundary-layer equations and gives surprisingly good agreement with experimental measurements, especially with regard to the trajectory of the jet.

No finite-difference algorithms will be provided in this section, since the procedures discussed in Section 7.3 can be adapted to free-shear flows in a straightforward manner. However, one numerical anomaly that sometimes occurs in the prediction of jets discharging to an ambient at rest is worth mentioning. For this case, some schemes are unable to correctly predict u to asymptotically approach the free stream velocity of zero. The problem is thought to be related to the treatment of the coefficients of the convective terms and the procedures used to locate the outer boundary. The difficulty is most evident if the coefficients are lagged. It is commonplace to overcome this problem in a practical manner by letting u_e be a small positive velocity of the order of 1-3% of the jet centerline velocity. Reports in the literature claim that this approximation does

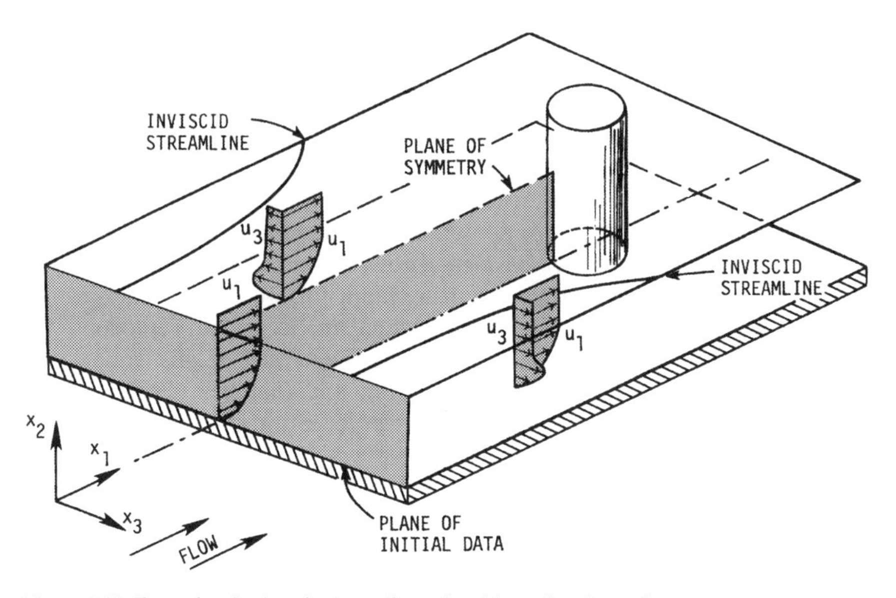

Figure 7.22 Example of subsonic three-dimensional boundary-layer flow.

not seriously degrade the accuracy of the calculations. Hornbeck (1973) shows that a satisfactory solution for $u_e = 0$ can be obtained with implicit methods by iteratively updating the coefficients.

7.7 THREE-DIMENSIONAL BOUNDARY LAYERS

7.7.1 Introduction

The majority of the flows that occur in engineering applications are 3-D. In this section we will consider finite-difference methods for those 3-D flows that are "thin" (i.e., with large velocity gradients) in only one coordinate direction. Such flows are sometimes referred to as "boundary sheets." Many flows occurring in applications are of this type. These are predominantly external flows. Examples include much of the viscous portion of the flow over wings and general aerodynamic bodies.

An example of a 3-D boundary-layer flow is illustrated in Fig. 7.22. The presence of the cylinder alters the pressure field, causing the inviscid flow streamlines to turn as indicated qualitatively in the figure. In accordance with the equations of motion, a component of the pressure gradient (responsible for the turning) is directed away from the center of curvature of the inviscid flow streamlines. Because the viscous layer is thin, this pressure gradient does not change in the direction normal to the surface. As a result, the velocity vector rotates toward the center of curvature of the inviscid streamlines as we move down within the boundary layer. This occurs because the pressure gradient remains fixed but the inertia of the fluid decreases as we move nearer the wall. This requires that the radius of streamline curvature decrease as we move in the normal direction toward the wall in the boundary layer. Thus the crossflow component of velocity will generally reach a maximum at some point within the boundary layer, as indicated in Fig. 7.22. This pressure-induced "cross flow" is referred to as a secondary flow in some applications and is responsible for such phenomena as the transport of sand toward the inside bank of a curved riverbed and the migration of tea leaves toward the center (near the bottom) of a stirred cup of tea.

Another interesting example of 3-D boundary-layer flow occurs on bodies of revolution at incidence. Such flows on a prolate spheroid, for example, have been studied extensively by several investigators, including Wang (1974, 1975), Blottner and Ellis (1973), Patel and Choi (1979), and Cebeci et al. (1979a).

The 3-D boundary-layer equations are not applicable to flows near the intersection of two surfaces (for example, near wing-body junctions and corners in channels) because stress gradients in two directions are important in those regions. Other reduced forms of the Navier-Stokes equations can be used to treat the flow near corners. These are discussed in Chapter 8.

The subject of 3-D boundary layers will not be covered in great detail here. Instead, we will outline the general numerical strategy required for the solution of the problem posed by these equations, making use of the material developed in earlier sections for the 2-D boundary-layer equations. Several new considerations arise with the 3-D problem, and these will be emphasized.

7.7.2 The Equations

The 3-D boundary-layer equations were presented in Chapter 5 in Cartesian coordinates [Eqs. (5.120)-(5.123)] and in body intrinsic orthogonal curvilinear coordinates [Eqs. (5.124)-(5.128)]. For certain special conditions (the laminar supersonic flow over a cone at incidence being one of them), the number of independent variables can be reduced from three to two. Special cases of this sort will not be discussed here.

The Cartesian coordinate system can be used for flows over developable surfaces (those that can be formed by bending a plane without stretching or shrinking), including of course, the special case of a flat surface. Curvilinear systems are required for flows over more general bodies. A few studies have been made using a curvilinear coordinate system coinciding with the inviscid streamlines (see Cebeci et al., 1973). However, most 3-D boundary-layer computations have been made with coordinate systems related to the geometry of the surface. Even with a body-oriented coordinate system, choices remain in the selection of the coordinate axes. Blottner (1975b) provides a review of the coordinate systems that have been used for 3-D flows.

The 3-D boundary-layer equations presented in Chapter 5 are singular at the origin at the x_1 coordinate. This singularity is of the same type as found at the leading edge of a flat plate in 2-D flow (see Section 7.3.7). Several investigators have satisfactorily used the equations in this form both for flows in the Cartesian coordinate system (Klinksiek and Pierce, 1973) and for more complex flows over axisymmetric bodies (Wang, 1972; Patel and Choi, 1979). These authors have generally used a separate procedure to generate a satisfactory stagnation point solution before utilizing their 3-D solution scheme.

It is also quite common to eliminate the singular behavior of the equations by use of a suitable dependent variable transformation. No single transformation has proven optimum for all flows. Blottner (1975b) discusses several that have been used for specific problems. An example transformation will be presented here that will remove the singularity at the origin of x_1 and will permit the stagnation point profiles to be obtained from the solution of the ODEs that remain when $x_1 = 0$. The transformed equations will permit the solution to be advanced from the stagnation point in a smooth and systematic manner. For laminar flows the boundary layer will tend to have a nearly uniform thickness in the transformed coordinates.

We first note that the boundary-layer equations revert to the Euler equations at the outer edge of the boundary layer, where the viscous terms vanish and $\partial u_1/\partial x_2$ and $\partial u_3/\partial x_2 \to 0$. This permits the components of the pressure gradient in Eqs. (5.126) and (5.127) to be written as

$$-\frac{1}{h_1}\frac{\partial p}{\partial x_1} = \frac{\rho u_{1,e}}{h_1}\frac{\partial u_{1,e}}{\partial x_1} + \frac{\rho u_{3,e}}{h_3}\frac{\partial u_{1,e}}{\partial x_3} + u_{1,e}u_{3,e}K_1 - \rho u_{3,e}^2 K_3 \quad (7.104)$$

$$-\frac{1}{h_3}\frac{\partial p}{\partial x_3} = \frac{\rho u_{1,e}}{h_1}\frac{\partial u_{3,e}}{\partial x_1} + \frac{\rho u_{3,e}}{h_3}\frac{\partial u_{3,e}}{\partial x_3} - u_{1,e}^2 K_1 + \rho u_{1,e}u_{3,e}K_3 \quad (7.105)$$

where $u_{1,e}(x_1, x_3)$ and $u_{3,e}(x_1, x_3)$ are given by the inviscid flow solution over the body. The subscript e denotes quantities evaluated at the outer edge of the boundary layer.

Let us assume that the Reynolds stresses for a turbulent flow will be evaluated by a viscosity model. That is, let

$$-\rho\overline{u_1'u_2'} = \mu_T\frac{\partial u_1}{\partial x_2} \qquad -\rho\overline{u_3'u_2'} = \mu_T\frac{\partial u_3}{\partial x_2}$$

$$-\rho c_p\overline{u_2'T'} = k_T\frac{\partial T}{\partial x_2} \qquad \frac{\mu_T c_p}{k_T} = \mathrm{Pr}_T$$

$$\bar{\mu} = \mu_T + \mu$$

The model for μ_T may be simple or complex; no assumption about the complexity of μ_T is made at this time. The equations remain valid for laminar flow where $\bar{\mu} = \mu$.

It is convenient to introduce nondimensional variables for velocity components defined by

$$F = \frac{u_1}{u_{1,e}} \qquad G = \frac{u_3}{W_e} \qquad I = \frac{H}{H_e}$$

where W_e will be designated below as either $u_{1,e}$ or $u_{3,e}$.

We now let $x = x_1$, $z = x_3$, and

$$\eta = \left[\frac{u_{1,e}}{x(\rho\mu)_e}\right]^{1/2} \int_0^{x_2} \rho \, dx_2$$

Using the chain rule for differentiation, derivatives with respect to the original independent variables can be replaced according to

$$\frac{\partial}{\partial x_1} = \frac{\partial}{\partial x} + \frac{\partial \eta}{\partial x}\frac{\partial}{\partial \eta}$$

$$\frac{\partial}{\partial x_2} = \frac{\partial \eta}{\partial x_2}\frac{\partial}{\partial \eta} = \left[\frac{u_{1,e}}{x(\rho\mu)_e}\right]^{1/2} \rho \frac{\partial}{\partial \eta}$$

$$\frac{\partial}{\partial x_3} = \frac{\partial}{\partial z} + \frac{\partial \eta}{\partial z}\frac{\partial}{\partial \eta}$$

Making the indicated substitutions permits Eqs. (5.125)-(5.128) to be written as follows:

continuity:

$$\frac{x}{h_1 h_3}\frac{\partial(h_3 F)}{\partial x} + \frac{F}{2h_1}(1+\beta_1) + \frac{\partial V}{\partial \eta} + \frac{1}{h_1 h_3[(\rho\mu)_e u_{1,e}/x]^{1/2}} \times \frac{\partial}{\partial z}\left\{\frac{h_1 W_e G}{u_{1,e}}[x u_{1,e}(\rho\mu)_e]^{1/2}\right\} = 0 \tag{7.106}$$

x momentum:

$$\frac{xF}{h_1}\frac{\partial F}{\partial x} + V\frac{\partial F}{\partial \eta} + \underbrace{\frac{xG}{h_3}\frac{\partial F}{\partial z}}_{(1)} + \underbrace{FGxK_1}_{(2)} - \underbrace{xG^2K_3}_{(3)}$$

$$= \beta_1\Big(\theta - \underbrace{F^2}_{(4)}\Big) + \beta_2\left(\frac{\theta u_{3,e}}{u_{1,e}} - \underbrace{FG}_{(5)}\right)$$

$$+ \theta\left(\frac{x u_{3,e} K_1}{u_{1,e}} - \frac{x u_{3,e}^2 K_3}{u_{1,e}^2}\right) + \frac{\partial}{\partial \eta}\left[\frac{\rho\bar{\mu}}{(\rho\mu)_e}\frac{\partial F}{\partial \eta}\right] \tag{7.107}$$

z momentum:

$$\frac{xF}{h_1}\frac{\partial G}{\partial x} + V\frac{\partial G}{\partial \eta} + \frac{xG}{h_3}\frac{\partial G}{\partial z} + xFGK_3 - xF^2K_1$$

$$= \theta\left(\beta_3 + \beta_4 + \frac{xK_3 u_{3,e}}{u_{1,e}} - xK_1\right) - \beta_5 GF - \beta_6 G^2 + \frac{\partial}{\partial \eta}\left[\frac{\rho\bar{\mu}}{(\rho\mu)_e}\frac{\partial G}{\partial \eta}\right] \tag{7.108}$$

energy:

$$\frac{xF}{h_1}\frac{\partial I}{\partial x} + V\frac{\partial I}{\partial \eta} + \frac{xGW_e}{h_3 u_{1,e}}\frac{\partial I}{\partial z}$$

$$= -\beta_7 FI - \frac{\beta_8 GW_e I}{u_{1,e}} + \frac{\partial}{\partial \eta}\left\{\left(\frac{\mu}{\text{Pr}} + \frac{\mu_T}{\text{Pr}_T}\right)\frac{\rho}{(\rho\mu)_e}\frac{\partial I}{\partial \eta}\right.$$

$$+ \frac{\rho u_{1,e}^2 F}{H_e(\rho\mu)_e}\left[\mu\left(1 - \frac{1}{\text{Pr}}\right) + \mu_T\left(1 - \frac{1}{\text{Pr}_T}\right)\right]\frac{\partial F}{\partial \eta}$$

$$\left. + \frac{\rho W_e^2 G}{H_e(\rho\mu)_e}\left[\mu\left(1 - \frac{1}{\text{Pr}}\right) + \mu_T\left(1 - \frac{1}{\text{Pr}_T}\right)\right]\frac{\partial G}{\partial \eta}\right\} \quad (7.109)$$

where

$$V = \rho\tilde{u}_2\left[\frac{x}{u_{1,e}(\rho\mu)_e}\right]^{1/2} + \frac{xF}{h_1}\frac{\partial \eta}{\partial x} + \frac{xGW_e}{h_3 u_{1,e}}\frac{\partial \eta}{\partial z}$$

$$\theta = \frac{\rho_e}{\rho} \qquad \beta_1 = \frac{x}{h_1 u_{1,e}}\frac{\partial u_{1,e}}{\partial x}$$

$$\beta_2 = \frac{x}{h_3 u_{1,e}}\frac{\partial u_{1,e}}{\partial z} \qquad \beta_3 = \frac{x}{h_1 W_e}\frac{\partial u_{3,e}}{\partial x}$$

$$\beta_4 = \frac{x u_{3,e}}{W_e h_3 u_{1,e}}\frac{\partial u_{3,e}}{\partial z} \qquad \beta_5 = \frac{x}{W_e h_1}\frac{\partial W_e}{\partial x}$$

The metrics and geodesic curvatures of the surface coordinate lines are as defined in Chapter 5.

$$\beta_6 = \frac{x}{h_3 u_{1,e}}\frac{\partial W_e}{\partial z} \qquad \beta_7 = \frac{x}{h_1 H_e}\frac{\partial H_e}{\partial x} \qquad \beta_8 = \frac{x}{h_3 H_e}\frac{\partial H_e}{\partial z}$$

An equation of state, $\rho = \rho(p, T)$, is needed to close the system of equations for a compressible flow. Several terms in Eq. (7.107) have been numbered for future reference.

The usual boundary conditions are

$\eta = 0$:

$$V = F = G = 0 \qquad I = I(x, 0, z)$$

or

$$\left.\frac{\partial I}{\partial \eta}\right)_{\eta=0} = Q(x, 0, z)$$

$\eta \to \infty$:

$$F = G = 1 \qquad G = \frac{u_{3,e}}{W_e}$$

where $Q(x, 0, z)$ is a specified function related to the wall heat flux. In addition, initial distributions of F, G, and I must be provided. Distributions of $u_{1,e}$, $u_{3,e}$, and H_e are also required.

The question of initial conditions requires careful consideration. Examination of the 3-D boundary-layer equations in the original orthogonal curvilinear coordinates prior to the transformation indicated above (or for that matter, in Cartesian coordinates) indicates that the roles of the x_1 and x_3 coordinates are interchangeable; that is, the equations are symmetric with respect to the interchange of x_1 and x_3 coordinates. As long as both u_1 and u_3 are positive, no single coordinate direction emerges as the obvious "marching" direction from considering the equations alone. Since first derivatives of u_1, u_3, and H appear with respect to both x_1 and x_3, it is expected that initial data should be provided in two intersecting planes to permit marching the dependent variables in both the x_1 and x_3 directions. The correct (permissible) marching direction is dictated by the zone-of-dependence principle, which will be discussed later. For now, we will proceed under the assumption that it is possible to march the solution in either the x_1 or x_3 directions and that initial data are needed on two intersecting planes. It is generally easy to determine a "main" flow direction from considerations of the body geometry and the direction of the oncoming stream. In defining η above, we have already assumed that the x or x_1 coordinates are in this main flow direction and the x_3 or z coordinates are in the crossflow direction. We will first discuss the determination of initial distributions of F, G, and I in the z, η plane, which will provide information appropriate for marching the solution in the x direction.

If the origin of the x coordinate is taken at the stagnation point (or line in some flows), the momentum and energy equations reduce to ODEs, which can be solved with the continuity equation to provide the necessary initial conditions in one plane. For the flow illustrated in Fig. 7.22, the appropriate form of the equations is obtained by simply neglecting all terms multiplied by x (which becomes equal to zero). This starting condition is similar to the flow at the leading edge of a sharp flat plate. On blunt bodies having a true stagnation point, $u_{1,e}$ and $u_{3,e}$ are known (Howarth, 1951) to vary linearly with x in the stagnation region. Thus, some of the terms that vanish for the sharp leading edge starting condition now have a nonzero limiting value as $x \to 0$ for blunt bodies. Blottner and Ellis (1973) discuss the stagnation point formulation in detail for incompressible flow.

In most 3-D boundary-layer flows, it is possible to compute initial distributions of F, G, and I (and u_1, u_3, and H when the untransformed curvilinear system is used) on a second intersecting plane by solving the PDEs on a *plane of symmetry*. The formulation of the plane-of-symmetry problem will be discussed below, but first, it is worth mentioning that in a few problems it is not possible to identify a plane of symmetry. An example of this is the sharp spinning cone considered by Dwyer (1971) and Dwyer and Sanders (1975). Controversy erupted over the question as to whether this flow can be treated as an initial value problem with the boundary-layer equations (Lin and Rubin,

1973a). It appears that use of difference schemes that lag the representation of the crossflow derivatives (Dwyer and Sanders, 1975; Kitchens et al., 1975) permits the solution to be marched away from a single initial data plane into those regions not forbidden by the zone-of-dependence principle. Such difference representations as well as the zone-of-dependence principle are discussed in Section 7.7.3.

The plane of symmetry is indicated in Fig. 7.22 for the flow over a flat plate with an attached cylinder. Flows over nonspinning bodies of revolution at incidence typically have both a windward and a leeward plane of symmetry, the former being most commonly used to develop the required second plane of initial data. Along the plane of symmetry,

$$G = \frac{\partial F}{\partial z} = \frac{\partial V}{\partial z} = \frac{\partial^2 G}{\partial z^2} = 0 \tag{7.110}$$

The inviscid flow and fluid properties are also symmetric about the plane of symmetry. Using Eq. (7.110), the x-momentum and energy equations reduce to 2-D form. The problem remains 3-D, however, because the cross-derivative term in the continuity equation does *not* vanish on the plane of symmetry. Expanding out the cross-derivative term in Eq. (7.106) and invoking the symmetry conditions, Eq. (7.110), permits the continuity equation to be written as

$$\frac{x}{h_1 h_3}\frac{\partial(h_3 F)}{\partial x} + \frac{F}{2h_1}(1+\beta_1) + \frac{\partial V}{\partial \eta} + \frac{xG_z}{h_3 u_{1,e}}\frac{\partial u_{3,e}}{\partial z} = 0 \tag{7.111}$$

where

$$G_z = \frac{\partial u_3/\partial z}{\partial u_{3,e}/\partial z}$$

The z-momentum equation in the form given by Eq. (7.108) provides no useful information because $G = 0$ everywhere in the plane of symmetry. However, differentiating the z-momentum equation with respect to z and again invoking the symmetry conditions provides an equation that can be solved for the required values of G_z:

$$\frac{xF}{h_1}\frac{\partial G_z}{\partial x} + V\frac{\partial G_z}{\partial \eta} + xFG_z K_3 = \beta_9(\theta - FG_z) + \beta_{10}(\theta - G_z^2) + x\theta K_3 + \frac{\partial}{\partial \eta}\left[\frac{\rho\bar{\mu}}{(\rho\mu)_e}\frac{\partial G_z}{\partial \eta}\right] \tag{7.112}$$

Defining $W_{e,z} = \partial u_{3,e}/\partial z$, we can express the parameters β_9 and β_{10} as

$$\beta_9 = \frac{x}{h_1 W_{e,z}}\frac{\partial W_{e,z}}{\partial x} \qquad \beta_{10} = \frac{xW_{e,z}}{h_3 u_{1,e}}$$

$W_{e,z}$ is to be obtained from the inviscid flow solution. Equation (7.112) for G_z has the same general form as the original z-momentum equation and can be solved by marching in the x direction along the plane of symmetry.

The arbitrary parameter W_e used in nondimensionalizing the crossflow velocity component is chosen to avoid singular behavior. We take $W_e = u_{3,e}$ at the stagnation point and along the plane of symmetry, and $W_e = u_{1,e}$ elsewhere.

7.7.3 Comments on Solution Methods for Three-Dimensional Flows

The 3-D boundary-layer problem involves several complicating features and considerations not present in the 2-D flows treated thus far. The inviscid flow solution required to provide the pressure gradient input for the boundary-layer solution is often considerably more difficult to obtain than for 2-D flows. Generation of the metrics and other information needed to establish the curvilinear-body-oriented coordinate system can also be a significant task for complex bodies. Turbulence models need to be extended to provide representation for the new apparent stress. In addition, the following features require special attention in the difference formulation: (1) implementation of the zone-of-dependence principle, and (2) representation of the crossflow convective derivatives in a manner to permit a stable solution for both positive and negative crossflow velocity components.

The 3-D boundary-layer equations have a hyperbolic character in the x-z plane, and the mathematical constraint that results is very much like the Courant-Friedrichs-Lewy condition discussed in connection with the wave equation. Major contributions to the formulation and interpretation of the zone of dependence principle for 3-D boundary layers can be found in the work of Raetz (1957), Der and Raetz (1962), Wang (1971), and Kitchens et al. (1975). The principle actually addresses both a zone of dependence and a zone of influence and is sometimes just referred to as the "influence" principle. The dependence part of the principle is the most relevant to the proper establishment of difference schemes and so has been given emphasis here.

If we consider the point labeled P in Fig. 7.23 within a 3-D boundary layer, the influence principle states that the influence of the solution at P is transferred instantaneously by diffusion to all points in the viscous flow on a line (labeled A-B in Fig. 7.23) normal to the surface passing through P and by convection downstream along all streamlines through that point. Normals to the body surface form the characteristic surfaces, and the speed of propagation is infinite in that direction. Disturbances anywhere along A-B are felt instantaneously along the whole line A-B and are carried downstream by *all* streamlines passing through A-B. The positions of the two outermost streamlines through A-B and extending downstream define the lateral extent of the wedge-shaped zone of influence for points on A-B. Events along A-B can influence the flow within the region bounded by the characteristics (lines normal to the wall) through these outermost streamlines. Typically, one outermost streamline is the limiting streamline at the wall and the other is the inviscid flow streamline. The flow along A-B is obviously influenced by the flow upstream, and a *zone of dependence* is defined by the characteristic lines passing through the two outermost streamlines extending upstream. Events at all points within this wedge-shaped

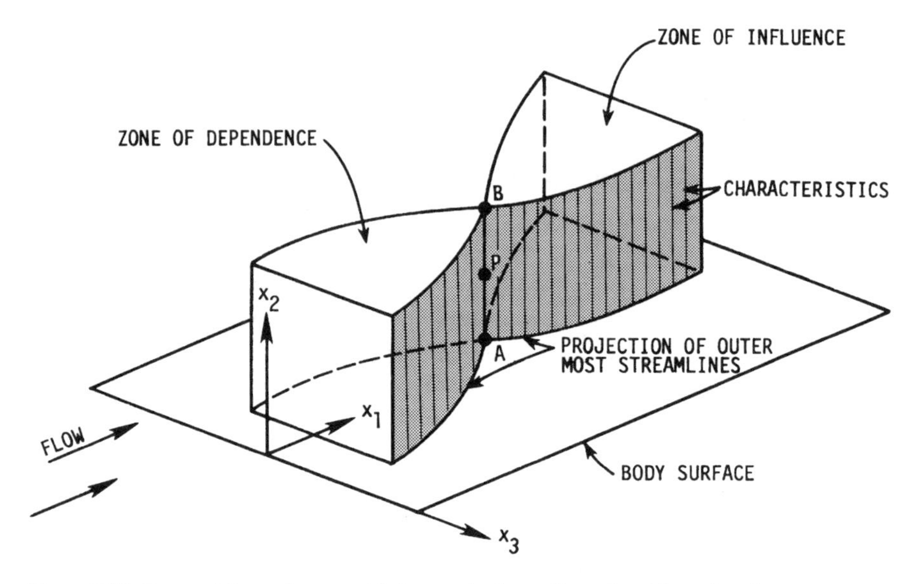

Figure 7.23 Zones of dependence and influence in three-dimensional boundary layers.

region upstream can influence events along A-B. The "outermost" streamlines are those having the maximum and minimum angular displacement from the constant x_3 (or z) surface passing through A-B. The zone of dependence then designates the minimum amount of initial data that must be supplied to determine the solution along A-B. These concepts apply to the PDEs. It is important for the difference molecule used along A-B not to exclude information in the zone of dependence; that is, the zone of dependence implied by the difference representation must be at least as large as the zone identified with the PDEs. This has also been shown previously for hyperbolic PDEs and identified as the CFL condition. The exact quantitative statement of the zone-of-dependence principle depends upon the difference molecule employed. For example, using a scheme that represents $\partial G/\partial z$ centrally (with Δz constant) at the n marching level as the solution is advanced to the $n+1$ level, the zone-of-dependence principle would require that

$$F > 0 \qquad \left| \frac{h_1\,\Delta x G}{h_3\,\Delta z F} \right| \leqslant 1 \tag{7.113}$$

Equation (7.113) indicates that the local angle made by the streamlines with the plane of constant z must be contained within the angle whose tangent is given by the mesh parameter $h_3\,\Delta z/(h_1\,\Delta z)$. We would like Eq. (7.113) to be satisfied at a given x level, with Δx being the increment back upstream. It would be unprofitable to iterate simply to establish the allowable step size, so the usual procedure is to utilize the most recently calculated values of G and F to establish the new step size using a safety factor to allow for anticipated changes in G and F over a Δx increment. In using Eq. (7.113) to establish the maximum allowable marching step increment, the inequality should be checked at each internal point at a given x level before Δx is established for the next step. With the use of certain difference schemes for flows in which G does not change sign, the zone-of-dependence constraint is automatically satisfied, as will be illustrated below for the 3-D Crank-Nicolson scheme.

Stability is also a concern in 3-D boundary-layer calculations. The presence of the additional convective derivative in the momentum equation generally influences the stability properties of the difference scheme. The stability constraint of a scheme is very likely to change as it is extended from 2-D to 3-D flow. The concept of stability is separate from the concept of the zone of dependence. This point is demonstrated very well in the work of Kitchens et al. (1975). For some schemes the constraint imposed by the zone-of-dependence principle will coincide with the stability constraint determined by the usual von Neumann analysis, but not always. Kitchens et al. (1975) show that for four difference schemes investigated, errors tend to grow whenever the zone-of-dependence principle was violated, but that for some schemes the solution remained very smooth and "stable" in appearance even though the errors were very significant. In other schemes, violation of the zone-of-dependence principle may trigger unstable behavior characterized by large oscillations even when such behavior is not predicted by a stability analysis. It is even possible to devise inherently unstable schemes that satisfy the zone-of-dependence constraint.

A few common difference schemes for 3-D boundary layers will be briefly described. In the following discussion, the indices n, j, k will be associated with the coordinate directions x_1, x_2, x_3 (or x, η, z). The solution is being advanced from the nth marching plane to the $n + 1$ plane. The solution at the $n + 1$ level will start at $k = 1$ (usually on the plane of symmetry), where the equations will be solved for the unknowns for all values of j. That is, fixing n and k, we obtain the solution along a line normal to the wall. Then the k index is advanced by 1, and the solution is obtained for another "column" of points along the surface normal. Thus the marching (or "sweeping") at the $n + 1$ level is in the crossflow direction. In difference representations below, the unknowns will be variables at the $n + 1, k$ levels.

Crank-Nicolson scheme. The 3-D extension of the Crank-Nicolson scheme has been used by several investigators. Its use is restricted to flows in which the crossflow component of velocity does not change sign owing to zone-of-dependence and stability considerations. The difference molecule is centered at $n + \frac{1}{2}, j, k - \frac{1}{2}$. Figure 7.24(a) illustrates the molecule as we look down on the flow (i.e., only points in the x-z plane are shown). The shaded area indicates the approximate maximum zone of dependence permitted by the molecule. The circled point indicates the location of the unknowns, and the x indicates the center of thc molecule. The presence of negative crossflow components of velocity causes this scheme to violate the zone-of-dependence principle because no information is contained in the molecule that would reflect flow conditions in

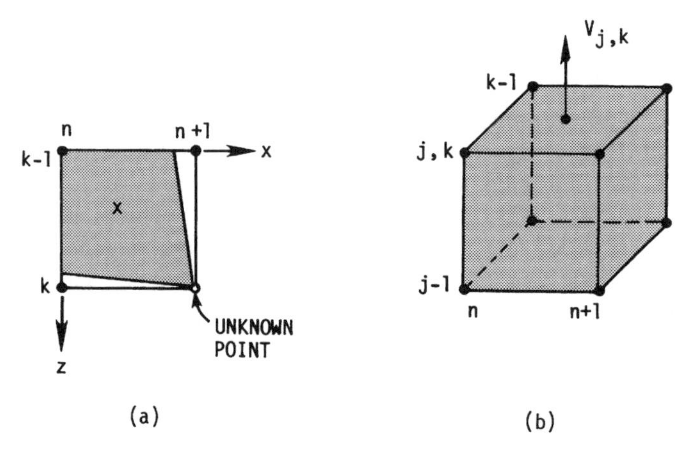

Figure 7.24 The Crank-Nicolson scheme. (a) The difference molecule projected on the x-z plane. (b) The control volume for the continuity equation.

the negative coordinate direction as the solution is advanced to the $n + 1, k$ level. On the other hand, the zone-of-dependence principle imposes no restriction on the size of Δx as long as $G \geqslant 0$, since the molecule spans all possible flow angles for which $F > 0$, $G \geqslant 0$.

More than one variation of the Crank-Nicolson scheme has been proposed. In the most frequently used version, terms of the form $\partial/\partial\eta(a\,\partial\phi/\partial\eta)$ are differenced as for the 2-D Crank-Nicolson scheme but *averaged* between $k - 1$ and k. Likewise, $\partial\phi/\partial x$ and $\partial\phi/\partial\eta$ terms are represented as for the 2-D Crank-Nicolson scheme but averaged over k and $k - 1$. Derivatives in the crossflow direction [as for example in the term labeled (1) in Eq. (7.107)] are represented by

$$\left.\frac{\partial\phi}{\partial z}\right)_{j,k-1/2}^{n+1/2} \cong \frac{\phi_{j,k}^{n+1} + \phi_{j,k}^{n} - \phi_{j,k-1}^{n+1} - \phi_{j,k-1}^{n}}{2\,\Delta z}$$

For flows over curved surfaces for which the curvature parameters K_1 and K_3 are nonzero, new terms of the form represented by terms labeled (2) and (3) in Eq. (7.107) must be represented. Similar terms appear in the untransformed equations in orthogonal curvilinear coordinates given in Chapter 5. Terms of this general form, not involving derivatives of the dependent variables, are considered source terms according to the definition given in Section 7.3.1. The terms labeled (4) and (5) in Eq. (7.107) are two additional source terms that arise due to the introduction of the F and G variables. These terms, (2)-(5) in Eq. (7.107), require linearization in the difference representations as do the convective terms. Any of the linearization techniques suggested in Section 7.3.3 can be used, although coupling of the equations is not commonly used. The source terms are represented at the center of the difference molecule

$(n + \frac{1}{2}, j, k - \frac{1}{2})$ by appropriate averages of variables at neighboring grid points. As an example, the term labeled (2) in Eq. (7.107) can be represented as

$$(FGxK_1)_{j,k-1/2}^{n+1} \simeq x^{n+1/2}(K_1)_{k-1/2}^{n+1/2}\left(F_{j,k}^{n} + F_{j,k-1}^{n} + F_{j,k}^{n+1} + F_{j,k-1}^{n+1}\right)$$
$$\times\left(G_{j,k}^{n} + G_{j,k-1}^{n} + G_{j,k-1}^{n+1} + \hat{G}_{j,k}^{n+1}\right)\Big/16 \qquad (7.114)$$

The only quantity treated algebraically as an unknown in Eq. (7.114) is $F_{j,k}^{n+1}$. The linearization is implemented by treating $\hat{G}_{j,k}^{n+1}$ algebraically as a known. The value of $\hat{G}_{j,k}^{n+1}$ can be determined by extrapolation, updated iteratively, or simply lagged, although lagging is not often used for the 3-D boundary-layer equations. Considerable flexibility exists in the way in which the various terms can be linearized. Other source terms appear on the right-hand side of Eq. (7.107), but these do not require linearization. The algebraic formulation for each momentum equation results in a simultaneous system of equations for the unknowns along the $n + 1, k$ column of points. The coefficient matrix is tridiagonal, so that the Thomas algorithm can be used.

Most current procedures for the 3-D boundary-layer equations solve the continuity equation separately for $V_{j,k}^{n+1}$ after F and G have been determined from the solution to the momentum equations. The difference representation for the continuity equation is usually established by considering a control volume centered about $(n + \frac{1}{2}, j - \frac{1}{2}, k - \frac{1}{2})$. Such a control volume is illustrated in Fig. 7.24(b). For F and G the average value of these quantities over a face of the control volume is established by taking the average of the quantities at the four corners of the face. Values of V are only needed at locations $n + \frac{1}{2}, j, k - \frac{1}{2}$ in the momentum equations. Thus, computational effort is normally saved by simply letting the value of V determined from the continuity equation be the value at the center of the x-z planes of the control volume. For computer storage the V physically considered to be located at $n + \frac{1}{2}, j, k - \frac{1}{2}$ is usually assigned the subscript $n + 1, j, k$. The location of $V_{j,k}^{n+1}$ is indicated in Fig. 7.24(b), where the labeling usually used for computer storage is employed. Grid schemes in which the dependent variables are evaluated at different locations in the computational domain are usually referred to as "staggered" grids. In this staggered grid, all variables except V are evaluated at regular grid points. Further examples of staggered grids arise in Chapter 8. The Crank-Nicolson scheme has the potential of being formally second-order accurate {(T. E. = $O[(\Delta x)^2, (\Delta \eta)^2, (\Delta z)^2]$}. The T.E. may be less favorable, depending on how linearizations and unequal mesh sizes are handled.

Krause zig-zag scheme. The Krause (1969) scheme has been widely used for flows in which the crossflow velocity component changes sign. The difference molecule is centered at $n + \frac{1}{2}, j, k$, and its projection on the x-z plane is given in Fig. 7.25(a). The shaded area again denotes the approximate maximum zone of dependence permitted by the molecule. We note that the molecule includes information in both z directions from point $n + 1, j, k$ so that, within limits,

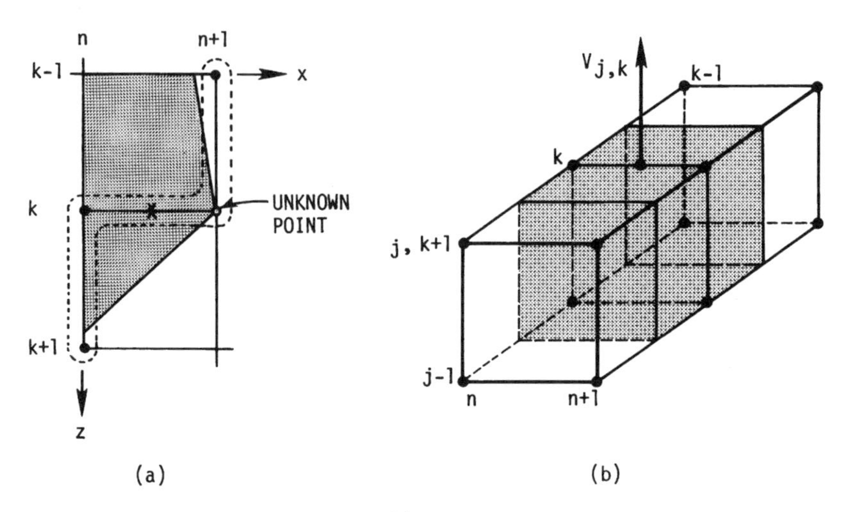

Figure 7.25 The Krause zig-zag scheme. (a) The difference molecule projected on the *x-z* plane. (b) The control volume for the continuity equation.

cross flow in both directions is permitted as long as the flow direction remains within the zone of dependence of the molecule. As for the Crank-Nicolson scheme, we observe that no mesh size constraint occurs when $F > 0$, $G \geqslant 0$. However, a constraint is observed when cross flow in the negative z direction occurs. The zone of dependence and the stability constraint for the Krause scheme can be stated as

$$F > 0 \qquad \frac{\Delta x h_1 G}{\Delta z h_3 F} \geqslant -1$$

It should be noted that the permitted flow direction can be altered by changing the aspect ratio, $\Delta z/\Delta x$, of the molecule.

The Krause difference representation is somewhat simpler algebraically than the Crank-Nicolson scheme, primarily because most difference representations are only averaged between $n+1$ and n, but *not* between two k levels. For the Krause scheme, terms of the form $\partial/\partial\eta(a\,\partial\phi/\partial\eta)$ and $\partial\phi/\partial x$ are differenced in the same manner as in the 2-D Crank-Nicolson scheme. Derivatives in the crossflow direction in the momentum equations are differenced using points in the zig-zag pattern denoted by the dashed lines in Fig. 7.25(a). For equal Δz increments this representation can be written

$$\left.\frac{\partial\phi}{\partial z}\right)_{j,k}^{n+1/2} \simeq \frac{\phi_{j,k+1}^{n} - \phi_{j,k}^{n} + \phi_{j,k}^{n+1} - \phi_{j,k-1}^{n+1}}{2\,\Delta z} \tag{7.115}$$

Since the sweep in the z direction is from columns $(n+1, k-1)$ to $(n+1, k)$, $\phi_{j,k}^{n+1}$ is the only unknown in Eq. (7.115). The problem of linearization of the algebraic representation is much the same as for the Crank-Nicolson scheme except that the molecule is more compact because quantities generally only need to be averaged between two grid points instead of four. For example, the term labeled (2) in Eq. (7.107) can be represented as

$$(FGxK_1)_{j,k}^{n+1/2} = \frac{x^{n+1/2}K_{1k}^{n+1/2}\left(F_{j,k}^n + F_{j,k}^{n+1}\right)\left(G_{j,k}^n + \hat{G}_{j,k}^{n+1}\right)}{4} \qquad (7.116)$$

A tridiagonal system of algebraic equations results from the Krause formulation, which can be solved by the Thomas algorithm.

The difference representation for the continuity equation is established by considering a control volume centered about $(n + \frac{1}{2}, j - \frac{1}{2}, k)$ as indicated in Fig. 7.25(b). The average value of F on an η-z face of the control volume can be determined from averaging only in the η direction, since the middle of the plane coincides with a k level. A zig-zag (or diagonal) average is used to represent G on an x-η plane. To illustrate this for equal Δz increments, we would represent a term of the form $\partial(aG)/\partial z$ in the Krause continuity equation as

$$\left.\frac{\partial(aG)}{\partial z}\right)_{j-1/2,k}^{n+1/2} \simeq \Big\{\Big[(aG)_{j,k+1}^n + (aG)_{j-1,k+1}^n + (aG)_{j,k}^{n+1} + (aG)_{j-1,k}^{n+1}\Big] - \Big[(aG)_{j,k}^n + (aG)_{j-1,k}^n + (aG)_{j,k-1}^{n+1} + (aG)_{j-1,k-1}^{n+1}\Big]\Big\}\Big/4\,\Delta z \qquad (7.117)$$

The V determined from the Krause continuity equation is located at the center of the upper x-z plane of the control volume (at $n + \frac{1}{2}, j, k$, but usually stored as $n + 1, j, k$). The storage index is the one indicated in the labeling of Fig. 7.25(b). The truncation error for the Krause scheme is the same as for the Crank-Nicolson scheme. Further details on the Crank-Nicolson and Krause schemes can be found in the work by Blottner and Ellis (1973).

Some variations. Two variations on the Krause scheme that have proven to be suitable for both positive and negative crossflow velocity components will be mentioned briefly. Wang (1973) has developed a second-order accurate two-step method that eliminates the need to linearize terms in the momentum equations. As with all multilevel methods, initial data must be provided at two marching levels. This is usually accomplished through the use of some other scheme for one or more steps. The projection of the two-step molecule on the x-z plane is shown in Fig. 7.26. The shaded area again indicates the approximate zone of dependence permitted by the molecule. Known data on the $n - 1$ and n levels are used to advance the solution. The method is implicit and centered at the point (n, j, k). Derivatives in the x and z direction are represented centrally about (n, j, k). Derivatives of the form

$$\frac{\partial}{\partial \eta}\left(a\frac{\partial \phi}{\partial \eta}\right)$$

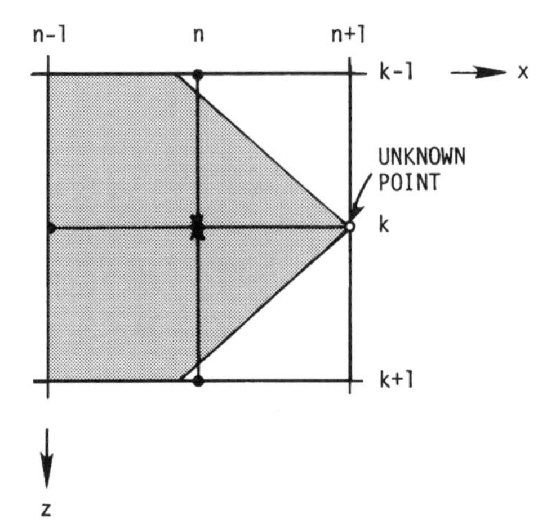

Figure 7.26 The two-step scheme.

are represented at $(n + 1, j, k)$ and $(n - 1, j, k)$ and averaged. The zone-of-dependence constraint is given by

$$F > 0 \qquad \left| \frac{\Delta x h_1 G}{\Delta z h_3 F} \right| \leqslant 1$$

No formal stability constraint is observed so long as $F > 0$.

Kitchens et al. (1975) compared the properties of four schemes for 3-D boundary layers and found that their scheme D had quite favorable error growth and stability properties. In addition, the results seemed relatively insensitive to violations in the zone-of-dependence constraint. The projection of this difference molecule on the x-z plane is shown in Fig. 7.27. The shaded area indicates the approximate zone of dependence for the method. The method is implicit. Derivatives in the x direction are represented centrally about $(n + \frac{1}{2}, j, k)$ with one very unique twist that converts an otherwise unstable scheme into a stable one. In the representation for $\partial\phi/\partial x$, values needed at n, j, k are replaced by the average of $\phi^n_{j,k+1}$ and $\phi^n_{j,k-1}$. Thus for equal increments we would use

$$\frac{\partial \phi}{\partial x} \simeq \frac{\phi^{n+1}_{j,k} - 0.5(\phi^n_{j,k+1} + \phi^n_{j,k-1})}{\Delta x}$$

Derivatives in the crossflow direction are represented centrally about (n, j, k). Derivatives of the form $\partial/\partial\eta(a\, \partial\phi/\partial\eta)$ are represented at $(n + 1, j, k)$ and (n, j, k) and averaged. The T.E. stated by Kitchens et al. (1975) is $O[\Delta x, (\Delta z)^2/\Delta x, (\Delta\eta)^2, (\Delta z)^2]$. The zone-of-dependence constraint for this method is the same as for the two-step method. In this case, the stability restriction is the same as the zone-of-dependence constraint.

Inverse methods and viscous-inviscid interaction. McLean and Randall (1979) reported the use of a viscous-inviscid interaction in 3-D for computing flows

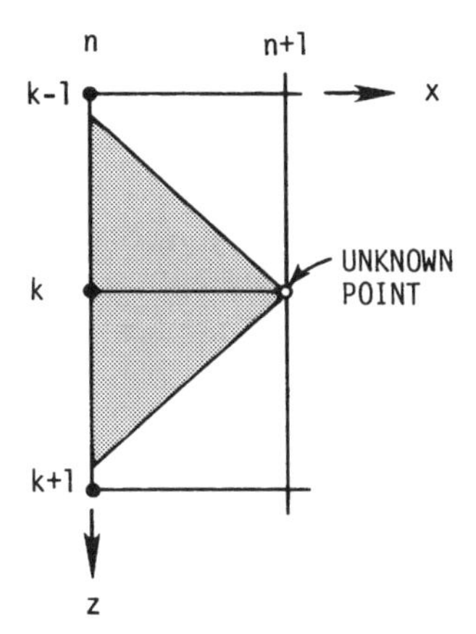

Figure 7.27 Scheme D (Kitchens et al., 1975).

over wings. The boundary-layer equations were solved in a direct manner in their work. The computation of the displacement surface needed to interface with the inviscid solution is somewhat more complex in 3-D. The simple Hilbert integral small-disturbance correction [Eq. (7.76)] for an inviscid flow has been extended to 3-D (Edwards, 1986), although the inviscid flow may also be recomputed in its entirety for each iterative pass. However, rather than recomputing the inviscid flow for the body modified by the displacement surface, it is frequently advantageous to maintain the same body in the inviscid calculation and represent the effects of the viscous flow by a distribution of sources and sinks (Lighthill, 1958). In the full inviscid potential flow solution, the sources and sinks (related to space derivatives of the displacement thicknesses) are represented as normal velocity boundary conditions (blowing or suction) at the body surface. This formulation offers an advantage when direct methods are used to solve the elliptic PDEs for the subsonic flow, in that the influence matrix and its inverse need not be recomputed for each interaction iteration.

A number of investigators (for example, Delery and Formery, 1983; Edwards and Carter, 1985; Edwards, 1986) have reported success in using an inverse procedure for solving the 3-D incompressible boundary-layer equations. Such a capability is the first step toward the development of 3-D viscous-inviscid interaction procedures that could permit computation of separated flows.

As in two dimensions, the 3-D inverse procedure treats the pressure as unknown. Since the direct solution of the 3-D boundary-layer system requires specification of two components of the pressure gradient, it is not surprising that an inverse solution method requires specification of two alternative functions. Edwards and Carter (1985) evaluated the suitability of specifying four different combinations of parameters and found that three of them worked satisfactorily but a fourth led to the formulation of an elliptic system that allowed departure (exponentially growing) solutions when solved in a forward marching manner. Examples of boundary condition combinations that worked included the specification of both integral parameters,

$$\delta_1 = \int_0^\infty \left(1 - \frac{u}{u_e}\right) dy \qquad \delta_2 = \int_0^\infty \left(\frac{w_e - w}{u_e}\right) dy$$

and δ_1 and the crossflow component of the edge velocity w_e. Three-dimensional viscous-inviscid interaction schemes based on an inverse treatment of the boundary-layer equations apparently have not been widely used in applications, although Edwards (1986) demonstrated that such a strategy is workable. This may be partly due to the inherent complexity of the two-solver approach and partly due to advances in methods based on the solution of a single set of governing equations, which have been enabled by rapid progress in computer technology.

7.7.4 Example Calculations

Here we briefly present some example computational results for the sample 3-D flow illustrated in Fig. 7.24. The results were obtained by application of the Krause scheme to Eqs. (7.106)–(7.108) for an incompressible laminar flow. The Crank-Nicolson scheme was used at the last z station to permit the calculation to end without requiring information from the $k + 1$ level. Computed results for this flow have been reported in the literature by several investigators (see, for example, Cebeci, 1975). For this flow, the inviscid velocity distribution is given by

$$u_{1,e} = u_\infty\left(1 + a^2\frac{\gamma_2}{\gamma_1^2}\right) \qquad u_{3,e} = -2u_\infty a^2\frac{\gamma_3}{\gamma_1^2}$$

where u_∞ is the reference free stream velocity and $\gamma_1 = (x - x_0)^2 + z^2$, $\gamma_2 = -(x - x_0)^2 + z^2$, and $\gamma_3 = (x - x_0)z$. The parameter x_0 is the distance of the cylinder axis from the leading edge, a is the cylinder radius, and x and z denote the distance measured from the leading edge and plane of symmetry, respectively. It is also useful to know $\partial u_{3,e}/\partial z$ along the plane of symmetry:

$$\left.\frac{\partial u_{3,e}}{\partial z}\right)_{z=0} = \frac{-2u_\infty a^2}{(x - x_0)^3}$$

Calculations were made for $u_\infty = 30.5$ m/s, $a = 0.061$ m, $x_0 = 0.457$ m using $\Delta x = 0.0061$ m, $\Delta\eta = 0.28$, and $\Delta z = 0.0061$ m. Typical velocity profiles for this flow are shown in Fig. 7.28. In particular, we note that the crossflow velocity component reaches a maximum within the inner one-third of the boundary layer. The variation in the flow angle (in x-z plane) with distance from the wall is shown in Fig. 7.29(a). The maximum skewing is observed to occur close to the wall. The velocity vector is seen to rotate through an angle of about 13° along the surface normal. This corresponds to the included angle made by the zone of dependence at this location (see Fig. 7.23). The variation in the skin-friction coefficient is shown in Fig. 7.29(b). The presence of the cylinder causes the flow

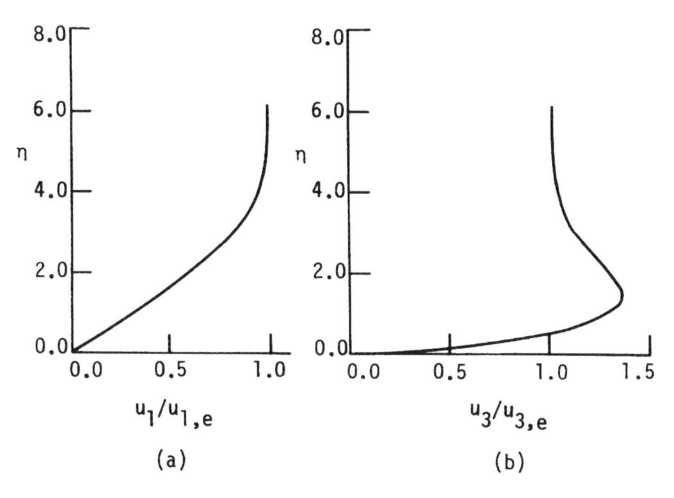

Figure 7.28 Velocity profiles at $x = 0.219$ m, $z = 0.079$ m, for an example three-dimensional flow over a flat plate with an attached cylinder. (a) Streamwise velocity distribution. (b) Crossflow velocity distribution.

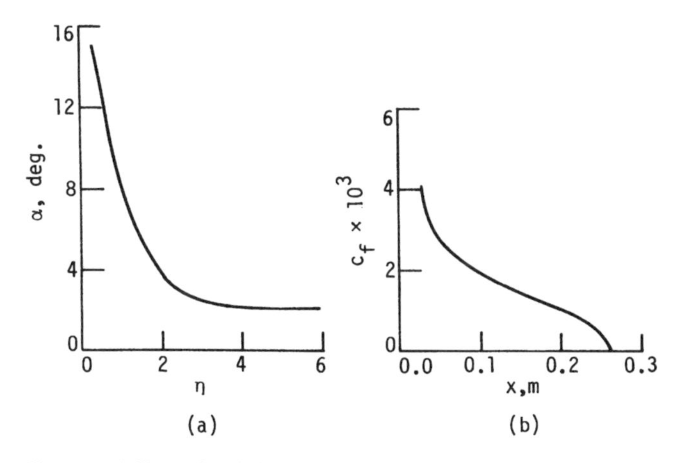

Figure 7.29 Example of three-dimensional boundary-layer flow over a flat plate with an attached cylinder. (a) Variation of flow angle (measured from x-η plane) along the surface normal at $x = 0.219$ m, $z = 0.079$ m. (b) Variation of skin-friction coefficient along plane of symmetry.

on the plane of symmetry to separate at $x \simeq 0.26$ m. Conventional boundary-layer calculation methods can proceed no farther along the plane of symmetry because both the x and z components of velocity have vanished.

7.7.5 Additional Remarks

Only a few representative difference schemes for 3-D boundary layers have been discussed in this chapter. Many other useful procedures have been suggested. Several of these are discussed by Wang (1974), Kitchens et al. (1975), and Blottner (1975b). Cebeci (1975) has extended the box scheme to 3-D flows, and Cebeci et al. (1979a) have implemented a zig-zag feature that permits the calculation of 3-D flows in which the crossflow velocity component changes sign. No single scheme has emerged to date as being superior for all flows. Several investigators have found the need to employ more than one scheme in order to cover all regions efficiently in some flows. The Krause zig-zag scheme is recommended as a reasonable starting point for the development of a 3-D boundary-layer finite-difference procedure. After the Krause procedure is well in hand, the user should be encouraged to explore the possible advantages offered by the several variations that have been suggested.

Turbulence modeling is certainly an important concern in 3-D flows. Most 3-D turbulent calculations to date have assumed that the turbulent viscosity is a scalar. Measurements tend to support the view that in the outer portion of the flow the apparent viscosity in a Boussinesq evaluation of the stress in the crossflow direction may be substantially less (by a factor ~ 0.4-0.7) than the viscosity for the apparent stress in the streamwise direction. Further research on turbulence modeling for 3-D flows would seem desirable.

The most successful application of 3-D boundary-layer theory has probably been for flows over wings. Reasonably refined computer programs have been documented for this application (Cebeci et al., 1977; McLean and Randall, 1979). Several papers and reports in the literature of a review or general nature should prove useful in obtaining a broad view of the status of predictions in 3-D boundary layers. The list includes Wang (1974, 1975), Bushnell et al. (1976), Blottner (1975b), and Kitchens et al. (1975).

7.8 UNSTEADY BOUNDARY LAYERS

It is frequently desirable to predict unsteady boundary-layer behavior, especially in the design of flight vehicles. Numerical aspects of this problem are reasonably well understood; however, challenging aspects of turbulence modeling remain. We will limit our discussion to 2-D unsteady boundary layers, although many of the concepts carry over to the 3-D case.

The unsteady 2-D boundary-layer equations appear as Eqs. (5.116)-(5.118) in Chapter 5. They differ from their steady flow counterparts only through the appearance of the term $\rho\, \partial u/\partial t$ in the momentum equation and $\partial \rho/\partial t$ in the continuity equation. The unsteady equations are also parabolic but with time as the marching parameter. Values of u, v, H, and fluid properties must be stored at grid points throughout the flow domain. Initial values of u, v, and H must be specified for all x and y. Boundary conditions may vary with time. The usual boundary conditions are as follows:

1. At $x = x_0$, $u(t, x_0, y)$ and $H(t, x_0, y)$ are prescribed for all y and t.
2. At $y = 0$, $u(t, x, 0) = v(t, x, 0) = 0$.
3. The $\lim_{y \to \infty} u(t, x, y) = u_e(t, x)$.

The main objective is to develop computational procedures that will provide accurate and stable solutions when flow reversal ($u < 0$) occurs. In this respect, the unsteady 2-D boundary-layer problem is similar to the 3-D steady problem, where the concern was to identify methods that would permit flow reversal in the crossflow direction. When flow reversal occurs in the unsteady problem, it is crucial to employ a difference representation that permits upstream influence. This principle has not been formulated in terms of a zone-of-dependence concept for the 2-D unsteady boundary-layer equations, but to ignore the possibility of information being convected in the flow direction is unacceptable. Furthermore, the steady boundary-layer equations are parabolic in the x direction, which again requires that information move in the direction of the x component of velocity; otherwise it would not be possible to achieve the correct steady-state solution from the transient formulation.

An adaptation of the zig-zag representation introduced by Krause for the 3-D boundary-layer equations has been frequently used to represent $\partial u/\partial x$ when flow reversal is present. This representation is illustrated in Fig. 7.30. Using the mesh notation introduced in the figure, the zig-zag representation of the streamwise derivative for equal Δx increments is

$$\frac{\partial u}{\partial x} \simeq \frac{u_{i,j}^{n+1} - u_{i-1,j}^{n+1} + u_{i+1,j}^{n} - u_{i,j}^{n}}{2\,\Delta x} \tag{7.118}$$

The j index is associated with the normal coordinate. The representation of Eq. (7.118) can be used with a difference scheme centered at $n + \frac{1}{2}, i, j$, which can

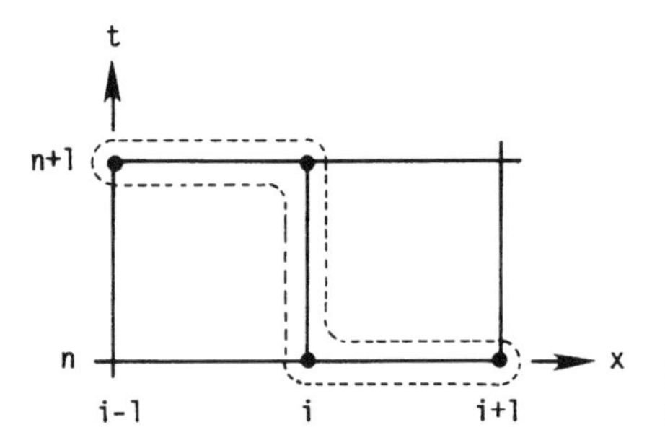

Figure 7.30 The zig-zag representation for streamwise derivatives in unsteady flows.

be thought of as the unsteady 2-D boundary-layer version of the Krause scheme for 3-D boundary layers.

Concepts from the zig-zag box scheme for 3-D boundary layers have also been used to develop a zig-zag box representation for unsteady boundary layers. This scheme also appears applicable when flow reversal is present (Cebeci et al., 1979a).

Other types of upwind differencing have been found to work satisfactorily with flow reversal by Telionis et al. (1973) and Murphy and Prenter (1981). The method of Murphy and Prenter (1981) also utilizes a fourth-order accurate discretization in the normal direction.

Blottner (1975a, 1975b) provides a helpful review of computational work on the unsteady boundary-layer equations. Other useful references include Telionis et al. (1973), Tsahalis and Telionis (1974), Telionis and Tsahalis (1976), Cebeci et al. (1979b), Phillips and Ackerberg (1973), Murphy and Prenter (1981), Telionis (1979), Kwon et al. (1988a, 1988b), and Ramin and Pletcher (1993).

PROBLEMS

7.1 Verify the stability constraints given in Section 7.3.2 for the two versions of the simple explicit procedure for the boundary-layer equations.

7.2 The term $(\partial u/\partial y)^2$ needs to be evaluated at the $n+1$ marching level, where u is an unknown. The marching coordinate is x, and y denotes the normal distance from the wall in a viscous flow problem. Utilize Newton linearization to obtain a difference representation for $(\partial u/\partial y)^2$ that could be used iteratively with the Thomas algorithm and that would be linear in the unknowns at each application of the algorithm.

7.3 Verify Eq. (7.24).

7.4 Generalize Eq. (7.24) to provide a second-order accurate representation when the mesh increments Δx and Δy are not constant.

7.5 Consider the following proposed implicit representation for the boundary-layer momentum equation:

$$u_j^n \frac{u_j^{n+1} - u_j^n}{\Delta x} + v_j^n \frac{u_j^{n+1} - u_{j-1}^{n+1}}{\Delta y} = \frac{\nu}{(\Delta y)^2}(u_{j+1}^{n+1} - 2u_j^{n+1} + u_{j-1}^{n+1})$$

Would you expect to find any mesh Reynolds number restrictions on the use of this representation when employing the Thomas algorithm for u and $v > 0$? Substantiate your answer.

7.6 Work Prob. 7.5 for the difference equation that results when the second term in the equation is replaced by

$$v_j^n \frac{u_{j+1}^{n+1} - u_j^{n+1}}{\Delta y}$$

7.7 Verify the stability constraint given by Eq. (7.30).

7.8 Establish that the algebraic system represented by Eqs. (7.38) and (7.39) is block tridiagonal with 2×2 blocks. Verify that it fits the format required by the modified tridiagonal elimination scheme given in Section 7.3.3 for solving the momentum and continuity equations in a coupled manner.

7.9 Write a computer program using an implicit method (either fully implicit, Crank-Nicolson, or the modified box scheme) to solve the incompressible laminar boundary-layer equations for flat plate flow in both physical (scheme A) and transformed (scheme B) [Eqs. (7.42) and (7.43)] coordinates. Linearize the difference equations either by lagging or extrapolating the coefficients v and u. Solve the momentum and continuity equations in an uncoupled manner. Use the tridiagonal elimination scheme to solve the system of simultaneous equations. Use $\Delta\eta = 0.3$ for scheme B and

$\rho u_\infty \Delta y/\mu = 60$ for scheme A. For scheme A the boundary layer will grow with increasing x, so that it will be necessary to add points to the computational domain as the calculation progresses. It will be possible to increase the marching step size in proportion to the boundary-layer thickness. It is suggested that the first step be established as $\Delta x = \rho u_\infty (\Delta y)^2/2\mu$ for scheme A.

Compare schemes A and B for ease of programming and accuracy. Consider the similarity solution tabulated by Schlichting (1979) as an "exact" solution for purposes of comparison. Calculate

$$C_f = \frac{\mu(\partial u/\partial y)_w}{\rho u_e^2/2}$$

from the solution. Determine $(\partial u/\partial y)_w$ by fitting a second-degree polynomial through the solution near the wall. Limit the downstream extent of the calculation to 75 streamwise steps. Investigate the sensitivity of methods to the streamwise step size. Perform the calculations for $\Delta x = 1\delta, 2\delta, 4\delta$. For scheme B, study the influence of the starting procedure on accuracy by first performing a streamwise calculation sweep by using $v = 0$ at $x = 0$ in the momentum equation and then repeating the calculation determining v at $x = 0$ iteratively through the use of the continuity equation.

7.10 Work Prob. 7.9 with the following changes. Select an implicit scheme and choose either physical or transformed coordinates in which to express the boundary-layer equations. Scheme A linearizes the coefficients by lagging, and scheme B implements the linearization through the Newton procedure with coupling of the continuity equation.

7.11 Work Prob. 7.9 using either physical or transformed coordinates. Let scheme A be an implicit scheme of your choice and scheme B be an explicit procedure such as DuFort-Frankel, hopscotch, or ADE.

7.12 Modify a difference scheme used in working Probs. 7.9 through 7.11 to permit the calculation of a boundary-layer flow in a pressure gradient. Verify your difference scheme by comparing the predicted velocity profiles with the results from the similarity solutions to the Faulkner-Skan equations (see Schlichting, 1979) for a potential flow given by $u_e(x) = u_1 x^m$ where u_1 and m are constants and x is the streamwise coordinate. Make your comparisons for $m = \frac{1}{3}$ and -0.0654. You may choose any convenient value for u_1.

7.13 Modify a difference scheme used in working Probs. 7.9 through 7.11 to permit the calculation of a boundary-layer flow with blowing or suction. Verify your difference scheme by comparing the predicted velocity profiles with the results obtained by Hartnett and Eckert (1957) for blowing and suction distributions given by

$$\frac{v_w(x)}{u_\infty}\sqrt{\text{Re}_x} = 0.25 \text{ and } -2.5$$

7.14 Develop a computer program to solve the 2-D incompressible constant property boundary-layer equations in transformed coordinates using the analytical transformation given in Section 7.3.7. Use either the Crank-Nicolson or fully implicit scheme and the Newton-linearized scheme with coupling. Validate your code by solving for the zero pressure gradient laminar flow on a flat plate at a plate Reynolds number of 800. Compare your predicted skin-friction coefficient with the analytical result: $C_f(\text{Re}_x)^{0.5} = 0.664$. Use a second-degree polynomial through points near the wall to compute the wall shear stress from the velocity distribution. Tabulate $C_f(\text{Re}_x)^{0.5}$ for at least 15 locations along the plate out to a location where the Re = 800. Use your scheme to estimate the separation point for a laminar flow where the free stream velocity distribution is given by $u_e = 100\text{–}300x$. Use at least 50 points across the boundary layer as separation is approached. Refine the $\Delta\xi$ step near separation until the separation point no longer changes.

7.15 Work Prob. 7.14 using the generalized coordinate approach instead of the analytical transformation. Establish the grid based on $\xi = x/L$, $\eta = (y/x)(\text{Re}_x)^{0.5}$.

7.16 Develop a finite-difference scheme for compressible laminar boundary-layer flow. Solve the energy equation in an uncoupled manner. Use the computer program to predict the skin-friction coefficient and Stanton number distributions for the flow of air over a flat plate at $M_e = 4$ and $T_w/T_\infty = 2$. Use the Sutherland equation (Section 5.1.4) to evaluate the fluid viscosity as a function of temperature. Assume constant values of Pr and c_p [Pr = 0.75, $c_p = 1 \times 10^3$ J/(kg K)]. Compare your predictions with the analytical results of van Driest (1952) [heat transfer results can be found in the work by Kays and Crawford (1993)].

7.17 Modify a difference scheme used in working Probs. 7.9 through 7.11 to permit calculation of an incompressible turbulent boundary layer on a flat plate. Use an algebraic turbulence model from Chapter 5. Use $u_\infty = 33$ m/s and $\nu = 1.51 \times 10^{-5}$ m^2/s. Compare your velocity profiles in law-of-the-wall coordinates with Fig. 5.7. Compare your predicted values of C_f with the measurements of Wieghardt and Tillmann (1951) tabulated below:

x, m	C_f
0.087	0.00534
0.187	0.00424
0.287	0.00386
0.387	0.00364
0.487	0.00345
0.637	0.00337
0.787	0.00317
0.937	0.00317
1.087	0.00308

7.18 Verify Eq. (7.72).

7.19 An inverse boundary-layer method is to be applied to steady incompressible 2-D flow. Use $\rho = 1$ kg/m^3, $\nu = 1$ m^2/s, and $\Delta y = \Delta x = 0.1$ m. The solution at the nth station is tabulated below. Use the simple explicit method to advance the solution to $n + 1$ (see Fig. P7.1). Use the secant method to determine the pressure gradient required to maintain a constant value of the displacement thickness as the solution is advanced to $n + 1$. That is, find the pressure gradient required to give $(\delta^*)^{n+1} = (\delta^*)^n$.

*n*th station

j	u	v
1	0	0
2	6	0
3	10	0
4	10	0

7.20 Work out the details of the terms Q_j^n and R_j^n in Eq. (7.90) for the DuFort-Frankel scheme for internal flows.

7.21 Using the boundary-layer equations, develop a finite-difference scheme for the incompressible laminar developing flow in a channel. Use the nondimensionalization that removes the Reynolds number from the equations (Eq. 7.80). Because of symmetry, you need only solve the equations to the channel centerline. Solve out to $x/(h \operatorname{Re}_h) = 0.02$, where h is the channel half-height. Suppose your friend needs the solution for Re = 500 from the inlet to a distance of 25 channel heights

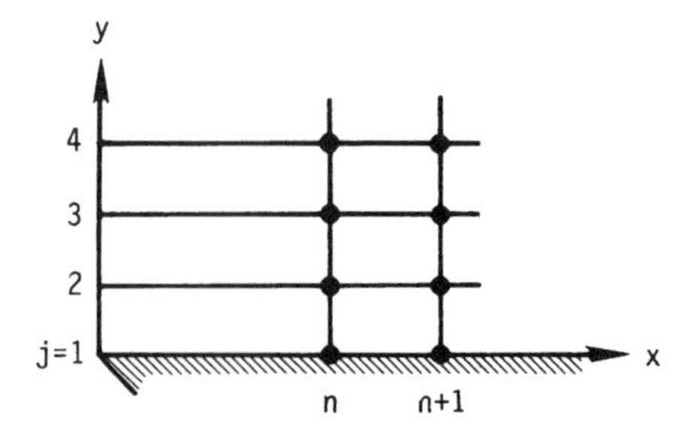

Figure P7.1

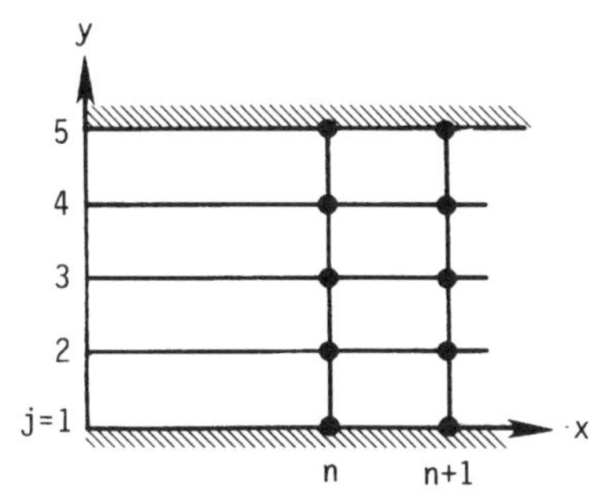

Figure P7.2

downstream. Explain how you would provide the information needed from your numerical solution. Is the flow fully developed after 25 channel heights? Compare the predicted centerline velocity development U_{CL}/U_{INLET} vs. $x/(h\,\mathrm{Re}_h)$ with the highest Reynolds number Navier-Stokes solutions reported by Chen and Pletcher (1991).

7.22 The steady viscous flow through a 2-D channel is to be determined by a solution to the boundary-layer equations. Use $\rho = 1$ kg/m^3, $\nu = 1$ m^2/s, $\Delta y = 0.1$ m, and $\Delta x = 0.025$ m. The solution at the nth station is tabulated below. Use the simple explicit method to advance the solution to the $n + 1$ station (see Fig. P7.2). Determine the pressure and the velocities.

nth station

$p = 100$ N/m^2

j	u	v
1	0	0
2	10	0
3	10	0
4	10	0
5	0	0

7.23 Work Prob. 7.22 using the fully implicit method with lagged coefficients.

7.24 Verify Eq. (7.93) for a fully implicit method.

7.25 Verify Eq. (7.96).

7.26 Verify Eq. (7.100).

7.27 Derive Eq. (7.112).

7.28 Specialize Eqs. (7.106)–(7.108) for an incompressible 3-D laminar flow in the Cartesian coordinate system. Write out the Crank-Nicolson representation for the equations. Explain your scheme for linearizing the algebraic equations.

7.29 Work Prob. 7.28 for the Krause zig-zag scheme.

7.30 Choose a suitable implicit finite-difference scheme to solve the 3-D laminar boundary-layer equations on the plane of symmetry for the example flow described in Section 7.7.4. Compare your predicted skin-friction coefficients with the results of Cebeci (1975) and/or Fig. 7.29(b).

7.31 Solve the example flow of Section 7.7.4 using the Crank-Nicolson scheme. Use the grid described in the example. Compare your results with those given by Cebeci (1975).

7.32 Write out a Krause-type difference scheme for the 2-D unsteady incompressible boundary-layer equations.

CHAPTER

EIGHT

NUMERICAL METHODS FOR THE "PARABOLIZED" NAVIER-STOKES EQUATIONS

The computational fluid dynamics (CFD) "frontier" has advanced from the simple to the complex. Generally, the simple methods taxed the available computational power when they occupied the frontier. The evolution proceeded from methods for various forms of the potential and boundary-layer equations to the Euler equations and then to various "parabolized" forms of the Navier-Stokes equations that are the subject of this chapter. Most of the schemes were developed at a time when the use of the full Navier-Stokes equations was prohibitive for many problems because of the large computer memory or CPU time required. If such parabolized schemes were considered economical of computer resources when they were introduced, they are still so. However, the need to save CPU time has diminished in a relative sense because of the incredible reduction in cost per operation experienced in recent times. Numerous numerical strategies will be discussed in this chapter. They all share the common characteristic that the steady form of the governing equations is employed, and the solution is "marched" in space. Some of the solution strategies to be described in this chapter share aspects in common with methods for the full Navier-Stokes equations discussed in Chapter 9.

8.1 INTRODUCTION

The boundary-layer equations can be utilized to solve many viscous flow problems, as discussed in Chapter 7. There are, however, a number of very important viscous flow problems that cannot be solved by using the boundary-layer equations. In these problems, the boundary-layer assumptions are not valid. For example, if the inviscid flow is fully merged with the viscous flow, the two flows cannot be solved independently of each other, as required by boundary-layer theory. As a result, it becomes necessary to solve a set of equations that is valid in both the inviscid and viscous flow regions.

Examples of viscous flow fields where the boundary-layer equations are not the appropriate governing equations are shown in Figs. 8.1(a)–(d). The hypersonic rarefied flow near the sharp leading edge of a flat plate [Fig. 8.1(a)] is a classic example of a viscous flow field that cannot be solved by the boundary-layer equations. In fact, very near the leading edge, the flow is not a continuum, so that the Navier-Stokes equations are invalid. In the merged-layer region, where the flow can first be considered a continuum, the shock layer and the viscous layer are fully merged and indistinguishable from each other. Further downstream, the shock layer coalesces into a discontinuity, and a distinct inviscid layer develops between the shock wave and the viscous layer. This is the beginning of the interaction region, which is further divided into the strong- and weak-interaction regions. The weak-interaction region eventually evolves into the classic Prandtl boundary-layer flow further downstream. Obviously, the boundary-layer equations cannot be used in the merged-layer region because the viscous layer and the shock layer are completely merged. At the beginning of the strong-interaction region, the viscous flow cannot be solved independently of the inviscid flow because of the strong interaction. In the weak-interaction region, it is possible to solve the inviscid and viscous portions of the flow separately, but this must be done in an iterative fashion, as discussed in Chapter 7. That is, the boundary-layer equations can be computed initially using approximate edge conditions. With the computed displacement thickness, the inviscid portion of the flow field can then be determined. This provides new edge conditions for the recomputation of the boundary layer. This procedure can be repeated until the solution for the entire flow field does not change between iterations. Unless the interaction is very weak, it has been observed that this iterative procedure is often inferior to solving a set of equations that is valid in both the inviscid and viscous flow regions (Davis and Rubin, 1980).

Figure 8.1(b) illustrates a mixing layer problem for which the boundary-layer (thin-shear-layer) equations are not applicable. Across the mixing layer, a strong normal pressure gradient exists. Consequently, the usual boundary-layer (thin-shear-layer) equations, which contain the normal momentum equation

$$\frac{\partial p}{\partial y} = 0 \tag{8.1}$$

are not valid. In this case, a more complete normal momentum equation is required. Another example of a flow field where the boundary-layer equations may not be applicable is the supersonic flow around a blunt body at high altitude, as seen in Fig. 8.1(c). In the region between the shock wave and the

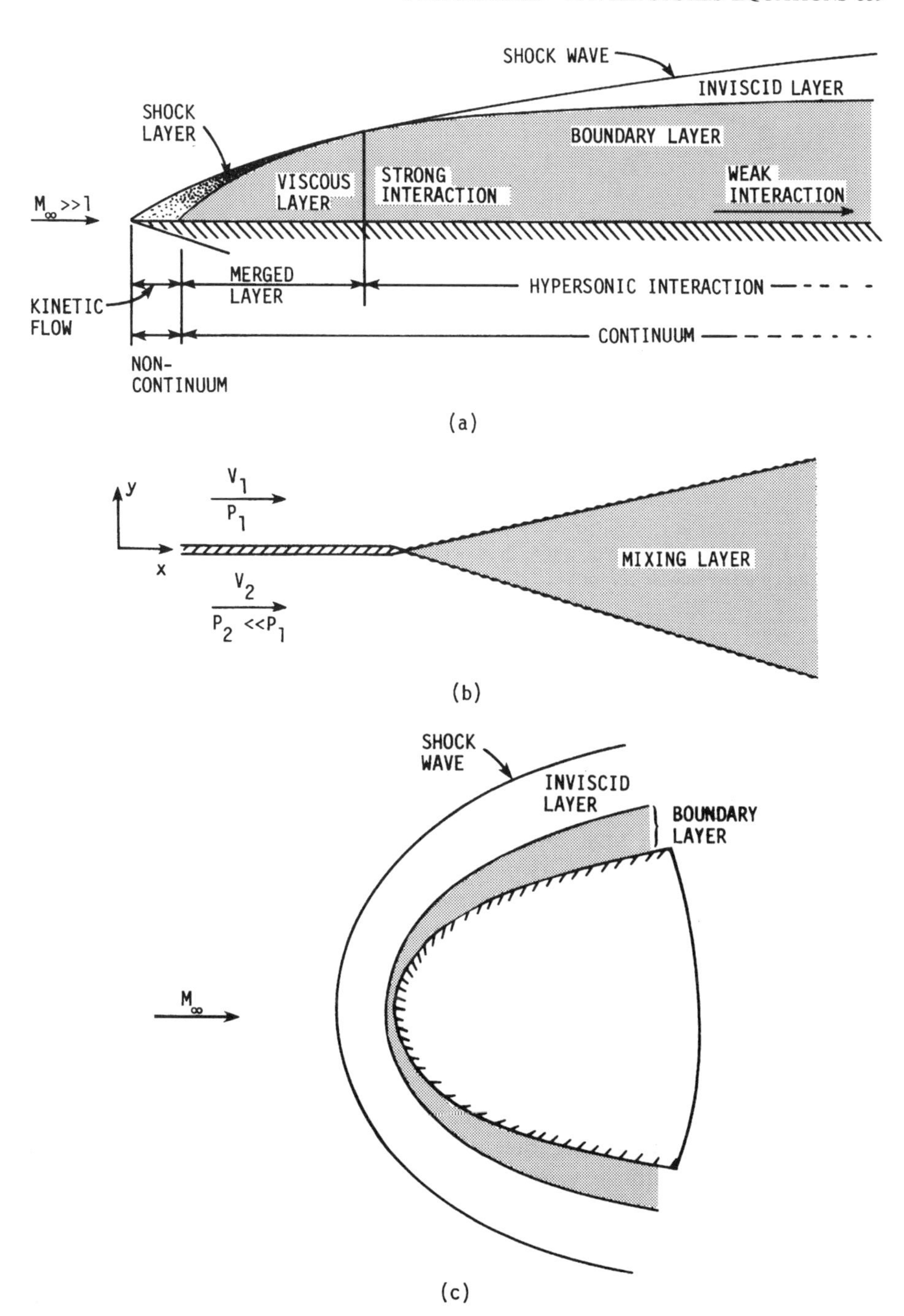

Figure 8.1 Examples of flow fields where the boundary-layer equations are not applicable. (a) Leading edge of a flat plate in a hypersonic rarefied flow. (b) Mixing layer with a strong transverse pressure gradient. (c) Blunt body in a supersonic flow at high altitude.

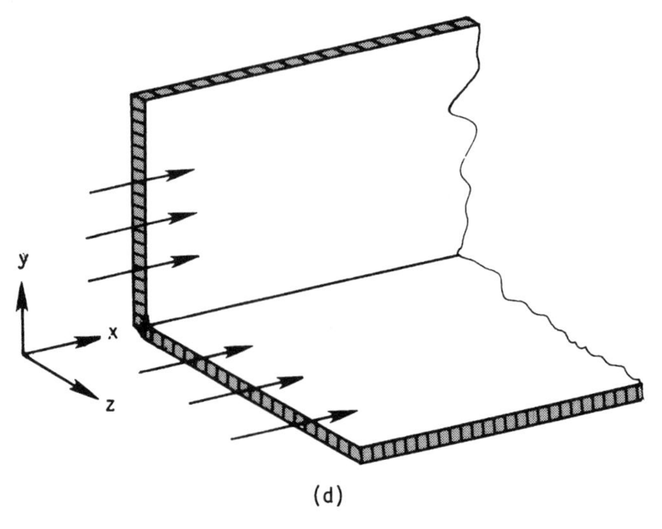

Figure 8.1 Examples of flow fields where the boundary-layer equations are not applicable (*Cont.*). (d) Flow along a streamwise corner.

body (i.e., the shock layer) there exists a strong interaction between the boundary layer and the inviscid flow region. As a result, sets of equations that are valid in both the inviscid and viscous regions are normally used to compute this type of flow field.

The flow along the corner formed by two intersecting surfaces, illustrated in Fig. 8.1(d), provides a final example of a flow for which the boundary-layer equations are not applicable. As pointed out in Chapter 7, the boundary-layer equations only include viscous derivatives with respect to a single "normal" coordinate direction. Very near the corner, viscous derivatives with respect to *both* "normal" directions will be important. Such a flow configuration occurs often in applications, as for example, near wing-body junctures and in rectangular channels.

The complete Navier-Stokes equations are an obvious set of equations that can be used to solve the flow fields in Fig. 8.1 as well as all other viscous flow fields for which the boundary-layer equations are not applicable. In some cases they are the only equations that apply. Unfortunately, the Navier-Stokes equations are very difficult to solve in their complete form. In general, a very large amount of computer time and storage is necessary to obtain a solution with these equations. This is particularly true for the compressible Navier-Stokes equations, which are a mixed set of elliptic-parabolic equations for a steady flow and a mixed set of hyperbolic-parabolic equations for an unsteady flow. The time-dependent solution procedure is normally used when a steady flow field is computed. That is, the unsteady Navier-Stokes equations are integrated in time until a steady-state solution is achieved. Thus, for a three-dimensional (3-D) flow field, a four-dimensional (4-D) (three space, one time) problem must be solved when the compressible Navier-Stokes equations are employed. Methods for solving the complete Navier-Stokes equations are discussed in Chapter 9.

Fortunately, for many of the viscous flow problems where the boundary-layer equations are not applicable, it is possible to solve a reduced set of equations that fall between the complete Navier-Stokes equations and the boundary-layer equations in terms of complexity. These reduced equations belong to a class of equations that is often referred to as the "thin-layer" or "parabolized" Navier-Stokes equations. There are several sets of equations that fall within this class:

1. thin-layer Navier-Stokes (TLNS) equations
2. parabolized Navier-Stokes (PNS) equations
3. reduced Navier-Stokes (RNS) equations
4. partially parabolized Navier-Stokes (PPNS) equations
5. viscous shock-layer (VSL) equations
6. conical Navier-Stokes (CNS) equations

The sets of equations in this class are characterized by the fact that they are applicable to both inviscid and viscous flow regions. In addition, the equations all contain a nonzero normal pressure gradient. This is a necessary requirement if viscous and inviscid regions are to be solved simultaneously. Finally, the equations in this class omit all viscous terms containing derivatives in the streamwise direction.

There are two very important advantages that result when these equations are used instead of the complete Navier-Stokes equations. First, there are fewer terms in the equations, which leads to some reduction in the required computation time. Second, and by far the most important advantage, is the fact that for a steady flow all of the equations in this class, except the TLNS equations, are a mixed set of hyperbolic-parabolic equations in the streamwise direction (provided that certain conditions are met). In other words, the Navier-Stokes equations are "parabolized" in the streamwise direction. As a consequence, the equations can be solved using a boundary-layer type of marching technique, so that a typical problem is reduced from four dimensions to three spatial dimensions. A substantial reduction in computation time and storage is thus achieved. In this chapter we will discuss the derivation of the equations in the "thin-layer Navier-Stokes" class and present a number of methods for solving them.

8.2 THIN-LAYER NAVIER-STOKES EQUATIONS

The unsteady boundary-layer equations can be formally derived from the complete Navier-Stokes equations by neglecting terms of the order of $1/(\mathrm{Re}_L)^{1/2}$ and smaller. As a consequence of this order-of-magnitude analysis, all viscous terms containing derivatives parallel to the body surface are dropped, since they are substantially smaller than viscous terms containing derivatives normal to the wall. In addition, the normal momentum equation is reduced to a simple equation [i.e., Eq. (8.1) for a Cartesian coordinate system] that indicates that the

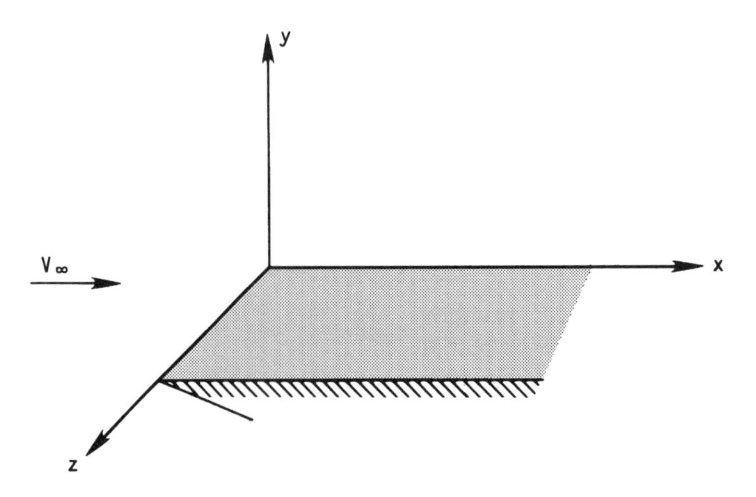

Figure 8.2 Flow over a flat plate.

normal pressure gradient is negligible. In the thin-layer approximation to the Navier-Stokes equations, the viscous terms containing derivatives in the directions parallel to the body surface are again neglected in the unsteady Navier-Stokes equations, but all other terms in the momentum equations are retained. One of the principal advantages of retaining the terms that are normally neglected in boundary-layer theory is that separated and reverse flow regions can be computed in a straightforward manner. Also, flows that contain a large normal pressure gradient, such as those shown in Fig. 8.1, can be readily computed.

The concept of the thin-layer approximation also arises from a detailed examination of typical high Reynolds number computations involving the complete Navier-Stokes equations (Baldwin and Lomax, 1978). In these computations, a substantial fraction of the available computer storage and time is expended in resolving the normal gradients in the boundary layer, since a highly stretched grid is required. As a result, the gradients parallel to the body surface are usually not resolved in an adequate manner even though the corresponding viscous terms are retained in the computations. Hence, for many Navier-Stokes computations it makes sense to drop those terms that are not being adequately resolved, provided that they are reasonably small. This naturally leads to the use of the thin-layer Navier-Stokes equations.

Upon simplifying the complete Navier-Stokes equations using the thin-layer approximation for the flow geometry shown in Fig. 8.2, the TLNS equations in Cartesian coordinates become as follows:

continuity:

$$\frac{\partial \rho}{\partial t} + \frac{\partial \rho u}{\partial x} + \frac{\partial \rho v}{\partial y} + \frac{\partial \rho w}{\partial z} = 0 \qquad (8.2)$$

x momentum:

$$\frac{\partial \rho u}{\partial t} + \frac{\partial}{\partial x}(p + \rho u^2) + \frac{\partial}{\partial y}\left(\rho u v - \mu \frac{\partial u}{\partial y}\right) + \frac{\partial}{\partial z}(\rho u w) = 0 \quad (8.3)$$

y momentum:

$$\frac{\partial \rho v}{\partial t} + \frac{\partial}{\partial x}(\rho u v) + \frac{\partial}{\partial y}\left(p + \rho v^2 - \frac{4}{3}\mu \frac{\partial v}{\partial y}\right) + \frac{\partial}{\partial z}(\rho v w) = 0 \quad (8.4)$$

z momentum:

$$\frac{\partial \rho w}{\partial t} + \frac{\partial}{\partial x}(\rho u w) + \frac{\partial}{\partial y}\left(\rho v w - \mu \frac{\partial w}{\partial y}\right) + \frac{\partial}{\partial z}(p + \rho w^2) = 0 \quad (8.5)$$

energy:

$$\frac{\partial E_t}{\partial t} + \frac{\partial}{\partial x}(E_t u + pu) + \frac{\partial}{\partial y}\left(E_t v + pv - \mu u \frac{\partial u}{\partial y} - \frac{4}{3}\mu v \frac{\partial v}{\partial y} - \mu w \frac{\partial w}{\partial y} - k \frac{\partial T}{\partial y}\right)$$
$$+ \frac{\partial}{\partial z}(E_t w + pw) = 0 \quad (8.6)$$

These equations are written for a laminar flow, but they can be readily modified to apply to a turbulent flow using the techniques of Section 5.4.

For more complicated body geometries it becomes necessary to map the body surface into a transformed coordinate surface in order to apply the thin-layer approximation. Suppose we apply the general transformation given by

$$\begin{aligned} \xi &= \xi(x, y, z, t) \\ \eta &= \eta(x, y, z, t) \\ \zeta &= \zeta(x, y, z, t) \\ t &= t \end{aligned} \quad (8.7)$$

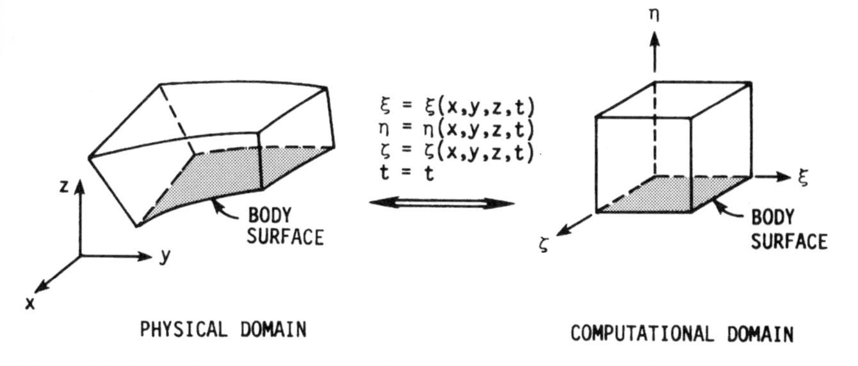

Figure 8.3 Generalized transformation. (a) Physical domain. (b) Computational domain.

to the complete Navier-Stokes equations (see Section 5.6.2) and let the body surface be defined as $\eta = 0$, as seen in Fig. 8.3. The transformed Navier-Stokes equations in strong conservation-law form become

$$\left(\frac{\mathbf{U}}{J}\right)_t + \left(\frac{\mathbf{U}\xi_t + \mathbf{E}\xi_x + \mathbf{F}\xi_y + \mathbf{G}\xi_z}{J}\right)_\xi + \left(\frac{\mathbf{U}\eta_t + \mathbf{E}\eta_x + \mathbf{F}\eta_y + \mathbf{G}\eta_z}{J}\right)_\eta$$
$$+\left(\frac{\mathbf{U}\zeta_t + \mathbf{E}\zeta_x + \mathbf{F}\zeta_y + \mathbf{G}\zeta_z}{J}\right)_\zeta = 0 \tag{8.8}$$

where J is the Jacobian of the transformation and **U**, **E**, **F**, and **G** are defined by Eqs. (5.44). We now apply the thin-layer approximation to the transformed Navier-Stokes equations. This approximation allows us to drop all viscous terms containing partial derivatives with respect to ξ and ζ. The resulting thin-layer equations may be written as (Pulliam and Steger, 1978):

$$\frac{\partial \mathbf{U}_2}{\partial t} + \frac{\partial \mathbf{E}_2}{\partial \xi} + \frac{\partial \mathbf{F}_2}{\partial \eta} + \frac{\partial \mathbf{G}_2}{\partial \zeta} = \frac{\partial \mathbf{S}_2}{\partial \eta} \tag{8.9}$$

where

$$\mathbf{U}_2 = \frac{\mathbf{U}}{J}$$

$$\mathbf{E}_2 = \frac{1}{J}\begin{bmatrix} \rho U \\ \rho u U + \xi_x p \\ \rho v U + \xi_y p \\ \rho w U + \xi_z p \\ (E_t + p)U - \xi_t p \end{bmatrix}$$

$$\mathbf{F}_2 = \frac{1}{J}\begin{bmatrix} \rho V \\ \rho u V + \eta_x p \\ \rho v V + \eta_y p \\ \rho w V + \eta_z p \\ (E_t + p)V - \eta_t p \end{bmatrix} \tag{8.10}$$

$$\mathbf{G}_2 = \frac{1}{J}\begin{bmatrix} \rho W \\ \rho u W + \zeta_x p \\ \rho v W + \zeta_y p \\ \rho w W + \zeta_z p \\ (E_t + p)W - \zeta_t p \end{bmatrix}$$

and all the viscous terms are contained in

$$\mathbf{S}_2 = \frac{1}{J}\begin{bmatrix} 0 \\ \mu\left(\eta_x^2 + \eta_y^2 + \eta_z^2\right)u_\eta + \dfrac{\mu}{3}(\eta_x u_\eta + \eta_y v_\eta + \eta_z w_\eta)\eta_x \\ \mu\left(\eta_x^2 + \eta_y^2 + \eta_z^2\right)v_\eta + \dfrac{\mu}{3}(\eta_x u_\eta + \eta_y v_\eta + \eta_z w_\eta)\eta_y \\ \mu\left(\eta_x^2 + \eta_y^2 + \eta_z^2\right)w_\eta + \dfrac{\mu}{3}(\eta_x u_\eta + \eta_y v_\eta + \eta_z w_\eta)\eta_z \\ \left(\eta_x^2 + \eta_y^2 + \eta_z^2\right)\left[\dfrac{\mu}{2}(u^2 + v^2 + w^2)_\eta + kT_\eta\right] \\ + \dfrac{\mu}{3}(\eta_x u + \eta_y v + \eta_z w)(\eta_x u_\eta + \eta_y v_\eta + \eta_z w_\eta) \end{bmatrix} \tag{8.11}$$

For compactness, Eqs. (8.10) are written in terms of the contravariant velocity components (U, V, W), which are defined by

$$\begin{aligned} U &= \xi_t + \xi_x u + \xi_y v + \xi_z w \\ V &= \eta_t + \eta_x u + \eta_y v + \eta_z w \\ W &= \zeta_t + \zeta_x u + \zeta_y v + \zeta_z w \end{aligned} \tag{8.12}$$

The contravariant velocity components U, V, W are in directions normal to constant ξ, η, ζ surfaces, respectively.

Although the TLNS equations are considerably less complicated than the complete Navier-Stokes equations, a substantial amount of computer effort is still required to solve these equations. The TLNS equations are a mixed set of hyperbolic-parabolic PDEs in time. As a consequence, the "time-dependent" approach can be applied in an identical manner to the procedure normally used to solve the compressible Navier-Stokes equations. Thus we will postpone our discussion of methods for solving the TLNS equations until Chapter 9, where the methods for solving the complete Navier-Stokes equations are discussed.

8.3 "PARABOLIZED" NAVIER-STOKES EQUATIONS

The "parabolized" Navier-Stokes (PNS) equations have steadily gained popularity because they can be used to predict complex 3-D steady supersonic viscous flow fields in an efficient manner. This efficiency is achieved because the equations can be solved using a space-marching technique as opposed to the time-marching technique that is normally employed for the complete Navier-Stokes equations. As a result, the computational effort required to solve the PNS equations for an entire supersonic flow field is similar to the effort required to solve either the inviscid portion of the flow field using the Euler equations or the viscous portion of the flow field using the boundary-layer equations. Furthermore, since the PNS equations are valid in both the inviscid and viscous portions of the flow field, the interaction between these regions of the flow field is automatically taken into account.

The term "parabolized" Navier-Stokes equations is somewhat of a misnomer, since the equations are actually a mixed set of hyperbolic-parabolic equations, provided that certain conditions are met. These conditions include the requirements that the inviscid outer region of the flow be supersonic and the streamwise velocity component be everywhere positive. Note that the last requirement excludes streamwise flow separation, but crossflow separation is permitted. An additional constraint is caused by the presence of the streamwise pressure gradient in the streamwise momentum equation. If this term is included everywhere in the flow field, then upstream influence can occur in the subsonic portion of the boundary layer and a single-pass space-marching method of solution is not well posed. This leads to exponentially growing solutions, which are often called *departure solutions*. Several techniques have been proposed to circumvent this difficulty, and they are discussed in Section 8.3.2.

8.3.1 Derivation of PNS Equations

The derivation of the PNS equations from the complete Navier-Stokes equations is, in general, not as rigorous as the derivation of the boundary-layer equations. Because of this, slightly different versions of the PNS equations have appeared in the literature. These versions differ in some cases because of the type of flow problem being considered. However, in all cases the normal pressure gradient term is retained, and the second derivative terms with respect to the streamwise direction are omitted.

One of the earliest studies involving the use of the PNS equations was by Rudman and Rubin (1968). In their study, the hypersonic laminar flow near the leading edge of a flat plate [see Fig. 8.1(a)] was computed using a set of PNS equations. Rudman and Rubin derived their PNS equations from the complete Navier-Stokes equations using a series expansion technique. This method for reducing the complexity of the Navier-Stokes equations is an alternative to the order-of-magnitude analysis used in Chapter 5 to derive the boundary-layer equations. In the series expansion method, the flow variables are first nondimensionalized with respect to local reference conditions in order to estimate the magnitude of the various terms in the Navier-Stokes equations. The flow variables are then expanded in an appropriate series. Rudman and Rubin assumed the following form:

$$\begin{aligned}
u &= V_\infty(u_0^* + \epsilon u_1^* + \cdots) \\
v &= V_\infty \delta^*(v_0^* + \epsilon v_1^* + \cdots) \\
p &= p_\infty p_{\text{ref}}^*(p_0^* + \epsilon p_1^* + \cdots) \\
\rho &= \rho_\infty \rho_{\text{ref}}^*(\rho_0^* + \epsilon \rho_1^* + \cdots) \\
T &= T_\infty T_{\text{ref}}^*(T_0^* + \epsilon T_1^* + \cdots)
\end{aligned} \tag{8.13}$$

$$\mu = \mu_\infty \mu^*_{\text{ref}}(\mu^*_0 + \epsilon\mu^*_1 + \cdots)$$
$$x = x^* L \qquad y = y^* \delta \qquad \delta = \delta^* L$$

where the superscript asterisk denotes a nondimensional quantity, the subscript ref represents the local reference value of a flow variable nondimensionalized with respect to the free stream value, L is the characteristic length in the x direction, and δ is the characteristic length in the y direction. The first term in the series expansion (denoted with a subscript zero) is used to obtain the zeroth-order solution, while both the first and second terms are needed to obtain the first-order solution. The relative magnitude of the coefficient ϵ is determined later in the analysis. For the relatively thin disturbed region shown in Fig. 8.1(a), the gradients normal to the surface are much greater than the gradients parallel to the surface, and δ^* can be assumed to be small.

When the expansions are substituted into the 2-D steady Navier-Stokes equations, the following nondimensional equations result (for convenience the subscript zero has been dropped):

continuity:

$$\frac{\partial \rho^* u^*}{\partial x^*} + \frac{\partial \rho^* v^*}{\partial y^*} = O(\epsilon) \tag{8.14}$$

x momentum:

$$\rho^* u^* \frac{\partial u^*}{\partial x^*} + \rho^* v^* \frac{\partial u^*}{\partial y^*} = -\Delta^2 \frac{\partial p^*}{\partial x^*} + \frac{1}{(\delta^*)^2 \operatorname{Re}_{\text{ref}}} \frac{\partial}{\partial y^*}\left(\mu^* \frac{\partial u^*}{\partial y^*}\right) + O\left[\epsilon, (\operatorname{Re}_{\text{ref}})^{-1}\right] \tag{8.15}$$

y momentum:

$$\rho^* u^* \frac{\partial v^*}{\partial x^*} + \rho^* v^* \frac{\partial v^*}{\partial y^*} = -\left(\frac{\Delta}{\delta^*}\right)^2 \frac{\partial p^*}{\partial y^*} + \frac{1}{(\delta^*)^2 \operatorname{Re}_{\text{ref}}}\left[\frac{4}{3}\frac{\partial}{\partial y^*}\left(\mu^* \frac{\partial v^*}{\partial y^*}\right) + \frac{\partial}{\partial x^*}\left(\mu^* \frac{\partial u^*}{\partial y^*}\right) - \frac{2}{3}\frac{\partial}{\partial y^*}\left(\mu^* \frac{\partial u^*}{\partial x^*}\right)\right] + O\left[\epsilon, (\operatorname{Re}_{\text{ref}})^{-1}\right] \tag{8.16}$$

energy:

$$\Delta^2\left[\rho^* u^* \frac{\partial T^*}{\partial x^*} + \rho^* v^* \frac{\partial T^*}{\partial y^*} + (\gamma - 1) p^* \left(\frac{\partial u^*}{\partial x^*} + \frac{\partial v^*}{\partial y^*}\right) - \frac{\gamma}{\Pr} \frac{1}{(\delta^*)^2 \operatorname{Re}_{\text{ref}}} \frac{\partial}{\partial y^*}\left(\mu^* \frac{\partial T^*}{\partial y^*}\right)\right]$$

$$= \frac{\gamma - 1}{(\delta^*)\,\mathrm{Re}_{\mathrm{ref}}} \left\{ \mu^* \left(\frac{\partial u^*}{\partial y^*} \right)^2 + (\delta^*)^2 \left[\frac{4}{3} \mu^* \left(\frac{\partial u^*}{\partial x^*} \right)^2 + \frac{4}{3} \mu^* \left(\frac{\partial v^*}{\partial y^*} \right)^2 \right. \right.$$

$$\left. - \frac{4}{3} \mu^* \frac{\partial u^*}{\partial x^*} \frac{\partial v^*}{\partial y^*} + 2\mu^* \frac{\partial v^*}{\partial x^*} \frac{\partial u^*}{\partial y^*} \right]$$

$$\left. + \epsilon \left[\mu_1^* \left(\frac{\partial u^*}{\partial y^*} \right)^2 + 2\mu^* \frac{\partial u^*}{\partial y^*} \frac{\partial u_1^*}{\partial y^*} \right] \right\}$$

$$+ O\left[\Delta^2 (\mathrm{Re}_{\mathrm{ref}})^{-1}, (\delta^*)^2 (\mathrm{Re}_{\mathrm{ref}})^{-1}, \frac{\epsilon^2 (\mathrm{Re}_{\mathrm{ref}})^{-1}}{(\delta^*)^2} \right] \tag{8.17}$$

In the above equations, $\mathrm{Re}_{\mathrm{ref}} = (\rho_\infty V_\infty L / \mu_\infty)(\rho^*_{\mathrm{ref}} / \mu^*_{\mathrm{ref}})$, $\Delta^2 = T^*_{\mathrm{ref}} / M_\infty^2 \gamma$, and a perfect gas is assumed.

The next step in the process is to determine which terms can be neglected in comparison to other terms in Eqs. (8.14)–(8.17). In order to do this, we need to obtain estimates for the magnitudes of $\mathrm{Re}_{\mathrm{ref}}$, Δ^2, and $(\Delta/\delta^*)^2$ in the various regions of the flow field. From our previous discussions on boundary layers, we know that for a thin viscous layer, $\mathrm{Re}_{\mathrm{ref}}$ is of order $1/(\delta^*)^2$. Also, near the edge of the viscous layer, Δ^2 is proportional to $(M_\infty^2)^{-1}$, since $T^*_{\mathrm{ref}} = 1$ in this region. From compressible boundary-layer theory (Schlichting, 1968) it is known that Δ^2 can achieve a maximum value of the order of $(\gamma - 1)/2A\gamma$, where A varies between $\mathrm{Pr}^{-1/2}$ for an adiabatic wall to about 4 in the cold wall limit. Hence for most cases, we can assume $\Delta^2 \ll 1$, provided that $M_\infty \geqslant 5$. Rudman and Rubin (1968) have shown that $(\Delta/\delta^*)^2$ is of order unity in the merged-layer region. Further downstream in the strong-interaction region, they have shown that $(\Delta/\delta^*)^2$ is very large near the wall but decreases in value to order unity at the edge of the boundary layer. Using the above information for the relative magnitudes of $\mathrm{Re}_{\mathrm{ref}}$, Δ^2, and $(\Delta/\delta^*)^2$ in the various regions of the flow field, we can now simplify Eqs. (8.14)–(8.17). For the set of equations valid to zeroth-order ($M_\infty \geqslant 5$), we can neglect terms of order $(\delta^*)^2$, Δ^2, and ϵ; but we must retain terms of order $(\Delta/\delta^*)^2$. As a result, the continuity equation and the y momentum equation cannot be reduced further. On the other hand, the x momentum equation is simplified, since the streamwise pressure gradient term can be dropped and the energy equation reduces to

$$\frac{\partial u^*}{\partial y^*} = 0 \tag{8.18}$$

If we combine Eq. (8.18) with the x momentum equation, we find that

$$u^* = \text{const} = 1 \tag{8.19}$$

or

$$u = V_\infty$$

Obviously, this is a trivial result (applicable only in the free stream), and we are forced to retain higher-order terms [i.e., $(\delta^*)^2$, Δ^2, and ϵ] in order to obtain a meaningful energy equation. Note that we can eliminate many of the higher-order terms by employing Eq. (8.19). The final forms of the zeroth-order equations in dimensional form become

continuity:

$$\frac{\partial \rho u}{\partial x} + \frac{\partial \rho v}{\partial y} = 0 \tag{8.20}$$

x momentum:

$$\rho u \frac{\partial u}{\partial x} + \rho v \frac{\partial u}{\partial y} = \frac{\partial}{\partial y}\left(\mu \frac{\partial u}{\partial y}\right) \tag{8.21}$$

y momentum:

$$\rho u \frac{\partial v}{\partial x} + \rho v \frac{\partial v}{\partial y} = -\frac{\partial p}{\partial y} + \frac{4}{3}\frac{\partial}{\partial y}\left(\mu \frac{\partial v}{\partial y}\right) + \frac{\partial}{\partial x}\left(\mu \frac{\partial u}{\partial y}\right) - \frac{2}{3}\frac{\partial}{\partial y}\left(\mu \frac{\partial u}{\partial x}\right) \tag{8.22}$$

energy:

$$\rho u c_v \frac{\partial T}{\partial x} + \rho v c_v \frac{\partial T}{\partial y} = -p\left(\frac{\partial u}{\partial x} + \frac{\partial v}{\partial y}\right) + \frac{\partial}{\partial y}\left(k \frac{\partial T}{\partial y}\right) + \mu\left(\frac{\partial u}{\partial y}\right)^2 + \frac{4}{3}\mu\left(\frac{\partial v}{\partial y}\right)^2 \tag{8.23}$$

The zeroth-order equations are valid for leading edge flow fields when $M_\infty \geqslant 5$, while the first-order equations are applicable when $M_\infty \geqslant 2$. The zeroth-order equations were derived by neglecting terms of order $(\delta^*)^2$, Δ^2, and ϵ. Since ϵ is the coefficient of the first-order terms, its order is given by the largest of $(\delta^*)^2$ and Δ^2. Rudman and Rubin have shown that in order for $(\delta^*)^2$ to be very small (i.e., $\leqslant 0.05$) the zeroth-order equations are not valid upstream of the point at which

$$\frac{\chi_\infty}{M_\infty^2} \cong 2$$

where χ_∞ is the strong-interaction parameter defined by

$$\chi_\infty = \left(\frac{\mu_{\text{wall}} T_\infty}{\mu_\infty T_{\text{wall}}}\right)^{1/2} \left(M_\infty^3 \operatorname{Re}_{x_\infty}\right)^{-1/2}$$

Consequently, an initial starting solution is required for the present leading edge problem. The same is true for all other problems that are solved using the PNS equations. For the present problem it is permissible to employ an approximate starting solution located very close to the leading edge because it will have a small effect on the flow field further downstream. This is because only a small amount of mass flow passes between the plate and the shock layer edge at this initial station as compared with the mass flow passing between the plate and the shock wave at stations further downstream. For other problems, however, the initial starting solution will have a definite effect on the downstream flow field, and in many cases, the starting solution must be determined accurately.

The set of PNS equations derived by Rudman and Rubin do not contain a streamwise pressure gradient term, so that there can be no upstream influence through the subsonic portion of the boundary layer. As a result, the equations behave in a strictly "parabolic" manner in the boundary-layer region. Because of this, Davis and Rubin (1980) refer to these equations as the parabolic Navier-Stokes equations instead of the "parabolized" Navier-Stokes equations. They use the latter name to refer to the sets of equations that do contain a streamwise pressure gradient term.

The PNS equations derived by Rudman and Rubin have been used to solve leading edge flows about both 2-D and 3-D geometries including flat plates, rectangular corners, cones, and wing tips (see Lin and Rubin, 1973b, for references). The 3-D equations are derived in a manner similar to the 2-D equations. The coordinates x, y, z are first nondimensionalized using L, δ_y, and δ_z, respectively. The velocities u, v, w are nondimensionalized using V_∞, $V_\infty \delta_y^*$, and $V_\infty \delta_z^*$, respectively, where $\delta_y^* = \delta_y/L$ and $\delta_z^* = \delta_z/L$. Terms of order $(\delta_z^*)^2$, $(\delta_y^*)^2$, $\delta_y^*\delta_z^*$, etc., are assumed small. After substituting the series expansions into the Navier-Stokes equations and neglecting higher-order terms, the 3-D zeroth-order equations become

continuity:

$$\frac{\partial \rho u}{\partial x} + \frac{\partial \rho v}{\partial y} + \frac{\partial \rho w}{\partial z} = 0 \tag{8.24}$$

x momentum:

$$\rho u \frac{\partial u}{\partial x} + \rho v \frac{\partial u}{\partial y} + \rho w \frac{\partial u}{\partial z} = \frac{\partial}{\partial y}\left(\mu \frac{\partial u}{\partial y}\right) + \frac{\partial}{\partial z}\left(\mu \frac{\partial u}{\partial z}\right) \tag{8.25}$$

y momentum:

$$\begin{aligned}
&\rho u \frac{\partial v}{\partial x} + \rho v \frac{\partial v}{\partial y} + \rho w \frac{\partial v}{\partial z} \\
&\quad = -\frac{\partial p}{\partial y} + \frac{4}{3}\frac{\partial}{\partial y}\left(\mu \frac{\partial v}{\partial y}\right) + \frac{\partial}{\partial z}\left(\mu \frac{\partial v}{\partial z}\right) + \frac{\partial}{\partial x}\left(\mu \frac{\partial u}{\partial y}\right) \\
&\qquad - \frac{2}{3}\frac{\partial}{\partial y}\left(\mu \frac{\partial u}{\partial x} + \mu \frac{\partial w}{\partial z}\right) + \frac{\partial}{\partial z}\left(\mu \frac{\partial w}{\partial y}\right)
\end{aligned} \tag{8.26}$$

z momentum:

$$\rho u \frac{\partial w}{\partial x} + \rho v \frac{\partial w}{\partial y} + \rho w \frac{\partial w}{\partial z}$$
$$= -\frac{\partial p}{\partial z} + \frac{4}{3}\frac{\partial}{\partial z}\left(\mu \frac{\partial w}{\partial z}\right) + \frac{\partial}{\partial y}\left(\mu \frac{\partial w}{\partial y}\right) + \frac{\partial}{\partial x}\left(\mu \frac{\partial u}{\partial z}\right)$$
$$-\frac{2}{3}\frac{\partial}{\partial z}\left(\mu \frac{\partial v}{\partial y} + \mu \frac{\partial u}{\partial x}\right) + \frac{\partial}{\partial y}\left(\mu \frac{\partial v}{\partial z}\right) \tag{8.27}$$

energy:

$$\rho u c_v \frac{\partial T}{\partial x} + \rho v c_v \frac{\partial T}{\partial y} + \rho w c_v \frac{\partial T}{\partial z}$$
$$= -p\left(\frac{\partial u}{\partial x} + \frac{\partial v}{\partial y} + \frac{\partial w}{\partial z}\right) + \frac{\partial}{\partial y}\left(k \frac{\partial T}{\partial y}\right)$$
$$+ \frac{\partial}{\partial z}\left(k \frac{\partial T}{\partial z}\right) + \mu\left[\left(\frac{\partial u}{\partial y}\right)^2 + \left(\frac{\partial u}{\partial z}\right)^2 + \left(\frac{\partial w}{\partial y} + \frac{\partial v}{\partial z}\right)^2\right]$$
$$+ \frac{4}{3}\mu\left[\left(\frac{\partial v}{\partial y}\right)^2 + \left(\frac{\partial w}{\partial z}\right)^2 - \frac{\partial v}{\partial y}\frac{\partial w}{\partial z}\right] \tag{8.28}$$

A set of PNS equations very similar to those of Rudman and Rubin's were derived independently by Cheng et al. (1970). Cheng et al. included a streamwise pressure gradient term in their equations.

The most common form of the PNS equations (Lubard and Helliwell, 1973, 1974) and the one that will be used for the rest of this chapter is obtained by assuming that the streamwise viscous derivative terms (including heat flux terms) are negligible compared to the normal and transverse viscous derivative terms. In other words, the streamwise viscous derivative terms are assumed to be of $O(1)$, while the normal and transverse viscous derivative terms are of $O(\mathrm{Re}_L^{1/2})$. Hence these PNS equations are derived by simply dropping all viscous terms containing partial derivatives with respect to the streamwise direction from the steady Navier-Stokes equations. The resulting set of equations for a Cartesian coordinate system (x is the streamwise direction) is given by

continuity:

$$\frac{\partial \rho u}{\partial x} + \frac{\partial \rho v}{\partial y} + \frac{\partial \rho w}{\partial z} = 0 \tag{8.29}$$

x momentum:

$$\rho u \frac{\partial u}{\partial x} + \rho v \frac{\partial u}{\partial y} + \rho w \frac{\partial u}{\partial z} = -\frac{\partial p}{\partial x} + \frac{\partial}{\partial y}\left(\mu \frac{\partial u}{\partial y}\right) + \frac{\partial}{\partial z}\left(\mu \frac{\partial u}{\partial z}\right) \tag{8.30}$$

y momentum:

$$\rho u \frac{\partial v}{\partial x} + \rho v \frac{\partial v}{\partial y} + \rho w \frac{\partial v}{\partial z} = -\frac{\partial p}{\partial y} + \frac{4}{3}\frac{\partial}{\partial y}\left(\mu \frac{\partial v}{\partial y}\right) + \frac{\partial}{\partial z}\left(\mu \frac{\partial v}{\partial z}\right) + \frac{\partial}{\partial z}\left(\mu \frac{\partial w}{\partial y}\right) - \frac{2}{3}\frac{\partial}{\partial y}\left(\mu \frac{\partial w}{\partial z}\right) \tag{8.31}$$

z momentum:

$$\rho u \frac{\partial w}{\partial x} + \rho v \frac{\partial w}{\partial y} + \rho w \frac{\partial w}{\partial z} = -\frac{\partial p}{\partial z} + \frac{4}{3}\frac{\partial}{\partial z}\left(\mu \frac{\partial w}{\partial z}\right) + \frac{\partial}{\partial y}\left(\mu \frac{\partial w}{\partial y}\right) + \frac{\partial}{\partial y}\left(\mu \frac{\partial v}{\partial z}\right) - \frac{2}{3}\frac{\partial}{\partial z}\left(\mu \frac{\partial v}{\partial y}\right) \tag{8.32}$$

energy

$$\begin{aligned}\rho u c_v \frac{\partial T}{\partial x} &+ \rho v c_v \frac{\partial T}{\partial y} + \rho w c_v \frac{\partial T}{\partial z} \\ &= -p\left(\frac{\partial u}{\partial x} + \frac{\partial v}{\partial y} + \frac{\partial w}{\partial z}\right) + \frac{\partial}{\partial y}\left(k \frac{\partial T}{\partial y}\right) \\ &\quad + \frac{\partial}{\partial z}\left(k \frac{\partial T}{\partial z}\right) + \mu\left[\left(\frac{\partial u}{\partial y}\right)^2 + \left(\frac{\partial u}{\partial z}\right)^2 + \left(\frac{\partial w}{\partial y} + \frac{\partial v}{\partial z}\right)^2\right] \\ &\quad + \frac{4}{3}\mu\left[\left(\frac{\partial v}{\partial y}\right)^2 + \left(\frac{\partial w}{\partial z}\right)^2 - \frac{\partial v}{\partial y}\frac{\partial w}{\partial z}\right]\end{aligned} \tag{8.33}$$

It is interesting to compare this set of PNS equations with the equations of Rudman and Rubin [Eqs. (8.24) – (8.28)]. We note that the continuity and energy equations are identical but the momentum equations are different. In particular, the present x momentum equation contains the streamwise pressure gradient term as discussed previously.

We now wish to express the PNS equations in terms of a generalized coordinate system. For the generalized transformation described in Section 5.6.2, the complete Navier-Stokes equations can be written as

$$\begin{aligned}\frac{\partial}{\partial t}\left(\frac{\mathbf{U}}{J}\right) &+ \frac{\partial}{\partial \xi}\left\{\frac{1}{J}\left[\xi_x(\mathbf{E}_i - \mathbf{E}_v) + \xi_y(\mathbf{F}_i - \mathbf{F}_v) + \xi_z(\mathbf{G}_i - \mathbf{G}_v)\right]\right\} \\ &+ \frac{\partial}{\partial \eta}\left\{\frac{1}{J}\left[\eta_x(\mathbf{E}_i - \mathbf{E}_v) + \eta_y(\mathbf{F}_i - \mathbf{F}_v) + \eta_z(\mathbf{G}_i - \mathbf{G}_v)\right]\right\} \\ &+ \frac{\partial}{\partial \zeta}\left\{\frac{1}{J}\left[\zeta_x(\mathbf{E}_i - \mathbf{E}_v) + \zeta_y(\mathbf{F}_i - \mathbf{F}_v) + \zeta_z(\mathbf{G}_i - \mathbf{G}_v)\right]\right\} = 0\end{aligned} \tag{8.34}$$

where

$$\mathbf{U} = \begin{bmatrix} \rho \\ \rho u \\ \rho v \\ \rho w \\ E_t \end{bmatrix}$$

$$\mathbf{E}_i = \begin{bmatrix} \rho u \\ \rho u^2 + p \\ \rho uv \\ \rho uw \\ (E_t + p)u \end{bmatrix} \qquad \mathbf{E}_v = \begin{bmatrix} 0 \\ \tau_{xx} \\ \tau_{xy} \\ \tau_{xz} \\ u\tau_{xx} + v\tau_{xy} + w\tau_{xz} - q_x \end{bmatrix}$$

$$\mathbf{F}_i = \begin{bmatrix} \rho v \\ \rho uv \\ \rho v^2 + p \\ \rho vw \\ (E_t + p)v \end{bmatrix} \qquad \mathbf{F}_v = \begin{bmatrix} 0 \\ \tau_{xy} \\ \tau_{yy} \\ \tau_{yz} \\ u\tau_{xy} + v\tau_{yy} + w\tau_{yz} - q_y \end{bmatrix} \tag{8.35}$$

$$\mathbf{G}_i = \begin{bmatrix} \rho w \\ \rho uw \\ \rho vw \\ \rho w^2 + p \\ (E_t + p)w \end{bmatrix} \qquad \mathbf{G}_v = \begin{bmatrix} 0 \\ \tau_{xz} \\ \tau_{yz} \\ \tau_{zz} \\ u\tau_{xz} + v\tau_{yz} + w\tau_{zz} - q_z \end{bmatrix}$$

and

$$\begin{aligned}
E_t &= \rho\left(e + \frac{u^2 + v^2 + w^2}{2}\right) \\
\tau_{xx} &= \tfrac{2}{3}\mu\big[2(\xi_x u_\xi + \eta_x u_\eta + \zeta_x u_\zeta) - (\xi_y v_\xi + \eta_y v_\eta + \zeta_y v_\zeta) \\
&\qquad - (\xi_z w_\xi + \eta_z w_\eta + \zeta_z w_\zeta)\big] \\
\tau_{yy} &= \tfrac{2}{3}\mu\big[2(\xi_y v_\xi + \eta_y v_\eta + \zeta_y v_\zeta) - (\xi_x u_\xi + \eta_x u_\eta + \zeta_x u_\zeta) \\
&\qquad - (\xi_z w_\xi + \eta_z w_\eta + \zeta_z w_\zeta)\big] \\
\tau_{zz} &= \tfrac{2}{3}\mu\big[2(\xi_z w_\xi + \eta_z w_\eta + \zeta_z w_\zeta) - (\xi_x u_\xi + \eta_x u_\eta + \zeta_x u_\zeta) \\
&\qquad - (\xi_y v_\xi + \eta_y v_\eta + \zeta_y v_\zeta)\big] \\
\tau_{xy} &= \mu(\xi_y u_\xi + \eta_y u_\eta + \zeta_y u_\zeta + \xi_x v_\xi + \eta_x v_\eta + \zeta_x v_\zeta) \\
\tau_{xz} &= \mu(\xi_z u_\xi + \eta_z u_\eta + \zeta_z u_\zeta + \xi_x w_\xi + \eta_x w_\eta + \zeta_x w_\zeta) \\
\tau_{yz} &= \mu(\xi_z v_\xi + \eta_z v_\eta + \zeta_z v_\zeta + \xi_y w_\xi + \eta_y w_\eta + \zeta_y w_\zeta)
\end{aligned} \tag{8.36}$$

$$q_x = -k(\xi_x T_\xi + \eta_x T_\eta + \zeta_x T_\zeta)$$
$$q_y = -k(\xi_y T_\xi + \eta_y T_\eta + \zeta_y T_\zeta)$$
$$q_z = -k(\xi_z T_\xi + \eta_z T_\eta + \zeta_z T_\zeta)$$

Note that the usual **E**, **F**, and **G** vectors have been split into an inviscid part (subscript i) and a viscous part (subscript v). The reason for doing this will become evident in Section 8.3.3, when we describe numerical procedures for solving the PNS equations. The PNS equations in generalized coordinates can now be obtained by simply dropping the unsteady terms and the viscous terms containing partial derivatives with respect to the streamwise direction ξ. The resulting equations become

$$\frac{\partial \mathbf{E}_3}{\partial \xi} + \frac{\partial \mathbf{F}_3}{\partial \eta} + \frac{\partial \mathbf{G}_3}{\partial \zeta} = 0 \tag{8.37}$$

where

$$\begin{aligned}
\mathbf{E}_3 &= \frac{1}{J}(\xi_x \mathbf{E}_i + \xi_y \mathbf{F}_i + \xi_z \mathbf{G}_i) \\
\mathbf{F}_3 &= \frac{1}{J}\left[\eta_x(\mathbf{E}_i - \mathbf{E}'_v) + \eta_y(\mathbf{F}_i - \mathbf{F}'_v) + \eta_z(\mathbf{G}_i - \mathbf{G}'_v)\right] \\
\mathbf{G}_3 &= \frac{1}{J}\left[\zeta_x(\mathbf{E}_i - \mathbf{E}'_v) + \zeta_y(\mathbf{F}_i - \mathbf{F}'_v) + \zeta_z(\mathbf{G}_i - \mathbf{G}'_v)\right]
\end{aligned} \tag{8.38}$$

and the prime is used to indicate that terms containing partial derivatives with respect to ξ have been omitted. Likewise, the shear stress and heat flux terms in Eqs. (8.36) reduce to

$$\begin{aligned}
\tau'_{xx} &= \tfrac{2}{3}\mu\left[2(\eta_x u_\eta + \zeta_x u_\zeta) - (\eta_y v_\eta + \zeta_y v_\zeta) - (\eta_z w_\eta + \zeta_z w_\zeta)\right] \\
\tau'_{yy} &= \tfrac{2}{3}\mu\left[2(\eta_y v_\eta + \zeta_y v_\zeta) - (\eta_x u_\eta + \zeta_x u_\zeta) - (\eta_z w_\eta + \zeta_z w_\zeta)\right] \\
\tau'_{zz} &= \tfrac{2}{3}\mu\left[2(\eta_z w_\eta + \zeta_z w_\zeta) - (\eta_x u_\eta + \zeta_x u_\zeta) - (\eta_y v_\eta + \zeta_y v_\zeta)\right] \\
\tau'_{xy} &= \mu(\eta_y u_\eta + \zeta_y u_\zeta + \eta_x v_\eta + \zeta_x v_\zeta) \\
\tau'_{xz} &= \mu(\eta_z u_\eta + \zeta_z u_\zeta + \eta_x w_\eta + \zeta_x w_\zeta) \\
\tau'_{yz} &= \mu(\eta_z v_\eta + \zeta_z v_\zeta + \eta_y w_\eta + \zeta_y w_\zeta) \\
q'_x &= -k(\eta_x T_\eta + \zeta_x T_\zeta) \\
q'_y &= -k(\eta_y T_\eta + \zeta_y T_\zeta) \\
q'_z &= -k(\eta_z T_\eta + \zeta_z T_\zeta)
\end{aligned} \tag{8.39}$$

For many applications (Schiff and Steger, 1979), the thin-layer approximation can also be applied to the PNS equations. With this additional assumption, the resulting equations are simply the steady form of the TLNS equations. For the generalized transformation described previously, these equations can be written as

$$\frac{\partial \mathbf{E}_2}{\partial \xi} + \frac{\partial \mathbf{F}_2}{\partial \eta} + \frac{\partial \mathbf{G}_2}{\partial \zeta} = \frac{\partial \mathbf{S}_2}{\partial \eta} \tag{8.40}$$

where $\mathbf{E}_2$, $\mathbf{F}_2$, $\mathbf{G}_2$, and $\mathbf{S}_2$ are defined by Eqs. (8.10) and (8.11).

8.3.2 STREAMWISE PRESSURE GRADIENT

The presence of the streamwise pressure gradient term in the streamwise momentum equation permits information to be propagated upstream through subsonic portions of the flow field such as a boundary layer. As a consequence, a single-pass space-marching method of solution is not well posed, and in many cases, exponentially growing solutions (departure solutions) are encountered. These departure solutions are characterized by either a separation-like increase in wall pressure or an expansion-like decrease in wall pressure. A similar behavior (Lighthill, 1953) is observed for the boundary-layer equations when the streamwise pressure gradient is not prescribed. The one difference, however, is that in the case of the PNS equations, the normal momentum equation allows a pressure interaction to occur between the critical subsonic boundary-layer region and the inviscid outer region.

In order to better understand why departure solutions occur, let us examine the influence of the streamwise pressure gradient term on the mathematical nature of the PNS equations. For simplicity, let us consider the 2-D PNS equations and assume a perfect gas with constant viscosity. With these assumptions, Eqs. (8.29)–(8.33) can be reduced to the following vector representation:

$$\frac{\partial \mathbf{E}}{\partial x} + \frac{\partial \mathbf{F}}{\partial y} = \frac{\partial \mathbf{F}_v}{\partial y} \tag{8.41}$$

where

$$\mathbf{E} = \begin{bmatrix} \rho u \\ \rho u^2 + \omega p \\ \rho u v \\ \left[\dfrac{\gamma}{\gamma - 1} p + \dfrac{\rho}{2}(u^2 + v^2)\right] u \end{bmatrix}$$

$$\mathbf{F} = \begin{bmatrix} \rho v \\ \rho u v \\ \rho v^2 + p \\ \left[\dfrac{\gamma}{\gamma - 1} p + \dfrac{\rho}{2}(u^2 + v^2)\right] v \end{bmatrix} \tag{8.42}$$

$$\mathbf{F}_v = \mu \begin{bmatrix} 0 \\ u_y \\ \frac{4}{3} v_y \\ u u_y + \frac{4}{3} v v_y + \frac{k}{\mu} T_y \end{bmatrix}$$

Note that in these equations a parameter ω has been inserted in front of the streamwise pressure gradient term in the x momentum equation. Thus if ω is set equal to zero, the streamwise pressure gradient term is omitted. On the other hand, if ω is set equal to 1, the term is retained completely.

If we first consider the inviscid limit ($\mu \to 0$), Eq. (8.41) reduces to the Euler equation

$$\frac{\partial \mathbf{E}}{\partial x} + \frac{\partial \mathbf{F}}{\partial y} = 0 \tag{8.43}$$

which is equivalent to

$$[A_1]\mathbf{Q}_x + [B_1]\mathbf{Q}_y = 0 \tag{8.44}$$

where

$$[A_1] = \begin{bmatrix} u & \rho & 0 & 0 \\ 0 & \rho u & 0 & \omega \\ 0 & 0 & \rho u & 0 \\ 0 & \rho u^2 + \frac{\gamma p}{\gamma - 1} & \rho u v & \frac{\gamma u}{\gamma - 1} \end{bmatrix} \qquad \mathbf{Q} = \begin{bmatrix} \rho \\ u \\ v \\ p \end{bmatrix} \tag{8.45}$$

$$[B_1] = \begin{bmatrix} v & 0 & \rho & 0 \\ 0 & \rho v & 0 & 0 \\ 0 & 0 & \rho v & 1 \\ 0 & \rho u v & \rho v^2 + \frac{\gamma p}{\gamma - 1} & \frac{\gamma v}{\gamma - 1} \end{bmatrix}$$

These equations are hyperbolic in x, provided that the eigenvalues of $[A_1]^{-1}[B_1]$ are real (see Section 2.5). The eigenvalues are

$$\begin{aligned} \lambda_{1,2} &= \frac{v}{u} \\ \lambda_{3,4} &= \frac{-b \pm \sqrt{b^2 - 4\bar{a}c}}{2\bar{a}} \end{aligned} \tag{8.46}$$

where

$$\bar{a} = [\gamma - \omega(\gamma - 1)]u^2 - \omega a^2$$
$$b = -uv[1 + \gamma - \omega(\gamma - 1)]$$
$$c = v^2 - a^2$$

and a is the speed of sound. If the streamwise pressure gradient is retained completely (i.e., $\omega = 1$), it is easy to show that the eigenvalues are all real, provided that

$$u^2 + v^2 \geqslant a^2$$

or

$$M \geqslant 1$$

This is the usual requirement that must be satisfied if the Euler equations are to be integrated using a space-marching technique. However, if only a fraction of the streamwise pressure gradient is retained (i.e., $0 \leqslant \omega \leqslant 1$), the eigenvalues will remain real even in subsonic regions, provided that

$$\omega \leqslant \frac{\gamma M_x^2}{1 + (\gamma - 1) M_x^2} \tag{8.47}$$

where $M_x = u/a$. This condition on the streamwise pressure gradient is derived by assuming that the normal component of velocity (v) is much smaller than the streamwise component (u).

We next consider the viscous limit by ignoring terms in Eq. (8.41) containing first derivatives with respect to y. The resulting equations can be written as

$$[A_2]\mathbf{Q}_x = [B_2]\mathbf{Q}_{yy} \tag{8.48}$$

where

$$[A_2] = \begin{bmatrix} u & \rho & 0 & 0 \\ u^2 & 2\rho u & 0 & \omega \\ uv & \rho v & \rho u & 0 \\ \dfrac{u(u^2 + v^2)}{2} & \dfrac{\gamma p}{\gamma - 1} + \dfrac{\rho(3u^2 + v^2)}{2} & \rho u v & \dfrac{\gamma u}{\gamma - 1} \end{bmatrix} \tag{8.49}$$

$$[B_2] = \mu \begin{bmatrix} 0 & 0 & 0 & 0 \\ 0 & 1 & 0 & 0 \\ 0 & 0 & \dfrac{4}{3} & 0 \\ \dfrac{-\gamma p}{(\gamma - 1)\rho^2 \Pr} & u & \dfrac{4}{3} v & \dfrac{\gamma}{(\gamma - 1)\rho \Pr} \end{bmatrix}$$

These equations are parabolic in the positive x direction if the eigenvalues of $[A_2]^{-1}[B_2]$ are real and positive (see Section 2.5). The eigenvalues must be positive in order for a positive viscosity to produce damping in the streamwise direction. The eigenvalues can be found from the following polynomial (assuming $u \neq 0$):

$$\lambda\left(\frac{\rho u}{\mu}\lambda - \frac{4}{3}\right)\left[\left(\frac{\rho u}{\mu}\lambda\right)^2 \{M_x^2[\gamma - \omega(\gamma - 1)] - \omega\} \right.$$
$$\left. + \left(\frac{\rho u}{\mu}\lambda\right)\left\{\left[\omega(\gamma - 1) - \gamma\left(\frac{1 + \Pr}{\Pr}\right)\right] M_x^2 + \frac{\omega}{\Pr}\right\} + \frac{\gamma M_x^2}{\Pr}\right] = 0 \tag{8.50}$$

Vigneron et al. (1978a) have shown that the eigenvalues determined from this equation will be real and positive if

$$u > 0 \tag{8.51}$$

and

$$\omega < \frac{\gamma M_x^2}{1 + (\gamma - 1)M_x^2} \tag{8.52}$$

Equation (8.51) prohibits reverse flows, while Eq. (8.52) places a restriction on the streamwise pressure gradient term in an identical manner to that given previously by Eq. (8.47). From this, we can conclude that the instability caused by the presence of the streamwise pressure gradient term in the PNS equations is actually an inviscid phenomenon.

Note that the right-hand side of Eq. (8.52), denoted by $f(M_x)$, is a function of the local streamwise Mach number (M_x) and becomes equal to 1 when $M_x = 1$ and is greater than 1 when $M_x > 1$ (see Fig. 8.4). Hence the streamwise pressure gradient term can be included fully when $M_x > 1$. However, when $M_x < 1$, only a fraction of this term (i.e., $\omega\, \partial p/\partial x$) can be retained if the

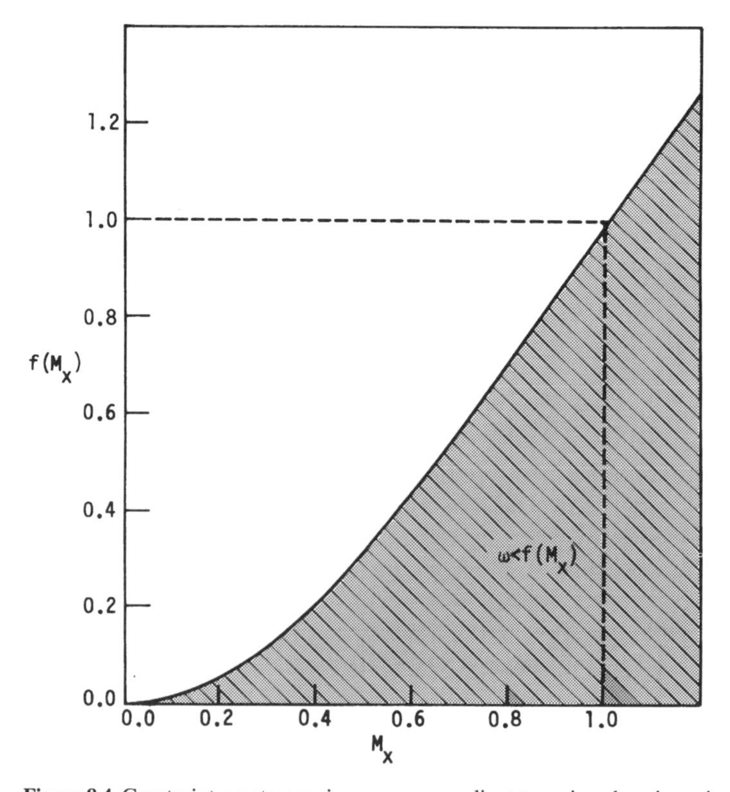

Figure 8.4 Constraint on streamwise pressure gradient term in subsonic regions.

eigenvalues are to remain real and positive. Note also that ω approaches zero close to a wall where $M_x = 0$. Thus we see that space-marched solutions of the PNS equations are subject to instabilities (departure solutions) when the streamwise pressure gradient term is retained fully in the subsonic portion of the boundary layer, since an "elliptic-like" behavior is introduced. A number of different techniques have been proposed to circumvent this difficulty, and they will now be discussed.

The obvious technique is to drop completely the streamwise pressure gradient term in subsonic regions. This will produce a stable marching scheme but will introduce errors in flow fields with large streamwise pressure gradients. It should be noted, however, that streamwise pressure variations will still exist in the numerical solution being evaluated through the y momentum equation and the energy equation. An alternative procedure is to specify the variation of the streamwise pressure gradient. Obviously, setting the pressure gradient equal to zero is just one of many ways that this can be done. If the streamwise pressure gradient is specified, we can remove this term from matrices $[A_1]$ and $[A_2]$ in Eqs. (8.44) and (8.48) and treat it as a source term in the eigenvalue analyses. As a consequence, the streamwise pressure gradient will not affect the mathematical character of the equations. For the solution of the boundary-layer equations, the streamwise pressure gradient is usually known either from the external inviscid flow or, for the case of internal flows, from the conservation-of-mass law. Unfortunately, for the flow fields normally computed with the PNS equations, the streamwise pressure gradient is not known a priori but must be computed as part of the solution.

In several studies the streamwise pressure gradient term has been retained in the subsonic viscous region by employing a backward-difference formula, which uses information from the previous marching step. For example, when the solution at the $i+1$ station is computed, $\partial p/\partial x$ can be evaluated from

$$\frac{\partial p}{\partial x} \cong \frac{p_i - p_{i-1}}{\Delta x} \tag{8.53}$$

which is a first-order backward-difference expression. Lubard and Helliwell (1973) studied the stability (departure behavior) of using a backward-difference formula for the streamwise pressure gradient term in both the momentum and energy equations. They applied a simple implicit differencing scheme to the PNS equations and used a Fourier stability analysis to show that an instability will occur if

$$\Delta x < (\Delta x)_{\min} \tag{8.54}$$

This stability condition is highly unusual, since we normally find from a Fourier stability analysis that an instability occurs when Δx is greater than some $(\Delta x)_{\max}$. When this analysis is applied to the 2-D PNS equations given by Eqs.

(8.48)–(8.49), $(\Delta x)_{\min}$ is given by

$$(\Delta x)_{\min} = \frac{\frac{1}{4}(\rho u/\mu)\left[(1/M_x^2) - 1\right](\Delta y)^2}{\gamma \sin^2(\beta/2)} \tag{8.55}$$

where β is the wave number $(k_m \, \Delta y)$. Lubard and Helliwell have also shown that if the streamwise pressure gradient term is differenced implicitly, like the rest of the terms in the PNS equations when simple implicit differencing is applied, the minimum allowable step size $(\Delta x)_{\min}$ is doubled. In order to explain these unusual stability conditions, Rubin (1981) has observed that $(\Delta x)_{\min}$ appears to represent the extent of the upstream elliptic interaction. If $(\Delta x) > (\Delta x)_{\min}$, the interaction is overstepped, and a forward marching procedure is stable. On the other hand, if $(\Delta x) < (\Delta x)_{\min}$, the numerical solution attempts to represent the elliptic interaction, and this leads to departure solutions, since upstream effects are not permitted by a forward-marched solution. Rubin and Lin (1980) have shown that the extent of the elliptic interaction region is of the order of the thickness of the subsonic region. Thus, if the subsonic region is relatively large, the minimum allowable Δx may be too large to permit accurate (or stable) calculations.

Another method that has been used to treat the streamwise pressure gradient term is called the "*sublayer approximation*" technique. This method was originally proposed by Rubin and Lin (1971) and later applied to the PNS equations by Schiff and Steger (1979). In the sublayer approximation technique, the pressure gradient term in the subsonic viscous region is calculated at a supersonic point outside of the sublayer region. This approximation is based on the fact that for a thin subsonic viscous layer, $\partial p/\partial y$ is negligible. Since the pressure gradient is specified in the subsonic region, it would appear that this technique would lead to stable space-marched solutions. However, it has been observed by Schiff and Steger that departure solutions still exist for some cases. This may be due to the pressure interaction between the supersonic and subsonic regions, which is permitted by the normal momentum equation and the energy equation.

A novel technique for handling the streamwise pressure gradient term was proposed by Vigneron et al. (1978a). In this approach, a fraction of the pressure gradient term $\omega(\partial p/\partial x)$ in the streamwise momentum equation is retained in the subsonic viscous region, and the remainder $(1 - \omega)(\partial p/\partial x)$ is usually omitted or is evaluated explicitly using a backward-difference formula or the "sublayer approximation" technique. For the Vigneron approach, Eq. (8.41) is rewritten as

$$\frac{\partial \mathbf{E}}{\partial x} + \frac{\partial \mathbf{P}}{\partial x} + \frac{\partial \mathbf{F}}{\partial y} = \frac{\partial \mathbf{F}_v}{\partial y} \tag{8.56}$$

where

$$\mathbf{P} = \begin{bmatrix} 0 \\ (1 - \omega)p \\ 0 \\ 0 \end{bmatrix} \tag{8.57}$$

and $\mathbf{E}$, $\mathbf{F}$, and $\mathbf{F}_v$ are defined in Eqs. (8.42). The parameter ω is computed using Eq. (8.47) with a safety factor σ applied:

$$\omega = \frac{\sigma \gamma M_x^2}{1 + (\gamma - 1) M_x^2} \tag{8.58}$$

Vigneron et al. (1978b) have used a Fourier stability analysis to study the "departure behavior" of this technique. They applied the simple implicit (Euler implicit) scheme to Eq. (8.56), with $\partial \mathbf{F}/\partial y$ omitted, and used a backward difference for $\partial \mathbf{P}/\partial x$. As expected, they found that if the "elliptic" pressure gradient term $\partial \mathbf{P}/\partial x$ is omitted, this technique will always lead to a stable space-marched solution, since the equations remain hyperbolic-parabolic. However, if this term is retained, an instability results if Δx is less than some $(\Delta x)_{\min}$. For $\omega = 0$, it was found that $(\Delta x)_{\min}$ is given by Eq. (8.55), which confirms the previous findings of Lubard and Helliwell. Thus it is obvious that in order to completely eliminate departure solutions, it is necessary to drop the term $(1 - \omega)(\partial p/\partial x)$ in subsonic regions when solving the PNS equations with a single marching sweep of the flow field. Other techniques for treating the streamwise pressure gradient term include those proposed by Lin and Rubin (1979), Buggeln et al. (1980), Yanenko et al. (1980), and Bhutta and Lewis (1985a).

For many flow problems the upstream elliptic effects are relatively small and the techniques described above will successfully prevent departure solutions while permitting an accurate solution to be computed with a single marching sweep through the flow field. For other problems where the upstream influence is significant (due to separation, wakes, shocks, etc.), the above techniques may prove to be inadequate. Either a departure solution results or the inconsistency introduced into the PNS equations to prevent the departure solution will lead to large errors. For these cases, a *global pressure relaxation procedure* (Rubin and Lin, 1980) can be used. In this procedure, an initial pressure distribution is used to determine the pressure gradient at each point in the elliptic region. The initial pressure distribution can be obtained by either setting the streamwise pressure gradient equal to zero, by using the "Vigneron" technique with $\partial \mathbf{P}/\partial x = 0$, or by taking a sufficiently large Δx. With the pressure gradient known, the PNS equations can be solved in a stable manner using a space-marching technique, provided that the pressure gradient term is differenced in an appropriate manner. The resulting solution will contain a new pressure distribution that can be used to determine the pressure gradient for the next sweep of the flow field. This iteration procedure is continued until the solution converges. In order for the elliptic character of the flow field to be properly modeled, the pressure gradient term must introduce downstream contributions. This can be accomplished by applying appropriate "upwind" differences to the "elliptic" and "hyperbolic" components of the pressure gradient term:

$$\frac{\partial p}{\partial x} = \underbrace{\omega \frac{\partial p}{\partial x}}_{\text{hyperbolic}} + \underbrace{(1 - \omega) \frac{\partial p}{\partial x}}_{\text{elliptic}} \tag{8.59}$$

For example, when the solution at the $i + 1$ station is computed, the streamwise pressure gradient term can be differenced on an equally spaced grid (Rakich, 1983) as

$$\left(\frac{\partial p}{\partial x}\right) \cong \omega\left(\frac{p_{i+1}^{n+1} - p_i^{n+1}}{\Delta x}\right) + (1 - \omega)\left(\frac{p_{i+2}^{n} - p_{i+1}^{n+1}}{\Delta x}\right)$$

where the superscript $n + 1$ indicates the current iteration level. This type of differencing is only possible when the global pressure relaxation procedure is used, since p_{i+2}^n is normally unknown.

The global pressure relaxation procedure can be used for problems where the upstream influence is significant. Although this procedure requires more computer time than a standard PNS calculation that employes one sweep of the flow field, it still offers advantages over the complete Navier-Stokes equations. One of the primary advantages is that only the pressure must be stored (if $u > 0$) during each sweep of the flow field. The global pressure relaxation procedure has been used by several investigators, including Rubin and Lin (1980), Rakich (1983), Barnett and Davis (1986), Khosla and Rubin (1987), Barnett and Power (1988), Power and Barber (1988), Power (1990), Rubin and Khosla (1990), and Miller et al. (1997).

The concept of splitting the streamwise pressure gradient term into its "elliptic" and "hyperbolic" components has led to the introduction of a new set of equations called the reduced Navier-Stokes (RNS) equations (Rubin, 1984). The RNS equations are derived from the complete unsteady Navier-Stokes equations by dropping the streamwise viscous terms, omitting the viscous terms in the normal momentum equation, and splitting the streamwise pressure gradient using Eq. (8.59). In most applications, the time-derivative terms are also omitted. For supersonic flows with embedded "elliptic" regions, the global pressure relaxation procedure (just described) can be used to solve the RNS equations. For subsonic flows the RNS equations are solved using the techniques described in Section 8.4.3 for the PPNS equations.

8.3.3 Numerical Solution of PNS Equations

As discussed previously, the PNS equations are a mixed set of hyperbolic-parabolic equations in the streamwise direction, provided that the following conditions are satisfied.

1. Inviscid flow is supersonic.
2. Streamwise velocity component is everywhere greater than zero.
3. Streamwise pressure gradient term in streamwise momentum equation is either omitted or the "departure behavior" is suppressed using one of the techniques described in the last section.

If these conditions are met, the PNS equations can be solved using methods similar to those employed for the parabolic boundary-layer equations. Thus the solution can be marched downstream in a stable manner from an initial data surface to the desired final station.

Early schemes. Some of the earliest solutions of the PNS equations were obtained using explicit finite-difference techniques (Rudman and Rubin, 1968; Boynton and Thomson, 1969; Rubin et al., 1969; Cresci et al., 1969; Cheng et al., 1970). Explicit schemes were used more for convenience than efficiency, since we have demonstrated in Chapter 7 that implicit methods are much more efficient for equations of this type. In later studies, the PNS equations have been primarily solved using implicit algorithms. Cheng et al. (1970) used the simple implicit scheme, Nardo and Cresci (1971) employed the Peaceman-Rachford alternating direction implicit (ADI) scheme, while Rubin and Lin (1972) and Lubard and Helliwell (1973) used similar iterative-implicit schemes. Rubin and Lin's predictor-corrector multiple iteration scheme is described in Section 4.5.10, where it is applied to the 3-D linear Burgers equation

$$u_x + cu_y + du_z = \mu(u_{yy} + u_{zz}) \tag{8.60}$$

The 3-D linear Burgers equation is a useful model equation for the PNS equations, but of course, it does not represent the nonlinear character of these equations. Thus, when the predictor-corrector multiple-iteration method is applied to the PNS equations, nonlinear terms such as $(u_{i+1,j,k}^{m+1})^2$ appear, where m is the iteration level, $x = i\,\Delta x$, $y = j\,\Delta y$, and $z = k\,\Delta z$. These nonlinear terms are linearized using a Newton-Raphson procedure (see Section 7.3.3). That is, if $f = f(x_1, x_2, \ldots, x_l)$ is a nonlinear term, then

$$f^{m+1} = f^m + \sum_{k=1}^{l} \left(\frac{\partial f}{\partial x_k}\right)^m (x_k^{m+1} - x_k^m) \tag{8.61}$$

where x_k denotes the dependent variables. Applying this formula to the nonlinear term $(u_{i+1,j,k}^{m+1})^2$ gives

$$\left(u_{i+1,j,k}^{m+1}\right)^2 = 2u_{i+1,j,k}^{m+1} u_{i+1,j,k}^{m} - (u_{i+1,j,k}^{m})^2 \tag{8.62}$$

After all the nonlinear terms are linearized in this manner, the resulting set of algebraic equations (at iteration level $m + 1$) can be solved using an efficient block tridiagonal solver. The iteration is continued until the solution converges at the $i + 1$ station. This method is implicit in the y direction, where the gradients are largest, but is explicit in the z direction (see Section 4.5.10), which leads to the following stability condition when applied to the 3-D PNS equations:

$$\Delta x \leqslant \Delta z \left|\frac{w}{u}\right| \tag{8.63}$$

Beam-Warming scheme. Until the latter part of the 1970s, the PNS equations were mainly solved using iterative, implicit finite-difference schemes like the ones described above. Vigneron et al. (1978a) were the first to employ a more efficient noniterative implicit approximate-factorization finite-difference scheme to solve the PNS equations. Their algorithm was adapted from the class of ADI schemes developed by Lindemuth and Killeen (1973), McDonald and Briley

(1975), and Beam and Warming (1978) to solve time-dependent equations such as the Navier-Stokes equations. Working independently of Vigneron et al. (1978a), Schiff and Steger (1979) developed a nearly identical algorithm except that the pressure gradient in the subsonic viscous layer was calculated at a supersonic point outside the layer using the "sublayer approximation" technique. In addition, a different linearization procedure was employed by Schiff and Steger.

In order to explain the Vigneron et al. algorithm, let us apply it to the 3-D PNS equations written in Cartesian coordinates (x is the streamwise direction) for a perfect gas. In this case, the generalized coordinates become

$$\begin{aligned} \xi &= x \\ \eta &= y \\ \zeta &= z \end{aligned} \tag{8.64}$$

and Eqs. (8.37)–(8.38) reduce to

$$\frac{\partial \mathbf{E}}{\partial x} + \frac{\partial \mathbf{F}}{\partial y} + \frac{\partial \mathbf{G}}{\partial z} = 0 \tag{8.65}$$

where

$$\begin{aligned} \mathbf{E} &= \mathbf{E}_i \\ \mathbf{F} &= \mathbf{F}_i - \mathbf{F}_v \\ \mathbf{G} &= \mathbf{G}_i - \mathbf{G}_v \end{aligned} \tag{8.66}$$

The vectors $\mathbf{E}_i$, $\mathbf{F}_i$, $\mathbf{G}_i$, $\mathbf{F}_v$, and $\mathbf{G}_v$ are given by Eqs. (8.35) and contain the following "parabolized" shear stress and heat flux terms:

$$\begin{aligned} \tau_{xx} &= \tfrac{2}{3}\mu(-v_y - w_z) \\ \tau_{yy} &= \tfrac{2}{3}\mu(2v_y - w_z) \\ \tau_{zz} &= \tfrac{2}{3}\mu(2w_z - v_y) \\ \tau_{xy} &= \mu u_y \\ \tau_{xz} &= \mu u_z \\ \tau_{yz} &= \mu(v_z + w_y) \\ q_x &= 0 \\ q_y &= -kT_y \\ q_z &= -kT_z \end{aligned} \tag{8.67}$$

In order to use the "Vigneron" technique for handling the streamwise pressure gradient, $\mathbf{E}$ can be replaced by $\mathbf{E}' + \mathbf{P}$, so that, Eq. (8.65) becomes

$$\frac{\partial \mathbf{E}'}{\partial x} + \frac{\partial \mathbf{P}}{\partial x} + \frac{\partial \mathbf{F}}{\partial y} + \frac{\partial \mathbf{G}}{\partial z} = 0 \tag{8.68}$$

where $\mathbf{E}'$ and $\mathbf{P}$ are given by

$$\mathbf{E}' = \begin{bmatrix} \rho u \\ \rho u^2 + \omega p \\ \rho uv \\ \rho uw \\ (E_t + p)u \end{bmatrix} \qquad \mathbf{P} = \begin{bmatrix} 0 \\ (1-\omega)p \\ 0 \\ 0 \\ 0 \end{bmatrix} \tag{8.69}$$

The solution of Eq. (8.65) is marched in x using the following difference formula suggested by Beam and Warming (1978):

$$\Delta^i \mathbf{E} = \frac{\theta_1 \, \Delta x}{1+\theta_2} \frac{\partial}{\partial x}(\Delta^i \mathbf{E}) + \frac{\Delta x}{1+\theta_2} \frac{\partial}{\partial x}(\mathbf{E}^i) + \frac{\theta_2}{1+\theta_2} \Delta^{i-1}\mathbf{E}$$
$$+ O\left[\left(\theta_1 - \frac{1}{2} - \theta_2\right)(\Delta x)^2 + (\Delta x)^3\right] \tag{8.70}$$

where

$$\Delta^i \mathbf{E} = \mathbf{E}^{i+1} - \mathbf{E}^i \tag{8.71}$$

and $x = i\,\Delta x$. This general difference formula, with the appropriate choice of the parameters θ_1 and θ_2, reproduces many of the standard difference schemes as seen in Table 8.1. For the PNS equations, either the first-order Euler implicit scheme ($\theta_1 = 1, \theta_2 = 0$) or the second-order, three-point backward scheme ($\theta_1 = 1, \theta_2 = \frac{1}{2}$) are normally used. As shown by Beam and Warming, the second-order trapezoidal differencing scheme ($\theta_1 = \frac{1}{2}, \theta_2 = 0$) will lead to unstable calculations when applied to parabolic equations. Note that the truncation error (T.E.) in Table 8.1 is for $\Delta^i \mathbf{E}$. When $\partial \mathbf{E}/\partial x$ is replaced by $\Delta^i \mathbf{E}/\Delta x$ in the numerical scheme, the T.E. is divided by Δx.

Substituting Eq. (8.65) into Eq. (8.70) yields

$$\Delta^i \mathbf{E} = -\frac{\theta_1 \, \Delta x}{1+\theta_2}\left[\frac{\partial}{\partial y}(\Delta^i \mathbf{F}) + \frac{\partial}{\partial z}(\Delta^i \mathbf{G})\right] - \frac{\Delta x}{1+\theta_2}\left[\frac{\partial}{\partial y}(\mathbf{F}^i) + \frac{\partial}{\partial z}(\mathbf{G}^i)\right]$$
$$+ \frac{\theta_2}{1+\theta_2}\Delta^{i-1}\mathbf{E} \tag{8.72}$$

with the T.E. term omitted. The difference formula is in the so-called "delta" form as discussed in Section 4.4.7. The delta terms $\Delta^i \mathbf{E}$, $\Delta^i \mathbf{F}$, and $\Delta^i \mathbf{G}$, which can

Table 8.1 Finite-difference schemes contained in Eq. (8.70)

θ_1	θ_2	Scheme	Truncation error in Eq. (8.70)
0	0	Euler explicit	$O[(\Delta x)^2]$
0	$-\frac{1}{2}$	Leap frog (explicit)	$O[(\Delta x)^3]$
$\frac{1}{2}$	0	Trapezoidal (implicit)	$O[(\Delta x)^3]$
1	0	Euler implicit	$O[(\Delta x)^2]$
1	$\frac{1}{2}$	Three-point backward (implicit)	$O[(\Delta x)^3]$

be written as

$$\begin{aligned} \Delta^i \mathbf{E} &= \Delta^i \mathbf{E}' + \Delta^i \mathbf{P} \\ \Delta^i \mathbf{F} &= \Delta^i \mathbf{F}_i - \Delta^i \mathbf{F}_v \\ \Delta^i \mathbf{G} &= \Delta^i \mathbf{G}_i - \Delta^i \mathbf{G}_v \end{aligned} \tag{8.73}$$

are linearized using truncated Taylor-series expansions. In order to linearize the inviscid delta terms $\Delta^i \mathbf{E}'$, $\Delta^i \mathbf{F}_i$, and $\Delta^i \mathbf{G}_i$, we make use of the fact that $\mathbf{E}'$, $\mathbf{F}_i$, and $\mathbf{G}_i$ are functions only of the $\mathbf{U}$ vector,

$$\mathbf{U} = \begin{bmatrix} \rho \\ \rho u \\ \rho v \\ \rho w \\ E_t \end{bmatrix} = \begin{bmatrix} U_1 \\ U_2 \\ U_3 \\ U_4 \\ U_5 \end{bmatrix} \tag{8.74}$$

For example, $\mathbf{F}_i$ can be expressed as

$$\mathbf{F}_i = \begin{bmatrix} U_3 \\ \dfrac{U_2 U_3}{U_1} \\ \dfrac{U_3^2}{U_1} + (\gamma - 1)\left(U_5 - \dfrac{U_2^2 + U_3^2 + U_4^2}{2U_1} \right) \\ \dfrac{U_3 U_4}{U_1} \\ \left[U_5 + (\gamma - 1)\left(U_5 - \dfrac{U_2^2 + U_3^2 + U_4^2}{2U_1} \right) \right] \dfrac{U_3}{U_1} \end{bmatrix} \tag{8.75}$$

As a consequence, we can readily expand $\mathbf{E}'$, $\mathbf{F}_i$, and $\mathbf{G}_i$ as

$$\begin{aligned} (\mathbf{E}')^{i+1} &= (\mathbf{E}')^i + \left(\frac{\partial \mathbf{E}'}{\partial \mathbf{U}} \right)^i \Delta^i \mathbf{U} + O[(\Delta x)^2] \\ (\mathbf{F}_i)^{i+1} &= (\mathbf{F}_i)^i + \left(\frac{\partial \mathbf{F}_i}{\partial \mathbf{U}} \right)^i \Delta^i \mathbf{U} + O[(\Delta x)^2] \\ (\mathbf{G}_i)^{i+1} &= (\mathbf{G}_i)^i + \left(\frac{\partial \mathbf{G}_i}{\partial \mathbf{U}} \right)^i \Delta^i \mathbf{U} + O[(\Delta x)^2] \end{aligned} \tag{8.76}$$

or

$$\begin{aligned} \Delta^i \mathbf{E}' &= [Q]^i \, \Delta^i \mathbf{U} + O[(\Delta x)^2] \\ \Delta^i \mathbf{F}_i &= [R]^i \, \Delta^i \mathbf{U} + O[(\Delta x)^2] \\ \Delta^i \mathbf{G}_i &= [S]^i \, \Delta^i \mathbf{U} + O[(\Delta x)^2] \end{aligned} \tag{8.77}$$

where $[Q]$, $[R]$, and $[S]$ are the Jacobian matrices $\partial \mathbf{E}'/\partial \mathbf{U}$, $\partial \mathbf{F}_i/\partial \mathbf{U}$, and $\partial \mathbf{G}_i/\partial \mathbf{U}$ given by

$$\frac{\partial \mathbf{E}'}{\partial \mathbf{U}} = \begin{bmatrix} 0 & 1 & 0 & 0 & 0 \\ \dfrac{\omega(\gamma-1)-2}{2}u^2 + \dfrac{\omega(\gamma-1)}{2}(v^2+w^2) & [2-\omega(\gamma-1)]u & -\omega(\gamma-1)v & -\omega(\gamma-1)w & \omega(\gamma-1) \\ -uv & v & u & 0 & 0 \\ -uw & w & 0 & u & 0 \\ \left[-\dfrac{\gamma E_t}{\rho} + (\gamma-1)(u^2+v^2+w^2)\right]u & \dfrac{\gamma E_t}{\rho} - (\gamma-1)\dfrac{3u^2+v^2+w^2}{2} & -(\gamma-1)uv & -(\gamma-1)uw & \gamma u \end{bmatrix} \tag{8.78}$$

$$\frac{\partial \mathbf{F}_i}{\partial \mathbf{U}} = \begin{bmatrix} 0 & 0 & 1 & 0 & 0 \\ -uv & v & u & 0 & 0 \\ \dfrac{\gamma-1}{2}(u^2+w^2) + \dfrac{\gamma-3}{2}v^2 & -(\gamma-1)u & (3-\gamma)v & -(\gamma-1)w & \gamma-1 \\ -vw & 0 & w & v & 0 \\ \left[-\dfrac{\gamma E_t}{\rho} + (\gamma-1)(u^2+v^2+w^2)\right]v & -(\gamma-1)uv & \dfrac{\gamma E_t}{\rho} - \dfrac{\gamma-1}{2}(u^2+3v^2+w^2) & -(\gamma-1)vw & \gamma v \end{bmatrix} \tag{8.79}$$

$$\frac{\partial \mathbf{G}_i}{\partial \mathbf{U}} = \begin{bmatrix} 0 & 0 & 0 & 1 & 0 \\ -uw & w & 0 & u & 0 \\ -vw & 0 & w & v & 0 \\ \dfrac{\gamma-1}{2}(u^2+v^2) + \dfrac{\gamma-3}{2}w^2 & -(\gamma-1)u & -(\gamma-1)v & (3-\gamma)w & \gamma-1 \\ \left[-\dfrac{\gamma E_t}{\rho} + (\gamma-1)(u^2+v^2+w^2)\right]w & -(\gamma-1)uw & -(\gamma-1)vw & \dfrac{\gamma E_t}{\rho} - \dfrac{\gamma-1}{2}(u^2+v^2+3w^2) & \gamma w \end{bmatrix} \tag{8.80}$$

The expression for the Jacobian $\partial \mathbf{E}'/\partial \mathbf{U}$ is derived by assuming ω to be locally independent of $\mathbf{U}$.

The viscous delta terms can be linearized using a method suggested by Steger (1977). In order to apply this linearization method, the coefficients of viscosity (μ) and the thermal conductivity (k) are assumed to be locally independent of $\mathbf{U}$ and the cross-derivative viscous terms are neglected. As a result of these assumptions, elements of $\mathbf{F}_v$ and $\mathbf{G}_v$ have the general form

$$\begin{aligned} f_k &= \alpha_k \frac{\partial}{\partial y}(\beta_k) \\ g_k &= \alpha_k \frac{\partial}{\partial z}(\beta_k) \end{aligned} \tag{8.81}$$

where α_k is independent of $\mathbf{U}$ and β_k is a function of $\mathbf{U}$. These elements are linearized in the following manner:

$$\begin{aligned} f^{i+1} &= f^i + \alpha_k^i \frac{\partial}{\partial y}\left[\sum_{l=1}^{5}\left(\frac{\partial \beta_k}{\partial U_l}\right)^i \Delta^i U_l\right] + O[(\Delta x)^2] \\ g^{i+1} &= g^i + \alpha_k^i \frac{\partial}{\partial z}\left[\sum_{l=1}^{5}\left(\frac{\partial \beta_k}{\partial U_l}\right)^i \Delta^i U_l\right] + O[(\Delta x)^2] \end{aligned} \tag{8.82}$$

so that we can write

$$\begin{aligned} \Delta^i \mathbf{F}_v &= [V]^i \, \Delta^i \mathbf{U} + O[(\Delta x)^2] \\ \Delta^i \mathbf{G}_v &= [W]^i \, \Delta^i \mathbf{U} + O[(\Delta x)^2] \end{aligned} \tag{8.83}$$

where $[V]$ and $[W]$ are the Jacobian matrices $\partial \mathbf{F}_v/\partial \mathbf{U}$ and $\partial \mathbf{G}_v/\partial \mathbf{U}$ given by

$$\frac{\partial \mathbf{F}_v}{\partial \mathbf{U}} = \mu \begin{bmatrix} 0 & 0 & 0 & 0 & 0 \\ -\partial_y\left(\frac{u}{\rho}\right) & \partial_y\left(\frac{1}{\rho}\right) & 0 & 0 & 0 \\ -\frac{4}{3}\partial_y\left(\frac{v}{\rho}\right) & 0 & \frac{4}{3}\partial_y\left(\frac{1}{\rho}\right) & 0 & 0 \\ -\partial_y\left(\frac{w}{\rho}\right) & 0 & 0 & \partial_y\left(\frac{1}{\rho}\right) & 0 \\ \begin{array}{c} -\partial_y\left(\frac{u^2}{\rho}\right) - \frac{4}{3}\partial_y\left(\frac{v^2}{\rho}\right) - \partial_y\left(\frac{w^2}{\rho}\right) \\ -\frac{\gamma}{\Pr}\partial_y\left[\frac{p}{(\gamma-1)\rho^2} - \frac{u^2+v^2+w^2}{2\rho}\right] \end{array} & \left(1-\frac{\gamma}{\Pr}\right)\partial_y\left(\frac{u}{\rho}\right) & \left(\frac{4}{3}-\frac{\gamma}{\Pr}\right)\partial_y\left(\frac{v}{\rho}\right) & \left(1-\frac{\gamma}{\Pr}\right)\partial_y\left(\frac{w}{\rho}\right) & \frac{\gamma}{\Pr}\partial_y\left(\frac{1}{\rho}\right) \end{bmatrix} \tag{8.84}$$

$$\frac{\partial \mathbf{G}_v}{\partial \mathbf{U}} = \mu \begin{bmatrix} 0 & 0 & 0 & 0 & 0 \\ -\partial_z\left(\frac{u}{\rho}\right) & \partial_z\left(\frac{1}{\rho}\right) & 0 & 0 & 0 \\ -\partial_z\left(\frac{v}{\rho}\right) & 0 & \partial_z\left(\frac{1}{\rho}\right) & 0 & 0 \\ -\frac{4}{3}\partial_z\left(\frac{w}{\rho}\right) & 0 & 0 & \frac{4}{3}\partial_z\left(\frac{1}{\rho}\right) & 0 \\ \begin{array}{c} -\partial_z\left(\frac{u^2}{\rho}\right) - \partial_z\left(\frac{v^2}{\rho}\right) - \frac{4}{3}\partial_z\left(\frac{w^2}{\rho}\right) \\ -\frac{\gamma}{\Pr}\partial_z\left[\frac{p}{(\gamma-1)\rho^2} - \frac{u^2+v^2+w^2}{2\rho}\right] \end{array} & \left(1-\frac{\gamma}{\Pr}\right)\partial_z\left(\frac{u}{\rho}\right) & \left(1-\frac{\gamma}{\Pr}\right)\partial_z\left(\frac{v}{\rho}\right) & \left(\frac{4}{3}-\frac{\gamma}{\Pr}\right)\partial_z\left(\frac{w}{\rho}\right) & \frac{\gamma}{\Pr}\partial_z\left(\frac{1}{\rho}\right) \end{bmatrix} \tag{8.85}$$

In these Jacobian matrices, ∂_y and ∂_z represent the partial derivatives $\partial/\partial y$ and $\partial/\partial z$.

We now substitute Eqs. (8.73), (8.77), and (8.83) into Eq. (8.72) to obtain

$$\left\{\left(\frac{\partial \mathbf{E}'}{\partial \mathbf{U}}\right)^i + \frac{\theta_1 \Delta x}{1+\theta_2}\left[\frac{\partial}{\partial y}\left(\frac{\partial \mathbf{F}_i}{\partial \mathbf{U}} - \frac{\partial \mathbf{F}_v}{\partial \mathbf{U}}\right) + \frac{\partial}{\partial z}\left(\frac{\partial \mathbf{G}_i}{\partial \mathbf{U}} - \frac{\partial \mathbf{G}_v}{\partial \mathbf{U}}\right)\right]^i\right\} \Delta^i \mathbf{U}$$
$$= -\frac{\Delta x}{1+\theta_2}\left[\frac{\partial}{\partial y}(\mathbf{F}^i) + \frac{\partial}{\partial z}(\mathbf{G}^i)\right] + \frac{\theta_2}{1+\theta_2}\Delta^{i-1}\mathbf{E} - \Delta^i \mathbf{P} \tag{8.86}$$

where the expression

$$\left[\frac{\partial}{\partial y}\left(\frac{\partial \mathbf{F}_i}{\partial \mathbf{U}} - \frac{\partial \mathbf{F}_v}{\partial \mathbf{U}}\right)\right]\Delta^i \mathbf{U}$$

implies

$$\frac{\partial}{\partial y}\left[\left(\frac{\partial \mathbf{F}_i}{\partial \mathbf{U}} - \frac{\partial \mathbf{F}_v}{\partial \mathbf{U}}\right)\Delta^i \mathbf{U}\right]$$

and the partial derivatives appearing in $\partial \mathbf{F}_v/\partial \mathbf{U}$ and $\partial \mathbf{G}_v/\partial \mathbf{U}$ are to be applied to all terms on their right, including $\Delta^i \mathbf{U}$. Note that in Eq. (8.86), all the implicit terms have been placed on the left-hand side of the equation, while all the explicit terms appear on the right-hand side. Included in the right-hand side of the equation is the pressure gradient term $\Delta^i \mathbf{P}$, which must be dropped in subsonic regions to avoid departure solutions when marching the solution with a single sweep of the flow field.

The left-hand side of Eq. (8.86) is approximately factored in the following manner:

$$\left\{\left[\left(\frac{\partial \mathbf{E}'}{\partial \mathbf{U}}\right)^i + \frac{\theta_1 \Delta x}{1+\theta_2}\frac{\partial}{\partial z}\left(\frac{\partial \mathbf{G}_i}{\partial \mathbf{U}} - \frac{\partial \mathbf{G}_v}{\partial \mathbf{U}}\right)^i\right]\left[\left(\frac{\partial \mathbf{E}'}{\partial \mathbf{U}}\right)^i\right]^{-1}\right.$$
$$\left.\times\left[\left(\frac{\partial \mathbf{E}'}{\partial \mathbf{U}}\right)^i + \frac{\theta_1 \Delta x}{1+\theta_2}\frac{\partial}{\partial y}\left(\frac{\partial \mathbf{F}_i}{\partial \mathbf{U}} - \frac{\partial \mathbf{F}_v}{\partial \mathbf{U}}\right)^i\right]\right\}\Delta^i \mathbf{U} = \text{RHS of Eq. (8.86)} \tag{8.87}$$

The order of accuracy of this factored expression can be determined by multiplying out the factored terms and comparing the result with the left-hand side of Eq. (8.86). Upon doing this, we obtain

$$\left\{\left(\frac{\partial \mathbf{E}'}{\partial \mathbf{U}}\right)^i + \frac{\theta_1 \Delta x}{1+\theta_2}\left[\frac{\partial}{\partial y}\left(\frac{\partial \mathbf{F}_i}{\partial \mathbf{U}} - \frac{\partial \mathbf{F}_v}{\partial \mathbf{U}}\right) + \frac{\partial}{\partial z}\left(\frac{\partial \mathbf{G}_i}{\partial \mathbf{U}} - \frac{\partial \mathbf{G}_v}{\partial \mathbf{U}}\right)\right]^i\right.$$
$$\left.+\left(\frac{\theta_1 \Delta x}{1+\theta_2}\right)^2 \frac{\partial}{\partial z}\left(\frac{\partial \mathbf{G}_i}{\partial \mathbf{U}} - \frac{\partial \mathbf{G}_v}{\partial \mathbf{U}}\right)^i\left[\left(\frac{\partial \mathbf{E}'}{\partial \mathbf{U}}\right)^i\right]^{-1}\frac{\partial}{\partial y}\left(\frac{\partial \mathbf{F}_i}{\partial \mathbf{U}} - \frac{\partial \mathbf{F}_v}{\partial \mathbf{U}}\right)^i\right\}\Delta^i \mathbf{U}$$
$$= \text{RHS of Eq. (8.86)} \tag{8.88}$$

so that

$$\text{LHS of Eq. (8.87)} = \text{LHS of Eq. (8.86)} + O[(\Delta x)^2] \tag{8.89}$$

As a consequence, the formal accuracy of the numerical algorithm is not affected by the approximate factorization. However, it has been observed that approximate factorization may lead to sizable errors in certain computations.

The partial derivatives $\partial/\partial y$ and $\partial/\partial z$ in Eq. (8.87) are approximated with second-order accurate central differences. For example, the inviscid term

$$\frac{\partial}{\partial y}\left(\frac{\partial \mathbf{F}_i}{\partial \mathbf{U}}\right)^i \Delta^i \mathbf{U}$$

is differenced as

$$\frac{\left[(\partial \mathbf{F}_i/\partial \mathbf{U})^i\, \Delta^i \mathbf{U}\right]_{j+1} - \left[(\partial \mathbf{F}_i/\partial \mathbf{U})^i\, \Delta^i \mathbf{U}\right]_{j-1}}{2\,\Delta y} \tag{8.90}$$

and each element of the viscous term,

$$\frac{\partial}{\partial y}\left(\frac{\partial \mathbf{F}_v}{\partial \mathbf{U}}\right)^i \Delta^i \mathbf{U}$$

which has the general form

$$\frac{\partial}{\partial y}\left[\alpha \frac{\partial}{\partial y}(\beta\, \Delta^i U_l)\right]$$

is differenced as

$$\begin{aligned}
&\frac{\{\alpha[\partial(\beta\,\Delta^i U_l)/\partial y]\}_{j+1/2} - \{\alpha[\partial(\beta\,\Delta^i U_l)/\partial y]\}_{j-1/2}}{\Delta y} \\
&\quad\cong \frac{\alpha_{j+1/2}[(\beta\,\Delta^i U_l)_{j+1} - (\beta\,\Delta^i U_l)_j] - \alpha_{j-1/2}[(\beta\,\Delta^i U_l)_j - (\beta\,\Delta^i U_l)_{j-1}]}{(\Delta y)^2} \\
&\quad\cong \frac{(\alpha_j + \alpha_{j+1})[(\beta\,\Delta^i U_l)_{j+1} - (\beta\,\Delta^i U_l)_j] - (\alpha_j + \alpha_{j-1})[(\beta\,\Delta^i U_l)_j - (\beta\,\Delta^i U_l)_{j-1}]}{2(\Delta y)^2}
\end{aligned} \tag{8.91}$$

The algorithm given by Eq. (8.87) is implemented in the following manner:

Step 1:

$$\left[\left(\frac{\partial \mathbf{E}'}{\partial \mathbf{U}}\right)^i + \frac{\theta_1\,\Delta x}{1+\theta_2}\frac{\partial}{\partial z}\left(\frac{\partial \mathbf{G}_i}{\partial \mathbf{U}} - \frac{\partial \mathbf{G}_v}{\partial \mathbf{U}}\right)^i\right]\Delta^i \mathbf{U}_1 = \text{RHS of Eq. (8.86)} \tag{8.92}$$

Step 2:
$$\Delta^i \mathbf{U}_2 = \left(\frac{\partial \mathbf{E}'}{\partial \mathbf{U}}\right)^i \Delta^i \mathbf{U}_1 \tag{8.93}$$

Step 3:
$$\left[\left(\frac{\partial \mathbf{E}'}{\partial \mathbf{U}}\right)^i + \frac{\theta_1 \, \Delta x}{1+\theta_2} \frac{\partial}{\partial y}\left(\frac{\partial \mathbf{F}_i}{\partial \mathbf{U}} - \frac{\partial \mathbf{F}_v}{\partial \mathbf{U}}\right)^i\right] \Delta^i \mathbf{U} = \Delta^i \mathbf{U}_2 \tag{8.94}$$

Step 4:
$$\mathbf{U}^{i+1} = \mathbf{U}^i + \Delta^i \mathbf{U} \tag{8.95}$$

In Step 1, $\Delta^i \mathbf{U}_1$ represents the vector quantity

$$\left[\left(\frac{\partial \mathbf{E}'}{\partial \mathbf{U}}\right)^i\right]^{-1} \left[\left(\frac{\partial \mathbf{E}'}{\partial \mathbf{U}}\right)^i + \frac{\theta_1 \, \Delta x}{1+\theta_2} \frac{\partial}{\partial y}\left(\frac{\partial \mathbf{F}_i}{\partial \mathbf{U}} - \frac{\partial \mathbf{F}_v}{\partial \mathbf{U}}\right)^i\right] \Delta^i \mathbf{U}$$

which is determined by solving the system of equations given by Eq. (8.92). This system of equations has the following block tridiagonal structure:

$$\begin{bmatrix} [B_1] & [C_1] & \cdots & & & 0 \\ [A_2] & [B_2] & [C_2] & & & \vdots \\ \vdots & [A_3] & [B_3] & [C_3] & & \vdots \\ \vdots & & \ddots & \ddots & \ddots & \vdots \\ \vdots & & & [A_{K-1}] & [B_{K-1}] & [C_{K-1}] \\ 0 & \cdots & & & [A_K] & [B_K] \end{bmatrix} \begin{bmatrix} [\Delta^i U_1]_1 \\ [\Delta^i U_1]_2 \\ [\Delta^i U_1]_3 \\ \vdots \\ [\Delta^i U_1]_{K-1} \\ [\Delta^i U_1]_K \end{bmatrix} = \begin{bmatrix} [\text{RHS}]_1 \\ [\text{RHS}]_2 \\ [\text{RHS}]_3 \\ \vdots \\ [\text{RHS}]_{K-1} \\ [\text{RHS}]_K \end{bmatrix} \tag{8.96}$$

where $[A]$, $[B]$, and $[C]$ are 5×5 matrices and $[\Delta^i \mathbf{U}_1]$ and [RHS] are column matrices whose elements are the components of the vectors $\Delta^i \mathbf{U}_1$ and the RHS of Eq. (8.86). This system of equations can be solved using the block tridiagonal solver given in Appendix B. Once $\Delta^i \mathbf{U}_1$ is determined, it is multiplied by $(\partial \mathbf{E}'/\partial \mathbf{U})^i$ in Step 2. As a result of this multiplication, the inverse matrix $[(\partial \mathbf{E}'/\partial \mathbf{U})^i]^{-1}$ does not have to be determined in the solution process. In Step 3, the block tridiagonal system of equations in the y direction is solved. Finally, in Step 4, the vector of unknowns at station $i+1$ (i.e., $\mathbf{U}^{i+1}$) is determined by simply adding $\Delta^i \mathbf{U}$ to the vector of unknowns at station i. The primitive variables can then be obtained from $\mathbf{U}^{i+1}$ in the following manner:

$$
\begin{aligned}
\rho^{i+1} &= U_1^{i+1} \\
u^{i+1} &= \frac{U_2^{i+1}}{U_1^{i+1}} \\
v^{i+1} &= \frac{U_3^{i+1}}{U_1^{i+1}} \\
w^{i+1} &= \frac{U_4^{i+1}}{U_1^{i+1}} \\
e^{i+1} &= \frac{U_5^{i+1}}{U_1^{i+1}} - \frac{(u^{i+1})^2 + (v^{i+1})^2 + (w^{i+1})^2}{2}
\end{aligned}
\tag{8.97}
$$

For centrally differenced algorithms like the present one, it is often necessary to add smoothing (artificial viscosity) in order to suppress high-frequency oscillations. This can easily be accomplished by adding a fourth-order explicit dissipation term of the form

$$
-\epsilon_e \left[(\Delta y)^4 \frac{\partial^4}{\partial y^4}(\mathbf{U}^i) + (\Delta z)^4 \frac{\partial^4}{\partial z^4}(\mathbf{U}^i) \right] \tag{8.98}
$$

to the right-hand side of Eq. (8.86). Since this is a fourth-order term, it does not affect the formal accuracy of the algorithm. The negative sign is required in front of the fourth derivatives in order to produce positive damping [see Eq. (4.21)]. The smoothing coefficient ϵ_e should be less than approximately $\frac{1}{16}$ for stability. The fourth-derivative terms can be evaluated using the following finite-difference approximations:

$$
\begin{aligned}
(\Delta y)^4 \frac{\partial^4}{\partial y^4}(\mathbf{U}^i) &\cong \mathbf{U}^i_{j+2,k} - 4\mathbf{U}^i_{j+1,k} + 6\mathbf{U}^i_{j,k} - 4\mathbf{U}^i_{j-1,k} + \mathbf{U}^i_{j-2,k} \\
(\Delta z)^4 \frac{\partial^4}{\partial z^4}(\mathbf{U}^i) &\cong \mathbf{U}^i_{j,k+2} - 4\mathbf{U}^i_{j,k+1} + 6\mathbf{U}^i_{j,k} - 4\mathbf{U}^i_{j,k-1} + \mathbf{U}^i_{j,k-2}
\end{aligned}
\tag{8.99}
$$

In the Schiff and Steger (1979) algorithm as well as that developed by Vigneron et al. (1978a), the solution is advanced using computational planes (i.e., solution surfaces) normal to the body axis. Most body shapes can be treated in this manner. However, for bodies with large surface slopes, the axial component of velocity in the inviscid part of the flow field may become subsonic, which prevents the computation from proceeding further. To alleviate this difficulty, Tannehill et al. (1982) have applied the numerical scheme (described previously) to the PNS equations written in general nonorthogonal coordinates, Eqs. (8.37)–(8.39). As a result, the orientation of each solution surface ($\xi =$ const) is left arbitrary, so that the most appropriate orientation can be selected for a given problem. In general, the optimum orientation occurs when the solution surface is nearly perpendicular to the local flow direction. In a similar manner, Helliwell et al. (1980) have incorporated a nonorthogonal coordinate system into the Lubard-Helliwell method to permit a more optimum orientation of the computational planes.

The PNS code developed by Schiff and Steger (1979) was further refined (Chaussee et al., 1981) and became the basis for the widely used AFWAL (Air Force Wright Aeronautical Laboratories) PNS code (Shanks et al., 1982; Stalnaker et al., 1986). Other implicit algorithms for solving the PNS equations have been developed by McDonald and Briley (1975) and Briley and McDonald (1980), who utilize a consistently split linearized block implicit (LBI) scheme, and by Li (1981a), who uses an iterative factored implicit scheme. The LBI scheme of McDonald and Briley has a linearized block implicit structure that is identical to the structure of the "delta" form of the Beam-Warming scheme.

The explicit-implicit scheme of MacCormack (1981) was applied to the solution of the PNS equations by Lawrence et al. (1984). This scheme requires the inversion of block bidiagonal systems rather than the block tridiagonal systems of the previous factored algorithms. The advent of high-speed vector-processing computers led Gielda and McRae (1986) to use the original explicit MacCormack (1969) scheme to solve the PNS equations. Since this scheme can be almost completely vectorized, it becomes competitive for certain classes of problems. In this algorithm the conservative streamwise flux vector $\mathbf{E}$ is solved at each ξ station, instead of the usual $\mathbf{U}$ vector. In another approach, Bhutta and Lewis (1985a) employed an implicit algorithm in conjunction with a pseudo-unsteady technique to solve the PNS equations.

Roe scheme. Nearly all of the previously described algorithms employ central differences for the derivatives in the crossflow (η, ζ) plane. A major difficulty for algorithms of this type is that the central differencing of fluxes across flow field discontinuities tends to introduce errors into the solution in the form of local flow property oscillations. In order to control these oscillations, some type of artificial dissipation is required. The correct magnitude of this added "smoothing" must be determined through a trial-and-error process. This has led to frustration on the part of many users of central-difference PNS codes. To alleviate this difficulty, Lawrence et al. (1986, 1987) developed an upwind implicit approximately factored finite-volume scheme based on Roe's approximate Riemann problem solver (Roe, 1981). With this upwind scheme, no user-specified smoothing coefficients are required when capturing discontinuities such as shock waves. The resulting upwind PNS code has been named UPS.

The development of the finite-volume scheme of Lawrence et al. begins with the integral form of the steady Navier-Stokes equations (Eq. 5.263):

$$\oiint_{\mathbf{S}} \overline{\mathbf{H}} \cdot d\mathbf{S} = 0 \tag{8.100}$$

The tensor $\overline{\mathbf{H}}$ can be expressed in terms of the Cartesian fluxes,

$$\overline{\mathbf{H}} = (\mathbf{E}_i - \mathbf{E}_v)\mathbf{i} + (\mathbf{F}_i - \mathbf{F}_v)\mathbf{j} + (\mathbf{G}_i - \mathbf{G}_v)\mathbf{k} \tag{8.101}$$

where the inviscid (subscript i) and viscous (subscript v) flux vectors are given by Eq. (8.35). The flow field is discretized using small but finite hexahedrons like the one shown in Fig. 8.5. Since the numerical solution is marched in the ξ direction, the flow field is discretized by successively adding slabs of thickness $\Delta\xi$ as the solution proceeds. The nth slab (n is the index for the ξ coordinate) is bounded by the two (η, ζ) systems of grid points at n and $n + 1$. Vinokur (1986) refers to these grid points as the primary grids. The vertices of each cell are located at mesh points of the primary grids and are connected by straight-line segments. The η and ζ coordinates are indexed using k and l, respectively. The present scheme is applied to area-averaged flow properties, which are assigned to the secondary grid points. The secondary grid points (see Fig. 8.5) are defined by averaging coordinates of the primary grids that define the constant ξ cell faces.

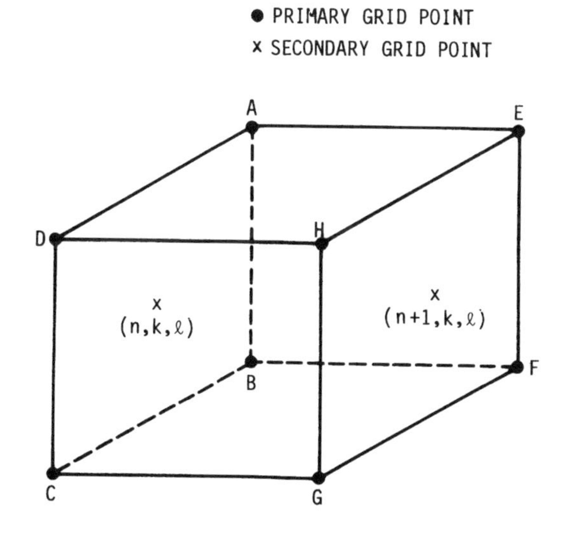

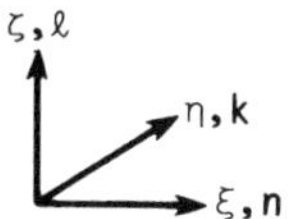

Figure 8.5 Finite-volume geometry.

After applying Eq. (8.100) to the structured-grid volume shown in Fig. 8.5, with constant properties on each cell face, the discretization becomes

$$\begin{aligned}\overline{\mathbf{H}}_{k,l}^{n+1}\cdot\mathbf{S}_{k,l}^{n+1}+\overline{\mathbf{H}}_{k+\frac{1}{2},l}^{n+\frac{1}{2}}\cdot\mathbf{S}_{k+\frac{1}{2},l}^{n+\frac{1}{2}}+\overline{\mathbf{H}}_{k,l+\frac{1}{2}}^{n+\frac{1}{2}}\cdot\mathbf{S}_{k,l+\frac{1}{2}}^{n+\frac{1}{2}}\\ -\overline{\mathbf{H}}_{k,l}^{n}\cdot\mathbf{S}_{k,l}^{n}-\overline{\mathbf{H}}_{k-\frac{1}{2},l}^{n+\frac{1}{2}}\cdot\mathbf{S}_{k-\frac{1}{2},l}^{n+\frac{1}{2}}-\overline{\mathbf{H}}_{k,l-\frac{1}{2}}^{n+\frac{1}{2}}\cdot\mathbf{S}_{k,l-\frac{1}{2}}^{n+\frac{1}{2}}=0\end{aligned}\tag{8.102}$$

with the cell-face-area vectors oriented in the positive coordinate directions. The cell-face-area vectors are indexed as follows:

$$\begin{aligned}&\mathbf{S}_{\mathrm{ABCD}}\Leftrightarrow\mathbf{S}_{k,l}^{n} && \mathbf{S}_{\mathrm{EHGF}}\Leftrightarrow\mathbf{S}_{k,l}^{n+1}\\ &\mathbf{S}_{\mathrm{DCGH}}\Leftrightarrow\mathbf{S}_{k-\frac{1}{2},l}^{n+\frac{1}{2}} && \mathbf{S}_{\mathrm{AEFB}}\Leftrightarrow\mathbf{S}_{k+\frac{1}{2},l}^{n+\frac{1}{2}}\\ &\mathbf{S}_{\mathrm{BFGC}}\Leftrightarrow\mathbf{S}_{k,l-\frac{1}{2}}^{n+\frac{1}{2}} && \mathbf{S}_{\mathrm{ADHE}}\Leftrightarrow\mathbf{S}_{k,l+\frac{1}{2}}^{n+\frac{1}{2}}\end{aligned}\tag{8.103}$$

and can be expressed with respect to a Cartesian coordinate system as

$$\begin{aligned}\mathbf{S}_{k,l}^{n+1}&=\left(\frac{\xi_x}{J}\right)_{k,l}^{n+1}\mathbf{i}+\left(\frac{\xi_y}{J}\right)_{k,l}^{n+1}\mathbf{j}+\left(\frac{\xi_z}{J}\right)_{k,l}^{n+1}\mathbf{k}\\ \mathbf{S}_{k+\frac{1}{2},l}^{n+\frac{1}{2}}&=\left(\frac{\eta_x}{J}\right)_{k+\frac{1}{2},l}^{n+\frac{1}{2}}\mathbf{i}+\left(\frac{\eta_y}{J}\right)_{k+\frac{1}{2},l}^{n+\frac{1}{2}}\mathbf{j}+\left(\frac{\eta_z}{J}\right)_{k+\frac{1}{2},l}^{n+\frac{1}{2}}\mathbf{k}\\ \mathbf{S}_{k,l+\frac{1}{2}}^{n+\frac{1}{2}}&=\left(\frac{\zeta_x}{J}\right)_{k,l+\frac{1}{2}}^{n+\frac{1}{2}}\mathbf{i}+\left(\frac{\zeta_y}{J}\right)_{k,l+\frac{1}{2}}^{n+\frac{1}{2}}\mathbf{j}+\left(\frac{\zeta_z}{J}\right)_{k,l+\frac{1}{2}}^{n+\frac{1}{2}}\mathbf{k}\\ \mathbf{S}_{k,l}^{n}&=-\left(\frac{\xi_x}{J}\right)_{k,l}^{n}\mathbf{i}-\left(\frac{\xi_y}{J}\right)_{k,l}^{n}\mathbf{j}-\left(\frac{\xi_z}{J}\right)_{k,l}^{n}\mathbf{k}\\ \mathbf{S}_{k-\frac{1}{2},l}^{n+\frac{1}{2}}&=-\left(\frac{\eta_x}{J}\right)_{k-\frac{1}{2},l}^{n+\frac{1}{2}}\mathbf{i}-\left(\frac{\eta_y}{J}\right)_{k-\frac{1}{2},l}^{n+\frac{1}{2}}\mathbf{j}-\left(\frac{\eta_z}{J}\right)_{k-\frac{1}{2},l}^{n+\frac{1}{2}}\mathbf{k}\\ \mathbf{S}_{k,l-\frac{1}{2}}^{n+\frac{1}{2}}&=-\left(\frac{\zeta_x}{J}\right)_{k,l-\frac{1}{2}}^{n+\frac{1}{2}}\mathbf{i}-\left(\frac{\zeta_y}{J}\right)_{k,l-\frac{1}{2}}^{n+\frac{1}{2}}\mathbf{j}-\left(\frac{\zeta_z}{J}\right)_{k,l-\frac{1}{2}}^{n+\frac{1}{2}}\mathbf{k}\end{aligned}\tag{8.104}$$

Equations (8.104) and (8.101) can then be inserted into Eq. (8.102) to yield

$$\begin{aligned}\left(\hat{\mathbf{E}}_i-\hat{\mathbf{E}}_v\right)_{k,l}^{n+1}+\left(\hat{\mathbf{F}}_i-\hat{\mathbf{F}}_v\right)_{k+\frac{1}{2},l}^{n+\frac{1}{2}}+\left(\hat{\mathbf{G}}_i-\hat{\mathbf{G}}_v\right)_{k,l+\frac{1}{2}}^{n+\frac{1}{2}}\\ -\left(\hat{\mathbf{E}}_i-\hat{\mathbf{E}}_v\right)_{k,l}^{n}-\left(\hat{\mathbf{F}}_i-\hat{\mathbf{F}}_v\right)_{k-\frac{1}{2},l}^{n+\frac{1}{2}}-\left(\hat{\mathbf{G}}_i-\hat{\mathbf{G}}_v\right)_{k,l-\frac{1}{2}}^{n+\frac{1}{2}}=0\end{aligned}\tag{8.105}$$

where

$$\left(\hat{\mathbf{E}}_i-\hat{\mathbf{E}}_v\right)=\left(\frac{\xi_x}{J}\right)(\mathbf{E}_i-\mathbf{E}_v)+\left(\frac{\xi_y}{J}\right)(\mathbf{F}_i-\mathbf{F}_v)+\left(\frac{\xi_z}{J}\right)(\mathbf{G}_i-\mathbf{G}_v)$$

$$\left(\hat{\mathbf{F}}_i - \hat{\mathbf{F}}_v\right) = \left(\frac{\eta_x}{J}\right)(\mathbf{E}_i - \mathbf{E}_v) + \left(\frac{\eta_y}{J}\right)(\mathbf{F}_i - \mathbf{F}_v) + \left(\frac{\eta_z}{J}\right)(\mathbf{G}_i - \mathbf{G}_v) \tag{8.106}$$

$$\left(\hat{\mathbf{G}}_i - \hat{\mathbf{G}}_v\right) = \left(\frac{\zeta_x}{J}\right)(\mathbf{E}_i - \mathbf{E}_v) + \left(\frac{\zeta_y}{J}\right)(\mathbf{F}_i - \mathbf{F}_v) + \left(\frac{\zeta_z}{J}\right)(\mathbf{G}_i - \mathbf{G}_v)$$

The metrics (i.e., components of the cell-face-area vectors) are evaluated using the formulas of Vinokur (1986), which were discussed in Section 5.7. For the three forward-facing sides of the finite volume shown in Fig. 8.5, these formulas yield

$$\begin{aligned}
\left(\frac{\xi_x}{J}\right)_{k,l}^{n+1} &= \frac{1}{2}[(y_D - y_B)(z_A - z_C) - (y_A - y_C)(z_D - z_B)] \\
\left(\frac{\xi_y}{J}\right)_{k,l}^{n+1} &= -\frac{1}{2}[(x_D - x_B)(z_A - z_C) - (x_A - x_C)(z_D - z_B)] \\
\left(\frac{\xi_z}{J}\right)_{k,l}^{n+1} &= \frac{1}{2}[(x_D - x_B)(y_A - y_C) - (x_A - x_C)(y_D - y_B)] \\
\left(\frac{\eta_x}{J}\right)_{k+\frac{1}{2},l}^{n+\frac{1}{2}} &= \frac{1}{2}[(y_E - y_D)(z_H - z_A) - (y_H - y_A)(z_E - z_D)] \\
\left(\frac{\eta_y}{J}\right)_{k+\frac{1}{2},l}^{n+\frac{1}{2}} &= -\frac{1}{2}[(x_E - x_D)(z_H - z_A) - (x_H - x_A)(z_E - z_D)] \qquad (8.107) \\
\left(\frac{\eta_z}{J}\right)_{k+\frac{1}{2},l}^{n+\frac{1}{2}} &= \frac{1}{2}[(x_E - x_D)(y_H - y_A) - (x_H - x_A)(y_E - y_D)] \\
\left(\frac{\zeta_x}{J}\right)_{k,l+\frac{1}{2}}^{n+\frac{1}{2}} &= \frac{1}{2}[(y_F - y_A)(z_E - z_B) - (y_E - y_B)(z_F - z_A)] \\
\left(\frac{\zeta_y}{J}\right)_{k,l+\frac{1}{2}}^{n+\frac{1}{2}} &= -\frac{1}{2}[(x_F - x_A)(z_E - z_B) - (x_E - x_B)(z_F - z_A)] \\
\left(\frac{\zeta_z}{J}\right)_{k,l+\frac{1}{2}}^{n+\frac{1}{2}} &= \frac{1}{2}[(x_F - x_A)(y_E - y_B) - (x_E - x_B)(y_F - y_A)]
\end{aligned}$$

Metrics calculated using these formulas satisfy the geometric conservation law

$$\oiint_{\mathbf{S}} d\mathbf{S} = 0 \tag{8.108}$$

which is obtained from Eq. (8.100) under uniform flow conditions. The description of the geometry is completed by determining the volume of the cell. The volume is only needed for evaluating the metrics in the viscous terms. Using the formula given in Section 5.7 by Vinokur (1986), the volume (i.e., Jacobian) is computed from

$$J = \tfrac{1}{3}\left(\mathbf{S}_{k,l}^{n} + \mathbf{S}_{k,l-\frac{1}{2}}^{n+\frac{1}{2}} + \mathbf{S}_{k+\frac{1}{2},l}^{n+\frac{1}{2}}\right) \cdot \left(\mathbf{r}_{k+\frac{1}{2},l-\frac{1}{2}}^{n} - \mathbf{r}_{k-\frac{1}{2},l+\frac{1}{2}}^{n+1}\right) \tag{8.109}$$

The parabolizing assumption is now applied to Eq. (8.105) dropping $\hat{\mathbf{E}}_v$ as well as derivatives with respect to ξ in $\hat{\mathbf{F}}_v$ and $\hat{\mathbf{G}}_v$. In addition, Vigneron's technique is used to suppress the ellipticity that is inherent in a space-marching procedure. This is accomplished by splitting the inviscid streamwise flux vector in the following manner:

$$\left(\hat{\mathbf{E}}_i\right)_{k,l}^n = \hat{\mathbf{E}}^*(\mathbf{S}_{k,l}^n, \mathbf{U}_{k,l}^n) + \hat{\mathbf{E}}^p(\mathbf{S}_{k,l}^n, \mathbf{U}_{k,l}^{n-1}) \tag{8.110}$$

where

$$\hat{\mathbf{E}}^* = \left[\rho\hat{U},\ \rho u\hat{U} + \left(\frac{\xi_x}{J}\right)\omega p,\ \rho v\hat{U} + \left(\frac{\xi_y}{J}\right)\omega p,\ \rho w\hat{U} + \left(\frac{\xi_z}{J}\right)\omega p, (E_t + p)\hat{U}\right]^T$$

$$\hat{\mathbf{E}}^p = (1-\omega)p\left[0, \left(\frac{\xi_x}{J}\right), \left(\frac{\xi_y}{J}\right), \left(\frac{\xi_z}{J}\right), 0\right]^T$$

and

$$\mathbf{U} = [\,\rho, \rho u, \rho v, \rho w, E_t]^T$$

$$\hat{U} = \left(\frac{\xi_x}{J}\right)u + \left(\frac{\xi_y}{J}\right)v + \left(\frac{\xi_z}{J}\right)w$$

The notation $\hat{\mathbf{E}}^*(\mathbf{S}_{k,l}^n, \mathbf{U}_{k,l}^n)$ indicates that $\hat{\mathbf{E}}^*$ is evaluated using geometrical properties at cell face $\mathbf{S}_{k,l}^n$ and flow variables from $\mathbf{U}_{k,l}^n$.

A change of dependent variable from $\hat{\mathbf{E}}^*$ to $\mathbf{U}$ is now made to avoid the difficult of extracting the flow properties from $\hat{\mathbf{E}}^*$ and also to simplify the application of the implicit algorithm. This is accomplished through the following linearization, which makes use of the homogeneous property of $\hat{\mathbf{E}}^*$:

$$\hat{\mathbf{E}}^*(\mathbf{S}^n, \mathbf{U}^n) = [\hat{A}^*]^{n-1}\mathbf{U}^n \tag{8.111}$$

where

$$[\hat{A}^*]^{n-1} = \frac{\partial \hat{\mathbf{E}}^*(\mathbf{S}^n, \mathbf{U}^{n-1})}{\partial \mathbf{U}^{n-1}}$$

After substituting Eqs. (8.110) and (8.111) into Eq. (8.105), the discretized conservation law becomes

$$\begin{aligned}
[\hat{A}^*]_{k,l}^n \Delta^n \mathbf{U}_{k,l} = &-\left([\hat{A}^*]_{k,l}^n - [\hat{A}^*]_{k,l}^{n-1}\right)\mathbf{U}_{k,l}^n - \left[\left(\hat{\mathbf{F}}_i - \hat{\mathbf{F}}_v\right)_{k+\frac{1}{2},l}^{n+\frac{1}{2}} - \left(\hat{\mathbf{F}}_i - \hat{\mathbf{F}}_v\right)_{k-\frac{1}{2},l}^{n+\frac{1}{2}}\right] \\
&-\left[\left(\hat{\mathbf{G}}_i - \hat{\mathbf{G}}_v\right)_{k,l+\frac{1}{2}}^{n+\frac{1}{2}} - \left(\hat{\mathbf{G}}_i - \hat{\mathbf{G}}_v\right)_{k,l-\frac{1}{2}}^{n+\frac{1}{2}}\right] \\
&-\left[\hat{\mathbf{E}}^p(\mathbf{S}_{k,l}^{n+1}, \mathbf{U}_{k,l}^n) - \hat{\mathbf{E}}^p(\mathbf{S}_{k,l}^n, \mathbf{U}_{k,l}^{n-1})\right]
\end{aligned} \tag{8.112}$$

where

$$\Delta^n \mathbf{U} = \mathbf{U}^{n+1} - \mathbf{U}^n$$

At this point the algorithm differs from a conventional finite-difference PNS solver only in the fact that the metrics are evaluated at cell interfaces rather than at grid points. A central-difference scheme is obtained by simply averaging the adjacent grid-point flow properties to obtain the cell-face numerical fluxes. Of course, this will lead to undesirable shock-capturing characteristics. To avoid this problem, Lawrence et al. determine the fluxes at the cell interfaces using

Roe's (1981) scheme, which is modified to make it applicable to a space-marching calculation. With Roe's scheme, the inviscid portions of the numerical fluxes are defined according to solutions of steady approximate Riemann problems. The fluxes $\hat{\mathbf{F}}_i$ and $\hat{\mathbf{G}}_i$ are determined separately by splitting the 2-D Riemann problem associated with the 3-D PNS equations into two 1-D Riemann problems. These 1-D problems have the generic form

$$\frac{\partial \hat{\mathbf{E}}^*}{\partial \xi} + [D]_{m+\frac{1}{2}} \frac{\partial \hat{\mathbf{E}}^*}{\partial \kappa} = 0 \tag{8.113}$$

with initial conditions

$$\hat{\mathbf{E}}^{*n} = \begin{cases} \hat{\mathbf{E}}^*\left(\mathbf{S}^n_{m+\frac{1}{2}}, \mathbf{U}_m\right) & \kappa < \kappa_{m+\frac{1}{2}} \\ \hat{\mathbf{E}}^*\left(\mathbf{S}^n_{m+\frac{1}{2}}, \mathbf{U}_{m+1}\right) & \kappa > \kappa_{m+\frac{1}{2}} \end{cases}$$

The coefficient matrix $[D]_{m+\frac{1}{2}}$ is defined by

$$[D]_{m+\frac{1}{2}} = \left(\frac{\kappa_x}{J}\right)_{m+\frac{1}{2}} \left(\frac{\partial \mathbf{E}_i}{\partial \hat{\mathbf{E}}^*}\right)_{m+\frac{1}{2}} + \left(\frac{\kappa_y}{J}\right)_{m+\frac{1}{2}} \left(\frac{\partial \mathbf{F}_i}{\partial \hat{\mathbf{E}}^*}\right)_{m+\frac{1}{2}} + \left(\frac{\kappa_z}{J}\right)_{m+\frac{1}{2}} \left(\frac{\partial \mathbf{G}_i}{\partial \hat{\mathbf{E}}^*}\right)_{m+\frac{1}{2}} \tag{8.114}$$

Although Eq. (8.113) is in nonconservative form, the local shock-capturing capabilities of the algorithm can be retained if the flow properties in $[D]_{m+\frac{1}{2}}$ are averaged between the grid points m and $m+1$, so that the following relation is satisfied:

$$\begin{aligned} &[D]_{m+\frac{1}{2}}\left[\hat{\mathbf{E}}^*\left(\mathbf{S}^n_{m+\frac{1}{2}}, \mathbf{U}_{m+1}\right) - \hat{\mathbf{E}}^*\left(\mathbf{S}^n_{m+\frac{1}{2}}, \mathbf{U}_m\right)\right] \\ &\quad = \left(\frac{\kappa_x}{J}\right)_{m+\frac{1}{2}} \Delta\mathbf{E}_i + \left(\frac{\kappa_y}{J}\right)_{m+\frac{1}{2}} \Delta\mathbf{F}_i + \left(\frac{\kappa_z}{J}\right)_{m+\frac{1}{2}} \Delta\mathbf{G}_i \end{aligned} \tag{8.115}$$

When the flow is supersonic, Roe's averaging (see Section 6.5.1) of the flow variables yields flow properties that satisfy Eq. (8.115).

The solution to the preceding approximate Riemann problem consists of four constant property regions separated by three surfaces of discontinuity (see Fig. 8.6) emanating from the cell edge $(\xi^n, \kappa_{m+\frac{1}{2}})$ and having slopes given by the eigenvalues of $[D]_{m+\frac{1}{2}}$. The resulting first-order accurate inviscid flux across the $m+\frac{1}{2}$ cell interface is given by

$$\begin{aligned} \mathbf{H}_{m+\frac{1}{2}} = &\left(\frac{\kappa_x}{J}\right)_{m+\frac{1}{2}} \frac{1}{2}[(\mathbf{E}_i)_m + (\mathbf{E}_i)_{m+1}] + \left(\frac{\kappa_y}{J}\right)_{m+\frac{1}{2}} \frac{1}{2}[(\mathbf{F}_i)_m + (\mathbf{F}_i)_{m+1}] \\ &+ \left(\frac{\kappa_z}{J}\right)_{m+\frac{1}{2}} \frac{1}{2}[(\mathbf{G}_i)_m + (\mathbf{G}_i)_{m+1}] \\ &- \frac{1}{2}[\operatorname{sgn} D]_{m+\frac{1}{2}}\left[\left(\frac{\kappa_x}{J}\right)_{m+\frac{1}{2}} \Delta\mathbf{E}_i + \left(\frac{\kappa_y}{J}\right)_{m+\frac{1}{2}} \Delta\mathbf{F}_i + \left(\frac{\kappa_z}{J}\right)_{m+\frac{1}{2}} \Delta\mathbf{G}_i\right] \end{aligned} \tag{8.116}$$

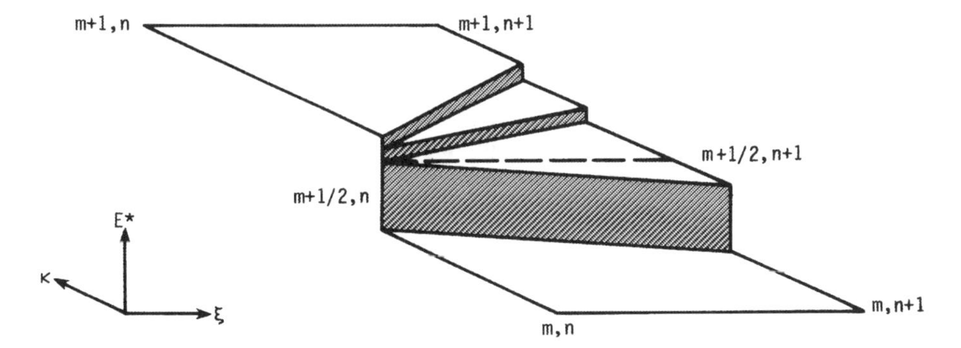

Figure 8.6 Approximate Riemann problem.

In this equation the matrix, [sgn D], is defined as

$$[\text{sgn } D] = [R][\text{sgn } \Lambda][R]^{-1}$$

where $[R]$ is the matrix of right eigenvectors of $[D]$ and [sgn Λ] is the diagonal matrix, which has elements related to the eigenvalues of $[D]$ by

$$\text{sgn } \lambda^i = \frac{\lambda^i}{|\lambda^i|}$$

First-order inviscid numerical fluxes in the η and ζ directions are then given by

$$\left(\hat{\mathbf{F}}_i\right)_{k+\frac{1}{2},l} = \mathbf{H}_{k+\frac{1}{2},l}$$

$$\left(\hat{\mathbf{G}}_i\right)_{k,l+\frac{1}{2}} = \mathbf{H}_{k,l+\frac{1}{2}}$$

In the definition of the flux in the η direction, $\mathbf{H}_{k+\frac{1}{2},l}$ is obtained by inserting η for κ and $k+\frac{1}{2}, l$ for $m+\frac{1}{2}$. Likewise, for the flux in the ζ direction, $\mathbf{H}_{k,l+\frac{1}{2}}$ is obtained by replacing κ with ζ and $m+\frac{1}{2}$ with $k, l+\frac{1}{2}$. Lawrence et al. have extended the algorithm to second-order accuracy in the crossflow directions by adapting the approach of Chakravarthy and Szema (1985) to the PNS equations. Details on the second-order fluxes can be found in the work by Lawrence (1987). Viscous stresses and heat transfer fluxes are evaluated in both crossflow directions using standard central differences.

The algorithm is made implicit by evaluating the first-order numerical flux at the $n+1$ marching station. The second-order flux terms are evaluated at the n marching station. The first-order flux is linearized using

$$(\mathbf{H})^{n+1}_{m+\frac{1}{2}} = (\mathbf{H})^{n}_{m+\frac{1}{2}} + \left(\frac{\partial(\mathbf{H})_{m+\frac{1}{2}}}{\partial \mathbf{U}_{m+1}}\right)^{n} \Delta^n \mathbf{U}_{m+1} + \left(\frac{\partial(\mathbf{H})_{m+\frac{1}{2}}}{\partial \mathbf{U}_{m}}\right)^{n} \Delta^u \mathbf{U}_m \quad (8.117)$$

The [sgn D] matrix that appears in $\mathbf{H}$ is assumed locally constant for the evaluation of the Jacobians. The viscous fluxes at $n+1$ are linearized in a similar manner using

$$\begin{aligned}
\left(\hat{\mathbf{F}}_v\right)^{n+1}_{k+\frac{1}{2},l} &= \left(\hat{\mathbf{F}}_v\right)^{n}_{k+\frac{1}{2},l} + \left(\frac{\partial\left(\hat{\mathbf{F}}_v\right)_{k+\frac{1}{2},l}}{\partial \mathbf{U}_{k+1,l}}\right)^n \Delta^n \mathbf{U}_{k+1,l} \\
&\quad + \left(\frac{\partial\left(\hat{\mathbf{F}}_v\right)_{k+\frac{1}{2},l}}{\partial \mathbf{U}_{k,l}}\right)^n \Delta^n \mathbf{U}_{k,l} \\
\left(\hat{\mathbf{G}}_v\right)^{n+1}_{k,l+\frac{1}{2}} &= \left(\hat{\mathbf{G}}_v\right)^{n}_{k,l+\frac{1}{2}} + \left(\frac{\partial\left(\hat{\mathbf{G}}_v\right)_{k,l+\frac{1}{2}}}{\partial \mathbf{U}_{k,l+1}}\right)^n \Delta^n \mathbf{U}_{k,l+1} \\
&\quad + \left(\frac{\partial\left(\hat{\mathbf{G}}_v\right)_{k,l+\frac{1}{2}}}{\partial \mathbf{U}_{k,l}}\right)^n \Delta^n \mathbf{U}_{k,l}
\end{aligned} \tag{8.118}$$

After substituting these linearized expressions into Eq. (8.112), the resulting block system of algebraic equations is approximately factored into two block tridiagonal systems. The algorithm can then be written as

$$\begin{aligned}
&\left[[\hat{A}^*]_{k,l} + \frac{\partial\left(\delta_\eta\left\{\hat{\mathbf{F}}_i - \hat{\mathbf{F}}_v\right\}\right)}{\partial \mathbf{U}_{k,l}} + \bar{\delta}_\eta\left(\frac{\partial\left\{\hat{\mathbf{F}}_i - \hat{\mathbf{F}}_v\right\}}{\partial \mathbf{U}}\right)\right]^n \left[[\hat{A}^*]^{-1}_{k,l}\right]^n \\
&\times \left[[\hat{A}^*]_{k,l} + \frac{\partial\left(\delta_\zeta\left\{\hat{\mathbf{G}}_i - \hat{\mathbf{G}}_v\right\}\right)}{\partial \mathbf{U}_{k,l}} + \bar{\delta}_\zeta\left(\frac{\partial\left\{\hat{\mathbf{G}}_i - \hat{\mathbf{G}}_v\right\}}{\partial \mathbf{U}}\right)\right]^n \Delta^n \mathbf{U}_{k,l} = (\mathbf{RHS})^n
\end{aligned} \tag{8.119}$$

where

$$\begin{aligned}
(\mathbf{RHS})^n = &-\left([\hat{A}^*]^n_{k,l} - [\hat{A}^*]^{n-1}_{k,l}\right)\mathbf{U}^n_{k,1} - \delta_\eta\left(\hat{\mathbf{F}}_i - \hat{\mathbf{F}}_v\right)^n - \delta_\zeta\left(\hat{\mathbf{G}}_i - \hat{\mathbf{G}}_v\right)^n \\
&-\left[\hat{\mathbf{E}}^p(\mathbf{S}^{n+1}_{k,l}, \mathbf{U}^n_{k,l}) - \hat{\mathbf{E}}^p(\mathbf{S}^n_{k,l}, \mathbf{U}^{n-1}_{k,l})\right]
\end{aligned} \tag{8.120}$$

and the difference operators are defined by

$$\delta_\kappa \Phi = \Phi_{m+\frac{1}{2}} - \Phi_{m-\frac{1}{2}}$$

$$\bar{\delta}_\kappa\left(\frac{\partial \Psi}{\partial \mathbf{U}}\Phi\right) = \frac{\partial \Psi_{m+\frac{1}{2}}}{\partial \mathbf{U}_{m+1}}\Phi_{m+1} - \frac{\partial \Psi_{m-\frac{1}{2}}}{\partial \mathbf{U}_{m-1}}\Phi_{m-1}$$

The system of equations can be solved using the same procedure as employed in the Beam-Warming scheme described previously. Further details of the algorithm can be found in the works by Lawrence et al. (1986, 1987) and Lawrence (1987, 1992).

Other schemes. The PNS equations have been solved using other upwind algorithms including the explicit finite-volume scheme of Korte and McRae (1988), which is based on Roe's approximate Riemann solver; the explicit finite-volume scheme of Gerbsch and Agarwal (1990), which is based on Osher's upwind method (Osher and Chakravarthy, 1983); and the scheme of Sturmayr et al. (1993), which is based on the ENO (essentially nonoscillatory) scheme of Yang (1991).

In addition, several upwind PNS algorithms have been developed that are derivatives of upwind time-dependent Navier-Stokes or TLNS solvers. These solvers are modified for space-marching using local time iterations at each streamwise step. Included in this group are the space-marched conservative supra-characteristic methods (CSCM-S) of Lombard et al. (1984) and Stookesberry and Tannehill (1987). The CSCM-S method is based on the CSCM for eigenvalue-based differencing developed by Lombard. Also included in this group is the algorithm of Newsome et al. (1987), which is based on the upwind Navier-Stokes scheme of Thomas and Walters (1987), and the algorithms of Ota et al. (1988), Thompson and Matus (1989), Molvik and Merkle (1989), and Matus and Bender (1990). An advantage of the time-iterative approach is that errors due to linearization and factorization can be reduced to negligible levels by iteration. An obvious disadvantage is that multiple iterations are required at each streamwise marching step.

8.3.4 Applications of PNS Equations

The PNS equations have been used to successfully compute the 3-D supersonic/hypersonic viscous flow over a variety of body shapes. For pointed bodies the initial starting solution is frequently obtained using the conical Navier-Stokes approximation (see Section 8.6). For blunt-nosed bodies, the initial starting solution is normally obtained using either a Navier-Stokes code or a viscous shock-layer (VSL) code.

Early studies involved the computation of flows over simple body shapes such as flat plates (Rudman and Rubin, 1968; Rubin et al., 1969; Cheng et al., 1970; Nardo and Cresci, 1971), corners (Cresci et al., 1969; Rubin and Lin, 1972), pointed cones (Rubin et al., 1969; Lin and Rubin, 1973; Lubard and Helliwell, 1974; Vigneron et al., 1978a, 1978b), and spinning cones (Lin and Rubin, 1974; Agarwal and Rakich, 1978).

Later, flows were computed over more complicated body shapes such as sphere-cones (Lubard and Rakich, 1975; Waskiewicz and Lewis, 1978; Rizk et al., 1981; Bhutta and Lewis, 1985a), hemisphere-cylinders (Schiff and Steger, 1979), ogive-cylinders (Rakich et al., 1979; Degani and Schiff, 1983), ogive-cylinder-boattails (Schiff and Sturek, 1980; Gielda and McRae, 1986), blunt biconics (Mayne, 1977; Helliwell et al., 1980, Chaussee et al., 1981; Kim and Lewis, 1982; Gnoffo, 1983; Neumann and Patterson, 1988), hyperboloids (Bhutta and Lewis, 1985b), sharp leading edge delta wings (Vigneron et al., 1978a, 1978b), blunt leading edge delta wings (Tannehill et al., 1982), and finned missiles (Rai et al., 1983).

More recently, flows have been computed over airplane-like vehicles such as the X-24C (Chaussee et al., 1981), Space Shuttle Orbiter (Li, 1981b; Rakich et al., 1984; Chaussee et al., 1984; Prabhu and Tannehill, 1984), generic fighter (Chaussee et al., 1985), hypersonic vehicles (Lawrence et al., 1987; Korte and McRae 1989; Bhutta and Lewis, 1989; Walker and Oberkampf, 1991), and generic versions of the National Aero-Space Plane (Buelow et al., 1990; Wadawadigi et al., 1994). See Fig. 8.7.

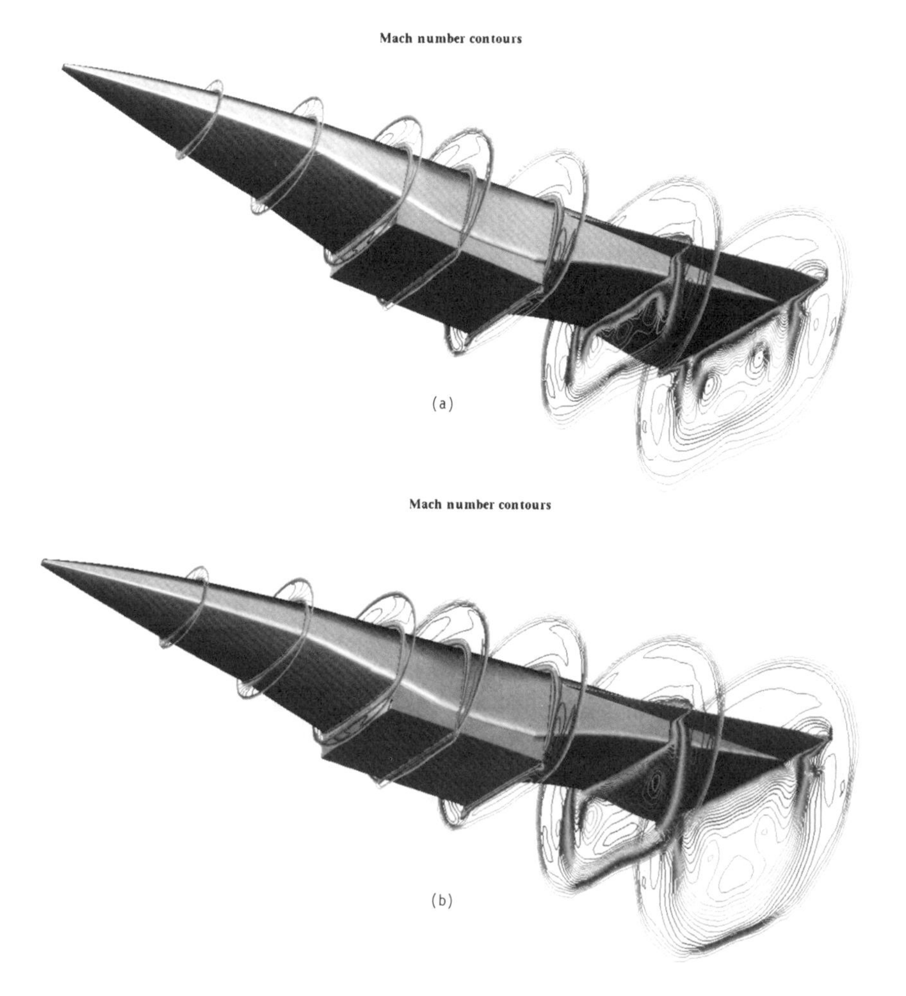

Figure 8.7 Comparison of Mach contours in various crossflow planes along the length of the Test Technology Demonstrator (TTD) configuration (Wadawadigi et al., 1994). (a) Scramjet engine off. (b) Scramjet engine on.

In addition to the computation of external flows, the PNS equations have been used to compute the supersonic/hypersonic flow in jets/plumes/wakes (Edelman and Weilerstein, 1969; Boynton and Thomson, 1969; Tannehill and Anderson, 1971; Dash and Wolf, 1984a, 1984b; Dash, 1985, 1989; Gielda and Agarwal, 1989; Sinha et al., 1990), ducts (Edelman and Weilerstein, 1969; Rubin et al., 1977; Sinha and Dash, 1986; Sinha et al., 1987, 1990), nozzles (Sinha and Dash, 1986; Chitsomboon and Northam, 1988; Gielda and Agarwal, 1989; Dang et al., 1989; Sinha et al., 1990), and inlets (Buggeln et al., 1980; Lawrence et al., 1986; Krawczyk and Harris, 1987; Gielda and Agarwal, 1989; Krawczyk et al., 1989).

The renewed interest in hypersonic aerothermodynamics has led to the development of several PNS codes that account for real-gas effects. PNS codes that include equilibrium chemistry have been written by Li (1981b), Gnoffo (1983), Prabhu and Tannehill (1984), Bhutta et al. (1985a), Banken et al. (1985), Molvik (1985), Stalnaker et al. (1986), Krawczyk and Harris (1987), Ota et al. (1988), Liou (1989), Tannehill et al. (1989), and Gerbsch and Agarwal (1991). For codes that utilize an upwind algorithm, it is also necessary to modify the numerical scheme to account for the equilibrium chemistry. Tannehill et al. (1989) have used both the approximate procedure of Grossman and Walters (1987) and the nearly exact procedure of Vinokur and Liu (1988) to modify the upwind implicit total variation diminishing (TVD) scheme in the UPS code to account for equilibrium chemistry.

For nonequilibrium (chemically reacting) flows, the species continuity equations must be solved in addition to the usual gas-dynamic equations. The gas-dynamic equations remain the same except for the additional diffusion term in the energy equation. The species continuity equations are parabolized by dropping the streamwise diffusion terms. The resulting nonequilibrium PNS equations have been solved using both fully coupled and loosely coupled approaches. In the fully coupled approach the gas-dynamic and species continuity equations are solved simultaneously using the same implicit algorithm. A drawback to this approach is that as the number of species increases, the size of the block matrices that must be inverted also increases. In the loosely coupled approach, the gas-dynamic and species continuity equations are solved separately, and the coupling between the two sets of equations is obtained through some type of iterative coupling. Nonequilibruim (finite-rate chemically reacting) PNS codes have been written by Bhutta et al. (1985b) Sinha and Dash (1986), Prabhu et al. (1987a, 1987b), Sinha et al. (1987), Tannehill et al. (1988) Gielda et al. (1988), Chitsomboon and Northam (1988), Gielda and Agarwal (1989), Molvik and Merkle (1989), Buelow et al. (1990), Kamath et al. (1991), Hugues and Vérant (1991), Wadawadigi et al. (1992), Ebrahimi and Gilbertson (1992), and White et al. (1993). A PNS code for both vibrational and chemical nonequilibrium flows has been developed by Miller et al. (1995). Other PNS codes have included the effects of ablation (Muramoto, 1993) and finite-catalytic walls (Miller et al., 1994).

8.4 PARABOLIZED AND PARTIALLY PARABOLIZED NAVIER-STOKES PROCEDURES FOR SUBSONIC FLOWS

Previous sections in this chapter dealt with flows that are predominantly supersonic. In this section we will discuss two computational strategies that are particularly useful for subsonic flows. These schemes can be categorized as either "once through" (fully parabolized) or multiple space-marching (partially parabolized). For both, the starting point is a form of the PNS equations. The approaches differ in the way that the pressure is treated.

8.4.1 Fully Parabolic Procedures

The fully parabolic procedure approach is applicable to a 3-D flow in which a predominant flow direction can be identified. The velocity component in this primary flow direction must generally be greater than zero, although some schemes may permit the existence of small regions of reversed flow by using the FLARE approximation in the primary flow direction. No restrictions are placed on the velocity components in the crossflow direction. As with all forms of the PNS equations, diffusion in the streamwise direction is neglected.

Unless further steps are taken, the PNS equations will permit transmission of influences in the streamwise direction through the pressure field for subsonic flows, as discussed in Section 8.3.2. In the present approach this elliptic behavior is suppressed in the streamwise direction by utilizing an approximation first suggested by Gosman and Spalding (1971). This approximation consists of representing the pressure gradient in the streamwise direction as an average over the flow cross section. A condition for using this approximation is that some method must exist for evaluating this pressure gradient. Possibilities include steady internal flows for which the pressure gradient can be determined with the aid of the global mass flow constraint and certain external flows for which the average pressure gradient can be taken as zero. Such external 3-D flows include flow in a corner and the subsonic jet discharging from a noncircular orifice into an ambient at rest or a co-flowing stream.

The computational strategy will be illustrated by considering flow through a straight rectangular channel. This permits use of the conservation equations in the Cartesian coordinate system. The same concepts are applicable to curved channels of constant cross-sectional area, but a different coordinate system must be used. The 3-D parabolic model has been extended to more general geometries by Briley and McDonald (1979).

The channel axis is in the x direction. Thus the y and z coordinates span planes perpendicular to the primary flow direction. The equations will be written in a form applicable to either laminar or turbulent flow. The variables are understood to represent time-mean quantities. This is the same convention as employed in Chapter 7. In developing the parabolized form of the Reynolds equations, diffusion in the streamwise direction by both molecular and turbulent mechanisms will be neglected. Furthermore, since only subsonic applications are to be considered, it will be assumed that $\overline{\rho' u'}/\bar{\rho}\bar{u}$, $\overline{\rho' v'}/\bar{\rho}\bar{v}$, and $\overline{\rho' w'}/\bar{\rho}\overline{w}$ are small, so that the difference between conventional and mass-weighted variables can be neglected. Terms involving pressure fluctuations in the energy equation will also be neglected. The symbol τ will denote the effective stress due to both molecular and turbulent mechanisms. Similarly, the symbol q will denote heat flux quantities from both molecular and turbulent mechanisms. Apart from the pressure gradient terms, which will be discussed below, the equations for the 3-D parabolic procedure follow from Eqs. (5.68), (5.73), and (5.84) after the simplifying assumptions given above are invoked:

continuity:

$$\frac{\partial \rho u}{\partial x} + \frac{\partial \rho v}{\partial y} + \frac{\partial \rho w}{\partial z} = 0 \tag{8.121a}$$

$$\int_A \rho u \, dA = \text{const} \qquad \text{(global)} \tag{8.121b}$$

x momentum:

$$\rho u \frac{\partial u}{\partial x} + \rho v \frac{\partial u}{\partial y} + \rho w \frac{\partial u}{\partial z} = -\frac{d\hat{p}}{dx} + \frac{\partial \tau_{xy}}{\partial y} + \frac{\partial \tau_{xz}}{\partial z} \tag{8.122}$$

y momentum:

$$\rho u \frac{\partial v}{\partial x} + \rho v \frac{\partial v}{\partial y} + \rho w \frac{\partial v}{\partial z} = -\frac{\partial p}{\partial y} + \frac{\partial \tau_{yy}}{\partial y} + \frac{\partial \tau_{yz}}{\partial z} \tag{8.123}$$

z momentum:

$$\rho u \frac{\partial w}{\partial x} + \rho v \frac{\partial w}{\partial y} + \rho w \frac{\partial w}{\partial z} = -\frac{\partial p}{\partial z} + \frac{\partial \tau_{zy}}{\partial y} + \frac{\partial \tau_{zz}}{\partial z} \tag{8.124}$$

energy:

$$\rho u c_p \frac{\partial T}{\partial x} + \rho v c_p \frac{\partial T}{\partial y} + \rho w c_p \frac{\partial T}{\partial z}$$

$$= \frac{\partial}{\partial y}(-q_y) + \frac{\partial}{\partial z}(-q_z) + \beta T u \frac{d\hat{p}}{dx} + \tau_{xy} \frac{\partial u}{\partial y} + \tau_{xz} \frac{\partial u}{\partial z} \tag{8.125}$$

state:

$$\rho = \rho(p, T) \tag{8.126}$$

In the pressure approximation of Gosman and Spalding (1971), a pressure $\hat{p}$ is defined for use in the x momentum equation which is assumed to vary *only in the x direction*. The pressure $\hat{p}$ will be determined with the aid of the global mass flow constraint much as for 2-D or axisymmetric channel flows computed through the thin-shear-layer equations. On the other hand, the p employed in the y and z momentum equations is permitted to vary across the channel cross section. The static pressure in the channel is assumed to be the sum of $\hat{p}$ and p.

The physical assumption in this decoupling procedure is that the pressure variations across the channel are so small that they would have a negligible effect if included in the streamwise momentum equation. Thus cross-plane pressure variations have been neglected in the streamwise momentum equation. On the other hand, these small pressure variations are included in the momentum equations in the y and z directions, since they play an important role in the distribution of the generally small components of velocity in the directions normal to the channel walls. The determination of $\hat{p}$ requires no information from downstream; $\hat{p}$ is a function of x only and can be uniquely determined at each cross section by employing the global mass flow constraint in combination with the momentum equations. This permits a "once through" calculation of the flow in a parabolic manner. On the other hand, since p varies with both y and z, the equations are elliptic (for subsonic flow) in the y-z plane. In fact, a Poisson equation can be developed for $p(y, z)$ in the cross plane from the y and z momentum equations. The overall calculation scheme then requires the use of procedures for elliptic equations in each cross plane, but the solution can be advanced in the x direction in a parabolic manner.

Using the Boussinesq approximation, the stresses (using summation notation) in the above equations can be evaluated from

$$\tau_{ij} = (\mu + \mu_T)\left(\frac{\partial u_i}{\partial x_j} + \frac{\partial u_j}{\partial x_i} - \frac{2}{3}\delta_{ij}\frac{\partial u_k}{\partial x_k}\right) - \frac{2}{3}\rho\bar{k}\delta_{ij} \tag{8.127}$$

With similar modeling assumptions, the heat flux quantities are normally represented by

$$q_y = -\left(k + \frac{\mu_T c_p}{\mathrm{Pr}_T}\right)\frac{\partial T}{\partial y}$$

$$q_z = -\left(k + \frac{\mu_T c_p}{\mathrm{Pr}_T}\right)\frac{\partial T}{\partial z}$$

Further simplifications to Eq. (8.127) are often found in specific applications, including the fully incompressible representation given by $\tau_{ij} = (\mu + \mu_T)\partial u_i/\partial x_j$. Suitable turbulence modeling for μ_T and Pr_T must be employed to close the system of equations. The usual boundary conditions for channel flow apply.

As was the case for the boundary-layer equations, the solution can proceed in either a coupled or sequential mode. The earliest and most widely used procedure followed the sequential strategy that will be outlined briefly here. We note that for a specified pressure field the momentum and energy equations would be entirely parabolic, and the solution could be marched in the primary flow direction using the x momentum equation to obtain u, the y momentum equation to obtain v, and the z momentum equation to obtain w. The energy equation provides T, and the density is obtained from the equation of state. However, for all but exactly the correct cross-plane pressure distribution, the velocity components will not satisfy the continuity equation. This, of course, is the crux of the problem—the momentum, energy, and state equations are a natural combination to use to advance the solution for the velocity components and density. The way in which the continuity and momentum equations can be used to determine the correct pressure distribution is less obvious. Workable procedures have been devised for correcting the pressure field, and these will be discussed next.

The computational strategy of solving the momentum equations in an uncoupled manner for the velocity components using a prescribed provisional pressure distribution and then using the continuity and momentum equations to correct the pressure field is known as the *pressure correction* or *segregated approach.*

The earliest solutions reported in the literature for the fully parabolic 3-D procedure followed the algebraic strategy outlined by Patankar and Spalding (1972) as the semi-implicit method for pressure-linked equations (SIMPLE) procedure. Some notable improvements in some of the solution steps have been suggested, and these are mentioned below. The Patankar and Spalding (1972) approach, in turn, draws heavily upon the earlier work of Harlow and Welch (1965), Amsden and Harlow (1970), and Chorin (1968). The segregated strategy proceeds as follows. The superscript $n + 1$ refers to the streamwise station being computed.

1. Employing suitable linearization for coefficients in Eq. (8.122), the pressure $\hat{p}^{n+1}$ can be determined in the same manner as for 2-D and axisymmetric channel flows solved by means of the boundary-layer equations (see Section 7.5), by making use of the global conservation-of-mass constraint. Then $u_{j,k}^{n+1}$ can be determined from the finite-difference solution of Eq. (8.122). The energy equation can be solved for $T_{j,k}^{n+1}$, and the equation of state used to determine $\rho_{j,k}^{n+1}$. An alternating direction implicit (ADI) scheme works very well for solving the momentum and energy equations.
2. Using an assumed pressure distribution in Eqs. (8.123) and (8.124), provisional values of v and w can be determined from a marching solution (an ADI scheme is recommended here too) to these momentum equations just as for the x momentum equations.
3. These provisional solutions for v and w in the cross plane will not generally satisfy the difference form of the continuity equation. By applying the continuity equation to the provisional solutions for the velocity components, mass sources (or sinks) can be computed at each grid point. We now seek a means for adjusting the pressure field in the cross plane so as to eliminate the mass sources. It is in the computation of the velocity and pressure corrections that the 3-D parabolic methods differ the most. Several investigators, including Briley (1974), Ghia et al. (1977b), and Ghia and Sokhey (1977) have followed the suggestion of Chorin (1968) and assumed that the corrective flow in the cross plane is irrotational, being driven by a pressure-like potential in such a manner as to annihilate the mass source. A Poisson equation can be developed for this potential from the continuity equation. Using p subscripts to denote provisional velocities and c subscripts to denote corrective quantities, we demand that

$$\frac{\partial \rho u}{\partial x} + \frac{\partial}{\partial y}\left[\rho(v_p + v_c)\right] + \frac{\partial}{\partial z}\left[\rho(w_p + w_c)\right] = 0 \qquad (8.128)$$

The streamwise derivative term and derivatives of the provisional velocities are known at the time the corrections are sought and can be incorporated into a single source term S_ϕ. Thus we can define a potential function $\hat{\phi}$ by $\rho v_c = \partial\hat{\phi}/\partial y$, $\rho w_c = \partial\hat{\phi}/\partial z$, and write Eq. (8.128) as

$$\frac{\partial^2\hat{\phi}}{\partial y^2} + \frac{\partial^2\hat{\phi}}{\partial z^2} = S_\phi \tag{8.129}$$

The required velocity corrections can then be computed from the $\hat{\phi}$ distribution resulting from the numerical solution of the Poisson equation in the cross plane. This approach preserves the vorticity of the original v_p and w_p velocity fields.

The original Patankar and Spalding proposal assumed that the velocity corrections were driven by pressure corrections in accordance with a very approximate form of the momentum equations in which the streamwise convective terms were equated to the pressure terms. This can be indicated symbolically by

$$\rho u \frac{\partial v_c}{\partial x} = -\frac{\partial p'}{\partial y} \tag{8.130}$$

$$\rho u \frac{\partial w_c}{\partial x} = -\frac{\partial p'}{\partial z} \tag{8.131}$$

In the above, p' can be viewed merely as a potential function (much like $\hat{\phi}$) used to generate velocity corrections that satisfy the continuity equation. In some schemes [as in the original Patankar and Spalding (1972) proposal], p' is viewed as an actual correction to be added to the provisional values of pressure. Since the velocity corrections can be assumed to be zero at the previous streamwise station, Eqs. (8.130) and (8.131) can be interpreted as

$$v_c = -A\frac{\partial p'}{\partial y} \tag{8.132}$$

$$w_c = -B\frac{\partial p'}{\partial z} \tag{8.133}$$

where A and B are coefficients that involve ρ, u, and Δx. The derivatives of p' are, of course, eventually to be represented on the finite-difference grid. The similarity between Eqs. (8.132) and (8.133) and the representation given earlier for the velocity corrections in terms of the potential $\hat{\phi}$ should be noted. Equations (8.132) and (8.133) can now be used in the continuity equation to develop a Poisson equation of the form

$$\frac{\partial^2 p'}{\partial y^2} + \frac{\partial^2 p'}{\partial z^2} = S_{p'} \tag{8.134}$$

The required velocity corrections can then be computed from the numerical solution of Eq. (8.134) using Eqs. (8.132) and (8.133). This approach is known as the p' procedure for obtaining velocity corrections. Improvements on this procedure have been suggested that attempt to employ a more complete form of the momentum equation in relating velocity corrections to p'. The paper by Raithby and Schneider (1979) describes several variations of the p' approach.

4. The next step is the pressure update. The velocity corrections just obtained have not been required to satisfy a complete momentum equation. It is now necessary to take steps to develop the improved pressure field in the cross plane that, when used in the complete momentum equations, will produce velocities that satisfy the continuity equation. Several procedures have been used. The corrected velocities can be employed in the difference form of the momentum equations to provide expressions for the pressure gradients that would be consistent with new velocities. We denote these symbolically by

$$\frac{\partial p}{\partial y} = F_1 \tag{8.135}$$

$$\frac{\partial p}{\partial z} = F_2 \tag{8.136}$$

One estimate of the "best" revised pressure field can be obtained by solving the Poisson equation that is developed from Eqs. (8.135) and (8.136):

$$\frac{\partial^2 p}{\partial y^2} + \frac{\partial^2 p}{\partial z^2} = \frac{\partial F_1}{\partial y} + \frac{\partial F_2}{\partial z} = S_p \tag{8.137}$$

The right-hand side of Eq. (8.137) is evaluated from the difference form of the momentum equations using the corrected velocities and is treated as a source term. Patankar (1980) has suggested a slightly different formulation, which also results in a Poisson equation to be solved for the updated pressure (the SIMPLER algorithm). SIMPLER stands for SIMPLE Revised. In all of these solutions of the Poisson equation, care must be taken in establishing the numerical representation of the boundary conditions. The differencing and solution strategy must ensure that the Gauss divergence theorem (see Section 3.3.7) is satisfied. A more detailed example of the boundary treatment for the Poisson equation for pressure is given in Section 8.4.3.

Raithby and Schneider (1979) have proposed a scheme for updating the pressure that does not require the solution of a second Poisson equation. Thcy refer to this as the procedure for pressure update from multiple path integration (PUMPIN). The idea is that the pressure change from grid point to grid point can be computed from integrating Eqs. (8.135) and (8.136) again using the corrected velocities in the momentum equations to evaluate F_1 and F_2. For exactly the correct velocities v and w, the pressure change computed by this procedure between any two points within the cross plane would be independent of path. If the velocities v and w are not exactly correct (they will only be correct as convergence is achieved), then each different path between two points will lead to a different result. We can fix one point as a reference and compute pressures at other points in the cross plane by averaging the pressures obtained by integrating over several different paths between the reference point and the grid point of interest. Raithby and Schneider (1979) reported good success at averaging pressures over only two paths, namely, from the reference point to the point of interest along constant y and then constant z, and also along constant z and then constant y.

The pressure can also be updated very simply by accepting the p' obtained in the velocity correction procedure of Patankar and Spalding (1972) [see Eq. (8.134)] as the correction to be added to the pressure.

5. Because the momentum and continuity equations have not been satisfied simultaneously in the procedures just described, steps 2–4 are normally repeated iteratively in sequence at each cross plane before the solution is advanced to the next marching station. Underrelaxation is commonly used for both the velocity and pressure corrections. That is, in moving from step 3 to 4, only a fraction of the computed velocity corrections may be added to the provisional v and w velocities. The fraction will vary from method to method. Likewise, it is common to only adjust the pressure by a fraction of the computed pressure correction before moving to step 2. Time-dependent forms of the governing equations are sometimes used to carry out this iterative process. Because steps 2–4 are to be repeated iteratively, it is common practice to terminate the intermediate Poisson equation solutions for velocity and pressure corrections (especially the latter) short of full convergence in early iterative passes through steps 2–4. The objective is to obtain an *improvement* in the pressure field with each iterative pass through steps 2–4. Until overall convergence is approached, there is little point in obtaining the best possible pressure field based on the wrong velocity distribution. The iterative sweeps through steps 2–4 are terminated when a pressure field has been established that will yield solutions to the momentum equations that satisfy the continuity equation within a specified tolerance; i.e., velocity corrections are no longer required.
6. After convergence to the specified degree is achieved, steps 1–5 are repeated for the next streamwise station.

Raithby and Schneider (1979) have reported on a comparative study of several of the methods described above for achieving the velocity and pressure corrections. The number of iterations through steps 2–5 above, required for convergence, was taken as the primary measure of merit. The computation time required for the various algorithms would be of interest but was not reported. Fixing the method for the pressure update, Raithby and Schneider observed that all of the methods given above for achieving the velocity corrections worked satisfactorily. There was very little difference between them in terms of the required number of iterations.

When the method for obtaining velocity corrections was fixed and several different methods for obtaining the pressure update were compared, the p'

method of Patankar and Spalding (1972) was observed to require notably more iterations for convergence than the other methods evaluated. The methods utilizing a Poisson equation and the PUMPIN procedure required only about half as many iterations as the p' method. The PUMPIN method required the fewest iterations by a slim margin. Use of the p' (Patankar and Spalding, 1972) method would not be recommended on the basis of the Raithby and Schneider (1979) study. This conclusion is confirmed by Patankar's (1980) recommendation that his SIMPLER algorithm that employs the Poisson equation formulation be used instead of the older p' method for updating the pressure. It is possible that the p' method may appear more competitive with the other methods when computation time rather than number of iterations is taken as the measure of merit.

Several investigators have reported calculations based on the 3-D parabolic model. These include the work of Patankar and Spalding (1972), Caretto et al. (1972), Briley (1974), Ghia et al. (1977b), Ghia and Sokhey (1977), and Patankar et al. (1974). For flows through channels of varying cross-sectional area, suggestions have been made to include an inviscid flow pressure (determined a priori) in the analysis to partially account for elliptic influences in the primary flow direction. Both regular and staggered grids have been used. The concepts of the mathematical model appear well established. The essential feature of the model is the replacement of the pressure in the streamwise momentum equation with an average pressure that can be determined in some manner (such as through application of global mass flow considerations) without consideration of the details of the downstream flow. Note that this pressure treatment is consistent with observations about solution methods for the PNS equations discussed earlier in this chapter for high-speed flows with embedded subsonic regions. The pressure decoupling discussed in this section is essential in order to avoid departure behavior when the solution is advanced in the streamwise direction. Aspects of this model, particularly the solution for the velocities and pressure in the cross plane, can be accomplished in a variety of ways, including through the use of a fully coupled procedure.

8.4.2 Parabolic Procedures for 3-D Free-Shear and Other Flows

As was mentioned above, the 3-D parabolic procedure is not restricted entirely to confined flows. The essential feature of the model was the decoupling of the pressure gradient terms in the primary flow and cross-flow directions. For confined flows the pressure gradient in the primary flow direction was determined with the aid of the global conservation-of-mass constraint. The main elements of the procedure can be used for other types of 3-D flows if the pressure gradient in the primary flow direction can be neglected or prescribed in advance. One such application occurs in the discharge of a subsonic free jet from a rectangular-shaped nozzle into a coflowing or quiescent ambient. The shape of such a jet gradually changes in the streamwise direction, eventually becoming round in cross section. For such flows, it is reasonable to neglect the streamwise pressure gradient. Small pressure variations in the cross plane must still be considered, as for 3-D confined flows. McGuirk and Rodi (1977) and Hwang and Pletcher (1978) have computed such flows by the 3-D parabolic procedure by setting $d\hat{p}/dx = 0$. An example of the 3-D parabolic procedure applied to free surface flows is given by Raithby and Schneider (1980).

8.4.3 Partially Parabolized (Multiple Space-Marching) Model

If the PNS equations are solved for a subsonic flow without making simplifying approximations regarding the pressure, the equations are only *partially* parabolized, leading to the terminology PPNS. The system remains elliptic overall because of the influence of the pressure field. The elliptic system can be solved by a direct method, although the memory requirements to do so are substantial for most applications. If the system of equations is solved in a marching mode, multiple streamwise sweeps are needed to resolve the elliptic effects transmitted by the pressure field. Thus, in developing the PPNS model from the Navier-Stokes equations, only certain diffusion processes are neglected, and no assumptions are made about the pressure. The neglected processes always include diffusion in the streamwise direction, but some schemes employed in two dimensions (Rubin and Reddy, 1983; Liu and Pletcher, 1986) may go a step further and neglect all diffusion terms in the normal momentum equation. The PPNS model has also been referred to as the semi-elliptic or reduced Navier-Stokes (RNS) formulation in the literature. The equations for the partially parabolized model are as given in Eqs. (8.121)–(8.126) with $d\hat{p}/dx$ replaced by $\partial p/\partial x$. The primary flow direction is assumed to be aligned with the x coordinate axis.

The solution strategies employed for the PPNS model can be distinguished as either pressure-correction (segregated) schemes or coupled approaches. Multiple streamwise sweeps are required in both approaches. The earliest procedures were of the pressure-correction type, and these will be discussed first.

Pressure-correction PPNS schemes. The PPNS model was first suggested by Pratap and Spalding (1976). Other partially parabolized procedures of the pressure-correction type have been proposed by Dodge (1977), Moore and Moore (1979), and Chilukuri and Pletcher (1980). The PPNS scheme was originally thought to be restricted to flows in which flow reversal in the primary direction does not occur. For these flows, 3-D storage is only required for the pressure (and the source term in the Poisson equation for pressure if the Poisson equation formulation is used) and not for the velocity components. This is the main computational advantage of the PPNS procedure compared to procedures for the full Navier-Stokes equations. Madavan and Pletcher (1982) have demonstrated that the PPNS model can be extended to 2-D applications in which reversal occurs in the component of velocity in the primary direction. This procedure requires that computer storage also be used for velocity components in and near the regions of primary flow reversal.

We will briefly describe how the PPNS strategy of Chilukuri and Pletcher (1980) can be applied to a steady incompressible 2-D laminar flow. The improvements suggested by Madavan and Pletcher (1982) will be included. For such a flow, the PPNS equations can be written

continuity:

$$\frac{\partial u}{\partial x} + \frac{\partial v}{\partial y} = 0 \tag{8.138}$$

x momentum:

$$u\frac{\partial u}{\partial x} + v\frac{\partial u}{\partial y} = -\frac{1}{\rho}\frac{\partial p}{\partial x} + \nu\frac{\partial^2 u}{\partial y^2} \tag{8.139}$$

y momentum:

$$u\frac{\partial v}{\partial x} + v\frac{\partial v}{\partial y} = -\frac{1}{\rho}\frac{\partial p}{\partial y} + \nu\frac{\partial^2 v}{\partial y^2} \tag{8.140}$$

A staggered grid (Harlow and Welch, 1965) is often used with pressure-correction schemes, and we will use it in the present 2-D example of the PPNS procedure. The idea is to define a different grid for each velocity component. This is illustrated in Fig. 8.8. To avoid confusion, only the grid location for the scalar variables (pressure and the velocity-correction potential $\hat{\phi}$, in this example)

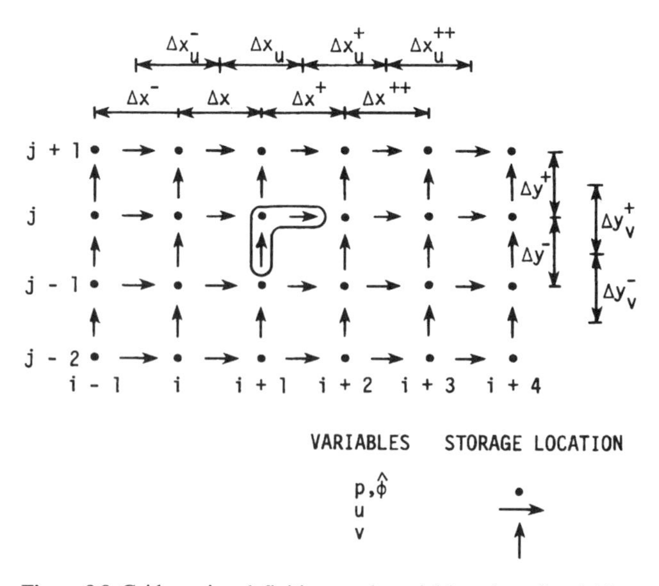

Figure 8.8 Grid spacing definitions and spatial location of variables on a staggered grid.

are denoted by solid symbols in the figure. Velocity components are calculated for "points" or locations on the faces of a control volume that could be drawn around the pressure points. The velocity components are located midway between pressure points, which means that for an unequally spaced grid, the pressure points are not necessarily in the geometric center of such a control volume. The locations of the velocity components are indicated by arrows in Fig. 8.8. Vertical arrows denote locations for v, and horizontal arrows indicate the locations for u. It is convenient to refer to the variables with a single set of grid indices, despite the fact that the variables are actually defined at different locations. Thus the designation $(i + 1, j)$ identifies a cluster of three distinct spatial locations as indicated by the boomerang-shaped enclosure in Fig. 8.8. In the staggered grid, $v_{i+1,j}^{n+1}$ is below $p_{i+1,j}$, and $u_{i+1,j}$ is to the right of $p_{i+1,j}$.

The staggered grid permits the divergence of the velocity field to be represented with second-order accuracy (for equally spaced grid points) at the solid grid points using velocity components at adjacent locations. Such a configuration ensures that the difference representation for this divergence has the conservative property. Also, the pressure difference between adjacent grid points becomes the natural driving force for velocity components located midway between the points. That is, a simple forward-difference representation for pressure derivatives is "central" relative to the location of velocity components. This permits the development of a Poisson equation for pressure that automatically satisfies the Gauss divergence theorem so long as care is taken in the treatment of the boundary conditions. Such boundary conditions are also more easily handled on the staggered grid. Patankar (1980) provides an excellent and more detailed discussion of the advantages of using a staggered grid for problems such as the present one.

The computational boundaries are most conveniently located along grid lines where components of velocity normal to the boundaries are located. This is illustrated in Fig. 8.9 for a lower boundary. Fictitious points are located outside of the physical boundary as necessary for imposing suitable boundary conditions. As an example, we will suppose that it is desired to impose no-slip boundary conditions at the lower boundary illustrated in Fig. 8.9. The v component of velocity is located at the physical boundary, and it is easy to simply specify $v_{i+1,1} = 0$. The treatment for the u component is not so obvious, since no u grid points are located on this boundary. Numerous possibilities exist. The main requirement is that the boundary formulation used must imply that the tangential component of velocity is zero at the location of the physical boundary. This can be achieved by developing a special difference form of the conservation equations for the control volume at the boundary, or by constraining the solution near the boundary such that an extrapolation to the boundary would satisfy the no-slip condition. A third and often used procedure is to employ a fictitious velocity point below the boundary with the constraint that $(u_{i+1,1} + u_{i+1,2})/2 = 0$. This is similar to the reflection technique for enforcing boundary conditions for inviscid flow, which was discussed in Chapter 6. The velocity at the fictitious point would then be used as required in the momentum equations in the

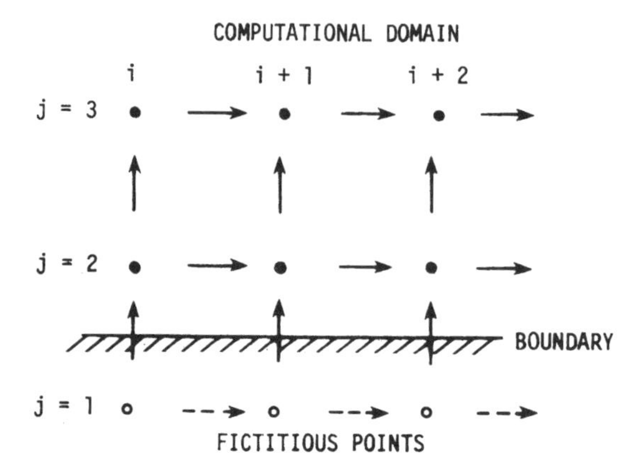

Figure 8.9 The staggered finite-difference grid near a boundary.

interior. Values of the potential function used to correct the velocities are also often obtained from points outside the physical boundaries. Values of pressure are not needed from such points for boundaries on which velocity boundary conditions are specified in the formulation described below. More details on the treatment of boundary conditions on a staggered grid can be found in the work of Amsden and Harlow (1970).

Several choices exist for the representation of the convective derivatives in the momentum equations. The scheme illustrated below makes use of three-point second-order accurate upwind representations for convective terms of the form $u\,\partial\phi/\partial x$. A hybrid scheme (see Section 7.3.3) will be used for terms of the form $v\,\partial\phi/\partial y$. These representations are linearized by extrapolating the coefficients based on values at the two adjacent upstream stations. When streamwise flow reversal is present, the direction of the "wind" changes, and this is taken into account in the representation used for the streamwise derivatives and in the extrapolation direction for the coefficients.

In the ensuing discussion, the following notation is adopted. The superscript $n+1$ denotes the current marching sweep; the subscript $i+1$ denotes the current streamwise step for which the solution is sought; and the subscript j denotes the grid points in the y direction. For the forward going flow, the following representation is used to obtain the extrapolated value of the coefficient $u_{i+1,j}^{n+1}$:

$$\hat{u}_{i+1,j}^{n+1} = \left(1 + \frac{\Delta x_u}{\Delta x_u^-}\right) u_{i,j}^{n+1} - \frac{\Delta x_u}{\Delta x_u^-} u_{i-1,j}^{n+1}$$

The circumflex indicates that $\hat{u}_{i+1,j}^{n+1}$ is a known quantity, determined by extrapolation. An extrapolation for $\hat{v}_{i+1,j}^{n+1}$ is obtained in a like manner using appropriate streamwise mesh increments. If $\hat{u}_{i+1,j}^{n+1}$ in the above expression becomes negative, the flow at $(i+1, j)$ is assumed to be reversed. When this occurs, the recommended procedure is to simply represent $\hat{u}^{n+1}_{i+1,j}$ by $u^n_{i+1,j}$ and $\hat{v}^{n+1}_{i+1,j}$ by $v^n_{i+1,j}$, making use of velocities from the previous iteration, which are stored for points in and near regions of reversed flow. As an alternative, extrapolation can also be used according to the expression

$$\hat{u}^{n+1}_{i+1,j} = \left(1 + \frac{\Delta x^+_u}{\Delta x^{++}_u}\right) u^n_{i+2,j} - \frac{\Delta x^+_u}{\Delta x^{++}_u} u^n_{i+3,j}$$

An expression for $\hat{v}^{n+1}_{i+1,j}$ can also be obtained by extrapolation in regions of reversed flow. The streamwise convective derivatives are then represented as follows. For the forward going flow,

$$\left(u\frac{\partial u}{\partial x}\right)^{n+1}_{i+1,j} \cong \hat{u}^{n+1}_{i+1,j}\left(\frac{\Delta x^-_u + 2\,\Delta x_u}{\Delta x_u(\Delta x^-_u + \Delta x_u)} u^{n+1}_{i+1,j} - \frac{\Delta x^-_u + \Delta x_u}{\Delta x^-_u\,\Delta x_u} u^{n+1}_{i,j} + \frac{\Delta x_u}{\Delta x^-_u(\Delta x^-_u + \Delta x_u)} u^{n+1}_{i-1,j}\right) \tag{8.141}$$

and for the reversed flow region,

$$\left(u\frac{\partial u}{\partial x}\right)^{n+1}_{i+1,j} \cong -\hat{u}^{n+1}_{i+1,j}\left(\frac{\Delta x^{++}_u + 2\,\Delta x^+_u}{\Delta x_u(\Delta x^{++}_u + \Delta x^+_u)} u^{n+1}_{i+1,j} + \frac{\Delta x^{++}_u + \Delta x^+_u}{\Delta x^{++}_u\,\Delta x^+_u} u^n_{i+2,j} - \frac{\Delta x^+_u}{\Delta x^{++}_u(\Delta x^{++}_u + \Delta x^+_u)} u^n_{i+3,j}\right) \tag{8.142}$$

The term $v\,\partial u/\partial y$ is represented by a hybrid scheme as follows:

$$\begin{aligned}\left(v\frac{\partial u}{\partial y}\right)^{n+1}_{i+1,j} \cong {} & \left[\hat{v}^{n+1}_{i+1,j}\left(u^{n+1}_{i+1,j} - u^{n+1}_{i+1,j-1}\right)\frac{\Delta y^+}{\Delta y^+ + \Delta y^-}\right. \\ & \left. + \hat{v}^{n+1}_{i+1,j+1}\frac{u^{n+1}_{i+1,j+1} - u^{n+1}_{i+1,j}}{\Delta y^+}\,\frac{\Delta y^-}{\Delta y^+ + \Delta y^-}\right] W \\ & + \hat{v}^{n+1}_{i+1,j}\frac{u^{n+1}_{i+1,j} - u^{n+1}_{i+1,j-1}}{\Delta y^-}(1 - W)A \\ & + \hat{v}^{n+1}_{i+1,j+1}\frac{u^{n+1}_{i+1,j+1} - u^{n+1}_{i+1,j}}{\Delta y^+}(1 - W)B\end{aligned} \tag{8.143}$$

The magnitudes of W, A, and B are determined as follows. Defining

$$R^+_m = \frac{\hat{v}^{n+1}_{i+1,j+1}\,\Delta y^-}{\nu}$$

$$R^-_m = \frac{\hat{v}^{n+1}_{i+1,j}\,\Delta y^+}{\nu}$$

and R_c as the critical mesh Reynolds number, equal to 1.9 (see Section 7.3.3), then when $R_m^+ > R_c$,

$$W = \frac{R_c}{R_m^+} \qquad A = 1 \qquad B = 0$$

When $R_m^+ < -R_c$,

$$W = \frac{R_c}{R_m^-} \qquad A = 0 \qquad B = 1$$

When $R_m^- < R_c < R_m^+$,

$$W = 1 \qquad A = 0 \qquad B = 0$$

This scheme is thus a weighted average of central and upwind differences for larger mesh Reynolds numbers and degenerates to central differencing for small mesh Reynolds numbers.

The second derivative term is represented by

$$\left(\frac{\partial^2 u}{\partial y^2}\right)_{i+1,j}^{n+1} \cong \frac{2}{\Delta y^+ + \Delta y^-}\left(\frac{u_{i+1,j+1}^{n+1} - u_{i+1,j}^{n+1}}{\Delta y^+} - \frac{u_{i+1,j}^{n+1} - u_{i+1,j-1}^{n+1}}{\Delta y^-}\right) \tag{8.144}$$

The pressure derivative in the streamwise momentum equation is represented by

$$\left.\frac{\partial p}{\partial x}\right)_{i+1,j}^{n} \cong \frac{p_{i+2,j}^n - p_{i+1,j}^n}{\Delta x^+} \tag{8.145}$$

The differencing of the pressure gradient term ensures that $u_{i+1,j}^{n+1}$ is influenced by the pressure downstream.

The y momentum equation is differenced in a similar manner. Because of the staggered grid being used, $v_{i+1,j}^{n+1}$ is not located at the same point in the flow as $u_{i+1,j}^{n+1}$. The evaluation of the coefficients in the difference representation of the y momentum equation should reflect this. For example, in representing the term $u\, \partial v/\partial x$, the coefficient should be formed using the average of u at two j levels. The pressure derivative utilizes pressure values on both sides of $v_{i+1,j}^{n+1}$:

$$\left.\frac{\partial p}{\partial y}\right)_{i+1,j}^{n} \cong \frac{p_{i+1,j}^n - p_{i+1,j-1}^n}{\Delta y^-} \tag{8.146}$$

As the momentum equations are solved, the best current estimate of the pressure field is used. Additional details on how this pressure field is determined will be discussed below. With the pressure fixed, the momentum equations are parabolic and are solved in a segregated manner, the x momentum equation for $u_{i+1,j}^{n+1}$ and the y momentum equation for $v_{i+1,j}^{n+1}$. The system of algebraic equations for the unknowns at the $i + 1$ level is tridiagonal and can be solved by employing the Thomas algorithm. As was observed for the 3-D parabolic procedure, the solution for the velocities will not satisfy the continuity equation until the correct pressure field is determined. Thus the velocities obtained from the solutions for the momentum equations are provisional. It is assumed that velocity corrections are driven by a potential $\hat{\phi}$ in such a manner that the continuity equation is satisfied by the corrected velocities. This requires that

$$\frac{\partial(u_p + u_c)}{\partial x} + \frac{\partial(v_p + v_c)}{\partial y} = 0 \tag{8.147}$$

where u_c and v_c are velocity corrections and u_p and v_p are the provisional velocities obtained from the solution to the momentum equations at marching level $i + 1$. Defining a potential function $\hat{\phi}$ by

$$u_c = \frac{\partial \hat{\phi}}{\partial x} \qquad v_c = \frac{\partial \hat{\phi}}{\partial y} \tag{8.148}$$

we obtain

$$\frac{\partial^2 \hat{\phi}}{\partial x^2} + \frac{\partial^2 \hat{\phi}}{\partial y^2} = -\frac{\partial u_p}{\partial x} - \frac{\partial v_p}{\partial y} = S_\phi \tag{8.149}$$

In difference form this becomes

$$\begin{aligned} &\frac{1}{\Delta x_u}\left(\frac{\hat{\phi}_{i+2,j} - \hat{\phi}_{i+1,j}}{\Delta x^+} - \frac{\hat{\phi}_{i+1,j} - \hat{\phi}_{i,j}}{\Delta x}\right) \\ &\quad + \frac{1}{\Delta y_v^+}\left(\frac{\hat{\phi}_{i+1,j+1} - \hat{\phi}_{i+1,j}}{\Delta y^+} - \frac{\hat{\phi}_{i+1,j} - \hat{\phi}_{i+1,j-1}}{\Delta y^-}\right) \\ &\quad = -2\frac{(u_p)_{i+1,j} - (u_p)_{i,j}}{\Delta x + \Delta x^+} - 2\frac{(v_p)_{i+1,j+1} - (v_p)_{i+1,j}}{\Delta y^+ + \Delta y^-} = (S_\phi)_{i+1,j} \end{aligned} \tag{8.150}$$

Such an algebraic equation can be written for each $\hat{\phi}$ grid point across the flow; $j = 2, 3, \ldots, NJ$, where $j = 2$ is the first $\hat{\phi}$ grid point above the lower boundary and $j = NJ$ denotes the $\hat{\phi}$ grid point just below the upper boundary. This results in a tridiagonal system of equations for the unknown $\hat{\phi}_{i+1,j}$ if $\hat{\phi}_{i,j}$ and $\hat{\phi}_{i+2,j}$ are known. The assumptions made to evaluate $\hat{\phi}_{i,j}$ and $\hat{\phi}_{i+2,j}$ are as follows.

Assumption 1: $\hat{\phi}_{i,j} = \hat{\phi}_{i+1,j}$

This implies that no corrective flow is present from the ith station where conservation of mass has already been established.

Assumption 2: $\hat{\phi}_{i+2,j} = 0$

This implies that $(v_c)_{i+2,j}$ is zero, which must be the case when convergence is achieved. Any other assumption regarding $\hat{\phi}_{i+2,j}$ would appear to be inconsistent with convergence. The boundary conditions used when solving the tridiagonal system of equations to determine $\hat{\phi}_{i+1,j}$ are chosen to be consistent with the prescribed velocity boundary conditions. For example, if velocities are prescribed along the top and bottom boundaries, v_c would be zero along these boundaries and the conditions used would be $\hat{\phi}_{i+1,1} = \hat{\phi}_{i+1,2}$ and $\hat{\phi}_{i+1,NJ} = \hat{\phi}_{i+1,NJ+1}$.

After the $\hat{\phi}_{i+1,j}$ are determined, velocity corrections are evaluated from the finite-difference representation of Eq. (8.148), namely,

$$(u_c)_{i+1,j} = -\frac{\phi_{i+1,j}}{\Delta x^+}$$

and

$$(v_c)_{i+1,j} = \frac{\phi_{i+1,j} - \phi_{i+1,j-1}}{\Delta y^-}$$

The corrected velocities now satisfy continuity at each grid point at the $i+1$ marching level, but unfortunately, until convergence, these velocities do not satisfy the momentum equations exactly.

The pressure is updated between marching sweeps by solving a Poisson equation for pressure using the method of SOR by points. The Poisson equation is formed from the difference representation of the momentum equations. That is, we can write

$$\frac{\partial p}{\partial x} = -\rho\left(u\frac{\partial u}{\partial x} + v\frac{\partial u}{\partial y} - \nu\frac{\partial^2 u}{\partial y^2}\right) = G1$$

$$\frac{\partial p}{\partial y} = -\rho\left(u\frac{\partial v}{\partial x} + v\frac{\partial v}{\partial y} - \nu\frac{\partial^2 v}{\partial y^2}\right) = G2$$

When the above equations are differenced, the G's are considered to be located midway between the pressure points used in representing the pressure derivatives on the left-hand side. Thus, $G1$ "points" are coincident with u locations, and $G2$ "points" are coincident with v locations. Then

$$\frac{\partial^2 p}{\partial x^2} + \frac{\partial^2 p}{\partial y^2} = \frac{\partial G1}{\partial x} + \frac{\partial G2}{\partial y} = S_p \tag{8.151}$$

where $G1$ and $G2$ are evaluated by using the *corrected* velocities that satisfy the continuity equation. The use of corrected velocities contributes to the development of a pressure field which will ultimately force the solutions to the momentum equations to conserve mass locally. The S_p terms are evaluated and stored as the marching integration sweep of the momentum equations proceeds. Normally, one successive overrelaxation (SOR) sweep of the pressure field is made during this marching procedure. It is easy to update the pressure by one line relaxation before advancing the velocity solution to the next i level. Several more SOR passes are made at the conclusion of the marching sweep. Overrelaxation factors of 1.7 have been successfully used, but the source term, S_p, is typically underrelaxed by a factor ranging from 0.2 to 0.65, the smaller factor being used for the earliest marching sweeps.

The boundary conditions for the Poisson equation for pressure are all Neumann conditions as derived from the momentum equations. The divergence theorem requires that

$$\iint S_p \, dx \, dy = \oint \frac{\partial p}{\partial n} \, dC$$

where C represents the boundary of the flow domain and $\partial p / \partial n$ is the magnitude of the Neumann boundary condition. The finite-difference equivalent of this constraint must be satisfied before the solution procedure for the Poisson equation will converge. With the staggered grid, this constraint can be satisfied by relating the boundary point pressures to the pressures in the interior through the specified derivative boundary conditions by an equation that is implicit with respect to iteration levels in the method of SOR by points. This step eliminates all dependence on the boundary pressures themselves (Miyakoda, 1962) when solving the Poisson equation for pressure. As long as the difference representation for S_p has the conservative property, the iterative procedure will converge. This boundary treatment is illustrated by writing Eq. (8.151) in difference form for a p point just inside the lower boundary

$$\begin{aligned} &\frac{1}{\Delta x_u}\left(\frac{p_{i+2,2}^{k} - p_{i+1,2}^{k+1}}{\Delta x^{+}} - \frac{p_{i+1,2}^{k+1} - p_{i,2}^{k+1}}{\Delta x}\right) \\ &\quad + \frac{1}{\Delta y_v^{+}}\left(\frac{p_{i+1,3}^{k+1} - p_{i+1,2}^{k+1}}{\Delta y^{+}} - \frac{p_{i+1,2}^{k+1} - p_{i+1,1}^{k+1}}{\Delta y^{-}}\right) \\ &\quad = \frac{G1_{i+1,2} - G1_{i,2}}{\Delta x_u} + \frac{G2_{i,3} - G2_{i,2}}{\Delta y_v^{+}} \end{aligned} \tag{8.152}$$

In the above, k refers to the iteration level in the SOR procedure for the Poisson equation, and $k+1$ denotes the level currently being computed. The boundary condition on the Poisson equation at the lower boundary is taken as

$$\left(\frac{\partial p}{\partial y}\right)_{\text{bdy}} = G2$$

That is, the boundary pressure derivative is evaluated from the momentum equation. In difference form this becomes

$$\frac{p_{i+1,2}^{k+1} - p_{i+1,1}^{k+1}}{\Delta y^{-}} = G2_{i+1,2} \tag{8.153}$$

where the pressures have been written implicitly at the present iteration level. The pressure below the lower boundary, $p_{i+1,1}^{k+1}$, can now be eliminated from the Poisson equation by substituting Eq. (8.153) into Eq. (8.152). This gives

$$\begin{aligned} &\frac{1}{\Delta x_u}\left(\frac{p_{i+2,2}^{k} - p_{i+1,2}^{k+1}}{\Delta x^{+}} - \frac{p_{i+1,2}^{k+1} - p_{i,2}^{k+1}}{\Delta x}\right) \\ &\quad + \frac{1}{\Delta y_v^{+}}\left(\frac{p_{i+1,3}^{k+1} - p_{i+1,2}^{k+1}}{\Delta y^{+}}\right) = \frac{G1_{i+1,2} - G1_{i,2}}{\Delta x_u} + \frac{G2_{i,3}}{\Delta y_v^{+}} \end{aligned} \tag{8.154}$$

An examination of the representation for S_p substantiates that the constraint imposed by the divergence theorem is satisfied by this procedure. An evaluation of $\int\int S_p \, dx \, dy$ leaves only terms involving $G1$ and $G2$ along the boundaries owing to cancellation of all other G. These boundary $G1$ and $G2$ values are exactly equal to $\phi(\partial p/\partial n) \, dC$ when the boundary conditions are expressed in terms of G, as illustrated by Eqs. (8.153) and (8.154).

The steps in the PPNS solution procedure are summarized below.

1. The momentum equations are solved for tentative velocity profiles at the $i + 1$ station using an estimated pressure filed. For the first streamwise sweep, this pressure field can be obtained by (a) assuming that $\partial p/\partial x = -\rho u_e(du_e/dx)$ and $\partial p/\partial y = 0$ or (b) assuming that $\partial p/\partial y = 0$ and using a secant procedure (see Section 7.4.3) to determine the value of $\partial p/\partial x$ that will conserve mass globally across the flow, much as is done when solving internal flows with boundary-layer equations. For sweeps beyond the first, a block adjustment can be added (or subtracted) to the downstream pressure through use of the secant procedure at each i station to ensure that mass is conserved globally across the flow. This forces the algebraic sum of the mass sources across the flow to be zero and appears to speed convergence in some cases. A noniterative scheme (Chiu and Pletcher, 1986) has also been used successfully to determine the block adjustment of pressure required to satisfy the global mass flow constraint. For the first streamwise sweep only, the FLARE approximation (see Section 7.4.2) is used to advance the solution through any regions of reversed flow.
2. The velocities are corrected to satisfy continuity locally using the potential function $\hat{\phi}$ as indicated above.
3. The pressure at $i + 1$ is now updated by one SOR pass across the flow at the $i + 1$ level. This is optional at this point, as all pressures are further improved at the end of the marching sweep.
4. Steps 1–3 are repeated for all streamwise stations until the downstream boundary is reached.
5. At the conclusion of the marching sweep, the pressures throughout the flow are updated by several iterations using the Poisson equation. This completes one global iteration. The next marching sweep then starts at the inflow boundary using the revised pressure field. The process continues until the velocity corrections become negligible; i.e., the pressure field obtained permits solutions to the momentum equations that also satisfy the continuity equation.

Sample computational results from the PPNS procedure are shown in Figs. 8.10 and 8.11. Chilukuri and Pletcher (1980) found that solutions to the PPNS equations for 2-D laminar channel inlet flows agreed well with solutions to the full Navier-Stokes equations for channel Reynolds numbers as low as 10. Velocity profiles predicted by the PPNS procedure are compared with the Navier-Stokes solutions obtained by McDonald et al. (1972) for a channel Reynolds number (Re $= u_\infty a/v$, where a is the channel half-width) equal to 75 in Fig. 8.10. For reference, solutions to the boundary-layer equations are also shown in the figure. The boundary-layer solutions fail to exhibit the velocity overshoots characteristic of solutions of the PPNS and Navier-Stokes equations. The PPNS scheme employed 32 grid points in the streamwise direction and 18 across the flow. The sum of the magnitudes of the mass sources at any streamwise station were reduced to less than 1% of the channel mass flow rate in seven streamwise marching sweeps.

PPNS results obtained by Madavan and Pletcher (1982) for a separated external flow are compared with numerical solutions to the Navier-Stokes equations obtained by Briley (1971) in Fig. 8.11. The flow separates under the influence of a linearly decelerating external stream. At a point downstream of separation, the external stream velocity becomes constant, causing the flow to reattach. Reversed flow exists over approximately one-third of the streamwise extent of the computational domain. In the PPNS calculation, 35 grid points were employed in the streamwise direction and 32 across the flow. Sixteen streamwise sweeps were required to reduce the sum of the magnitudes of the mass sources at any streamwise station to less than 1% of the mass flow rate. The computation was continued for a total of 43 streamwise sweeps, at which

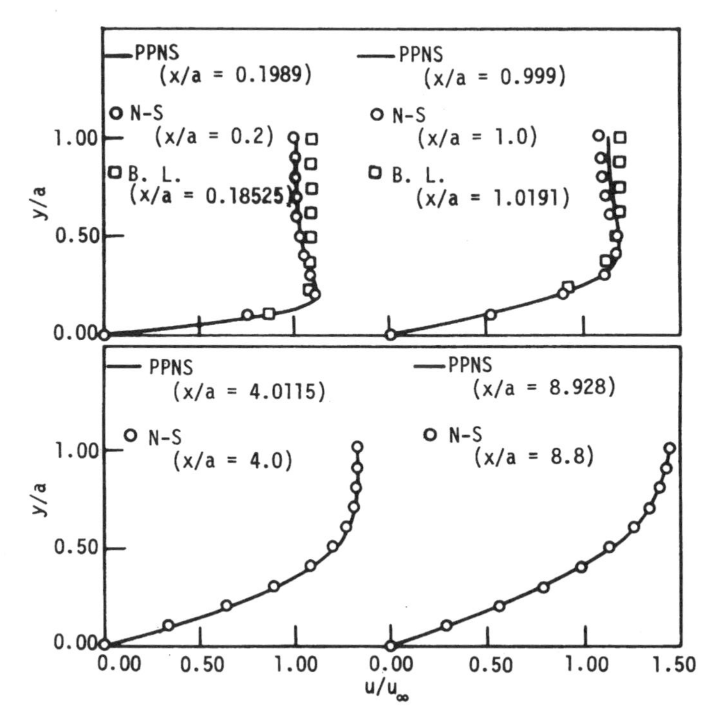

Figure 8.10 Comparison of velocity profiles predicted by the PPNS procedure (Chilukuri and Pletcher, 1980) with the Navier-Stokes (N-S) solutions of McDonald et al. (1972) and with boundary-layer (B.L.) solutions obtained using the method of Nelson and Pletcher (1974), Re = 75.

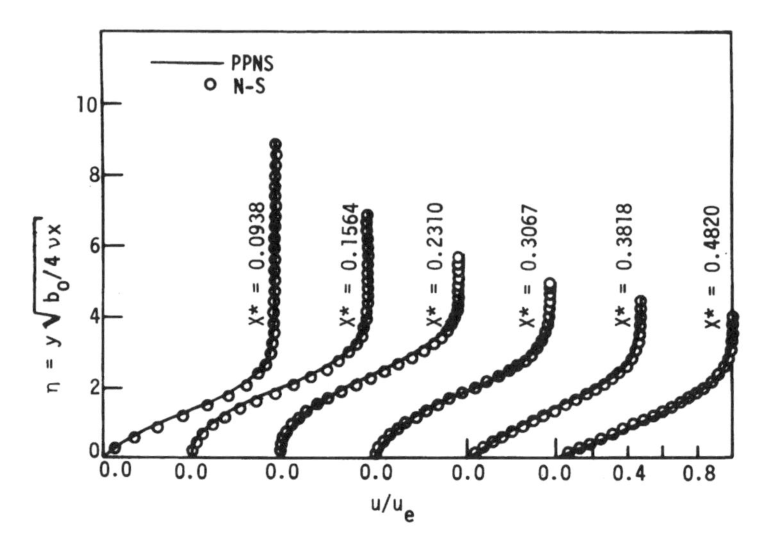

Figure 8.11 Comparison of velocity profiles predicted by the PPNS procedure (Madavan and Pletcher, 1982) with the Navier-Stokes (N-S) solutions of Briley (1971) for a laminar separating and reattaching flow; $x^* = b_1 x / b_0$, $b_0 = 30.48$ m/s, $b_1 = 300\ \text{s}^{-1}$.

time the sum of the magnitudes of the mass sources at any streamwise station was less than 0.05%.

Coupled PPNS schemes. Although the multiple space-marching strategy for coupled systems does not require that the pressure be entirely fixed for each streamwise marching sweep, certain constraints do exist. Specifically, stability requires that the downstream value of pressure in the difference expression for the streamwise pressure gradient be fixed. To see this, we will consider solving the following 2-D system by a coupled space-marching scheme.

continuity:

$$\frac{\partial(\rho^* u^*)}{\partial x^*} + \frac{\partial(\rho^* v^*)}{\partial y^*} = 0 \tag{8.155}$$

x momentum:

$$\rho^* u^* \frac{\partial u^*}{\partial x^*} + \rho^* v^* \frac{\partial u^*}{\partial y^*} = -\frac{\partial p^*}{\partial x^*} + \frac{1}{\text{Re}} \frac{\partial}{\partial y^*}\left(\mu \frac{\partial u^*}{\partial y^*}\right) \tag{8.156}$$

y momentum:

$$\rho^* u^* \frac{\partial v^*}{\partial x^*} + \rho^* v^* \frac{\partial v^*}{\partial y^*} = -\frac{\partial p^*}{\partial y^*} + \underbrace{\frac{1}{\text{Re}} \frac{\partial}{\partial y^*}\left(\mu \frac{\partial v^*}{\partial y^*}\right)} \tag{8.157}$$

energy:

$$T^* + \tfrac{1}{2}(u^{*2} + v^{*2}) = H_0 \tag{8.158}$$

state:

$$p^* = \frac{\gamma - 1}{\gamma}\rho^* T^* \tag{8.159}$$

In the above equations, asterisks denote nondimensional variables. Length coordinates have been nondimensionalized by a characteristic length L, the velocities by a reference velocity u_∞, the density by ρ_∞, the pressure by $\rho_\infty u_\infty^2$, the temperature by u_∞^2/c_{p_∞} and the viscosity by μ_∞. Note that a simplified version of the energy equation, a statement of constant total enthalpy, is used here for convenience. The Reynolds number is given by $\text{Re} = \rho_\infty u_\infty L/\mu_\infty$ and $\gamma = 1.4$. The term marked off by a brace in Eq. (8.157) is often omitted on the basis of order-of-magnitude arguments (Rubin, 1984). Without the viscous term in the y momentum equation, the system represents the composite of the traditional interacting boundary-layer flow model in which the boundary-layer equations are solved in viscous regions using a pressure gradient determined by a solution of the Euler equations. Most often, such a system has been called the RNS equations, although more general formulations (Ramakrishnan and Rubin, 1987) have also included the time term in the RNS system.

It is instructive to consider the conditions under which Eqs. (8.155)–(8.159) can be solved by a space-marching procedure. Following Liu and Pletcher (1986), we first use the equations of state and energy to rewrite the continuity equation in the form

$$\left(1 + \frac{u^{*2}}{T^*}\right)\frac{\partial u^*}{\partial x^*} + \frac{u^* v^*}{T^*}\frac{\partial v^*}{\partial x^*} + \frac{u^*}{p^*}\frac{\partial p^*}{\partial x^*} + \frac{u^* v^*}{T^*}\frac{\partial u^*}{\partial y^*} + \left(1 + \frac{v^{*2}}{T^*}\right)\frac{\partial v^*}{\partial y^*} + \frac{v^*}{p^*}\frac{\partial p^*}{\partial y^*} = 0 \tag{8.160}$$

We let $\mathbf{Z}$ be the vector of primitive variables u, v, p and write Eqs. (8.156), (8.157), and (8.160) in matrix-vector form as

$$[A]\frac{\partial \mathbf{Z}}{\partial x^*} + [B]\frac{\partial \mathbf{Z}}{\partial y^*} = [C]\frac{\partial}{\partial y^*}\left(\mu\frac{\partial \mathbf{Z}}{\partial y^*}\right) + \mathbf{d} \tag{8.161}$$

where

$$[A] = \begin{bmatrix} \rho^* u^* & 0 & 1 \\ 0 & \rho^* u^* & 0 \\ 1 + \dfrac{u^{*2}}{T^*} & \dfrac{u^* v^*}{T^*} & \dfrac{u^*}{p^*} \end{bmatrix} \qquad [B] = \begin{bmatrix} \rho^* v^* & 0 & 0 \\ 0 & \rho^* v^* & 1 \\ \dfrac{u^* v^*}{T^*} & 1 + \dfrac{v^{*2}}{T^*} & \dfrac{v^*}{p^*} \end{bmatrix}$$

$$[C] = \begin{bmatrix} \dfrac{1}{\text{Re}} & 0 & 0 \\ \dfrac{1}{\text{Re}} & 0 & 0 \\ 0 & 0 & 0 \end{bmatrix} \qquad \mathbf{d} = \begin{bmatrix} 0 \\ 0 \\ 0 \end{bmatrix} \tag{8.162}$$

As indicated in Chapter 2, a marching procedure in the x^* direction can be used to solve Eq. (8.161) if the eigenvalues of $[A]^{-1}[C]$ are nonnegative and real. Wang et al. (1981) have shown that the necessary condition is actually that the roots of the characteristic equation

$$\det |[C] - \lambda[A]| = 0 \tag{8.163}$$

be real and nonnegative. The roots (eigenvalues) of Eq. (8.163) and the eigenvalues of $[A]^{-1}[C]$ are identical as long as the inverse of $[A]$ exists. The three eigenvalues of $[A]^{-1}[C]$ are $\{0, 0, [(u^* + p^* v^*/\rho^*)/\text{Re}\, p^*(M^2 - 1)]\}$. When all viscous terms are neglected in the y momentum equation, the expression simplifies to $\{0, 0, [u^*/\text{Re}\, p^*(M^2 - 1)]\}$. In both cases, when the velocities are positive, the eigenvalues will be nonnegative only for supersonic flow. Thus for subsonic flow the system is not well posed for a marching solution in x^*.

To enable space-marching for subsonic flow, the system can be rearranged by treating the pressure gradient in the x momentum equation as a source term depending at most on the unknown pressure. That is, in the discretization of the pressure gradient, the pressure at one streamwise location must be fixed, for example, by using a value from the previous global sweep. Stability considerations as well as arguments based on characteristics dictate that it is the downstream pressure that should be fixed. This can be incorporated into the system by modifying $[A]$ and $\mathbf{d}$ in Eq. (8.162) as follows:

$$[A] = \begin{bmatrix} \rho^* u^* & 0 & 0 \\ 0 & \rho^* u^* & 0 \\ 1 + \dfrac{u^{*2}}{T^*} & \dfrac{u^* v^*}{T^*} & \dfrac{u^*}{p^*} \end{bmatrix} \qquad \mathbf{d} = \begin{bmatrix} -\dfrac{\partial p^{*+}}{\partial x^*} \\ 0 \\ 0 \end{bmatrix}$$

where it is understood that the pressure gradient in $\mathbf{d}$ is to be discretized such that the downstream value is fixed and the other pressure in the difference is treated as an unknown. The eigenvalues for the system so modified are $\{0, 0, [1/\text{Re}\, \rho^* u^*)]\}$ with or without the viscous term included in the y momentum equation. Thus for $u^* > 0$, the system can be solved by marching in the positive x^* direction. In some applications of interest, local regions of flow reversal are embedded in a flow that is predominantly in the positive x^* direction. Under these conditions ($u^* < 0$, $M^2 < 1$), type-dependent differencing can be used for the streamwise convective terms in the momentum equations with the values at the upstream point in the difference being fixed using values from a previous global sweep. The FLARE approximation can be used for the very first global sweep. In the context of marching the solution in the positive x^* direction, this amounts to treating the streamwise convective term as a source term as was done for the pressure gradient in the streamwise momentum equation. The eigenvalues for the system treated as indicated above in embedded regions of reversed flow are $(0, 0, 0)$.

In the coupled-marching approach, values of u, v, and p (in 2-D) are computed at each global sweep. For flows that are fully attached ($u > 0$), as was the case for the pressure-correction methods, the entire history of the flow from one global sweep to the next is recorded in the pressure field. Storage for the entire velocity field is not required. If flow reversal occurs, velocities in the reversed flow region must be stored for use in the type-dependent differencing of convective terms in the next global sweep.

Rubin and Lin (1981), Rubin and Reddy (1983), and Israeli and Lin (1985) were among the first to report coupled space-marching methods for reduced forms of the Navier-Stokes (RNS) equations for incompressible or subsonic flow. Other coupled approaches for reduced equations include those reported by Liu and Pletcher (1986) and TenPas (1990). A discussion of recent coupled approaches is included in the review article by Rubin and Tannehill (1992).

A variety of discretizations have been demonstrated for coupled space-marching schemes. Some have maintained the conservation-law form of the equations and have employed generalized coordinates. Implicit methods have been most often used. In two dimensions, a block tridiagonal elimination scheme has worked well. Three-dimensional applications have been demonstrated by Reddy and Rubin (1988). Ramakrishnan and Rubin (1987) describe a time-consistent version of the 2-D PPNS formulation (all viscous terms in the y momentum equation neglected).

Many aspects of the discretization for the coupled PPNS approaches are similar to procedures discussed previously for the boundary-layer equations, PNS equations, and pressure-correction PPNS procedures. Two key features of the approach will be discussed in some detail for clarity. The most unique feature of the coupled multiple space-marching procedure is the treatment of the pressure term in the streamwise momentum equation. In order to accurately accommodate a wide range of flow Mach numbers, it is desirable (Rubin, 1988) to split the representation of the pressure gradient in the streamwise momentum equation into positive and negative flux contributions (as described in Section 8.3.2). This representation is given by

$$\left.\frac{\partial p^*}{\partial x^*}\right|_{i+1} \cong \omega_{i+1/2}\frac{p_{i+1}^* - p_i^*}{\Delta x^*} + (1 - \omega_{i+3/2})\frac{p_{i+2}^* - p_{i+1}^*}{\Delta x^*}$$

where ω is the Vigneron parameter, $0 \leqslant \omega \leqslant \min(1, \omega_m)$ and $\omega_m = \gamma M_x^2/[1 + (\gamma - 1)M_x^2]$ except that $\omega = 0$ is used in regions of reversed flow. In the above, M_x is the local Mach number of the flow in the x direction. Rubin (1988) reports that splitting the streamwise pressure gradient appearing in the energy equation is optional. For incompressible flows, $\omega = 0$ is clearly appropriate. For subsonic compressible flows the optimum value of ω is problem dependent, but satisfactory results have been observed using $\omega = 0$ over a wide range of low-speed subsonic flows.

To further illustrate, a simple two-point difference will be used in the Cartesian coordinate system for an incompressible or very low speed flow ($\omega = 0$). The solution is being advanced from the i to the $i + 1$ marching station, as is always the case for global sweeps of a space-marching procedure.

Thus the pressure term can be represented as

$$\left.\frac{\partial p^*}{\partial x^*}\right|_{i+1} \cong \frac{p^{*k}_{i+2} - p^{*k+1}_{i+1}}{\Delta x^*}$$

where the superscript k indicates the global marching level. It is important to notice that the value of p^* at the $i + 2$ position is being specified using the solution from a previous sweep, while the pressure at location $i + 1$ is being treated as an unknown. Note that when splitting is employed for subsonic flows, it is still only the pressure at $i + 2$ that is fixed. Of course, when pressure splitting is employed and the local flow is supersonic, $\omega = 1$ and the influence of the downstream pressure vanishes. Rubin and Reddy (1983) point out that the best agreement with known solutions is observed if the pressure computed at $i + 1$ for incompressible flow is interpreted as the pressure at the upstream station (station i for $u^* > 0$).

The other, somewhat subtle, point is the representation of streamwise convective terms in embedded regions of reversed flow. The procedure normally used is to treat the downwind value in the difference stencil as the unknown and the upwind value as known. In our example we shall let the value being computed (downwind value) be located at the $i + 1$ station. Thus the upwind station would be at level i for $u^* > 0$ and at level $i + 2$ for $u^* < 0$. Due to the existence of a predominant flow direction, at each streamwise location some of the flow is moving in the positive x^* direction. In that region a convective derivative (choosing an incompressible flow for simplicity) might be represented as

$$\left.\frac{\partial u^*\phi^*}{\partial x^*}\right|_{i+1} \cong \frac{u^{*k}_{i+1}\overline{\phi^{*k+1}_{i+1}} - u^{*k+1}_{i}\phi^{*k+1}_{i} + \overline{u^{*k+1}_{i+1}}\phi^{*k}_{i+1} - u^{*k}_{i+1}\phi^{*k}_{i+1}}{\Delta x^*}$$

where the expression has been linearized by a Newton method expanding about the values obtained at the previous global sweep. The values treated as unknowns in the algebraic formulation are marked by overbars for emphasis. The symbol ϕ indicates a variable such as u^* or v^*. In the reversed flow portion of the flow at $i + 1$ the representation would be

$$\left.\frac{\partial u^*\phi^*}{\partial x^*}\right|_{i+1} \cong \frac{u^{*k}_{i+2}\phi^{*k}_{i+2} - \overline{u^{*k+1}_{i+1}}\phi^{*k}_{i+1} - u^{*k}_{i+1}\overline{\phi^{*k+1}_{i+1}} + u^{*k}_{i+1}\phi^{*k}_{i+1}}{\Delta x^*}$$

where the unknown being determined is again marked by overbars.

Boundary condition specification may vary somewhat depending on the specific discretization employed, problem being solved, and whether or not a diffusion term is used in the y momentum equation. Most frequently for incompressible or subsonic flows, velocities or one velocity and a streamwise velocity derivative and a thermal variable (when an energy equation is included) would be specified at the inflow boundary. The pressure at inflow is extrapolated from the interior but specified at the outflow boundary. No-slip conditions are specified at solid boundaries, and one velocity and the pressure or a pressure derivative is specified at free stream boundaries. When the viscous term is included in the y momentum equation, one more variable, usually the normal component of velocity, must be specified at one of the boundaries. Liu and Pletcher (1986) describe a systematic way to couple the boundary conditions on both sides of a 2-D domain, which provides flexibility for choosing the variables to be specified.

A number of improvements to the basic multiple space-marching coupled procedure have been suggested for subsonic applications. These have been motivated largely by the observation that the influence of the downstream pressure propagates upstream only one grid point per global iteration. Thus to improve convergence, some investigators (TenPas, 1990; Bentson and Vradis, 1987) have employed a backsweep of an approximate Poisson equation between global marching sweeps to rapidly transmit pressure changes in the upstream direction and thereby accelerate the global convergence.

It will be pointed out in the next chapter that space-marching procedures can be employed to solve the steady flow version of the full Navier-Stokes equations. The only difference between the PPNS equations and the steady-flow Navier-Stokes equations is the omission of streamwise diffusion terms in the former. Such terms can be included in a multiple space-marching scheme if values from a previous global sweep are used at the $i + 2$ level in the representation of the streamwise diffusion terms. The effects of streamwise diffusion are usually only significant for flows at low Reynolds numbers. Figure 8.12, taken from TenPas (1990), shows the streamwise development of the axial component of velocity at the center of a 2-D channel for several Reynolds numbers. A uniform velocity distribution was specified at the inlet. Solutions to the full Navier-Stokes equations and the PPNS equations are shown, revealing that at the lowest Re, 0.5, the differences are significant. Differences can still be discerned at Re = 10 but become insignificant at Re = 75.

8.5 VISCOUS SHOCK-LAYER EQUATIONS

The viscous shock-layer (VSL) equations are a more approximate set of equations than the PNS equations. In terms of complexity, they fall between the PNS equations and the boundary-layer equations. The major advantage of the VSL equations is that they remain hyperbolic-parabolic in both the streamwise and crossflow directions. Thus the VSL equations can be solved using a marching procedure in both directions very similar to techniques employed for the 3-D boundary-layer equations. This is in contrast to the PNS equations, which must be solved simultaneously over the entire crossflow plane. As a consequence, the VSL equations can be solved (in most cases) with less computer time than the PNS equations. An additional advantage of the VSL equations is that they can be used to compute the viscous flow in the subsonic blunt-nose region, where the PNS equations are not applicable. Thus, for bodies with blunt noses, the VSL equations can be solved to provide a starting solution for a subsequent PNS

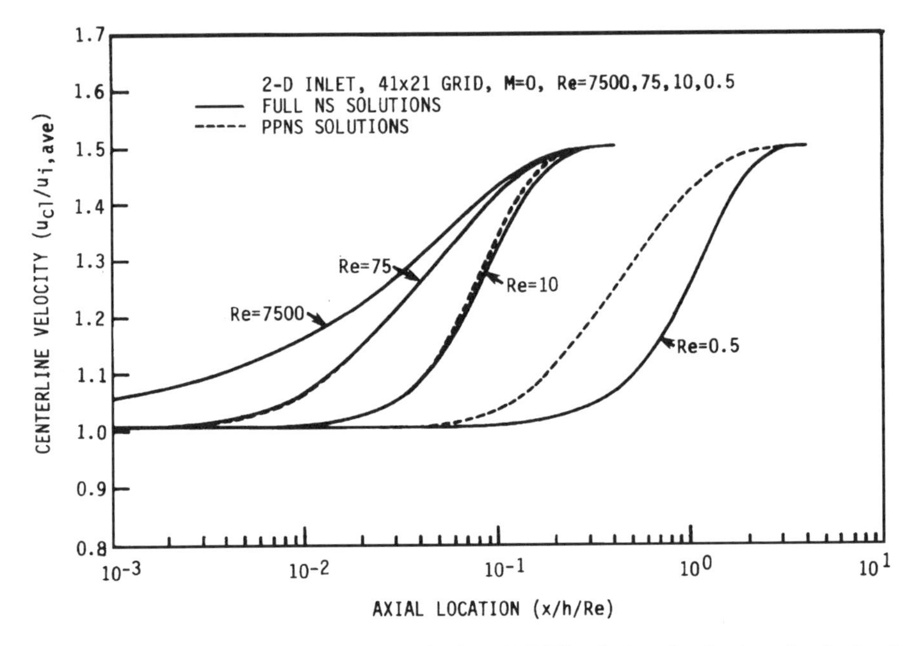

Figure 8.12 Comparison of Navier-Stokes (NS) and PPNS solutions for laminar developing two-dimensional flow in a channel (TenPas, 1990).

computation. The major disadvantage of the VSL equations is that they cannot be used to compute flow fields with crossflow separations. This is a direct result of the fact that the VSL equations are not elliptic in the crossflow plane.

The concept of using a set of equations such as the VSL equations to solve for the high Mach number flow past a blunt body had its origins in the work of Cheng (1963) and Davis and Flügge-Lotz (1964). As mentioned previously, the solution of a set of equations like the VSL equations avoids the need to explicitly determine the second-order boundary-layer effects of vorticity and displacement thickness. Furthermore, it eliminates the difficulty encountered in matching the viscous and inviscid solutions, when the boundary layer is significantly merged with the outer inviscid flow.

Of all the early studies involving the use of the VSL equations, the method of Davis (1970) was the most successful. He solved the axisymmetric VSL equations in order to determine the hypersonic laminar flow over a hyperboloid. The VSL equations used by Davis are derived by first nondimensionalizing the Navier-Stokes equations with variables that are of order 1 in the boundary layer for large Reynolds numbers. In a similar manner, another set of equations is formed by nondimensionalizing the Navier-Stokes equations with variables that are of order 1 in the inviscid region of the flow field. In both sets of equations, terms up to second order in ϵ are retained:

$$\epsilon = \left(\frac{\mu_{\text{ref}}}{\rho_\infty V_\infty r_{\text{nose}}} \right)^{1/2} \tag{8.164}$$

where μ_{ref} is the coefficient of viscosity evaluated at the reference temperature,

$$T_{\text{ref}} = \frac{V_\infty^2}{c_{p_\infty}} \tag{8.165}$$

The two sets of equations are then compared and combined into a single set of equations, which is valid to second order from the body to the shock. For a 2-D ($m = 0$) or axisymmetric ($m = 1$) body intrinsic coordinate system (see Fig. 5.3), the VSL equations in nondimensional form become as follows:

continuity:

$$\frac{\partial}{\partial \xi^*}\left[(r^* + \eta^* \cos\phi)^m \rho^* u^*\right] + \frac{\partial}{\partial \eta^*}\left[(1 + K^*\eta^*)(r^* + \eta^* \cos\phi)^m \rho^* v^*\right] = 0 \tag{8.166}$$

ξ momentum:

$$\rho^* \left(\frac{u^*}{1 + K^*\eta^*} \frac{\partial u^*}{\partial \xi^*} + v^* \frac{\partial u^*}{\partial \eta^*} + \frac{K^* u^* v^*}{1 + K^*\eta^*} \right) + \frac{1}{1 + K^*\eta^*} \frac{\partial p^*}{\partial \xi^*}$$
$$= \frac{\epsilon^2}{(1 + K^*\eta^*)^2 (r^* + \eta^* \cos\phi)^m} \frac{\partial}{\partial \eta^*}\left[(1 + K^*\eta^*)^2 (r^* + \eta^* \cos\phi)^m \tau^*\right] \tag{8.167}$$

where

$$\tau^* = \mu^* \left(\frac{\partial u^*}{\partial \eta^*} - \frac{K^* u^*}{1 + K^*\eta^*} \right)$$

η momentum:

$$\rho^* \left(\frac{u^*}{1 + K^*\eta^*} \frac{\partial v^*}{\partial \xi^*} + v^* \frac{\partial v^*}{\partial \eta^*} - \frac{K^* (u^*)^2}{1 + K^*\eta^*} \right) + \frac{\partial p^*}{\partial \eta^*} = 0 \tag{8.168}$$

energy:

$$\rho^* \left(\frac{u^*}{1 + K^*\eta^*} \frac{\partial T^*}{\partial \xi^*} + v^* \frac{\partial T^*}{\partial \eta^*} \right) - \frac{u^*}{1 + K^*\eta^*} \frac{\partial p^*}{\partial \xi^*} - v^* \frac{\partial p^*}{\partial \eta^*}$$
$$= \frac{\epsilon^2 (\tau^*)^2}{\mu^*} + \frac{\epsilon^2}{(1 + K^*\eta^*)(r^* + \eta^* \cos\phi)^m}$$
$$\times \frac{\partial}{\partial \eta^*}\left[(1 + K^*\eta^*)(r^* + \eta^* \cos\phi)^m \frac{\mu^*}{\text{Pr}} \frac{\partial T^*}{\partial \eta^*}\right] \tag{8.169}$$

These equations have been nondimensionalized in the following manner:

$$\xi^* = \frac{\xi}{r_{\text{nose}}} \qquad \eta^* = \frac{\eta}{r_{\text{nose}}} \qquad r^* = \frac{r}{r_{\text{nose}}} \qquad K^* = \frac{K}{r_{\text{nose}}}$$
$$u^* = \frac{u}{V_\infty} \qquad v^* = \frac{v}{V_\infty} \qquad T^* = \frac{T}{T_{\text{ref}}} \tag{8.170}$$
$$p^* = \frac{p}{\rho_\infty V_\infty^2} \qquad \rho^* = \frac{\rho}{\rho_\infty} \qquad \mu^* = \frac{\mu}{\mu_{\text{ref}}}$$

By assuming a thin shock layer, the normal momentum equation reduces to

$$\frac{\partial p^*}{\partial \eta^*} = \frac{K^* \rho^* (u^*)^2}{1 + K^* \eta^*} \tag{8.171}$$

The above equations can be readily converted to a 2-D Cartesian coordinate system by setting

$$\begin{aligned} m &= 0 \\ K^* &= 0 \\ x^* &= \xi^* \\ y^* &= \eta^* \end{aligned} \tag{8.172}$$

The resulting VSL equations in Cartesian coordinates can then be compared directly with the PNS equations given previously by Eqs. (8.29)–(8.33). This comparison shows that the continuity and x momentum equations are the same but the y momentum and energy equations in the VSL set of equations are simpler than the corresponding PNS equations.

In the original solution technique of Davis, the VSL equations were normalized with variable values behind the shock. This was done to permit the same grid in the normal direction to be used over the entire body. An initial global solution was obtained by utilizing the thin shock-layer assumption. This assumption makes the VSL equations totally parabolic and permits the use of standard boundary-layer solution algorithms. Subsequent global iterations retained the complete normal momentum equation. Also, for the first global iteration the shock was assumed to be concentric with the body. This assumption was possible because only hyperboloid body shapes were considered because of the difficulties associated with curvature discontinuities in body shapes such as sphere-cones. The shock angles for the second iteration were determined from the shock-layer thicknesses computed during the first iteration.

The marching procedure was initiated from an approximate stagnation streamline solution. This stagnation streamline solution was obtained from the VSL equations, which reduce to ODEs along $\xi = 0$. The solution at each subsequent ξ station was obtained by solving the VSL equations individually in the following order:

1. energy
2. ξ momentum
3. continuity
4. η momentum

The original method of Davis was not entirely satisfactory because of several limitations. First, the method was restricted to analytic body shapes such as hyperboloids. This difficulty was circumvented by Miner and Lewis (1975), who computed the flow around a sphere-cone body. They started with an initial shock shape from an inviscid blunt-body solution and used a transition function near the sphere-cone juncture in order to obtain a smooth distribution of curvature. Later, Srivastava et al. (1978) overcame this same difficulty by applying special difference formulas to the jump conditions across surface discontinuities.

Another difficulty associated with the original Davis method was the poor convergence of the shock shape when the shock layer became thick. This problem was resolved by Srivastava et al. (1978, 1979), who noted that the relaxation process associated with the shock shape was similar to the interaction between displacement thickness and the outer inviscid flow in supersonic interacting boundary-layer theory. As a result of this observation, they were able to solve the shock-shape divergence problem by adapting the ADI method of Werle and Vatsa (1974) for interacting boundary layers. Another problem with the original Davis method was that it was not able to solve the flow far downstream on slender bodies. This difficulty was traced to the fact that the VSL equations were being solved in an uncoupled manner. In particular, the two first-order equations (continuity and normal momentum) introduced instabilities that grew in the streamwise direction. By solving the continuity and normal momentum equations in a coupled fashion, Waskiewicz et al. (1978) were able to eliminate this stability problem. In a similar manner, Hosny et al. (1978) overcame the problem by completely coupling the VSL equations through a quasi-linearization technique. More recently, Gordon and Davis (1992) coupled the VSL equations with an additional equation for the shock standoff distance and have thereby eliminated the need for local iterations. They also developed a new global iteration procedure that uses Vigneron's technique (Vigneron et al., 1978a, 1978b) to split the differencing of the streamwise derivatives, $\partial p/\partial \xi$ and $\partial v/\partial \xi$, into forward and backward parts. In addition, Gupta et al. (1992) developed a solution procedure for the VSL equations in which global iterations are required only in the nose region of a blunt body.

The VSL equations have been successfully applied to a large number of different blunt-body flow fields. Murray and Lewis (1978) were the first to solve the flow around general 3-D body shapes at angle of attack. Since then, the VSL equations have been applied to a variety of body shapes, including sphere-cones (Murray and Lewis, 1978; Gogineni et al., 1980; Thompson et al., 1987), ellipsoids (Szema and Lewis, 1981), sphere-cone-cylinder-flares (Kim and Lewis, 1982), space shuttle geometry (Szema and Lewis, 1981; Shinn et al., 1982; Kim et al., 1983; Thompson, 1987), and aeroassist orbital transfer vehicles (Shinn and Jones, 1983; Carlson and Gally, 1991). In addition, VSL codes have been enhanced to account for turbulence (Anderson et al., 1976; Szema and Lewis, 1980; Thareja et al., 1982; Gupta et al., 1990), equilibrium chemistry (Thareja et al., 1982; Swaminathan et al., 1982; Gupta, 1987), nonequilibrium chemistry (Moss, 1974; Miner and Lewis, 1975; Swaminathan et al., 1983; Shinn et al., 1982; Kim et al., 1983; Song and Lewis, 1986; Thompson, 1987; Gupta et al., 1987; Zoby et al., 1989; Bhutta and Lewis, 1991), ablation (Thompson et al., 1983; Song and Lewis, 1986; Bhutta et al., 1989; Gupta et al., 1990), slip effects (Swaminathan et al., 1984; Lee et al., 1990), and catalytic walls (Shinn et al., 1982; Kim et al., 1983; Thompson, 1987).

8.6 "CONICAL" NAVIER-STOKES EQUATIONS

The conical flow assumption for inviscid flows makes use of the fact that a significant length scale is missing in the conical direction for a flow field surrounded by conical boundaries. As a result, no variations in flow properties in the radial direction can occur, and a 3-D inviscid flow problem is reduced to a 2-D problem. This leads to a self-similar solution, which is the same for all constant radius surfaces but scales linearly with the radius. The concept of conical flow is strictly valid only for inviscid flows. However, the viscous portions of the same flow fields have been observed in experiments to be strongly dominated by the outer inviscid conical flow. For these flow fields, Anderson (1973) suggested that a quick estimate of the heat transfer and skin friction could be obtained by solving the unsteady Navier-Stokes equations in a time-dependent fashion on the unit sphere with all derivatives in the radial direction set equal to zero. Thus the Navier-Stokes equations are solved subject to a local conical approximation. We will refer to the equations solved in this manner as the "conical" Navier-Stokes (CNS) equations. The local Reynolds number is determined by the radial position where the solution is computed. As a result, the solution is not self-similar in the sense of inviscid conical flow, but is scaled through the local Reynolds number, which remains in the resulting set of equations.

The CNS equations were originally used by McRae (1976) to compute the laminar flow over a cone at high angle of attack. Since then, Vigneron et al. (1978a), Bluford (1978), McMillin et al. (1987), and Ruffin and Murman (1988) have computed the laminar flow over a delta wing, and Tannehill and Anderson (1980) have computed the flow in a 3-D axial corner. Also, McRae and Hussaini (1978) have employed an eddy viscosity model in conjunction with the CNS equations to compute the turbulent flow over a cone at high angle of attack. In all of the above cases (except one, where the inviscid flow was not completely conical) the computed inviscid and viscous portions of the flow field agree surprisingly well with the available experimental data.

The CNS equations have also proved quite useful in computing starting solutions for PNS calculations of flows over conical (or pointed) body shapes. This is the primary use of the CNS equations today. Schiff and Steger (1979) and others have incorporated a marching step-back method in their PNS codes, which is equivalent to solving the CNS equations using the time-dependent approach described above. In these codes the flow variables are initially set equal to their free stream values, and the equations are marched from $x = x_0$ to $x = x_0 + \Delta x$ using the same implicit scheme as used to solve the PNS equations but with $\partial p / \partial x = 0$. After each marching step, the solution is scaled back to $x = x_0$. The computation is repeated until no change in flow variables occurs.

The CNS equations are derived from the complete Navier-Stokes equations,

$$\frac{\partial \mathbf{U}^*}{\partial t^*} + \frac{\partial \mathbf{E}^*}{\partial x^*} + \frac{\partial \mathbf{F}^*}{\partial y^*} + \frac{\partial \mathbf{G}^*}{\partial z^*} = 0 \tag{8.173}$$

where $\mathbf{U}^*$, $\mathbf{E}^*$, $\mathbf{F}^*$, and $\mathbf{G}^*$ are the nondimensional vectors defined by Eqs. (5.46). The following conical transformation

$$\begin{aligned} \alpha &= \left[(x^*)^2 + (y^*)^2 + (z^*)^2 \right]^{1/2} \\ \beta &= \frac{y^*}{x^*} \\ \gamma &= \frac{z^*}{x^*} \\ \tau &= t^* \end{aligned} \tag{8.174}$$

is initially applied to these equations. The resulting transformed equations can be written in the following strong conservation-law form:

$$\frac{\partial}{\partial \tau}\left(\frac{\alpha^2}{\lambda^3} \mathbf{U}^* \right) + \frac{\partial}{\partial \alpha}\left[\frac{\alpha^2}{\lambda^4} (\mathbf{E}^* + \beta \mathbf{F}^* + \gamma \mathbf{G}^*) \right] + \frac{\partial}{\partial \beta}\left[\frac{\alpha}{\lambda^2} (-\beta \mathbf{E}^* + \mathbf{F}^*) \right] + \frac{\partial}{\partial \gamma}\left[\frac{\alpha}{\lambda^2} (-\gamma \mathbf{E}^* + \mathbf{G}^*) \right] = 0 \tag{8.175}$$

where

$$\lambda = (1 + \beta^2 + \gamma^2)^{1/2}$$

The assumption of local conical self-similarity requires that

$$\begin{aligned} \frac{\partial \mathbf{E}^*}{\partial \alpha} &= 0 \\ \frac{\partial \mathbf{F}^*}{\partial \alpha} &= 0 \\ \frac{\partial \mathbf{G}^*}{\partial \alpha} &= 0 \end{aligned} \tag{8.176}$$

which reduces Eq. (8.175) to

$$\frac{\partial}{\partial \tau}\left(\frac{\alpha^2}{\lambda^3}\mathbf{U}^*\right) + \frac{2\alpha}{\lambda^4}(\mathbf{E}^* + \beta\mathbf{F}^* + \gamma\mathbf{G}^*) + \frac{\partial}{\partial \beta}\left[\frac{\alpha}{\lambda^2}(-\beta\mathbf{E}^* + \mathbf{F}^*)\right] + \frac{\partial}{\partial \gamma}\left[\frac{\alpha}{\lambda^2}(-\gamma\mathbf{E}^* + \mathbf{G}^*)\right] = 0 \tag{8.177}$$

The solution is computed on a spherical surface whose nondimensional radius ($r^* = r/L$) is equal to 1. On this computational surface, $\alpha = 1$, since

$$r^* = \left[(x^*)^2 + (y^*)^2 + (z^*)^2\right]^{1/2} = \alpha$$

As a result, Eq. (8.177) can be rewritten as

$$\frac{\partial \mathbf{U}_4}{\partial \tau} + \frac{\partial \mathbf{F}_4}{\partial \beta} + \frac{\partial \mathbf{G}_4}{\partial \gamma} + \mathbf{H}_4 = 0 \tag{8.178}$$

where

$$\begin{aligned} \mathbf{U}_4 &= \frac{\mathbf{U}^*}{\lambda^3} \\ \mathbf{F}_4 &= \frac{-\beta\mathbf{E}^* + \mathbf{F}^*}{\lambda^2} \\ \mathbf{G}_4 &= \frac{-\gamma\mathbf{E}^* + \mathbf{G}^*}{\lambda^2} \\ \mathbf{H}_4 &= \frac{2(\mathbf{E}^* + \beta\mathbf{F}^* + \gamma\mathbf{G}^*)}{\lambda^4} \end{aligned} \tag{8.179}$$

The partial derivatives appearing in the viscous terms of $\mathbf{E}^*$, $\mathbf{F}^*$, and $\mathbf{G}^*$ are readily transformed using

$$\begin{aligned} \frac{\partial}{\partial x^*} &= -\beta\lambda\frac{\partial}{\partial \beta} - \gamma\lambda\frac{\partial}{\partial \gamma} \\ \frac{\partial}{\partial y^*} &= \lambda\frac{\partial}{\partial \beta} \\ \frac{\partial}{\partial z^*} &= \lambda\frac{\partial}{\partial \gamma} \end{aligned} \tag{8.180}$$

Thus the shear-stress and heat flux terms, given in Eqs. (5.47), become

$$\tau_{xx}^* = \frac{2\mu^*}{3\,\mathrm{Re}_L}(-2\beta\lambda u_\beta^* - 2\gamma\lambda u_\gamma^* - \lambda v_\beta^* - \lambda w_\gamma^*)$$

$$\tau_{yy}^* = \frac{2\mu^*}{3\,\mathrm{Re}_L}(2\lambda v_\beta^* + \beta\lambda u_\beta^* + \gamma\lambda u_\gamma^* - \lambda w_\gamma^*)$$

$$\tau_{zz}^* = \frac{2\mu^*}{3\,\mathrm{Re}_L}(2\lambda w_\gamma^* + \beta\lambda u_\beta^* + \gamma\lambda u_\gamma^* - \lambda v_\beta^*)$$

$$
\begin{aligned}
\tau_{xy}^* &= \frac{\mu^*}{\mathrm{Re}_L}(\lambda u_\beta^* - \beta\lambda v_\beta^* - \gamma\lambda v_\gamma^*) \\
\tau_{xz}^* &= \frac{\mu^*}{\mathrm{Re}_L}(\lambda u_\gamma^* - \beta\lambda w_\beta^* - \gamma\lambda w_\gamma^*) \\
\tau_{yz}^* &= \frac{\mu^*}{\mathrm{Re}_L}(\lambda v_\gamma^* + \lambda w_\beta^*) \\
q_x^* &= \frac{\mu^*}{(\gamma-1)M_\infty^2 \,\mathrm{Re}_L \,\mathrm{Pr}}(-\beta\lambda T_\beta^* - \gamma\lambda T_\gamma^*) \\
q_y^* &= \frac{\mu^*}{(\gamma-1)M_\infty^2 \,\mathrm{Re}_L \,\mathrm{Pr}}\lambda T_\beta^* \\
q_z^* &= \frac{\mu^*}{(\gamma-1)M_\infty^2 \,\mathrm{Re}_L \,\mathrm{Pr}}\lambda T_\gamma^*
\end{aligned}
\tag{8.181}
$$

Note that the Reynolds number Re_L remains in the expressions for shear stress and heat flux. This Reynolds number is evaluated using

$$
\mathrm{Re}_L = \frac{\rho_\infty V_\infty L}{\mu_\infty} \tag{8.182}
$$

where L is the radius of the spherical surface where the solution is computed. As a consequence, solutions of the CNS equations depend directly on the position ($r = L$) where they are computed. This is different from inviscid solutions, which are independent of r and, thus, truly conical.

An analysis by Rasmussen and Yoon (1990) has shown that the boundary layer computed on a circular cone with the CNS equations is thinner than would be computed with the complete Navier-Stokes equations. Nevertheless, the CNS equations are very useful in providing approximate starting solutions for PNS calculations of flows over pointed bodies.

The CNS equations can be solved using the same "time-dependent" algorithms that are applied in Chapter 9 to the 2-D compressible Navier-Stokes equations. Thus we will postpone our discussion of numerical schemes for the CNS equations until then. In closing, it should be remembered that the CNS equations are a very approximate form of the complete Navier-Stokes equations, and as such, they should not be used for flow problems where a high degree of accuracy is required.

PROBLEMS

8.1 Verify Eq. (8.8).

8.2 Derive Eqs. (8.9)–(8.11).

8.3 Reduce the thin-layer equations in Cartesian coordinates to the set of boundary equations that are valid at a no-slip wall ($y = 0$). Assume the wall is held at a constant temperature of T_w.

8.4 Reduce the thin-layer equations written in the transformed coordinate system [Eqs. (8.9)–(8.11)] to the set of boundary equations that are valid at a no-slip wall ($\eta = 0$). Assume the wall is held at a constant temperature of T_w.
8.5 Obtain Eq. (8.15) from Eq. (5.19).
8.6 Obtain Eq. (8.16) from Eq. (5.19).
8.7 Obtain Eq. (8.17) from Eq. (5.31).
8.8 Obtain Eq. (8.23) from Eq. (8.17).
8.9 Derive the compressible laminar boundary-layer equations starting with Eqs. (8.14)–(8.17). Note that in the boundary-layer region, $\Delta^2 \sim O(1)$ and $(\Delta/\delta^*)^2 \gg 1$.
8.10 Apply the thin-layer approximation to Eqs. (8.37)–(8.39) and show that they are equivalent to Eqs. (8.40), (8.10), and (8.11).
8.11 Verify that Eq. (8.44) is equivalent to Eq. (8.43).
8.12 Show that the eigenvalues of Eq. (8.44) are given by Eq. (8.46).
Hint: $|\lambda[I] - [A_1]^{-1}[B_1]| = |[A_1]^{-1}||\lambda[A_1] - [B_1]|$.
8.13 Derive Eq. (8.47).
8.14 Verify Eqs. (8.48) and (8.49).
8.15 Derive Eq. (8.50).
8.16 For the flow conditions,

$$M_x = 0.6$$

$$\frac{\text{Re}}{L} = \frac{\rho u}{\mu} = \frac{1000}{m}$$

$$\gamma = 1.4$$

$$\text{Pr} = 0.72$$

solve Eq. (8.50), and show that all the roots will be real and positive if $\omega = 0.4$, which satisfies Eq. (8.52).
8.17 Repeat Prob. 8.16 with $\omega = 0.5$, and show that at least one root of Eq. (8.50) will *not* be real and positive.
8.18 If all the eigenvalues of Eq. (8.50) are real, show that these eigenvalues are positive, provided that the conditions given by Eqs. (8.51) and (8.52) are satisfied.
8.19 Place an ω in front of the streamwise pressure gradient term in both the streamwise momentum equation and the energy equation, and evaluate the condition that must be satisfied in order for Eq. (8.44) to remain hyperbolic if $\omega < 1$. You may assume that $v \ll u$.
8.20 Linearize the following terms using Eq. (8.61):

(*a*) $u^{m+1}_{i+1,j,k} v^{m+1}_{i+1,j,k}$

(*b*) $(u^{m+1}_{i+1,j,k})^2 v^{m+1}_{i+1,j,k}$

(*c*) $(u^{m+1}_{i+1,j,k})^3 v^{m+1}_{i+1,j,k}$

(*d*) $u^{m+1}_{i+1,j,k} v^{m+1}_{i+1,j,k} w^{m+1}_{i+1,j,k}$

(*e*) $u^{m+1}_{i+1,j,k} (v^{m+1}_{i+1,j,k})^2 w^{m+1}_{i+1,j,k}$

8.21 Derive the expression for the Jacobian $\partial \mathbf{E}^*/\partial \mathbf{U}$ given by Eq. (8.78).
8.22 Derive the expression for the Jacobian $\partial \mathbf{F}/\partial \mathbf{U}$ given by Eq. (8.79).
8.23 Derive the expression for the Jacobian $\partial \mathbf{G}/\partial \mathbf{U}$ given by Eq. (8.80).
8.24 If ω can be approximated by

$$\omega \simeq \gamma M_x^2$$

derive the expression for the Jacobian $\partial \mathbf{E}^*/\partial \mathbf{U}$ that results when ω is no longer assumed independent of $\mathbf{U}$.
8.25 Derive the expression for the Jacobian $\partial \mathbf{F}_v/\partial \mathbf{U}$ given by Eq. (8.84).
8.26 Derive the expression for the Jacobian $\partial \mathbf{G}_v/\partial \mathbf{U}$ given by Eq. (8.85).
8.27 The elements of the matrix $[C]_k$ in Eq. (8.96) can be represented by $(c_{lm})_k$, where $l = 1, 2, \ldots, 5$ and $m = 1, 2, \ldots, 5$. Determine the element $(c_{24})_k$.
8.28 Determine the element $(c_{32})_k$ in Prob. 8.27.

8.29 Determine the element $(c_{43})_k$ in Prob. 8.27.
8.30 The elements of the matrix $[B]_k$ in Eq. (8.96) can be represented by $(b_{lm})_k$, where $l = 1, 2, \ldots, 5$ and $m = 1, 2, \ldots, 5$. Determine the element $(b_{24})_k$.
8.31 Determine the element $(b_{43})_k$ in Prob. 8.30.
8.32 Determine the elements $(a_{33})_k$, $(b_{33})_k$, and $(c_{33})_k$ of the matrices $[A]_k$, $[B]_k$, and $[C]_k$ in Eq. (8.96).
8.33 Apply the difference formula given in Eq. (8.70) to the 2-D PNS equation,

$$\frac{\partial \mathbf{E}^*}{\partial x} + \frac{\partial \mathbf{P}}{\partial x} + \frac{\partial \mathbf{F}}{\partial y} = 0$$

and develop a solution algorithm like that given by Eqs. (8.92)–(8.95) for the 3-D PNS equation.
8.34 Derive the expressions for the cell-face-area vectors given in Eq. (8.104).
8.35 Verify Eqs. (8.105) and (8.106).
8.36 Starting with the x momentum Navier-Stokes equation [Eq. (5.19)], show how this equation is reduced and write out the resulting x momentum equation for the following simplified sets of fluid dynamic equations. Assume that the streamwise flow is in the x direction, while the body surface is in the x-z plane.

(*a*) thin-layer Navier-Stokes (TLNS) equations
(*b*) reduced Navier-Stokes (RNS) equations
(*c*) PNS equations
(*d*) PNS equations with thin-layer approximation
(*e*) three-dimensional steady boundary-layer equations
(*f*) Euler equations

8.37 Repeat Prob. 8.36 for the y momentum Navier-Stokes equation.
8.38 Repeat Prob. 8.36 for the Navier-Stokes energy equation [Eq. (5.25)]. You may ignore all body forces and external heat transfer.
8.39 Work out the details for a velocity correction procedure for the 3-D parabolic procedure for an incompressible flow in a rectangular channel. Use both the $\hat{\phi}$ potential and p' methods. Employ a staggered grid.
8.40 Write the y momentum equation in finite-difference form for the PPNS model following the strategy outlined in Section 8.4.3 for the x momentum equation.
8.41 Prove that the formulation described for the Poisson equation for pressure in the PPNS model satisfies the constraint

$$\iint S_p \, dx\, dy = \oint \frac{\partial p}{\partial n} \, dC$$

8.42 Suggest a way that the PPNS procedure might be extended to 3-D flows.
8.43 Explain how the velocity boundary conditions can be implemented for a boundary that is a line of symmetry (such as the centerline of a 2-D channel) when a staggered grid is used for the PPNS momentum equations. Explain in terms of the Thomas algorithm.
8.44 The incompressible partially parabolized Navier-Stokes momentum equations are given below for a flow predominantly in the positive x direction:

$$\underbrace{u\frac{\partial u}{\partial x}}_{\text{term C}} + v\frac{\partial u}{\partial y} = -\underbrace{\frac{1}{\rho}\frac{\partial p}{\partial x}}_{\text{term A}} + \nu\frac{\partial^2 u}{\partial y^2}$$

$$\underbrace{u\frac{\partial v}{\partial x}}_{\text{term B}} + v\frac{\partial v}{\partial y} = -\frac{1}{\rho}\frac{\partial p}{\partial y} + \nu\frac{\partial^2 v}{\partial y^2}$$

The equations are to be solved by a coupled space-marching fully implicit scheme. Assume that the unknowns are to be solved at the $i + 1$ marching station, $k + 1$ marching sweep.

(*a*) Explain how term A is to be treated in the discretization. Give an appropriate difference representation.

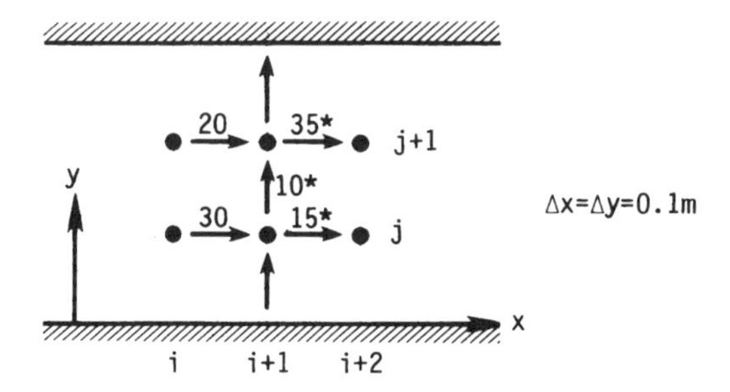

Figure P8.1

(*b*) Use Newton linearization to obtain a representation for term B that is linear in the unknown values of u and v.

(*c*) Give a difference representation for term C that is appropriate for use in a small region of recirculation where $u < 0$. Apply Newton linearization.

8.45 The partially parabolized Navier-Stokes equations are being solved on a uniform staggered grid for a steady laminar incompressible flow by the method described in the text. Provisional velocities (in m/s) have been obtained at the $i + 1$ marching level, and these are indicated by an asterisk in Fig. P8.1. Use the velocity potential method to determine velocity corrections that must be added to the three provisional velocities so that mass is conserved. Note that velocities on the upper and lower boundaries are fixed by the boundary conditions to be zero (no slip), so that no corrections are needed for these.

8.46 Suggest an implicit discretization scheme to implement a coupled space-marching procedure for the PPNS equations given by Eqs. (8.155)–(8.159) for subsonic applications. Use Newton linearization. Explain how the linear equations can be solved.

8.47 Suggest appropriate boundary conditions for solving the PPNS equations given by Eqs. (8.155)–(8.159) by an implicit scheme on a rectangular domain for the following problems. State the conditions needed when the viscous term in the y momentum equation is (1) included; (2) omitted.

(*a*) developing flow in a 2-D channel

(*b*) flow over a flat plate of finite length

8.48 Apply the transformation

$$\alpha = x^*$$

$$\beta = \frac{y^*}{x^*}$$

$$\gamma = \frac{z^*}{x^*}$$

$$\tau = t^*$$

to Eq. (8.173) and derive the "conical" Navier-Stokes equations that can be used to compute a solution at the station $x = L$, where $\alpha = x^* = 1$.

CHAPTER

NINE

NUMERICAL METHODS FOR THE NAVIER-STOKES EQUATIONS

9.1 INTRODUCTION

For certain viscous flow problems, it is not possible to obtain an accurate solution using the simplified flow equations discussed in Chapters 6–8. Examples of such flow problems include shock–boundary-layer interactions, leading edge flows, certain wake flows, and other flows that involve strong viscous-inviscid interactions with large separated flow regions. For these cases, it becomes necessary to solve the complete set of Navier-Stokes (N-S) equations (or the Reynolds averaged form of these equations). Unfortunately, these equations are very complex and require a substantial amount of computer time and storage in order to obtain a solution. However, if the flow is incompressible, the equations can be simplified, and the required computer time is decreased accordingly.

Numerical schemes for solving the N-S equations are based on the same methods described in Chapter 6 for the Euler equations. Since the N-S equations consist of the Euler equations plus shear-stress and heat flux terms, the only change that is required is to discretize the additional terms in an appropriate manner. Because of the dissipative nature of the viscous terms, they are almost always discretized using central differences. One of the major differences that occurs when solving the N-S equations, as compared to the Euler equations, is the need to use fine meshes in order to properly resolve viscous layers. In many cases, this requirement will lead to meshes (cells) with large aspect ratios in the viscous regions of the flow field. When the aspect ratio of the cells becomes large, a typical numerical scheme produces larger truncation errors (T.E.s), and the rate of convergence is decreased. Another consideration that must be taken into account when solving the N-S equations is the amount of numerical dissipation that is present in the numerical method. This numerical dissipation may be inherent in the scheme or could be explicitly added (i.e., smoothing terms). Obviously, the numerical dissipation should be substantially less than the actual physical dissipation in order to obtain an accurate solution.

The unsteady compressible N-S equations are a mixed set of hyperbolic-parabolic equations in time, while the unsteady incompressible N-S equations are a mixed set of elliptic-parabolic equations. As a consequence, different numerical techniques have been used in the past to solve the N-S equations in the compressible and incompressible flow regimes. These techniques are discussed in this chapter beginning with the techniques for solving the compressible N-S equations. Recently, methods have been developed to efficiently solve the compressible N-S equations at very low Mach numbers. With these methods, it becomes possible to solve both compressible and incompressible flows using the same approach. These methods are discussed in Section 9.2.6.

9.2 COMPRESSIBLE NAVIER-STOKES EQUATIONS

The compressible N-S equations in Cartesian coordinates without body forces or external heat addition can be written (see Section 5.1.6) as

$$\frac{\partial \mathbf{U}}{\partial t} + \frac{\partial \mathbf{E}}{\partial x} + \frac{\partial \mathbf{F}}{\partial y} + \frac{\partial \mathbf{G}}{\partial z} = 0 \tag{9.1}$$

where **U**, **E**, **F**, and **G** are vectors given by

$$\mathbf{U} = \begin{bmatrix} \rho \\ \rho u \\ \rho v \\ \rho w \\ E_t \end{bmatrix} \tag{9.2}$$

$$\mathbf{E} = \begin{bmatrix} \rho u \\ \rho u^2 + p - \tau_{xx} \\ \rho u v - \tau_{xy} \\ \rho u w - \tau_{xz} \\ (E_t + p)u - u\tau_{xx} - v\tau_{xy} - w\tau_{xz} + q_x \end{bmatrix} \tag{9.3}$$

$$\mathbf{F} = \begin{bmatrix} \rho v \\ \rho u v - \tau_{xy} \\ \rho v^2 + p - \tau_{yy} \\ \rho v w - \tau_{yz} \\ (E_t + p)v - u\tau_{xy} - v\tau_{yy} - w\tau_{yz} + q_y \end{bmatrix} \tag{9.4}$$

$$\mathbf{G} = \begin{bmatrix} \rho w \\ \rho u w - \tau_{xz} \\ \rho v w - \tau_{yz} \\ \rho w^2 + p - \tau_{zz} \\ (E_t + p)w - u\tau_{xz} - v\tau_{yz} - w\tau_{zz} + q_z \end{bmatrix} \tag{9.5}$$

and the components of the shear-stress tensor and heat flux vector are given by

$$\begin{aligned}
\tau_{xx} &= \frac{2}{3}\mu\left(2\frac{\partial u}{\partial x} - \frac{\partial v}{\partial y} - \frac{\partial w}{\partial z}\right) \\
\tau_{yy} &= \frac{2}{3}\mu\left(2\frac{\partial v}{\partial y} - \frac{\partial u}{\partial x} - \frac{\partial w}{\partial z}\right) \\
\tau_{zz} &= \frac{2}{3}\mu\left(2\frac{\partial w}{\partial z} - \frac{\partial u}{\partial x} - \frac{\partial v}{\partial y}\right) \\
\tau_{xy} &= \mu\left(\frac{\partial u}{\partial y} + \frac{\partial v}{\partial x}\right) = \tau_{yx} \\
\tau_{xz} &= \mu\left(\frac{\partial w}{\partial x} + \frac{\partial u}{\partial z}\right) = \tau_{zx} \\
\tau_{yz} &= \mu\left(\frac{\partial v}{\partial z} + \frac{\partial w}{\partial y}\right) = \tau_{zy} \\
q_x &= -k\frac{\partial T}{\partial x} \\
q_y &= -k\frac{\partial T}{\partial y} \\
q_z &= -k\frac{\partial T}{\partial z}
\end{aligned} \tag{9.6}$$

These equations can be expressed in terms of a generalized orthogonal curvilinear coordinate system (x_1, x_2, x_3) using the formulas in Section 5.1.8. In addition, the compressible N-S equations can be written in terms of a generalized nonorthogonal curvilinear coordinate system (ξ, η, ζ) using the general transformation described in Section 5.6.2:

$$\begin{aligned} \xi &= \xi(x, y, z) \\ \eta &= \eta(x, y, z) \\ \zeta &= \zeta(x, y, z) \end{aligned} \tag{9.7}$$

The transformed equations are given in Chapter 8 as Eqs. (8.34)–(8.36).

The thin-layer approximation to the compressible N-S equations is discussed in Section 8.2. This approximation allows one to drop a number of terms from the complete N-S equations. However, the mathematical character of the resulting equations is identical to that of the complete N-S equations, and as a result, the two sets of equations are normally solved in the same manner. The thin-layer N-S equations are given in Chapter 8 for a Cartesian coordinate system [Eqs. (8.2)–(8.6)] and for a general nonorthogonal coordinate system [Eqs. (8.9)–(8.12)].

For turbulent flows, it is convenient to use the Reynolds averaged equations instead of the N-S equations. Employing the Boussinesq approximation (see Section 5.4.2), the N-S equations can be changed to a modeled form of the Reynolds averaged equations by replacing the coefficient of viscosity μ with

$$\mu + \mu_T$$

and by also replacing the coefficient of thermal conductivity k with

$$k + k_T$$

where μ_T is the eddy viscosity and k_T is the turbulent thermal conductivity. The turbulent thermal conductivity can be expressed in terms of the eddy viscosity using the turbulent Prandtl number Pr_T:

$$k_T = \frac{c_p \mu_T}{\mathrm{Pr}_T} \tag{9.8}$$

Techniques for determining μ_T are described in detail in Section 5.4.

As mentioned previously, the unsteady compressible N-S equations are a mixed set of hyperbolic-parabolic equations in time. If the unsteady terms are dropped from these equations, the resulting equations become a mixed set of hyperbolic-elliptic equations, which are difficult to solve because of the differences in numerical techniques required for hyperbolic and elliptic types of equations. As a consequence, most solutions of the compressible N-S equations have employed the unsteady form of the equations. The steady-state solution is obtained by marching the solution in time (or pseudo-time) until convergence is achieved. This procedure is called the *time-dependent approach* and is the method that will be discussed in this chapter for solving the compressible N-S equations.

Both explicit and implicit schemes have been used with the time-dependent approach to solve the compressible N-S equations. Nearly all of these methods are at least second-order accurate in space and are either first- or second-order accurate in "time." If an accurate time evolution of the flow is required, the numerical scheme should at least be second-order accurate in time. On the other hand, if only the steady-state solution is desired, it is often advantageous to employ a scheme that is not time accurate, since the steady-state solution can usually be achieved with fewer time steps (iterations). Because of the added complexity, only a handful of third-order (or higher) time-accurate methods have appeared in the literature to solve the compressible N-S equations. Many feel that a second-order method is the optimum choice, since higher-order accuracy is at the expense of more computer time. For a complete review of nearly all papers that report solutions to the compressible N-S equations prior to 1976, the reader is urged to consult the excellent survey paper of Peyret and Viviand (1975). More recent reviews are given by MacCormack (1985, 1993). We will now begin our detailed discussion of methods for solving the compressible N-S equations.

9.2.1 Explicit MacCormack Method

When the original MacCormack (1969) scheme is applied to the 3-D compressible N-S equations given by Eq. (9.1), the following algorithm results.

Predictor:

$$\mathbf{U}_{i,j,k}^{\overline{n+1}} = \mathbf{U}_{i,j,k}^{n} - \frac{\Delta t}{\Delta x}(\mathbf{E}_{i+1,j,k}^{n} - \mathbf{E}_{i,j,k}^{n}) - \frac{\Delta t}{\Delta y}(\mathbf{F}_{i,j+1,k}^{n} - \mathbf{F}_{i,j,k}^{n}) - \frac{\Delta t}{\Delta z}(\mathbf{G}_{i,j,k+1}^{n} - \mathbf{G}_{i,j,k}^{n}) \tag{9.9}$$

Corrector:

$$\mathbf{U}_{i,j,k}^{n+1} = \frac{1}{2}\left[\mathbf{U}_{i,j,k}^{n} + \mathbf{U}_{i,j,k}^{\overline{n+1}} - \frac{\Delta t}{\Delta x}\left(\mathbf{E}_{i,j,k}^{\overline{n+1}} - \mathbf{E}_{i-1,j,k}^{\overline{n+1}}\right) - \frac{\Delta t}{\Delta y}\left(\mathbf{F}_{i,j,k}^{\overline{n+1}} - \mathbf{F}_{i,j-1,k}^{\overline{n+1}}\right) - \frac{\Delta t}{\Delta z}\left(\mathbf{G}_{i,j,k}^{\overline{n+1}} - \mathbf{G}_{i,j,k-1}^{\overline{n+1}}\right)\right] \tag{9.10}$$

where $x = i\,\Delta x$, $y = j\,\Delta y$, and $z = k\,\Delta z$. This explicit scheme is second-order accurate in both space and time. In the present form of this scheme, forward differences are used for all spatial derivatives in the predictor step, while backward differences are used in the corrector step. The forward and backward differencing can be alternated between predictor and corrector steps as well as between the three spatial derivatives in a sequential fashion. This eliminates any bias due to the one-sided differencing. An example of a suitable sequence is given in Table 9.1.

The derivatives appearing in the viscous terms of **E**, **F**, and **G** must be differenced correctly in order to maintain second-order accuracy. This is accomplished in the following manner. The x derivative terms appearing in **E**

Table 9.1 Differencing sequence for MacCormack scheme

	Predictor			Corrector		
Step	x Derivative	y Derivative	z Derivative	x Derivative	y Derivative	z Derivative
1	F	F	F	B	B	B
2	B	B	F	F	F	B
3	F	F	B	B	B	F
4	B	F	B	F	B	F
5	F	B	F	B	F	B
6	B	F	F	F	B	B
7	F	B	B	B	F	F
8	B	B	B	F	F	F
9	F	F	F	B	B	B
.	.	.	.	.	.	.
.	.	.	.	.	.	.

F, forward difference; B, backward difference.

are differenced in the opposite direction to that used for $\partial \mathbf{E}/\partial x$, while the y derivatives and the z derivatives are approximated with central differences. Likewise, the y derivative terms appearing in $\mathbf{F}$ and the z derivative terms appearing in $\mathbf{G}$ are differenced in the opposite direction to that used for $\partial \mathbf{F}/\partial y$ and $\partial \mathbf{G}/\partial z$, respectively, while the cross-derivative terms in $\mathbf{F}$ and $\mathbf{G}$ are approximated with central differences. For example, consider the following term in $\mathbf{F}$, which corresponds to the x-momentum equation:

$$F_2 = \rho uv - \mu \frac{\partial u}{\partial y} - \mu \frac{\partial v}{\partial x} \tag{9.11}$$

In the predictor step, given by Eq. (9.9), this term in $\mathbf{F}^n_{i,j,k}$ is differenced as

$$(F_2)^n_{i,j,k} = (\rho uv)^n_{i,j,k} - \mu^n_{i,j,k} \frac{u^n_{i,j,k} - u^n_{i,j-1,k}}{\Delta y} - \mu^n_{i,j,k} \frac{v^n_{i+1,j,k} - v^n_{i-1,j,k}}{2\,\Delta x} \tag{9.12}$$

while in the corrector step, given by Eq. (9.10), this term in $\mathbf{F}^{\overline{n+1}}_{i,j-1,k}$ is differenced as

$$(F_2)^{\overline{n+1}}_{i,j-1,k} = (\rho uv)^{\overline{n+1}}_{i,j-1,k} - \mu^{\overline{n+1}}_{i,j-1,k} \frac{u^{\overline{n+1}}_{i,j,k} - u^{\overline{n+1}}_{i,j-1,k}}{\Delta y}$$
$$- \mu^{\overline{n+1}}_{i,j-1,k} \frac{v^{\overline{n+1}}_{i+1,j-1,k} - v^{\overline{n+1}}_{i-1,j-1,k}}{2\,\Delta x} \tag{9.13}$$

Because of the complexity of the compressible N-S equations, it is not possible to obtain a closed-form stability expression for the MacCormack scheme applied to these equations. However, the following empirical formula (Tannehill et al., 1975) can normally be used:

$$\Delta t \leqslant \frac{\sigma(\Delta t)_{\mathrm{CFL}}}{1 + 2/\mathrm{Re}_\Delta} \tag{9.14}$$

where σ is the safety factor ($\cong 0.9$), $(\Delta t)_{\mathrm{CFL}}$ is the inviscid Courant-Friedrichs-Levy (CFL) condition (MacCormack, 1971)

$$(\Delta t)_{\mathrm{CFL}} \leqslant \left(\frac{|u|}{\Delta x} + \frac{|v|}{\Delta y} + \frac{|w|}{\Delta z} + a\sqrt{\frac{1}{(\Delta x)^2} + \frac{1}{(\Delta y)^2} + \frac{1}{(\Delta z)^2}} \right)^{-1} \tag{9.15}$$

Re_Δ is the minimum mesh Reynolds number given by

$$\mathrm{Re}_\Delta = \min(\mathrm{Re}_{\Delta x}, \mathrm{Re}_{\Delta y}, \mathrm{Re}_{\Delta z}) \tag{9.16}$$

where

$$\begin{aligned} \mathrm{Re}_{\Delta x} &= \frac{\rho|u|\,\Delta x}{\mu} \\ \mathrm{Re}_{\Delta y} &= \frac{\rho|v|\,\Delta y}{\mu} \\ \mathrm{Re}_{\Delta z} &= \frac{\rho|w|\,\Delta z}{\mu} \end{aligned} \tag{9.17}$$

and a is the local speed of sound,

$$a = \sqrt{\frac{\gamma p}{\rho}}$$

Before each step, Δt can be computed for each grid point using Eq. (9.14). The smallest value of Δt is then used to advance the solution over the entire mesh. If only the steady-state solution is desired, Li (1973) has suggested that the solution at each point be advanced using the maximum possible Δt, as computed from Eq. (9.14), in order to accelerate the convergence of the solution. This procedure is referred to as *local time stepping*. In addition, multigrid procedures (see Section 4.3.5) can also be used to accelerate the convergence of N-S calculations.

After each predictor or corrector step, the primitive variables (ρ, u, v, w, e, p, T) can be found by "decoding" the $\mathbf{U}$ vector,

$$\mathbf{U} = \begin{bmatrix} \rho \\ \rho u \\ \rho v \\ \rho w \\ E_t \end{bmatrix} = \begin{bmatrix} U_1 \\ U_2 \\ U_3 \\ U_4 \\ U_5 \end{bmatrix} \tag{9.18}$$

in the following manner

$$\begin{aligned}
\rho &= U_1 \\
u &= \frac{U_2}{U_1} \\
v &= \frac{U_3}{U_1} \\
w &= \frac{U_4}{U_1} \\
e &= \frac{U_5}{U_1} - \frac{u^2 + v^2 + w^2}{2} \\
p &= p(\rho, e) \\
T &= T(\rho, e)
\end{aligned} \tag{9.19}$$

MacCormack (1971) modified his original method by incorporating time splitting into the scheme. This revised method, which was applied to the viscous Burgers equation in Section 4.5.8, "splits" the original MacCormack scheme into a sequence of one-dimensional (1-D) operations. As a result, the stability condition is based on a 1-D scheme that is less restrictive than the original 3-D scheme. Thus it becomes possible to advance the solution in each direction with the maximum possible time step. This is particularly advantageous if the allowable time steps $(\Delta t_x, \Delta t_y, \Delta t_z)$ are much different because of large differences in the mesh spacings $(\Delta x, \Delta y, \Delta z)$. In order to apply this algorithm to Eq. (9.1), we define the 1-D difference operators $L_x(\Delta t_x)$, $L_y(\Delta t_y)$, and $L_z(\Delta t_z)$ in the following manner. The $L_x(\Delta t_x)$ operator applied to $\mathbf{U}^*_{i,j,k}$,

$$\mathbf{U}^{**}_{i,j,k} = L_x(\Delta t_x)\mathbf{U}^*_{i,j,k} \tag{9.20}$$

is equivalent to the two-step formula

$$\begin{aligned}
U^{\overline{**}}_{i,j,k} &= \mathbf{U}^*_{i,j,k} - \frac{\Delta t_x}{\Delta x}(\mathbf{E}^*_{i+1,j,k} - \mathbf{E}^*_{i,j,k}) \\
\mathbf{U}^{**}_{i,j,k} &= \frac{1}{2}\left[\mathbf{U}^*_{i,j,k} + \mathbf{U}^{\overline{**}}_{i,j,k} - \frac{\Delta t_x}{\Delta x}\left(\mathbf{E}^{\overline{**}}_{i,j,k} - \mathbf{E}^{\overline{**}}_{i-1,j,k}\right)\right]
\end{aligned} \tag{9.21}$$

These expressions make use of the dummy time indices * and **. The $L_y(\Delta t_y)$ and $L_z(\Delta t_z)$ operators are defined in a similar manner. That is, the $L_y(\Delta t_y)$ operator applied to $\mathbf{U}^*_{i,j,k}$,

$$\mathbf{U}^{**}_{i,j,k} = L_y(\Delta t_y)\mathbf{U}^*_{i,j,k} \tag{9.22}$$

is equivalent to

$$\begin{aligned} \mathbf{U}_{i,j,k}^{\overline{**}} &= \mathbf{U}_{i,j,k}^{*} - \frac{\Delta t_y}{\Delta y}(\mathbf{F}_{i,j+1,k}^{*} - \mathbf{F}_{i,j,k}^{*}) \\ \mathbf{U}_{i,j,k}^{**} &= \frac{1}{2}\left[\mathbf{U}_{i,j,k}^{*} + \mathbf{U}_{i,j,k}^{\overline{**}} - \frac{\Delta t_y}{\Delta y}\left(\mathbf{F}_{i,j,k}^{\overline{**}} - \mathbf{F}_{i,j-1,k}^{\overline{**}}\right)\right] \end{aligned} \tag{9.23}$$

and the $L_z(\Delta t_z)$ operator applied to $\mathbf{U}_{i,j,k}^{*}$,

$$\mathbf{U}_{i,j,k}^{**} = L_z(\Delta t_z)\mathbf{U}_{i,j,k}^{*} \tag{9.24}$$

is equivalent to

$$\begin{aligned} \mathbf{U}_{i,j,k}^{\overline{**}} &= \mathbf{U}_{i,j,k}^{*} - \frac{\Delta t_z}{\Delta z}(\mathbf{G}_{i,j,k+1}^{*} - \mathbf{G}_{i,j,k}^{*}) \\ \mathbf{U}_{i,j,k}^{**} &= \frac{1}{2}\left[\mathbf{U}_{i,j,k}^{*} + \mathbf{U}_{i,j,k}^{\overline{**}} - \frac{\Delta t_z}{\Delta z}\left(\mathbf{G}_{i,j,k}^{\overline{**}} - \mathbf{G}_{i,j,k}^{\overline{**}}\right)\right] \end{aligned} \tag{9.25}$$

As mentioned in Section 4.5.8, a sequence of operators is consistent if the sums of the time steps for each of the operators are equal and is second-order accurate if the sequence is symmetric. A sequence that satisfies these criteria and is applicable to Eq. (9.1) is given by

$$\mathbf{U}_{i,j,k}^{n+2} = L_x(\Delta t_x)L_y(\Delta t_y)L_z(\Delta t_z)L_z(\Delta t_z)L_y(\Delta t_y)L_x(\Delta t_x)\mathbf{U}_{i,j,k}^{n} \tag{9.26}$$

Another sequence that satisfies these criteria, and is applicable when $\Delta y \ll \min(\Delta x, \Delta z)$, is given by

$$\mathbf{U}_{i,j,k}^{n+2} = L_x(\Delta t_x)\left[L_y\left(\frac{\Delta t_y}{m}\right)\right]^m L_z(\Delta t_z)L_z(\Delta t_z)\left[L_y\left(\frac{\Delta t_y}{m}\right)\right]^m L_x(\Delta t_x)\mathbf{U}_{i,j,k}^{n} \tag{9.27}$$

where m is an integer.

The algorithms resulting from a sequence of operators such as Eqs. (9.26) and (9.27) are stable if the time step size in the argument of each does not exceed the maximum allowed for that operator. Since it is not possible to analyze the stability of each operator applied to the complete N-S equations, a 1-D form of the empirical stability formula, given by Eq. (9.14), can be used for each operator:

$$\begin{aligned} \Delta t_x &\leqslant \frac{\sigma\,\Delta x}{(|u| + a)(1 + 2/\mathrm{Re}_{\Delta x})} \\ \Delta t_y &\leqslant \frac{\sigma\,\Delta y}{(|v| + a)(1 + 2/\mathrm{Re}_{\Delta y})} \\ \Delta t_z &\leqslant \frac{\sigma\,\Delta z}{(|w| + a)(1 + 2/\mathrm{Re}_{\Delta z})} \end{aligned} \tag{9.28}$$

where σ is the safety factor and a is the local speed of sound.

Computations involving the compressible N-S equations sometimes become unstable (i.e., "blow up") because of numerical oscillations. These oscillations are the result of inadequate mesh refinement in regions of large gradients such as shock waves and are accentuated when central differences are used for the spatial derivatives, as in the present MacCormack scheme. In many cases, it is impractical to refine the mesh in these regions, particularly if they are far removed from the region of interest. For such situations, MacCormack and Baldwin (1975) have devised a "product" fourth-order smoothing scheme, which is an alternative to the fourth-order type of smoothing given by Eq. (8.98). In the MacCormack type of smoothing, dissipation terms are added to each operator. For example, they are added to the $L_x(\Delta t_x)$ operator in the following manner:

$$\begin{aligned} \mathbf{U}^{\overline{**}}_{i,j,k} &= \mathbf{U}^{*}_{i,j,k} - \frac{\Delta t_x}{\Delta x}\left(\mathbf{E}^{*}_{i+1,j,k} + \mathbf{S}^{*}_{i+1,j,k} - \mathbf{E}^{*}_{i,j,k} - \mathbf{S}^{*}_{i,j,k}\right) \\ \mathbf{U}^{**}_{i,j,k} &= \frac{1}{2}\left[\mathbf{U}^{*}_{i,j,k} + \mathbf{U}^{\overline{**}}_{i,j,k} - \frac{\Delta t_x}{\Delta x}\left(\mathbf{E}^{\overline{**}}_{i,j,k} + \mathbf{S}^{\overline{**}}_{i,j,k} - \mathbf{E}^{\overline{**}}_{i-1,j,k} - \mathbf{S}^{\overline{**}}_{i-1,j,k}\right)\right] \end{aligned} \tag{9.29}$$

where

$$\begin{aligned} \mathbf{S}^{*}_{i,j,k} &= \epsilon_e\left[\left(|u^{*}_{i,j,k}| + a^{*}_{i,j,k}\right)\frac{|\delta_x^2 p^{*}_{i,j,k}|}{\left(p^{*}_{i+1,j,k} + 2p^{*}_{i,j,k} + p^{*}_{i-1,j,k}\right)}\left(\mathbf{U}^{*}_{i,j,k} - \mathbf{U}^{*}_{i-1,j,k}\right)\right] \\ \mathbf{S}^{\overline{**}}_{i,j,k} &= \epsilon_e\left[\left(|u^{\overline{**}}_{i,j,k} + a^{\overline{**}}_{i,j,k}\right)\frac{|\delta_x^2 p^{\overline{**}}_{i,j,k}|}{\left(p^{\overline{**}}_{i+1,j,k} + 2p^{\overline{**}}_{i,j,k} + p^{\overline{**}}_{i-1,j,k}\right)}\left(\mathbf{U}^{\overline{**}}_{i+1,j,k} - \mathbf{U}^{\overline{**}}_{i,j,k}\right)\right] \end{aligned} \tag{9.30}$$

and $0 \leqslant \epsilon_e \leqslant 0.5$ for stability. Thus an artificial viscosity term of the form

$$\epsilon_e(\Delta x)^4 \frac{\partial}{\partial x}\left(\frac{|u| + a}{4p}\left|\frac{\partial^2 p}{\partial x^2}\right|\frac{\partial \mathbf{U}}{\partial x}\right) \tag{9.31}$$

has been added to the N-S equations. This smoothing term has a very small magnitude except in regions of pressure oscillations, where the T.E. is already producing erroneous results.

The explicit MacCormack algorithm is a suitable method for solving both steady and unsteady flows at moderate to low Reynolds numbers. However, it is not a satisfactory method for solving high Reynolds numbers flows, where the viscous regions become very thin. For these flows, the mesh must be highly refined in order to accurately resolve the viscous regions. This leads to small time steps and, subsequently, long computing times if an explicit scheme such as the MacCormack method is used. In order to explain this further, let us consider the 2-D flow over a flat plate at high Reynolds number. In this case, a very fine mesh is required near the flat plate in order to resolve the boundary layer, but a coarser grid can be used in the inviscid portion of the flow field, as illustrated in Fig. 9.1. In the coarse grid region, the MacCormack time-split scheme can be

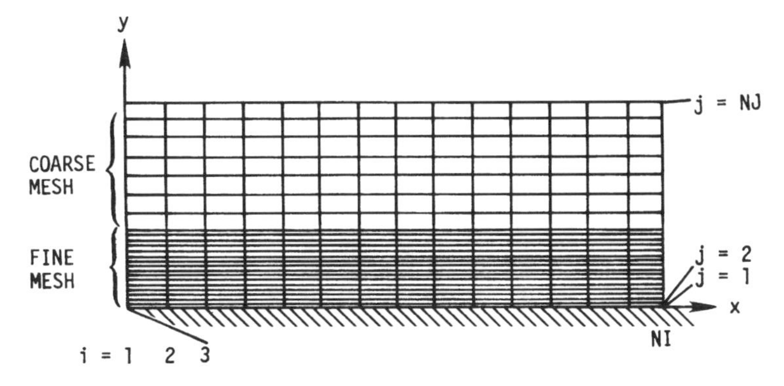

Figure 9.1 Mesh for high Reynolds number flow over a flat plate.

applied in the following manner:

$$\mathbf{U}_{i,j}^{n+1} = L_x\left(\frac{\Delta t}{2}\right) L_y(\Delta t) L_x\left(\frac{\Delta t}{2}\right) \mathbf{U}_{i,j}^{n} \tag{9.32}$$

where

$$\Delta t \leqslant \min\left(2\,\Delta t_x, \Delta t_y\right)_{\text{coarse mesh}} \tag{9.33}$$

In the fine grid region the following sequence of operators can be used:

$$\mathbf{U}_{i,j}^{n+1} = \left[L_y\left(\frac{\Delta t}{2m}\right) L_x\left(\frac{\Delta t}{m}\right) L_y\left(\frac{\Delta t}{2m}\right)\right]^m \mathbf{U}_{i,j} \tag{9.34}$$

where m is the smallest integer such that

$$\frac{\Delta t}{m} \leqslant \min\left(\Delta t_x, 2\,\Delta t_y\right)_{\text{fine mesh}} \tag{9.35}$$

For high Reynolds numbers, the fine-grid region becomes very thin, requiring Δy to be very small. This causes Δt_y in the L_y operator to be very small and the integer m to be very large. Consequently, a substantial amount of calculation time is required in the fine-grid region. To overcome this difficulty, MacCormack (1976) developed a hybrid version of his scheme, which is known as the MacCormack rapid solver method. This hybrid method is part explicit and part implicit. For the flat-plate problem described above, the rapid solver method is implemented by replacing the $L_y(\Delta t/2m)$ operator in Eq. (9.34) with

$$L_{y_H}\left(\frac{\Delta t}{2m}\right) L_{y_P}\left(\frac{\Delta t}{2m}\right)$$

where the L_{y_H} operator is applied to the inviscid (hyperbolic) portion of the N-S equations, i.e.,

$$\frac{\partial \mathbf{U}}{\partial t} + \frac{\partial \mathbf{F}_H}{\partial y} = 0 \tag{9.36}$$

with $\mathbf{F}_H$ defined as

$$\mathbf{F}_H = \begin{bmatrix} \rho v \\ \rho uv \\ \rho v^2 + p \\ (E_t + p)v \end{bmatrix} \tag{9.37}$$

The L_{H_P} operator is applied to the viscous (parabolic) portion of the N-S equations:

$$\frac{\partial \mathbf{U}}{\partial t} + \frac{\partial \mathbf{F}_P}{\partial y} = 0 \tag{9.38}$$

where $\mathbf{F}_P = \mathbf{F} - \mathbf{F}_H$. The L_{y_H} operator solves Eq. (9.36) using either the method of characteristics or the original MacCormack scheme (Li, 1977; Shang, 1977). The L_{y_P} operator solves Eq. (9.38) using an implicit scheme such as the Crank-Nicolson or Laasonen schemes. Thus it is possible to solve Eq. (9.36) and (9.38) using a time step that is not limited by the viscosity stability constraint. The rapid solver method has proved to be from 10 to 100 times faster than the time-split scheme for high Reynolds number flows. However, because of its complexity, more traditional implicit schemes (see Sections 9.2.3 and 9.2.4) are preferred when solving high Reynolds number flows.

9.2.2 Other Explicit Methods

In addition to the MacCormack scheme, other explicit methods that have been used to solve the compressible N-S equations include the following:

1. Hopscotch method (Section 4.2.13)
2. Leap frog/DuFort-Frankel method (Section 4.5.2)
3. Brailovskaya method (Section 4.5.3)
4. Allen-Cheng method (Section 4.5.4)
5. Lax-Wendroff method (Section 4.5.5)
6. Runge-Kutta method (Section 4.1.13)

These methods were discussed in earlier sections (as indicated), where they were applied to model equations. When these methods are applied to the more complicated compressible N-S equations, certain difficulties can arise, as we have seen before. For example, the mixed-derivative terms create a problem for the hopscotch method. If these terms are differenced in the usual manner by applying Eq. (3.51), the hopscotch method is no longer explicit, since a matrix solution is required. This problem can be circumvented by lagging the mixed-derivative terms (i.e., evaluating them at the previous time level).

All of the above methods, except the Lax-Wendroff and Runge-Kutta schemes, are first-order accurate in time, so that they cannot be used to accurately compute the time evolution of a flow field. In addition, all of the methods have a stability restriction that limits the maximum time step. However, the stability conditions for the hopscotch and Allen-Cheng methods are independent of the viscosity, which gives them an advantage over the other methods. The allowable time step for the hopscotch method is given by the inviscid CFL condition, which for a 2-D problem becomes

$$(\Delta t)_{\mathrm{CFL}} \leqslant \frac{\Delta x}{|u| + |v| + 2\sqrt{a}} \tag{9.39}$$

if $\Delta x = \Delta y$. An advantage of the Brailovskaya method is that the viscous terms need to be computed only once during the two-step procedure.

Of the six explicit methods listed above, only the Runge-Kutta method is widely used to solve the compressible N-S equations. The Runge-Kutta method, developed by Jameson et al. (1981, 1983) to solve the Euler equations, has been extended to the N-S equations by Swanson and Turkel (1985, 1987), Martinelli et al. (1986), Martinelli and Jameson (1988), and Turkel et al. (1991). This method utilizes a blend of second- and fourth-order damping terms and employs local time stepping, implicit residual smoothing, and multigrid to accelerate the convergence to steady state. The Runge-Kutta method is employed in the computer code TLNS3D (thin layer N-S program for 3-D flows) developed at NASA Langley Research Center (Vatsa and Wedan, 1989).

9.2.3 Beam-Warming Scheme

Prior to the mid-1970s, the numerical methods available to solve the compressible N-S equations were nearly all explicit and were limited in time step size by the CFL condition. As a consequence, it was difficult to compute high Reynolds number flows because of the fine meshes required to resolve the viscous regions. This difficulty was overcome with the application of noniterative implicit methods to the N-S equations. Briley and McDonald (1974) and Beam and Warming (1978) were the first to apply this type of scheme to solve the compressible N-S equations. We will describe the Beam-Warming scheme in this section.

The Beam-Warming numerical scheme for solving the compressible N-S equations belongs to the same class of alternating direction implicit (ADI) schemes developed by Lindemuth and Killeen (1973) and McDonald and Briley (1975). Under certain conditions, these schemes can be shown to be equivalent. The Briley-McDonald scheme is discussed in Section 4.5.7, where it is applied to the viscous Burgers equation.

For simplicity, we will apply the Beam-Warming finite-difference scheme to the 2-D compressible N-S equations, which can be written in the following vector form:

$$\frac{\partial \mathbf{U}}{\partial t}+\frac{\partial \mathbf{E}(\mathbf{U})}{\partial x}+\frac{\partial \mathbf{F}(\mathbf{U})}{\partial y}=\frac{\partial \mathbf{V}_1(\mathbf{U},\mathbf{U}_x)}{\partial x}+\frac{\partial \mathbf{V}_2(\mathbf{U},\mathbf{U}_y)}{\partial x}+\frac{\partial \mathbf{W}_1(\mathbf{U},\mathbf{U}_x)}{\partial y}+\frac{\partial \mathbf{W}_2(\mathbf{U},\mathbf{U}_y)}{\partial y} \tag{9.40}$$

where

$$\mathbf{U}=\begin{bmatrix}\rho\\ \rho u\\ \rho v\\ E_t\end{bmatrix}\qquad \mathbf{E}(\mathbf{U})=\begin{bmatrix}\rho u\\ \rho u^2+p\\ \rho uv\\ (E_t+p)u\end{bmatrix}\qquad \mathbf{F}(\mathbf{U})=\begin{bmatrix}\rho v\\ \rho uv\\ \rho v^2+p\\ (E_t+p)v\end{bmatrix}$$

$$\mathbf{V}_1+\mathbf{V}_2=\begin{bmatrix}0\\ \frac{2}{3}\mu(2u_x-v_y)\\ \mu(u_y+v_x)\\ \mu v(u_y+v_x)+\frac{2}{3}\mu u(2u_x-v_y)+kT_x\end{bmatrix} \tag{9.41}$$

$$\mathbf{W}_1+\mathbf{W}_2=\begin{bmatrix}0\\ \mu(u_y+v_x)\\ \frac{2}{3}\mu(2v_y-u_x)\\ \mu u(u_y+v_x)+\frac{2}{3}\mu v(2v_y-u_x)+kT_y\end{bmatrix}$$

In the Beam-Warming scheme the solution is marched in time using the following difference formula:

$$\Delta^n\mathbf{U}=\frac{\theta_1\,\Delta t}{1+\theta_2}\frac{\partial}{\partial t}(\Delta^n\mathbf{U})+\frac{\Delta t}{1+\theta_2}\frac{\partial}{\partial t}(\mathbf{U}^n)+\frac{\theta_2}{1+\theta_2}\Delta^{n-1}\mathbf{U}+O\left[\left(\theta_1-\frac{1}{2}-\theta_2\right)(\Delta t)^2+(\Delta t)^3\right] \tag{9.42}$$

where $\Delta^n\mathbf{U}=\mathbf{U}^{n+1}-\mathbf{U}^n$. This general difference formula, with the appropriate choice of the parameters θ_1 and θ_2, represents many of the standard difference schemes, as we have seen from Section 8.3.3. For the compressible N-S equations, either the Euler implicit scheme ($\theta_1=1, \theta_2=0$), which is first-order accurate in time, or the three-point backward implicit scheme ($\theta_1=1, \theta_2=\frac{1}{2}$), which is second-order accurate in time, is normally used.

After substituting Eq. (9.40) into Eq. (9.42), we obtain

$$\Delta^n \mathbf{U} = \frac{\theta_1 \, \Delta t}{1+\theta_2}\left[\frac{\partial}{\partial x}(-\Delta^n \mathbf{E} + \Delta^n \mathbf{V}_1 + \Delta^n \mathbf{V}_2) + \frac{\partial}{\partial y}(-\Delta^n \mathbf{F} + \Delta^n \mathbf{W}_1 + \Delta^n \mathbf{W}_2)\right]$$
$$+ \frac{\Delta t}{1+\theta_2}\left[\frac{\partial}{\partial x}(-\mathbf{E}^n + \mathbf{V}_1^n + \mathbf{V}_2^n) + \frac{\partial}{\partial y}(-\mathbf{F}^n + \mathbf{W}_1^n + \mathbf{W}_2^n)\right]$$
$$+ \frac{\theta_2}{1+\theta_2}\Delta^{n-1}\mathbf{U} + O\left[\left(\theta_1 - \frac{1}{2} - \theta_2\right)(\Delta t)^2 + (\Delta t)^3\right] \tag{9.43}$$

This difference formula is in the so-called delta form, which is discussed in Section 4.4.7. The delta terms are linearized using truncated Taylor-series expansions. For example, $\Delta^n \mathbf{E}$ is linearized using

$$\mathbf{E}^{n+1} = \mathbf{E}^n + \left(\frac{\partial \mathbf{E}}{\partial \mathbf{U}}\right)^n (\mathbf{U}^{n+1} - \mathbf{U}^n) + O[(\Delta t)^2] \tag{9.44}$$

which can be rewritten as

$$\Delta^n \mathbf{E} = [A]^n \, \Delta^n \mathbf{U} + O[(\Delta t)^2] \tag{9.45}$$

where $[A]$ is the Jacobian matrix $\partial \mathbf{E}/\partial \mathbf{U}$ given by

$$[A] = -\begin{bmatrix} 0 & -1 & 0 & 0 \\ \dfrac{3-\gamma}{2}u^2 + \dfrac{1-\gamma}{2}v^2 & (\gamma-3)u & (\gamma-1)v & (1-\gamma) \\ uv & -v & -u & 0 \\ \dfrac{\gamma E_t u}{\rho} + (1-\gamma)u(u^2+v^2) & -\dfrac{\gamma E_t}{\rho} + \dfrac{\gamma-1}{2}(3u^2+v^2) & (\gamma-1)uv & -\gamma u \end{bmatrix} \tag{9.46}$$

and γ is the ratio of specific heats. This Jacobian matrix is derived assuming a perfect gas. In a like manner, $\Delta^n \mathbf{F}$ can be linearized as

$$\Delta^n \mathbf{F} = [B]^n \, \Delta^n \mathbf{U} + O[(\Delta t)^2] \tag{9.47}$$

where $[B]$ is the Jacobian matrix $\partial \mathbf{F}/\partial \mathbf{U}$ given by

$$[B] = -\begin{bmatrix} 0 & 0 & -1 & 0 \\ uv & -v & -u & 0 \\ \dfrac{3-\gamma}{2}v^2 + \dfrac{1-\gamma}{2}u^2 & (\gamma-1)u & (\gamma-3)v & 1-\gamma \\ \dfrac{\gamma E_t v}{\rho} + (1-\gamma)v(u^2+v^2) & (\gamma-1)uv & -\dfrac{\gamma E_t}{\rho} + \dfrac{\gamma-1}{2}(3v^2+u^2) & -\gamma v \end{bmatrix} \tag{9.48}$$

The viscous delta term $\Delta^n \mathbf{V}_1(\mathbf{U}, \mathbf{U}_x)$ is linearized by writing

$$\begin{aligned}
\Delta^n \mathbf{V}_1 &= \left(\frac{\partial \mathbf{V}_1}{\partial \mathbf{U}}\right)^n \Delta^n \mathbf{U} + \left(\frac{\partial \mathbf{V}_1}{\partial \mathbf{U}_x}\right)^n \Delta^n \mathbf{U}_x + O[(\Delta t)^2] \\
&= [P]^n \, \Delta^n \mathbf{U} + [R]^n \, \Delta^n \mathbf{U}_x + O[(\Delta t)^2] \\
&= ([P] - [R_x])^n \, \Delta^n \mathbf{U} + \frac{\partial}{\partial x}([R]^n \, \Delta^n \mathbf{U}) + O[(\Delta t)^2] \qquad (9.49)
\end{aligned}$$

where $[P]$ is the Jacobian $\partial \mathbf{V}_1 / \partial \mathbf{U}$, $[R]$ is the Jacobian $\partial \mathbf{V}_1 / \partial \mathbf{U}_x$, and $[R_x] = \partial [R] / \partial x$. These matrices can be written as

$$[P] - [R_x] = -\frac{1}{\rho} \left[\begin{array}{c:c:c:c}
0 & 0 & 0 & 0 \\
-u\left(\frac{4}{3}\mu\right)_x & \left(\frac{4}{3}\mu\right)_x & 0 & 0 \\
-v\mu_x & 0 & \mu_x & 0 \\
-u^2\left(\frac{4}{3}\mu\right)_x - v^2\mu_x & u\left(\frac{4}{3}\mu\right)_x & v\mu_x & 0
\end{array} \right] \qquad (9.50)$$

$$[R] = \frac{1}{\rho} \left[\begin{array}{c:c:c:c}
0 & 0 & 0 & 0 \\
-\frac{4}{3}\mu u & \frac{4}{3}\mu & 0 & 0 \\
-\mu v & 0 & \mu & 0 \\
-\left(\frac{4}{3}\mu - \frac{k}{c_v}\right)u^2 - \left(\mu - \frac{k}{c_v}\right)v^2 - \frac{k}{c_v}\frac{E_t}{\rho} & \left(\frac{4}{3}\mu - \frac{k}{c_v}\right)u & \left(\mu - \frac{k}{c_v}\right)v & \frac{k}{c_v}
\end{array} \right] \qquad (9.51)$$

The matrix for $[P] - [R_x]$ is obtained by assuming that μ and k are locally independent of $\mathbf{U}$. In a like manner, $\Delta^n \mathbf{W}_2(\mathbf{U}, \mathbf{U}_y)$ is linearized as

$$\Delta^n \mathbf{W}_2 = ([Q] - [S_y])^n \, \Delta^n \mathbf{U} + \frac{\partial}{\partial y}([S]^n \, \Delta^n \mathbf{U}) + O[(\Delta t)^2] \qquad (9.52)$$

where

$$[Q] - [S_y] = -\frac{1}{\rho} \left[\begin{array}{c:c:c:c}
0 & 0 & 0 & 0 \\
-u\mu_y & \mu_y & 0 & 0 \\
-v\left(\frac{4}{3}\mu\right)_y & 0 & \left(\frac{4}{3}\mu\right)_y & 0 \\
-v^2\left(\frac{4}{3}\mu\right)_y - u^2\mu_y & u\mu_y & v\left(\frac{4}{3}\mu\right)_y & 0
\end{array} \right] \qquad (9.53)$$

and

$$[S] = \frac{1}{\rho}\left[\begin{array}{c|c|c|c} 0 & 0 & 0 & 0 \\ -\mu u & \mu & 0 & 0 \\ -\frac{4}{3}\mu v & 0 & \frac{4}{3}\mu & 0 \\ -\left(\frac{4}{3}\mu - \frac{k}{c_v}\right)v^2 - \left(\mu - \frac{k}{c_v}\right)u^2 - \frac{k}{c_v}\frac{E_t}{\rho} & \left(\mu - \frac{k}{c_v}\right)u & \left(\frac{4}{3}\mu - \frac{k}{c_v}\right)v & \frac{k}{c_v} \end{array}\right] \tag{9.54}$$

The cross-derivative terms can be evaluated without loss of accuracy by noting that

$$\begin{aligned} \Delta^n \mathbf{V}_2 &= \Delta^{n-1}\mathbf{V}_2 + O[(\Delta t)^2] \\ \Delta^n \mathbf{W}_1^n &= \Delta^{n-1}\mathbf{W}_1 + O[(\Delta t)^2] \end{aligned} \tag{9.55}$$

for a uniform time step Δt. By evaluating the cross-derivative terms in this manner, the block tridiagonal form of the final equations is maintained. The Steger method (Steger, 1977) for linearizing viscous terms, described in Section 8.3.3, can be used in place of the linearizations given by Eqs. (9.49) and (9.52). The Steger form of linearization is particularly useful when coordinate transformations have been applied to the N-S equations.

Substituting Eqs. (9.45), (9.47), (9.49), (9.52), and (9.55) into Eq. (9.43) yields

$$\begin{aligned} &\left\{[I] + \frac{\theta_1 \Delta t}{1+\theta_2}\left[\frac{\partial}{\partial x}([A] - [P] + [R_x])^n - \frac{\partial^2}{\partial x^2}[R]^n \right.\right. \\ &\qquad \left.\left. + \frac{\partial}{\partial y}([B] - [Q] + [S_y])^n - \frac{\partial^2}{\partial y^2}[S]^n\right]\right\}\Delta^n \mathbf{U} \\ &= \frac{\Delta t}{1+\theta_2}\left[\frac{\partial}{\partial x}(-\mathbf{E} + \mathbf{V}_1 + \mathbf{V}_2)^n + \frac{\partial}{\partial y}(-\mathbf{F} + \mathbf{W}_1 + \mathbf{W}_2)^n\right] \\ &\quad + \frac{\theta_1 \Delta t}{1+\theta_2}\left[\frac{\partial}{\partial x}(\Delta^{n-1}\mathbf{V}_2) + \frac{\partial}{\partial y}(\Delta^{n-1}\mathbf{W}_1)\right] + \frac{\theta_2}{1+\theta_2}\Delta^{n-1}\mathbf{U} \\ &\quad + O\left[\left(\theta_1 - \frac{1}{2} - \theta_2\right)(\Delta t)^2, (\Delta t)^3\right] \end{aligned} \tag{9.56}$$

where $[I]$ is the unity matrix. In Eq. (9.56), expressions such as

$$\left[\frac{\partial}{\partial x}([A] - [P] + [R_x])^n\right]\Delta^n \mathbf{U}$$

should be interpreted as

$$\frac{\partial}{\partial x}\left[([A] - [P] + [R_x])^n \Delta^n \mathbf{U}\right]$$

The left-hand side (LHS) of Eq. (9.56) is approximately factored in the following manner:

$$\left\{[I] + \frac{\theta_1 \Delta t}{1+\theta_2}\left[\frac{\partial}{\partial x}([A] - [P] + [R_x])^n - \frac{\partial^2}{\partial x^2}[R]^n\right]\right\}$$

$$\times\left\{[I] + \frac{\theta_1 \Delta t}{1+\theta_2}\left[\frac{\partial}{\partial y}([B] - [Q] + [S_y])^n - \frac{\partial^2}{\partial y^2}[S]^n\right]\right\}\Delta^n \mathbf{U}$$

$$= \text{LHS of Eq. (9.56)} + O[(\Delta t)^3] \tag{9.57}$$

and the final form of the Beam-Warming algorithm becomes

$$\text{LHS of Eq. (9.57)} = \text{RHS of Eq. (9.56)} \tag{9.58}$$

It should be noted that the approximate factorization introduced by Eq. (9.57) may limit the size of the allowable time step because of the added T.E. The partial derivatives in Eq. (9.57) are evaluated using second-order accurate central differences.

The Beam-Warming algorithm is implemented in the following manner:

Step 1:

$$\left\{[I] + \frac{\theta_1 \Delta t}{1+\theta_2}\left[\frac{\partial}{\partial x}([A] - [P] + [R_x])^n - \frac{\partial^2}{\partial x^2}[R]^n\right]\right\}\Delta^n \mathbf{U}_1$$

$$= \text{RHS of Eq. (9.56)} \tag{9.59}$$

Step 2:

$$\left\{[I] + \frac{\theta_1 \Delta t}{1+\theta_2}\left[\frac{\partial}{\partial y}([B] - [Q] + [S_y])^n - \frac{\partial^2}{\partial y^2}[S]^n\right]\right\}\Delta^n \mathbf{U} = \Delta^n \mathbf{U}_1 \tag{9.60}$$

Step 3:

$$\mathbf{U}^{n+1} = \mathbf{U}^n + \Delta^n \mathbf{U} \tag{9.61}$$

In Step 1, $\Delta^n \mathbf{U}_1$ represents the remaining terms on the LHS of Eq. (9.57). Equations (9.59) and (9.60) represent systems of equations that have the same block tridiagonal structure as shown in Eq. (8.96) except that for the 2-D compressible N-S equations, the blocks are 4×4 matrices.

Warming and Beam (1977) have studied the stability of their algorithm by applying it to both the 2-D wave equation,

$$u_t + c_1 u_x + c_2 u_y = 0 \tag{9.62}$$

and the diffusive equation,

$$u_t = a u_{xx} + b u_{xy} + c u_{yy} \tag{9.63}$$

The latter equation is parabolic if $b^2 < 4ac$ and $(a, c) > 0$. They found that the algorithm is unconditionally stable when applied to Eq. (9.62), provided that $\theta_2 > 0$. When applied to Eq. (9.63), the algorithm is unconditionally stable, provided that $\theta_2 \geqslant 0.385$. Note that neither the leap frog scheme ($\theta_1 = 0, \theta_2 = -\frac{1}{2}$) nor the trapezoidal scheme ($\theta_1 = \frac{1}{2}, \theta_2 = 0$) is unconditionally stable when applied to Eq. (9.63). However, the three-point backward scheme ($\theta_1 = 1, \theta_2 = \frac{1}{2}$) is unconditionally stable and can be used when second-order temporal accuracy is desired.

In order to suppress oscillations that will occur near flow field discontinuities as a result of the central differences that are used for the spatial derivatives, it is necessary to add damping (artificial viscosity) to the Beam-Warming scheme. This can be accomplished by adding a fourth-order explicit dissipation term of the form given by Eq. (8.98) to the RHS of Eq. (9.56). In addition, if only the steady-state solution is of interest, a second-order implicit smoothing term can also be added to the LHS of Eq. (9.56). This latter smoothing term can be second order, since it has no effect on the steady-state solution where $\Delta^n \mathbf{U} = 0$. After the smoothing terms are added, the final differenced form of the algorithm becomes as follows:

Step 1:

$$\left\{[I] + \frac{\theta_1 \Delta t}{1+\theta_2}\left[\bar{\delta}_x([A] - [P] + [R_x])^n - \delta_x^2[R]^n - \epsilon_i \delta_x^2\right]\right\} \Delta^n \mathbf{U}_1$$
$$= \text{RHS of Eq. (9.56)} - \epsilon_e\left(\delta_x^4 + \delta_y^4\right)\mathbf{U}^n \tag{9.64}$$

Step 2:

$$\left\{[I] + \frac{\theta_1 \Delta t}{1+\theta_2}\left[\bar{\delta}_y([B] - [Q] + [S_y])^n - \delta_y^2[S]^n - \epsilon_i \delta_y^2\right]\right\} \Delta^n \mathbf{U} = \Delta^n \mathbf{U}_1 \tag{9.65}$$

Step 3:

$$\mathbf{U}^{n+1} = \mathbf{U}^n + \Delta^n \mathbf{U} \tag{9.66}$$

where $\bar{\delta}$, δ^2, and δ^4 are the usual central-difference operators and ϵ_e and ϵ_i are the coefficients of the explicit and implicit smoothing terms, respectively. Using a Fourier stability analysis, it can be shown that the coefficient of the explicit smoothing term must be in the range

$$0 \leqslant \epsilon_e \leqslant \frac{1+2\theta_2}{8(1+\theta_2)} \tag{9.67}$$

to ensure stability.

Désidéri et al. (1978) have investigated the possibility of maximizing the rate of convergence of the time-dependent solution by using the proper ratio of the coefficients of the smoothing terms. They found that when the Beam-Warming scheme (with Euler implicit differencing) is applied to the Euler equations, the rate of convergence is optimized when

$$\frac{\epsilon_i}{\epsilon_e} = 2 \tag{9.68}$$

Beam and Warming have pointed out that their algorithm can be simplified considerably if μ is assumed locally constant. In this case, $(\mu_x, \mu_y) = 0$ and Eqs. (9.50) and (9.53) reduce to

$$\begin{aligned} [P] - [R_x] &= 0 \\ [Q] - [S_y] &= 0 \end{aligned} \tag{9.69}$$

If only the steady-state solution is desired, Tannehill et al. (1978) have suggested that all the viscous terms on the LHS of the algorithm (i.e., $[P], [R_x], [R], [Q], [S_y], [S]$) can be set equal to zero, provided that implicit smoothing ($\epsilon_i > 0$) is retained. This takes advantage of the fact that the LHS of Eq. (9.57) approaches zero as the steady-state solution is approached. With this simplification, the complexity of the Beam-Warming algorithm is greatly reduced, particularly if a non-Cartesian coordinate system is employed. It is believed that this simplifying technique can be used in all moderate to high Reynolds number computations, since tests confirm that the convergence rate is not affected for these cases. To reduce computation time further, Chaussee and Pulliam (1981) have transformed the coupled set of thin-layer N-S equations into an uncoupled diagonal form.

The Beam-Warming scheme is employed in the widely used ARC3D code (Pulliam and Steger, 1980) developed at NASA Ames Research Center. This code has recently been incorporated into the OVERFLOW code (Buning et al., 1994), which is an outgrowth of both the ARC3D code and the flux-vector splitting F3D code (Steger et al., 1986). In addition, the Beam-Warming scheme is used in the Transonic Navier-Stokes (TNS) code developed by Holst et al. (1987).

9.2.4 Other Implicit Methods

MacCormack (1981) developed an implicit analog of his explicit finite-difference method. This method consists of two stages. The first stage uses the original MacCormack scheme, while the second stage employs an implicit scheme to eliminate any stability restrictions. The resulting matrix equations are either upper or lower block bidiagonal equations, which can be solved in an easier fashion than the usual block tridiagonal systems. A major disadvantage of this scheme is due to the difficulties encountered in applying non-Dirichlet boundary conditions.

Obayashi and Kuwahara (1986) modified the Beam-Warming scheme by applying lower-upper (LU) factorization (see Section 4.3.4) in conjunction with flux-vector splitting (see Section 6.4.1). As a result, each ADI operator is decomposed into the product of lower and upper bidiagonal matrices, which are easier to solve. This technique is referred to as LU-ADI factorization and is similar to that used in the "implicit" MacCormack scheme described above.

The lower-upper symmetric Gauss-Seidel (LU-SGS) implicit scheme was developed by Yoon and Jameson (1987, 1988) to obtain steady-state solutions of the unsteady Euler and N-S equations. The LU-SGS method employs an approximate Newton iteration procedure that permits scaler diagonal inversions as opposed to the block matrix inversions required in the conventional line Gauss-Seidel (LGS) methods. The LU-SGS method ensures that the matrix is diagonally dominant without resorting to flux splitting. Rieger and Jameson (1988) extended the LU-SGS scheme to three dimensions. This scheme is widely used to solve the compressible N-S equations. It can be combined with an upwind scheme (see next section) to eliminate oscillations near flow field discontinuities.

9.2.5 Upwind Methods

The central-difference schemes described previously for solving the compressible N-S equations almost always require additional dissipation for numerical stability. Upwind schemes, on the other hand, inherently possess the needed dissipation to control these numerical instabilities. Upwind schemes were initially applied to the Euler equations in the early 1980s, as described in Chapter 6. Shortly thereafter, in the mid-1980s, they were applied to the compressible N-S equations. The extension to the compressible N-S equations is straightforward, since the additional shear-stress and heat flux terms are centrally differenced. However, an important consideration that must be taken into account when solving viscous flows with upwind schemes is whether they will produce excessive dissipation that will swamp the natural dissipation in boundary-layer regions. This problem is effectively eliminated by using higher-order upwind schemes.

Some of the earliest applications of upwind schemes to the compressible N-S equations were by Lombard et al. (1983), Coakley (1983b), MacCormack (1985), Chakravarthy et al. (1985), and Thomas and Walters (1985). Lombard et al. (1983) employed an implicit, upwind flux-difference splitting algorithm called the conservative supracharacteristics method (CSCM). Coakley (1983b) and MacCormack (1985) used implicit upwind finite-volume schemes similar to the flux-vector splitting method of Steger and Warming (1981). MacCormack applied LGS and Newton iteration procedures to solve the resulting matrix equations.

Chakravarthy et al. (1985) employed a family of high-order accurate total variation diminishing (TVD) schemes (Chakravarthy and Osher, 1985) based on Roe's approximate Riemann solver to model the convection terms in the compressible N-S equations. (See Section 4.4.12 for a discussion of TVD schemes.) This work has led to a series of widely used, unified computer programs (the USA-series of codes) that were developed at Rockwell International Science Center. These codes can be used for a wide variety of flow situations, including steady and unsteady flows; low-speed, subsonic, transonic, supersonic, and hypersonic flows; internal and external flows; and perfect-gas and real-gas (equilibrium and nonequilibrium chemistry) flows (Palaniswamy et al., 1989).

Thomas and Walters (1985) initially used the flux-vector splitting method developed by van Leer and co-workers (van Leer et al., 1982; Anderson et al., 1985) to solve the thin-layer compressible N-S equations. This implicit upwind finite-volume method includes third-order accurate spatial differencing along with either approximate factorization or LGS relaxation to solve the resulting matrix equations. Since it has been shown (van Leer et al., 1987) that flux-vector splitting schemes will produce excessive dissipation in boundary layers, as compared with Roe and Osher's approximate Riemann solvers, Thomas and Walters have incorporated Roe's flux-difference splitting into their algorithm. This work has led to the widely used Computational Fluids Laboratory 3-D (CFL3D) code developed at NASA Langley Research Center (Vasta et al., 1987).

Since the initial applications of upwind schemes to the compressible N-S equations, numerous investigators have refined these procedures and have applied them to ever more complicated problems. For example, flow problems involving finite-rate chemistry and thermal nonequilibrium have been successfully computed using the compressible N-S equations. Included in this latter category is the work of Gnoffo (1986, 1989), who developed the Langley aerothermodynamic upwind relaxation algorithm (LAURA) code. This code was developed primarily to solve 3-D external hypersonic flows in chemical and thermal nonequilibrium. It uses an implicit upwind finite-volume algorithm based on Roe's scheme with second-order TVD corrections (Yee, 1985a, 1985b). The fluids, chemistry, and thermodynamics are fully coupled in the code.

Candler and MacCormack (1988) extended the upwind algorithm of MacCormack (1985) to account for chemical and thermal nonequilibrium processes. Molvik and Merkle (1989) developed the TUFF code to solve 3-D external reacting hypersonic flows. The TUFF code uses an implicit upwind finite-volume algorithm and employs a temporal Riemann solver that fully accounts for the multicomponent mixture of gases. Higher-order accuracy is obtained using Chakravarthy and Osher's (1985) TVD scheme. The code has been enhanced to permit the calculation of the internal reacting flow in scramjet engines (Molvik et al., 1993).

9.2.6 Compressible Navier-Stokes Equations at Low Speeds

Until recently, most algorithms designed for compressible flows were observed to become very inefficient or inaccurate at low Mach numbers. The traditional remedy was to solve the incompressible form of the equations for problems requiring solutions in the low-speed regime. This appears unreasonable. The incompressible equations are merely a subset of the compressible equations, and it is well known that the physics itself is usually no more complex merely because the Mach number is low. For most flows, no important changes would be observed if the Mach number were reduced from, say, 0.2 to 0.01 if all other dimensionless parameters of the flow remained the same. If no significant changes in flow structure are noted as the Mach number drops from about 0.2 toward zero, any difficulties encountered must be due to an inappropriate construction of the numerical algorithm itself. The computational difficulties are believed to arise from two separate mechanisms: (1) an ill-conditioned algebraic problem and (2) round-off errors due to a disparity between magnitudes of variables. The first mechanism is the most troublesome. These difficulties have been addressed by several investigators, including Turkel (1987, 1992), Feng and Merkle (1990), Choi and Merkle (1991), and Peyret and Viviand (1985).

The key issues can be outlined sufficiently by using the 1-D N-S equations as they apply to an ideal gas. Nondimensional variables are defined as

$$\begin{gathered} t=\frac{\tilde{t}}{L_{\text{ref}}/u_{\text{ref}}} \qquad x=\frac{\tilde{x}}{L_{\text{ref}}} \qquad u=\frac{\tilde{u}}{u_{\text{ref}}} \\ p=\frac{\tilde{p}}{(\rho_{\text{ref}}u_{\text{ref}}^2)} \qquad T=\frac{\tilde{T}}{T_{\text{ref}}} \qquad \mu=\frac{\tilde{\mu}}{\mu_{\text{ref}}} \\ R=\frac{\tilde{R}}{(u_{\text{ref}}^2/T_{\text{ref}})}=\frac{1}{\gamma M^2} \qquad c_p=\frac{\tilde{c}_p}{(u_{\text{ref}}^2/T_{\text{ref}})}=\frac{1}{(\gamma-1)M^2} \end{gathered} \tag{9.70}$$

where the tildes denote the dimensional variables, subscript ref denotes dimensional reference quantities, and the Mach number M is based on reference quantities and the gas constant:

$$M=\frac{u_{\text{ref}}}{\sqrt{\gamma\tilde{R}T_{\text{ref}}}}$$

We will solve for the primitive variables p, u, and T rather than the "conserved" variables ρ, ρu, and E_t for two reasons. First, we know that the primitive variables are appropriate for low-speed flows because they are widely used in incompressible formulations. If we elect to solve the compressible equations for the same primitive variables, it should be possible to detect what, if anything, is causing the numerical difficulty as M approaches zero and the equations reduce to a variable-property version of the incompressible flow equations. Second, it is possible to identify and remedy the source of numerical difficulties more quickly if we choose to solve for the primitive variables rather than the conserved variables.

Substituting for density by using the ideal gas equation of state and utilizing primitive variables p, u, and T, the conservation equations for mass, momentum, and energy can be written

$$\frac{\partial \mathbf{Q}(\mathbf{q})}{\partial t}+\frac{\partial \mathbf{E}(\mathbf{q})}{\partial x}-\frac{\partial \mathbf{E}_v(\mathbf{q})}{\partial x}=0 \tag{9.71}$$

where

$$\mathbf{q} = \begin{bmatrix} p \\ u \\ T \end{bmatrix} \qquad \mathbf{Q} = \begin{bmatrix} \dfrac{p}{RT} \\ \dfrac{pu}{RT} \\ \dfrac{p}{\gamma R} + M^2 \dfrac{(\gamma - 1)}{2} \dfrac{pu^2}{RT} \end{bmatrix}$$

$$\mathbf{E} = \begin{bmatrix} \dfrac{pu}{RT} \\ \dfrac{pu^2}{RT} + p \\ \dfrac{pu}{R} + M^2 \dfrac{(\gamma - 1)}{2} \dfrac{pu^3}{RT} \end{bmatrix} \qquad \mathbf{E}_v = \begin{bmatrix} 0 \\ \dfrac{4\mu}{3\,\mathrm{Re}} \dfrac{\partial u}{\partial x} \\ \dfrac{4(\gamma - 1)M^2 \mu u}{3\,\mathrm{Re}} \dfrac{\partial u}{\partial x} + \dfrac{\mu}{\mathrm{Re}\,\mathrm{Pr}} \dfrac{\partial T}{\partial x} \end{bmatrix} \tag{9.72}$$

The Reynolds and Prandtl numbers are defined as

$$\mathrm{Re} = \frac{\rho_{\mathrm{ref}} u_{\mathrm{ref}} L_{\mathrm{ref}}}{\mu_{\mathrm{ref}}} \qquad \mathrm{Pr} = \frac{\tilde{c}_p \tilde{\mu}}{\tilde{k}}$$

The Prandtl number of the fluid is assumed to be constant. The viscosity and thermal conductivity can be evaluated from Sutherland's equation. Note that the equations are in the strong conservation-law form even though primitive variables are used. Some of the computational fluid dynamics (CFD) literature gives the impression that "conserved" variables must be used to achieve the favorable properties of the conservation-law form of the equations. This is simply not true. It will be indicated below how the conservation-law form of the discretized equations can be satisfied using primitive variables.

First, we will consider the consequences of the reference Mach number approaching zero. Note that $p/RT = \rho/\rho_{\mathrm{ref}}$ and $R = 1/(\gamma M^2)$. We observe that the combination p/RT approaches a perfectly acceptable finite limit as M goes to zero. Note that $\mathbf{E}$ and $\mathbf{E}_v$ will reduce to a form appropriate for a fluid whose density may be a function of temperature but for which the density cannot be altered by changes in pressure alone, i.e., $\partial\rho/\partial p)_T = 0$. This is the sense in which the compressible N-S equations reduce to an incompressible form by virtue of the zero Mach number limit. Solutions to this system under isothermal conditions can be made to behave as close to the incompressible limit $[(1/\rho)(D\rho/Dt) = 0]$ as desired by choosing the reference Mach number to be sufficiently low.

To consider the mathematical properties of the 1-D conservation equations, it is convenient to write Eq. (9.71) as

$$[A_t]\frac{\partial \mathbf{q}}{\partial t} + [A_x]\frac{\partial \mathbf{q}}{\partial x} = \frac{\partial \mathbf{E}_v}{\partial x} \tag{9.73}$$

where $[A_t]$ and $[A_x]$ are Jacobian matrices to be evaluated at the most recent iteration level and $\mathbf{q}$ is the vector of unknown primitive variables given in Eq. (9.72). The Jacobian matrices can be written as

$$[A_t] = \begin{bmatrix} \dfrac{\gamma M^2}{T} & 0 & -\dfrac{p}{RT^2} \\ \dfrac{\gamma M^2 u}{T} & \dfrac{p}{RT} & -\dfrac{pu}{RT^2} \\ M^2 + \gamma M^4 u^2 \dfrac{(\gamma-1)}{2T} & \gamma M^4 p \dfrac{(\gamma-1)u}{T} & -\gamma M^4 p u^2 \dfrac{(\gamma-1)}{2T^2} \end{bmatrix} \tag{9.74}$$

$$[A_x] = \begin{bmatrix} \dfrac{\gamma M^2 u}{T} & \dfrac{p}{RT} & -\dfrac{pu}{RT^2} \\ \dfrac{\gamma M^2 u^2}{T} + 1 & \dfrac{2pu}{RT} & -\dfrac{pu^2}{RT^2} \\ \gamma M^2 u + \gamma M^4 \dfrac{(\gamma-1)u^3}{2T} & \dfrac{p}{R} + 3\gamma M^4 p u^2 \dfrac{(\gamma-1)}{2T} & -\gamma M^4 p u^3 \dfrac{(\gamma-1)}{2T^2} \end{bmatrix} \tag{9.75}$$

In the above, R has been replaced by $1/(\gamma M^2)$ in those terms that will vanish as M goes to zero. Notice that the variable property incompressible form (in the sense of $\partial\rho/\partial p)_T = 0$) of the equations is recovered as M goes to zero. Notice also that as M goes to zero, the terms containing the time derivative of pressure tend toward zero unless a vanishingly small time step is used. We shall show next that singular behavior of the coupled time-dependent system accompanies the vanishing of the pressure-time derivatives.

The mathematical nature of the time-marching problem can be established by writing the system as

$$\frac{\partial \mathbf{q}}{\partial t} + [A_t]^{-1}[A_x]\frac{\partial \mathbf{q}}{\partial x} = [A_t]^{-1}\frac{\partial \mathbf{E}_v}{\partial x} \tag{9.76}$$

and considering the eigenvalues of $[A_t]^{-1}[A_x]$, which in this case are u, $u + a$, and $u - a$ in terms of nondimensional variables. The parameter a is the nondimensional local speed of sound. The problem can be solved by a marching method, since the eigenvalues are real. As M becomes small, $[A_t]$ becomes ill conditioned. That is, the determinant of $[A_t]$ becomes small, and errors are expected to arise in computing $[A_t]^{-1}$. In the limit as M goes to zero, the inverse of $[A_t]$ does not exist (the first column and third row vanish), and the system is singular.

The magnitudes of the eigenvalues of the matrix product $[A_t]^{-1}[A_x]$ also provide information about the properties of the system. As M is decreased in the subsonic regime, the eigenvalues of the matrix product $[A_t]^{-1}[A_x]$ begin to differ more and more in magnitude, the ratio of the smallest eigenvalue to the largest being approximately the ratio of the convective speed to the acoustic speed. The condition and degree of "stiffness" of the system can be related to the relative magnitudes of the eigenvalues. When the eigenvalues differ greatly in magnitude, convergence to a steady-state solution is usually slow, or for time-dependent solutions, the allowable time step becomes very small. This occurs because greatly varying signal speeds appear in the equations and the traditional solution schemes attempt to honor all of them, creating a "stiff" system. Since M has almost no influence on the physical characteristics of the flow at very low values of M (it effectively "cancels out" of the physics), it must be possible to devise a solution scheme in which M has very little influence on the convergence rate ("cancels out") of the numerical scheme over the range in which it is an unimportant physical parameter.

Another observation is that the pressure is eliminated as an unknown in the time derivative as M approaches zero. This indicates that the solution to the incompressible equations carry no direct pressure history and thus the pressure field is established anew at each time step.

To compute flows at very low M, one could of course, adopt a fully incompressible scheme, as has often been the case in the past. On the other hand, there are advantages to maintaining the variable-property fully coupled arrangement, and we will indicate that it is possible to modify the compressible formulation so that it will work well for vanishingly small Mach numbers with virtually no sacrifice in accuracy or efficiency.

To overcome the awkward mathematical situation that arises with the unsteady form of the coupled compressible equations at low M, it is necessary to make changes in the formulation of the time terms. At least two alternatives exist. The existing time terms can be modified to permit efficient solutions to be obtained to the steady-flow equations, or an efficient scheme can be obtained by adding a pseudo-time (artificial time) term that vanishes at convergence at each physical time step. This latter approach is recommended because it permits time-accurate solutions to the N-S equations to be obtained when they are needed.

Although the pseudo-time term can take many different forms and still be effective, a suitable arrangement can be obtained by simply adding a term to each equation having the same form as the physical time term but with M removed from the coefficients to the unknown pressure that cause the fatal ill conditioning. The equations then become

$$[A_p]\frac{\partial \mathbf{q}}{\partial \tau} + [A_t]\frac{\partial \mathbf{q}}{\partial t} + [A_x]\frac{\partial \mathbf{q}}{\partial x} = \frac{\partial \mathbf{E}_v}{\partial x} \tag{9.77}$$

where τ is a pseudo-time and the preconditioning matrix $[A_p]$ is given by

$$[A_p] = \begin{bmatrix} \dfrac{1}{T} & 0 & -\dfrac{p}{RT^2} \\ \dfrac{u}{T} & \dfrac{p}{RT} & -\dfrac{pu}{RT^2} \\ \dfrac{1}{\gamma} + M^2u^2\dfrac{(\gamma-1)}{2T} & \gamma M^4 pu\dfrac{(\gamma-1)}{T} & -\gamma M^4 pu^2\dfrac{(\gamma-1)}{2T^2} \end{bmatrix} \tag{9.78}$$

Note that $[A_p]$ is formed from $[A_t]$ by simply dividing the first column of $[A_t]$ by γM^2 (equivalent to multiplying by the nondimensional gas constant R). Presumably, dividing by only M^2 would have the same effect. It is also quite likely that the preconditioning matrix can be simplified somewhat by setting some of the off-diagonal entries in the second and third columns equal to zero. The form above has the conceptual advantage of being easily developed from $[A_t]$ (with minimal change) using commonsense logic.

The hyperbolic system is solved by advancing in pseudo-time until no changes occur at each physical time step. At that point, the time-accurate equations are satisfied. Obviously, this involves "subiterations," but that is consistent with the observation that for a completely incompressible flow, the pressure field must be established at each physical time step with no direct dependence on a previous pressure field. The addition of the pseudo-time term changes the eigenvalues of the hyperbolic system, so that they are clustered closer together in magnitude, effectively at speeds closer to the convective speed. The hyperbolic system being solved is in τ and x, and it is now the eigenvalues of the matrix product $[A_p]^{-1}[A_x]$ that characterize the system. This is similar to marching in pseudo-time to a "steady" solution at each physical time step. The eigenvalues of $[A_p]^{-1}[A_x]$ are

$$u, \tfrac{1}{2}\left[u(1+\gamma M^2) \pm \sqrt{u^2(1-\gamma M^2)^2 + 4\gamma T}\right]$$

If we assume a value of 1.4 for the ratio of specific heats and evaluate u and T at the reference values, the above expressions give 2.275 for the ratio of the magnitudes of the largest to the smallest eigenvalues in the limit of zero Mach number. Without preconditioning, this ratio approaches infinity at the zero Mach number limit. For $M = 0.3$ the ratio is 2.89 with preconditioning.

The second and minor source of difficulty with the coupled compressible formulation at low M is related to the differences in the relative magnitudes of the nondimensional dependent variables. The magnitudes of the velocity and temperature remain of order 1, but the nondimensional pressure tends to increase without limit as M decreases. This permits round-off errors to become significant. Again, this is not a problem of physical origin. Using double-precision arithmetic, no difficulties will usually be observed until M decreases to about

10^{-5} or 10^{-6}. To completely eliminate the problem, a "gauge" or relative pressure can be introduced, so that differences in pressure (originating from pressure derivatives) become differences in the gauge pressure. This procedure is very easy to implement. The nondimensional thermodynamic pressure p is replaced by the sum of p_c, a constant, and p_g, the variable part of the pressure. The constant part is selected to be as large as possible for the problem at hand.

Although the discussion above utilized primitive variables, the results can be interpreted in terms of the more traditional conserved variables (Feng and Merkle, 1990). However, there are no particular disadvantages to using the primitive variables for computations, and the required form of preconditioning is much easier to develop in terms of primitive variables. We shall indicate how the preconditioned system can be solved by an implicit scheme while maintaining the conservation-law form of the discretized equations. Putting Eq. (9.77) in conservation-law form except for the pseudo-time term gives

$$[A_p]\frac{\partial \mathbf{q}}{\partial \tau} + \frac{\partial \mathbf{Q}(\mathbf{q})}{\partial t} + \frac{\partial \mathbf{E}(\mathbf{q})}{\partial x} - \frac{\partial \mathbf{E}_v(\mathbf{q})}{\partial x} = 0 \tag{9.79}$$

The vectors $\mathbf{Q}$, $\mathbf{E}$, and $\mathbf{E}_v$ can be linearized by iterating at each pseudo-time step using a Newton method:

$$\mathbf{Q} = \tilde{\mathbf{Q}} + \left[\tilde{A}_t\right]\Delta\mathbf{q} \qquad \mathbf{E} = \tilde{\mathbf{E}} + \left[\tilde{A}_x\right]\Delta\mathbf{q} \qquad \mathbf{E}_v = \tilde{\mathbf{E}}_v + \left[\tilde{A}_v\right]\Delta\mathbf{q}$$

where the Jacobian matrices $[A_t]$ and $[A_x]$ have been defined previously and

$$[A_v] = \begin{bmatrix} 0 & 0 & 0 \\ 0 & \dfrac{4\mu}{3\,\mathrm{Re}}\dfrac{\partial}{\partial x} & 0 \\ 0 & \dfrac{4(\gamma-1)M^2\mu}{3\,\mathrm{Re}}\dfrac{\partial u}{\partial x} & \dfrac{\mu}{\mathrm{Re}\,\mathrm{Pr}}\dfrac{\partial}{\partial x} \end{bmatrix} \tag{9.80}$$

In the above, the tilde indicates evaluation at the most recently determined values (from the previous iteration) and the Δ indicates changes from the previous pseudo-time iterations. Equation (9.79) can then be written as

$$[A_p]\frac{\partial \Delta\mathbf{q}}{\partial \tau} + \frac{\partial\left(\left[\tilde{A}_t\right]\Delta\mathbf{q}\right)}{\partial t} + \frac{\partial\left(\left[\tilde{A}_x\right]\Delta\mathbf{q}\right)}{\partial x} - \frac{\partial\left(\left[\tilde{A}_v\right]\Delta\mathbf{q}\right)}{\partial x} = -\left(\frac{\partial \tilde{\mathbf{Q}}}{\partial t} + \frac{\partial \tilde{\mathbf{E}}}{\partial x} - \frac{\partial \tilde{\mathbf{E}}_v}{\partial x}\right) \tag{9.81}$$

Further details will depend somewhat on the discretization scheme selected. For example, if a centrally differenced fully implicit discretization is employed, the algebraic system can be solved using the block tridiagonal algorithm given in Appendix B. The algebraic system is solved for $\Delta\mathbf{q}$, the change between pseudo-time iterations. When the changes vanish at each physical step, the LHS goes to zero. The RHS is the discretized original partial differential equation

(PDE) (the residual) *in conservation-law form* evaluated using $\mathbf{q}$ from the most recent pseudo-time iteration. Thus the conservation-law form of the equation is satisfied at each physical time step when iterative (pseudo-time) convergence is achieved.

The procedure readily extends to the 2-D and 3-D N-S equations. For example, in 2-D the matrix $[A_p]$ becomes

$$[A_p] = \begin{bmatrix} \dfrac{1}{T} & 0 & 0 & -\dfrac{p}{RT^2} \\ \dfrac{u}{T} & \dfrac{p}{RT} & 0 & -\dfrac{pu}{RT^2} \\ \dfrac{v}{T} & 0 & \dfrac{p}{RT} & -\dfrac{pv}{RT^2} \\ \dfrac{1}{\gamma} + \dfrac{u^2+v^2}{2c_pT} & \dfrac{pu}{Rc_pT} & \dfrac{pv}{Rc_pT} & -\dfrac{p(u^2+v^2)}{2c_pRT^2} \end{bmatrix} \tag{9.82}$$

Further details on implementing this procedure for the 2-D N-S equations can be found in the work by Pletcher and Chen (1993).

9.3 INCOMPRESSIBLE NAVIER-STOKES EQUATIONS

From the above discussion on solving the compressible N-S equations at low speeds, it should be evident that the preconditioned compressible form of the N-S equations can be solved efficiently at vanishingly small M. For low-speed, nearly "incompressible" gas flows where property variations may be important or where the presence of heat transfer requires a solution to the energy equation, solving the compressible form of the equations may be the best choice.

When heat transfer or significant property variations are not present, the traditional incompressible form of the N-S equations is usually selected for numerical solution. Panton (1984) provides a useful discussion of the limiting forms of the N-S equations and the range of applicability of the incompressible form of the equations.

The incompressible N-S equations for a constant property flow without body forces or external heat addition are given by (see Chapter 5)

continuity:

$$\nabla \cdot \mathbf{V} = 0 \tag{9.83}$$

momentum:

$$\rho \frac{D\mathbf{V}}{Dt} = -\nabla p + \mu \nabla^2 \mathbf{V} \tag{9.84}$$

energy:

$$\rho c_v \frac{DT}{Dt} = k \nabla^2 T + \Phi \tag{9.85}$$

These equations (one vector, two scalar) are a mixed set of elliptic-parabolic equations that contain the unknowns $(\mathbf{V}, p, T)$. Note that the temperature appears directly only in the energy equation, so that we can uncouple this equation from the continuity and momentum equations. This uncoupling is exact if the fluid properties are independent of temperature. For many applications, the temperature changes are either insignificant or unimportant, and it is not necessary to solve the energy equation. However, if we wish to find the temperature distribution, this can be easily accomplished, since the unsteady energy equation is a parabolic PDE, provided that $\mathbf{V}$ has already been computed. With this in mind, we will focus our attention on methods for solving the continuity and momentum equations during the remainder of this chapter.

The 2-D incompressible N-S equations written in Cartesian coordinates (without the energy equation) are

continuity:

$$\frac{\partial u}{\partial x} + \frac{\partial v}{\partial y} = 0 \tag{9.86}$$

x momentum:

$$\frac{\partial u}{\partial t} + u\frac{\partial u}{\partial x} + v\frac{\partial u}{\partial y} = -\frac{1}{\rho}\frac{\partial p}{\partial x} + \nu\left(\frac{\partial^2 u}{\partial x^2} + \frac{\partial^2 u}{\partial y^2}\right) \tag{9.87}$$

y momentum:

$$\frac{\partial v}{\partial t} + u\frac{\partial v}{\partial x} + v\frac{\partial v}{\partial y} = -\frac{1}{\rho}\frac{\partial p}{\partial y} + \nu\left(\frac{\partial^2 v}{\partial x^2} + \frac{\partial^2 v}{\partial y^2}\right) \tag{9.88}$$

where ν is the kinematic viscosity μ/ρ. These equations are written in the *primitive-variable* form, where p, u, v are the primitive variables. Incompressible flows have been solved successfully by using both primitive and derived (such as vorticity and stream function) variables. Both approaches will be discussed here. We start with the vorticity–stream function technique in Section 9.3.1. Methods for solving Eqs. (9.86)–(9.88) for the primitive variables will follow in Section 9.3.2.

9.3.1 Vorticity–Stream Function Approach

The vorticity–stream function approach has been one of the most popular methods for solving the 2-D incompressible N-S equations. In this approach, a change of variables is made that replaces the velocity components with the vorticity ζ and the stream function ψ. The vorticity vector $\boldsymbol{\zeta}$ was defined in Chapter 5 as

$$\boldsymbol{\zeta} = \nabla \times \mathbf{V} \tag{9.89}$$

The magnitude of the vorticity vector is

$$|\boldsymbol{\zeta}| = |\nabla \times \mathbf{V}| \tag{9.90}$$

while the scaler value of vorticity for 2-D flows can be written as

$$\zeta = \frac{\partial v}{\partial x} - \frac{\partial u}{\partial y} \tag{9.91}$$

for a 2-D Cartesian coordinate system. Also in this coordinate system, the stream function ψ is defined by the equations

$$\begin{aligned} \frac{\partial \psi}{\partial y} &= u \\ \frac{\partial \psi}{\partial x} &= -v \end{aligned} \tag{9.92}$$

Using these new dependent variables, the two momentum equations [Eqs. (9.87) and (9.88)] can be combined (thereby eliminating pressure) to give

$$\frac{\partial \zeta}{\partial t} + u\frac{\partial \zeta}{\partial x} + v\frac{\partial \zeta}{\partial y} = \nu\left(\frac{\partial^2 \zeta}{\partial x^2} + \frac{\partial^2 \zeta}{\partial y^2}\right) \tag{9.93}$$

or

$$\frac{D\zeta}{Dt} = \nu\,\nabla^2 \zeta \tag{9.94}$$

This parabolic PDE is called the *vorticity transport equation*. The 1-D form of this equation,

$$\frac{\partial \zeta}{\partial t} + u\frac{\partial \zeta}{\partial x} = \nu\frac{\partial^2 \zeta}{\partial x^2} \tag{9.95}$$

is the 1-D advection-diffusion equation, which is often used as a model equation. In addition, the nonlinear Burgers equation can be used to model the vorticity transport equation. In fact, the numerical techniques described in Section 4.5 to solve the nonlinear Burgers equation can be directly applied to the vorticity transport equation.

An additional equation involving the new dependent variables ζ and ψ can be obtained by substituting Eqs. (9.92) into Eq. (9.91), which gives

$$\frac{\partial^2 \psi}{\partial x^2} + \frac{\partial^2 \psi}{\partial y^2} = -\zeta \tag{9.96}$$

or

$$\nabla^2 \psi = -\zeta \tag{9.97}$$

This elliptic PDE is the *Poisson equation*. Methods for solving equations of this type are discussed in Section 4.3.

As a result of the change of variables, we have been able to separate the mixed elliptic-parabolic 2-D incompressible N-S equations into one parabolic equation (the vorticity transport equation) and one elliptic equation (the Poisson equation). These equations are normally solved sequentially using a time-marching procedure, which is described by the following steps:

1. Specify initial values for ζ and ψ at time $t = 0$.
2. Solve the vorticity transport equation for ζ at each interior grid point at time $t + \Delta t$.
3. Iterate for new ψ values at all points by solving the Poisson equation using new ζ at interior points.
4. Find the velocity components from $u = \psi_y$ and $v = -\psi_x$.
5. Determine values of ζ on the boundaries using ψ and ζ values at interior points.
6. Return to Step 2 if the solution is not converged.

An alternative sequential procedure has been used successfully by Mallinson and de Vahl Davis (1973) that is based on a pseudo-transient representation of Eq. (9.96):

$$\frac{\partial \psi}{\partial \tau} - \left(\frac{\partial^2 \psi}{\partial x^2} + \frac{\partial^2 \psi}{\partial y^2} + \zeta \right) = 0$$

The pseudo-time step is an additional parameter in the scheme that can be varied in order to accelerate convergence. Upon convergence, the pseudo-time term vanishes, and Eq. (9.96) is satisfied.

The vorticity–stream function system can also be solved efficiently in a coupled rather than sequential manner. Rubin and Khosla (1981) solved the 2×2 coupled system for ψ and ζ using the modified strongly implicit procedure. The coupled procedure was also used in combination with multigrid acceleration by Ghia et al. (1982).

The pressure does not appear explicitly in the vorticity–stream function formulation. However, in those applications where the pressure is of interest, it can be readily determined from the velocity solution by solving an additional Poisson equation. This equation is derived by differentiating Eq. (9.87) with respect to x:

$$\frac{\partial}{\partial t}\left(\frac{\partial u}{\partial x} \right) + \left(\frac{\partial u}{\partial x} \right)^2 + u \frac{\partial^2 u}{\partial x^2} + \frac{\partial v}{\partial x}\frac{\partial u}{\partial y} + v \frac{\partial^2 u}{\partial x \partial y} = -\frac{1}{\rho}\frac{\partial^2 p}{\partial x^2} + \nu \frac{\partial}{\partial x}(\nabla^2 u) \tag{9.98}$$

differentiating Eq. (9.88) with respect to y,

$$\frac{\partial}{\partial t}\left(\frac{\partial v}{\partial y} \right) + \left(\frac{\partial v}{\partial y} \right)^2 + v \frac{\partial^2 v}{\partial y^2} + \frac{\partial u}{\partial y}\frac{\partial v}{\partial x} + u \frac{\partial^2 v}{\partial x \partial y} = -\frac{1}{\rho}\frac{\partial^2 p}{\partial y^2} + \nu \frac{\partial}{\partial y}(\nabla^2 v) \tag{9.99}$$

and adding the results to obtain

$$\frac{\partial}{\partial t}\left(\frac{\partial u}{\partial x}+\frac{\partial v}{\partial y}\right)+\left(\frac{\partial u}{\partial x}\right)^2+\left(\frac{\partial v}{\partial y}\right)^2+2\left(\frac{\partial v}{\partial x}\right)\left(\frac{\partial u}{\partial y}\right)+u\left(\frac{\partial^2 u}{\partial x^2}+\frac{\partial^2 v}{\partial x\,\partial y}\right)$$
$$+v\left(\frac{\partial^2 u}{\partial x\,\partial y}+\frac{\partial^2 v}{\partial y^2}\right)=-\frac{1}{\rho}\nabla^2 p+\nu\left[\frac{\partial}{\partial x}(\nabla^2 u)+\frac{\partial}{\partial y}(\nabla^2 v)\right] \tag{9.100}$$

Using the continuity equation, Eq. (9.100) can be reduced to

$$\nabla^2 p = 2\rho\left(\frac{\partial u}{\partial x}\frac{\partial v}{\partial y}-\frac{\partial u}{\partial y}\frac{\partial v}{\partial x}\right) \tag{9.101}$$

In terms of the stream function this equation can be rewritten as

$$\nabla^2 p = S \tag{9.102}$$

where

$$S = 2\rho\left[\left(\frac{\partial^2\psi}{\partial x^2}\right)\left(\frac{\partial^2\psi}{\partial y^2}\right)-\left(\frac{\partial^2\psi}{\partial x\,\partial y}\right)^2\right] \tag{9.103}$$

Thus we have obtained a Poisson equation for pressure that is analogous to Eq. (9.97). In fact, all the methods discussed in Section 4.3 for solving Eq. (9.97) will also apply to Eq. (9.102) if S is differenced in an appropriate manner. A suitable second-order difference representation is given by

$$S_{i,j} = 2\rho_{i,j}\left[\left(\frac{\psi_{i+1,j}-2\psi_{i,j}+\psi_{i-1,j}}{(\Delta x)^2}\right)\left(\frac{\psi_{i,j+1}-2\psi_{i,j}+\psi_{i,j-1}}{(\Delta y)^2}\right)\right.$$
$$\left.-\left(\frac{\psi_{i+1,j+1}-\psi_{i+1,j-1}-\psi_{i-1,j+1}+\psi_{i-1,j-1}}{4\,\Delta x\,\Delta y}\right)^2\right] \tag{9.104}$$

For a steady flow problem, the Poisson equation for pressure is only solved once, i.e., after the steady-state values of ζ and ψ have been computed. If only the wall pressures are desired, it is not necessary to solve the Poisson equation over the entire flow field. Instead, a simpler equation can be solved for the wall pressures. This equation is obtained by applying the tangential momentum equation to the fluid adjacent to the wall surface. For a wall located at $y = 0$ in a Cartesian coordinate system (see Fig. 9.2), the steady tangential momentum equation (x momentum equation) reduces to

$$\left.\frac{\partial p}{\partial x}\right)_{\text{wall}} = \left.\mu\frac{\partial^2 u}{\partial y^2}\right)_{\text{wall}} \tag{9.105}$$

or

$$\left.\frac{\partial p}{\partial x}\right)_{\text{wall}} = \left.-\mu\frac{\partial \zeta}{\partial y}\right)_{\text{wall}} \tag{9.106}$$

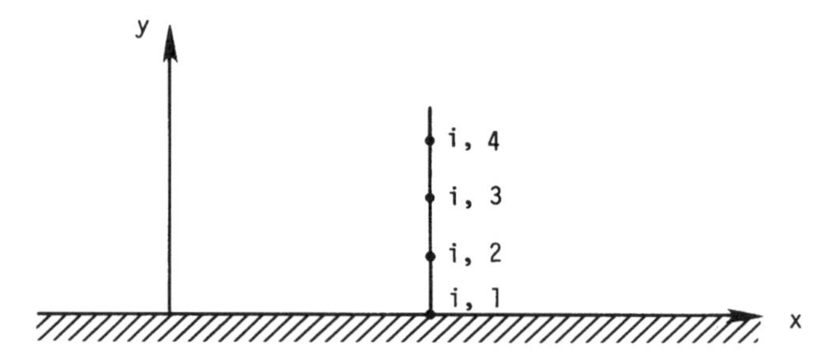

Figure 9.2 Grid points normal to a flat plate at $y = 0$.

which can be differenced as

$$\frac{p_{i+1,1} - p_{i-1,1}}{2\,\Delta x} = -\mu\left(\frac{-3\zeta_{i,1} + 4\zeta_{i,2} - \zeta_{i,3}}{2\,\Delta y}\right) \tag{9.107}$$

In order to apply Eq. (9.107) the pressure must be known for at least one point on the wall surface. The pressure at the adjacent point can be determined using a first-order one-sided difference expression for $\partial p/\partial x$ in Eq. (9.107). Thereafter, Eq. (9.107) can be used to find the pressure at all other wall points. For a body intrinsic coordinate system, Eq. (9.106) becomes

$$\left.\frac{\partial p}{\partial s}\right)_{\text{wall}} = -\mu\left.\frac{\partial \zeta}{\partial n}\right)_{\text{wall}} \tag{9.108}$$

where s is measured along the body surface and n is normal to it.

The time-marching procedure described earlier for solving the vorticity transport equation and the Poisson equation requires that appropriate expressions for ψ and ζ be specified at the boundaries. The specification of these boundary conditions is extremely important, since it directly affects the stability and accuracy of the solution. Let us examine the application of boundary conditions on a wall located at $y = 0$. At the wall surface, ψ is a constant that is usually set equal to zero. In order to find ζ at the wall surface, we expand ψ using a Taylor series about the wall point $(i, 1)$:

$$\psi_{i,2} = \psi_{i,1} + \left.\frac{\partial \psi}{\partial y}\right)_{i,1}\Delta y + \frac{1}{2}\left.\frac{\partial^2 \psi}{\partial y^2}\right)_{i,1}(\Delta y)^2 + \cdots \tag{9.109}$$

Since

$$\left.\frac{\partial \psi}{\partial y}\right)_{i,1} = u_{i,1} = 0$$

$$\left.\frac{\partial^2 \psi}{\partial y^2}\right)_{i,1} = \left.\frac{\partial u}{\partial y}\right)_{i,1} \tag{9.110}$$

and

$$\zeta_{i,1} = \cancelto{0}{\frac{\partial v}{\partial x}}\Bigg)_{i,1} - \frac{\partial u}{\partial y}\Bigg)_{i,1} = -\frac{\partial^2 \psi}{\partial y^2}\Bigg)_{i,1} \tag{9.111}$$

we can rewrite Eq. (9.109) as

$$\psi_{i,2} = \psi_{i,1} - \tfrac{1}{2}\zeta_{i,1}(\Delta y)^2 + O\left[(\Delta y)^3\right]$$

or

$$\zeta_{i,1} = \frac{2(\psi_{i,1} - \psi_{i,2})}{(\Delta y)^2} + O(\Delta y) \tag{9.112}$$

This first-order expression for $\zeta_{i,1}$ often gives better results than higher-order expressions, which are susceptible to instabilities at higher Reynolds numbers. For example, the following second-order expression, which was first used by Jensen (1959), leads to unstable calculations at moderate to high Reynolds numbers:

$$\zeta_{i,1} = \frac{7\psi_{i,1} - 8\psi_{i,2} + \psi_{i,3}}{2(\Delta y)^2} + O\left[(\Delta y)^2\right] \tag{9.113}$$

Briley (1970) explained the instability by noting that the polynomial expression for ψ, assumed in the derivation of Eq. (9.113), is inconsistent with the evaluation of $u = \partial\psi/\partial y$ at $(i, 2)$ using a central difference. By evaluating u at $(i, 2)$ using the following expression, which is consistent with Eq. (9.113),

$$u_{i,2} = \frac{\partial \psi}{\partial y}\Bigg)_{i,2} = \frac{-5\psi_{i,1} + 4\psi_{i,2} + \psi_{i,3}}{4\,\Delta y} + O\left[(\Delta y)^2\right] \tag{9.114}$$

Briley found his computations to be stable even at high Reynolds numbers.

A classical problem that has wall boundaries surrounding the entire computational region is the driven cavity problem illustrated in Fig. 9.3. In this problem the incompressible viscous flow in the cavity is driven by the uniform translation of the upper surface (lid). The boundary conditions for this problem are indicated in Fig. 9.3. The driven cavity problem is an excellent test case for comparing methods that solve the incompressible N-S equations. A standard test condition of $\text{Re}_l = 100$ is frequently chosen in these comparisons, where

$$\text{Re}_l = \frac{Ul}{\nu} \tag{9.115}$$

and l is the width of the cavity. Two-dimensional computational results are available from numerous investigators, including Burggraf (1966), Bozeman and Dalton (1973), Rubin and Harris (1975), and Ghia et al. (1982). The Ghia et al. results are among the most detailed, having been computed on a 257 × 257 grid. Unsteady 2-D driven cavity results can be found in the works by Soh and Goodrich (1988) and Pletcher and Chen (1993). Three-dimensional computational results have been reported by Iwatsu et al. (1993) and Freitas et

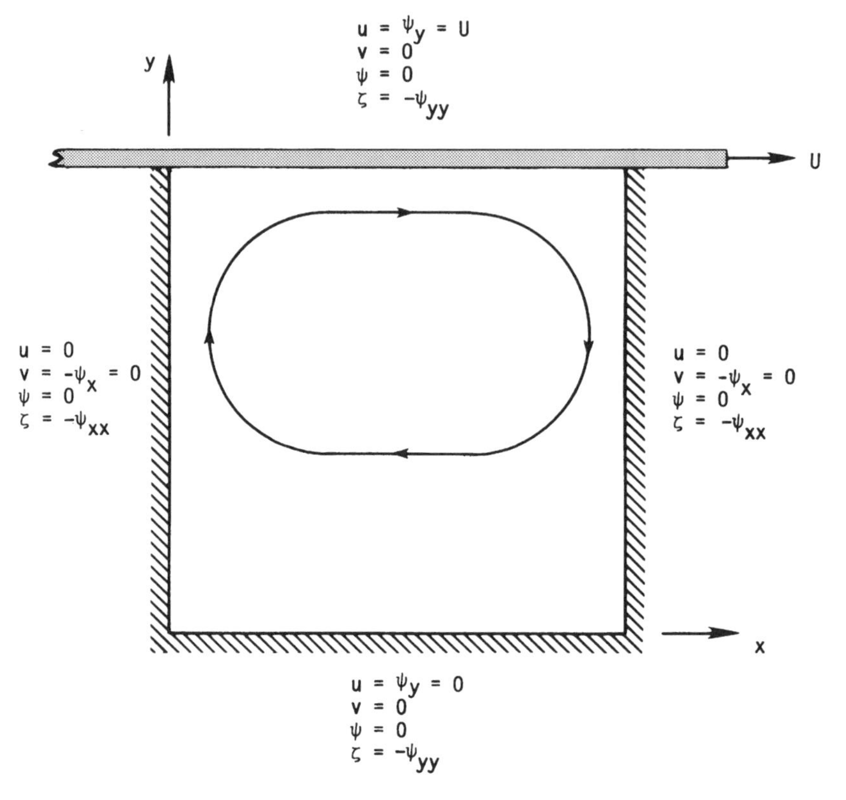

Figure 9.3 Driven cavity problem.

al. (1985). Experimental data are available in the works by Mills (1965), Pan and Acrivos (1967), and Koseff and Street (1984).

The specification of appropriate values for ζ and ψ at other types of boundaries, such as symmetry lines, upper surfaces, inflow and outflow planes, and slip lines, is extremely important, and care must be taken to ensure that the physics of the problem is correctly modeled. An excellent discussion on how to treat these various boundaries can be found in the works by Roache (1972) and Fletcher (1988).

An alternate way of solving the incompressible N-S equations written in the vorticity–stream function formulation, involves using the steady form of the vorticity transport equation

$$u\frac{\partial \zeta}{\partial x} + v\frac{\partial \zeta}{\partial y} = \nu \nabla^2 \zeta \tag{9.116}$$

This equation is elliptic and can be solved using methods similar to those employed for the Poisson equation. This approach has been successfully used by several investigators, but it appears to be susceptible to instabilities. For this reason, the transient approach is recommended over this steady-state method.

The extension of the vorticity–stream function approach to 3-D problems is complicated by the fact that a stream function does not exist for a truly 3-D flow. However, the vorticity–stream function approach can be generalized to 3-D by making use of a *vector potential.* Several variations have been used. The earliest method, generally referred to as the *vorticity-potential method* (see, for example, Aziz and Hellums, 1967), expressed the velocity as the curl of a vector potential:

$$\boldsymbol{\psi} = \psi_x \mathbf{i} + \psi_y \mathbf{j} + \psi_z \mathbf{k} \tag{9.117}$$

which satisfies the continuity equation

$$\nabla \cdot \mathbf{V} = 0 \tag{9.118}$$

so that

$$\mathbf{V} = \nabla \times \boldsymbol{\psi} \tag{9.119}$$

and

$$u = \frac{\partial \psi_z}{\partial y} - \frac{\partial \psi_y}{\partial z}$$

$$v = -\frac{\partial \psi_z}{\partial x} + \frac{\partial \psi_x}{\partial z}$$

$$w = \frac{\partial \psi_y}{\partial x} - \frac{\partial \psi_x}{\partial y}$$

After inserting Eq. (9.119) into Eq. (9.89), we obtain

$$\nabla \times (\nabla \times \boldsymbol{\psi}) = \boldsymbol{\zeta} \tag{9.120}$$

Since the vector potential can be arbitrarily chosen to satisfy

$$\nabla \cdot \boldsymbol{\psi} = 0$$

we can simplify Eq. (9.120) to yield

$$\nabla^2 \boldsymbol{\psi} = -\boldsymbol{\zeta} \tag{9.121}$$

This vector Poisson equation represents three scalar Poisson equations that must be solved after each time step. Likewise, the vorticity transport equation for a 3-D problem is a vector equation, which must be separated into three scalar parabolic equations:

$$\begin{aligned}
\frac{\partial \zeta_x}{\partial t} + u\frac{\partial \zeta_x}{\partial x} + v\frac{\partial \zeta_x}{\partial y} + w\frac{\partial \zeta_x}{\partial z} - \zeta_x\frac{\partial u}{\partial x} - \zeta_y\frac{\partial u}{\partial y} - \zeta_z\frac{\partial u}{\partial z} &= \nu\,\nabla^2\zeta_x \\
\frac{\partial \zeta_y}{\partial t} + u\frac{\partial \zeta_y}{\partial x} + v\frac{\partial \zeta_y}{\partial y} + w\frac{\partial \zeta_y}{\partial z} - \zeta_x\frac{\partial v}{\partial x} - \zeta_y\frac{\partial v}{\partial y} - \zeta_z\frac{\partial v}{\partial z} &= \nu\,\nabla^2\zeta_y \\
\frac{\partial \zeta_z}{\partial t} + u\frac{\partial \zeta_z}{\partial x} + v\frac{\partial \zeta_z}{\partial y} + w\frac{\partial \zeta_z}{\partial z} - \zeta_x\frac{\partial w}{\partial x} - \zeta_y\frac{\partial w}{\partial y} - \zeta_z\frac{\partial w}{\partial z} &= \nu\,\nabla^2\zeta_z
\end{aligned} \tag{9.122}$$

to find the three components $(\zeta_x, \zeta_y, \zeta_z)$ of the vorticity vector, although only two of the above vorticity transport equations are actually independent. Given two components of the vorticity vector, the third can be found from a linear combination of derivatives of the other two. In practice, this approach requires the solution of three parabolic and three elliptic equations at each time level. Although the vorticity-potential method would appear to require more computational effort than the primitive-variable formulation, Aziz and Hellums (1967) reported that the vorticity-potential method was faster and more accurate than a method based on the primitive-variable approach.

More recent studies utilizing derived variables have made use of a variation of the vorticity-potential method known as the *dual-potential method*. This variation was motivated by the discovery by Hirasaki and Hellums (1970) that the boundary conditions for problems with inflow and outflow can be simplified if the velocity is composed of the sum of a scalar potential and a vector potential. This formulation is based on the Helmholtz decomposition theorem, which states that any vector field can be split into a curl-free and a divergence-free part. Thus the dual-potential approach represents the velocity as

$$\mathbf{V} = \nabla\phi + \nabla \times \boldsymbol{\psi} \tag{9.123}$$

where $\nabla\phi$ is the curl-free part and $\nabla \times \boldsymbol{\psi}$ is the divergence-free (or *solenoidal*) part. In addition, it is possible to select $\boldsymbol{\psi}$, such that $\nabla \cdot \boldsymbol{\psi} = 0$ as before. Because $\nabla \times \boldsymbol{\psi}$ is divergence free, the continuity equation requires that

$$\nabla^2\phi = 0 \tag{9.124}$$

The relation between the vector potential and vorticity is obtained by taking the curl of Eq. (9.123),

$$\nabla^2\boldsymbol{\psi} - \nabla(\nabla \cdot \boldsymbol{\psi}) = -\boldsymbol{\zeta} \tag{9.125}$$

Because the vector potential has been selected to be solenoidal, it follows that

$$\nabla^2\boldsymbol{\psi} = -\boldsymbol{\zeta} \tag{9.126}$$

The solution procedure for the dual-potential formulation is nearly the same as given above for the vector-potential formulation, except that in the general case, an additional Laplace equation, Eq. (9.124), needs to be solved for the scalar potential ϕ. The velocity components are then determined from the scalar and vector potentials as

$$\begin{aligned} u &= \frac{\partial\phi}{\partial x} + \frac{\partial\psi_z}{\partial y} - \frac{\partial\psi_y}{\partial z} \\ v &= \frac{\partial\phi}{\partial y} - \frac{\partial\psi_z}{\partial x} + \frac{\partial\psi_x}{\partial z} \\ w &= \frac{\partial\phi}{\partial z} + \frac{\partial\psi_y}{\partial x} - \frac{\partial\psi_x}{\partial y} \end{aligned} \tag{9.127}$$

Notice that the part of the velocity components that is derived from the vector-potential function is the same as observed earlier for the vector-potential formulation. In fact, when the dual-potential formulation is applied to a viscous problem in which there is no throughflow, i.e., a flow in which the boundary conditions on all boundaries are the no-slip conditions (such as the driven cavity problem), a trivial solution of $\phi = 0$ satisfies the Laplace equation for the scalar potential as well as the required boundary conditions. Thus, for such problems, the scalar potential is not needed. It is only when throughflow exists that the merits of including the scalar potential become evident. Accommodating throughflow boundary conditions with the vector potential alone can be done but is overly complex (Hiraski and Hellums, 1970). Further details on the treatment of boundary conditions for the dual-potential method can be found in the works of Aregbesola and Burley (1977), Richardson and Cornish (1977), Morino (1986), and Gegg et al. (1989).

Before moving on to a discussion of the primitive-variable approach, we will briefly describe an approach that can be considered a hybrid of the stream function–vorticity approach and the primitive-variable approach. In this hybrid vorticity-velocity approach, the dependent variables are the vorticity components $(\zeta_x, \zeta_y, \zeta_z)$ and the velocity components (u, v, w). The vorticity components are obtained by solving Eq. (9.122), and the velocity components are determined from

$$\nabla^2 \mathbf{V} = -\nabla \times \boldsymbol{\zeta} \tag{9.128}$$

This vector equation is derived by taking the vector cross product of the del operator with the definition of the vorticity vector and then simplifying the resulting expression,

$$\nabla \times (\nabla \times \mathbf{V}) = \nabla \times \boldsymbol{\zeta} \tag{9.129}$$

using the appropriate vector identity. Agarwal (1981) states that this hybrid vorticity-velocity approach avoids the necessity of using a staggered-grid arrangement, which is required in some primitive-variable approaches. Other applications of the vorticity-velocity approach can be found in the works of Dennis et al. (1979), Gastski et al. (1982), Fasel and Booz (1984), Farouk and Fusegi (1985), Osswald et al. (1987), and Guj and Stella (1988).

9.3.2 Primitive-Variable Approach

General. The approaches based on derived variables such as the vorticity–stream function and dual-potential methods lose some of their attractiveness when applied to a 3-D flow, as discussed in the last section. Consequently, the incompressible N-S equations are most often solved in their primitive-variable form (u, v, w, p) for 3-D problems. Even for 2-D problems, the use of primitive variables is quite common.

The incompressible N-S equations written in nondimensional primitive-variable form for a Cartesian coordinate system are given by the following.

continuity:

$$\frac{\partial u^*}{\partial x^*} + \frac{\partial v^*}{\partial y^*} + \frac{\partial w^*}{\partial z^*} = 0 \tag{9.130}$$

x momentum:

$$\frac{\partial u^*}{\partial t^*} + u^* \frac{\partial u^*}{\partial x^*} + v^* \frac{\partial u^*}{\partial y^*} + w^* \frac{\partial u^*}{\partial z^*} = -\frac{\partial p^*}{\partial x^*} + \frac{1}{\text{Re}_L}\left(\frac{\partial^2 u^*}{\partial x^{*2}} + \frac{\partial^2 u^*}{\partial y^{*2}} + \frac{\partial^2 u^*}{\partial z^{*2}}\right) \tag{9.131}$$

y momentum:

$$\frac{\partial v^*}{\partial t^*} + u^* \frac{\partial v^*}{\partial x^*} + v^* \frac{\partial v^*}{\partial y^*} + w^* \frac{\partial v^*}{\partial z^*} = -\frac{\partial p^*}{\partial y^*} + \frac{1}{\text{Re}_L}\left(\frac{\partial^2 v^*}{\partial x^{*2}} + \frac{\partial^2 v^*}{\partial y^{*2}} + \frac{\partial^2 v^*}{\partial z^{*2}}\right) \tag{9.132}$$

z momentum:

$$\frac{\partial w^*}{\partial t^*} + u^* \frac{\partial w^*}{\partial x^*} + v^* \frac{\partial w^*}{\partial y^*} + w^* \frac{\partial w^*}{\partial z^*} = -\frac{\partial p^*}{\partial z^*} + \frac{1}{\text{Re}_L}\left(\frac{\partial^2 w^*}{\partial x^{*2}} + \frac{\partial^2 w^*}{\partial y^{*2}} + \frac{\partial^2 w^*}{\partial z^{*2}}\right) \tag{9.133}$$

These equations are nondimensionalized using

$$\begin{aligned} u^* &= \frac{u}{V_\infty} & x^* &= \frac{x}{L} & p^* &= \frac{p}{\rho_\infty V_\infty^2} \\ v^* &= \frac{v}{V_\infty} & y^* &= \frac{y}{L} & t^* &= \frac{tV_\infty}{L} \\ w^* &= \frac{w}{V_\infty} & z^* &= \frac{z}{L} & \text{Re}_L &= \frac{V_\infty L}{\nu_\infty} \end{aligned} \tag{9.134}$$

Notice that no time derivative of pressure appears in these equations. For an incompressible fluid, pressure waves propagate at infinite speed. The pressure is determined through the governing equations and boundary conditions, but it also can be shown to be governed by an elliptic PDE.

Methods for solving the incompressible N-S equations in primitive variables can be grouped into two broad categories. The first we shall refer to as the *coupled approach*. In this approach the discretized conservation equations are solved, treating all dependent variables as simultaneous unknowns. For the time-dependent N-S equations this is implemented in what is known as the *artificial compressibility* (also known as the *pseudo-compressibility*) *method*. An artificial-time derivative of pressure is added to the continuity equation to permit the solution to the coupled hyperbolic system to be advanced in time. Without the addition of such a time derivative, the algebraic system of equations resulting from a coupled time-dependent discretization is singular. This occurs in a manner similar to the singular behavior noted in Section 9.2.6 for the compressible time-dependent formulation taken to the incompressible limit. The addition of the artificial-time term plays a role similar to that of preconditioning applied to the compressible formulation. Upon convergence, the artificial-time term vanishes, as will be explained below. On the other hand, it will be seen that the steady-flow incompressible equations can be solved by a coupled discretization without the addition of artificial terms.

The second strategy often employed for the incompressible N-S equations will be referred to as the *pressure correction* approach. Such methods are also known as pressure-based, uncoupled, sequential, or segregated methods. The distinguishing feature of this approach is the use of a derived equation to determine the pressure. Typically, the momentum equations are solved for the velocity components in an uncoupled manner. The x momentum equation is solved for the x component of velocity, the y momentum equation is solved for the y component of velocity, etc. In doing so, the equations are linearized by using values lagged in iteration level for the other unknowns, including pressure. The velocity components have thus been computed without using the continuity equation as a constraint. Usually, a Poisson equation is developed for the pressure, or changes in the pressure, that will alter the velocity field in a direction such as to satisfy the continuity equation. Such an equation for pressure can be derived from the conservation equations in a rigorous manner. However, several well-known schemes use an approximate formulation for the pressure equation, which is justified as long as the iterative procedure produces a solution that satisfies all of the discretized conservation equations, including the continuity equation.

The literature on numerical schemes for the incompressible N-S equations is quite extensive. A great many numerical schemes, each differing from others in some detail but falling within the two basic approaches defined above, have been proposed for solving the incompressible N-S equations. Only a few specific examples will be given here.

Coupled approach: The method of artificial compressibility. One of the early techniques proposed for solving the incompressible N-S equations in primitive-variable form was the artificial compressibility method of Chorin (1967). In this method, the continuity equation is modified to include an artifical compressibility term that vanishes when the steady-state solution is reached. With the addition of this term to the continuity equation, the resulting N-S equations are a mixed set of hyperbolic-parabolic equations, which can be solved using a standard time-dependent approach. In order to explain this method, let us apply it to Eqs. (9.130)–(9.133). The continuity equation is replaced by

$$\frac{\partial \tilde{\rho}^*}{\partial \tilde{t}^*} + \frac{\partial u^*}{\partial x^*} + \frac{\partial v^*}{\partial y^*} + \frac{\partial w^*}{\partial z^*} = 0 \tag{9.135}$$

where $\tilde{\rho}^*$ is an artificial density and $\tilde{t}^*$ is a fictitious time that is analogous to real time in a compressible flow. The artificial density is related to the pressure by the artificial equation of state,

$$p^* = \beta \tilde{\rho}^* \tag{9.136}$$

where β is the artificial compressibility factor to be determined later. Note that at steady-state the solution is independent of $\tilde{\rho}^*$ and $\tilde{t}^*$, since $\partial\tilde{\rho}^*/\partial\tilde{t}^* \to 0$. After replacing t^* with $\tilde{t}^*$ in Eqs. (9.131)–(9.133) and substituting Eq. (9.136) into Eq. (9.135), we can apply a suitable numerical technique to the resulting equations and march the solution in $\tilde{t}^*$ to obtain a final steady-state incompressible solution. Obviously, this technique is applicable only to steady-flow problems, since it is not time accurate.

In order to facilitate the application of the numerical scheme, Eqs. (9.130)–(9.133) and Eqs. (9.135)–(9.136) can be combined into the following vector form:

$$\frac{\partial \mathbf{u}^*}{\partial \tilde{t}^*} + \frac{\partial \mathbf{e}^*}{\partial x^*} + \frac{\partial \mathbf{f}^*}{\partial y^*} + \frac{\partial \mathbf{g}^*}{\partial z^*} = \frac{1}{\mathrm{Re}_L}\left(\frac{\partial^2}{\partial x^{*2}} + \frac{\partial^2}{\partial y^{*2}} + \frac{\partial^2}{\partial z^{*2}}\right)[D]\mathbf{u}^* \tag{9.137}$$

where

$$\mathbf{u}^* = \begin{bmatrix} p^* \\ u^* \\ v^* \\ w^* \end{bmatrix} \qquad \mathbf{e}^* = \begin{bmatrix} \beta u^* \\ p^* + (u^*)^2 \\ u^* v^* \\ u^* w^* \end{bmatrix}$$

$$\mathbf{f}^* = \begin{bmatrix} \beta v^* \\ u^* v^* \\ p^* + (v^*)^2 \\ v^* w^* \end{bmatrix} \qquad \mathbf{g}^* = \begin{bmatrix} \beta w^* \\ u^* w^* \\ v^* w^* \\ p^* + (w^*)^2 \end{bmatrix} \tag{9.138}$$

$$[D] = \begin{bmatrix} 0 & 0 & 0 & 0 \\ 0 & 1 & 0 & 0 \\ 0 & 0 & 1 & 0 \\ 0 & 0 & 0 & 1 \end{bmatrix}$$

Note that since Eq. (9.136) represents an artificial equation of state, then $\beta^{1/2}$ plays the role of an artificial sound speed. Defining Jacobians,

$$[A] = \frac{\partial \mathbf{e}^*}{\partial \mathbf{u}^*} \qquad [B] = \frac{\partial \mathbf{f}^*}{\partial \mathbf{u}^*} \qquad [C] = \frac{\partial \mathbf{g}^*}{\partial \mathbf{u}^*} \tag{9.139}$$

the LHS of Eq. (9.137) can be rewritten as

$$\frac{\partial \mathbf{u}^*}{\partial \tilde{t}^*} + [A]\frac{\partial \mathbf{u}^*}{\partial x^*} + [B]\frac{\partial \mathbf{u}^*}{\partial y^*} + [C]\frac{\partial \mathbf{u}^*}{\partial z^*} \tag{9.140}$$

where

$$[A] = \begin{bmatrix} 0 & \beta & 0 & 0 \\ 1 & 2u^* & 0 & 0 \\ 0 & v^* & u^* & 0 \\ 0 & w^* & 0 & u^* \end{bmatrix} \qquad [B] = \begin{bmatrix} 0 & 0 & \beta & 0 \\ 0 & v^* & u^* & 0 \\ 1 & 0 & 2v^* & 0 \\ 0 & 0 & w^* & v^* \end{bmatrix}$$

$$[C] = \begin{bmatrix} 0 & 0 & 0 & \beta \\ 0 & w^* & 0 & u^* \\ 0 & 0 & w^* & v^* \\ 1 & 0 & 0 & 2w^* \end{bmatrix}$$

The eigenvalues of $[A]$, $[B]$, and $[C]$ are

$$\left(u, u, u \pm \sqrt{u^2 + \beta}\right), \quad \left(v, v, v \pm \sqrt{v^2 + \beta}\right), \quad \left(w, w, w \pm \sqrt{w^2 + \beta}\right)$$

respectively. This suggests that the magnitude of β should be close to that of the convective velocities to avoid the stiffness associated with a disparity in the magnitudes of the eigenvalues. Although the artificial equation of state suggests that $\beta^{1/2}$ is an artificial speed of sound, the eigenvalues above indicate that the effective acoustic wave speeds are really the quantities under the radicals in the eigenvalues above ($\sqrt{u^2 + \beta}$, for example), which are functions of the velocity components as well as β. The optimum value of β may be somewhat problem dependent. Kwak et al. (1986) suggest that a value of β in the range 0.1–10.0 will work well for most problems. On the high side, the problem is one of stiffness, which retards the convergence rate. On the low side, the value of $\beta \Delta \tilde{t}^*$ should be large enough to permit pressure waves (which actually should move at infinite speed in the incompressible limit) to propagate far enough to reasonably balance viscous effects during the artificial transient, or the pseudo-time iterations will tend not to converge.

In the original paper of Chorin, the leap frog/DuFort-Frankel finite-difference scheme (see Section 4.5.2) was used. Since that time, a variety of numerical schemes have been used to solve the hyperbolic system of equations, including the multistage Runge-Kutta explicit method, approximate-factorization implicit schemes, the LU-SGS implicit scheme, and the coupled strongly implicit scheme. Generally, it is believed that any numerical solution strategy that is appropriate for solving the time-dependent compressible N-S equations as a coupled system will also work for solving the discretized equations resulting from the artificial compressibility formulation. On balance, it seems that implicit formulations have been favored over explicit methods for incompressible viscous flow applications.

Over the years, a number of investigators have reported good success with the artificial compressibility method in a number of impressive applications using a variety of algorithms. Among these are the works of Steger and Kutler (1976), Choi and Merkle (1985), Kwak et al. (1986), Hartwich and Hsu (1987), and Hartwich et al. (1988). Initially, such methods were considered to be only suitable for obtaining steady solutions because the solutions had to be iterated to time convergence for the artificial term to vanish. It has now been demonstrated (Merkle and Athavale, 1987; Pan and Chakravarthy, 1989; Rogers et al., 1989; Chen and Pletcher, 1993) that this approach can be made time accurate by considering the time terms in the momentum equations to be the real, physical time terms, and the time term added to the continuity equation to be in pseudo-time. The solution is then iterated to pseudo-time convergence at each real (physical) time step. Pseudo-time terms can also be added to the momentum equations (while leaving physical time terms intact) as an option to aid in maintaining diagonal dominance of the algebraic system. When the pseudo-time term (or terms) vanish, the solution obtained satisfies the complete time-dependent N-S equations. This approach employing pseudo-time terms in all equations will be discussed next in some detail.

To obtain time-accurate solutions to the N-S equations by the artificial compressibility method, Eq. (9.137) is modified by retaining the physical time terms but adding pseudo-time terms to give

$$[A_p]\frac{\partial \mathbf{u}^*}{\partial \tilde{t}^*} + [A_t]\frac{\partial \mathbf{u}^*}{\partial t^*} + [A]\frac{\partial \mathbf{u}^*}{\partial x^*} + [B]\frac{\partial \mathbf{u}^*}{\partial y^*} + [C]\frac{\partial \mathbf{u}^*}{\partial z^*} = \frac{1}{\mathrm{Re}_L}\left(\frac{\partial^2}{\partial x^{*2}} + \frac{\partial^2}{\partial y^{*2}} + \frac{\partial^2}{\partial z^{*2}}\right)[D]\mathbf{u}^* \tag{9.141}$$

where $\mathbf{u}^*$, $[A]$, $[B]$, $[C]$, and $[D]$ are as defined previously and

$$[A_p] = \begin{bmatrix} a & 0 & 0 & 0 \\ 0 & b & 0 & 0 \\ 0 & 0 & c & 0 \\ 0 & 0 & 0 & d \end{bmatrix} \qquad [A_t] = \begin{bmatrix} 0 & 0 & 0 & 0 \\ 0 & 1 & 0 & 0 \\ 0 & 0 & 1 & 0 \\ 0 & 0 & 0 & 1 \end{bmatrix}$$

The above formulation works well with the parameters a, b, c, d set equal to 1, although other values may enhance the convergence rate or robustness of some algorithms. At each physical time step, the computations are advanced in pseudo-time until convergence (no further changes in the variables are observed). The equations are often solved in conservation-law form with the linearization achieved by a Newton method, whereby the Jacobians are evaluated using the most recently computed values. Notice that numerical errors associated with the linearization can be driven to zero during the pseudo-time iteration cycle, and at iterative convergence, the conservation-law form of the equations is satisfied. This strategy follows the pattern described in Section 9.2.6 for solving the compressible N-S equations in primitive variables. The eigenvalues of the system are the eigenvalues of $[A_p]^{-1}[A]$, $[A_p]^{-1}[B]$, and $[A_p]^{-1}[C]$. The eigenvalues of $[A_p]^{-1}[A]$ are $u^*/c, u^*/d, (u^*/b) \pm \sqrt{(u^{*2}/b^2) + (\beta/ab)}$. The other eigenvalues are obtained from the above expression by replacing u^* with v^* to obtain the eigenvalues of $[A_p]^{-1}[B]$ and by replacing u^* with w^* to obtain the eigenvalues of $[A_p]^{-1}[C]$. For $a = b = c = d = 1$ the eigenvalues are seen to be the same as for the original steady-flow pseudo-compressibility formulation of Chorin.

Some flexibility exists in establishing $[A_p]$, and for some choices, the eigenvalues are altered somewhat. For example, we can add a pseudo-time derivative of pressure to each momentum equation. Taking some liberties with the form of the coefficients of that term, we find the following candidate preconditioning matrix:

$$[A_p] = \begin{bmatrix} 1 & 0 & 0 & 0 \\ \dfrac{au}{\beta} & 1 & 0 & 0 \\ \dfrac{av}{\beta} & 0 & 1 & 0 \\ \dfrac{aw}{\beta} & 0 & 0 & 1 \end{bmatrix}$$

where a is an arbitrary constant. Using this last form for $[A_p]$, we compute the eigenvalues of $[A_p]^{-1}[A]$ to be $u, u, u - \frac{1}{2}(au \pm \sqrt{a^2u^2 - 4au^2 + 4u^2 + 4\beta})$. Notice that this result reduces to the eigenvalues for the original Chorin scheme if $a = 0$. However, if $a = 2$, the eigenvalues become $u, u, \pm\beta^{1/2}$, and the effective acoustic speed becomes independent of the velocity and equal to $\beta^{1/2}$. We close this discussion with the reminder that adding the pseudo-time terms to the momentum equations is not essential for time accuracy if pseudo-time convergence is achieved at each physical time step (see Pan and Chakravarthy et al. 1989).

Note that the strategy employed in the time-accurate version of the artificial compressibility scheme bears some resemblance to that suggested in Section 9.2.6 for solving the preconditioned compressible N-S equations. In fact, if isothermal conditions are assumed, the application of the preconditioned compressible formulation to low-speed flows becomes effectively an artificial compressibility scheme to the extent that percentage changes in the density remain small.

Coupled approach: Space marching. It is possible to solve the steady form of both the compressible and incompressible N-S equations by coupled space-marching methods. Because the steady-flow system of equations is elliptic for subsonic flows, repeated calculation sweeps are made from inflow to outflow until convergence is achieved. Examples of coupled space-marching methods applied to the steady N-S equations can be found in the works of Bentson and Vradis (1987), TenPas and Pletcher (1987, 1991), Vradis et al. (1992), and

TenPas and Hancock (1992). Following the formulation of TenPas and Hancock (1992), the 2-D incompressible continuity and momentum equations are written in the form

$$\frac{\partial \mathbf{e}^*}{\partial x^*} + \frac{\partial \mathbf{f}^*}{\partial y^*} = \frac{1}{\mathrm{Re}_L}\left(\frac{\partial^2}{\partial x^{*2}} + \frac{\partial^2}{\partial y^{*2}}\right)[D]\mathbf{u}^* \tag{9.142}$$

where

$$\mathbf{u}^* = \begin{bmatrix} p^* \\ u^* \\ v^* \end{bmatrix} \qquad \mathbf{e}^* = \begin{bmatrix} u^* \\ u^{*2} + p^* \\ u^* v^* \end{bmatrix} \qquad \mathbf{f}^* = \begin{bmatrix} v^* \\ u^* v^* \\ v^{*2} + p \end{bmatrix} \qquad [D] = \begin{bmatrix} 0 & 0 & 0 \\ 0 & 1 & 0 \\ 0 & 0 & 1 \end{bmatrix}$$

The solution is to be advanced by marching in the "streamwise" direction, which we shall assume to be the positive x direction. For an implicit space-marching strategy to work successfully, attention must be given to two details. The first is the manner in which the streamwise pressure gradient is treated. It is essential in the marching sweep to treat the *downstream* value of pressure as given either from an initial estimate or from the value computed from the most recent sweep. Note that this is the downstream and not the downwind value, in that the downstream value is fixed regardless of the local flow direction. For example, if the marching solution is being advanced from marching level i to $i + 1$, the pressure gradient term in the x momentum equation would be forward differenced (either first order or second order), treating the value of pressure at $i + 1$ as unknown. For a first-order representation, this would give

$$\left.\frac{\partial p}{\partial x}\right|_{i+1} \cong \frac{p_{i+2} - p_{i+1}}{\Delta x}$$

where p_{i+2} would be evaluated from the previous sweep (i.e., "fixed") as the solution at $i + 1$ is computed, but p_{i+1} would be an unknown. The transverse pressure gradient term in the y momentum equation is discretized at the unknown $i + 1$ level using forward (rather than central) second-order differences to prevent even-odd decoupling. The pressure is fixed by the boundary conditions at the downstream boundary. The fixing of the downstream pressure at each marching step is consistent with the physical and mathematical nature of the steady incompressible flow problem, in that the local solution should be influenced by information coming from all directions.

The second concern in the space-marching procedure is to ensure that the difference stencil honors the appropriate zone of dependence in a manner that maintains diagonal dominance in the implicit solution algorithm. To achieve this, the streamwise convective derivative in the momentum equations is invariably upwinded. In regions where the flow is reversed, as behind a step in a rearward-facing step flow, the differencing direction changes. The upwind value is treated as known, and the downwind value is unknown, which is the value at $i + 1$ in the present discussion. The streamwise second-derivative terms are differenced centrally, with the $i + 1$ values treated as unknowns, and the upstream and downstream values lagged to the most recently computed values.

The continuity equation is differenced in a form equivalent to a finite-volume approximation for a control volume shifted upstream between i and $i + 1$. This allows for global mass conservation to be ensured at each marching sweep. The streamwise derivative uses second-order central differences, while the derivative in the transverse direction is approximated by a second-order backward difference. The need to march in the predominant flow direction restricts somewhat the available options for grid construction when generalized coordinates are used. For most flow configurations, the use of space marching is limited to "H" grids (see Thompson et al., 1985).

Multiple marching sweeps are required in space-marching schemes in order to develop the final pressure distribution because the downstream value of pressure in the streamwise momentum equation is always fixed at the value determined at the previous sweep until convergence is achieved. It has been observed that convergence can be accelerated by "correcting" the pressure between streamwise sweeps. Presumably, this accelerates the rate at which information travels upstream. The pressure correction usually takes the form of a single sweep from downstream to upstream of an approximate Poisson equation for the pressure. More details on space-marching schemes for the complete N-S equations can be found in the work of TenPas (1990). Such schemes have been demonstrated for computing developing 2-D and 3-D flows in a channel, flows over a rearward-facing step, and flow over a cylinder at low Reynolds numbers. Space-marching procedures have also been employed for the compressible N-S equations in the subsonic regime (TenPas and Pletcher, 1991; Pappalexis and TenPas, 1993).

Pressure-correction approach: General. The general pressure-correction approach is characterized by a formulation in which the momentum equations are solved sequentially for the velocity components using the best available estimate for the pressure distribution. Such a procedure will not yield a velocity field that satisfies the continuity equation unless the correct pressure distribution is employed. If mass sources exist, the pressure is improved in a separate step in a manner that will eliminate the mass sources (satisfy continuity). If the pressure changes, the solution to the momentum equations will change, and particularly, in implicit schemes, the sequence is repeated iteratively until a divergence-free velocity field is established. Such schemes have often been called "pressure-based" schemes, as contrasted with coupled-solution schemes (such as the artificial compressibility scheme) which have been referred to as "density based." This terminology (pressure or density based) is being abandoned here because even in coupled approaches, it is increasingly common to employ primitive variables (u, v, p, T, for example), where the density no longer appears as a variable. This is especially evident in schemes employing low Mach number preconditioning. Thus the terminology "pressure correction" is being suggested as being more descriptive than "pressure based," since both coupled and sequential schemes are likely to employ pressure as a dependent variable.

Pressure-correction methods have been widely used for solving the incompressible N-S equations. The methods differ primarily in the algorithms used to solve the component equations and the strategy employed to develop an equation to be solved for an improved pressure. Such an equation is most often a Poisson equation. Early forms of these methods employed staggered grids to avoid even-odd decoupling of the pressure. More recently, regular (colocated) grids have been employed satisfactorily.

Some of the most commonly used variations of the pressure-correction method are the marker-and-cell (MAC) method of Harlow and Welch (1965), the SIMPLE and SIMPLER methods of Caretto et al. (1972) and Patankar (1980), the fractional-step method (Chorin, 1968; Yanenko, 1971; Marchuk, 1975), and the primitive-variable implicit split operator (PISO) method of Issa (1986). A sampling of these methods will now be discussed.

Pressure-correction approach: Marker-and-cell method. Perhaps the earliest pressure-correction scheme for solving the incompressible N-S equations was the MAC method introduced by Harlow and Welch (1965) and Welch et al. (1966). The scheme was based on a staggered grid similar to that introduced in Chapter 8. The marker-and-cell terminology arose because the method had the capability to resolve time-dependent free surface flows by tracing the paths of fictitious massless marker particles introduced on the free surface. The solution was advanced in time by solving the momentum equations for velocity components using the best current estimate of the pressure distribution, much as indicated in Section 8.4.3. Such a solution initially would not satisfy the continuity equation unless the correct pressure distribution was used. The pressure is improved by numerically solving a Poisson equation derived in the same manner as Eq. (9.100). In nondimensional form this equation can be written as

$$\nabla^2 p^* = S_p^* - \frac{\partial D^*}{\partial t^*} \tag{9.143}$$

where

$$\begin{aligned} S_p^* = {} & \frac{d}{dx^*}\left[-(u^* u_x^* + v^* u_y^* + w^* u_z^*) + \frac{1}{\mathrm{Re}_L}(u_{xx}^* + u_{yy}^* + u_{zz}^*)\right] \\ & + \frac{d}{dy^*}\left[-(u^* v_x^* + v^* v_y^* + w^* v_z^*) + \frac{1}{\mathrm{Re}_L}(v_{xx}^* + v_{yy}^* + v_{zz}^*)\right] \\ & + \frac{d}{dz^*}\left[-(u^* w_x^* + v^* w_y^* + w^* w_z^*) + \frac{1}{\mathrm{Re}_L}(w_{xx}^* + w_{yy}^* + w_{zz}^*)\right] \end{aligned}$$

and D^* is the local dilatation term given by

$$D^* = u_x^* + v_y^* + w_z^*$$

and terms such as u_x^* denote $\partial u^* / \partial x^*$. The value of S_p^* is determined from the solution of the momentum equations using the provisional values of pressure.

counter and $D_{i,j,k}^{m+1}$ is set equal to zero. That is, the correction in pressure is required to compensate for the nonzero dilation at the m iteration level. The Poisson equation is then solved for the revised pressure field. Depending on the details of the discretization method employed (explicit, implicit, etc.), the improved pressure may then be used in the momentum equation for a better solution at the present time step. If the dilation (divergence of velocity field) is not zero, the cyclic process of solving the momentum equations and the Poisson equation is repeated until the velocity field is divergence free.

A point that needs careful attention in all schemes requiring the solution of a Poisson equation is the proper application of boundary conditions. These boundary conditions are Neumann as derived from the momentum equations. When a staggered grid is used, the pressure at the boundary itself is not required for the solution of the momentum equations. However, in representing the Neumann boundary condition, the pressure at a point below the boundary (a fictitious point) is called for. It is shown in Chapter 8 (Section 8.4.3) that this pressure cancels out of the representation of the boundary condition if it is evaluated implicitly at the current iteration level. To achieve convergence of the Poisson equation, the solution must satisfy the integral constraint,

$$\iint_A \nabla^2 p \, dA = \oint_C \frac{\partial p}{\partial n} \, ds \tag{9.144}$$

where C is the closed boundary of the solution domain of area A and ds is a differential length along C. On a staggered grid (see Chapter 8) this is satisfied automatically. When a nonstaggered mesh is used, the possible inconsistency arising from Eq. (9.144) can be circumvented (Ghia et al., 1977a, 1979, 1981; Briley, 1974) by reducing the source term of the Poisson equation [RHS of Eq. (9.143)] at every point by the same fixed amount computed as required to satisfy the global constraint of Eq. (9.144). With the nonstaggered mesh, the pressure gradient at the wall is required. A technique for its computation will be illustrated for a wall located at $y = 0$ as shown in Fig. 9.4. Note that a fictitious row of grid points for pressure has been added below the wall surface in this

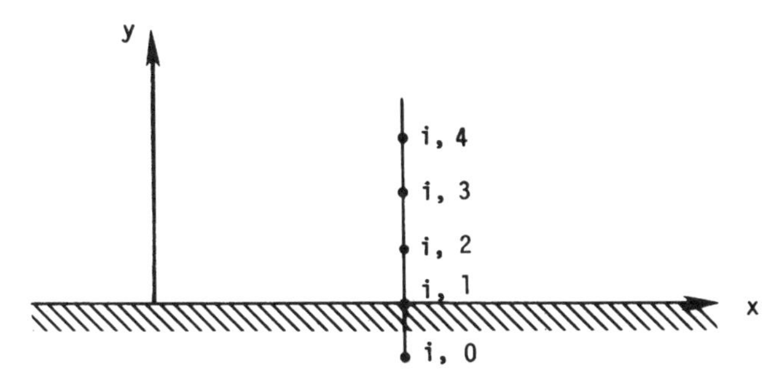

Figure 9.4 Grid points for determination of pressure boundary condition.

nonstaggered grid. At the wall surface, the y momentum equation [Eq. (9.132)] reduces to

$$\left.\frac{\partial p^*}{\partial y^*}\right)_{i,1} = \frac{1}{\mathrm{Re}_L} \left.\frac{\partial^2 v^*}{\partial y^{*2}}\right)_{i,1} \tag{9.145}$$

This equation can be differenced using the familiar second-order accurate central-difference expressions

$$\frac{p^*_{i,2} - p^*_{i,0}}{2\,\Delta y^*} = \frac{1}{\mathrm{Re}_L}\left(\frac{v^*_{i,2} - 2\cancelto{0}{v^*_{i,1}} + v^*_{i,0}}{(\Delta y^*)^2}\right) \tag{9.146}$$

where $v^*_{i,0}$ is the value of v^* at the fictitious point. An expression for $v^*_{i,0}$ can be determined from the continuity equation, which reduces to

$$\left.\frac{\partial v^*}{\partial y^*}\right)_{i,1} = 0 \tag{9.147}$$

at the wall. Using a third-order accurate finite-difference expression for this reduced form of the continuity equation,

$$\left.\frac{\partial v^*}{\partial y^*}\right)_{i,1} = \frac{-2v^*_{i,0} - 3\cancelto{0}{v^*_{i,1}} + 6v^*_{i,2} - v^*_{i,3}}{6(\Delta y^*)} + O\left[(\Delta y^*)^3\right] = 0 \tag{9.148}$$

allows us to compute $v^*_{i,0}$ and retain second-order accuracy in Eq. (9.146). Similar techniques can be used to find the pressure gradient at other boundaries in order to solve the Poisson pressure equation.

A very large number of methods can trace their ancestry back to the MAC method. In 1970, Amsden and Harlow (1970) introduced a simpler MAC procedure (SMAC), which employed a second Poisson equation for a velocity potential that would drive a corrective velocity in order to satisfy the continuity equation. Such a procedure is described in Chapter 8 in connection with the partially parabolized procedures for subsonic flows. The original MAC and SMAC schemes employed an explicit time-marching procedure. Implicit discretizations have been widely used in more recent variations of these schemes (Deville, 1974; Briley, 1974; Ghia et al., 1979).

Pressure-correction approach: Projection (fractional step) methods. A key feature of the MAC method is the splitting of the solution into two distinct steps, one of which is the solution of a Poisson equation for the pressure. A great many variations to this "splitting" of the solution procedure have been suggested. One such variation proposed by Chorin (1968) and Temam (1969) is known as the *projection method*, or the *method of fractional steps*. The projection method was originally formulated on a regular rather than a staggered grid. Another difference is in the way the pressure Poisson equation is developed. In the original formulation proposed by Chorin, the pressure gradient terms are omitted from the momentum equations in the first step. The unsteady equations are advanced in time to obtain a provisional nondimensional velocity $\mathbf{V}^*$. In a second step, we wish to correct this provisional velocity by accounting for the pressure gradient and the continuity equation. This is achieved by considering

$$\frac{\mathbf{V}^{n+1} - \mathbf{V}^*}{\Delta t} + \nabla p^{n+1} = 0 \tag{9.149}$$

subject to the continuity constraint $\nabla \cdot \mathbf{V}^{n+1} = 0$. By taking the divergence of Eq. (9.149) subject to the continuity constraint above, we obtain the Poisson equation:

$$\nabla^2 p^{n+1} = \frac{\nabla \cdot \mathbf{V}^*}{\Delta t} \tag{9.150}$$

The solution procedure consists of first computing $\mathbf{V}^*$ from the momentum equations while neglecting the pressure gradient terms. The pressure Poisson equation is then solved for the pressure field, after which the velocities are computed from Eq. (9.149). More recently, investigators have found that the procedure also works well if a provisional pressure distribution is used in the momentum equations in the first step, where provisional velocities are determined. Then the p in Eq. (9.149) becomes a pressure correction, which can be determined from the solution to the Poisson equation.

The projection method has been implemented on both regular and staggered grids. On a staggered grid the scheme is very similar to the MAC method. Both explicit and implicit formulations have been employed. A rather detailed discussion of the projection method that points out similarities between it and the MAC scheme is provided by Peyret and Taylor (1983).

Pressure-correction approach: SIMPLE family of methods. The semi-implicit method for pressure linked equations (SIMPLE) algorithm of Caretto et al. (1972) and Patankar and Spalding (1972) which is introduced in Section 8.4.1 for the solution of the partially parabolized N-S equations can also be applied to the incompressible N-S equations (see Caretto et al., 1972; Patankar, 1975, 1981). This procedure is based on a cyclic series of guess-and-correct operations to solve the governing equations. The velocity components are first calculated from the momentum equations using a guessed pressure field. The pressures and velocities are then corrected, so as to satisfy continuity. This procedure continues until the solution converges. The main distinction between this method and the MAC and projection methods is in the way in which the pressure and velocity corrections are achieved.

In this procedure, the actual pressure p is written as

$$p = p_0 + p' \tag{9.151}$$

where p_0 is the estimated (or intermediate) value of pressure and p' is the pressure correction. Likewise, the actual velocity components (for a 2-D flow) are written as

$$\begin{aligned} u &= u_0 + u' \\ v &= v_0 + v' \end{aligned} \tag{9.152}$$

where u_0, v_0 are the estimated (or intermediate) values of velocity and u', v' are the velocity corrections. The pressure corrections are related to the velocity corrections by approximate forms of the momentum equations:

$$\begin{aligned} \rho \frac{\partial u'}{\partial t} &= -\frac{\partial p'}{\partial x} \\ \rho \frac{\partial v'}{\partial t} &= -\frac{\partial p'}{\partial y} \end{aligned} \tag{9.153}$$

Since the velocity corrections can be assumed to be zero at the previous iteration step, the above equations can be written as

$$\begin{aligned} u' &= -A \frac{\partial p'}{\partial x} \\ v' &= -A \frac{\partial p'}{\partial y} \end{aligned} \tag{9.154}$$

where A is a fictitious time increment divided by density. After combining Eqs. (9.152) and (9.154) and substituting the result into the continuity equation, we obtain

$$\cancelto{0}{\left(\frac{\partial u}{\partial x} + \frac{\partial v}{\partial y} \right)} - \left(\frac{\partial u_0}{\partial x} + \frac{\partial v_0}{\partial y} \right) + A \left(\frac{\partial^2 p'}{\partial x^2} + \frac{\partial^2 p'}{\partial y^2} \right) = 0 \tag{9.155}$$

or

$$\nabla^2 p' = \frac{1}{A} (\mathbf{\nabla} \cdot \mathbf{V}_0) \tag{9.156}$$

where $\mathbf{V}_0$ is the estimated velocity vector. This Poisson equation can be solved for the pressure correction. Note that if the estimated velocity vector satisfies continuity at every point, then the pressure correction is zero at every point. In the actual SIMPLE algorithm, an equivalent differenced form of Eq. (9.156) is used as shown by Raithby and Schneider (1979).

The SIMPLE procedure can now be described by the following steps:

1. Guess the pressure (p_0) at each grid point.
2. Solve the momentum equations to find the velocity components (u_0, v_0). A staggered grid in conjunction with a block-iterative method is recommended by Patankar and Spalding.

3. Solve the pressure-correction equation [i.e., Eq. (9.156)] to find p' at each grid point.
4. Correct the pressure and velocity using Eqs. (9.151) and (9.154):

$$p = p_0 + p'$$

$$u = u_0 - \frac{A}{2\,\Delta x}(p'_{i+1,j} - p'_{i-1,j})$$

$$v = v_0 - \frac{A}{2\,\Delta y}(p'_{i,j+1} - p'_{i,j-1})$$

5. Replace the previous intermediate values of pressure and velocity (p_0, u_0, v_0) with the new corrected values (p, u, v), and return to Step 2. Repeat this process until the solution converges.

The SIMPLE procedure has been used successfully to solve a number of incompressible flow problems. However, in certain cases it is found that the rate of convergence is not satisfactory. This is due to the fact that the pressure-correction equation tends to overestimate the value of p' even though the corresponding velocity corrections are reasonable. Because of this, Eq. (9.151) is often replaced with

$$p = p_0 + \alpha_p p'$$

where α_p is an underrelaxation constant. For the same reason, underrelaxation is also employed in the solution of the momentum equations. In the present formulation, underrelaxation can be accommodated by varying the parameter A in Eqs. (9.154) and (9.156).

A large number of variations of the SIMPLE strategy have been proposed for the purpose of improving the convergence rate of the scheme. Among them are the SIMPLE revised (SIMPLER) scheme (Patankar, 1981) and the SIMPLEC scheme (van Doormal and Raithby, 1984).

The SIMPLER algorithm appears to make use of a feature contained in the fractional-step method, whereby provisional velocities are defined from a momentum equation in which the pressure gradient is absent. Patankar refers to these as pseudo-velocities. The algorithm starts with a guessed velocity field. With this, coefficients are computed for the momentum equations, and pseudo-velocities are obtained by solving a form of the momentum equations in which the pressure gradient is missing. These velocities are treated much as the provisional velocities of the SIMPLE algorithm to obtain a Poisson equation for pressure, making use of the continuity equation. The Poisson equation is solved for the pressure, which is then used to obtain a solution to the momentum equations. Velocity corrections are then computed as required to satisfy the continuity equation using the p' (or an alternative) procedure of the SIMPLE algorithm. This usually will require the solution of another Poisson-like equation for the p' field that will drive the corrections. Only the velocities are corrected with the p' solution. This revised scheme makes use of a more exact (less approximate) procedure for revising the pressure. Looking more closely at the algorithm, it should be evident that if the first velocity field is the correct one and continuity is satisfied, then the pressure field computed with the help of the pseudo-velocities will also be the correct one. Thus, when the momentum equations are solved with this pressure field, a mass-conserving velocity field will be computed, and no further iterations will be required.

The SIMPLEC procedure developed by van Doormal and Raithby (1984) attempts to improve on the convergence rate of SIMPLE by using a more complete or consistent approximation to the momentum equations to compute p'. The scheme attempts to approximate the effects of some terms in the momentum equations neglected in the SIMPLE algorithm for p'. This is equivalent to modifying the A in Eq. (9.154). With this modification, van Doormal and Raithby reported that it was no longer necessary to underrelax the pressure correction. They also observed that SIMPLEC performed more efficiently than both SIMPLE and SIMPLER for the several test cases they considered. Other suggested improvements to SIMPLE and SIMPLER were included in their paper, which should be consulted for further details.

Pressure-correction approach: SIMPLE on nonstaggered grids. Until the early 1980s the SIMPLE family of methods was generally only employed on staggered grids. It is not straightforward to implement staggered grid schemes on general nonorthogonal curvilinear coordinate systems. However, use of nonstaggered (colocated) grids with the SIMPLE family of methods was observed to result in decoupling of the velocity and pressure fields, yielding "wiggles" in solutions. This provided motivation for numerical experimentation, and in 1981, several investigators (Hsu, 1981; Prakash, 1981; Rhie, 1981) reported success in implementing pressure-correction schemes on a regular grid. In effect, fourth-order pressure smoothing or dissipation was added to the continuity equation to achieve this result. The description given here follows the report of Rhie and Chow (1983).

The colocated scheme of Rhie and Chow (1983) follows the SIMPLE sequence. However, all variables are located at the same grid points. The momentum equations are solved using the best guess for the pressure field. Until the correct pressure field is established, this velocity field will not satisfy the continuity equation. The "trick" comes into play in computing the mass sources needed to correct the velocities and pressure. It is required in the SIMPLE strategy to compute the mass sources in each computational control volume or cell. On a colocated grid in which the grid points are on the interior of each control volume, this means that velocities need to be interpolated to the cell faces. Assuming a uniform grid in the Cartesian coordinate system, a simple linear interpolation would ordinarily be considered, giving for example, in two dimensions

$$\tilde{u}_{i+1/2,j} = \frac{\tilde{u}_{i,j} + \tilde{u}_{i+1,j}}{2}$$

where $\tilde{u}$ is the provisional x component of velocity obtained from solving the x momentum equation with a provisional pressure field. Inspection of the momentum equations used to compute the velocities at i and $i+1$ reveals that the cell-face velocity indicated above depends on velocities computed using centrally differenced pressure gradient terms formed from differences over two grid spacings. The Rhie-Chow interpolation scheme can be interpreted as an attempt to estimate the cell-face velocity that would have been computed if the resultant pressure gradient influencing the cell-face velocity were $(p_{i+1,j} - p_{i,j})/\Delta x$ (adjacent nodes only) instead of $(p_{i+2,j} - p_{i,j} + p_{i+1,j} - p_{i-1,j})/4\,\Delta x$, which follows from the momentum equations. The problem with the second expression is that it is insensitive to "checkerboard" oscillations because the differences are between every second node. The interpolation scheme to obtain cell-face velocities for computing the mass sources, as in Eq. (9.155), can be written as

$$\tilde{u}_{i+1/2,j} = \tilde{u}_{\mathrm{li}} + B_{\mathrm{li}}\left(\frac{p_{i+1,j} - p_{i,j}}{\Delta x} - \left.\frac{\partial p}{\partial x}\right|_{\mathrm{li}}\right) \tag{9.157}$$

where the subscript li denotes a value linearly interpolated to the cell face at $i + 1/2$ and B is the coefficient of the pressure gradient term in the momentum equation after it has been rearranged to isolate $u_{i,j}$ on the LHS. Thus Eq. (9.157) can be thought of as correcting the linearly interpolated velocity by providing a local value of the pressure gradient instead of the one resulting from the centrally differenced forms used in the momentum equations. For the uniform Cartesian grid example initiated above, Eq. (9.157) can be written as

$$\tilde{u}_{i+1/2,j} = \frac{\tilde{u}_{i,j} + \tilde{u}_{i+1,j}}{2} + B_{\mathrm{li}}\left(\frac{p_{i+1,j} - p_{i,j}}{\Delta x} - \frac{p_{i+2,j} - p_{i,j} + p_{i+1,j} - p_{i-1,j}}{4\,\Delta x}\right) \tag{9.158}$$

which can be further rearranged into the form

$$\tilde{u}_{i+1/2,j} = \frac{\tilde{u}_{i,j} + \tilde{u}_{i+1,j}}{2} + \frac{B_{\mathrm{li}}}{4\,\Delta x}(3p_{i+2,j} - 3p_{i,j} + p_{i+1,j} - p_{i-1,j}) \tag{9.159}$$

The term in parentheses can be recognized as the fourth difference commonly used in fourth-order dissipation, which formed the basis for the earlier observation that fourth-order dissipation of pressure is being added to the continuity equation. The cell-face values of velocities computed as indicated above are then used to compute the mass source term. To enforce mass conservation, velocity and pressure corrections are introduced as in SIMPLE. The pressure Poisson equation is solved for the pressure corrections, and then the velocity corrections at the nodes can be computed from a simplified relationship between nodal velocities and the pressure gradient having the form of Eq. (9.154). These correction relationships follow the discretization used in the momentum equations and do not make further use of the special interpolation formula for cell-face velocities discussed above.

Several investigators have compared the accuracy and computational efficiency of the colocated and staggered-grid version of the SIMPLE family of methods. Among these studies are the works of Burns et al. (1986), Peric et al. (1988), and Melaaen (1992). Generally, the accuracy and convergence rate of both formulations have been found comparable. Examples have been cited in which each of the two schemes has been slightly more accurate than the other. Generally, however, the difference between the two results has been less than the estimated numerical error in the calculations of either scheme. There seems to be general agreement that the colocated approach is more convenient for working in curvilinear nonorthogonal coordinate systems and in three dimensions. Implementation of multigrid is also more straightforward with the colocated arrangement.

The SIMPLE family of methods has also been applied to compressible flows. For details, refer to the works of van Doormaal et al. (1987), Karki and Patankar (1989), McGuirk and Page (1989), and Shyy et al. (1992).

Pressure-correction approach: PISO (pressure-implicit with splitting of operators) method. A predictor-corrector strategy was proposed by Issa (1985) for solving the discretized time-dependent N-S equations in a sequential uncoupled manner. The scheme is applicable to both the incompressible and compressible forms of the equations and has been implemented on both colocated and staggered grids. The scheme is largely implicit, and various strategies for solving the simultaneous algebraic equations can be employed. The splitting strategy will be outlined here for incompressible flow using symbolic operator notation.

One predictor step and two corrector steps are utilized. Let the asterisks denote intermediate values computed during the splitting process. The calculation proceeds in the following steps.

1. *Predictor step.* The pressure field prevailing at time level n is used in the implicit solution of the momentum equations. This step is identical to the first step in the SIMPLE algorithm when the latter is applied to a time-dependent flow:

$$\frac{\rho}{\Delta t}(u_i^* - u_i^n) = H(u_i^*) - \Delta_i p^n + S_i \tag{9.160}$$

   Index notation is employed in Eq. (9.160), and the operator H stands for the finite-difference representation of the spatial convective and diffusive fluxes of momentum. The operator Δ_i is the finite-difference equivalent of $\partial/\partial x_i$. This velocity field will not generally satisfy the continuity equation.
2. *First corrector step.* In this step a new pressure field p^* is sought along with a revised velocity field u_i^{**} that will satisfy conservation of mass. Treating the velocities explicitly, the momentum equation is considered in the form

$$\frac{\rho}{\Delta t}(u_i^{**} - u_i^n) = H(u_i^*) - \Delta_i p^* + S_i \tag{9.161}$$

Requiring that $\Delta_i u_i^{**} = 0$ and taking the divergence of Eq. (9.161) gives a discretized Poisson equation:

$$\Delta_i^2 p^* = \Delta_i H(u_i^*) + \Delta_i S_i + \frac{\rho}{\Delta t} \Delta_i u_i^n \tag{9.162}$$

Note that the Poisson equation can be immediately solved for pressure p^*, since the RHS contains quantities already determined in the predictor step. This pressure field can then be used in Eq. (9.161) to compute u_i^{**}, which by design should satisfy the continuity equation.

3. *Second corrector step*. This step is essentially a recorrection, using the strategy outlined above. The momentum equation is considered in the form

$$\frac{\rho}{\Delta t}(u_i^{***} - u_i^n) = H(u_i^{**}) - \Delta_i p^{**} + S_i \tag{9.163}$$

Taking the divergence of Eq. (9.163) and requiring the continuity equation to be satisfied with the new velocity field, $\Delta_i u_i^{***} = 0$, gives an equation that can be solved for an updated pressure field:

$$\Delta_i^2 p^{**} = \Delta_i H(u_i^{**}) + \Delta_i S_i + \frac{\rho}{\Delta t} \Delta_i u_i^n \tag{9.164}$$

The pressure field is first determined from Eq. (9.164) and then used in the momentum equation, Eq. (9.163), where the updated velocity field, u_i^{***}, is computed. Following this format, more recorrections can be made, but Issa (1985) suggests that the two correction steps should be sufficient for most purposes.

Issa (1985) discusses the errors associated with the method and argues that the splitting errors are sufficiently small that time-accurate solutions can be obtained without iterative application of the algorithm using time steps dictated only by the accuracy of the difference scheme. The favorable features of the scheme were demonstrated by Issa et al. (1986) in a paper that considered time-dependent and subsonic compressible flows. A successive overrelaxation by lines procedure was used to solve the simultaneous linearized algebraic equations resulting from a finite-volume discretization on a staggered grid.

PROBLEMS

9.1 Show how all the terms in the 2-D y momentum equation are differenced when the explicit MacCormack (1969) method is applied to the compressible N-S equations.
9.2 Repeat Prob. 9.1 for the 2-D energy equation.
9.3 Apply the explicit MacCormack scheme to the N-S equations written in cylindrical coordinates (see Section 5.1.8), and show how all the terms in the r momentum equation are differenced.
9.4 Apply the Allen-Cheng method instead of the explicit MacCormack method in Prob. 9.1.
9.5 Derive the Jacobian matrix $[A]$ given by Eq. (9.46).
9.6 Derive the Jacobian matrix $[B]$ given by Eq. (9.48).
9.7 Derive the Jacobian matrix $[R]$ given by Eq. (9.51).

9.8 Derive the Jacobian matrix $[S]$ given by Eq. (9.54).
9.9 Derive the matrix $[P] - [R_x]$ given by Eq. (9.50).
9.10 Derive the matrix $[Q] - [S_y]$ given by Eq. (9.53).
9.11 Determine the amplification factor G for the explicit MacCormack scheme applied to the linearized Burgers equation. Does Eq. (4.313) satisfy $|G| \leqslant 1$ for all values of β when $\nu = \frac{1}{2}$ and $r = \frac{1}{4}$?
9.12 Repeat Prob. 9.11 for $\nu = 1$ and $r = \frac{1}{2}$.
9.13 Use the explicit MacCormack method to solve the linearized Burgers equation for the initial condition

$$u(x,0) = 0 \qquad 0 \leqslant x \leqslant 1$$

and the boundary conditions

$$u(0,t) = 100$$
$$u(1,t) = 0$$

on a 21 grid point mesh. Find the steady-state solution for the conditions

$$r = 0.5$$
$$\nu = 0.5$$

and compare the numerical solution with the exact solution.
9.14 Derive the Jacobian matrix $[A_t]$ given by Eq. (9.74).
9.15 Derive the Jacobian matrix $[A_x]$ given by Eq. (9.75).
9.16 Formulate a preconditioned implicit scheme for solving the 2-D compressible N-S equations at low Mach numbers using primitive variables. Derive all necessary Jacobian matrices. Explain your work.
9.17 Explain how preconditioning can be applied to the 2-D compressible N-S equations when conserved variables are employed in an implicit formulation.
9.18 Obtain Eq. (9.124).
9.19 Solve the square driven cavity problem for $\text{Re}_l = 50$. Use the forward-time, centered-space (FTCS) method (Section 4.5.1) to solve the vorticity transport equation and the successive overrelaxation method to solve the Poisson equation. Employ a first-order evaluation of the vorticity at the wall, and use an 8×8 grid.
9.20 Repeat Prob. 9.19 for $\text{Re}_l = 100$ and a 15×15 grid.
9.21 Derive the vorticity transport equations for a 3-D Cartesian coordinate system.
9.22 Use the artificial compressibility method to solve the square driven cavity problem for $\text{Re}_l = 100$. Apply the leap frog/DuFort-Frankel finite-difference scheme to the governing equations on a 15×15 grid. Determine pressure at the wall using a suitable finite-difference representation of the normal momentum equation applied at the wall.
9.23 Use the method of artificial compressibility to solve the steady square driven cavity problem for $\text{Re}_l = 100$. Use a coupled implicit scheme. Compare your solution from a 21×21 grid with the results of Ghia et al. (1982).
9.24 Add a pseudo-time term to the method of Prob. 9.23, and compute the driven cavity problem for $\text{Re}_l = 100$ in a time-accurate manner, starting the lid impulsively from rest. Plot the x component of velocity at the center of the cavity as a function of time.
9.25 Use a preconditioned compressible formulation for the 2-D N-S equations to solve the steady driven cavity problem for $\text{Re}_l = 100$. Obtain solutions at Mach numbers of 0.2, 0.1, 0.01, and 0.001. Compare the convergence histories of the scheme for the Mach numbers specified.

CHAPTER

TEN

GRID GENERATION

10.1 INTRODUCTION

One of the first steps in computing a numerical solution to the equations that describe a physical process is the construction of a grid. The physical domain must be covered with a mesh, so that discrete volumes or elements are identified where the conservation laws can be applied. A well-constructed grid greatly improves the quality of the solution, and conversely, a poorly constructed grid is a major contributor to a poor result. In many applications, difficulties with numerical simulations can be traced to poor grid quality. For example, the lack of convergence to a desired level is often a result of poor grid quality. In this chapter, techniques for generating grids using both structured and unstructured approaches will be discussed. Since grid generation is a very large field, only the basic ideas of a limited number of methods will be presented.

Structured grid generation can be thought of as being composed of three categories:

1. Complex variable methods
2. Algebraic methods
3. Differential equation techniques

Complex variable techniques have the advantage that the transformations used are analytic or partially analytic, as opposed to those methods that are entirely numerical. Unfortunately, they are restricted to two dimensions. For this reason, the technique has limited applicability and will not be covered here. For details of the application of complex variable methods, the works by Churchill (1948), Moretti (1979), Davis (1979), and Ives (1982) should be consulted. Algebraic and differential equation techniques can be used on complicated three-dimensional (3-D) problems. Of the structured methods, these have received the most use and will be discussed in this chapter. Unstructured methods are primarily based on using triangular or prismatic elements, although recently, randomly shaped cells with arbitrary connectivity have been increasingly used in flow simulations. Unstructured grid generation schemes may be thought of as being divided into three groups:

1. Point insertion schemes
2. Advancing front methods
3. Domain decomposition techniques

Details of point insertion methods will be discussed. However, only the procedure for constructing grids satisfying the Delaunay (1934) criterion using point insertion will be given. The detailed description of advancing front schemes or those using arbitrarily shaped cells is beyond the scope of this text. Domain decomposition techniques rely on recursively subdividing domains to provide a cell structure that may be used to complete a field calculation. These methods will not be discussed in detail in this chapter, but some information on the basic idea will be included.

Early work using finite-difference methods was restricted to problems where suitable coordinate systems could be selected in order to solve the governing equations in that base system. As experience in computing solutions for complex flow fields was gained, general mappings were employed to transform the physical plane into a computational domain. Numerous advantages accrue when this procedure is followed. For example, the body surface can be selected as a boundary in the computational plane, permitting easy application of surface boundary conditions. In general, transformations are used that lead to a uniformly spaced grid in the computational plane, while points in physical space may be unequally spaced. This situation is shown in Fig. 10.1. When this procedure is used, it is necessary to include the metrics of the mapping in the differential equations.

In applying finite-volume methods to the solution of physical problems, the direct application of the conservation statement to elements in the physical plane can be made without transforming the original differential equations. As we have seen in previous chapters, the discrete equations may be generated with this procedure and the metrics that appear with the generalized mapping now appear through the direct use of the volumes and the surface areas of the cell faces. With either approach, the physical domain is divided into volumes or cells. Techniques for creating the cell or element structure forms the basis of grid generation.

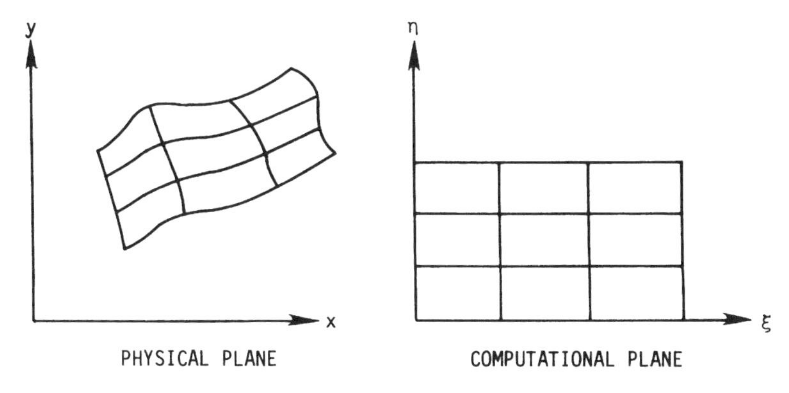

Figure 10.1 Mapping to computational space.

10.2 ALGEBRAIC METHODS

In Section 5.6 we used algebraic expressions to cluster grid points near solid boundaries to provide adequate resolution of the viscous boundary layer. In another example, a domain-normalizing transformation was used in order to align the grid lines with the body and shock wave in physical space. These are examples of simple algebraic mappings. To generate computational grids using this technique, known functions are used in one, two, or three dimensions to take arbitrarily shaped physical regions into a rectangular computational domain. Although the computational domain is not required to be rectangular, the usual procedure uses a rectangular region for simplicity.

The simplest procedure available that may be used to produce a boundary fitted computational mesh is the normalizing transformation discussed in Section 5.6.1. Suppose a body fitted mesh is desired in order to solve for the flow in a diverging nozzle. The geometry of the nozzle is shown in Fig. 10.2, and the describing function for the nozzle is given as

$$y_{\max} = x^2 \qquad 1.0 \leqslant x \leqslant 2.0 \tag{10.1}$$

In this example, a computational grid can easily be generated by choosing equally spaced increments in the x direction and using uniform division in the y direction. This may be described as

$$\begin{aligned} \xi &= x \\ \eta &= \frac{y}{y_{\max}} \end{aligned} \tag{10.2}$$

where $y_{\max}$ denotes the y coordinate of the nozzle wall. In this case the values of x and y for a given ξ and η are easily recovered. The mesh generated in the physical domain is shown in Fig. 10.3.

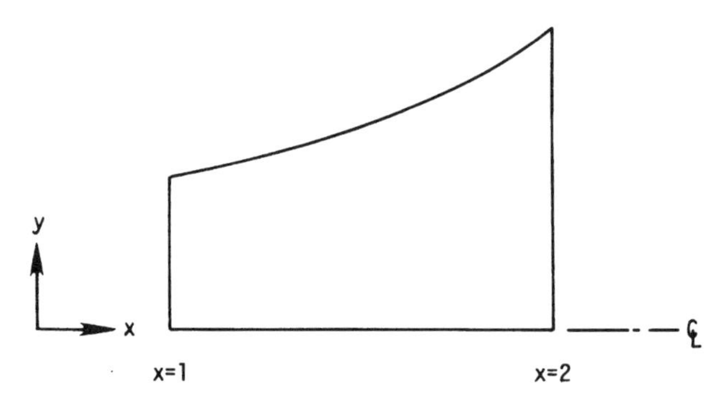

Figure 10.2 Nozzle geometry.

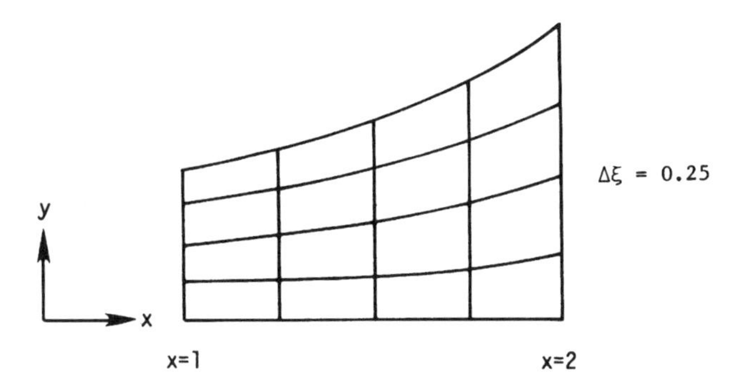

Figure 10.3 Computational mesh in physical space.

Care must be exercised when the metrics of the transformation are derived. In particular, the η_x derivative of Eq. (10.2) is written

$$\eta_x = -\frac{y}{y_{\max}^2}\frac{dy_{\max}}{dx} = -\frac{2\eta}{\xi} \tag{10.3}$$

and

$$\eta_y = \frac{1}{y_{\max}} = \frac{1}{\xi^2} \tag{10.4}$$

In the example just completed, the transformation was analytic and the point distribution was obtained through the given mapping. The same normalizing transformation could have been constructed by assigning points in the physical plane along constant ξ and constant η lines and numerically computing the metrics by using second-order central differences. This has the advantage of permitting assignment of grid points in the physical plane where desired. The disadvantage is that all metrics must be determined using numerical techniques. In this case the transformation would be numerical and not algebraic.

If numerical methods are used to generate the required transformation, the terms x_ξ, x_η, y_ξ, and y_η are determined using finite differences. The quantities ξ_x, ξ_y, η_x, and η_y appear in the differential equation, which must be solved. These quantities are obtained from the expressions

$$
\begin{aligned}
\xi_x &= \frac{y_\eta}{\mathrm{I}} \\
\xi_y &= -\frac{x_\eta}{\mathrm{I}} \\
\eta_x &= -\frac{y_\xi}{\mathrm{I}} \\
\eta_y &= \frac{x_\xi}{\mathrm{I}}
\end{aligned}
\tag{10.5}
$$

where the inverse Jacobian (I) is given by

$$\mathrm{I} = (J)^{-1} = x_\xi y_\eta - y_\xi x_\eta$$

More details will be presented in the section treating mappings governed by differential equations.

Example 10.1 Compare the metrics for the simple normalizing transformation just discussed by computing them analytically and also by using a finite-difference approximation.

Solution We select the point (1.75, 2.2969) in the nozzle of Fig. 10.3 to compare the metrics. From Eq. (10.3), the analytic evaluation is

$$\eta_x = -\frac{2(0.75)}{1.75} = -0.85714$$

The numerical calculation is performed by using Eq. (10.5). The inverse Jacobian (I) is evaluated first as

$$\mathrm{I} = \overset{1}{\cancel{x_\xi}} y_\eta - \overset{0}{\cancel{y_\xi}} x_\eta = \frac{3.0625 - 1.53125}{2(0.25)} = 3.06250$$

Next, the y_ξ term is computed as

$$y_\xi = \frac{3 - 1.6875}{0.5} = 2.6250$$

Thus

$$\eta_x = -\frac{2.6250}{3.0625} = -0.85714$$

In this example, the metrics computed by analytical and numerical methods give equally good results. Of course, this is not true for many problems.

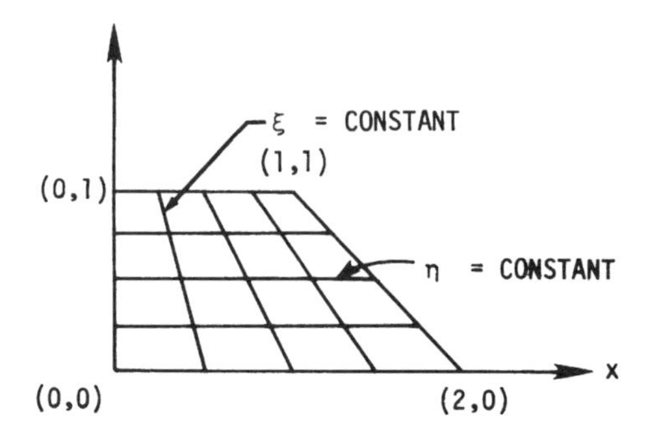

Figure 10.4 Trapezoid to rectangle mapping.

Example 10.2 The trapezoidal region shown in Fig. 10.4 is mapped into a corresponding rectangular region by the equations given by

$$x = \left(\frac{1+\xi}{2}\right)\left(\frac{3-\eta}{2}\right)$$
$$y = \frac{\eta+1}{2} \tag{10.6}$$

In this example the physical domain is mapped into a rectangular region centered at the origin. This demonstrates the use of a normalizing transformation in one direction along with a simple translation.

This choice of parameterization will give a different grid even for the case where Lagrange interpolation is used. While the preceding examples show acceptable results for computational grids, it is not always possible to construct a satisfactory grid without a more systematic approach. In particular, the application of general interpolation techniques provides a more formal approach toward the generation of grids using algebraic methods.

Smith and Weigel (1980) developed a flexible method of directly providing grids using interpolation between surfaces. In this method, the domain of interest is defined by an upper and a lower boundary in the physical plane. As shown in Fig. 10.5(a), these boundaries are denoted by

$$\mathbf{r}_1 = \mathbf{r}_{B_1}(\xi) \tag{10.7}$$

$$\mathbf{r}_2 = \mathbf{r}_{B_2}(\xi) \tag{10.8}$$

The upper and lower boundaries have been parameterized using the computational coordinate as the parameter. These curves may also be written in terms of the scalar coordinates

$$\begin{aligned} x_{B_1} &= x_1(\xi) \\ y_{B_1} &= y_1(\xi) \\ x_{B_2} &= x_2(\xi) \\ y_{B_2} &= y_2(\xi) \end{aligned} \tag{10.9}$$

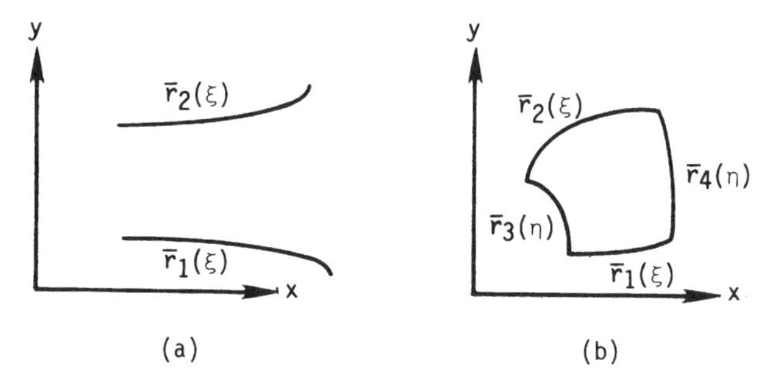

Figure 10.5 Algebraic interpolation domains. (a) Open domain for two-surface methods. (b) Closed domain for transfinite interpolation.

The range on ξ in the computational plane is

$$0 \leqslant \xi \leqslant 1$$

and the transformation is defined so that at $\eta = 0$,

$$\begin{aligned} x_{B_1} &= x_1(\xi) = x(\xi, 0) \\ y_{B_1} &= y_1(\xi) = y(\xi, 0) \end{aligned} \tag{10.10}$$

and at $\eta = 1$,

$$\begin{aligned} x_{B_2} &= x_2(\xi) = x(\xi, 1) \\ y_{B_2} &= y_2(\xi) = y(\xi, 1) \end{aligned} \tag{10.11}$$

A function defined on $0 \leqslant \eta \leqslant 1$ with parameters on the two boundaries completes the algebraic relation. This is chosen to be of the form

$$\begin{aligned} x = x(\xi, \eta) &= F\left(x_1, \frac{dx_1}{d\eta}, \ldots, x_2, \frac{dx_2}{d\eta}, \ldots\right) \\ y = y(\xi, \eta) &= G\left(y_1, \frac{dy_1}{d\eta}, \ldots, y_2, \frac{dy_2}{d\eta}, \ldots\right) \end{aligned} \tag{10.12}$$

If linear variations across the domain are selected, the grid is determined on the basis of Lagrange interpolation, and the form of the expression for (x, y) is

$$\mathbf{r}(\xi, \eta) = (1 - \eta)\mathbf{r}_1(\xi) + \eta\mathbf{r}_2(\xi) \tag{10.13}$$

or in scalar form,

$$x = (1 - \eta)x_1(\xi) + \eta x_2(\xi)$$

$$y = (1 - \eta)y_1(\xi) + \eta y_2(\xi)$$

Example 10.3 To demonstrate the use of this approach, suppose we wish to map the trapezoid defined by the equations

$$x = 0$$

$$x = 1$$

$$y = 0$$

$$y = 1 + x$$

into the computational plane. In this case, the upper and lower boundaries may be written

$$x_{B_1} = x_1(\xi) = \xi$$

$$y_{B_1} = y_1(\xi) = 0$$

$$x_{B_2} = x_2(\xi) = \xi$$

$$y_{B_2} = y_2(\xi) = 1 + \xi$$

This produces the mapping required in Eq. (10.13) and is of the form

$$x = \xi$$

$$y = (1 + \xi)\eta$$

This parameterization produces the simple normalizing transformation discussed earlier in this section. In this example, both the right and left boundaries are also correctly mapped. This is coincidental and will not occur in more general problems. A different point distribution can be obtained by choosing a nonlinear function for the boundary parameterization. For example, if

$$x_1 = \xi^2$$

$$x_2 = \xi^2$$

then

$$x = \xi^2$$

$$y = \eta(1 + \xi^2)$$

In this case, the nonlinear boundary parameterization produces some clustering of the grid points. However, the ability to cluster points is limited to the influence of the boundary point distribution on the interior through interpolation.

Additional control of the grid point distribution can be attained if higher-order interpolation polynomials are used. Hermite interpolation is often employed, since the derivatives specifying the initial slope of the constant coordinate curves leaving the boundaries are also included in the interpolation. The expression for the interpolated grid coordinates is of the form

$$\mathbf{r}_{\text{inter}}(\xi, \eta) = \mathbf{r}_1(\xi) f_1(\eta) + \mathbf{r}_2(\xi) f_2(\eta) + \frac{d\mathbf{r}_1(\xi)}{d\eta} f_3(\eta) + \frac{d\mathbf{r}_2(\xi)}{d\eta} f_4(\eta) \tag{10.14}$$

where the functions are given by

$$\begin{aligned} f_1(\eta) &= 2\eta^3 - 3\eta^2 + 1 \\ f_2(\eta) &= -2\eta^3 + 3\eta^2 \\ f_3(\eta) &= \eta^3 - 2\eta^2 + \eta \\ f_4(\eta) &= \eta^3 - \eta^2 \end{aligned} \tag{10.15}$$

This additional flexibility can be used to produce orthogonality at the upper and lower boundaries [see Kowalski (1980) for details].

In most problems, the boundaries are not analytic functions but are simply prescribed as a set of data points. In this case, the boundary must be approximated by a curve fitting procedure to employ algebraic mappings. Eiseman and Smith (1980) discuss possible methods of accomplishing this and particularly recommend tension splines. Tension splines are suggested because higher-order approximations including cubic splines tend to produce wiggles in the boundary. The tension parameter in the tension spline allows control of this phenomenon.

With the simple interpolation schemes investigated thus far, only two boundaries are matched. Unfortunately, many problems include the necessity of matching four boundaries enclosing the physical domain. Gordon and Hall (1973) describe transfinite interpolation, and Rizzi and Eriksson (1981) provide application details for generating boundary conforming grids with algebraic interpolation. To understand transfinite interpolation, consider the geometry of the physical domain given in Fig. 10.5(b). The simple interpolation scheme of Eq. (10.13) does not produce a grid that matches the boundaries denoted by 3 and 4 in Fig. 10.5(b). In fact, this simple interpolation produces an error at these boundaries that may be specifically identified. The error at boundaries denoted by 3 and 4 may be written

$$\begin{aligned} \mathbf{e}_3 &= \mathbf{r}_3(\eta) - \mathbf{r}_{\text{inter}}(0, \eta) \\ \mathbf{e}_4 &= \mathbf{r}_4(\eta) - \mathbf{r}_{\text{inter}}(1, \eta) \end{aligned} \tag{10.16}$$

where the maximum and minimum values of ξ are taken to be 0 and 1, respectively, to define the left and right boundaries of the domain and the subscript, inter, indicates the interpolated values of (x, y). If Lagrange interpolation is used, these errors are given by

$$\begin{aligned} \mathbf{e}_3 &= \mathbf{r}_3(\eta) - (1 - \eta)\mathbf{r}_1(0) - \eta\mathbf{r}_2(0) \\ \mathbf{e}_4 &= \mathbf{r}_4(\eta) - (1 - \eta)\mathbf{r}_1(1) - \eta\mathbf{r}_2(1) \end{aligned} \tag{10.17}$$

To eliminate these errors, we interpolate them onto the domain, so the errors at both the left and right boundaries are eliminated. The expression for the interpolated error may be written

$$\mathbf{e}(\xi, \eta) = (1 - \xi)\mathbf{e}_3(\eta) + \xi\mathbf{e}_4(\eta) \tag{10.18}$$

The expression for the interpolated computational coordinates is written

$$\mathbf{r}(\xi, \eta) = \mathbf{r}_{\text{inter}} + \mathbf{e}(\xi, \eta) \tag{10.19}$$

and the final result becomes

$$\begin{aligned} \mathbf{r}(\xi, \eta) = (1 - \eta)\mathbf{r}_1(\xi) + \eta\mathbf{r}_2(\xi) + (1 - \xi)[\mathbf{r}_3(\eta) - (1 - \eta)\mathbf{r}_1(0) - \eta\mathbf{r}_2(0)] \\ + \xi[\mathbf{r}_4(\eta) - (1 - \eta)\mathbf{r}_1(1) - \eta\mathbf{r}_2(1)] \end{aligned} \tag{10.20}$$

This result shows that the transfinite interpolation (TFI) is composed of interpolations between the corresponding edges and an interpolation from each of the corner points. This interpolation will match all of the edge data required by a fixed domain. If Hermite interpolation is used, a different final result is obtained that matches the domain boundaries and also the initial slope requirements. In this case, it is possible to control the orthogonality of the mesh at the boundaries, and details can be found in the work of Chawner and Anderson (1991). The TFI method can also be used when only three boundaries are matched. This can be accomplished by noting that the error need not vanish on the fourth boundary, but may create a fourth boundary that assumes any shape. This is similar to using a TFI scheme to create a grid on an open domain. When transfinite interpolation is used, it must be remembered that no control over the Jacobian of the transformation is maintained. Grid crossing can and will occur when different parameterizations of the boundaries are used or when the derivatives at the boundaries (Hermite) are given certain values. These behavior traits can be seen through the applications provided by the problems at the end of this chapter.

The basic idea behind the most common algebraic methods used to construct grids was presented in this section. With this basic understanding of the TFI scheme, the method can be developed with considerably more rigor. However, the intent here is to introduce the most common methods of grid generation, and a more sophisticated presentation will not be included.

10.3 DIFFERENTIAL EQUATION METHODS

In the previous section, algebraic methods were presented that can be used to produce usable grids. Any procedure that results in an acceptable grid is a valid one. One of the most frequently used and most highly developed procedures is the differential equation method. If a partial differential equation is used to generate a grid, we can exploit the properties of the solution of the grid-generating equation in producing the mesh. All three classes of partial differential equations have been used to produce grid construction methods, and a short discussion of each is presented in this section.

10.3.1 Elliptic Schemes

Elliptic partial differential equations (PDEs) have the property that the solutions are generally very smooth. This smoothness can be used to advantage, and for this reason, Laplace's equation is a good choice. To better understand the choice of Laplace's equation, consider the solution of a steady heat conduction problem in two dimensions with Dirichlet boundary conditions. The solution of this problem produces isotherms that are smooth (class C^{II} properties) and are nonintersecting. The number of isotherms in a given region can be increased by adding a source term. If the isotherms are used as grid lines, they will be smooth, nonintersecting, and can be densely packed in any region by controlling the source term.

One of the attractive features of using Laplace's equation is that the Jacobian is guaranteed to be positive as a result of the maximum principle for harmonic functions. Unfortunately, this theorem only applies to the analytic equations and solution (Thompson et al., 1985). When the differential equation is discretized, the truncation errors may lead to grid crossing even though the maximum principle holds for the solution of the analytic equation. This point must be clearly understood. If the numerical formulation of the differential equation satisfies the consistency condition, then the maximum principle will be satisfied in the limit of vanishing mesh size. However, no promises can be made for finite mesh sizes. In some cases, an estimate can be made to determine when mesh crossing will occur (see Prob. 10.14).

This idea of using elliptic differential equations is based on the work of Crowley (1962) and Winslow (1966) and transforms the physical domain into the computational plane, where the mapping is controlled by a Poisson equation. Thompson et al. (1974) have worked extensively on using elliptic PDEs to generate grids. When the Poisson grid generators are used, the mapping is constructed by specifying the desired grid points (x, y) on the boundary of the physical domain with the interior point distribution determined through the solution of the equations

$$\begin{aligned} \xi_{xx} + \xi_{yy} &= P(\xi, \eta) \\ \eta_{xx} + \eta_{yy} &= Q(\xi, \eta) \end{aligned} \tag{10.21}$$

where (ξ, η) represent the coordinates in the computational domain and P and Q are terms that control the point spacing on the interior of D. Equations (10.21) are then transformed to computational space by interchanging the roles of the independent and dependent variables. This yields a system of two elliptic equations of the form

$$\begin{aligned} \alpha x_{\xi\xi} - 2\beta x_{\xi\eta} + \gamma x_{\eta\eta} &= -\mathrm{I}^2(P x_\xi + Q x_\eta) \\ \alpha y_{\xi\xi} - 2\beta y_{\xi\eta} + \gamma y_{\eta\eta} &= -\mathrm{I}^2(P y_\xi + Q y_\eta) \end{aligned} \tag{10.22}$$

where

$$\begin{aligned} \alpha &= x_\eta^2 + y_\eta^2 \\ \beta &= x_\xi x_\eta + y_\xi y_\eta \\ \gamma &= x_\xi^2 + y_\xi^2 \\ \mathrm{I} &= \frac{\partial(x, y)}{\partial(\xi, \eta)} = x_\xi y_\eta - x_\eta y_\xi \end{aligned}$$

This system of equations is solved on a uniformly spaced grid in the computational plane. This provides the (x, y) coordinates of each point in physical space. For simply connected regions, Dirichlet boundary conditions can be used

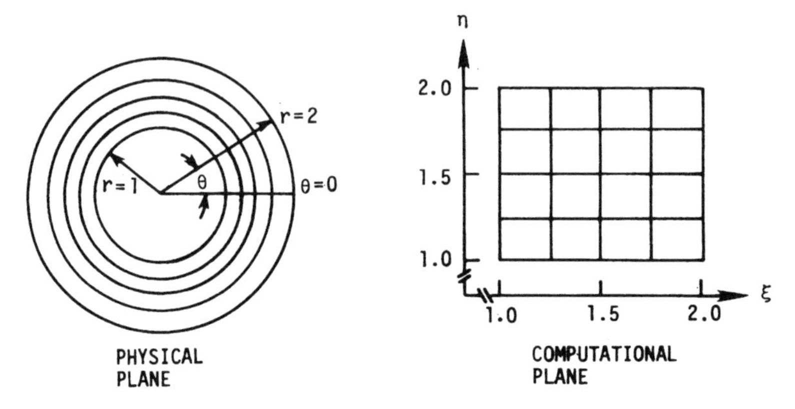

Figure 10.6 Application of Thompson scheme.

at all boundary points. The advantages of using this technique to generate a computational mesh are many. The resulting grid is smooth, the transformation is one to one, and complex boundaries are easily treated. Of course, there are some disadvantages. Specification of P and Q is not an easy task, grid point control on the interior is difficult to achieve, and boundaries may be changing with time. In the latter case, the grid must be computed after each time step. This can consume large amounts of computer time.

A simple example demonstrating the application of the Thompson scheme is shown in Fig. 10.6. The region between two concentric circles is mapped into the computational domain, and the resulting constant ξ and η surfaces in physical space are shown. The inner circle is of radius r_0, and the outer circle is of radius r_1. For this problem, the circle is cut at $\theta = 0$ and mapped into the region between 1 and $\xi_{\max}$ and 1 and $\eta_{\max}$ in computational space. In this problem the mapping is determined by a solution of two Laplace's equations,

$$\nabla^2 \xi = 0$$
$$\nabla^2 \eta = 0$$

subject to boundary conditions

$$\begin{aligned} r &= r_0 & \eta &= 1 \\ r &= r_1 & \eta &= \eta_{\max} \\ \theta &= 0 & \xi &= 1 \\ \theta &= 2\pi & \xi &= \xi_{\max} \end{aligned}$$

The solution is of the form

$$x = R \cos \phi$$
$$y = R \sin \phi$$

where

$$R = r_0 \left(\frac{r_1}{r_0} \right)^{(\eta - 1)/(\eta_{\max} - 1)}$$

$$\phi = \left(\frac{\xi - 1}{\xi_{\max} - 1} \right) 2\pi$$

This solution is interesting, in that a uniform grid in the physical domain is not achieved in this case. The distribution in the radial direction is a series of concentric circles. To obtain the mapping with a series of uniformly spaced concentric circles, $P = 0$ and $Q = 1/\eta$ (see Prob. 10.9).

As previously noted, one of the difficulties with this scheme is point control on the interior of the domain. This requires that methods for developing P and Q be devised in order to obtain the desired point distribution. Middlecoff and Thomas (1979) have developed a method that provides approximate control of point spacing by evaluating P and Q according to the desired point distribution on the boundary.

In order to demonstrate this idea, we suppose that a solution of Eq. (10.21) is required subject to Dirichlet boundary conditions. We elect to write P and Q in the form

$$\begin{aligned} P &= \phi(\xi, \eta)\left(\xi_x^2 + \xi_y^2\right) \\ Q &= \psi(\xi, \eta)\left(\eta_x^2 + \eta_y^2\right) \end{aligned} \tag{10.23}$$

where ϕ and ψ will be specified through the boundary conditions. With this convention, our original system [Eq. (10.22)] may be written

$$\begin{aligned} \alpha(x_{\xi\xi} + \phi x_\xi) - 2\beta x_{\xi\eta} + \gamma(x_{\eta\eta} + \psi x_\eta) &= 0 \\ \alpha(y_{\xi\xi} + \phi y_\xi) - 2\beta y_{\xi\eta} + \gamma(y_{\eta\eta} + \psi y_\eta) &= 0 \end{aligned} \tag{10.24}$$

Middlecoff and Thomas (1979) proposed writing these equations along either constant ξ or η surfaces corresponding to the boundaries of the domain, assuming that the grid was orthogonal at the boundaries and that the opposite family of lines had zero curvature at the intersection. If we are interested in finding the values of ϕ along a constant η boundary, it is assumed that the constant ξ curves intersecting this boundary have no curvature at the intersection point and that the two are orthogonal. If S represents arc length along the constant η boundary, then the expression relating this arc length to the grid control function ϕ is

$$S_{\xi\xi} + \phi S_\xi = 0 \tag{10.25}$$

In a similar fashion, if arc length along the constant ξ boundaries is denoted by N, the equation for the relationship between N and the grid control parameter ψ is given by

$$N_{\eta\eta} + \psi N_\eta = 0 \tag{10.26}$$

Since S and N represent arc length along the boundaries, the values of (x, y) specified on the domain boundaries permit S and N to be determined. Finite-difference forms of the above two equations may be used to find the values of ϕ and ψ that are needed to determine the interior grid from the Thomas and Middlecoff (TM) form of the Thompson scheme. The interior values for ϕ and ψ are found by interpolating the boundary values onto the interior. A simple Lagrange interpolation is usually adequate.

The interior point distribution or clustering is determined by either P and Q in the Thompson formulation or by ϕ and ψ in the TM formulation of the Poisson grid generation equations. In order to control grid point distribution on the interior of the domain, it is important to understand how the construction of these grid control functions influences grid point location. In the original TM formulation, the values of ϕ and ψ were determined from the boundary and interpolated to determine the interior distribution. Next we discuss why the values of ϕ and ψ found from the approximations at the boundaries result in control of the grid points.

Anderson (1987) examined the TM form of the Poisson grid generation equations written along the constant coordinate lines without the assumption of orthogonality and zero curvature of the intersecting family. The resulting equations show that

$$S_{\xi\xi} + S_{\xi}\left[\phi - (\mu_{\xi} - 2\nu_{\xi}) \cot \theta - \frac{S_{\xi}\nu_{\xi}}{N\eta \sin \theta}\right] = 0 \tag{10.27}$$

and

$$N_{\eta\eta} + N_{\eta}\left[\psi + (\nu_{\eta} - 2\mu_{\eta}) \cot \theta + \frac{\mu_{\xi}N_{\eta}}{S_{\xi} \sin \theta}\right] = 0 \tag{10.28}$$

In these expressions, the first terms inside the square brackets are the same as the TM terms that are associated with the orthogonality and local curvature of the grid. The values of ν and μ represent the local inclination of constant ξ and η lines, respectively, and θ is the angle of intersection between the two families of curves. If the grid control parameters are sufficiently large in comparison with the other terms, the grid will be determined primarily by the values of ϕ and ψ. The governing equations for the arc lengths are then consistent with the TM formulation. The expressions given by Eqs. (10.27) and (10.28) are equidistribution laws, and the values of the grid control parameters are related to weight functions for this equidistribution. Consider the equidistribution of a weight function w in the discrete form

$$(\Delta S)w = \text{const} = C \tag{10.29}$$

where ΔS is the distance between any two mesh points along a constant η curve. If ΔS is large, w is small, and vice versa. This shows that control of the mesh spacing can be attained by correctly formulating the weight function. The continuous equivalent of the discrete equidistribution law may be written

$$S_{\xi}w = C \tag{10.30}$$

where the arc length derivative is now controlled by the weight function. If this equation is differentiated, we obtain

$$S_{\xi\xi} + S_{\xi} w_{\xi}/w = 0 \tag{10.31}$$

This is similar to the form of the original TM equation and shows that the TM method of finding the correct values of ϕ and ψ is an approximate equidistribution law with

$$\phi = w_{\xi}/w \tag{10.32}$$

The grid spacing control described above shows why control can be exercised by proper construction of the weight functions, or equivalently, the values of ϕ and ψ. Geometric functions that provide clustering near points or lines have been developed and are generally written in the form of an exponential (Thompson, 1975, 1980). A function that clusters near the line $\eta = \eta_j$ is of the form

$$\psi(\xi, \eta) = -A \operatorname{sgn}(\eta - \eta_j) e^{(-B|\eta - \eta_j|)} \tag{10.33}$$

where A and B are positive constants. To cluster near a point (ξ_j, η_j), the function has a correction to the distance and is of the form

$$\psi(\xi, \eta) = -A \operatorname{sgn}(\eta - \eta_j) e^{[-B\sqrt{(\xi - \xi_j)^2 + (\eta - \eta_j)^2}]} \tag{10.34}$$

where the constants A and B are taken to be positive. A corresponding expression may be written for ϕ.

Other techniques for the control of interior grid point locations with control of the orthogonality at the boundaries have been developed. Sorenson and Steger (1983) and Hilgenstock (1988) have presented methods for the control of the orthogonality at boundaries and the spacing of the first mesh interval on the interior. These procedures use an iteration scheme to attain orthogonality at the boundaries and satisfy the specified spacing. The orthogonality constraint is typically allowed to attenuate into the interior to prevent overspecification of the problem. If orthogonality at the boundary is a critical issue for a given application, these methods are very effective.

Other variations on the use of elliptic differential equations may be found in the literature. One of the interesting variations has been presented by Winslow (1981). In the original Poisson grid generation equations, the control of the arc lengths requires that two grid control parameters (in two dimensions) be specified. A simpler approach might only require the specification of the grid cell area or volume. This would necessitate prescription of only one parameter, regardless of the number of dimensions in the problem. Winslow (1981) called this parameter the diffusion and wrote the governing equations in the form

$$\nabla \cdot (D\, \nabla \xi) = 0 \tag{10.35a}$$

$$\nabla \cdot (D\, \nabla \eta) = 0 \tag{10.35b}$$

The parameter D may be specified to control the spacing of the computational coordinates, as can be seen if these equations are integrated over an arbitrary control volume. Anderson (1990) has shown analytically that the diffusion parameter is approximately proportional to the Jacobian of the transformation. Consequently, to specify the cell area or volume, the diffusion is set equal to the desired volume multiplied by a scaling factor. Of course, the simplicity of this approach must be traded off against the loss of ability to control anything except the cell volume.

General construction of orthogonal grids using elliptic methods is also of great interest, especially if the mesh spacing is also controlled. This may be accomplished in 2-D problems, and the works of Eiseman (1982), Arina (1986), and Sharp and Anderson (1991) are recommended reading.

Many other researchers have contributed to the state of the art in elliptic grid generation, and the interested reader is encouraged to consult the many conference publications on grid generation and the recent review paper by Thompson (1996).

10.3.2 Hyperbolic Schemes

Hyperbolic systems can also be used to generate grids. The advantage in using this type of partial differential equation is that the grid may be generated by solving the governing equations only once. This type of grid generation scheme is usually applied to problems with open domains consistent with the type of PDE describing the physical problem. The initial point distribution is specified along an initial data line with appropriate boundary conditions, and the solution is marched outward. The outer boundary at the end of the computation must be accepted wherever it occurs, with the shape that has resulted from the calculation. Steger and Sorenson (1980) described a method using a system of hyperbolic equations to generate a mesh. They have proposed an arc length orthogonality scheme and a volume orthogonality method. Only the latter will be presented in detail here.

In a 2-D problem, the Jacobian of the transformation controls the magnification of area elements between the physical and computational planes. If we imagine that mesh spacing in computational space is given by $\Delta\xi = \Delta\eta = 1$, then the area elements are also one unit in size. The inverse of the Jacobian,

$$x_\xi y_\eta - y_\xi x_\eta = \mathrm{I} \tag{10.36}$$

then represents the area in physical space for a given area element in computational space. If I is specified as a function of position, then Eq. (10.36) can be used as a single equation specifying grid control in the physical plane. A second equation is obtained by requiring that the grid lines be orthogonal at the boundary in physical space. Along a boundary where $\xi(x, y) = \text{const}$, we may write

$$d\xi = 0 = \xi_x\, dx + \xi_y\, dy$$

or

$$\left.\frac{dy}{dx}\right)_{\xi=\text{const}} = -\frac{\xi_x}{\xi_y} = \frac{y_\eta}{x_\eta} \tag{10.37}$$

Along an η = const surface

$$\left.\frac{dy}{dx}\right)_{\eta=\text{const}} = -\frac{\eta_x}{\eta_y} = \frac{y_\xi}{x_\xi} \tag{10.38}$$

If we require that ξ and η surfaces be perpendicular, the slopes must be negative reciprocals. This requirement becomes

$$x_\xi x_\eta + y_\xi y_\eta = 0 \tag{10.39}$$

The system given by Eqs. (10.36) and (10.39) is linearized by expanding about a known state denoted by the tilde. Using this convention, we may linearize one of the terms in Eq. (10.39) as

$$\begin{aligned} x_\xi y_\eta &= (\tilde{x} + x - \tilde{x})_\xi(\tilde{y} + y - \tilde{y})_\eta \\ &= \tilde{x}_\xi \tilde{y}_\eta + \tilde{y}_\eta(x_\xi - \tilde{x}_\xi) + \tilde{x}_\xi(y_\eta - \tilde{y}_\eta) + O(\Delta^2) \\ &= \tilde{y}_\eta x_\xi + \tilde{x}_\xi y_\eta - \tilde{x}_\xi \tilde{y}_\eta + O(\Delta^2) \end{aligned} \tag{10.40}$$

If the other terms are linearized in a similar manner, we obtain

$$[A]\mathbf{w}_\xi + [B]\mathbf{w}_\eta = \mathbf{f} \tag{10.41}$$

where

$$\mathbf{w} = \begin{bmatrix} x \\ y \end{bmatrix}$$

$$[A] = \begin{bmatrix} \tilde{x}_\eta & \tilde{y}_\eta \\ \tilde{y}_\eta & -\tilde{x}_\eta \end{bmatrix} \qquad [B] = \begin{bmatrix} \tilde{x}_\xi & \tilde{y}_\xi \\ -\tilde{y}_\xi & \tilde{x}_\xi \end{bmatrix} \qquad \mathbf{f} = \begin{bmatrix} 0 \\ \mathrm{I} + \tilde{\mathrm{I}} \end{bmatrix} \tag{10.42}$$

The eigenvalues of $[B]^{-1}[A]$ must be real if the system is hyperbolic in the η direction. These eigenvalues are

$$\lambda_{1,2} = \pm\sqrt{\frac{\tilde{x}_\eta^2 + \tilde{y}_\eta^2}{\tilde{x}_\xi^2 + \tilde{y}_\xi^2}} \tag{10.43}$$

This shows that Eq. (10.41) is hyperbolic in the η direction and can be marched in η so long as $\tilde{x}_\xi^2 + \tilde{y}_\xi^2 \neq 0$.

The procedure to use in generating a grid with this scheme is to assume the body is the $\eta = 0$ surface and specify the distribution of points along the body. Next, the inverse Jacobian I in Eq. (10.36) is computed. Steger and Sorenson suggest that I be determined by laying out a straight line with length equal to that of the body surface (l) and distribute the body points on this line. Next, a line parallel to the first is drawn at an η = const surface as desired. Once this is done, the quantity I is easily determined by estimating the area elements of the

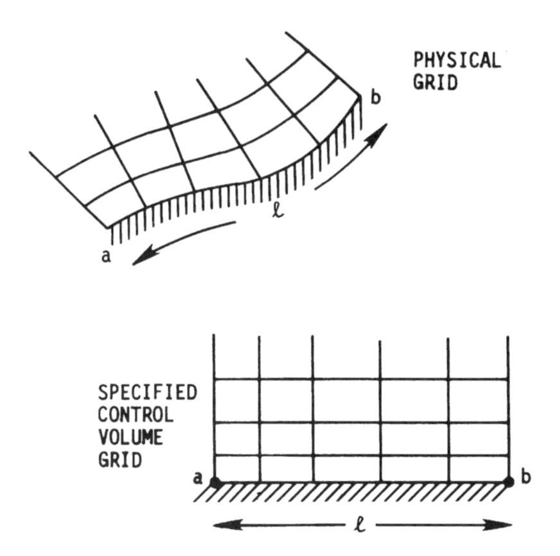

Figure 10.7 Area element computation.

grid. This procedure is illustrated in Fig. 10.7. The system of governing equations given by Eq. (10.41) is now solved using any standard method for solving systems of hyperbolic PDEs.

Since we specify I in this scheme, a smoothly varying grid is obtained if I is well chosen. However, poor selection of the I variation leads to possible "shocks" or discontinuous propagation of this information through the mesh. It is also true that discontinuous boundary data are propagated in the mesh. On the other hand, the mesh is orthogonal and is generated very rapidly. Figure 10.8 shows the grid generated about a typical airfoil shape. In this case, points have been clustered near the body in order to permit resolution of the viscous boundary layer.

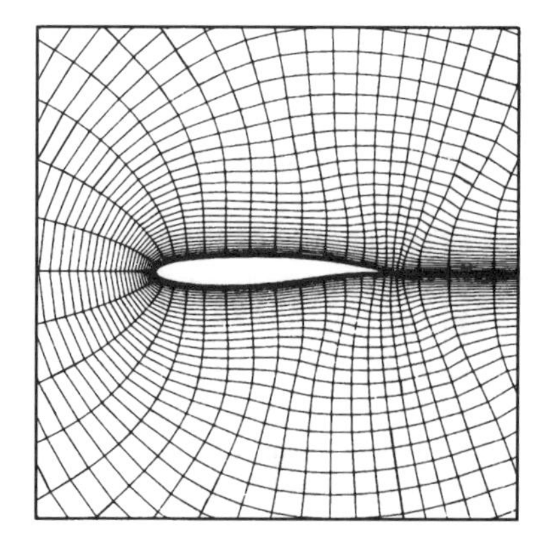

Figure 10.8 Grid for an airfoil configuration.

10.3.3 Parabolic Schemes

Parabolic PDEs are also solved by advancing the solution away from an initial data surface while satisfying boundary conditions at the ends of the domain. As was true for hyperbolic grid generators, parabolic generators should be useful in producing grids using a single-pass strategy. Nakamura (1982) and Edwards (1985) developed the basic ideas used in parabolic grid generation, and these techniques provide another way of producing acceptable grids.

The idea of parabolic grid generation is based on using the Laplace or Poisson grid generator and specially treating the parts of the equation that control the elliptic behavior. In order to understand the basic idea, consider Laplace's equation as the fundamental generating equation. The geometry of the domain is assumed to be consistent with that shown in Fig. 10.7 for the hyperbolic case. The initial data are given as the coordinates of all points along the $\eta = 0$ surface. The idea is to advance the solution for the grid outward from this surface subject to the boundary conditions along the minimum and maximum ξ edges.

If either Laplace's or Poisson's equation is used, the problem is elliptic, and the solution cannot be advanced in the η direction because the central differencing requires that information from the advanced $(j + 1)$ level be used. To illustrate this problem, consider the differencing of the second derivative,

$$\frac{\partial^2 \mathbf{r}}{\partial \eta^2}$$

If a second-order central difference is used, this is represented as

$$\frac{\mathbf{r}_{j+1} - 2\mathbf{r}_j + \mathbf{r}_{j-1}}{(\Delta\eta)^2}$$

If the integration of the equation is started with given data at the location indicated by $j - 1$, the unknown level is then indicated by j. However, the difference equations show that information from the next level at $j + 1$ is needed. We supply this information by assuming that this can be approximated by replacing any value at $j + 1$ by the outer boundary value, as originally suggested by Nakamura. It is also necessary to use this idea in evaluating the cross-derivative terms and first-derivative terms. When this approach is selected, the grid generation equation in discrete form may be solved as a marching problem, with the unknowns at the jth level constituting the values to be determined. At each step, the $j + 1$ level information is supplied by simply continuing to use the values on the outer boundary of the domain. This method creates a technique that allows a solution to the elliptic equation to be computed via a marching scheme. It has the conceptual advantage of producing a grid in a single pass.

The original parabolic methods of Nakamura (1982) and Edwards (1985) used the outer boundary to evaluate the necessary $j + 1$ point data in solving Laplace's equation. Other methods may be used to approximate the information needed at the advanced levels. One way is to use a reference grid (Noack, 1986), where an initial grid is constructed using any simple method, usually an algebraic scheme, and the reference grid point locations are used to supply the needed advanced point information for the solution of Laplace's equation. If, after the initial solution is computed, the reference grid is taken to be the first iterative pass from a Laplace or Poisson equation solution, an additional pass through the grid solver is made, and this process is repeated until a satisfactory grid is produced. This effectively becomes a solution to the elliptic equation that has not been completely converged. Hodge et al. (1987) also extended the idea of parabolic grid generation by using the Poisson equation in place of Laplace's equation and also provided some latitude in the selection of the direction that the equations could be parabolized.

At this point, no information has been given to suggest a means to control the grid spacing. In the works of Nakamura (1982) and Edwards (1985), grid control was accomplished by using nonuniform spacing in the computational domain. This variation in the cell sizes in the computational domain was used in solving a Laplace grid generation equation, providing control of mesh spacing in the physical domain. Some control of orthogonality was also provided by altering the location of the outer boundary points. This effectively is accomplished by altering the source terms that appear in the difference equations. The reference grid was used by Noack (1985) as a means of controlling the space of the grid points. Hodge et al. (1987) has given some guidance in the selection of the source terms in the Poisson equation for parabolic grid generation.

Parabolic grid generation has the advantage that no grid shocks occur as is possible in the hyperbolic case. In this sense, we expect grids to be relatively smooth. However, the effort required to set up the reference grid, or the outer boundary, as well as select a variable step size to control the grid point locations is time consuming. As with any method, there are advantages and disadvantages. However, if sufficient familiarity with these techniques is gained through experience, parabolic grid generation can be very effective.

10.4 VARIATIONAL METHODS

Variational methods have recently gained in popularity as a grid generation tool. When a function is minimized, several measures of mesh quality can be included. Brackbill and Saltzman (1980) and Brackbill (1982) have developed a technique for constructing an adaptive grid using a variational approach. In their scheme, a function that contains a measure of grid smoothness, orthogonality, and volume variation is minimized using variational principles. The smoothness of the transformation is represented by the integral

$$I_s = \int_D \left[(\nabla \xi)^2 + (\nabla \eta)^2 \right] dV \tag{10.44}$$

A measure of orthogonality is provided by

$$I_0 = \int_D (\nabla\xi \cdot \nabla\eta)^2 \mathrm{I}^3 \, dV \tag{10.45}$$

and the volume measure is given as

$$I_v = \int_D w \mathrm{I} \, dV \tag{10.46}$$

where w is a given weighting function.

The transformation relating the physical and computational domains is determined by minimizing a linear combination of the above three integrals. This linear combination with coefficient multipliers λ_v and λ_0 is written

$$I_t = I_s + \lambda_v I_v + \lambda_0 I_0 \tag{10.47}$$

In order to minimize I_t, the Euler-Lagrange equations must be formed (Weinstock, 1952). As an example, the smoothness measure, Eq. (10.44), may be written

$$I_s = \iint \left(\frac{x_\xi^2 + x_\eta^2 + y_\xi^2 + y_\eta^2}{\mathrm{I}} \right) d\xi \, d\eta \tag{10.48}$$

when the variables are interchanged and the integration is performed in computational space. If we construct the Euler-Lagrange equations corresponding to I_s, they are of the form

$$\begin{aligned} \left(\frac{\partial}{\partial x} - \frac{\partial}{\partial \xi}\frac{\partial}{\partial x_\xi} - \frac{\partial}{\partial \eta}\frac{\partial}{\partial x_\eta} \right) \left(\frac{x_\xi^2 + x_\eta^2 + y_\xi^2 + y_\eta^2}{\mathrm{I}} \right) &= 0 \\ \left(\frac{\partial}{\partial y} - \frac{\partial}{\partial \xi}\frac{\partial}{\partial y_\xi} - \frac{\partial}{\partial \eta}\frac{\partial}{\partial y_\eta} \right) \left(\frac{x_\xi^2 + x_\eta^2 + y_\xi^2 + y_\eta^2}{\mathrm{I}} \right) &= 0 \end{aligned} \tag{10.49}$$

If the differentiation is performed, these expressions may be written

$$\begin{aligned} A(\alpha x_{\xi\xi} - 2\beta x_{\xi\eta} + \gamma x_{\eta\eta}) - B(\alpha y_{\xi\xi} - 2\beta y_{\xi\eta} + \gamma y_{\eta\eta}) &= 0 \\ -B(\alpha x_{\xi\xi} - 2\beta x_{\xi\eta} + \gamma x_{\eta\eta}) + C(\alpha y_{\xi\xi} - 2\beta y_{\xi\eta} + \gamma y_{\eta\eta}) &= 0 \end{aligned} \tag{10.50}$$

The coefficients A, B, C, α, β, and γ are functions of the metrics, and their evaluation is left as an exercise (see Prob. 10.13). If

$$B^2 - AC \neq 0$$

these equations may be written as

$$\begin{aligned} \alpha x_{\xi\xi} - 2\beta x_{\xi\eta} + \gamma x_{\eta\eta} &= 0 \\ \alpha y_{\xi\xi} - 2\beta y_{\xi\eta} + \gamma y_{\eta\eta} &= 0 \end{aligned} \tag{10.51}$$

This is the form of the original mapping given by Winslow and is also the basic system of equations for Thompson's work. If I_t as defined in Eq. (10.47) is minimized, each of the integrals, I_v and I_0, contribute terms to a significantly more complicated set of Euler-Lagrange equations than those given in Eq. (10.51).

The use of a variational approach provides a solid mathematical basis for the grid but also entails additional effort in solving more PDEs. The Euler-Lagrange equations must be solved in addition to those governing the fluid motion. In the example shown here, the adaptive grid is constructed by implementing a new mesh after each iteration or time step and computing the grid speed by using a backward difference. The variational approach clearly offers a powerful method for constructing computational grids. The disadvantage is that a considerable effort must be expended in solving the equations that govern the grid generation. If a linear combination of the integrals of Eq. (10.47) is used, the λ's must also be selected. However, some remarkable results have been obtained with the proper choice of these coefficients.

The book by Knupp and Steinberg (1993) is a good source for a comprehensive treatment of the application of variational methods to the grid generation problem. Examples of the application of direct methods may be found in the literature, and typical of this is the work of Kennon and Dulikravich (1985) and Carcaillet (1986). Future applications of the variational approach will likely involve more work on direct minimization of integrals as opposed to the construction of the Euler-Lagrange equations. This simplifies the work by eliminating the laborious construction of the governing differential equations by using additional CPU time. Integrals representing a measure of desired qualities in a grid can be minimized with a number of well-proven methods that are readily available in the literature. The Euler-Lagrange equations can in practice be obtained with symbolic manipulators that also remove much of the difficulty in application if this classical approach is used. Variational techniques are a powerful way to formulate measures of grid quality and provide guidance in the construction of grid generation schemes. With continuing improvements in CPU power and inexpensive storage, more extensive use will be made of these methods.

10.5 UNSTRUCTURED GRID SCHEMES

Unstructured grid generation schemes have gained in popularity in recent years for a number of reasons. The increase in computer power and the reduction in memory costs have been major factors. One of the attractive features of unstructured mesh generation schemes is the promise they seemingly hold of ultimately providing a method that automates the grid generation process. In constructing grids using a structured approach, the grid must be segmented into blocks due to the topology of the domain and the configuration of interest, with the logical structure defined to provide appropriate connectivity. The flow solver must also be written to interpret and use the data format produced by the grid generator. This process of generating a structured mesh is a time-intensive task for engineers and scientists working in the field. Although good progress has been made in attempting to automate the blocking and subdivision for structured grids (Dannenhoffer, 1991, 1995, 1996), interactive grid generation is still used for the majority of structured mesh problems. When an unstructured approach is employed, defining the configuration of interest forms the most complex portion of the problem for the user, and the unstructured grid generator is employed to create the grid automatically. This is the case at least in concept, although in reality, the ability to generate grids automatically, in general, is still beyond the state of the art. For unstructured grids, the connectivity information stored is cell-to-cell as opposed to block-to-block, so additional storage is necessary when compared to the structured approach. However, the increase in available CPU power and memory makes the trade-off between CPU time and engineering hours favor the unstructured approach. There are other factors that may play an equally important role. One consideration is the solver efficiency. Due to the problem of random cell location and connectivity, unstructured solvers are usually not as computationally efficient as their structured counterparts. One must also try to construct cells where the volumes are as nearly equal or change very smoothly to avoid the problem of introducing errors in the solutions that are grid induced. This problem of the smoothly changing volume size is common to both techniques. Unstructured mesh schemes must also be monitored to reduce the thin or so-called high aspect ratio cells that are created in the generation process, since these cells contribute to increased errors.

Other considerations are of importance in the construction of grids for solution of flow problems. The grid point or cell densities that give adequate resolution for flow problems create difficulties for both structured and unstructured grids. For example, in the boundary layer, the use of structured mesh schemes naturally suggests a cell shape that is elongated in the flow direction. This configuration is consistent with the boundary layer assumptions, in that more cells appear in the normal direction as compared to the flow direction, where only small changes in the flow may occur. On the other hand, the use of unstructured grids, for example, triangles in a 2-D problem and tetrahedra in 3-D, requires a higher cell density in the boundary layer because the cells need to be as nearly equilateral (analogous to orthogonality in structured meshes) as possible in order to avoid grid-induced errors in the solution. The storage requirements are much larger for the unstructured grid. This can be visualized by imagining that a 2-D structured mesh is used as a base and the mesh is then triangulated by simply inserting the diagonal in each cell. In this example, the number of cells produced is larger by a factor of 2 for the unstructured result. In 3-D problems, the number of cells produced using this procedure is at least a factor of 5 larger. In addition, the cells produced may be long and narrow (high aspect ratio), and mesh refinement is then needed to reduce this aspect ratio.

In this section, the procedure for construction of a Delaunay (1931) mesh will be outlined using the Bowyer (1981) insertion scheme. This is intended to provide an introduction to some of the concepts associated with the logic for constructing unstructured grids.

10.5.1 Connectivity Information

As a starting point, consider the connected triangles shown in Fig. 10.9. We must determine what information is necessary to completely identify the cell and all of the neighbors of that cell in the computational mesh. In generating an unstructured mesh, the point locations are arbitrary, and we may choose to place them at any desired position. As in the structured case, each point must be identified. We consider a point insertion scheme where each point is independently inserted and the cell connectivity resulting from this insertion is determined. This suggests that points be identified sequentially as they are inserted. If 35 points have been inserted into the mesh, the next point that needs to be inserted is identified as number 36. In addition to the identification of the grid point number, the coordinates of this point must be known and stored as $[x(36), y(36)]$.

After a grid point is inserted into an existing mesh, logic for establishing the new connectivity is employed. Data that identify the grid points that form a given cell are needed. As each cell is formed, the cell is numbered, and the forming points for that cell are also stored. For example, the convention can be taken that the forming points for this 2-D example are numbered in a counterclockwise direction around each triangle. We number these as forming point one, fp1(ncell), and continue around the triangular cell including all three points. This is illustrated in Fig. 10.9. In this figure, the three triangles that constitute the cell structure are formed by using five points, which are numbered on the exterior of each triangle vertex. The number assigned to each cell is shown in parentheses on the interior, and the forming point convention shows the local identification of the forming points as 1, 2, 3 on the interior of each cell near the vertices where the forming points are located.

In addition, the neighbor cell information is needed. Cells are considered to be neighbors if they share a common face. As a convenient convention, we may

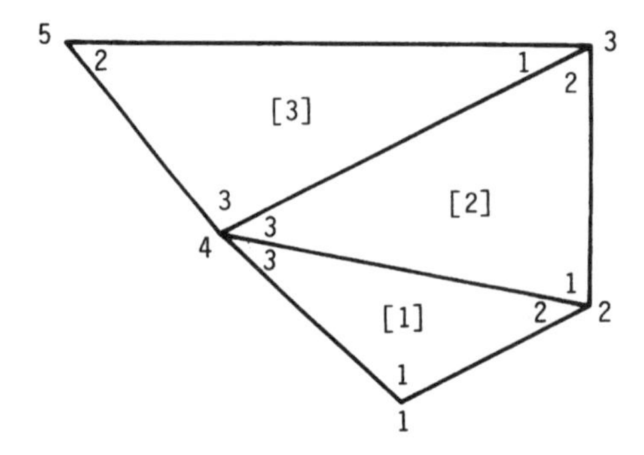

Figure 10.9 Notation for unstructured cells.

identify the first neighbor of a cell as that cell opposite to forming point 1. For example, if cell 2 is given as shown in Fig. 10.9, the second neighbor is identified as cell 1, and the first neighbor is cell 3. The neighbor information for this figure is given as

$$\text{nei1}(1) = 2$$

$$\text{nei1}(2) = 3, \qquad \text{nei2}(2) = 1$$

$$\text{nei2}(3) = 2$$

In general, each triangular cell will have three neighbors. In this example, the first and third cells have only one and the second cell has two neighboring cells. The forming point information for these cells would be stored as

$$\text{fp1}(1) = 1 \qquad \text{fp2}(1) = 2 \qquad \text{fp3}(1) = 4$$

$$\text{fp1}(2) = 2 \qquad \text{fp2}(2) = 3 \qquad \text{fp3}(2) = 4$$

$$\text{fp1}(3) = 3 \qquad \text{fp2}(3) = 5 \qquad \text{fp3}(3) = 4$$

The data contained in the cell-numbering scheme, the neighbors, and the forming points are sufficient to establish any of the parameters that are needed in the mesh or in a computational fluid dynamics (CFD) code using the cell structure derived from this mesh.

The discussion in this section regarding the information storage for a triangular mesh is also applicable to a 3-D tetrahedral cell structure except that the convenience of some of the numbering conventions identifying the neighbors may not apply. The same data are required when rectangular cells are mixed with triangles in a 2-D case. Of course, when mixed-cell hybrid grids are used, the flow solver must be written to accept any cell structure and any arbitrary connectivity.

10.5.2 Delaunay Triangulation

When an unstructured grid is constructed, the task is simplified if a fixed set of rules is followed, leading to a grid that has certain attractive properties. The Delaunay triangulation provides a grid where a fixed set of rules applies to the construction, and the grid properties include the following:

1. Given a set of points, the triangulation is unique.
2. The triangulation produces the most equilateral mesh for the given point set.
3. The grid point generation and the triangulation are decoupled.

The origins of this approach go back to the work of Dirichlet (1850), where a technique for decomposing a given domain into a set of convex polygons was studied. The geometric dual of this construction is called the Delaunay triangulation. The Delaunay triangulation has a number of implementations and includes the diagonal swapping (Cendes et al., 1985), the Bowyer insertion scheme (Bowyer, 1981), and the sweepline method (Fortune, 1987). While this approach has the advantages enumerated above, there are disadvantages as well.

These include the following:

1. Lack of uniqueness when four points lie on a circle and the counterpart in 3-D
2. The complex logic required to preserve boundaries
3. The lack of uniqueness resulting from the numerical implementation of the analytical theory of the triangulation
4. The solution errors associated with high aspect ratio or elongated cells (slivers)

These issues will become clear as the details of the triangulation emerge.

Given a point set $P = p_i(x_i)$ that is not colinear and does not have four points that lie on a circle, the set of points that is closer to vertex v_i than any other vertex is called the Voronoi polygon (Voronoi, 1908). This is illustrated in Fig. 10.10, where the Voronoi polygons are shown for a finite set of points. The dashed lines are the Voronoi polygons formed by constructing cells with sides corresponding to the perpendicular bisectors of the line segments in the triangulation. The vertices of the polygons are formed from the intersection of the perpendicular bisectors of the lines connecting the points, $P = p_i(x_i)$. As the mesh grows, more cells are added due to the addition of more line segments connecting the points in the triangulation. As the tesselation continues, the boundary polygons are those on the convex hull of the domain. The complete set of polygons including those closed on the interior and those open on the boundary of the domain is referred to as the Voronoi tesselation of the domain.

When the nuclei (point p_i contained in the polygon) of the Voronoi polygons are connected to the two nearest neighbors, the resulting structure is called the Delaunay triangulation or Delaunay tesselation. This is also shown in Fig. 10.10. In CFD, the cell structure used for a finite-volume solution of a flow problem may be applied to either the Voronoi polygons or the Delaunay tesselation. For the 2-D discussion presented here, the triangular cells always have three cell faces, while the Voronoi cells, sometimes called the mesh dual, may have a random number of edges. This suggests that a flow solver that uses the Delaunay triangulation for control volumes may have simpler logic and be easier to construct.

Although the discussion has centered on 2-D space, the ideas are also applicable to 3-D. In that case, the edges of the cell are planes, and the cells are tetrahedra or polyhedra. The increase in complexity of the grid generation problem in going from 2-D to 3-D is dramatic. Consequently, only 2-D cases will be considered here.

The circumcircle test is the simplest method to construct the Delaunay mesh and determine the connectivity of a set of points. For the planar case, three points determine a circle. For a triangular cell, the cell is a valid cell if no other point falls within the circle defined by the forming points of the circle. This is the standard test used in the Bowyer algorithm to complete the connections for the Delaunay tesselation. Figure 10.11 shows four points that

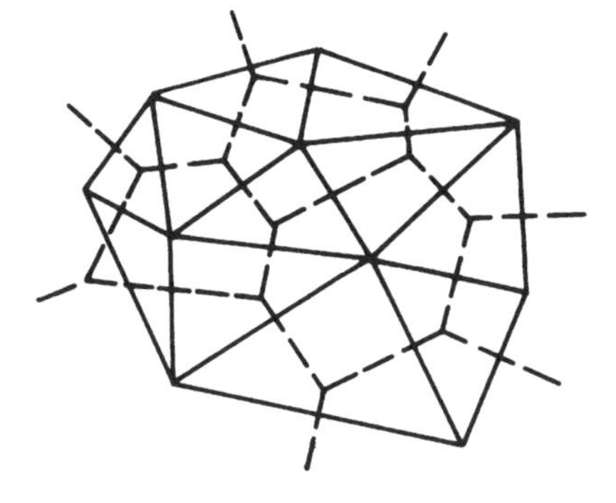

Figure 10.10 Voronoi (dashed lines) and Delaunay (solid lines) tesselations.

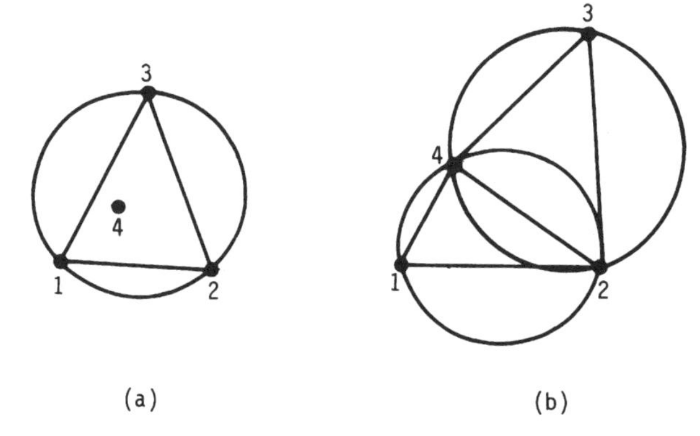

Figure 10.11 Circumcircle test. (a) Incorrect connectivity. (b) Correct connectivity.

are to be connected but the proper connections must be established. The first connection shows that the circle formed by connecting points 1, 2, and 3 encloses point 4. This violates the circle criterion, and other connections must be made. The proper connections and the associated circumcircles are also shown where points 1, 2, and 4 form one cell, and points 2, 3, and 4 form the second cell.

10.5.3 Bowyer Algorithm

Bowyer (1981) developed a scheme that can be used to triangulate a set of points. This approach is usually termed the "Bowyer insertion algorithm" because the scheme is based on inserting points into a valid Delaunay mesh and retriangulating the mesh. The basic technique relies on the circle test and a series of data tree searches to determine the new connectivity. The search can be efficiently carried out and the method can be used to refine the grid by simply inserting additional points, and finding the new connectivity as each point is inserted. The Bowyer algorithm consists of a number of steps as described below.

Step 1. Generate a set of grid point locations that are desired for the domain of interest. This set should include the boundary points and all of the interior points. The points can be generated a number of ways including the following:
   a. A random number generator
   b. Structured grid generator such as TFI or elliptic generation
   c. A self-adjusting method that determines the largest cell in the mesh, the highest aspect ratio, or some other characteristic of the generated grid and inserts a point at the circumcenter of the circle or some other predetermined location to refine the mesh to the desired level.
   d. Methods based on domain decomposition (discussed in more detail below)

Step 2. Create an initial supertriangle that completely encloses the entire domain. This may be any valid triangulation and the simplest geometry is a supertriangle or a rectangle that is triangulated.

Step 3. Insert a mesh point from the list established in Step 1 in the existing triangulation, and delete the first triangle that fails the circumcircle test. This will be the cell where the point is inserted.

Step 4. Initiate a search of the neighbors of the first deleted cell to determine if any other neighbor cells have violated circumcircles. If a neighbor cell is deleted, the common face between that neighbor and the first deleted cell must be removed, and the search proceeds through the neighbors of this cell. The tree search continues until the complete list of deleted cells is compiled.

Step 5. Establish the new connectivity by connecting the newly inserted point with the boundary points of the cavity created by the deleted cells. Add each of the new cells to the list of valid triangles.

Step 6. Repeat this procedure, starting with Step 3, until all the grid points generated in Step 1 have been inserted.

The Bowyer insertion technique described above provides a correctly triangulated mesh for convex domains. Unfortunately, most of the domains that surround practical shapes are not convex. However, the unstructured grid can be constructed by beginning with the superstructure and filling the entire superstructure as well as the body interior and including the boundary of the physical domain. This valid triangulation must undergo a postprocessing phase to remove all triangles interior to the body and those triangles between the outer boundary of the superstructure and the outer boundary of the domain.

One of the problems that must be addressed is that of preserving the integrity of the body surface when a given set of points is triangulated. When the set of points for a domain is compiled, the outer boundary of the domain and the boundary of the body are usually the first points selected to insert into the initial superstructure. The interior points between these two boundaries may be determined by any of the methods noted in step 1. However, the integrity of the body surface must be preserved, and without some special checks, this cannot be

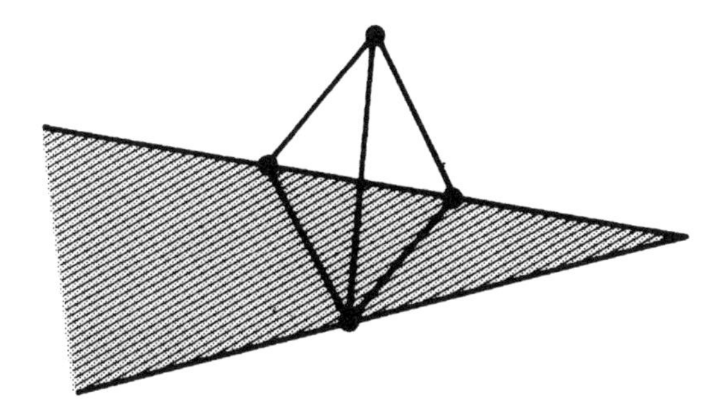

Figure 10.12 Surface fidelity violation near an airfoil trailing edge.

guaranteed. Figure 10.12 shows a connection across a body surface near an airfoil trailing edge where an improper connection has been made. This must be corrected as part of the postprocessing phase of the grid generation. Two popular ways of dealing with this problem are diagonal swapping routines or point insertion schemes that delete and reform the triangles so that the surface segment forms a cell edge.

An example of a grid constructed for an NACA 0012 airfoil is given in Fig. 6.14. This grid shows the mesh density increasing near the body in order to provide the resolution desired for accurate pressure calculations using an Euler code. The process of constructing this grid follows the steps given above, and a good reference source for additional details on constructing unstructured grids using the Bowyer scheme is that of Holmes and Snyder (1988).

Baker (1987) studied the Bowyer insertion scheme and has shown that the method is based upon two theorems.

Theorem 1 Given a Delaunay triangulation T of a planar set of points S, introduce a new point $p \in S$ and remove all triangles that fail the Delaunay circle test. All the edges of the Delaunay cavity are visible from point p.

Theorem 2 The retriangulation of a Delaunay cavity, by joining the point p to each of the boundary points of the cavity is Delaunay.

One additional issue addressed by Baker deals with the problem of precision in applying the circle test. Since the circle test is performed with a computer with finite accuracy, the precision of the test will determine whether or not the circle test is satisfied. As a consequence, if care is not exercised, the test may actually hinge on the round-off error of the machine. Baker has stated the following theorem, regarding the precision of this test.

Theorem 3 Let p_i be a finite point set, and let $d(p_i, p_j)$ represent the Euclidian metric. If L represents $\max[d(p_i, p_j)]$ and ϵ represents $\min[d(p_i, p_j)]$, the precision of the floating point accuracy used in the circle test with Delaunay triangulation must be greater than ϵ^2/L^2.

This places a substantial restriction on the precision of the test procedure. When standard engineering workstations are used, it is imperative that double-precision arithmetic be employed.

In addition to the point insertion scheme provided by the Bowyer approach, a Delaunay mesh may be constructed by using the sweepline algorithm first suggested by Fortune (1987). This is an advancing front method that builds Delaunay cells as the front proceeds over the domain including the configuration that is the object of the study. For details on the application of this scheme, the work of Fang et al. (1993) for the 2-D case and Fang (1995) for the 3-D case is recommended.

10.6 OTHER APPROACHES

In Section 10.5.3, the Bowyer insertion scheme was outlined as a technique for constructing a Delaunay mesh. Other important methods of constructing unstructured grids have been developed using advancing fronts. While these schemes forego the Delaunay criterion, they have been used with good success in a variety of applications (Löhner and Baum, 1990; Löhner and Parikh, 1988). With this approach, the grid is advanced by adding cells at the front as it advances into the domain. These fronts are usually started from known structures such as a body or other boundary and may be composed of either structured or unstructured cells. When advancing fronts collide, rules are needed for treating the collisions and constructing cells under such circumstances. Unfortunately, these rules are constructed to treat individual exceptions, and general theorems providing construction rules are difficult to identify. However, advancing front

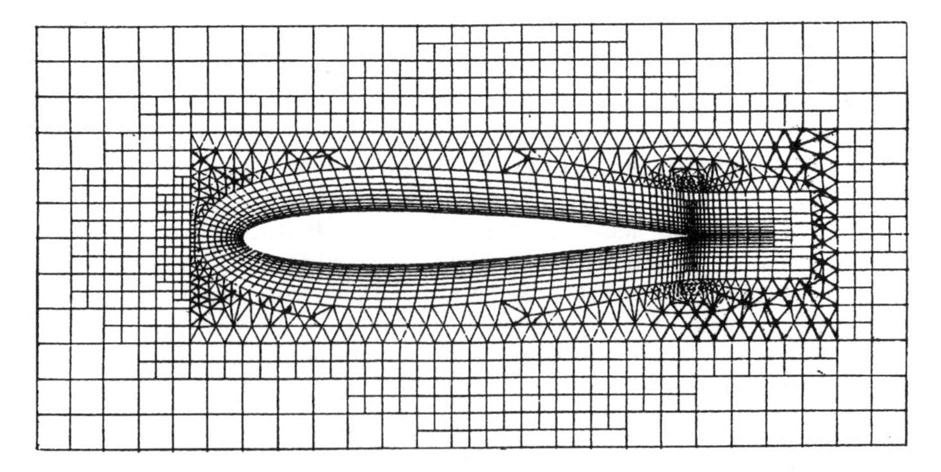

Figure 10.13 Hybrid grid for NACA 0012 airfoil.

schemes have been used to grid very complex configurations, and these grids have been used with success in solving very difficult fluid mechanics problems.

The concept of hybrid grids is also of great interest in applications in CFD. The works of Kallinderis and Ward (1993), Kallinderis (1996), and Noack et al. (1996) are representative of the state of the art. These methods are mixed, in the sense that combinations of structured and unstructured grids are used to completely cover the domain. Regions around bodies are usually grided with a structured body-conforming scheme, and the zones away from walls are covered with unstructured sections. The interface between these different zones requires logic to provide the connectivity to close the problem. This approach shows great promise as a technique to simplify the automatic grid generation problem. As an example, Fig. 10.13 shows a hybrid grid around an NACA 0012 airfoil.

The use of rectangular grids has also been of interest for some time in the CFD field. These schemes are based on using quadtree or octree data structures (Yerry and Shephard, 1983, 1984). Rectangular grid schemes have the promise of completely automating the grid generation process. The idea involves recursive subdivision of a domain until the body surface is identified at the highest refinement level in the mesh. After the refinement level is satisfied, the body-surface cells are then specially treated by considering the way the body slices these cells (Karman, 1995a, 1995b; Coirer and Powell, 1995). This is a very natural scheme to consider and forms an automated way to grid a domain once the logic for the sliced cells is complete. However, the problem of storage and data management must be carefully considered. The use of domain subdivision methods requires large storage, and usually, long computation times are necessary. The problem becomes clear when considering the resolution required to solve for flow in the turbulent boundary layer of a typical vehicle. In many cases, the refinement in the boundary layer may be extreme to achieve the desire level of refinement. A sketch of the idea used with rectangular cell

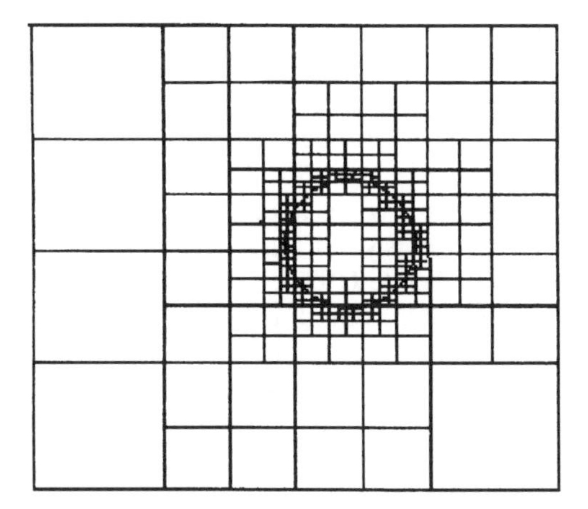

Figure 10.14 Domain subdivision using rectangular cells.

subdivision is shown in Fig. 10.14. The body has a nearly circular geometry, and note that the subdivision is completed so that only one level of refinement is accomplished at any cell boundary. This is desirable from both a logic and a solver accuracy point of view. The effectiveness of this approach is apparent when one considers the simplicity of the concept applied to very complex objects. Again, the major difficulty is in defining logic to produce correct body-surface cells. This idea of subdivision is not restricted to rectangular cells but can also be applied using other geometric structures as a base.

As is true in most of the hybrid and unstructured schemes, the issue is one of deciding what trade-off of labor hours versus CPU time is a good one. If an automated approach using any of these successful schemes can be utilized to completely solve for the flow around a vehicle in a matter of days, this becomes an attractive option. This is especially true when the calculation can be completed on an engineering workstation. Of course, the flow solver must be written to take advantage of the special features of these grid generation schemes. Future research will define the optimum use of these techniques.

10.7 ADAPTIVE GRIDS

Techniques for generating grids as a prelude to numerically solving a PDE were presented in the previous section. One problem in solving a PDE with this approach is that the grid is constructed and points are distributed in the physical domain before details of the solution are known. As a consequence, the grid may not be the best one for the particular problem.

Adaptive methods for solving PDEs have been developed to aide in increasing the accuracy of computed solutions. These methods have been classified into three categories. These categories naturally appear if one views the adaptivity as a means of reducing some measure of the global error in the solution.

In finite-element theory (Oden, 1988), the adaptive method is referred to as an h method if mesh refinement is used, an r method if the number of grid points is fixed but is redistributed, and a p method if the order of the solution scheme is increased. In finite-difference and finite-volume applications, the h and r ideas are the most popular due to the way these methods are constructed.

The adaptive grid strategies that are followed when a fixed number of points are redistributed to improve the solution are usually based on an application of the equidistribution scheme outlined previously in this chapter. Early applications of this idea include the works of White (1982), Dwyer et al. (1979, 1980), and Gnoffo (1980). These authors applied the equidistribution idea in one dimension in solving a variety of problems in fluid mechanics and heat transfer. The application of the equidistribution idea to multidimensional problems has been accomplished in several ways. The simplest to understand are the Poisson grid generators with control functions based on equidistribution of a weight function as given in Eq. (10.32), or using Eqs. (10.35*a*) and (10.35*b*) with the diffusion set equal to a constant times the desired cell volume. Other approaches that have been applied with success are the spring analogy of Nakhashi and Diewert (1986) and the application of the strict equidistribution law to multidimensional problems by Anderson (1983) and Eiseman (1983). Variational methods as outlined in Section 10.4 based on the original work of Brackbill and Saltzman (1982) are also useful in constructing adaptive grids. These methods have been applied to structured meshes in most cases. Mesh redistribution schemes have also been applied to unstructured meshes. The difficulty is that the connectivity must be altered for these cases if the mesh point movement is very large. While the mechanics of changing this connectivity are automatically accounted for in the grid generation algorithm, the associated redistribution of the flux terms for the fluid dynamic variables may not be as easily accomplished when compared with the structured grid approaches. Other ways of r adaption of unstructured grids based on measures of solution quality can be cited. Hagmeijer and Kok (1996), Catherall (1996), Carpenter and McRae (1996), and Riemslagh and Vierendeels (1996) give representative results using these methods.

Adaptive grid construction is applied to both steady flow problems and to time-accurate flow calculations. For adapting grids in a steady flow problem, the grid is adapted or refined after a predetermined number of iterations or time steps have been taken. When the solution converges, the grid will stop adjusting to the changes that occur and will reflect the properties that appear in the solution that have been used to calculate the grid motion and refinement. In the time-accurate case, the grid point motion and refinement are performed in conjunction with the time-accurate solution of a physical problem. This requires the time-accurate coupling of the PDEs of the physical problem and those describing the grid movement or the mesh refinement.

Grid movement schemes can produce substantial improvements in solution quality. However, mesh refinement methods promise significantly better results because no limitations exist that define the limit on grid resolution that can be attained. The Bowyer scheme for generating an unstructured mesh was presented in Section 10.5.3 and can be used as a simple technique to refine a grid to the desired level. The idea of mesh refinement can be applied without limitation to any grid. The idea works if one starts with structured, unstructured, or hybrid grids that are formed of arbitrarily shaped cells. Of course, the use of grid refinement necessitates storage of information as if the grid was unstructured even though the original grid, before refinement, was structured. The rectangular grid schemes that use subdomain division are based on refinement of the mesh until a desired cell size is achieved. The division of cells using either triangular or rectangular shapes (2-D case) relies on splitting an edge or edges of existing cells. When triangular grids are used, the splitting results in construction of new cells where connecting nodes produce either four new cells when each edge is divided or two new cells when only one edge is split. In the case of rectangular cells, the subdivision of cells based on edge splitting leads to extra nodes that appear in the center of edge segments when a cell is divided on one side. This does not create problems, since the numerical method is assumed to be cell based and the flux terms can be associated with the parent cells and redistributed. The major issue for refinement is correctly managing the database associated with the changing mesh and selecting an appropriate criterion to use to determine the need for additional subdivision. Numerous recent papers show results that illustrate the use of these methods. Examples of recent work include Schneiders (1996), Kallinderis et al. (1993, 1995, 1996), Noack et al. (1996), and Smith and Johnson (1996).

PROBLEMS

10.1 Verify the equations for the transformation metrics given in Eq. (10.5).
10.2 Suppose that a physical domain is defined on the interval $0 \leqslant x \leqslant 1$ with an upper boundary given by

$$y_{\text{upper}} = 1 + 0.2 \sin(\pi x)$$

and a lower boundary given by

$$y_{\text{lower}} = 0.1 \cos(\pi x)$$

Devise a transformation that provides a uniform distribution of mesh points between the upper and lower boundaries. Use a simple normalizing transformation.
10.3 In Prob. 10.2 the interval was defined by two $x = \text{const}$ lines. If the left boundary is defined as

$$y_{\text{L}} = 10x$$

and the right boundary is defined by

$$y_{\text{R}} = 4(x - 1)$$

with the same upper and lower boundaries, determine a normalizing transformation to provide equal grid spacing in the physical plane. Why does this become so much more complicated than the transformation of Prob. 10.2?
10.4 Work Prob. 10.2 using the algebraic method demonstrated in Example 10.3. Use linear functions to verify your results and then use cubic functions.
10.5 Work Prob. 10.3 using linear functions with the method given in Example 10.3.
10.6 Suppose that you are required to solve a system of PDEs in (t, x, y) on the rectangular domain

$$0 \leqslant x \leqslant 1$$
$$0 \leqslant y \leqslant 1$$

A surface $F(t, x, y) = 0$ is to be tracked similar to a shock and computed as part of the solution. Devise a transformation that converts the physical plane into two rectangular computational domains joined at the boundary $F(t, x, y) = 0$. Assume that the surface is smooth and always intersects the left and right boundaries in physical space.
10.7 Verify the transformation given in Eq. (10.14) and the associated f_i functions.
10.8 The Thompson scheme for generating grids is based upon Eq. (10.21). Derive the computational domain equations given in Eq. (10.22).
10.9 Show that the mapping governed by the differential equations

$$\nabla^2 \xi = 0$$

$$\nabla^2 \eta = \frac{1}{\eta}$$

maps uniformly spaced circles in physical space into a uniform rectangular grid in the computational plane.
10.10 Show that a solution of the Cauchy-Riemann equations is a solution of Laplace's equation but the reverse is not necessarily true.

10.11 Repeat Prob. 10.3 and use the Thompson technique to obtain the mapping using the method of Middlecoff and Thomas [Eq. (10.25) and Eq. (10.26)] to effectively determine P and Q. Discuss your result, and point out any difficulties encountered in establishing your choice in selecting ϕ and ψ.
10.12 Construct the Euler-Lagrange equations that result when a mesh is obtained using a minimization of the orthogonality measure given by Eq. (10.45).
10.13 Complete the differentiation indicated in Eq. (10.49), and determine the coefficients identified in Eq. (10.50).
10.14 Consider the 1-D form for the Poisson equation. Use a central difference for the second derivative and estimate the maximum value of the control function that may be used before grid crossing occurs.
10.15 The equations given in Prob. 10.2 define the upper and lower boundaries of a physical domain. If the right and left boundaries are straight lines connecting the end points of these defining equations, use transfinite interpolation with Lagrange polynomials to construct a grid covering this domain.
10.16 Work Prob. 10.15 using Hermite polynomials. Show that the proper choice of the coordinate line slopes at the boundaries must be made to prevent grid crossing.
10.17 Develop the TFI expression that may be used to grid an open domain where the outer boundary is not prescribed. Construct a numerical example illustrating this application.
10.18 You have been assigned the task of constructing a grid for a NACA 0012 airfoil. Use parabolic grid generation with Laplace's equation to construct this grid. Select the outer boundary to be uniformly two chord lengths from the body, and use the outer boundary as the forward point in the difference approximation.
10.19 Construct an algorithm using the Bowyer insertion scheme to correctly triangulate a given set of points. Assume the initial Delaunay triangulation is given by a single triangle and insert a total of 10 points to verify your work.
10.20 Using the computer code from Prob. 10.19, insert points in the supertriangle defining a rectangular outer boundary enclosing a NACA 0012 airfoil.
10.21 With the code developed in Prob. 10.19, insert points between the airfoil boundary and the outer boundary to provide adequate resolution to complete a flow field computation. Base the refinement on selecting the largest triangle, and insert a point at the circumcenter of this triangle. Perform this exercise for several grid point densities. Discuss your results and include any problems you identify with this technique.
10.22 Devise a method to eliminate cells interior to the airfoil and exterior to the rectangular outer boundary using the code from the previous problem. Be sure to reorder the cell structure as a continuous list for ease of use with a flow solver.

APPENDIX

A

SUBROUTINE FOR SOLVING A TRIDIAGONAL SYSTEM OF EQUATIONS

Subroutine SY solves a tridiagonal system of equations following the Thomas algorithm described in Chapter 4. To use the subroutine, the equations must be of the form

$$\begin{bmatrix} D_{IL} & A_{IL} & & \\ B_I & D_I & A_I & \\ & \ddots & \ddots & \ddots \\ & & B_{IU} & D_{IU} \end{bmatrix} \begin{bmatrix} U_{IL} \\ U_I \\ \vdots \\ U_{IU} \end{bmatrix} = \begin{bmatrix} C_{IL} \\ C_I \\ \vdots \\ C_{IU} \end{bmatrix} \tag{A.1}$$

The call statement for subroutine SY is of the form

$$\text{CALL SY}(IL, IU, B, D, A, C)$$

where B, D, A, and C are the array names for the singly subscripted real variables $B(I)$, $D(I)$, $A(I)$, $C(I)$. The variables IL and IU are unsubscripted integer variables. The arrays must be defined for subscripts ranging from IL to IU according to

B, Coefficient behind (to the left of) the main diagonal
D, Coefficient on the main diagonal
A, Coefficient ahead (to the right of) the main diagonal
C, Element in the constant vector

The equations in the system are ordered according to the value of the subscript. The variable *IL* corresponds to the subscript of the first equation in the system and *IU* corresponds to the subscript of the last equation in the system. The number of equations in the system is $IU - IL + 1$. *The solution vector*, **U**, *is returned to the calling program in the* C *array*. That is, the constant vector **C** is overwritten in the subroutine with the solution. The *D* array is also altered by the subroutine. *A* and *B* remain unchanged.

LISTING OF SUBROUTINE SY

```
C...
      SUBROUTINE SY(IL,IU,BB,DD,AA,CC)
      DIMENSION AA(1),BB(1),CC(1),DD(1)
C...
C...SUBROUTINE SY SOLVES TRIDIAGONAL SYSTEM BY ELIMINATION
C...IL = SUBSCRIPT OF FIRST EQUATION
C...IU = SUBSCRIPT OF LAST EQUATION
C...BB = COEFFICIENT BEHIND DIAGONAL
C...DD = COEFFICIENT ON DIAGONAL
C...AA = COEFFICIENT AHEAD OF DIAGONAL
C...CC = ELEMENT OF CONSTANT VECTOR
C...
C...ESTABLISH UPPER TRIANGULAR MATRIX
C...
      LP = IL+1
      DO 10 I = LP,IU
      R = BB(I)/DD(I-1)
      DD(I) = DD(I)-R*AA(I-1)
   10 CC(I) = CC(I)-R*CC(I-1)
C...
C...BACK SUBSTITUTION
C...
      CC(IU) = CC(IU)/DD(IU)
      DO 20 I = LP,IU
      J = IU-I+IL
   20 CC(J) = (CC(J)-AA(J)*CC(J+1))/DD(J)
C...
C...SOLUTION STORED IN CC
C...
      RETURN
      END
```

APPENDIX

B

SUBROUTINES FOR SOLVING BLOCK TRIDIAGONAL SYSTEMS OF EQUATIONS

The subroutines described here for solving block tridiagonal systems of equations were provided by Sukumar R. Chakravarthy of Rockwell International Science Center. Subroutine NBTRIP solves a block tridiagonal system of equations of the form

$$\begin{bmatrix} B_{IL} & C_{IL} & & \\ A_I & B_I & C_I & \\ & \ddots & \ddots & \ddots \\ & & A_{IU} & B_{IU} \end{bmatrix} \begin{bmatrix} X_{IL} \\ X_I \\ \vdots \\ X_{IU} \end{bmatrix} = \begin{bmatrix} D_{IL} \\ D_I \\ \vdots \\ D_{IU} \end{bmatrix} \tag{B.1}$$

Subroutine PBTRIP solves a periodic block tridiagonal system of equations in the form

$$\begin{bmatrix} B_{IL} & C_{IL} & & A_{IL} \\ A_I & B_I & C_I & \\ & \ddots & \ddots & \ddots \\ C_{IU} & & A_{IU} & B_{IU} \end{bmatrix} \begin{bmatrix} X_{IL} \\ X_I \\ \vdots \\ X_{IU} \end{bmatrix} = \begin{bmatrix} D_{IL} \\ D_I \\ \vdots \\ D_{IU} \end{bmatrix} \tag{B.2}$$

The block matrices A, B, and C are $N \times N$ matrices at every point I with N being an integer greater than 1. Note that for $N = 1$, the Thomas algorithm of Appendix A can be employed. The right-hand side vector D has length N at each point I. The total number of I points at which the matrices are defined (denoted by NI) is given by

$$NI = (IU - IL + 1) \tag{B.3}$$

The matrices A, B, and C are dimensioned as

$$A(N, N, NI)$$
$$B(N, N, NI)$$
$$C(N, N, NI)$$

while the vector D is dimensioned as

$$D(N, NI)$$

The call statement for subroutine NBTRIP is

CALL NBTRIP($A, B, C, D, IL, IU, ORDER$)

with arguments defined by

A, Subdiagonal block matrix
B, Diagonal block matrix
C, Superdiagonal block matrix
D, Right-hand side vector
IL, Lower value of I for which matrices are defined
IU, Upper value of I for which matrices are defined
$ORDER$, N (order can be any integer greater than 1)

The solution (X) is returned to the calling program by overwriting the D vector with the X vector. The calling statement for subroutine PBTRIP is

CALL PBTRIP($A, B, C, D, IL, IU, ORDER$)

with the same arguments as subroutine NBTRIP. However, if $ORDER$ is greater than 5 a dimension statement must be changed in this subroutine (see listing of subroutine).

Subroutines NBTRIP and PBTRIP employ no pivoting strategy in their elimination schemes. It should be noted that a specialized subroutine for solving a block tridiagonal system of equations can be written for each value of N which will be faster than the general subroutines given here.

LISTING OF SUBROUTINE NBTRIP

```
C...
C...SUBROUTINE TO SOLVE NON-PERIODIC BLOCK TRIDIAGONAL
C...SYSTEM OF EQUATIONS WITHOUT PIVOTING STRATEGY
C...WITH THE DIMENSIONS OF THE BLOCK MATRICES BEING
C...N x N (N IS ANY NUMBER GREATER THAN 1).
```

```
C...
      SUBROUTINE NBTRIP(A,B,C,D,IL,IU,ORDER)
      INTEGER ORDER,ORDSQ
      DIMENSION A(1),B(1),C(1),D(1)

C...
C...A = SUB DIAGONAL MATRIX
C...B =     DIAGONAL MATRIX
C...C = SUP DIAGONAL MATRIX
C...D = RIGHT HAND SIDE VECTOR
C...IL = LOWER VALUE OF INDEX FOR WHICH MATRICES ARE DEFINED
C...IU = UPPER VALUE OF INDEX FOR WHICH MATRICES ARE DEFINED
C...     (SOLUTION IS SOUGHT FOR BTRI(A,B,C)*X = D
C...     FOR INDICES OF X BETWEEN IL AND IU (INCLUSIVE).
C...     SOLUTION WRITTEN IN D VECTOR (ORIGINAL CONTENTS
C...     ARE OVERWRITTEN)).
C...ORDER = ORDER OF A,B,C MATRICES AND LENGTH OF D VECTOR
C...     AT EACH POINT DENOTED BY INDEX I
C...     (ORDER CAN BE ANY INTEGER GREATER THAN 1).
C...
C...THE MATRICES AND VECTORS ARE STORED IN SINGLE SUBSCRIPT FORM
C...
      ORDSQ = ORDER**2
C...
C...FORWARD ELIMINATION
C...
      I = IL
      I0MAT = 1+(I-1)*ORDSQ
      I0VEC = 1+(I-1)*ORDER
      CALL LUDECO(B(I0MAT),ORDER)
      CALL LUSOLV(B(I0MAT),D(I0VEC),D(I0VEC),ORDER)
      DO 100 J=1,ORDER
      I0MATJ = I0MAT+(J-1)*ORDER
      CALL LUSOLV(B(I0MAT),C(I0MATJ),C(I0MATJ),ORDER)
  100 CONTINUE
  200 CONTINUE
      I = I+1
      I0MAT = 1+(I-1)*ORDSQ
      I0VEC = 1+(I-1)*ORDER
      I1MAT = I0MAT-ORDSQ
      I1VEC = I0VEC-ORDER
      CALL MULPUT(A(I0MAT),D(I1VEC),D(I0VEC),ORDER)
      DO 300 J=1,ORDER
      I0MATJ = I0MAT+(J-1)*ORDER
      I1MATJ = I1MAT+(J-1)*ORDER
      CALL MULPUT(A(I0MAT),C(I1MATJ),B(I0MATJ),ORDER)
  300 CONTINUE
      CALL LUDECO(B(I0MAT),ORDER)
```

```
      CALL LUSOLV(B(IOMAT),D(IOVEC),D(IOVEC),ORDER)
      IF(I.EQ.IU) GO TO 500
      DO 400 J=1,ORDER
      IOMATJ = IOMAT+(J-1)*ORDER
      CALL LUSOLV(B(IOMAT),C(IOMATJ),C(IOMATJ),ORDER)
  400 CONTINUE
      GO TO 200
  500 CONTINUE
C...
C...BACK SUBSTITUTION
C...
      I = IU
  600 CONTINUE
      I = I-1
      IOMAT = 1+(I-1)*ORDSQ
      IOVEC = 1+(I-1)*ORDER
      I1VEC = IOVEC+ORDER
      CALL MULPUT(C(IOMAT),D(I1VEC),D(IOVEC),ORDER)
      IF (I.GT.IL) GO TO 600
C...
      RETURN
      END
```

LISTING OF SUBROUTINE PBTRIP

```
C...
C...SUBROUTINE TO SOLVE PERIODIC BLOCK TRIDIAGONAL
C...SYSTEM OF EQUATIONS WITHOUT PIVOTING STRATEGY.
C...EACH BLOCK MATRIX MAY BE OF DIMENSION N WITH
C...N ANY NUMBER GREATER THAN 1.
C...
      SUBROUTINE PBTRIP(A,B,C,D,IL,IU,ORDER)
      INTEGER ORDER,ORDSQ
      DIMENSION A(1),B(1),C(1),D(1)
      DIMENSION AD(25),CD(25)
C...
C...A = SUB DIAGONAL MATRIX
C...B =     DIAGONAL MATRIX
C...C = SUP DIAGONAL MATRIX
C...D = RIGHT HAND SIDE VECTOR
C...IL = LOWER VALUE OF INDEX FOR WHICH MATRICES ARE DEFINED
C...IU = UPPER VALUE OF INDEX FOR WHICH MATRICES ARE DEFINED
C...     (SOLUTION IS SOUGHT FOR BTRI(A,B,C)*X = D
C...     FOR INDICES OF X BETWEEEN IL AND IU (INCLUSIVE).
C...     SOLUTION WRITTEN IN D VECTOR (ORIGINAL CONTENTS
C...     ARE OVERWRITTEN)).
C...ORDER = ORDER OF A,B,C MATRICES AND LENGTH OF D VECTOR
C...     AT EACH POINT DENOTED BY INDEX I
C...     (ORDER CAN BE ANY INTEGER GREATER THAN 1)
```

```
C...      (ARRAYS AD AND CD MUST BE AT LEAST OF LENGTH ORDER**2)
C...      (CURRENT LENGTH OF 25 ANTICIPATES MAXIMUM ORDER OF 5).
C...
      IS = IL+1
      IE = IU-1
      ORDSQ = ORDER**2
      IUMAT = 1+(IU-1)*ORDSQ
      IUVEC = 1+(IU-1)*ORDER
      IEMAT = 1+(IE-1)*ORDSQ
      IEVEC = 1+(IE-1)*ORDER
C...
C...FORWARD ELIMINATION
C...
      I = IL
      IOMAT = 1+(I-1)*ORDSQ
      IOVEC = 1+(I-1)*ORDER
      CALL LUDECO(B(IOMAT),ORDER)
      CALL LUSOLV(B(IOMAT),D(IOVEC),D(IOVEC),ORDER)
      DO 10 J=1,ORDER
      IOMATJ = IOMAT+(J-1)*ORDER
      CALL LUSOLV(B(IOMAT),C(IOMATJ),C(IOMATJ),ORDER)
      CALL LUSOLV(B(IOMAT),A(IOMATJ),A(IOMATJ),ORDER)
   10 CONTINUE
C...
      DO 200 I = IS,IE
      IOMAT = 1+(I-1)*ORDSQ
      IOVEC = 1+(I-1)*ORDER
      I1MAT = IOMAT-ORDSQ
      I1VEC = IOVEC-ORDER
      DO 20 J=1,ORDSQ
      IOMATJ = J-1+IOMAT
      IUMATJ = J-1+IUMAT
      AD(J) = A(IOMATJ)
      CD(J) = C(IUMATJ)
      A(IOMATJ) = 0.0
      C(IUMATJ) = 0.0
   20 CONTINUE
      CALL MULPUT(AD,D(I1VEC),D(IOVEC),ORDER)
      DO 22 J=1,ORDER
      IOMATJ = IOMAT+(J-1)*ORDER
      I1MATJ = I1MAT+(J-1)*ORDER
      CALL MULPUT(AD,C(I1MATJ),B(IOMATJ),ORDER)
      CALL MULPUT(AD,A(I1MATJ),A(IOMATJ),ORDER)
   22 CONTINUE
      CALL LUDECO(B(IOMAT),ORDER)
      CALL LUSOLV(B(IOMAT),D(IOVEC),D(IOVEC),ORDER)
      DO 24 J=1,ORDER
      IOMATJ = IOMAT+(J-1)*ORDER
      CALL LUSOLV(B(IOMAT),C(IOMATJ),C(IOMATJ),ORDER)
```

```
      CALL LUSOLV(B(IOMAT),A(IOMATJ),A(IOMATJ),ORDER)
   24 CONTINUE
      CALL MULPUT(CD,D(I1VEC),D(IUVEC),ORDER)
      DO 26 J=1,ORDER
      IUMATJ = IUMAT+(J-1)*ORDER
      I1MATJ = I1MAT+(J-1)*ORDER
      CALL MULPUT(CD,A(I1MATJ),B(IUMATJ),ORDER)
      CALL MULPUT(CD,C(I1MATJ),C(IUMATJ),ORDER)
   26 CONTINUE
  200 CONTINUE
C...
      DO 30 J=1,ORDSQ
      IUMATJ = J-1+IUMAT
      AD(J) = A(IUMATJ)+C(IUMATJ)
   30 CONTINUE
      CALL MULPUT(AD,D(IEVEC),D(IUVEC),ORDER)
      DO 32 J=1,ORDER
      IUMATJ = IUMAT+(J-1)*ORDER
      IEMATJ = IEMAT+(J-1)*ORDER
      CALL MULPUT(AD,C(IEMATJ),B(IUMATJ),ORDER)
      CALL MULPUT(AD,A(IEMATJ),B(IUMATJ),ORDER)
   32 CONTINUE
      CALL LUDECO(B(IUMAT),ORDER)
      CALL LUSOLV(B(IUMAT),D(IUVEC),D(IUVEC),ORDER)
C...
C...BACK SUBSTITUTION
C...
      DO 40 IBAC = IL,IE
      I = IE-IBAC+IL
      IOMAT = 1+(I-1)*ORDSQ
      IOVEC = 1+(I-1)*ORDER
      I1VEC = IOVEC+ORDER
      CALL MULPUT(A(IOMAT),D(IUVEC),D(IOVEC),ORDER)
      CALL MULPUT(C(IOMAT),D(I1VEC),D(IOVEC),ORDER)
   40 CONTINUE
C...
      RETURN
      END

C...
C...SUBROUTINE TO CALCULATE L-U DECOMPOSITION
C...OF A GIVEN MATRIX A AND STORE RESULT IN A
C...(NO PIVOTING STRATEGY IS EMPLOYED)
C...
      SUBROUTINE LUDECO(A,ORDER)
      INTEGER ORDER
      DIMENSION A(ORDER,1)

C...
      DO 8 JC=2,ORDER
```

```
    8 A(1,JC) = A(1,JC)/A(1,1)
      JRJC = 1
   10 CONTINUE
      JRJC = JRJC+1
      JRJCM1 = JRJC-1
      JRJCP1 = JRJC+1
      DO 14 JR=JRJC,ORDER
      SUM = A(JR,JRJC)
      DO 12 JM=1,JRJCM1
   12 SUM = SUM-A(JR,JM)*A(JM,JRJC)
   14 A(JR,JRJC) = SUM
      IF (JRJC.EQ.ORDER) RETURN
      DO 18 JC = JRJCP1,ORDER
      SUM = A(JRJC,JC)
      DO 16 JM=1,JRJCM1
   16 SUM = SUM-A(JRJC,JM)*A(JM,JC)
   18 A(JRJC,JC) = SUM/A(JRJC,JRJC)
      GO TO 10
      END

C...
C...SUBROUTINE TO MULTIPLY A VECTOR B BY A MATRIX A
C...SUBTRACT RESULT FROM ANOTHER VECTOR C AND STORE
C...RESULT IN C.  THUS VECTOR C IS OVERWRITTEN.
C...
      SUBROUTINE MULPUT(A,B,C,ORDER)
      INTEGER ORDER
      DIMENSION A(1),B(1),C(1)

C...
      DO 200 JR=1,ORDER
      SUM = 0.0
      DO 100 JC=1,ORDER
      IA = JR+(JC-1)*ORDER
  100 SUM = SUM+A(IA)*B(JC)
  200 C(JR) = C(JR)-SUM
C...
      RETURN
      END

C...
C...SUBROUTINE TO SOLVE LINEAR ALGEBRAIC SYSTEM OF
C...EQUATIONS A*C=B AND STORE RESULTS IN VECTOR C.
C...MATRIX A IS INPUT IN L-U DECOMPOSITION FORM.
C...(NO PIVOTING STRATEGY HAS BEEN EMPLOYED TO
C...COMPUTE THE L-U DECOMPOSITION OF THE MATRIX A).
C...
      SUBROUTINE LUSOLV(A,B,C,ORDER)
      INTEGER ORDER
      DIMENSION A(ORDER,1),B(1),C(1)
```

```
C...
C...FIRST L(INV)*B
C...
      C(1) = C(1)/A(1,1)
      DO 14 JR=2,ORDER
      JRM1 = JR-1
      SUM = B(JR)
      DO 12 JM=1,JRM1
   12 SUM = SUM-A(JR,JM)*C(JM)
   14 C(JR) = SUM/A(JR,JR)
C...
C...NEXT U(INV) OF L(INV)*B
C...
      DO 18 JRJR=2,ORDER
      JR = ORDER-JRJR+1
      JRP1 = JR+1
      SUM = C(JR)
      DO 16 JMJM = JRP1,ORDER
      JM = ORDER-JMJM+JRP1
   16 SUM = SUM-A(JR,JM)*C(JM)
   18 C(JR) = SUM
C...
      RETURN
      END
```

APPENDIX

C

THE MODIFIED STRONGLY IMPLICIT PROCEDURE

This appendix describes the Modified Strongly Implicit (MSI) procedure (Schneider and Zedan, 1981) for solving a class of elliptic PDE's. The overall strategy of this procedure was described in Chapter 4. This appendix supplies further details. Schneider and Zedan (1981) presented the procedure as a means for solving the algebraic equations arising from the finite-difference representation of the elliptic equation

$$\frac{\partial}{\partial x}\left(k_x \frac{\partial u}{\partial x}\right) + \frac{\partial}{\partial y}\left(k_y \frac{\partial u}{\partial y}\right) = q(x, y) \tag{C.1}$$

which governs two-dimensional steady-state heat conduction when u is the temperature. In the above, k_x and k_y are thermal conductivities for heat flow in the x and y directions, respectively, and $q(x, y)$ is a source term accounting for possible heat generation. It should be clear that a wide variety of problems are governed by equations of the form given by Eq. (C.1). With $k_x = k_y =$ constant and $q(x, y) \neq 0$, Eq. (C.1) becomes the Poisson equation. With $k_x = k_y =$ constant and $q(x, y) = 0$, Eq. (C.1) reduces to the Laplace equation. Only numerical examples for the solution to the Laplace equation were presented in Schneider and Zedan (1981). Examples presented employed Dirichlet, Neumann, and Robins (convective) boundary conditions.

The algorithm is developed to handle a nine-point finite-difference representation of Eq. (C.1) and treats the five-point representation as a special case. A nine-point [see Eq. (4.114)] representation of Eq. (C.1) can be written in the general form

$$A^1_{i,j}u_{i,j+1} + A^2_{i,j}u_{i+1,j+1} + A^3_{i,j}u_{i+1,j} + A^4_{i,j}u_{i+1,j-1} + A^5_{i,j}u_{i,j-1} + A^6_{i,j}u_{i-1,j-1}$$
$$+ A^7_{i,j}u_{i-1,j} + A^8_{i,j}u_{i-1,j+1} + A^9_{i,j}u_{i,j} = q_{i,j} \tag{C.2}$$

The i, j subscript refers to location within the grid network rather than the matrix row-column designation. Note that superscripts are used to identify the coefficients in the difference equation written for the general point (i, j). The five-point representation becomes a special case in which

$$A^2_{i,j} = A^4_{i,j} = A^6_{i,j} = A^8_{i,j} = 0$$

The equations can be written in the form

$$[A]\mathbf{u} = \mathbf{C} \tag{C.3}$$

where the coefficient matrix has the form

$$[A] = \begin{bmatrix} A^9_{i,j} & A^3_{i,j} & & A^8_{i,j} & A^1_{i,j} & A^2_{i,j} & & \\ A^7_{i,j} & * & \ddots & & & * & \ddots & \\ & \ddots & \ddots & \ddots & & & \ddots & \ddots \\ A^6_{i,j} & A^5_{i,j} & A^4_{i,j} & \ddots & \ddots & & & \\ & \ddots & * & \ddots & \ddots & \ddots & & \\ & & A^6_{i,j} & A^5_{i,j} & A^4_{i,j} & & A^7_{i,j} & A^9_{i,j} \end{bmatrix}$$

For reference, the diagonals corresponding to grid points having the same value for the i index (same grid column) and are identified by an asterisk. We now construct a matrix

$$[B] = [A + P]$$

such that $[B]$ can be decomposed into upper and lower triangular matrices, $[L]$ and $[U]$. We require that the original nine coefficients ($A^1_{i,j}$ through $A^9_{i,j}$) remain unchanged as $[A + P]$ is constructed. The $[L]$ and $[U]$ matrices have the form

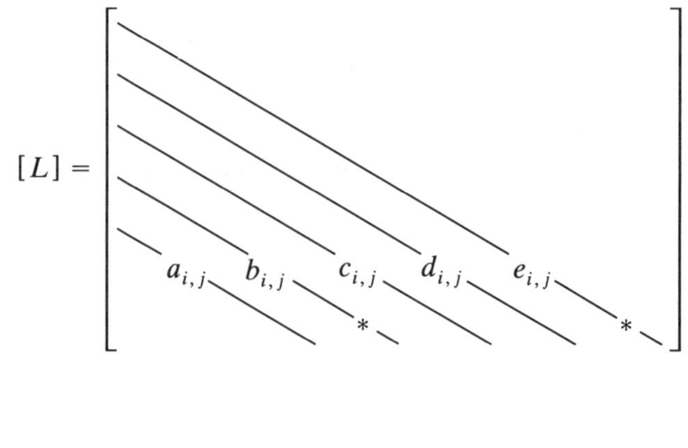

$$[U] = \left[\quad 1 \quad f_{i,j} \quad g_{i,j} \quad h_{i,j} \quad s_{i,j} \quad \right]$$

(diagonals marked $*$ under $f_{i,j}$ and $s_{i,j}$)

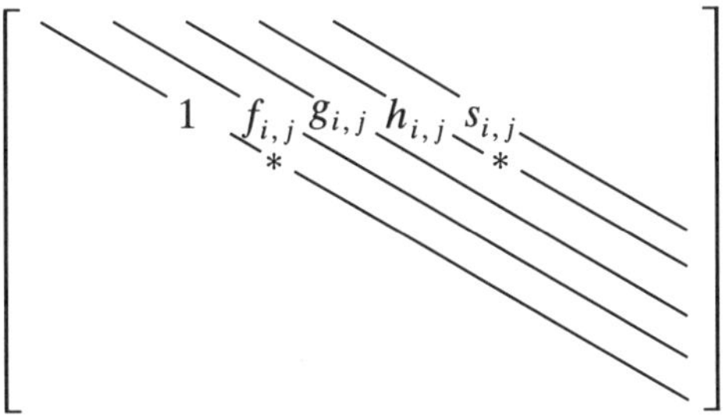

Again the asterisk is used to identify diagonals corresponding to grid points having the same value for the i index.

The equations to be used to determine the coefficients of $[L]$ and $[U]$ such that the original nine coefficients in $[A]$ remain unchanged in $[B]$ are

$$a_{i,j} = A^6_{i,j} \tag{C.3a}$$

$$a_{i,j} f_{i-1,j-1} + b_{i,j} = A^5_{i,j} \tag{C.3b}$$

$$b_{i,j} f_{i,j-1} + C_{i,j} = A^4_{i,j} \tag{C.3c}$$

$$a_{i,j} h_{i-1,j-1} + b_{i,j} g_{i,j-1} + d_{i,j} = A^7_{i,j} \tag{C.3d}$$

$$a_{i,j} s_{i-1,j-1} + b_{i,j} h_{i,j-1} + c_{i,j} g_{i+1,j-1} + d_{i,j} f_{i-1,j} + e_{i,j} = A^9_{i,j} \tag{C.3e}$$

$$b_{i,j} s_{i,j-1} + c_{i,j} h_{i+1,j-1} + e_{i,j} f_{i,j} = A^3_{i,j} \tag{C.3f}$$

$$d_{i,j} h_{i-1,j} + e_{i,j} g_{i,j} = A^8_{i,j} \tag{C.3g}$$

$$d_{i,j} s_{i-1,j} + e_{i,j} h_{i,j} = A^1_{i,j} \tag{C.3h}$$

$$e_{i,j} s_{i,j} = A^2_{i,j} \tag{C.3i}$$

The modified coefficient matrix $[B] = [A + P]$ has the form

$$[B] = \left[\begin{array}{l} \qquad\qquad\qquad \phi^4_{i,j} \quad A^8_{i,j} \quad A^1_{i,j} \quad A^2_{i,j} \\ A^6_{i,j} \quad A^5_{i,j} \quad A^4_{i,j} \quad \phi^1_{i,j} \quad \phi^2_{i,j} \quad A^7_{i,j} \quad A^9_{i,j} \quad A^3_{i,j} \quad \phi^3_{i,j} \end{array} \right]$$

(diagonals marked $*$ under $A^1_{i,j}$, $A^5_{i,j}$ and $A^3_{i,j}$)

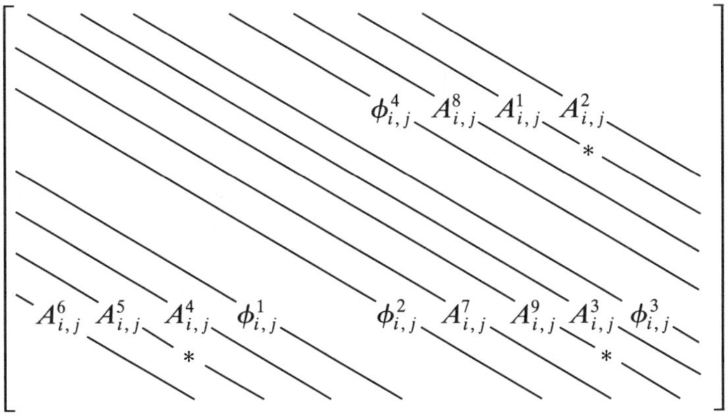

where the asterisk has the same meaning as before.

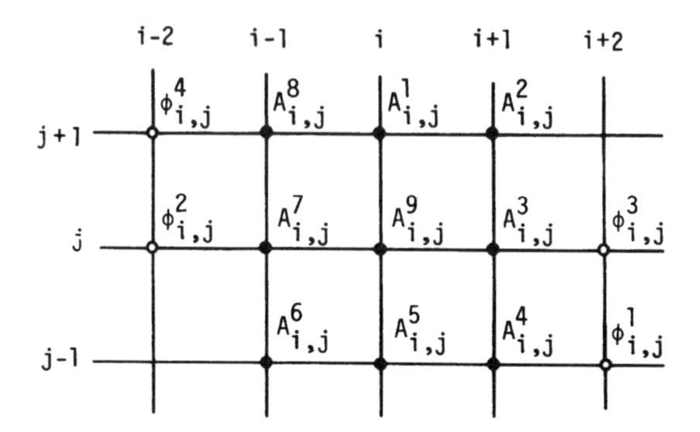

Figure C.1 The numerical molecule for the MSI procedure for a nine-point formulation, points labeled $A^2_{i,j}$, $A^4_{i,j}$, $A^6_{i,j}$, $A^8_{i,j}$, $\phi^2_{i,j}$, $\phi^3_{i,j}$ are eliminated when a five-point formulation is used.

The elements in $[B]$ denoted by $\phi^1_{i,j}$, $\phi^2_{i,j}$, $\phi^3_{i,j}$, and $\phi^4_{i,j}$ are determined from

$$\phi^1_{i,j} = c_{i,j}f_{i+1,j-1} \tag{C.4a}$$

$$\phi^2_{i,j} = a_{i,j}g_{i-1,j-1} \tag{C.4b}$$

$$\phi^3_{i,j} = c_{i,j}s_{i+1,j-1} \tag{C.4c}$$

$$\phi^4_{i,j} = d_{i,j}g_{i-1,j} \tag{C.4d}$$

The numerical molecule associated with the modified matrix $[B]$ is shown schematically in Fig. C.1.

Schneider and Zedan (1981) employed Taylor-series expansions to obtain values of $u_{i-2,j}$, $u_{i+2,j}$, $u_{i+2,j-1}$, and $u_{i-2,j+1}$ in terms of u's in the original nine-point molecule to partially cancel the influence of the additional $(\phi_{i,j})$ terms in the $[B]$ matrix. These are

$$u_{i-2,j} = -u_{i,j} + 2u_{i-1,j} \tag{C.5a}$$

$$u_{i+2,j} = -u_{i,j} + 2u_{i+1,j} \tag{C.5b}$$

$$u_{i+2,j-1} = -2u_{i,j} + 2u_{i+1,j} + u_{i,j-1} \tag{C.5c}$$

$$u_{i-2,j+1} = -2u_{i,j} + 2u_{i-1,j} + u_{i,j+1} \tag{C.5d}$$

Other "extrapolation" schemes for obtaining values outside the original molecule may work equally well. The use of such approximations affects only the approach to convergence of the iterative sequence and not the final converged solution.

An iterative parameter α is employed to implement partial cancellation of the influence of the $\phi_{i,j}$ terms appearing in $[B]$. This is done by using a modified representation for the nine-point scheme in the form

$$A^5_{i,j}u_{i,j-1} + A^7_{i,j}u_{i-1,j} + A^9_{i,j}u_{i,j} + A^1_{i,j}u_{i,j+1} + A^3_{i,j}u_{i+1,j}$$
$$+ A^6_{i,j}u_{i-1,j-1} + A^8_{i,j}u_{i-1,j+1} + A^2_{i,j}u_{i+1,j+1}$$
$$+ A^4_{i,j}u_{i+1,j-1} + \phi^1_{i,j}\left[u_{i+2,j-1} - \alpha(-2u_{i,j} + 2u_{i+1,j} + u_{i,j-1})\right]$$

$$+ \phi_{i,j}^{2}\left[u_{i-2,j} - \alpha(-u_{i,j} + 2u_{i-1,j})\right] + \phi_{i,j}^{3}\left[u_{i+2,j} - \alpha(-u_{i,j} + 2u_{i+1,j})\right]$$
$$+ \phi_{i,j}^{4}\left[u_{i-2,j+1} - \alpha(-2u_{i,j} + 2u_{i-1,j} + u_{i,j+1})\right] = q_{i,j} \tag{C.6}$$

Equations (C.3) and (C.4) are modified to include the partial cancellation indicated in Eq. (C.6) and rearranged to permit the explicit evaluation of the elements of $[L]$ and $[U]$:

$$a_{i,j} = A_{i,j}^{6} \tag{C.7a}$$

$$b_{i,j} = \frac{A_{i,j}^{5} - a_{i,j} f_{i-1,j-1} - \alpha A_{i,j}^{4} f_{i+1,j-1}}{1 - \alpha f_{i,j-1} f_{i+1,j-1}} \tag{C.7b}$$

$$c_{i,j} = A_{i,j}^{4} - b_{i,j} f_{i,j-1} \tag{C.7c}$$

$$d_{i,j} = \frac{A_{i,j}^{7} - a_{i,j} h_{i-1,j-1} - b_{i,j} g_{i,j-1} - 2\alpha a_{i,j} g_{i-1,j-1}}{1 + 2\alpha g_{i-1,j}} \tag{C.7d}$$

$$e_{i,j} = A_{i,j}^{9} - a_{i,j} s_{i-1,j-1} - b_{i,j} h_{i,j-1} - c_{i,j} g_{i+1,j-1} - d_{i,j} f_{i-1,j}$$
$$+ \alpha\left(2\phi_{i,j}^{1} + \phi_{i,j}^{2} + \phi_{i,j}^{3} + 2\phi_{i,j}^{4}\right) \tag{C.7e}$$

$$f_{i,j} = \frac{A_{i,j}^{3} - b_{i,j} s_{i,j-1} - c_{i,j} h_{i+1,j-1} - 2\alpha\left(\phi_{i,j}^{1} + \phi_{i,j}^{3}\right)}{e_{i,j}} \tag{C.7f}$$

$$g_{i,j} = \frac{A_{i,j}^{8} - d_{i,j} h_{i-1,j}}{e_{i,j}} \tag{C.7g}$$

$$h_{i,j} = \frac{A_{i,j}^{1} - d_{i,j} s_{i-1,j} - \alpha\phi_{i,j}^{4}}{e_{i,j}} \tag{C.7h}$$

$$s_{i,j} = \frac{A_{i,j}^{2}}{e_{i,j}} \tag{C.7i}$$

The $\phi_{i,j}$'s appearing in the above are evaluated as indicated in Eqs. (C.4) using the values of a, b, c, d, f, g, and s obtained from Eqs. (C.7). Note that the $\phi_{i,j}$'s are needed in Eqs. (C.7) and should be evaluated as soon as the evaluation of $d_{i,j}$ is complete. The results obtained by Schneider and Zedan (1981) indicate that the MSI procedure is not extremely sensitive to the choice of α. Values of α between 0.3 and 0.6 worked well in their calculations.

It is important to observe that when the MSI procedure is used for the five-point difference representation,

$$A_{i,j}^{2} = A_{i,j}^{4} = A_{i,j}^{6} = A_{i,j}^{8} = 0 \tag{C.8}$$

and, as a result,

$$a_{i,j} = s_{i,j} = \phi_{i,j}^{2} = \phi_{i,j}^{3} = 0 \tag{C.9}$$

The iterative sequence is developed as follows. Adding $[P]\mathbf{u}$ to both sides of Eq. (C.3) gives

$$[A + P]\mathbf{u} = \mathbf{C} + [P]\mathbf{u} \tag{C.10}$$

We evaluate the unknowns on the right-hand side at the n iteration level to write

$$[A + P]\mathbf{u}^{n+1} = \mathbf{C} + [P]\mathbf{u}^n \tag{C.11}$$

Decomposing $[A + P]$ into the $[L]$ and $[U]$ matrices gives

$$[L][U]\mathbf{u}^{n+1} = \mathbf{C} + [P]\mathbf{u}^n \tag{C.12}$$

Defining an intermediate vector $\mathbf{V}^{n+1}$ by

$$\mathbf{V}^{n+1} = [U]\mathbf{u}^{n+1} \tag{C.13}$$

we can employ the two-step process

Step 1: $$[L]\mathbf{V}^{n+1} = \mathbf{C} + [P]\mathbf{u}^n \tag{C.14a}$$

Step 2: $$[U]\mathbf{u}^{n+1} = \mathbf{V}^{n+1} \tag{C.14b}$$

The elements of $[P]$ are simply the $\phi^1, \phi^2, \phi^3, \phi^4$ (only ϕ^1 and ϕ^4 when the five-point scheme is used) values determined from Eqs. (C.4).

Alternatively, we can define a difference vector

$$\boldsymbol{\delta}^{n+1} = \mathbf{u}^{n+1} - \mathbf{u}^n \tag{C.15}$$

and a residual vector

$$\mathbf{R}^n = [A]\mathbf{u}^n - \mathbf{C} \tag{C.16}$$

so that Eq. (C.11) becomes

$$[A + P]\boldsymbol{\delta}^{n+1} = -\mathbf{R}^n \tag{C.17}$$

Replacing $[A + P]$ by the $[L][U]$ product gives

$$[L][U]\boldsymbol{\delta}^{n+1} = -\mathbf{R}^n$$

Defining an intermediate vector $\mathbf{W}^{n+1}$ by

$$\mathbf{W}^{n+1} = [U]\boldsymbol{\delta}^{n+1} \tag{C.18}$$

the solution procedure can again be written as a two-step process:

Step 1: $$[L]\mathbf{W}^{n+1} = -\mathbf{R}^n \tag{C.19a}$$

Step 2: $$[U]\boldsymbol{\delta}^{n+1} = \mathbf{W}^{n+1} \tag{C.19b}$$

The processes represented by Eqs. (C.14) and (C.19) consist of a forward substitution to determine $\mathbf{V}^{n+1}$ or $\mathbf{W}^{n+1}$ followed by a backward substitution to obtain $\mathbf{u}^{n+1}$ or $\boldsymbol{\delta}^{n+1}$. The coefficients remain unchanged for the iterative process. The right-hand side of the Step 1 equation is then updated and the procedure is repeated.

APPENDIX

D

FINITE-VOLUME DISCRETIZATION FOR GENERAL CONTROL VOLUMES

The finite-volume method enforces conservation principles in integral form to fixed regions in space known as control volumes. The purpose of this appendix is to illustrate how this can be carried out for general control volumes for which the boundaries may not necessarily intersect in an orthogonal manner. The main points can be readily illustrated by considering two examples in two dimensions, that of mass conservation in steady, incompressible flow and thermal energy conservation.

The approach is largely the same regardless of the shape of the control volume. The grid may be structured or unstructured. We will utilize the generally nonorthogonal but structured grid illustrated in Fig. D.1. in which control volumes are quadrilaterals. The boundaries of control volumes are placed approximately halfway between grid points. More precisely, the coordinates of the corners of the control volumes (points a, b, c, d) are taken as the average of the coordinates of the four surrounding grid points. That is,

$$x_a = (x_{i,j-1} + x_{i+1,j-1} + x_{i+1,j} + x_{i,j})/4$$

$$y_a = (y_{i,j-1} + y_{i+1,j-1} + y_{i+1,j} + y_{i,j})/4$$

The coordinates of points b, c, d are located in a similar manner. The corners are connected by straight lines to form the control volume. This illustrates a cell vertex or nodal point scheme, the procedure employed in Chapter 3 (Section 3.4.4). In passing we note that other choices for the establishment of control volumes could have been made. For example, we could have considered the

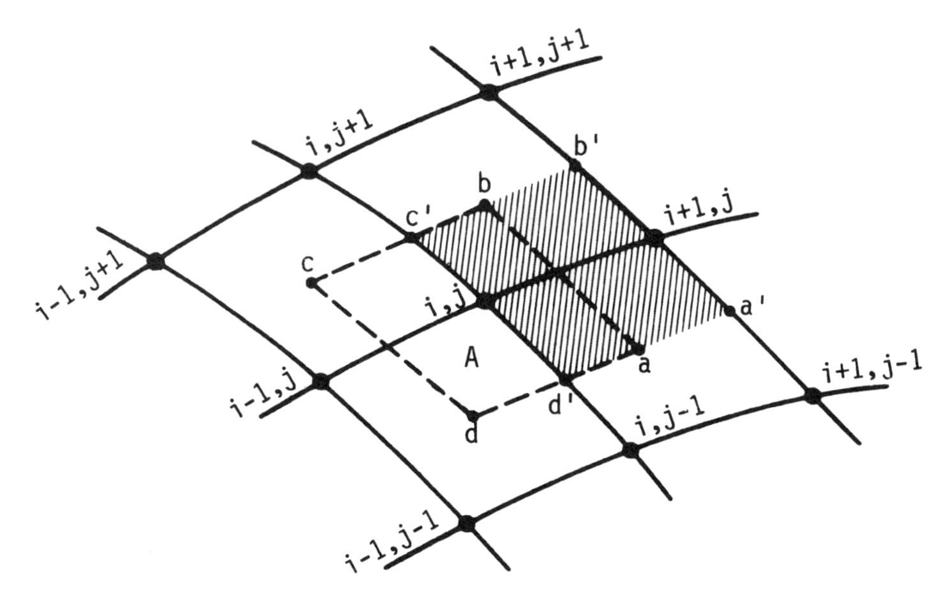

Figure D.1 Control volume for finite-volume discretization; shaded area is secondary volume used in representation of boundary derivatives.

intersecting solid lines in Fig. D.1 to be the control volume boundarics and placed "grid" points (where variables will be evaluated) in the center of the volumes. This would have been a cell-centered scheme, the scheme utilized in Chapter 5 (Section 5.7).

For a control volume, conservation of mass in steady, incompressible, two-dimensional flow requires

$$\oiint_S \mathbf{V} \cdot \mathbf{n}\, dS = 0 \tag{D.1}$$

where $\mathbf{n}$ is a unit vector normal to the control volume (positive when pointing outward) and $\mathbf{V}$ is the velocity vector, $\mathbf{V} = u\mathbf{i} + v\mathbf{j}$. The integral represents the net volume rate of flow out of the volume S. For two-dimensional flow, the "volume" is formed by including a unit depth normal to the x-y plane. This unit depth will be omitted in the equations that follow. Equation (D.1) serves as a model for conservation statements that require evaluation of surface fluxes that can be represented in terms of simple functions of the dependent variables themselves.

In two dimensions, we can represent $\mathbf{n}dS$ as $\mathbf{i}dy - \mathbf{j}dx$ (omitting the unit depth) for an integration path around the boundary in a *counter-clockwise* direction. Thus, Eq. (D.1) can be written as

$$\oiint_S (u\,dy - v\,dx) = 0 \tag{D.2}$$

To evaluate this integral for the control volume A in Fig. D.1, we need to represent the velocity components on the boundaries of the control volume. One choice is to use the average of the velocity components at nodes on either side of the boundary. Thus, for boundary a-b we could use

$$u_{ab} = u_{i+1/2,j} \cong (u_{i+1,j} + u_{i,j})/2 \qquad v_{ab} = v_{i+1/2,j} \cong (v_{i+1,j} + v_{i,j})/2$$

Thus,

$$\begin{aligned}\oiint_S (u\,dy - v\,dx) \\ &\cong u_{i+1/2,j}\Delta y_{ab} - v_{i+1/2,j}\Delta x_{ab} + u_{i,j+1/2}\Delta y_{bc} - v_{i,j+1/2}\Delta x_{bc} \\ &\quad + u_{i-1/2,j}\Delta y_{cd} - v_{i-1/2,j}\Delta x_{cd} + u_{i,j-1/2}\Delta y_{da} - v_{i,j-1/2}\Delta x_{da}\end{aligned} \tag{D.3}$$

where the increments in the coordinates indicated above must be evaluated very carefully as $\Delta x_{ab} = x_b - x_a$, $\Delta y_{ab} = y_b - y_a$, for example. Some of the increments will be positive and some will be negative as the integration proceeds around the control volume. Note that if the control volume is a rectangle with sides aligned with the Cartesian coordinate system, one of the coordinate increments will be zero along each boundary and some of the velocity components will cancel resulting in the central representation

$$\oiint_S (u\,dy - v\,dx) \cong (u_{i+1,j} - u_{i-1,j})\Delta y_{ab} + (v_{i,j+1} - v_{i,j-1})\Delta x_{da} \tag{D.4}$$

Many conservation statements required in applications govern unsteady phenomena and others contain fluxes that depend on derivatives. The integral form of the 2-D heat equation will be used as a model to illustrate how such terms can be handled for a control volume of arbitrary shape. The differential form of the 2-D heat equation is given by Eq. (3.92a)

$$\rho c \frac{\partial T}{\partial t} = \frac{\partial}{\partial x}\left(k\frac{\partial T}{\partial x}\right) + \frac{\partial}{\partial y}\left(k\frac{\partial T}{\partial y}\right) = \boldsymbol{\nabla}\cdot(k\boldsymbol{\nabla}T)$$

The corresponding integral form is

$$\iiint_R \rho c \frac{\partial T}{\partial t}\,dR + \oiint_S \mathbf{q}\cdot\mathbf{n}\,dS = 0 \tag{D.5}$$

where $\mathbf{q} = -k\dfrac{\partial T}{\partial x}\mathbf{i} - k\dfrac{\partial T}{\partial y}\mathbf{j}$. Following the strategy employed with the surface integral that appeared in the previous example, we can write Eq. (D.5) as

$$\iiint_R \rho c \frac{\partial T}{\partial t}\,dR + \oiint_S \left(-k\frac{\partial T}{\partial x}\,dy + k\frac{\partial T}{\partial y}\,dx\right) = 0 \tag{D.6}$$

The term on the left containing the time derivative can be evaluated by assuming that the temperature at point (i, j) is the mean value for the volume and then using a forward-time difference to obtain

$$\rho c \frac{\left(T_{i,j}^{n+1} - T_{i,j}^{n}\right)}{\Delta t} R_{abcd} \tag{D.7}$$

where R_{abcd} is the volume of the control volume. In this 2-D example, we will take the volume to be the area of the 2-D control volume in the x-y plane, A_{abcd}, times a unit depth normal to that plane. As before, the unit depth will be omitted in the equations to follow. We will determine the area of the quadrilateral region A_{abcd} as one-half the magnitude of the cross products of its diagonals,

$$A_{abcd} = 0.5|(\Delta x_{db} \Delta y_{ac} - \Delta y_{db} \Delta x_{ac})|$$

Note that for a rectangular region with boundaries aligned with the Cartesian coordinate system, this results in $A_{abcd} = \Delta x \Delta y$.

To evaluate the surface integral in Eq. (D.6) we will first discretize by approximating the integrand on the boundaries. When doing this, a decision must be made as to the time level at which the boundary fluxes are to be evaluated. This determines whether the scheme will be explicit or implicit. Reasonable choices include time levels n, $n + 1$ or an average of the two. Selecting time level n,

$$\begin{aligned} \oiint_S \left(-k \frac{\partial T}{\partial x}\, dy + k \frac{\partial T}{\partial y}\, dx \right) &\cong -k \frac{\partial T}{\partial x}\bigg)_{i+1/2,j}^{n} \Delta y_{ab} \\ &+ k \frac{\partial T}{\partial y}\bigg)_{i+1/2,j}^{n} \Delta x_{ab} - k \frac{\partial T}{\partial x}\bigg)_{i,j+1/2}^{n} \Delta y_{bc} \\ &+ k \frac{\partial T}{\partial y}\bigg)_{i,j+1/2}^{n} \Delta x_{bc} - k \frac{\partial T}{\partial x}\bigg)_{i-1/2,j}^{n} \Delta y_{cd} \\ &+ k \frac{\partial T}{\partial y}\bigg)_{i-1/2,j}^{n} \Delta x_{cd} - k \frac{\partial T}{\partial x}\bigg)_{i,j-1/2}^{n} \Delta y_{da} + k \frac{\partial T}{\partial y}\bigg)_{i,j-1/2}^{n} \Delta x_{da} \end{aligned} \tag{D.8}$$

For rectangular volumes whose boundaries align with the Cartesian coordinate system, the derivatives at boundaries are readily represented by central differences utilizing neighboring nodal values of temperature. This is demonstrated in Chapter 3 (Section 3.4.4). However, it is possible to develop appropriate representations for derivatives at control volume boundaries by integral methods in a manner that is not restricted to Cartesian or even orthogonal grids. The method makes use of results that can be obtained from application of the divergence theorem,

$$\iiint_R \nabla \cdot \mathbf{V}\, dS = \oiint_S \mathbf{V} \cdot \mathbf{n}\, dS$$

Letting $\mathbf{V} = T\mathbf{i}$ and applying the divergence theorem gives

$$\iiint_R \frac{\partial T}{\partial x}\, dR = \oiint_S T dy \tag{D.9}$$

where again we are making use of the fact that $\mathbf{n}\, dS$ can be represented by $\mathbf{i} dy - \mathbf{j} dx$. In a like manner, letting $\mathbf{V} = T\mathbf{j}$, we observe

$$\iiint_R \frac{\partial T}{\partial y}\, dR = -\oiint_S T dx \tag{D.10}$$

Further, by multiplying the first expression by $\mathbf{i}$ and the second by $\mathbf{j}$ and adding, we can see that

$$\iiint_R \nabla T dR = \oiint_S T\mathbf{n}\, dS \tag{D.11}$$

These results provide a way in which derivatives of temperature on the control volume boundary can be represented by integrating the temperature itself around the boundaries of a suitable volume. Suppose, for example, that we wish to represent $\left.\frac{\partial T}{\partial x}\right)_{i+1/2,j}^{n}$ and $\left.\frac{\partial T}{\partial y}\right)_{i+1/2,j}^{n}$ for an interior control volume such as volume A in Fig. D.1. We first establish a secondary volume (area in this 2-D example) in such a manner that the point at which the representation is desired, $i + \frac{1}{2}, j$ in this case, is approximately in the center. The shaded area in Fig. D.1 will serve that purpose. This procedure assumes that $\left.\frac{\partial T}{\partial x}\right)_{i+1/2,j}^{n}$ and $\left.\frac{\partial T}{\partial y}\right)_{i+1/2,j}^{n}$ are mean values for the secondary area $A_{a'b'c'd'}$. Thus, it follows from the results above that

$$\left.\frac{\partial T}{\partial x}\right)_{i+1/2,j}^{n} \cong \frac{1}{A_{a'b'c'd'}} \oiint_S T dy \quad \text{and} \quad \left.\frac{\partial T}{\partial y}\right)_{i+1/2,j}^{n} \cong -\frac{1}{A_{a'b'c'd'}} \oiint_S T dx$$

The coordinates of the secondary volume are established by employing suitable averages of the coordinates of neighboring points. The determination of the coordinates of points a, b, c, d has already been discussed. The coordinates are found to be averages of the coordinates of the four neighboring nodal points. The location of a' can be determined by averaging the coordinates of points $(i + 1, j)$ and $(i + 1, j - 1)$. The coordinates of point b' can be determined by averaging the coordinates of points $(i + 1, j)$ and $(i + 1, j + 1)$. The coordinates of points c' and d' are determined as averages in a similar manner. We next approximate the line integrals as, for example,

$$\left.\frac{\partial T}{\partial x}\right)_{i+1/2,j}^{n} \cong \frac{1}{A_{a'b'c'd'}} \oint_S T dy \cong T_{i+1,j}^{n} \Delta y_{a'b'} + T_b^n \Delta y_{b'c'} + T_{i,j}^{n} \Delta y_{c'd'} + T_a^n \Delta y_{d'a'} \tag{D.12}$$

where T_a^n and T_b^n are determined as averages of the four neighboring nodal temperatures. Notice that for a rectangular volume with boundaries that align with the Cartesian coordinate system, Eq. (D.12) reduces to

$$\frac{T_{i+1,j}^{n} - T_{i,j}^{n}}{\Delta x}$$

To complete the discretization of Eq. (D.8), seven more equations similar to Eq. (D.12) must be developed. This requires that a different secondary area be established for approximating derivatives on each of the four sides of the original control volume labeled A in Fig. D.1. This is computationally intensive, but the result is general and not restricted to orthogonal grids.

NOMENCLATURE

a	speed of sound
A	area
A	$\partial F / \partial u$
$\mathbf{A}$	area vector
c	wave speed
c	specific heat
c_f	skin-friction coefficient
c_i	species mass fraction
c_p	specific heat at constant pressure
c_v	specific heat at constant volume
C_p	pressure coefficient
dA	differential area
dl	differential length
$d\mathbf{r}$	differential length vector along a line
dR	differential volume
dS	differential surface area
dt	differential time
dx, dy, dz	differential lengths in Cartesian system
dv	differential volume
dV	differential volume
$d\mathscr{V}$	differential volume
D	diameter
$\mathscr{D}_{\text{im}}$	multi-component diffusion coefficient
e	error

e	internal energy per unit mass
E	shift operator
Ec	Eckert number
E_t	total energy per unit volume, [$= \rho(e + V^2/2)$ if only internal energy and kinetic energy are included]
f	function
f	numerical flux
$\mathbf{f}$	body force per unit mass
f_x, f_y, f_z	components of body force per unit mass in a Cartesian system
F	denotes function, nondimensional velocity variable
$\mathbf{F}$	flux
$\mathbf{g}$	gravity vector
G	amplification factor
G	nondimensional velocity variable
h	height
h	enthalpy per unit mass ($= e + p/\rho$)
h	Δx
h	heat transfer coefficient
h_i°	enthalpy of formation for species i
h_1, h_2, h_3	scale factors in an orthogonal curvilinear coordinate system
H	total enthalpy ($= h + V^2/2$)
i	$\sqrt{-1}$
$\mathbf{i}_1, \mathbf{i}_2, \mathbf{i}_3$	unit vectors in a generalized curvilinear coordinate system
$\mathbf{i}, \mathbf{j}, \mathbf{k}$	unit vectors in a Cartesian coordinate system
I	nondimensional enthalpy variable
I	inverse Jacobian
I	transfer operator
J	Jacobian
k	Δy
k	coefficient of thermal conductivity
$\bar{k}$	kinetic energy of turbulence
k_m	wave number
k_T	turbulent thermal conductivity
K	local body curvature
K	$\Delta y_+/\Delta y_-$
l	length
l	mixing length
l_ϵ	dissipation length
$l_{1,2}$	scalar components of $\mathbf{L}$
L	difference operator
L	reference length
$\mathbf{L}$	eigenvector
$\dot{m}$	mass flow rate
M	Mach number
M_x	local streamwise Mach number

$\mathscr{M}$	molecular weight of mixture
$\mathscr{M}_i$	molecular weight of species i
n	time level
n	normal distance
$\mathbf{n}$	unit normal
N	total number of time steps
p	pressure
Pr	Prandtl number
Pr_T	turbulent Prandtl number
q	intensity of line source or sink
q	magnitude of heat flux vector
$\mathbf{q}$	heat flux vector
Q	external heat addition per unit volume
r	$\alpha\,\Delta t/(\Delta x)^2$
r	radius, radial distance
$\mathbf{r}$	position vector
r_x	$\alpha\,\Delta t/(\Delta x)^2$
r_y	$\alpha\,\Delta t/(\Delta y)^2$
r_z	$\alpha\,\Delta t/(\Delta z)^2$
R	explicit operator
R	residual
R	radius of curvature
R	volume
R	gas constant
Re	Reynolds number
Re_L	freestream Reynolds number based on length $L, (= \rho_\infty V_\infty L/\mu_\infty)$
$\mathrm{Re}_{\Delta x}$	mesh Reynolds number ($= c\,\Delta x/\mu$ for Burgers' equation)
$\mathscr{R}$	universal gas constant
s	entropy per unit mass
s	arc length
S	source term
$\mathbf{S}$	surface area vector
S_{ij}	mean strain tensor
t	time
T	temperature
TV	total variation
u	unknown
u^+	nondimensional velocity used in turbulent flow
u, v, w	velocity components in a Cartesian system
u_1, u_2, u_3	velocity components in a generalized coordinate system
u_r, u_θ, u_z	velocity components in a cylindrical coordinate system
u_r, u_θ, u_ϕ	velocity components in a spherical coordinate system
U, V, W	contravariant velocity components
U	freestream velocity in x direction
$\mathbf{U}_i$	species diffusion velocity

v	unknown
v_T	characteristic velocity of the turbulence
$\mathbf{V}$	velocity vector
V	magnitude of velocity vector
w	weight function
w	unknown
$\mathbf{w}$	primitive variable vector
x, y, z	Cartesian coordinates
x_1, x_2, x_3	generalized curvilinear coordinates
y^+	nondimensional distance used in turbulent flow
α	thermal diffusivity
α, β, γ	conical coordinates
β	volumetric expansion coefficient
β	grid aspect ratio $(= \Delta x / \Delta y)$
β	artificial compressibility factor
β	stretching parameter
β	$k_m \Delta x$
β	$\sqrt{M_\infty^2 - 1}$
β	pressure gradient parameter, $[= (x/u_e)\, du_e/dx]$
β_x	$k_m \Delta x$
β_y	$k_m \Delta y$
γ	ratio of specific heats
Γ	finite-difference operator, circulation
δ	characteristic length in y direction
δ	boundary-layer thickness
δ	central-difference operator defined by Eq. (3.14)
δ_u	represents change in u between two iterations
$\bar{\delta}$	central-difference operator defined by Eq. (3.13)
$\hat{\delta}$	central-difference operator defined by Eqs. (4.100)
δ^*	displacement thickness
δ_{ij}	Kronecker delta
Δ	forward-difference operator defined by Eq. (3.9)
Δx_+	$x_{j+1} - x_j$
Δx_-	$x_j - x_{j-1}$
Δy_+	$y_{j+1} - y_j$
Δy_-	$y_j - y_{j-1}$
$\Delta^n(\ )$	$(\ )^{n+1} - (\ )^n$
η	nondimensional distance variable
ϵ	turbulence dissipation rate
ϵ	round-off error
ϵ_i	coefficient of implicit smoothing term
ϵ_e	coefficient of explicit smoothing term
$\boldsymbol{\zeta}$	vorticity $(= \nabla \times \mathbf{V})$
ζ	magnitude of vorticity vector

θ	nondimensional thermal variable defined by Eq. (5.95)
θ	angle measured in circumferential direction
θ	parameter controlling type of difference scheme
θ	momentum thickness
θ_1, θ_2	parameters controlling type of difference scheme
κ	coefficient of bulk viscosity
κ	von Kármán constant
λ	eigenvalue
λ	generalized diffusion coefficient
$\overline{\lambda}$	$\lambda_T + \lambda$
μ	viscous coefficient in Burgers' equation
μ	coefficient of viscosity
μ	averaging operator defined by Eq. (3.16)
$\overline{\mu}$	$\mu + \mu_T$
μ'	second coefficient of viscosity
μ_T	eddy viscosity
ξ, η, ζ	transformed coordinates
π	3.14159...
$\mathbf{\Pi}_{ij}$	stress tensor
ρ	density
ρ	artificial density
ρ_i	species density
σ	eigenvalue
σ	shock angle
τ	parameter
τ	computational time or pseudo time
τ	shear stress
$\boldsymbol{\tau}_{ij}$	viscous stress tensor
ν	kinematic viscosity ($= \mu/\rho$)
ν	$c\,\Delta t/\Delta x$
ϕ	velocity potential
ϕ	angle in spherical coordinate system
ϕ	phase angle
ϕ	generalized variable
ϕ	limiter function
ϕ	grid control function
ϕ, ψ	boundary point clustering function
Φ	dissipation function
χ	strong-interaction parameter
χ	pressure gradient parameter, $[= (-1/\rho)\,dp/dx]$
ψ	stream function
$\boldsymbol{\psi}$	vector potential
ψ	limiter function
ψ	grid control function
ω	fraction of streamwise pressure gradient term

ω	damping function or damping parameter
ω_{i}	rate of production of species i
$\omega, \overline{\omega}$	overrelaxation parameters
∇	backward-difference operator defined by Eq. (3.11)
$\boldsymbol{\nabla}$	vector differential operator
∇^2	Laplacian operator $(= \boldsymbol{\nabla} \cdot \boldsymbol{\nabla})$

SUBSCRIPTS

b	body value
bdy	boundary value
B	boundary value
CFL	Courant-Friedrichs-Lewy condition
e	exact value
e	edge of boundary layer
f	frozen
i	inlet
i	inner
i	inviscid term
i, j, k	grid locations in x, y, z directions
inv	denotes inviscid value
lam	laminar-like in form
l	lower
l	left value
L	left value
min	minimum
max	maximum
n	normal or normal component
nose	nose value
o	intermediate (or estimated) value
o	initial value
o	outer
opt	optimum value
r	right value
ref	reference conditions
R	right value
s	shock value
stag	stagnation value
t	tangential or tangential component
t	thermal
t	partial differentiation with respect to time
T	tangential
T	turbulent
turb	turbulent quantity

u upper
v viscous term
wall wall value
x partial differentiation with respect to x
y partial differentiation with respect to y
z partial differentiation with respect to z
x, y, z differences in x, y, z directions
x, y, z components in x, y, z directions
1 conditions in front of shock
2 conditions behind shock
∞ freestream value

SUPERSCRIPTS

i index in marching direction
k iteration level
m iteration level
n index in marching direction
n time level
$*$ dummy time index
$**$ dummy time index
$*$ denotes a nondimensional quantity
$*$ sonic conditions
$+$ positive state
$-$ negative state
$'$ denotes fluctuation in turbulent flow, conventionally-averaged variables
$'$ perturbation quantity
$'$ correction term
$''$ denotes fluctuation in turbulent flow, mass-averaged variables

OVERBARS

$\bar{}$ denotes averaged quantity or time-averaged quantity
$\tilde{}$ dimensional variables
$\tilde{}$ denotes mass-averaged variables [see Eq. (5.64)]
$\hat{}$ Roe-averaged quantity
$\hat{}$ denotes value of variable from previous iteration

REFERENCES

Abbett, M. J. (1973). Boundary Condition Calculation Procedures for Inviscid Supersonic Flow Fields, *Proc. AIAA Computational Fluid Dynamics Conference*, Palm Springs, California, pp. 153–172.

Acirvos, A. and Schrader, M. L. (1982). Steady Flow in a Sudden Expansion at High Reynolds Numbers, *Phys. Fluids*, vol. 25, pp. 923–930.

Adams Jr., J. C. and Hodge, B. K. (1977). The Calculation of Compressible, Transitional, Turbulent, and Relaminarizational Boundary Layers Over Smooth and Rough Surfaces Using an Extended Mixing Length Hypothesis, AIAA Paper 77-682, Albuquerque, New Mexico.

Agarwal, R. K. (1981). A Third-Order-Accurate Upwind Scheme for Navier-Stokes Solutions in Three Dimensions, *Proc. ASME/AIAA Conference on Computers in Flow Predictions and Fluid Dynamics Experiments*, Washington, D.C., pp. 73–82.

Agarwal, R. K. and Rakich, J. V. (1978). Hypersonic Laminar Viscous Flow Past Spinning Cones at Angle of Attack, AIAA Paper 78-65, Huntsville, Alabama.

Akselvoll, K. and Moin, P. (1993). Large-Eddy Simulation of a Backward-Facing Step Flow, *Second International Symposium on Engineering Turbulence Modeling and Measurements* (W. Rodi and M. Martelli, eds.), Elsevier, New York.

Aldama, A. A. (1990). Filtering Techniques for Turbulent Flow Simulation, *Lect. Notes Eng.*, vol. 56, Springer-Verlag, New York.

Allen, D. N. de G. (1954). *Relaxation Methods*, McGraw-Hill, New York.

Allen, D. and Southwell, R. V. (1955). Relaxation Methods Applied to Determine the Motion, in Two Dimensions, of a Viscous Fluid Past a Fixed Cylinder, *Q. J. Mech. Appl. Math.*, vol. 8, pp. 129–145.

Allen, J. S. and Cheng, S. I. (1970). Numerical Solutions of the Compressible Navier-Stokes Equations for the Laminar Near Wake, *Phys. Fluids*, vol. 13, pp. 37–52.

Ames Research Staff (1953). Equations, Tables, and Charts for Compressible Flow, NACA Report 1135.

Ames, W. F. (1977). *Numerical Methods for Partial Differential Equations*, 2d ed., Academic, New York.

Amsden, A. A. and Harlow, F. H. (1970). The SMAC Method: A Numerical Technique for Calculating Incompressible Fluid Flows, Los Alamos Scientific Laboratory Report LA-4370, Los Alamos, New Mexico.

Anderson, C., Moss, J. N., and Sutton, K. (1976). Turbulent Viscous Shock-Layer Solutions with Strong Vorticity Interaction, AIAA Paper 76-120, Washington, D.C.

Anderson, D. A. (1973). Private Communication to W. L. Hankey.

Anderson, D. A. (1983). Adaptive Grid Methods for Partial Differential Equations, *Advances in Grid Generation* (K. N. Ghia and U. Ghia, eds.), FED-vol. 5, ASME Applied Mechanics, Bioengineering, and Fluids Engineering Conference, Houston, Texas.

Anderson, D. A. (1987). Equidistribution Schemes, Poisson Generators, and Adaptive Grids, *Appl. Math. Comput.*, vol. 24, pp. 211–277.

Anderson, D. A. (1990). Grid Cell Volume Control with an Adaptive Grid Generator, *Appl. Math. Comput.*, vol. 35, pp. 209–217.

Anderson, D. A., Tannehill, J. C., and Pletcher, R. H. (1984). *Computational Fluid Mechanics and Heat Transfer*, McGraw-Hill/Hemisphere, New York.

Anderson Jr., J. D. (1982). *Modern Compressible Flow*, McGraw-Hill, New York.

Anderson Jr., J. D. (1989). *High Hypersonic and Temperature Gas Dynamics*, McGraw-Hill, New York.

Anderson Jr., J. D. (1995). *Computational Fluid Dynamics*, McGraw-Hill, New York.

Anderson, W. K. (1986). Implicit Multigrid Algorithms for the Three-Dimensional Flux-Split Euler Equations, Ph.D. dissertation, Mississippi State University, Mississippi State.

Anderson, W., Thomas, J., and van Leer, B. (1985). A Comparison of Finite-Volume Flux-Vector Splittings for the Euler Equations, AIAA Paper 85-0122, Reno, Nevada.

Anon. (1982). Compendium of Unsteady Aerodynamic Measurements, AGARD R-702m.

Appel, K. and Haken, W. (1976). Every Planar Map is 4-Colorable, *Bull. Am. Math. Soc.*, vol. 82, pp. 711–712.

Aregbesola, Y. A. S. and Burley, D. M. (1977). The Vector and Scalar Potential Method for the Numerical Solution of Two- and Three-dimensional Navier-Stokes Equations, *J. Comput. Phys.*, vol. 24, pp. 398–415.

Arina, R. (1986). Orthogonal Grids with Adaptive Control, *Numerical Grid Generation in Computational Field Simulations* (B. K. Soni, J. F. Thompson, J. Hauser, P. Eiseman, eds.), NSF Engineering Research Center for Computational Field Simulation, Mississippi State University, Mississippi State, pp. 113–124.

Aziz, K. and Hellums, J. D. (1967). Numerical Solution of the Three-Dimensional Equations of Motion for Laminar Natural Convection, *Phys. Fluids*, vol. 10, pp. 314–324.

Bailey, F. R. and Ballhaus, W. F. (1972). Relaxation Methods for Transonic Flow about Wing-Cylinder Combinations and Lifting Swept Wings, Proc. Third Int. Conf. Num. Methods Fluid Mech., *Lect. Notes Phys.*, vol. 19, Springer-Verlag, New York, pp. 2–9.

Bailey, H. E. (1967). Programs for Computing Equilibrium Thermodynamic Properties of Gases, NASA TN D-3921.

Baker, A. J. (1983). *Finite Element Computational Fluid Mechanics*, McGraw-Hill, New York.

Baker, R. J. and Launder, B. E. (1974). The Turbulent Boundary Layer with Foreign Gas Injection: II—Predictions and Measurements in Severe Streamwise Pressure Gradients, *Int. J. Heat Mass Transfer*, vol. 17, pp. 293–306.

Baker, T. J. (1987). Three-Dimensional Mesh Generation by Triangulation of Arbitrary Point Set, AIAA Paper 87-1124, Honolulu, Hawaii.

Baldwin, B. S. and Barth, T. J. (1990). A One-Equation Turbulence Transport Model for High Reynolds Number Wall-Bounded Flows, NASA TM-102847.

Baldwin, B. S. and Lomax, H. (1978). Thin Layer Approximation and Algebraic Model for Separated Turbulent Flows, AIAA Paper 78-257, Huntsville, Alabama.

Bank, R. E. (1977). Marching Algorithms for Elliptic Boundary Value Problems: II—The Variable Coefficient Case, *SIAM J. Numer. Anal.*, vol. 5, pp. 950–970.

Banken, G. J., Roberts, D. W., Holcomb, J. E., and Birch, S. F. (1985). An Investigation of Film Cooling on a Hypersonic Vehicle Using a PNS Flow Analysis Code, AIAA Paper 85-1591, Cincinnati, Ohio.

Barakat, H. Z. and Clark, J. A. (1966). On the Solution of the Diffusion Equations by Numerical Methods, *J. Heat Transfer*, vol. 87–88, pp. 421–427.

Barbin, A. R. and Jones, J. B. (1963). Turbulent Flow in the Inlet Region of a Smooth Pipe, *Trans. ASME J. Basic Eng.*, vol. 85, pp. 29–34.

Barnett, M. and Davis, R. T. (1986). Calculation of Supersonic Flows with Strong Viscous-Inviscid Interaction, *AIAA J.*, vol. 24, pp. 1949–1955.

Barnett, M. and Power, G. D. (1988). An Efficient Algorithm for Strong Viscous/Inviscid Interaction in Hypersonic Flows, AIAA Paper 88-0712, Reno, Nevada.

Bartels, R. and Rothmayer, A. P. (1994). An Efficient Interaction Boundary Layer Method for Supersonic Flow in the Transonic Regime, *Comput. Fluids*, vol. 23, pp. 55–62.
Barth, T. J. (1987). Analysis of Implicit Local Linearization Techniques for TVD and Upwind Algorithms, AIAA Paper 87-0595, Reno, Nevada.
Barth, T. J. (1991). A 3-D Upwind Euler Solver for Unstructured Meshes, AIAA Paper 91-1548-CP, Honolulu, Hawaii.
Beam, R. M. and Warming, R. F. (1976). An Implicit Finite-Difference Algorithm for Hyperbolic Systems in Conservation Law Form, *J. Comput. Phys.*, vol. 22, pp. 87–110.
Beam, R. M. and Warming, R. F. (1978). An Implicit Factored Scheme for the Compressible Navier-Stokes Equations, *AIAA J.*, vol. 16, pp. 393–401.
Beckwith, I. E. and Gallagher, J. J. (1961). Local Heat Transfer and Recovery Temperatures on a Yawed Cylinder at a Mach Number of 4.15 and High Reynolds Numbers, NASA TR R-104.
Bender, E. E. and Khosla, P. K. (1988). Application of Sparse Matrix Solvers and Newton's Method to Fluid Flow Problems, AIAA Paper 88-3700, Cincinnati, Ohio.
Benton, E. R. and Platzman, G. W. (1972). A Table of Solutions of the One-Dimensional Burgers Equation, *Q. Appl. Math.*, vol. 30, pp. 195–212.
Bentson, J. and Vradis, G. (1987). A Two-Stage Pressure Correction Technique for the Incompressible Navier-Stokes Equations, AIAA Paper 87-0545, Reno, Nevada.
Bhutta, B. A. and Lewis, C. H. (1985a). Prediction of Three-Dimensional Hypersonic Reentry Flows Using a PNS Scheme, AIAA Paper 85-1604, Cincinnati, Ohio.
Bhutta, B. A. and Lewis, C. H. (1985b). Low Reynolds Number Flows Past Complex Multiconic Geometries, AIAA Paper 85-0362, Reno, Nevada.
Bhutta, B. A. and Lewis, C. H. (1989). Prediction of Nonequilibrium Viscous Hypersonic Flows over Lifting Configurations, AIAA Paper 89-1696, Buffalo, New York.
Bhutta, B. A. and Lewis, C. H. (1991). Recent Improvements in the Nonequilibrium VSL Scheme for Hypersonic Blunt-Body Flows, AIAA Paper 91-0469, Reno, Nevada.
Bhutta, B. A., Lewis, C. H., and Kautz II, F. A. (1985a). Hypersonic Equilibrium-Air Flows Using an Implicit Non-Iterative Parabolized Navier-Stokes Scheme, AIAA Paper 85-0169, Reno, Nevada.
Bhutta, B. A., Lewis, C. H., and Kautz II, F. A. (1985b). A Fast Fully Iterative Parabolized Navier-Stokes Scheme for Chemically Reacting Reentry Flows, AIAA Paper 85-0926, Reno, Nevada.
Bhutta, B. A., Song, D. J., and Lewis, C. H. (1989). Nonequilibrium Viscous Hypersonic Flows over Ablating Teflon Surfaces, AIAA Paper 89-0314, Reno, Nevada.
Birch, S. F. (1976). A Critical Reynolds Number Hypothesis and Its Relation to Phenomenological Turbulence Models, *Proc. 1976 Heat Transfer and Fluid Mechanics Institute*, Stanford University Press, Stanford, California, pp. 152–164.
Birkhoff, G., Varga, R. S., and Young, D. (1962). Alternating Direction Implicit Methods, *Advances in Computers*, vol. 3, Academic, New York, pp. 189–273.
Blottner, F. G. (1974). Variable Grid Scheme Applied to Turbulent Boundary Layers, *Comput. Methods Appl. Mech. Eng.*, vol. 4, pp. 179–194.
Blottner, F. G. (1975a). Investigation of Some Finite-Difference Techniques for Solving the Boundary Layer Equations, *Comput. Methods Appl. Mech. Eng.*, vol. 6, pp. 1–30.
Blottner, F. G. (1975b). Computational Techniques for Boundary Layers, *AGARD Lecture Series No. 73 on Computational Methods for Inviscid and Viscous Two- and Three-Dimensional Flowfields*, pp. (3-1)–(3-51).
Blottner, F. G. (1977). Numerical Solution of Slender Channel Laminar Flows, *Comput. Methods Appl. Mech. Eng.*, vol. 11, pp. 319–339.
Blottner, F. G. and Ellis, M. A. (1973). Finite-Difference Solution of the Incompressible Three-Dimensional Boundary Layer Equations for a Blunt Body, *Comput. Fluids*, vol. 1, Pergamon, Oxford, pp. 133–158.
Blottner, F. G., Johnson, M., and Ellis, M. (1971). Chemically Reacting Viscous Flow Program for Multi-Component Gas Mixtures, Sandia Labs., Report SC-RR-70-754, Albuquerque, New Mexico.
Bluford, G. S. (1978). Navier-Stokes Solution of Supersonic and Hypersonic Flow Field Around Delta Wings, AIAA Paper 78-1136, Seattle, Washington.

Boris, J. and Book, D. (1973). Flux-Corrected Transport: I SHASTA, A Fluid Transport Algorithm That Works, *J. Comput. Phys.*, vol. 11, pp. 38–69.

Boussinesq, J. (1877). Essai Sur La Théorie Des Eaux Courantes, *Mem. Présentés Acad. Sci.*, vol. 23, Paris, p. 46.

Bowyer, A. (1981). Computing Dirichlet Tesselations, *Comput. J.*, vol. 24, pp. 162–167.

Boynton, F. P. and Thomson, A. (1969). Numerical Computation of Steady, Supersonic, Two-Dimensional Gas Flow in Natural Coordinates, *J. Comput. Phys.*, vol. 3, pp. 379–398.

Bozeman, J. D. and Dalton, C. (1973). Numerical Study of Viscous Flow in a Cavity, *J. Comput. Phys.*, vol. 12, pp. 348–363.

Brackbill, J. U. (1982). Coordinate System Control: Adaptive Meshes, *Numerical Grid Generation, Proceedings of a Symposium on the Numerical Generation of Curvilinear Coordinate Systems and Their Use in the Numerical Solution of Partial Differential Equations* (J. F. Thompson, ed.), Elsevier, New York, pp. 277–294.

Brackbill, J. U. and Saltzman, J. (1980). An Adaptive Computation Mesh for the Solution of Singular Perturbation Problems, *Numerical Grid Generation Techniques*, NASA Conference Publication 2166, pp. 193–196.

Bradshaw, P., Ferriss D. H., and Altwell, N. D. (1967). Calculation of Boundary Layer Development Using the Turbulent Energy Equation, *J. Fluid Mech.*, vol. 28, pp. 593–616.

Bradshaw, P., Dean, R. B., and McEligot, D. M. (1973). Calculation of Interacting Turbulent Shear Layers: Duct Flow, *J. Fluids Eng.*, vol. 95, pp. 214–219.

Brailovskaya, I. (1965). A Difference Scheme for Numerical Solution of the Two-Dimensional Nonstationary Navier-Stokes Equations for a Compressible Gas, *Sov. Phys. Dokl.*, vol. 10, pp. 107–110.

Brandt, A. (1972). Multilevel Adaptive Technique for Fast Numerical Solution to Boundary Value Problems, Proceedings of the 3rd International Conference on Numerical Methods in Fluid Dynamics, *Lect. Notes Phys.*, vol. 18, Springer-Verlag, New York, pp. 82–89.

Brandt, A. (1977). Multilevel Adaptive Solutions to Boundary Value Problems, *Math. Comput.*, vol. 31, pp. 333–390.

Briggs, W. L. (1987). *A Multigrid Tutorial*, Society for Industrial and Applied Mathematics, Lancaster Press, Philadelphia, Pennsylvania.

Briley, W. R. (1970). A Numerical Study of Laminar Separation Bubbles using the Navier-Stokes Equations, United Aircraft Research Laboratories, Report J110614-1, East Hartford, Connecticut.

Briley, W. R. (1971). A Numerical Study of Laminar Separation Bubbles using the Navier-Stokes Equations, *J. Fluid Mech.*, vol. 47, pp. 713–736.

Briley, W. R. (1974). Numerical Method for Predicting Three-Dimensional Steady Viscous Flow in Ducts, *J. Comput. Phys.*, vol. 14, pp. 8–28.

Briley, W. R. and McDonald, H. (1974). Solution of the Three-Dimensional Compressible Navier-Stokes Equations by an Implicit Technique, Proc. Fourth Int. Conf. Num. Methods Fluid Dyn., Boulder Colorado, *Lect. Notes Phys.*, vol. 35, Springer-Verlag, New York, pp. 105–110.

Briley, W. R. and McDonald, H. (1975). Numerical Prediction of Incompressible Separation Bubbles, *J. Fluid Mech.*, vol. 69, pp. 631–656.

Briley, W. R. and McDonald, H. (1979). Analysis and Computation of Viscous Subsonic Primary and Secondary Flows, AIAA Paper 79-1453, Williamsburg, Virginia.

Briley, W. R. and McDonald, H. (1980). On the Structure and Use of Linearized Block Implicit Schemes, *J. Comput. Phys.*, vol. 34, pp. 54–73.

Brown, S. N. and Stewartson, K. (1969). Laminar Separation, *Annu. Rev. Fluid Mech.*, vol. 1, Annual Reviews, Inc., Palo Alto, California, pp. 45–72.

Buelow, P. E., Tannehill, J. C., Ievalts, J. O., and Lawrence, S. L. (1990). A Three-Dimensional Upwind Parabolized Navier-Stokes Code for Chemically Reacting Flows, AIAA Paper 90-0394, Reno, Nevada.

Buelow, P. E., Tannehill, J. C., Ievalts, J. O., and Lawrence, S. L. (1991). Three-Dimensional, Upwind, Parabolized Navier-Stokes Code for Chemically Reacting Flows, *J. Thermophys. Heat Transfer*, vol. 5, pp. 274–283.

Buggeln, R. C., McDonald, H., Kreskovsky, J. P., and Levy, R. (1980). Computation of Three-Dimensional Viscous Supersonic Flow in Inlets, AIAA Paper 80-0194, Pasadena, California.
Buneman, O. (1969). A Compact Non-Iterative Poisson Solver, Institute for Plasma Research SUIPR Report 294, Stanford University, California.
Buning, P. G., Chiu, I. T., Obayashi, S., Rizk, Y. M., and Steger, J. L. (1988). Numerical Simulation of the Integrated Space Shuttle Vehicle in Ascent, AIAA Paper 88-4359-CP, Minneapolis, Minnesota.
Buning, P. G., Chen, W. M., Renze, K. J., Sondak, D., Chiu, I.-T., and Slotnick, J. P. (1994). OVERFLOW/F3D User's Manual, Version 1.6 ap, NASA Ames Research Center, Moffett Field, California.
Burgers, J. M. (1948). A Mathematical Model Illustrating the Theory of Turbulence, *Adv. Appl. Mech.*, vol. 1, pp. 171–199.
Burggraf, O. R. (1966). Analytical and Numerical Studies of the Structures of Steady Separated Flows, *J. Fluid Mech.*, vol. 24, pp. 113–151.
Burggraf, O. R., Werle, M. J., Rizzetta, D., and Vatsa, V. N. (1979). Effect of Reynolds Number on Laminar Separation of a Supersonic Stream, *AIAA J.*, vol. 17, pp. 336–343.
Burns, A. D., Wilkes, N. S., Jones, I. P., and Kightley, J. R. (1986). FLOW3D: Body-Fitted Coordinates, Atomic Energy Research Establishment, Harwell, U.K., Report No. AERE-R 12262.
Burstein, S. Z. and Mirin, A. A. (1970). Third Order Difference Methods for Hyperbolic Equations, *J. Comput. Phys.*, vol. 5, pp. 547–571.
Bushnell, D. M., Cary Jr., A. M., and Holley, B. B. (1975). Mixing Length in Low Reynolds Number Compressible Turbulent Boundary Layers, *AIAA J.*, vol. 13, pp. 1119–1121.
Bushnell, D. M., Cary Jr., A. M., and Harris, J. E. (1976). Calculation Methods for Compressible Turbulent Boundary Layers, von Kármán Institute for Fluid Dynamics, *Lecture Series 86 on Compressible Turbulent Boundary Layers*, vol. 2, Rhode-St.-Genese, Belgium.
Buzbee, B. L., Golub, G. H., and Nielson, C. W. (1970). On Direct Methods for Solving Poisson's Equations, *SIAM J. Numer. Anal.*, vol. 7, pp. 627–656.
Candler, G. V. and MacCormack, R. W. (1988). The Computation of Hypersonic Ionized Flows in Chemical and Thermal Nonequilibrium, AIAA Paper 88-0511, Reno, Nevada.
Carcaillet, R. (1986). Optimization of Three-Dimensional Computational Grids and Generation of Flow Adaptive Computational Grids, AIAA Paper 86-0156, Reno, Nevada.
Caretto, L. S., Gosman, A. D., Patankar, S. V., and Spalding, D. (1972). Two Calculation Procedures for Steady, Three-Dimensional Flows with Recirculation, Proc. Third. Int. Conf. Num. Methods Fluid Mech., *Lect. Notes Phys.*, vol. 19, Springer-Verlag, New York, pp. 60–68.
Carlson, L. A. and Gally, T. A. (1991). Nonequilibrium Chemical and Radiation Coupling Phenomena in AOTV Flowfields, AIAA Paper 91-0569, Reno, Nevada.
Carnahan, B., Luther, H. A., and Wilkes, J. O. (1969). *Applied Numerical Methods*, John Wiley, New York.
Carpenter, J. G. and McRae, D. S. (1996). Adaption of Unstructured Meshes Using Node Movement, *Numerical Grid Generation in Computational Field Simulations* (B. K. Soni, J. F. Thompson, J. Hauser, and P. Eiseman, eds.), NSF Engineering Research Center for Computational Field Simulation, Mississippi State University, Mississippi State, pp. 269–278.
Carter, J. E. (1971). Numerical Solutions of the Supersonic, Laminar Flow over a Two-Dimensional Compression Corner, Ph.D. thesis, Virginia Polytechnic Institute and State University, Blacksburg.
Carter, J. E. (1978). A New Boundary-Layer Interaction Technique for Separated Flows, NASA TM-78690.
Carter, J. E. (1981). Viscous-Inviscid Interaction Analysis of Transonic Turbulent Separated Flow, AIAA Paper 81-1241, Palo Alto, California.
Carter, J. E. and Wornom, S. F. (1975). Forward Marching Procedure for Separated Boundary Layer Flows, *AIAA J.*, vol. 13, pp. 1101–1103.
Carter, J. E., Edwards, D. E., and Werle, M. J. (1980). A New Coordinate Transformation for Turbulent Boundary Layer Flows, *Numerical Grid Generation Techniques*, NASA Conference Publication 2166, pp. 197–212.

Catherall, D. (1996). Adaptivity via Mesh Movement with Three-Dimensional Block-Structured Grids, *Numerical Grid Generation in Computational Field Simulations* (B. K. Soni, J. F. Thompson, J. Hauser, and P. Eiseman, eds.), NSF Engineering Research Center for Computational Field Simulation, Mississippi State University, Mississippi State, pp. 57–66.

Cebeci, T. (1975). Calculation of Three-Dimensional Boundary Layers—II. Three-Dimensional Flows in Cartesian Coordinates, *AIAA J.*, vol. 13, pp. 1056–1064.

Cebeci, T. (1976). Separated Flows and Their Representation by Boundary Layer Equations, Report ONR-CR215-234-2, Office of Naval Research, Arlington, Virginia.

Cebeci, T. and Chang, K. C. (1978). A General Method for Calculating Momentum and Heat Transfer in Laminar and Turbulent Duct Flows, *Numer. Heat Transfer*, vol. 1, pp. 39–68.

Cebeci, T. and Smith, A. M. O. (1974). *Analysis of Turbulent Boundary Layers*, Academic, New York.

Cebeci, T., Kaups, K., Mosinskis, G. J., and Rehn, J. A. (1973). Some Problems of the Calculation of Three-Dimensional Boundary-Layer Flows on General Configurations, NASA CR-2285.

Cebeci, T., Kaups, K., and Ramsey, J. A. (1977). A General Method for Calculating Three-Dimensional Compressible Laminar and Turbulent Boundary Layers on Arbitrary Wings, NASA CR-2777.

Cebeci, T., Khattab, A. A., and Stewartson, K. (1979a). Prediction of Three-Dimensional Laminar and Turbulent Boundary Layers on Bodies of Revolution at High Angles of Attack, *Proc. Second Symposium on Turbulent Shear Flows*, Imperial College, London, pp. 15.8–15.13.

Cebeci, T., Carr, L. W., and Bradshaw, P. (1979b). Prediction of Unsteady Turbulent Boundary Layers with Flow Reversal, *Proc. Second Symposium on Turbulent Shear Flows*, Imperial College, London, pp. 14.23–14.28.

Cendes, Z. J., Shenton, D. N., and Shahnasser, H. (1983). Magnetic Field Computations using Delaunay Triangulations and Complementary Finite Element Methods, *IEEE Trans. Magnetics*, vol. 19, pp. 2551–2554.

Chakravarthy, S. R. (1979). The Split-Coefficient Matrix Method for Hyperbolic Systems of Gasdynamic Equations, Ph.D. dissertation, Department of Aerospace Engineering, Iowa State University, Ames.

Chakravarthy, S. R. (1983). Euler Equation—Implicit Schemes and Boundary Conditions, *AIAA J.*, vol. 21, pp. 699–706.

Chakravarthy, S. R. and Osher, S. (1983). High Resolution Application of the Osher Upwind Scheme for the Euler Equations, AIAA Paper 83-1943, Danvers, Massachusetts.

Chakravarthy, S. R. and Osher, S. (1985). A New Class of High Accuracy TVD Schemes for Hyperbolic Conservation Laws, AIAA Paper 85-0363, Reno, Nevada.

Chakravarthy, S. R. and Szema, K. Y. (1985). An Euler Solver for Three-Dimensional Supersonic Flows with Subsonic Pockets, AIAA Paper 85-1703, Cincinnati, Ohio.

Chakravarthy, S. R., Anderson, D. A, and Salas, M. D. (1980). The Split-Coefficient Matrix Method for Hyperbolic Systems of Gasdynamic Equations, AIAA Paper 80-0268, Pasadena, California.

Chakravarthy, S. R., Szema, K. Y., Goldberg, U. C., Gorski, J. J., and Osher, S. (1985). Application of a New Class of High Accuracy TVD Schemes to the Navier-Stokes Equations, AIAA Paper 85-0165, Reno, Nevada.

Chambers, T. L. and Wilcox, D. C. (1976). A Critical Examination of Two-Equation Turbulence Closure Models, AIAA Paper 76-352, San Diego, California.

Chan, Y. Y. (1972). Compressible Turbulent Boundary Layer Computations Based on an Extended Mixing Length Approach, *Can. Aeronaut. Space Inst. Trans.*, vol. 5, pp. 21–27.

Chang, S.-C. (1987). On the Validity of the Modified Equation Approach to the Stability Analysis of Finite-Difference Methods, AIAA Paper 87-1120, Honolulu, Hawaii.

Chapman, A. J. (1974). *Heat Transfer*, 3d. ed., Macmillan, New York.

Chapman, D. R. (1975). Introductory Remarks, NASA SP-347, pp. 4–7.

Chapman, D. R. (1979). Computational Aerodynamics Development and Outlook, *AIAA J.*, vol. 17, pp. 1293–1313.

Chaussee, D. S. and Pulliam, T. H. (1981). Two-Dimensional Inlet Simulation Using a Diagonal Implicit Algorithm, *AIAA J.*, vol. 19, pp. 153–159.

Chaussee, D. S., Patterson, J. L., Kutler, P., Pulliam, T. H., and Steger, J. L. (1981). A Numerical Simulation of Hypersonic Viscous Flows over Arbitrary Geometrics at High Angle of Attack, AIAA Paper 81-0050, Pasadena, California.

Chaussee, D. S., Rizk, Y. M., and Buning, P. G. (1984). Viscous Computation of a Space Shuttle Flowfield, *Ninth International Conference on Numerical Methods in Fluid Dynamics*, vol. 218, Springer-Verlag, Berlin, pp. 148–153.

Chaussee, D. S., Blom, G., and Wai, J. C. (1985). Numerical Simulation of Viscous Supersonic Flow over a Generic Fighter Configuration, NASA TM-86823.

Chawner, J. R. and Anderson, D. A. (1991). Development of an Algebraic Grid Generation Method with Orthogonality and Clustering Control, *Numerical Grid Generation in Computational Field Simulation and Related Fields* (A. S. Arcilla, J. Hauser, P. R. Eiseman, and J. F. Thompson, eds.), North Holland, New York, pp. 107–114.

Chen, K.-H., and Pletcher, R. H. (1991). Primitive Variable, Strongly Implicit Calculation Procedure for Viscous Flows at All Speeds, *AIAA J.* vol. 29, pp. 1241–1249.

Chen, K.-H. and Pletcher, R. H. (1993). Simulation of Three-Dimensional Liquid Sloshing Flows Using a Strongly Implicit Calculation Procedure, *AIAA J.*, vol. 31, pp. 901–910.

Cheng, H. K. (1963). The Blunt-Body Problem in Hypersonic Flow at Low Reynolds Number, Cornell Aeronautical Laboratory, AF-1285-A-10, Buffalo, New York.

Cheng, H. K., Chen, S. Y., Mobley, R., and Huber, C. R. (1970). The Viscous Hypersonic Slender-Body Problem: A Numerical Approach Based on a System of Composite Equations, The Rand Corporation, RM-6193-PR, Santa Monica, California.

Chien, K.-Y. (1982). Predictions of Channel and Boundary-Layer Flows with a Low-Reynolds-Number Turbulence Model, *AIAA J.*, vol. 20, pp. 33–38.

Chilukuri, R. and Pletcher, R.H. (1980). Numerical Solutions to the Partially Parabolized Navier-Stokes Equations for Developing Flow in a Channel, *Numer. Heat Transfer*, vol. 3, pp. 169–188.

Chitsomboon, T. and Northam, G. B. (1988). A 3D-PNS Computer Code for the Calculation of Supersonic Combusting Flows, AIAA Paper 88-0438, Reno, Nevada.

Chiu, I.-T. and Pletcher, R. H. (1986). Prediction of Heat Transfer in Laminar Flow over a Rearward-Facing Step Using the Partially Parabolized Navier-Stokes Equations, *Heat Transfer 1986, Proceedings of the Eighth International Heat Transfer Conference* (C. L. Tien, V. P. Carey, and J. K. Ferrell, eds.), Hemisphere, Washington, D.C., pp. 414–420.

Choi, D. and Merkle, C. L. (1985). Application of Time-Iterative Schemes to Incompressible Flow, *AIAA J.*, vol. 23, pp. 1518–1524.

Choi, D. and Merkle, C. L. (1991). Time Derivative Preconditioning for Viscous Flows, AIAA Paper 91-1652, Honolulu, Hawaii.

Choi, D. and Merkle, C. L. (1993). The Application of Preconditioning in Viscous Flows, *J. Comput. Phys.*, vol. 105, pp. 203–223.

Chorin, A. J. (1967). A Numerical Method for Solving Incompressible Viscous Flow Problems, *J. Comput. Phys.*, vol. 2, pp. 12–26.

Chorin, A. J. (1968). Numerical Solution of the Navier-Stokes Equations, *Math. Comput.*, vol. 22, pp. 745–762.

Chow, C.-Y. (1979). *An Introduction to Computational Fluid Mechanics*, John Wiley, New York.

Chow, L. C. and Tien, C. L. (1978). An Examination of Four Differencing Schemes for Some Elliptic-Type Convection Equations, *Numer. Heat Transfer*, vol. 1, pp. 87–100.

Christoph, G. H. and Pletcher, R. H. (1983). Prediction of Rough-Wall Skin Friction and Heat Transfer, *AIAA J.*, vol. 21, pp. 509–515.

Christov, C. I. (1982). Orthogonal Coordinate Meshes with Manageable Jacobian, *Numerical Grid Generation* (J. F. Thompson, ed.), Elsevier, New York, pp. 885–892.

Chung, T. J. (1978). *Finite Element Analysis in Fluid Dynamics*, McGraw-Hill, New York.

Churchill, R. V. (1941). *Fourier Series and Boundary Value Problems*, McGraw-Hill, New York.

Churchill, R. V. (1948). *Introduction to Complex Variables*, McGraw-Hill, New York.

Churchill, R. V. (1960). *Introduction to Complex Variables and Applications*, 2d. ed., McGraw-Hill, New York.

Churchill, S. W. (1974). *The Interpretation and Use of Rate Data: The Rate Concept*, Hemisphere, Washington, D.C.

Coakley, T. J. (1983a). Turbulence Modeling Methods for the Compressible Navier-Stokes Equations, AIAA Paper 83-1693, Danvers, Massachusetts.

Coakley, T. L. (1983b). Implicit Upwind Methods for the Compressible Navier-Stokes Equations, AIAA Paper 83-1958, Danvers, Massachusetts.

Coirer, W. J. and Powell, K. G. (1995). A Cartesian, Cell-Based Approach for Adaptively Refined Solutions of the Euler and Navier-Stokes Equations, AIAA Paper 95-0566, Reno, Nevada.

Cole, J. D. and Messiter, A. F. (1957). Expansion Procedures and Similarity Laws for Transonic Flow, *Z. Angew. Math. Phys.*, vol. 8, pp. 1–25.

Coles, D. E. (1953). Measurements in the Boundary Layer on a Smooth Flat Plate in Supersonic Flow, III. Measurements in a Flat Plate Boundary Layer at the Jet Propulsion Laboratory, Jet Propulsion Laboratory Report 20-71, California Institute of Technology, Pasadena, California.

Courant, R. and Friedrichs, K. O. (1948). *Supersonic Flow and Shock Waves*, Interscience Publishers, New York.

Courant, R., Friedrichs, K. O., and Lewy, H. (1928). Über die Partiellen Differenzengleichungen der Mathematischen Physik, *Math. Ann.*, vol. 100, pp. 32–74. (Translated to: On the Partial Difference Equations of Mathematical Physics, *IBM J. Res. Dev.*, vol. 11, pp. 215–234, 1967.)

Crank, J. and Nicolson, P. (1947). A Practical Method for Numerical Evaluation of Solutions of Partial Differential Equations of the Heat-Conduction Type, *Proc. Cambridge Philos. Soc.*, vol. 43, pp. 50–67.

Crawford, M. E. and Kays, W. M. (1975). STAN5—A Program for Numerical Computation of Two-Dimensional Internal/External Boundary Layer Flows, Report No. HMT-23, Thermosciences Division, Department of Mechanical Engineering, Stanford University, California.

Cresci, R. J., Rubin, S. G., Nardo, C. T., and Lin, T. C. (1969). Hypersonic Interaction along a Rectangular Corner, *AIAA J.*, vol. 7, pp. 2241–2246.

Crowley, W. P. (1962). Internal Memorandum, Lawrence Radiation Laboratory, Livermore, California.

Crowley, W. P. (1967). Second-Order Numerical Advection, *J. Comput. Phys.*, vol. 1, pp. 471–484.

Daly, B. J. and Harlow, F. H. (1970). Transport Equations in Turbulence, *Phys. Fluids*, vol. 13, pp. 2634–2649.

Dancey, C. L. and Pletcher, R. H. (1974). A Boundary Layer Finite Difference Method for Calculating Through the Separation Point and into the Region of Recirculation in Incompressible Laminar Flow, Engineering Research Institute Technical Report 74103/HTL-2, Iowa State University, Ames.

Dang, A. L., Kehtarnavaz, H., and Coats, E. E. (1989). The Use of Richardson Extrapolation in PNS Solutions of Rocket Nozzle Flow, AIAA Paper 89-2895, Monterey, California.

Dannenhoffer, J. F. (1991). Computer-Aided Block-Structuring through the Use of Optimization and Expert System Techniques, AIAA Paper-91-1585, Honolulu, Hawaii.

Dannenhoffer, J. F. (1995). Automatic Blocking for Complex Three-Dimensional Configurations, *Surface Modeling, Grid Generation, and Related Issues in Computational Fluid Dynamics Workshop: Workshop Proceedings*, NASA Conference Publication 3291, NASA Lewis Research Center, Cleveland, Ohio.

Dannenhoffer, J. F. (1996). Automatic Generation of Block Structures—Progress and Challenges, *Numerical Grid Generation in Computational Fluid Simulations* (B. K. Soni, J. F. Thompson, J. Hauser, and P. Eiseman, eds.), NSF Engineering Research Center for Computational Field Simulation, Mississippi State University, Mississippi State, pp. 403–412.

Dash, S. M. (1985). Recent Developments in the Modeling of High Speed Jets, Plumes and Wakes, AIAA Paper 85-1616, Cincinnati, Ohio.

Dash, S. M. (1989). Recent Advances in Parabolized and Full Navier-Stokes Solvers for Analyzing Hypersonic Chemically Reacting Flowfield Problems, AIAA Paper 89-1698, Buffalo, New York.

Dash, S. M. and Wolf, D. E. (1984a). Interactive Phenomena in Supersonic Jet Mixing Problems, Part I: Phenomenology and Numerical Modeling Techniques, *AIAA J.*, vol. 22, pp. 905–913.

Dash, S. M. and Wolf, D. E. (1984b). Iterative Phenomena in Supersonic Jet Mixing Problems, Part II: Numerical Studies, *AIAA J.*, vol. 22, pp. 1395–1404.

Davis, R. T. (1963). Laminar Compressible Flow Past Axisymmetric Blunt Bodies (Results of a Second Order Theory), Ph.D. dissertation, Stanford University, California.
Davis, R. T. (1970). Numerical Solution of the Hypersonic Viscous Shock-Layer Equations, *AIAA J.*, vol. 8, pp. 843–851.
Davis, R. T. (1979). Numerical Methods for Coordinate Generation Based on Schwarz-Christoffel Transformations, AIAA Paper 79-1463, Williamsburg, Virginia.
Davis, R. T. and Flügge-Lotz, I. (1964). Second-Order Boundary-Layer Effects in Hypersonic Flow Past Axisymmetric Blunt Bodies, *J. Fluid Mech.*, vol. 20, pp. 593–623.
Davis, R. T. and Rubin, S. G. (1980). Non-Navier-Stokes Viscous Flow Computations, *Comput. Fluids*, vol. 8, pp. 101–131.
Davis, S. F. (1984). A Rotationally Biased Upwind Difference Scheme for the Euler Equations, *J. Comput. Phys.*, vol. 56, pp. 65–92.
Daywitt, J. E. and Anderson, D. A. (1974). Analysis of a Time-Dependent Finite-Difference Technique for Shock Interaction and Blunt-Body Flows, Engineering Research Institute Technical Report 74074, Iowa State University, Ames.
Deardorff, J. W. (1970). A Numerical Study of Three-Dimensional Turbulent Channel Flow at Large Reynolds Numbers, *J. Fluid Mech.*, vol. 41, pp. 453–480.
Deconick, H., Struijs, R., Bourgois, G., Paillaere, H., and Roe, P. L. (1992). Multidimensional Upwind Methods for Unstructured Grids, AGARD-R-787.
Degani, D. and Schiff, L. B. (1983). Computation of Supersonic Viscous Flows around Pointed Bodies at Large Incidence, AIAA Paper 83-0034, Reno, Nevada.
Delaunay, B. (1934). Sur la sphere vide, *Bul. Acad. Sci.*, *URSS*, Class. Sci. Nat., pp. 793–800.
Delery, J. M. and Formery, M. J. (1983). A Finite Difference for Inverse Solutions of 3-D Turbulent Boundary-Layer Flow, AIAA Paper 83-0301, Reno, Nevada.
De Neef, T. and Moretti, G. (1980). Shock Fitting For Everybody, *Comput. Fluids*, vol. 8, pp. 327–334.
Dennis, S. C. R., Ingham, D. B. and Cook, R. N. (1979). Finite-Difference Methods for Calculating Steady Incompressible Flows in Three Dimensions, *J. Comput. Phys.*, vol. 33, pp. 325–339.
Der Jr., J. and Raetz, G. S. (1962). Solution of General Three-Dimensional Laminar Boundary Layer Problems by an Exact Numerical Method, Institute of the Aerospace Sciences, Paper 62-70, New York.
Désidéri, J.-A. and Tannehill, J. C. (1977a). Over-Relaxation Applied to the MacCormack Finite-Difference Scheme, *J. Comput. Phys.*, vol. 23, pp. 313–326.
Désidéri, J.-A. and Tannehill, J. C. (1977b). Time-Accuracy of the Over-Relaxed MacCormack Finite-Difference Scheme, ERI Report 77251, Iowa State University, Ames.
Désidéri, J.-A., Steger, J. L., and Tannehill, J. C. (1978). On Improving the Iterative Convergence Properties of an Implicit Approximate-Factorization Finite Difference Algorithm, NASA Technical Memorandum 78495.
Diewert, S. and Green M. (1986). Computational Aerothermodynamics, 33rd Annual Meeting of the American Astronautical Association, Boulder, Colorado.
Dirichlet, G. L. (1850). Uber die Reduction der Positiven Quadratischen Formen mit Drei Undestimmten Ganzen Zahlen, *Z. Reine Angew. Math.*, vol. 40, pp. 209–227.
Dodge, P. R. (1977). Numerical Method for 2D and 3D Viscous Flows, *AIAA J.*, vol. 15, pp. 961–965.
Donaldson, C. duP. (1972). Calculation of Turbulent Shear Flows for Atmospheric and Vortex Motions, *AIAA J.*, vol. 10, pp. 4–12.
Donaldson, C. duP. and Rosenbaum, H. (1968). Calculation of Turbulent Shear Flows Through Closure of the Reynolds Equations by Invariant Modeling, Aero. Res. Assoc. of Princeton Report 127.
Dorrance, W. H. (1962). *Viscous Hypersonic Flow*, McGraw-Hill, New York.
Douglas Jr., J. (1955). On the Numerical Integration of $\partial^2 u/\partial x^2 + \partial^2 u/\partial y^2 = \partial u/\partial t$ by Implicit Methods, *J. Soc. Ind. Appl. Math.*, vol. 3, pp. 42–65.
Douglas, J. and Gunn, J. E. (1964). A General Formulation of Alternating Direction Methods—Part I. Parabolic and Hyperbolic Problems, *Numer. Math.*, vol. 6, pp. 428–453.
Douglas, J. and Rachford, H. H. (1956). On the Numerical Solution of Heat Conduction Problems in Two and Three Space Variables, *Trans. Am. Math. Soc.*, vol. 82, pp. 421–439.

DuFort, E. C. and Frankel, S. P. (1953). Stability Conditions in the Numerical Treatment of Parabolic Differential Equations, *Math. Tables Other Aids Comput.*, vol. 7, pp. 135–152.

Dwyer, H. A. (1971). Hypersonic Boundary Layer Studies on a Spinning Sharp Cone at Angle of Attack, AIAA Paper 71-57, New York, New York.

Dwyer, H. A. and Sanders, B. R. (1975). A Physically Optimum Difference Scheme for Three-Dimensional Boundary Layers, Proc. Fourth Int. Conf. Num. Methods Fluid Dyn., *Lect. Notes Phys.*, vol. 35, Springer-Verlag, New York, pp. 144–150.

Dwyer, H. A., Kee, R. J., and Sanders, B. R. (1979). An Adaptive Grid Method for Problems in Fluid Mechanics and Heat Transfer, AIAA Paper 79-1464, Williamsburg, Virginia.

Dwyer, H. A., Raiszadeh, F., and Otey, G. (1980). A Study of Reactive Diffusion Problems with Stiff Integrators and Adaptive Grids, Proc. Seventh Int. Conf. Num. Methods Fluid Dyn., *Lect. Notes Phys.*, vol. 141, Springer-Verlag, New York, pp. 170–175.

Ebrahimi, H. B. and Gilbertson, M. (1992). Two- and Three-Dimensional Parabolized Navier-Stokes Code for Scramjet Combustor, Nozzle and Film Cooling Analysis, AIAA Paper 92-0391, Reno, Nevada.

Edelman, R. and Weilerstein, G. (1969). A Solution of the Inviscid-Viscid Equations with Applications to Bounded and Unbounded Multicomponent Reacting Flows, AIAA Paper 69-83, New York.

Edwards, D. E. (1986). Analysis of Three-Dimensional Separated Flow Using Interacting Boundary-Layer Theory, *Boundary-Layer Separation* (F. T. Smith and S. N. Brown, eds.), Springer-Verlag, New York.

Edwards, D. and Carter, J. (1985). Analysis of Three-Dimensional Separated Flow with the Boundary-Layer Equations, AIAA Paper 85-1499-CP, Cincinnati, Ohio.

Edwards, T. A. (1985). Noniterative Three-Dimensional Grid Generation Using Parabolic Partial Differential Equations, AIAA Paper 85-0485, Reno, Nevada.

Eiseman, P. R. (1979). A Multi-Surface Method of Coordinate Generation, *J. Comput. Phys.*, vol. 33, pp. 118–150.

Eiseman, P. R. (1982). Orthogonal Grid Generation, *Numerical Grid Generation* (J. F. Thompson, ed.), Elsevier, New York, pp. 193–230.

Eiseman, P. R. (1983). Alternating Direction Adaptive Grid Generation, AIAA Paper 83-1937, Danvers, Massachusetts.

Eiseman, P. R. and Smith, R. E. (1980). Mesh Generation Using Algebraic Techniques, *Numerical Grid Generation Techniques*, NASA Conference Publication 2166, pp. 73–120.

Eisenstadt, S. C., Gursky, M. C., Schultz, M. H., and Sherman, A. H. (1977). Yale Sparse Matrix Package: II The Nonsymmetric Codes, Research Report No. 114, Yale University, New Haven.

Emery, A. F. and Gessner, F. B. (1976). The Numerical Prediction of the Turbulent Flow and Heat Transfer in the Entrance Region of a Parallel Plate Duct, *J. Heat Transfer*, vol. 98, pp. 594–600.

Enquist, B. and Osher, S. (1980). Stable and Entropy Satisfying Approximations for Transonic Flow Calculations, *Math. Comput.*, vol. 34, pp. 45–75.

Enquist, B. and Osher, S. (1981). One-Sided Difference Approximations for Nonlinear Conservation Laws, *Math. Comput.*, vol. 36, pp. 321–352.

Erlebacher, G., and Hussaini, M. Y., Speziale, C. G., and Zang, T. A. (1992). Toward the Large-Eddy Simulation of Compressible Turbulent Flows, *J. Fluid Mech.*, vol. 238, pp. 155–185.

Evans, M. E. and Harlow, F. H. (1957). The Particle-in-Cell Method for Hydrodynamic Calculations, Los Alamos Scientific Laboratory Report LA-2139, Los Alamos, New Mexico.

Fang, F. (1991). Sweepline Algorithm for Unstructured Grid Generation, Ph.D. dissertation, University of Texas at Arlington.

Fang, J., Parpia, I., and Kennon, S. (1993). Sweepline Algorithm for Unstructured-Grid Generation on Two-Dimensional Non-Convex Domains, *Int. J. Numer. Eng.*, vol. 36, pp. 2761–2778.

Farouk, B. and Fusegi, T. (1985). A Coupled Solution of the Vorticity-Velocity Formulation of the Incompressible Navier-Stokes Equations, *Int. J. For Numerical Methods in Fluids*, vol. 5, pp. 1017–1034.

Fasel, H. and Booz, O. (84). Numerical Investigation of Supercritical Taylor-Vortex Flow for a Wide Gap, *J. Fluid Mech.*, vol. 138, pp. 21–52.

Favre, A. (1965). Equations des Gaz Turbulents Compressibles: 1. Formes Générales, *J. Mec.*, vol. 4, pp. 361–390.
Fedorenko, R. P. (1962). A Relaxation Method for Solving Elliptic Equations, *USSR Comput. Math. Math. Phys.*, vol. 1, pp. 1092–1096.
Fedorenko, R. P. (1964). The Speed of Convergence of an Iterative Process, *USSR Comput. Math. Math. Phys.*, vol. 4, pp. 227–235.
Feng, J. and Merkle, C. L. (1990). Evaluation of Preconditioning Methods for Time-Marching Systems, AIAA Paper 90-0016, Reno, Nevada.
Fletcher, C. A. J. (1988). *Computational Techniques for Fluid Dynamics*, vols. I and II, Springer-Verlag, New York.
Flores, J., Holst, T. L. and Gundy, K. L. (1987). Transonic Navier-Stokes Solution for a Fighter-like Configuration, AIAA Paper 87-0032, Reno, Nevada.
Forsythe, G. E. and Wasow, W. (1960). *Finite Difference Methods for Partial Differential Equations*, Wiley, New York.
Fortune, S. (1987). A Sweepline Algorithm for Voronoi Diagrams, *Algorithmica*, vol. 2. pp. 153–174.
Frankel, S. P. (1950). Convergence Rates of Iterative Treatments of Partial Differential Equations, *Math. Tables Other Aids Comput.*, vol. 4, pp. 65–75.
Freitas, C. J., Street, R. L., Findikakis, A. N., and Koseff, J. R. (1985). Numerical Simulation of Three-Dimensional Flow in a Cavity, *Int. J. Numer. Methods Fluids*, vol. 5, pp. 561–575.
Freidrich, C. M. and Forstall Jr., W. (1953). A Numerical Method for Computing the Diffusion Rate of Coaxial Jets, *Proceedings of the Third Midwestern Conference on Fluid Mechanics*, Univ. of Minnesota Inst. of Technol., Minneapolis, Minnesota, pp. 635–649.
Fröberg, C. (1969). *Introduction to Numerical Analysis*, 2d ed., Addison-Wesley, Reading, Massachusetts, pp. 21–28.
Fromm, J. E. (1968). A Method for Reducing Dispersion in Convective Difference Schemes, *J. Comput. Phys.*, vol. 3, pp. 176–189.
Gabutti, B. (1983). On Two Upwind Finite-Difference Schemes for Hyperbolic Equations in Nonconservative Form, *Comput. Fluids*, vol. 11, pp. 207–230.
Garabedian, P. R. (1964). *Partial Differential Equations*, Wiley, New York.
Gardner, W. D. (1982). The Independent Inventor, *Datamation*, vol. 28, pp. 12–22.
Gary, J. (1962). Numerical Computation of Hydrodynamic Flows Which Contain a Shock, Courant Institute of Mathematical Sciences Report NYO 9603, New York University.
Gary, J. (1969). The Numerical Solution of Partial Differential Equations, National Center for Atmospheric Research, NCAR Manuscript 69-54, Boulder, Colorado.
Gatski, T. B., Grosch, C. E., and Rose, M. E. (1982). A Numerical Study of the Two-Dimensional Navier-Stokes Equations in Vorticity-Velocity Variables, *J. Comput. Phys.*, vol. 48, pp. 1–22.
Gault, D. E. (1955). An Experimental Investigation of Regions of Separated Laminar Flow, NACA TN-3505.
Gegg, S. G., Pletcher, R. H., and Steger, J. L. (1989). A Dual Potential Formulation of the Navier-Stokes Equations, Report HTL-51, CFD-21, ISU-ERI-Ames-90030, College of Engineering, Iowa State University, Ames.
Gerbsch, R. A. and Agarwal, R. K. (1990). Solution of the Parabolized Navier-Stokes Equations Using Osher's Upwind Scheme, AIAA Paper 90-0392, Reno, Nevada.
Gerbsch, R. A. and Agarwal, R. K. (1991). Solution of the Parabolized Navier-Stokes Equations for Three-Dimensional Real Gas Flows Using Osher's Upwind Scheme, AIAA Paper 91-0248, Reno, Nevada.
Germano, M., Piomelli, U., Moin, P., and Cabot, W. H. (1991). A Dynamic Subgrid-Scale Eddy Viscosity Model, *Phys. Fluids, A*, vol. 3, pp. 1760–1765.
Ghia, K. N. and Sokhey, J. S. (1977). Laminar Incompressible Viscous Flow in Curved Ducts of Regular Cross-Sections, *J. Fluids Eng.*, vol. 99, pp. 640–648.
Ghia, K. N., Hankey Jr., W. L., and Hodge, J. K. (1977a). Study of Incompressible Navier-Stokes Equations in Primitive Variables Using Implicit Numerical Technique, AIAA Paper 77-648, Albuquerque, New Mexico.
Ghia, K. N., Hankey Jr., W. L., and Hodge, J. K. (1979). Use of Primitive Variables in the Solution of Incompressible Navier-Stokes Equations, *AIAA J.*, vol. 17, pp. 298–301.

Ghia, U., Ghia, K. N., and Struderus, C. J. (1977b). Three-Dimensional Laminar Incompressible Flow in Straight Polar Ducts, *Comput. Fluids*, vol. 5, pp. 205–218.

Ghia, U., Ghia, K. N., Rubin, S. G., and Khosla, P. K. (1981). Study of Incompressible Flow Separation Using Primitive Variables, *Comput. Fluids*, vol. 9, pp. 123–142.

Ghia, U., Ghia, K. N., and Shin, C. T. (1982). High-Re Solutions for Incompressible Flow Using the Navier-Stokes Equations and a Multigrid Method, *J. Comput. Phys.*, vol. 48, pp. 387–411.

Gielda, T. P. and Agarwal, R. K. (1989). Efficient Finite-Volume Parabolized Navier-Stokes Solutions for Three-Dimensional, Hypersonic, Chemically Reacting Flowfields, AIAA Paper 89-0103, Reno, Nevada.

Gielda, T. P. and McRae, D. S. (1986). An Accurate, Stable, Explicit, Parabolized Navier-Stokes Solver for High-Speed Flows, AIAA Paper 86-1116, Atlanta, Georgia.

Gielda, T. P., Hunter, L. G., and Chawner, J. R. (1988). Efficient Parabolized Navier-Stokes Solutions of Three-Dimensional, Chemically Reacting Scramjet Flowfields, AIAA Paper 88-0096, Reno, Nevada.

Gnoffo, P. A. (1980). Complete Supersonic Flowfields over Blunt Bodies in a Generalized Orthogonal Coordinate System, NASA TM 81784.

Gnoffo, P. A. (1983). Hypersonic Flows over Biconics Using a Variable-Effective-Gamma Parabolized Navier-Stokes Code, AIAA Paper 83-1666, Danvers, Massachusetts.

Gnoffo, P. A. (1986). Application of Program LAURA to Three-Dimensional AOTV Flow Fields, AIAA Paper 86-0565, Reno, Nevada.

Gnoffo, P. A. (1989). Upwind-Biased, Point-Implicit Relaxation Strategies for Viscous Hypersonic Flow, AIAA Paper 89-1972-CP, Buffalo, New York.

Godunov, S. K. (1959). Finite-Difference Method for Numerical Computation of Discontinuous Solutions of the Equations of Fluid Dynamics, *Mat. Sb.*, vol. 47, pp. 271–306.

Gogineni, P. R., Lewis, C. H., and Denysyk, B. (1980). Three-Dimensional Viscous Hypersonic Flows over General Bodies, AIAA Paper 80-0029, Pasadena, California.

Goldstein, S. (1948). On Laminar Boundary Layer Flow near a Position of Separation, *Q. J. Mech. Appl. Math.*, vol. 1, pp. 43–69.

Golub, G. H. and van Loan, C. F. (1989). *Matrix Computations*, 2d ed., John Hopkins University Press, Baltimore, Maryland.

Gordon, P. (1969). The Diagonal Form of Quasi-Linear Hyperbolic Systems as a Basis for Difference Equations, General Electric Company Final Report, Naval Ordinance Laboratory Contract No. N60921-7164, pp. II.D-1, II.D-22.

Gordon, R. and Davis, R. T. (1992). Improved Method for Solving the Viscous Shock Layer Equations, *AIAA J.*, vol. 30, pp. 1770–1779.

Gordon, S. and McBride, B. J. (1971). Computer Program for Calculation of Complex Chemical Equilibrium Compositions, Rocket Performance, Incident and Reflected Shocks, and Chapman-Jouguet Detonations, NASA SP-273.

Gordon, W. and Hall, C. (1973). Construction of Curvilinear Coordinate Systems and Application to Mesh Generation, *Int. J. Numer. Methods Eng.*, vol. 7, pp. 461–477.

Gosman, A. D. and Spalding, D. B. (1971). The Prediction of Confined Three-Dimensional Boundary Layers, *Salford Symposium on Internal Flows*, Paper 19, Inst. Mech. Engrs., London.

Gottlieb, J. J. and Groth, C. P. T. (1988). Assessment of Riemann Solvers for Unsteady One-Dimensional Inviscid Flows of Perfect Gases, *J. Comput. Phys.*, vol. 78, pp. 437–458.

Greenspan, D. (1961). *Introduction to Partial Differential Equations*, McGraw-Hill, New York.

Grossman, B. (1979). Numerical Procedure for the Computation of Irrotational Conical Flows, *AIAA J.*, vol. 17, pp. 828–837.

Grossman, B. and Siclari, M. J. (1980). The Nonlinear Supersonic Potential Flow over Delta Wings, AIAA Paper 80-0269, Pasadena, California.

Grossman, B. and Walters, R. W. (1987). An Analysis of Flux-Split Algorithms for Euler's Equations with Real Gases, AIAA Paper 87-1117-CP, Honolulu, Hawaii.

Guj, G. and Stella, F. (1988). Numerical Solutions of High Re Recirculating Flows in Vorticity-Velocity Form, *Int. J. for Numerical Methods in Fluids*, vol. 8, pp. 405–416.

Gupta, R. N. (1987). Navier-Stokes and Viscous Shock-Layer Solutions for Radiating Hypersonic Flows, AIAA Paper 87-1576, Honolulu, Hawaii.

Gupta, R. N., Lee, K, P., Zoby, E. V., Moss, J. N., and Thompson, R. A. (1990). Hypersonic Viscous Shock-Layer Solutions over Long Slender Bodies, Part I: High Reynolds Number Flows, *J. Spacecraft Rockets*, vol. 27, pp. 175–184.

Gupta, R. N., Lee, K. P., and Zoby, E. V. (1992). Enhancements to Viscous-Shock-Layer Technique, AIAA Paper 92-2897, Nashville, Tennessee.

Hackbusch, W. (1985). *Multigrid Methods and Applications*, Springer Ser. Comput. Math., Springer, New York.

Hackbusch, W. and Trottenberg, U. (eds.) (1982). Multigrid Methods, *Lect. Notes Math.*, vol. 960. Springer-Verlag, New York.

Hadamard, J. (1952). *Lectures on Cauchy's Problem in Linear Partial Differential Equations*, Dover, New York.

Hafez, M., South, J., and Murman, E. (1979). Artificial Compressibility Methods for Numerical Solutions of Transonic Full Potential Equation, *AIAA J.*, vol. 17, pp. 838–844.

Hageman, P. and Young, D. M. (1981). *Applied Iterative Methods*, Academic Press, New York.

Hagmeijer, R. and Kok, J. C. (1996). Adaptive 3D Single-Block Grids, for the Computation of Viscous Flows around Wings, *Numerical Grid Generation in Computational Field Simulations* (B. K. Soni, J. F. Thompson, J. Hauser, and P. Eiseman, eds.), NSF Engineering Research Center for Computational Field Simulation, Mississippi State University, Mississippi State, pp. 47–56.

Hall, M. G. (1981). Computational Fluid Dynamics—A Revolutionary Force in Aerodynamics, AIAA Paper 81-1014, Palo Alto, California.

Hanel, D. and Schwane, R. (1989). An Implicit Flux-Vector Splitting Scheme for the Computation of Viscous Hypersonic Flow, AIAA Paper 89-0274, Reno, Nevada.

Hanel, D., Schwane, R., and Seider, G. (1987). On the Accuracy of Upwind Schemes for the Solution of the Navier-Stokes Equations, AIAA Paper 87-1104, Honolulu, Hawaii.

Hanjalić, K. and Launder, B. E. (1972). A Reynolds Stress Model of Turbulence and Its Application to Asymmetric Shear Flows, *J. Fluid Mech.*, vol. 52, pp. 609–638.

Hansen, A. G. (1964). *Similarity Analyses of Boundary Value Problems in Engineering*, Prentice-Hall, Englewood Cliffs, New Jersey.

Harlow, F. H. and Fromm, J. E. (1965). Computer Experiments in Fluid Dynamics, *Sci. Am.*, vol. 212, pp. 104–110.

Harlow, F. H. and Nakayama, P. I. (1968). Transport of Turbulence Energy Decay Rate, Los Alamos Scientific Laboratory Report LA-3854, Los Alamos, New Mexico.

Harlow, F. H. and Welch, J. E. (1965). Numerical Calculation of Time-Dependent Viscous Incompressible Flow of Fluids with Free Surface, *Phys. Fluids*, vol. 8, pp. 2182–2189.

Harris, J. E. (1971). Numerical Solution of the Equations for Compressible Laminar, Transitional, and Turbulent Boundary Layers and Comparisons with Experimental Data, NASA TR-R 368.

Harten, A. (1978). The Artificial Compression Method for Computation of Shocks and Contact Discontinuities: III. Self-Adjusting Hybrid Schemes, *Math. Comput.*, vol. 32, pp. 363–389.

Harten, A. (1983). High-Resolution Schemes for Hyperbolic Conservation Laws, *J. Comput. Phys.*, vol. 49, pp. 357–385.

Harten, A. (1984). On a Class of High-Resolution Total Variation Stable Finite Difference Schemes, *SIAM J. Numer. Anal.*, vol. 21, pp. 1–23.

Harten, A. and Hyman, J. M. (1983). Self-Adjusting Grid Methods for One-Dimensional Hyperbolic Conservation Laws, *J. Comput. Phys.*, vol. 50, pp. 235–269.

Harten, A. and Zwas, G. (1972). Self-Adjusting Hybrid Schemes for Shock Computations, *J. Comput. Phys.*, vol. 9, pp. 568–583.

Hartnett, J. P. and Eckert, E. R. G. (1957). Mass-Transfer Cooling in a Laminar Boundary Layer with Constant Fluid Properties, *Trans. ASME*, vol. 79, 247–254.

Hartwich, P.-M. and Hsu, C.-H. (1987). High-Resolution Upwind Schemes for the Three-Dimensional Incompressible Navier-Stokes Equations, AIAA Paper 87-0547, Reno, Nevada.

Hartwich, P.-M., Hsu, C.-H., and Liu, C. H. (1988). Vectorizable Implicit Algorithms for the Flux-Difference Split, Three-Dimensional Navier-Stokes Equations, *J. Fluids Eng.*, vol. 110, pp. 297–305.

Hayes, W. D. (1966). La Seconde Approximation Pour les Écoulements Transsoniques Non Visqueux, *J. Méc*., vol. 5, pp. 163–206.
Hayes, W. D. and Probstein, R. F. (1966). *Hypersonic Flow Theory*, 2d ed., Academic, New York.
Healzer, J. M., Moffat, R. J., and Kays, W. M. (1974). The Turbulent Boundary Layer on a Rough Porous Plate: Experimental Heat Transfer with Uniform Blowing, Thermosciences Division, Report No. HMT-18, Department of Mechanical Engineering, Stanford University, California.
Helliwell, W. S., Dickinson, R. P., and Lubard, S. C. (1980). Viscous Flow over Arbitrary Geometries at High Angle of Attack, AIAA Paper 80-0064, Pasadena, California.
Hellwig, G. (1977). *Partial Differential Equations: An Introduction*, B. G. Teubner, Stuttgart.
Herring, H. J. and Mellor, G. L. (1968). A Method of Calculating Compressible Turbulent Boundary Layers, NASA CR-1144.
Hess, J. L. and Smith, A. M. O. (1967). Calculation of Potential Flow about Arbitrary Bodies, *Prog. Aeronaut. Sci*., vol. 8, pp. 1–138.
Hildebrand, F. B. (1956). *Introduction to Numerical Analysis*, McGraw-Hill, New York.
Hilgenstock, A. (1988). A Fast Method for the Elliptic Generation of Three-Dimensional Grids with Full Boundary Control, *Numerical Grid Generation in Computational Fluid Mechanics*, Pineridge Press, Swansea, U.K., pp. 137–146.
Hindman, R. G. (1981). Geometrically Induced Errors and Their Relationship to the Form of the Governing Equations and the Treatment of Generalized Mappings, AIAA Paper 81-1008, Palo Alto, California.
Hindman, R. G. and Spencer, J. (1983). A New Approach to Truly Adaptive Grid Generation, AIAA Paper 83-0450, Reno, Nevada.
Hindman, R. G., Kutler, P., and Anderson, D. A. (1979). A Two-Dimensional Unsteady Euler Equation Solver for Flow Regions with Arbitrary Boundaries, AIAA Paper 79-1465, Williamsburg, Virginia.
Hindman, R. G., Kutler, P., and Anderson, D. A. (1981). Two-Dimensional Unsteady Euler-Equation Solver for Arbitrarily Shaped Flow Regions, *AIAA J*., vol. 19, pp. 424–431.
Hinze, J. O. (1975). *Turbulence*, 2d ed., McGraw-Hill, New York.
Hirasaki, G. J. and Hellums, J. D. (1970). Boundary Conditions on the Vector and Scalar Potentials in Viscous Three-Dimensional Hydrodynamics, *Q. Appl. Math*., vol. 28, pp. 293–296.
Hirsch, C. (1988). *Numerical Computation of Internal and External Flows*, vol. 1, *Fundamentals of Numerical Discretization*, John Wiley, New York.
Hirsch, C. (1990). *Numerical Computation of Internal and External Flows*, vol. 2, *Computational Methods for Inviscid and Viscous Flows*, John Wiley, New York.
Hirschfelder, J. O., Curtiss, C. F., and Bird, R. B. (1954). *Molecular Theory of Gases and Liquids*, Wiley, New York.
Hirsh, R. S. and Rudy, D. H. (1974). The Role of Diagonal Dominance and Cell Reynolds Number in Implicit Methods for Fluid Mechanics Problems, *J. Comput. Phys*., vol. 16, pp. 304–310.
Hirt, C. W. (1968). Heuristic Stability Theory for Finite-Difference Equations, *J. Comput. Phys*., vol. 2, pp. 339–355.
Hixon, R. and Sankar, L. N. (1992). Application of a Generalized Minimal Residual Method to 2D Unsteady Flows, AIAA Paper 92-0422, Reno, Nevada.
Hockney, R. H. and Jesshope, C. R. (1981). *Parallel Computers: Architecture, Programming and Algorithms*, Adam Hilger, Bristol.
Hockney, R. W. (1965). A Fast Direct Solution of Poisson's Equation Using Fourier Analysis, *J. Assoc. Comput. Mach*., vol. 12, pp. 95–113.
Hockney, R. W. (1970). The Potential Calculation and Some Applications, *Methods Comput. Phys*., vol. 9, pp. 135–211.
Hodge, J. K., Leone, S. A., McCarty, R. L. (1987). Non-Iterative Parabolic Grid Generation for Parabolized Equations, *AIAA J*., vol. 25, pp. 542–549.
Hoffman, K. A. (1989). *Computational Fluid Dynamics for Engineers*, Engineering Education System, Austin, Texas.
Holmes, D. and Snyder, D. (1988). The Generation of Unstructured Triangular Meshes Using Delaunay Triangulation, *Numerical Grid Generation in Computational Fluid Mechanics*, Pineridge Press, Swansea, U.K., pp. 643–652.

Holst, T. L. (1979). Implicit Algorithm for the Conservative Transonic Full-Potential Equation Using an Arbitrary Mesh, *AIAA J.*, vol. 17, pp. 1038–1045.

Holst, T. L. (1980). Fast, Conservative Algorithm for Solving the Transonic Full-Potential Equation, *AIAA J.*, vol. 18, pp. 1431–1439.

Holst, T. L. and Ballhaus, W. F. (1979). Fast, Conservative Schemes for the Full Potential Equation Applied to Transonic Flows, *AIAA J.*, vol. 17, pp. 145–152.

Holst, T. L., Kaynak, U., Gundy, K. L., Thomas, S. D., Flores, J., and Chaderjian, N. M. (1987). Transonic Wing Flows Using an Euler Navier-Stokes Zonal Approach, *J. Aircraft*, vol. 24, pp. 17–24.

Holst, T. L., Salas, M., and Claus, R. (1992). The NASA Computational Aerosciences Program toward Teraflop Computing, AIAA Paper 92-0558, Reno, Nevada.

Holt, M. (1977). *Numerical Methods in Fluid Dynamics*, Springer-Verlag, New York.

Hong, S. W. (1974). Laminar Flow Heat Transfer in Ordinary and Augmented Tubes, Ph.D. dissertation, Iowa State University, Ames.

Hornbeck, R. W. (1963). Laminar Flow in the Entrance Region of a Pipe, *Appl. Sci. Res., Sec. A*, vol. 13, pp. 224–232.

Hornbeck, R. W. (1973). *Numerical Marching Techniques for Fluid Flows with Heat Transfer*, NASA SP-297.

Horstman, C. C. (1977). Turbulence Model for Non-Equilibrium Adverse Pressure Gradient Flows, *AIAA J.*, vol. 15, pp. 131–132.

Hosny, W. M., Davis, R. T., and Werle, M. J. (1978). Improvements to the Solution of the Viscous Shock Layer Equations, Department of Aerospace Engineering and Applied Mechanics, Report AFL 78-11-45, University of Cincinnati, Ohio.

Houwink, R. and Veldman, A. E. P. (1984). Steady and Unsteady Separated Flow Computations for Transonic Airfoils, AIAA Paper 84-1618, Snowmass, Colorado.

Howarth, L. (1938). On the Solution of the Laminar Boundary Layer Equations, *Proc. R. Soc. London, Ser. A*, vol. 164, pp. 547–579.

Howarth, L. (1951). The Boundary Layer in Three-Dimensional Flow. Part I: Derivation of the Equations for Flow along a General Curved Surface, *Philos. Mag.*, vol. 42, pp. 239–243.

Hsu, C. (1981). A Curvilinear-Coordinate Method for Momentum, Heat and Mass Transfer in Domains of Irregular Geometry, Ph.D. dissertation, University of Minnesota, Minneapolis.

Hugues, E. C. and Vérant, J. L. (1991). Nonequilibrium Parabolized Navier-Stokes Code with an Implicit Finite Volume Method, AIAA Paper 91-0470, Reno, Nevada.

Hwang, S. S. and Pletcher, R. H. (1978). Prediction of Turbulent Jets and Plumes in Flowing Ambients, Engineering Research Institute Technical Report 79003/HTL-15, Iowa State University, Ames.

Israeli, M. and Lin, T. (1985). Iterative Numerical Solutions and Boundary Conditions for the Parabolized Navier-Stokes Equations, *Comput. Fluids*, vol. 13, pp. 397–410.

Issa, R. I. (1985). Solution of the Implicitly Discretized Fluid Flow Equations by Operator-Splitting, *J. Comput. Phys.*, vol. 62, pp. 40–65.

Issa, R. I., Gosman, A. D., and Watkins, A. P. (1986). The Computation of Compressible and Incompressible Recirculating Flows by a Non-iterative Implicit Scheme, *J. Comput. Phys.*, vol. 62, pp. 66–82.

Ives, D. C. (1982). Conformal Grid Generation, *Numerical Grid Generation, Proceedings of a Symposium on the Numerical Generation of Curvilinear Coordinate Systems and Their Use in the Numerical Solution of Partial Differential Equations* (J. F. Thompson, ed.), Elsevier, New York, pp. 107–130.

Iwatsu, R., Hyun, J. M., and Kuwahara, K. (1993). Numerical Simulations of Three-Dimensional Flows in a Cubic Cavity with an Oscillating Lid, *J. Fluids Eng.*, vol. 115, pp. 680–686.

Jacquotte, O. P. (1996). Recent Progress in Mesh Generation in Europe, *Grid Generation in Computational Field Simulations* (B. K. Soni, J. F. Thompson, J. Hauser, and P. Eiseman, eds.), NSF Engineering Research Center for Computational Field Simulation, Mississippi State University, Mississippi State, pp. 1121–1130.

James, M. L., Smith, G. M., and Wolford, J. C. (1967). *Applied Numerical Methods for Digital Computation with FORTRAN*, International Textbook Company, Scranton, Pennsylvania.

Jameson, A. (1974). Iterative Solution of Transonic Flows over Airfoils and Wings Including Flows at Mach 1, *Commun. Pure Appl. Math.*, vol. 27, pp. 283–309.
Jameson, A. (1975). Transonic Potential Flow Calculations using Conservation Form, *Proc. AIAA 2nd Computational Fluid Dynamics Conference*, Hartford, Connecticut, pp. 148–161.
Jameson, A. (1982). Steady State Solution of the Euler Equations for Transonic Flow, Transonic Shock, and Multidimensional Flows, *Adv. Sci. Comput.*, no. 47, Academic Press, New York.
Jameson, A. (1983). Solution of the Euler Equations by a Multigrid Method, *Appl. Math. Comput.*, vol. 13, pp. 327–356.
Jameson, A. (1986). Current Status of Future Directions of Computational Transonics, *Computational Mechanics—Advances and Trends* (A. K. Noor, ed.), Publication AMD 75, ASME, New York.
Jameson, A. (1987). Successes and Challenges in Computational Aerodynamics, AIAA Paper 87-1184, Honolulu, Hawaii.
Jameson, A. and Baker, T. J. (1983). Solution of Euler Equations for Complex Configurations, AIAA Paper 83-1929, Danvers, Massachusetts.
Jameson, A. and Lax, P. (1984). Conditions for the Construction of Multi-Point Total Variation Diminishing Difference Schemes, Department of Mechanical and Aerospace Engineering, Princeton University, MAE Report 1650.
Jameson, A., Schmidt, W., and Turkel, E. (1981). Numerical Solutions of the Euler Equations by Finite Volume Methods using Runge-Kutta Time-Stepping Schemes, AIAA Paper 81-1259, Palo Alto, California.
Jameson, A., Baker, T., and Weatherill, N. (1986). Calculation of Inviscid Flow over a Complete Aircraft, AIAA Paper 86-0103, Reno, Nevada.
Jeffrey, A. and Taniuti, T. (1964). *Nonlinear Wave Propagation with Applications to Physics and Magneto-Hydrodynamics*, Academic, New York.
Jensen, V. G. (1959). Viscous Flow Round a Sphere at Low Reynolds Numbers ($\leqslant$ 40), *Proc. R. Soc. London, Ser. A*, vol. 249, pp. 346–366.
Jesperson, D. C. and Pulliam, T. H. (1983). Flux Vector Splitting and Approximate Newton Methods, AIAA Paper 83-1899, Danvers, Massachusetts.
Jobe, C. E. (1974). The Numerical Solution of the Asymptotic Equations of Trailing Edge Flow, Technical Report AFFDL-TR-74-46, Air Force Flight Dynamics Laboratory, Dayton, Ohio.
Jobe, C. E. and Burggraf, O. R. (1974). The Numerical Solution of the Asymptotic Equations of Trailing Edge Flow, *Proc. R. Soc. London, Ser. A*, vol. 340, pp. 91–111.
Johnson, D. A. and Coakley, T. J. (1990). Improvements to a Nonequilibrium Algebraic Turbulence Model, *AIAA J.*, vol. 28, pp. 2000–2003.
Johnson, D. A. and King, L. S. (1985). A Mathematically Simple Turbulent Closure Model for Attached and Separated Turbulent Boundary Layers, *AIAA J.*, vol. 23, pp. 1684–1692.
Johnson, F. and Rubbert, P. (1975). Advanced Panel-Type Influence Coefficient Methods Applied to Subsonic Flows, AIAA Paper 75-50, Pasadena, California.
Johnson, G. M. (1980). An Alternative Approach to the Numerical Simulation of Steady Inviscid Flow, Proc. Seventh Int. Conf. Num. Methods Fluid Dyn., *Lect. Notes in Phys.*, vol. 141, Springer-Verlag, New York, pp. 236–241.
Jones, W. P. and Launder, B. E. (1972). The Prediction of Laminarization with a Two-Equation Model of Turbulence, *Int. J. Heat Mass Transfer*, vol. 15, pp. 301–314.
Kallinderis, Y. (1996). Discretization of Complex 3-D Flow Domains with Adaptive Hybrid Grids, *Numerical Grid Generation in Computational Field Simulations* (B. K. Soni, J. F. Thompson, J. Hauser, and P. Eiseman, eds.), NSF Engineering Research Center for Computational Field Simulation, Mississippi State University, Mississippi State, pp. 505–516.
Kallinderis, Y. and Vidwans, A. (1994). Generic Parallel Adaptive-Grid Navier-Stokes Algorithm, *AIAA J.*, vol. 32, pp. 54–61.
Kallinderis, Y. and Ward, S. (1993). Prismatic Grid Generation for Three-Dimensional Complex Geometries, *AIAA J.*, vol. 31, pp. 1850–1858.
Kamath, P. S., Mao, M. M., and McClinton, C. R. (1991). Scramjet Combustor Analysis with the Ship3D PNS Code, AIAA Paper 91-5090, Orlando, Florida.
Karamcheti, K. (1966). *Principles of Ideal-Fluid Aerodynamics*, Wiley, New York.

Karki, K. C. and Patankar, S. V. (1989). Pressure Based Calculation Procedure for Viscous Flows at All Speeds in Arbitrary Configurations, *AIAA J.*, vol. 27, pp. 1167–1174.

Karman, S. L. (1995a). SPLITFLOW: A 3-D Unstructured Cartesian/Prismatic Grid CFD Code for Complex Geometries, AIAA Paper 95-0343, Reno, Nevada.

Karman, S. L. (1995b). Unstructured Cartesian/Prismatic Grid Generation for Complex Geometries, *Proceedings of the Surface Modeling, Grid Generation and Related Issues in Computational Fluid Dynamics Workshop*, NASA Conference Publication 3291, NASA Lewis Research Center, Cleveland, Ohio, pp. 251–257.

Kays, W. M. (1972). Heat Transfer to the Transpired Turbulent Boundary Layer, *Int. J. Heat Mass Transfer*, vol. 15, pp. 1023–1044.

Kays, W. M. and Crawford, M. E. (1980). *Convective Heat and Mass Transfer*, 2d ed., McGraw-Hill, New York.

Kays, W. M. and Crawford, M. E. (1993). *Convective Heat and Mass Transfer*, 3rd ed., McGraw-Hill, New York.

Kays, W. M. and Moffat, R. J. (1975). The Behavior of Transpired Turbulent Boundary Layers, *Studies in Convection: Theory, Measurement, and Applications*, vol. 1, Academic, New York, pp. 223–319.

Keller, H. B. (1970). A New Difference Scheme for Parabolic Problems, *Numerical Solutions of Partial Differential Equations*, vol. 2 (J. Bramble, ed.), Academic, New York.

Keller, H. B. and Cebeci, T. (1972). Accurate Numerical Methods for Boundary-Layer Flows. II: Two-Dimensional Turbulent Flows, *AIAA J.*, vol. 10, pp. 1193–1199.

Kennon, S. R. and Dulikravich, G. S. (1985). A Posteriori Optimization of Computational Grids, AIAA Paper 85-0483, Reno, Nevada.

Kentzer, C. P. (1970). Discretization of Boundary Conditions on Moving Discontinuities, Proc. Second Int. Conf. Num. Methods Fluid Dyn., *Let. Notes Phys.*, vol. 8, Springer-Verlag, New York, pp. 108–113.

Khosla, P. K. and Rubin, S. G. (1987). Consistent Strongly Implicit Iterative Procedures for Two-Dimensional Unsteady and Three-Dimensional Space Marching Flow Calculations, *Comput. Fluids*, vol. 15, pp. 361–377.

Kim, J., Moin, P., and Moser, R. (1987). Turbulence Statistics in Fully Developed Channel Flow at Low Reynolds Number, *J. Fluid Mech.*, vol. 177, pp. 133–166.

Kim, M. D. and Lewis, C. H. (1982). Computation of Hypersonic Viscous Flow Past General Bodies at Angle-of-Attack and Yaw, AIAA Paper 82-0225, Orlando, Florida.

Kim, M. D., Swaminathan, S., and Lewis, C. H. (1983). Three-Dimensional Nonequilibrium Viscous Shock-Layer Flow over the Space Shuttle Orbiter, AIAA Paper 83-0487, Reno, Nevada.

Kitchens Jr., C. W., Sedney, R., and Gerber, N. (1975). The Role of the Zone of Dependence Concept in Three-Dimensional Boundary-Layer Calculations, *Proc. AIAA 2nd Computational Fluid Dynamics Conference*, Hartford, Connecticut, pp. 102–112.

Klineberg, J. M. and Steger, J. L. (1974). On Laminar Boundary Layer Separation, AIAA Paper 74-94, Washington, D.C.

Klinksiek, W. F. and Pierce, F. J. (1973). A Finite-Difference Solution of the Two- and Three-Dimensional Incompressible Turbulent Boundary Layer Equations, *J. Fluids Eng.*, vol. 95, pp. 445–458.

Klopfer, G. H. and McRae, D. S. (1981a). The Nonlinear Modified Equation Approach to Analyzing Finite-Difference Schemes, AIAA Paper 81-1029, Palo Alto, California.

Klopfer, G. H. and McRae, D. S. (1981b). Nonlinear Analysis of the Truncation Errors in Finite-Difference Schemes for the Full System of Euler Equations, AIAA Paper 81-0193, St. Louis, Missouri.

Klunker, E. (1971). Contributions to Methods for Calculating the Flow about Thin Lifting Wings at Transonic Speeds—Analytical Expression for the Fair Field, NASA TN D-6530.

Knechtel, E. D. (1959). Experimental Investigation at Transonic Speeds of Pressure Distributions over Wedge and Circular Arc Sections and Evaluation of Perforated-Wall Interference, NASA TN D-15.

Kontinos, D. A. and McRae, D. S. (1991). An Explicit, Rotated Upwind Algorithm for Solution of the Euler/Navier-Stokes Equations, AIAA Paper 91-1531, Honolulu, Hawaii.

Kontinos, D. A. and McRae, D. S. (1994). Rotated Upwind Strategies for Solution of the Euler Equations, AIAA Paper 94-0079, Reno, Nevada.

Korkegi, R. H. (1956). Transition Studies and Skin-Friction Measurements on an Insulated Flat Plate at a Mach Number of 5.8, *J. Aeronaut. Sci.*, vol. 25, pp. 97–192.

Korte, J. J. and McRae, D. S. (1988). Explicit Upwind Algorithm for the Parabolized Navier-Stokes Equations, AIAA Paper 88-0716, Reno, Nevada.

Korte, J. J. and McRae, D. S. (1989). Numerical Simulation of Flow over a Hypersonic Aircraft Using an Explicit Upwind PNS Solver, AIAA Paper 89-1829, Buffalo, New York.

Koseff, J. R. and Street, R. L. (1984). Visualization Studies of a Shear Driven Three-Dimensional Recirculation Flow, *J. Fluids Eng.*, vol. 106, pp. 21–29.

Kowalski, E. J. (1980). Boundary-Fitted Coordinate Systems for Arbitrary Computational Regions, *Numerical Grid Generation Techniques*, NASA Conference Publication 2166, pp. 331–353.

Krause, E. (1969). Comment on "Solution of a Three-Dimensional Boundary-Layer Flow with Separation," *AIAA J.*, vol. 7, pp. 575–576.

Krause, E. (1985). Computational Fluid Dynamics: Its Present Status and Future Direction, *Comput. Fluids*, vol. 13, pp. 239–269.

Krawczyk, W. J. and Harris, T. B. (1987). Parabolized Navier-Stokes Analysis of Two-Dimensional Scramjet Inlet Flow Fields, AIAA Paper 87-1899, San Diego, California.

Krawczyk, W. J., Harris, T. B., Rajendran, N., and Carlson, D. R. (1989). Progress in the Development of Parabolized Navier-Stokes Technology for External and Internal Supersonic Flows, AIAA Paper 89-1828, Buffalo, New York.

Kreskovsky, J. P., Shamroth, S. J., and McDonald, H. (1974). Parametric Study of Relaminarization of Turbulent Boundary Layers on Nozzle Walls, NASA CR-2370.

Kutler, P. (1983). A Perspective of Theoretical and Applied Computational Fluid Dynamics, AIAA Paper 83-0037, Reno, Nevada.

Kutler, P. (1993). Multidisciplinary Computational Aerosciences, *Proceedings of the 5th International Symposium on Computational Fluid Dynamics*, vol. II, Springer-Verlag, Berlin, pp. 109–119.

Kutler, P. and Lomax, H. (1971). The Computation of Supersonic Flow Fields about Wing-Body Combinations by "Shock-Capturing" Finite Difference Techniques, Proc. Second Int. Conf. Num. Methods Fluid Dyn., *Lect. Notes Phys.*, vol. 8, Springer-Verlag, New York, pp. 24–29.

Kutler, P., Warming, R. F., and Lomax, H. (1973). Computation of Space Shuttle Flowfields using Noncentered Finite-Difference Schemes, *AIAA J.*, vol. 11, pp. 196–204.

Kutler, P., Steger, J. L., and Bailey, F. R. (1987). Status of Computational Fluid Dynamics in the United States, AIAA Paper 87-1135-CP, Honolulu, Hawaii.

Kwak, D., Chang, J. L. C., Shanks, S. P., and Chakravarthy, S. K. (1986). A Three-Dimensional Incompressible Navier-Stokes Solver Using Primitive Variables, *AIAA J.*, vol. 24, pp. 390–396.

Kwon, O. K. and Pletcher, R. H. (1979). Prediction of Incompressible Separated Boundary Layers Including Viscous-Inviscid Interaction, *J. Fluids Eng.*, vol. 101, pp. 466–472.

Kwon, O. K. and Pletcher, R. H. (1981). Prediction of the Incompressible Flow over a Rearward-Facing Step, Engineering Research Institute Technical Report 82019/HTL-26, Iowa State University, Ames.

Kwon, O. K. and Pletcher, R. H. (1986a). A Viscous-Inviscid Interaction Procedure, Part 1: Method for Computing Two-Dimensional Separated Channel Flows, *J. Fluids Eng.*, vol. 108, pp. 64–70.

Kwon, O. K. and Pletcher, R. H. (1986b). A Viscous-Inviscid Interaction Procedure, Part 2: Application to Turbulent Flow over a Rearward-Facing Step, *J. Fluids Eng.*, vol. 108, pp. 71–75.

Kwon, O. K., Pletcher, R. H., and Lewis, J. P. (1984). Prediction of Sudden Expansion Flows Using the Boundary-Layer Equations, *J. Fluids Eng.*, vol. 106, pp. 285–291.

Kwon, O. K., Pletcher, R. H., and Delaney, R. A. (1988a). Solution Procedure for Unsteady Two-Dimensional Boundary Layers, *J. Fluids Eng.*, vol. 110, pp. 69–75.

Kwon, O. K., Pletcher, R. H., and Delaney, R. A. (1988b). Calculation of Unsteady Turbulent Boundary Layers, *J. Turbomach.*, vol. 110, pp. 195–201.

Laasonen, P. (1949). Über eine Methode zur Lösung der Wärmeleitungsgleichung, *Acta Math.*, vol. 81, pp. 309–317.

Lam, C. K. G. and Bremhorst, K. A. (1981). Modified Form of k-ϵ Model for Predicting Wall Turbulence, *J. Fluids Eng.*, vol. 103, pp. 456–460.

Lapidus, L. and Pinder, G. F. (1982). *Numerical Solution of Partial Differential Equations in Science and Engineering*, Wiley-Interscience, New York.

Larkin, B. K. (1964). Some Stable Explicit Difference Approximations to the Diffusion Equation, *Math. Comput.*, vol. 18, pp. 196–202.

Launder, B. E. (1979). Stress-Transport Closures: Into the Third Generation, *Proc. First Symposium on Turbulent Shear Flows*, Springer-Verlag, New York.

Launder, B. E. and Sharma, B. I. (1974). Application of the Energy Dissipation Model of Turbulence to the Calculation of Flow near a Spinning Disc, *Lett. Heat Mass Transfer*, vol. 1, pp. 131–138.

Launder, B. E. and Spalding, D. B. (1972). *Mathematical Models of Turbulence*, Academic, New York.

Launder, B. E. and Spalding, D. B. (1974). The Numerical Computation of Turbulent Flows, *Comput. Methods Appl. Mech. Eng.*, vol. 3, pp. 269–289.

Launder, B. E., Reece, G. J., and Rodi, W. (1975). Progress in the Development of a Reynolds-Stress Turbulence Closure, *J. Fluid Mech.*, vol. 68, pt. 3, pp. 537–566.

Lawrence, S. L. (1987). Application of an Upwind Algorithm to the Parabolized Navier-Stokes Equations, Ph.D. dissertation, Iowa State University, Ames.

Lawrence, S. L. (1992). Parabolized Navier-Stokes Methods for Hypersonic Flows, Von Karman Institute for Fluid Dynamics Lecture Series 1992-06, Rhode-St.-Genese, Belgium.

Lawrence, S. L., Tannehill, J. C., and Chaussee, D. S. (1984). Application of the Implicit MacCormack Scheme to the Parabolized Navier-Stokes Equations, *AIAA J.*, vol. 22, pp. 1755–1763.

Lawrence, S. L., Tannehill, J. C., and Chaussee, D. S. (1986). An Upwind Algorithm for the Parabolized Navier-Stokes Equations, AIAA Paper 86-1117, Atlanta, Georgia.

Lawrence, S. L., Chaussee, D. S., and Tannehill, J. C. (1987). Application of an Upwind Algorithm to the Three-Dimensional Parabolized Navier-Stokes Equations, AIAA Paper 87-1112-CP, Honolulu, Hawaii.

Lax, P. D. (1954). Weak Solutions of Nonlinear Hyperbolic Equations and their Numerical Computation, *Commun. Pure Appl. Math.*, vol. 7, pp. 159–193.

Lax, P. D. (1973). *Hyperbolic Systems of Conservation Laws and the Mathematical Theory of Shock Waves*, Society for Industrial and Applied Mathematics, Philadelphia, PA.

Lax, P. D. and Wendroff, B. (1960). Systems of Conservation Laws, *Commun. Pure Appl. Math.*, vol. 13, pp. 217–237.

LeBail, R. C. (1972). Use of Fast Fourier Transforms for Solving Partial Differential Equations in Physics, *J. Comput. Phys.*, vol. 9, pp. 440–465.

Le Balleur, J. C. (1984). A Semi-Implicit and Unsteady Numerical Method of Viscous-Inviscid Interaction for Transonic Separated Flows, *Rech. Aerospatiale*, no. 1984-1, pp. 15–37.

Leck, C. L. and Tannehill, J. C. (1993). A New Rotated Upwind Difference Scheme for the Euler Equations, AIAA Paper 93-0066, Reno, Nevada.

Lee, D. and Pletcher, R. H. (1988). Simultaneous Viscous-Inviscid Interaction Calculation Procedure for Transonic Turbulent Flows, *AIAA J.*, vol. 25, pp. 1354–1362.

Lee, K. P., Gupta, R. N., Zoby, E. V., and Moss, J. N. (1990). Hypersonic Viscous Shock Layer Solutions over Long Slender Bodies, Part II: Low Reynolds Number Flows, *J. Spacecraft Rockets*, vol. 26, pp. 221–228.

Leonard, A. (1974). Energy Cascade in Large-Eddy Simulations of Turbulent Fluid Flows, *Advances in Geophysics*, vol. 18A, pp. 237–248.

Leonard, B. P. (1979a). A Stable and Accurate Convective Modelling Procedure Based on Quadratic Upstream Interpolation, *Comput. Methods Appl. Mech. Eng.*, vol. 19, pp. 59–98.

Leonard, B. P. (1979b). A Survey of Finite Differences of Opinion on Numerical Muddling of the Incomprehensible Defective Confusion Equation, *Finite Element Methods for Convective Dominated Flows*, AMD-vol. 34, The American Society of Mechanical Engineers.

Levine, R. D. (1982). Supercomputers, *Sci. Am.*, vol. 246, pp. 118–135.

Levy, D. W., Powell, K. G., and van Leer, B. (1993). Use of a Rotated Riemann Solver for the Two-Dimensional Euler Equations, *J. Comput. Phys.*, vol. 106, pp. 201–214.

Lewis, J. P. and Pletcher, R. H. (1986). Limitations of the Boundary-Layer Equations for Predicting Laminar Symmetric Sudden Expansion Flows, *J. Fluids Eng.*, vol. 108, pp. 208–213.

Li, C. P. (1973). Numerical Solution of Viscous Reacting Blunt Body Flows of a Multicomponent Mixture, AIAA Paper 73-202, Washington, D.C.

Li, C. P. (1977). A Numerical Study of Separated Flows Induced by Shock-Wave/Boundary-Layer Interaction, AIAA Paper 77-168, Los Angeles, California.

Li, C. P. (1981a). Application of an Implicit Technique to the Shock-Layer Flow around General Bodies, AIAA Paper 81-0191, St. Louis, Missouri.

Li, C. P. (1981b). Numerical Simulation of Re-entry Flow around the Shuttle Orbiter Including Real Gas Effects, *Computers in Flow Predictions and Fluid Dynamics Experiments* (K. Ghia, T. Mueller, and B. Patel, eds.), ASME, New York, pp. 141–149.

Liebmann, L. (1918). Die Angenäherte Ermittelung Harmonischer Funktionen und Konformer Abbildungen, *Sitzungsber. Math. Phys. Kl. Bayer. Akad. Wiss.*, vol. 3, p. 385.

Liepmann, H. W. and Roshko, A. (1957). *Elements of Gasdynamics*, Wiley, New York.

Lighthill, M. J. (1953). On Boundary Layers and Upstream Influence. II. Supersonic Flows without Separation, *Proc. R. Soc. London, Ser. A*, vol. 217, pp. 478–507.

Lighthill, M. J. (1958). On Displacement Thickness, *J. Fluid Mech.*, vol. 4, pp. 383–392.

Lin, T. C. and Rubin, S. G. (1973a). Viscous Flow over Spinning Cones at Angle of Attack, *AIAA J.*, vol. 12, pp. 975–985.

Lin, T. C. and Rubin, S. G. (1973b). Viscous Flow over a Cone at Moderate Incidence: I. Hypersonic Tip Region, *Comput. Fluids*, vol. 1, pp. 37–57.

Lin, T. C. and Rubin, S. G. (1974). Viscous Flow over Spinning Cones at Angle of Attack, *AIAA J.*, vol. 12, pp. 975–985.

Lin, T. C. and Rubin, S. G. (1979). A Numerical Model for Supersonic Viscous Flow over a Slender Reentry Vehicle, AIAA Paper 79-0205, New Orleans, Louisiana.

Lin, T. C., Rubin, S. G., and Widhopf, G. F. (1981). A Two-Layer Model for Coupled Three Dimensional Viscous and Inviscid Flow Calculations, AIAA Paper 81-0118, St. Louis, Missouri.

Lindemuth, I. and Killeen, J. (1973). Alternating Direction Implicit Techniques for Two Dimensional Magnetohydrodynamics Calculations, *J. Comput. Phys.*, vol. 13, pp. 181–208.

Liou, M. F. (1989). Three-Dimensional PNS Solutions of Hypersonic Internal Flows with Equilibrium Chemistry, AIAA Paper 89-0002, Reno, Nevada.

Liou, N.-S. and Steffan, C. (1991). A New Flux Splitting Scheme, NASA TM-104404.

Liu, X. and Pletcher, R. H. (1986). A Coupled Marching Procedure for the Partially Parabolized Navier-Stokes Equations, *Numer. Heat Transfer*, vol. 10, pp. 539–556.

Lock, R. C. (1970). Test Cases for Numerical Methods in Two-Dimensional Transonic Flows, AGARD Report 575.

Löhner, R. (1995). Mesh Adaption in Fluid Mechanics, *Eng. Frac. Mech.*, vol. 50, pp. 819–847.

Löhner, R. and Baum, J. D. (1990). Numerical Simulation of Shock Interaction with Complex Geometry Three-Dimensional Structures Using a New Adaptive H-Refinement Scheme on Unstructured Grids, AIAA Paper 90-0700, Reno, Nevada.

Löhner, R. and Parikh, P. (1988). Generation of Three-Dimensional Unstructured Grids by the Advancing Front Method, AIAA Paper 88-0515, Reno, Nevada.

Lomax, H., Kutler, P., and Fuller, F. B. (1970). The Numerical Solution of Partial Differential Equations Governing Convection, AGARDograph 146.

Lombard, C. K., Bardina, J., Venkatapathy, E., and Oliger, J. (1983). Multi-Dimensional Formulation of CSCM: An Upwind Flux Difference Eigenvector Split Method for the Compressible Navier-Stokes Equations, AIAA Paper 83-1895, Danvers, Massachusetts.

Lombard, C. K., Venkatapathy, E., and Bardina, J. (1984). Universal Single Level Implicit Algorithm for Gas Dynamics, AIAA Paper 84-1533, Snowmass, Colorado.

Lubard, S. C. and Helliwell, W. S. (1973). Calculation of the Flow on a Cone at High Angle of Attack, R & D Associates Technical Report, RDA-TR-150, Santa Monica, California.

Lubard, S. C. and Helliwell, W. S. (1974). Calculation of the Flow on a Cone at High Angle of Attack, *AIAA J.*, vol. 12, pp. 965–974.

Lubard, S. C. and Rakich, J. V. (1975). Calculation of the Flow on a Blunted Cone at a High Angle of Attack, AIAA Paper 75-149, Pasadena, California.

Ludford, G. (1951). The Behavior at Infinity of the Potential Function of a Two-Dimensional Subsonic Compressible Flow, *J. Math. Phys.*, vol. 30, pp. 131–159.

Lugt, H. J. and Ohring, S. (1974). Efficiency of Numerical Methods in Solving the Time-Dependent, Two-Dimensional Navier-Stokes Equations, *Numerical Methods in Fluid Dynamics*, Peutech, London.

Lumley, J. L. (1970). Toward a Turbulent Constitutive Equation, *J. Fluid Mech.*, vol. 41, pp. 413–434.

Lumley, J. L. (1978). Computational Modeling of Turbulent Flows, *Adv. Appl. Mech.*, vol. 18, pp. 123–176.

Luther, H. A. (1966). Further Explicit Fifth-Order Runge-Kutta Formulas, *SIAM Rev.*, vol. 8, pp. 374–380.

Macagno, E. O. (1965). Some New Aspects of Similarity in Hydraulics, *La Houille Blanche*, vol. 20, pp. 751–759.

Macagno, E. O. and Hung, T.-K. (1967). Computational and Experimental Study of a Captive Annular Eddy, *J. Fluid Mech.*, vol. 28, pp. 43–67.

MacCormack, R. W. (1969). The Effect of Viscosity in Hypervelocity Impact Cratering, AIAA Paper 69-354, Cincinnati, Ohio.

MacCormack, R. W. (1971). Numerical Solution of the Interaction of a Shock Wave with a Laminar Boundary Layer, Proc. Second Int. Conf. Num. Methods Fluid Dyn., *Lect. Notes Phys.*, vol. 8, Springer-Verlag, New York, pp. 151–163.

MacCormack, R. W. (1976). At Efficient Numerical Method for Solving the Time-Dependent Compressible Navier-Stokes Equations at High Reynolds Number, NASA TM X-73,-129.

MacCormack, R. W. (1981). A Numerical Method for Solving the Equations of Compressible Viscous Flow, AIAA Paper 81-0110, St. Louis, Missouri.

MacCormack, R. W. (1985). Current Status of Numerical Solutions of the Navier-Stokes Equations, AIAA Paper 85-0032, Reno, Nevada.

MacCormack, R. W. (1993). A Perspective on a Quarter Century of CFD Research, AIAA Paper 93-3291-CP, Honolulu, Hawaii.

MacCormack, R. W. and Baldwin, B. S. (1975). A Numerical Method for Solving the Navier-Stokes Equations with Application to Shock-Boundary Layer Interactions, AIAA Paper 75-1, Pasadena, California.

MacCormack, R. W. and Paullay, A. J. (1972). Computational Efficiency Achieved by Time Splitting of Finite Difference Operators, AIAA Paper 72-154, San Diego, California.

Madavan, N. K. and Pletcher, R. H. (1982). Prediction of Incompressible Laminar Separated Flows Using the Partially Parabolized Navier-Stokes Equations, Engineering Research Institute Technical Report 82127/HTL-27, Iowa State University, Ames.

Madni, I. K. and Pletcher, R. H. (1975a). A Finite-Difference Analysis of Turbulent, Axisymmetric, Buoyant Jets and Plumes, Engineering Research Institute Technical Report 76096/HTL-10, Iowa State University, Ames.

Madni, I. K. and Pletcher, R. H. (1975b). Prediction of Turbulent Jets in Coflowing and Quiescent Ambients, *J. Fluids Eng.*, vol. 97, pp. 558–567.

Madni, I. K. and Pletcher, R. H. (1977a). Prediction of Turbulent Forced Plumes Issuing Vertically into Stratified or Uniform Ambients, *J. Heat Transfer*, vol. 99, pp. 99–104.

Madni, I. K. and Pletcher, R. H. (1977b). Buoyant Jets Discharging Nonvertically into a Uniform Quiescent Ambient—A Finite-Difference Analysis and Turbulence Modeling, *J. Heat Transfer*. vol. 99, pp. 641–647.

Malik, M. R. and Pletcher, R. H. (1978). Computation of Annular Turbulent Flows with Heat Transfer and Property Variations, *Heat Transfer 1978, Proc. Sixth Int. Heat Transfer Conference*, vol. 2, Hemisphere, Washington, D.C., pp. 537–542.

Malik, M. R. and Pletcher, R. H. (1981). A Study of Some Turbulence Models for Flow and Heat Transfer in Ducts of Annular Cross-Section, *J. Heat Transfer*, vol. 103, pp. 146–152.

Mallinson, G. D. and de Vahl Davis, G. (1973). The Method of the False Transient for the Solution of Coupled Elliptic Equations, *J. Comput. Phys.*, vol. 12, pp. 435–461.

Mani, M., Willhite, P., and Ladd, J. (1995). Performance of One-Equation Turbulence Models in CFD Applications, AIAA Paper 95-2221, San Diego, CA.

Marchuk, G. M. (1975). *Methods of Numerical Mathematics*, Springer-Verlag, Berlin.

Marconi, F. (1980). Supersonic, Inviscid, Conical Corner Flowfields, *AIAA J.*, vol. 18, pp. 78–84.

Martin, E. D. and Lomax, H. (1975). Rapid Finite-Difference Computation of Subsonic and Slightly Supercritical Aerodynamic Flows, *AIAA J.*, vol. 13, pp. 579–586.

Martinelli, L. and Jameson, A. (1988). Validation of a Multigrid Method for Reynolds Averaged Equations, AIAA Paper 88-0414, Reno, Nevada.

Martinelli, L., Jameson, A., and Grasso, F. (1986). A Multigrid Method for the Navier-Stokes Equations, AIAA Paper 86-0208, Reno, Nevada.

Matus, R. J. and Bender, E. E. (1990). Application of a Direct Solver for Space Marching Solutions, AIAA Paper 90-1442, Seattle, Washington.

Mayne Jr., A. W. (1977). Calculation of the Laminar Viscous Shock Layer on a Blunt Biconic Body at Incidence to Supersonic and Hypersonic Flow, AIAA Paper 77-88, Los Angeles, California.

McBride, B. J., Heimel, S., Ehlers, J. G., and Gordon, S. (1963). Thermodynamic Properties to 6000°K for 210 Substances Involving the First 18 Elements, NASA SP-3001.

McDonald, H. (1970). Mixing Length and Kinematic Eddy Viscosity in a Low Reynolds Number Boundary Layer, United Aircraft Research Laboratory Report J2 14453-1, East Hartford, Connecticut.

McDonald, H. (1978). Prediction of Boundary Layers in Aircraft Gas Turbines, *The Aerothermodynamics of Aircraft Gas Turbine Engines*, Air Force Aero Propulsion Laboratory Report AFAPL-TR-78-52, Wright-Patterson Air Force Base, Ohio.

McDonald, H. and Briley, W. R. (1975). Three-Dimensional Supersonic Flow of a Viscous or Inviscid Gas, *J. Comput. Phys.*, vol. 19, pp. 150–178.

McDonald, H. and Camerata, F. J. (1968). An Extended Mixing Length Approach for Computing the Turbulent Boundary Layer Development, *Proc. Computation of Turbulent Boundary Layers—1968 AFOSR-IFP-Stanford Conference*, vol. 1, Stanford University, California, pp. 83–98.

McDonald, H. and Fish, R. W. (1973). Practical Calculation of Transitional Boundary Layers, *Int. J. Heat Mass Transfer*, vol. 16, pp. 1729–1744.

McDonald, H. and Kreskovsky, J. P. (1974). Effect of Free Stream Turbulence on the Turbulent Boundary Layer, *Int. J. Heat Mass Transfer*, vol. 17, pp. 705–716.

McDonald, J. W., Denny, V. E., and Mills, A. F. (1972). Numerical Solutions of the Navier-Stokes Equations in Inlet Regions, *J. Appl. Mech.*, vol. 39, pp. 873–878.

McEligot, D. M., Smith, S. B., and Bankston, C. A. (1970). Quasi-Developed Turbulent Pipe Flow with Heat Transfer, *J. Heat Transfer*, vol. 92, pp. 641–650.

McGuirk, J. and Page, G. (1989). Shock Capturing Using a Pressure-Correction Method, AIAA Paper 89-0561, Reno, Nevada.

McGuirk, J. J. and Rodi, W. (1977). The Calculation of Three-Dimensional Turbulent Free Jets, *Proc. Symposium on Turbulent Shear Flows*, Pennsylvania State University, University Park.

McLean, J. D. and Randall, J. L. (1979). Computer Program to Calculate Three-Dimensional Boundary Layer Flows over Wings with Wall Mass Transfer, NASA CR-3123.

McMillin, S. N., Thomas, J. L., and Murman, E. M. (1987). Euler and Navier-Stokes Solutions for the Leeside Flow over Delta Wings at Supersonic Speeds, AIAA Paper 87-2270-CP, Monterey, California.

McRae, D. S. (1976). A Numerical Study of Supersonic Cone Flow at High Angle of Attack, AIAA Paper 76-97, Washington, D.C.

McRae, D. S. and Hussaini, M. Y. (1978). Numerical Simulation of Supersonic Cone Flow at High Angle of Attack, ICASE Report 78-21.

Melaaen, M. C. (1992). Calculation of Fluid Flows with Staggered and Nonstaggered Curvilinear Nonorthogonal Grids—A Comparison, *Numerical Heat Transfer, Part B*, vol. 21, pp. 21–39.

Merkle, C. L. and Athavale, M. (1987). Time-Accurate Unsteady Incompressible Flow Algorithm Based on Artificial Compressibility, AIAA Paper 87-1137-CP, Honolulu, Hawaii.
Middlecoff, J. F. and Thomas, P. D. (1979). Direct Control of the Grid Point Distribution in Meshes Generated by Elliptic Equations, AIAA Paper 79-1462, Williamsburg, Virginia.
Miller, J. H., Tannehill, J. C., Wadawadigi, G., Edwards, T. A., and Lawrence, S. L. (1994). Computation of Hypersonic Flows with Finite-Catalytic Walls, AIAA Paper 94-2354, Colorado Springs, Colorado.
Miller, J. H., Tannehill, J. C., Lawrence, S. L., and Edwards, T. A. (1995). Development of an Upwind PNS Code for Thermo-Chemical Nonequilibrium Flows, AIAA Paper 95-2009, San Diego, California.
Miller, J. H., Tannehill, J. C., and Lawrence, S. L. (1997). Computation of Supersonic Flows with Embedded Separated Regions Using an Efficient PNS Algorithm, AIAA Paper 97-1942, Snowmass, Colorado.
Mills, R. D. (1965). Numerical Solutions of the Viscous Flow Equations for a Class of Closed Flows, *J. R. Aeronaut. Soc.*, vol. 69, pp. 714–718.
Minaie, B. N. and Pletcher, R. H. (1982). A Study of Turbulence Models for Predicting Round and Plane Heated Jets, *Heat Transfer 1982, Proc. Seventh Int. Heat Transfer Conference*, vol. 3, Hemisphere, Washington, D.C., pp. 383–388.
Miner, E. W. and Lewis, C. H. (1975). Hypersonic Ionizing Air Viscous Shock-Layer Flows over Sphere Cones, *AIAA J.*, vol. 13, pp. 80–88.
Mitchell, A. R. and Griffiths, D. F. (1980). *The Finite Difference Method in Partial Differential Equations*, Wiley, Chichester.
Miyakoda, K. (1962). Contribution to the Numerical Weather Prediction—Computation with Finite Difference, *Jpn. J. Geophys.*, vol. 3, pp. 75–190.
Moin, P. and Kim, J. (1982). Numerical Investigation of Turbulent Channel Flow, *J. Fluid Mech.*, vol. 118, pp. 341–377.
Moin, P., Squires, K., Cabot, W. and Lee, S. (1991). A Dynamic Subgrid-Scale Model for Compressible Turbulent and Scalar Transport, *Phys. Fluids A*, vol. 3, pp. 2746–2757.
Mollenhoff, C. R. (1988). *Atanasoff: Forgotten Father of the Computer*, Iowa State University Press, Ames.
Molvik, G. A. (1985). A Parabolized Navier-Stokes Code with Real Gas Effects, Parabolized N-S Code Workshop, Wright-Patterson Air Force Base, Dayton, Ohio.
Molvik, G. A. and Merkle, C. L. (1989). A Set of Strongly Coupled, Upwind Algorithms for Computing Flows in Chemical Nonequilibrium, AIAA Paper 89-0199, Reno, Nevada.
Molvik, G. A., Bowles, J. V., and Huynh, L. C. (1993). Analysis of a Hypersonic Research Vehicle with a Hydrocarbon Scramjet Engine, AIAA Paper 93-0509, Reno, Nevada.
Moore, J. and Moore, J. G. (1979). A Calculation Procedure for Three-Dimensional Viscous, Compressible Duct Flow, Parts I and II, *J. Fluids Eng.*, vol. 101, pp. 415–428.
Moretti, G. (1969). Importance of Boundary Conditions in the Numerical Treatment of Hyperbolic Equations, *Phys. Fluids*, Supplement II, vol. 12, pp. 13–20.
Moretti, G. (1971). Complicated One-Dimensional Flows, Polytechnic Institute of New York, PIBAL Report No. 71-25.
Moretti, G. (1974). On the Matter of Shock Fitting, Proc. Fourth Int. Conf. Num. Methods Fluid Dyn., Boulder, Colorado, *Lect. Notes Phys.*, vol. 35, Springer-Verlag, New York, pp. 287–292.
Moretti, G. (1975). A Circumspect Exploration of a Difficult Feature of Multidimensional Imbedded Shocks, *Proc. AIAA 2nd Computational Fluid Dynamics Conference*, Hartford, Connecticut, pp. 10–16.
Moretti, G. (1978). An Old-Integration Scheme for Compressible Flows Revisited, Refurbished and Put to Work, Polytechnic Institute of New York, M/AE Report No. 78-22.
Moretti, G. (1979). Conformal Mappings for the Computation of Steady Three-Dimensional Supersonic Flows, *Numerical/Laboratory Computer Methods in Fluid Mechanics* (A. A. Pouring and V. I. Shah, eds.), ASME, New York, pp. 13–28.
Moretti, G. and Abbett, M. (1966). A Time-Dependent Computational Method for Blunt Body Flows, *AIAA J.*, vol. 4, pp. 2136–2141.

Moretti, G. and Bleich, G. (1968). Three-Dimensional Inviscid Flow about Supersonic Blunt Cones at Angle of Attack, Sandia Laboratories Report SC-RR-68-3728, Albuquerque, New Mexico.
Moretti, P. M. and Kays, W. M. (1965). Heat Transfer to a Turbulent Boundary Layer with Varying Free-Stream Velocity and Varying Surface Temperature—An Experimental Study, *Int. J. Heat Mass Transfer*, vol. 8, pp. 1187–1202.
Morino, L. (1986). Helmholtz Decomposition Revisited: Vorticity Generation and Trailing Edge Condition, *Comput. Mech.*, vol. 1, pp. 65–90.
Moss, J. N. (1974). Reacting Viscous Shock-Layer Solutions with Multicomponent Diffusion and Mass Injection, NASA TR-R-411.
Muramoto, K. K. (1993). The Prediction of Viscous Nonequilibrium Hypersonic Flows about Ablating Configurations Using an Upwind Parabolized Navier-Stokes Code, AIAA Paper 93-2998, Orlando, Florida.
Murman, E. M. and Cole, J. D. (1971). Calculation of Plane Steady Transonic Flows, *AIAA J.*, vol. 9, pp. 114–121.
Murphy, J. D. and Prenter, P. M. (1981). A Hybrid Computing Scheme for Unsteady Turbulent Boundary Layers, *Proc. Third Symposium on Turbulent Shear Flows*, University of California, Davis, pp. 8.26–8.34.
Murray, A. L. and Lewis, C. H. (1978). Hypersonic Three-Dimensional Viscous Shock-Layer Flows over Blunt Bodies, *AIAA J.*, vol. 16, 1279–1286.
Nakahashi, K. and Diewert, G. S. (1986). Three-Dimensional Adaptive Grid Method, *AIAA J.*, vol. 24, pp. 948–954.
Nakamura, S. (1982). Marching Grid Generation Using Parabolic Partial Differential Equations, *Numerical Grid Generation* (J. F. Thompson, ed.), Elsevier, New York, pp. 79–105.
Napolitano, M., Werle, M. J., and Davis, R. T. (1978). A Numerical Technique for the Triple-Deck Problem, AIAA Paper 78-1133, Seattle, Washington.
Nardo, C. T. and Cresci, R. J. (1971). An Alternating Directional Implicit Scheme for Three-Dimensional Hypersonic Flows, *J. Comput. Phys.*, vol. 8, pp. 268–284.
NASA (1972). Free Turbulent Shear Flows, vol. 1, in Proceedings of the Langley Working Conference on Free Turbulent Shear Flows, NASA SP-321.
Nee, V. W. and Kovasznay, L. S. G. (1968). The Calculation of the Incompressible Turbulent Boundary Layer by a Simple Theory, *Phys. Fluids*, vol. 12, pp. 473–484.
Nelson, R. M. and Pletcher, R. H. (1974). An Explicit Scheme for the Calculation of Confined Turbulent Flows with Heat Transfer, *Proc. 1974 Heat Transfer and Fluid Mechanics Institute*, Stanford University Press, Stanford, California, pp. 154–170.
Neumann, R. D. and Patterson, J. L. (1988). Results of an Industry Representative Study of Code to Code Validation of Axisymmetric Configurations at Hypervelocity Flight Condition, AIAA Paper 88-2691, San Antonio, Texas.
Newsome, R. W., Walters, R. W., and Thomas, J. L. (1987), An Efficient Iteration Strategy for Upwind/Relaxation Solutions to the Thin Layer Navier-Stokes Equations, AIAA Paper 87-1113, Honolulu, Hawaii.
Ng, K. H. and Spalding, D. B. (1972). Turbulence Model for Boundary Layers Near Walls, *Phys. Fluids*, vol. 15, pp. 20–30.
Noack, R. W. (1985). Inviscid Flow Field Analysis of Maneuvering Hypersonic Vehicles Using the SCM Formulation and Parabolic Grid Generation, AIAA Paper 85-1682, Cincinnati, Ohio, 1985.
Noack, R. W., Steinbrenner, J. P., and Bishop, D. G. (1996). A Three-Dimensional Hybrid Grid Generation Technique with Application to Bodies in Relative Motion, *Numerical Grid Generation in Computational Field Simulations* (B. K. Soni, J. F. Thompson, J. Hauser, and P. Eiseman, eds.), NSF Engineering Research Center for Computational Field Simulation, Mississippi State University, Mississippi State, pp. 547–556.
Obayashi, S. and Kuwahara, K. (1986). An Approximate LU Factorization Method for the Compressible Navier-Stokes Equations, *J. Comput. Phys.*, vol. 63, pp. 157–167.
Obayashi, S., Fujii, K., and Takanashi, S. (1987). Toward the Navier-Stokes Analysis of Transport Aircraft Configurations, AIAA Paper 87-0428, Reno, Nevada.

O'Brien, G. G., Hyman, M. A., and Kaplan, S. (1950). A Study of the Numerical Solution of Partial Differential Equations, *J. Math. Phys.*, vol. 29, pp. 223–251.

Oden, J. T. (1988). Adaptive Finite Element Methods for Problems in Solid and Fluid Mechanics, chap. 12, *Element Theory and Application Overview* (R. Voight, ed.), Springer-Verlag, New York.

Oleinik, O. A. (1957). Discontinuous Solutions of Nonlinear Differential Equations, *Uspekhi Mat. Nauk*, vol. 12, pp. 95–172.

Oran, E. S. and Boris, J. P. (1987). *Numerical Simulation of Reactive Flows*, Elsevier Science Publishing, New York.

Orszag, S. A. and Israeli, M. (1974). Numerical Simulation of Viscous Incompressible Flows, *Annu. Rev. Fluid Mech.*, vol. 6, Annual Reviews, Inc., Palo Alto, California, pp. 281–318.

Osher, S. (1984). Riemann Solvers, Entropy Conditions and Difference Approximations, *SIAM J. Numer. Anal.*, vol. 21, pp. 217–235.

Osher, S. and Chakravarthy, S. R. (1983). Upwind Schemes and Boundary Conditions with Applications to Euler Equations in General Coordinates, *J. Comput. Phys.*, vol. 50, pp. 447–481.

Osswald, G. A., Ghia, K. N., and Ghia, U. (1984). Unsteady Navier-Stokes Simulation of Internal Separated Flows over Plane and Axisymmetric Sudden Expansions, AIAA Paper 84-1584, Snowmass, Colorado.

Osswald, G. A., Ghia, K. N., and Ghia, U. (1987). A Direct Algorithm for Solution of Incompressible Three-Dimensional Unsteady Navier-Stokes Equations, AIAA Paper 87-1139, Honolulu, Hawaii.

Ota, D. K., Chakravarthy, S. R., and Darling, J. C. (1988). An Equilibrium Air Navier-Stokes Code for Hypersonic Flows, AIAA Paper 88-0419, Reno, Nevada.

Owczarek, J. A. (1964). *Fundamentals of Gas Dynamics*, International Textbook Company, Scranton, Pennsylvania.

Palaniswamy, S., Chakravarthy, S. R., and Ota, D. K. (1989). Finite Rate Chemistry for USA-Series Codes: Formulation and Applications, AIAA Paper 89-0200, Reno, Nevada.

Palumbo, D. J. and Rubin, E. L. (1972). Solution of the Two-Dimensional, Unsteady Compressible Navier-Stokes Equations Using a Second-Order Accurate Numerical Scheme, *J. Comput. Phys.*, vol. 9, pp. 466–495.

Pan, F. and Acrivos, A. (1967). Steady Flows in Rectangular Cavities, *J. Fluid Mech.*, vol. 28, pp. 643–655.

Pan, D. and Chakravarthy, S. R. (1989). Unified Formulation for Incompressible Flows, AIAA Paper 89-0122, Reno, Nevada.

Panton, R. L. (1984). *Incompressible Flow*, Wiley Interscience, New York.

Pappalexis, J. N. and TenPas, P. W. (1993). Effect of Wall Temperature Ratio on Laminar Forced Convection Heat Transfer Behind a Rearward-Facing Step, General Papers on Convection, HTD-vol. 256, ASME, New York, pp. 29–35.

Park, C. (1990). *Nonequilibrium Hypersonic Aerothermodynamics*, John Wiley, New York.

Parpia, I. (1994). Personal Communication.

Parpia, I. and Michalek, D. J. (1993). Grid-Independent Upwind Scheme for Multidimensional Flow, *AIAA J.*, vol. 31, pp. 646–656.

Patankar, S. V. (1975). Numerical Prediction of Three-Dimensional Flows, *Studies in Convection: Theory, Measurement, and Applications* (B. E. Launder, ed.), vol. 1, Academic, New York, pp. 1–78.

Patankar, S. V. (1980). *Numerical Heat Transfer and Fluid Flow*, Hemisphere, Washington, D.C.

Patankar, S. V. (1981). A Calculation Procedure for Two-Dimensional Elliptic Situations, *Numer. Heat Transfer*, vol. 4, pp. 409–425.

Patankar, S. V. and Spalding, D. B. (1970). *Heat and Mass Transfer in Boundary Layers*, 2d ed., Intertext Books, London

Patankar, S. V. and Spalding, D. B. (1972). A Calculation Procedure for Heat, Mass and Momentum Transfer in Three-Dimensional Parabolic Flows, *Int. J. Heat Mass Transfer*, vol. 15, pp. 1787–1806.

Patankar, S. V., Pratap, V. S., and Spalding, D. B. (1974). Prediction of Laminar Flow and Heat Transfer in Helically Coiled Pipes, *J. Fluid Mech.*, vol. 62, pp. 539–551.

Patankar, S. V., Basu, D. K., and Alpay, S. A. (1977). Prediction of the Three-Dimensional Velocity Field of a Deflected Turbulent Jet, *J. Fluids Eng.*, vol. 99, pp. 758–762.

Patankar, S. V., Ivanović, M., and Sparrow, E. M. (1979). Analysis of Turbulent Flow and Heat Transfer in Internally Finned Tubes and Annuli, *J. Heat Transfer*, vol. 101, pp. 29–37.

Patel, V. C. and Choi, D. H. (1979). Calculation of Three-Dimensional Laminar and Turbulent Boundary Layers on Bodies of Revolution at Incidence, *Proc. Second Symposium on Turbulent Shear Flows*, Imperial College, London, pp. 15.14–15.24.

Peaceman, D. W. and Rachford, H. H. (1955). The Numerical Solution of Parabolic and Elliptic Differential Equations, *J. Soc. Ind. Appl. Math.*, vol. 3, pp. 28–41.

Peng, T. C. and Pindroh, A. L. (1962). An Improved Calculation of Gas Properties at High Temperatures: Air, Boeing Co., Report D2-11722, Seattle, Washington.

Peric, M., Kessler, R., and Scheuerer, G. (1988). Comparison of Finite-Volume Numerical Methods with Staggered and Colocated Grids, *Comput. Fluids*, vol. 16, pp. 389–403.

Peyret, R. and Taylor, T. D. (1983). *Computational Methods for Fluid Flow*, Springer-Verlag, New York.

Peyret, R. and Viviand, H. (1975). Computation of Viscous Compressible Flows Based on the Navier-Stokes Equations, AGARD-AG-212.

Peyret, R. and Viviand, H. (1985). Pseudo-Unsteady Methods for Inviscid or Viscous Flow Computations, *Recent Advances in the Aerospace Sciences* (C. Casi, ed.), Plenum, New York, pp. 312–343.

Phillips, J. H. and Ackerberg, R. C. (1973). A Numerical Method for Integrating the Unsteady Boundary-Layer Equations When There are Regions of Backflow, *J. Fluid Mech.*, vol. 58, pp. 561–579.

Piomelli, U. (1988). *Models for Large-Eddy Simulations of Turbulent Channel Flows Including Transpiration*, Ph.D. dissertation, Stanford University, Stanford, California.

Pletcher, R. H. (1969). On a Finite-Difference Solution for the Constant-Property Turbulent Boundary Layer, *AIAA J.*, vol. 7, pp. 305–311.

Pletcher, R. H. (1970). On a Solution for Turbulent Boundary Layer Flows with Heat Transfer, Pressure Gradient, and Wall Blowing or Suction, *Heat Transfer 1970, Proc. Fourth Int. Heat Transfer Conference*, vol. 1, Elsevier, Amsterdam.

Pletcher, R. H. (1971). On a Calculation Method for Compressible Boundary Layers with Heat Transfer, AIAA Paper 71-165, New York.

Pletcher, R. H. (1974). Prediction of Transpired Turbulent Boundary Layers, *J. Heat Transfer*, vol. 96, pp. 89–94.

Pletcher, R. H. (1976). Prediction of Turbulent Boundary Layers at Low Reynolds Numbers, *AIAA J.*, vol. 14, pp. 696–698.

Pletcher, R. H. (1978). Prediction of Incompressible Turbulent Separating Flow, *J. Fluids Eng.*, vol. 100, pp. 427–433.

Pletcher, R. H. and Chen, K.-H. (1993). On Solving the Compressible Navier-Stokes Equations for Unsteady Flows at Very Low Mach Numbers, AIAA Paper 93-3368-CP, Orlando, Florida.

Pletcher, R. H. and Dancey, C. L. (1976). A Direct Method of Calculating Through Separated Regions in Boundary Layer Flow, *J. Fluids Eng.*, vol. 98, pp. 568–572.

Polezhaev, V. I. (1967). Numerical Solution of the System of Two-Dimensional Unsteady Navier-Stokes Equations for a Compressible Gas in a Closed Region, *Fluid Dyn.*, vol. 2, pp. 70–74.

Power, G. D. (1990). A Novel Approach for Analyzing Supersonic High Reynolds Number Flows with Separation, AIAA Paper 90-0764, Reno, Nevada.

Power, G. D. and Barber, T. J. (1988). Analysis of Complex Hypersonic Flows with Strong Viscous/Inviscid Interaction, *AIAA J.*, vol. 26, pp. 832–840.

Prabhu, D. K. and Tannehill, J. C. (1984). Numerical Solution of Space Shuttle Orbiter Flowfield Including Real Gas Effects, AIAA Paper 84-1747, Snowmass, Colorado.

Prabhu, D. K., Tannehill, J. C., and Marvin, J. G. (1987a). A New PNS Code for Chemical Nonequilibrium Flows, AIAA Paper 87-0248, Reno, Nevada.

Prabhu, D. K., Tannehill, J. C., and Marvin, J. G. (1987b). A New PNS Code for Three-Dimensional Chemically Reacting Flows, AIAA Paper 87-1472, Honolulu, Hawaii.

Prakash, C. (1981). A Finite Element Method for Predicting Flow through Ducts with Arbitrary Cross Sections, Ph.D. dissertation, University of Minnesota, Minneapolis.

Prandtl, L. (1926). Ueber die ausgebildete Turbulenz, *Proceedings of the 2nd International Congress for Applied Mechanics*, Zürich, pp. 62–74.

Pratap, V. S. and Spalding, D. B. (1976). Fluid Flow and Heat Transfer in Three-Dimensional Duct Flows, *Int. J. Heat Mass Transfer*, vol. 19, pp. 1183–1188.

Pulliam, T. H. (1985). Efficient Solution Methods for the Navier-Stokes Equations, Lecture Notes for the von Karman Institute for Fluid Dynamics Lecture Series, *Numerical Techniques for Viscous Flow Computation in Turbomachinery Blading*, van Karman Institute, Rhode-St.-Genese, Belgium.

Pulliam, T. H. and Chaussee, D. S. (1981). A Diagonal Form of an Implicit Approximate Factorization Algorithm, *J. Comput. Phys.*, vol. 39, pp. 347–363.

Pulliam, T. H. and Steger, J. L. (1978). On Implicit Finite-Difference Simulations of Three Dimensional Flow, AIAA Paper 78-10, Huntsville, Alabama.

Pulliam, T. H. and Steger, J. L. (1980). Implicit Finite Difference Simulations of Three-Dimensional Compressible Flow, *AIAA J.*, vol. 18, pp. 159–167.

Raetz, G. S. (1957). A Method of Calculating Three-Dimensional Laminar Boundary Layers of Steady Compressible Flow, Report No. NAI-58-73 (BLC-114), Northrop Corporation.

Rai, M. M. (1982). A Philosophy for Construction of Solution Adaptive Grids, Ph.D. dissertation, Iowa State University, Ames.

Rai, M. M. (1987). Navier-Stokes Simulation of Blade Vortex Interaction Using High-Order Accurate Upwind Schemes, AIAA Paper 87-0543, Reno, Nevada.

Rai, M. M. and Anderson, D. A. (1980). Grid Evolution in Time Asymptotic Problems, *Numerical Grid Generation Techniques*, NASA Conference Publication 2166, pp. 409–430.

Rai, M. M. and Anderson, D. A. (1981). The Use of Adaptive Grids in Conjunction with Shock Capturing Methods, AIAA Paper 81-1012, St. Louis, Missouri.

Rai, M. M. and Anderson, D. A. (1982). Application of Adaptive Grids to Fluid Flow Problems with Asymptotic Solutions, *AIAA J.*, vol. 20, pp. 496–502.

Rai, M. M. and Chaussee, D. S. (1994). New Implicit Boundary Procedures—Theory and Application, *AIAA J.*, vol. 22, pp. 1094–1168.

Rai, M. M., Chaussee, D. S., and Rizk, Y. M. (1983). Calculation of Viscous Supersonic Flows over Finned Bodies, AIAA Paper 83-1667, Danvers, Massachusetts.

Raithby, G. D. (1976). Skew Upstream Differencing Schemes for Problems Involving Fluid Flow, *Comput. Methods Appl. Mech. Eng.*, vol. 9, pp. 153–164.

Raithby, G. D. and Schneider, G. E. (1979). Numerical Solution of Problems in Incompressible Fluid Flow: Treatment of the Velocity-Pressure Coupling, *Numer. Heat Transfer*, vol. 2, pp. 417–440.

Raithby, G. D. and Schneider, G. E. (1980). The Prediction of Surface Discharge Jets by a Three-Dimensional Finite-Difference Model, *J. Heat Transfer*, vol. 102, pp. 138–145.

Raithby, G. D. and Torrance, K. E. (1974). Upstream-Weighted Differencing Schemes and Their Application to Elliptic Programs Involving Fluid Flow, *Comput. Fluids*, vol. 2, pp. 191–206.

Rakich, J. V. (1978). Computational Fluid Mechanics—Course Notes, Department of Mechanical and Aerospace Engineering, North Carolina State University, Raleigh.

Rakich, J. V. (1983). Iterative PNS Method for Attached Flows with Upstream Influence, AIAA Paper 83-1955, Danvers, Massachusetts.

Rakich, J. V., Vigneron, J. C., and Agarwal, R. (1979). Computation of Supersonic Viscous Flows over Ogive-Cylinders at Angle of Attack, AIAA Paper 79-0131, New Orleans, Louisiana.

Rakich, J. V., Venkatapathy, E. Tannehill, J. C., and Prabhu, D. K. (1984). Numerical Solution of Space Shuttle Orbiter Flowfield, *J. Spacecraft Rockets*, vol. 21, pp. 9–15.

Ralston, A. (1965). *A First Course in Numerical Analysis*, McGraw-Hill, New York.

Ramakrishnan, S. V. and Rubin, S. G. (1987). Time-Consistent Pressure Relaxation Procedure for Compressible Reduced Navier-Stokes Equations, *AIAA J.*, vol. 25, pp. 905–913.

Ramin, T. H. and Pletcher, R. H. (1993). Developments on an Unsteady Boundary-Layer Analysis: Internal and External Flows, *Numer. Heat Transfer, Part B*, vol. 23, pp. 289–307.

Rasmussen, M. L. and Yoon, B.-H. (1990). Boundary Layers for the Conical Navier-Stokes Equations, *AIAA J.*, vol. 28, pp. 752–754.

Reddy, D. R. and Rubin, S. G. (1988). Consistent Boundary Conditions for RNS Scheme Applied to Three-Dimensional Viscous Flows, *J. Fluids Eng*, vol. 10, pp. 306–314.

Reyhner, T. A. (1968). Finite-Difference Solution of the Compressible Turbulent Boundary Layer Equations, *Proc. Computation of Turbulent Boundary Layers—1968 AFOSR-IFP-Stanford Conference*, vol. 1, Stanford University, California, pp. 375–383.

Reyhner, T. A. and Flügge-Lotz, I. (1968). The Interaction of a Shock Wave with a Laminar Boundary Layer, *Int. J. Non-Linear Mech.*, vol. 3, pp. 173–199.

Reynolds, A. J. (1975). The Prediction of Turbulent Prandtl and Schmidt Numbers, *Int. J. Heat Mass. Transfer*, vol. 18, pp. 1055–1069.

Rhie, C. M. (1981). A Numerical Study of the Flow Past an Isolated Airfoil with Separation, Ph.D. dissertation, University of Illinois, Urbana-Champaign.

Rhie, C. M. and Chow, W. L. (1983). Numerical Study of the Turbulent Flow Past an Airfoil with Trailing Edge Separation, *AIAA J.*, vol. 21, pp. 1525–1532.

Richardson, L. F. (1910). The Approximate Arithmetical Solution by Finite Differences of Physical Problems Involving Differential Equations, with an Application to the Stresses in a Masonry Dam, *Philos, Trans. R. Soc. London, Ser. A*, vol. 210, pp. 307–357.

Richardson, S. M. and Cornish, A. R. H. (1977). Solution of Three-Dimensional Incompressible Flow Problems, *J. Fluid Mech.*, vol. 82, pp. 309–319.

Richtmyer, R. D. (1957). *Difference Methods for Initial-Value Problems*, Interscience Publishers, New York.

Richtmyer, R. D. and Morton, K. W. (1967). *Difference Methods for Initial-Value Problems*, 2d ed., Interscience Publishers, Wiley, New York.

Rieger, H. and Jameson, A. (1988). Solution of the Three-Dimensional Compressible Euler and Navier-Stokes Equations by an Implicit LU Scheme, AIAA Paper 88-0619, Reno, Nevada.

Riemslagh, K. and Vierendeels, J. (1996). Grid Generation for Complex Shaped Moving Domains, *Numerical Grid Generation in Computational Field Simulations* (B. K. Soni, J. F. Thompson, J. Hauser, and P. Eiseman, eds.), NSF Engineering Research Center for Computational Field Simulation, Mississippi State University, Mississippi State, pp. 569–578.

Rizk, Y. M., Chaussee, D. S., and McRae, D. S. (1981). Computation of Hypersonic Viscous Flow around Three-Dimensional Bodies at High Angles of Attack, AIAA Paper 81-1261, Palo Alto, California.

Rizzi, A. and Eriksson, L. E. (1981). Transfinite Mesh Generation and Damped Euler Equation Algorithm for Transonic Flow around Wing-Body Configurations, *Proc. AIAA 5th Computational Fluid Dynamics Conference*, Palo Alto, California, pp. 43–69.

Roache, P. J. (1972). *Computational Fluid Dynamics*, Hermosa, Albuquerque, New Mexico.

Roberts, G. O. (1971). Computational Meshes for Boundary Layer Problems, Proc. Second Int. Conf. Num. Methods Fluid Dyn., *Lect. Notes Phys.*, vol. 8, Springer-Verlag, New York, pp. 171–177.

Rodi, W. (1975). A Review of Experimental Data of Uniform Density Free Turbulent Boundary Layers, *Studies in Convection: Theory, Measurement, and Applications*, vol. 1 (B. E. Launder, ed.), Academic, New York.

Rodi, W. (1976). A New Algebraic, Relation for Calculating Reynolds Stresses, *ZAMM*, vol. 56, pp. 219–221.

Roe, P. L. (1980). The Use of the Riemann Problem in Finite-Difference Schemes, *Lect. Notes Phys.*, vol. 141, Springer-Verlag, New York, pp. 354–359.

Roe, P. L. (1981). Approximate Riemann Solvers, Parameter Vectors and Difference Schemes, *J. Comput. Phys.*, vol. 43, pp. 357–372.

Roe, P. L. (1984). Generalized Formulation of TVD Lax-Wendroff Schemes, ICASE Report 84-53.

Roe, P. L. (1985). Some Contributions to the Modeling of Discontinuous Flows, Proc. *1983* AMS-SIAM Summer Seminar on Large Scale Computing in Fluid Mechanics, *Lect. Appl. Math.*, vol. 22, pp. 163–193.

Roe, P. L. and Pike, J. (1985). Efficient Construction and Utilization of Approximate Riemann Solutions, *Comput. Methods in Applied Sciences and Engineering* (R. Glowinske and J. L. Lions, eds.), North Holland, Amsterdam.

Rogers, S. E. and Kwak, D. (1988). An Upwind Differencing Scheme for the Time-Accurate Incompressible Navier-Stokes Equations, AIAA Paper 88-2583, Williamsburg, Virginia.

Rogers, S. E., Kwak, D., and Kiris, C. (1989). Numerical Solution of the Incompressible Navier-Stokes Equations for Steady-State and Time-Dependent Problems, AIAA Paper 89-0463, Reno, Nevada.

Rotta, J. (1951). Statistische Theorie nichthomogener Turbulenz, *Z. Phys.*, vol. 129, pp. 547–572.

Rouleau, W. T. and Osterle, J. F. (1955). The Application of Finite Difference Methods to Boundary-Layer Type Flows, *J. Aeronaut. Sci.*, vol. 22, pp. 249–254.

Rubbert, P. E. and Saaris, G. R. (1972). Review and Evaluation of a Three-Dimensional Lifting Potential Flow Analysis Method for Arbitrary Configurations, AIAA Paper 72-188, San Diego, California.

Rubesin, M. W. (1976). A One-Equation Model of Turbulence for Use with the Compressible Navier-Stokes Equations, NASA TM X-73-128.

Rubesin, M. W. (1977). Numerical Turbulence Modeling, *AGARD Lecture Series No. 86 on Computational Fluid Dynamics*, pp. 3-1 to 3-37.

Rubin, S. G. (1981). A Review of Marching Procedures for Parabolized Navier-Stokes Equations, *Proceedings of Symposium on Numerical and Physical Aspects of Aerodynamic Flows*, Springer-Verlag, New York, pp. 171–186.

Rubin, S. G. (1984). Incompressible Navier-Stokes and Parabolized Navier-Stokes Formulations and Computational Techniques, *Computational Methods in Viscous Flows*, vol. 3 (W. G. Habash, ed.), Pineridge, Swansea, U.K., pp. 53–99.

Rubin, S. G. (1988). RNS/Euler Pressure Relaxation and Flux Vector Splitting, *Comput. Fluids*, vol. 16, pp. 485–490.

Rubin, S. G. and Harris, J. E. (1975). Numerical Studies of Incompressible Viscous Flow in a Driven Cavity, NASA SP-378.

Rubin, S. G. and Khosla, P. K. (1981). Navier-Stokes Calculations with a Coupled Strongly Implicit Method—I. Finite-Difference Solutions, *Comput. Fluids*, vol. 9, pp. 163–180.

Rubin, S. G. and Khosla, P. K. (1990). A Review of Reduced Navier-Stokes Computations for Compressible Viscous Flows, *J. Comput. Syst. Eng.*, vol. 1, pp. 549–562.

Rubin, S. G. and Lin, T. C. (1971). Numerical Methods for Two- and Three-Dimensional Viscous Flow Problems: Application to Hypersonic Leading Edge Equations, Polytechnic Institute of Brooklyn, PIBAL Report 71-8, Farmingdale, New York.

Rubin, S. G. and Lin, T. C. (1972). A Numerical Method for Three-Dimensional Viscous Flow: Application to the Hypersonic Leading Edge, *J. Comput. Phys.*, vol. 9, pp. 339–364.

Rubin, S. G. and Lin, A. (1980). Marching with the PNS Equations, *Proceedings of 22nd Israel Annual Conference on Aviation and Astronautics*, Tel Aviv, Israel, pp. 60–61. See *Isr. J. Technol.*, vol. 18, 1980.

Rubin, S. G. and Lin, A. (1981). Marching with the PNS Equations, *Isr. J. Technol.*, vol. 18, pp. 21–31.

Rubin, S. G. and Reddy, D. R. (1983). Analysis of Global Pressure Relaxation for Flows with Strong Interaction and Separation, *Comput. Fluids*, vol. 11, pp. 281–306.

Rubin, S. G., Lin. T. C., Pierucci, M., and Rudman, S. (1969). Hypersonic Interactions near Sharp Leading Edges, *AIAA J.*, vol. 7, pp. 1744–1751.

Rubin, S. G., Khosla, P. K., and Saari, S. (1977). Laminar Flow in Rectangular Channels, *Comput. Fluids*, vol. 5, pp. 151–173.

Rubin, S. and Tannehill, J. (1992). Parabolized/Reduced Navier-Stokes Computational Techniques, *Annu. Rev. Fluid Mech.*, vol. 24, Annual Reviews, Inc., Palo Alto, California, pp. 117–144.

Rudman, S. and Rubin, S. G. (1968). Hypersonic Viscous Flow over Slender Bodies with Sharp Leading Edges, *AIAA J.*, vol. 6, pp. 1883–1889.

Ruffin, S. M. and Murman, E. M. (1988). Solutions for Hypersonic Viscous Flow over Delta Wings, AIAA Paper 88-0126, Reno, Nevada.

Rusanov, V. V. (1970). On Difference Schemes of Third Order Accuracy for Nonlinear Hyperbolic Systems, *J. Comput. Phys.*, vol. 5, pp. 507–516.

Saad, Y. and Schultz, M. H. (1986). GMRES, A Generalized Minimum Residual Algorithm for Solving Nonsymmetric Linear Systems, *SIAM J. Sci. Stat. Comput.*, vol. 7, pp. 856–869.

Saffman, P. G. and Wilcox, D. C. (1974). Turbulence Model Predictions for Turbulence Boundary Layers, *AIAA J.*, vol. 12, pp. 541–546.

Salas, M. D. (1975). The Anatomy of Floating Shock Fitting, *Proc. AIAA 2nd Computational Fluid Dynamics Conference*, Hartford, Connecticut, pp. 47–54.

Salas, M. D. (1979). Flow Properties for a Spherical Body at Low Supersonic Speeds, presented at the Symposium on Computers in Aerodynamics, Twenty-Fifth Anniversary of the Aerodynamics Laboratories, Polytechnic Institute of New York.

Sanders, R. N. and Prendergast, K. H. (1974). The Possible Relation of the Three-Kiloparsec Arm to Explosions in the Galactic Nucleus, *Astrophys. J.* vol. 188, pp. 231–238.

Saul'yev, V. K. (1957). On a Method of Numerical Integration of a Diffusion Equation, *Dokl. Akad. Nauk SSSR*, vol. 115, pp. 1077–1079. (In Russian)

Schiff, L. B. and Steger, J. L. (1979). Numerical Simulation of Steady Supersonic Viscous Flow, AIAA Paper 79-0130, New Orleans, Louisiana.

Schiff, L. B. and Sturek, W. B. (1980). Numerical Simulation of Steady Supersonic Flow over an Ogive-Cylinder-Boattail Body, AIAA Paper 80-0066, Pasadena, California.

Schlichting, H. (1968). *Boundary-Layer Theory*, 6th ed., translated by J. Kestin, McGraw-Hill, New York.

Schlichting, H. (1979). *Boundary-Layer Theory*, 7th ed., translated by J. Kestin, McGraw-Hill, New York.

Schneider, G. E. and Zedan, M. (1981). A Modified Strongly Implicit Procedure for the Numerical Solution of Field Problems, *Numer. Heat Transfer*, vol. 4, pp. 1–19.

Schneiders, R. (1996). Refining Quadrilateral and Hexahedral Element Meshes, *Numerical Grid Generation in Computational Field Simulations* (B. K. Soni, J. F. Thompson, J. Hauser, and P. Eiseman, eds.), NSF Engineering Research Center for Computational Field Simulation, Mississippi State University, Mississippi State, pp. 679–688.

Schubauer, G. B. and Tchen, C. M. (1959). Section B of *Turbulent Flows and Heat Transfer*, vol. 5, in High Speed Aerodynamics and Jet Propulsion, Princeton University Press, New Jersey.

Schumann, U. (1977). Realizability of Reynolds Stress Turbulence Models, *Phys. Fluids*, vol. 20, pp. 721–725.

Schumann, U. (1980). Fast Elliptic Solvers and their Application in Fluid Dynamics, *Computational Fluid Dynamics*, Hemisphere, Washington, D.C., pp. 402–430.

Schwartztrauber, P. N. and Sweet, R. A. (1977). The Direct Solution of the Discrete Poisson Equation on a Disc, *SIAM J. Numer. Anal.*, vol. 5, pp. 900–907.

Shaanan, S., Ferziger, J. H., and Reynolds, W. C. (1975). Numerical Simulation of Turbulence in the Presence of Shear, Report TF-6, Department of Mechanical Engineering, Stanford University, Stanford, California.

Shah, R. K. and London, A. L. (1978). *Laminar Flow Forced Convection in Ducts*, Academic, New York.

Shang, J. S. (1977). An Implicit-Explicit Method for Solving the Navier-Stokes Equations, AIAA Paper 77-646, Albuquerque, New Mexico.

Shang, J. S. and Hankey Jr., W. L. (1975). Supersonic Turbulent Separated Flows Utilizing the Navier-Stokes Equations, *Flow Separation*, AGARD-CCP-168.

Shang, J. S. and Scherr, S. J. (1985). Navier-Stokes Solution of the Flow Field around a Complete Aircraft, AIAA Paper 85-1509, Cincinnati, Ohio.

Shang, J. S. and Scherr, S. J. (1986). Numerical Simulation of X24C-10D, Air Force Wright Aeronautical Laboratories Report AFWAL-TR-86-3072.

Shankar, V. (1981). Treatment of Conical and Nonconical Supersonic Flows by an Implicit Marching Scheme Applied to the Full Potential Equation, *Proc. ASME/AIAA Conference on Computers in Flow Predictions and Fluid Dynamics Experiments*, Washington, D.C., pp. 163–170.

Shankar, V. and Chakravarthy, S. (1981). An Implicit Marching Procedure for the Treatment of Supersonic Flow Fields using the Conservative Full Potential Equation, AIAA Paper 81-1004, Palo Alto, California.

Shankar, V. and Osher, S. (1982). An Efficient Full Potential Implicit Method Based on Characteristics for Analysis of Supersonic Flows, AIAA Paper 82-0974, St. Louis, Missouri.

Shankar, V., Ide, H., Gorski, J., and Osher, S. (1985). A Fast, Time-Accurate Unsteady Full Potential Scheme, AIAA Paper 85-1512, Cincinnati, Ohio.

Shanks, S. P., Srinivasan, G. R., and Nicolet, W. E. (1982). AFWAL Parabolized Navier-Stokes Code: Formulation and User's Manual, AFWAL TR-82-3034.

Shapiro, A. H. (1953). *The Dynamics and Thermodynamics of Compressible Fluid Flow*, vol. I, Ronald Press, New York.

Sharpe, H. and Anderson, D. A. (1991). Orthogonal Adaptive Grid Generation with Fixed Internal Boundaries for Oil Reservoir Simulation, *Numerical Grid Generation in Computational Field Simulation and Related Fields* (A. S. Arcilla, J. Hauser, P. R. Eiseman, and J. F. Thompson, eds.), North Holland, New York, pp. 405–417.

Shih, T.-H., Zhu, J., and Lumley, J. L. (1994). A New Reynolds Stress Algebraic Equation Model, NASA Technical Memorandum 106644.

Shinn, J. L. and Jones, J. J. (1983). Chemical Nonequilibrium Effects on Flowfields for Aeroassist Orbital Transfer Vehicles, AIAA Paper 83-0214, Reno, Nevada.

Shinn, J. L., Moss, J. N., and Simmonds, A. L. (1982). Viscous Shock-Layer Heating Analysis for the Shuttle Windward-Symmetry Plane with Surface Finite Catalytic Recombination Rates, AIAA Paper 82-0842, St. Louis, Missouri.

Shyy, W., Chen, M.-H., and Sun, C.-S. (1992). A Pressure-Based FMG/FAS Algorithm for Flow at All Speeds, AIAA Paper 92-0548, Reno, Nevada.

Sichel, M. (1963). Structure of Weak Non-Hugoniot Shocks, *Phys. Fluids*, vol. 6, pp. 653–663.

Simon, H. D. (1995). Seven Years of Parallel Computing at NAS (1987–1994): What Have We Learned? AIAA Paper 95-0219, Reno, Nevada.

Sinha, N. and Dash, S. M. (1986). Parabolized Navier-Stokes Analysis of Ducted Turbulent Mixing Problems with Finite-Rate Chemistry, AIAA Paper 86-0004, Reno, Nevada.

Sinha, N., Dash, S. M., and Krawczyk, W. J. (1987). Inclusion of Chemical Kinetics into Beam-Warming Based PNS Model for Hypersonic Propulsion Applications, AIAA Paper 87-1898, San Diego, California.

Sinha, N., Dash, S. M., and Lee, R. A. (1990). 3-D PNS Analysis of Scramjet Combustor/Nozzle and Exhaust Plume Flowfields, AIAA Paper 90-0094, Reno, Nevada.

Smagorinsky, J. (1963). General Circulation Experiments with the Primitive Equations, I: The Basic Experiment, *Mon. Weather Rev.*, vol. 91, pp. 99–164.

Smith, R. E. and Weigel, B. L. (1980). Analytic and Approximate Boundary Fitted Coordinate Systems for Fluid Flow Simulation, AIAA Paper 80-0192, Pasadena, California.

Smith, R. J. and Johnson, L. J. (1996). Automatic Grid Generation and Flow Solution for Complex Geometries, *AIAA J.*, vol. 34, pp. 1120–1125.

Sod, G. (1985). *Numerical Methods in Fluid Dynamics*, Cambridge University Press, Cambridge.

Soh, W. Y. and Goodrich, J. W. (1988). Unsteady Solutions of Incompressible Navier-Stokes Equations, *J. Comput. Phys.*, vol. 79, pp. 113–134.

Song, D. J. and Lewis, C. H. (1986). Hypersonic Finite-Rate Chemically Reacting Viscous Flows over an Ablating Carbon Surface, *J. Spacecraft Rockets*, vol. 23, pp. 47–54.

Sorenson, R. L. and Steger, J. L. (1983). Grid Generation in Three Dimensions by Poisson Equations with Control of Cell Size and Skewness at Boundary Surfaces, *Advances in Grid Generation* (K. N. Ghia and U. Ghia, eds.), FED-vol. 5, ASME Applied Mechanics, Bioengineering, and Fluids Engineering Conference, Houston, Texas.

Southwell, R. V. (1940). *Relaxation Methods in Engineering Science*, Oxford University Press, London.

Spalart, P. R., and Allmaras, S. R. (1992). A One-Equation Turbulence Model for Aerodynamic Flows, AIAA Paper 92-0439, Reno, Nevada.

Spalding, D. B. (1972). A Novel Finite-Difference Formulation for Differential Expressions Involving Both First and Second Derivatives, *Int. J. Numer. Methods Eng.*, vol. 4, pp. 551–559.

Speziale, C. G. (1985). Galilean Invariance of Subgrid-Scale Stress Models in the Large-Eddy Simulation of Turbulence, *J. Fluid Mech.*, vol. 156, pp. 55–62.

Speziale, C. G. (1987). On Nonlinear *k-l* and *k-ε* Models of Turbulence, *J. Fluid Mech.*, vol. 178, pp. 459–475.

Speziale, C. G. (1991). Analytical Methods for the Development of Reynolds-Stress Closures in Turbulence, *Annu. Rev. Fluid Mech.*, vol. 23, Annual Reviews, Inc., Palo Alto, California, pp. 107–157.

Srinivasan, S. and Tannehill, J. C. (1987). Simplified Curve Fits for the Transport Properties of Equilibrium Air, NASA CR-178411.

Srinivasan, S., Tannehill, J. C., and Weilmuenster, K. J. (1987). Simplified Curve Fits for the Thermodynamic Properties of Equilibrium Air, NASA RP-1181.

Srivastava, B. N., Werle, M. J., and Davis, R. T. (1978). Viscous Shock-Layer Solutions for Hypersonic Sphere Cones, *AIAA J.*, vol. 16, pp. 137–144.

Srivastava, B. N., Werle, M. J., and Davis, R. T. (1979). Numerical Solutions of Hypersonic Viscous Shock-Layer Equations, *AIAA J.*, vol. 17, pp. 107–110.

Stalnaker, J. F., Nicholson, L. A., Hanline, D. S., and McGraw, E. H. (1986). Improvements to the AFWAL PNS Code Formulation, AFWAL-TR-86-3076.

Steger, J. L. (1977). Implicit Finite-Difference Simulation of Flow about Arbitrary Geometries with Application to Airfoils, AIAA Paper 77-665, Albuquerque, New Mexico.

Steger, J. L. (1978). Coefficient Matrices for Implicit Finite-Difference Solution of the Inviscid Fluid Conservation Law Equations, *Comput. Methods Appl. Mech. Eng.*, vol. 13, pp. 175–188.

Steger, J. L. (1981). A Preliminary Study of Relaxation Methods for the Inviscid Conservative Gasdynamics Equations using Flux Splitting, NASA Contractor Report 3415.

Steger, J. L. and Kutler, P. (1976). Implicit Finite-Difference Procedures for the Computation of Vortex Wakes, AIAA Paper 76-385, San Diego, California.

Steger, J. L. and Sorenson, R. L. (1980). Use of Hyperbolic Partial Differential Equations to Generate Body Fitted Coordinates, *Numerical Grid Generation Techniques*, NASA Conference Publication 2166, pp. 463–478.

Steger, J. L. and Warming, R. F. (1979). Flux Vector Splitting of the Inviscid Gasdynamic Equations with Application to Finite-Difference Methods, NASA TM D-78605.

Steger, J. L. and Warming R. F. (1981). Flux Vector Splitting of the Inviscid Gas Dynamics Equations with Application to Finite Difference Methods, *J. Comput. Phys.*, vol. 40, pp. 263–293.

Steger, J. L., Ying, S. X., and Schiff, L. B. (1986). A Partially Flux-Split Algorithm for Numerical Simulations of Compressible Inviscid and Viscous Flows, *Proceedings of the Workshop on Computational Fluid Dynamics*, Institute of Nonlinear Sciences, University of California, Davis.

Steinhoff, J. and Jameson, A. (1981). Multiple Solutions of the Transonic Potential Flow Equation, *Proc. AIAA 5th Computational Fluid Dynamics Conference*, Palo Alto, California, pp. 347–353.

Stephenson, P. L. (1976). A Theoretical Study of Heat Transfer in Two-Dimensional Turbulent Flow in a Circular Pipe and Between Parallel and Diverging Plates, *Int. J. Heat Mass Transfer*, vol. 19, pp. 413–423.

Stewartson, K. (1974). Multistructured Boundary Layers on Flat Plates and Related Bodies, *Adv. Appl. Mech.*, vol. 14, Academic, New York, pp. 145–239.

Stonc, II. L. (1968). Iterative Solution of Implicit Approximations of Multidimensional Partial Equations, *SIAM J. Numer. Anal.*, vol. 5, pp. 530–558.

Stookesberry, D. C. and Tannehill, J. C. (1987). Computation of Separated Flow Using the Space-Marching Conservative Supra-Characteristics Method, *AIAA J.*, vol. 25, pp. 1063–1070.

Sturmayr, A. M., Moschetta, J.-M., and Lafon, A. (1993). Comparison of ENO and TVD Schemes for the Parabolized Navier-Stokes Equations, AIAA Paper 93-2970, Orlando, Florida.

Svehla, R. A. (1962). Estimated Viscosities and Thermal Conductivity of Gases at High Temperatures, NASA TR R-132.

Swaminathan, S., Kim, M. D., and Lewis, C. H. (1982). Real Gas Flows over Complex Geometries at Moderate Angles of Attack, AIAA Paper 82-0392, Orlando, Florida

Swaminathan, S., Kim, M. D., and Lewis, C. H. (1983). Three-Dimensional Nonequilibrium Viscous Shock-Layer Flows over Complex Geometries, AIAA Paper 83-0212, Reno, Nevada.

Swaminathan, S., Song, D. J., and Lewis, C. H. (1984). High Altitude Effects on Three-Dimensional Nonequilibrium Viscous Shock Layer Flows, AIAA Paper 84-0304, Reno, Nevada.

Swanson, R. C. and Turkel, E. (1985). A Multistage Time-Stepping Scheme for the Navier-Stokes Equations, AIAA Paper 85-0035, Reno, Nevada.

Swanson, R. C. and Turkel, E. (1987). Artificial Dissipation and Central Difference Schemes for the Euler and Navier-Stokes Equations, AIAA Paper 87-1107, Honolulu, Hawaii.

Swartztrauber, P. N. (1977). The Methods of Cyclic Reduction, Fourier Analysis and the FACR Algorithm for the Discrete Solution of Poisson's Equation on a Rectangle, *SIAM Rev.*, vol. 19, pp. 490–501.

Sweby, P. K. (1984). High Resolution Schemes Using Flux Limiter for Hyperbolic Conservation Laws, *SIAM J. Num. Anal.*, vol. 21, pp. 995–1011.

Szema, K. Y. and Lewis, C. H. (1980). Three-Dimensional Hypersonic Laminar, Transitional and/or Turbulent Shock-Layer Flows, AIAA Paper 80-1457, Snowmass, Colorado.

Szema, K. Y. and Lewis, C. H. (1981). Three-Dimensional Viscous Shock-Layer Flows over Lifting Bodies at High Angles of Attack, AIAA Paper 81-1146, Palo Alto, California.

Tannehill, J. C. and Anderson, E. W. (1971). Intermediate Altitude Rocket Exhaust Plumes, *J. Spacecraft Rockets*, vol. 8, pp. 1052–1057.

Tannehill, J. C. and Anderson, D. A. (1980). Computation of Three-Dimensional Supersonic Viscous Flows in Internal Corners, Technical Report AFWAL-TR-80-3017.

Tannehill, J. C., Holst, T. L., and Rakich, J. V. (1975). Numerical Computation of Two-Dimensional Viscous Blunt Body Flows with an Impinging Shock, AIAA Paper 75-154, Pasadena, California.

Tannehill, J. C., Vigneron, Y. C., and Rakich, J. V. (1978). Numerical Solution of Two-Dimensional Turbulent Blunt Body Flows with an Impinging Shock, AIAA Paper 78-1209, Seattle, Washington.

Tannehill, J. C., Venkatapathy, E., and Rakich, J. V. (1982). Numerical Solution of Supersonic Viscous Flow over Blunt Delta Wings, *AIAA J.*, vol. 20, 203–210.

Tannehill, J. C., Ievalts, J. O., Prabhu, D. K., and Lawrence, S. L. (1988). An Upwind Parabolized Navier-Stokes Code for Chemically Reacting Flows, AIAA Paper 88-2614, San Antonio, Texas.

Tannehill, J. C., Buelow, P. E., Ievalts, J. O., and Lawrence, S. L. (1989). Three-Dimensional Upwind Parabolized Navier-Stokes Code for Real Gas Flows, AIAA Paper 89-1651, Buffalo, New York.

Taylor, A. E. (1955). *Advanced Calculus*, Ginn and Company, Boston.

Telionis, D. P. (1979). REVIEW—Unsteady Boundary Layers Separated and Attached, *J. Fluids Eng.*, vol. 101, pp. 29–43.

Telionis, D. P. and Tsahalis, D. Th. (1976). Unsteady Turbulent Boundary Layers and Separation, *AIAA J.*, vol. 14, pp. 468–474.

Telionis, D. P., Tsahalis, D. Th., and Werle, M. J. (1973). Numerical Investigation of Unsteady Boundary-Layer Separation, *Phys. Fluids*, vol. 16, pp. 968–973.

TenPas, P. W. (1990). Numerical Solution of the Steady, Compressible, Navier-Stokes Equations in Two and Three Dimensions by a Coupled Space-Marching Method, Ph.D. dissertation, Iowa State University, Ames.

TenPas, P. W. and Hancock, P. D. (1992). Numerical Simulation of Laminar Flow and Heat Transfer in Channels with Symmetric and Asymmetric Sudden Expansions, *ASME Topics in Heat Transfer*, vol. 1, HTD-vol. 206-1, ASME, New York.

TenPas, P. W. and Pletcher, R. H. (1991). Coupled Space-Marching Method for the Navier-Stokes Equations for Subsonic Flows, *AIAA J.*, vol. 29, 219–226.

Thareja, R., Szema, K. Y., and Lewis, C. H. (1982). Effects of Chemical Equilibrium on Three-Dimensional Viscous Shock-Layer Analysis of Hypersonic Laminar on Turbulent Flows, AIAA Paper 82-0305, Orlando, Florida.

Thom, A. and Apelt, C. J. (1961). *Field Computations in Engineering and Physics*, C. Van Nostrand, Princeton, New Jersey.

Thomas, J. L. and Walters, R. W. (1985). Upwind Relaxation Algorithms for the Navier-Stokes Equations, AIAA Paper 85-1501, Cincinnati, Ohio.

Thomas, J. L. and Walters, R. W. (1987). Upwind Relaxation Algorithms for the Navier-Stokes Equations, *AIAA J.*, vol. 25, pp. 527–534.

Thomas, L. H. (1949). Elliptic Problems in Linear Difference Equations over a Network, *Watson Sci. Comput. Lab. Rept.*, Columbia University, New York.

Thomas, P. D. and Lombard, C. K. (1978). The Geometric Conservation Law—A Link between Finite-Difference and Finite-Volume Methods of Flow Computation on Moving Grids, AIAA Paper 78-1208, Seattle, Washington.

Thomas, P. D., Vinokur, M., Bastianon, R. A., and Conti, R. J. (1972). Numerical Solution for Three-Dimensional Inviscid Supersonic Flow, *AIAA J.*, vol. 10, pp. 887–894.

Thommen, H. U. (1966). Numerical Integration of the Navier-Stokes Equations, *Z. Angew. Math. Phys.*, vol. 17, pp. 369–384.

Thompson, D. S. and Matus, R. J. (1989). Conservation Errors and Convergence Characteristics of Iterative Space-Marching Algorithms, AIAA Paper 89-1935-CP, Buffalo, New York.

Thompson, J. F. (1980). Numerical Solution of Flow Problems using Body Fitted Coordinate Systems, *Computational Fluid Dynamics*, vol. 1 (W. Kollmann, ed.), Hemisphere, Washington, D.C.

Thompson, J. F. (1996). A Reflection on Grid Generation in the 90's: Trends, Needs, and Influences, *Numerical Grid Generation in Computational Field Simulations* (B. K. Soni, J. F. Thompson, J. Hauser, and P. Eiseman, eds.), NSF Engineering Research Center for Computational Field Simulation, Mississippi State University, Mississippi State, pp. 1029–1110.

Thompson, J. F., Thames, F. C., and Mastin, C. W. (1974). Automatic Numerical Generation of Body-Fitted Curvilinear Coordinate System for Field Containing any Number of Arbitrary Two-Dimensional Bodies, *J. Comput. Phys.*, vol. 15, pp. 299–319.

Thompson, J. F., Thames, F. C., Mastin, C. W., and Shanks, S. P. (1975). Use of Numerically Generated Body-Fitted Coordinate Systems for Solution of the Navier-Stokes Equations, *Proc. AIAA 2nd Computational Fluid Dynamics Conference*, Hartford, Connecticut, pp. 68–80.

Thompson, J. F., Warsi, Z. U. A., and Matson, C. W. (1985). *Numerical Grid Generation*, Elsevier, New York.

Thompson, R. A. (1987). Comparison of Nonequilibrium Viscous-Shock-Layer Solutions with Windward Surface Shuttle Heating Data, AIAA Paper 87-1473, Honolulu, Hawaii.

Thompson, R. A., Lewis, C. H., and Kautz II, F. A. (1983). Comparison of Techniques for Predicting 3-D Viscous Flows over Ablated Shapes, AIAA Paper 83-0345, Reno, Nevada.

Thompson, R. A., Zoby, E. V., Wurster, K. E., and Gnoffo, P. A. (1987). An Aerothermodynamic Study of Slender Conical Vehicles, AIAA Paper 87-1475, Honolulu, Hawaii.

Ting, L. (1965). On the Initial Conditions for Boundary Layer Equations, *J. Math. Phys.*, vol. 44, pp. 353–367.

Truman, C. R., Shirazi, S. A., and Blottner, F. G. (1993). Noniterative Solution for Pressure in Parabolic Flows, *J. Fluids Eng.*, vol. 115, pp. 627–630.

Tsahalis, D. Th. and Telionis, D. P. (1974). Oscillating Laminar Boundary Layers and Unsteady Separation, *AIAA J.*, vol. 12, pp. 1469–1475.

Turkel, E. (1987). Preconditioned Methods for Solving the Incompressible and Low Speed Compressible Equations, *J. Comput. Phys.*, vol. 72, pp. 227–298.

Turkel, E. (1992). Review of Preconditioning Methods for Fluid Dynamics, ICASE Report 92-47, Institute for Computer Applications in Science and Engineering, NASA Langley Research Center, Langley, Virginia.

Turkel, E., Swanson, R. C., Vasta, V. N., and White, J. A. (1991). Multigrid for Hypersonic Viscous Two- and Three-Dimensional Flows, AIAA Paper 91-1572-CP, Honolulu, Hawaii.

van Albada, G. D., van Leer, B., and Roberts, W. W. (1982). A Comparative Study of Computational Methods in Cosmic Gas Dynamics, *Astron. Astrophysics*, vol. 108, pp. 76–84.

van Doormal, J. P. and Raithby, G. D. (1984). Enhancements of the SIMPLE Method for Predicting Incompressible Fluid Flows, *Numer. Heat Transfer*, vol. 7, pp. 147–163.

van Doormaal, J. P., Raithby, G. D., and McDonald, B. H. (1987). The Segregated Approach to Predicting Viscous Compressible Fluid Flows, *J. Turbomachinery*, vol. 109, pp. 268–277.

van Driest, E. R. (1951). Turbulent Boundary Layer in Compressible Fluids, *J. Aeronaut. Sci.*, vol. 18, pp. 145–160.

van Driest, E. R. (1952). Investigation of Laminar Boundary Layer in Compressible Fluids using the Crocco Method, NACA TN-2597.

van Driest, E. R. (1956). On Turbulent Flow Near a Wall, *J. Aeronaut. Sci.*, vol. 23, pp. 1007–1011.

van Dyke, M. (1969). Higher-Order Boundary-Layer Theory, *Annu. Rev. Fluid Mech.*, vol. 1, Annual Reviews, Inc., Palo Alto, California, pp. 265–292.

Vanka, S. P. (1986). Block-Implicit Multigrid Calculation of Navier-Stokes Equations in Primitive Variables, *J. Comput. Phys.*, vol. 65, pp. 138–156.

van Leer, B. (1974). Towards the Ultimate Conservation Difference Scheme, II: Monotonicity and Conservation Combined in a Second-Order Scheme, *J. Comput. Phys.*, vol. 14, pp. 361–370.

van Leer, B. (1979). Towards the Ultimate Conservative Difference Scheme, V: A Second-Order Sequel to Godunov's Method, *J. Comput. Phys.*, vol. 32, pp. 101–136.

van Leer, B. (1982). Flux Vector Splitting for the Euler Equations, Proceedings of the 8th International Conference on Numerical Methods in Fluid Dynamics, *Lect. Notes Phys.*, vol. 170, Springer-Verlag, New York, pp. 507–512.

van Leer, B. (1984). On the Relation Between the Upwind Differencing Schemes of Godunov, Engquist-Osher, and Roe, *SIAM J. Sci. Stat. Computing*, vol. 5, pp. 1–20.

van Leer, B. (1990). Flux Vector Splitting for the 1990's, *Proceedings of the Computational Fluid Dynamics Symposium on Aeropropulsion* NASA CP-3078, pp. 203–214.

van Leer, B. (1992). Progress in Multi-Dimensional Upwind Differencing, Proc. of the 13th International Conference on Numerical Methods in Fluid Mechanics, Rome 1992, *Lect. Notes Phys.*, vol. 414, Springer-Verlag, New York, pp. 1–26.

van Leer, B., Thomas, J. L., Roe, P. L., and Newsome, R. W. (1987). A Comparison of Numerical Flux Formulas for the Euler and Navier-Stokes Equations, AIAA Paper 87-1104-CP, Honolulu, Hawaii.

Varga, R. S. (1962). *Matrix Iterative Numerical Analysis*, Wiley, New York.

Vasta, V. N. and Wedan, B. W. (1989). Development of an Efficient Multigrid Code for 3-D Navier-Stokes Equations, AIAA Paper 89-1791, Buffalo, New York.

Vasta, V. N., Thomas, J. L., and Wedan, B. W. (1987). Navier-Stokes Computations of Prolate Spheroids at Angle of Attack, AIAA Paper 87-2627-CP, Honolulu, Hawaii.

Venkatakrishnan, V. and Barth, T. J. (1989). Application of Direct Solvers to Unstructured Meshes for the Euler and Navier-Stokes Equations Using Upwind Schemes, AIAA Paper 89-0364, Reno, Nevada.

Venkatakrishnan, V. and Mavriplis, D. J. (1991). Implicit Solvers for Unstructured Meshes, ICASE Report 91-40.

Viegas, J. R. and Horstman, C. C. (1979). Comparison of Multi-Equation Turbulence Models for Several Shock Boundary-Layer Interaction Flows, *AIAA J.*, vol. 17, pp. 811–820.

Viegas, J. R., Rubesin, M. W., and Horstman, C. C. (1985). On the Use of Wall Functions as Boundary Conditions for Two-Dimensional Separated Compressible Flows, AIAA Paper 85-0180, Reno, Nevada.

Vigneron, Y. C., Rakich, J. V., and Tannehill, J. C. (1978a). Calculation of Supersonic Viscous Flow over Delta Wings with Sharp Subsonic Leading Edges, AIAA Paper 78-1137, Seattle, Washington.

Vigneron, Y. C., Rakich, J. V., and Tannehill, J. C. (1978b). Calculation of Supersonic Viscous Flow over Delta Wings with Sharp Subsonic Leading Edges, NASA TM-78500.

Vincenti, W. G. and Kruger, C. H. (1965). *Introduction to Physical Gas Dynamics*, John Wiley, New York.

Vinokur, M. (1974). Conservation Equations of Gas-Dynamics in Curvilinear Coordinate Systems, *J. Comput. Phys.*, vol. 14, pp. 105–125.

Vinokur, M. (1986). An Analysis of Finite-Difference and Finite-Volume Formulation of Conservation Laws, NASA CR-177416.

Vinokur, M. and Liu, Y. (1988). Equilibrium Gas Flow Computations, II: An Analysis of Numerical Formulations of Conservation Laws, AIAA Paper 88-0127, Reno, Nevada.

Viviand, H. (1974). Conservative Forms of Gas Dynamic Equations, *Rech. Aerosp.*, No. 1974-1, pp. 65–68.

von Neumann, J. and Richtmyer, R. D. (1950). A Method for the Numerical Calculation on Hydrodynamic Shocks, *J. Appl. Phys.*, vol. 21, pp. 232–237.

Vornoi, G. (1908). Nouvelles applications des parametres continus a la theorie des formes quadratiques. Recherches sur les parallelloedres primitifs, *J. Reine Angew. Math.*, vol. 134.

Vradis, G., Zalak, V., and Bentson, J. (1992). Simultaneous Variable Solutions of the Incompressible Steady Navier-Stokes Equations in General Curvilinear Coordinate Systems, *J. Fluids Eng.*, vol. 114, pp. 299–305.

Wachspress, E. L. (1966). *Iterative Solution of Elliptic Systems*, Prentice-Hall, Englewood Cliffs, New Jersey.

Wadawadigi, G., Tannehill, J. C., Buelow, P. E., and Lawrence, S. L. (1992). A Three-Dimensional Upwind PNS Code for Chemically Reacting Scramjet Flowfields, AIAA Paper 92-2898, Nashville, Tennessee.

Wadawadigi, G., Tannehill, J. C., Lawrence, S. L., and Edwards, T. A. (1994). Three-Dimensional Computation of the Integrated Aerodynamic and Propulsive Flowfields of a Generic Hypersonic Space Plane, AIAA Paper 94-0633, Reno, Nevada.

Walker, M. A. and Oberkampf, W. L. (1991). Joint Computational/Experimental Aerodynamics Research on a Hypersonic Vehicle, Part 2: Computational Results, AIAA Paper 91-0321, Reno, Nevada.

Wang, K. C. (1971). On the Determination of the Zones of Influence and Dependence for Three-Dimensional Boundary-Layer Equations, *J. Fluid Mech.*, vol. 48, pt. 2, pp. 397–404.

Wang, K. C. (1972). Separation Patterns of Boundary Layer over an Inclined Body of Revolution, *AIAA J.*, vol. 10, pp. 1044–1050.

Wang, K. C. (1973). Three-Dimensional Laminar Boundary Layer over a Body of Revolution at Incidence, Part VI: General Methods and Results of the Case at High Incidence, MML TR 73-02c, Martin Marietta Laboratories, Baltimore, Maryland.

Wang, K. C. (1974). Boundary Layer over a Blunt Body at High Incidence with an Open-Type of Separation, *Proc. R. Soc. London, Ser. A*, vol. 340, pp. 33–55.

Wang, K. C. (1975). Boundary Layer over a Blunt Body at Low Incidence with Circumferential Reversed Flow, *J. Fluid Mech.*, vol. 72, pt. 1, pp. 49–65.

Wang, R. Q., Jiao, L. Q., and Liu, X. Z. (1981). Numerical Methods for the Solution of the Simplified Navier-Stokes Equations, Proceedings of the Seventh International Conference on Numerical Methods in Fluid Dynamics, *Lect. Notes Phys.*, vol. 141, Springer-Verlag, New York, pp. 423–427.

Wang, W.-P. and Pletcher, R. H. (1995a). Evaluation of Some Coupled Algorithms for Large Eddy Simulation of Turbulent Flow Using a Dynamic SGS Model, AIAA Paper 95-2244, San Diego, California.

Wang, W.-P. and Pletcher, R. H. (1995b). Large Eddy Simulation of a Low Mach Number Channel Flow with Property Variations, *Proceedings of the 10th Symposium on Turbulent Shear Flows*, Pennsylvania State University, .

Warming, R. F. and Beam, R. M (1975). Upwind Second-Order Difference Schemes and Applications in Unsteady Aerodynamic Flows, *Proc. AIAA 2nd Computational Fluid Dynamics Conference*, Hartford, Connecticut, pp. 17–28.

Warming, R. F. and Beam, R. M. (1977). On the Construction and Application of Implicit Factored Schemes for Conservation Laws, Symposium on Computational Fluid Dynamics, New York. See *SIAM-AMS Proceedings*, vol. 11, 1978, pp. 85–129.

Warming, R. F. and Hyett, B. J. (1974). The Modified Equation Approach to the Stability and Accuracy Analysis of Finite-Difference Methods, *J. Comput. Phys.*, vol. 14, pp. 159–179.

Warming, R. F., Kutler, P., and Lomax, H. (1973). Second- and Third-Order Noncentered Difference Schemes for Nonlinear Hyperbolic Equations, *AIAA J.*, vol. 11, pp. 189–196.

Waskiewicz, J. D. and Lewis, C. H. (1978). Hypersonic Viscous Flows over Sphere-Cones at High Angle of Attack, AIAA Paper 78-64, Huntsville, Alabama.

Waskiewicz, J. D., Murray, A. L., and Lewis, C. H. (1978). Hypersonic Viscous Shock-Layer Flow over a Highly Cooled Sphere, *AIAA J.*, vol. 16, pp. 189–192.

Weatherill, N. (1992). Delaunay Triangulation in Computational Fluid Dynamics, *Math. Appl.*, vol. 24, pp. 129–150.

Weinberger, H. F. (1965). *A First Course in Partial Differential Equations*, John Wiley, New York.

Weinstock, R. (1952). *Calculus of Variations with Applications to Physics and Engineering*, McGraw-Hill, New York.

Welch, J. E., Harlow, F. H., Shannon, J. P., and Daly, B. J. (1966). The MAC Method, Los Alamos Scientific Laboratory Report LA-3425, Los Alamos, New Mexico.

Werle, M. J. and Bertke, S. D. (1972). A Finite-Difference Method for Boundary Layers with Reverse Flow, *AIAA J.*, vol. 10, pp. 1250–1252.

Werle, M. J. and Dwoyer, D. L. (1972). Laminar Hypersonic Interacting Boundary Layers: Subcritical Branching in the Strong Interaction Regime, ARL 72-0011, Wright-Patterson Air Force Base, Dayton, Ohio.

Werle, M. J. and Vasta, V. N. (1974). A New Method for Supersonic Boundary Layer Separations, *AIAA J.*, vol. 12, pp. 1491–1497.

Werle, M. J. and Verdon, J. M. (1979). Solutions for Supersonic Trailing Edges Including Separation, AIAA Paper 79-1544, Williamsburg, Virginia.

Werle, M. J., Polak, A., and Bertke, S. D. (1973). Supersonic Boundary-Layer Separation and Reattachment—Finite Difference Solutions, Report No. AFL 72-12-1, Department of Aerospace Engineering, University of Cincinnati, Ohio.

White, Jr., A. B. (1982). On the Numerical Solution of Initial/Boundary-Value Problems in One Space Dimension, *SIAM J. Numer. Anal.*, vol. 19, pp. 683–697.

White, F. M. (1974). *Viscous Fluid Flow*, McGraw-Hill, New York.

White, F. M. (1991). *Viscous Fluid Flow*, 2d ed., McGraw-Hill, New York.

White, J. A., Korte, J. J., and Gaffney Jr., R. L., (1993), Flux-Difference Split Parabolized Navier-Stokes Algorithm for Nonequilibrium Chemically Reacting Flows, AIAA Paper 93-0534, Reno, Nevada.

Whitham, G. B. (1974). *Linear and Nonlinear Waves*, Wiley, New York.

Whittaker, E. T. and Watson, G. N. (1927). *A Course in Modern Analysis*, 4th ed. (reprinted 1962, Cambridge University Press).

Wieghardt, K. and Tillman, W. (1951). On the Turbulent Friction Layer for Rising Pressure, NACA TM-1314.

Wigton, L. B. and Holt, M. (1981). Viscous-Inviscid Interaction in Transonic Flow, AIAA Paper 81-1003, Palo Alto, California.

Wigton, L. B., Yu, N. J., and Young, D. P. (1985). GMRES Acceleration of Computational Fluid Dynamics Codes, AIAA Paper 85-1494-CP, Cincinnati, Ohio.

Wilcox, D. C. (1988). Reassessment of the Scale Determining Equation for Advanced Turbulence Models, *AIAA J.*, vol. 26, pp. 1299–1310.

Wilcox, D. C. (1993). *Turbulence Modeling for CFD*, DCW Industries, La Canada, California.

Wilcox, D. C. and Traci, R. M. (1976). A Complete Model of Turbulence, AIAA Paper 76-351, San Diego, California.

Wilke, C. R. (1950). A Viscosity Equation for Gas Mixtures, *J. Chem. Phys.*, vol. 18, pp. 517–519.

Williams, J. C. (1977). Incompressible Boundary Layer Separation, *Annu. Rev. Fluid Mech.*, vol. 9, Annual Reviews, Inc., Palo Alto, California, pp. 113–144.

Winslow, A. (1966). Numerical Solution of the Quasi-linear Poisson Equation, *J. Comput. Phys.*, vol. 1, pp. 149–172.

Winslow, A. (1981). Adaptive Zoning by the Equipotential Method, UCID-19062, U. of California, Livermore.

Wolfstein, M. (1969). The Velocity and Temperature Distribution in One-Dimensional Flow with Turbulence Augmentation and Pressure Gradient, *Int. J. Heat Mass Transfer*, vol. 12, pp. 301–318.

Wornom, S. F. (1977). A Critical Study of Higher-Order Numerical Methods for Solving the Boundary-Layer Equations, *Proc. AIAA 3rd Computational Fluid Dynamics Conference*, Albuquerque, New Mexico.

Wu, J. C. (1961). On the Finite-Difference Solution of Laminar Boundary Layer Problems, *Proc. 1961 Heat Transfer and Fluid Mechanics Institute*, Stanford University Press, Stanford, California.

Wylie Jr., C. R. (1951). *Advanced Engineering Mathematics*, McGraw-Hill, New York.

Yakhot, V. and Orszag, S. A. (1986). Renormalization Group Analysis of Turbulence, I: Basic Theory, *J. Sci. Comput.*, vol. 1, pp. 3–51.

Yanenko, N. N. (1971). *The Method of Fractional Steps: The Solution of Problems of Mathematical Physics in Several Variables* (M. Holt, ed.), Springer-Verlag, New York.

Yanenko, N. N., Kovenya, V. M., Tarnavsky, G. A., and Cherny, S. G. (1980). Economical Methods for Solving the Problems of Gas Dynamics, Proc. Seventh Int. Conf. Num. Methods Fluid Dyn., *Lect. Notes Phys.*, vol. 141, Springer-Verlag, New York, pp. 448–453.

Yang, J. Y. (1991). Third-Order Non-Oscillatory Schemes for the Euler Equations, *AIAA J.*, vol. 29, pp. 1611–1618.

Yang, K.-S. and Ferziger, J. H. (1993). Large-Eddy Simulation of Turbulent Flow with a Surface-Mounted Two-Dimensional Obstacle Using a Dynamic Subgrid-Scale Model, AIAA Paper 93-542, Reno, Nevada.

Yee, H. C. (1981). Numerical Approximations of Boundary Conditions with Applications to Inviscid Gas Dynamics, NASA Report TM-81265.

Yee, H. C. (1985a). On Symmetric and Upwind TVD Schemes, NASA TM-86842.

Yee, H. C. (1985b). On Symmetric and Upwind TVD Schemes, *Proceedings of the 6th GAMM Conference on Numerical Methods in Fluid Mechanics*, Broanschweig, Vieweg, pp. 399–407.

Yee, H. C. (1986). Linearized Form of Implicit TVD Schemes for the Multidimensional Euler and Navier-Stokes Equations, *Comput. Math. Appl.*, vol. 12A, pp. 413–432.

Yee, H. C. (1987). Upwind and Symmetric Shock Capturing Methods, NASA TM-89464.

Yee, H. C. (1989). A Class of High-Resolution Explicit and Implicit Shock-Capturing Methods, NASA TM-101088.

Yee, H. C. and Harten, A. (1985). Implicit TVD Shock-Capturing Schemes for Hyperbolic Conservation Laws in Curvilinear Coordinates, AIAA Paper 85-1513-CP, Cincinnati, Ohio.

Yerry, M. A. and Shephard, M. S. (1983). A Modified-Quadtree Approach to Finite Element Mesh Generation, *IEEE Comput. Graphics Appl.*, vol. 3, pp. 39–46.

Yerry, M. A. and Shephard, M. S. (1984). Automatic Three-Dimensional Mesh Generation by the Modified-Octree Technique, *Int. J. Numer. Methods Eng.*, vol. 20, pp. 1965–1990.

Yoon, S. and Jameson, A. (1987). Lower-Upper Implicit Schemes with Multiple Grids for the Euler Equations, *AIAA J.*, vol. 25, pp. 929–935.

Yoon, S. and Jameson, A. (1988). Lower-Upper Symmetric-Gauss-Seidel Method for the Euler and Navier-Stokes Equations, *AIAA J.*, vol. 26, pp. 1025–1026.

Yoon, S. and Kwak, D. (1993). Multigrid Convergence of an Implicit Symmetric Relaxation Scheme, AIAA Paper 93-3357, Orlando, Florida.

Young, D. (1954). Iterative Methods for Solving Partial Difference Equations of Elliptic Type, *Trans. Am. Math. Soc.*, vol. 76, pp. 92–111.

Yu, N. J., Chen, H. C., Kusunose, K., and Sommerfield, D. M. (1987). Flow Simulation of a Complex Airplane Configuration Using Euler Equations, AIAA Paper 87-0454, Reno, Nevada.

Yu, S. T., Tsai, Y.-L. P., and Hsieh, K. C. (1992). Runge-Kutta Methods Combined with Compact Difference Schemes for the Unsteady Euler Equations, AIAA Paper 92-3210, Nashville, Tennessee.

Zachmanoglou, E. C. and Thoe, D. W. (1976). *Introduction to Partial Differential Equations with Applications*, Williams & Wilkins, Baltimore, Maryland.

Zedan, M. and Schneider, G. E. (1985). A Coupled Strongly Implicit Procedure for Velocity and Pressure Calculations in Fluid Flow Problems, *Numer. Heat Transfer*, vol. 8, pp. 537–557.

Zha, G. C. and Bilgen, E. (1993). Numerical Solutions of Euler Equations by Using a New Flux Vector Splitting Scheme, *Int. J. Numer. Methods Fluids*, vol. 17, pp. 115–144.

Zoby, E. V., Lee, K. P., Gupta, R. N., Thompson, R. A., and Simmonds, A. L. (1989). Viscous Shock-Layer Solutions with Nonequilibrium Chemistry for Hypersonic Flows Past Slender Bodies, *J. Spacecraft Rockets*, vol. 26, pp. 221–228.

INDEX